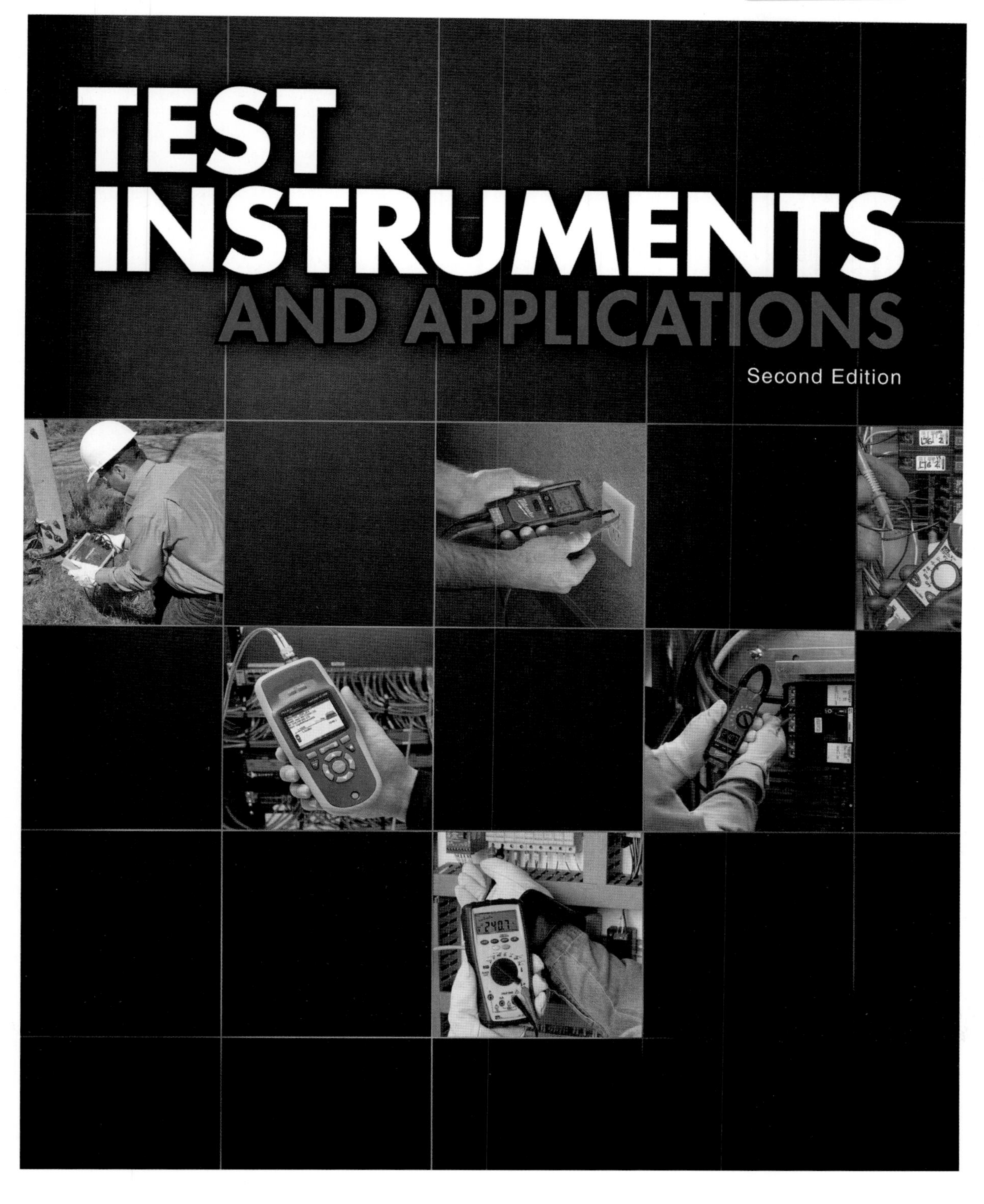

TEST INSTRUMENTS AND APPLICATIONS

Second Edition

AMERICAN TECHNICAL PUBLISHERS
Orland Park, Illinois

Glen A. Mazur

2 3 4 5 6 7 8 9 – 19 – 9 8 7

Printed in the United States of America

ISBN 978-0-8269-1332-6

This book is printed on recycled paper.

ACKNOWLEDGMENTS

The author and publisher are grateful for the technical information and assistance provided by the following companies and organizations.

- ABB Inc.
- ABB Motors and Drives
- AEMC® Instruments
- Amprobe/Advanced Test Products
- ASI Robicon
- B&K Precision
- Dranetz-BMI
- Electrical Apparatus Service Association, Inc.
- Extech by FLIR
- FLIR Systems
- Fluke Corporation
- Fluke Networks
- GE Lighting
- GE Panametrics
- Greenlee Textron, Inc.
- Hach
- Hubbell Power Systems, Inc.
- IDEAL Industries, Inc.
- Infineon
- Mastercool® Inc.
- McQuay International
- Megger Group Limited
- Meriam Process Technologies
- Milwaukee Tool Corporation
- MSA Safety, Inc.
- Robinair
- Rockwell Automation-Allen Bradley Company Inc.
- Salisbury by Honeywell
- Siemens
- Vacuum Interrupters, Inc./Group CBS
- Yellow Jacket Div. Ritchie Engineering Co., Inc.

CONTENTS

CONTENTS

TROUBLESHOOTING 377

Troubleshooting • Troubleshooting Levels • Troubleshooting Methods • Troubleshooting Procedures • Equipment Reliability • Troubleshooter Responsibilities

TOPICS COVERED

APPLICATIONS APPENDIX 397

GENERAL APPENDIX 627

GLOSSARY 643

INDEX 653

INTRODUCTION

Industry demand for safety, quality, and productivity requires the routine use of test instruments by electricians, maintenance technicians, and other industry personnel. As electrical devices have become more technologically complex, taking accurate measurements has become more critical. Increased test instrument capabilities necessitate training for safe and efficient measurement of specific electrical quantities. *Test Instruments and Applications* provides an overview of typical electrical test instruments used for installation tests, process equipment operation, quality control, and troubleshooting activities.

Test Instruments and Applications begins with an introduction to safety and test instruments, and then focuses on test instruments used for specific applications found in industry. Common electrical measurements are introduced with an overview of the measurement principles and test instrument procedures required. Personal protective equipment depicted is based on requirements specified in National Fire Protection Association standard *NFPA 70E, Standard for Electrical Safety in the Workplace*. Required safety practices and common industrial applications are emphasized throughout the book.

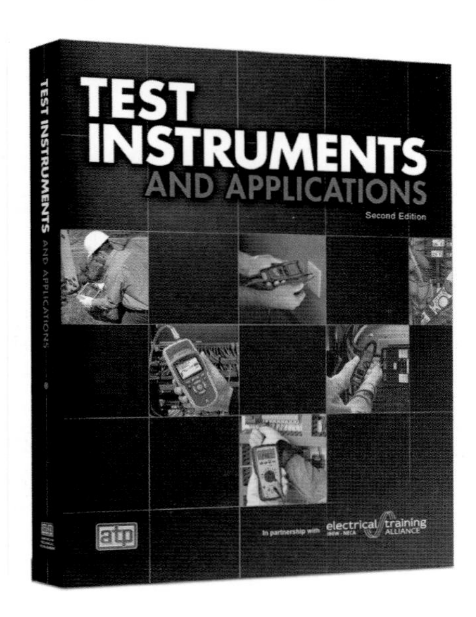

Full-color illustrations are used to supplement concise text. Procedures for the various measurements are detailed with sequenced steps. Factoids, Technical Tips, and Safety Tips throughout complement the content presented. An extensive Glossary and comprehensive Appendix offer additional reference material.

Test Instruments and Applications is one of many high-quality training products available from American Technical Publishers. To obtain information about related training products, visit the American Technical Publishers web site at www.atplearning.com.

The Publisher

FEATURES

Chapter introductions provide an overview of key content in the chapter

Full-color photos supplement text and illustrations

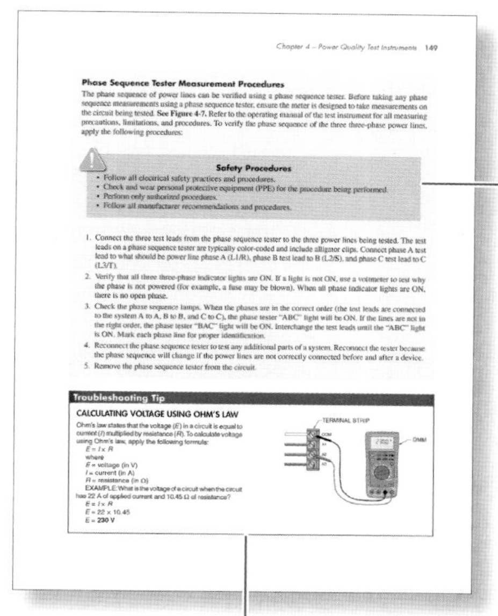

Safety information is included throughout the book

Trouble shooting tips offer supplementary topics to enhance troubleshooting efficiency

Illustrated step-by-step procedures depict common test instrument procedures

MEGOHMMETER INSULATION STEP VOLTAGE TEST MEASUREMENT PROCEDURES

SAFETY CONSIDERATIONS

Reviews test for comprehension of content covered

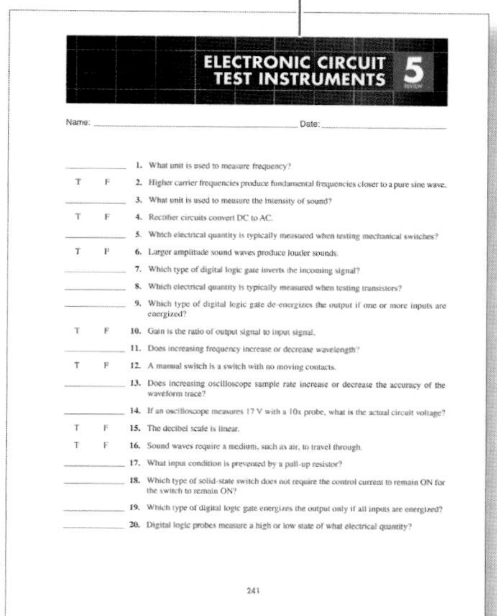

LEARNER RESOURCES

Test Instruments and Applications includes access to online Learner Resources that reinforce content and enhance learning. These online resources can be accessed using either of the following methods:

- Key ATPeResources.com/QuickLinks into a web browser and enter QuickLinks™ Access code 842525.
- Use a Quick Response (QR) reader app to scan the QR Code with a mobile device.

The Learner Resources include the following:

- **Quick Quizzes™** that provide interactive questions for each chapter, with embedded links to highlighted content within the textbook and to the Illustrated Glossary
- **Illustrated Glossary** that serves as a helpful reference to commonly used terms, with selected terms linked to textbook illustrations
- **Flash Cards** that provide a self-study/review of common terms and their definitions
- **Interactive Simulations** that are interactive activities to help the learner apply proper procedures with commonly used test instruments
- **Media Library** that consists of videos and animations that reinforce textbook content
- **Internet Resources** that provide access to additional online resources that support continued learning

To obtain information on other related training material, visit the American Technical Publishers website at www.atplearning.com.

The Publisher

1 INTRODUCTION TO TEST INSTRUMENTS

All test instruments and meters are used where measurements of electrical properties are required. All test instruments and meters have OSHA, manufacturer, facility, and application safety procedures (National Fire Protection Association (NFPA) 70E, Standard for Electrical Safety in the Workplace) that must be followed at all times.

IDEAL Industries, Inc.

INTRODUCTION TO TEST INSTRUMENTS

Electricity is a relatively new science when compared to other scientific endeavors. In 600 BC, the Greeks discovered a substance, amber (fossilized tree sap), that when rubbed against wool, causes other substances to be attracted to the amber. The Greek word for amber is the source of our word electricity. In modern terminology, the attraction is the result of static electricity and opposite polarities attracting one another. The foundation of electrical science is the ability to measure the various phenomena caused by electricity. Science, by definition, is the ability to reduce any physical event to a quantitative, numerical value.

In order for the electrical industry to develop and grow to where electricity is so important in our lives today, test instruments were developed that provide numerical information about how electricity behaves. Because the scientists could not see electricity, the scientists were dependent upon electrical properties that had been defined. Instruments developed by scientists measured only those electrical properties that people understood. With the need for new electrical properties to describe the characteristics of electricity, all the properties were defined in the physical terms of the time. **See Figure 1-1.**

Milwaukee Tool Corporation

Test instruments allow technicians to identify poorly performing equipment, circuit faults, and hazardous conditions.

EARLY ELECTRICAL MEASUREMENTS

Megger Group Limited

Figure 1-1. Knowledge about electricity evolved with measurement capabilities.

There are two parts to any measurement. The first part is the actual measurement. The second part is a standard to compare the measurement against. Any measurement of value is relative to something else — and this relationship can be expressed in a mathematical formula. Without a standard, there is no logical means of communicating the value of a measurement with any degree of accuracy, or in a meaningful way. Without the ability to accurately communicate values of measurement, commerce cannot exist or scientific investigations take place. Modern electrical measurements are taken and displayed during normal circuit operation to provide a visual indication of circuit and component performance. **See Figure 1-2.** Measurements are also taken during troubleshooting to indicate any problems in a circuit or help predict possible future problems.

MODERN ELECTRICAL MEASUREMENTS

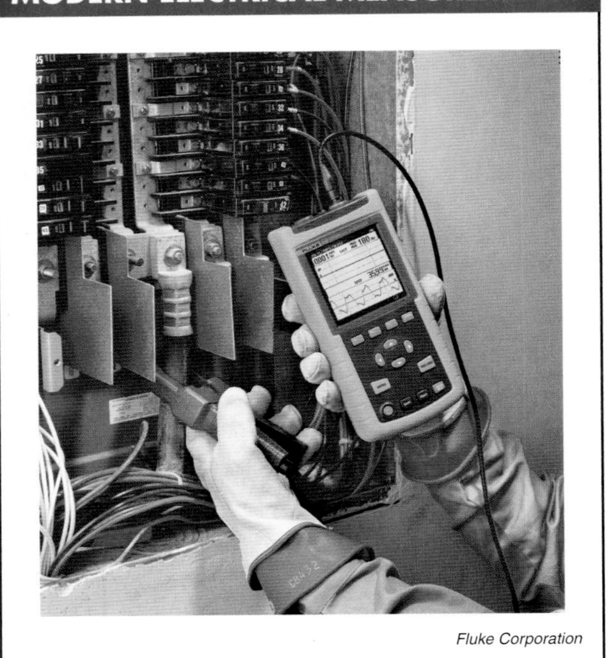

Fluke Corporation

Figure 1-2. Modern electrical measurements are displayed numerically and graphically.

Technical Tip

Voltmeters that were used in the early 20th century consisted of an analog meter (with one scale) with two screw terminals used to connect it in series with the circuit. A separate multiplier would be added to the circuit if the voltage to be measured was greater than the full-scale deflection of the instrument.

HISTORY OF MEASUREMENTS

Archeologists have determined from exploration and the limited information of recorded history that people had a strong interest in measuring length, weight, time, and temperature. Not much has changed through the millennia. Modern science still relies on accurately measuring length, weight, time, and temperature. The only additional contribution by science has been to more accurately define, accept, and maintain a standard for each measurement. All other units of measure are defined and compared to the four basic measurements.

Length

The earliest attempt of man to convey length can be described in terms of distance traveled. The term "day's journey" is still used today. However, "day's journey" is not an accurate standard because, when walking is the means of travel, people travel different distances in the same amount of time. When a person runs instead of walking, the distance traveled is greater. When a horse, coach, car, train, or jet aircraft is used, the difference in distances traveled is tremendous. The Romans refined the distance traveled to a unit of length, the mile. The Latin words "milia passuum" mean thousands of paces. A pace consisted of two steps, or approximately 5′. A thousand paces approximately equals 5000′, or about one mile (5280′). **See Figure 1-3.**

METRIC AND U.S. UNITS OF LENGTH

METRIC UNITS OF LENGTH	U.S. UNITS OF LENGTH
meter (m) 1 decimeter (dm) = 0.1 m 1 centimeter (cm) = 0.01 m 1 millimeter (mm) = 0.001 m	yard (yd) 1 foot = ⅓ yd = 12″ 1 inch = 1/36 yd = 1/12′ —— —— ——
1 m = 10 dm = 100 cm = 1000 mm	1 yd = 3′ = 36″
kilometer (km) 1 kilometer = 1000 m	1 rod = 5.5 yd mile = 1760 yd

METRIC AND U.S. UNITS OF LENGTH RELATIONSHIP

Length	Meter (m)	Millimeter (mm)	Foot (ft)	Inch
1 meter	1	1,000	3.281	39.37
1 millimeter	0.001	1	3.281×10^{-3}	0.03937
1 foot	0.3048	304.8	1	12
1 inch	0.0254	25.4	0.0833	1

Figure 1-3. Measurements of length were first defined by parts of the body, for example using shoulder-to-hand or elbow-to-hand distances.

One of the earliest standards for the measure of length was the cubit, which was developed in ancient Egypt. The cubit was used as a standard of length as long ago as 3000 BC. The length of the cubit was defined as the length of the forearm from the elbow to the tip of the middle finger (approximately 21″). Cubit sticks used to make measurements of length were compared to the royal cubit at regular intervals. The royal cubit stone used to develop the standard had distinct markings that divide the length of the stone into smaller units. The basic subunit of the royal cubit stone is the digit. There are 28 digits to a cubit. Four digits equal a palm, with five digits equaling a hand. Twelve digits, or three palms, equal a small span, and 24 digits equal a large span. The digit is also subdivided on the stone. The 14th digit of the royal cubit stone is subdivided into 16 equal parts. The 15th digit is divided into 15 parts, and so on, to the 28th digit, which is divided into two equal parts. The smallest division of 1/16 of a digit is equal to 1/448 of the royal cubit. The accuracy of the cubit system of measurement is attested to in the Great Pyramid of Giza. Although thousands of cubit sticks were used in the construction of the Great Pyramid, the sides of the pyramid vary no more than 0.05% from the mean length of 230.364 m. **See Figure 1-4.**

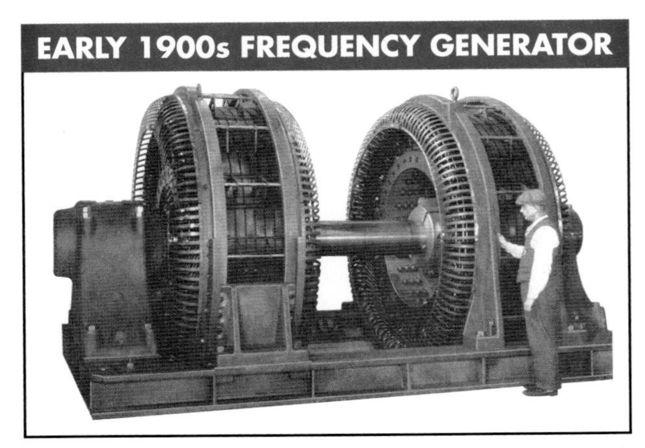

Figure 1-4. By the early 1900s the accuracy of length measurement had progressed to the point of allowing electrical equipment to be created.

One of the most significant outcomes of the French Revolution was the establishment of the metric system (SI). The meter was established as the basic unit for measuring length. The meter was defined as $\frac{1}{10,000,000}$ of the distance of the meridian passing through Paris from the North Pole to the Equator. France adopted the metric system in June, 1799. A platinum-iridium bar was cast and cut to the exact length required for use as the standard. In 1840, due to the superiority of the metric system over older systems, the metric system was given official sanction.

The United States adopted the metric system in 1875 and was issued a Prototype #27 bar from the International Bureau of Weights and Measure. Even though the United States used the customary British system of weights and measure, length was legally defined in reference to the length of the standard meter. For example, the yard is equal to 0.9144 times the length of the meter, and an inch equals 2.54 centimeters. With the unit of length defined, defining area and volume was possible.

Area is a multiple of length and width. The standard unit of area is the square meter (m^2). When an area has a side equal to 50 m and another side (at 90°) equal to 40 m, the area equals 2000 square meters ($50 \text{ m} \times 40 \text{ m} = 2000 \text{ m}^2$). In the case of volume, the defined area is three-dimensional. The standard unit for volume is the cubic meter (m^3). A box measuring 2 meters in length, width, and height contains 8 cubic meters ($2 \text{ m} \times 2 \text{ m} \times 2 \text{ m} = 8 \text{ m}^3$). **See Figure 1-5.**

British units of area and capacity such as the acre and the quart can be referenced mathematically to the metric system. Land area in the metric system is measured with the hectare (ha), which is defined as 10,000 m^2. One acre equals 0.404686 ha. The unit of measure of capacity in the metric system is the liter (l). The liter is equal to 1 cubic decimeter (0.001 m^3). The quart is equal to 0.946353 l.

Modern length measurements are taken during all phases of electrical equipment installation. Estimated measurements are taken when running and cutting wire or conduit, estimating the cost of a project, and other applications in which an exact measurement is not always required. Accurate measurements are required when bending conduit, mounting motors, and installing electrical equipment. Accurate measurements are measurements that provide information for use in the real world, but are not mathematically exact. **See Figure 1-6.**

Weight

The oldest known weight standards date to 4000 BC, which used stones used to balance scales for the measurement of gold in Egyptian temples. Babylonians used seeds or grain as weight standards. The United States still uses a unit called the grain to measure the weight of drugs and precious metals.

A system of weights and measures was developed in China that was completely separate from the Mediterranean-European systems. The Chinese system exhibits all the same characteristics as the western systems such as using parts of the body as the source of the units. The Chinese system was chaotic in that there was no relationship between various types of units such as length and volume. Shih Huang Ti, who became emperor of China in 221 BC, provided the first standards. The basic weight, the "shili" or the "tan," was fixed at approximately 132 lb. A noteworthy contribution of the Chinese system was an allowance for a decimal notation on foot rulers as long ago as the sixth century BC.

METRIC AND U.S. UNITS OF AREA

METRIC UNITS OF AREA	U.S. UNITS OF AREA
square meter (m^2)	square yard (yd^2) = 9 ft^2
square decimeter (dm^2) = 0.01 m^2	square foot (ft^2) = 144 in^2
square centimeter (cm^2) = 0.0001 m^2	square inch (in^2)
square millimeter (mm^2) = 0.000001 m^2	square rod = 30.25 yd^2
square kilometer (km^2) = 1,000,000 m^2	acre = 4840 yd^2
hectare = 10,000 m^2	square mile = 640 $acres^2$

METRIC AND U.S. UNITS OF AREA RELATIONSHIP

Area	m^2	cm^2	ft^2	in^2
1 square meter	1	10^4	10.764	1549.9
1 square centimeter	10^{-4}	1	0.00108	0.155
1 square foot	0.0929	929	1	144
1 square inch	6.452×10^{-4}	6.452	0.00694	1

Figure 1-5. Once standard length measurements had been defined, two-dimensional area and three-dimensional volume measurements were defined.

LASER DISTANCE ESTIMATORS

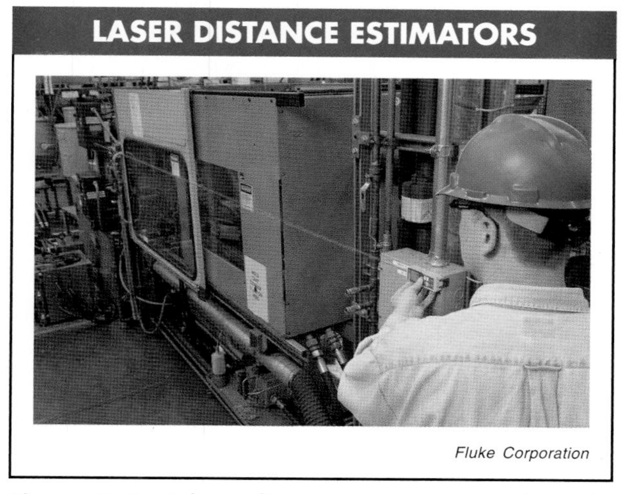

Fluke Corporation

Figure 1-6. A laser distance estimator is used to take measurements without the use of a tape measure.

A unique characteristic of the Chinese system was the inclusion of an acoustical measurement. A standard vessel used for measuring grain or wine was defined by the pitch of the vessel when the vessel was struck. The vessel having the proper weight would also have the proper pitch when struck. The ancient Chinese words to describe the acoustical standards still mean "wine bowl," "grain measure," and "bell" in the Chinese language today.

Western culture received the unit of weight from the Romans. The unit "libra" acquired a Germanic name in parts of northern Europe, but still retained a Roman identity in the English abbreviation of pound (lb). The troy pound, for weighing precious metals, was only 12 oz (troy). In 1963 the British accepted the metric system as the standard for the country. **See Figure 1-7.**

In the metric system, the kilogram is the unit of measure for mass. A kilogram equals the weight of 1000 cm^3 of water at 4°C. Water density is the greatest at a temperature of 4°C (39.2°F). One pound equals 0.453592 kg. A kilogram equals 2.204624 lb. Mass is used as the standard instead of weight. Weight differs with the pull of gravity. For example, an astronaut who weighs 165 lb on Earth has no weight when in space. Weight is equal to mass multiplied by the acceleration due to gravity. The acceleration due to gravity is measured in meters per second squared (m/s^2). The pull of gravity varies slightly across the surface of the Earth according to latitude.

Gravity is an accelerating unit. When a body is dropped in a vacuum, the body falls 16.08 feet the first second and is traveling 32.16 ft/s at that instant. Weight measurements are specified for electrical equipment to aid in the installation of the equipment. For example, luminaires are mounted using one of several different methods, depending on the weight of the fixture. **See Figure 1-8.**

METRIC AND U.S. UNITS OF WEIGHT (MASS)				
Metric Units of Mass			**U.S. Units of Mass**	
1 kilogram (kg) = 1000 g			1 pound (lb) = 16 oz	
1 gram (g) = 0.001 kg			1 ounce (oz) = 1/16 lb	
1 metric ton (t) = 1000 kg			1 slug = 32.17 lb 1 short ton = 2000 lb 1 long ton = 2240 lb	
Metric and U.S. Units of Weight Relationship				
Mass	**g**	**kg**	**lb**	**sl**
1 gram	1	0.001	0.002205	6.852×10^{-5}
1 kilogram	1000	1	2.2045	0.06852
1 pound	453.6	0.4536	1	0.03108
1 slug	14,590	14.590	32.17	1

Figure 1-7. The standard unit of weight in the metric system is the gram. For the U.S. system the standard unit is the pound.

Time

In earliest human history, time was probably measured in terms of the rising and setting of the sun. The calendars of the ancient Egyptians and Babylonians are some of the oldest artifacts showing the interest of these cultures in the passing of time. Later, sundials and water clocks were used to measure shorter periods of time. The Babylonians had a number system based on 60. The Babylonians believed the year consisted of 360 days. The 360-day year led to fractions such as 1/60 which are still used in astronomy and for the minutes and seconds of a day. The Babylonians determined time by the relationship of the movements of the Earth to the stars. However, the speed of the Earth changes during the year, causing a discrepancy between intervals. A standard time measurement was required for modern science. **See Figure 1-9.**

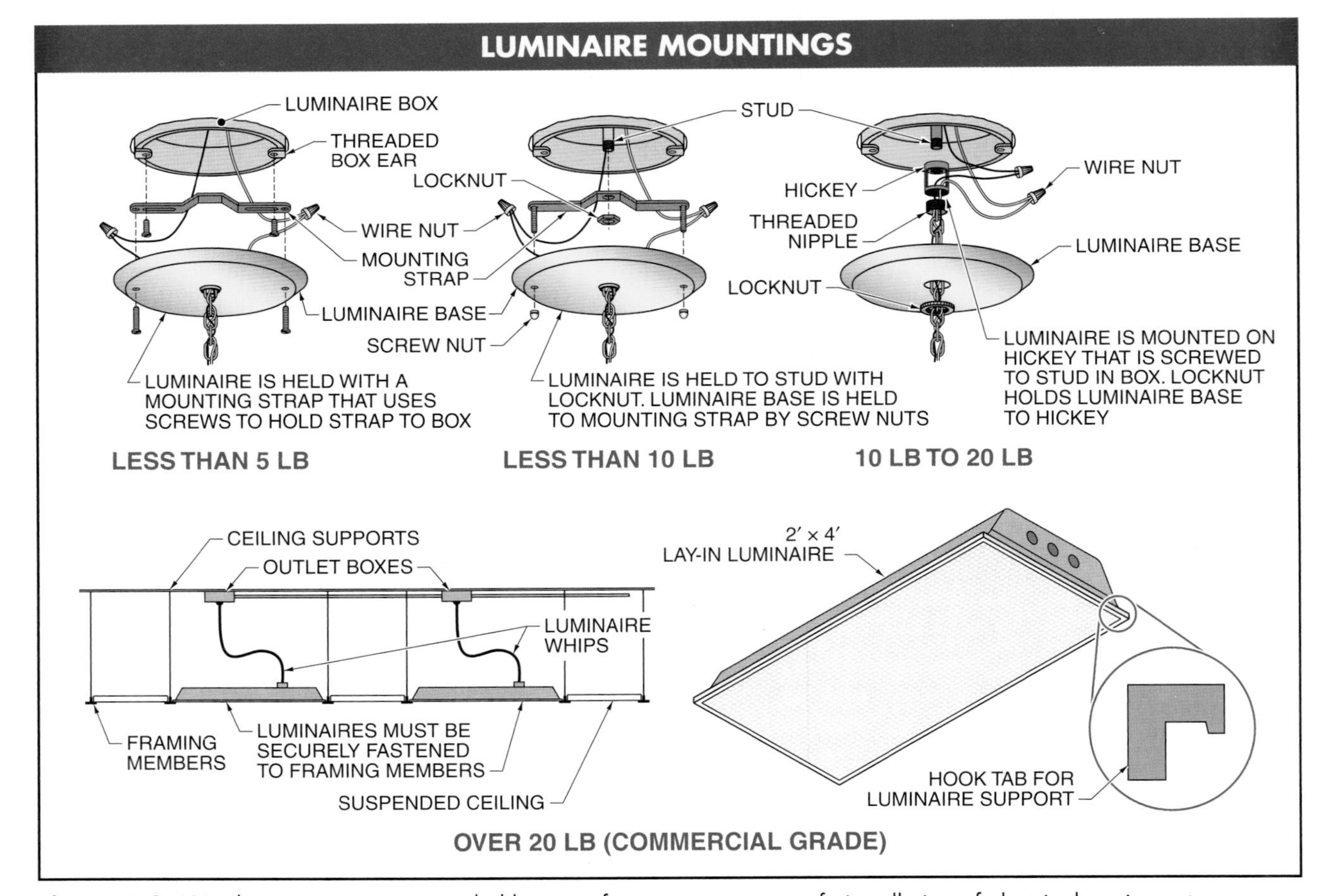

Figure 1-8. Weight capacities are provided by manufacturers to ensure safe installation of electrical equipment.

METRIC AND U.S. UNITS OF TIME	
Units	**Relationship**
second (s) minute (min) hour (h) day (d) year (y)	1 min = 60 s 1 h = 60 min 1 d = 24 h 1 y = 365.26 d

Figure 1-9. The standard unit of time in the metric and U.S. systems of measurement is the second.

Troubleshooting Tip

TIME MEASUREMENT ACCURACY

Cesium-beam atomic clocks measure frequency with an accuracy of 2 to 3 parts in 10^{14}. This time measurement is accurate to 2 nanoseconds per day or one second in 1,400,000 years. Currently, it is the most accurate time measurement that has been achieved.

Scientists believed time to be a fundamental quantity that can be measured accurately like length and mass. Albert Einstein realized that time measurements are affected by the relative motion between two objects. Because of the work of Einstein, time became known as the fourth dimension. The metric system adopted the second as the standard unit of measurement for time. Today, time is part of some electrical measurements to provide an understanding of how a circuit is operating and to determine possible problems. For example, voltage measurements together with time are used to define momentary power interruptions (0 V for 0.5 cycles to 3 seconds), temporary power interruptions (0 V for 3 seconds to 1 minute), and sustained power interruptions (0 V for more than 1 minute). **See Figure 1-10.**

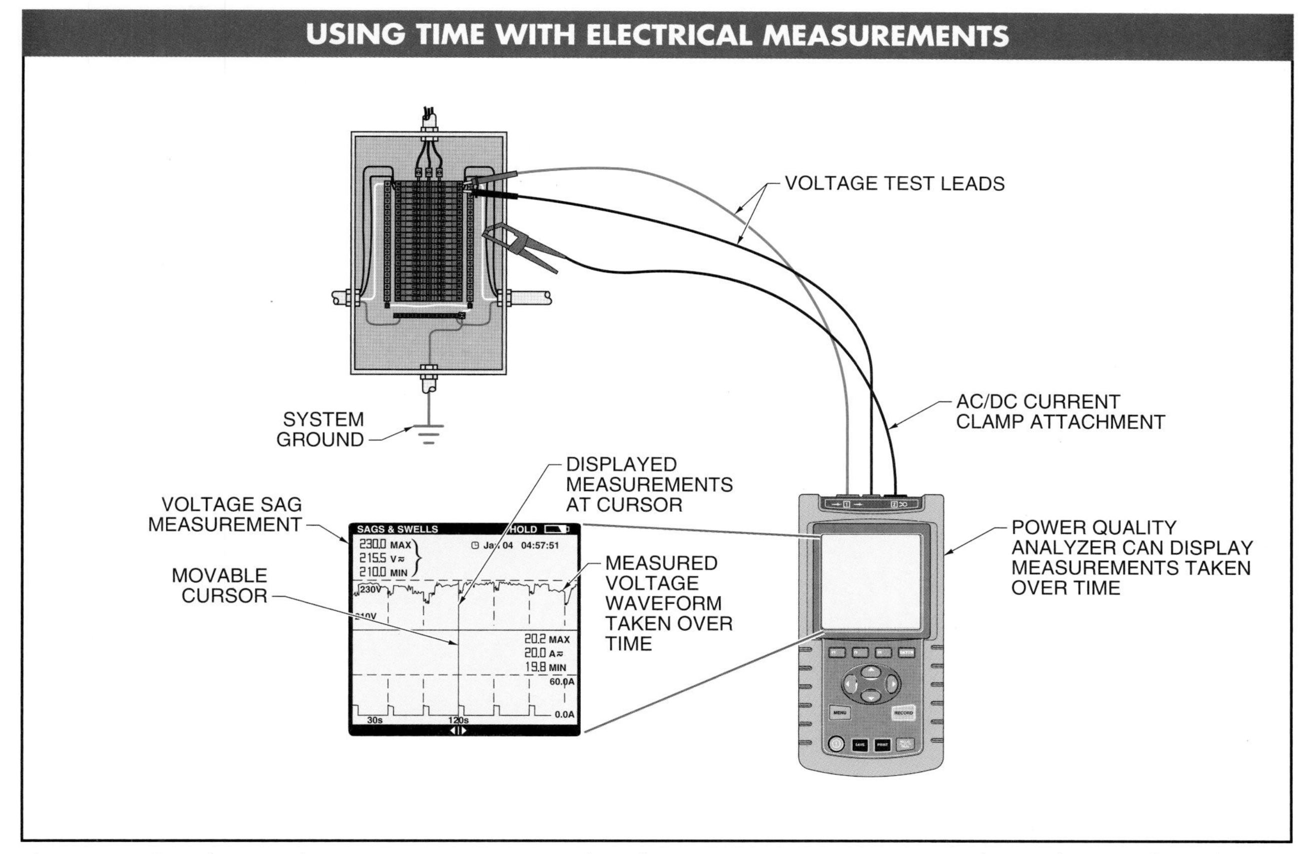

Figure 1-10. Voltage and current measurements can be taken over time using a power quality analyzer to show changes in the circuit.

Temperature

Measurements of temperature appeared late in history. Galileo is credited with the invention of the first thermometer, the "thermoscope," in 1593, but the thermoscope was not very accurate. An accurate thermometer was invented in 1641 using alcohol. Thermometers were continually improved until 1714 when Gabriel Daniel Fahrenheit built the mercury thermometer, which is used today. The centigrade (Celsius) scale came about in 1742 through the work of Andrew Celsius. Both thermometers have scales based on the freezing and boiling points of water. The Fahrenheit temperature scale uses 32° as the freezing point of water and 212° as the boiling point of water. The Celsius temperature scale uses 0° as the freezing point of water and 100° as the boiling point of water. **See Figure 1-11.**

Lord Kelvin of England developed another thermometer. This was a gas thermometer with its 0° mark set at absolute zero. The metric system uses the Kelvin scale as the standard for the measurement of temperature. *Absolute zero* is the lowest theoretical temperature that a substance can reach, which corresponds to –273.16° on the Celsius scale and –459.67° on the Fahrenheit scale. The Kelvin scale is based on the fact that a gas at 0°C loses 1⁄273 of its volume when the temperature of the gas is lowered to –1°C. In theory, the gas will disappear at –273.16°C, or at absolute zero. At 0 K, all molecules are at absolute rest with no motion, and a substance possesses no heat. In fact, all gases change to liquids and then to solids prior to reaching absolute zero. Helium 3 is used to get within a fraction of a degree of absolute zero before helium 3 solidifies. The lowest temperature ever achieved has been within a few millionths of a degree of absolute zero.

CELSIUS AND FAHRENHEIT TEMPERATURE SCALE EQUIVALENTS

°C	°F	°C	°F	°C	°F	°C	°F	°C	°F	°C	°F	°C	°F
10,000	18,032	500	932	140	284	72	161.6	41	105.8	10	50.0	–21	–5.8
9500	17,132	480	896	130	266	71	159.8	40	104.0	9	48.2	–22	–7.6
9000	16,232	460	860	120	248	70	158.0	39	102.2	8	46.4	–23	–9.4
8500	15,332	440	824	110	230	69	156.2	38	100.4	7	44.6	–24	–11.2
8000	14,432	420	788	100	212	68	154.4	37	98.6	6	42.8	–25	–13.0
7500	13,532	400	752	99	210.2	67	152.6	36	96.8	5	41.0	–26	–14.8
7000	12,632	390	734	98	208.4	66	150.8	35	95.0	4	39.2	–27	–16.6
6500	11,732	380	716	97	206.6	65	149.0	34	93.2	3	37.4	–28	–18.4
6000	10,832	370	698	95	203.0	64	147.2	33	91.4	2	35.6	–29	–20.2
5500	9932	360	680	94	201.2	63	145.4	32	89.6	1	33.8	–30	–22.0
5000	9032	350	662	93	199.4	62	143.6	31	87.8	0	32.0	–31	–23.8
4500	8132	340	644	92	197.6	61	141.8	30	86.0	–1	30.2	–32	–25.6
4000	7232	330	626	91	195.8	60	140.0	29	84.2	–2	28.4	–33	–27.4
3500	6332	320	608	90	194.0	59	138.2	28	82.4	–3	26.6	–34	–29.2
3000	5432	310	590	89	192.2	58	136.4	27	80.6	–4	24.8	–35	–31.0
2500	4532	300	572	88	190.4	57	134.6	26	78.8	–5	23.0	–36	–32.8
2000	3632	290	554	87	188.6	56	132.8	25	77.0	–6	21.2	–37	–34.6
1500	2732	280	536	86	186.8	55	131.0	24	75.2	–7	19.4	–38	–36.4
1000	1832	270	518	85	185.0	54	129.2	23	73.4	–8	17.6	–39	–38.2
950	1742	260	500	84	183.2	53	127.4	22	71.6	–9	15.8	–40	–40.0
900	1652	250	482	83	181.4	52	125.6	21	69.8	–10	14.0	–50	–58.0
850	1562	240	464	82	179.6	51	123.8	20	68.0	–11	12.2	–60	–76.0
800	1472	230	446	81	177.8	50	122.0	19	66.2	–12	10.4	–70	–94.0
750	1382	220	428	80	176.0	49	120.2	18	64.4	–13	8.6	–80	–112.0
700	1292	210	410	79	174.2	48	118.4	17	62.6	–14	6.8	–90	–130.0
650	1202	200	392	78	172.4	47	116.6	16	60.8	–15	3.2	–100	–148.0
600	1112	190	374	77	170.6	46	114.8	15	59.0	–16	3.2	–125	–193.0
580	1076	180	356	76	168.8	45	113.0	14	57.2	–17	1.4	–150	–238.0
560	1040	170	338	75	167.0	44	111.2	13	55.4	–18	–0.4	–200	–328.0
540	1004	160	320	74	165.2	43	109.4	12	53.6	–19	–2.2	–250	–418.0
520	968	150	302	73	163.4	42	107.6	11	51.8	–20	–4.0	–273	–459.4

Figure 1-11. Both the Celsius and Fahrenheit temperature scales base temperature measurements on the boiling point and freezing point of water.

At absolute zero, any volume of gas will disappear. Temperatures on an absolute zero scale are called "degrees absolute". The absolute zero scale in centigrade is often called the Kelvin scale, and degrees of temperature are kelvins (K). When using the Fahrenheit scale, absolute temperatures are degrees Rankine (°R). Absolute temperatures are determined on the Celsius scale by adding 273° to the reading. For example, at 0°C the absolute value is 273 K. For the Fahrenheit scale, add 460°. At 32°F, the freezing point of water, the absolute value is 492°R. Modern temperature measurements are required when troubleshooting or when performing preventive maintenance tasks. **See Figure 1-12.**

KELVIN AND RANKINE TEMPERATURE SCALES

	Kelvin	Rankine
Water boils	373	672
50°F	284	511
Water freezes	273	492
–75°F	214	385
Absolute	0	0

Figure 1-12. Both Kelvin and Rankine scales are absolute zero temperature scales.

SAFETY LABELS

A safety label is a label that indicates areas or tasks that can pose a hazard to personnel and/or equipment. Safety labels appear several ways on equipment and in equipment manuals. Safety labels use signal words to communicate the severity of a potential problem. The three most common signal words are danger, warning, and caution. **See Figure 1-13.**

Danger Signal Word

Danger is a signal word that indicates an imminently hazardous situation which, if not avoided, will result in death or serious injury. The information indicated by a danger signal word indicates the most extreme type of potential situation, and must be followed. The danger symbol is an exclamation mark enclosed in a triangle followed by the word danger written boldly in a red box.

Warning Signal Word

Warning is a signal word that indicates a potentially hazardous situation which, if not avoided, could result in death or serious injury. The information indicated by a warning signal word indicates a potentially hazardous situation and must be followed. The warning symbol is an exclamation mark enclosed in a triangle followed by the word warning written boldly in an orange box.

Caution Signal Word

Caution is a signal word that indicates a potentially hazardous situation which, if not avoided, may result in minor or moderate injury. The information indicated by a caution signal word indicates a potential situation that may cause a problem to people and/or equipment. A caution signal word also warns of problems due to unsafe work practices. The caution symbol is an exclamation mark enclosed in a triangle followed by the word caution written boldly in a yellow box.

Other signal words may also appear with the danger, warning, and caution signal words used by manufacturers. ANSI Z535.4, *Product Safety Signs and Labels*, provides additional information concerning safety labels. Additional signal words may be used alone or in combination on safety labels. Additional signal words used by electric motor drive manufacturers are electrical warning and explosion warning.

Technical Tip

Information on safety labels may be communicated using symbols or may be preceded by the word "warning" written in bold type.

SAFETY LABELS AND SIGNAL WORDS

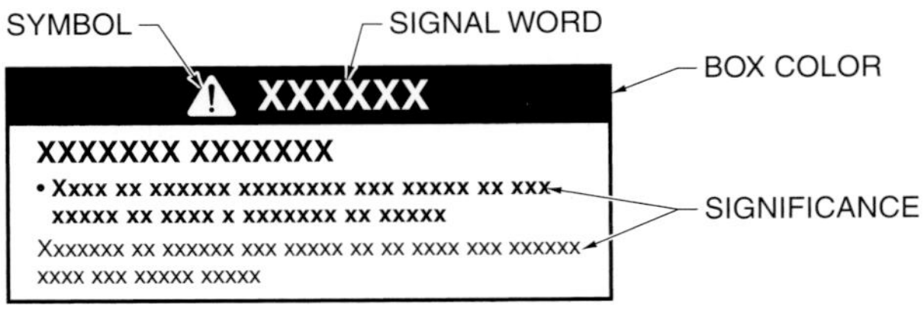

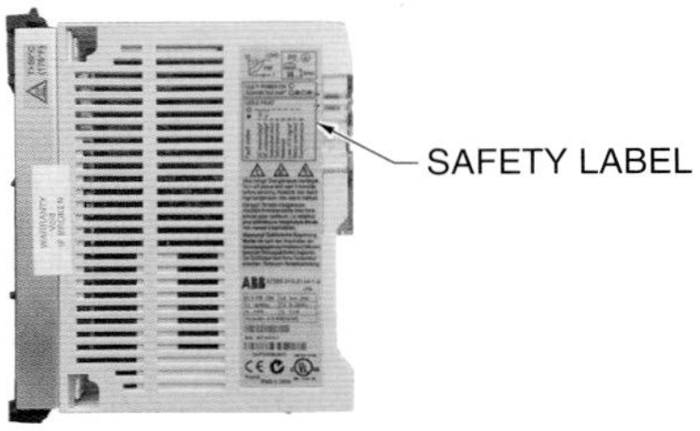

Safety Label	Box Color	Symbol	Significance
⚠ DANGER HAZARDOUS VOLTAGE • Ground equipment using screw provided. Electric motor drive panel must be properly grounded before applying power. • Do not use metallic conduits as a ground conductor. Failure to observe these precautions will cause shock or burn, resulting in severe personal injury or death!	RED		**DANGER** – INDICATES AN IMMINENTLY HAZARDOUS SITUATION WHICH, IF NOT AVOIDED, WILL RESULT IN DEATH OR SERIOUS INJURY
⚠ WARNING MEASUREMENT HAZARD When taking measurements inside the electric motor drive, make sure that only the test lead tips touch internal metal parts. Keep hands behind the protective finger guard provided on the test leads.	ORANGE		**WARNING** – INDICATES A POTENTIALLY HAZARDOUS SITUATION WHICH, IF NOT AVOIDED, COULD RESULT IN DEATH OR SERIOUS INJURY
⚠ CAUTION MOTOR OVERHEATING This drive controller does not provide direct thermal protection for the motor. Use of a thermal sensor in the motor may be required for protection at all speeds and loading conditions. Consult motor manufacturer for thermal capability of motor when operated over desired speed range. Failure to observe this precaution can result in equipment damage.	YELLOW		**CAUTION** – INDICATES A POTENTIALLY HAZARDOUS SITUATION WHICH, IF NOT AVOIDED, MAY RESULT IN MINOR OR MODERATE INJURY, OR DAMAGE TO EQUIPMENT. MAY ALSO BE USED TO ALERT AGAINST UNSAFE WORK PRACTICES
 WARNING Disconnect electrical supply before working on this equipment.	ORANGE		**ELECTRICAL WARNING** – INDICATES A HIGH VOLTAGE LOCATION AND CONDITIONS THAT COULD RESULT IN DEATH OR SERIOUS INJURY FROM AN ELECTRICAL SHOCK
 WARNING Do not operate the meter around explosive gas, vapor, or dust.	ORANGE		**EXPLOSION WARNING** – INDICATES LOCATION AND CONDITIONS WHERE EXPLODING ELECTRICAL PARTS MAY CAUSE DEATH OR SERIOUS INJURY

Figure 1-13. Safety labels are used to identify situations for different degrees of possible injury to personnel or death.

Electrical Warning

An *electrical warning* is a warning used to indicate a high voltage location and conditions that could result in death or serious personal injury from an electrical shock if proper precautions are not taken. The electrical warning safety label is usually placed where there is a potential for coming in contact with live electrical wires, terminals, or parts. The electrical warning symbol is a lightning bolt enclosed in a triangle. The safety label may be shown with no words or may be preceded by the word warning written boldly.

Explosion Warning

An *explosion warning* is a warning used to indicate locations and conditions where exploding parts may cause death or serious personal injury if proper precautions and procedures are not followed. The explosion warning symbol is an explosion enclosed in a triangle. The safety label may be shown with no words or may be preceded by the word warning written boldly.

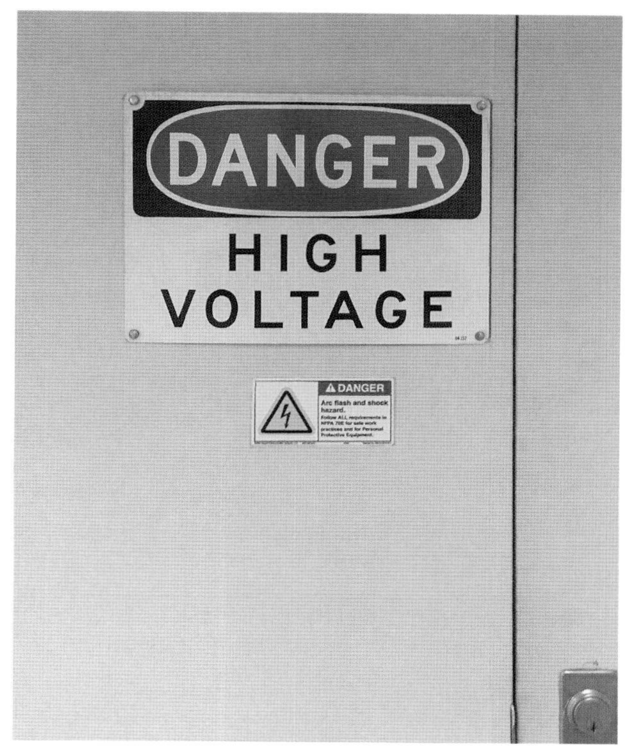

Danger signal words indicate hazardous conditions which, if not avoided, can result in serious injury or death.

SAFETY AND TEST INSTRUMENT PROCEDURES

Proper equipment and procedures are required when taking measurements with test instruments. To take and interpret a displayed measurement, the test instrument must be properly set to the correct measuring position and properly connected into the circuit to be tested, and the measurement read. There is always the possibility that the test instrument or meter will not be properly set to the correct function or to the correct range, and/or misread. Test instruments not properly connected to a circuit increase the likelihood of an improper measurement or meter reading, or creating an unsafe condition.

Function Switch and Selector Switch Settings

Ensure that the function switch of a meter matches the desired measurement and the connection of the test leads to the circuit. Before taking a measurement, a meter must be set to the correct function switch position for the measurement to be taken. For example, the function switch of a meter must be set to the correct voltage position (VAC, VDC, or mVDC) when measuring voltage, and the correct current position (A or mA) when measuring current. Some meters may require that the function switch be set to voltage or current, and a selector switch is used to select the desired range. The test leads of the test instrument or meter must also be moved to the proper jack positions to take a voltage measurement or a current measurement. **See Figure 1-14.**

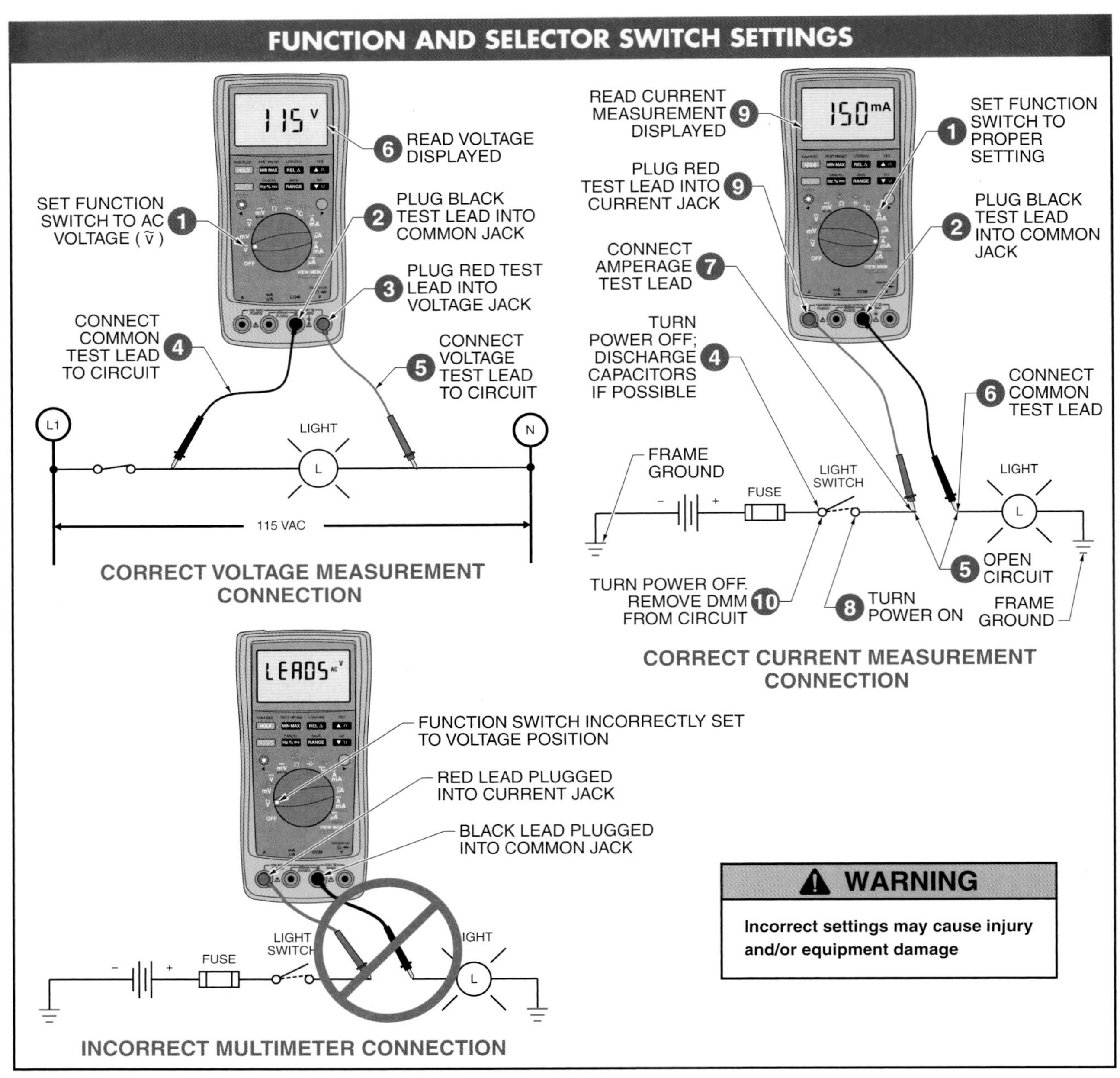

Figure 1-14. A voltage or current measurement requires the proper function switch position with test leads moved to the jacks required for that measurement.

Always ensure that the test leads are in the jacks that correspond to the setting of the meter function and selector switch before taking a measurement. The importance of verifying that test lead jack position corresponds to the settings of the meter before taking any measurement cannot be overemphasized. When the test leads of a test instrument or meter are connected to the amperage jacks (A, mA, or µA), the resistance of the meter is very low (typically less than 1 Ω), and when connected to an energized circuit for voltage measurement, the test instrument or meter creates a short circuit.

Connecting the test leads of a test instrument or meter to the amperage jacks and setting the function and selector switches to a voltage measurement position for measurement is the main reason meter fuses blow (when the meter has fuses) and the end of the test leads become burnt. When a test instrument or meter does not have fused amperage jacks, the meter circuits become the fuse and personal injury can result. In addition to fusing the current jacks, some test instruments and meters also include an audible warning to help prevent improper jack usage for the meter setting. Some meters emit a sound (constant audible warning) when the test leads are connected in the current jacks and a noncurrent mode is selected. Audible warnings also help protect against the common mistake of leaving the test leads plugged into the current input jacks and attempting a voltage measurement. Prevent the nuisance of blowing DMM fuses due to operator error and reduce the risk of damage to the DMM and equipment.

Meter Fuses

Never assume that the fuses of a test instrument or meter are good. Most test instruments and meters have fuses that are part of the current-measuring circuits. The fuses prevent possible injury and/or meter damage when the meter is not properly connected to a circuit. Meter fuses must be tested prior to taking any current measurements. A blown (open) fuse prevents the measurement of any circuit current, even when current is present in the circuit. A dangerous situation is created in which the person taking the measurement believes that there is no current in the circuit (or that the circuit is not energized) because the meter is not displaying a current measurement. Most test instrument and meter manufacturers specify procedures for testing fuses. When a fuse-testing procedure is not provided, the fuses must be removed and tested with a separate meter or the fuses can be tested by using the meter on a low-current (few milliamps) test circuit known to be operating. A typical fuse-testing procedure provided by the manufacturer lists basic steps. **See Figure 1-15.** To test the fuses of a test instrument or meter, apply the following general procedures:

1. Set the function switch of the test instrument or meter to resistance.
2. Plug the red test lead into the voltage jack of the meter.
3. Touch the tip of the red test lead to the high current terminal (A) of the test instrument or meter. Move the test lead tip of the test lead probe around until a reading is obtained. A low resistance reading should be displayed. Consult the user's manual for exactly what the reading should be. When the meter reads OL, replace the fuse and retest.
4. Touch the red test lead tip to the DMM low current terminal (mA/μA). Move the test lead tip within the terminal until a reading is obtained. A resistance reading should be displayed (and usually will be higher than the "A" jack reading). Consult the user's manual for exactly what the reading should be. If the DMM reads OL, replace the fuse and retest.
5. Turn the meter OFF and remove the test lead.

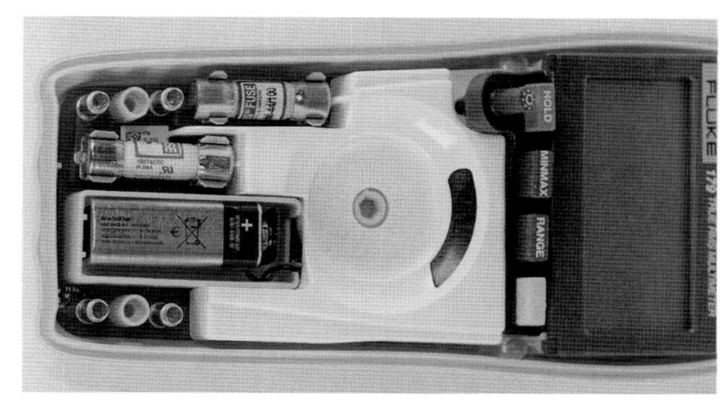

Replacement fuses used in test instruments must be the type specified by the manufacturer for safe operation of the test instrument.

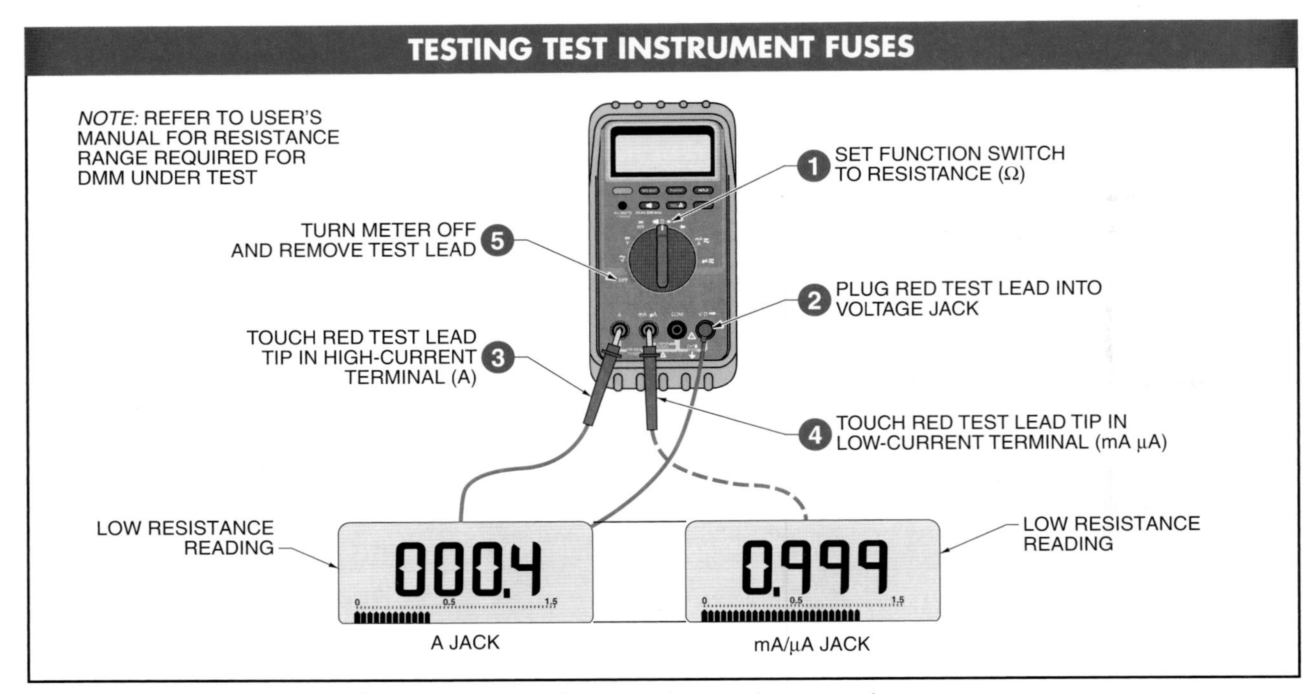

Figure 1-15. For maximum safety, test instrument fuses must be tested prior to taking any current measurements.

Meter Prefixes and Symbols

Understanding meter prefixes and meter symbols used as part of the reading displayed on a test instrument or meter is a must. Test instruments use prefixes and symbols when displaying measurements. There are big differences among displayed measurements of 12 VDC, 12 VAC, 12 mVDC, 0.12 kVDC, 0.12 VAC, –12 VDC, or –12 mVDC. Always consider every prefix and symbol included in the display when interpreting a measurement. **See Figure 1-16.**

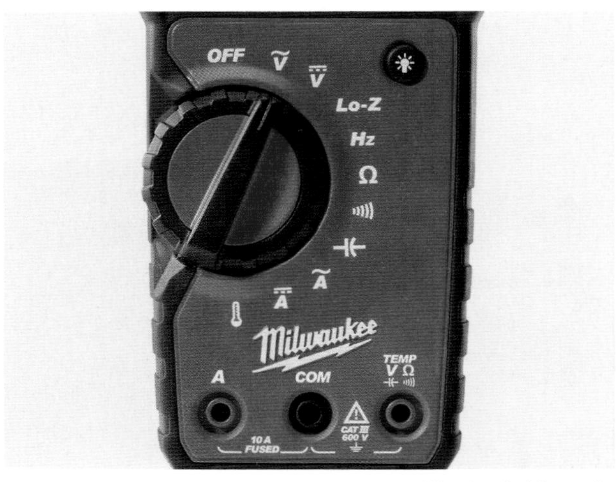

Milwaukee Tool Corporation

Abbreviations and symbols identify the types of measurements selectable with multimeter function switches.

Proper Test Instrument Applications

Choose the correct test instrument for the application. Any test instrument is designed for taking measurements on some type of application, but not for all applications. Using the proper meter in an application ensures that an accurate measurement is taken. Using an incorrect meter in an application may mean that the displayed measurement is not correct, and can be over 100% in error. For example, only a meter marked "TRUE RMS" can be used when taking a measurement on a circuit that includes solid-state components that distort the shape of AC sine waves (motor drives, PLCs, computers, and printers). **See Figure 1-17.**

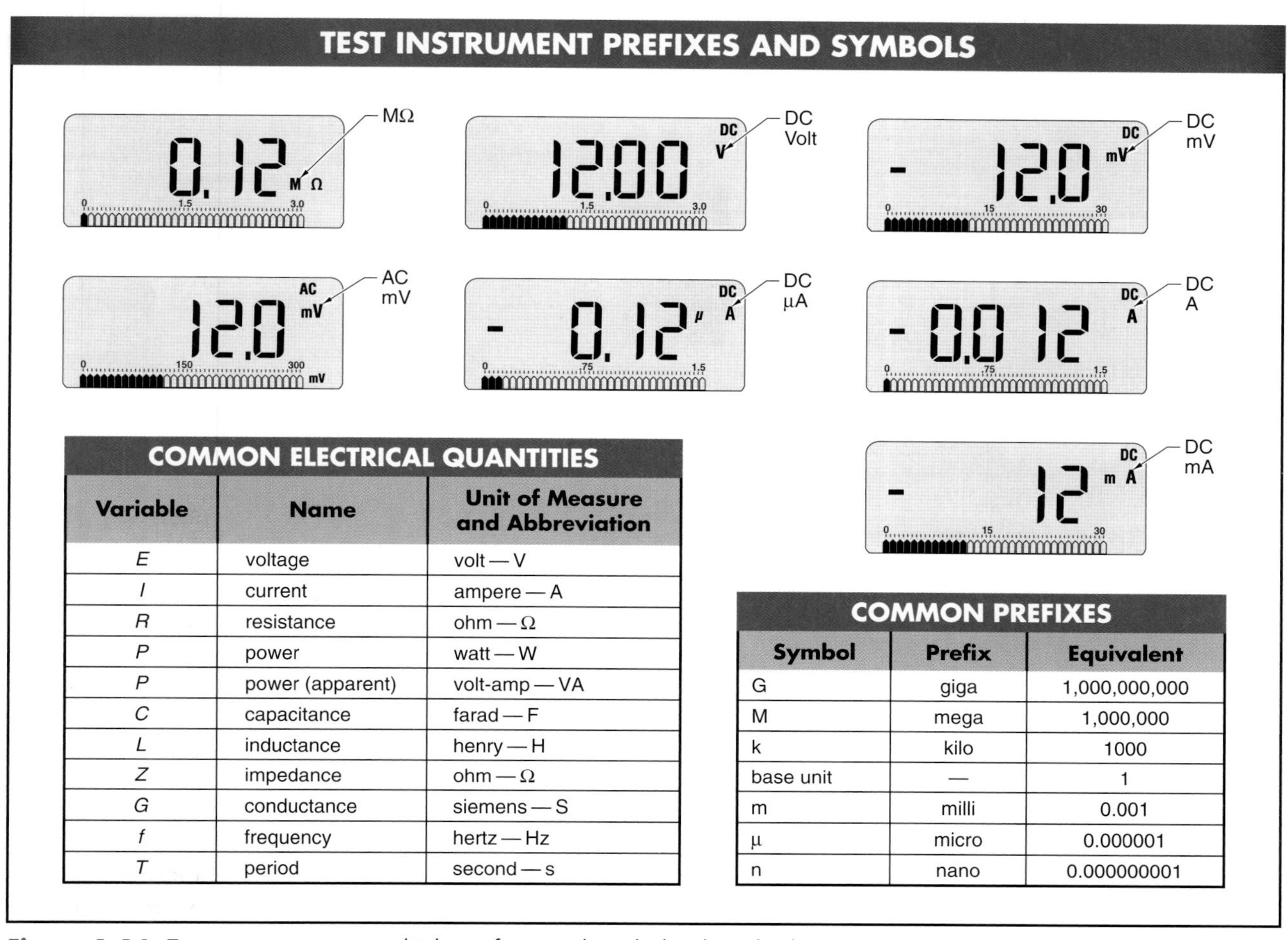

COMMON ELECTRICAL QUANTITIES

Variable	Name	Unit of Measure and Abbreviation
E	voltage	volt — V
I	current	ampere — A
R	resistance	ohm — Ω
P	power	watt — W
P	power (apparent)	volt-amp — VA
C	capacitance	farad — F
L	inductance	henry — H
Z	impedance	ohm — Ω
G	conductance	siemens — S
f	frequency	hertz — Hz
T	period	second — s

COMMON PREFIXES

Symbol	Prefix	Equivalent
G	giga	1,000,000,000
M	mega	1,000,000
k	kilo	1000
base unit	—	1
m	milli	0.001
µ	micro	0.000001
n	nano	0.000000001

Figure 1-16. Test instruments use multiple prefixes and symbols when displaying a measurement.

Test Instrument Limits

Understanding the range limits (specifications) of a test instrument is critical to acquiring accurate measurements. All test instruments and meters have specified measuring limits and ranges. For example, a meter that includes a frequency-measuring position (selection) can be used to measure the frequency (Hz) of an AC voltage (or current) measurement. However, in some applications such as AC drives, the frequency measurement range of a meter may not be accurate or even make sense because the meter cannot accurately measure the power line frequencies that the motor is receiving (0 Hz to about 100 Hz) because all AC drives produce a carrier frequency on the motor power lines. *Carrier frequency* is the frequency that controls the number of times the solid-state switches in the inverter section of a pulse width modulated (PWM) variable frequency drive turn ON and turn OFF per second. The carrier frequency is typically between 2 kHz and 20 kHz. **See Figure 1-18.** Only a meter designed for taking frequency measurements on variable frequency drives can be used to take an accurate frequency measurement at the motor (or out of the drive). The meter is designed to distinguish the difference between the power frequency and the carrier frequency and will only display the power frequency. Always verify the measuring limits of a meter and the meter type when a measurement does not appear correct. Verify that the proper meter is being used for each application at all times.

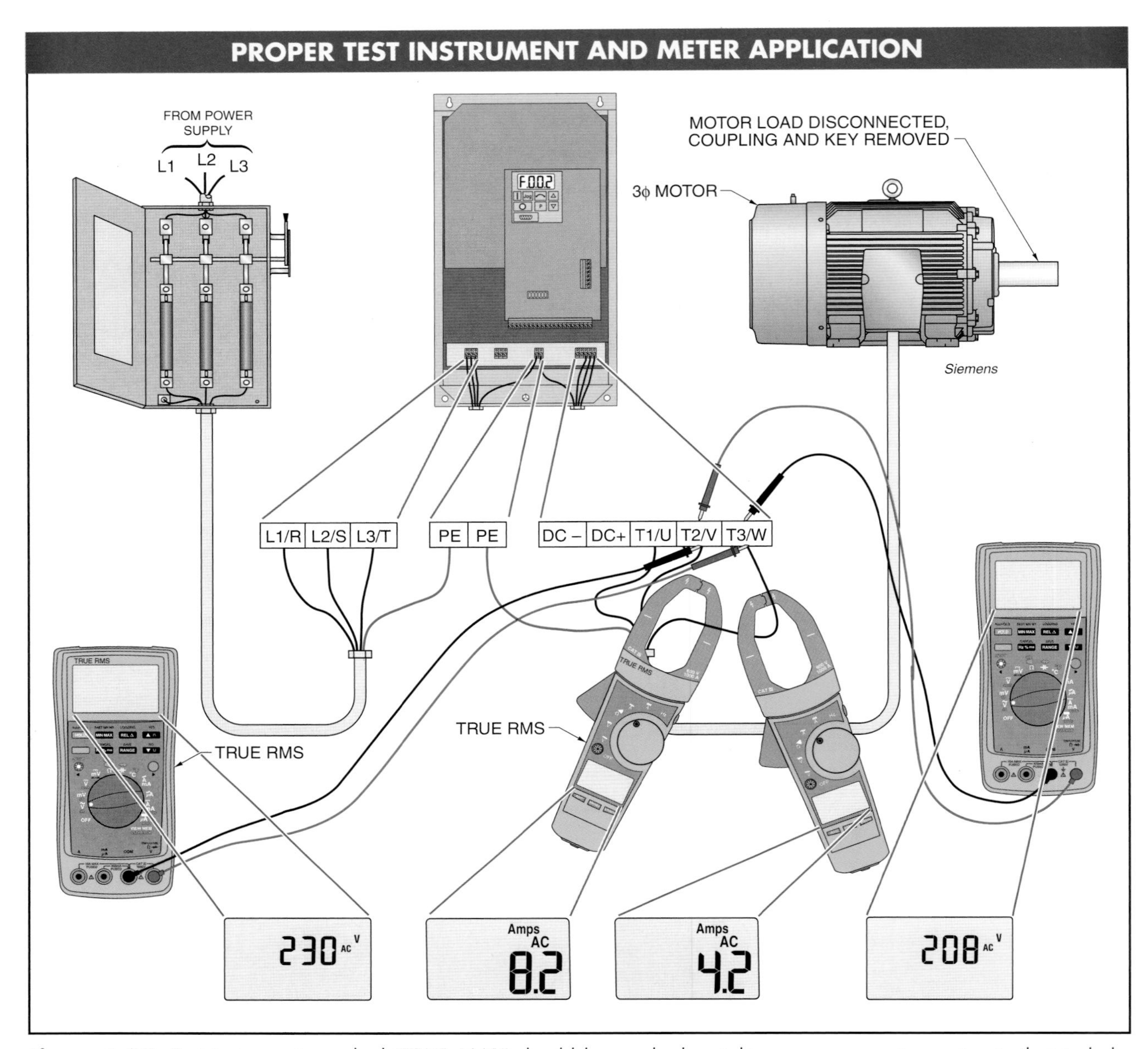

Figure 1-17. Test instruments marked "TRUE RMS" should be used when taking measurements on circuits that include solid-state components.

Analog versus Digital Test Instruments

Both analog and digital test instruments and meters are available for taking basic electrical measurements such as voltage, current, and resistance and other nonelectrical measurements such as temperature and pressure. Analog test instruments and meters were in use for years before digital meters were invented. However, the use of analog test instruments and meters is declining as manufacturers produce digital meters. The main reason for the increase in the use of digital electronic technology is that digital meters have far greater measuring capabilities than analog meters, in a much smaller space.

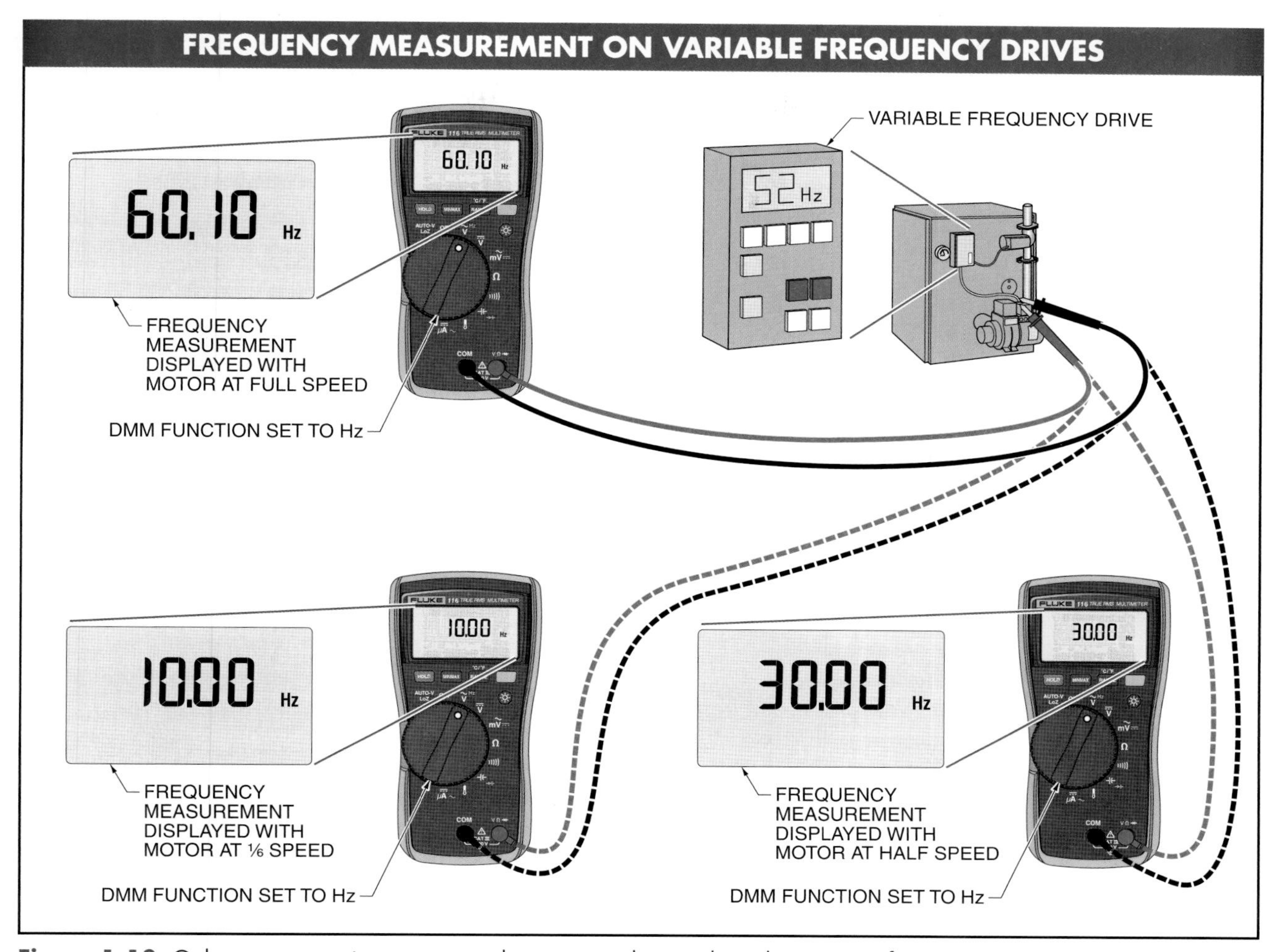

Figure 1-18. Only true rms test instruments and meters are designed to take accurate frequency measurements on variable frequency drives.

Analog and Digital Test Instrument Advantages and Disadvantages

There are advantages and disadvantages to both analog and digital test instruments and meters. In some cases the function of a test instrument or meter can be an advantage or disadvantage, depending upon the application an electrician is using the meter with. For example, most digital multimeters have an "autoranging" capability. Autoranging allows a meter to automatically adjust to a higher range setting if the set range is not high enough for the application. The advantage of autoranging is that only the measuring function must be selected (voltage, current, or resistance) and the meter will automatically select the best range (400 mV, 4 V, 40 V, 400 μA, or 4000 mA range). The disadvantage of autoranging is that when a meter is set to a selected measurement such as VAC, the meter can display a measurement before being connected to the circuit (ghost voltage). **See Figure 1-19.** The test instrument or meter is programmed to search for any voltage, no matter how small, and therefore the meter automatically selects the most sensitive mV range (400 mV). The displayed millivolt value may be misinterpreted as an actual voltage measurement. Ghost voltages are also to blame when a meter is connected to a circuit that has no power and the meter displays readings such as 108 mV, 213 mV, or 455 mV. The display of ghost voltages can be eliminated on some meters by selecting the manual range (RANGE button) mode on the meter. The meter requires that range be manually set to a higher setting, such as the 400 V range or the 4000 V range.

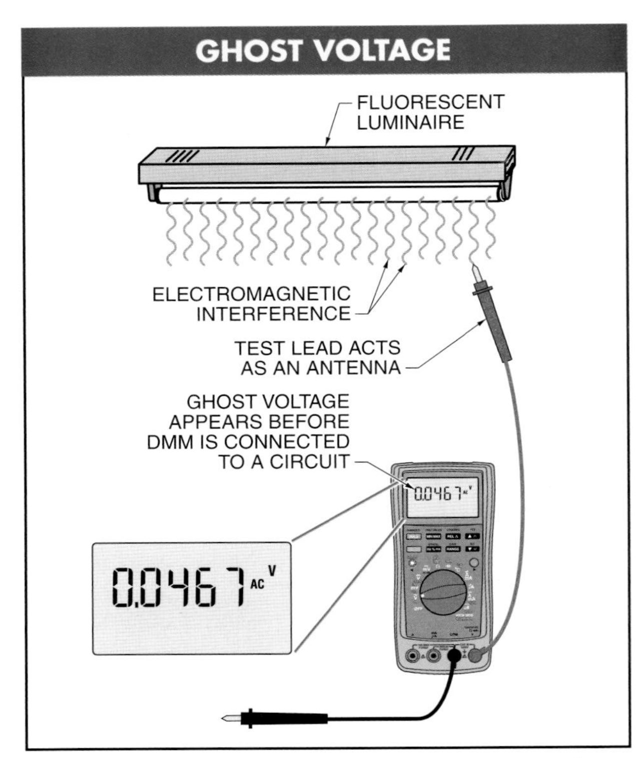

Figure 1-19. Digital test instruments display ghost voltages caused by electromagnetic interference.

Analog test instruments and meters have other advantages over digital meters such as the following:

- Analog test instruments and meters are less susceptible to noise (magnetic field coupling) and thus do not display ghost readings as easily as digital meters.
- Analog test instruments and meters have a lower input impedance (resistance), which reduces false readings from induced ghost voltages.
- Analog test instruments and meters have been around much longer and some electricians are more comfortable with what has worked in the past.

Analog test instruments and meters also have disadvantages compared to digital meters such as the following:

- Analog test instrument and meter scales can be misread, especially when a meter has multiple settings (1 V, 10 V, 25 V, 50 V, 250 V) used with the same display scale.
- Analog test instruments and meters are less accurate (for example, typical voltage measurements are within 1% to 5%, depending on meter specifications).
- Analog test instruments and meters have lower input impedance (resistance), which can load down sensitive circuits when taking measurements.
- Analog test instruments and meters offer only basic protection using fuses. There is no protection such as an audible input alert (available with digital meters) that warns when a selector switch position does not match the position of the test leads (for example, the meter is set to measure voltage, but the test leads are plugged into the amperage jacks).
- Analog test instruments and meters cannot offer features that require a digital electronic circuit, such as autoranging, MIN MAX function, Recording function, Measurement Hold function, or Relative function.

Digital test instruments and meters have advantages over analog meters such as the following:

- Digital test instruments and meters display measurements as numerical numbers, not a scale position, and are less likely to be misread unless the prefixes and symbols accompanying numerical values are misapplied.
- Most digital test instruments and meters are autoranging. Once a measuring function such as VAC is selected, the meter will automatically select the best meter range for taking the measurement (400 mV, 4 V, 40 V, 400 V, or 4000 V).
- Digital test instruments and meters are more accurate (typical voltage measurements are within 0.01% to 1.5%, depending on meter specifications).

Technical Tip

In some situations, digital test instruments are more forgiving of operator error than analog test instruments. When a digital meter shows a negative value, typically the probes are backwards. Disconnect the probes from the device being tested and reconnect in reverse. Probes connected backwards on analog test instruments may show no reading and may even damage the movements of the instrument.

- Digital test instruments and meters have a high input impedance (resistance), which will not load down sensitive circuits when taking a measurement.
- Digital test instruments and meters offer special functions such as Autoranging, MIN MAX function, Recording function, Measurement Hold function, and Relative function.
- Some digital test instruments and meters come with downloading capabilities, which when used with software allow for measurement data recording and manipulation.

Digital test instruments and meters also have disadvantages compared to analog meters such as the following:

- Digital test instruments and meters can display ghost readings (mV or V readings) when the meter is in the automatic range setting mode.
- Some digital test instrument and meter models offer so many features that the meters can be intimidating to use before the measurement functions are all understood and used a few times.
- Digital test instruments and meters require time (seconds) to acquire a reading when the source is changing.

Greenlee Textron, Inc.

Digital test instruments can provide greater accuracy than analog test instruments.

Prior to using electrical test equipment, refer to the user's manual for proper operating procedures, safety precautions, and limits.

ELECTRICAL MEASUREMENT SAFETY

Electrical measurements are taken with electrical test instruments designed for specific tasks. Test instruments such as voltage testers, analog and digital multimeters, clamp-on ammeters, and power quality meters measure electrical quantities such as voltage, current, power, and frequency. Each meter has specific features and limits. The user's manual details specifications and features, proper operating procedures, safety precautions, warnings, and allowed applications.

Conditions can change quickly as voltage and current levels vary in individual circuits. General safety precautions required when using test instruments include the following:

- When a circuit does not have to be energized (as when taking a resistance measurement or checking diodes and capacitors in a circuit), lockout and tagout all equipment and circuits to be tested.
- Never assume a test instrument is operating correctly. Check the test instrument that will be measuring voltage on a known (energized) voltage source before taking a measurement on an unknown voltage source. After taking a measurement on an unknown voltage source, retest the test instrument on a known source to verify the meter still operates properly.
- Always assume that equipment and circuits are energized until positively identified as de-energized by taking proper measurements.

- Never work alone when working on or near exposed energized circuits that may cause an electrical shock.
- Always wear personal protective equipment (safety glasses, insulating gloves, cover gloves, arc flash protection) appropriate for the test area. **See Figure 1-20.**

PROCEDURAL SAFETY

⚠ WARNING

- Follow all electrical safety practices and procedures
- Check and wear personal protective equipment (PPE) for the procedure being performed
- Perform only authorized procedures
- Follow all manufacturer recommendations and procedures

Figure 1-20. The four warning procedures used in procedural lists are typically used to encompass many aspects of safety for an electrician.

- Ensure that the function switch of a test instrument or meter is set to the proper range and function before applying test leads to the circuit. A test instrument set to the wrong function can be damaged. For example, damage can occur when test leads contact an AC power source while the meter is set to measure resistance or continuity.
- Ensure that the test leads of the test instrument or meter are connected properly. Test leads that are not connected to the correct jacks can be dangerous. For example, attempting to measure voltage while the test leads are in the amperage jack produces a short circuit. For additional safety, use a meter that is self-protected with a high-energy fuse. Follow the operating instructions in the user's manual or the directions on the graphic display for proper test lead connections and operating information. **See Figure 1-21.**
- When using a test instrument or meter, ensure that the function switch is set to the proper range and function before connecting test leads to a circuit. When using a graphic display meter, verify that the proper screen (voltage or current measurement and power measurement) is selected.
- Start with the highest range when measuring unknown values. Using a range that is too low can damage the meter. Attempting a voltage or current measurement above the rated limit can be dangerous.
- Connect the ground test lead (black) first, the voltage test lead (red) next. Disconnect the voltage test lead (red) first, and the ground test lead (black) next after taking measurements.
- Whenever possible, connect test leads to the output side (load side) of a circuit breaker or fuse to provide better short circuit protection.
- Never assume that a circuit is de-energized or equipment is fully discharged. Capacitors can hold a charge for a long time — several minutes or more. Always check for the presence of voltage before taking any other measurements.
- Check test leads for frayed or broken insulation. Electrical shock can occur from accidental contact with live components. Electrical test equipment should have double-insulated test leads, recessed input jacks on the meter, shrouds on the test lead plugs, and finger guards on test probes.
- Use meters that conform to the IEC 1010 category in which they will be used. For example, to measure 480 V in an electrical distribution feeder panel, a meter rated at CAT III-600 V or CAT III-1000 V is used.
- Avoid taking measurements in humid or damp environments.
- Ensure that no atmospheric hazard such as flammable dust or vapor is present in the area.
- Use one hand when working on a live circuit to reduce the chance of an electrical shock passing through the heart and lungs.

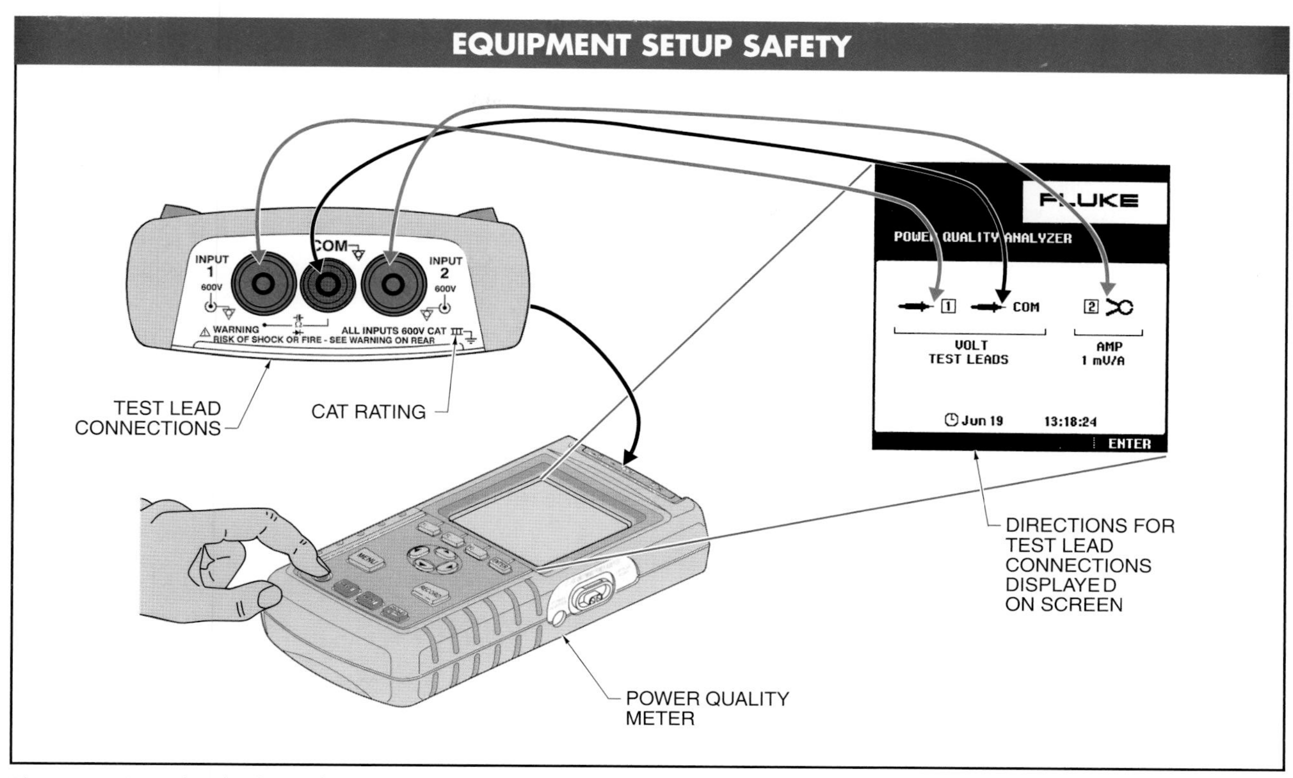

Figure 1-21. The displays of some test instruments prompt the user for correct connection of test leads and probes.

PERSONAL PROTECTIVE EQUIPMENT

Personal protective equipment (PPE) is clothing and/or equipment worn by a technician to reduce the possibility of injury in the work area. The use of personal protective equipment is required whenever work may occur on or near energized exposed electrical circuits. The National Fire Protection Association standard *NFPA 70E, Standard for Electrical Safety in the Workplace*, addresses "electrical safety requirements for employee workplaces that are necessary for the safeguarding of employees in pursuit of gainful employment."

NFPA 70E

For maximum safety, personal protective equipment and safety requirements for test instrument procedures must be followed as specified in NFPA 70E, OSHA Standard Part 1910 *Subpart I—Personal Protective Equipment* (1910.132 through 1910.138), and other applicable safety mandates. Personal protective equipment includes protective clothing, head protection, eye protection, ear protection, hand and foot protection, back protection, knee protection, and rubber insulated matting. **See Figure 1-22.**

Electricians performing test instrument procedures must follow all 70E requirements for personal protective clothing, test instruments, and equipment for maximum safety.

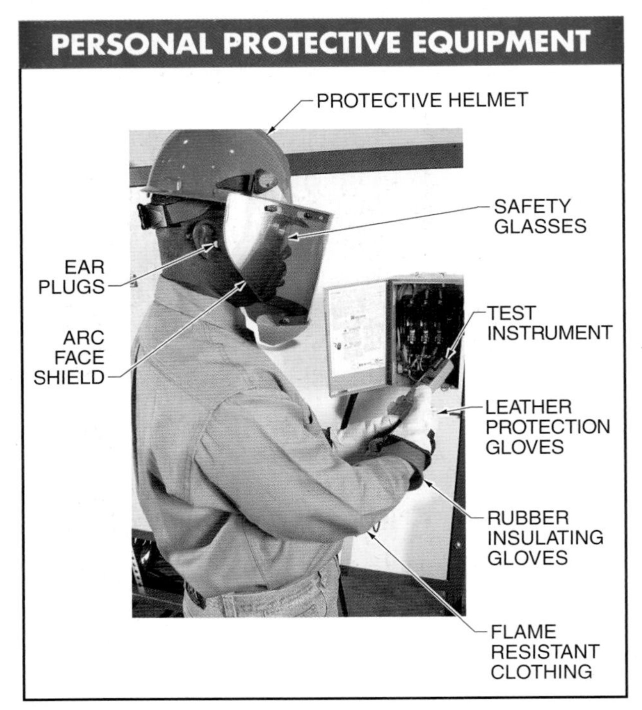

Figure 1-22. Personal protective equipment includes protective clothing, head protection, eye protection, ear protection, hand and foot protection, back protection, knee protection, and rubber insulated matting.

Rubber Insulating Gloves

Rubber insulating gloves are an important article of personal protective equipment for electrical workers. Safety requirements for the usage of rubber insulating gloves and leather protectors must be followed at all times. The primary purpose of rubber insulating gloves and leather protectors is to insulate hands and lower arms from possible contact with live conductors. Rubber insulating gloves offer a high resistance to current flow to help prevent an electrical shock and the leather protectors protect the rubber glove and add additional insulation. Rubber insulating gloves are rated and labeled for maximum voltage allowed.

WARNING

Rubber insulating gloves are designed for specific applications. Leather protectors are required for protecting rubber insulating gloves. Rubber insulating gloves offer the highest resistance and greatest insulation. Serious injury or death can result from improper use of rubber insulating gloves, or from using outdated and/or the wrong type of rubber insulating gloves for an application.

Per NFPA 70E, "Only qualified persons shall perform testing on or near live parts operating at 50 V or more." All personal protective equipment and tools are selected to be appropriate to the operating voltage (or higher) of the equipment or circuits being worked on or near. Equipment, devices, tools, and test instruments and meters must be suited for the work to be performed. In most cases, voltage-rated gloves and tools are required. Voltage-rated gloves and tools are rated and tested for the maximum line-to-line voltage upon which work will be performed. Protective gloves must be inspected or tested as required for maximum safety before each task.

Technical Tip

Always consider the worst possible operating conditions when selecting personal protective equipment.

Inspection of Rubber Insulating Gloves. Rubber insulating gloves must be field tested (by visual inspection and air test) prior to each use. Rubber insulating gloves must also be laboratory tested by an approved laboratory every six months. Rubber insulating gloves have a color-coded label for applicable voltage ratings. **See Figure 1-23.** Gloves must also be air-tested when there is cause to suspect any damage. The entire surface must be inspected by rolling the cuff tightly toward the palm in such a manner that air is trapped inside the glove, or by using a mechanical inflation device. When using a mechanical inflation device, care must be taken to avoid overinflation. The glove is examined for punctures and other defects. Puncture detection may be enhanced by listening for escaping air. Some brands of rubber insulating gloves are available with two color layers. When one color layer becomes visible, the color serves as notification to the user that the gloves must be replaced.

RUBBER INSULATING GLOVE RATING*		
Class	Maximum Use	Label Color
00	500 V	Beige
0	1 kV (1000 V)	Red
1	7.5 kV (7500 V)	White
2	17 kV (17,000 V)	Yellow
3	26.5 kV (26,500 V)	Green
4	36 kV (36,000 V)	Orange

* Refer to ASTM D 120-02 Standard Specification for Rubber Insulating Gloves.

Figure 1-23. Rubber insulating gloves have a color-coded label representing voltage ratings for proper application.

Visual inspection of rubber insulating gloves is performed by stretching a small area (particularly fingertips) and checking for defects such as the following:

- punctures or pin holes
- embedded or foreign material
- deep scratches or cracks
- cuts or snags
- deterioration caused by oil, heat, grease, insulating compounds, or any other substances which may harm rubber

When visual inspection is completed, an air test is performed using the following procedures:

- Grasp the gauntlet of the glove between thumb and forefinger.
- Spin the glove around rapidly to fill with air.
- Trap the air by squeezing the gauntlet with one hand, while using the other hand to squeeze the palm, fingers, and thumb while looking for defects.
- Hold the glove to your face or ear to detect escaping air. Gloves failing the air test must be tagged "unsafe" and returned to a supervisor, or properly disposed of.

Proper care of leather protector gloves is essential for user safety. Leather protector gloves are checked for cuts, tears, holes, abrasions, defective or worn stitching, oil contamination, and any other condition that might prevent the glove from adequately protecting rubber gloves. Any substance that can physically damage rubber gloves must be removed before testing.

Insulating gloves and protector gloves found to be defective shall not be discarded or destroyed in the field, but shall be tagged "unsafe" and returned to a supervisor.

Flame-Resistant (FR) Clothing

Sparks from an electrical circuit can cause a fire. Approved flame-resistant (FR) clothing must be worn in conjunction with rubber insulating gloves for protection from electrical arcs when performing certain operations on or near energized equipment or circuits. FR clothing must be kept as clean and sanitary as practical and must be inspected prior to each use. Defective clothing must be removed from service immediately and replaced. Defective FR clothing must be tagged "unsafe" and returned to a supervisor.

Eye Protection

Eye protection must be worn to prevent eye or face injuries caused by flying particles, contact arcing, and radiant energy. Eye protection must comply with OSHA 29 CFR 1910.133, *Eye and Face Protection*. Eye protection standards are specified in ANSI Z87.1, *Occupational and Educational Eye and Face Protection*. Eye protection includes safety glasses, face shields, and arc blast hoods. **See Figure 1-24.** *Safety glasses* are an eye protection device with special impact-resistant glass or plastic lenses, reinforced frames, and side shields. Plastic frames are designed to keep the lenses secured in the frame when an impact occurs to minimize the shock hazard when working with electrical equipment. Side shields provide additional protection from flying objects. Tinted-lens safety glasses protect against low voltage arc hazards.

Technical Tip

NFPA codes, standards, recommended practices, and guides, such as 70E, were developed through a consensus of standards and approved by the American National Standards Institute. 70E Standard for Electrical Safety in the Workplace consists of: Safety-Related Work Practices, Safety Related Maintenance Requirements, Safety Requirements for Special Equipment, and Installation Safety Requirements.

A *face shield* is any eye and face protection device that covers the entire face with a plastic shield, and is used for protection from flying objects. Tinted face shields protect against low voltage arc hazards. Goggles are an eye protection device with a flexible frame that is secured on the face with an elastic headband. Goggles fit snugly against the face to seal the areas around the eyes, and may be used over prescription glasses. Goggles with clear lenses protect against small flying particles or splashing liquids. Tinted goggles are sometimes used to protect against low-voltage arc hazards.

Safety glasses, face shields, and goggle lenses must be properly maintained to provide protection and clear visibility. Lens cleaners are available that clean without risk of lens damage. Pitted, scratched, and crazed lenses (crazing is a defect caused by exposure to aggressive solvents, chemicals, or heat which leaves microscopic cracks within the lenses) reduce vision and may cause lenses to fail on impact.

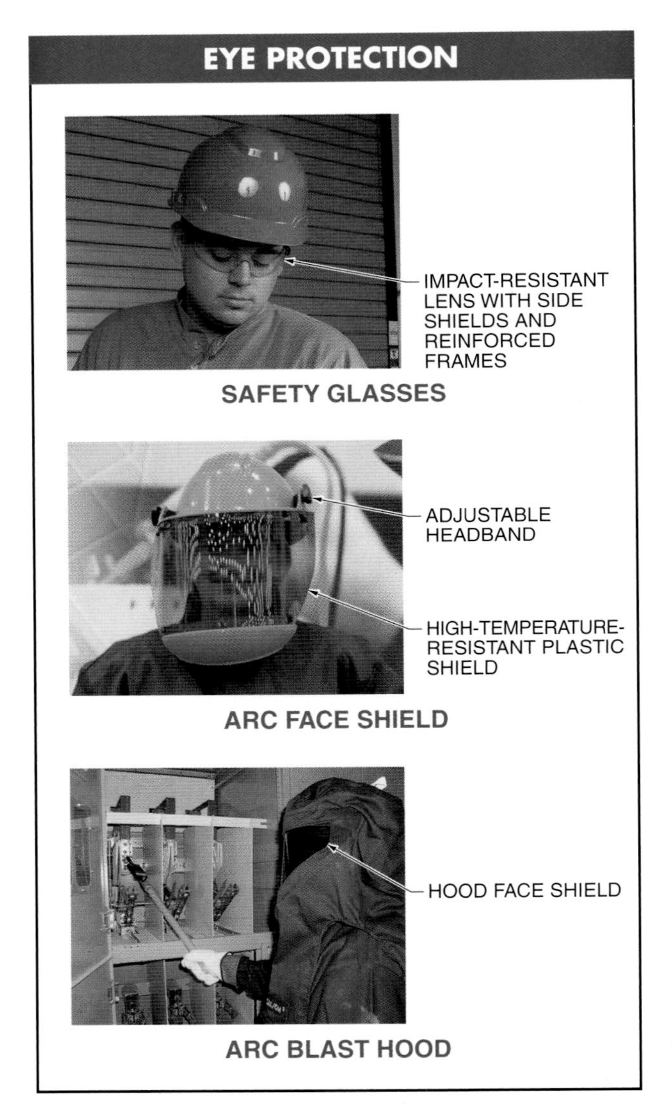

Figure 1-24. Eye protection must be worn to prevent eye or face injuries caused by flying particles, contact arcing, or radiant energy.

TEST INSTRUMENT SAFETY STANDARDS

The *International Electrotechnical Commission (IEC)* is an organization that develops international safety standards for electrical equipment. The IEC standards reduce safety hazards that can occur from unpredictable circumstances when using electrical test equipment such as test instruments and meters. For example, voltage surges on a power distribution system can cause a safety hazard when a test instrument is being used in an electrical system.

A *voltage surge* is a higher-than-normal voltage that temporarily exists on one or more power lines. Voltage surges vary in voltage amount and time present on the power lines. One type of voltage surge is a transient voltage. A *transient voltage (voltage spike)* is a temporary, undesirable voltage in an electrical circuit. Transient voltages typically exist for a very short time, but are large in magnitude and very erratic. Transient voltages occur due to lightning strikes, unfiltered electrical equipment, and power being switched ON and OFF.

Troubleshooting Tip

SAFETY PROCEDURES

Failure to properly control hazardous energy sources during maintenance and testing procedures accounts for an average of 217 nonfatal injuries per year.

High transient voltages can reach several thousand volts. A transient voltage on a 120 V power line can reach 1000 V (1 kV) or more. High transient voltages exist close to a lightning strike or when large (high-current) loads are switched OFF. **See Figure 1-25.** For example, when a large motor (100 HP) is turned OFF, a transient voltage moves down the power distribution system. When a test instrument or meter is connected to a point along the system in which a high transient voltage is present, an arc can be created inside the meter. Once started, the arc can cause a high-current short in the power distribution system even after the original high transient voltage is gone. A high-current short can turn into an arc blast.

An *arc blast* is an explosion that occurs when the air surrounding electrical equipment becomes ionized and conductive. The amount of current drawn and the potential damage caused depend on the specific location of the arc blast in the power distribution system. All power distribution systems have current limits set by fuses and circuit breakers along the system. The current rating (size) of the fuses and circuit breakers decreases further away from the main distribution panel. The further away from the main distribution panel, the less likely a high transient voltage is to cause damage.

CAUTION

Other than a laser, an electric arc is the hottest heat source on Earth. Electric arcs are capable of producing temperatures up to 10,000°F. Temperatures of such intensity are capable of producing serious burns at distances up to 20′ and can be fatal at distances up to 8′.

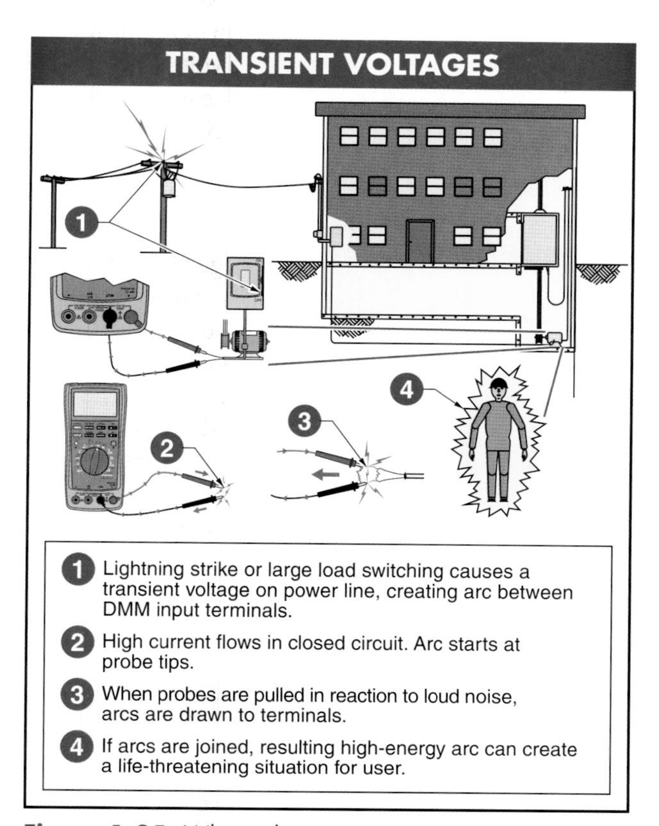

Figure 1-25. When taking measurements in an electrical circuit, transient voltages can cause electrical shock and/or damage to test instruments and meters.

CAT Ratings

IEC 1010 classifies the applications in which test instruments may be used into four overvoltage installation categories (Category I – Category IV). The four categories are typically abbreviated as CAT I, CAT II, CAT III, and CAT IV. The CAT ratings determine what magnitude of transient voltage a test instrument or other electrical appliance can withstand when used on a power distribution system. For example, a test instrument or other electrical measurement tool specified for use in a CAT III installation must withstand a 6000 V transient (2 ms rise time with a 50 ms, 50% duration) voltage without resulting in a hazard. When a test instrument or other meter is operated on voltages above 600 V, the test instrument must be capable of withstanding an 8000 V transient voltage. Also, a test instrument or meter that is designed to withstand a transient voltage can be damaged but the transient cannot result in a hazard to the technician or the facility. To protect technicians from transient voltages, protection must be built into all test equipment.

Safety standards such as IEC 61010-1 2nd edition, the harmonized North America standard, and the UL standard 61010-1 vary, but are closely matched. The requirements of the standards are used to rate test equipment for minimizing hazards such as shock, fire, and arc blast among other concerns. A test instrument designed to these standards offers a high level of protection. A measurement category rating such as CAT III or CAT IV indicates acceptable usage on three-phase permanently installed loads and three phase distribution panels in a building or facility. All exposed electrical installations and the power panels of a facility are considered high-voltage areas. Measurement categories such as CAT III and CAT IV ratings are important criteria for test instruments and meters used in industrial applications. **See Figure 1-26.**

Power distribution systems are divided into categories based on the magnitude of transient voltage test instruments must withstand when used on the power distribution system. Dangerous high-energy transient voltages such as a lightning strike are attenuated (lessened) or dampened as the transient travels through the impedance (AC resistance) of the system and system grounds. Within an IEC 1010 standard category, a higher voltage rating denotes a higher transient voltage withstanding rating. For example, a CAT III-1000 V (steady-state) rated test instrument has better protection compared to a CAT III-600 V (steady-state) rated test instrument. Between categories, a higher voltage rating (steady-state) might not provide higher transient voltage protection. For example, a CAT III-600 V test instrument has better transient protection compared to a CAT II-600 V test instrument. A test instrument must be chosen based on the IEC overvoltage installation category first and voltage second.

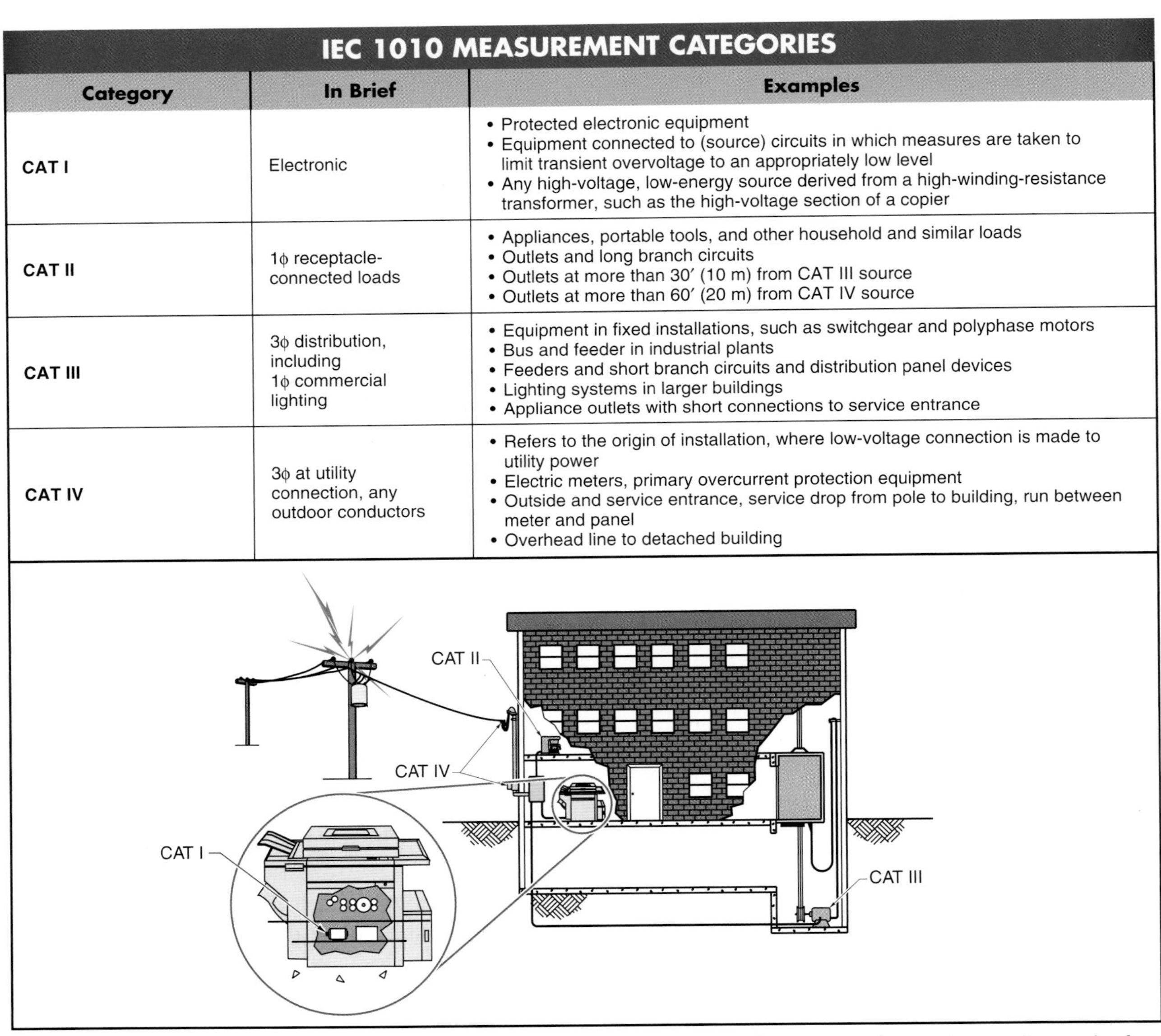

IEC 1010 MEASUREMENT CATEGORIES

Category	In Brief	Examples
CAT I	Electronic	• Protected electronic equipment • Equipment connected to (source) circuits in which measures are taken to limit transient overvoltage to an appropriately low level • Any high-voltage, low-energy source derived from a high-winding-resistance transformer, such as the high-voltage section of a copier
CAT II	1ϕ receptacle-connected loads	• Appliances, portable tools, and other household and similar loads • Outlets and long branch circuits • Outlets at more than 30′ (10 m) from CAT III source • Outlets at more than 60′ (20 m) from CAT IV source
CAT III	3ϕ distribution, including 1ϕ commercial lighting	• Equipment in fixed installations, such as switchgear and polyphase motors • Bus and feeder in industrial plants • Feeders and short branch circuits and distribution panel devices • Lighting systems in larger buildings • Appliance outlets with short connections to service entrance
CAT IV	3ϕ at utility connection, any outdoor conductors	• Refers to the origin of installation, where low-voltage connection is made to utility power • Electric meters, primary overcurrent protection equipment • Outside and service entrance, service drop from pole to building, run between meter and panel • Overhead line to detached building

Figure 1-26. IEC 1010 defines the applications in which test instruments and meters can be used according to the four categories.

Independent Testing Organizations

National, state, and local codes and standards are used to protect people and property from electrical dangers. A *code* is a regulation or minimum requirement. A *standard* is an accepted reference or practice. Codes and standards ensure that electrical equipment is built and installed safely and every effort is made to protect people from electrical shock. The IEC sets standards but does not test or inspect for code and standard compliance.

A test instrument with a symbol and listing number of an independent testing lab such as Underwriters Laboratories Inc.® (UL), Canadian Standards Association (CSA), or other recognized testing organization indicates compliance with the standards of the organization. A manufacturer can claim to "design to" a standard with no independent verification. To be UL listed or CSA certified, a manufacturer must employ the services of an approved agency to test a product's compliance with a standard. **See Figure 1-27.** For example, UL3111-1 or CSA C22.2 No. 1010.1-92 indicates that the IEC 1010 standard is met.

To help prevent electrical shock, proper personal protective equipment must be worn at all times.

WARNING

Before using any electrical test instruments or meters, always refer to the user's manual for proper operating procedures, safety precautions, and technical limits. Conditions can change quickly as voltage and current levels vary in individual circuits.

Electrical Shock

According to the National Safety Council, over 1000 people are killed by electrical shock in the United States each year. Electricity is the number one cause of fires. More than 100,000 people are killed in electrical fires each year. Improper electrical wiring or misuse of electricity causes destruction of equipment and fire damage to property.

Safe working habits are required when troubleshooting an electrical circuit or component because the electrical parts that are normally enclosed are exposed during troubleshooting.

An electrical shock results any time a body becomes part of an electrical circuit. Electrical shock varies from a mild shock to fatal current. The severity of an electrical shock depends on the amount of electric current (in mA) that flows through the body, the length of time the body is exposed to the current flow, the path the current takes through the body, the physical size and condition of the body through which the current passes, and the amount of body area exposed to the electric contact. The amount of current that passes through a circuit depends on the voltage and resistance of the circuit. During an electrical shock, the body of an electrician becomes part of the circuit. The resistance a body offers to the flow of current varies. Sweaty hands have less resistance than dry hands. A wet floor has less resistance than a dry floor. The lower the resistance, the greater the current flow. The greater the current flow, the greater the severity of a shock. **See Figure 1-28.**

RECOGNIZED TESTING LABORATORIES (RTLs) AND STANDARDS ORGANIZATIONS*	
Underwriters Laboratories, Inc.® (UL)	333 Pfingsten Rd., Northbrook, IL 60062 USA Tel: 847-272-8800 www.ul.com
American National Standards Institute (ANSI)	1899 L Street, NW, 11th Floor Washington DC 20036 Tel: 202-293-8020 www.ansi.org
British Standards Institution (BSI)	389 Chiswick High Road, London W4 4AL United Kingdom Tel: +44 345 086 9001 www.bsigroup.com
CENELEC European Committee for Electrotechnical Standardization	Rue de la Science, 23 B-1040 Brussels Tel: +32 2 550 08 19 www.cenelec.eu
Canadian Standards Association (CSA)	Central Office 178 Rexdale Blvd. Etobicoke (Toronto), Ont. M9W 1R3 Tel: 416-747-4000 www.csagroup.org
Verband der Elektrotechnik und Informationstechnik (VDE)	Frankfurt am main Germany www.vde.com
Japanese Standards Association (JSA)	Mita Mt Bldg, 3-13-12 Mita Minato-ku Tokyo 108-0073 Japan www.jsa.or.jp
International Electrotechnical Commission (IEC)	3, rue de Varembé PO Box 131 1211 Geneva 20 Switzerland Tel: +41 22 919 02 11 www.iec.ch
The Institute of Electrical and Electronic Engineers, Inc. (IEEE)	3 Park Ave., 17th Floor New York, NY 10017 Tel: 212-419-7900 www.ieee.org
National Institute of Standards and Technology (NIST)	100 Bureau Dr., Gaithersburg, MD 20899 Tel: 301-975-2000 www.nist.gov
National Electrical Manufacturers Association (NEMA)	1300 N. 17th St. STE 900 Arlington, VA 22209 Tel: 703-841-3200 www.nema.org
International Standards Organization (ISO)	BIBC 11 Cheminde Blandonnet 8 CP 401 1214 Vernier Geneva Switzerland Tel: +41 22 749 01 11 www.iso.org
OSHA Region 1 Regional Office	JFK Federal Building, 25 New Sudbury St. Room E340 Boston, MA 02203 Tel: 617-565-9860 www.osha.gov
TÜV Rheinland of North America, Inc.	12 Commerce Rd., Newton, CT 06470 Tel: 203-426-0888 www.tuv.com

* Partial Listing

Figure 1-27. Test instruments have symbols listing the nationally recognized testing laboratories and standards organizations that the meters are in compliance with.

When handling a victim of an electrical shock accident, apply the following procedures:

1. Break the circuit to free the victim immediately and safely. Never touch any part of a victim's body when in contact with the circuit. When the circuit cannot be turned OFF, use any nonconducting device to free the victim. Resist the temptation to touch the victim when power is not turned OFF.
2. After the victim is free from the circuit, send for help and determine if the victim is breathing. When there is no breathing or pulse, start CPR. Always get medical attention for a victim of electrical shock.
3. When the victim is breathing and has a pulse, check for burns and cuts. Burns are caused by contact with the live circuit, and are found at the points that the electricity entered and exited the body. Treat the entrance and exit burns as thermal burns and get medical help immediately.

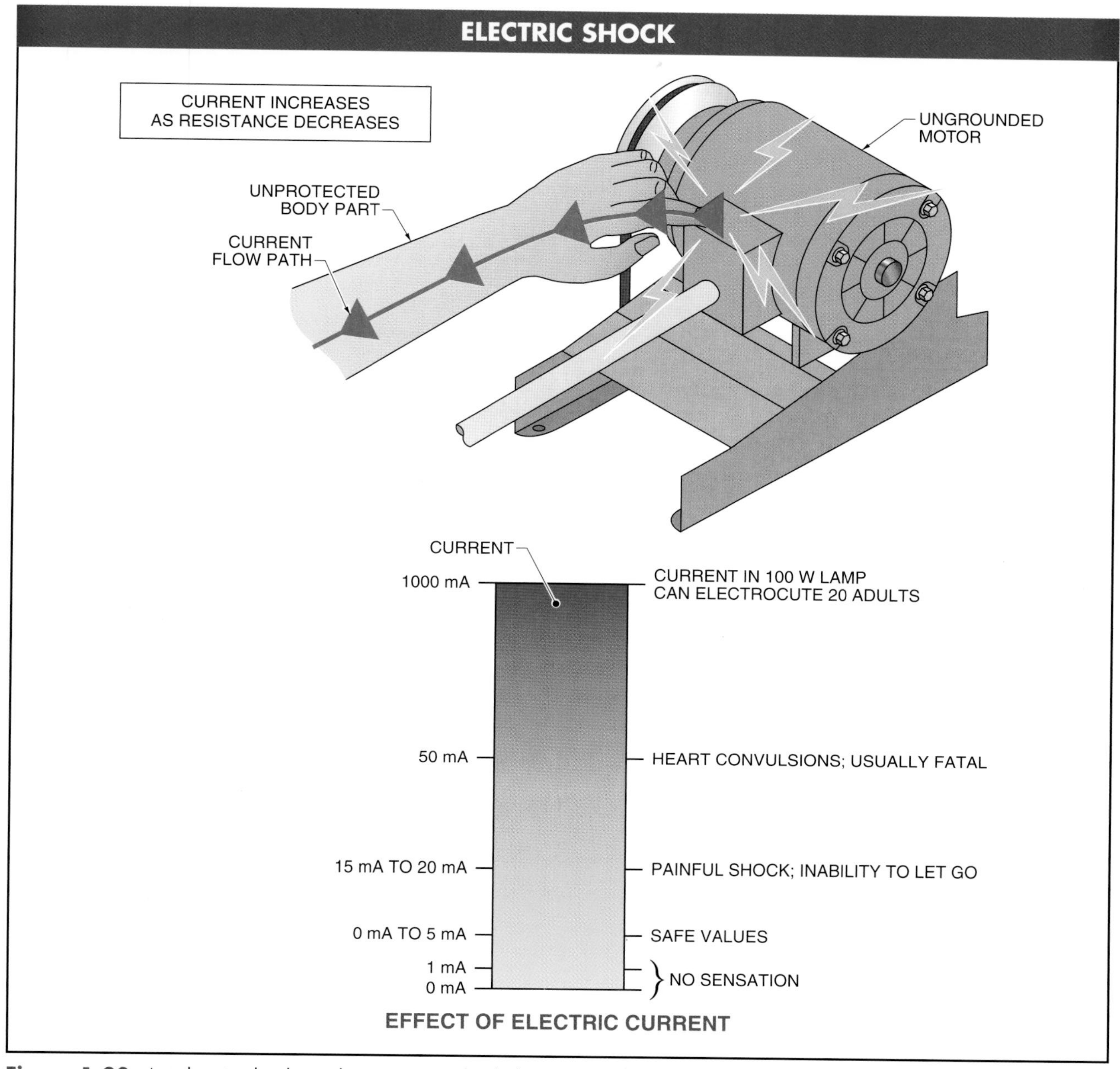

Figure 1-28. An electric shock results any time a body becomes part of an electrical circuit.

Rubber Insulating Matting

Rubber insulating matting is a personal protective device that provides electricians protection from electrical shock when working on energized electrical circuits. Dielectric black fluted rubber matting is specifically designed for use in front of open cabinets or high voltage equipment. Matting is used to protect electricians when voltages are over 50 V. Two types of matting that differ in chemical and physical characteristics are designated as Type I and Type II matting. **See Figure 1-29.**

RUBBER INSULATING MATTING RATINGS					
Safety Standard	Material Thickness		Material Width (in.)	Test Voltage	Maximum Working Voltage
	Inches	Millimeters			
BS921*	0.236	6	36	11,000	450
BS921*	0.236	6	48	11,000	450
BS921*	0.354	9	36	15,000	650
BS921*	0.354	9	48	15,000	650
VDE0680†	0.118	3	39	10,000	1000
ASTM D178‡	0.236	6	24	25,000	17,000
ASTM D178‡	0.236	6	30	25,000	17,000
ASTM D178‡	0.236	6	36	25,000	17,000
ASTM D178‡	0.236	6	48	25,000	17,000

* BSI–British Standards Institute
† VDE–Verband Deutscher Elektrotechniker Testing and Certification Institute
‡ ASTM International

Figure 1-29. Rubber insulating matting provides protection from electrical shock when working on energized electrical circuits.

Grounding

Electrical circuits are grounded to safeguard equipment and personnel against the hazards of electrical shock. Proper grounding of electrical tools, machines, equipment, and delivery systems is one of the most important factors in preventing hazardous conditions. *Grounding* is the connection of all exposed non-current-carrying metal parts to the earth. Grounding provides a direct path for unwanted (fault) current to the earth without causing harm to persons or equipment. Grounding is accomplished by connecting the circuit to a metal underground pipe, a metal frame of a building, a concrete-encased electrode, or a ground ring. **See Figure 1-30.**

Non-current-carrying metal parts that are connected to ground include all metal boxes, raceways, enclosures, and equipment. Unwanted current exists because of insulation failure or because a current-carrying conductor makes contact with a non-current-carrying part of the system. In a properly grounded system, the unwanted current flow trips fuses or circuit breakers. Once the fuse or circuit breaker is tripped, the circuit is opened and no additional current flows.

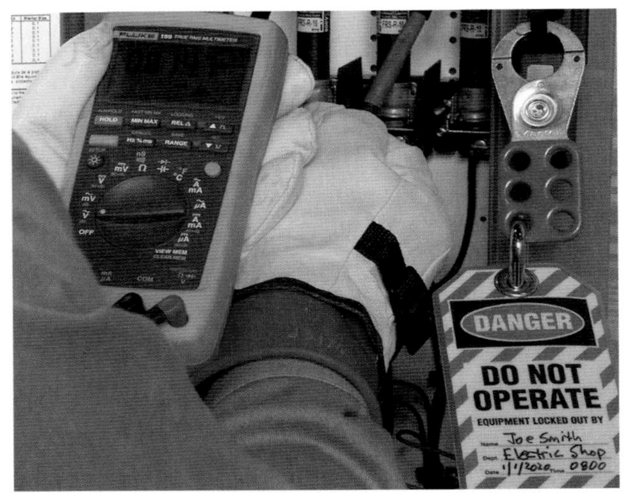

Electricians sometimes use a lockout/tagout hasp to ensure that no one person can accidentally energize a dangerous circuit or system while another is still working on it.

LOCKOUT/TAGOUT

Lockout is the process of removing the source of electrical power and installing a lock that prevents the power from being turned ON. To ensure the safety of personnel working with equipment, all electrical, pneumatic, and hydraulic power is removed and the equipment must be locked out and tagged out. *Tagout* is the process of placing a danger tag on the source of electrical power, which indicates that the equipment may not be operated until the danger tag is removed. Per OSHA standards, equipment is locked out and tagged out before any installation or preventive maintenance is performed. **See Figure 1-31.**

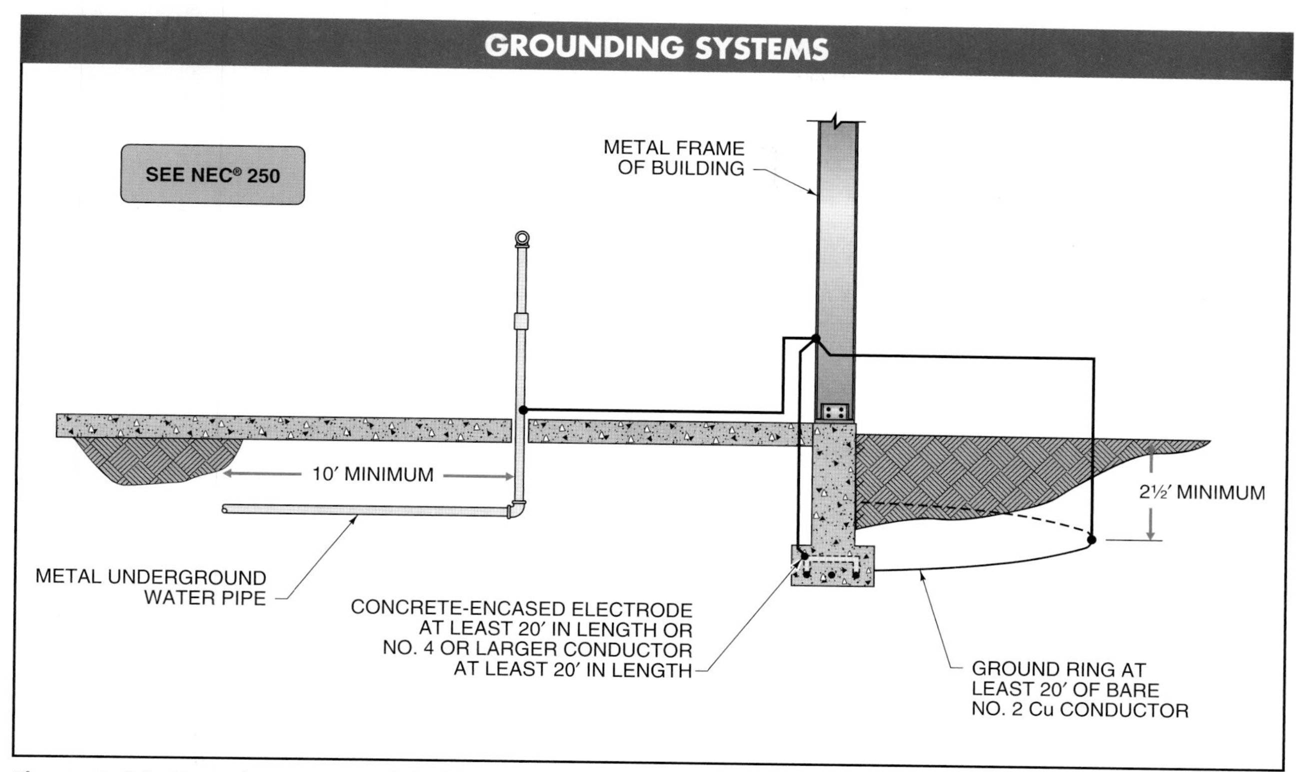

Figure 1-30. Grounding is accomplished by connecting a circuit to a metal underground pipe, a metal frame of a building, a concrete-encased electrode, or a grounding ring.

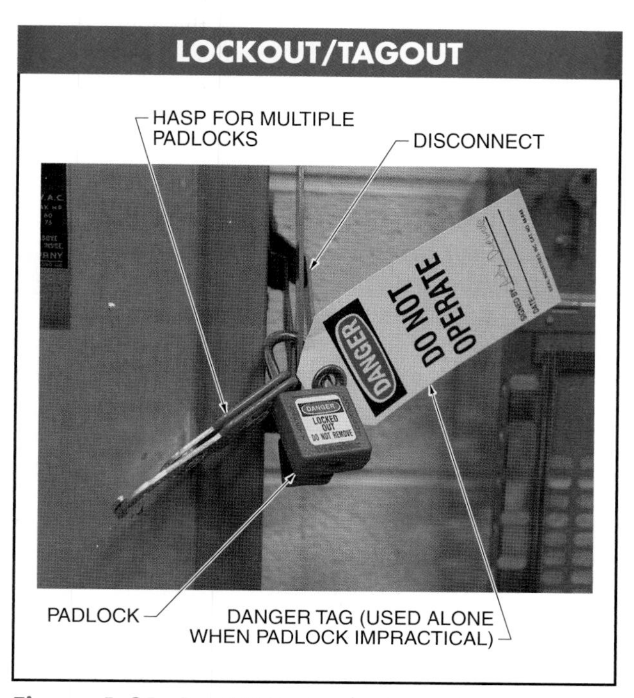

Figure 1-31. Per OSHA standards, equipment must be locked out and tagged out before any installation or preventive maintenance is performed.

A danger tag has the same importance and purpose as a lock and is used alone only when a lock does not fit the disconnect device. A danger tag must be attached at the disconnect device with a tag tie or equivalent and must have space for the technician's name, craft, and other company-required information. A danger tag must withstand the elements and expected atmosphere for the maximum period of time that exposure is expected. **See Figure 1-32.**

Lockout/tagout is used in the following situations:

- Power is not required to be on to a piece of equipment to perform a task
- When removing or bypassing machine guards or other safety devices
- The possibility exists of being injured or caught in moving machinery
- When jammed equipment is being cleared
- The danger exists of being injured if equipment power is turned ON

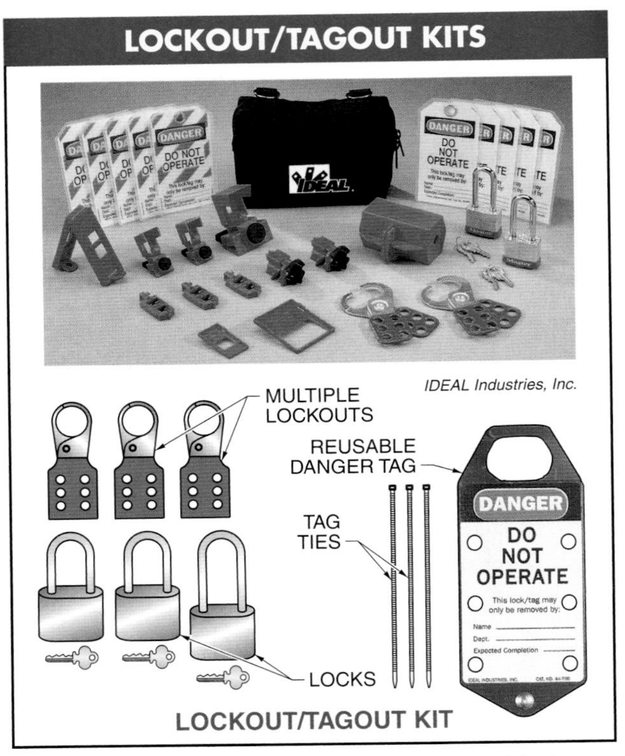

Figure 1-32. Lockout/tagout kits contain reusable danger tags, tag ties, multiple lockouts, magnetic signs, and information on lockout/tagout procedures.

Lockout and tagouts do not by themselves remove power from a machine or its circuitry. OSHA provides a standard procedure for equipment lockout/tagout. Lockout is performed and tags are attached only after the equipment is turned OFF and tested.

Typical company lockout/tagout procedures are as follows:

1. Notify all affected persons that a lockout/tagout is required. Notification must include the reason for the lockout/tagout, and the expected duration.
2. When the equipment is operating, shut it down using normal procedures.
3. Operate the energy-isolating device(s) so that the equipment is isolated from all energy sources. Stored energy in springs, elevated machine members, and capacitors must be dissipated or restrained by blocking, discharging, or other appropriate methods.
4. Lockout/tagout the energy-isolating devices with assigned locks and danger tags. **See Figure 1-33.**
5. After ensuring that no personnel are exposed, operate the normal operating controls, verifying that the equipment is inoperable and that all energy sources have been isolated.
6. Inspect and test the equipment with appropriate test instruments to verify that all energy sources are disconnected. Multiphase electrical power requires that each phase be tested. The equipment is now locked out and tagged out.

A lockout/tagout must not be removed by any person other than the authorized person who installed the lockout/tagout, except in an emergency. In an emergency, only supervisory personnel may remove a lockout/tagout, and only upon notification of the lockout/tagout person. A list of company rules and procedures is given to authorized personnel and any person who may be affected by a lockout/tagout.

When more than one electrician is required to perform a task on a piece of equipment, each electrician must place a lockout /tagout on the energy-isolating device(s). A multiple lockout/tagout device (hasp) must be used because energy-isolating devices typically cannot accept more than one lockout/tagout at one time. A *hasp* is a multiple lockout/tagout device.

Troubleshooting Tip

LOCKOUT/TAGOUT REQUIREMENTS

Lockout/tagout procedures must conform to OSHA 29 CFR 1910.147 – The *Control of Hazardous Energy (Lockout/Tagout)* and company rules and procedures. A lockout/tagout shall not be removed by any person other than the person who installed it, except in an emergency. In an emergency, only authorized personnel (persons who have been trained in proper lockout/tagout procedures) may remove the lockout/tagout. Tagouts must be attached by hand, easy to read, self-locking, and resistant to accidental removal. Lockouts and tagouts must be tough and must resist damage from environmental and working conditions. Written procedures must be established for each piece of equipment in the facility.

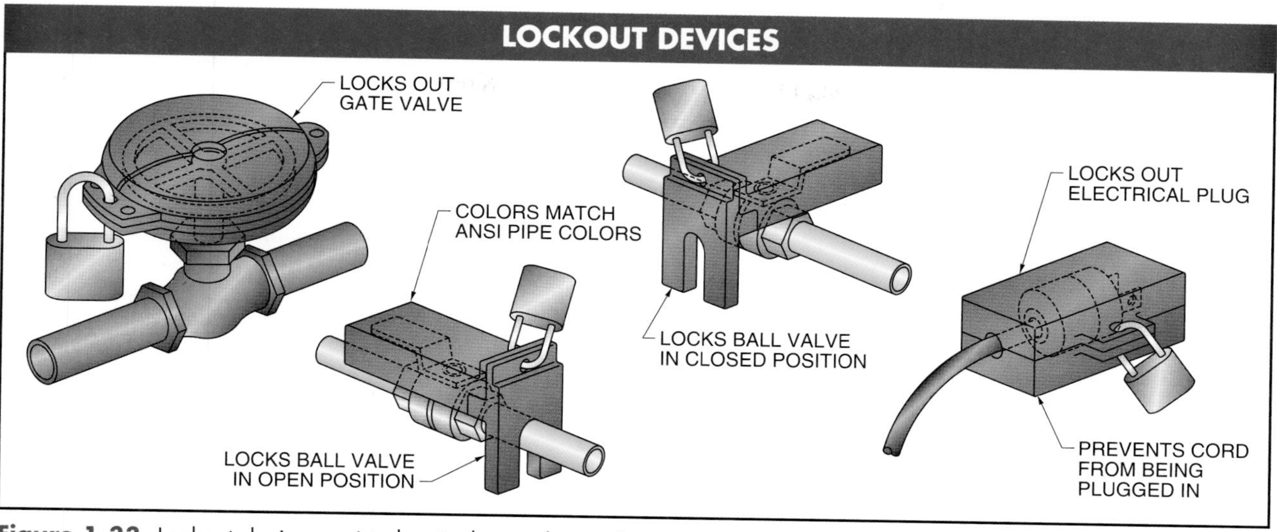

Figure 1-33. Lockout devices resist chemicals, cracking, abrasion, and temperature changes, and are available in colors to match ANSI pipe colors. Lockout devices are sized to fit standard industry control device sizes.

ARC BLAST SAFETY

An *electric arc* is a discharge of electric current across an air gap. Arcs are caused by excessive voltage ionizing an air gap between two conductors, or by accidental contact and reseparation between two conductors. When an electric arc occurs, there is the possibility of "arc flash" or "arc blast." *Arc flash* is an extremely high-temperature discharge produced by an electrical fault in the air. Arc flash temperatures reach 35,000°F. An *arc blast* is an explosion that occurs when the air surrounding electrical equipment becomes ionized and conductive. Arc blasts are a threat to electrical systems of 480 V and higher. Arc blasts are possible in systems of lesser voltage, but arc blasts are not likely to be as destructive as in a high-voltage system.

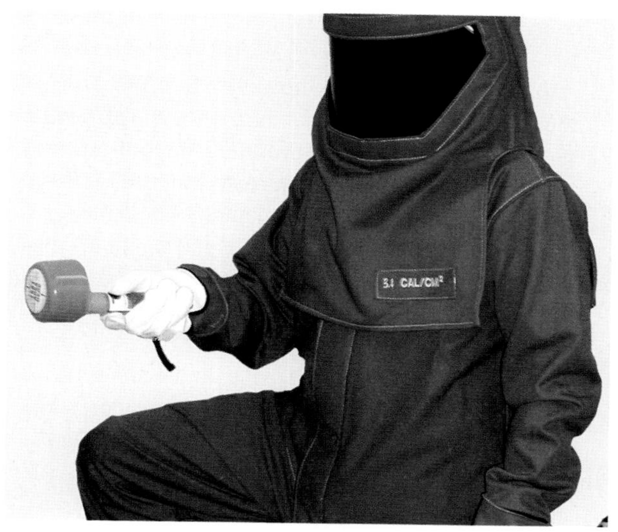

Salisbury by Honeywell

Electricians working with energized high voltage circuits may be required to wear additional protective clothing to guard against arc flash.

Arc flash and arc blast are always a possibility when working with electrical test instruments. Arc flash can occur when using a voltmeter or DMM to measure voltage in a 480 V or higher electrical system when there happens to be a power line transient, such as a lightning strike or power surge, at the same time. Some voltmeters and DMMs indicate a CAT rating specifying the tolerance limit for overvoltage transients and subsequent safety provisions. A potential cause for arc flash and arc blast is improper test instrument and meter use. For example, an arc blast can occur when connecting an ammeter across two points of a circuit that is energized with a higher voltage than the rating of the meter. To avoid causing arc blast or arc flash, an electrical system needs to be de-energized, locked out, and tagged out prior to performing work. Only qualified electricians are allowed to work on energized circuits of 50 V or higher.

Flash Protection Boundary

The *flash protection boundary* is the distance at which PPE is required to prevent burns when an arc occurs. **See Figure 1-34.** Per NFPA 70E, systems of 600 V and less require a flash protection boundary of 4′, based on the time of the circuit breaker to act. While a circuit that is being worked on should always be de-energized, the possibility exists that nearby circuits are still energized and within the flash point boundary. Barriers such as insulation blankets, along with the proper PPE, must be used to protect against flashing from nearby energized circuits.

Personal Protective Equipment (PPE) for Arc Blast Protection. Proper PPE must always be worn when working with energized electrical circuits. Clothing made of synthetic materials such as nylon, polyester, or rayon, alone or combined with cotton, must never be worn as synthetic materials burn and melt to the skin. Per NFPA 70E, the type of PPE worn depends on the type of work being performed. The minimum PPE requirement for electrical work is an untreated natural material long-sleeve shirt, long pants, safety glasses with side shields, and rubber insulating shoes or boots. Additional PPE includes flame-retardant (FR) coveralls, FR long-sleeve shirts and pants, a hard hat with an FR liner, hearing protection, and double-layer flash suit jacket and pants. Flash suits are similar to firefighter turnout gear and must be worn when working near a Category 4 Hazard/Risk area. **See Figure 1-35.**

CODES AND STANDARDS

To protect people and property from electrical dangers, national, state, and local codes and standards are used. A code is a regulation or minimum requirement. A standard is an accepted reference or practice. Codes and standards ensure electrical equipment is built and installed safely and every effort is made to protect electricians from electrical shock.

National Fire Protection Association (NFPA)

NFPA is a national organization that provides guidance in assessing the hazards of the products of combustion. The NFPA publishes the National Electrical Code® (NEC®). The purpose of the NEC® is the practical safeguarding of persons and property from the hazards arising from the use of electricity. The NEC® is updated every three years. Many city, county, state, and federal agencies use the NEC® to set requirements for electrical installations. The primary concern of the NEC® is to protect lives and property.

Technical Tip

According to U.S. government statistics, electrical injuries are not the most common but have a high fatality rate when they do occur. The biggest electrical hazard is a live power line, and the injury rates are highest in construction, manufacturing, and services industries. These industries account for 44% of electrical injuries but only 7% of the workforce.

APPROACH BOUNDARIES TO ENERGIZED PARTS FOR SHOCK PREVENTION

Nominal System Voltage, Range, Phase to Phase*	Limited Approach Boundary		Restricted Approach Boundary (Allowing for Accidental Movement)	Prohibited Approach Boundary
	Exposed Movable Conductor	Exposed Fixed-Circuit Part		
0 to 50	N/A	N/A	N/A	N/A
51 to 300	10′-0″	4′-0″	Avoid contact	Avoid contact
301 to 750	10′-0″	4′-0″	1′-0″	0′-1″
751 to 15,000	10′-0″	4′-0″	2′-2″	0′-7″

* in volts

Figure 1-34. The approach boundary is the distance at which PPE is required while working on energized circuits to prevent burns if an arc occurs.

FLAME-RESISTANT PROTECTIVE EQUIPMENT REQUIREMENTS				
Flame-Retardant Clothing Type	**Category Number (1 = Least Hazardous)**			
	1	**2**	**3**	**4**
Flash suit jacket				X
Flash suit pants				X
Head protection				
Hard hat	X	X	X	X
Flame-retardant hard hat liner			X	X
Safety glasses w/side shields or goggles	X	X	X	X
Face protection (2-layer hood)			X	X
Hearing protection (ear canal inserts)			X	X
Rubber gloves w/leather protectors	X	X	X	X
Leather shoes w/rubber soles	X	X	X	X

Figure 1-35. Per NFPA 70E, the type of PPE required depends on the voltage and where work is being performed.

Occupational Safety and Health Administration

The Occupational Safety and Health Administration (OSHA) is a federal government agency established under the Occupational Safety and Health Act of 1970, which requires all employers to provide a safe environment for their employees. The Act requires that all employers provide work areas free from recognized hazards likely to cause serious harm.

National Electrical Manufacturers Association (NEMA)

NEMA is a national organization that assists with information and standards concerning proper selection, ratings, construction, testing, and performance of electrical equipment. NEMA standards are used as guidelines for the manufacture and use of electrical equipment.

CARE OF TEST INSTRUMENTS

All test instruments and meters are designed for taking measurements and should last virtually forever when used correctly. The life expectancy of a test instrument or meter depends upon how the meter is used, how the meter is taken care of, and how well the meter is made. When a test instrument or meter is used correctly (within its range, specifications, and operating environment) and physically taken care of, the meter will last a long time. Test instruments that are built better will last longer when handled roughly or used all the time because they are stronger and more durable. Test instruments must be handled with reasonable care.

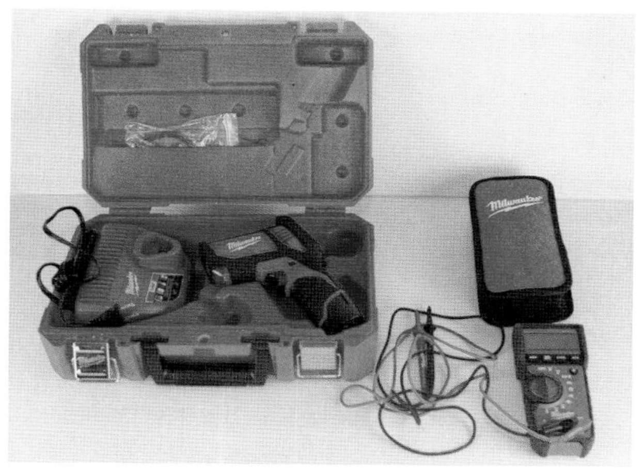

When not in use, guard test instruments against accidental damage by storing them in their protective bags or cases, or as the manufacturer recommends.

Carrying Cases

When using a test instrument or meter, ensure the meter is placed in such a manner as to prevent the meter from being accidentally dropped or damaged. Being dropped is the number one reason why many test instruments and meters are damaged. When a meter is likely to be dropped, use a meter that includes a heavy outer protective case designed for rough (commercial and industrial) use. Keep test instruments and meters in protective carrying cases when in transit or not in use. Most manufacturers offer a variety of carrying holsters, soft carrying cases, and hard carrying cases. A carrying holster, typically connected to the belt or tool pouch, prevents dropping a test instrument or meter. Soft carrying cases prevent damage during normal storage and transportation. Hard carrying cases prevent damage when a test tool is kept or transported along with heavier tools and other items that can damage softer carrying cases. **See Figure 1-36.**

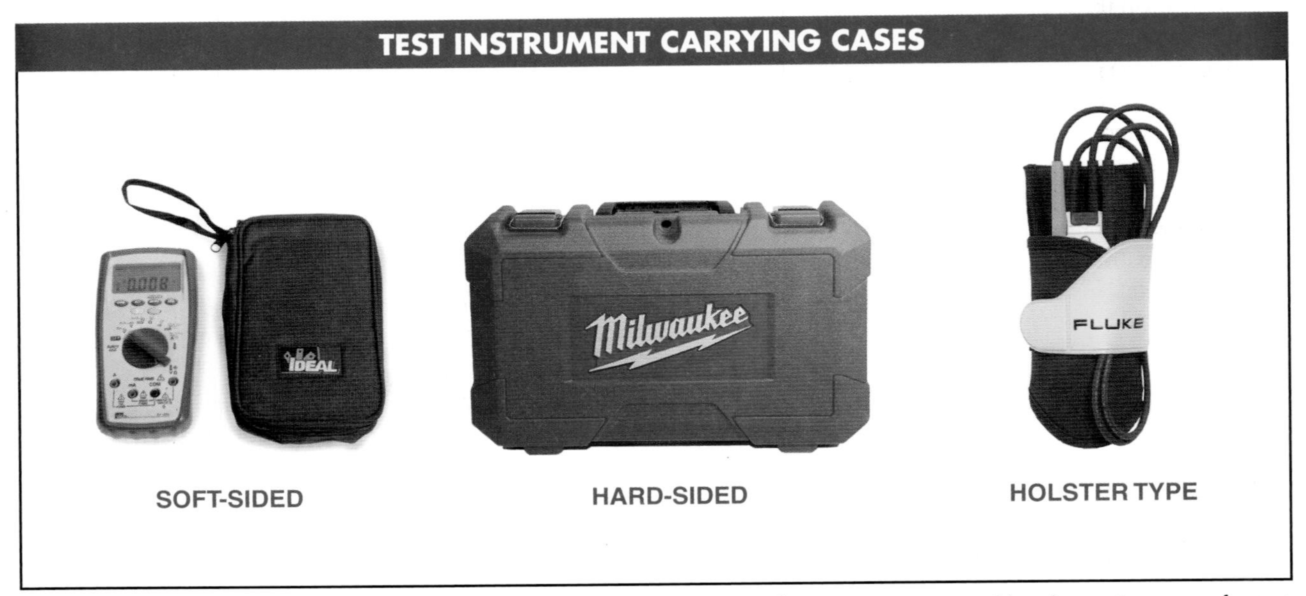

Figure 1-36. Most manufacturers offer a variety of carrying holsters, soft carrying cases, and hard carrying cases for test instruments and meters.

Troubleshooting Tip

CARRYING CASE DESIGN

Test instrument carrying cases and equipment cases are designed to protect contents from impact, shock, vibration, rain, and dust. Case applications include carrying, shipping, containment, and military and aerospace environments. Test instrument cases are available in a variety of materials and configurations, including EMI/RFI shielding, telescopic covers and handles, and provisions for rack mount equipment.

Test Instrument and Meter Care Practices

Extra care must be provided to analog test instruments and meters because analog meter movements are susceptible to damage from shock. The ability of a test instrument or meter to take an accurate measurement is impaired when a meter is dropped. Hard carrying cases are the best when transporting analog instruments. Electricians must understand how to properly handle and operate any test instrument used.

Because many of the components may be of precision quality, repairing test instruments and meters is expensive. Some test instruments are unforgiving when exposed to overvoltages/currents and hostile conditions. Some manufacturers protect test instruments and meters with fast-acting fuses. When a fuse blows (or an overload trips), remove the instrument from the circuit before replacing the fuse or resetting the overload. Fuses must be replaced with the specified types and values indicated in the owner's manual. Failure to use the specified fuses can result in serious damage to the instrument. Even though a fuse fits and appears to be the same, the fuse may not perform the same. For example, a 10 A rated automobile fuse (designed to open at 10 A in a 12 VDC circuit) is not the same as a 10 A rated meter fuse (designed to open at 10 A in an application in which there may be several hundred volts AC or DC). Even when an automobile fuse fits into the fuse holder of a meter, only the correct fuse recommended by the manufacturer can be used.

Test leads must be stored carefully and checked periodically for excessive wear.

Test leads are one of the most important parts of any test instrument because test leads are the part that connects the test instrument to the circuit or component being tested and are typically the part of the test instrument an electrician is holding. Test leads associated with any test instrument or meter must be checked on a regular basis. When test leads are used with test instruments that measure dangerous currents or voltages, the test leads must be checked each time they are used. Look for cuts, burned areas, deterioration, or other damage that may reduce the strength of the insulation. Test leads use stranded wire to make the leads flexible, and the individual strands break over time during normal operation. As strands break, the diameter of the test lead wire is reduced and the resistance of the test lead increases.

Test leads are tested by connecting the test leads to a meter that measures resistance and touching the metal tips of the test leads together. The resistance of test leads must be less than 1 Ω. When a test lead resistance is more than 1 Ω, the test lead is probably damaged. The higher the resistance, the more damaged the leads. However, before assuming that a reading of more than 1 Ω means damaged test leads, ensure the meter taking the resistance measurement is zeroed. Analog meters have a zeroing knob or screw for setting the needle to zero, and digital meters have a Relative button (REL) to zero the digital display.

Do not operate test instruments and meters for prolonged periods of time in areas with extreme temperatures and humidity. Avoid use in areas that have excessive dust or corrosive fumes. Test instruments should not be exposed to strong electrical or electromagnetic interference. When a liquid is spilled on a test instrument or meter, immediately clean the spill from the instrument and wipe the meter dry. In the case of corrosive liquids, use a suitable cleaner to neutralize the corrosive action. When a spill occurs on the plastic window of a meter, be careful not to scratch the display window by using an abrasive cleaner.

Place all switches of a test instrument or meter in the OFF position when testing is completed. When a test instrument or meter is to be stored, place the meter in a protective case. Store the instrument in a clean area, free of vibration, extreme temperatures, and humidity. Remove batteries from test instruments and meters that will be stored for an extended period of time. Batteries can rupture and spill corrosive materials into the instrument, causing serious damage. For instruments in constant use, check the batteries for evidence of electrical and mechanical failure at least several times a year. Replace batteries of a test instrument or meter every year, even when the instrument appears to be operating properly.

All test instruments must be tested for accuracy on a periodic basis. Some manufacturers recommend monthly tests with an annual calibration at an authorized service center. At a minimum, test instrument measurement must be tested against a known measurement (fixed voltage, current, or resistance) value and against another instrument of the same type. Any discrepancy can signal that the test instrument requires additional testing. Only qualified personnel or the manufacturer should perform repairs to test instruments. Replacement parts must be electrically and mechanically identical to the original parts. In many cases, the parts are only available from the manufacturer of the instrument. The use of improper substitute parts can result in an inaccurate device and can cause a dangerous situation for the next user.

INTRODUCTION TO TEST INSTRUMENTS

REVIEW

Name: ______________________________ Date: ____________________

______________ **1.** What is the abbreviation for milliamperes?

T F **2.** Digital meters are more popular than analog meters because digital meters typically have greater measuring capabilities.

T F **3.** The caution signal word indicates an imminently hazardous situation that, if not avoided, results in death or serious injury.

______________ **4.** What is the symbol for ohms?

______________ **5.** Which type of meter, analog or digital, is more likely to be affected by electrical noise?

T F **6.** Tagout physically prevents energy sources from being energized.

______________ **7.** What unit is used to measure capacitance?

______________ **8.** Which type of meter, analog or digital, is more likely to warn the user if the function switch is improperly set?

______________ **9.** What is the maximum allowable resistance for test leads?

______________ **10.** What unit is used to measure frequency?

T F **11.** A high CAT rating indicates that a test instrument can be used in a circuit with high power, great possibility of short circuits, and high energy transients.

______________ **12.** Which unit prefix indicates one million times the base unit?

T F **13.** Test instruments should be periodically tested for accuracy.

______________ **14.** Which unit prefix indicates one thousandth of the base unit?

______________ **15.** What type of additional personal protective equipment must be worn when working around electrical systems with a risk of arc flash?

______________ **16.** What electrical unit is used to rate rubber insulating gloves?

T F **17.** Leather protector gloves are used to protect wearers from electrical shock.

______________ **18.** Which organization publishes the National Electrical Code®?

______________ **19.** Which two temperature scales are most commonly used by test instruments?

T F **20.** Both the function switch and the connection of the test leads must match the desired measurement before testing a circuit.

21. Why do some meters include fuses?

22. Why is it often more convenient to use unit prefixes to express measurements?

23. What are ghost voltages?

24. What is grounding?

25. What is the difference between lockout and tagout?

2 GENERAL USE TEST INSTRUMENTS

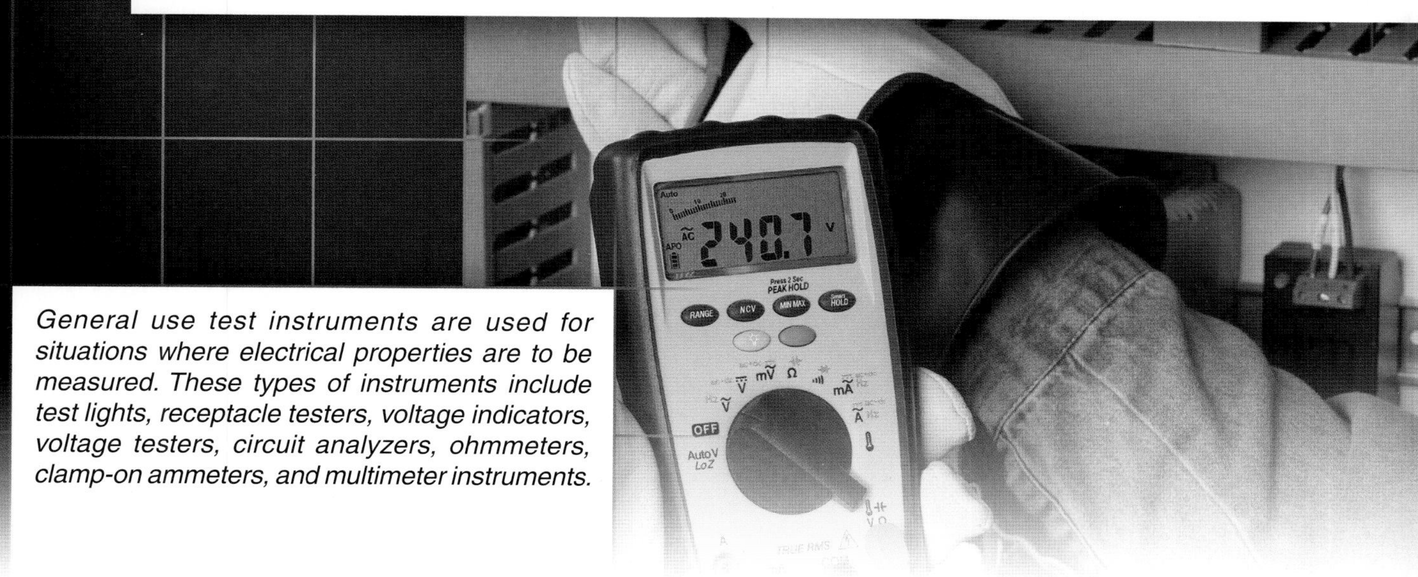

General use test instruments are used for situations where electrical properties are to be measured. These types of instruments include test lights, receptacle testers, voltage indicators, voltage testers, circuit analyzers, ohmmeters, clamp-on ammeters, and multimeter instruments.

IDEAL Industries, Inc.

GENERAL USE TEST INSTRUMENTS

A *test instrument* is a device used to measure electrical properties such as voltage, current, resistance, frequency, power, and conductivity. The most common electrical properties measured with test instruments are voltage, current, and resistance. Test instruments are also used to measure nonelectrical quantities such as temperature, pressure, and speed (rpm). **See Figure 2-1.** Test instruments are used when installing or troubleshooting a circuit or component.

Technical Tip

Test instruments are warranted to be free from defects in materials and workmanship for the instruments' lifetime. Warranties do not cover fuses; disposable batteries; damage from neglect, misuse, contamination, alteration, accident, or abnormal conditions of operation or handling, including overvoltage failures caused by use outside the specified ratings; or normal wear and tear of the test instrument.

Test instruments are portable or permanent devices. A *portable test instrument (meter)* is a device that is used to take temporary measurements and is typically powered by batteries. A *temporary electrical measurement* is a measurement taken when a test instrument is briefly connected to a circuit or to a component and then removed. A *permanent test instrument* is a device that is installed in a process or at a bench to continually measure and display quantities and is powered by a 115 V receptacle. Test instruments are either single-function or multifunction devices. A *single-function test instrument,* such as an ammeter, is a device capable of measuring and displaying only one electrical property. A *multifunction test instrument (multimeter)* is a device that is capable of measuring two or more electrical properties.

Each test instrument has specific features and limits. The user's manual of a test instrument lists safety precautions and warnings, specifications and features, and recommended operating procedures.

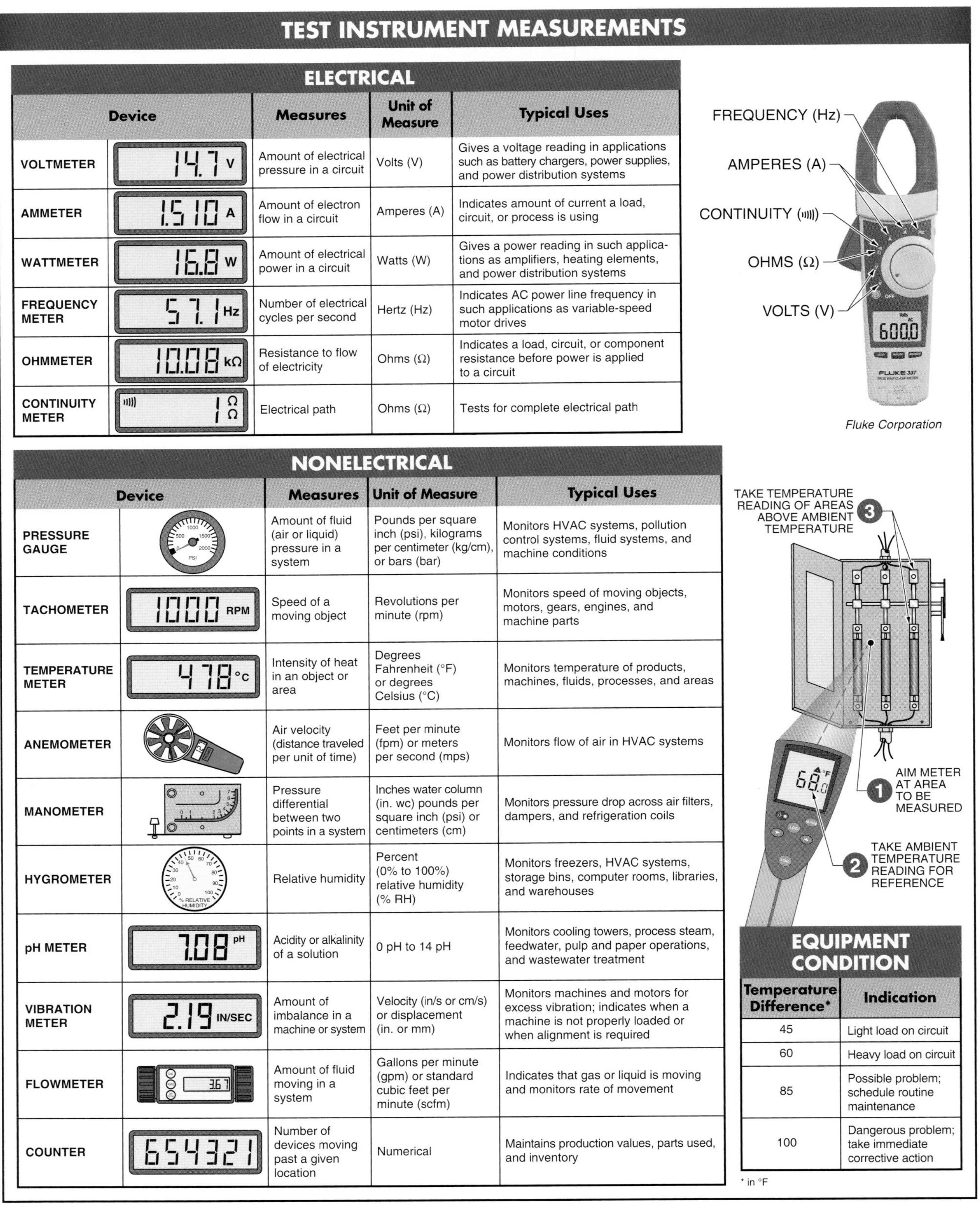

TEST INSTRUMENT MEASUREMENTS

ELECTRICAL

Device	Display	Measures	Unit of Measure	Typical Uses
VOLTMETER	14.7 V	Amount of electrical pressure in a circuit	Volts (V)	Gives a voltage reading in applications such as battery chargers, power supplies, and power distribution systems
AMMETER	1.510 A	Amount of electron flow in a circuit	Amperes (A)	Indicates amount of current a load, circuit, or process is using
WATTMETER	16.8 W	Amount of electrical power in a circuit	Watts (W)	Gives a power reading in such applications as amplifiers, heating elements, and power distribution systems
FREQUENCY METER	57.1 Hz	Number of electrical cycles per second	Hertz (Hz)	Indicates AC power line frequency in such applications as variable-speed motor drives
OHMMETER	10.08 kΩ	Resistance to flow of electricity	Ohms (Ω)	Indicates a load, circuit, or component resistance before power is applied to a circuit
CONTINUITY METER	1 Ω	Electrical path	Ohms (Ω)	Tests for complete electrical path

NONELECTRICAL

Device	Display	Measures	Unit of Measure	Typical Uses
PRESSURE GAUGE		Amount of fluid (air or liquid) pressure in a system	Pounds per square inch (psi), kilograms per centimeter (kg/cm), or bars (bar)	Monitors HVAC systems, pollution control systems, fluid systems, and machine conditions
TACHOMETER	1000 RPM	Speed of a moving object	Revolutions per minute (rpm)	Monitors speed of moving objects, motors, gears, engines, and machine parts
TEMPERATURE METER	478 °C	Intensity of heat in an object or area	Degrees Fahrenheit (°F) or degrees Celsius (°C)	Monitors temperature of products, machines, fluids, processes, and areas
ANEMOMETER		Air velocity (distance traveled per unit of time)	Feet per minute (fpm) or meters per second (mps)	Monitors flow of air in HVAC systems
MANOMETER		Pressure differential between two points in a system	Inches water column (in. wc) pounds per square inch (psi) or centimeters (cm)	Monitors pressure drop across air filters, dampers, and refrigeration coils
HYGROMETER		Relative humidity	Percent (0% to 100%) relative humidity (% RH)	Monitors freezers, HVAC systems, storage bins, computer rooms, libraries, and warehouses
pH METER	7.08 pH	Acidity or alkalinity of a solution	0 pH to 14 pH	Monitors cooling towers, process steam, feedwater, pulp and paper operations, and wastewater treatment
VIBRATION METER	2.19 IN/SEC	Amount of imbalance in a machine or system	Velocity (in/s or cm/s) or displacement (in. or mm)	Monitors machines and motors for excess vibration; indicates when a machine is not properly loaded or when alignment is required
FLOWMETER	36.7	Amount of fluid moving in a system	Gallons per minute (gpm) or standard cubic feet per minute (scfm)	Indicates that gas or liquid is moving and monitors rate of movement
COUNTER	654321	Number of devices moving past a given location	Numerical	Maintains production values, parts used, and inventory

EQUIPMENT CONDITION

Temperature Difference*	Indication
45	Light load on circuit
60	Heavy load on circuit
85	Possible problem; schedule routine maintenance
100	Dangerous problem; take immediate corrective action

* in °F

Figure 2-1. Test instruments are used to measure electrical properties such as voltage, current, resistance, frequency, power, lumens, and conductivity; and to measure nonelectrical quantities such as temperature, pressure, and speed (rpm).

General Use Test Instrument Abbreviations

General use test instruments commonly use standard abbreviations to represent a quantity or term for quick recognition. An *abbreviation* is a letter or combination of letters that represents a word. The specific abbreviations used are dependent on which language is being used. Generally, abbreviations that spell a word are followed by a period. For example, electrical properties such as voltage and current are identified with the abbreviations V for volts and A for amps, respectively. Abbreviations can be used individually or in combination with prefixes as in mV for millivolt or kV for kilovolt. **See Figure 2-2.**

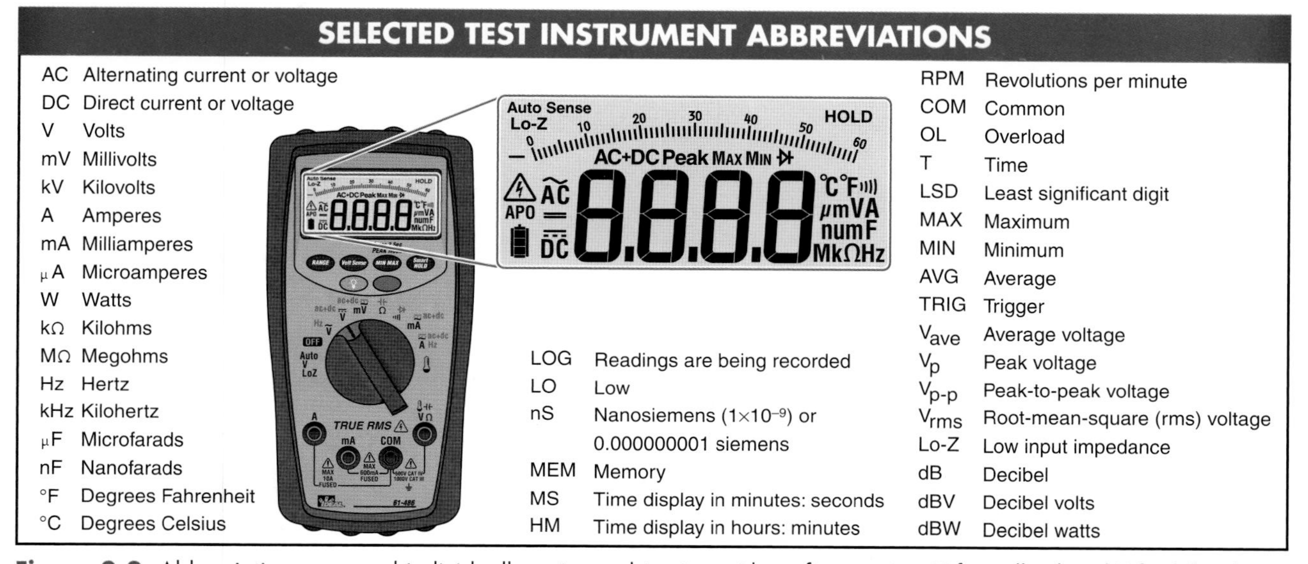

SELECTED TEST INSTRUMENT ABBREVIATIONS

Abbreviation	Meaning
AC	Alternating current or voltage
DC	Direct current or voltage
V	Volts
mV	Millivolts
kV	Kilovolts
A	Amperes
mA	Milliamperes
μA	Microamperes
W	Watts
kΩ	Kilohms
MΩ	Megohms
Hz	Hertz
kHz	Kilohertz
μF	Microfarads
nF	Nanofarads
°F	Degrees Fahrenheit
°C	Degrees Celsius
LOG	Readings are being recorded
LO	Low
nS	Nanosiemens (1×10^{-9}) or 0.000000001 siemens
MEM	Memory
MS	Time display in minutes: seconds
HM	Time display in hours: minutes
RPM	Revolutions per minute
COM	Common
OL	Overload
T	Time
LSD	Least significant digit
MAX	Maximum
MIN	Minimum
AVG	Average
TRIG	Trigger
V_{ave}	Average voltage
V_p	Peak voltage
V_{p-p}	Peak-to-peak voltage
V_{rms}	Root-mean-square (rms) voltage
Lo-Z	Low input impedance
dB	Decibel
dBV	Decibel volts
dBW	Decibel watts

Figure 2-2. Abbreviations are used individually or in combination with prefixes as in mV for millivolt or kV for kilovolt.

General Use Test Instrument Symbols

A *symbol* is a graphic element that represents a quantity, unit, or component. Symbols provide quick recognition and are independent of language because a symbol can be interpreted regardless of the language a person speaks. For example, standard symbols commonly used on DMMs can represent electrical components (diode, battery, and capacitor), terms (AC, DC, and ohm), or a message to the user (warning). Some test instrument functions are represented by two symbols that are combined. For example, a resistance measurement is represented by the symbol for ohms (Ω) and audio beeper or continuity (·))). **See Figure 2-3.**

Technical Tip

Standardized symbols and colors are used on electrical test instruments such as multimeters, megohmmeters, and oscilloscopes to efficiently denote electrical functions, with red being hot and black being common. Some test instruments indicate condition with color (green for low voltage and red for high voltage).

General Use Test Instrument Terminology

All test instruments use specific terms to describe functions and displayed information. The terms may be abbreviated or shown as a symbol. In general, test instrument terms and symbols have standard definitions that are used consistently throughout the electrical industry. **See Figure 2-4. See Appendix.**

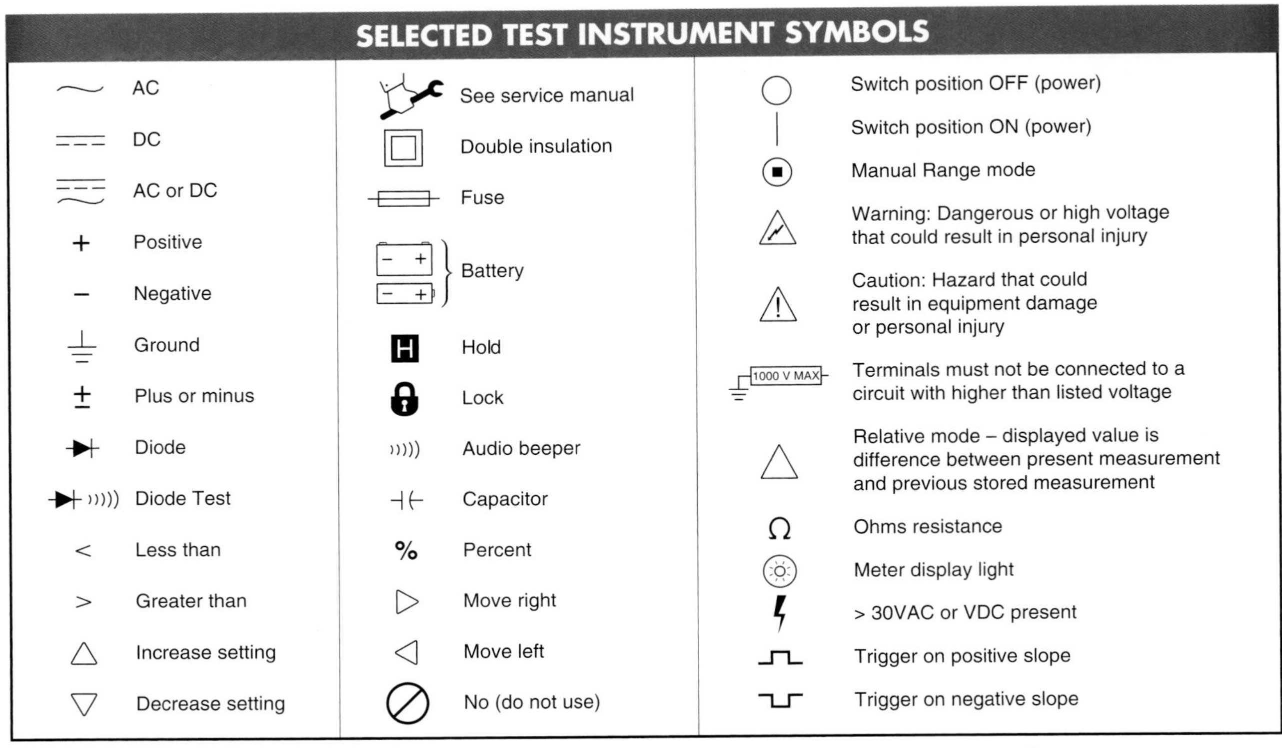

SELECTED TEST INSTRUMENT SYMBOLS

Symbol	Meaning	Symbol	Meaning	Symbol	Meaning
~	AC		See service manual	○	Switch position OFF (power)
	DC		Double insulation	|	Switch position ON (power)
	AC or DC		Fuse		Manual Range mode
+	Positive		Battery		Warning: Dangerous or high voltage that could result in personal injury
–	Negative				Caution: Hazard that could result in equipment damage or personal injury
	Ground	H	Hold	1000 V MAX	Terminals must not be connected to a circuit with higher than listed voltage
±	Plus or minus		Lock	△	Relative mode – displayed value is difference between present measurement and previous stored measurement
	Diode	)))))	Audio beeper	Ω	Ohms resistance
)))))	Diode Test		Capacitor		Meter display light
<	Less than	%	Percent		> 30VAC or VDC present
>	Greater than	▷	Move right		Trigger on positive slope
△	Increase setting	◁	Move left		Trigger on negative slope
▽	Decrease setting	⊘	No (do not use)		

Figure 2-3. Symbols provide quick recognition and interpretation regardless of the language spoken.

Troubleshooting Tip

TEST INSTRUMENT COLOR CODING

Test instruments use color coding to help identify proper meter jack and lead connections for usage. Test instruments were developed by engineers in the electronic world, where the color red identifies positive (+) and the color black identifies negative (–) or ground (⏚). Test instrument color coding is different from color coding used for electrical wiring, where black and red conductors (wires) are considered ungrounded and white conductors are grounded.

Although the actual color of a test lead or conductor does not affect the measurement and displayed meter value, color coding does affect the safety of a technician when taking the measurement. For example, when troubleshooting a circuit with a ground (115 VAC residential circuit), the black meter lead should be connected to the white (grounded) conductor of the circuit first, and then the red lead of the meter is connected to the ungrounded (black or red) conductor of the circuit to take the measurement. By connecting the black meter lead first, the red meter lead is not ungrounded until the lead is connected to the circuit.

If the red meter lead is connected to the grounded (black or red) conductor of a circuit first, the black meter lead that is not connected can cause an electrical shock if the lead is accidentally touched by a technician who is grounded. The meter is in series with the technician and the full circuit voltage is divided between the meter and the technician. Depending on whether the meter or the technician has the higher resistance and therefore the higher voltage drop, the meter could be destroyed or the technician could be seriously injured.

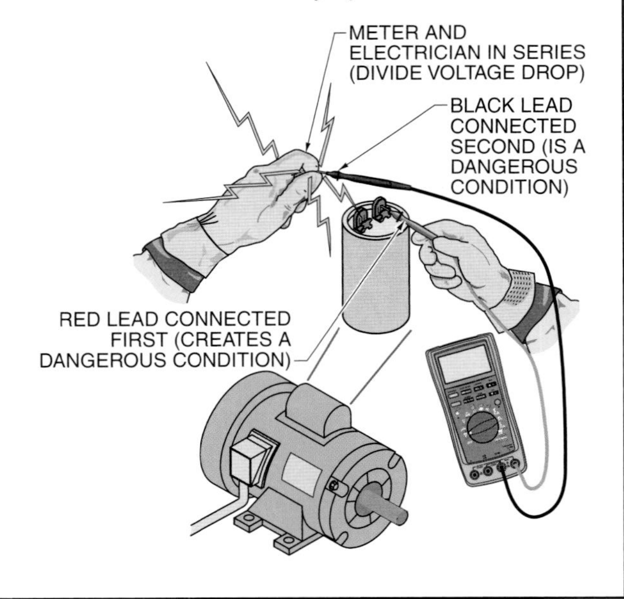

TEST INSTRUMENT TERMINOLOGY		
Term	**Symbol**	**Definition**
AC		Continually changing current that reverses direction at regular intervals; standard U.S. frequency is 60 Hz
AC COUPLING		Signal that passes an AC signal and blocks a DC signal; used to measure AC signals that are riding on a DC signal
AC/DC		Indicates ability to read or operate on alternating and direct current
ACCURACY ANALOG METER		Largest allowable error (in percent of full scale) made under normal operating conditions; the reading of a meter set on the 250 V range with an accuracy rating of ±2% could vary ±5 V; analog meters have greater accuracy when readings are taken on the upper half of the scale
DIODE	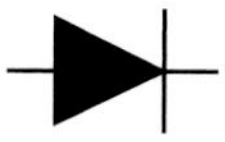	Semiconductor that allows current to flow in only one direction
DISCHARGE		Removal of an electric charge
DUAL TRACE	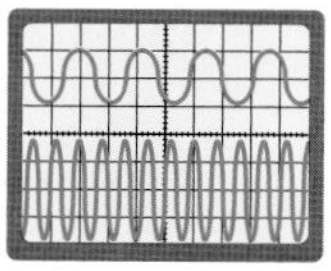	Feature that allows two separate waveforms to be displayed simultaneously
EARTH GROUND		Reference point that is directly connected to ground
FREQUENCY		Number of complete cycles occurring per second
FUNCTION SWITCH	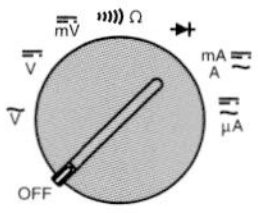	Switch that selects the function (AC voltage, DC voltage, etc.) that a meter is to measure
GLITCH	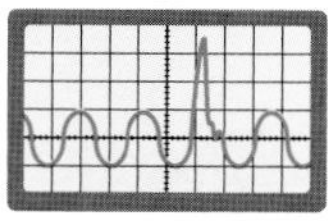	Momentary spike in a waveform
GLITCH DETECT		Function that increases the meter sampling rate to maximize the detection of the glitch(es)
TRIGGER		Device which determines the starting point of a measurement
WAVEFORM		Pattern defined by an electrical signal
ZOOM	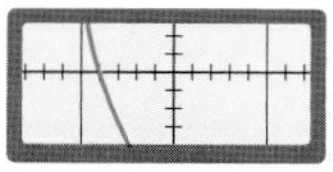	Function that allows a waveform (or part of waveform) to be magnified

Figure 2-4. Test instruments use standard industry terms to describe instrument functions and displayed information.

VOLTMETERS

A *voltmeter* is a test instrument that measures voltage. Voltage is either direct current (DC) or alternating current (AC). *DC voltage* is voltage in a circuit that has current that flows in one direction only. *AC voltage* is voltage in a circuit that has current that reverses its direction of flow at regular intervals. AC voltage is stated and measured as peak, average, or rms values. **See Figure 2-5.** The *peak value (V_{max} or V_p)* of a sine wave is the maximum instantaneous value of either the positive or negative alternations. The *average value (V_{ave})* of a sine wave is the mathematical mean of all instantaneous voltage values in ½ of the sine wave. The average value is equal to 0.637 of the peak value of a standard sine wave. The *root-mean-square (effective) value (V_{rms})* of an AC sine wave is the value that produces the same amount of heat in a pure resistive circuit as DC of the same value. The rms value is equal to 0.707 of the peak value of a sine wave.

Test Lights

A *test light* is a test instrument with a bulb that is connected to two test leads to give a visual indication when voltage is present in a circuit. The most common test light is a neon test light. **See Figure 2-6.** A *neon test light* has a bulb that is filled with neon gas and uses two electrodes to ionize the gas (excite the atoms). Neon test lights are preferred because neon bulbs have extended lifetimes compared to other bulbs. The long life of neon bulbs is attributed to the bulbs having a very high resistance so neon bulbs draw very little current when taking a measurement. The bulb of a test light illuminates when voltage is present in the circuit being tested.

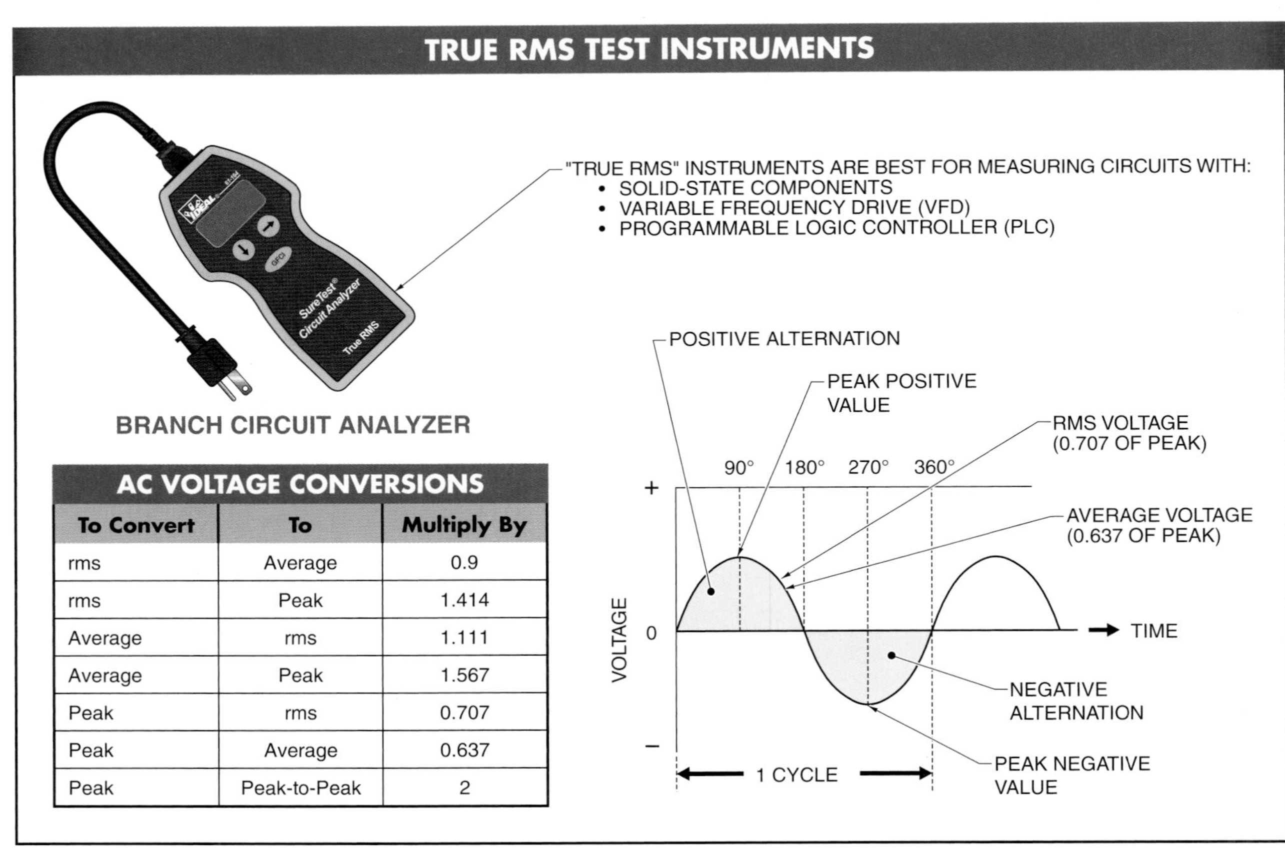

AC VOLTAGE CONVERSIONS

To Convert	To	Multiply By
rms	Average	0.9
rms	Peak	1.414
Average	rms	1.111
Average	Peak	1.567
Peak	rms	0.707
Peak	Average	0.637
Peak	Peak-to-Peak	2

Figure 2-5. True RMS test instruments use an rms value that is equal to 0.707 of the peak value of a sine wave, and are the best test instruments to use when measuring circuits with solid-state components.

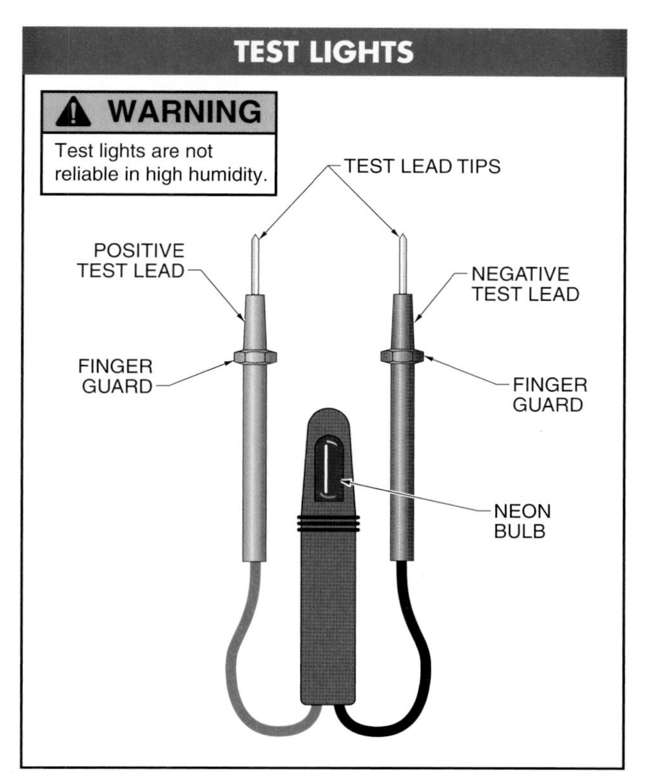

Figure 2-6. Test lights provide a visual indication when voltage is present in a circuit, but do not indicate the amount of voltage.

Test Light Advantages/ Disadvantages

The advantage of using test lights is that test lights are inexpensive, small enough to carry in a pocket, and easy to use. The disadvantage of test lights is that test lights have a limited voltage-indicating range and cannot determine the actual voltage of a circuit, only that voltage is present in a circuit. Test lights that have a wider voltage range are better than test lights that have only one voltage rating. For example, a neon test light rated for 90 VAC to 600 VAC is better than a test light rated for only 120 VAC. Another disadvantage of neon test lights is that neon test lights must not be used to test ground fault circuit interrupters (GFCIs) or ground fault interrupters (GFIs), because the neon bulb does not draw enough current to trip the GFCI when connected between the hot side of the receptacle and ground.

Test Light Applications

Test lights are primarily used to determine when voltage is present in a circuit (the circuit is energized), such as when testing receptacles. When testing a receptacle, the test light bulb illuminates when the receptacle is energized. **See Figure 2-7.**

When a receptacle is properly wired, a test light bulb will illuminate when the test light leads are connected from the neutral slot to the hot slot. A test light bulb will also illuminate when the leads are connected from the ground slot to the hot slot. If the test light illuminates when the leads are connected from the neutral slot to the ground slot, the hot (black) and neutral (white) wires are wired in reverse polarity. The situation of having the hot and neutral wires reversed is a safety hazard and must be corrected.

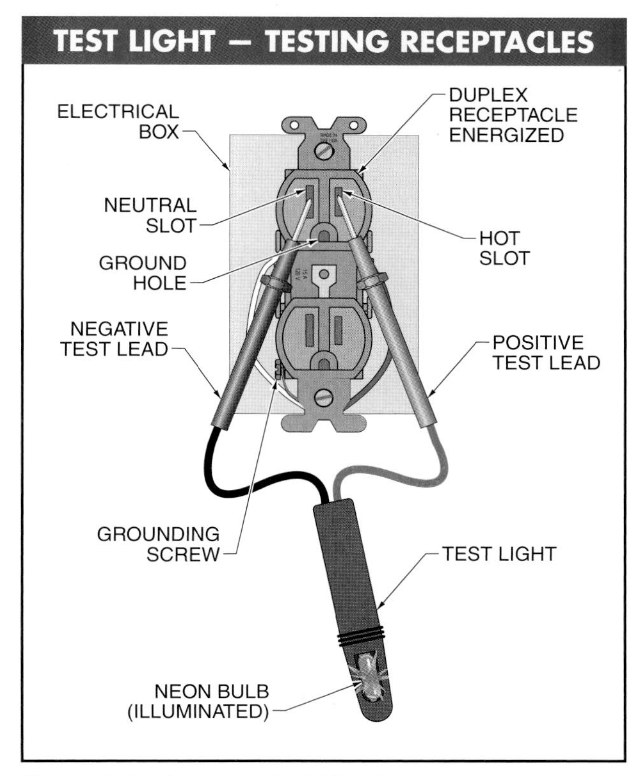

Figure 2-7. A test light (bulb) illuminates when a receptacle is energized.

If a test light illuminates when the test leads are connected to the neutral slot and hot slot but does not light when connected to the ground slot and the hot slot, the receptacle is not grounded. When a test light illuminates, but is dimmer than when connected between the neutral slot and hot slot, the receptacle has an improper ground (having higher resistance). Improper grounds are also a safety hazard and must be corrected.

Testing receptacles is also possible with a receptacle tester. A *receptacle tester* is a device that is plugged into a standard receptacle to determine if the receptacle is properly wired and energized. **See Figure 2-8.** Some receptacle tester models include a ground fault circuit interrupter or ground fault interrupter test button that allows the receptacle tester to be used on GFCI or GFI receptacles.

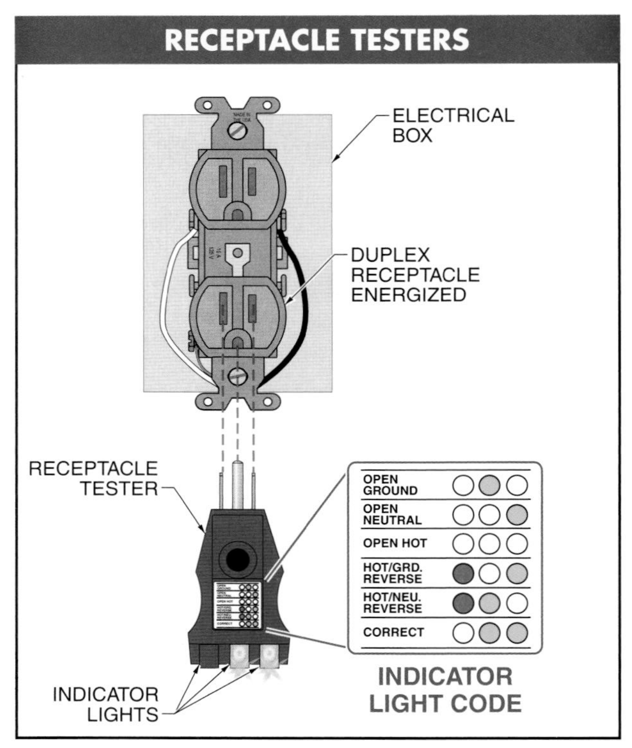

Figure 2-8. Receptacle testers are plugged into standard receptacles to determine if the receptacle is properly wired and energized.

WARNING

Always wear proper protective equipment when working around energized circuits. Exercise caution when testing voltages over 24 V.

Test Light Measurement Procedures

Before using a test light or any voltage measuring instrument, always check the test light on a known energized circuit that is within the test light's rating to ensure that the test light is operating correctly.

Before taking any measurements using a test light, ensure the test light is designed to take measurements on the circuit being tested. Refer to the operating manual of the test instrument for all measuring precautions, limitations, and procedures. **See Figure 2-9.** To test for voltage using a test light, apply the following procedures:

Safety Procedures

- Follow all electrical safety practices and procedures.
- Check and wear personal protective equipment (PPE) for the procedure being performed.
- Perform only authorized procedures.
- Follow all manufacturer recommendations and procedures.

1. Verify that the test light has a voltage rating higher than the highest potential voltage in the circuit. Care must be taken to guarantee that the exposed metal tips of the test light leads do not touch fingers or any metal parts not being tested.
2. Connect one test lead of the test light to one side of the circuit or ground. When testing a circuit that has a neutral or ground, connect to the neutral or ground side of the circuit first.

3. Connect the other test lead of the test light to the other side (hot side) of the circuit. Voltage is present when the test light bulb illuminates. Voltage is less than the rating of the test light when the test light is dimly lit and is higher than the rating of the test light when the test light glows brighter than normal. Voltage is not present in a circuit or present at a very low level when a test light does not illuminate.

4. Remove the test light from the circuit.

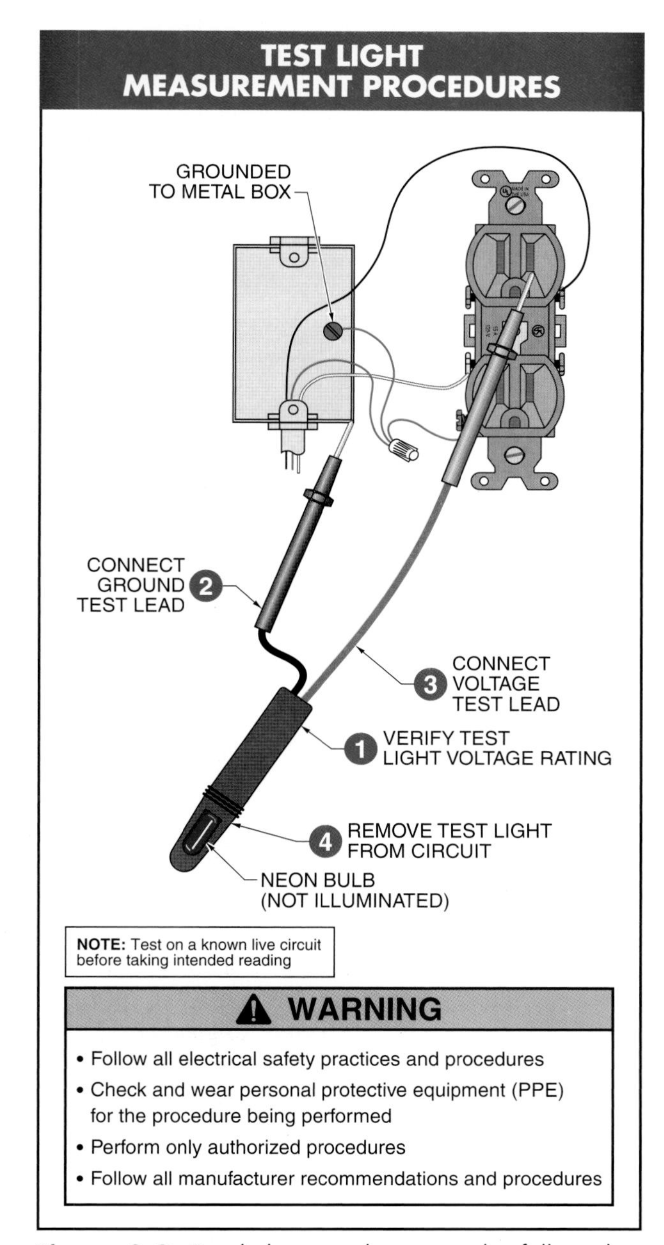

Figure 2-9. Test light procedures must be followed to ensure the safety of electricians, test lights, and circuits being tested.

WARNING

When a test light does not illuminate, a voltage can still be present that could cause an electrical shock. A test light can be damaged during testing by too high a voltage, so always retest a test light on a known energized circuit to verify that the bulb of a test light that indicated no voltage is still operating correctly.

Voltage Indicators

A *voltage indicator* is a test instrument that indicates the presence of voltage when the test tip touches, or is near, an energized hot conductor or energized metal part. The tip glows and/or the device creates a sound when voltage is present at the test point. Voltage indicators are used to test receptacles, fuses/circuit breakers, breaks in cables, and other applications in which the presence of voltage must be detected. **See Figure 2-10.**

Voltage indicators are available in various voltage ranges (a few volts to hundreds of volts) and in the different voltage types (AC, DC, AC/DC) for testing various types of circuits. Voltage indicators rated for 90 VAC to 600 VAC are used to test power-supply circuits. Voltage indicators rated for 24 VAC to 90 VAC are used to test low-voltage control circuits. Some models of voltage indicators are available to test for magnetic fields (AC, DC, or permanent magnets) or are used to test solenoids, transformers, and other types of coils for voltage.

Technical Tip

Ninety-nine percent of voltage test lights are for AC use only. Test lights can lie (false positive) due to resistance values built into the testers by manufacturers. Test lights provide visual (LED) indicators that are difficult to see in direct sunlight.

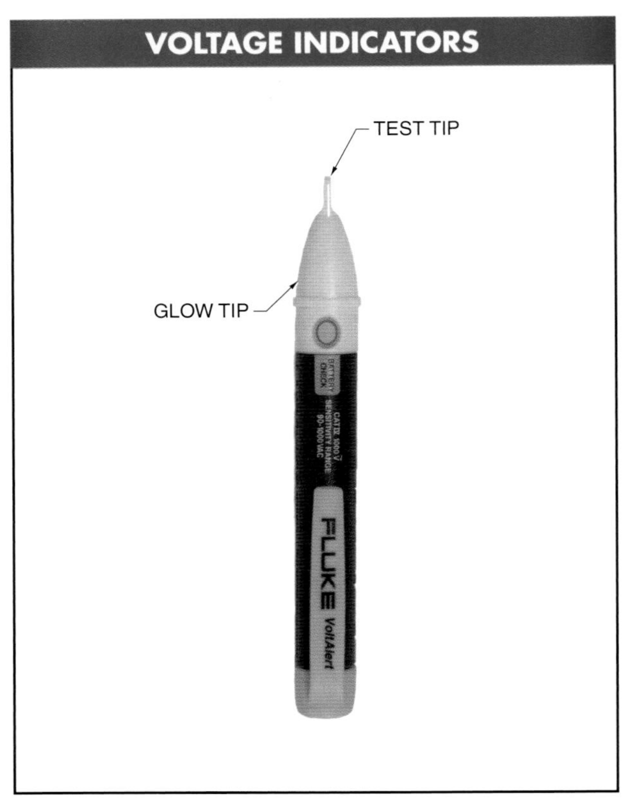

Figure 2-10. A voltage indicator glows to indicate the presence of voltage when the voltage indicator touches, or is near, an energized hot conductor or energized metal part.

Voltage indicators are pocket test instruments that are easy to use and provide a quick visual indication when voltage is present by glowing red.

Voltage Indicator Advantages/ Disadvantages

The advantages of voltage indicators are that voltage indicators are inexpensive, are small enough to carry in a pocket, are easy to use, are nonconductive, and indicate a voltage without touching any live parts of the circuit, even through conductor insulation. One disadvantage of voltage indicators is that voltage indicators only indicate that voltage is present but do not indicate the actual voltage amount. Another disadvantage of voltage indicators is that voltage indicators may not provide an indication that voltage is present, even when voltage is present, as when the wire being tested is shielded.

Voltage Indicator Applications

Voltage indicators are used for many applications, such as finding a break in an extension cord or wire, or determining when a receptacle is hot (energized). One important application of a voltage indicator is in making a preliminary test to determine if any metal parts are hot before beginning service on a circuit or component. A dangerous condition occurs when any hot (energized) conductor touches a metal part that is not grounded. Such conditions occur because of a nick in conductor insulation, or when an exposed metal terminal screw contacts another metal part when a switch or receptacle is loose. An ungrounded situation occurs any time a ground conductor is loose or not connected properly. **See Figure 2-11.**

Before touching any metal parts of a circuit, use a voltage indicator to indicate whether any conductors or metal components are hot (energized). Voltage indicators glow and/or emit sounds when a wire or metal component is probably energized. The wire or component must be tested with a voltmeter to determine the actual voltage amount. When a voltage indicator does not glow and/or sound, the wire or metal component is probably not hot (energized), but it may be energized when shielded conductors are being used. Voltage indicators must only be used as a quick precheck to determine if a wire or metal component is energized.

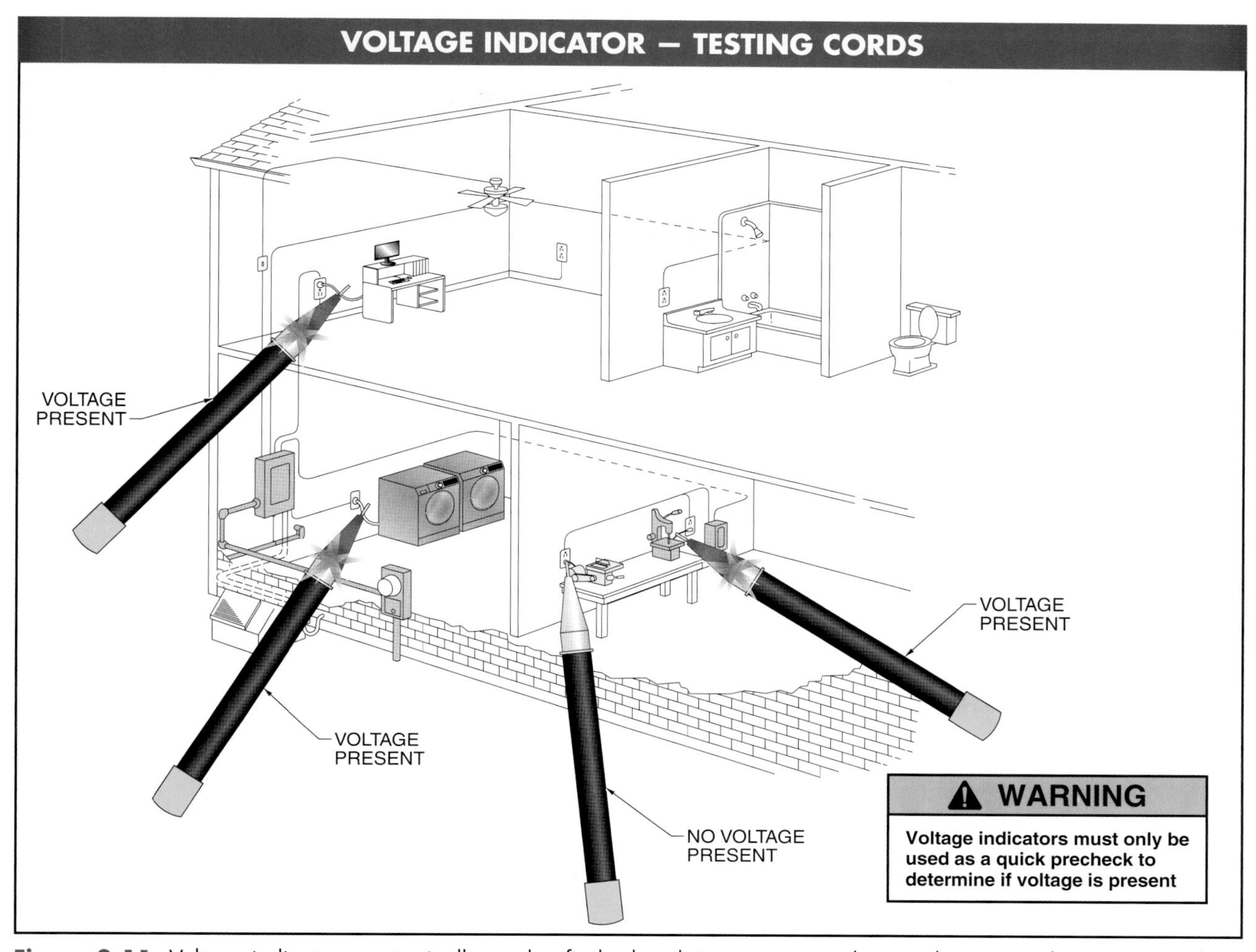

Figure 2-11. Voltage indicators are typically used to find a break in a power cord, or to determine when a receptacle is hot (energized).

Voltage Indicator Measurement Procedures

Before using a voltage indicator or any voltage measuring instrument, always check the voltage indicator on a known energized circuit that is within the voltage rating of the voltage indicator to verify proper operation. A voltage indicator is operating properly when the indicator glows or emits a sound when the tip is placed next to a hot (energized) conductor. The hot side (short slot or black wire) of a standard receptacle is typically used to test voltage indicators. **See Figure 2-12.**

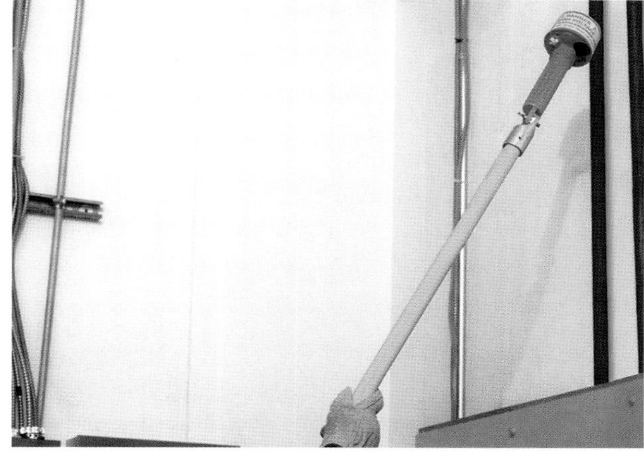

AEMC® Instruments

Noncontact voltage indicators can detect voltages from 220 V to 275 kV.

Before taking any measurements using a voltage indicator, ensure the indicator is designed to take measurements on the circuit being tested. Refer to the operating manual of the test instrument for all measuring precautions, limitations, and procedures. Always wear required personal protective equipment and follow all safety rules when taking the measurement. To test for voltage using a voltage indicator, apply the following procedures:

1. Verify that the voltage indicator has a voltage rating higher than the highest potential voltage in the circuit being tested. When circuit voltage is unknown, slowly bring the voltage indicator near the conductor being tested. The voltage indicator will glow and/or sound when voltage is present. The brighter a voltage indicator glows, the higher the voltage or the closer the voltage indicator is to the voltage source.
2. Place the tip of the voltage indicator on or near the wire or device being tested. When testing an extension cord for a break, test several points along the wire. Expect the voltage tester to turn ON and OFF when moved along a cord that has twisted wire conductors, because the hot wire will change position along the cord.
3. Remove voltage indicator from test area.
4. When the voltage indicator does not indicate the presence of voltage by glowing or making a sound, do not assume that there is no voltage and start working on exposed components of a circuit. Always take a second test instrument (voltmeter or multimeter) and measure for the presence of voltage before working around or on exposed wires and electrical components.

WARNING

When a voltage indicator does not glow or sound, a voltage that can cause an electrical shock may still be present. Always retest a voltage indicator on a known energized circuit after use to verify that the voltage indicator is operating properly.

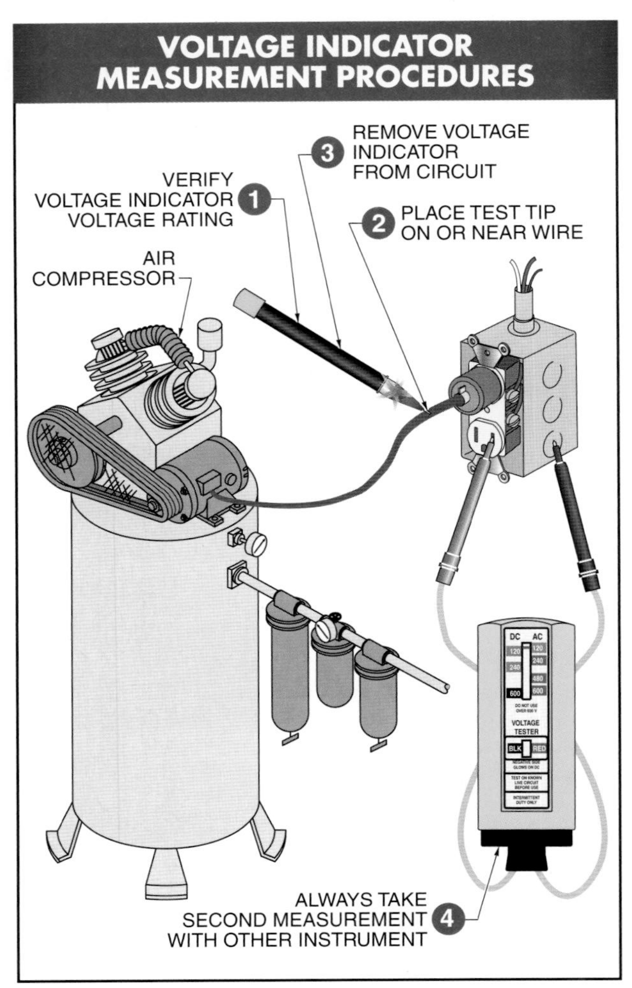

Figure 2-12. A second test instrument is used to verify when voltage is present.

Voltage Testers

A *voltage tester* is an electrical test instrument that indicates the approximate voltage amount and type of voltage (AC or DC) in a circuit by the movement of a pointer (and vibration on some models). When a voltage tester includes a solenoid, the solenoid vibrates when the tester is connected to AC voltage. Some voltage testers include a colored plunger or other indicator such as a light that indicates the polarity of the test leads as positive or negative when measuring a DC circuit. **See Figure 2-13.**

Voltage Tester Advantages/Disadvantages

Advantages of voltage testers are that electricians can concentrate on placing the test leads instead of reading the tester and the meters have a lower impedance (resistance) than voltage indicators or digital multimeters. GFCIs are designed to trip at approximately 6 mA (0.006 A). A disadvantage of voltage testers is that electronic equipment measurements are affected by solenoid voltage testers. Voltage indicators and multimeters cannot be used to trip GFCIs during a test (when connected from the hot receptacle slot to ground). The lower impedance of voltage testers allows voltage testers to be used for testing GFCIs.

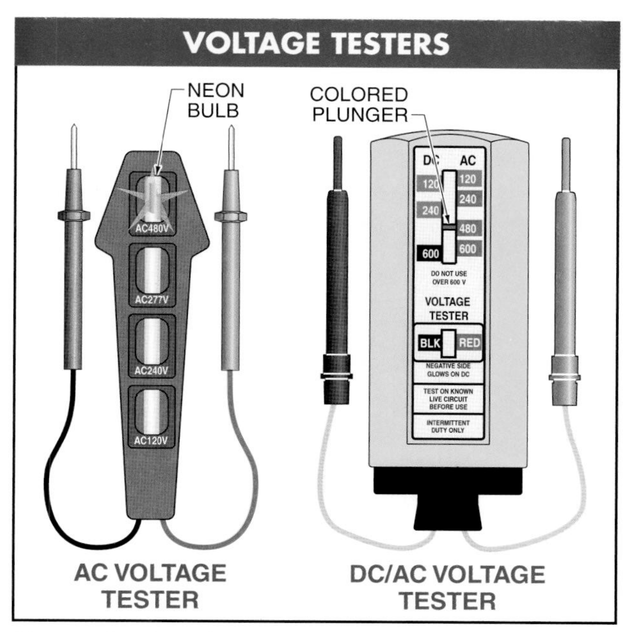

Figure 2-13. Voltage testers indicate the approximate voltage amount and type of voltage (AC or DC) in a circuit.

Voltage Tester Applications

Voltage testers are used to take voltage measurements any time the voltage of a circuit being tested is within the rating of the tester and an exact voltage measurement is not required. Exact voltage measurements are not required to determine when a receptacle is hot (energized), a system is grounded, fuses or circuit breakers are good or bad, or when a system is a 115 VAC, 230 VAC, or 460 VAC circuit. Because test lights, voltage indicators, and voltmeters do not draw enough current to trip a GFCI receptacle, voltage testers are considered the best test instrument for testing GFCI receptacles. Properly wired GFCI receptacles trip when the test button on the receptacle is pressed. See Figure 2-14.

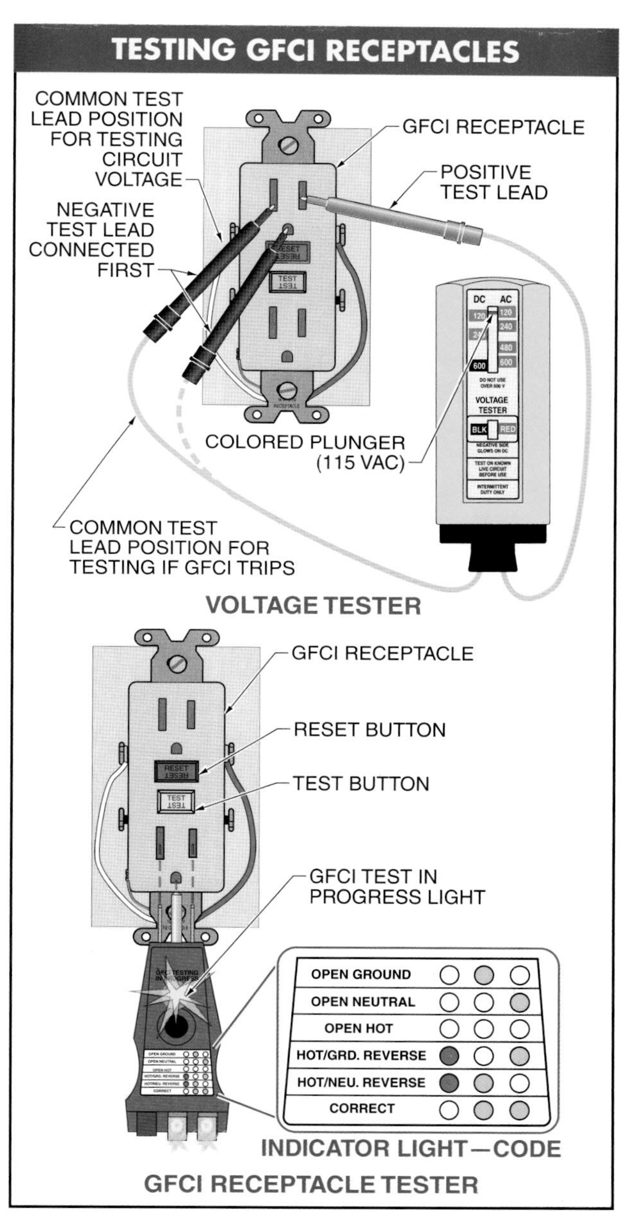

Figure 2-14. Voltage testers and GFCI receptacle testers are the best test instruments to use when testing GFCI receptacles.

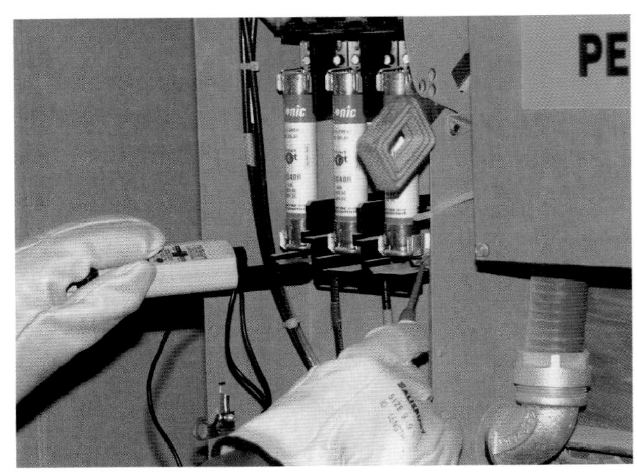

Voltage testers are test instruments commonly used when testing the fuses in a 3ϕ disconnect.

Voltage Tester Measurement Procedures

Before using a voltage tester or any voltage measuring instrument, always check the voltage tester on a known energized circuit that is within the voltage rating of the voltage tester to verify proper operation. **See Figure 2-15.**

Before taking any voltage measurements using a voltage tester, ensure the tester is designed to take measurements on the circuit being tested. Refer to the operating manual of the test instrument for all measuring precautions, limitations, and procedures. Always wear required personal protective equipment and follow all safety rules when taking the measurement. To take a voltage measurement with a voltage tester, apply the following procedures:

1. Verify that the voltage tester has a voltage rating higher than the highest potential voltage in the circuit being tested.
2. Connect the common test lead to the point of testing (neutral or ground).
3. Connect the voltage test lead to the point of testing (ungrounded conductor). The pointer of the voltage tester indicates a voltage reading and vibrates when the current in the circuit is AC. The indicator shows a voltage reading and does not vibrate when the current in the circuit is DC.
4. Observe the voltage measurement displayed.
5. Remove the voltage tester from the circuit.

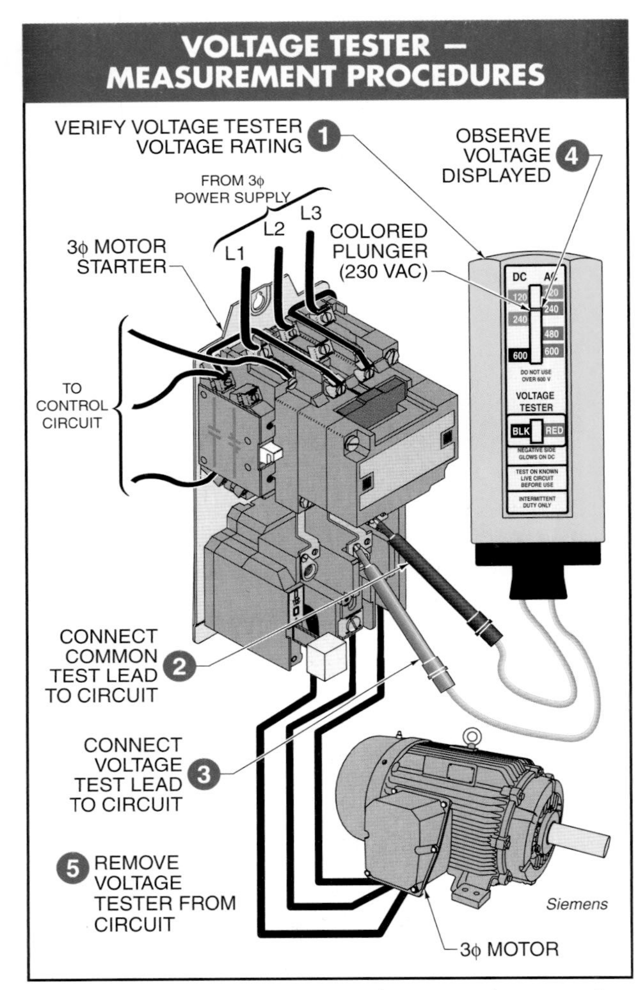

Figure 2-15. A voltage tester shows a voltage reading and vibrates if the circuit current is AC. The indicator shows a voltage reading, but does not vibrate if the current in the circuit is DC.

WARNING

When a voltage tester does not indicate a voltage, a voltage that can cause an electrical shock may still be present. A voltage tester can be damaged during testing by excessive high voltages, so always retest a voltage tester on a known energized circuit to verify that the tester that did not indicate a voltage is still operating correctly.

Circuit Analyzers

Most loads found in households, offices, construction sites, schools, and stores are loads that are designed to be connected to receptacles (outlets) for power. Voltage testers and GFCI receptacle testers are used to verify that GFCI receptacles are wired correctly and operating at the proper voltage level, and that the GFCI trips properly.

Circuit analyzers are used to test multiple circuit operating functions and potential problems. A *circuit analyzer* is a receptacle plug and meter that determines circuit wiring faults (reverse polarity or open ground), tests for proper operation of GFCIs and arc fault breakers (AFCIs), and displays important circuit measurements (hot/neutral/ground voltages, impedance, and line frequency). Circuit analyzers are used to test a circuit for the following:

- Proper receptacle wiring. Circuit analyzers detect open grounds, reverse polarity (hot and neutral are reversed), when hot and ground are reversed, and when there is an open hot or neutral wire.
- Proper GFCI receptacle wiring and operation. Circuit analyzers test GFCIs not only by tripping a GFCI but also by displaying the amount of current (in mA) that tripped the GFCI and the time (in ms) the GFCI took to trip.
- Proper AFCI wiring and operation.
- Proper rms circuit voltages. Voltage measurements taken between ground and neutral provide an approximation of how much load is on a circuit and when there is an illegal ground to neutral connection. Zero volts indicates that there is a neutral to ground connection and a measurement over 5 V typically indicates that the circuit is overloaded.
- Proper line frequency.
- Proper line impedances. High conductor impedance indicates the existence of too much load on the circuit, undersized conductors, or a high-resistance connection in the circuit.
- Proper peak voltage. By comparing peak voltage to the rms voltage of a circuit, an indication of the number and amplitude of harmonics present on a line can be obtained. When an AC sine wave is pure (has no distortion), the peak voltage is equal to the rms voltage multiplied by 1.41. The greater the difference from the expected calculated value (rms times 1.41), the more distorted the AC sine wave.

Circuit Analyzer Measurement Procedures

Before taking any circuit measurements using a circuit analyzer, ensure the analyzer is designed to take measurements on the circuit being tested. **See Figure 2-16.** Refer to the operating manual of the test instrument for all measuring precautions, limitations, and procedures. Always wear required personal protective equipment and follow all safety rules when taking the measurement. To test a receptacle using a circuit analyzer, apply the following procedures:

1. Plug the circuit analyzer into the receptacle (branch circuit) being tested.
2. Allow the analyzer to take circuit measurements.
3. Read and record measurements while scrolling through values recorded.
4. Compare the measurements to normal expected values. **See Appendix.**
5. Remove the circuit analyzer from the receptacle (branch circuit).

Voltage Measurement Analysis

Typically, all AC voltage sources vary from fluctuations in the AC voltage over the power distribution system. When voltage measurements are different from what was expected, the voltage measurement is more likely to be lower than normal. Voltage measurements taken from AC power systems must be within –10% and +5% of the rating of the load(s) in the circuit. Voltage measurements taken from DC power systems must be within –10% and +10% of the rating of the load(s) in the circuit. **See Figure 2-17.**

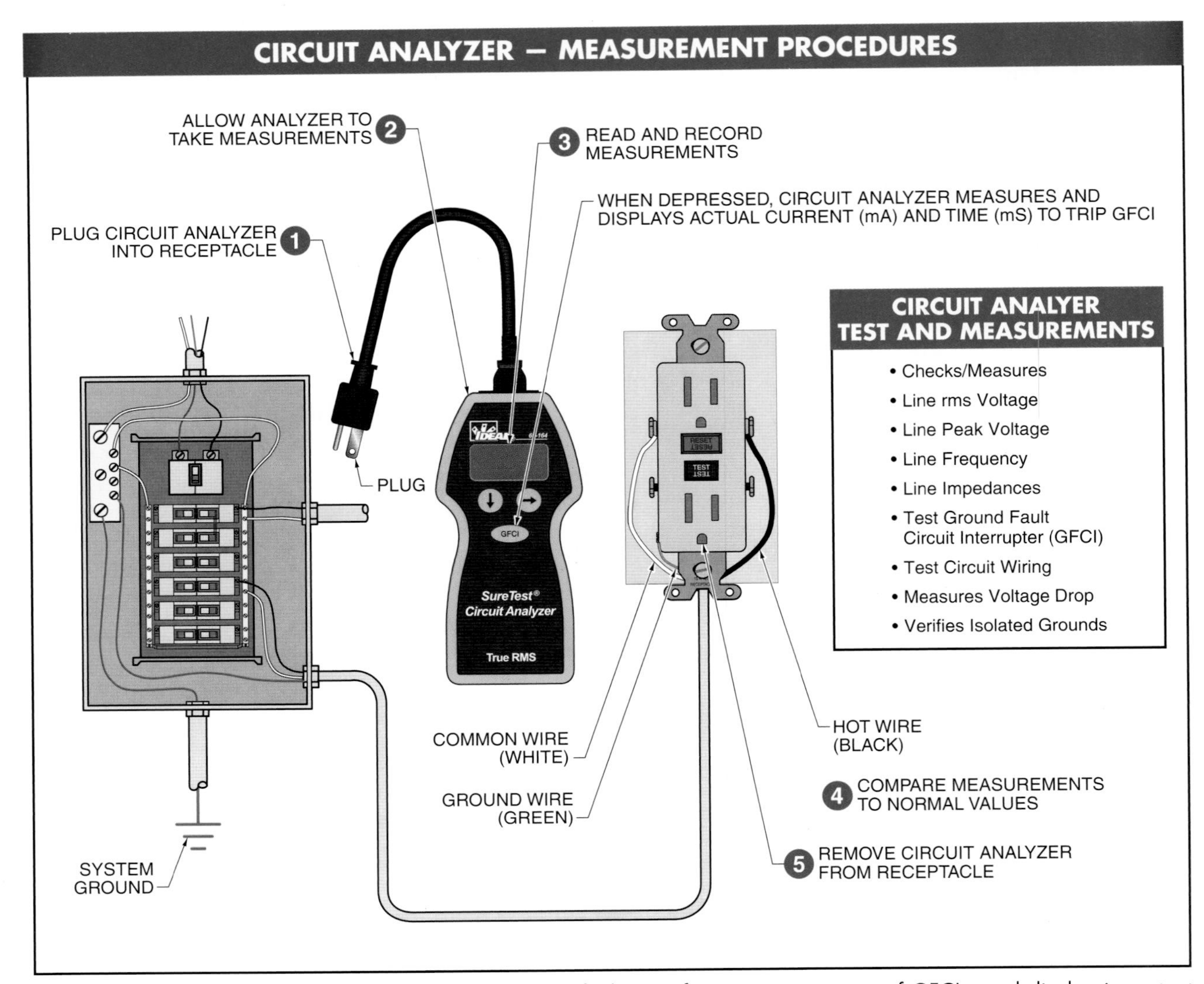

Figure 2-16. Circuit analyzers determine circuit wiring faults, test for proper operation of GFCIs, and display important circuit measurements.

SYSTEM VOLTAGE RANGES*

Supply	Service Range		Point of Use Range	
	Satisfactory	Acceptable	Satisfactory	Acceptable
120, 1ϕ	114 – 126	114 – 126	110 – 126	106 – 127
120/240, 1ϕ	114/228 – 126/252	114/228 – 126/252	110/220 – 126/252	106/212 – 127/254
120/208, 3ϕ	114/197 – 126/218	114/197 – 126/218	110/191 – 126/218	106/184 – 127/220
120/240, 3ϕ	114/228 – 126/252	114/228 – 126/252	110/220 – 126/252	106/212 – 127/254
277/480, 3ϕ	236/456 – 291/504	236/456 – 291/504	254/440 – 291/504	254/424 – 293/508

* in volts

Figure 2-17. Voltage measurements taken from AC power systems should be within –10% and +5% of the rating of the load(s) in the circuit, and DC power systems should be within –10% and +10% of the rating of the load(s) in the circuit.

Voltage Unbalance

Voltage unbalance (imbalance) is the unbalance that occurs when the voltages at the terminals of a motor or other 3ϕ load are not equal. When voltage to a 3ϕ motor is unbalanced, one or two windings overheat, causing thermal deterioration of the windings. Voltage unbalance also results in a current unbalance. Line (L1, L2, and L3) voltages must be tested for voltage unbalance periodically and during all service calls. Voltage unbalance should not be more than 1%. Whenever there is a voltage unbalance of 2% or greater, the following steps must be taken:

1. Check the surrounding power system for excessive loads connected to one line (L1, L2, or L3). Use a power quality meter to measure the true power, apparent power, reactive power, power factor, and displacement power factor of a circuit.
2. Adjust the load or motor rating by reducing the load on the motor or by oversizing the motor if the voltage unbalance cannot be corrected.
3. Notify the local power company.

The primary source of voltage unbalances of less than 2% is too many 1ϕ loads on one phase of a three-phase distribution system. High voltage unbalances are commonly caused by a blown fuse in one phase of a three-phase capacitor bank.

Voltage unbalance is found by applying the following standard procedures. **See Figure 2-18.**

1. Measure the voltage between each incoming power line. Take measurements from L1 to L2, L1 to L3, and L2 to L3.
2. Add the voltages.
3. Find the voltage average by dividing by 3.
4. Find the voltage deviation by subtracting the voltage average from the voltage with the largest measurement.
5. Find the voltage unbalance by applying the formula:

$$V_u = \frac{V_d \times 100}{V_a}$$

where

V_u = voltage unbalance (in %)

V_d = voltage deviation (in V)

V_a = voltage average (in V)

100 = constant

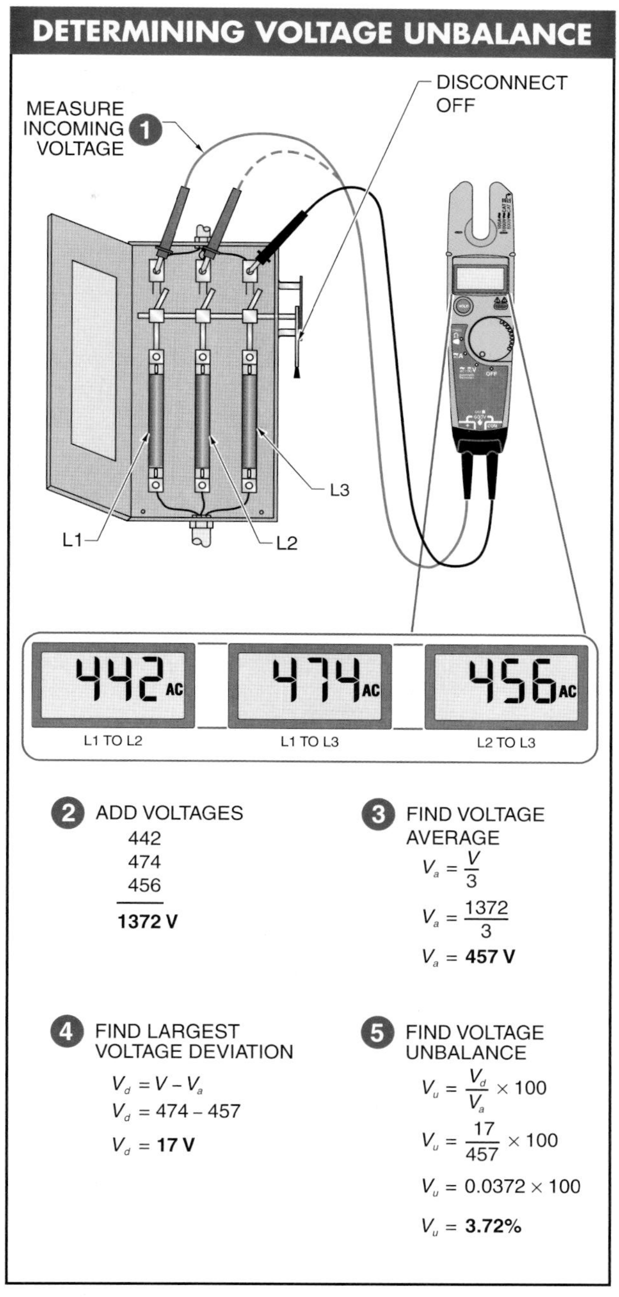

Figure 2-18. Voltage unbalance should not be more than 1%. Voltage unbalances that are 2% or more must be eliminated.

When a 3ϕ motor fails due to voltage unbalance, one or two of the stator windings become blackened. **See Figure 2-19.** The darkest winding is the winding with the largest voltage unbalance.

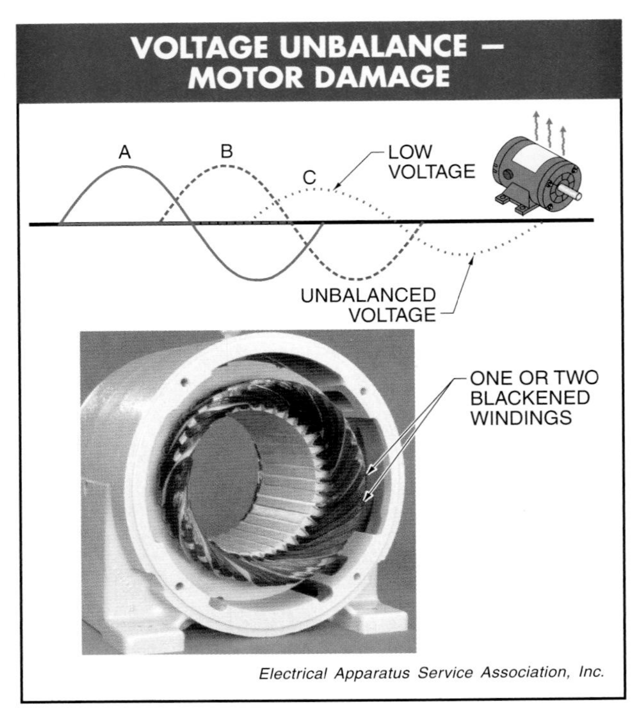

Figure 2-19. One or two of the stator windings of a motor become blackened when a 3f motor fails due to voltage unbalance.

OHMMETERS

An *ohmmeter* is a test instrument that measures resistance. **See Figure 2-20.** *Resistance (R)* is the opposition to the flow of electrons in a circuit. Resistance is measured in ohms (Ω). Higher resistance measurements are expressed using prefixes, as in kilohms (kΩ) and megohms (MΩ). Resistance measurements on permanent test instruments are displayed as Ω, kΩ, and MΩ. Prefixes are used to simplify the measurement displayed.

All test instruments that include an ohmmeter setting measure the amount of resistance in a de-energized circuit or component. Circuits or components are not required to be energized because the test instrument (set for resistance measurement) has an internal battery that is used to supply voltage to the test instrument leads and the component being tested. All resistance measurements must be taken with a de-energized circuit. When a circuit includes a capacitor, the capacitor must be discharged before taking any resistance readings of the circuit.

Resistance measurements are normally taken to indicate the condition of a circuit or component. The higher the resistance in a circuit, the lower the current flow through the circuit. Likewise, the lower the resistance in a circuit, the higher the current flow through the circuit. Components designed to insulate, such as rubber or plastic, have very high resistance values. Components designed to conduct, such as conductors or component contacts, have very low resistance values.

When insulators are damaged by moisture and/or overheating, resistance decreases. When conductors are damaged by burning and/or corrosion, resistance increases. Other components such as heating elements and resistors have a fixed resistance value. Any significant change in the fixed resistance value typically indicates a problem. Some components such as resistors include on the component a resistance value along with a tolerance. When a tolerance is indicated, the measured resistance value must be within the specified resistance range.

WARNING

To avoid injury to personnel or damage to equipment, always verify that a circuit is de-energized before taking any measurements.

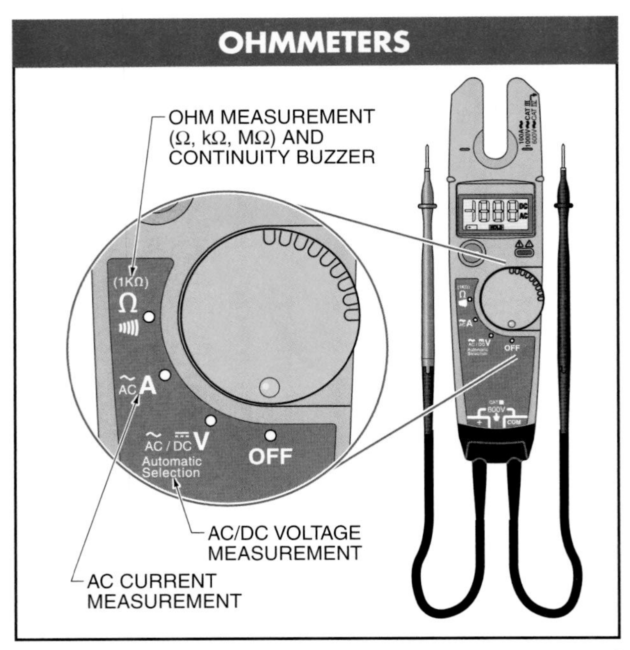

Figure 2-20. The amount of resistance (ohms) is measured in de-energized circuits or components.

Continuity Testers

A *continuity tester* is a test instrument that tests for a complete path for current to flow. For example, a closed switch that is operating properly has continuity. An open switch does not have continuity. **See Figure 2-21.** Some test instruments test for continuity using a continuity test mode. The continuity test mode is commonly used to test components such as switches, fuses, electrical connections, and individual conductors. The test instrument emits an audible response (beeps) when there is a complete path. Indication of a complete path can be used to determine the condition of a component as open or closed. For example, a good fuse should have continuity, whereas a bad fuse does not have continuity.

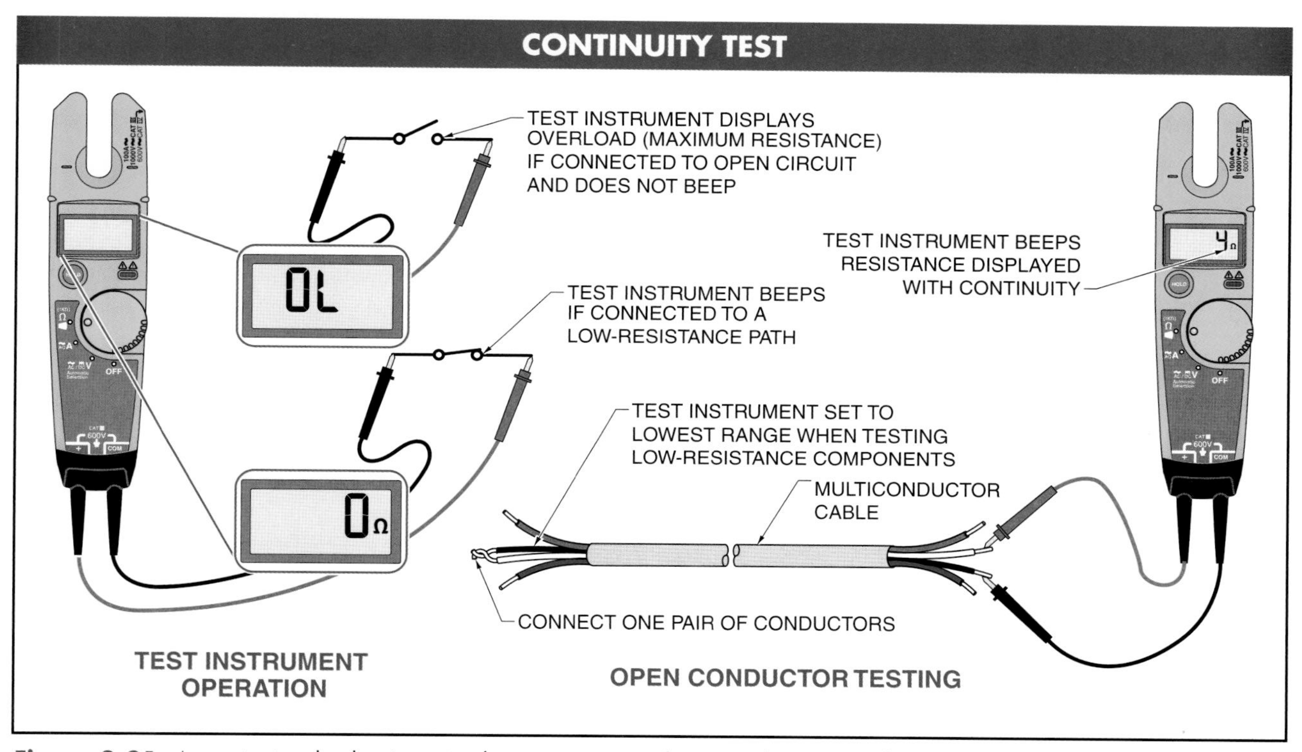

Figure 2-21. A continuity checker is a simple test instrument that tests de-energized circuits or components for a complete path for current to flow.

Continuity Tester Advantages/Disadvantages

The main advantage of using the continuity test mode of a test instrument is that an audible response is sometimes more desirable than reading a resistance measurement. An audible response allows the electrician to concentrate on the testing procedures without looking at the display. The main disadvantage of using a continuity tester is that continuity testers only indicate continuity by sound on a circuit or component. Continuity testers only operate with circuits that have very low resistance (typically 40 Ω or lower) and will not indicate the actual resistance measurement of the circuit or component being tested.

Continuity Tester Applications

Continuity testers are used to test for a complete flow path (open, broken, or closed) in any de-energized low-resistance device. A continuity tester is a common test tool to use when testing single-pole, 3-way, and 4-way switches. **See Figure 2-22.** For example, a 3-way switch has one terminal that is the "common" terminal. Depending upon the manufacturer, the common terminal can be located in any position on a 3-way switch. The common terminal is often identified on a 3-way switch by a different color. On specialty switches, such as automobile switches, the common terminal is not identified. Continuity testers are used to determine the common terminal and proper switch operation.

Continuity Tester Measurement Procedures

Continuity is tested with a test instrument set on the continuity test mode. **See Figure 2-23.**

 **WARNING**

A continuity tester must only be used on de-energized circuits or components. Any voltage applied to a continuity tester causes damage to the test instrument and/or harm to the electrician. Always test a circuit for voltage before taking a continuity test.

Before taking any continuity measurements using a continuity tester, ensure the meter is designed to take measurements on the circuit being tested. Refer to the operating manual of the test instrument for all measuring precautions, limitations, and procedures. Always wear required personal protective equipment and follow all safety rules when taking the measurement. To take continuity measurements with a continuity tester, apply the following procedures:

1. Set the continuity tester function switch to continuity test mode as required. Most test instruments have the continuity test mode and resistance mode sharing the same function switch position.

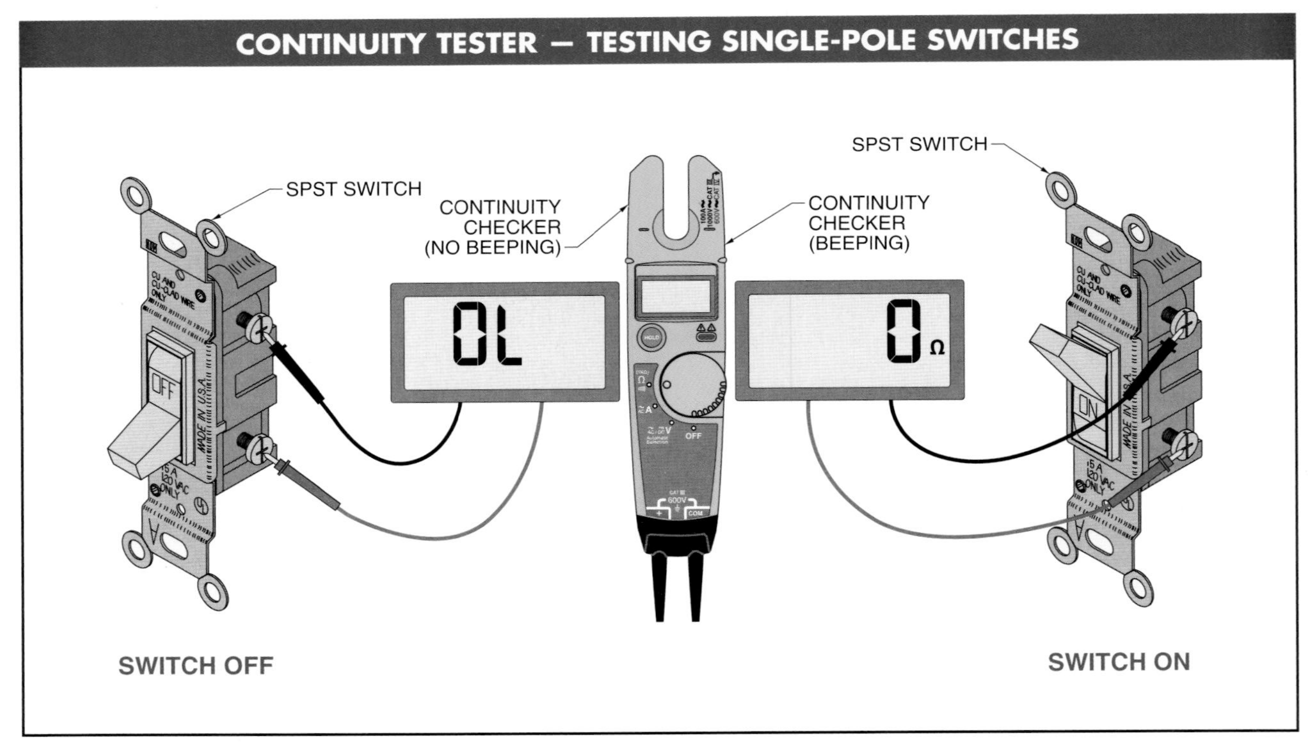

Figure 2-22. Single-pole switches (SPST) have continuity (buzzing) when closed (ON) and no continuity (no buzzing) when open (OFF).

2. With the circuit de-energized, connect the test leads across the component being tested. The position of the test leads is arbitrary.
3. When there is a complete path (continuity), the continuity tester beeps. When there is no continuity (open circuit), the continuity tester does not beep.
4. After completing all continuity tests, remove the continuity tester from the circuit or component being tested and turn the instrument OFF to prevent battery drain.

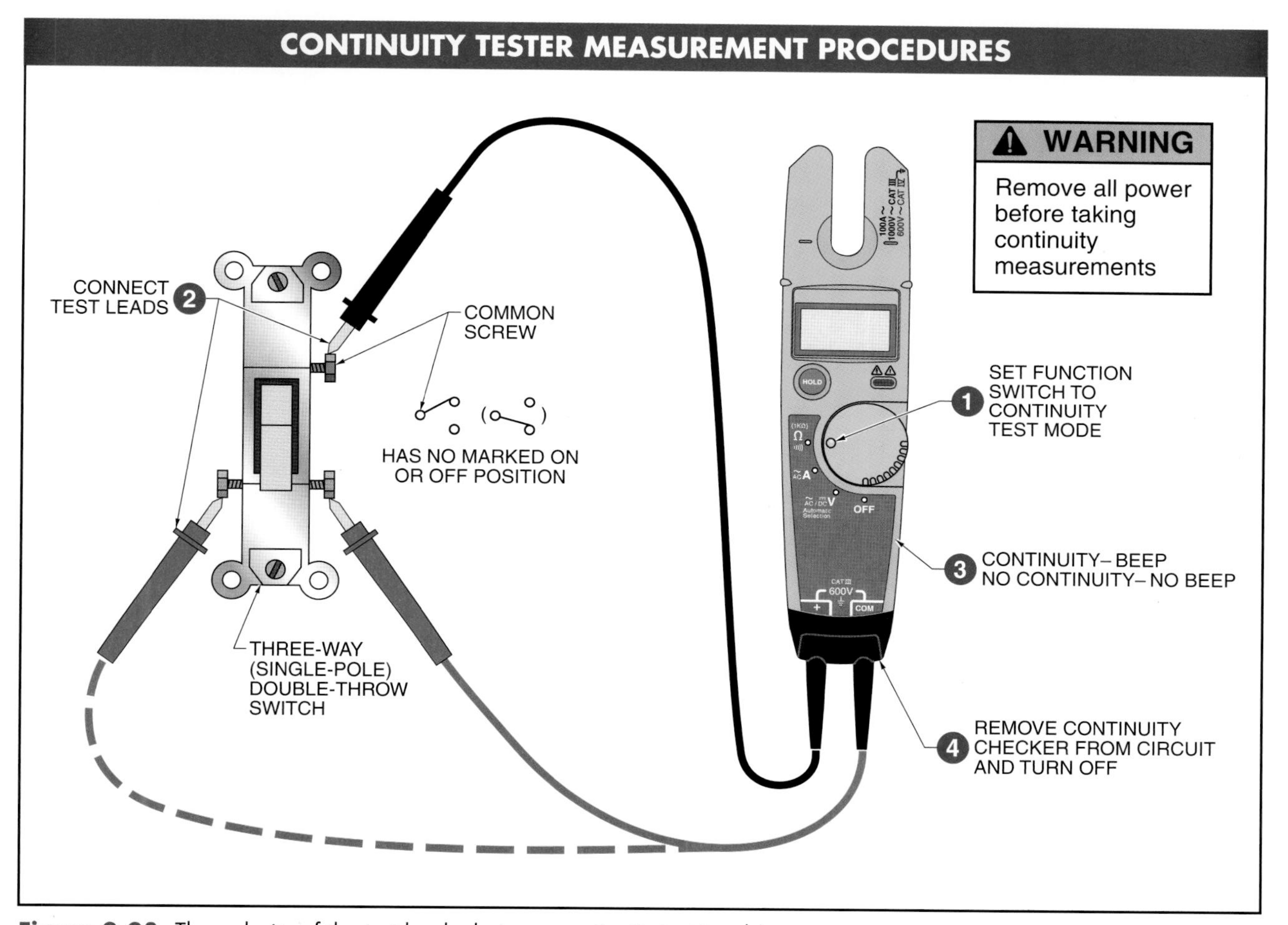

Figure 2-23. The polarity of the test leads during a continuity test is arbitrary.

Technical Tip

PC board trace continuity testers test the paths of traces by touching the tips of the tester to the components on the circuit board. An audible beep verifies the continuity of a specific circuit. Trace continuity testers use a low voltage that will not bias semiconductor devices into conduction or harm any sensitive components on a PC board.

Ohmmeter Resistance Measurement

Ohmmeter resistance measurements are taken to determine the resistance (condition) of de-energized (having all power OFF) components or circuits. Analog ohmmeters or digital ohmmeters are used to take resistance measurements. **See Figure 2-24.**

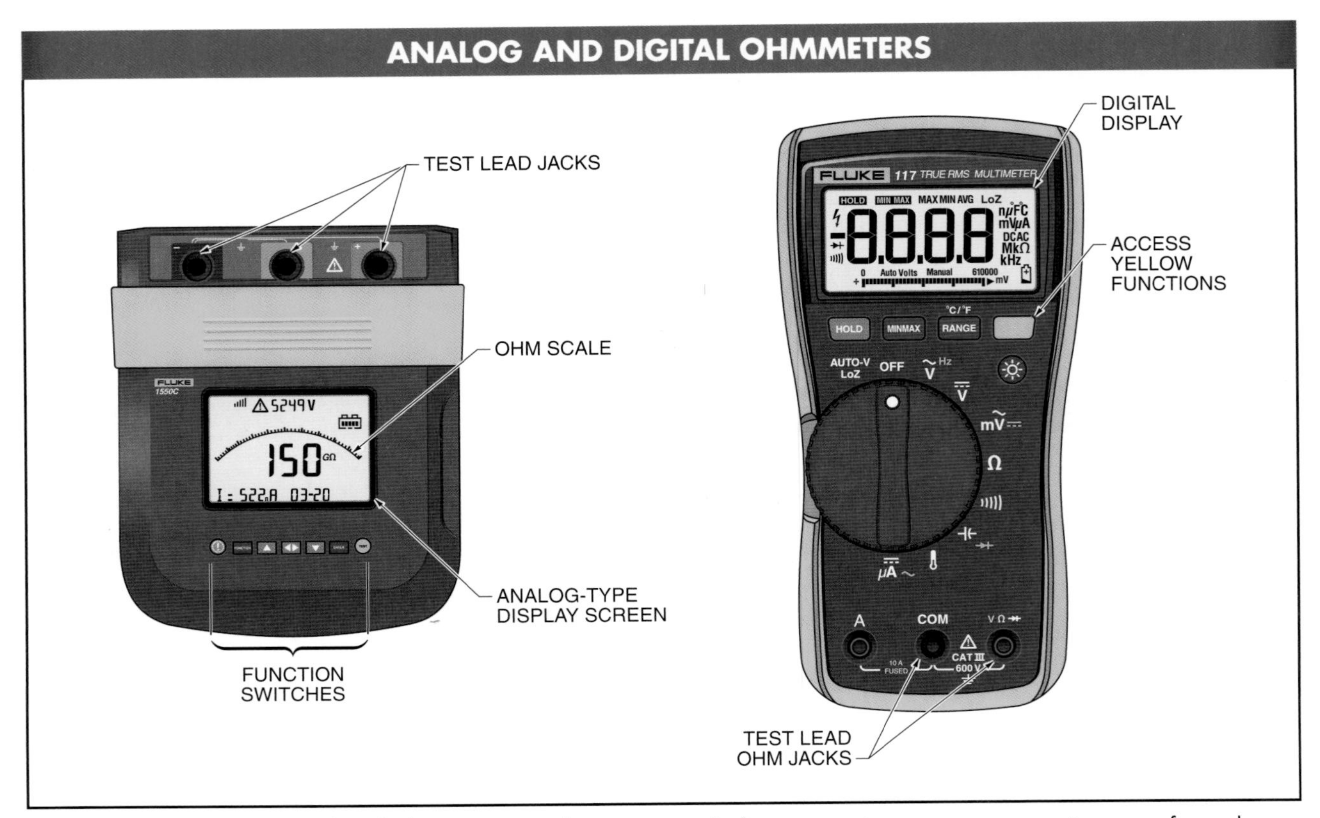

Figure 2-24. Analog and digital ohmmeters must be set to zero before any resistance measurements are performed.

Analog ohmmeters have a function switch that must be set to ohms and a selector switch that must be set to a range position (Ω, R × 1, R × 100, R × 1K, or R × 10,000). Because resistance measurements are taken using the internal batteries of an ohmmeter, the pointer of the meter must be zeroed for the condition of the batteries before taking any resistance measurements. The pointer of an analog ohmmeter is zeroed by connecting the two test leads together and adjusting the zero adjuster control knob until the pointer is on 0 Ω. Zeroing the pointer must be performed each time the ohmmeter function switch or selector switch positions are changed to a different range because battery condition affects resistance measurements.

Digital ohmmeters have a function switch that must be set to the resistance measurement (Ω) mode. As with analog ohmmeters, digital ohmmeters also use internal batteries to take resistance measurements. Digital ohmmeters are zeroed (meter reads resistance of test leads) by setting the function switch of the meter to the resistance measurement (Ω) position and touching the two test leads of the meter together. Some digital ohmmeters read the resistance of the test leads, while other ohmmeters automatically set to zero. When the resistance of the test leads is displayed, the displayed resistance of the test leads is eliminated, zeroing the meter, by pressing the relative mode (REL) button on the digital ohmmeter.

Ohmmeter Applications

Ohmmeter measurements not only are used to determine the resistance of a circuit or component, but are also used to determine the type of wiring of a component being tested. For example, an ohmmeter being used for a resistance test can be used to determine if a 9-lead, 3ϕ motor is wye or delta connected and if the windings of the motor are in good condition. **See Figure 2-25.**

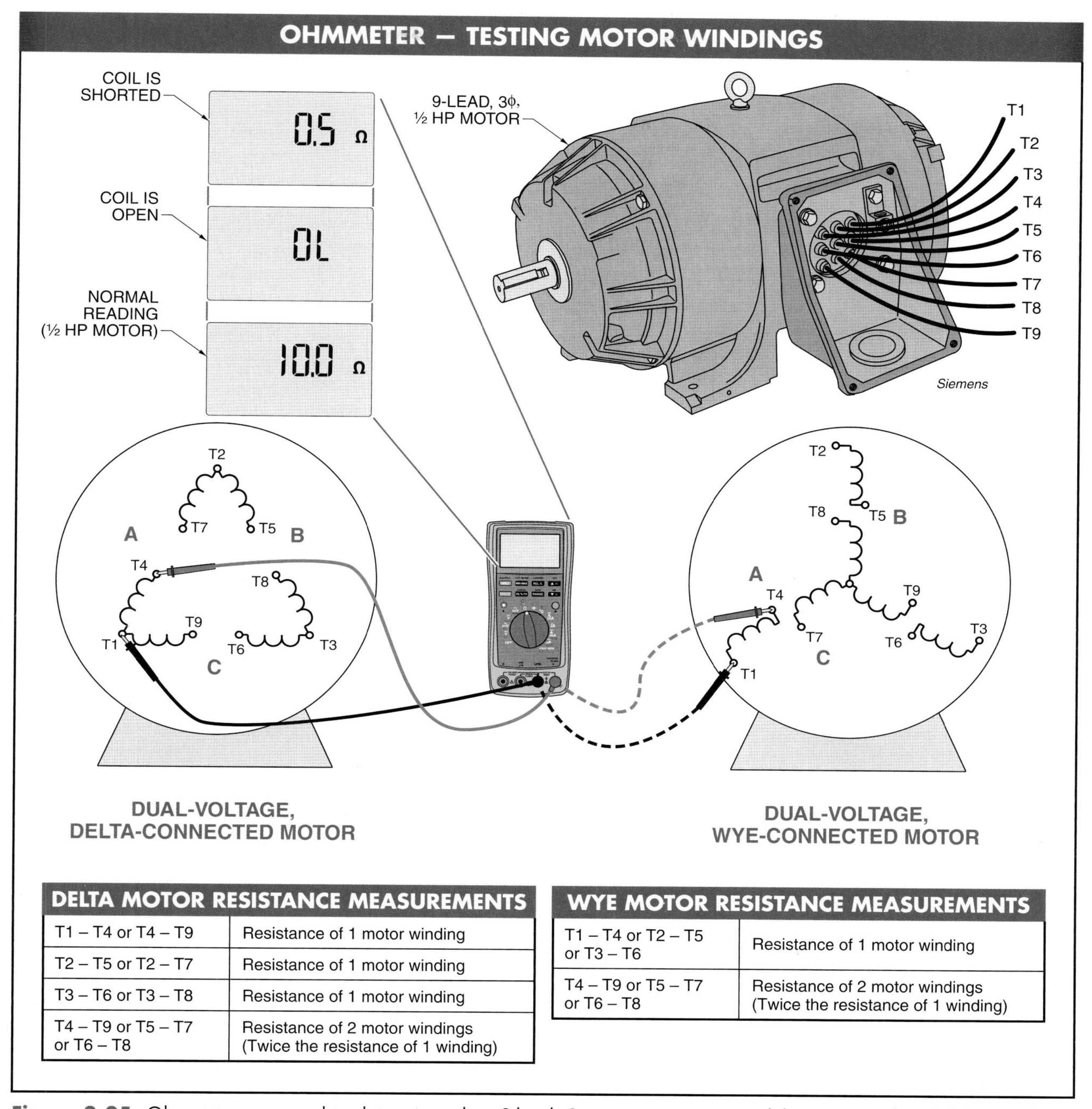

DELTA MOTOR RESISTANCE MEASUREMENTS

T1 – T4 or T4 – T9	Resistance of 1 motor winding
T2 – T5 or T2 – T7	Resistance of 1 motor winding
T3 – T6 or T3 – T8	Resistance of 1 motor winding
T4 – T9 or T5 – T7 or T6 – T8	Resistance of 2 motor windings (Twice the resistance of 1 winding)

WYE MOTOR RESISTANCE MEASUREMENTS

T1 – T4 or T2 – T5 or T3 – T6	Resistance of 1 motor winding
T4 – T9 or T5 – T7 or T6 – T8	Resistance of 2 motor windings (Twice the resistance of 1 winding)

Figure 2-25. Ohmmeters are used to determine when 9-lead, 3ϕ motors are wye or delta connected and if the windings are open or shorted.

Ohmmeters are used to test for open and short motor windings, and/or to test the resistance of the windings. An open winding of a motor is displayed as an overload (OL). A shorted motor winding is displayed as 0 Ω or as a lower than normal winding resistance; this applies for T1 to T4 on both wye- and delta-connected nine-lead motors.

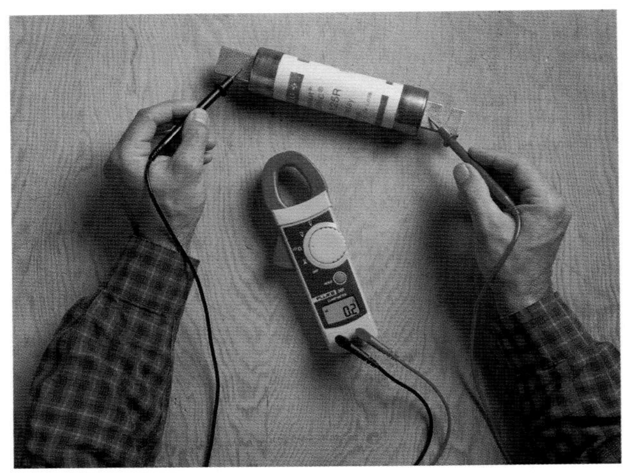

Fluke Corporation

Continuity testers beep, but some multimeters set to measure continuity emit an acoustic signal that is proportional in pitch to the resistance measured in a circuit.

WARNING

Ohmmeters measure resistance on de-energized circuits or components. Low voltages from circuits applied to ohmmeters set to measure resistance cause inaccurate readings and possible meter damage. High voltages from circuits applied to ohmmeters set to measure resistance cause ohmmeter damage, even when the DMM has internal protection. Always verify that circuits or components do not have voltage before taking any resistance measurements.

Ohmmeter Measurement Procedures

Before taking any resistance measurements using an ohmmeter, ensure the meter is designed to take measurements on the circuit being tested. **See Figure 2-26.** Refer to the operating manual of the test instrument for all measuring precautions, limitations, and procedures. Always wear required personal protective equipment and follow all safety rules when taking the measurement. Resistance is measured with test instruments that include a resistance measurement mode by applying the following procedures:

1. Verify that all power is OFF to the circuit and remove the component being tested from the circuit.
2. Set the ohmmeter function switch to the resistance mode as required. Test instruments display OL and the ohm (Ω) symbol when set to the resistance mode.
3. Plug the black test lead into the common jack when required.
4. Plug the red test lead into the resistance jack when required.
5. Ensure that the batteries of the test instrument are in good condition. Most digital meters display the battery symbol when the batteries are low.
6. Zero the ohmmeter. When the test leads of the ohmmeter are connected together the display of the meter should read 0 Ω. However, the resistance of the test leads (typically 0.2 Ω to 0.5 Ω) may be displayed and can affect a very low resistance measurement when the ohmmeter does not have a relative mode button to set the meter to zero.
7. Connect the test leads across the circuit or component being tested. Ensure that the contact between the test leads and the circuit or component is correct. Dirt, solder flux, oil, and other foreign substances greatly affect resistance readings. Also, any contact with the metal ends of the test leads by the fingers of the electrician during the resistance test affects the resistance measurement.
8. Read the resistance measurement displayed on the ohmmeter.
9. After completing all resistance measurements, remove the ohmmeter from the circuit or component and turn the meter OFF to prevent battery drain.

Technical Tip

Only batteries approved by the manufacturer of a test instrument may be used in a test instrument. After removing the old batteries, the contacts of the new batteries and the battery holders of the test instrument must be cleaned. Ensure the correct polarity when inserting new batteries into holders.

Follow the procedure for battery replacement:

- Always replace all of the batteries at one time.
- Ensure that the polarity of the new batteries matches the test instrument.
- Dispose of used batteries properly, in an environmentally friendly manner.

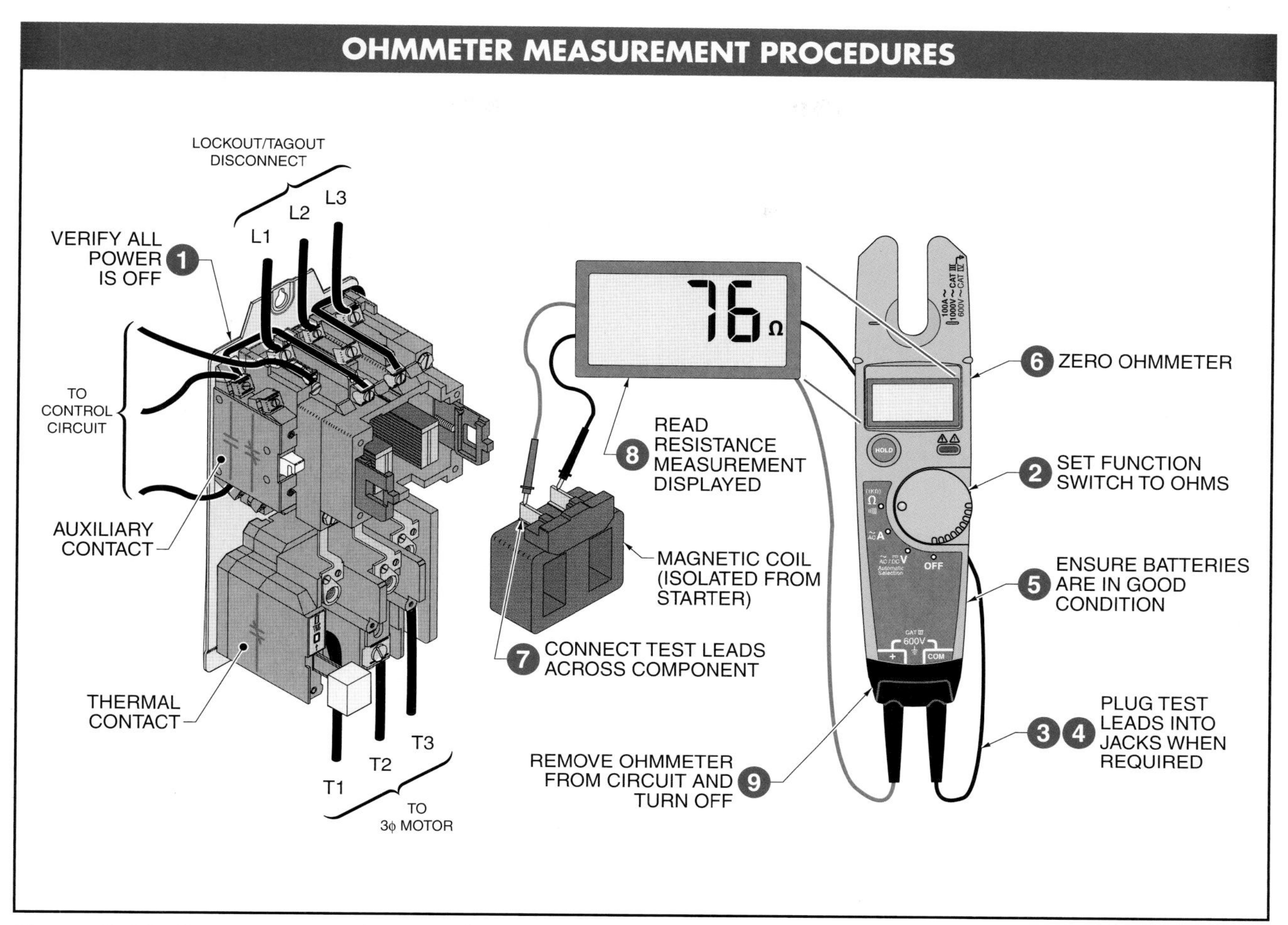

Figure 2-26. Electricians must always verify that circuits or components do not have voltage before taking any resistance measurements.

Resistance Measurement Analysis

The significance of the information obtained from a resistance measurement depends on the component being tested. Typically, the resistance of a component varies over time, and resistance varies slightly from component to component. Slight resistance changes are not considered critical, but changes may indicate a pattern that must be noted. For example, as the resistance of a heating element increases over time, the current passing through the heating element decreases, and the heat produced by the heating element also decreases.

Typically, components used in control circuits, such as switches and relay contacts, have very low resistance when new, but increase in resistance over time from causes such as wear, dirt, and heat. Loads such as motors and solenoids decrease in resistance over time from insulation breakdown and moisture in the ambient air.

The resistance measurement displayed by ohmmeters is the total resistance through all possible paths found between the test leads. Electricians must be aware of the possible series and parallel connections involved when taking resistance measurements across a component that is part of a circuit. The resistance of all components connected in parallel with the component being tested affects, and typically lowers, the resistance reading. Always check the circuit schematic for parallel and series flow paths. **See Figure 2-27.**

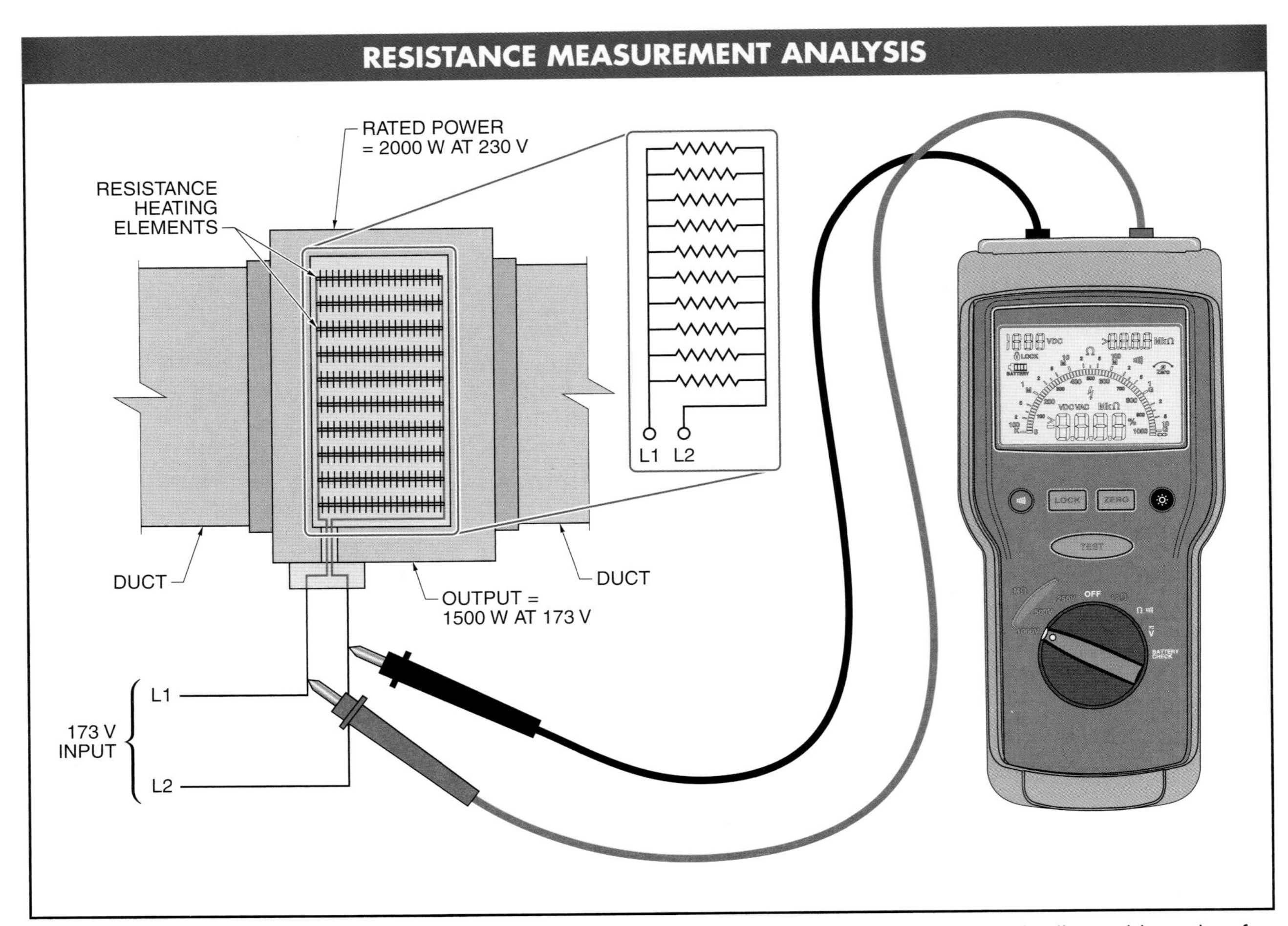

Figure 2-27. The resistance measurement displayed by ohmmeters is the total resistance through all possible paths of a circuit, found between the test leads.

Megohmmeters/Insulation Testers

A *megohmmeter* is a high-resistance ohmmeter used to measure insulation deterioration on various wires by measuring high resistance values during high voltage test conditions. Megohmmeter test voltages range from 50 V to 5000 V. A megohmmeter detects insulation failure or potential failure of insulation caused by excessive moisture, dirt, heat, cold, corrosive substances, vibration, and aging. **See Figure 2-28.** Megohmmeters are sometimes referred to as insulation testers.

Megohmmeter Applications

Some insulation, such as that found on conductors used to wire branch circuits, is thick insulation that is hard to damage or break down. Other insulation, such as the insulation used on motor windings, is very thin (to save weight and space) and breaks down more easily. Megohmmeters are used to test for insulation breakdown in long wire runs, motor windings, and transformers. **See Figure 2-29.**

The minimum resistance of motor windings depends on the voltage rating of the motor. The minimum acceptable resistance measurements are typically found in manufacturer recommended resistance tables. **See Figure 2-30.**

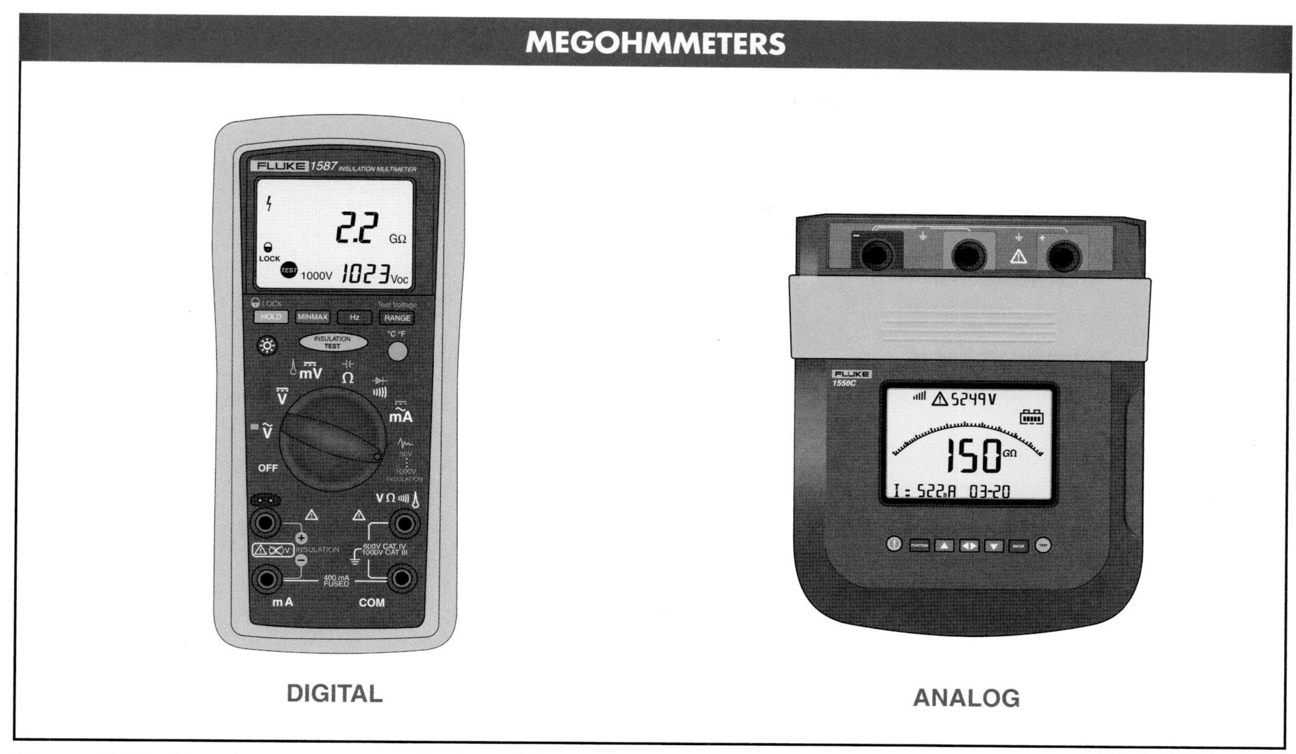

Figure 2-28. Megohmmeters are available in different designs with a variety of functions and features.

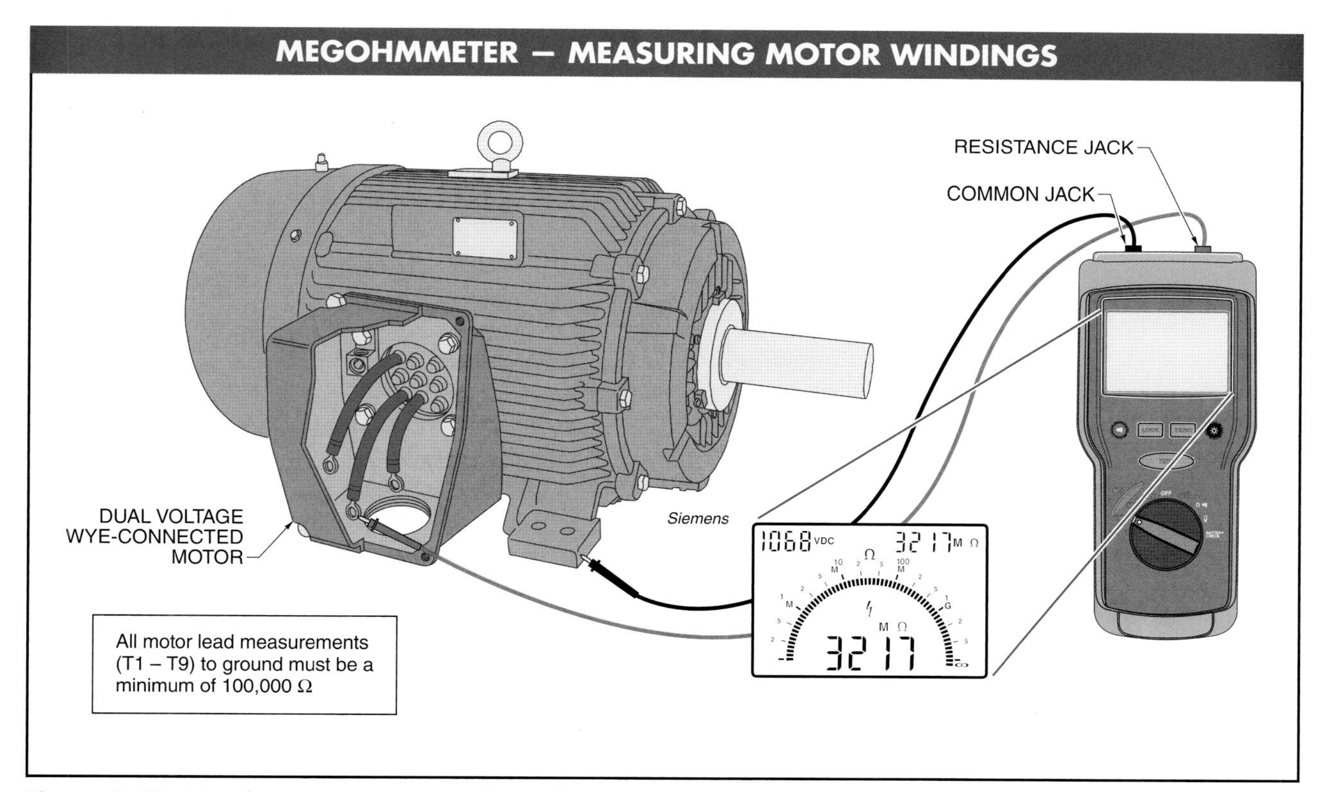

Figure 2-29. Megohmmeters measure insulation deterioration on conductors, motor windings, and transformers.

RECOMMENDED MINIMUM RESISTANCE*	
Motor Voltage Rating (From Nameplate)	**Minimum Acceptable Resistance†**
Less than 208	100,000 Ω
208 – 240	200,000 Ω
240 – 600	300,000 Ω
600 – 1000	1 MΩ
2400 – 5000	2 MΩ
1000 – 2400	3 MΩ

* values for motor windings at 40°C
† in ohms

Figure 2-30. The minimum resistance of motor windings depends on the voltage rating of the motor as recommended by the manufacturer.

Insulation allows conductors to stay separated from each other and from earth ground. Insulation must have a high resistance to prevent current from leaking through the insulation. All insulation has a resistance value that is less than infinity, which allows some current leakage to occur. *Leakage current* is current that leaves the normal path of current flow (hot to neutral) and flows through a ground wire. Under normal operating conditions, the amount of leakage current is so small—measured in microamperes—that the leaking current has no effect on the operation or safety of a circuit. The total leakage current through insulation is the combination of conductive leakage current, capacitive leakage current, and surface leakage current.

Conductive Leakage Current

Conductive leakage current is the small amount of current that normally flows through the insulation of a conductor. Conductive leakage current flows from conductor to conductor or from a hot conductor to ground. **See Figure 2-31.** Ohm's law is used to determine the amount of conductive leakage current.

Insulation decreases in resistance as insulation ages and is exposed to damaging elements. The resistance of conductor insulation decreases as conductive leakage current increases. An increase in conductive leakage current results in additional insulation deterioration. Conductive leakage current is kept to a minimum by keeping insulation clean and dry.

Capacitive Leakage Current

Capacitive leakage current is leakage current that flows through conductor insulation due to a capacitive effect. A *capacitor* is an electronic device used to store an electric charge. A capacitor is created by separating two plates with a dielectric material. Two conductors that run next to each other act as a low-level capacitor. The insulation between the conductors is the dielectric and the conductors are the plates. **See Figure 2-32.**

Conductors carrying DC voltages typically produce little capacitive leakage current because the leakage current lasts for an extremely short time period and then stops. AC voltages produce continuous capacitive leakage current but the capacitive leakage current can be kept to a minimum by separating or twisting the conductors along the run.

Troubleshooting Tip

OHM'S LAW

Ohm's law is the relationship between voltage, current, and resistance in an electrical circuit.

$E = I \times R$

where

E = voltage (in V)
I = current (in A)
R = resistance (in Ω)

$E = I \times R$
Voltage = current × resistance

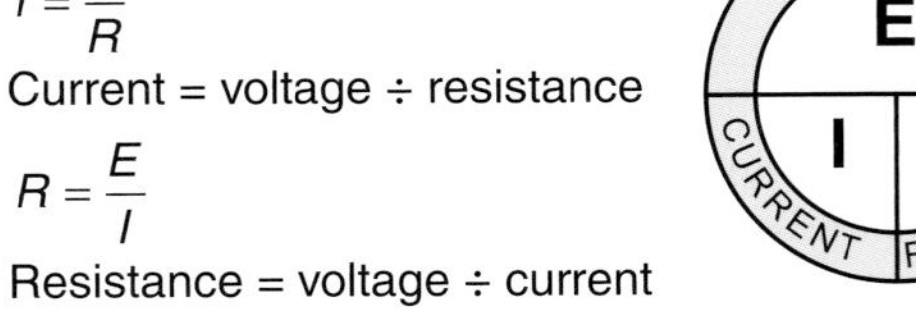

$I = \frac{E}{R}$
Current = voltage ÷ resistance

$R = \frac{E}{I}$
Resistance = voltage ÷ current

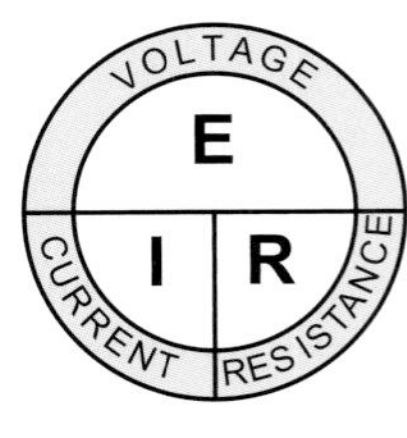

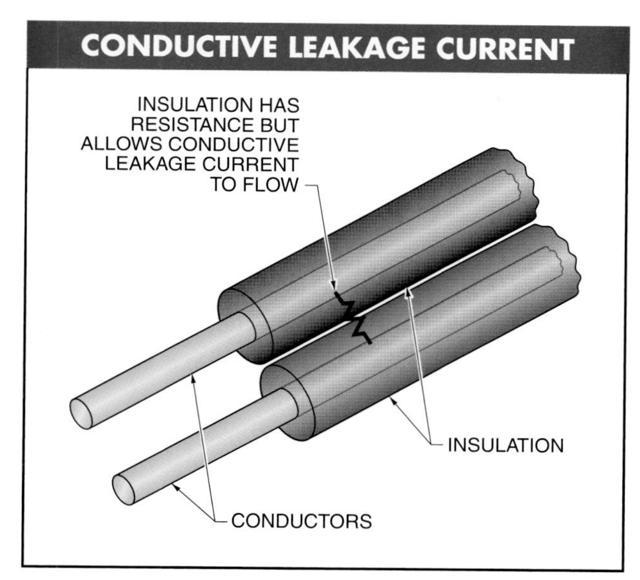

Figure 2-31. Conductive leakage current is a small amount of current that flows through the insulation of a conductor, from conductor to conductor, or from a hot conductor to ground.

Technical Tip

According to the International Electrotechnical Commission Safety Standard 950, double-insulated meters can have a maximum leakage current of 0.25 mA, grounded handheld meters can have a maximum leakage current of 0.75 mA.

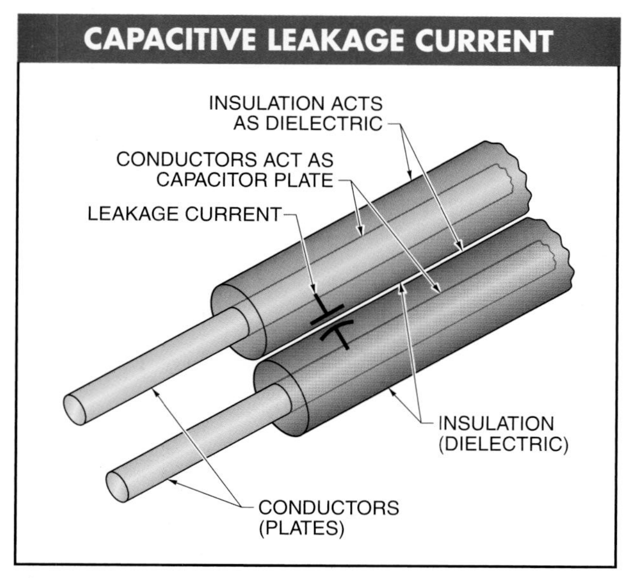

Figure 2-32. Capacitive leakage current is a small amount of current that flows around conductor insulation due to a capacitive effect, because the insulation acts as a dielectric, and the conductors act as plates.

Surface Leakage Current

Surface leakage current is current that flows from areas on conductors where insulation has been removed to allow electrical connections. Conductors are terminated with wire nuts, splices, push-on connectors, terminal posts, and other fastening devices at different points along an electrical circuit. The point at which insulation is removed from a wire provides a low resistance path for surface leakage current, with dirt and moisture allowing additional surface leakage current to occur. Surface leakage current results in increased heat at the point of connection. Increased heat contributes to an increase in insulation deterioration, which makes the conductor brittle. Surface leakage current is kept to a minimum by making all connections clean and tight. **See Figure 2-33.**

Conductors carrying AC voltage have the voltage varying between 0 V and peak voltage due to the alternating characteristic of AC voltage. Surface leakage current flows continuously as AC voltage alternates.

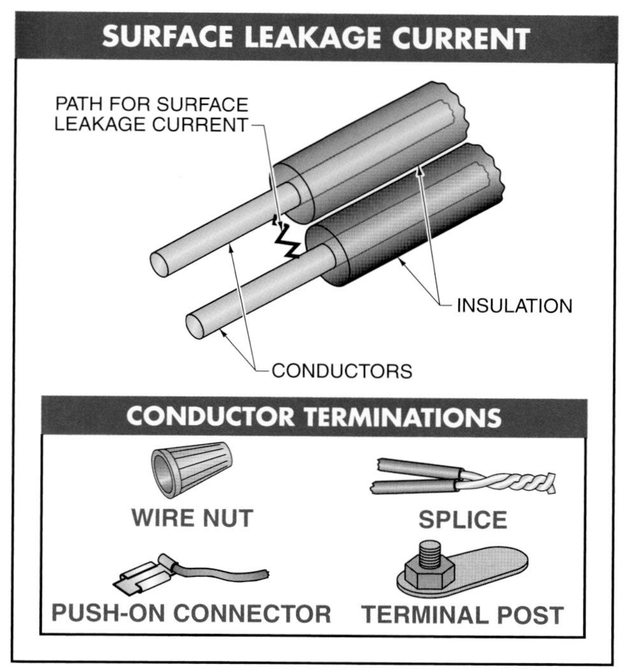

Figure 2-33. Surface leakage current is current that flows from areas on conductors where insulation has been removed creating increased heat, which contributes to insulation deterioration.

Megohmmeter Measurement Procedures

Megohmmeters send high voltages into conductors or motor windings to test the insulation. **See Figure 2-34.**

Before taking any resistance measurements using a megohmmeter, ensure the meter is designed to take measurements on the circuit being tested. Refer to the operating manual of the test instrument for all measuring precautions, limitations, and procedures. Always wear required personal protective equipment and follow all safety rules when taking the measurement. To take a resistance measurement with a megohmmeter, apply the following procedures:

1. Ensure that all power is OFF in the circuit or component being tested. Test the circuit for voltage using a voltmeter.
2. Set the megohmmeter function switch to voltage (the proper voltage range for the circuit to be tested). The test voltage used to test a conductor or motor winding must be as high as or higher than the highest voltage to which the conductor or motor winding being tested is exposed.
3. Plug the black test lead into the negative (earth) jack.
4. Plug the red test lead into the positive (line) jack.

WARNING

Verify that voltage is not present in a circuit or component being tested before taking any resistance measurements with a megohmmeter. Ensure that no body part contacts the high-voltage section of the megohmmeter test leads.

5. Ensure that the batteries are in good condition. A megohmmeter does not contain batteries if the meter includes a hand crank or if the meter plugs into a standard 115 V receptacle.
6. Connect the line test lead (red) to the conductor being tested.
7. Connect the earth test lead (black) to a second conductor in the circuit or to earth ground.
8. Press the test button or turn the hand crank on the megohmmeter and read the resistance displayed (can take 10 sec to 20 sec). Change the megohmmeter resistance or voltage range if necessary.
9. Consult the equipment manufacturer or megohmmeter manufacturer for the minimum recommended resistance values. The insulation is in good condition when the megohmmeter reading is equal to or higher than the minimum values indicated by the manufacturer.
10. Remove the megohmmeter from the conductors.
11. Discharge circuit or conductors being tested.
12. Turn megohmmeter OFF.

Troubleshooting Tip

CALCULATING RESISTANCE USING OHM'S LAW

Ohm's law states that the resistance (R) in a circuit is equal to voltage (E) divided by current (I). To calculate resistance using Ohm's law, apply the formula:

$$R = \frac{E}{I}$$

where

R = resistance (in Ω)

E = voltage (in V)

I = current (in A)

EXAMPLE: What is the resistance of wire insulation, when the wires have 1000 V applied and 20 mA of current result?

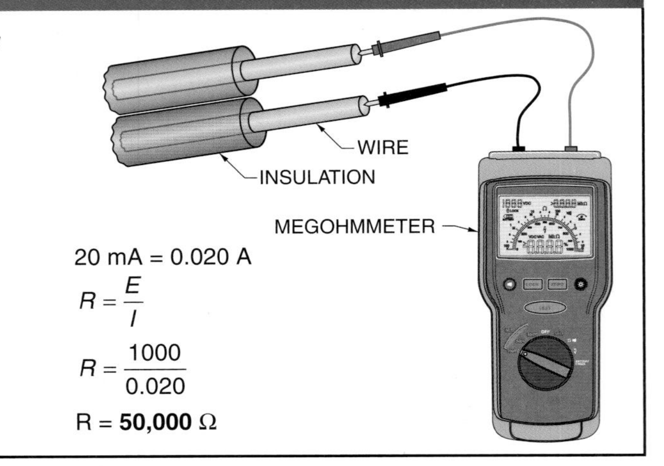

20 mA = 0.020 A

$$R = \frac{E}{I}$$

$$R = \frac{1000}{0.020}$$

R = **50,000 Ω**

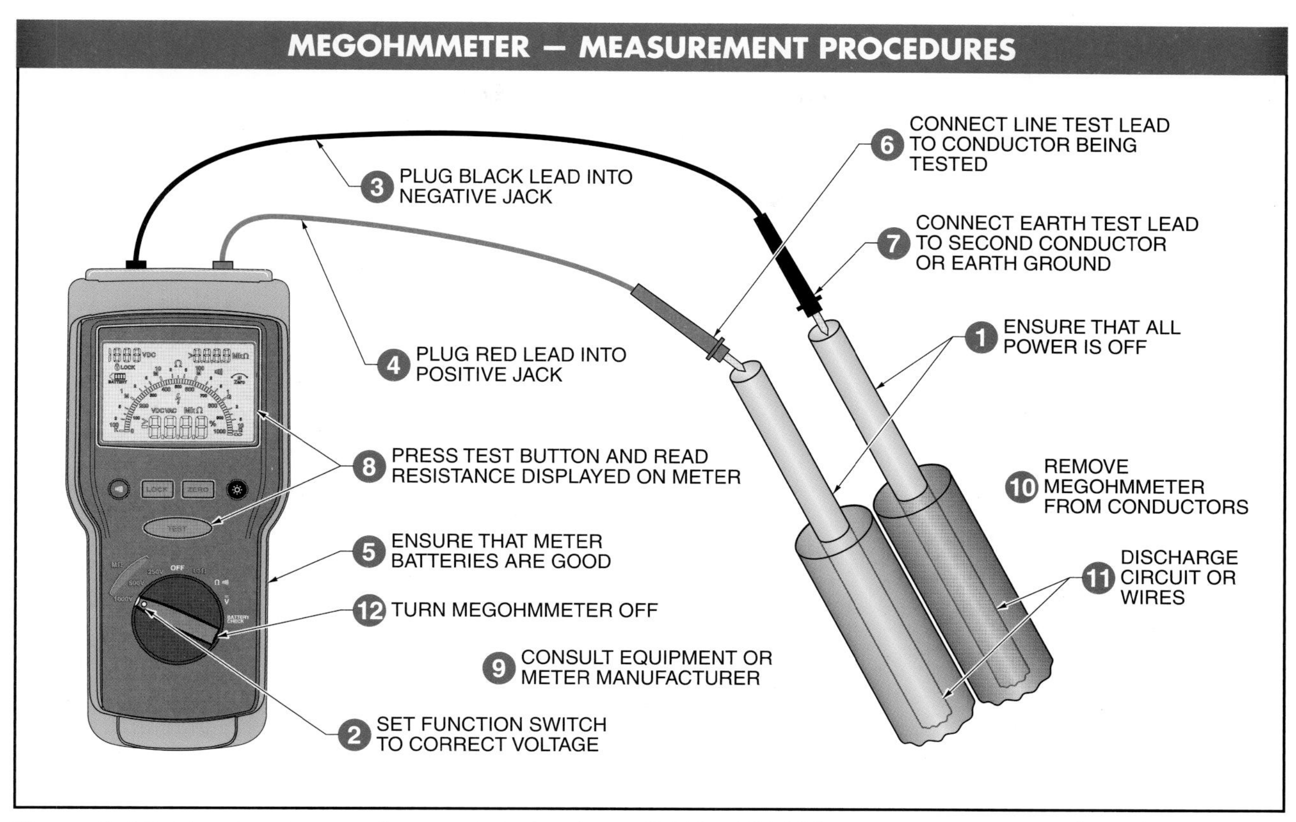

Figure 2-34. Megohmmeters send extremely high voltages (50 V to 5000 V) into wires or motor windings during testing.

AMMETERS

An *ammeter* is a test instrument that measures the amount of current in an electrical circuit. *Current* is the amount of electrons flowing through a conductor, component, or circuit. Current is measured in amperes. An *ampere* is the number of electrons passing a given point in 1 sec. Large amounts of current are measured in amperes (A). Small amounts of current are measured in milliamperes (mA) or microamperes (μA). Current flows through a closed circuit when a power source is connected to a load that uses electricity.

Amperage measurements are used to determine the amount of circuit loading or the condition of an electrical component (load). Every load (light bulb, motor, heating element, speaker) that converts electrical energy into some other form of energy (light, rotating motion, heat, sound) uses current. The more electrical energy required, the higher the current usage. Every time a load is switched ON or a new load is added to a circuit, the power source must provide more current through the circuit for the new loads. **See Figure 2-35.**

Current is typically measured using clamp-on ammeters or multimeters with clamp-on current probe accessories. Small amounts of current (as in electronics) can be measured using a multimeter connected as an in-line ammeter. A *clamp-on ammeter* is a test instrument that measures current in a circuit by measuring the strength of the magnetic field around a single conductor. The jaws of clamp-on ammeters are placed around conductors to measure current. Clamp-on ammeters are available for measuring AC only, or for measuring both AC and DC. Most clamp-on ammeters also include features that allow for the measurement of AC voltage, DC voltage, and resistance.

SYSTEM VOLTAGE RANGES*				
Watts	Volts Single-Phase		Volts Three-Phase	
	120	240	240	480
100	0.83	0.42	0.24	0.13
150	1.25	0.63	0.36	0.18
200	1.67	0.83	0.49	0.25
250	2.08	1.04	0.61	0.30
300	2.50	1.25	0.73	0.37
400	3.33	1.67	0.97	0.49
500	4.17	2.08	1.20	0.60
600	5.00	2.50	1.45	0.73
700	5.83	2.92	1.70	0.85
800	6.67	3.33	1.93	0.97
900	7.50	3.75	2.17	1.09
1000	8.33	4.17	2.41	1.21
1200	10.0	5.00	2.90	1.45
1500	12.5	6.25	3.62	1.82
2000	16.7	8.33	4.82	2.41
2500	20.8	10.4	6.10	3.05
5000	41.7	20.8	12.1	6.10
8000	66.7	33.3	19.3	9.65
10,000	83.3	41.7	24.1	12.1

* in amps

Figure 2-35. Amperage measurements are used to determine the amount of circuit loading or the condition of an electrical component (load).

When a clamp-on current probe accessory is used with a multimeter, the multimeter functions as a clamp-on ammeter. Clamp-on current probe accessories are available for multimeters that allow for the measurement of AC and DC current. A multimeter connected as an in-line ammeter measures small amounts of current in a circuit by inserting the test leads of the ammeter in series with the component(s) being tested. In-line ammeters require the circuit to be opened so the ammeter can be inserted into the circuit to acquire a reading. In-line ammeters are available for DC only, or AC and DC. **See Figure 2-36.**

The advantage of clamp-on ammeters or multimeters with clamp-on current probe accessories is that current measurements are taken without opening the circuit. Clamp-on ammeters and multimeters with current probes are commonly used to measure currents from 1 A or less to 3000 A. The main advantage of using multimeters with clamp-on current probes instead of clamp-on ammeters is the features available on multimeters such as MIN MAX recording mode (MIN MAX), relative mode (REL), and a bar graph display. Various multimeter features are helpful when troubleshooting circuits.

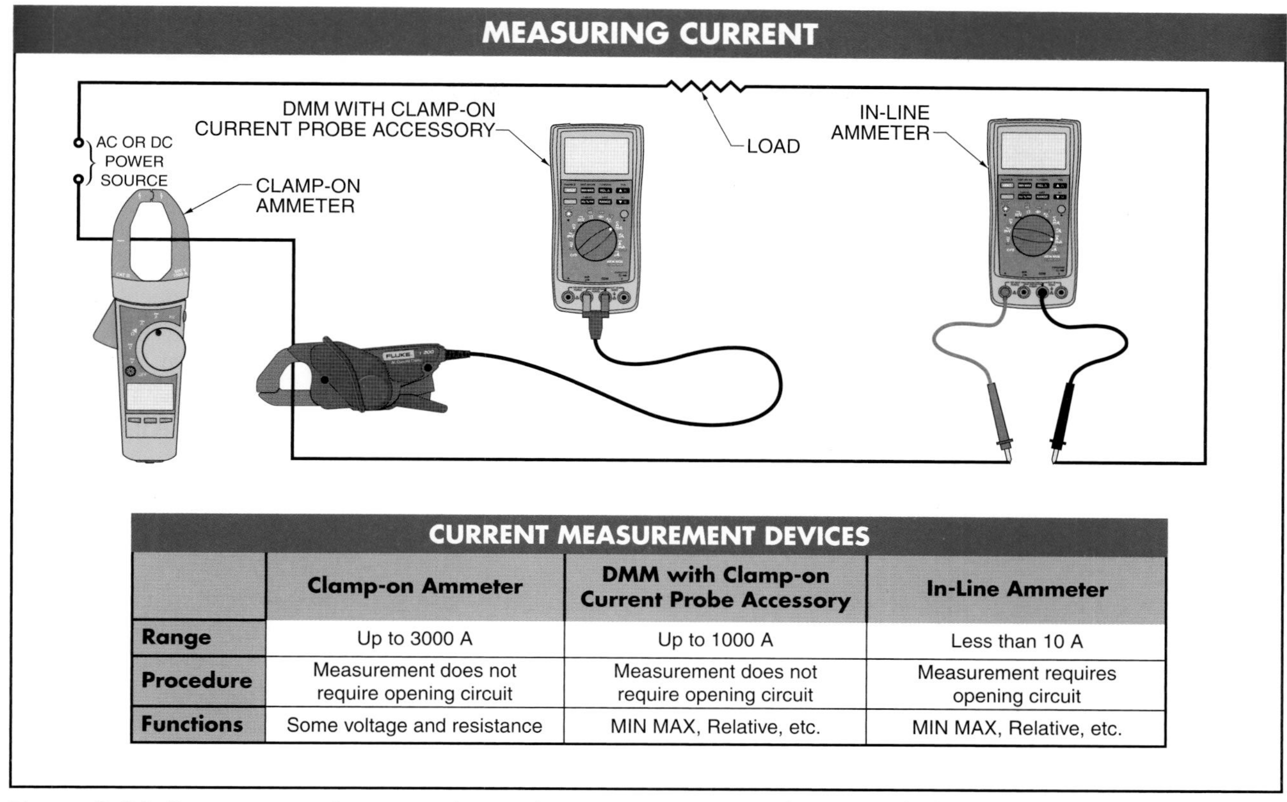

CURRENT MEASUREMENT DEVICES

	Clamp-on Ammeter	DMM with Clamp-on Current Probe Accessory	In-Line Ammeter
Range	Up to 3000 A	Up to 1000 A	Less than 10 A
Procedure	Measurement does not require opening circuit	Measurement does not require opening circuit	Measurement requires opening circuit
Functions	Some voltage and resistance	MIN MAX, Relative, etc.	MIN MAX, Relative, etc.

Figure 2-36. Current is typically measured using clamp-on ammeters or multimeters with clamp-on current probe accessories or by in-line ammeters for small amounts of current.

In-line ammeters are used to measure small amounts of current—normally less than 1 A. The maximum current that can be measured is determined by the function switch position and the placement of the test leads in the current jacks of the ammeter. A multimeter typically has two current jacks. One jack is used for measuring small currents (typically up to 300 mA or 400 mA). The other jack is used for measuring larger currents (typically up to 1 A). In-line current measurements must be limited to circuits that can be easily opened and circuits known to have currents less than 1 A.

Typically, the first choice for measuring current in a circuit is a clamp-on ammeter or a multimeter with a clamp-on current probe accessory. When a clamp-on ammeter cannot be used, or an in-line measurement is required, steps must be taken to prevent possible injury to the electrician, and/or damage to the test instrument and equipment.

Clamp-on Ammeters

A clamp-on ammeter is a portable test instrument that measures current in a circuit by measuring the strength of the magnetic field around a single conductor. A clamp-on ammeter is typically used to measure current in a circuit with over 1 A of current and in applications in which current can be measured by easily placing the jaws of the ammeter around one of the conductors. Most clamp-on ammeters can also measure voltage and resistance. To measure voltage and resistance, the clamp-on ammeter must include test leads and a voltage and resistance mode. In addition to basic measuring functions (current, voltage, and resistance), some ammeters also include special functions such as a hold button (HOLD) for maintaining a displayed measurement, a min/max button (MIN MAX) for recording the minimum and maximum measured values, and other special functions. **See Figure 2-37.**

CLAMP-ON AMMETERS

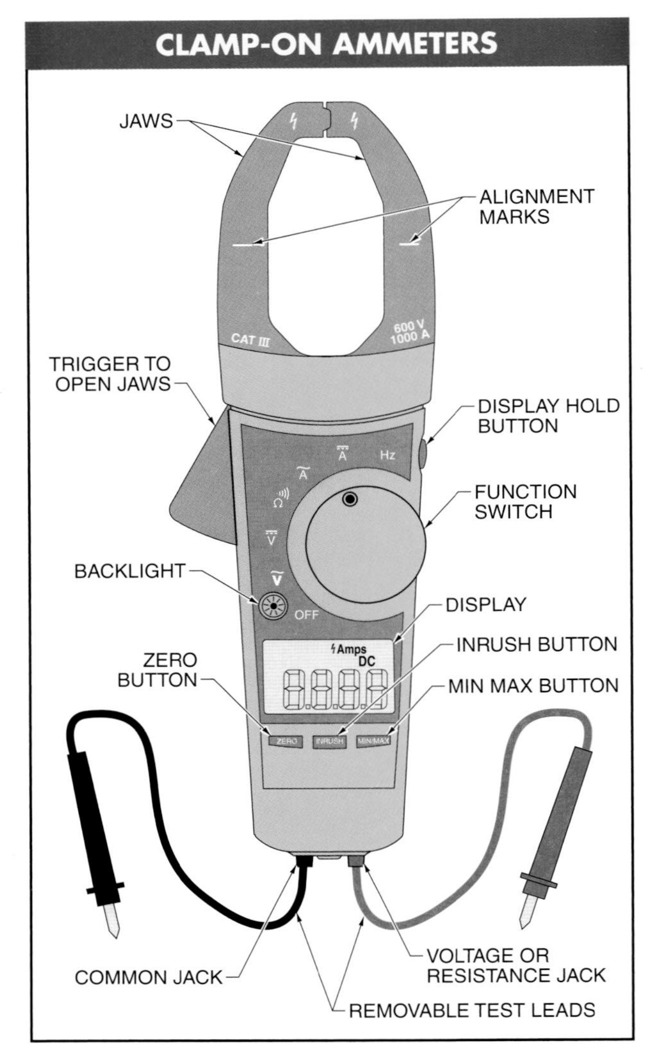

Figure 2-37. Clamp-on ammeters measure current in a circuit by measuring the strength of the magnetic field around a single conductor.

Technical Tip

Clamp-on ammeters are typically used where voltages are less than 600 V with a frequency under 2 kHz.

Clamp-on Ammeter Applications

Current is a common troubleshooting measurement, because only a current measurement can be used by an electrician to determine how much a circuit is loaded (or working). Current measurements vary because current can vary at different points in parallel or series-parallel circuits. Current in series circuits is constant throughout the circuit. When loads are added to parallel circuits, the supply voltage remains the same, but the current must increase with each load that is added.

The largest amount of current in a parallel circuit is at a point closest to the power source; current decreases as the system distributes current to each of the parallel loads. On an individual leg of a combination series-parallel circuit, the current going to a load (motor, heating element, or light bulb) is the same as the current on the return line from the load. A slight measured difference—up to 10%—between lines (supply and return) is possible when using a clamp-on ammeter because of the placement of the meter, or when low current measurements are taken on high ammeter ranges (measuring 2 A on a 400 A setting). Any variation that is excessive must be investigated because the current measurement may indicate that a partial short exists on one of the lines and current is flowing to ground.

Troubleshooting Tip

CALCULATING CURRENT USING OHM'S LAW

Ohm's law states that the current (I) in a circuit is equal to voltage (V) divided by resistance (R). To calculate current using Ohm's law, apply the formula:

$$I = \frac{V}{R}$$

where

I = current (in A)

V = voltage (in V)

R = resistance (in Ω)

EXAMPLE: What is the current through a wire, when the wire has 5750 Ω of resistance and 115 V applied?

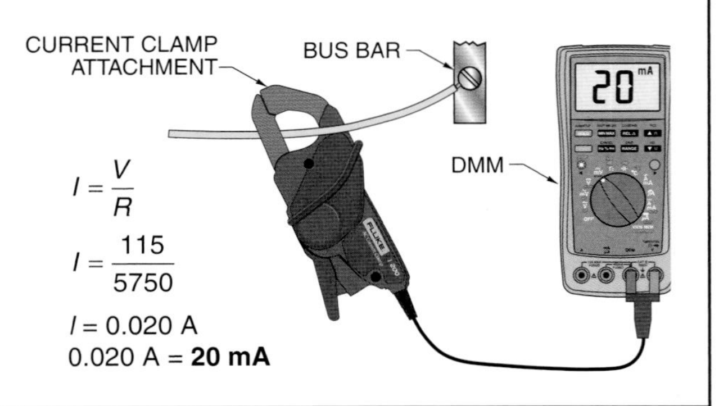

$$I = \frac{V}{R}$$

$$I = \frac{115}{5750}$$

I = 0.020 A

0.020 A = **20 mA**

When measuring current along a circuit that feeds several loads, the current may not be the same on each line. When current measurements are taken at the main power panel or from an electric motor drive, the current on the hot and neutral conductors of a 115 V or 208 V single-phase circuit, and all three phases of a 3ϕ circuit, should basically be the same. **See Figure 2-38.**

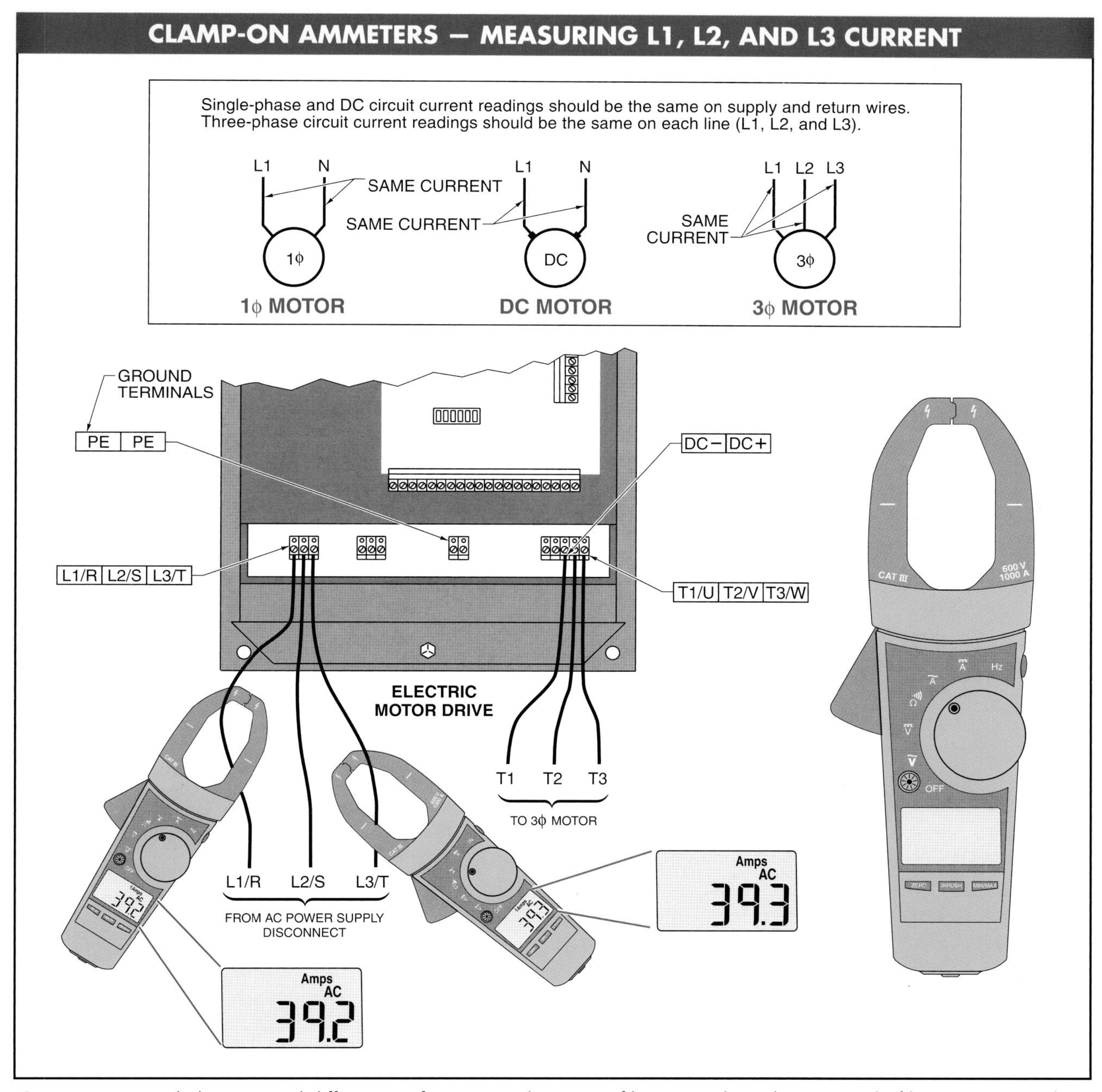

Figure 2-38. A slight measured difference, of up to 10% between 1f lines (supply and return) and 3f lines (L1, L2, and L3) is possible when the conductor is not centered in the jaws of a clamp-on ammeter.

IDEAL Industries, Inc.

Clamp-on ammeters are used to measure the current (magnetic field) flowing through one isolated conductor.

Clamp-on Ammeter Measurement Procedures

Clamp-on ammeters measure current in a circuit by measuring the strength of the magnetic field around a single conductor. Electricians must ensure that clamp-on ammeters do not pick up stray magnetic fields by separating conductors being tested from other conductors as much as possible during testing. When stray magnetic fields may be affecting a measurement, several measurements at various locations along the same conductor must be taken.

AC or DC current measurements performed with clamp-on ammeters or multimeters with clamp-on current probe accessories must be performed following standard procedures. **See Figure 2-39.**

Before taking any current measurements using a clamp-on ammeter, ensure the meter is designed to take measurements on the circuit being tested. Refer to the operating manual of the test instrument for all measuring precautions, limitations, and procedures. Always wear required personal protective equipment and follow all safety rules when taking the measurement. To take a current measurement with a clamp-on ammeter, apply the following procedures:

1. Determine if the current in the circuit to be measured is AC or DC.
2. Select a clamp-on ammeter required to measure the circuit current (AC or DC). When both AC and DC measurements are required, select an ammeter that can measure both AC and DC current.
3. Ensure that the current range of the clamp-on ammeter is high enough to measure the maximum current that exists in the circuit being tested. When the clamp-on ammeter range is not high enough, select a range that is high enough, or select an ammeter with a higher range capability.
4. Set the clamp-on ammeter function switch to the highest current range (600 A, 200 A, 10 A or 400 mA). Select a setting as high as or higher than the highest possible circuit current when there is more than one test position or if the circuit current is unknown.
5. When required, plug the clamp-on current probe accessory into the multimeter. The black test lead of the clamp-on current probe accessory is plugged into the common jack. The red test lead is plugged into the mA jack for current measurement accessories that produce a current output. The red test lead is plugged into the voltage (V) jack for current measurement accessories that produce a voltage output. The current measurement accessories that produce current output are designed to measure AC only and generally deliver 1 mA to the meter for every 1 A of measured current (1 mA/A). Current accessories that produce a voltage output are designed to measure AC or DC current and deliver 1 mV to the meter for every 1 A of measured current (1 mV/A).
6. Open the clamp-on meter or probe accessory jaws by pressing against the trigger.
7. Enclose one conductor in the center of the jaws. Ensure that the jaws are completely closed before taking any current measurements.
8. Read the current measurement displayed on the clamp-on ammeter or multimeter with clamp-on probe accessory.
9. Remove the clamp-on ammeter or clamp-on accessory from the circuit.

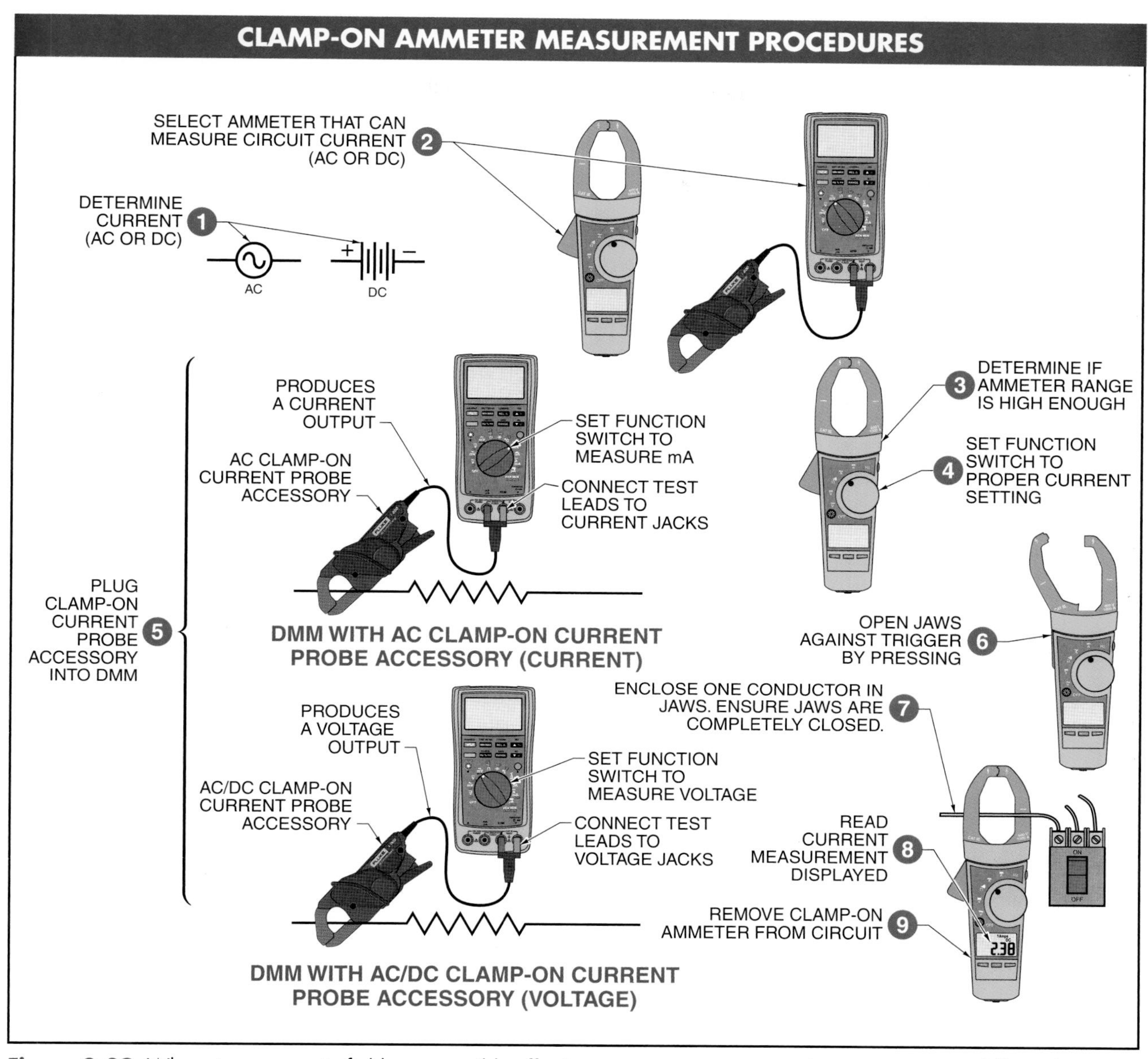

Figure 2-39. When stray magnetic fields are possibly affecting a measurement, several measurements at different locations along the same conductor must be taken.

In-Line Ammeters

In-line ammeters provide the most accurate current measurements when measuring currents less than 1 A. In-line ammeters are the most accurate because the meter is connected in series with the load and every milliamp of current that flows through the load must flow through the meter.

Some applications require that the in-line method of current measurement be used because in-line is the only way or best way to test a circuit or component. The in-line method of current measurement is typically used only for applications requiring very accurate current measurements and having less than 1 A of current.

WARNING

In-line current measurements are the most dangerous measurement to take of any electrical measurement (voltage, resistance, clamp-on current measurements). Any in-line current measurement requires that extreme caution be used.

In-Line Ammeter Applications

Forklift electrical circuits draw large amounts of current because forklift circuits operate at low voltages (the lower the voltage, the higher the required current for any given power). For example, a 100 W safety strobe light designed for a 36 V system requires 2.77 A to deliver 100 W of power.

$$I = \frac{P}{E}$$

where

I = current (in amps)
P = power (in watts)
E = voltage (in volts)

$$I = \frac{P}{E}$$

$$I = \frac{100}{36}$$

$$\mathbf{I = 2.77\ A}$$

However, when a forklift or automobile is turned OFF and all accessories (hydraulic valves, lights, radio) are OFF, the battery should have very little current flow (a little current can be drawn by digital clocks, computers, and security systems). The small amount of current that is drawn should not drain a battery that is in good shape for several weeks. However, when a load is faulted (as when a hydraulic valve solenoid stays energized, a glove box lamp remains on, or water is in the safety light or tail-lamp holder) or insulation is damaged, the battery can be drained when the forklift or automobile is OFF in a couple of hours. To test the drain (current draw) on a battery, an in-line ammeter is used. **See Figure 2-40.**

To test the current draw on a battery, the ignition switch is turned to the OFF position and all accessories are turned OFF, the positive battery post is disconnected from the battery, and an in-line ammeter is connected between the battery post and the disconnected terminal.

In-line ammeters accurately measure the small amount of current drawn from batteries. To isolate a problem or test each circuit, the fuses of the forklift or automobile can be removed one at a time.

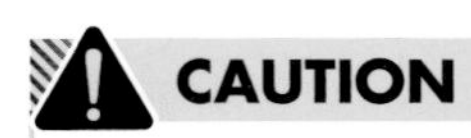

CAUTION

Only fused in-line ammeters set to the highest current measuring range can be used.

In-Line Ammeter Measurement Procedures

Care is required to protect the in-line ammeter, circuit, and electrician when measuring AC or DC current with an in-line ammeter. Standard safety precautions must be followed when using an in-line ammeter. The following are the standard procedures for taking an AC current measurement with an in-line ammeter:

1. Test the ammeter fuses following manufacturer recommended procedures.
2. Verify that the expected circuit current measurement is less than the current rating (setting) of the in-line ammeter. Start with the highest current-measuring range of the in-line ammeter when the circuit current is unknown. When the circuit current can exceed the rating of the in-line ammeter, use a clamp-on ammeter, or do not take the current measurement.
3. Set the function switch of the in-line ammeter to the proper setting for measuring AC current and set the function or selector switch to the proper current rating because most ammeters include more than one current setting (A, mA, and μA).
4. Verify that the test leads of the in-line ammeter are connected to the proper jacks for measuring the correct amount of AC current. Most ammeters include more than one current jack.

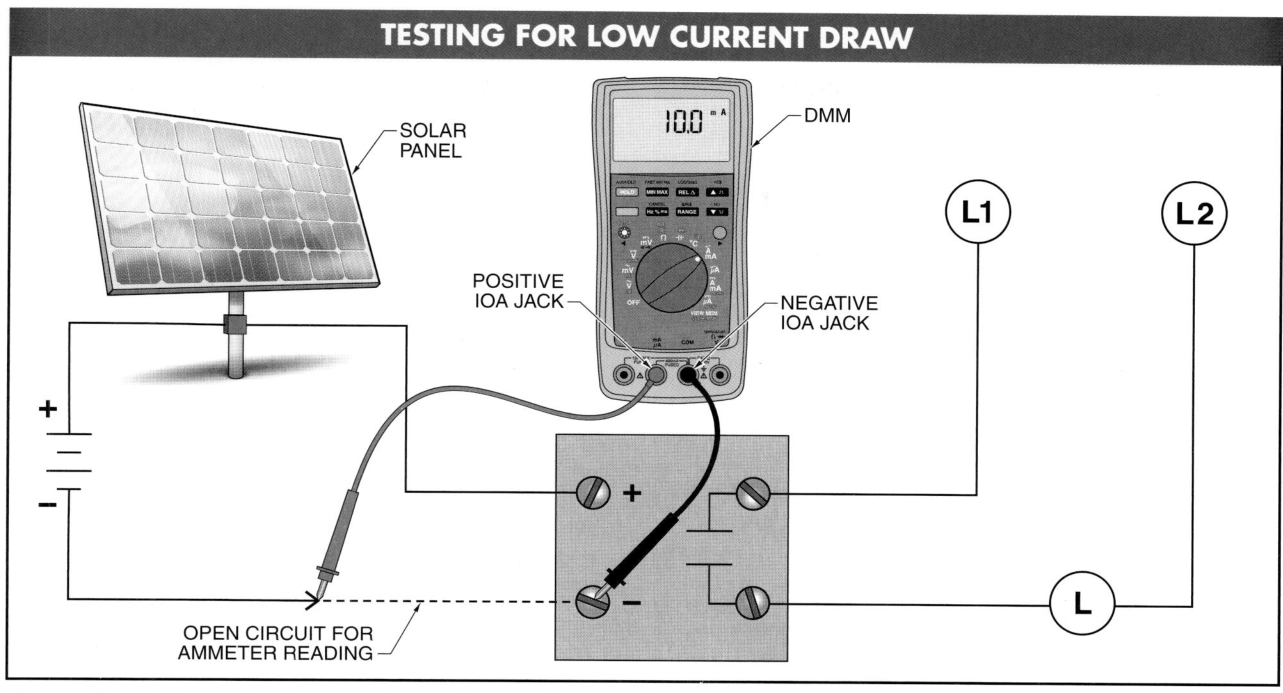

Figure 2-40. To test the drain an electrical system has on a battery, an in-line ammeter is used.

WARNING

Always ensure that the function switch position matches the jack connections of the test leads. In-line ammeters are damaged when the test leads are connected to measure current and the function switch is set for a different measurement such as voltage or resistance. Some meters have an input warning function that provides a constant audible warning (beep) when the test leads are connected in the current jacks and a noncurrent mode is selected by the function switch.

5. Verify that the power to the circuit being tested is OFF before connecting and disconnecting the in-line ammeter test leads. When necessary, take voltage measurement to ensure that the circuit is de-energized.
6. Do not change the function switch position on an in-line ammeter while the meter is connected to an energized circuit.
7. Turn power to the test circuit and in-line ammeter OFF before changing any ammeter settings.
8. Connect the in-line ammeter in series with the load(s) to be measured. Never connect an ammeter in parallel (as with a voltmeter) with the load(s) to be measured.
9. Turn the power ON to the circuit being tested.
10. Read the AC current measurement displayed.
11. Turn the power OFF to the circuit and in-line ammeter and remove the in-line ammeter from the circuit.

Many test instruments include a fuse in the current-measuring circuit to prevent damage caused by excessive current. Before using an in-line ammeter, verify that the meter is fused on the current range being used. Most in-line ammeters are marked as fused or not fused at the test lead current terminals. When an in-line ammeter is not marked as fused, consider the meter unusable for measuring current. Use a clamp-on ammeter or use a clamp-on current probe accessory instead of an unfused meter. DC current is measured with an in-line ammeter using standard procedures. **See Figure 2-41.**

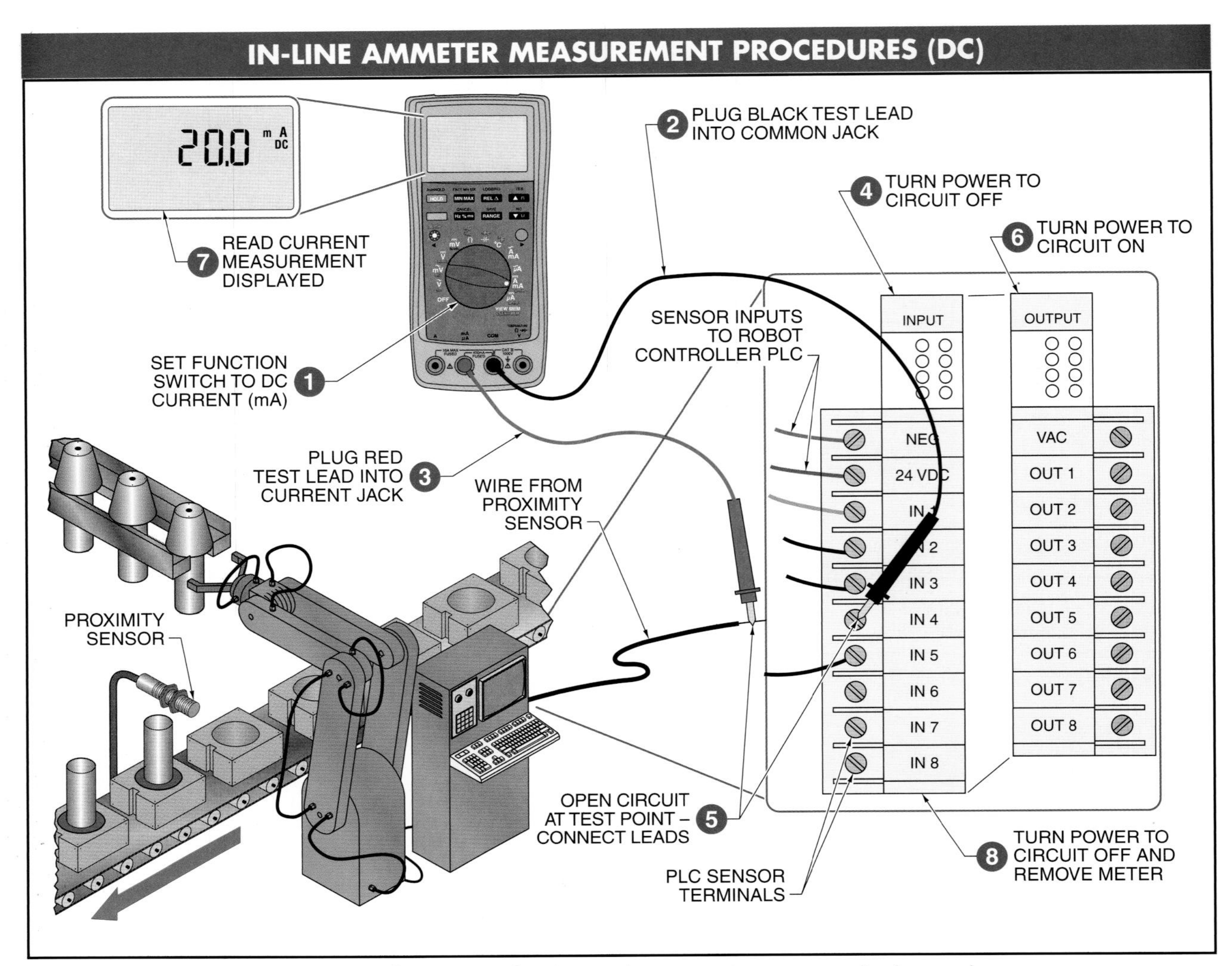

Figure 2-41. AC or DC current can be measured with an in-line ammeter using standard procedures.

WARNING

Ensure that no body parts contact any part of an energized circuit, including metal contact points at the tip of the in-line ammeter test leads.

WARNING

In-line ammeters not marked as fused, must be considered unusable for measuring current.

Before taking any DC current measurements using an in-line ammeter, ensure the meter is designed to take measurements on the circuit being tested. Refer to the operating manual of the test instrument for all measuring precautions, limitations, and procedures. Always wear required personal protective equipment and follow all safety rules when taking the measurement. To take a DC current measurement with an in-line ammeter, apply the following procedures:

1. Set the function switch of the in-line ammeter to the proper position for measuring DC current (A, mA, or µA). Select a setting with a high enough rating to measure the highest possible circuit current when the ammeter has more than one position.
2. Plug the black test lead into the common jack.

3. Plug the red test lead into the current jack. The current jack may be marked A or mA/μA.
4. Turn the power OFF to the circuit or component being tested and discharge all capacitors when possible.
5. Open the circuit at the test point and connect the test leads to each side of the opening. For DC current, the black (negative) test lead is connected to the negative side of the opening and the red (positive) test lead is connected to the positive side of the opening. Reverse the black and red test leads when a negative sign appears to the left of the measurement displayed.
6. Turn the power ON to the circuit being tested and the in-line ammeter.
7. Read the current measurement displayed.
8. Turn the circuit power and the in-line ammeter OFF and remove the ammeter from the circuit.

Current Measurement Analysis

Knowing the measured current value of a system, circuit, or component is very useful when troubleshooting. Current measurements of a motor can be compared with the motor nameplate current rating, allowing an electrician to determine if the motor is operating correctly.

For example, an electric motor in a car wash system is working at maximum capacity (fully loaded), and the motor draws the rated nameplate current (8 A in this example). The motor is overloaded when it draws more than the rated nameplate current, and the motor is underloaded when it draws less than the rated nameplate current. **See Figure 2-42.**

The size of a motor may be increased or the load on the motor decreased when overloading is a problem. Current measurements taken on a power distribution system are used to indicate total system loading.

Typically, higher than rated currents indicate a problem. Higher than normal current in a circuit produces higher temperatures that cause the insulation on conductors and motors to break down, leading to component failure.

Excessive temperatures are a major cause of failure for transistors, integrated circuits (chips), and other solid-state electronic components. For maximum efficiency, always take current measurements when equipment is first installed and during normal operation. Measurements taken when a circuit is operating properly are used to provide a baseline comparison when troubleshooting a problem with the circuit in the future.

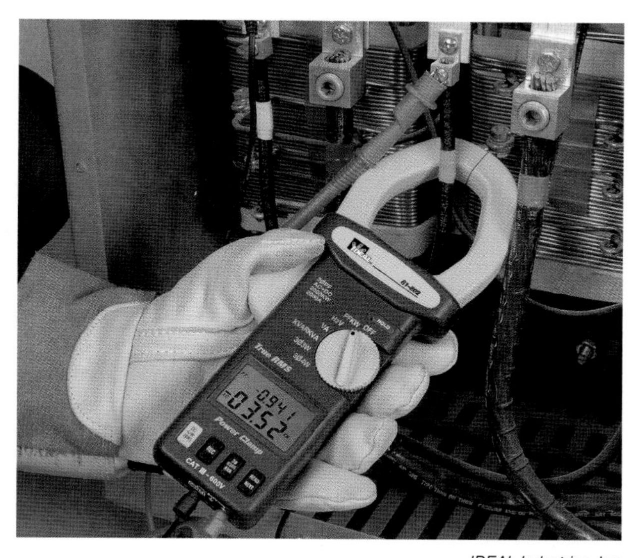

IDEAL Industries, Inc.

Some clamp-on ammeters are capable of determining power by measuring current.

Current Unbalance

Current unbalance (imbalance) is the unbalance that occurs when current on each of the three power lines of a three-phase power supply to a 3ϕ motor or to other 3ϕ loads is not equal. The problem with small voltage unbalances is that small voltage unbalances cause high current unbalances. High current unbalances cause excessive heat, resulting in insulation breakdown. Typically, voltage unbalances cause current unbalances at a rate of about 8:1. For example, a 2% voltage unbalance creates a 16% current unbalance.

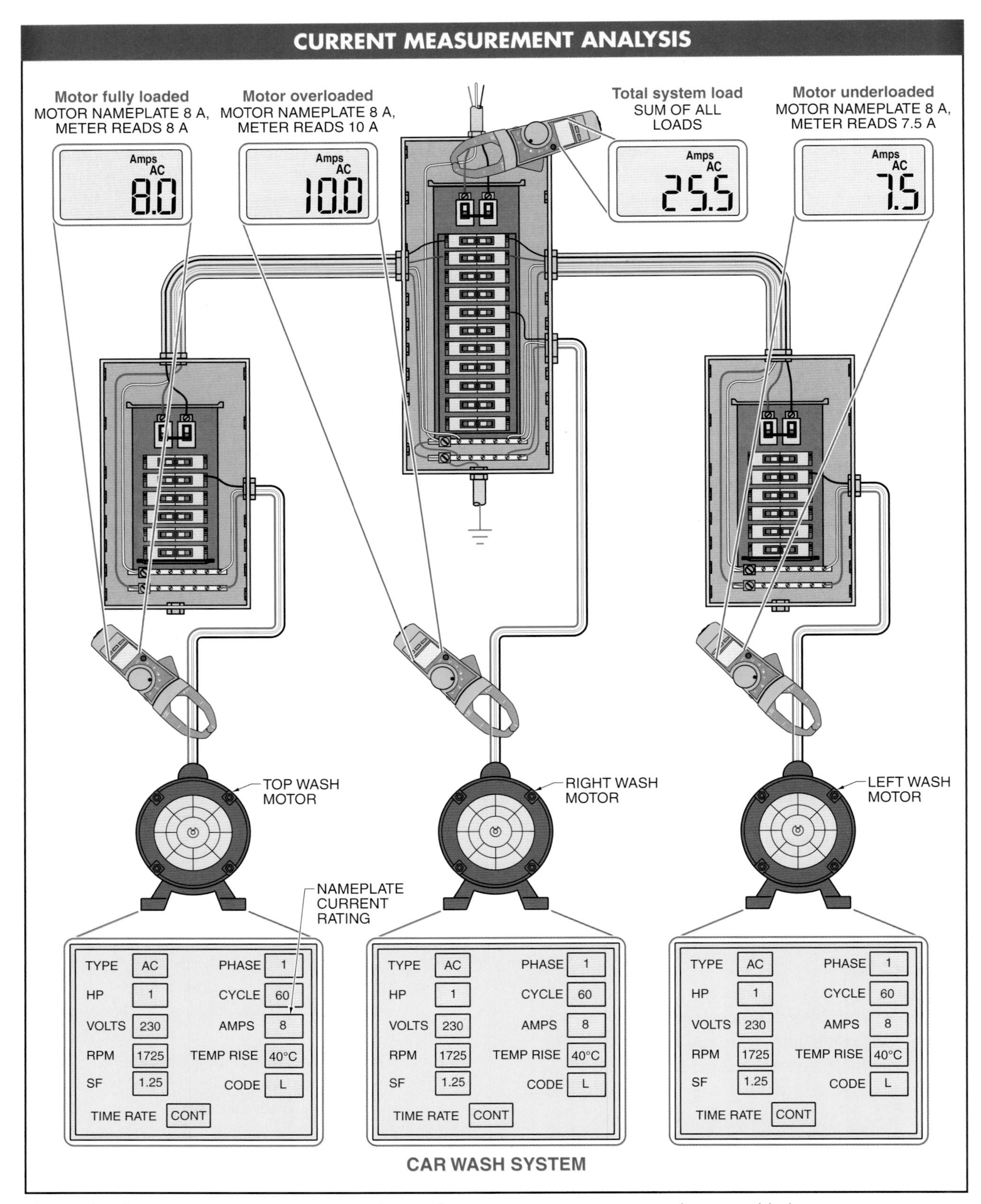

Figure 2-42. Current measurements are used to verify motor operation in a system during troubleshooting.

Current unbalances should never exceed 10%. Any time a current unbalance exceeds 10%, electricians must test for voltage unbalance. Likewise, any time there is a voltage unbalance of more than 1%, electricians must test for current unbalance. Current unbalance is found the same way as voltage unbalance is found, except current measurements are taken. **See Figure 2-43.** Current unbalance is found by applying the following procedure:

1. Measure the current on each incoming power line (L1, L2, and L3) entering the circuit disconnect.
2. Add the currents.
3. Find the current average by dividing the sum by 3.
4. Find the current deviation by subtracting the current average from the current with the largest deviation.
5. To find current unbalance, apply the following formula:

$$I_u = \frac{I_d}{I_a} \times 100$$

where

I_u = current unbalance (in %)

I_d = current deviation (in V)

I_a = current average (in V)

100 = constant

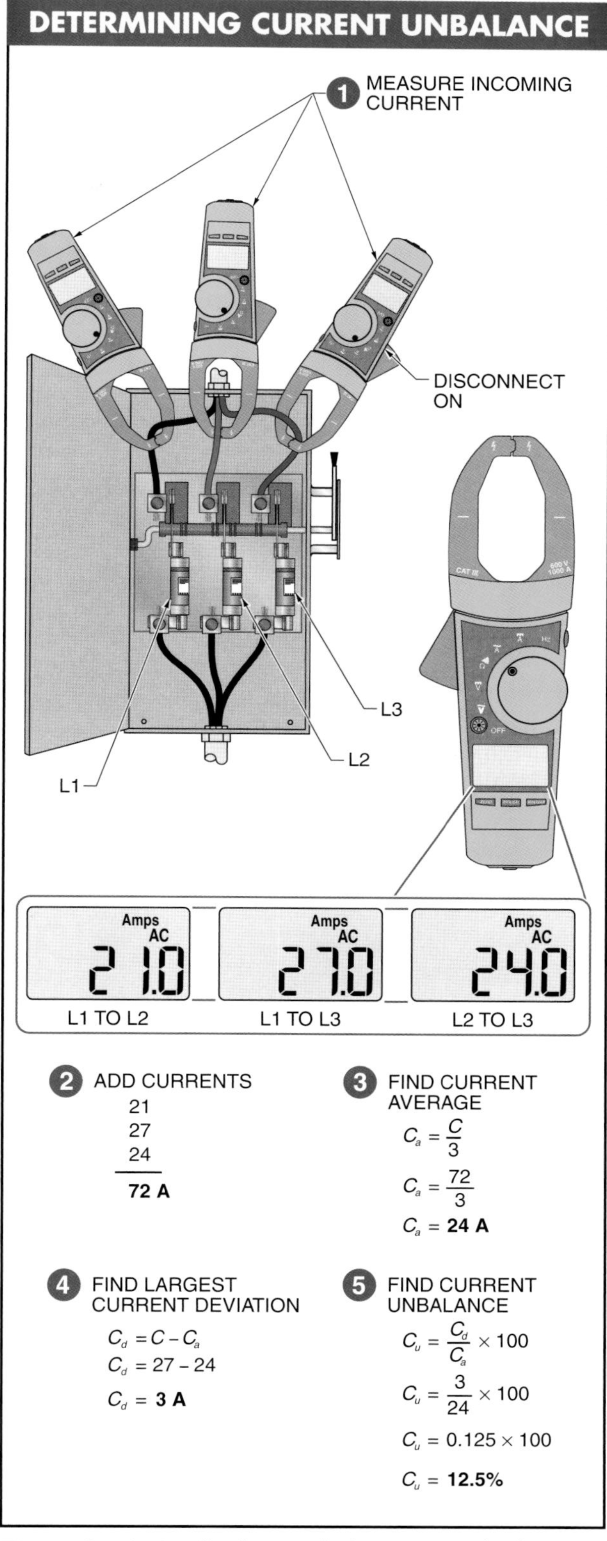

Figure 2-43. Small voltage unbalances cause high current unbalances. Typically, voltage unbalances cause current unbalances at a rate of about 8:1.

MULTIMETERS

A *multimeter* is a portable test instrument that is capable of measuring two or more electrical properties. Multimeters are either analog or digital. An analog multimeter is a portable test instrument that uses electromechanical components to display measured values. A digital multimeter is a portable test instrument that uses electrical components to display measured values. **See Figure 2-44.**

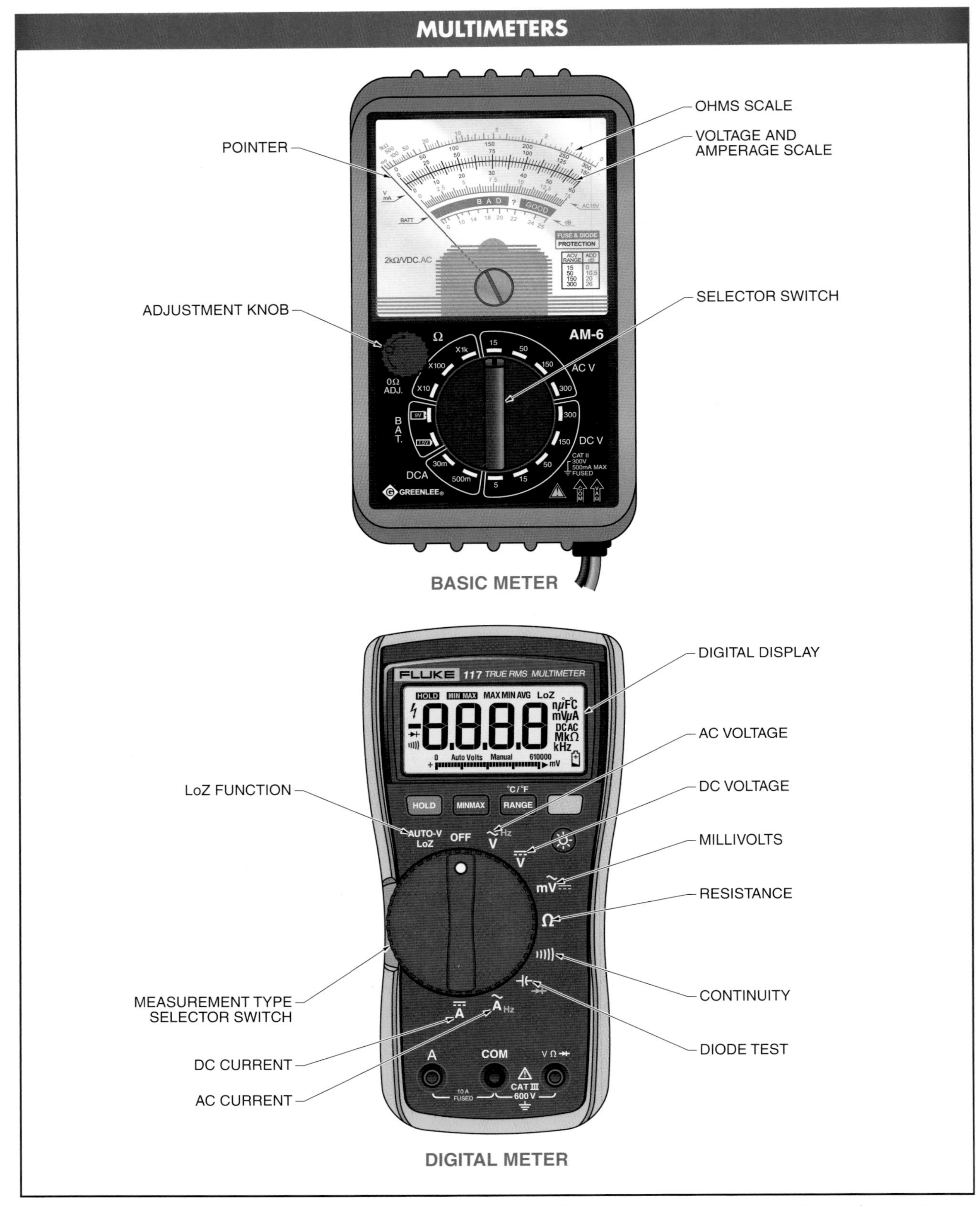

Figure 2-44. Multimeters are portable test instruments that are capable of measuring two or more electrical properties.

Analog Multimeters

An *analog multimeter* is a meter that can measure two or more electrical properties and displays the measured properties along calibrated scales using a needle. Most analog multimeters have several calibrated scales which correspond to the different selector switch settings (AC, DC, and R) and jack usage (mA jack and 10 A jack). An important part of reading an analog multimeter is verifying that the correct scale is used when reading the measurement. The most common measurements made with analog multimeters are voltage, resistance, and current. Many analog multimeters also include scales and ranges for measuring decibels (dB) and checking batteries. The *decibel scale* is a scale that indicates the comparison (ratio) between two or more signal powers or voltages. The battery tester function allows for displaying the charged condition (25%, 50%, 75%, or 100%) of a battery being tested.

Reading Analog Multimeter Displays

An *analog display* is an electromechanical device that indicates readings by the mechanical motion of a pointer. Analog displays use linear and nonlinear scales to display measured values. A *linear scale* is a scale that is divided into equally spaced segments. A *nonlinear scale* is a scale that is divided into unequally spaced segments. A mirror on the analog display helps to reduce measurement errors from parallax (distorted view of pointer caused by angle other than head-on view). **See Figure 2-45.**

Analog scales are divided using primary divisions, secondary divisions, and subdivisions. A *primary division* is a division with a listed value. A *secondary division* is a division that divides primary divisions into halves, thirds, fourths, fifths, etc. A *subdivision* is a division that divides secondary divisions into halves, thirds, fourths, fifths, etc. Secondary divisions and subdivisions do not have listed numerical values. When reading an analog scale, add the primary, secondary, and subdivision readings. **See Figure 2-46.**

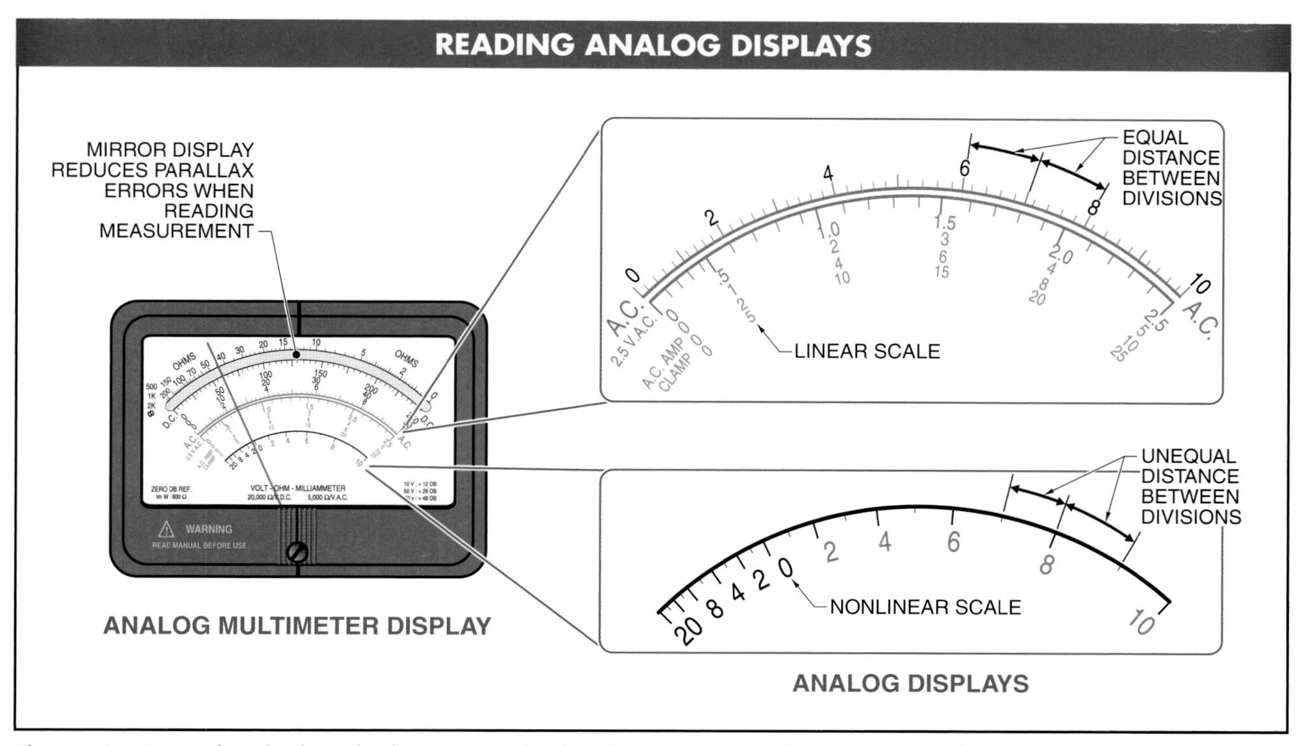

Figure 2-45. Analog displays display measured values by using the mechanical motion of a pointer on a scale.

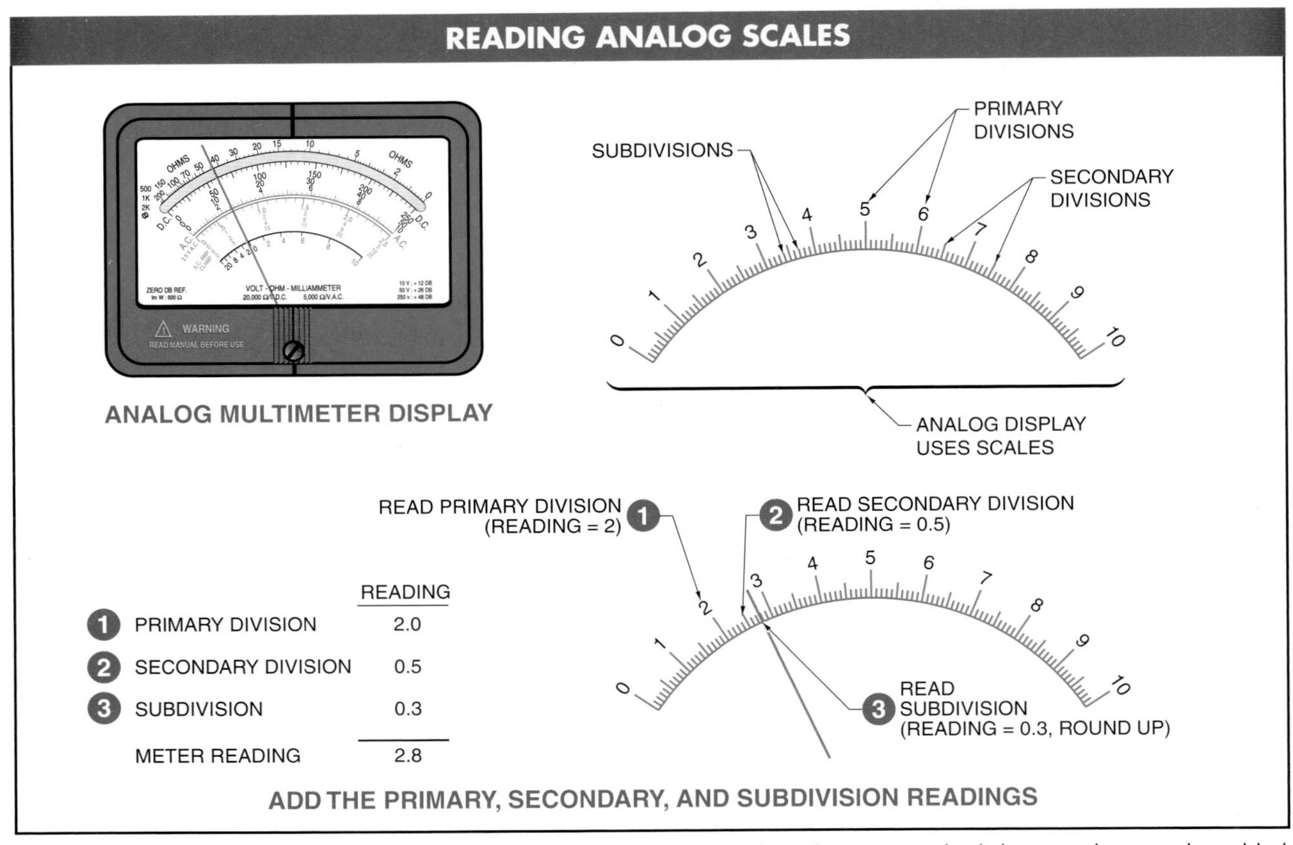

Figure 2-46. Analog scales are divided into primary divisions, secondary divisions, and subdivisions that must be added together to acquire a reading.

Reading Analog Display Procedures

To read an analog display, apply the following procedures:

1. Read the primary divisions.
2. Read the secondary divisions when the pointer moves past a secondary division. *Note:* May not occur with very low readings.
3. Read the subdivisions when the pointer is not directly on a primary or secondary division. Round the reading to the nearest subdivision when the pointer is not directly on a subdivision. Round the reading to the next highest subdivision when rounding to the nearest subdivision is unclear.
4. Add the primary division, secondary division, and subdivision readings to obtain the analog reading.

Digital Multimeters

A *digital multimeter (DMM)* is a meter that can measure two or more electrical properties and displays the measured properties as numerical values. Basic digital multimeters measure voltage, current, and resistance. Advanced digital multimeters include special functions such as measuring capacitance and temperature. The main advantage of a digital multimeter over an analog multimeter is the ability of a digital meter to record measurements in addition to making it easier to read the displayed values.

Technical Tip

To avoid parallax errors, view the pointer head on, or use a display that includes a mirror that allows viewing the measurement from an angle to the pointer.

Reading Digital Displays

A *digital display* is an electronic device that displays measurements as numerical values. Digital displays help eliminate human error when taking measurements by displaying exact values measured. Errors occur when reading a digital display when displayed prefixes, symbols, and decimal points are not properly applied.

Digital displays display values using either light-emitting diodes (LEDs) or liquid crystal displays (LCDs). LED displays are easier to read but use more power than LCD displays. Most portable digital test instruments use LCD displays. The exact value on digital displays is determined from the numbers displayed and the position of the decimal point. Selector switches (range switches) determine the placement of the decimal point on a digital display.

Typical voltage ranges on a digital display are 3 V, 30 V, and 300 V; or simply V and mV. The highest possible reading with the range switch set on 3 V is 2.99 V. The highest possible reading with the range switch set on 30 V is 29.99 V. The highest possible reading with the range switch set on 1 V is 0.999 V. **See Figure 2-47.**

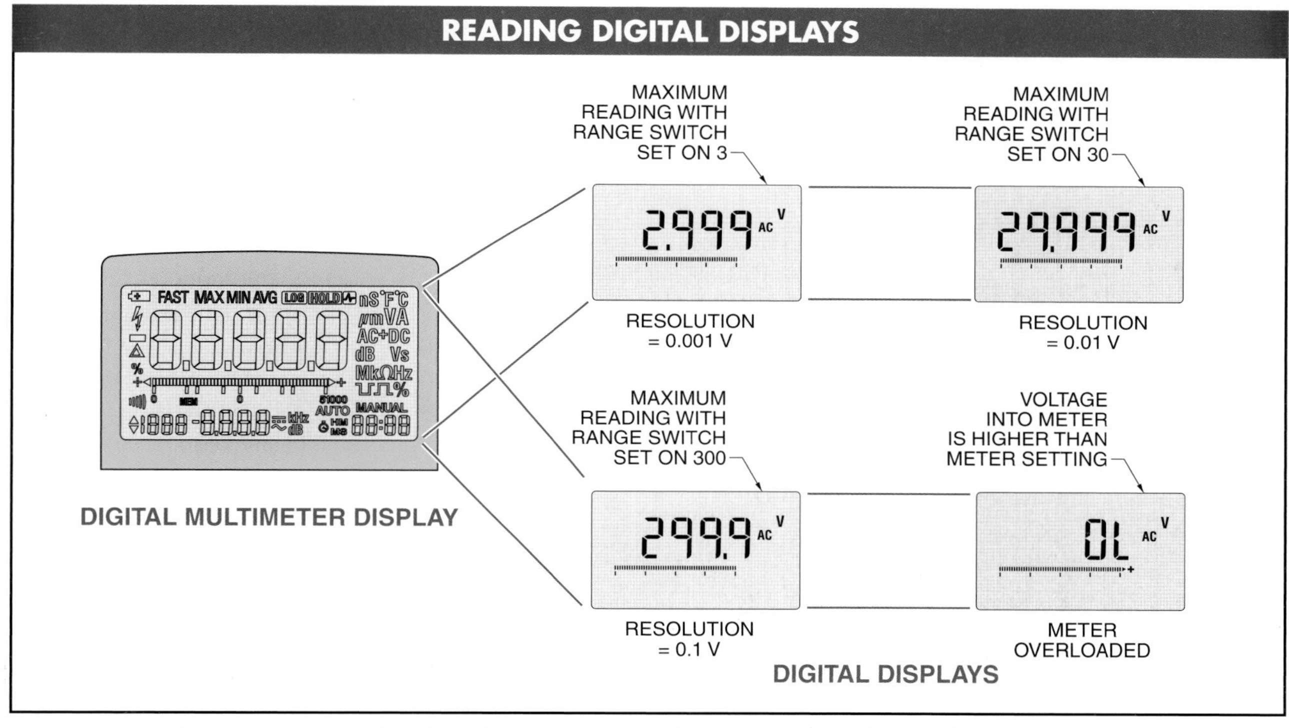

Figure 2-47. The exact value on digital displays is determined from the numbers displayed and the position of the decimal point set by the selector switch of the meter.

Bar Graphs

Most digital displays include a bar graph to show changes and trends in a circuit. A *bar graph* is a graph composed of segments that function as an analog pointer. The displayed bar graph segments increase as the measured value increases and decrease as the measured value decreases. Reverse the polarity of the test leads when a negative sign is displayed at the beginning of the bar graph. **See Figure 2-48.**

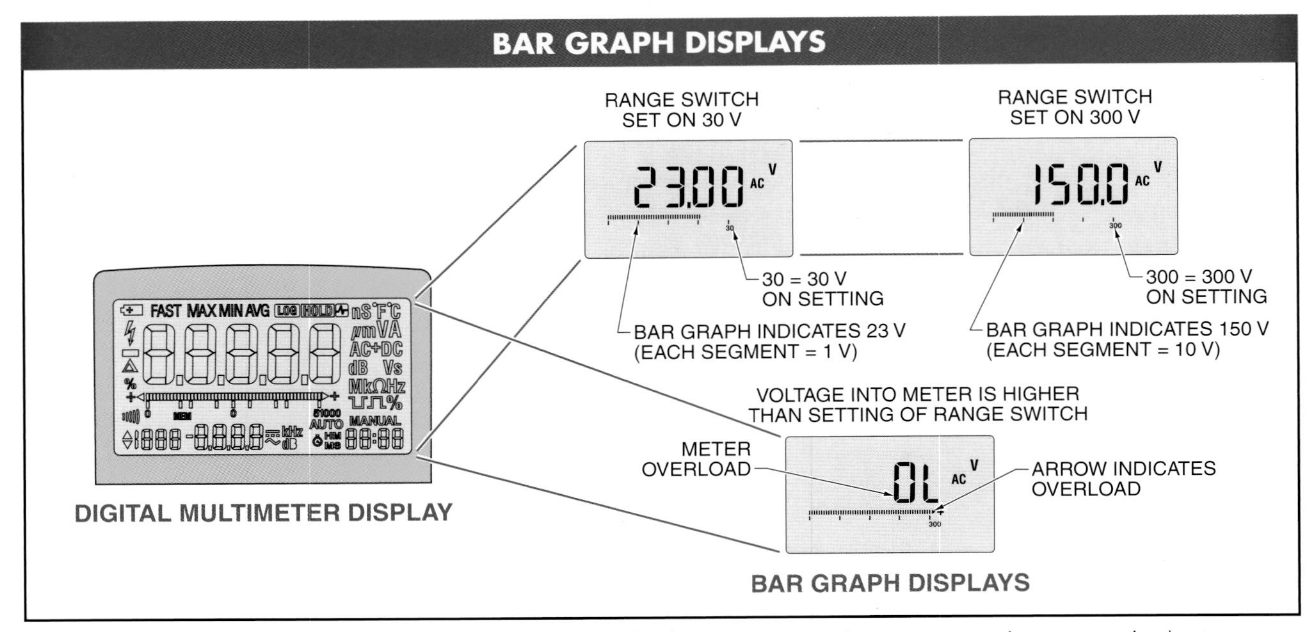

Figure 2-48. Bar graphs function as an analog pointer by displaying segments that increase as the measured value increases and decrease as the measured value decreases.

DMM Advanced Features

A DMM is useful when troubleshooting basic voltage, current, and resistance problems. Common troubleshooting problems include testing for power loss because of blown fuses, excessive current levels from overloaded circuits, and improper resistance from damaged insulation or components. However, when troubleshooting complex problems, a DMM with advanced features such as RANGE, HOLD, MIN MAX, and REC mode must be used. DMM advanced features are helpful when troubleshooting problems such as improper frequencies, overheated circuit neutral wires, and intermittent problems. Advanced features vary by manufacturer and the specific DMM model.

DMM Range Mode

A DMM has the best self-protection when left in the autorange mode when taking most measurements. Using the manual range mode (RANGE) requires that extra caution be used by an electrician as a DMM in the manual range mode does not automatically correct for values outside the set range. For example, when a DMM is set to the 3.000 V range, any value exceeding 3.000 V is displayed as OL. The OL reading prevents the display of any measurement over 3.000 V and may lead to confusion regarding the circuit being tested.

DMMs are sometimes changed from autorange mode to manual range mode when the value to be measured is known. For example, DMMs used to measure voltages on standard 115 VAC residential wall receptacles are set to the 300.0 VAC range to greatly reduce or eliminate ghost voltages. When left in the autorange mode, DMMs typically default to the 300.0 mV range during power-up. The lower voltage range makes DMMs much more sensitive during voltage measurements and may result in ghost voltages.

DMM Measurement Hold Mode

Measurement hold mode (HOLD) is a DMM mode that captures the measurement, beeps, and locks the meter measurement on the display for later viewing. Measurement hold mode automatically updates with each new measurement and allows the electrician to concentrate on test probe placement to reduce the risk of accidental circuit damage or electrical shock. Measurement hold mode is typically used when taking voltage, current, and resistance measurements. The HOLD button is pressed to select the measurement hold mode and is pressed again to exit. **See Figure 2-49.**

Technical Tip

Multimeters require general maintenance. Periodically wipe the case of the meter with a damp cloth and mild detergent. Do not use abrasives or cleaners (solvents). Dirt or moisture in the jacks can affect measurements taken and can falsely activate meter alarms. To clean the jacks do the following:

1. Turn meter OFF and remove test leads.
2. Shake out any dirt that may be in the jacks.
3. Soak a swab in alcohol. Move the swab around each jack.

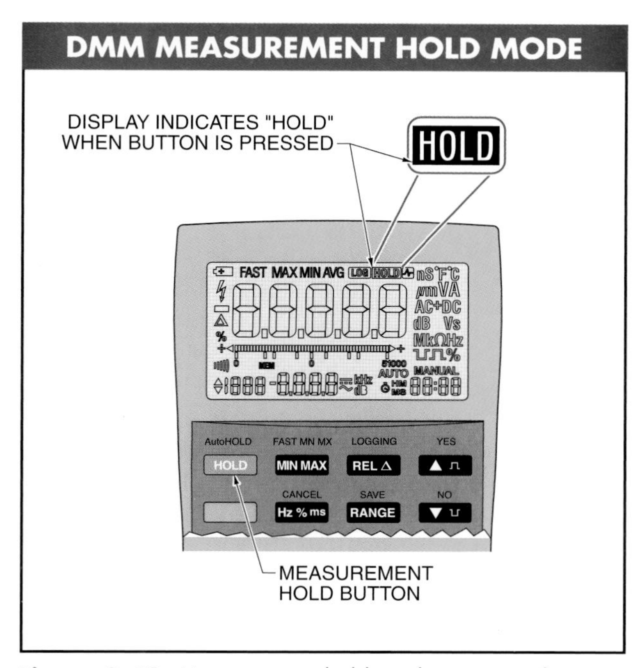

Figure 2-49. Measurement hold mode captures the measurement, beeps, and locks the meter measurement on the display for later viewing.

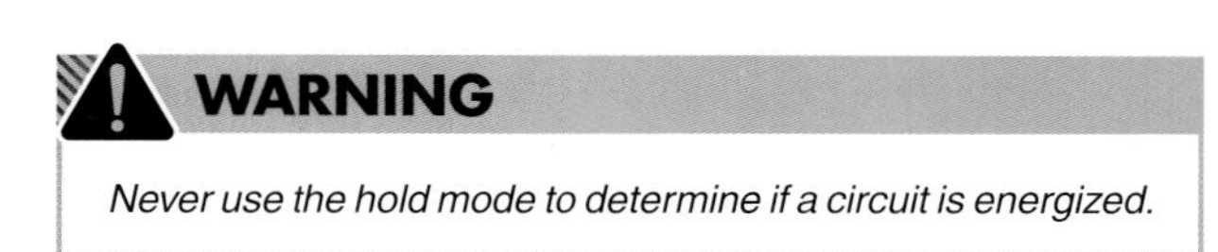

DMM MIN MAX Recording Mode

The *MIN MAX recording mode* is a DMM mode that captures and stores the lowest and highest measurements for later display. The MIN MAX recording mode is used with DMMs when taking voltage, current, and resistance measurements. The MIN MAX recording mode is also used with DMM accessories to record temperature, pressure, and vacuum measurements. When troubleshooting, the MIN MAX recording mode is especially useful for immediately recording the effect of any adjustments.

Minimum or maximum measured values are indicated by pressing the MIN MAX button after a DMM is connected to a circuit. The abbreviation MIN or MAX is displayed with the value to identify the measured value. Most DMMs sound (beep) each time a new minimum or maximum value is recorded. A continuous beep indicates that a DMM is overloaded.

DMM MIN MAX Recording Mode Applications

Voltage measurements using the MIN MAX recording mode are useful in identifying low-voltage conditions caused by large motors starting up, by temporary voltage losses, or by circuit overloading or high-voltage surges. Transformers have a fixed power rating (VA or kVA rating) that is listed on the nameplate of the transformer. When the total load connected to a transformer is within the rating of the transformer, the transformer delivers the correct secondary voltage. For example, most industrial control circuits are powered by a transformer that provides the proper voltage (24 V is common) to the control circuit. However, when the total load connected to the transformer exceeds the rating of the transformer, the voltage output of the transformer decreases. The greater the load, the greater the voltage decrease. A low-voltage condition causes circuit problems such as relays and motor starters dropping out (acting as if de-energized). **See Figure 2-50.** Low-voltage conditions are a difficult problem to detect because voltage measurements are often taken when all or almost all loads in a circuit are OFF.

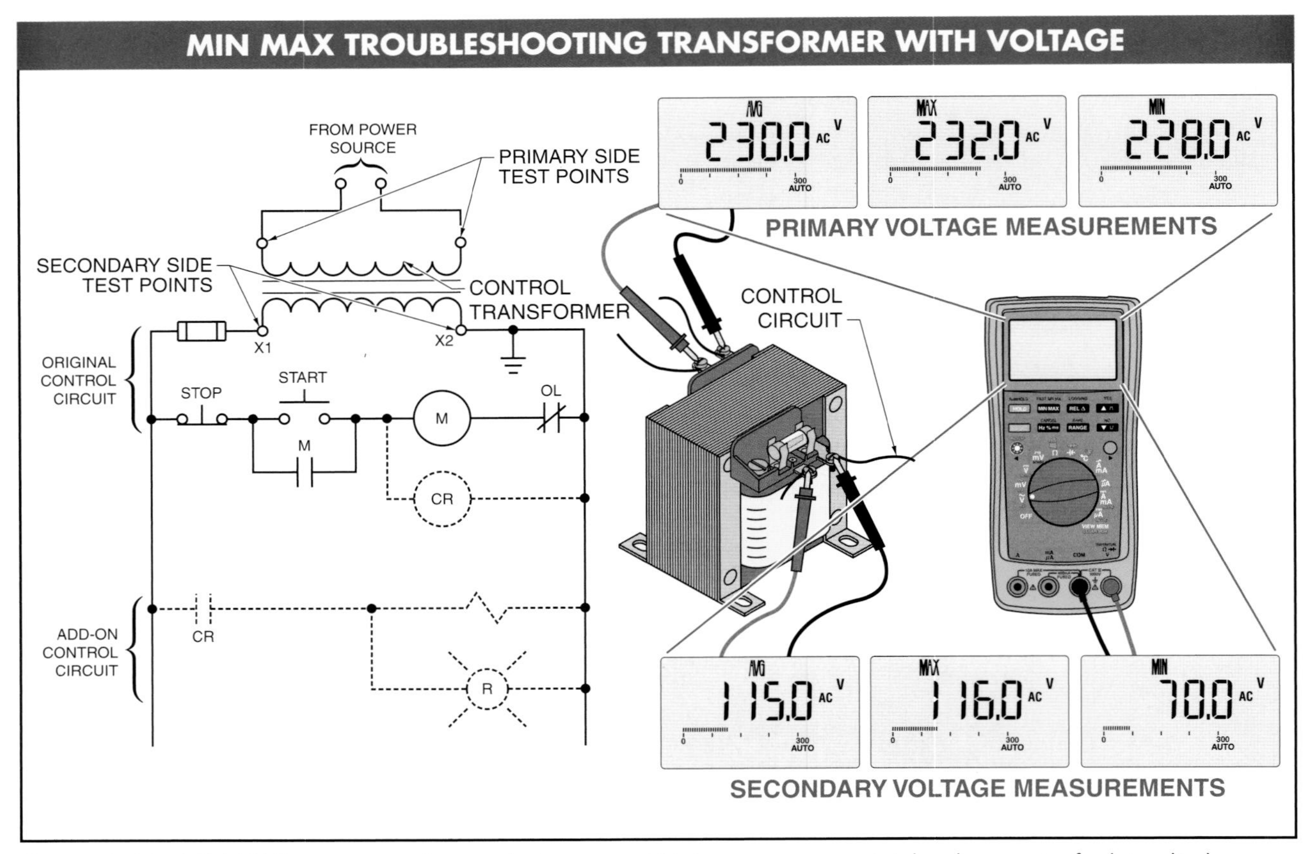

Figure 2-50. MIN MAX captures and stores the lowest and highest measurements taken by a meter for later display.

Because the voltage output of a transformer appears to be within specifications, the cause of the problem is often overlooked. A DMM set to monitor voltage with the MIN MAX recording mode on all loads during typical system operating conditions can detect a low-voltage condition caused by circuit overloading. A low-voltage condition is corrected by reducing the load on the transformer or increasing the size of the transformer.

Current measurements with the MIN MAX recording mode are useful in determining when circuits are overloaded or if there is capacity for additional loads to be added to the circuit. When all loads in a circuit are OFF, there is no current draw. When there is current draw with all loads OFF, typically there is a leakage current problem. Leakage current problems are caused by a partial short in the system from problems such as insulation breakdown. With all loads ON, the total current draw must be within the specifications for the system. When the measured current is less than the maximum current capacity of the circuit, the circuit has room for additional loads. **See Figure 2-51.**

Resistance measurements with the MIN MAX recording mode are useful in isolating problems from loose connections, corrosion, or shorts. With all power OFF, the DMM test leads are connected across a connection, splice, load, or circuit. The resistance of the connection or circuit being tested is displayed. When a DMM is set to measure resistance using the MIN MAX recording mode, the measurement should not change. When a new high-resistance maximum (MAX) measurement is recorded, there is probably a loose connection (open) in the circuit. For a connection with a fixed-resistance value (0.05 Ω), a new high-resistance maximum (150.0 Ω) measurement indicates there is probably a bad connection. A DMM can be left in MIN MAX recording mode for long-term measurements. A DMM records measurements as long as the batteries are in good condition (typically 1000 hr). **See Figure 2-52.**

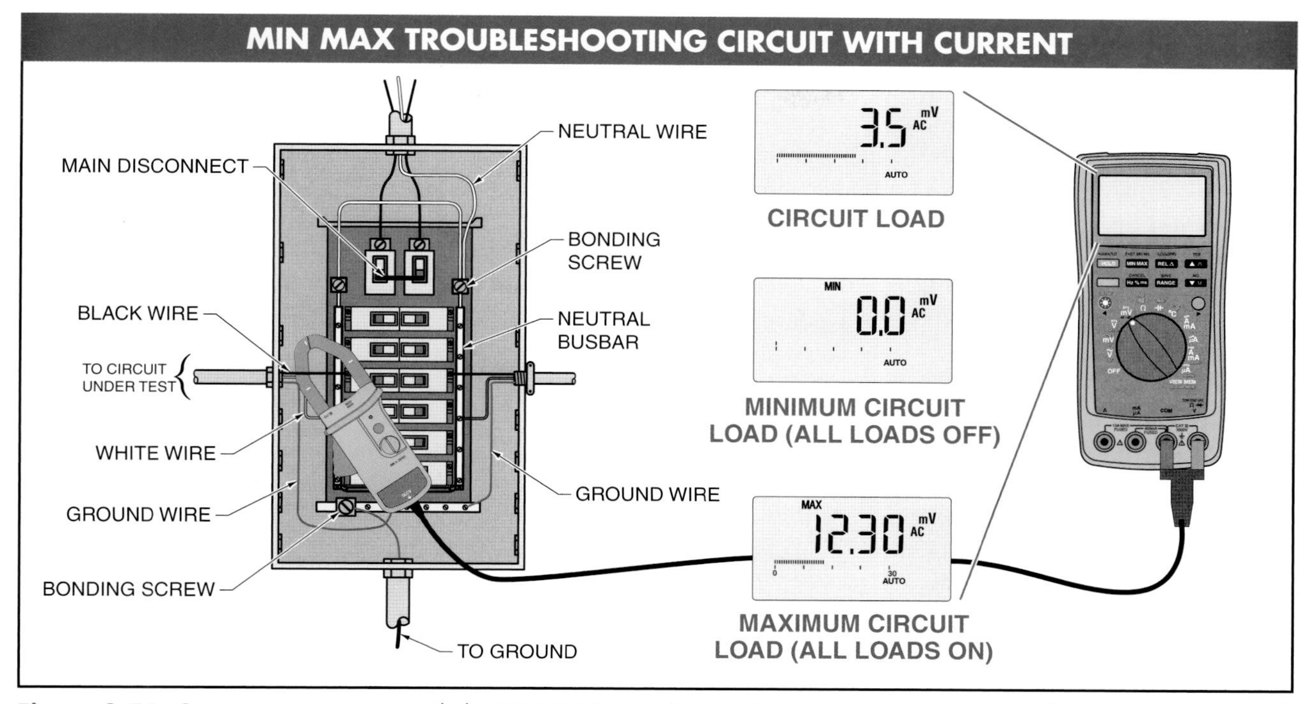

Figure 2-51. Current measurements with the MIN MAX recording mode are useful in determining if circuits are overloaded or if there is capacity for additional loads to be added to the circuit.

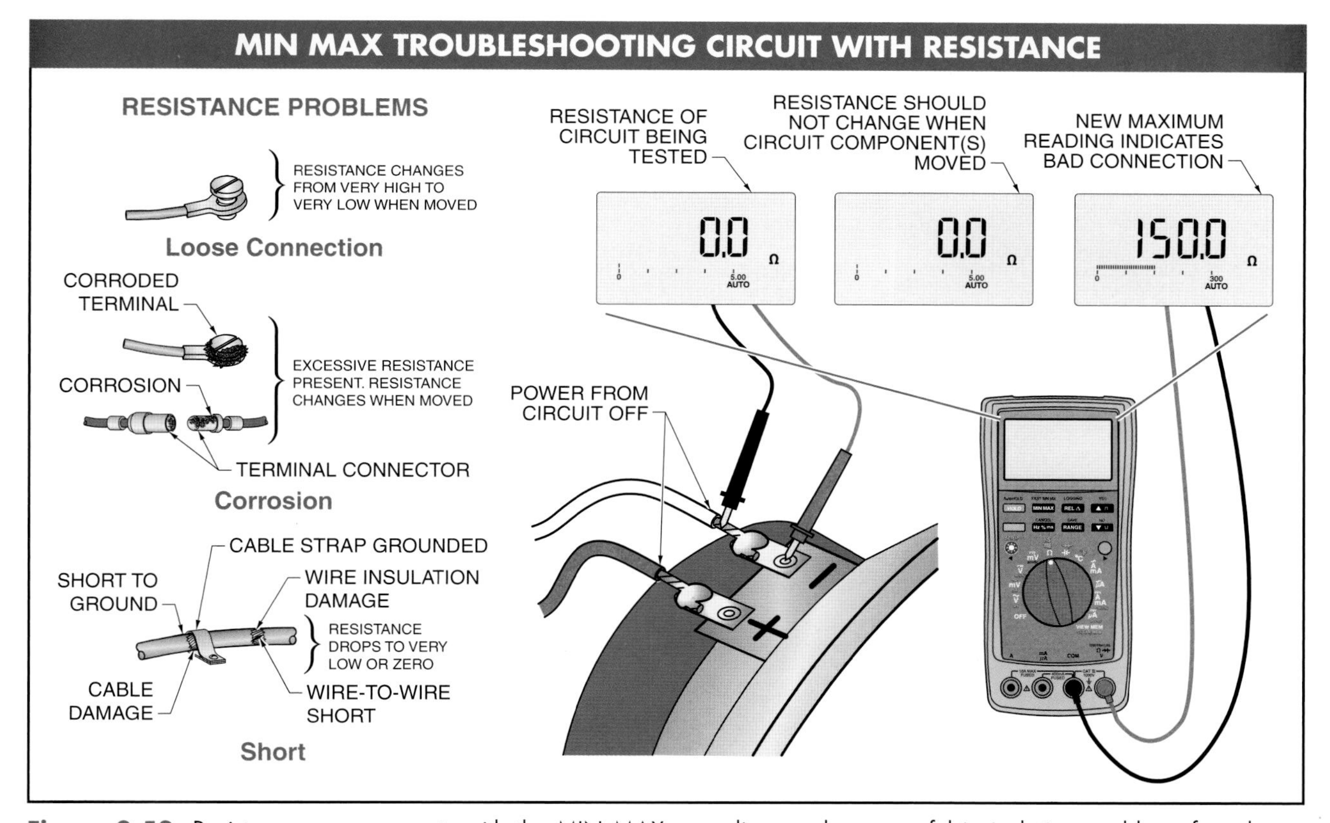

Figure 2-52. Resistance measurements with the MIN MAX recording mode are useful in isolating problems from loose connections, corrosion, or shorts.

Relative Mode

Relative mode (REL) is a DMM mode that records a measurement and displays the difference between that reading and subsequent measurements. The relative mode is typically used to show measurements above or below a specific value, or to zero out the baseline reading to eliminate the resistance detected in the test leads of the meter.

To use the relative mode, a DMM is connected to the circuit being tested. The REL button is pressed and the DMM displays a zero reading. The measured value at the time the REL button is pressed is stored as a reference value. In relative mode, any measurement value displayed on the DMM is the difference between the measured test value and the stored reference value.

Relative mode is used with voltage, current, and resistance measurements. For example, the relative mode is used when testing a forklift electrical system for the amount of current drawn by a load or circuit. A DMM set to measure DC using a DC clamp-on current probe accessory can measure total system current at the battery or any individual circuit current at the fuse box. **See Figure 2-53.**

Relative mode can also be used to zero out any current that is constantly drawn by an electrical system during normal operation. For example, the memory circuits in computers, digital clocks, and cellular telephones constantly draw a small amount of operating current. Pressing the REL button zeros out the DMM by subtracting the memory circuit current from the measurement being displayed. As an individual load is turned ON, current measurements are displayed for that load only. The relative mode allows for greater DMM accuracy when taking measurements.

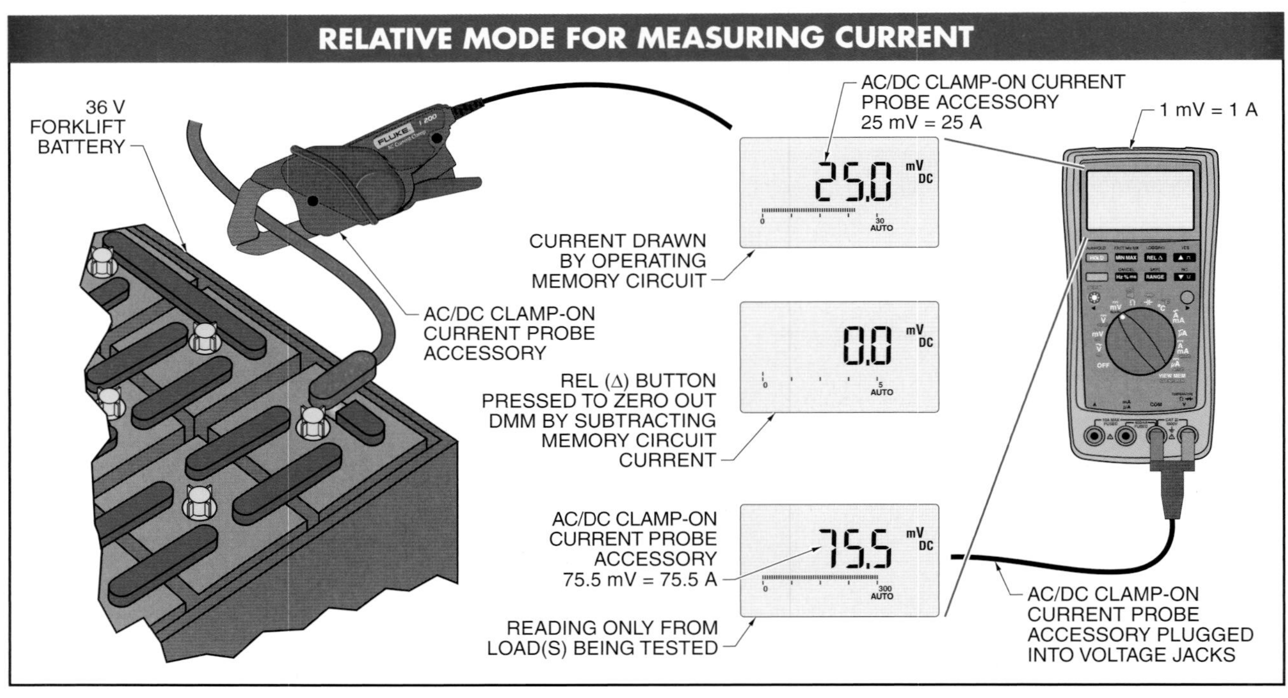

Figure 2-53. Relative mode records a measurement and displays the difference between that reading and subsequent measurements taken by a meter.

Diode Test Mode

Diode test mode is a DMM mode used to test the condition of a diode. A *diode* is an electronic device that allows current to flow in one direction only. Diodes are commonly used in electrical and electronic circuits to control current flow. A DMM in diode test mode produces a small voltage between the test leads that is used to measure the voltage drop across a diode. Values are compared with specifications acquired from the manufacturer for proper diode operation. Typically, a diode will measure 0.2 V to 0.8 V when the test meter leads are connected in the forward-bias direction and OL (overload) when the test meter leads are reversed and the diode is reversed biased. A shorted diode will allow AC to pass in both directions. An open diode will measure OL in both directions. See **Figure 2-54.**

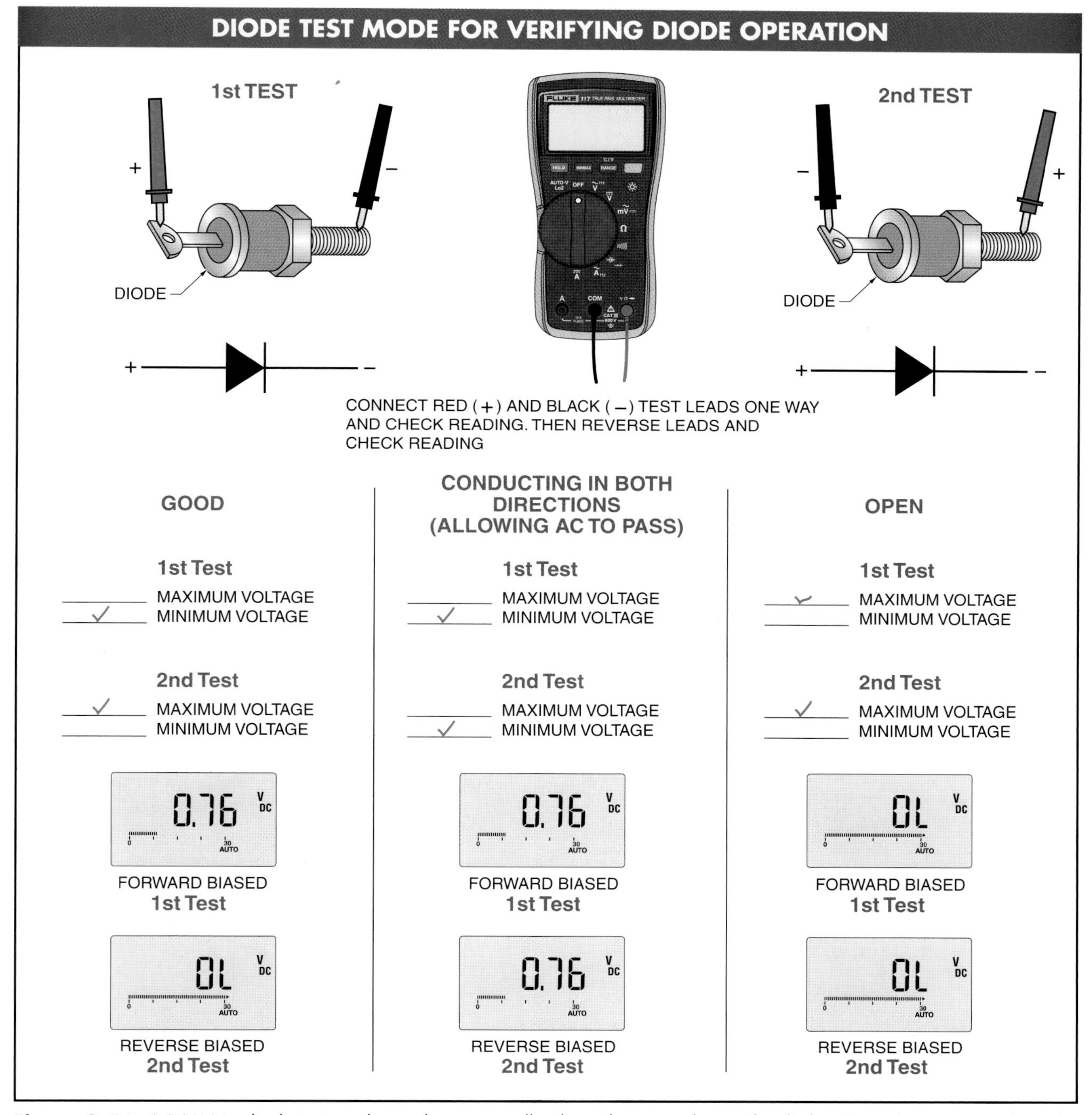

Figure 2-54. A DMM in diode test mode produces a small voltage between the test leads that is used to measure the voltage drop across a diode.

Capacitance Measurement Mode

Capacitance measurement mode is a DMM mode used to measure capacitance or test the condition of a capacitor. A capacitor is a device used to store an electrical charge. When in capacitance measurement mode, a DMM charges the capacitor with a known current, measures the resultant voltage, and calculates the capacitance. Values are compared with specifications acquired from the manufacturer for proper capacitor operation. Capacitance measurement mode is commonly used to test the condition of capacitors used on capacitor-start motors. **See Figure 2-55.**

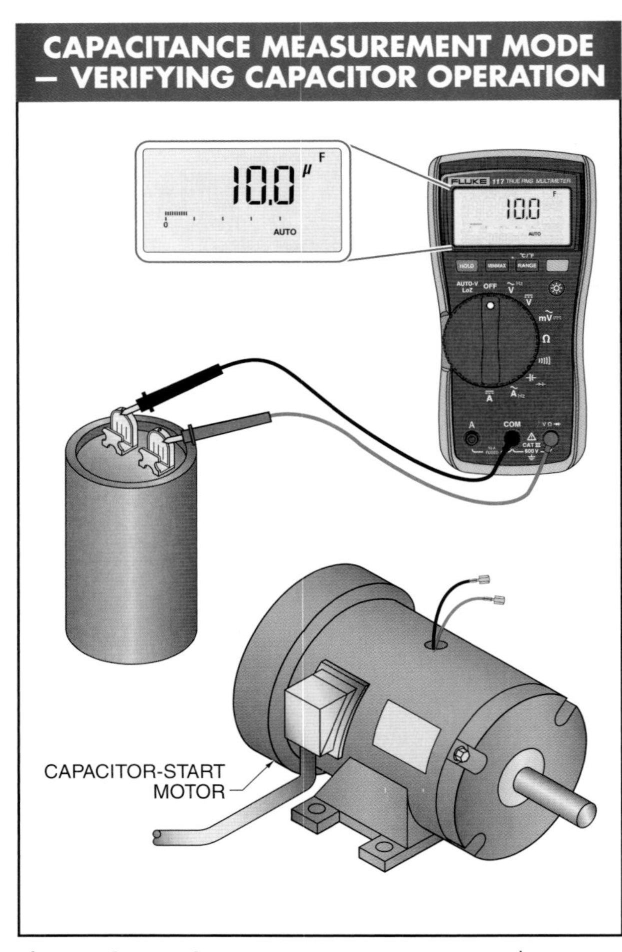

Figure 2-55. Capacitance measurement mode is commonly used to test the condition of capacitors used on capacitor-start motors.

Capacitor-start motors are used in applications where a motor must be brought to up to 75% of maximum speed quickly and cycled on and off rapidly.

Selecting a Test Instrument or Meter

Portable and permanent test instruments are available in a wide range of styles and prices. Cost is dependent on specifications (measuring range or accuracy), added features (such as carrying case, stand, or clamp-on current probe), built-in safety features and ratings (such as silicon test leads and CAT rating), warranties, intended usage (contractor grade or industrial), and other features (such as size of display and sleep mode to extend battery life).

GENERAL USE TEST INSTRUMENTS

2

REVIEW

Name: ______________________ Date: ______________________

______ **1.** What type of voltage test instrument does not require direct contact with the circuit?

______ **2.** Would a quality control testing lab be more likely to use permanent or portable test instruments?

T F **3.** A 1% current unbalance can create an 8% voltage unbalance.

______ **4.** Would a maintenance technician who troubleshoots production equipment be more likely to use permanent or portable test instruments?

T F **5.** In-line ammeters are used to measure small amounts of current.

______ **6.** Which common multimeter mode records a measurement and displays the difference between the reading and subsequent measurements?

______ **7.** Which type of ammeter is used when the circuit to be measured cannot be opened?

______ **8.** Which multimeter test mode is commonly used to test components such as switches, fuses, electrical connections, and individual conductors?

T F **9.** The measurement hold mode is a multimeter mode that captures and stores the lowest and highest measurements for later display.

______ **10.** Which electrical property equals current multiplied by resistance?

T F **11.** Loads such as motors and solenoids decrease in resistance over time.

______ **12.** Which electrical property is measured to indicate total circuit loading?

______ **13.** What color is the test lead that typically indicates negative or ground?

T F **14.** Reverse the polarity of the test leads when a negative sign is displayed at the beginning of a multimeter bar graph.

______ **15.** Which common multimeter mode would be most useful in measuring the change in current when a load is turned ON?

______ **16.** Which common multimeter mode would be most useful in recording high- or low-voltage conditions over a period of time?

______ **17.** What type of electronic component is tested by measuring the voltage drop across it in two directions?

T F **18.** Surface leakage current is current that flows from areas on conductors where insulation has been removed to allow for electrical connections.

______________ **19.** Which type of AC voltage measurement is useful because it equates the heat produced by an AC sine wave to DC resistive circuit?

______________ **20.** Which part of an analog multimeter display is determined by the position of the range switch?

21. Why is a small amount (microamperes) of leakage current unavoidable?

22. Why are neon test lights not used to test ground fault circuit interrupters (GFCIs)?

23. Why is it important to test insulation on conductors?

24. Why must megohmmeters, and not regular ohmmeters, be used to test insulation?

25. How is the ohmmeter resistance reading of one component affected by other components connected in parallel?

3 VOICE-DATA-VIDEO (VDV) TEST INSTRUMENTS

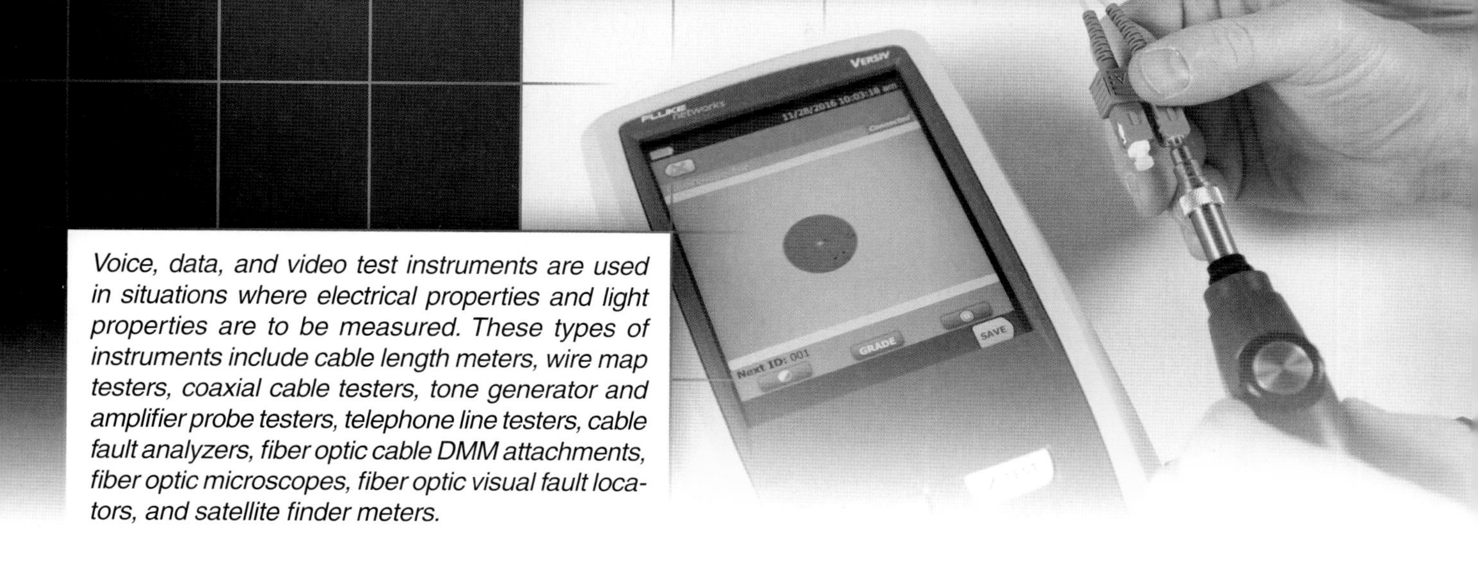

Voice, data, and video test instruments are used in situations where electrical properties and light properties are to be measured. These types of instruments include cable length meters, wire map testers, coaxial cable testers, tone generator and amplifier probe testers, telephone line testers, cable fault analyzers, fiber optic cable DMM attachments, fiber optic microscopes, fiber optic visual fault locators, and satellite finder meters.

VOICE-DATA-VIDEO COMMUNICATION

People have always communicated by one means or another. Voice-data-video communication has evolved over the years. In the past, voice communication was as simple as people conversing or giving speeches, data communication was as simple as recording quantities and measurements, and video communication involved the use of photographs and visual aids.

Modern requirements for voice-data-video (VDV) communication still include the most basic forms of communication but also include the need to communicate over great distances as accurately and quickly as possible. The four biggest needs in modern communication are the following:

- To move large amounts of data at one time. Entire volumes that once took years to manually copy may now be electronically copied, electronically moved over great distances, and electronically reproduced in print or video.
- To move large amounts of data at faster speeds. Where a second was once considered a very short time period for almost any type of transaction, modern operating speeds are milliseconds (ms), microseconds (μs), and nanoseconds (ns). VDV transmission speed has moved from kilohertz (kHz), to megahertz (MHz), to gigahertz (GHz).
- To maintain proper system operation, prevent communication and storage mistakes, and prevent file corruption.
- To be able to move VDV communication to points along the network using methods other than copper wires. VDV communication using means other than copper wires has allowed the growth in the use of fiber optic cable and wireless communication. **See Figure 3-1.**

Technical Tip

While a DVD disc is the same diameter as a compact disc (CD), a DVD disc has smaller physical pits beneath the disc surface, allowing it to store more information than an ordinary CD.

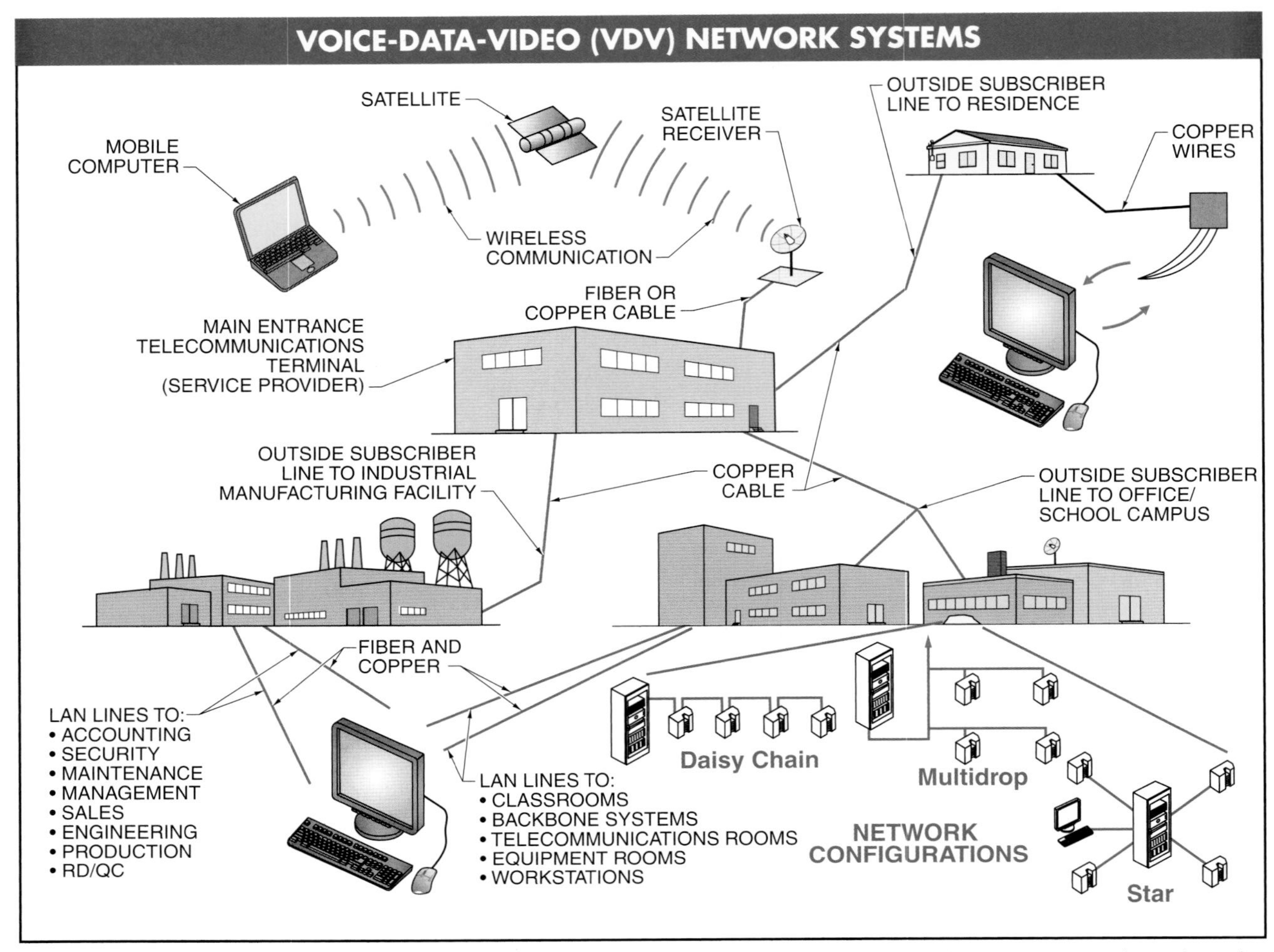

Figure 3-1. VDV communication networks utilize copper and fiber optic cables, and wireless connectivity throughout the entire system.

The ability to move large amounts of information at high speeds, without mistakes or file corruption, continues to be provided by advances in technology. For example, using fiber optics to transmit VDV signals allows for much faster operating speeds than copper wire signal transmission. Wireless transmission of VDV signals allows VDV communication from almost any location. Proper communication is ensured by testing newly installed cables and lines using VDV test instruments. After the cables and lines are operational, VDV test instruments are used to locate potential problems before the potential problems cause data or communication losses.

Technical Tip

The maximum distance of a cable run from a telecommunications closet to a work area is 295 ft (90 m).

VDV Testing

Electrical circuits are tested to ensure that the circuits will work properly when placed in full operation. However, not all electrical circuits are tested 100% before being placed in operation. Some circuits, such as 120 VAC branch circuits, are spot tested, or tested only when a problem exists. All VDV circuits must be tested and certified as meeting minimum industry standards prior to being placed in service.

Knowing which test to perform, and understanding the test results, is an important part of installing and maintaining all VDV systems. All problems must be identified and corrected before there is an interruption in service. Testing includes utilizing test instruments to perform basic tests as well as advanced tests that are required to meet industry specifications.

VDV circuit testing starts with the same tests that have been performed on copper phone lines since the beginning of voice communication. Typical tests include ensuring that the lines do not include any basic problems such as open circuits, short circuits, incorrect wiring, and basic wire map problems. Advanced electronic testing parameters are performed on all VDV cables to ensure minimum industry standards are met upon complete installation.

Advanced tests are required because problems that may not be a factor with voice transmission can cause major problems during data transmission. For example, crosstalk on phone lines is when a line hears part of another phone conversation. *Crosstalk* is any unwanted reception of signals induced on a communication line from another communication line or from an outside source. Crosstalk causes problems in data lines. Crosstalk problems are amplified as the frequency is increased. As data moves at faster speeds, crosstalk problems will only increase. Data that once was transmitted at kilohertz operating speeds now moves at megahertz and gigahertz speeds.

- 1 cycle in 1 sec = 1 Hz (hertz)
- 60 cycles in 1 sec = 60 Hz
- 1000 cycles in 1 sec = 1 kHz
- 1,000,000 (one million) cycles in 1 sec = 1 MHz
- 1,000,000,000 (one billion) cycles in 1 sec = 1 GHz

Most VDV tests are based on a pass or fail indication. However, when a cable system fails a test, there is a reason. Understanding the test and test results is important so corrective action can be taken to fix the problem. Problems that occur only for a short time duration or intermittently require special test instruments that can detect and record intermittent signal problems.

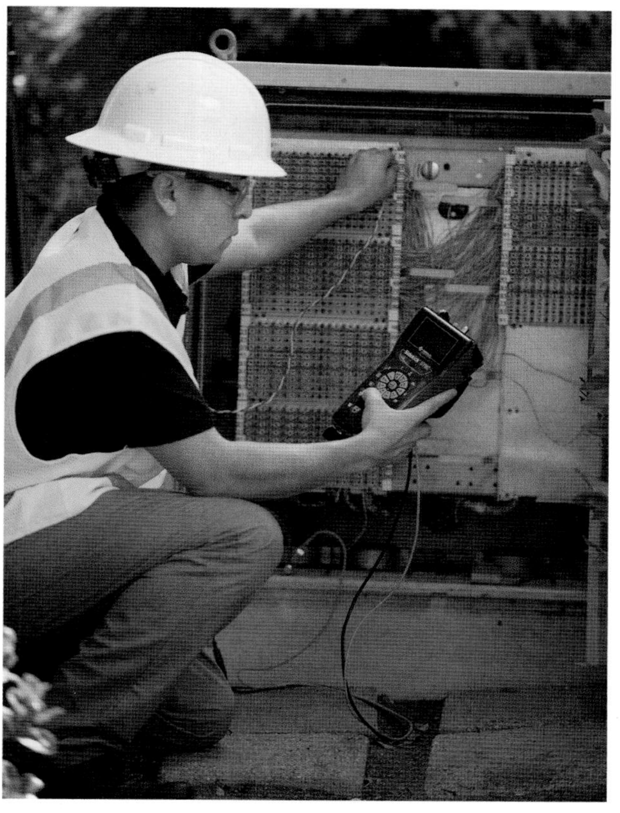

Greenlee Textron, Inc.

Even simple voice networks require a variety of test instruments to verify, install, troubleshoot, and repair telephone conductors and cables.

VDV Terminology

All technical fields, including the electrical and electronic fields, use acronyms and abbreviations. The VDV communication field is a technical area within the electrical and electronic fields that uses terminology with many acronyms and abbreviations to shorten a word or phrase. An *acronym* is a word formed from the first letters of a compound term (voice, data, video = VDV). An *abbreviation* is a letter or combination of letters that represent a word (hertz = Hz).

Understanding VDV acronyms and abbreviations is important when working within the VDV field. Understanding VDV acronyms and abbreviations is also important when using test instruments because VDV test instruments use acronyms and abbreviations to indicate the tests performed and the information displayed. **See Figure 3-2.**

VDV TEST INSTRUMENT ACRONYMS

Acronym	Meaning
AC	alternating current
ACR	attenuation to crosstalk ratio
A/D	analog to digital
ADSL	asymmetric DSL
AF	audio frequency
AM	amplitude modulation
ANSI	American National Standards Institute
AWG	American Wire Gauge
BER	bit error rate
BICSI	Building Industry Consulting Services International
BNC	Bayonet Neill-Concelman
b/s	bits per second
CAN	campus area network
CATV	community access television
CBC	coupled bonding conductor
CCTV	closed circuit television
CM	communication cable
CMR	communication riser cable
CO	central office
CPE	customer premises equipment
CSA	Canadian Standards Association
CT	cable tray
D/A	digital to analog
dB (DB)	decibels
DC	direct current
DD	distribution device
DLC	digital loop carrier
DS	digital signal
DSL	digital subscriber line
DTMF	dual tone multiple frequency
EF	entrance facility
ELFEXT	equal level far end crosstalk
EMC	electromagnetic compatibility
EMI	electromagnetic interference
EMT	electrical metal tubing
ER	equipment room
FCC	Federal Communications Commission
FDDI	fiber distributed data interface
FEXT	far end crosstalk
ft	feet
FM	frequency modulation
FTP	foiled twisted pair
Gb/s	gigabits per second
GRD	ground
HC	horizontal cross-connect
HF	high frequency
Hub	a networking device that allows attached devices to receive information/data
HVAC	heating, ventilation, air conditioning
Hz	hertz
IC	intermediate cross-connect
IEEE	Institute of Electrical & Electronic Engineers
ISDN	integrated service digital network
ISO	International Organization for Standardization
ISP	internet service provider
Kb/s	kilobits per second
Km	kilometer
KTS	key telephone system
LAN	local area network
m	meter
MAN	metropolitan area network
MAX	maximum
Mbps	megabits per second
MC	main cross-connect
MDF	main distribution frame
MHz	megahertz
MIN	minimum
MP	multipurpose cable
MPP	multipurpose plenum cable
MPR	multipurpose riser cable
MUX	multiplex
NEC	National Electrical Code
NEG	negative
NEMA	National Electrical Manufacturers Association
NEXT	near end crosstalk
NFPA	National Fire Protection Association
NIC	network interface card
NM	non-metallic
NOC	network operations center
NOS	network operating system
OC	optical carrier
OSI	open system interconnection
PBX	private branch exchange
POS	positive
POTS	plain old telephone service
PSACR	power sum attenuation to crosstalk ratio
PSELFEXT	power sum equal level far end crosstalk
PSTN	public switched telephone network
RF	radio frequency
RFI	radio frequency interference
RUS	rural utility service
ScTP	screened twisted pair
SNMP	simple network management protocol
SNR	signal to noise ratio
SONET	synchronous optical network
STP	shielded twisted pair
TC	telecommunications closet
TDR	time domain reflectometer
TGB	telecommunications grounding bar
TO	telecommunications outlet
TP	twisted pair
UDC	universal data connector
UL	Underwriters Laboratories
UTP	unshielded twisted pair
VDSL	very-high bit rate DSL
VDV	voice-data-video
WA	work area
WAN	wide area network
X-Box	cross-connection box
Z	impedance

Figure 3-2. Knowledge of VDV acronyms is required when working with test instruments.

For example, a data communication tester may be listed as having the following capabilities:

- Certified up to CAT 7 — ISO E limits to 600 MHz
- Tests UTP cables
- Tests ScTP cables
- Tests STP cables
- Tests LAN cables
- TDR capability for Cu and fiber measurements
- Measures signal loss greater than 0.3 dB
- Includes DSL filtering

VDV CONDUCTORS AND CABLE

Voice, data, and video signals must be transmitted from a transmitter to a receiver. In between a transmitter and receiver the signals may be amplified, filtered, stored, redistributed, and/or changed from one form to another (analog to digital, digital to analog) as required.

Voice is transmitted from the transmitter of one phone to the receiver of another phone. Voice or music is transmitted from a source (radio station, CD, or DVD) to a receiver that converts and amplifies the signal back into an audible signal that is heard through speakers. Likewise, video signals are transmitted from cameras to receivers that convert and amplify the signal back into a visual signal that is seen on a CRT or LCD screen. Data is transmitted from computers, bar code readers, and transducers (pressure and temperature) to be received by other computers, recording devices, and other types of displays.

VDV signals are transmitted using any combination of hard wiring (cable), fiber optic cables, or wireless (laser and satellite) devices. Hard-wired circuits have been the standard method of transmitting and receiving VDV signals. Hard-wired circuits use copper wire or fiber optics to send and receive signals between various locations. Wireless transmission allows signals to be transmitted and received without hard wiring between locations.

VDV Conductors

Electrical circuits and components are connected using conductors. A *conductor* is a low resistance metal that carries electricity to various parts of a circuit. VDV conductors include several different cable types. **See Figure 3-3.**

Conductors are available as individual wires or in groups, such as cable and cord. A *wire* is an individual conductor. A *cable* is two or more conductors grouped together within a common protective cover and used to connect individual components. A *cord* is two or more very flexible conductors grouped together and used to deliver power to a load by means of a plug.

Most individual conductors are enclosed in an insulated cover to protect the conductor, increase safety, and meet code requirements. Some individual conductors, such as a ground wire, may be bare. In coaxial cable, an outer braided conductor is wrapped around an insulated inner conductor. The braided conductor is used to shield the inner conductor from outside electromagnetic interference (noise).

Technical Tip

Refer to TIA-569,Pathways and Spaces Standard, prior to performing commercial VDV installation projects.

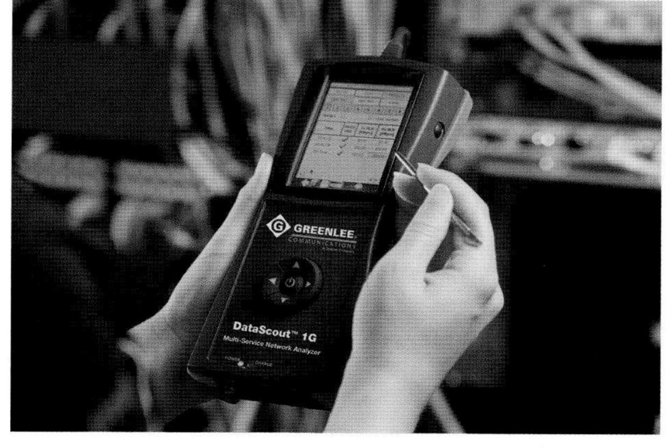

Greenlee Textron, Inc.

VDV test instruments ensure low voltage conductors and other types of cables have no physical damage and will work correctly when installed.

Communication Cable

Cable is used to deliver electrical power or signals from one location to another location within an electrical system. Power cables are used to deliver high voltage and current throughout an electrical system. Power cables commonly deliver 90 VDC, 115 VAC, 180 VDC, 230 VAC, or 460 VAC to residential, commercial, and industrial loads. However, power cables also deliver high currents at lower voltage loads as in automobile, marine, or aviation applications.

VDV CABLE TYPES		
4 UTP	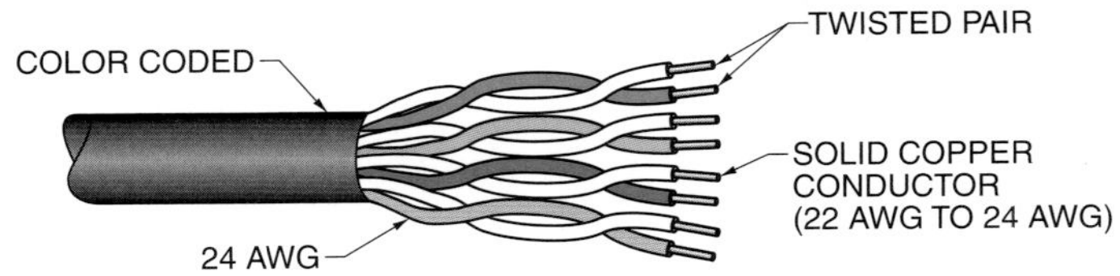 COLOR CODED; TWISTED PAIR; SOLID COPPER CONDUCTOR (22 AWG TO 24 AWG); 24 AWG	TWISTED-PAIR CABLE USED FOR VOICE AND DATA TRANSMISSION
25 UTP	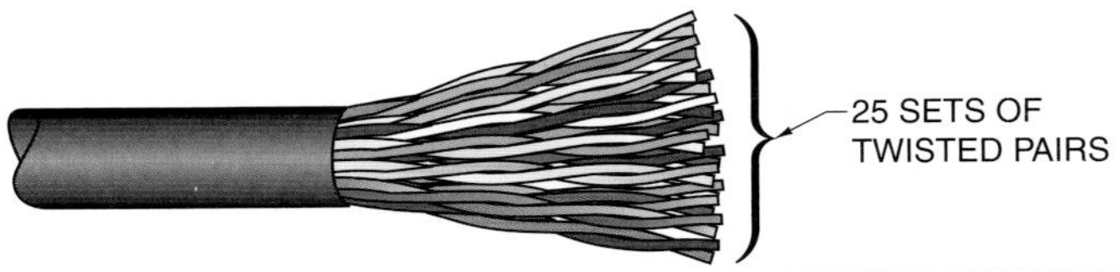 25 SETS OF TWISTED PAIRS	CABLE USED FOR DATA TRANSMISSION
COAXIAL CABLES	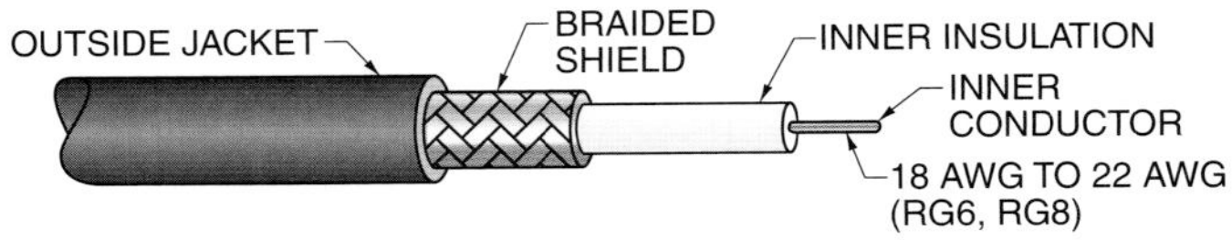 OUTSIDE JACKET; BRAIDED SHIELD; INNER INSULATION; INNER CONDUCTOR; 18 AWG TO 22 AWG (RG6, RG8)	CONNECTS CABLE TELEVISION AND VCR TO TELEVISION: ALSO USED FOR DATA
CAT 3/FTP/ 100PR/24 AWG	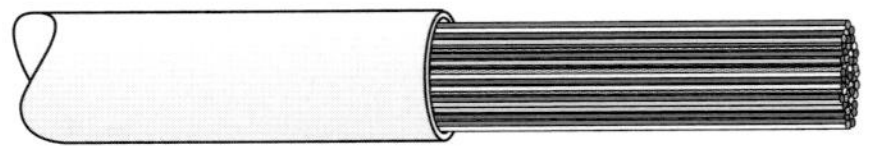	CAT 3 ISO 11801/TIA 568A COMPLIANT. USED AS BACKBONE CONNECTION FOR DATA SYSTEMS
ALUMINUM ARMOR OSP CABLES	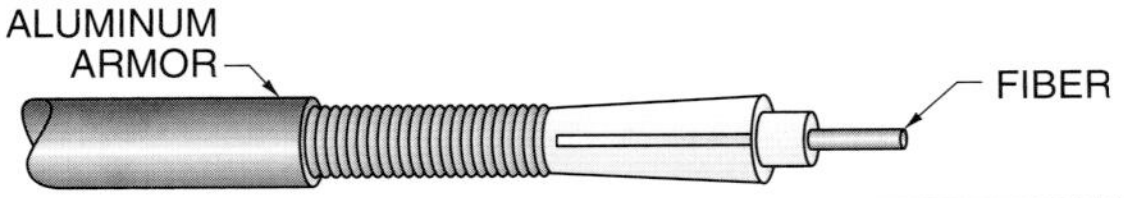 ALUMINUM ARMOR; FIBER	CAT 3 ISO 11801/TIA 568A COMPLIANT. USED AS BACKBONE CONNECTION FOR DATA SYSTEMS
UTP CABLES	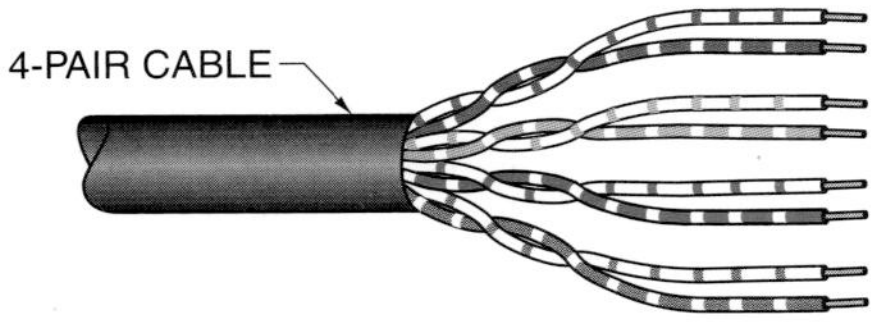 4-PAIR CABLE	TWISTED-PAIR CABLE USED FOR VOICE, DATA, AND VIDEO TRANSMISSION
STP CABLES	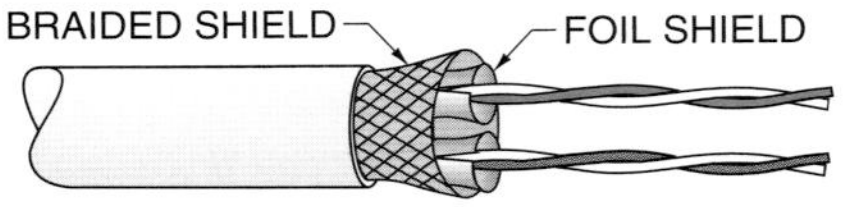 BRAIDED SHIELD; FOIL SHIELD	USED FOR PROTECTION FROM ELECTROMAGNETIC INTERFERANCE
ScTP CABLES	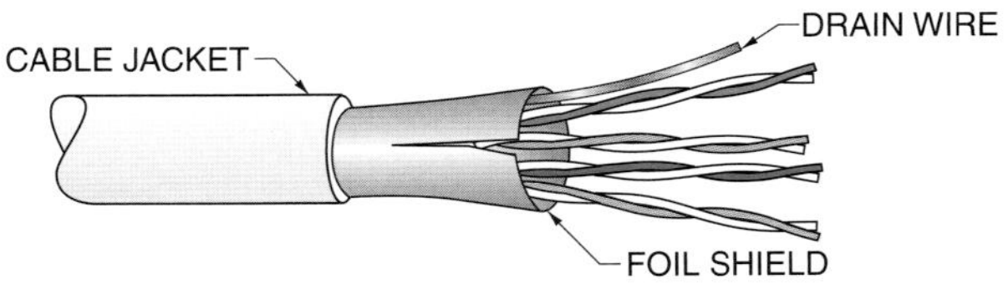 DRAIN WIRE; CABLE JACKET; FOIL SHIELD	USED FOR PROTECTION FROM ELECTROMAGNETIC INTERFERENCE; HAS LESS RESISTANCE THAN STP
FIBER-OPTIC CABLES	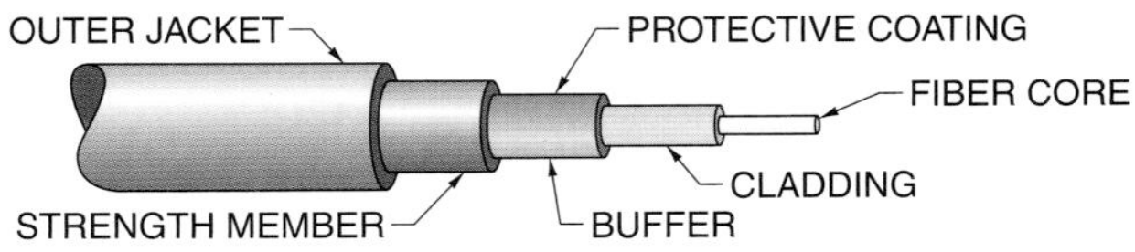 OUTER JACKET; PROTECTIVE COATING; FIBER CORE; CLADDING; STRENGTH MEMBER; BUFFER	FIBER-OPTIC CABLE USED OUTSIDE FOR DIRECT BURIAL AND LONG RUNS
HYBRID FIBER COAX (HFC)	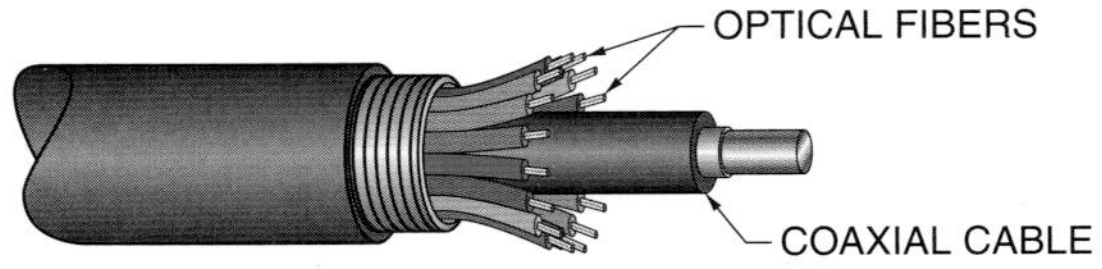 OPTICAL FIBERS; COAXIAL CABLE	

Figure 3-3. VDV cable types include several different cable types.

Cables are also used to deliver communication signals within a VDV system. Communication cable is cable (typically copper or fiber optic) used to transmit data from one location or device within a system to another location. Communication cable is also commonly referred to as low voltage cable. Typical AWG #22, AWG #24, and AWG #26 wire (solid or stranded) is used, but the CAT (category) rating of the cable is more important than the wire size. The following types of communication cable are used to interconnect various systems:

- Audio or voice cable is used to interconnect telephones, speaker systems, intercoms, and digital surround sound systems. Voice wiring typically uses a four-pair, 100 Ω UTP (unshielded twisted pair) cable. A CAT 3 or CAT 5e rated cable must be used for audio or voice communication.
- Video cable (coaxial) is used to interconnect security systems, television, cable television, digital cable, DVDs, projection screens, and multimedia displays. Video hard wiring uses 75 Ω coaxial cable.
- Data cable is used to interconnect computer networks, fax machines, and printers. Data wiring uses a four-pair, 100 Ω UTP cable. A CAT 5e or CAT 6 rated cable must be used for data communication.

Communication cables are designed for aerial, protected (behind walls), or underground applications. *Aerial cable* is cable suspended in the air on poles or other overhead structures. Aerial cable typically includes a steel cable used for strength. Communication cable designed to be buried underground, also known as outside plant cable, includes additional outer layers for moisture protection. Underground communication cable is used to interconnect stations, servers, terminals, telecommunication rooms, entrance facilities, and backbone cables. *Backbone cables* are conductors (copper or fiber optic) used between telecommunication closets, or floor distribution terminals and central offices (equipment rooms) within a building. Communication cable is often labeled using the acronym of the cable, such as unshielded twisted pair (UTP), screened twisted pair (ScTP), shielded twisted pair (STP), or shielded twisted pair-A (STP-A).

Communication Cable Category (CAT) Rating

Communication cable is rated by a category (CAT) rating number. CAT ratings standards are referenced from ANSI/TIA 568-C.2. New CAT ratings are continually created as manufacturers achieve faster operating speeds, and old CAT ratings such as CAT 1, 2, 3, and 5 are phased out. **See Figure 3-4.**

The CAT rating of a cable determines the conductor bandwidth rating. *Bandwidth* is the width of a range of frequencies that have been specified as performance limits within which a meter can be used. The higher the bandwidth, the higher the data transfer capacity. Cable CAT ratings include the following:

- CAT 1—A rating used for older wire and cable used in POTS ("plain old telephone service"). CAT 1 cables are of various gauges and are used in inside applications with frequencies less than 1 MHz. Examples include UTPSTP backbone voice applications and doorbells.
- CAT 2—A rating used for inside wire and cable systems for voice applications and data transmission up to 4 MHz. Cable sizes are AWG #22 or AWG #24 and are typically used in UTPSTP backbone voice applications.
- CAT 3—A rating used for inside wire and cable systems for voice applications and data transmission up to 16 MHz. High-speed data/LAN systems use CAT 3. CAT 3 cable size is AWG #24, which is used in UTPSTP backbone voice applications that span a relatively small area.
- CAT 4—A rating used for inside wire and cable systems for voice applications and data transmission up to 20 MHz. Cable sizes are various gauges.
- CAT 5—A rating used for inside wire and cable systems for voice and data transmission up to 100 MHz. Cable sizes are various gauges. Early on, CAT 5 cable was more common in LAN systems because CAT 5 offered extended frequency transmission, above that of CAT 3 and CAT 4 cable.

COMMUNICATION CABLE CATEGORY RATINGS

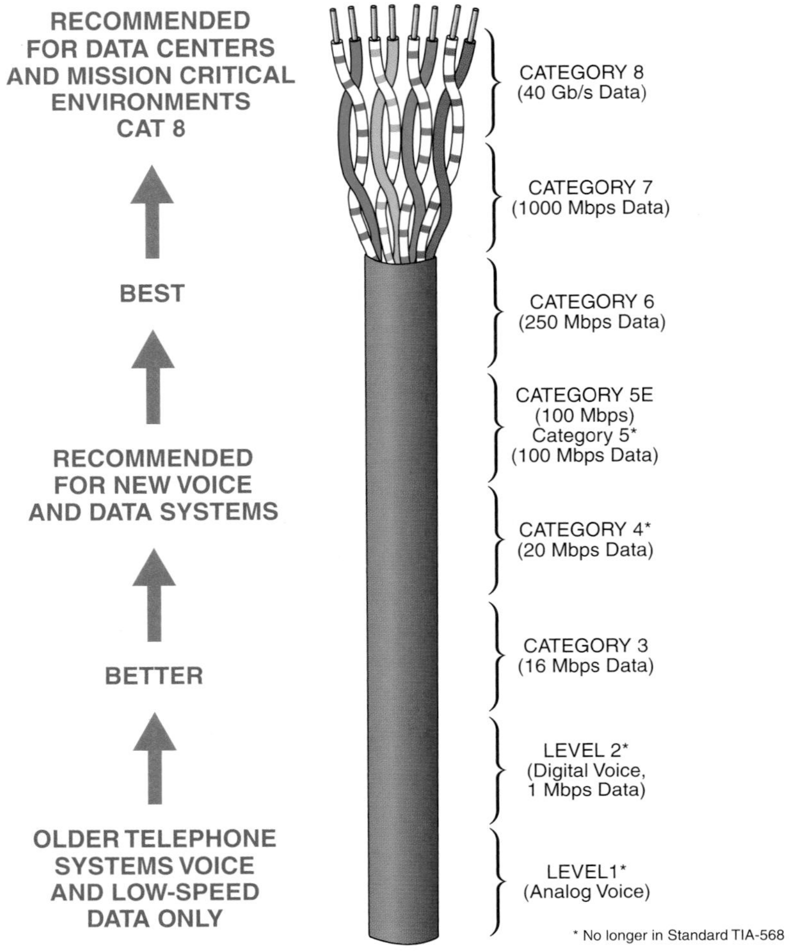

VDV CABLE CLASSIFICATION SYSTEM

TIA-568a Categories	Comments
Level 1	Used with POTS cable installations; suitable for voice only applications; <1 MHz frequency capacity.
Level 2	Used with IBM token ring cable installations; 4 MHz frequency capacity; 1 Mbps data.
CAT 3	Used with 10Base-T cable installations; Original Ethernet; 16 MHz frequency capacity; 16 Mbps data.
CAT 4	Used with enhanced token ring instalations; 20 MHz frequency capacity; 20 Mbps data.
CAT 5	Originally described as "the last cable you'll ever need"; 100 MHz frequency capacity; 100 Mbps data.
CAT 5E	Enhanced Cat 5 or Cat 5e; eventually replaced Cat 5 entirely; 100 MHz frequency capacity at distances up 328′ (100 m); 100 Mbps data.
CAT 6	Used with gigabit Ethernet, 1000Base-T network cable installations; 250 MHz frequency capacity; twice the bandwidth capacity of Cat 5e cable; 250 Mbps data.
CAT 7	Used with 10 gigabit Ethernet cable installations; 600 MHz frequency capacity; mainly used in Europe; 1000 Mbps data.
CAT 8	Used for server-to-access-switch interconnect applications; 40 Gb/s frequency capacity; 1 MHz to 2000 MHz data.

Figure 3-4. Levels of VDV cable performance were developed by ANSI/TIA/EIA.

- CAT 5e—A rating used for inside wire and cable systems for voice and data transmission up to 100 MHz. CAT 5e cable allows for higher bandwidth and less crosstalk than CAT 5 because the wire is packaged tighter, with greater electrical balancing between pairs. Cable sizes are various gauges.
- CAT 6—A rating used for inside wire and cable systems for voice and data transmissions of up to 250 MHz. Cable sizes are AWG #22 for STP and AWG #23 or AWG #24 for UTP.
- CAT 7—A rating used for inside wire and cable systems for voice and data transmissions of up to 600 MHz. Cable sizes are AWG #22 for STP and AWG #23 or AWG #24 for UTP.
- CAT 8—A rating used for inside wire and cable systems for voice and data transmission with performance requirements extended up to 1.2 GHz. Cable sizes are AWG #22 for STP and AWG #23 or AWG #24 for UTP.

Communication Cable Color Codes and Pin Designations

Communication cable is arranged in twisted pairs of conductors in a common protective cover. *Twisted conductors* are conductors that are intertwined at regular intervals. The number of pairs typically ranges from four to hundreds of pairs. A standard color code identification system identifies each pair of conductors, regardless of the number of pairs in the cable. **See Figure 3-5.**

Each conductor pair has a tip conductor and a ring conductor. A *tip conductor* is the first wire in a pair of wires. A *ring conductor* is the second wire in a pair of wires. Tip and ring conductor pair colors complement each other for easy identification. For example, the first conductor of the pair (tip conductor) is colored white with a blue stripe and the second conductor (ring conductor) is blue with a white stripe.

The terms "tip" and "ring" conductor come from the cordboard plugs of early telephone system switchboards. The wire at the end of the plug was called the tip and wire at the connecting point of the jack was called the ring. **See Figure 3-6.**

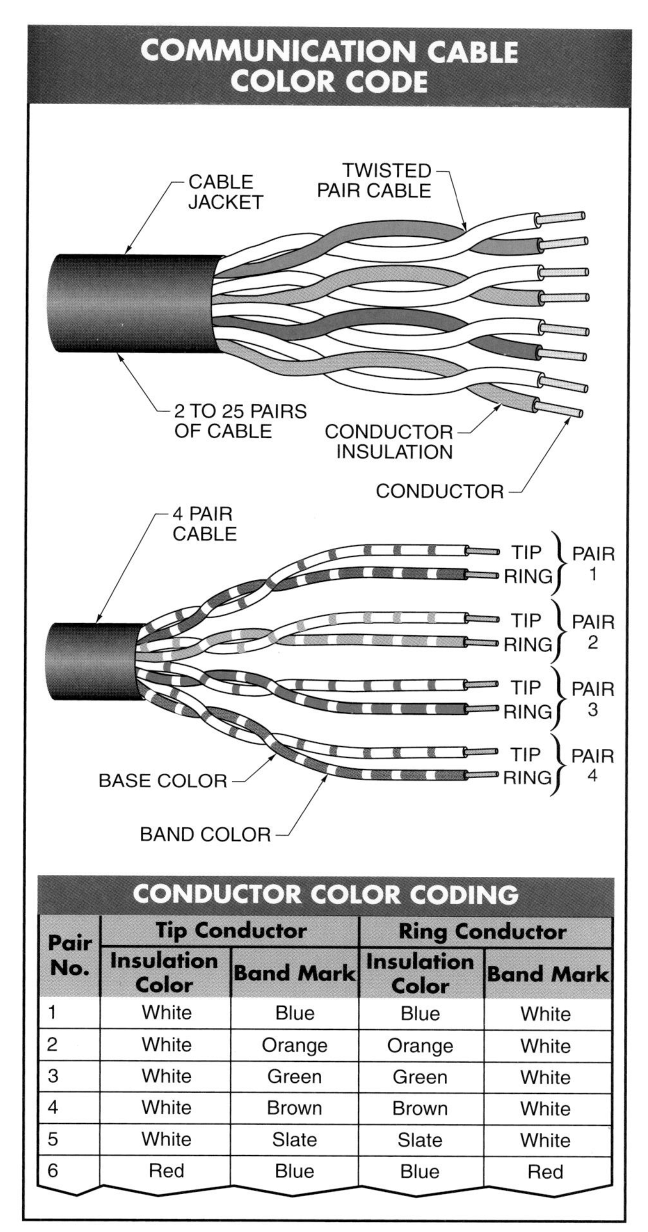

CONDUCTOR COLOR CODING

Pair No.	Tip Conductor		Ring Conductor	
	Insulation Color	Band Mark	Insulation Color	Band Mark
1	White	Blue	Blue	White
2	White	Orange	Orange	White
3	White	Green	Green	White
4	White	Brown	Brown	White
5	White	Slate	Slate	White
6	Red	Blue	Blue	Red

Figure 3-5. A standard color code system identifies each pair of conductors, regardless of the number of pairs in the cable.

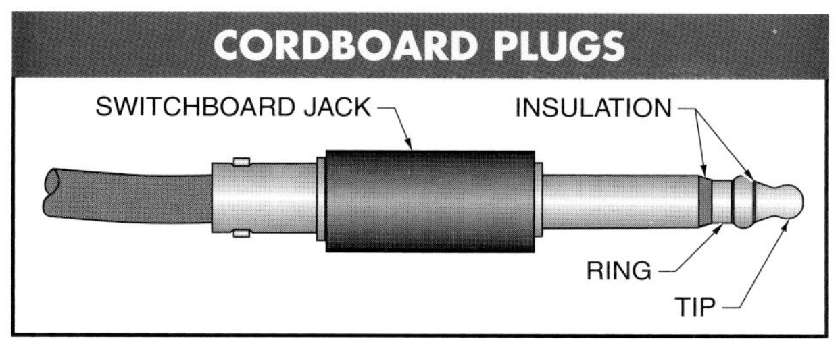

Figure 3-6. Tip and ring derive their names from the early telephone operator's cordboard plug.

Communication Cable Bundle Color Code Designations

VDV systems often require communication cable that is bundled into hundreds of individual pairs of conductors. Communication cables that include large numbers of conductor pairs are arranged with conductor pairs grouped into sets of 25. Each set of 25 conductor pairs is wrapped with a color-coded binder (string) that identifies the grouping. The binder color code system follows the same color code as the 25 pair conductor color code (1 = blue, 2 = orange, and 3 = green). Binders are color-coded using one or two colors. When a one-color binder is used, the color of the binder follows the color of the ring conductor band color code (blue, orange, green, brown). When a two-color binder is used, the color of the binder follows the color of the tip conductor insulation band color code (white-blue, white-orange, white-green). **See Figure 3-7.**

In addition to the color-coded binders used to band and identify conductor sets, a cable may also include a ripcord. A ripcord is a cord included in a cable that aids in removing the outer jacket.

Communication Cable Mechanical Arrangement Numbering System

Conductor pairs are terminated using connection blocks, cable outlets, plugs, couplers, and other types of quick-connect, quick-disconnect connections. Connectors are designed to provide good mechanical strength and electrical conductivity. The wiring of connectors must also separate the conductors for identification of each twisted pair.

Conductors are terminated using connectors designed to hold one pair of conductors or multiple pairs of conductors. Standard connectors are designed for holding one, four, or 25 pairs of conductors. Connectors follow a numbering system that allows standardization when wiring conductor pairs. Some recommended communication cable installation tips include the following:

- Use as few connections as possible because excessive connections lower system performance.
- Wire to the highest projected data rate.
- Allow at least 18″ of spare wire at all connection points.
- Keep connectors clean during installation and use dust covers on fiber optic cables and ferrules (connectors).
- Never splice wires together on communication cable runs.
- Never run communication cables in parallel with power conductors.
- Keep VDV cables and wires away from heat sources.
- Use nonmetallic staples when securing cable or wiring. Do not staple tightly over VDV cables.
- Install communication jacks at the same height as electrical power receptacles.
- Install communication jacks near electrical power receptacles, since both powers are typically required.
- Use inner walls for cable runs to avoid problems with outer wall insulation.
- Test connections for open or short circuits, and grounds.

COMMUNICATION PROBLEMS

Communication problems occur in many forms. The most common causes include improper cable installations, electromagnetic interference (EMI), radio frequency interference (RFI), pulling too hard (25 ft/lb), exceeding minimum bend radius, and using staples on CAT 5e and CAT 6 cables. Test instruments are used to troubleshoot problems and determine when new equipment must be installed according to industry and design specifications.

Troubleshooting Tip

CABLE MANAGEMENT

Cable management is a method by which cable users, both end users and service providers, keep track of their cable systems. Cable management entails the actual route a cable or patch cord takes to get from one point to another. End users physically manage their cables with horizontal and vertical devices such as racks, trays, drawers, and enclosures that route the cables. ANSI/TIA/EIA 606 covers the proper procedure for administering cable.

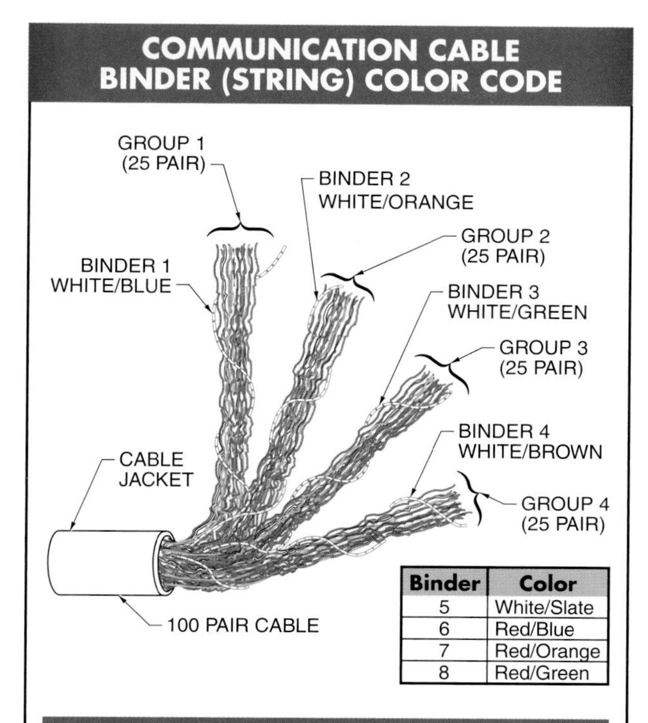

Binder	Color
5	White/Slate
6	Red/Blue
7	Red/Orange
8	Red/Green

25 PAIR COLOR CODES

Type	Color	Type	Color
Pair 1	White/Blue Blue	Pair 14	Black/Brown Brown
Pair 2	White/Orange Orange	Pair 15	Black/Slate Slate
Pair 3	White/Green Green	Pair 16	Yellow/Blue Blue
Pair 4	White/Brown Brown	Pair 17	Yellow/Orange Orange
Pair 5	White/Slate Slate	Pair 18	Yellow/Green Green
Pair 6	Red/Blue Blue	Pair 19	Yellow/Brown Brown
Pair 7	Red/Orange Orange	Pair 20	Yellow/Slate Slate
Pair 8	Red/Green Green	Pair 21	Violet/Blue Blue
Pair 9	Red/Brown Brown	Pair 22	Violet/Orange Orange
Pair 10	Red/Slate Slate	Pair 23	Violet/Green Green
Pair 11	Black/Blue Blue	Pair 24	Violet/Brown Brown
Pair 12	Black/Orange Orange	Pair 25	Violet/Slate Slate
Pair 13	Black/Green Green		

Figure 3-7. With communication cable, each set of 25 conductor pairs is wrapped with a color-coded binder that identifies the grouping.

Technical Tip

Improperly installed cables and conductors that must be removed and replaced with new cables and conductors are the main cause of cost and budget overages at most job sites.

Causes of Cable Failure

Cable problems are the number one cause of problems within a VDV system. Estimates indicate that over 50% of reported problems with VDV systems can be traced back to some type of cable problem. When a VDV system is designed correctly and includes the correct components and parts, the interconnections of the cabling system are where problems are most likely to occur. **See Figure 3-8.**

Each error, no matter how small, increases the amount of attenuation (decrease in signal, power, or light wave strength) in the cable, and in the system. For example, removing just 1″ of cable sheathing at a terminated end can increase signal loss (attenuation) by about 1 dB. Leaving ends untwisted by ½″ increases signal loss by about 1.5 dB, and leaving ends untwisted for 6″ increases signal loss by 10 dB. The major causes of cable problems are the following:

- Cable ends not terminated properly to terminals, plugs, and jacks— for example, not maintaining the required pair-twist when making a connection (adding an additional twist to the pair when terminating helps maintain proper line impedance). Another common mistake is to remove too much of the cable jacketing.
- Cable not properly installed changes the impedance or ability of the cable to carry the signal at the required operating speed. Common installation problems include overtightened cable ties, staples driven too tight, twists or knots in the cable, cables bent tighter than the minimum bend radius, and overfilled cable supports (or trays and runner systems).
- Cable runs not properly routed. VDV cables must be separated from power cables to prevent induced magnetic fields from causing noise problems.
- Cables not properly identified, marked, and tested, or not having the necessary documentation.
- Cables run too close to high heat sources.

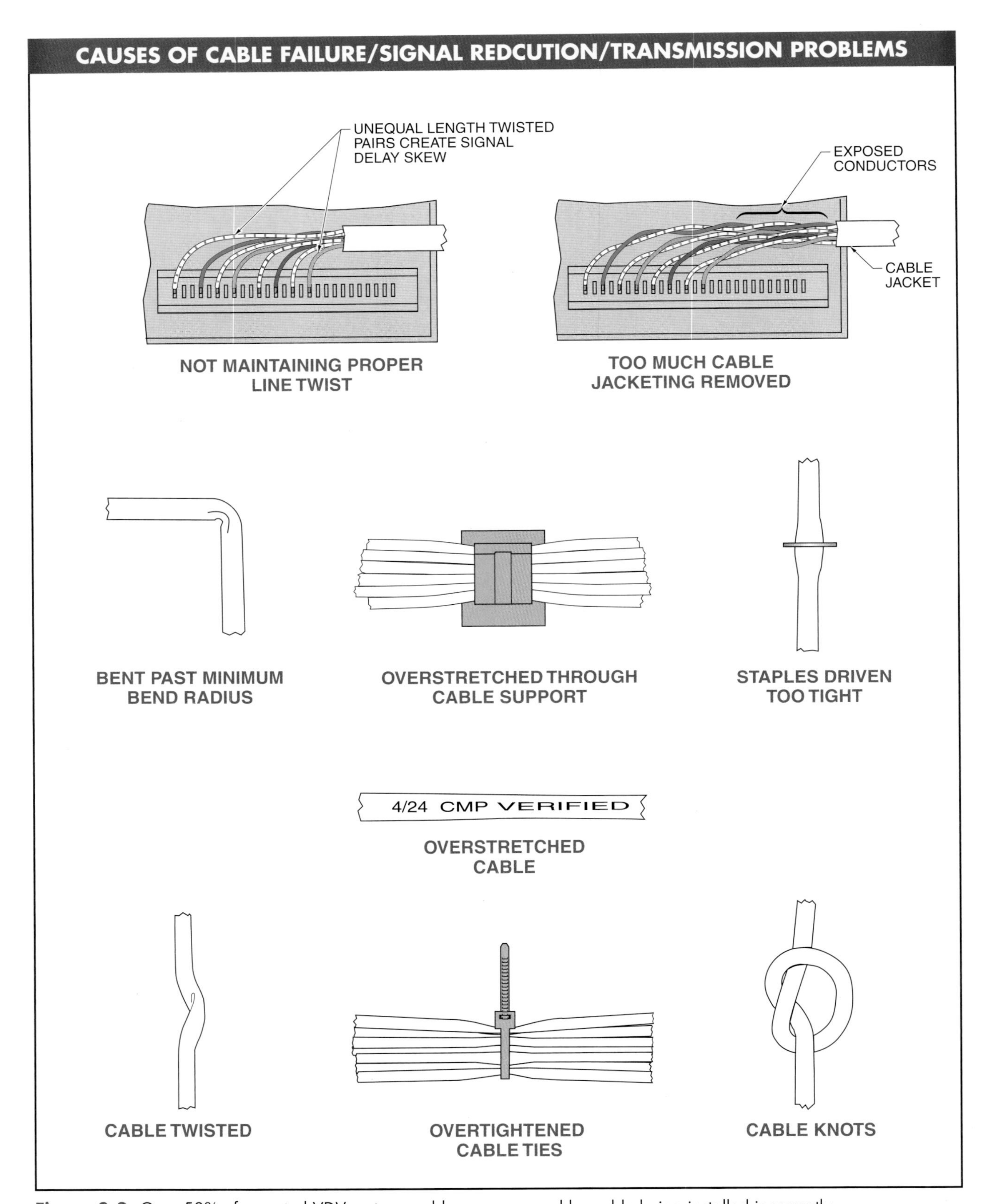

Figure 3-8. Over 50% of reported VDV system problems are caused by cable being installed incorrectly.

EMI/RFI Interference

Electromagnetic interference (EMI) is an unwanted signal (noise) induced on electrical power or VDV cables through magnetic coupling of adjacent wires or magnetic field-producing devices such as motors and transformers. *Radio frequency interference (RFI)* is an unwanted signal (noise) induced on electrical power and VDV cables from transmitted radio frequencies such as AM radio and analog cellular. **See Figure 3-9.**

Induced EMI and RFI interference produce false signals in VDV cables which lead to processing errors, incorrect data transfer, and printing errors. EMI and RFI problems are reduced by using shielded cables and proper separation of VDV cables from power cables and other electrical loads that produce EMI.

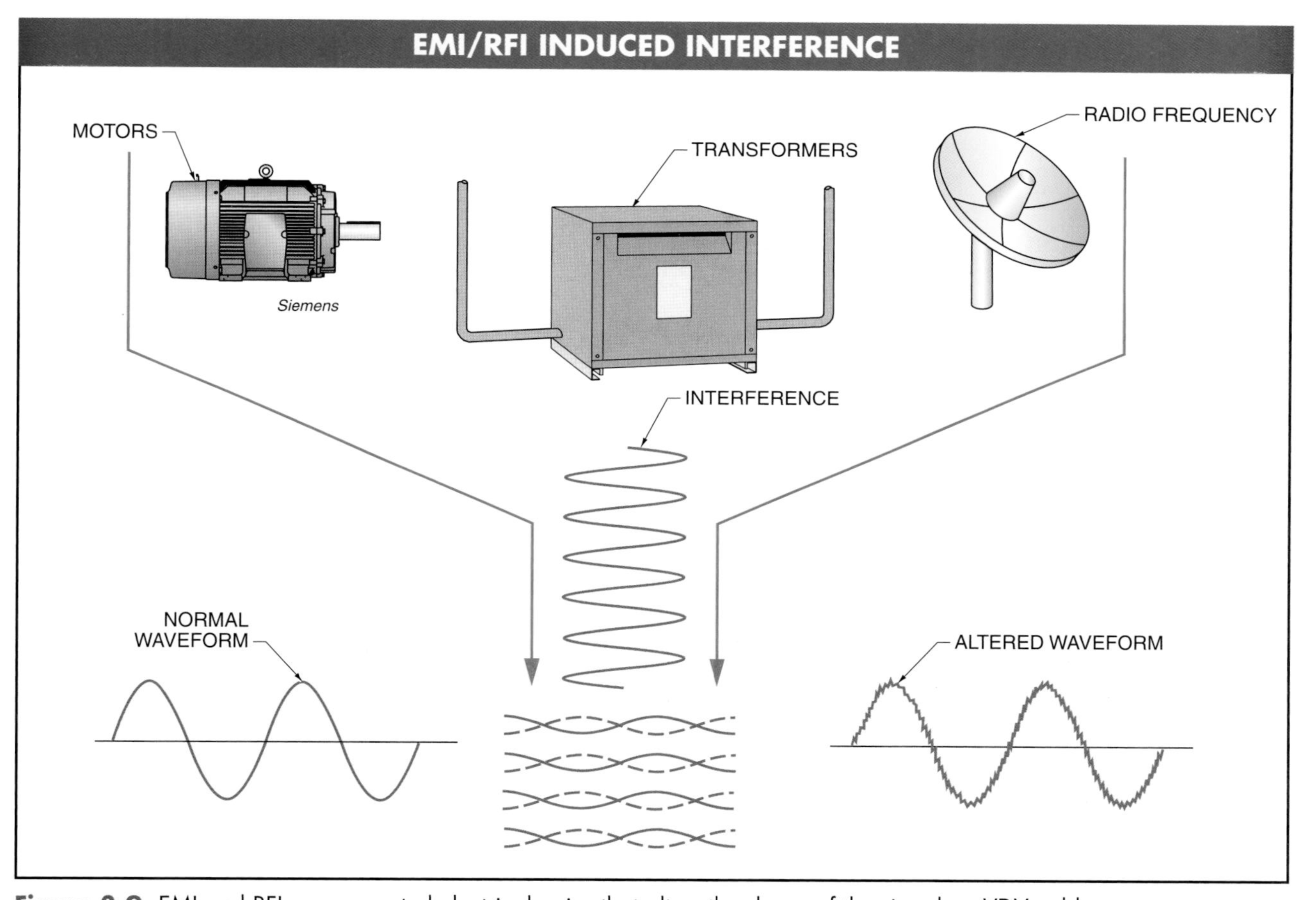

Figure 3-9. EMI and RFI are unwanted electrical noise that alters the shape of the signal on VDV cable.

VDV System Testing

Test instruments are typically thought of as being used for troubleshooting. However, test instruments are used any time cable is installed, moved, or replaced. VDV test instruments are used after cable installation to verify correct VDV cable operation. Every cable must be tested and the results documented. However, in addition to using VDV test instruments during certification and troubleshooting, VDV testers are also used during all stages of cable installation and operation.

When VDV cable is installed as part of new construction, testing and documentation becomes even more important. Testing the cable, and documenting the results, after cable installation identifies any damaged cables. VDV cables are installed after the walls are closed. VDV cables must be kept separate from other wiring and use routes through a building only for VDV cable. Any VDV cabling found to be defective by testing is replaced.

When a VDV cable is installed and certified, continued cable testing must be part of a preventive maintenance program. As loads are added to a system and new cables installed, cables that appear to be working properly can be damaged to the point that the cable is unusable. Problems never occur at a time when lost data does not create a problem. Routine testing of VDV cable will indicate signal losses (attenuation) and other potential problems.

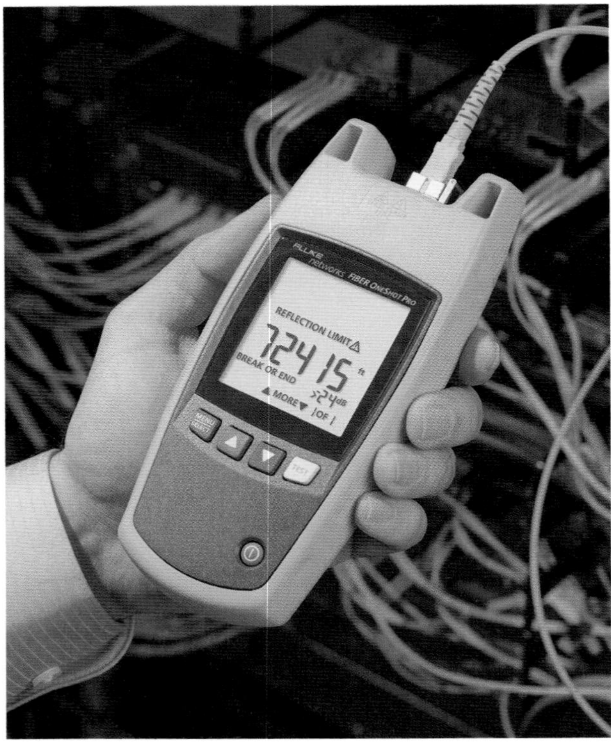

Fluke Networks

Cable length testers measure cable length and distance to faults and usually work with a variety of cable types. Testing cable length is important for ensuring clean VDV signals.

VDV Test Measurements

A number of different tests can be performed on VDV cables and systems. Some tests, such as measuring the length of cable on a roll of wire, are performed prior to installing any cable. Other tests, such as a wire map test, are performed after the cable is installed and before the cable is placed in service. Some tests, such as a TDR (time-domain reflectometer) test, can be performed after a cable is installed, as part of a preventive maintenance program, or when troubleshooting a problem. VDV tests include a number of different types of tests.

Cable Length Test. A *cable length test* is a test that measures the length of conductors (in feet or meters). The test is used to measure the length of conductors still on the rolls, previously cut from the roll, or as part of installed cables. The test also verifies that cables do not exceed the set limits, such as the 295′ (90 m) limit on horizontal cabling from the transmitter/receiver to the telecommunication outlet. Knowing the length of VDV cable is important because VDV cable, like all conductors, has resistance that reduces the amount of power passed through the cable. The greater the length of a VDV cable, the greater the power reduction (attenuation) created by the cable. Because all VDV transmitted signals are at a relatively low power, any reduction in power caused by a cable must be kept to a minimum.

Resistance Test. A multimeter set to measure resistance can be used to take resistance measurements of switches and connectors, verify ground wire connections, and test splices. However, when using a continuity tester or ohmmeter on VDV cable and equipment, the measurements have limited meaning. The usefulness of the measurements is limited because continuity testers and ohmmeters are only able to indicate a low or high resistance. A low resistance might appear fine, but low resistance does not mean that a VDV signal can still pass through as required or specified. Thus, actual resistance measurements are helpful, but a resistance measurement must only be used as a basic measurement that identifies a low or high resistance at the measuring point.

Wire Map Test. A *wire map test* is a basic test for testing circuits for open or short circuits, crossed pairs, reversed wires, and split pairs. Wire map tests are performed on all VDV cables after the cable is installed. Wire map tests are used to test four-pair cables, 25-pair cables, and other cables. A wire map test is used to ensure that a cable is properly terminated to the correct terminals and that there are not any wires left unconnected or connected to the wrong terminal.

Tone and Probe Test. A *tone and probe test* is a test that identifies specific pairs of wires within a system. A tone generator is used to produce a signal (tone) at one end of a pair of wires and a probe (inductive amplifier) is used to identify the wire carrying the signal at the other end of the cable. Tone and probe testing is a method of tracing individual wire pairs in locations that include numerous cable pairs.

Attenuation Test. An *attenuation test* is a test that measures the attenuation (power loss) of each cable pair. Power loss occurs from one end of a cable to the other end because of the resistance of the wire and leakage current (power) through the insulation of the cable. Attenuation is measured in decibels (dB). The lower the dB loss measurement, the more efficient the cable. A cable with an attenuation of 5 dB performs better than a cable with an attenuation of 10 dB. **See Figure 3-10.**

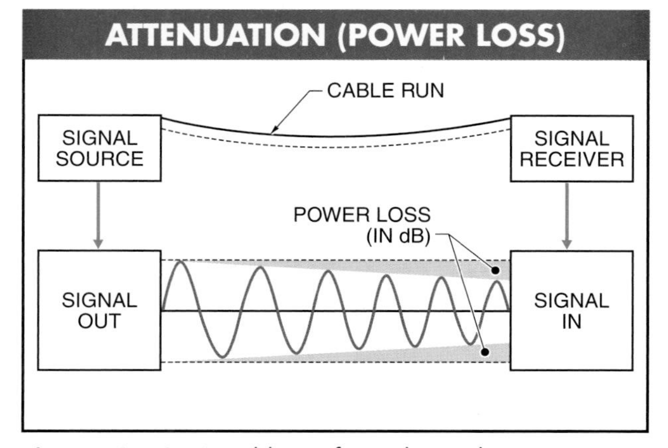

Figure 3-10. A cable performs best when attenuation (power loss) is kept to a minimum.

Telephone Test. A *telephone test* is a test that is used to test the voice carrying capability of circuits and to identify circuits. Telephone test sets are used to test phone systems, but are also used to perform basic tests on other VDV cables. Testing a communication cable to verify that the cable can carry a voice signal isolates a cable problem from other equipment problems, even on cables designed to carry video and data information. For example, when a cable cannot carry a voice signal, the problem is probably in the cable. When a cable can carry a voice signal but there are data transmission problems, the problem is probably with the transmission or receiving equipment.

CAUTION

Both AC and DC voltages are used in telephone circuits. A telephone jack is wired to operate using 48 VDC and 75 VAC to 90 VAC for the ringer voltage. Because the AC voltage for the ringer is only present when a call is coming in, and most individuals come in contact with the lines only when DC voltage is present, it may seem that a telephone line cannot produce a shock. When contact is made with the lines while AC voltage for the ringer is present, an electrical shock is possible.

NEXT Test. A *NEXT test* is a test that checks cable for near end crosstalk (in dB). Crosstalk is unwanted induced noise on a pair of cables. An example of crosstalk on voice lines is when unrelated conversations are heard in the background over the phone line being tested. The source of crosstalk is typically from another pair of wires in the same group or from another adjacent group of wires. Since crosstalk increases as frequency increases, crosstalk problems increase as VDV signals are operated at faster speeds (higher frequencies). NEXT (near end crosstalk) is the amount of coupling loss (in dB) that occurs when a transmitted signal sent out on one cable pair is picked up as crosstalk on another cable pair. The name NEXT comes from the fact that the test (measurement of crosstalk) is conducted at the end of the cable pair closest to the transmitter. **See Figure 3-11.**

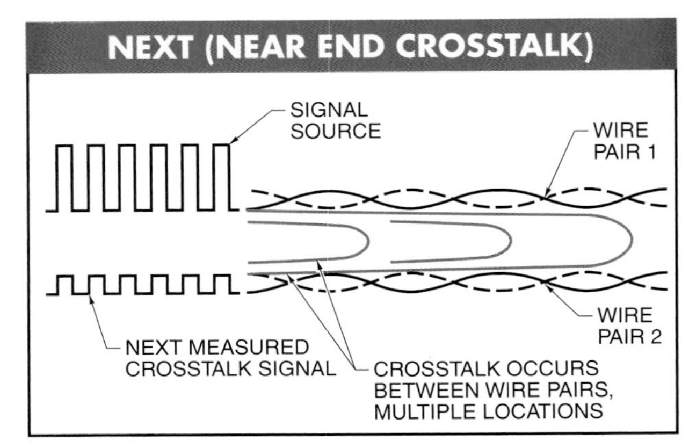

Figure 3-11. NEXT (near end crosstalk) is the amount of coupling loss (in dB) that occurs when a transmitted signal sent on one cable pair is picked up as crosstalk on another cable pair, and can occur in multiple locations.

Propagation Delay Test. A *propagation delay test* is a test that measures the time (in ns) required for a signal to travel the length of a cable pair. Propagation losses limit the length of a cable because high losses cause signal and data reception problems. Causes for signal losses can be runs that are too long, bad splices, low-quality cable, and cable drops. Propagation losses also increase as a function of frequency and can create a bigger problem for high-speed transmissions. Applying a high-frequency test signal at one end of a cable and measuring the propagation loss at the other end determines the propagation delay of the cable.

Delay Skew Test. A *delay skew test* is a test that determines the difference in propagation delay between cable pairs. The difference in propagation is important because in some systems, more than one pair of cables are used to send simultaneous signals over different pairs in order to move large amounts of data at fast speeds. Simultaneous signals require that the transmitted data reach the receiver points at the same time.

Impedance Test. An *impedance test* is a test that measures the impedance of each cable pair. *Impedance (Z)* is the total opposition that any combination of resistance (R), inductive reactance (X_L) and capacitive reactance (X_C) offers to the flow of alternating current. Impedance, like resistance, is measured in ohms and limits current flow, which limits the total power that can be sent through a cable. An impedance mismatch is correct when conductors of a pair of wires do not have the same impedance (resistance). A damaged conductor, bad splice, or improper conductor termination typically causes impedance mismatches. Properly terminated cable ends typically eliminate impedance mismatches.

RL Test. A *return loss (RL) test* is a test that measures the signal loss due to signal reflection (return loss) in a cable. RL tests indicate how well the impedance characteristics of a cable match the rated impedance over a range of frequencies. In a VDV cable, some of the transmitted signal is reflected back to the transmitter, never traveling to the receiver. In a voice system, RL is heard as an echo of the original sounds.

TDR Test. A *time domain reflectometer (TDR) test* is a test used for measuring cable length and locating faults (short or open circuits, poor connections) on cables (twisted pair and coaxial). A TDR tester transmits pulses into one end of a cable. The pulses travel along at a speed determined by the propagation of the pulses. When the pulses reach an impedance change in the cable, a fault, or the end of the cable, a reflected signal is sent back to the TDR. The tester reads the size, shape, and return time of reflected pulses to identify the type of problem. **See Figure 3-12.**

Troubleshooting Tip

SIGNAL-TO-NOISE RATIO

Signal-to-noise ratio is the measurement of the ratio of usable signals to the undesired noise on a cable or circuit and is usually measured in decibels. The ratio is a measure of the quality of transmission. The only way to overcome unwanted signal noise is to increase the signal strength or to reduce the volume of noise so the receiver can "hear" the signal.

VDV COPPER WIRE TEST INSTRUMENTS

All conductors are designed to carry power from one point in a circuit to another point. The amount of power that a conductor can carry depends on the applied voltage (5 VDC, 12 VDC, 115 VAC), the voltage and current type (AC or DC), the frequency of the voltage and current (60 Hz, 20,000 Hz), the conductor size (AWG #14 or AWG #22), the conductor material (copper or aluminum), and the insulation used on the conductors. Conductors used in high power circuits operating at lower frequencies (115 VAC, 60 Hz) can generally carry power with few or no problems, even when the installation is less than acceptable (poor splices or connections). Problems that do occur may not be known for some time, such as a low voltage condition (caused by a bad splice) that is causing a lamp to burn at less than full brightness or a heating element that produces less than the rated heat. However, conductors carrying low power VDV signals operate at high frequencies. Because of the higher frequencies, any installation less than standard causes problems. Problems typically show up immediately or at an undesired time, such as when transmitting large amounts of data. In order to ensure proper cable installation and uninterrupted service, VDV test instruments are used.

Technical Tip

Cable assemblies are wires or cables bundled together with connectors on at least one end, and are made by over 300 companies.

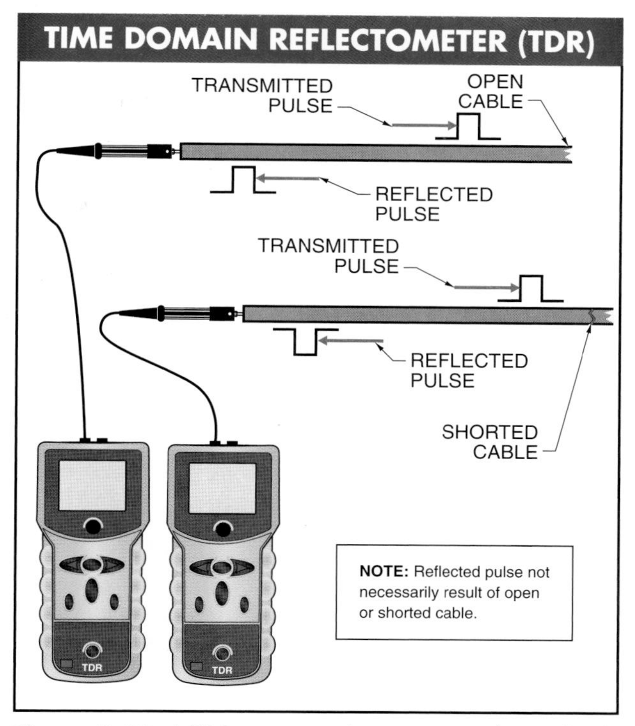

Figure 3-12. A TDR measures distance, impedance, levels of RFI/EMI, connector/terminator problems, and the presence of open and short circuits in network systems that use cable.

Low power VDV wires operating at high frequencies will have transmission problems if installation is less than standard.

Cable Length Meter

A *cable length meter* is a handheld test instrument that measures the length of cables that are still on rolls, a cut length of cable, or an already installed cable. A cable length meter can be used with power cable, telecommunication cable, data cable, twisted pair cable, coaxial cable, and other copper or aluminum cable. Most cable length meters can measure cable lengths of several thousand feet or more. **See Figure 3-13.** Cable length meters are not only used for determining cable length, but also aid in troubleshooting. When a cable is open, the cable length meter will display the distance to the open section. Cable length meters are relatively accurate, depending upon the model used and length of cable being tested.

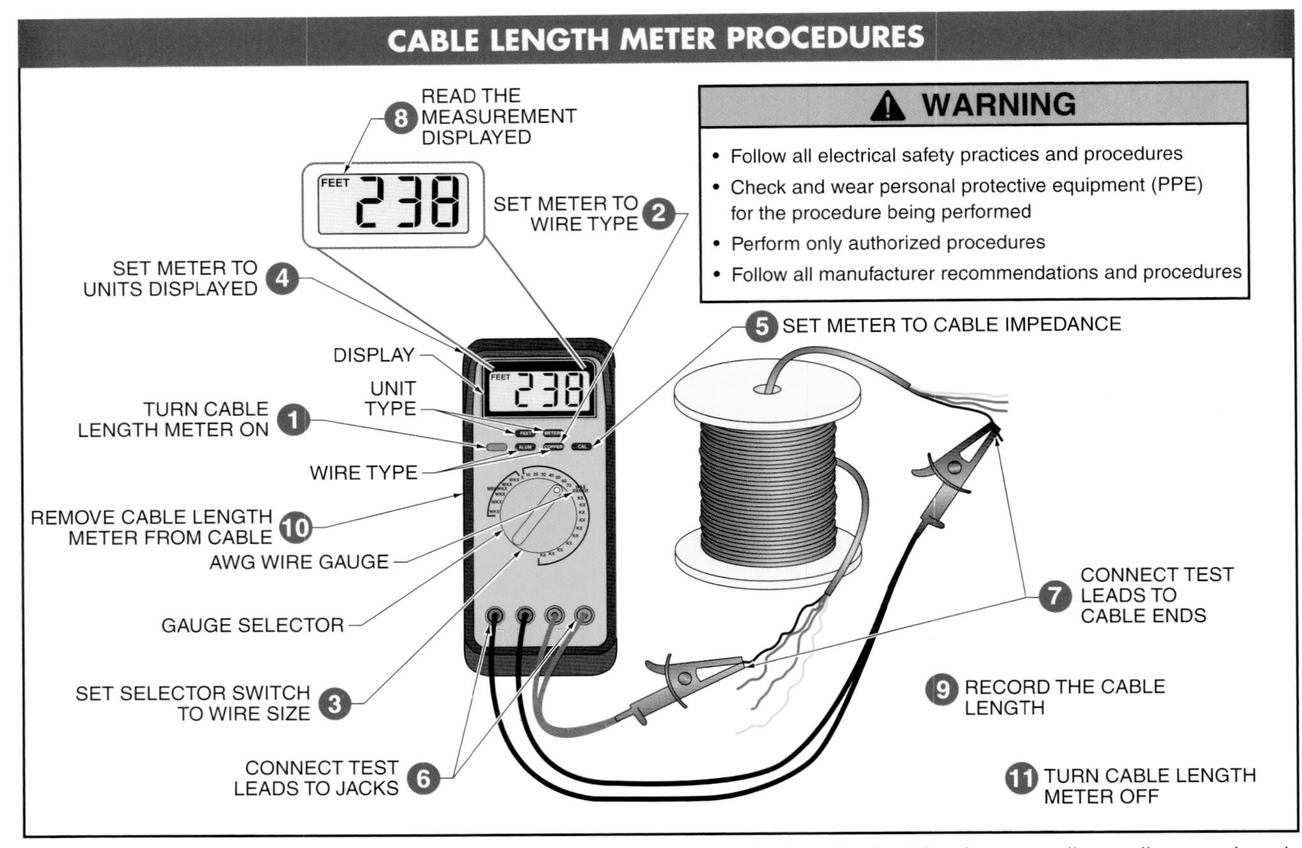

Figure 3-13. A cable length meter is a handheld unit that measures the length of cables that are still on rolls, a cut length of cable, or an already installed cable.

Refer to the operating manual of the test instrument for all measuring precautions, limitations, and procedures. To measure the length of a cable, apply the following procedures:

Safety Procedures

- Follow all electrical safety practices and procedures.
- Check and wear personal protective equipment (PPE) for the procedure being performed.
- Perform only authorized procedures.
- Follow all manufacturer recommendations and procedures.

1. Turn the cable length meter ON.
2. Set the function switch of the cable length meter to the type of wire to be measured (copper or aluminum).
3. Set the selector switch of the cable length meter to the size of wire to be measured (such as AWG #18 or AWG #22).
4. Set the selector switch of the cable length meter to the desired measurement display (feet or meters).

5. When a cable tester meter calls for setting the impedance of a typical cable, set the selector switch of the meter for the impedance of the cable being tested. Typically, the setting should be 50 Ω for power cables, 75 Ω for coaxial cable, and 100 Ω for twisted pair.
6. Connect test leads to cable length meter jacks.
7. Attach the cable length meter to both ends of the cable being tested.
8. Read the cable length measurement displayed on the meter.
9. Record the cable length.
10. Remove the cable length meter from the cable being measured.
11. Turn the cable length meter OFF.

Expect a 2% to 5% difference in length measurement between twisted pairs. The variance exists because there is a difference in the twist rate (twists per foot or TPF) between the pairs of cable. The twisting of each pair at a different rate reduces problems from crosstalk and reduces the strength of the electromagnetic field radiated from the twisted pair. The more energy that remains in a circuit, the farther the signal can travel without signal boosting. **See Figure 3-14.**

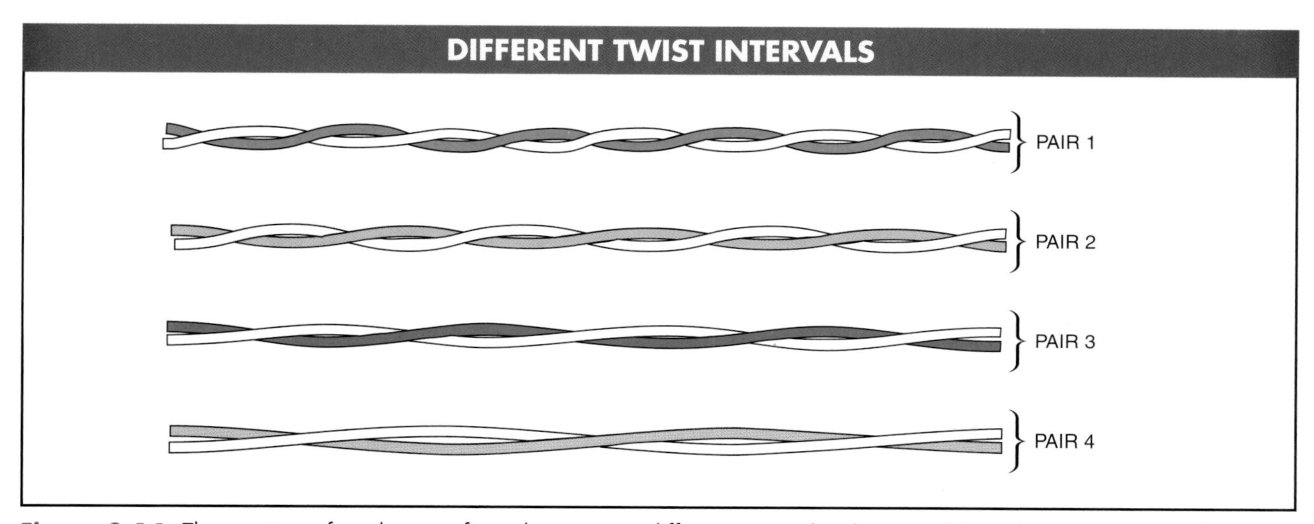

Figure 3-14. The twisting of each pair of conductors at a different interval reduces problems from crosstalk and reduces the strength of the electromagnetic field radiated from the circuit.

Cable Fault Finder (Wire Map Tester)

A *cable fault finder* (wire map tester or cable map tester) is a test instrument used to find open circuits, short circuits, and improper cable wiring. A wire map tester typically displays a PASS or FAIL indication after each test. A wire map tester tests and displays the wire conditions between the tester ends of the cable and the far ends of the cable on all four pairs. **See Figure 3-15.**

CAUTION

Always turn a VDV test instrument ON before connecting the VDV test instrument to a circuit. Many VDV test instruments include an automatic input protection circuit that is activated when the test instrument is turned ON.

Technical Tip

Electrical metal tubing (EMT) is also commonly known as conduit. EMT acts as a fire barrier for telecommunications systems. Installers, designers, and technicians should refer to local municipal codes to determine when EMT is a requirement for routing VDV cables and circuits.

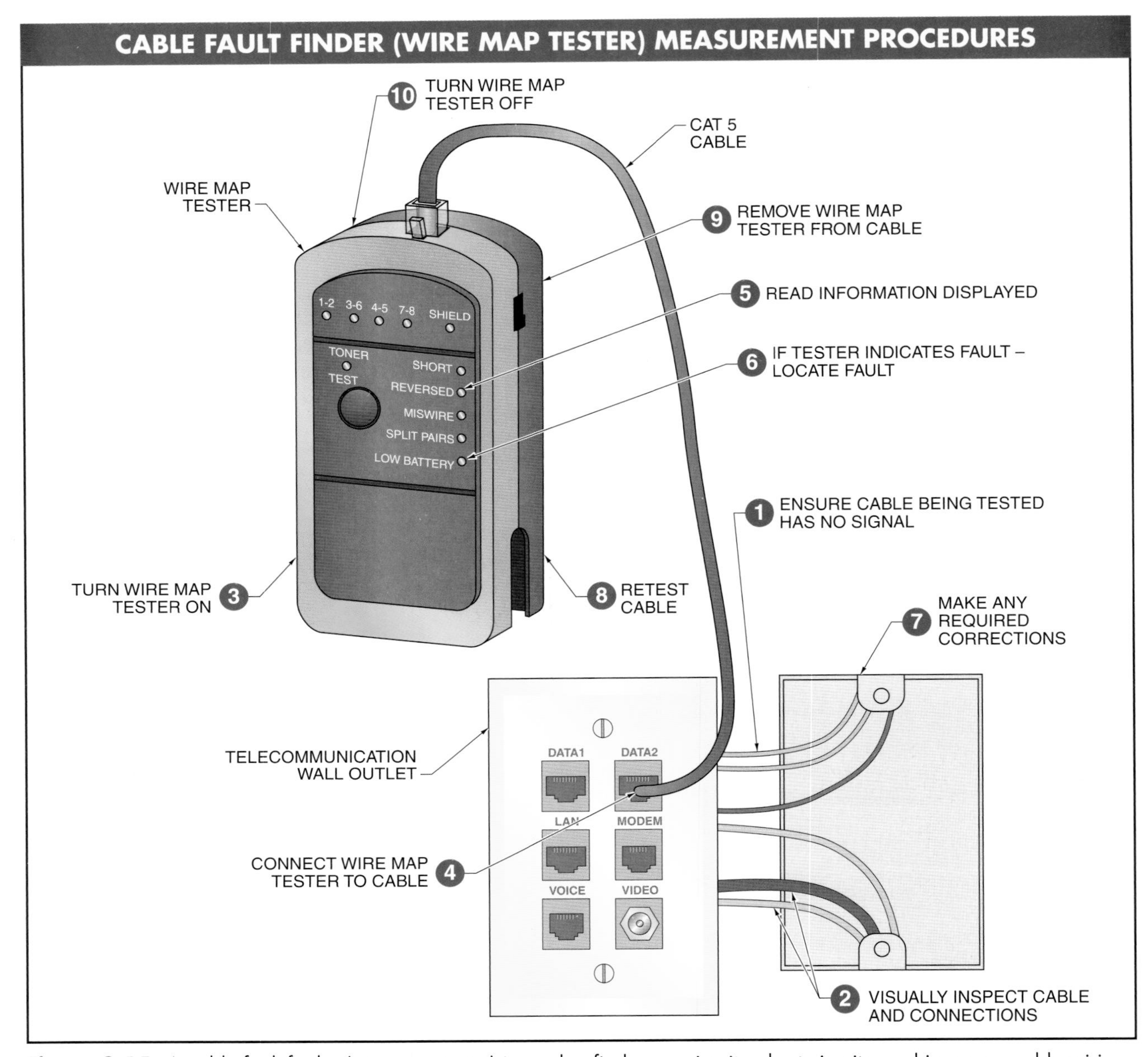

Figure 3-15. A cable fault finder (wire map tester) is used to find open circuits, short circuits, and improper cable wiring.

Before taking any measurements using a wire map tester, ensure the tester is designed to take measurements on the cables being tested. Refer to the operating manual of the test instrument for all measuring precautions, limitations, and procedures. Always wear required personal protective equipment and follow all safety rules when taking the measurement. To find a fault using a wire map tester, apply the following procedures:

1. Ensure the cables being tested are not carrying any signals.
2. Perform a visual inspection of the cable and cable connections, and the area in which the cable is located. Look for signs of damage, loose wires, and pinched cables.
3. Turn the wire map tester ON.

4. Connect the wire map tester to the cable being tested.
5. Read the information displayed on the wire map tester. The display may use PASS or FAIL test lights, or a problem may be displayed.
6. When a wire map tester indicates a fault, check for the fault in the cable.
7. Make any required corrections.
8. Retest the cable.
9. Remove the wire map tester from the cable being tested.
10. Turn the wire map tester OFF.

Cable Fault Analysis

A wire map tester indicates when a cable passes or fails a given test. When a cable fails a test, understanding the fault will allow for correcting the problem and retesting the cable. Although there can be any number of wiring mistakes and problems, there are several typical ones. **See Figure 3-16.**

Correct Wiring. When wiring is performed properly, a wire map tester displays a PASS condition. Due to the PASS condition on the display, the wire map tester will not display a FAIL condition.

Shorted Wire. A *shorted wire* (short) is any conductor (cable or wire) that has an unwanted low resistance path between two conductors or to ground. Typically a short takes place between wire pairs, but can also take place between cables.

Open Wire. An *open wire* is any conductor (cable or wire) that does not have continuity (low resistance path) between the two ends of the conductor. Typically an open takes place in only one wire, but there may be more than one when a cable is severely overpulled.

Crossed (Transposed) Cable Pair. A crossed (transposed) cable pair occurs when two common cable pairs are connected at the terminals of the other. The tip end of one pair must be switched with the tip end of the other pair or the ring end of one pair must be switched with the ring of the other pair.

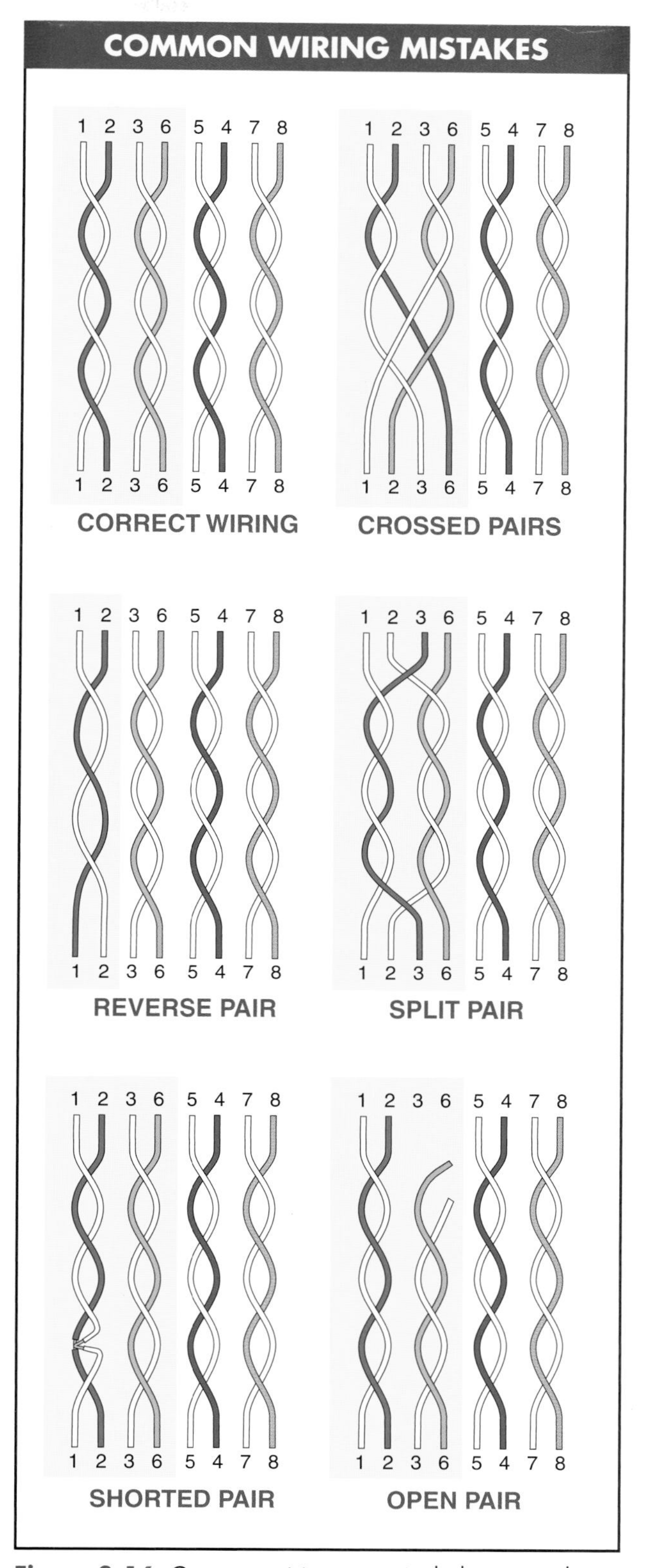

Figure 3-16. Common wiring errors include reversed wire pairs, crossed cable pairs, split wire pairs, open wires, and shorted wires.

Reversed Wire Pair (Same Pair). A *reversed wire pair* is a common cable pair in which the tip and ring conductors are reversed with each other. When ring conductors are reversed, the condition is called a tip/ring reversal.

Split Wire Pair. A split wire pair occurs when any two wires from different pairs are each wired into the position of the other.

Troubleshooting Tip

CO-LOCATION OPERATIONS

A co-locator is a carrier other than the service provider who puts their equipment in the service provider's facility and then joins their equipment to the service provider's equipment for a fee. Co-location is done to provide better service, solve technical problems faster, have better interconnections, and save money.

Coaxial Cable Testers

A *coaxial cable tester* is a handheld test instrument that is used to test for open or short circuits on coaxial cable. A *coaxial cable* is a cable that has a center conductor surrounded by an insulating layer with a braided metal jacket around the outside of the insulating layer. The braided metal jacket provides a high degree of protection against EMI and RFI interference. **See Figure 3-17.**

CAUTION

When taking standard voltage and current measurements on power lines (115 VAC, 230 VAC, etc.), the testing does not interrupt service or cause any system problems. However, VDV service can be interrupted or data lost or damaged by improper use of some types of test instruments on a communication line that is transmitting information. This is because some testers can cause impedance changes in the circuit. A VDV test instrument should never be connected to an active network unless the instrument is designed to safely monitor network activity.

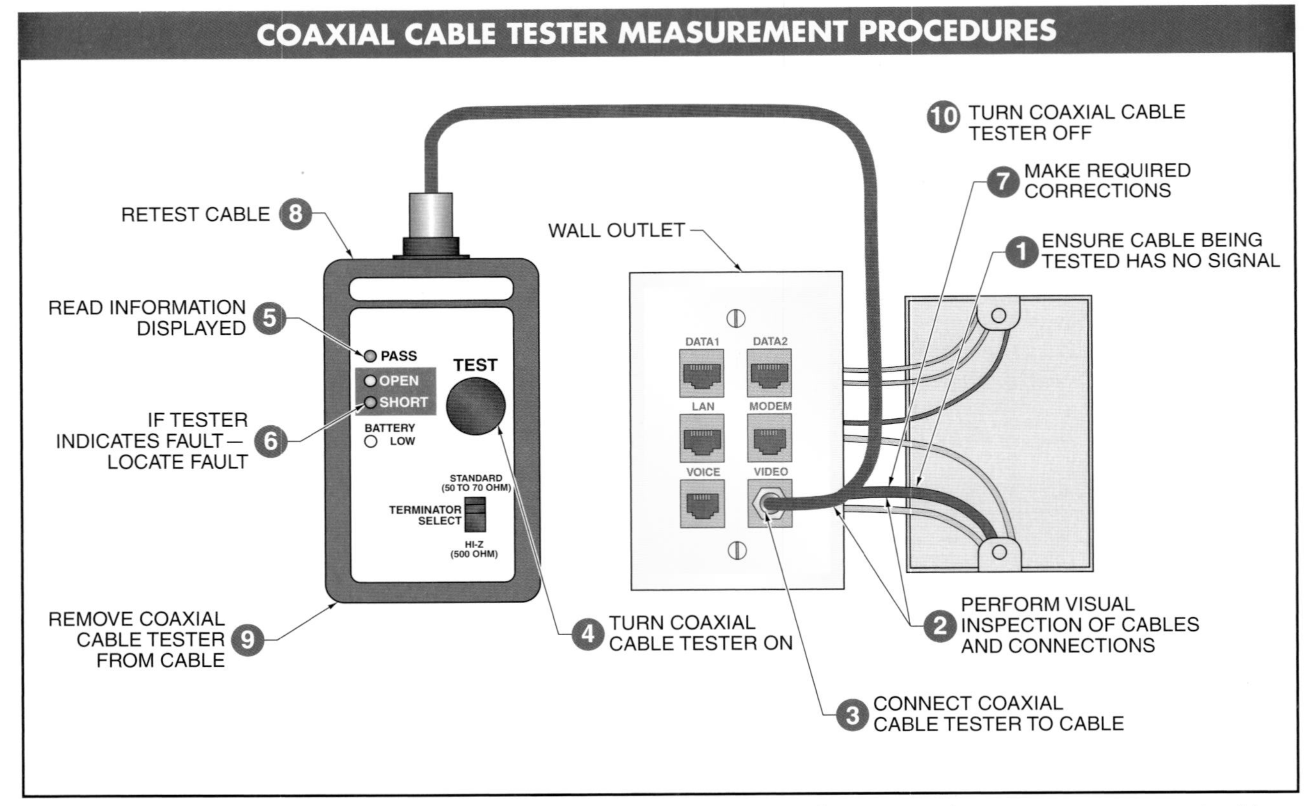

Figure 3-17. A coaxial cable tester is a handheld tester that is used to test for open or short circuits on a coaxial cable.

Before taking any measurements using a coaxial cable tester, ensure the tester is designed to take measurements on the coaxial cables being tested. Refer to the operating manual of the test instrument for all measuring precautions, limitations, and procedures. Always wear required personal protective equipment and follow all safety rules when taking the measurement. To test a coaxial cable using a basic coaxial cable tester, apply the following procedures:

1. Ensure the cables being tested are not carrying any signals.
2. Perform a visual inspection of the cable, cable connections, and area in which the cable is located. Look for signs of damage, loose wires, and pinched cables. At the end of a coaxial cable, look for damage to the inner connector (the small, thin wire sticking out the farthest).
3. Connect the coaxial cable tester to the cable being tested.
4. Turn the coaxial cable tester ON.
5. Read the information displayed on the coaxial cable tester. The display may use PASS or FAIL test lights, or a problem may be displayed.
6. When the coaxial cable tester indicates a fault, check for the fault in the cable.
7. Make any required corrections.
8. Retest the cable.
9. Remove the coaxial cable tester from the cable being tested.
10. Turn the coaxial cable tester OFF.

Advanced Testing. When a coaxial cable tester includes additional test capabilities such as measuring cable impedance, cable length, and TDR, each test should always be performed as specified by the manufacturer. When a coaxial cable tester indicates a fault, the electrician should check for the fault in the cable, make any required corrections, and retest the cable.

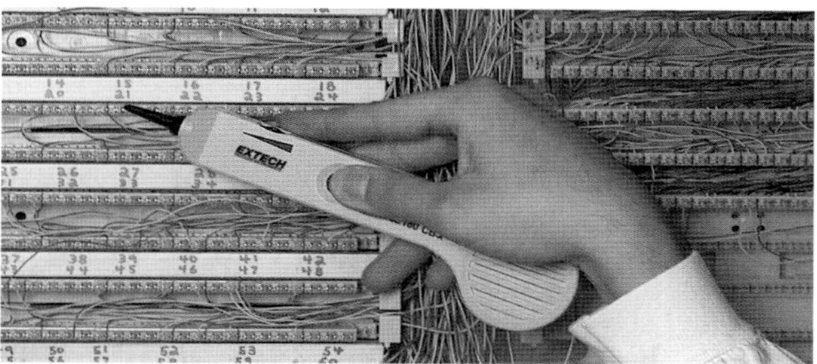

Extech by FLIR

Amplifier probes work with tone generators to identify individual circuits when there are a large number of circuits or when the runs are too long to trace manually.

Tone Generator and Amplifier Probe Testers

Electrical systems are comprised of individual electrical circuits. The number of individual electrical circuits varies from a few to thousands of circuits. VDV wiring systems can include hundreds of individual wires and cables or bundles of wires and cables. When the wires and cables are not marked, finding the correct wire and/or troubleshooting with the unmarked wires is time consuming. To trace and identify individual wires and cables, a tone generator and amplifier probe are used.

A tone generator is a test instrument that is used to place a tone on a cable that can be received by an amplifier probe at the other end of the cable. Tone generators and amplifier probes are used for cable identification with standard electrical wire, security wire, telephone wire, and coaxial wire. Tone generators are used with amplifier probes or can be included as part of other VDV test instruments. **See Figure 3-18.**

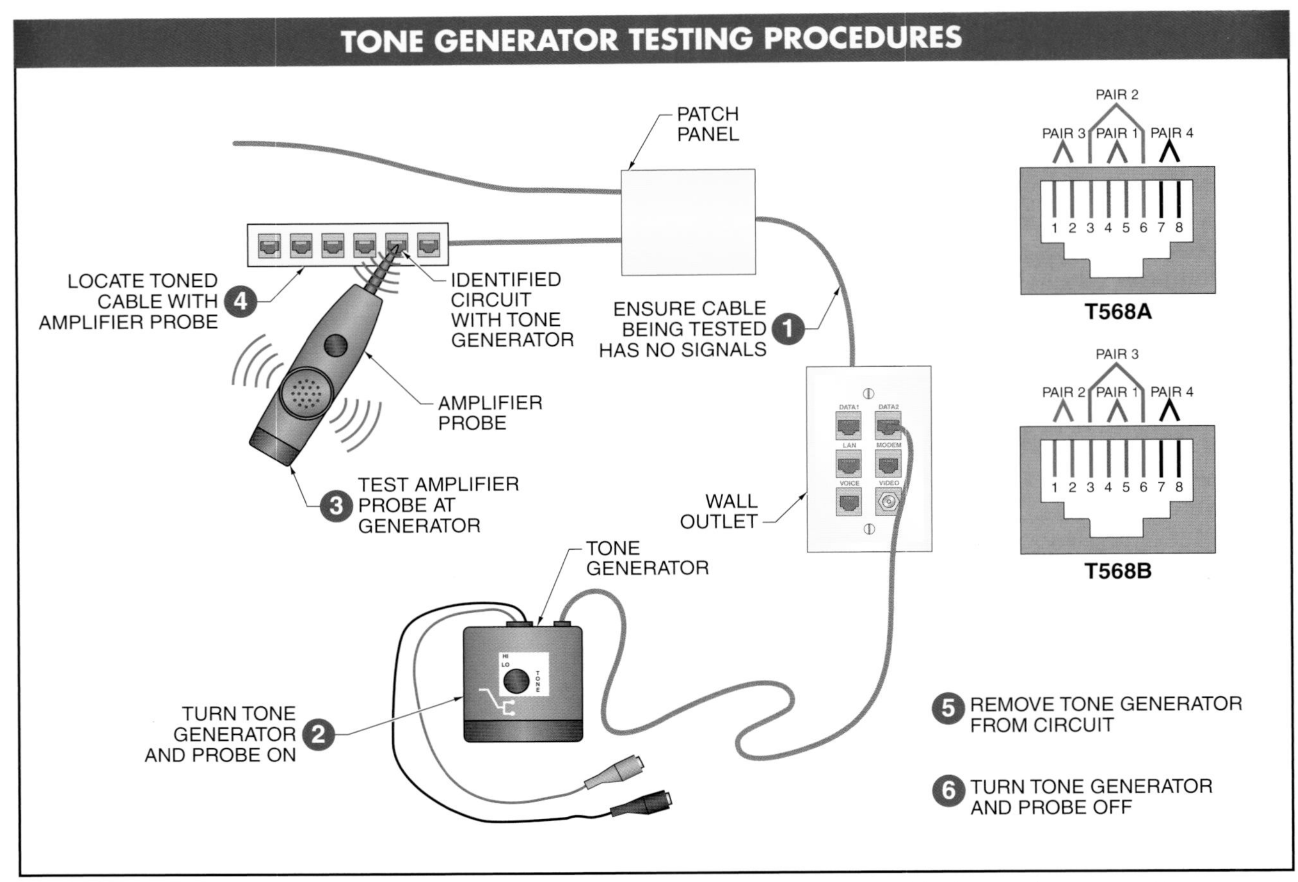

Figure 3-18. A tone generator is a testing device that is used to place a tone on a cable that can be received by an amplifier probe on the other end, and is used mostly for cable identification of standard electrical wire, security wire, telephone wire, and coaxial wire.

Before taking any measurements using a tone generator and amplifier probe, ensure the tester is designed to take measurements on the cable being tested. Refer to the operating manual of the test instrument for all measuring precautions, limitations, and procedures. Always wear required personal protective equipment and follow all safety rules when taking the measurement. To test a cable using a tone generator and amplifier probe, apply the following procedures:

1. Ensure the cable being tested is not carrying any signals.
2. Turn the tone generator and amplifier probe ON.
3. Prior to connecting the tone generator to the cable (circuit) being tested, verify that the amplifier probe is working by placing the amplifier next to the tone generator.
4. Use the amplifier probe to locate the cable (circuit) connected to the tone generator. When there is no tone, the cable being tested is not the correct cable to be identified. When the wire being tested causes the amplifier probe to sound, the correct wire has been identified.
5. Remove the tone generator from the cable or circuit being tested.
6. Turn the tone generator and amplifier probe OFF.

Troubleshooting Tip

REQUIRED TESTING

The tests required for 100 Ω twisted pair cabling as required by ANSI/TIA/EIA 568-B are wire map, cable length, insertion loss (insertion loss is an updated term for attenuation), near end crosstalk (NEXT), power sum NEXT (PSNEXT) for category 5e and higher, equal level far end crosstalk (ELFEXT) for category 6 and higher, power sum equal level far end cross talk (PSELFEXT) for category 5e and higher, delay and delay skew for category 5e and higher, and return loss for category 6 and higher.

Telephone Line Testers

Hard-wired telephone systems and telephones used to be the primary method of voice communication over distance. Telephones are still the primary method of voice communication, but are now hard-wired or wireless. Hard-wired telephone lines that once carried only voice are now also used to carry data and video signals. When voice, data, or video problems occur on telephone lines, the telephone lines must be tested. Testing a telephone line for proper voice transmission and reception is important because when a line cannot satisfactorily carry a voice signal, the line will not be able to carry video and data signals. To test telephone lines, a telephone line tester is used. A *telephone line tester* is a test instrument that is used to simulate the telephone of a caller so the telephone equipment and line can be tested. A telephone line tester is used to identify telephone line problems (troubleshoot) on the phone line and equipment. Typical telephone line testers can detect, measure, or perform the following:

- low voltage condition
- high voltage condition
- low current (mA) condition
- tone generation for tracing wires with a tone probe
- line polarity
- capturing and identifying dial-out numbers
- inbound caller ID testing
- outgoing caller ID testing

Telephone line testers range from basic testers that are used to identify only a few basic circuit conditions, such as whether a line has a signal, to complete line/signal analyzing functions. The basic tester is primarily used to verify that a line is powered. The analyzer is used to isolate and identify any problem with the line and/or certify the line was installed correctly. **See Figure 3-19.**

Before taking any measurements using a telephone line tester, ensure the tester is designed to take measurements on the telephone line being tested. Refer to the operating manual of the test instrument for all measuring precautions, limitations, and procedures. Always wear required personal protective equipment and follow all safety rules when taking the measurement. To test a telephone line using a telephone line tester, apply the following procedures:

1. Turn the telephone line tester ON.
2. Connect the telephone line tester to the circuit being tested.
3. Observe any warning lights or tones that indicate a circuit problem, such as an overcurrent line condition.
4. Remove the tester if any warning lights are displayed and correct the problem on the line prior to reconnecting the tester.
5. With the phone tester connected to the circuit being tested, perform all required line or system tests, such as trying to send and receive a clear voice (phone) signal.
6. When a problem is identified, make any required circuit corrections.
7. Retest the telephone line to verify that the line and equipment are working properly.
8. Remove the telephone line tester from the circuit.
9. Turn the telephone line tester OFF.

Technical Tip

Insertion loss is the combined signal loss of each of the components in a permanent link. Total insertion loss = (IL × number of connectors) + (IL × total length of cable) + (IL × total length of patch cords). Total loss should be 3 dB or less at any frequency.

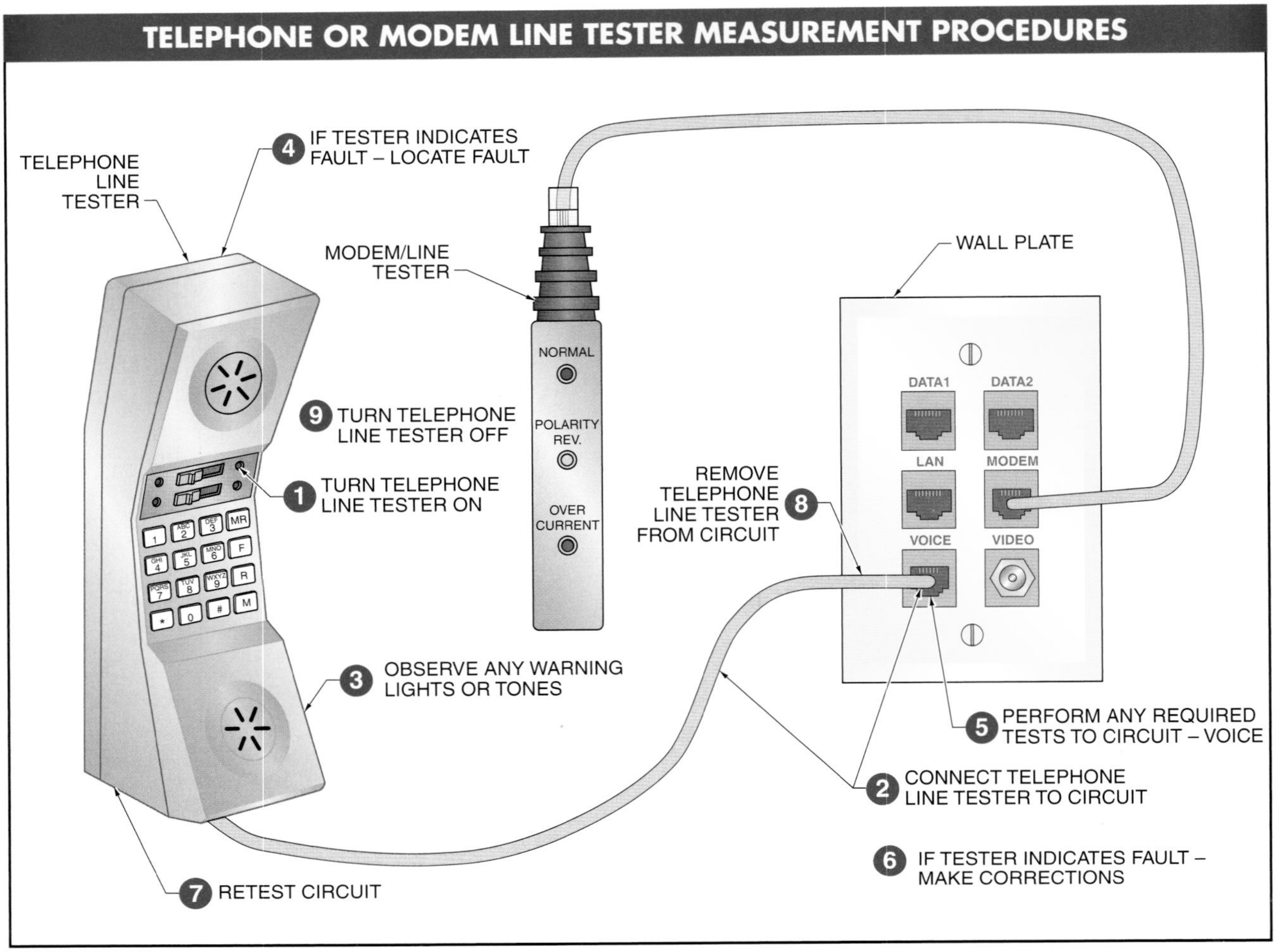

TELEPHONE OR MODEM LINE TESTER MEASUREMENT PROCEDURES

Figure 3-19. Telephone line testers and modem/line testers are used to troubleshoot telephone lines for unknown problems.

Cable Fault Analyzers

There are several tests that can be performed to evaluate the condition or quality of twisted pair wires or coaxial cables. Cable fault analyzers are typically used for the following tests and functions:

- cable length
- wire map
- attenuation
- NEXT (near-end crosstalk)
- propagation delay
- delay skew
- impedance
- RL (return loss)
- TDR (time domain reflectometer)

The types of tests an analyzer can perform depends upon the model used (models also vary in cost). Certain tests require only one unit that connects at one end of a cable while other tests require a receiver connected at one end of the cable with a main unit at the other end. Cable fault analyzers can perform various tests such as wire length, wire map, attenuation, opens, and shorts. Cable fault analyzers display information in numerical values, in words (PASS or FAIL), or with graphic displays. **See Figure 3-20.**

Technical Tip

Fault tolerance is the ability of a system to provide normal operation despite failure of hardware, software, or cabling.

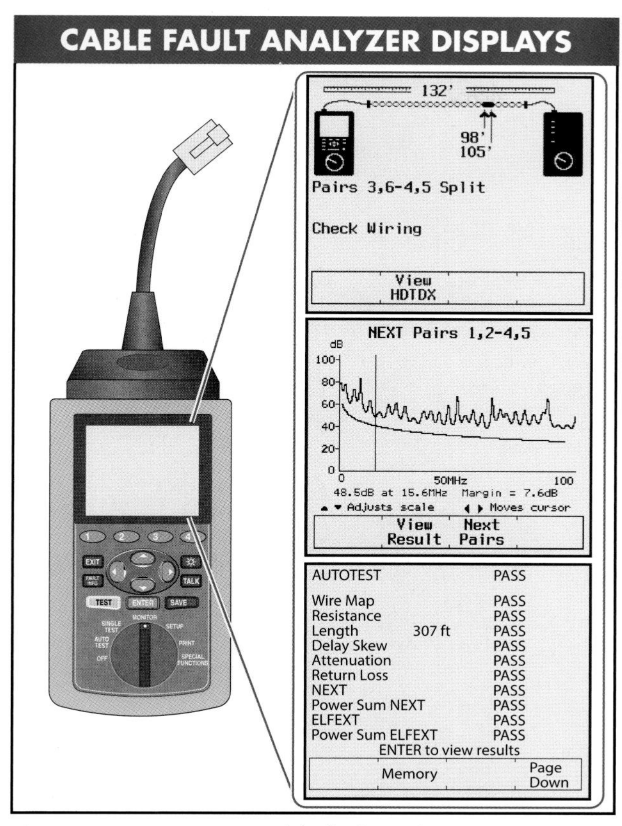

Figure 3-20. Copper cable fault analyzers typically use graphic displays to indicate information but can have numerical or light indicator displays.

Cable Fault Analyzer Measurement Procedures

Cable fault analyzers are used to perform all basic tests on a twisted pair or coaxial cable and certify that the cable meets set standards. **See Figure 3-21.** Before taking any measurements using a cable fault analyzer, ensure the fault analyzer is designed to take measurements on the wires or cables being tested. Refer to the operating manual of the test instrument for all measuring precautions, limitations, and procedures. Always wear required personal protective equipment and follow all safety rules when taking the measurement. To test twisted pair wires or cables using a cable fault analyzer, apply the following procedures:

1. Ensure that the cables being tested are not carrying any signals.
2. Turn the cable fault analyzer, main unit, and receiver unit ON (if used).
3. Input any required or optional setup information (cable type, user name, printer type, time, and date) into the cable fault analyzer.
4. Run any self-test the cable fault analyzer includes.
5. Set the reference level by following the procedures outlined in the manual of the cable fault analyzer. Setting the reference level typically requires that the main unit be connected to the receiver unit. A reference level establishes a reference point so the cable fault analyzer can measure attenuation changes in the signal strength once the analyzer is connected to the cable being tested.
6. Disconnect the main unit from the receiver.
7. Connect the main unit to one end of the cable being tested.
8. Connect the receiver unit to the other end of the cable being tested.
9. Use the directions provided with the analyzer to perform all required tests (wire map, NEXT, attenuation).
10. Record, or save in the memory of the cable fault analyzer, all important or required measurements.
11. Make any required corrections to faults in the twisted pair or cable.
12. Retest the twisted pair or cable after any changes are made.
13. Remove the main unit and the receiver unit from the cable being tested.
14. Turn the main unit and receiver unit of the cable fault analyzer OFF.

CAUTION

Some cable fault analyzers must be turned ON prior to being connected to a circuit. Fault analyzers must be on to activate internal protection circuits.

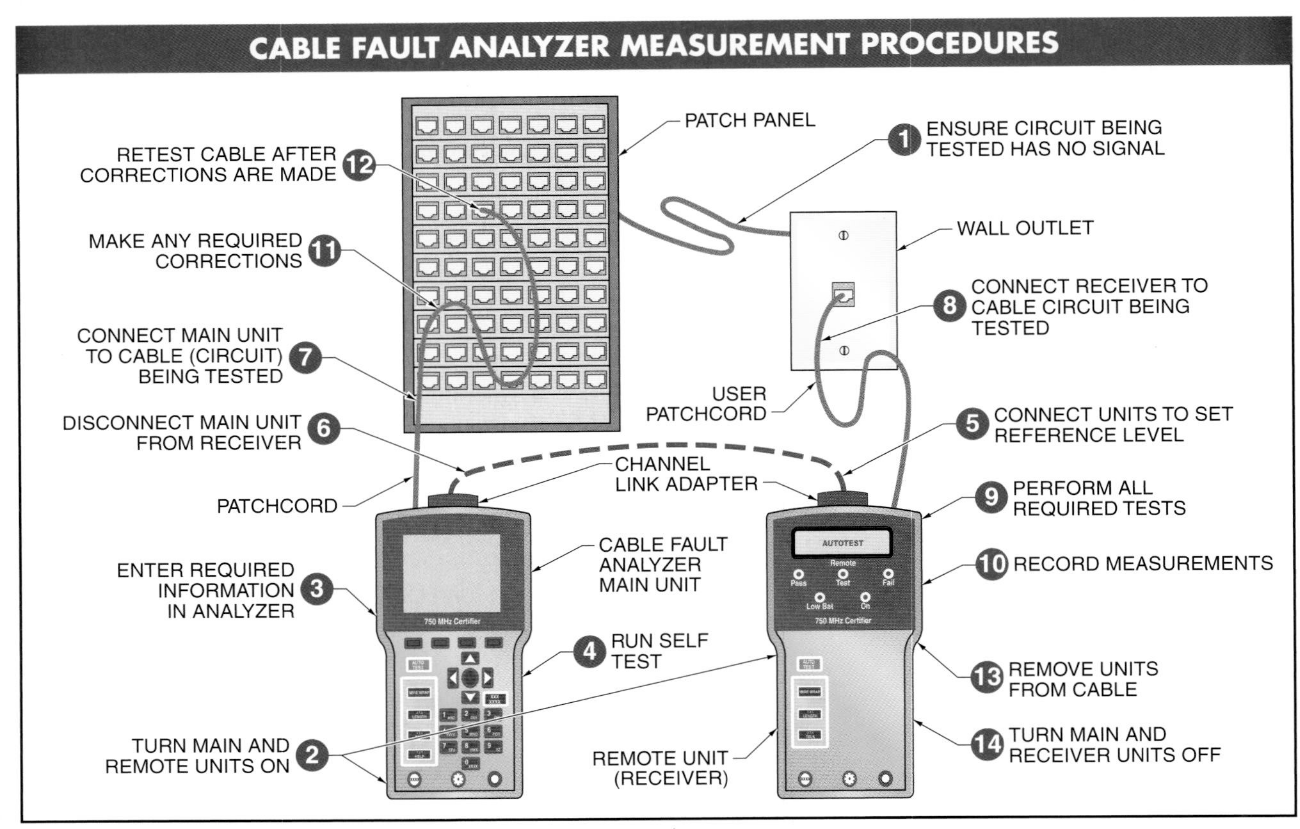

Figure 3-21. A cable fault analyzer is used to perform all basic tests on a twisted pair and coaxial cable and certify that the cable meets set standards while displaying pertinent information.

VDV FIBER OPTIC TEST INSTRUMENTS

Various test instruments are used for the proper installation and troubleshooting of fiber optic cable and related hardware in telecommunication systems. Most testing is performed with digital multimeters (DMMs) with fiber optic test attachments. Other tests require fiber optic microscopes, fiber optic fault analyzers, or optical time domain reflectometers (OTDR).

Technical Tip

Approximately 88% of all optical fiber service lines are single mode (SM). Single-mode fiber has a core diameter of 8 μm to 8-9 μm and a cladding diameter of 125 μm. Advantages of single-mode fiber include high bandwidth and low attenuation rates. Public networks use single-mode fiber for applications such as telephony, cable television, and backbone networks.

Fiber Optic Cable

Similar to copper conductors and cable, fiber optic conductors and cable are used to transmit data from one location to another location. *Fiber optics* is the method of using light to transport information from one location to another location through thin filaments of glass or plastic. In a fiber optic system, electrical information (voice, data, or video) is converted to light energy by a fiber optic transceiver, transmitted to a receiver over optical fibers, and then converted back into electrical information by a fiber optic transceiver.

Optical fibers are either glass or plastic. Because of higher quality, glass is more suitable for optic data transmission over long distances but is more expensive and fragile than plastic. Due to glass being more fragile than plastic, plastic is used in applications requiring short transmission lines (50′ or less).

Fiber optic cables are lightweight and offer the highest bandwidth and speeds possible. Fiber optics also uses smaller diameter cables, which are not affected by electromagnetic interference (EMI) or radio frequency interference (RFI). Fiber optic cables are used in large backbone systems and at individual workstations. Fiber optic connectors must be protected from physical damage and moisture. To protect the fibers, protective cladding and coatings are used along with strength members. *Cladding* is a layer of glass or other transparent material surrounding the fiber core that acts as insulation.

There are various types of fiber optic cables available to meet varying cable requirements. Fiber optic cables typically consist of more than one fiber-carrying strand. In addition to standard fiber optic cable, two other types of fiber optic cable are the tight-buffered fiber optic cable and the loose tube fiber optic cable. **See Figure 3-22.**

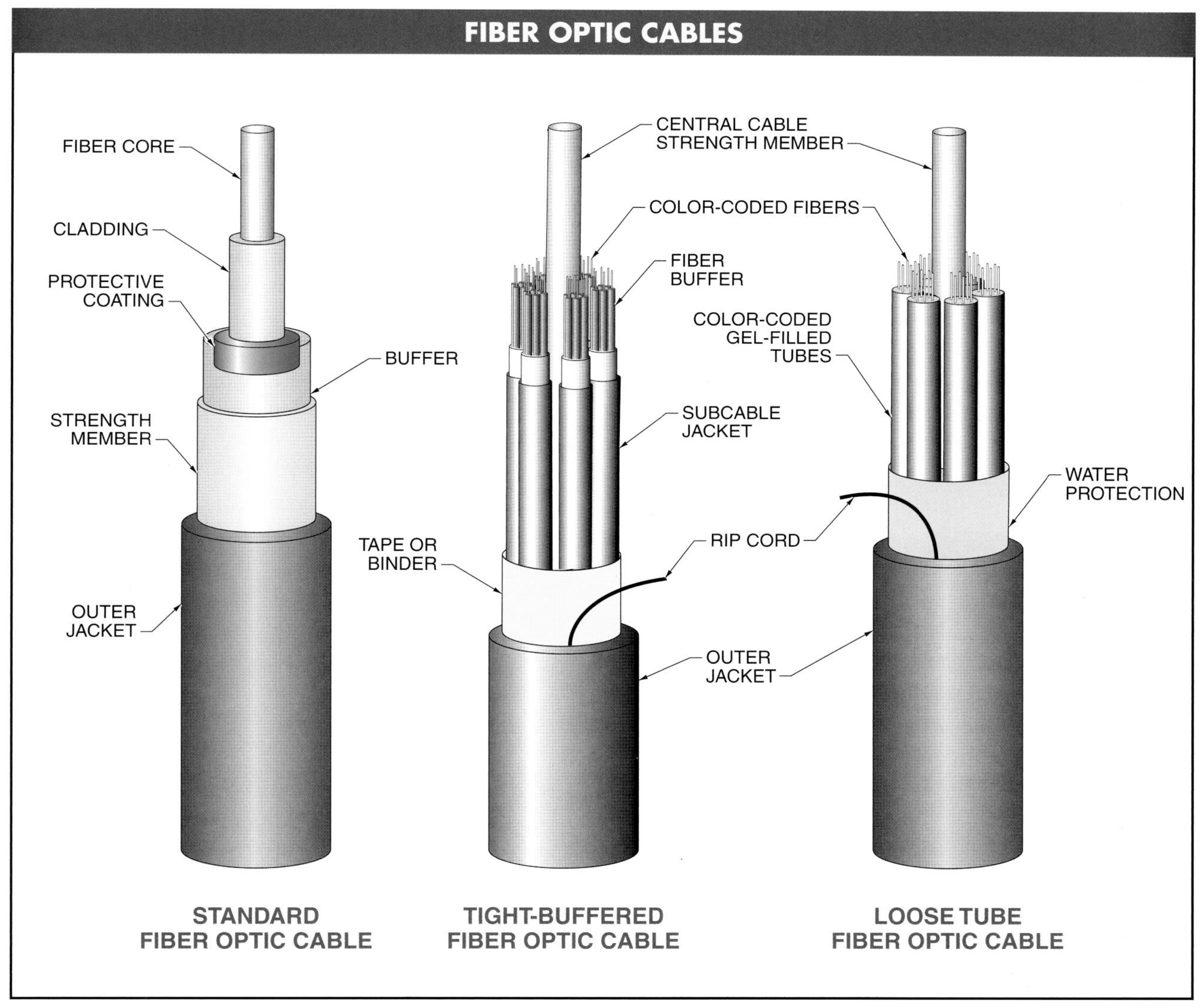

Figure 3-22. In addition to standard fiber optic cable, two other basic types of fiber optic cable are tight-buffered fiber optic cable and loose tube fiber optic cable.

Tight-buffered fiber cable has a more restricted fiber movement. Loose tube fiber cable has fiber that is not restricted and can move freely inside the housing. Although both types use air or gel-filled tubes to allow fiber movement, tight-buffered fiber cable typically uses gel-filled tubes. The movement of fiber cable allows for expansion and contraction during temperature changes. Loose tube fiber cable is stronger and is used in applications requiring long-distance cable runs. Although signal losses occur because of the length of the fiber cable, losses from splices and connectors often exceed the losses from the actual cable length. Common signal losses occur from axial and angular misalignment, excessive separation between cables, and rough ends. **See Figure 3-23.**

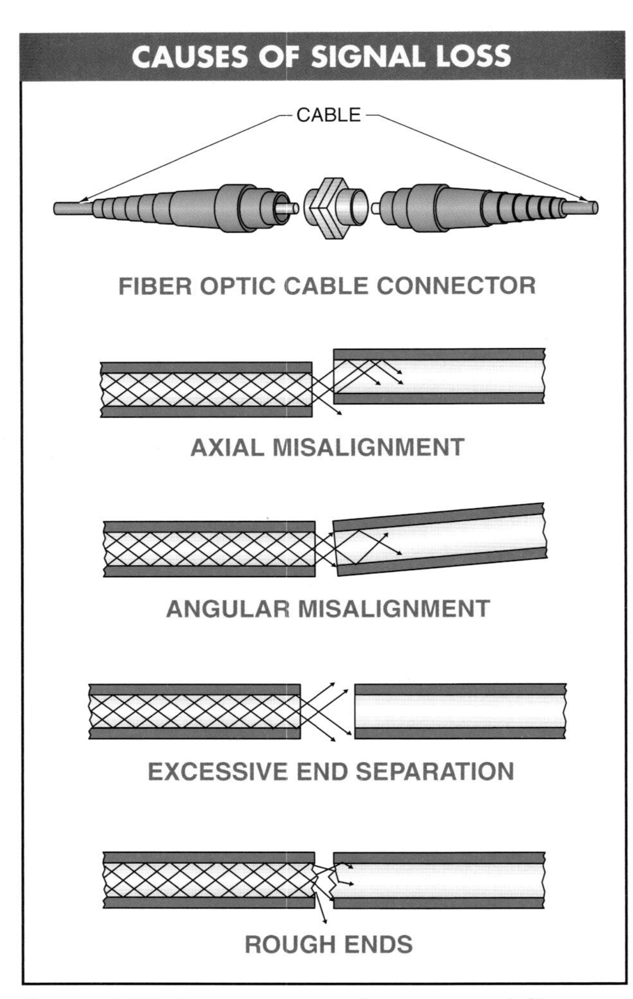

Figure 3-23. Common causes of signal loss with fiber optic cable include axial and angular misalignment, excessive end separation, and rough (unpolished) ends.

Technical Tip

The minimum long-term bend radius of most optical cable is the cable's diameter × 10.

Fiber Optic Cable DMM Test Attachments

A *fiber optic cable DMM test attachment* is a DMM accessory used to test fiber optic cable and fittings. When transmitting light through a fiber optic cable, light loss must be kept to a minimum. Fiber optic cable DMM test attachments are used to test fiber optic cable and fittings for any losses. **See Figure 3-24.**

DMM fiber optic attachments plug into any DMM that can measure and display mV DC. The fiber optic attachments are used to measure the signal loss (attenuation) of light on any fiber optic cable. An independent fiber optic test light source is required to supply a light source of a known strength for testing. The fiber optic attachments output 1 mV DC per 1 dB to the DMM.

Before taking any measurements using a fiber optic cable loss (attenuation) attachment, ensure the attachment and DMM are designed to take measurements on the fiber cables being tested. Refer to the operating manual of the test instrument for all measuring precautions, limitations, and procedures. Always wear required personal protective equipment and follow all safety rules when taking the measurement. To test the attenuation of a fiber optic cable using a fiber optic attenuation attachment, apply the following procedures:

1. Ensure the fiber optic cable being tested is not carrying any signals.
2. Connect the fiber optic test light source to one end of the fiber optic cable being tested.
3. Connect the fiber optic attenuation attachment to the voltage inputs of the DMM.

4. Set the DMM to measure mV DC.
5. Connect the fiber optic attenuation attachment to the fiber optic cable (end opposite the fiber optic test light source).
6. Read the displayed mV DC, which is the amount of loss (attenuation) through the fiber optic cable and splices. A reading of 0 mV DC indicates no loss, 5 mV equals a 5 dB loss, and 20 mV equals a 20 dB loss.
7. Reverse the position of the fiber optic attenuation attachment with the position of the fiber optic test light source.
8. Read the displayed mV DC, which is the amount of loss (attenuation) through the fiber optic cable and splices, but in the opposite direction. The highest reading (greatest dB loss) is used to evaluate the cable system.
9. Record the attenuation and compare the measurement to established standards.
10. Turn the fiber optic attenuation attachment and fiber optic test light source OFF.
11. Remove the fiber optic attenuation attachment and fiber optic test light source from the fiber optic cable.

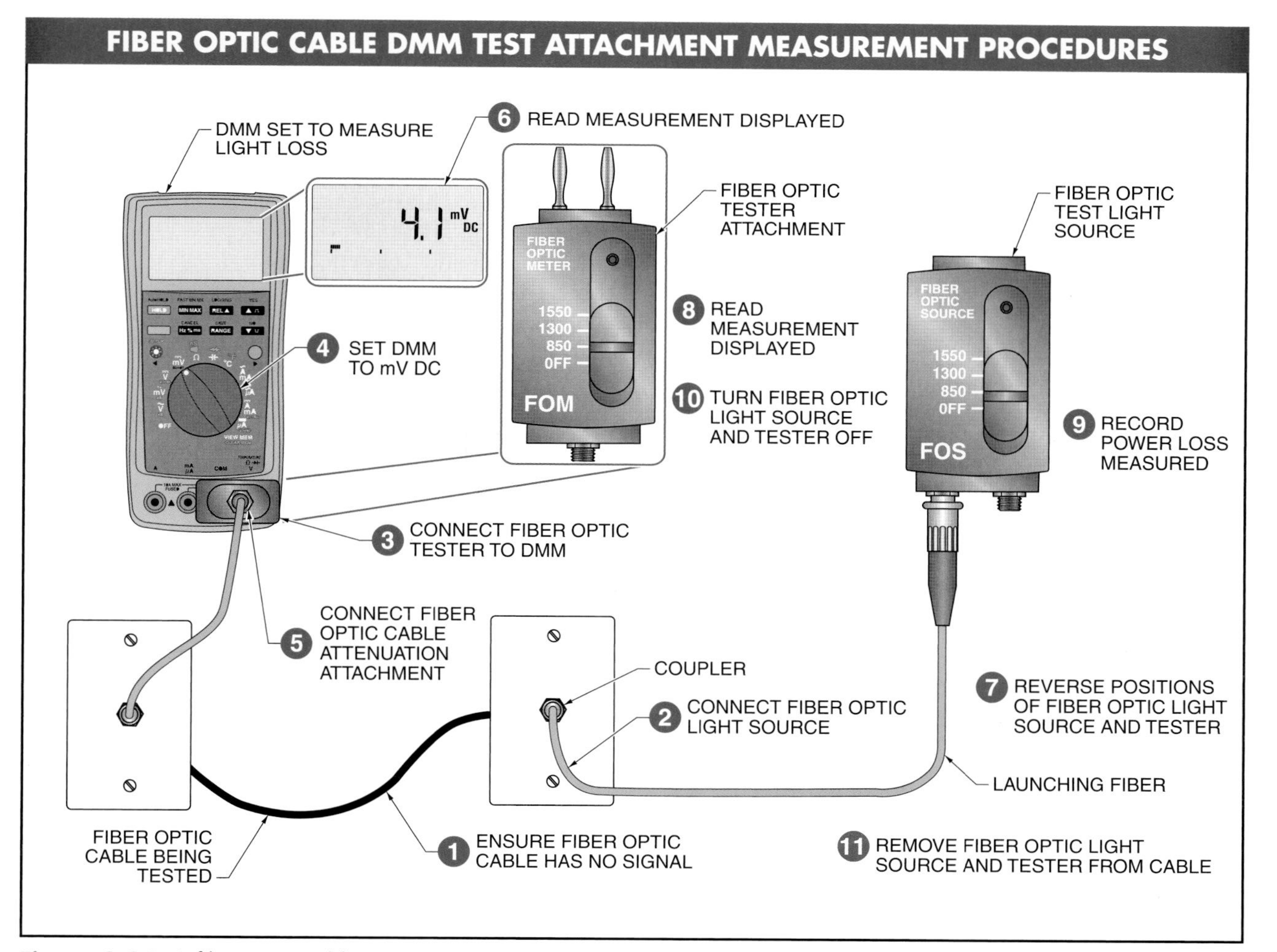

Figure 3-24. A fiber optic cable DMM test attachment is a DMM accessory used to test fiber optic cable and fittings.

Fiber Optic Fault Analyzers

Fiber optics is a technology that uses light signals transmitted through flexible fibers of glass or plastic. Fiber optic cable is used to connect two electronic circuits together. Electronic circuits convert electrical signals into light signals at the fiber optic transmitter or convert the light signals back into electronic signals at the fiber optic receiver. The light signal (beam) must be transmitted to the receiver without excessive loss of power or signal distortion.

There are several tests and measurements that can be taken to evaluate the condition or quality of a fiber cable. The four most common tests performed on fiber cable are the following:

- continuity test of fiber optic cable
- measurement of the amount of optical power (strength of signal)
- measurement of attenuation (loss of signal) through the fiber
- measurement of light signal dispersion over the length of the fiber optic cable

Continuity Tests. One of the most common tests is a continuity test. A continuity test checks the power going through a fiber. The light that actually carries the VDV signals cannot be seen. Thus, a white light or red laser light is typically used to test continuity visually. The light source is connected to one end of the fiber optic cable being tested and the other end of the cable is tested to verify that light is coming through the cable. When there is a complete path (continuity) for the light to travel through the fiber (i.e., there are no breaks), the next test is to test the actual quantity and quality of the light source delivered to the receiver from the transmitter by the fiber optic cable.

Optical Power Measurements. As in any electrical system, there must be enough power to perform the required work. A fiber optic transmission system has a light source that must have enough power to transmit the signal through the cable and connections. Optical power strength is measured in dBm (decibels per milliwatt), with 0 dBm equal to 1 milliwatt.

The decibel scale is logarithmic, which means that for every 3 dB increase in strength the power level must be doubled, and for every 3 dB decrease in the strength the power level is cut in half.

Light Strength	Electrical Power
0 dBm =	1 milliwatt or 1000 microwatts
–3 dBm =	0.5 milliwatts or 500 microwatts
–10 dBm =	0.1 milliwatts or 100 microwatts
–20 dBm =	0.01 milliwatts or 10 microwatts
–30 dBm =	0.001 milliwatts or 1 microwatt

Technical Tip

The minimum short-term bend radius (during installation procedures) of most optical cable is the cable's diameter × 20.

Power Attenuation Measurements. When a power source has enough power to operate a load, the power must still get from the power source to the load. Power attenuation is one of the most important measurements taken on a fiber cable. Power attenuation is measured in dB.

POWER LOSS	POWER LOST	POWER RECEIVED
0 dB	0%	100%
3 dB	50%	50%
10 dB	90%	10%
20 dB	99%	1%
30 dB	99.9%	0.1%
40 dB	99.99%	0.01%
50 dB	99.999%	0.001%

An *optical time domain reflectometer (OTDR)* is a test instrument that is used to measure cable attenuation. An OTDR uses a laser light source that sends out short pulses into a fiber and analyzes the light scattered back. The light source decays with fiber attenuation. Attenuation is produced by reflections from splices, connectors, and other areas in the cable that cause problems. Based on the amount of reflected back signal, the type and location of a fault is displayed. **See Figure 3-25.**

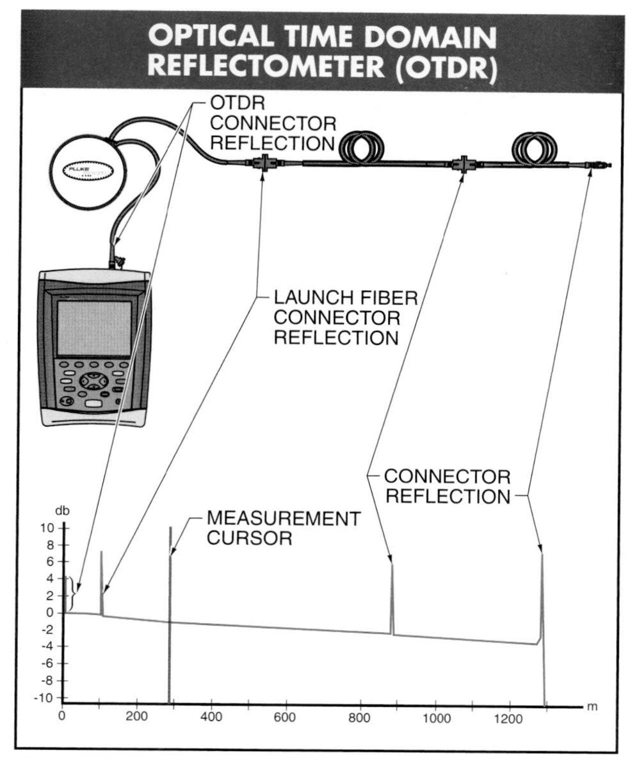

Figure 3-25. An OTDR is used to measure fiber optic signal loss (attenuation), length, and fiber endface integrity.

Troubleshooting Tip

ADVANTAGES OF FIBER OPTIC TRANSMISSION

Although fiber optic technology is more recent than copper as well as initially more expensive, fiber does have several cost-saving advantages over copper connectivity. These advantages include the following:

- Long distance transmission without the use of repeaters or regenerators; fiber can be routed 10,000′ versus 328′ for copper
- Higher capacity; one fiber cable can replace several copper cables.
- Reduced system cost, once installed
- Reduced maintenance cost
- Ease of upgrade
- Lower life cycle cost
- Smaller size for same bandwidth
- Lighter weight
- Dielectric nature; light waves do not conduct electric current.
- Immunity to EMI/RFI

Light Dispersion Measurements. In a perfect electrical circuit, a signal (sine wave, square wave, or digital pulse) is the same shape at the fiber optic receiver as at the fiber optic transmitter. However, in all electrical circuits (copper wire or fiber optic), the received signal is always distorted to some extent. Light dispersion is the main source of distortion in a fiber optic cable. **See Figure 3-26.** Light dispersion lengthens the wavelength of the transmitted signal as the light travels down a fiber. Light dispersion is what limits the length of a fiber cable, because light dispersion increases proportionally with cable length.

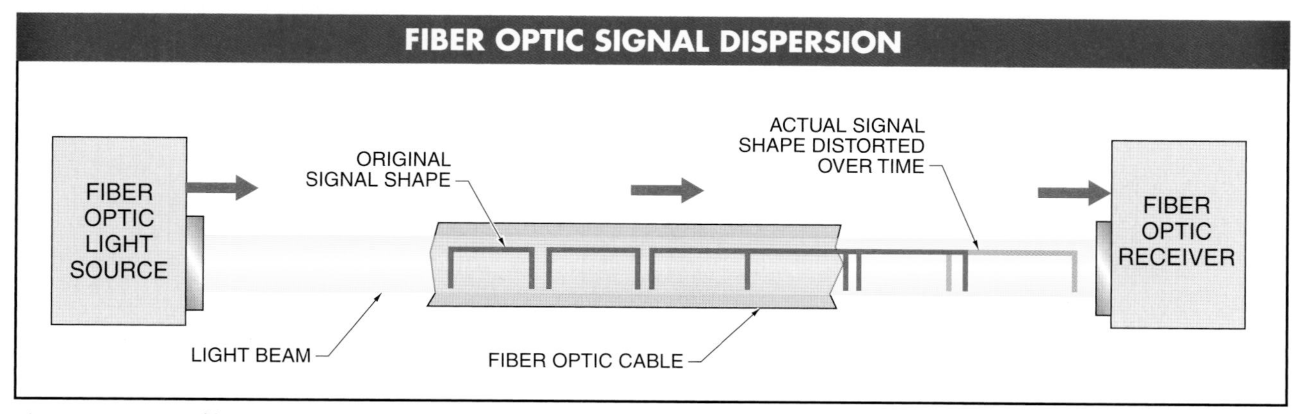

Figure 3-26. A fiber optic signal will become distorted from its original shape as it moves along the length of the cable from the signal source to the receiver.

Power losses and light dispersion are increased by impurities in the cable such as bubbles in the fibers, and by dirty connections, bends in the cable, increased length of cable, and coupling losses. Even the oil from the hand of an electrician left on a fiber end causes losses. Due to the sensitivity of fiber optic cable to contaminants, fiber optic cable as well as copper cable installations must be performed in the correct way, with the correct tools and by qualified electricians. A special type of fiber cable is zero-dispersion fiber optic cable. Zero-dispersion fiber optic cable is thinner and has fewer impurities than standard fiber optic cable. While being of the highest quality, zero-dispersion fiber optic cable is very expensive and is only used for new installations, rather than upgrades.

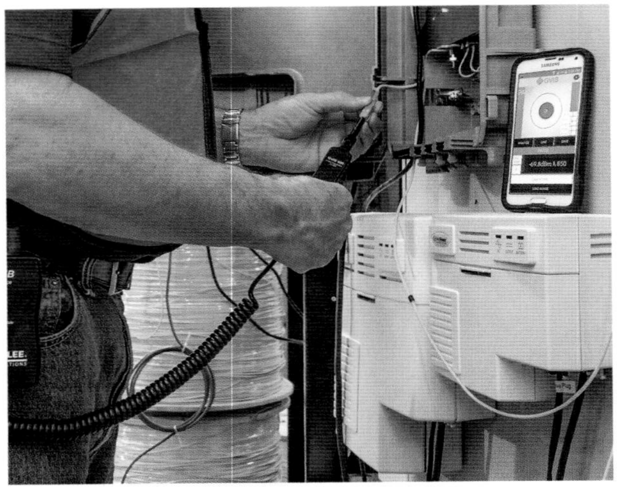

Greenlee Textron, Inc.

Simple fiber link problems can be diagnosed with a visual fault locator, which can locate fibers, verify continuity and polarity, and help find breaks in cables, connectors, and splices.

Fiber Optic Visual Fault Locators

A fiber optic fault analyzer can be used to perform all basic tests on a fiber cable and certify that the cable meets industry specifications. While a standard fiber optic fault analyzer will provide in-depth information as to the cause of a problem, a portable fiber optic fault locator (also referred to as a fault finder) locates fractures, breaks, cracks, and other faults in the glass fiber by using a red beam as a light source. The red beam will only be visible at the point of the fault. Most technicians utilize handheld models of visual fault locators to initially locate a problem.

Fiber Optic Visual Fault Locator Testing Procedures

Before testing a fiber optic cable using a fiber optic visual fault locator, ensure the visual fault locator is designed to take measurements on the fiber cables being tested. Refer to the operating manual of the test instrument for all measuring precautions, limitations, and procedures. Always wear required personal protective equipment and follow all safety rules when taking the measurement. **See Figure 3-27.** To test for faults in a fiber optic cable using a standard fiber optic visual fault locator, apply the following procedures:

1. Ensure the fiber optic cables being tested are not carrying any signals.
2. Connect the fiber optic visual fault locator to the fiber optic cable.
3. Turn the visual fault locator ON.
4. Press the FLASH button to toggle between continuous and flash mode.
5. Locate the red beam of light along the fiber optic cable, which indicates a fault area.
6. Make any required corrections to any faults.
7. Retest the cable after any changes are made.
8. Turn the fiber optic visual fault locator OFF.
9. Remove the fiber optic visual fault locator from the fiber optic cable.

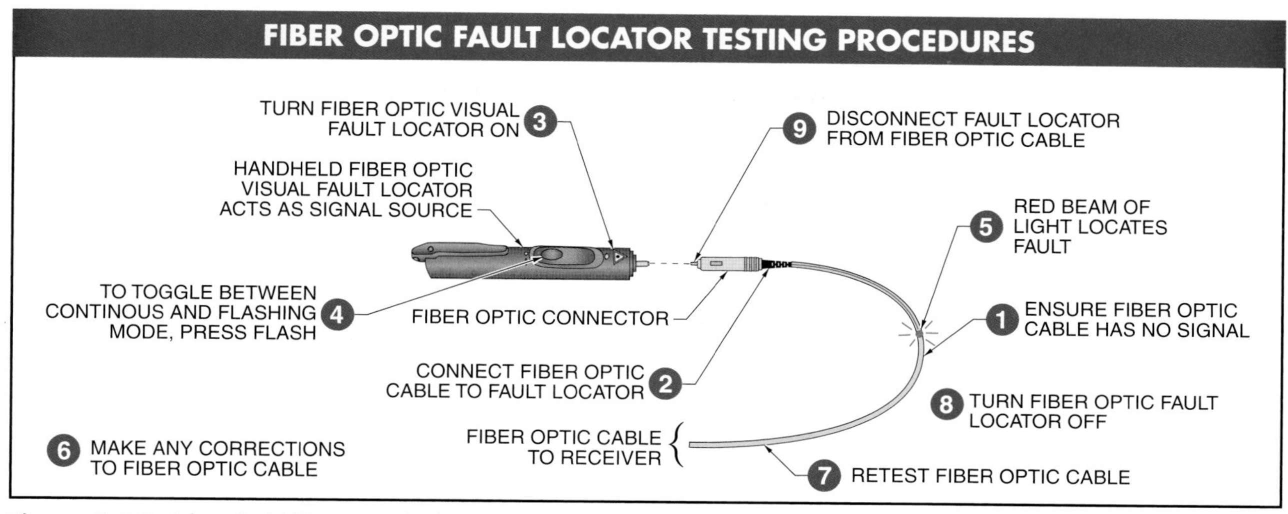

Figure 3-27. A handheld fiber optic fault analyzer initially locates a problem in the glass fiber by using a red beam of light that is visible at the point of the fault.

FIBER-OPTIC TESTING SAFETY

VDV technicians must always follow safety practices when testing fiber-optic cables. Safety practices include wearing appropriate eye protection and other PPE as required. VDV technicians must follow all federal, state, local, and job site safety regulations. They must verify that the light source is turned off at both ends of a cable and that the cable is disconnected at both ends. A VDV technician must never look directly into a fiber-optic connector or directly into lasers on communication or test equipment. Finally, they must verify that all unmated, fiber-optic connectors have a protective dust cap installed over the ferrule for each connector.

Troubleshooting Tip

PREPOLISHED CONNECTORS

A prepolished fiber optic connector is polished to specifications in the factory. No field (hand) polish is required. The cable fiber is inserted through an indexing gel behind the connector's ferrule and crimped to the connector in the field, saving time and cost.

WARNING

Never look directly at the red beam. Permanent eye damage could result.

CAUTION

Always turn a fiber optic fault analyzer ON before connecting the analyzer to a circuit. Fault analyzers must be on to activate internal protection circuits.

INSPECTING AND CLEANING FIBER-OPTIC CONNECTORS

VDV technicians must inspect and clean a fiber-optic connector when necessary before inserting it into a port or joining it with another connector. *Note:* This applies to all fiber-optic connectors including test jumpers, patch cords, launch cords, and receive cords. It is equally important to inspect the mating port or mating connector involved. A small amount of dirt, dust, or other contaminants on the face of a connector or in the interior of a port will block the cable's light signal. Contaminants on the face of a connector can be transferred to a port or to another connector. Sources of contamination include dust, lint, and skin oils. Contamination is the leading cause of fiber-optic cable failures.

It is impossible to know if the face of a connector is clean unless it is inspected. The face of a fiber-optic connector or port can be inspected with a fiber-inspection scope. The two main types of fiber-inspection scopes are optical fiber-inspection scopes and video fiber-inspection scopes. **See Figure 3-28.**

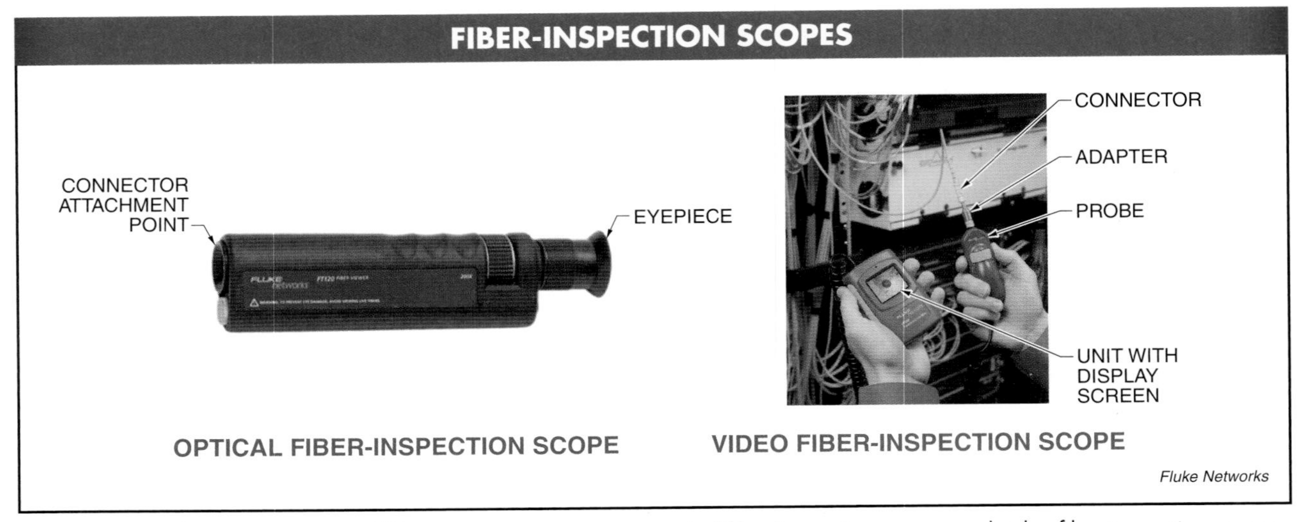

Figure 3-28. The two main types of fiber-inspection scopes are optical fiber-inspection scopes and video fiber-inspection scopes.

Optical fiber-inspection scopes are handheld devices that use a lens to magnify the connector face image. The connector is attached to one end of the fiber inspection scope, and the VDV technician examines the connector face through an eyepiece. Adapters allow different types of connectors to be inspected. An optical fiber-inspection scope cannot be used to inspect ports.

WARNING: An optical fiber-inspection scope must only be used with a cable that is not connected to a light source, such as a loose cable. Viewing a cable that is connected to a light source with an optical fiber-inspection scope even once can cause permanent damage or blindness to the eye. Never use this type of instrument if the cable connections cannot be verified.

Video fiber-inspection scopes are handheld devices that consist of a probe and a separate unit with a display screen. The fiber-optic connector or port attaches to the probe, and a video image of the connector or port face is displayed on the screen. Adapters allow different types of connectors and connector ports to be inspected. Advanced video fiber-inspection scopes can evaluate a connector face, compare it to known standards, and provide a "pass" or "fail" rating. This decreases the amount of time necessary to inspect the connector. This instrument is the most common type of instrument used to inspect fiber as it is safer than an optical fiber-inspection scope.

Troubleshooting Tip

VISUAL FAULT LOCATORS

Many issues with individual fibers used in VDV installations can be identified with a small, handheld test instrument known as a visual fault locator (VFL). A VFL can be inserted at the end of the connecter located on a cable. The VFL transmits a laser beam through the glass, which can pinpoint any breaks or macrobends in the fiber.

WIRELESS COMMUNICATION

Wireless communication is the sending and receiving of voice, data, and video signals without using wires or cables. The term "wireless" is actually an old term created when radio receivers first came out at a time when wired telegraph was the primary way of sending signals from one point to another. Today, cellular telephones and wireless communication by laptop computers are easily recognized forms of wireless communication. Wireless communication is one of the fastest growing segments of the communication field in many areas other than cellular phones and laptop computers. For example, wireless communication between tagged parts and data systems of companies allows real-time continuous monitoring and tracking of everything in a warehouse, on the job site, and during transportation. Construction and transportation companies now know the exact location of every piece of equipment owned by the company at any time of the day or night.

The transmission of signals is accomplished using radio frequencies (RF), infrared (IR) waves, laser light beams, or microwave signals. Most wireless communication systems use low-power radio (about 3 kHz to 300 GHz) or infrared frequencies (about 3 THz to 430 THz). Microwave and laser transmission and reception are sometimes used between two fixed points, such as two buildings in a campus setting.

The advantage of a microwave link is that a microwave link allows reliable communication regardless of rain, fog, or other atmospheric conditions that can distort other wireless signals (laser and radio waves). Satellites and satellite technology now allow wireless communication signals to be transmitted and received almost anywhere in the world. With the combination of satellite communication and Global Positioning System (GPS), worldwide communication and positioning make it possible to not only transmit and receive information worldwide, but also to know the exact location that signals are coming from and going to. Aligning satellite receivers to the best position for receiving signals is accomplished by using satellite finder meters.

Satellite Finder Meters

A *satellite finder meter (satellite strength meter)* is a test instrument that is used to locate a satellite signal and measure and display the signal strength. Satellite finder meters are used to align satellite-receiving dishes (dish receivers) to the best position to receive the strongest signal.

Technical Tip

Topology is the description of connections in a network. Logical topology is one station transmitting to many receivers. Physical topology is any two stations connected to a telecommunications closet where all nodes can be cross-connected.

Satellite Finder Meter Measurement Procedures

Before finding the location of a satellite using a satellite finder meter, ensure the satellite finder meter is designed to take measurements in the area being tested. **See Figure 3-29.** Refer to the operating manual of the test instrument for all measuring precautions, limitations, and procedures. Always wear required personal protective equipment and follow all safety rules when taking the measurement. To use a satellite finder meter to locate the strongest satellite signal, apply the following procedures:

1. Open the line connecting the satellite dish to the satellite receiver.
2. Connect the satellite finder meter in series with the open line.
3. Turn the satellite finder meter ON.
4. Observe the displayed signal strength and position the satellite dish so that the highest signal strength is displayed.
5. Remove the satellite finder meter from the dish and receiver circuit.
6. Reconnect the cable from the satellite dish to the satellite receiver.
7. Check that the system is operating properly.

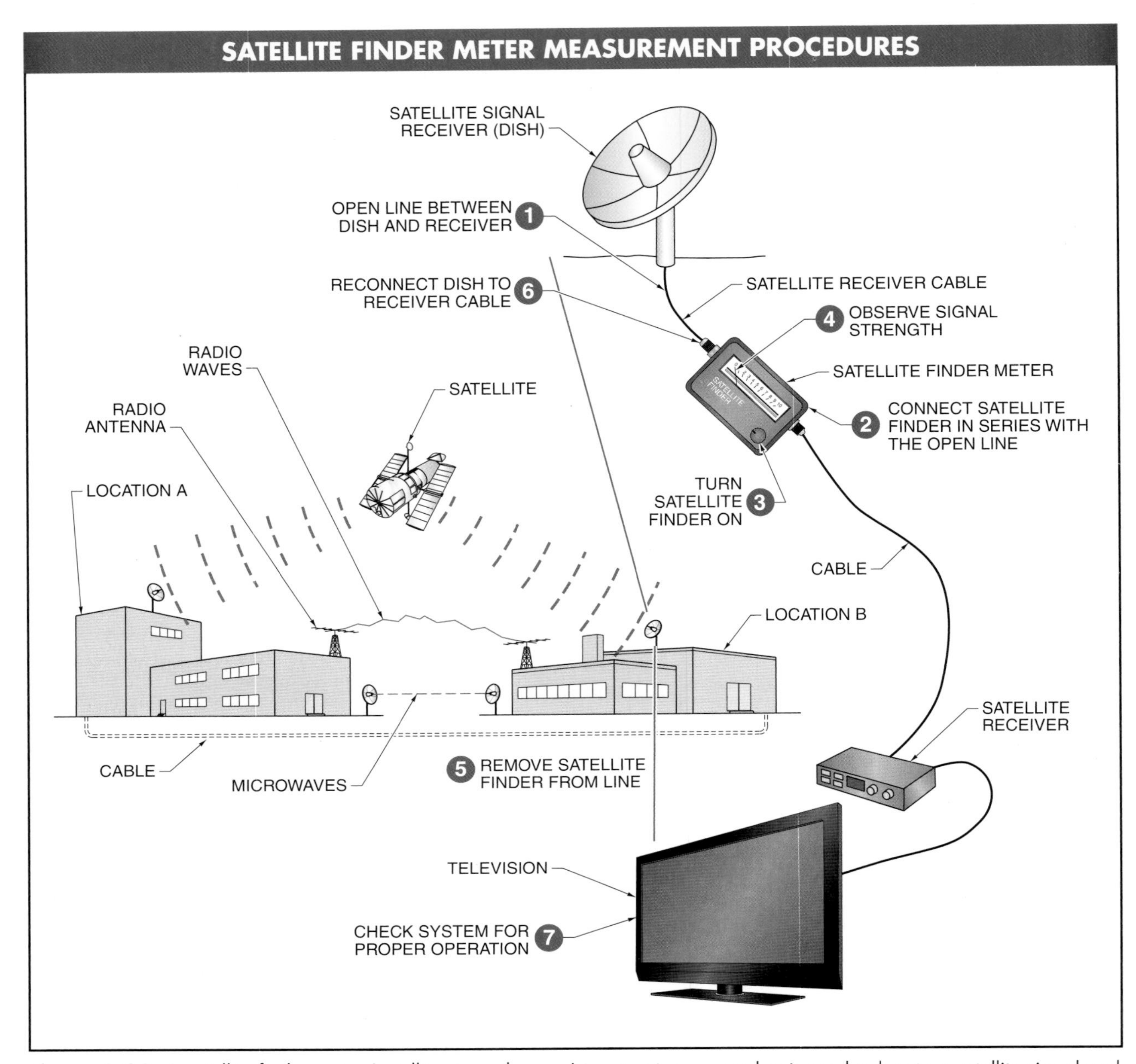

Figure 3-29. A satellite finder meter (satellite strength meter) is a test instrument that is used to locate a satellite signal and measure/display the signal strength.

POWER OVER ETHERNET

Ethernet is a cabling technology used for network connectivity. *Power over Ethernet (PoE)* is a network system technology that is used to safely pass electric power over Ethernet cabling. PoE is used to power VDV equipment such as IP telephones, wireless access points (WAPs), security systems, access control systems, and notification systems typically within the same building or campus. Ethernet operates over twisted pair, coaxial, fiber optic cabling and/or through the air.

Network technicians who have implemented or are considering implementing PoE and installers offering this technology as part of their service should be aware of potential impacts on the networking testing process. The installation of some types of PoE equipment will require recertification of the existing cabling plant. Existing PoE equipment on the network may require changes in test procedures to ensure accurate testing of the data network. Additional tests on new PoE equipment will help verify and document its proper operation.

All network appliances require both data connectivity and a power supply. The advantage of PoE is that it requires only one set of wires to fulfill both of these needs, reducing installation time and cost and saving space. PoE also makes it easier to move an appliance, since it can simply be plugged into a PoE-enabled network jack.

All network appliances require both data connectivity and a power supply equipment (PSE). The two basic types of PSEs used with PoE are end-span and mid-span. **See Figure 3-30.** There are significant differences between these PSEs in their effect on network testing. End-span PSEs use an Ethernet switch with an embedded power supply to deliver power and data. Mid-span PSEs are positioned between legacy switches or routers and the powered devices. Mid-span PSEs can be individual devices or they may be integrated into a patch panel or switch.

An important difference between power insertion patch panels and end-span PSEs is that while power insertion patch panels become part of the permanent link, end-span devices do not. Since the performance of the connection and the quality of the termination of power insertion patch panels are major contributors to permanent link performance, each permanent link needs to be recertified whenever a power insertion patch panel is installed. End-span devices are not in the link when it is certified and have minimal effect on network testing.

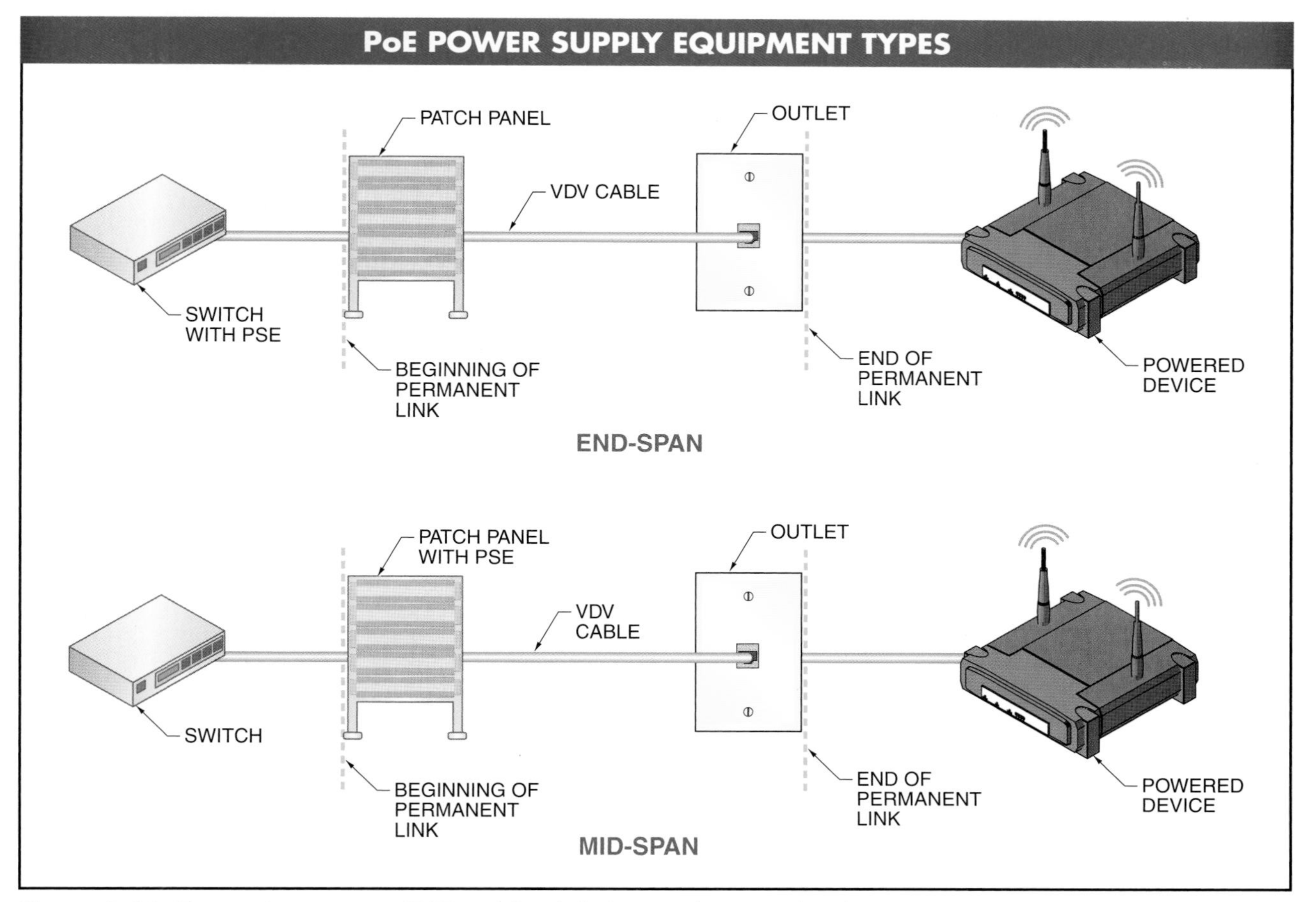

Figure 3-30. The two basic types of PSEs used with PoE are end-span and mid-span.

Testing PoE Systems

As PSEs are prepared for installation or for troubleshooting of the various powered devices (PDs), such as VoIP phones, security cameras, wireless access points, and badge scanners, it is important to calculate the overall power requirement for all PDs that are to be connected to a given PSE to ensure it does not get oversubscribed.

Although it is acceptable to test the PoE voltage directly at the PSE, the best practice is to test for the wattage or voltage level at the wall jack where a PD is to be connected. This is important because PoE will dissipate as it traverses the cable. The power at the wall jack must be verified because it is what will be required for the PD. Low-voltage network analyzers are used to test and troubleshoot PoE systems. **See Figure 3-31.** A low-voltage network analyzer is used to accurately locate problems within the system as well as provide information on each system port.

PoE is subject to the same cable distance limitation as standard CAT 5e cable, which is 100 m (328′). If the physical cable is out of specification and longer than the TIA standard, power may be too weak by the time it reaches the PD. Common PoE variables to review when testing PoE include the following:

- PoE is subject to the same distance limitations as standard network cable runs (100 m/328′).
- There is incompatibility between the PD and PSE.
- The switch is oversubscribed for the PoE.
- The switch provisioning of PoE is incorrect.
- Each port has power limitations.
- Cable faults are present.

The increasing use of PoE means that network technicians must understand its impact on the certification process. Each variety of PoE equipment has varying effects. By adjusting test procedures accordingly, network technicians can easily validate the performance of PoE systems along with the underlying Ethernet infrastructure.

LOW-VOLTAGE NETWORK ANALYZERS

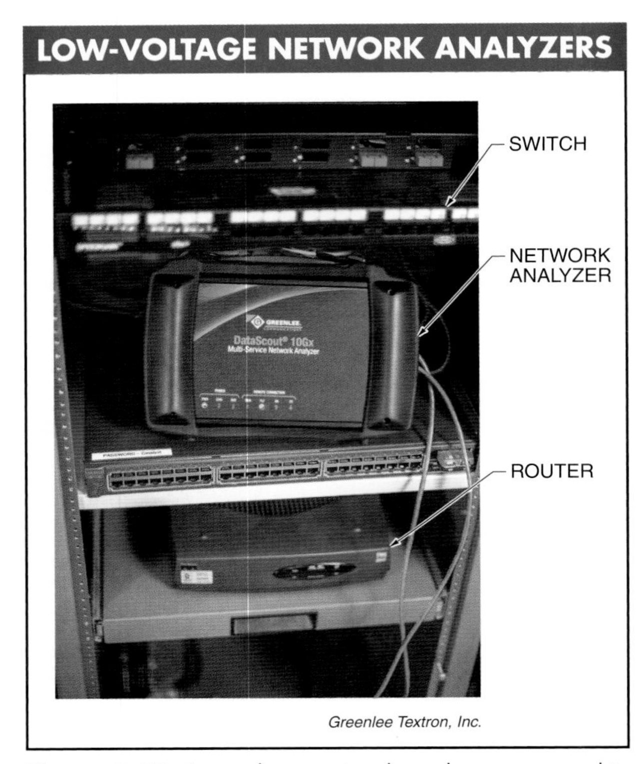

Figure 3-31. Low-voltage network analyzers are used to accurately locate problems within a PoE system as well as to provide information on each system port.

VOICE-DATA-VIDEO (VDV) TEST INSTRUMENTS

3 REVIEW

Name: ______________________________ Date: ____________________

______________	**1.**	What does UTP stand for?
______________	**2.**	What is the function of an outer braided conductor in a cable?
______________	**3.**	What does the CAT rating of a conductor indicate?
T F	**4.**	Fiber optic cable uses light signals to transmit communications data.
______________	**5.**	Which VDV test measures the power loss of a signal over a cable pair?
______________	**6.**	Which type of communication signal does not require hard wiring between locations?
T F	**7.**	Signal losses from splices and connectors often exceed losses from cable length.
______________	**8.**	What are the two conductors in a twisted pair called?
______________	**9.**	Does a lower CAT rating indicate a higher or lower data transfer capacity?
______________	**10.**	Which VDV test measures the time required for a signal to travel the length of the cable pair?
T F	**11.**	The color code for the first pair in a communication cable typically includes red.
______________	**12.**	Should communications cables be run in parallel with power conductors?
______________	**13.**	Which VDV test uses voice signals to verify communication circuits?
T F	**14.**	Fiber optic signals are susceptible to EMI and RFI.
T F	**15.**	Coaxial cable typically carries video signals.
______________	**16.**	Which VDV cable problem results in signals in one circuit inducing unwanted signals in another circuit?
______________	**17.**	Which test instrument can be used to visually examine contaminants or faults in fiber optic cables?
______________	**18.**	Which VDV test determines if wires in communications circuits are crossed, reversed, open, shorted, or split?
T F	**19.**	10 dB of power loss is equivalent to 10% loss.
______________	**20.**	Which VDV test measures the total opposition to alternating current on a cable?

21. What are some possible sources of electromagnetic or radio frequency interference?

22. What is the advantage to twisting cable pairs?

23. Explain how a tone generator and amplifier probe work together to test communication wiring.

24. What advantages do fiber optic networks have over copper networks?

25. How does a time domain reflectometer (TDR) measure the distance to a cable fault or the end of a cable?

4 POWER QUALITY TEST INSTRUMENTS

All test instruments and meters are used where measurements of electrical properties are required. All test instruments and meters have OSHA, manufacturer, facility, and application safety procedures (National Fire Protection Association (NFPA) 70E, Standard for Electrical Safety in the Workplace) that must be followed at all times.

Fluke Corporation

POWER QUALITY PROBLEMS

An *electrical power system* is a system that produces, transmits, distributes, and delivers electrical power to satisfactorily operate electrical loads designed for connection to the system. An electrical power system may be small and simple or large and complex. For example, a portable generator used by company electricians or by an emergency crew of a utility company is a small, self-contained electrical power system. A large utility company servicing a metropolitan area is a very large electrical power system.

Regardless of the size of an electrical power system, power must be supplied that allows loads to operate satisfactorily. Damage to electrical equipment occurs when electrical power is not supplied at the proper level (voltage), amount (current), type (single-phase or three-phase, AC or DC), or condition (purity).

When electrical power is properly supplied to a load, the load should operate for years without a problem. When problems do occur with a load, the load must be serviced or replaced. Servicing or replacing a load is only a short-term solution when the problem is with the incoming power from the utility company. Problems with power systems or loads result in production equipment damage, costly downtime, and/or safety hazards.

Before a load is connected or when a load is serviced, the quality of the incoming power must be tested to ensure proper system and load operation. Power quality must also be tested as part of a preventive maintenance program. A *preventive maintenance program* is a combination of unscheduled and scheduled maintenance work required to maintain equipment in peak operating condition. Preventive maintenance is performed to keep electrical systems and loads operating with little or no downtime.

When troubleshooting power quality problems, an electrician must identify the source of the power problem. The problem might come from an outside source such as a lightning strike on the incoming power lines of the facility. The electrical distribution system of the facility or building can be the source of problems because of improper grounding and/or undersized conductors. Power quality problems are also caused by one or all of the loads connected to a power distribution system. Loads connected to a power distribution system can cause problems such as harmonics on the power lines. Most power quality problems show up when loads are connected to or disconnected from a power distribution system. **See Figure 4-1.**

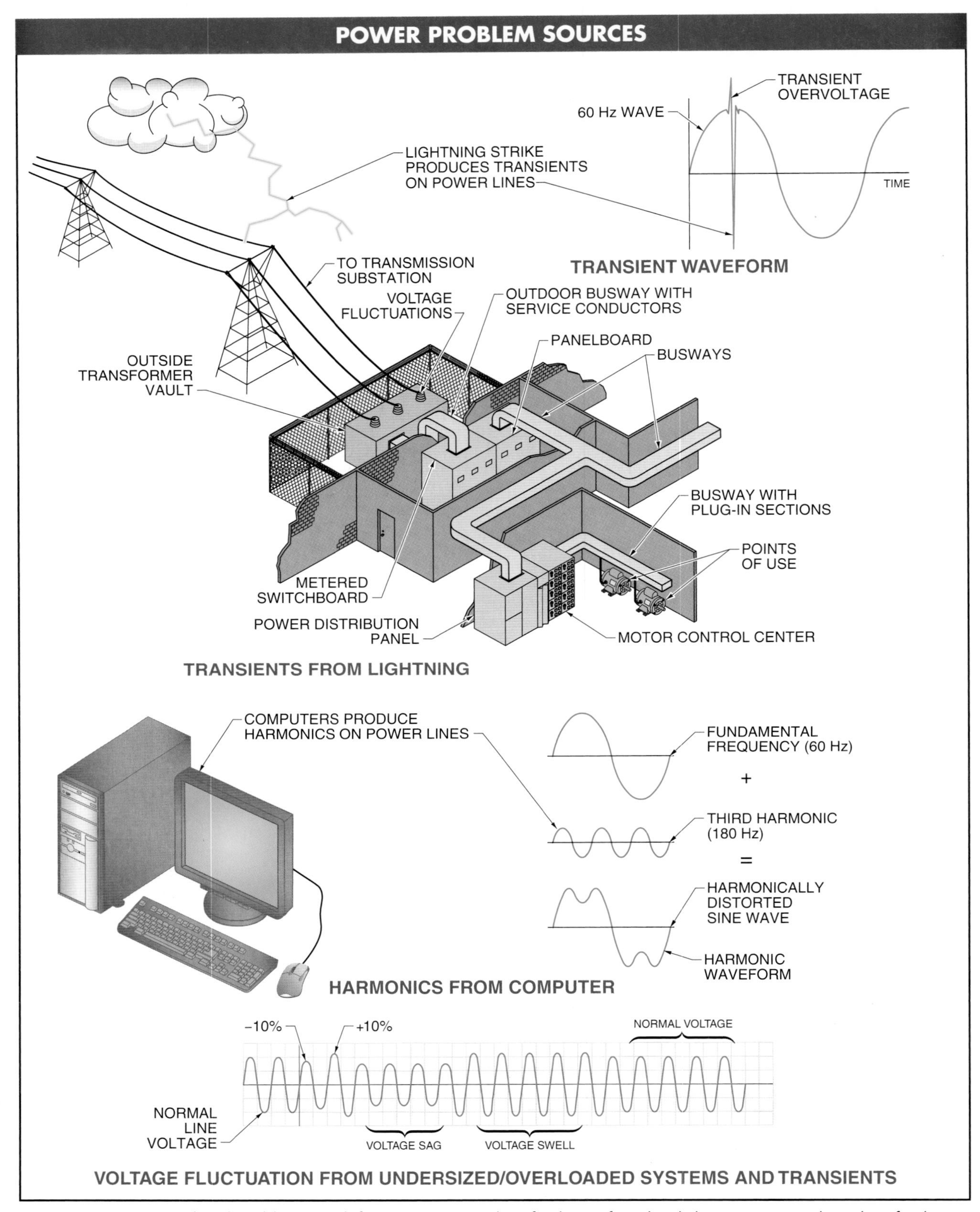

Figure 4-1. Power-related problems result from causes outside a facility or from loads being connected inside a facility.

Troubleshooting Tip

ISOLATED AND ELECTRICALLY FLOATING

Isolated and electrically floating are terms used to describe measurements in which the common test lead of a power quality meter is connected to a voltage other than earth ground.

To troubleshoot power quality problems in power distribution systems, measurements must be taken. Some measurements are taken and interpreted immediately, such as those for phase sequence, low or high voltages, or overloaded circuits. Other measurements (transient voltages, voltage sags, or voltage swells) require time in which to acquire a good understanding of the system and circuit problems, and time for acquiring usable data. Acquiring measurements over time is preferred because power quality problems vary and are only detectable at certain times of the day or during a certain sequence of production operations. Power quality problems are found using test instruments such as oscilloscopes, clamp-on ammeters, phase detectors, or multimeters, or by using specialized test instruments such as harmonic analyzers and power quality analyzers. **See Figure 4-2.**

POWER QUALITY PROBLEM TEST INSTRUMENTS

Power Quality Problem	Test Instrument Used to Detect Problem
Improper Phase Sequence, 3ϕ Lines	Phase Detector Tester
Phase Unbalance	Three-Phase Power Analyzer Meter
Voltage Unbalance, 3ϕ Lines	Digital Multimeter
Single Phasing	Clamp-On Ammeter
Current Unbalance, 3ϕ Lines	Clamp-On Ammeter
Overcurrent Problems	Clamp-On Ammeter
Overheated Equipment	Temperature Meter
Transients	Power Analyzer Meter
Harmonics	Harmonic Meter – Power Analyzer Meter
Power Interruptions	Voltmeter with MIN MAX – Power Analyzer Meter
Voltage Problems (Sags/Swells)	Voltmeter with MIN MAX – Power Analyzer Meter
Power Factor	Power Analyzer Meter
Noise	Oscilloscope – Power Analyzer Meter

Figure 4-2. Power quality problems are found using test instruments such as oscilloscopes, clamp-on ammeters, multimeters, temperature meters, and 3ϕ power quality analyzers.

POWER QUALITY TROUBLESHOOTING

Troubleshooting is a logical, step-by-step process used to find a problem in an electrical power system or process as quickly and easily as possible. Some power quality problems are found by taking only one or two measurements, such as measuring a low voltage condition on a branch circuit or receptacle (outlet). Other power quality problems require that several different measurements (voltage, harmonic, and transient) be taken, often at different locations and at various times. Following common troubleshooting steps helps organize the troubleshooting processes to determine the type of fault and the location of the electrical problem. **See Figure 4-3.** When troubleshooting a power quality problem, the following steps are performed:

- Obtain information
- Observe system and load operation
- Isolate problem(s)
- Test fault
- Document findings

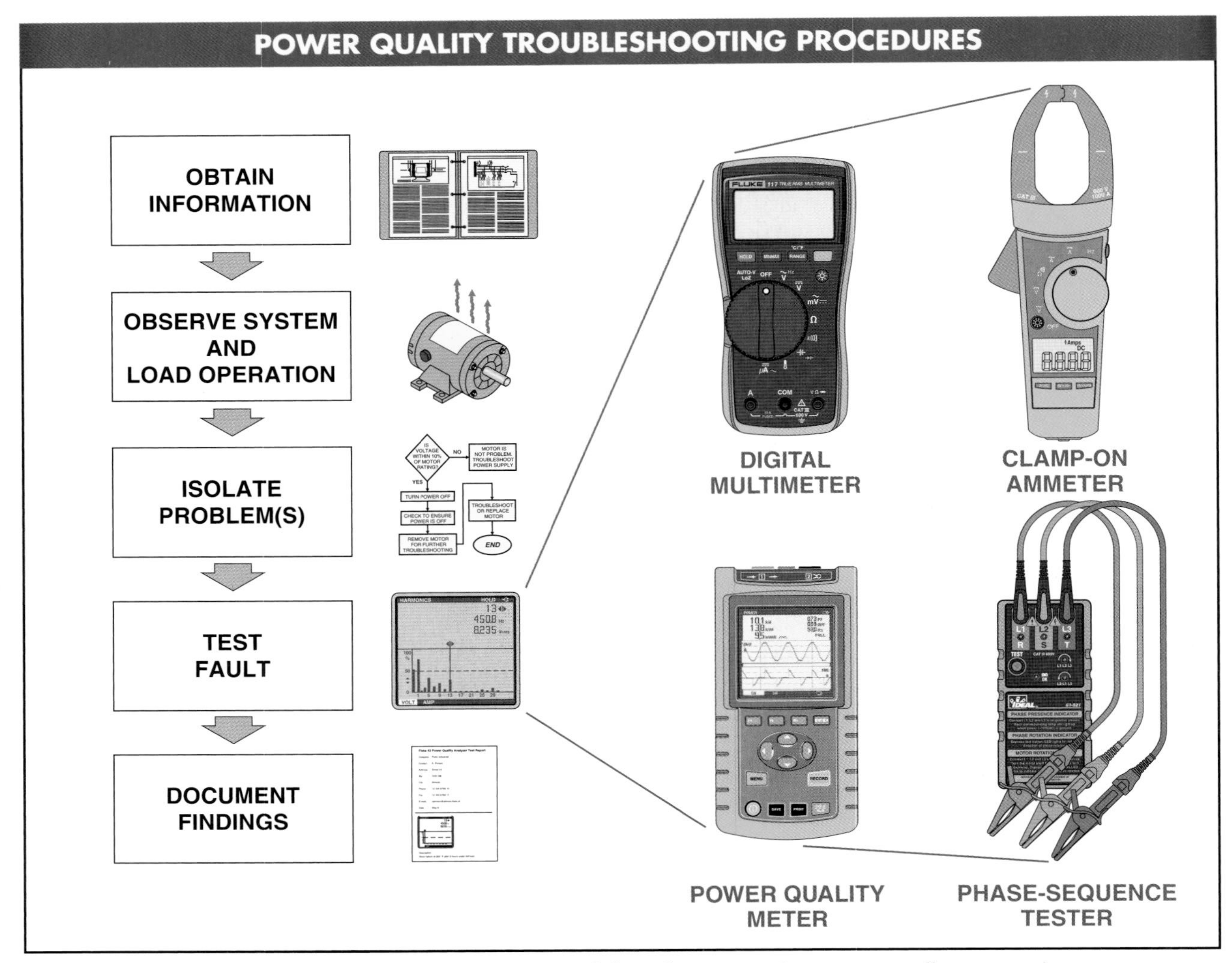

Figure 4-3. Troubleshooting a power distribution system follows basic steps for maximum efficiency and accuracy.

Obtain Information

Before taking any measurements or taking any corrective action, obtain as much information about the problem as possible or practical. System and circuit information can be obtained by applying the following procedures:

1. Walk through the building or facility and observe the type and number of linear loads (motors, heating elements, incandescent lamps, magnetic motor starters) and nonlinear loads (computers, printers, electric motor drives, programmable controllers, copy machines) used in the building. Look for possible sources of problems such as new loads being added to existing circuits, or new installations (circuits), as well as the type and quality of wiring. Nonlinear loads are a common source of harmonic problems. New installations can overload power system components (transformers and wiring) if additional power usage has not been allowed for or reserve power is not available. The quality of wiring to loads may be correct, or may be poor because of extension cord use and overloading. An electrical system that is poorly wired is more likely to have power quality problems. An electrical system that has undersized wires and transformers, plus distribution panels that are not grounded properly, will have power quality problems.

2. Gather technical information, such as prints of the power distribution system and electric bills from the utility company. When electrical prints are not available, make basic sketches and take notes about the condition of the system in various locations. Electric bills detail such information as any penalties (such as for poor power factor and demand charges) that the utility has imposed for poor power usage. Electric bills can also show when there was high power usage. An undersized electrical system will have more problems during high (peak) energy usage.
3. Collect as much information as possible about the nature of the problem being reported or the suspected problem. Ask questions such as, What is the nature of the problem? When does the problem occur? How often does the problem occur? What loads are affected? What actions have caused the problem in the past? An operator may not know anything about electricity or proper wiring, but an operator can often supply an electrician with valuable information about a problem. Always take notes when collecting information.

Observe System and Load Operation

Observe the system and load operation during startup, partial operation, full operation, and shutdown. Starting large 3ϕ motors or equipment causes voltage sags. Proper operation of electrical equipment during partial operation of a facility but not during full operation typically indicates that system transformers are overloaded. Observe if the entire system or only one circuit has a problem. Observations help determine the starting point of the troubleshooting process. Troubleshooting a problem with an entire system usually requires starting at the main power source and working to the loads. Troubleshooting a load problem usually requires starting at the load and working to the main power source.

Isolate Problems

Isolate the area or section of an electrical system that is suspected of having a problem. Narrow down the sources of the problem based on information obtained. Look for potential secondary causes. For example, overhead service entrances can be subjected to more problems than lateral underground services because overhead service entrances are affected by close lightning strikes and other weather-related problems. The more an electrician troubleshoots, the easier it will be for the electrician to isolate the problem area of a system. Often, power systems have primary and secondary problem areas.

Troubleshooting Tip

GENERAL MEASUREMENT GUIDELINES

When taking electrical power quality measurements, keep the following general limits in mind.

- **Voltage Range:** Within +5% to −10% of loads rating.
- **Voltage Unbalance:** Not more than 2% between the three motor leads.
- **Current Unbalance:** Not more than 10% between the three motor leads.
- **Voltage Total Harmonic Distortion:** Not more than 5%.
- **Current Total Harmonic Distortion:** Not more than 20%.

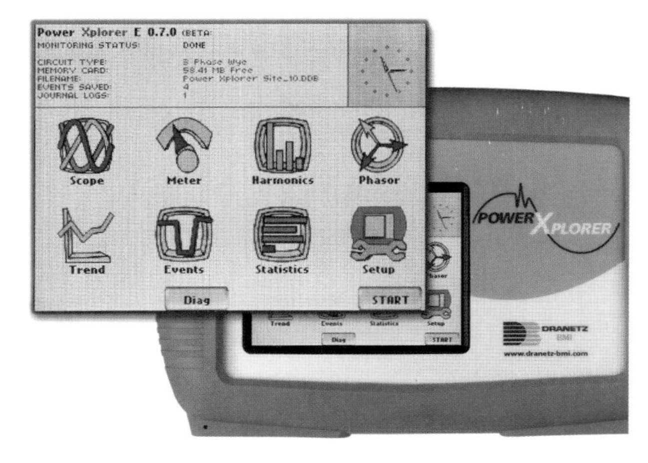

Dranetz-BMI

Some power quality meters have a graphic interface and are Wi-Fi enabled for quick access to power quality measurements.

Test Fault

Start at the potential source of a problem and take measurements. Taking measurements is the only way, other than visual observation, of understanding how an electrical system is or is not operating. Start with basic test instruments (voltmeter, current clamp, temperature meter) to acquire general information and use more advanced meters (power analyzer meter, harmonic meter) to acquire specific information. Take as many temperature, voltage, current, harmonic, and other power measurements as practical with various test instruments. Test instruments that include several functions are useful when taking various measurements. For example, when using a power analyzer meter, voltage measurements for sags and swells, harmonic measurements, power measurements, and measurements for transients can be taken. Record all measurements taken and displayed information (waveforms).

Document Findings

Even when documentation is not required, documentation is important because documentation records what work was done to a system and can be used as a reference for future problems. Documentation also helps identify problems that lead to failure over time. Record all malfunctions or problems identified and list all the components and parts of the system tested. List suggestions for correcting or eliminating each problem. Print out stored measurements and waveforms when possible. Compare the findings with data from the manufacturer of other, similar equipment that is operating properly.

THREE-PHASE POWER LINE PROBLEMS

When an electrical device fails or malfunctions, or an overcurrent protection device such as a fuse or circuit breaker removes power from a circuit, a problem exists in the electrical system. The problem can be caused by insulation breakdown, improper application, circuit overloading, or problems within the power distribution system. Power distribution system problems that can cause problems for loads include:

- improper phase sequence
- voltage unbalance
- current unbalance
- single phasing

Voltage unbalance and single phasing problems are found by taking voltage measurements with multimeters that display actual circuit voltage. Test instruments such as test lights and voltage indicators are not used because test lights and voltage indicators only approximate the voltage level in a circuit. Current unbalance is found by using clamp-on ammeters. Improper phase sequence is found by using phase sequence testers.

Technical Tip

Phase monitor relays are used to protect against phase loss, phase reversal, phase unbalance, undervoltage, and overvoltage.

Improper Phase Sequence

Improper phase sequence (phase reversal) is the changing of the sequence of any two phases in a three-phase system or circuit. Improper phase sequence reverses motor rotation. Reversing motor rotation can damage driven machinery and/or injure personnel. **See Figure 4-4.**

Phase reversals typically occur when modifications are made to power distribution systems or when maintenance is performed on electrical conductors or switching equipment. The National Electrical Code® (NEC®) requires phase reversal protection on all personnel transportation equipment such as moving walkways, escalators, and ski lifts. Phase reversal protection is also applied in applications in which an accidental phase reversal will damage equipment, such as in a pump application where the pump motor can operate in only one direction without causing damage to the pump.

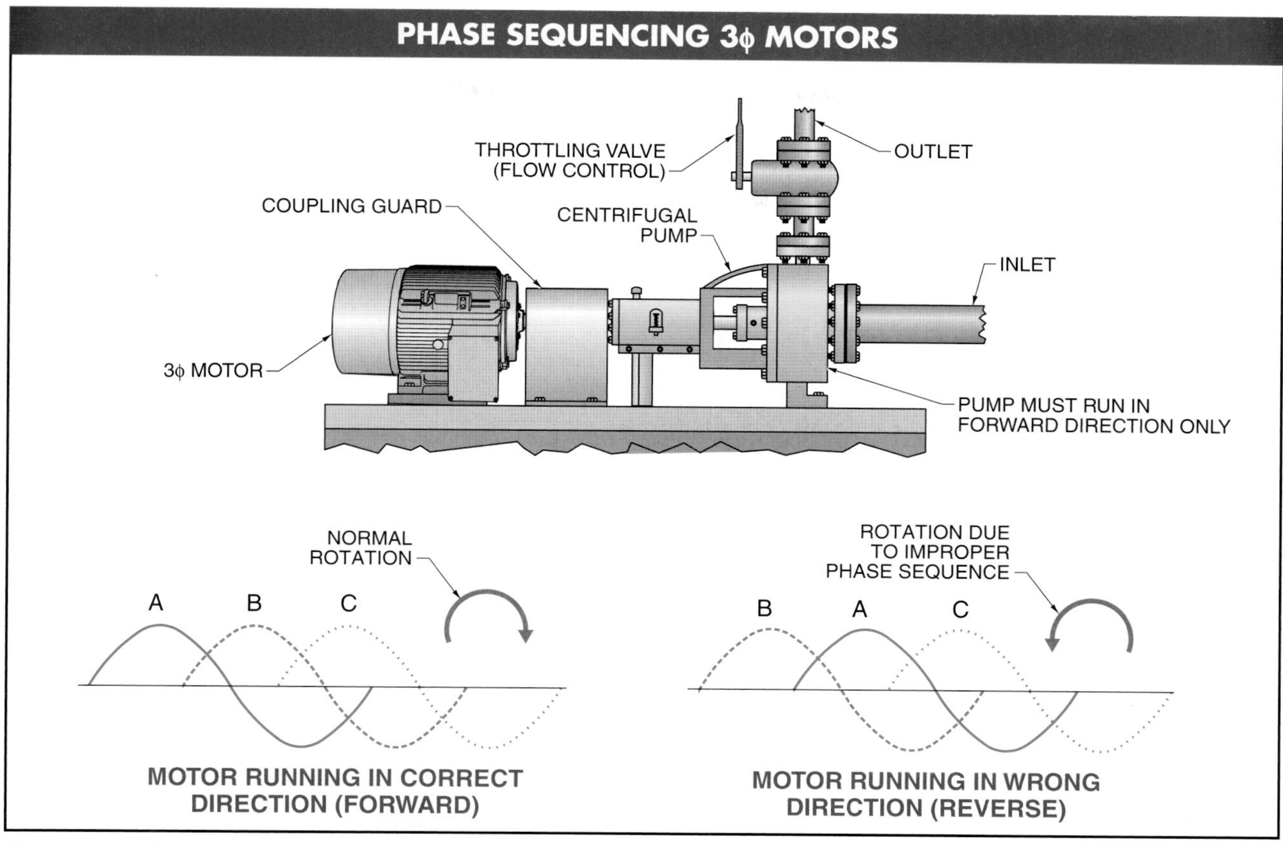

Figure 4-4. Phase sequencing of a 3ϕ motor requires the changing of any two phases (phase reversal) of the power supply to the motor.

Phase Sequence and Motor Rotation Tester

A *phase sequence tester* is a test instrument used to determine which of the three-phase power lines are powered and which power line is phase A, which is phase B, and which is phase C. The phase sequence is important because phase sequence determines the direction a motor shaft will rotate. Phase sequence testers are available in two types—phase sequence tester only, or phase sequence tester and motor rotation tester combined.

A phase sequence tester is connected to the supply lines, and lights on the tester indicate if voltage is present and if the power lines are in the correct order (phase A is A, phase B is B, and phase C is C). When the phases are not in the correct order, the phase sequence tester indicates that the phases are not in ABC order.

A motor rotation tester (when included as part of a phase sequence tester) is connected to the T1, T2, and T3 motor terminals before the motor is connected to power or mechanical equipment. Lights are used to indicate the ABC condition of the motor and circuit. Three lights indicate the presence of line power and two lights indicate motor shaft rotation. The clockwise or counterclockwise rotation of the motor shaft is determined by viewing the end of the motor shaft.

Technical Tip

Phase indicators and motor rotation testers vary in cost, depending on the desired frequency range.

Phase Sequence and Motor Rotation Applications

Three-phase circuits include three individual ungrounded (hot) power lines. The three power lines are referred to as phases A (L1, R), B (L2, S), and C (L3, T). Phases A, B, and C must be connected to switchboards and panelboards per NEC® requirements. The phases shall be arranged A, B, C from front to back, top to bottom, and left to right as viewed from the front of the switchboard or panelboard. **See Figure 4-5.**

If present, the high voltage leg of a three-phase circuit shall be considered phase B and colored orange (or clearly marked) per the NEC® when the switchboard or panelboard is fed from a 120/240 V, 3ϕ, 4-wire or delta-connected service. The color orange and the B designation marking are required because there is approximately 208 V between phase B (high voltage leg) and the neutral of the system. One hundred and ninety-five volts is considered an unreliable source of power because 195 V is too high for standard 115 V loads and too low for standard 230 V loads.

Conductors (wires) are covered with an insulating material that is available in various colors. The advantage of using various colors on conductors is that the identification (function) of each conductor can easily be determined. Some colors have definite meaning. For example, the color white is used as a neutral conductor. Other colors may have more than one meaning depending on the circuit. For example, a red conductor may be used to identify a hot wire in a 230 V circuit or a switched wire in a 115 V circuit.

Conductor color-coding helps conductor identification when balancing loads among the three phases and aids in the troubleshooting process. Conductor color-coding is also used in applications that do not require every conductor to be color-coded because standard circuit conductor color-coding is always used where applicable. Green or green with a yellow stripe is the standard color for grounding conductors. A solid green conductor is the most common color used for grounding conductors regardless of the voltage level or circuit type. A *grounding conductor* is a conductor that does not normally carry current, except during a fault (short circuit). The grounding conductor must be sized to carry full fault current.

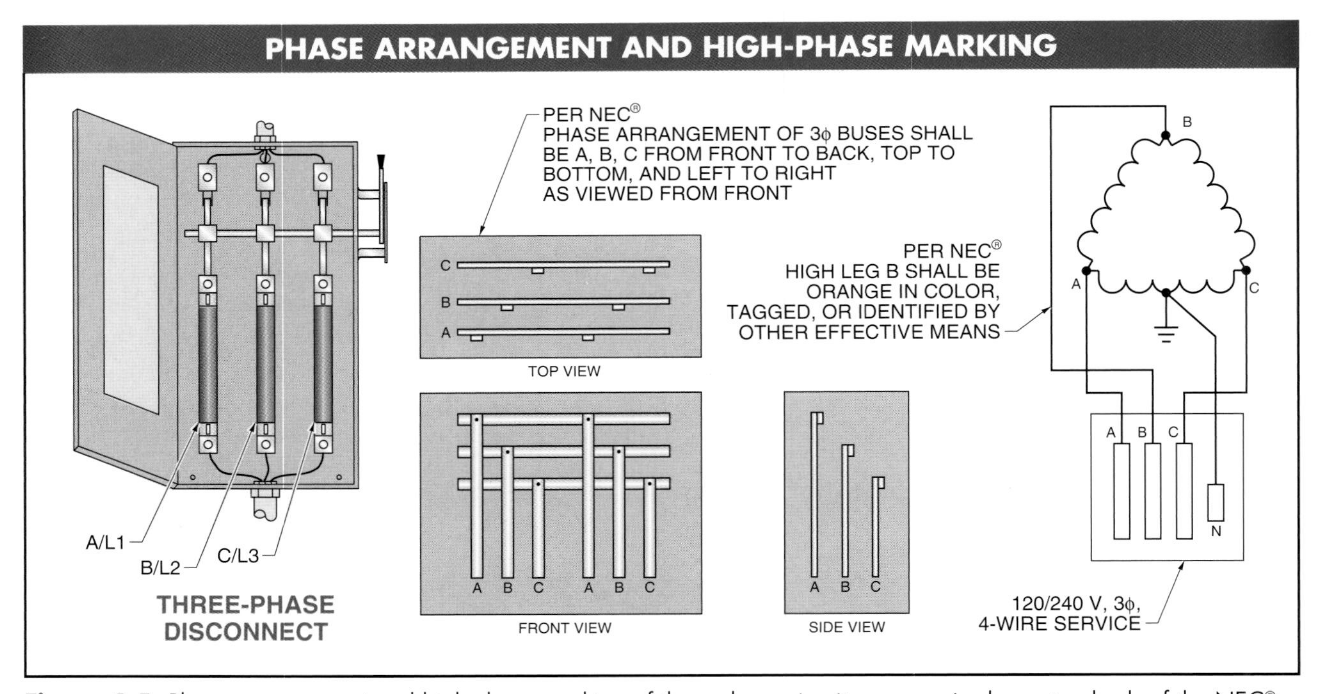

Figure 4-5. Phase arrangement and high-phase marking of three-phase circuits are required per standards of the NEC®.

The color white or natural gray is used for the neutral (grounded circuit) conductor. A *neutral conductor* is a grounded current-carrying conductor that carries current from the load back to the power source. A good ground reference exists at the power source. If the green wire and ground rod were removed, the circuit would still operate but would create a very dangerous condition. Neutral conductors are connected directly from the power source to the loads and never connected through fuses, circuit breakers, or switches except in cases of circuits involving motor fuel dispensing. The conductor colors green (or green with a yellow stripe) and white (or natural gray) are required per the NEC®.

Electrical circuits include ungrounded (hot) conductors in addition to neutral and grounding conductors. An *ungrounded conductor* is a current-carrying conductor that is connected to loads through fuses, circuit breakers, and switches. Ungrounded conductors can be any color other than white, natural gray, green, or green with a yellow stripe. Black is the most common color used for ungrounded conductors. Red, blue, orange, brown, and yellow are also used for ungrounded conductors. Although such colors as red, blue, orange, and yellow are used to indicate a hot conductor, the exact color used to indicate various typical hot conductors (A, B, C) of systems may vary. **See Figure 4-6.**

Hot conductors in a three-phase system may be identified with other types of markings in addition to color-coding. For example, a three-phase system may use three black conductors for each of the three hot conductors (A, B, and C), and tape band markings for each conductor. One band of colored identification tape may be placed around the line one (L1) black conductor when marking three-phase hot conductors. Likewise, two bands of colored identification tape may be placed around the line two (L2) black conductor, and three bands of colored identification tape may be placed around the line three (L3) black conductor.

Technical Tip

Article 250 of the National Electrical Code® covers all requirements for bonding and grounding of electrical installations.

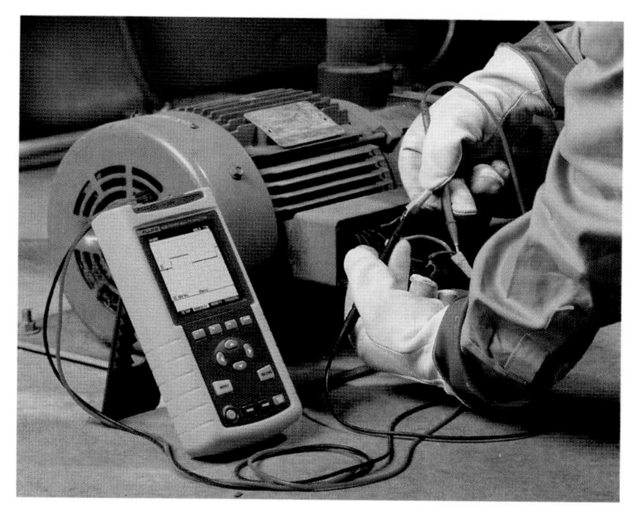

Fluke Corporation

Power quality meters and phase testers are used to identify the wiring phases of a 3ϕ motor.

Wire markers are typically used in place of tape bands. A wire marker is a preprinted peel-off marker designed to adhere when wrapped around a conductor. Wire markers resist moisture, dirt, and oil, and are used to identify conductors of the same color that have different meanings. For example, the three hot black conductors (L1, L2, and L3) of a three-phase system are marked with different numbered wire markers. Wire markers are used even when different colored conductors are used. Using wire markers in addition to color-coding conductors further clarifies the meaning of all conductors.

Conductors (wires) marked phase A, B, or C, should be phase A, B, and C throughout a building or facility. As wire is pulled through a system, phase sequence testers are used to guarantee that all three phases are properly marked. Phase A must be phase A everywhere in a power distribution system.

The clockwise or counterclockwise rotation of a 3ϕ motor depends upon the sequence of the power lines. Even when the sequence of the power lines is known, a motor may not operate in the direction an electrician thought because not all motors are wired the same way. Motor rotation testers are used to test the rotation of a motor before power is applied.

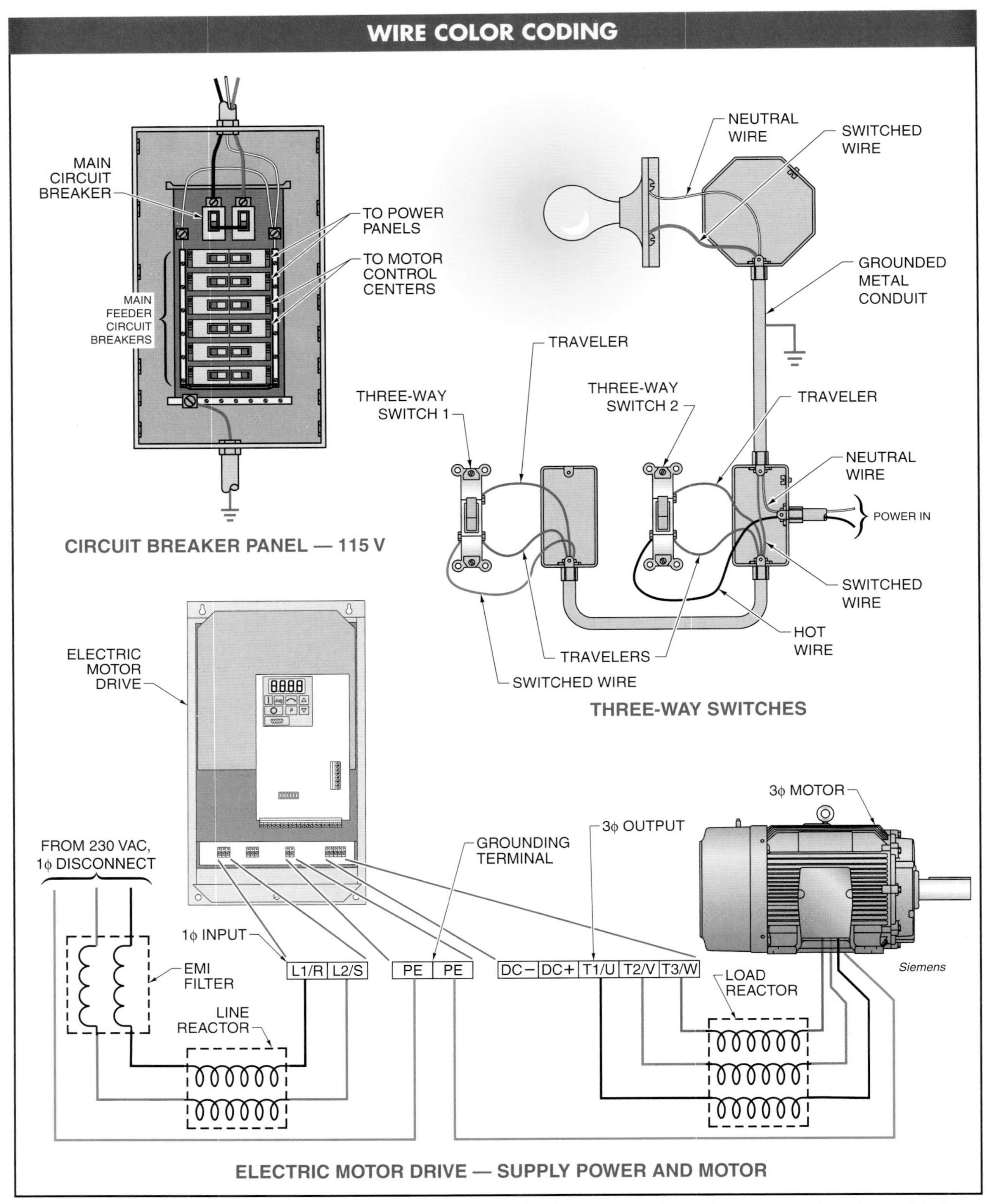

Figure 4-6. Conductor color coding allows for the balancing of loads among the three phases and aids in the troubleshooting process.

Phase Sequence Tester Measurement Procedures

The phase sequence of power lines can be verified using a phase sequence tester. Before taking any phase sequence measurements using a phase sequence tester, ensure the meter is designed to take measurements on the circuit being tested. **See Figure 4-7.** Refer to the operating manual of the test instrument for all measuring precautions, limitations, and procedures. To verify the phase sequence of the three three-phase power lines, apply the following procedures:

Safety Procedures

- Follow all electrical safety practices and procedures.
- Check and wear personal protective equipment (PPE) for the procedure being performed.
- Perform only authorized procedures.
- Follow all manufacturer recommendations and procedures.

1. Connect the three test leads from the phase sequence tester to the three power lines being tested. The test leads on a phase sequence tester are typically color-coded and include alligator clips. Connect phase A test lead to what should be power line phase A (L1/R), phase B test lead to B (L2/S), and phase C test lead to C (L3/T).
2. Verify that all three three-phase indicator lights are ON. If a light is not ON, use a voltmeter to test why the phase is not powered (for example, a fuse may be blown). When all phase indicator lights are ON, there is no open phase.
3. Check the phase sequence lamps. When the phases are in the correct order (the test leads are connected to the system A to A, B to B, and C to C), the phase tester "ABC" light will be ON. If the lines are not in the right order, the phase tester "BAC" light will be ON. Interchange the test leads until the "ABC" light is ON. Mark each phase line for proper identification.
4. Reconnect the phase sequence tester to test any additional parts of a system. Reconnect the tester because the phase sequence will change if the power lines are not correctly connected before and after a device.
5. Remove the phase sequence tester from the circuit.

Troubleshooting Tip

CALCULATING VOLTAGE USING OHM'S LAW

Ohm's law states that the voltage (E) in a circuit is equal to current (I) multiplied by resistance (R). To calculate voltage using Ohm's law, apply the following formula:

$E = I \times R$

where

E = voltage (in V)

I = current (in A)

R = resistance (in Ω)

EXAMPLE: What is the voltage of a circuit when the circuit has 22 A of applied current and 10.45 Ω of resistance?

$E = I \times R$

$E = 22 \times 10.45$

$E =$ **230 V**

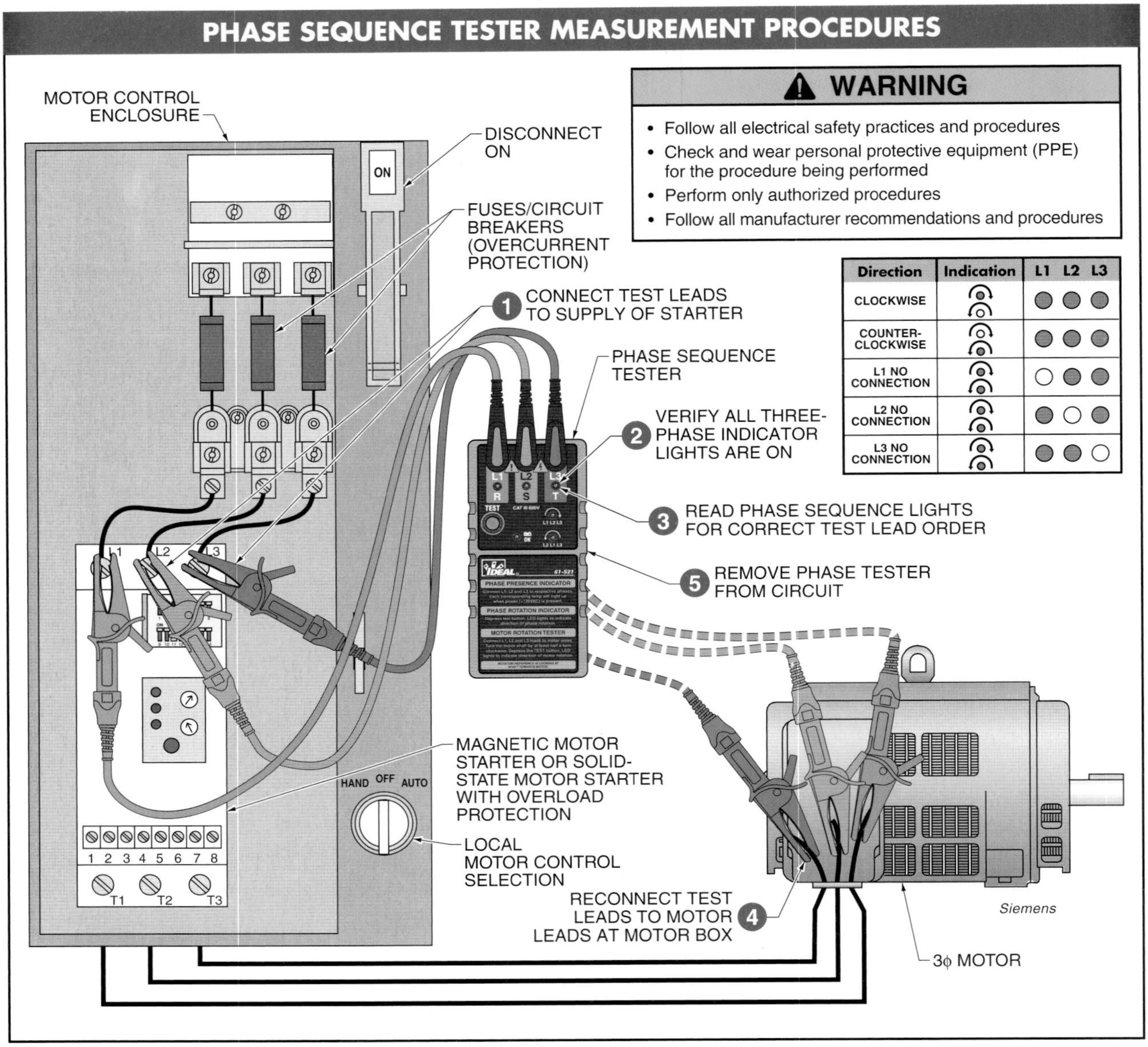

PHASE SEQUENCE TESTER MEASUREMENT PROCEDURES

⚠ WARNING

- Follow all electrical safety practices and procedures
- Check and wear personal protective equipment (PPE) for the procedure being performed
- Perform only authorized procedures
- Follow all manufacturer recommendations and procedures

Direction	Indication	L1 L2 L3
CLOCKWISE		● ● ●
COUNTER-CLOCKWISE		● ● ●
L1 NO CONNECTION		○ ● ●
L2 NO CONNECTION		● ○ ●
L3 NO CONNECTION		● ● ○

Figure 4-7. The phase sequence of power lines can be verified using test instruments such as phase sequence testers.

Motor Rotation Tester Measurement Procedures

The rotation of a motor can be tested using a motor rotation tester before the motor is connected to an electrical circuit or mechanical system. **See Figure 4-8.**

Before taking any frequency measurements using a digital multimeter, ensure the meter is designed to take measurements on the motor being tested. Refer to the operating manual of the test instrument for all measuring precautions, limitations, and procedures. Always wear required personal protective equipment and follow all safety rules when taking the measurement. To test the direction of motor rotation of a 3ϕ motor, apply the following procedures:

1. When the motor is not disconnected from the power lines, verify that the circuit power to the motor is OFF using a voltmeter (first test the voltmeter on a known energized circuit). Some motor rotation test instruments have a button that is pressed to test that all power in the system is OFF. When a system is powered, a warning light will turn ON.
2. Connect the test leads of the motor rotation tester to the motor input terminals T1 (U), T2 (V), and T3 (W).
3. Rotate the motor shaft clockwise by hand. When the clockwise light is lit, the motor will run in the clockwise direction when T1 is connected to L1, T2 to L2, and T3 to L3. When the counterclockwise light is lit, the motor will run in the counterclockwise direction when connected T1 to L1, T2 to L2, and T3 to L3.
4. When the motor must run in the opposite direction, interchange two power (or motor) lines. Interchanging 1 and 3 is the industrial standard.
5. Remove the motor rotation tester from the motor.

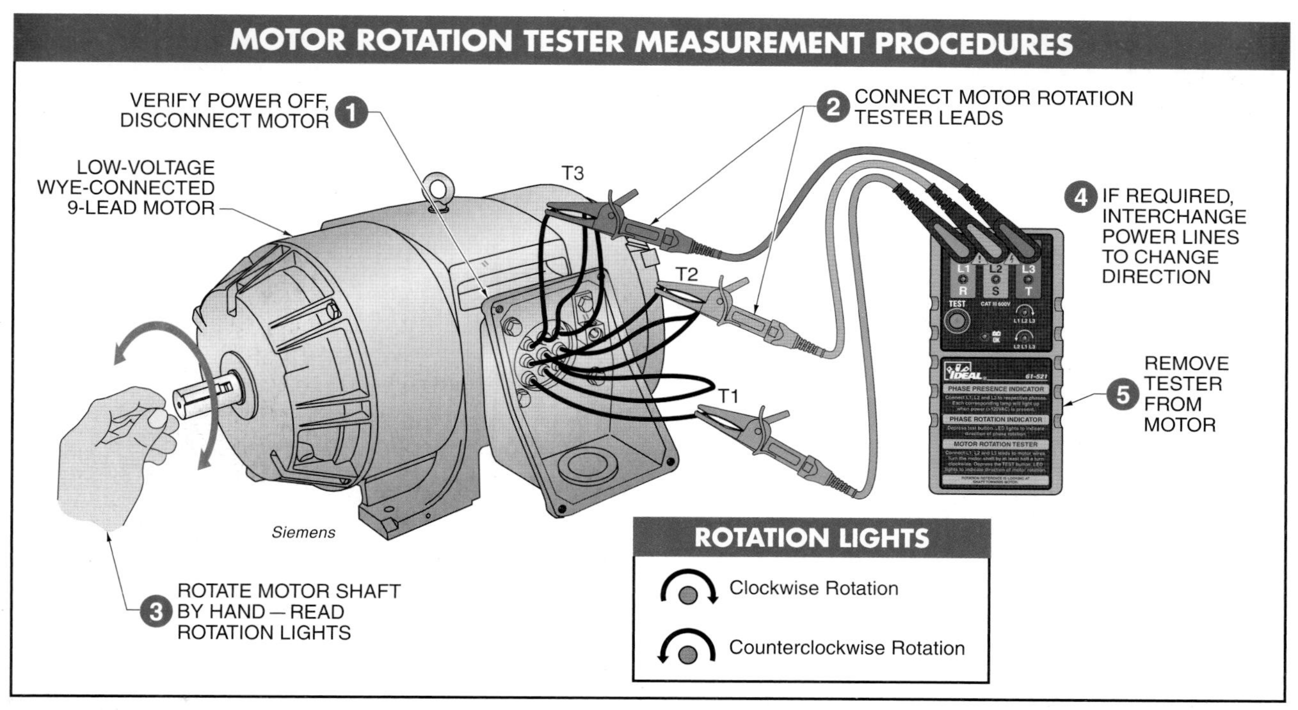

Figure 4-8. Motor rotation must be checked before connecting a motor to an electrical circuit or mechanical system.

Phase Unbalance

When three-phase power is generated and distributed, the three power lines are electrically 120° out of phase with each other. *Phase unbalance (imbalance)* is the unbalance that occurs when three-phase power lines are more or less than 120° out of phase. Phase unbalance of a three-phase power system occurs when single-phase loads are applied, causing one or two of the lines to carry more or less of the load. Electricians must balance the loads on three-phase power systems during installation. Phase unbalance begins to occur as additional 1ϕ loads are added to the system. The unbalance causes the three-phase lines to move out of phase so the power lines are no longer 120 electrical degrees apart. **See Figure 4-9.**

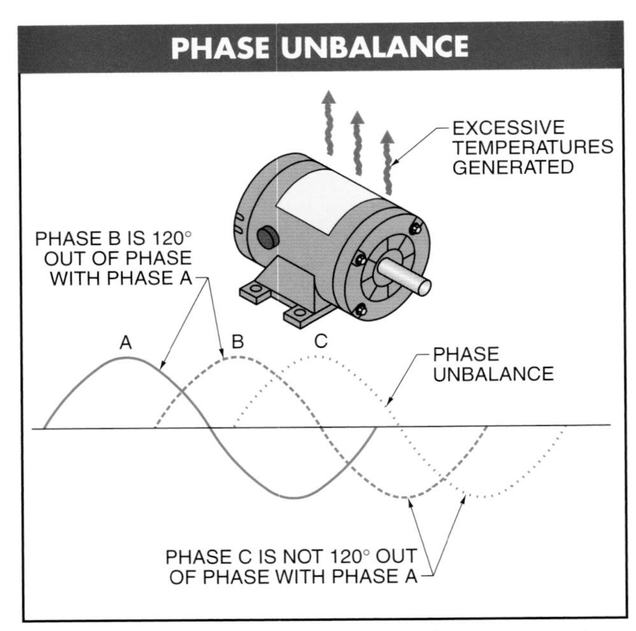

Figure 4-9. Phase unbalance occurs when 3ϕ power lines move out of phase as single-phase loads are added or connected to a system.

Phase unbalance causes 3ϕ motors to operate at temperatures higher than nameplate ratings. The greater the phase unbalance, the greater the temperature rise. High temperatures produce insulation breakdown and other related problems. A 3ϕ motor operating in an unbalanced circuit cannot deliver its rated horsepower. For example, a phase unbalance of 3% causes a motor to work at 90% of its rated power and the motor must be de-rated.

Phase unbalance in an electrical system can be thought of like the timing system in an automobile. A little off on the timing of an automobile causes a much larger power loss problem because the system is operating out of sequence.

Technical Tip

When leads from a three-phase generator that is to operate in parallel with other generators are being connected to the generator switch, the circuits must be phased out. The leads must be connected so that when the generator switch is thrown, each lead from the generator will connect to the corresponding lead from the other generators. This is done to prevent damage from an interchange of current when the machines are in parallel.

Three-Phase Power Analyzer Meter Measurement Procedures

Phase unbalance is tested using a three-phase power analyzer meter. A three-phase power analyzer meter is connected to a circuit so that the voltage and current on each of the three power lines are simultaneously monitored and compared. To measure voltage and current, a power analyzer meter is set to a screen or function setting that displays the three-phase sine waves and the phase relationship of each phase. **See Figure 4-10.**

Before taking any phase balance measurements using a power analyzer meter, ensure the meter is designed to take measurements on the circuit being tested. Refer to the operating manual of the test instrument for all measuring precautions, limitations, and procedures. Always wear required personal protective equipment and follow all safety rules when taking the measurement. To test the phase balance (or unbalance) on three-phase power lines, apply the following procedures:

1. Set the power analyzer meter to the screen or function setting with which phase sequence measurements are taken.
2. Connect the test leads of the power analyzer meter to the meter jacks as required.
3. Connect the power analyzer voltage test leads to phase A (L1, R), phase B (L2, S), and phase C (L3, T) as indicated by the meter.
4. Connect the power analyzer clamp-on ammeter lead(s) to phase A (L1, R), phase B (L2, S), and phase C (L3, T) as indicated by the meter.
5. Read the graphic display and information provided on the power analyzer meter display. If the meter is not displaying information on all three power lines, one or more of the power lines is not powered and/or the meter leads are not connected correctly. Verify that the test leads are properly connected. When the test leads are properly connected, test the power lines for voltage. To test the voltage on the three power lines, use a standard voltmeter, or the power analyzer can be used as a voltmeter by changing the function of the meter (not the meter connections).

6. When the power lines are more than 3% out of phase, there is a problem. Voltage unbalance can only be eliminated by balancing the loads on the power lines. To balance the power lines, take current measurements on all the single-phase power lines (use clamp-on ammeter). When one or two power lines are overloaded, remove the load from the overloaded line and place the load on a line that is not overloaded. After balancing the loads, re-test the phases for proper phase balance.
7. Remove the power analyzer meter from the circuit.

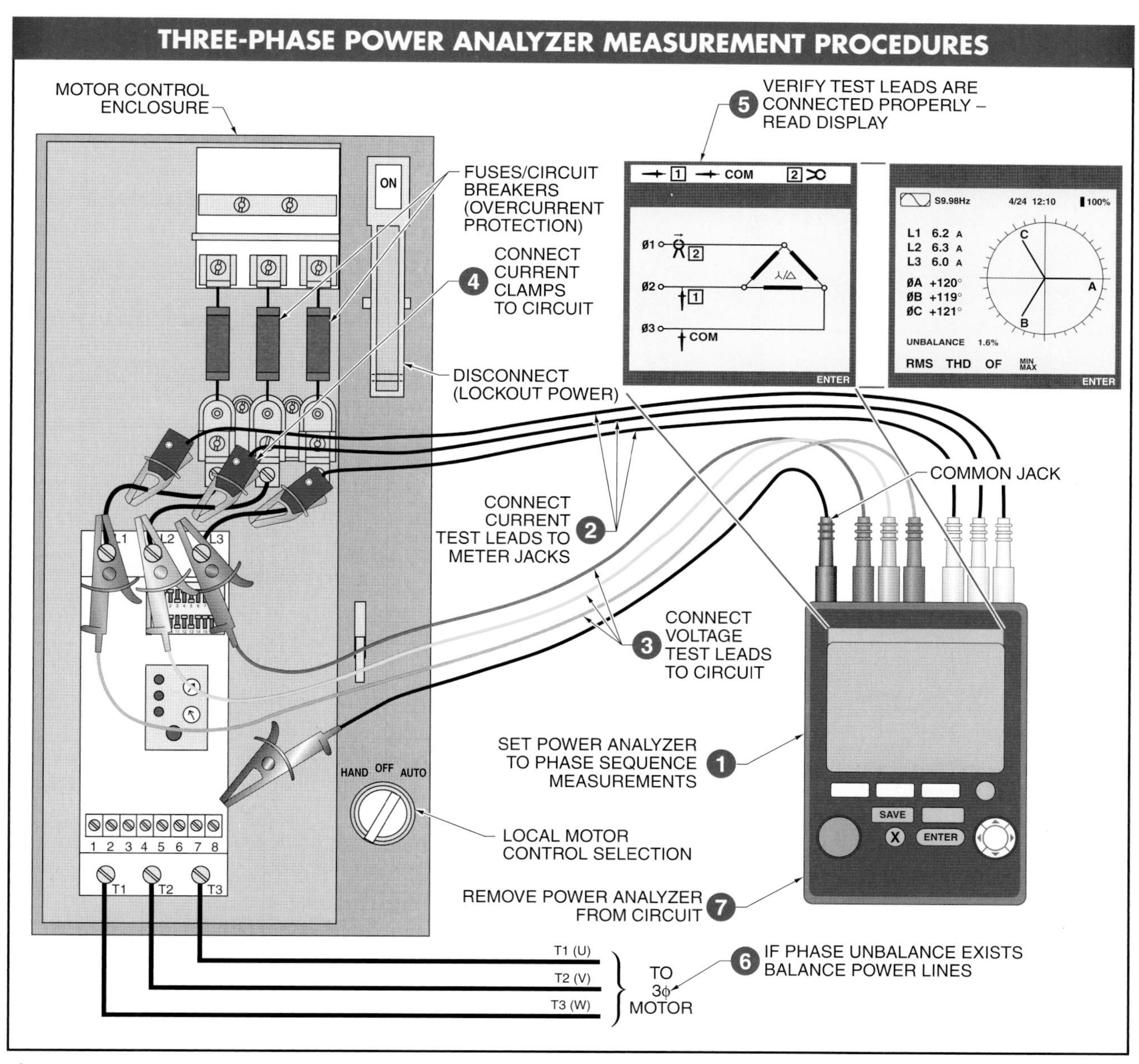

Figure 4-10. Power analyzers are connected to circuits so that voltage and current on each of the three phase lines are measured simultaneously and compared to check for phase unbalance.

Voltage Unbalance

Voltage unbalance (imbalance) is the unbalance that occurs when voltages at the terminals of a motor or other three-phase load are not equal. Voltage unbalance causes motor windings to overheat, resulting in thermal deterioration of the windings. When a 3ϕ motor fails due to voltage unbalance, one or two of the stator windings become blackened. The darkest stator winding is the winding with the largest voltage unbalance. Voltage unbalance creates current unbalance. Typically, for every 1% of voltage unbalance, current unbalance is 4% to 8%. **See Figure 4-11.**

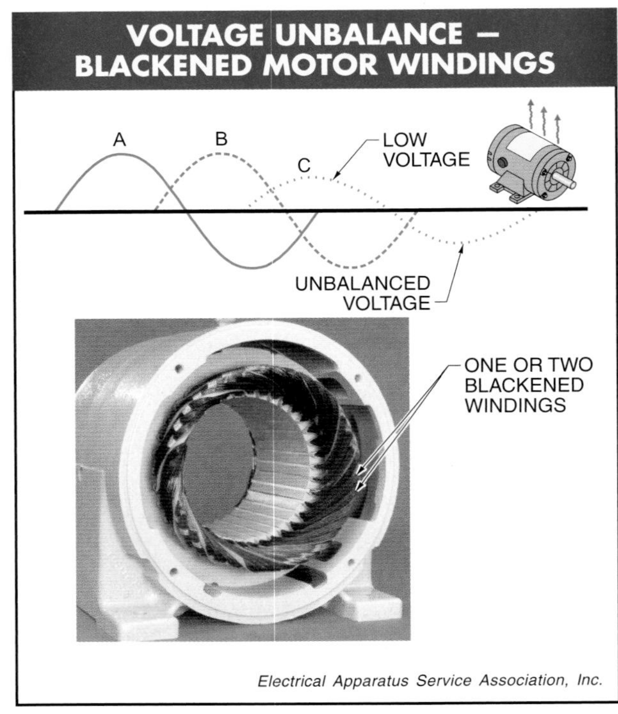

Figure 4-11. When examining blackened motor stator windings, the darkest winding is the winding with the largest voltage unbalance.

Technical Tip

When unbalanced voltages are detected, a blown fuse may be the cause. Often, a service or feeder fuse opens and goes undetected with the exception of a slight unbalance of line-to-line voltage. Voltage unbalance is a common occurrence on installations such as wastewater treatment plants, industrial manufacturing facilities, and warehouses where large pump, vacuum, and blower motors operate continuously.

Voltage unbalance of the three power lines must be tested during all service calls and during preventive maintenance. Typically, voltage unbalance must not be more than 1%. Whenever there is a 2% or greater voltage unbalance, the following steps must be taken:

1. Test the surrounding power system for excessive loads connected to one of the phases. Use a power analyzer meter to measure true power, apparent power, reactive power, power factor, and displacement power factor of the circuit. All transformers have a voltamp (VA) power rating. Circuits on transformers that are overloaded are a source of problems for loads.
2. Adjust the load on a motor by reducing the load or adjust the rating of the motor by oversizing the motor when voltage unbalance cannot be corrected.
3. Notify the utility company when voltage unbalance appears at the main service entrance. Testing for voltage unbalance at the main service entrance with loads ON and with loads OFF can isolate utility problems from outside or within a facility. When voltage unbalance improves as loads are removed, the problem is inside the facility, but do not eliminate an undersized utility transformer or feed conductor as the problem (have the utility company check transformers and feed conductors).

Voltage Unbalance Measurement Procedures

The primary source of voltage unbalance that is less than 2% is single-phase loads on a three-phase circuit. **See Figure 4-12.** A blown fuse can cause higher voltage unbalances on one phase of a three-phase capacitor bank. Voltage unbalance is found by applying the following procedures:

1. Measure the voltage between each incoming power line. Take measurements from L1 to L2, L1 to L3, and L2 to L3.
2. Add the voltages.
3. Find the voltage average by dividing by 3.

4. Find the voltage deviation by subtracting the voltage average from the voltage with the largest deviation.
5. Find the voltage unbalance by applying the following formula:

$$V_u = \frac{V_d}{V_a} \times 100$$

where

V_u = voltage unbalance (in %)

V_d = voltage deviation (in V)

V_a = voltage average (in V)

100 = constant

Current Unbalance

Current unbalance (imbalance) is the unbalance that occurs when current on each of the three power lines of a three-phase power supply to a 3ϕ motor or to other 3ϕ loads is not equal. Current unbalances or even high current unbalances can be created by small voltage unbalances. High current unbalance results in excessive heat and in turn insulation breakdown. A 2% voltage unbalance typically results in an 8% to 16% current unbalance.

Current unbalances must not exceed 10%. Any time current unbalance exceeds 10%, the system must be tested for voltage unbalance. Any time a voltage unbalance of more than 1% exists, the system must be tested for current unbalance.

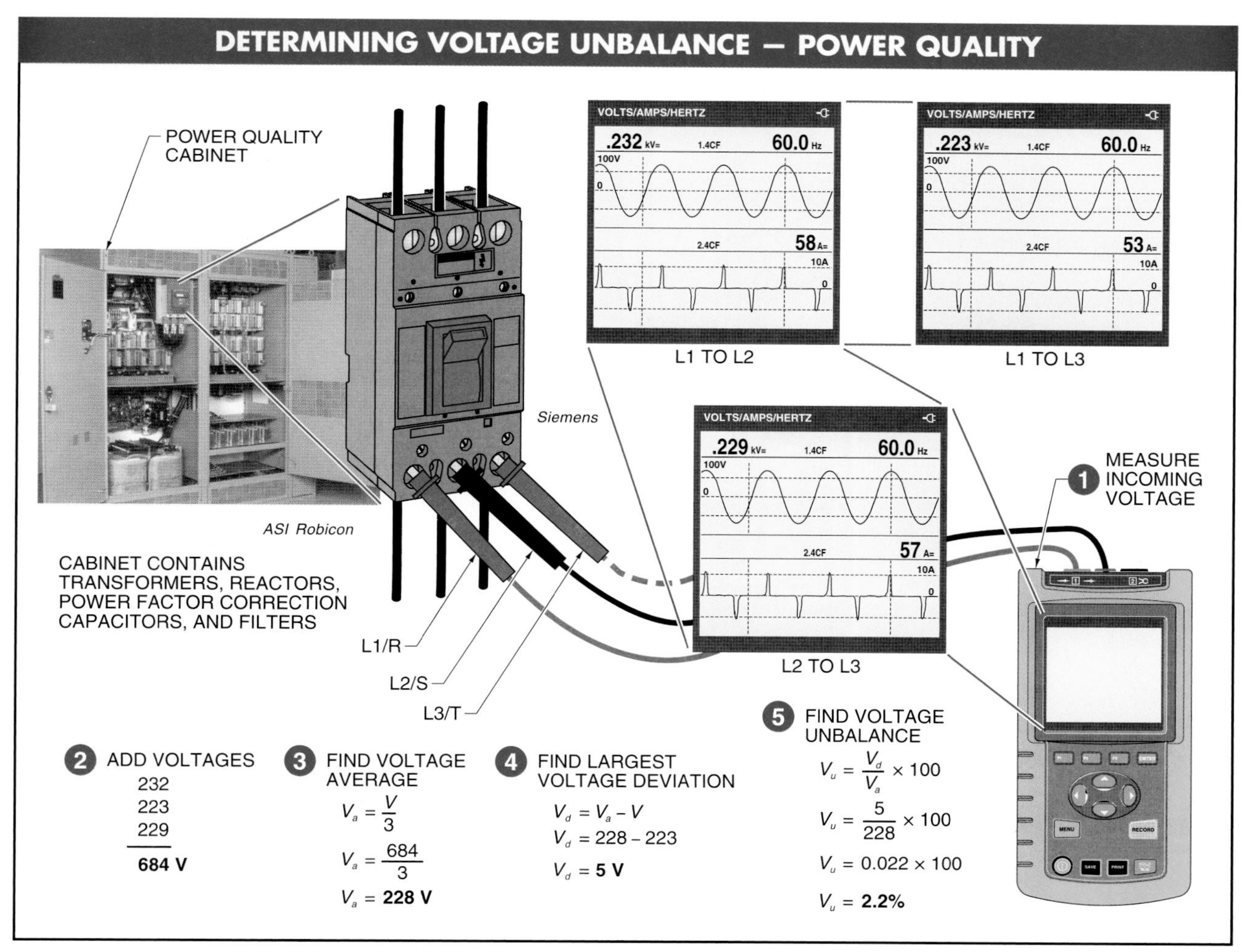

Figure 4-12. The primary source of voltage unbalances that are less than 2% is single-phase loads on 3ϕ circuits.

Current Unbalance Measurement Procedures

Current unbalance is found using the same method as finding voltage unbalance, except current measurements are taken. **See Figure 4-13.** Current unbalance is found by applying the following procedures:

1. Measure the current on each of the incoming power lines. Take current measurements of L1, L2, and L3.
2. Add the currents.
3. Find the current average by dividing the sum by 3.
4. Find the current deviation by subtracting the current average from the current with the largest deviation.
5. To find current unbalance, apply the following formula:

$$I_u = \frac{I_d}{I_a} \times 100$$

where

I_u = current unbalance (in %)

I_d = current deviation (in A)

I_a = current average (in A)

100 = constant

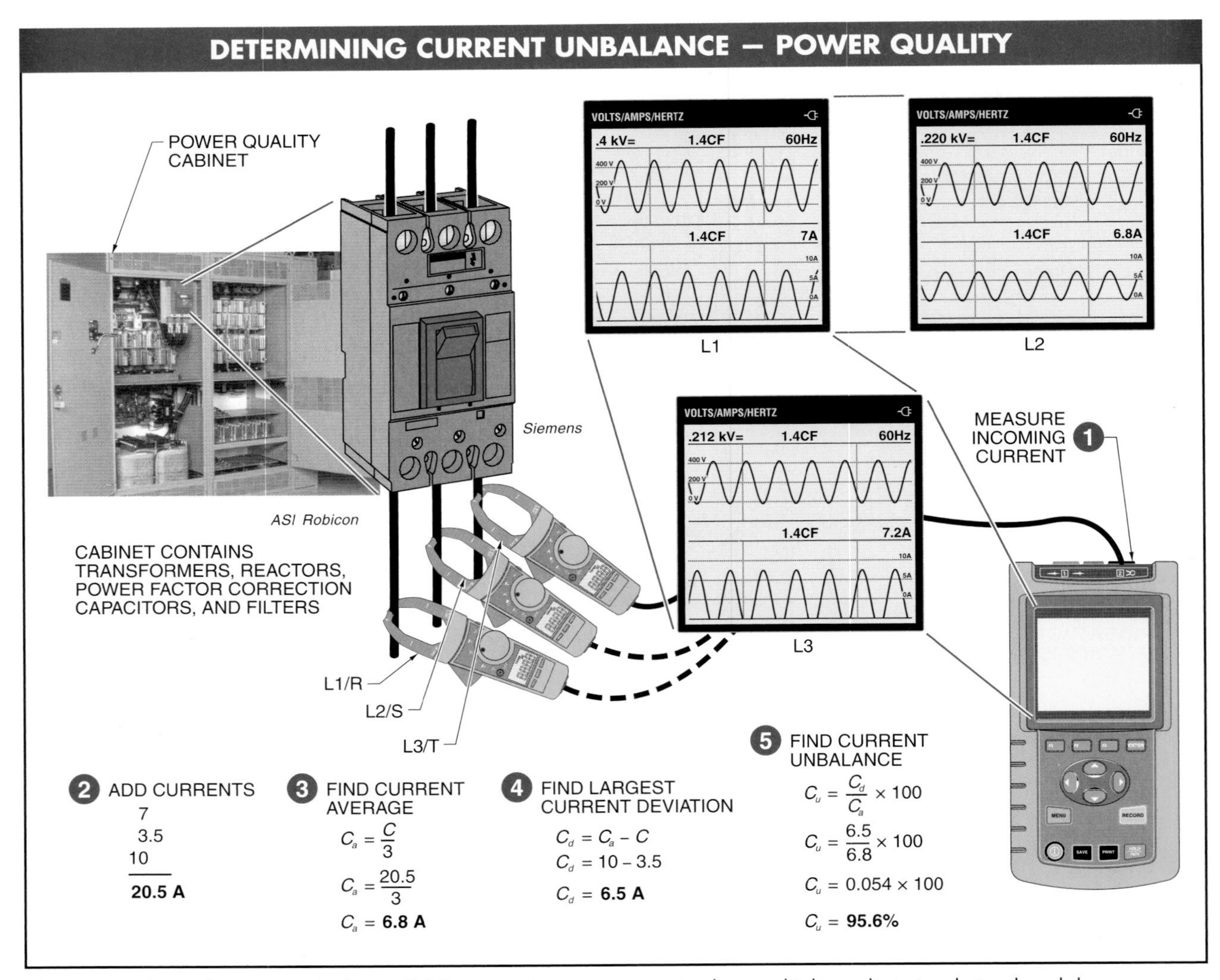

Figure 4-13. High current unbalance (10% or more) causes excessive heat, which results in insulation breakdown.

Single Phasing

Single phasing is the operation of a three-phase load on two phases due to one phase being lost. For example, single phasing occurs when one of the three-phase lines to a 3ϕ motor (T1, T2, or T3) does not deliver voltage to the motor. Single phasing is the maximum condition of voltage unbalance. **See Figure 4-14.**

Single phasing occurs when one phase opens on either the primary or secondary power distribution system. Common causes of single phasing include blown fuses, mechanical failure within switching equipment, or a lightning strike on one of the power lines. Single phasing can go undetected on systems because 3ϕ motors continue to operate on two phases in most applications. A motor will not start on two phases, but will continue operating if a phase is lost when the motor is already operating. When a motor is not properly protected, single phasing will cause a motor to burn out because the motor draws all required current from the two remaining powered lines.

Measuring voltage at a motor does not always detect a single phasing condition because the open winding in the motor generates a voltage that is almost equal to the phase voltage that is lost. The open winding of the motor acts as the secondary side of a transformer, and the two windings that are still connected to the power source act as the primary side of a transformer. To test for a motor or load that is single phasing, both voltage and current measurements must be taken.

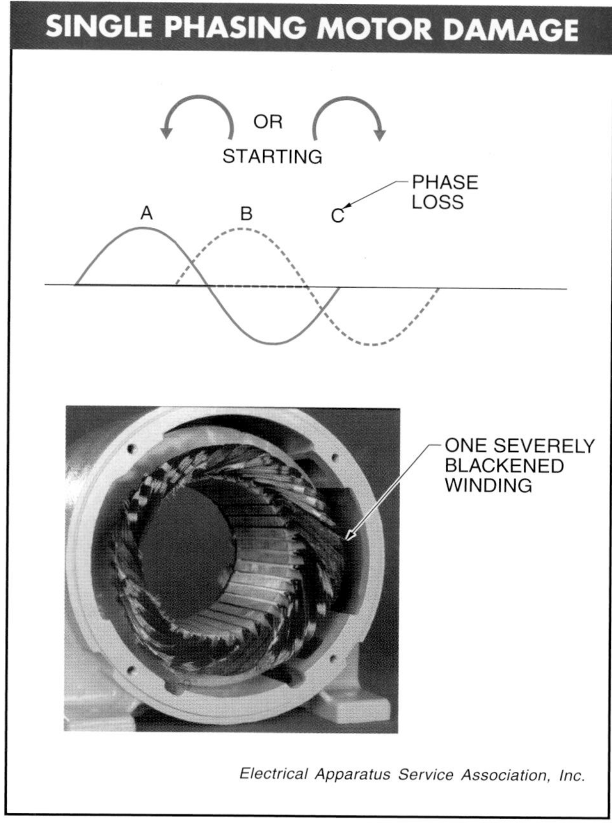

Figure 4-14. Single phasing occurs when one phase of a 3ϕ motor supply does not deliver voltage to the motor.

Troubleshooting Tip

POWER QUALITY METERS

Power quality meters measure and display active power (W), apparent power (VA), reactive power (VAR), power factor (PF), displacement power factor (DPF), and frequency.

Single Phasing Measurement Procedures

Single phasing measurements are best performed when a motor is running. An operating motor allows voltage and current measurements to be taken and checked against motor nameplate ratings. A motor that is operating within nameplate ratings is not required to be shut down unless there is some other obvious problem (noise, vibration, or overheating). **See Figure 4-15.**

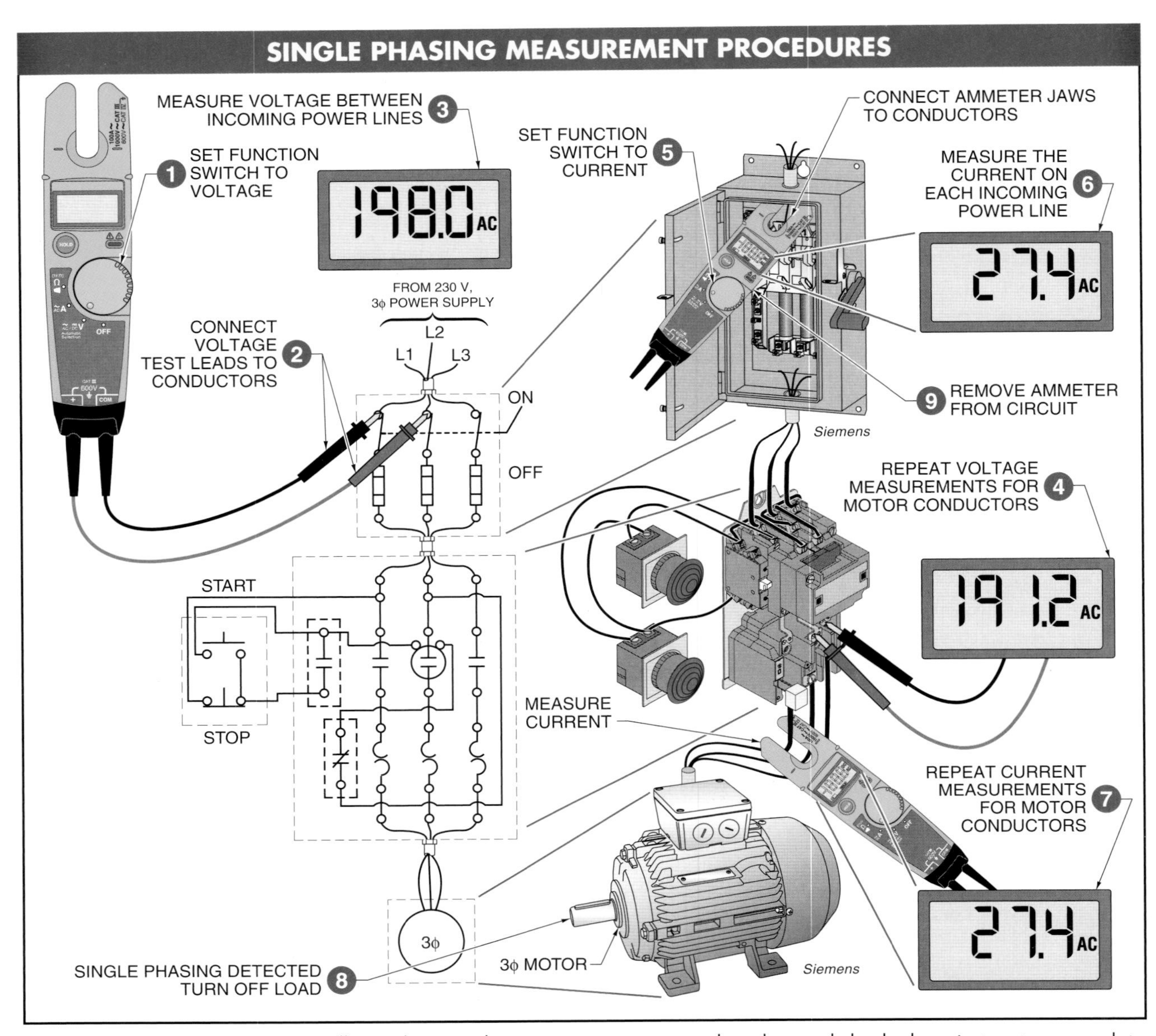

Figure 4-15. Operating motors allow voltage and current measurements to be taken and checked against motor nameplate ratings for single phasing.

Before taking any phase voltage and current measurements using a digital multimeter, ensure the meter is designed to take measurements on the circuit being tested. Refer to the operating manual of the test instrument for all measuring precautions, limitations, and procedures. Always wear required personal protective equipment and follow all safety rules when taking the measurement. Single phasing is found by applying the following procedures:

1. Set the function switch of the meter to AC voltage.
2. Connect the voltage test leads to the circuit being tested.

3. Measure the voltage between each incoming power line (L1, L2, and L3) and/or motor terminal (T1, T2, and T3), depending upon what part of the circuit is being tested. Take measurements from L1 to L2, L1 to L3, and L2 to L3. When the measured voltages are within 1%, there is no problem. When the measured voltage is within 2% to 5%, there is a voltage unbalance problem. When the measured voltage is greater than 5%, there is probably a single phasing problem. A single phasing problem on a 230 V system with L2 lost will show readings as follows:

   L1 to L2 = approximately 190 V
   L2 to L3 = approximately 190 V
   L1 to L3 = 230 V

   The two power lines (L1 and L3) that have the normal voltage readings are the good lines. The ability to determine a single phase condition by using voltage measurements is helpful when a current meter is not available.
4. Repeat voltage measurements for motor conductors.

   T1 to T2 = approximately 190 V
   T2 to T3 = approximately 190 V
   T1 to T3 = 230 V
5. Set function switch of the meter to AC current.
6. Measure the current at each incoming power line (L1, L2, and L3). The three phases of current supplying power to a motor must be within 10% of each other and within the nameplate current rating of the motor. When a motor is single phasing, the current on two of the lines will be high and the current on the third line will be zero. The two lines with the higher current are the good lines and the current is higher because the two good lines have to carry all the working current of the motor. The line with no current is the open line.
7. Repeat current measurements for motor conductors (T1, T2, and T3).
8. When a single-phase condition is detected, turn the load OFF and test for a blown fuse, open switch, burned contact, or lost connection.
9. Remove the ammeter from the circuit.

LOAD TYPES

Many types of loads are connected to electrical circuits. A *load* is any electrical device that converts electrical energy into some other form of energy, such as light, heat, sound, or mechanical. Mechanical energy is movement by linear motion (produced by solenoids) or rotary motion (produced by motors). There are two basic types of electrical loads, linear loads and nonlinear loads. Linear loads have been around since the first electrical circuits were used. Nonlinear loads started to become popular in the 1980s and have grown in numbers and percentage of total loads ever since. Nonlinear loads cause power quality problems because nonlinear loads produce harmonics.

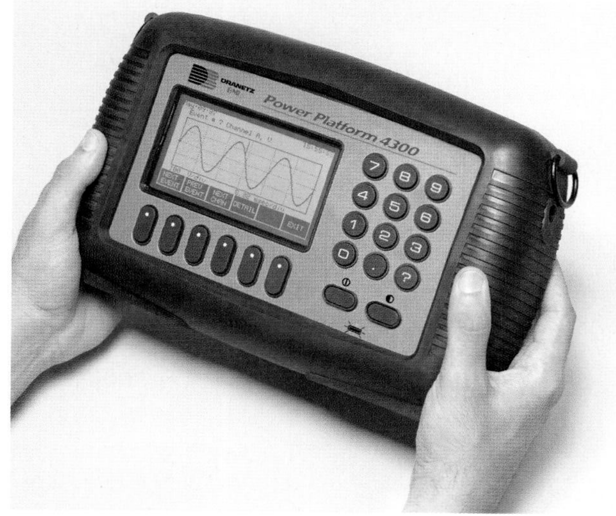

Dranetz-BMI

Linear loads such as incandescent lamps and heating elements produce sinusoidal waveforms.

Electrical problems caused by linear loads are typically found by testing a circuit with standard voltmeters and ammeters. Electrical problems caused by nonlinear loads often require power analyzers to determine the exact electrical problem. Power analyzers display voltage and current waveforms in addition to measuring voltage, current, power, harmonics, and other circuit characteristics.

Linear Loads

A *linear load* is any load in which current increases proportionately as voltage increases and current decreases proportionately as voltage decreases. In an AC circuit that includes a linear load, voltage and current are both sinusoidal even when both are out of phase. A *sinusoidal waveform* is a waveform that is consistent with a pure sine wave. Pure resistance, inductance, and capacitance loads, such as incandescent lamps, heating elements, motors, alarms (bells and horns), solenoids, and relay coils are linear. Linear loads cause problems to a power distribution system when shorted out, when oversized for the power distribution system, or when they are not operating properly. Linear loads do not cause problems like harmonic distortion to a power distribution system. **See Figure 4-16.**

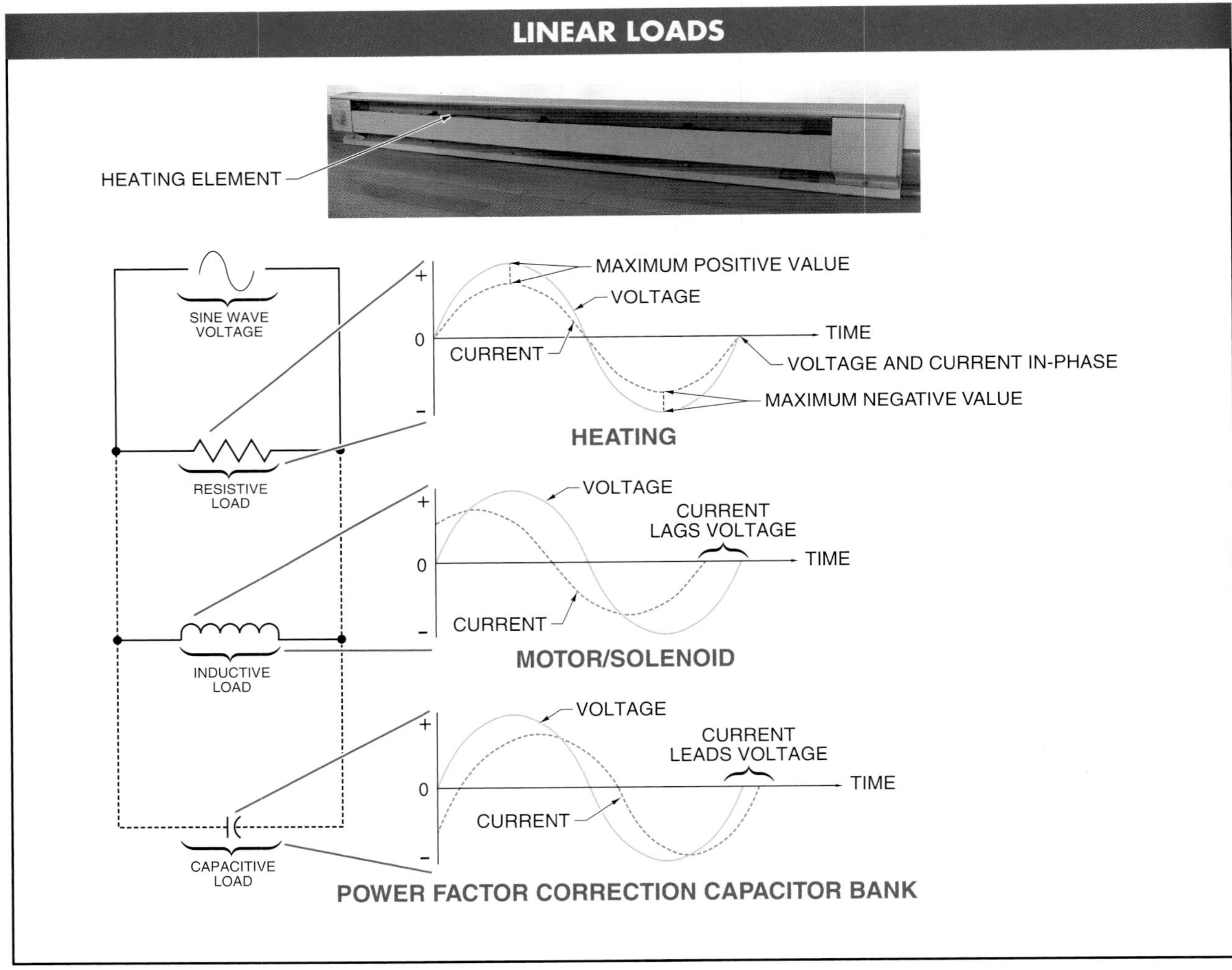

Figure 4-16. Linear loads have current increasing proportionally as voltage increases, and current decreasing proportionally as voltage decreases.

Nonlinear Loads

A *nonlinear load* is any load in which the instantaneous load current is not proportional to the instantaneous voltage. Voltage and current are not proportional because nonlinear loads draw current in short pulses, even when the source voltage is a pure sine wave. **See Figure 4-17.**

Technical Tip

Nonsinusoidal waveforms are current and voltage variations that vary with time but not sinusoidally. Nonlinear devices such as PLCs, personal computers, electric motor drives, halide lighting, and transistors usually cause these variations. Circuits with nonsinusoidal waveforms are analyzed by breaking the waveform into a series of sinusoidal waves of different frequencies known as a Fourier series. Each frequency circuit is then analyzed by typical AC circuit techniques and the results are combined to give the total response.

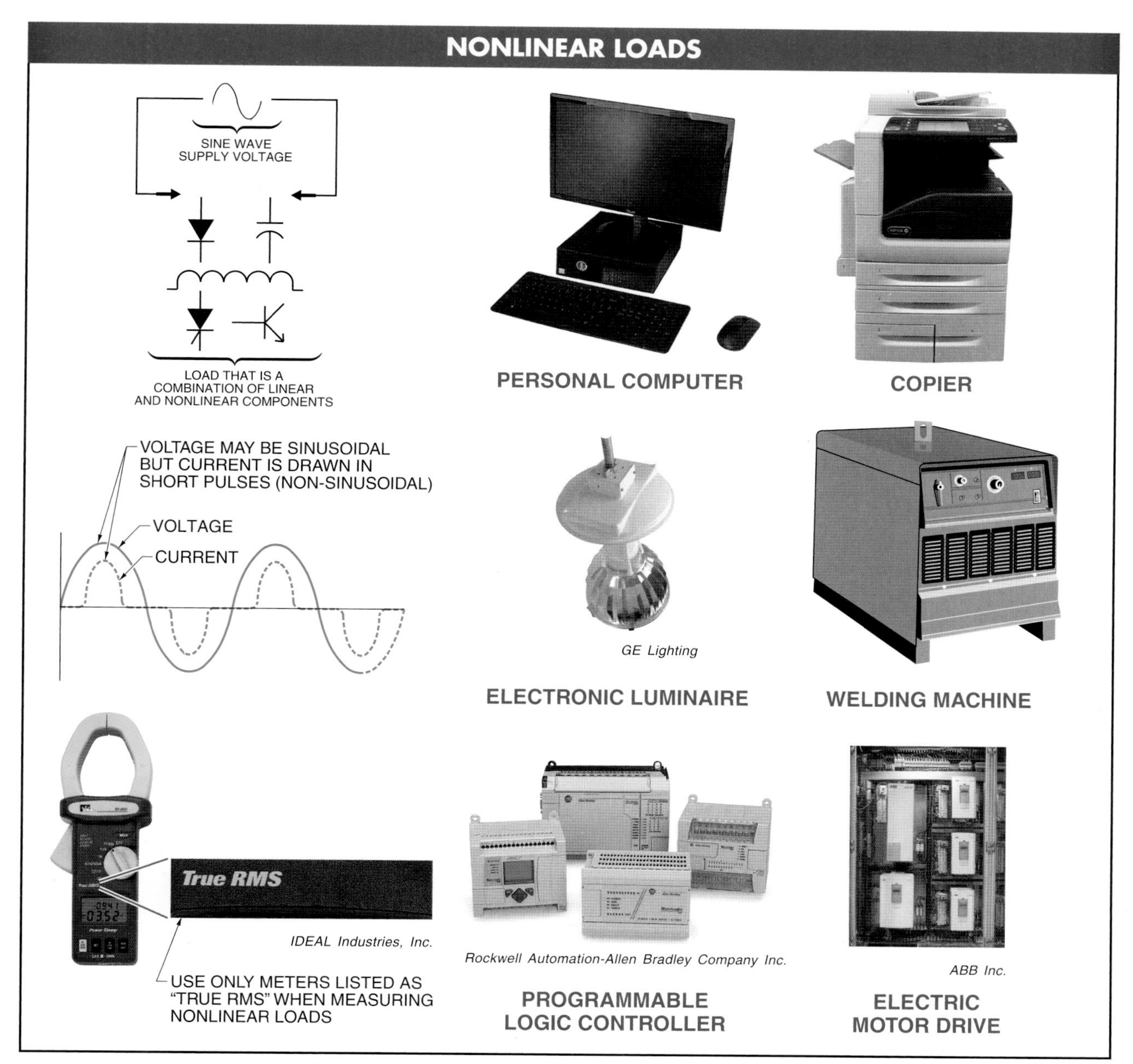

Figure 4-17. Nonlinear loads do not have the instantaneous current proportional to the instantaneous voltage.

There are both single-phase nonlinear loads and three-phase nonlinear loads. Single-phase nonlinear loads include personal computers, photocopiers, printers, televisions, electronic lighting ballasts, energy-efficient lamps, electric welding equipment, most medical equipment, and programmable logic controllers (PLCs). Three-phase nonlinear loads include solid-state electric motor drives and uninterruptible power supplies (UPSs). All nonlinear loads affect the power distribution system because nonlinear loads produce harmonics on the three three-phase power lines. Harmonics cause problems such as neutral conductor overheating, transformer overloading, and equipment failure.

The problem created by single-phase nonlinear loads is in the power supply used to convert AC voltage into DC voltage. The first section of a single-phase power supply is typically a combination of a diode bridge (used to change AC voltage into DC voltage) and capacitors (used to smooth out and store the rectified DC voltage). Any nonlinear load, such as a computer, draws power from the power-supply capacitors. As the load draws power, the charged capacitors start to decrease in charge. To stay charged, capacitors draw power only from the peak of the voltage sine wave. Drawing power only at the peaks creates gaps, leaving the current to be drawn in pulses. Because the current is drawn in pulses, the current waveform is no longer sinusoidal.

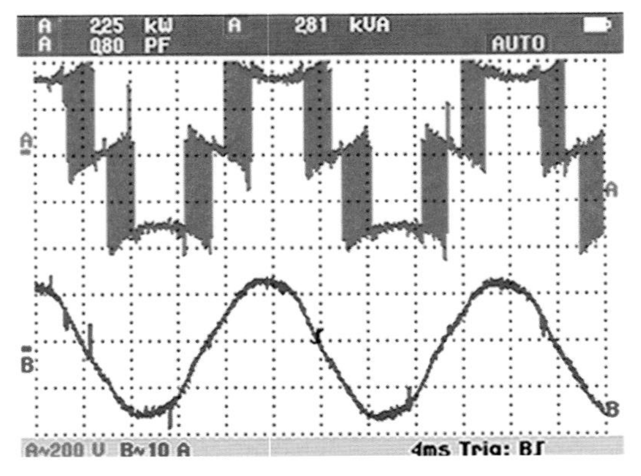

Fluke Corporation

Graphic display meters provide real-time information (graphic and numerical) on the operation of power distribution systems.

When troubleshooting nonlinear loads, electricians must always use test instruments marked "TRUE RMS." True RMS test instruments can measure nonsinusoidal waveforms (distorted sine waves) produced by nonlinear loads.

WARNING

Never use the current probes of a power quality analyzer on circuits rated higher than 600 V in overvoltage according to Category III (CATIII) of EN/ICEC61010-1.

GRAPHIC DISPLAYS

In addition to measuring and reading electrical properties (voltage and current) with basic test instruments, electrical properties are also measured and displayed using meters with graphic displays such as oscilloscopes and power analyzers. An oscilloscope is an instrument that graphically displays an instantaneous voltage (trace). An oscilloscope is used to display the shape of a voltage waveform when bench testing electronic circuits. *Bench testing* is testing performed when equipment being tested is brought to a designated service area. Meters with graphic displays are typically a combination oscilloscope and multimeter (voltage/current/resistance). Graphic display meters are designed for use as portable test instruments or as recording meters that are connected and left in place so electrical values can be recorded over time. **See Figure 4-18.**

A graphic display screen contains horizontal and vertical axes. The horizontal (x) axis represents time. The vertical (y) axis represents amplitude of the measured waveform (voltage or current). Values on the two axes from the voltage or current measurements form the waveform displayed.

To properly view the waveform, the amplitude (vertical axis) and the number of cycles (horizontal axis) must be adjusted. Depending on the meter, amplitude and number of cycles can be adjusted by the user or automatically adjusted. In addition to displaying electrical property waveforms, graphic display meters typically display numerical data also.

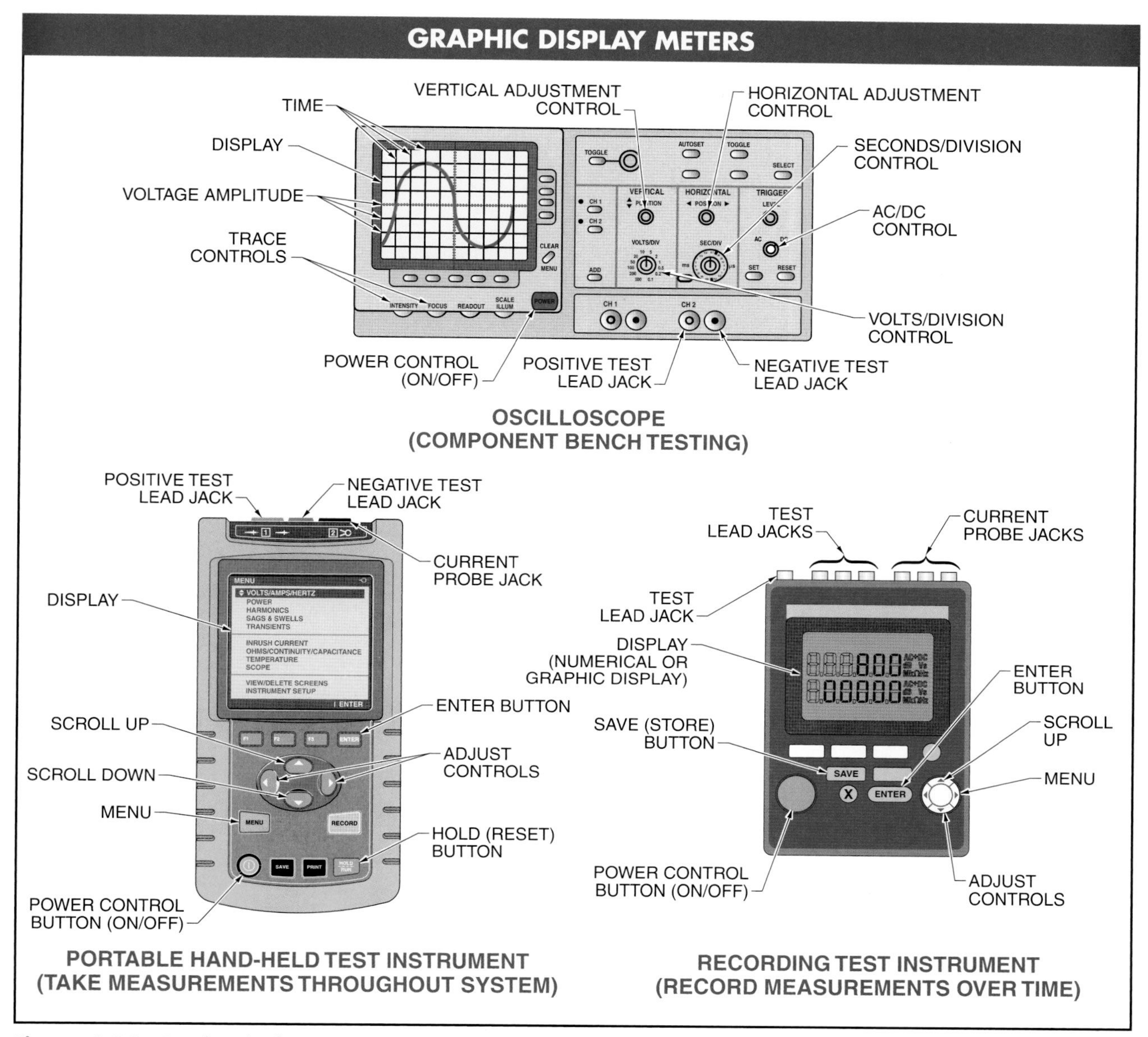

Figure 4-18. Graphic display meters include meters such as oscilloscopes, power analyzers, and three-phase recording analyzers.

After analyzing the shape and numerical data of a waveform, the value (RMS voltage, maximum voltage, minimum voltage) and condition (sinusoidal or distorted) of the power being tested can be determined. The shape of the waveform and measured values are often used to obtain more valuable troubleshooting data than just a standard voltage and/or current measurement. For example, the measurement of a voltage tester captures a value only at a particular moment. A voltmeter with a MIN MAX recording mode captures high and low measured values, but depending on the manufacturer, can or cannot indicate when a high or low value was captured or the number of values captured. Power analyzer meters capture and record waveform and measurement values over long periods of time (minutes, hours, or days). The ability to store data over time allows detection and analysis of power problems within a system. The measurements and displays can be reviewed and printed out as necessary. **See Figure 4-19.**

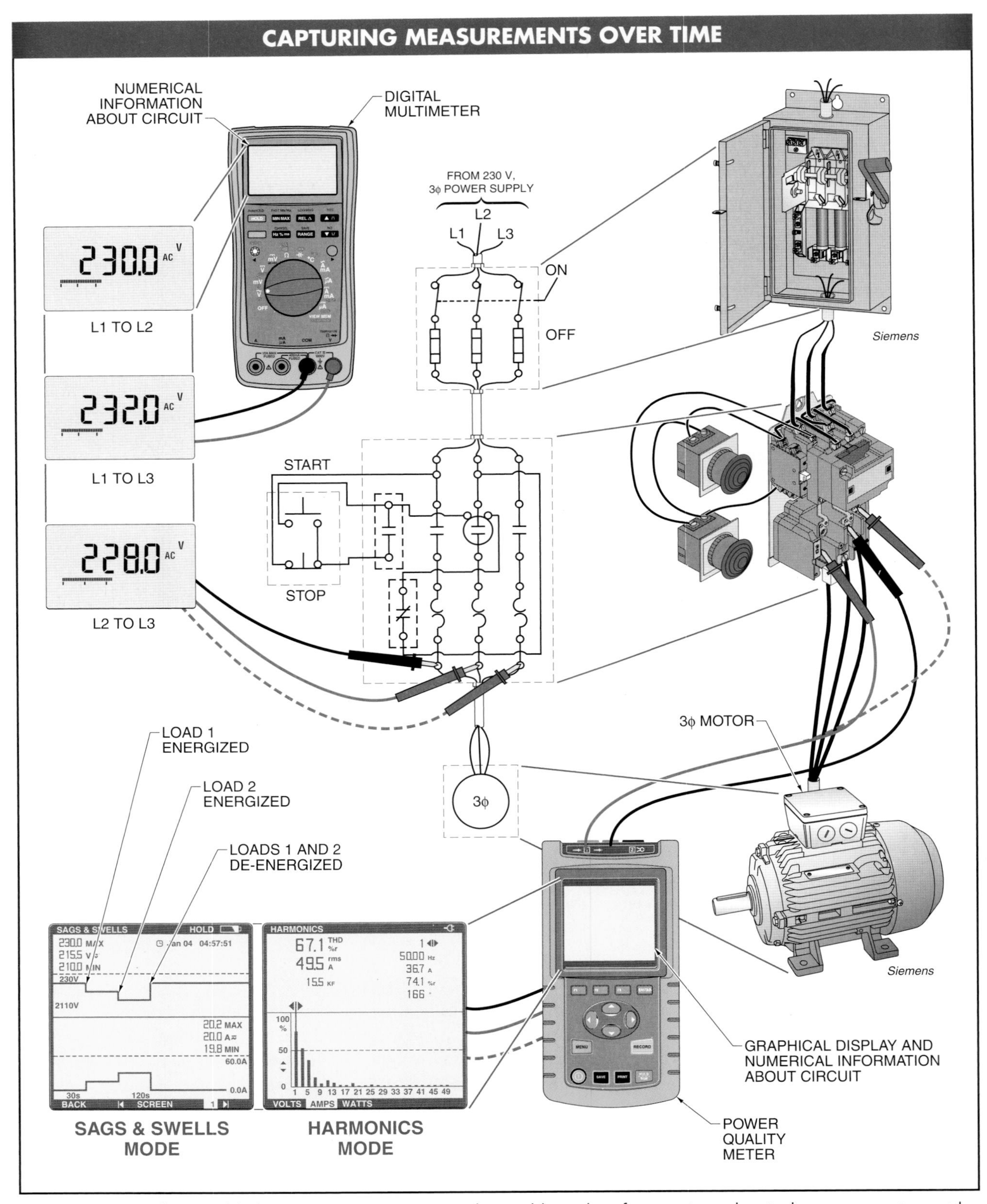

Figure 4-19. When troubleshooting a motor power quality problem, data from various electrical measurements must be analyzed.

AC Voltage Measurement Procedures (Graphic Display)

A meter with a graphic display of circuit voltage provides more usable information than numerical values from a voltmeter. Any additional information an electrician can acquire is especially important when troubleshooting problems such as voltage sags, voltage swells, transients, or other voltage-related problems. A graphic display meter such as a power analyzer meter improves troubleshooting accuracy and efficiency. Graphic display meter voltage measurement procedures, such as method of connecting test leads, menu usage, and how data is displayed on the screen, vary by model and manufacturer. **See Figure 4-20.**

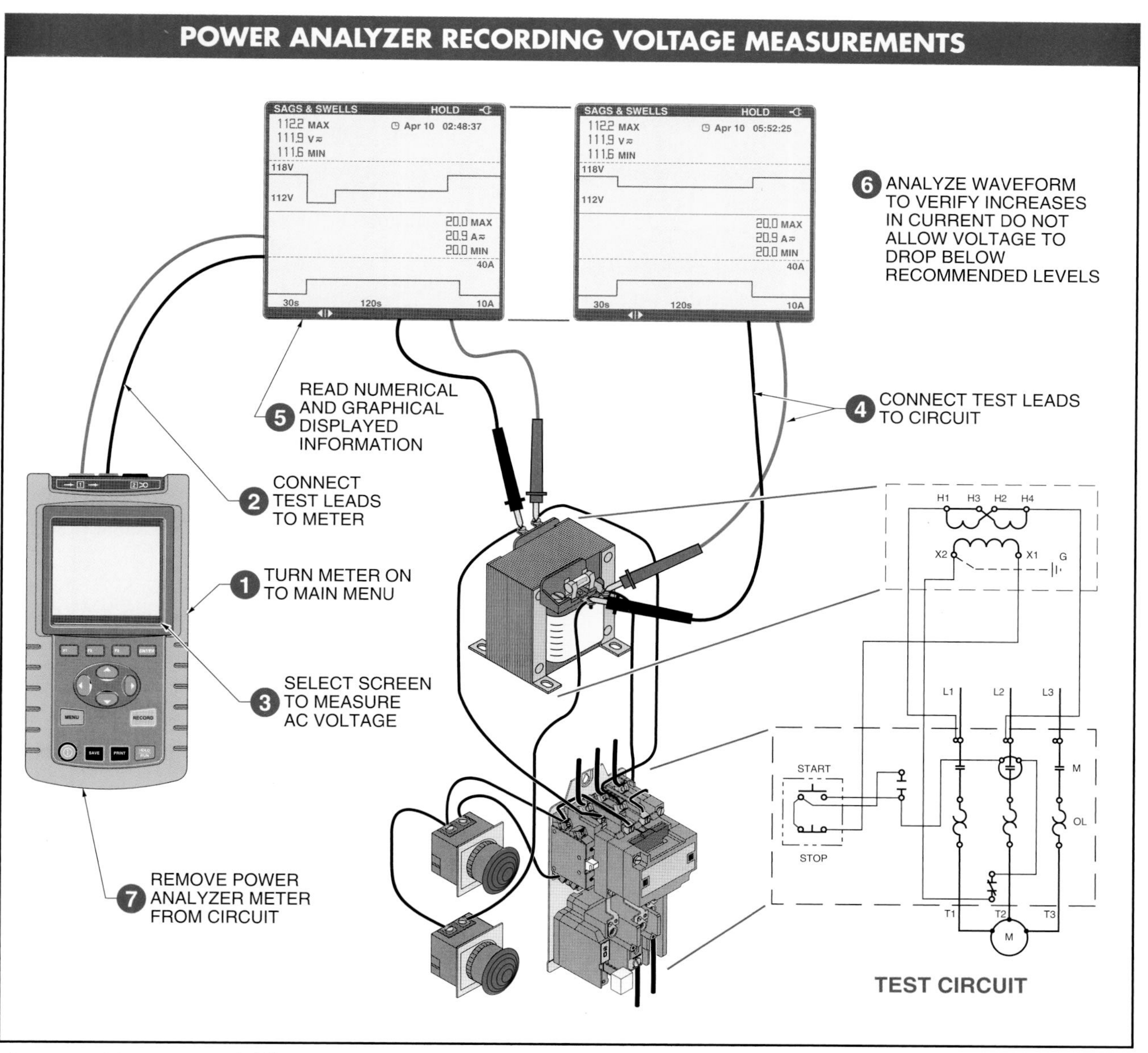

Figure 4-20. Because of differing procedures used by various manufacturers, and variations among models, always refer to specific instructions in the user's manual or directions shown on the display of a meter.

Before taking any AC voltage measurements using a power analyzer meter, ensure the meter is designed to take measurements on the circuit being tested. Always wear required personal protective equipment and follow all safety rules when taking the measurement. To measure AC voltages with a graphic display meter, apply the following procedures:

1. Turn the graphic display meter ON to display the main menu screen.
2. Connect the test leads of the power analyzer meter to the meter jacks as required.
3. Select the screen used to measure AC voltage. When the meter requires a specific voltage range setting, set the meter to a range high enough to measure the expected voltage. When the voltage measurement level is not known, set the meter to the highest range.
4. Connect the voltage test leads to the circuit being tested.
5. Read the voltage measurement and waveform displayed.
6. The waveform should be sinusoidal. When the waveform is not sinusoidal, additional measurements and displays must be made.
7. Remove the power analyzer meter from the circuit.

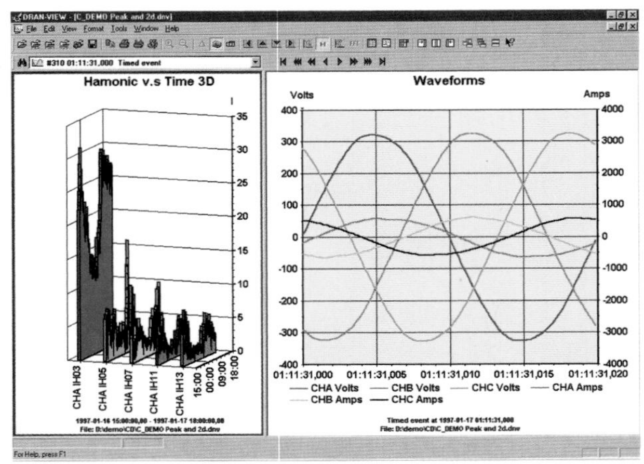

Dranetz-BMI

Graphic display meters store information that is downloadable or that can be viewed at a later time.

AC Current Measurement Procedures (Graphic Display)

Graphic display meter current measurement procedures, such as method of connecting test leads, menu usage, and how data is displayed on the screen vary by model and manufacturer. Because of the various procedures used, always refer to specific instructions in the user's manual or directions shown on the display of the meter. **See Figure 4-21.**

Before taking any AC voltage and current measurements using a three-phase graphic display meter, ensure the meter is designed to take measurements on the circuit being tested. Always wear required personal protective equipment and follow all safety rules when taking the measurement. To measure AC current with a three-phase graphic display meter, apply the following procedures:

1. Turn the graphic display meter ON to display the main menu screen.
2. Connect the voltage test leads to the meter as required. Connect the clamp-on current probe accessory to the meter as required. The voltmeter leads are not required for measuring current, but when used, provide voltage measurements in waveform for comparison with current measurements.
3. Select the screen used to measure AC voltage and AC current.
4. Connect ground test lead.
5. Connect the voltage test leads to the circuit being tested.
6. Connect the clamp-on current probe accessory or accessories to the circuit being tested.
7. Read the current measurement and waveform displayed.
8. The waveform should be sinusoidal for linear loads. When the waveform is not sinusoidal, additional measurements and displays must be made.
9. Remove the three-phase graphic display meter from the circuit.

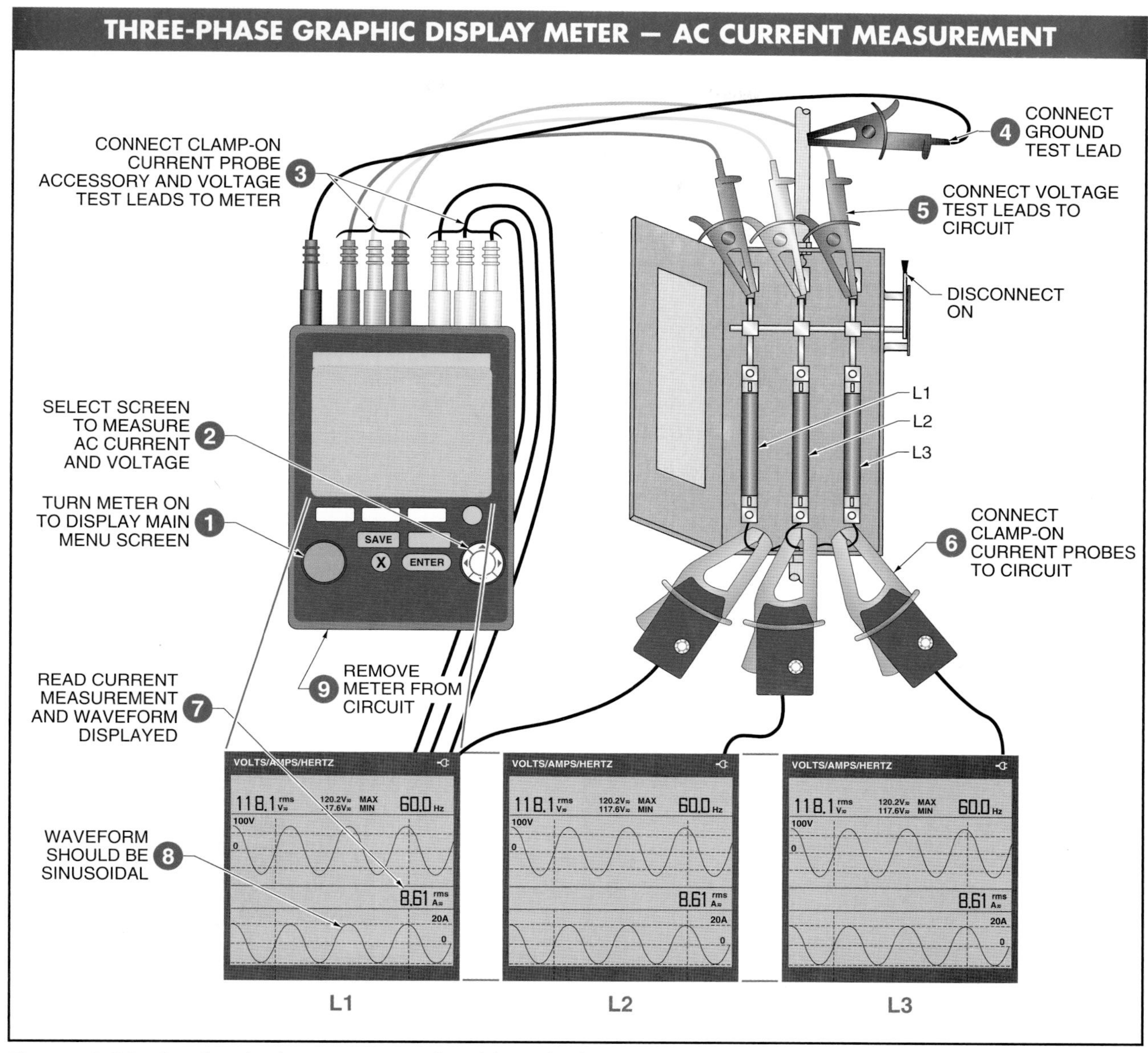

Figure 4-21. Graphic display meters provide additional information compared to numeric displays that is useful when troubleshooting.

TEMPERATURE PROBLEMS

Heat is produced in all electrical systems. The heat may be deliberate, such as in heating elements and heat lamps, or the heat may be the result of electricity flowing through electrical distribution equipment such as conductors (wire), transformers, and switching gear. Heat is produced when current flows through a resistance. The higher the current, the greater the temperature produced at the resistance. The higher the resistance of the conductors or switching gear, the greater the temperature produced. Temperature measurements identify problems or potential problems in electrical systems by identifying the heat produced by undersized conductors, switching gear, and transformers.

Temperature measurements are important because power quality problems, such as harmonics, produce extra heat in electrical systems even when current flow is within acceptable limits. Harmonics in an electrical system produce additional heat in all transformers. Due to the heat created by harmonics, transformers have K ratings. K-rated transformers have larger neutral conductors and special windings to reduce efficiency losses caused by harmonics. K ratings such as K-4 or K-13 are listed on transformers and represent the ability of transformers to operate correctly at higher temperatures produced by harmonics. K factor measurements are typically taken using power analyzer meters. As the amount of harmonics increases in an electrical system, the measured K factor will also increase.

Temperature measurements are taken using contact and noncontact temperature test instruments. When testing for power quality problems, a noncontact temperature test instrument is preferred. Noncontact temperature instruments are typically of the infrared type or thermal imaging type. An *infrared meter* is a meter that measures heat energy by measuring the infrared energy emitted by a material and displays the temperature as a numerical value. All materials emit infrared energy in proportion to the temperature of the material. A *thermal imaging camera* is a meter that measures heat energy by measuring the infrared energy emitted by a material and displays the temperature as a color-coded thermal picture. **See Figure 4-22.**

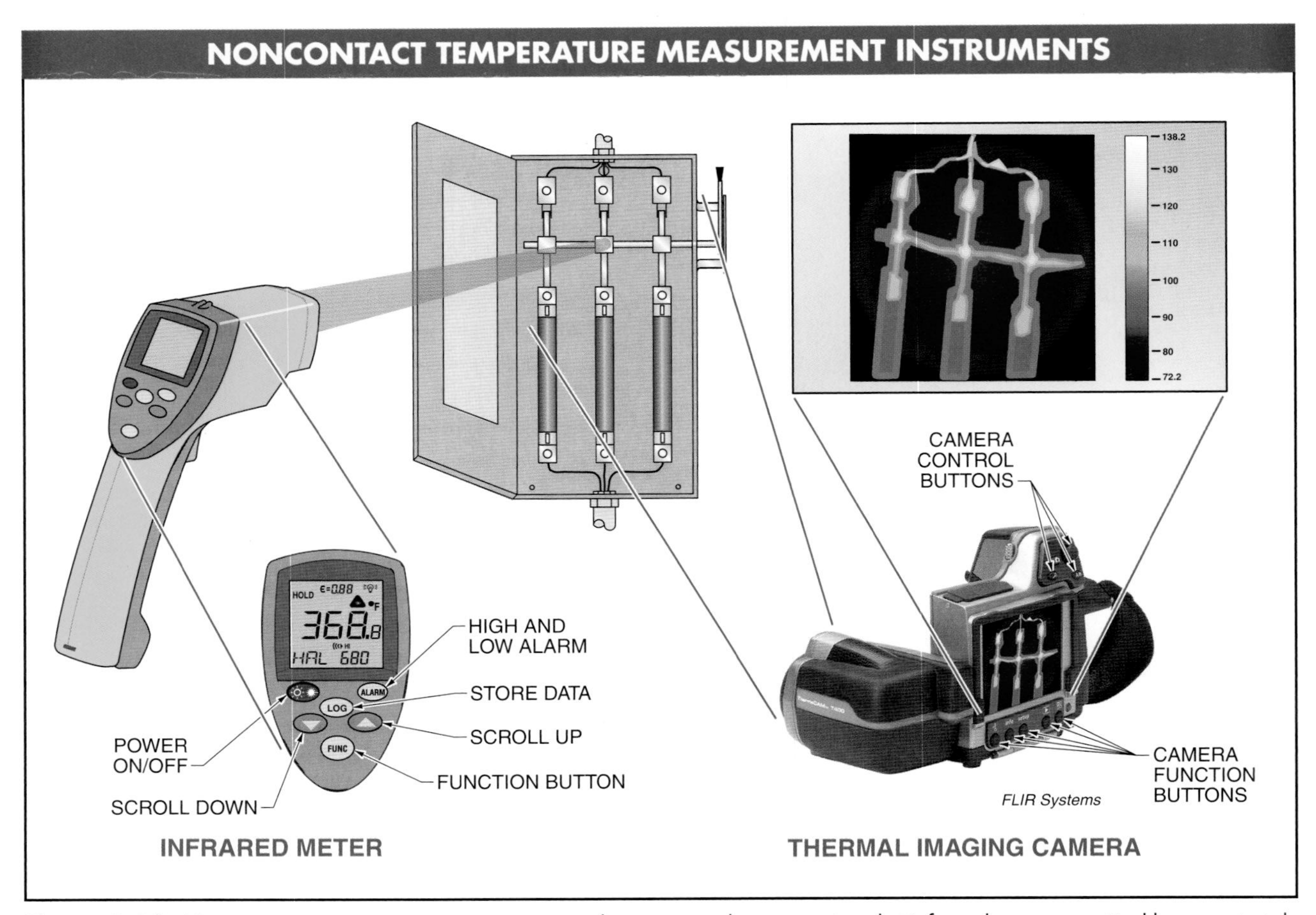

Figure 4-22. Noncontact temperature meters measure heat energy by measuring the infrared energy emitted by a material, and display the temperature as a numerical value.

Infrared Meters

Infrared meters are able to take temperature measurements without touching an object but can only measure the surface temperature of a small area at one time. Infrared meters are also less expensive than thermal imaging cameras. Thermal imaging cameras are able to measure the temperature of large areas and are great troubleshooting and preventive maintenance test instruments.

Infrared meters are able to detect small amounts of heat variances in electrical distribution systems and loads. A poor electrical connection is a typical problem that causes a small to moderate heat increase. The worse the connection, the higher the resistance and the higher the heat created. Loose, corroded, or dirty electrical connections generate unwanted resistance and heat. The temperature rise at a connection depends on the current flowing through the connection and the resistance of the connection. A temperature rise above ambient temperature is expected on all electrical equipment carrying current, but the higher the temperature, the greater the possibility of equipment failure or the system overloading.

Maintenance must be performed when electrical equipment is operating at higher than normal temperatures to avoid equipment failure. Temperature rises of 50°F (28°C) must be investigated. Temperature rises of 100°F (56°C) or more require immediate action (immediate shutdown of system and repair of fault).

Infrared Temperature Meter Measurement Procedures

Most of the resistance in a normal circuit occurs at the load. Resistance in a circuit that has poor connections, corrosion, or other high resistance paths has resistance that occurs at points other than at the load. Infrared temperature meter measurements prevent problems by locating areas of unwanted heat before the heat causes component failure. **See Figure 4-23.**

To measure temperature using an infrared meter, apply the following procedures:

1. Aim the infrared temperature meter at the area being measured. The meter is focused based on the distance between the object and the meter.
2. Take an ambient temperature reading for reference.
3. Take temperature readings of any areas suspected to have temperatures above ambient temperature.
4. Turn meter OFF.

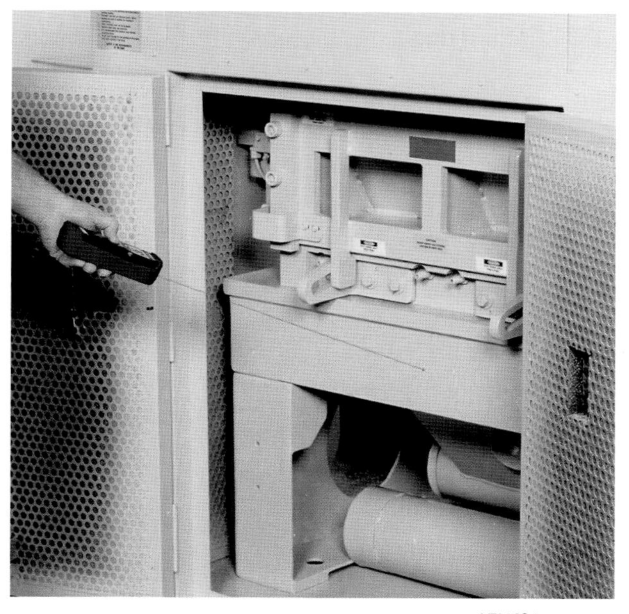

AEMC® Instruments

Temperature instruments may be explosionproof, waterproof, or designed for specific applications.

To determine temperature readings in such areas, subtract the ambient temperature reading from the reading obtained on the infrared temperature meter. The difference in temperature is the amount of heat produced by the electrical component, circuit, or equipment being tested.

Technical Tip

Temperature measurements are used to indicate a problem or severity of a problem.

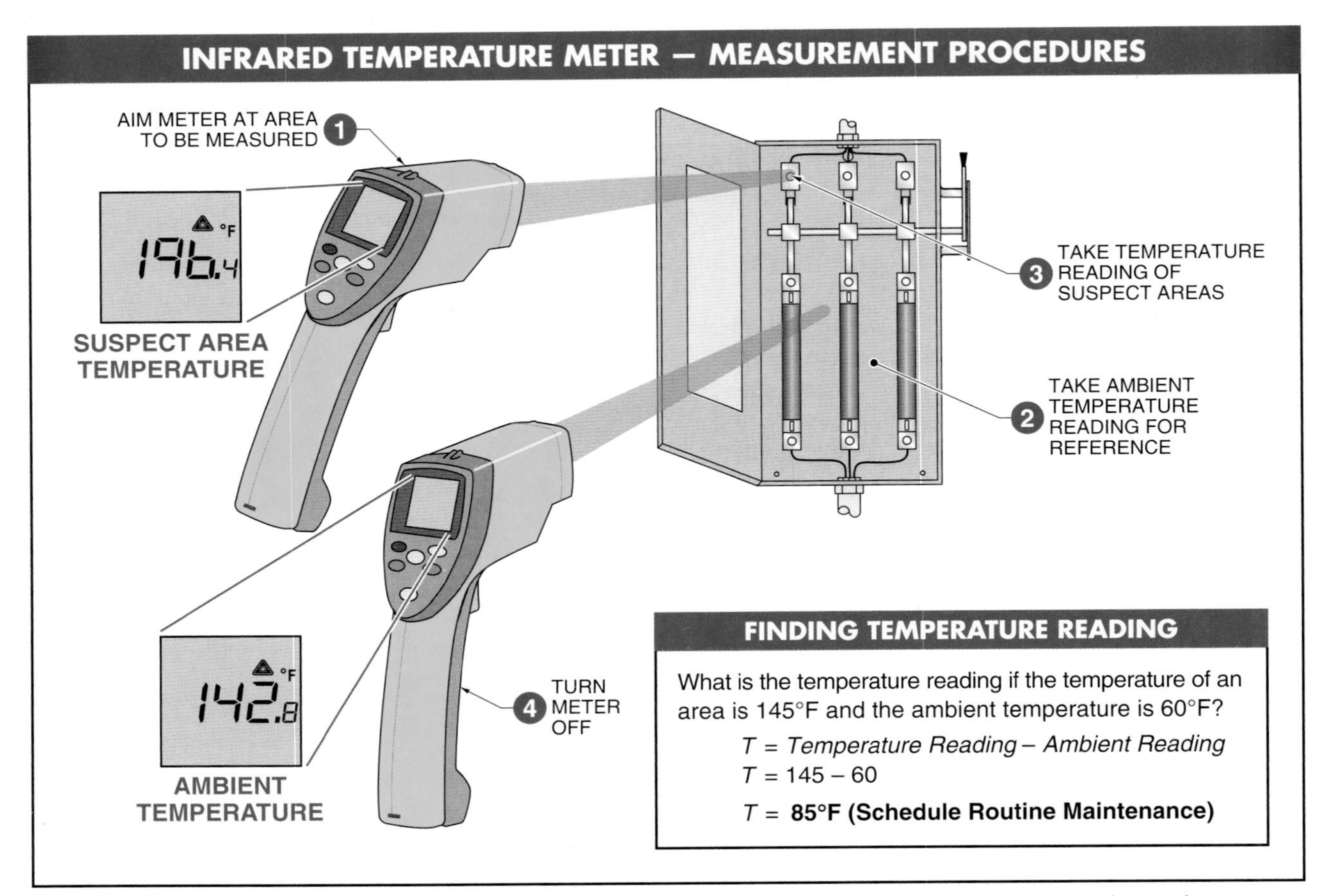

Figure 4-23. Infrared meter temperature measurements prevent problems by locating unwanted heat in electrical equipment before the heat causes failure.

POWER INTERRUPTIONS

All electrical and electronic equipment is rated for operation at a specific voltage. The rated voltage is accepted as a voltage range. The typical range was ±10%, but when components of an electrical system are derated to save energy and operating cost, the typical range is +5% to –10%. The +5% to –10% range is used because overvoltages are more damaging to loads than undervoltages. Equipment manufacturers, utility companies, and regulating agencies must routinely deal with changes in system voltage. Power interruptions are classified into standard industry categories. **See Figure 4-24.**

Momentary Power Interruptions

A *momentary power interruption* is a decrease to 0 V on one or more power lines lasting from 0.5 cycles up to 3 sec. All power distribution systems encounter momentary power interruptions during normal operation. Momentary power interruptions are caused when lightning strikes (strikes nearby), by utility grid switching during a problem (short on one line), or during open circuit transition switching. *Open circuit transition switching* is a process in which power is momentarily disconnected when switching a circuit from one voltage supply or level to another voltage supply or level.

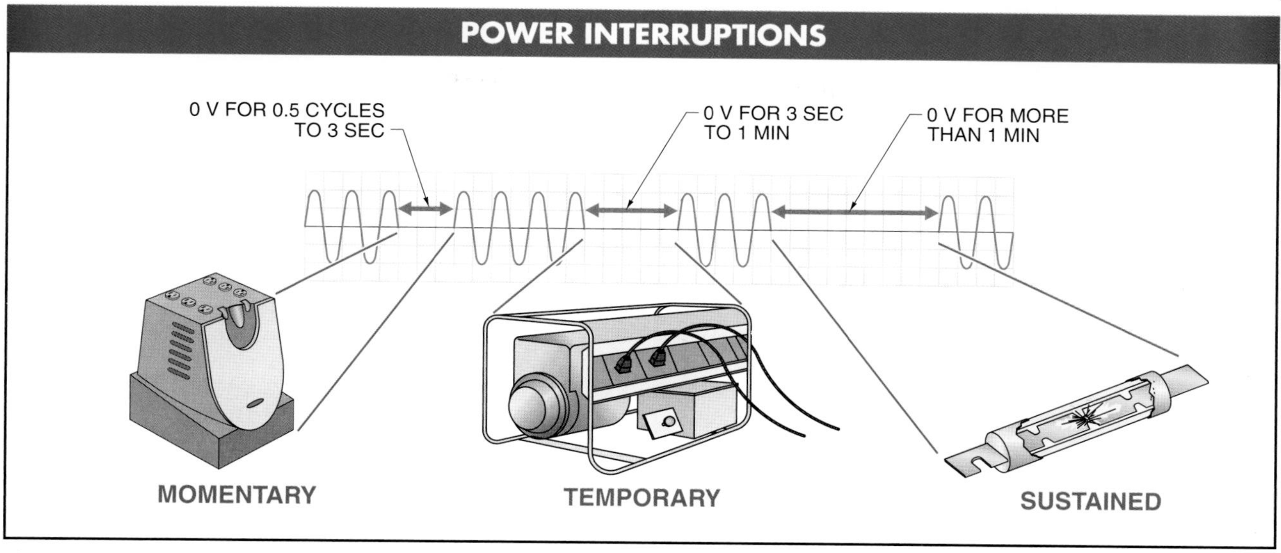

Figure 4-24. Power interruptions to an electrical system can be momentary, temporary, or sustained.

Temporary Power Interruptions

A *temporary power interruption* is a decrease to 0 V on one or more power lines lasting for more than 3 sec up to 1 min. Automatic circuit breakers and other circuit protection equipment protect power distribution systems by removing faults. An automatic circuit breaker typically requires 20 cycles to about 5 sec to restore power after a temporary interruption. A temporary power interruption can also be caused by a gap in time between a power interruption and when a back-up power supply (generator) takes over, or if someone accidentally opens a circuit by switching the wrong circuit breaker or switch gear. When power is not restored, a temporary power interruption becomes a sustained power interruption.

Sustained Power Interruptions

A *sustained power interruption* is a decrease in voltage to 0 V on all power lines for a period of more than 1 min. Even the best power distribution systems have a complete loss of power at some time. Sustained power interruptions (outages) are commonly the result of storms or circuit breakers tripping due to damaged equipment.

Power Quality Meter Characteristics

Several different test instruments are used to detect power interruptions. The specific test instrument used depends upon the application, required results, and cost considerations. A voltage tester can be used to test for loss of power at any time. Voltmeters with MIN MAX recording functions are used to record power interruptions over time, but cannot indicate when or for how long an interruption took place. Recording meters are used to show exactly when power was interrupted and for what length of time. **See Figure 4-25.**

Power Interruption Measurement Applications

As utility companies are bought and sold, parts of the distribution system are subcontracted out to private companies, leaving greater demand on older, undersized systems and equipment. Utility companies typically charge penalties for poor building, facility, or plant power factors. Analyzing the reasons for power interruptions and poor power factors using test instruments is an economical way for both the supplier (utility company) and user (customer) to document what is causing the poor power factor or what caused a power interruption.

POWER QUALITY METER CHARACTERISTICS		
Test Instrument Used to Check Power Interruption	**Advantages**	**Disadvantages**
Voltage Tester	1. Inexpensive and usually available	1. Cannot record measurements over time
Voltmeter with MIN MAX Recording Mode	1. Can record circuit minimum and maximum voltages over time 2. Can display a minimum and maximum voltage that was recorded	1. Cannot tell when a power interruption occurred or for how long
Portable (Handheld) Power Analyzer Meter	1. Can be carried like a standard handheld meter and record circuit conditions over time with numerical and graphical displays and download software	1. Basic power analyzers cost several times more than voltmeters with MIN MAX recording, advanced feature meters cost 10 times more
Recording and Logging Meter	1. Can be permanently mounted to panels along a system 2. Designed for single function by measuring just voltage, current, watts, harmonics, or transients (less expensive)	1. Made of less rugged design than other meters due to permanent mounting 2. Because of single function, several are required if various circuit properties are to be measured

Figure 4-25. Power quality meters may be portable (handheld) or permanently mounted; may be inexpensive or cost thousands of dollars; and may or may not be able to take measurements over time.

Power Interruption Measurement Procedures

Power interruptions can be a problem for a section of a power distribution system inside a facility or for individual branch circuits. To test for loose connections or the tripping of automatic overloads, a voltmeter with a MIN MAX recording function is used. Tripping of automatic overloads is a problem when the overloads automatically reset. During the time of the overload trip a power interruption occurs, but after the overloads automatically reset, the short power interruption may not be seen as a problem. To test and record power interruptions over time, use a power analyzer meter with recording function. **See Figure 4-26.**

Before taking any power interruption measurements using a digital multimeter and power analyzer meter, ensure the meters are designed to take measurements on the system being tested. Refer to the operating manual of the test instrument for all measuring precautions, limitations, and procedures. Always wear required personal protective equipment and follow all safety rules when taking the measurement. To test for power interruptions, apply the following procedures:

1. Connect the test leads of the voltmeter and power analyzer meter to the meter jacks as required.
2. Set the voltmeter to the MIN MAX recording mode, and the power analyzer meter to voltage recording mode.
3. Connect the voltage test leads of the meters to the system being tested.
4. Allow meters to record voltage measurements as loads are switched ON and OFF. Check for loose connections by wiggling equipment and/or conductors.
5. Read and record any findings of measured (recorded) power interruptions.
6. Take additional voltage and/or current measurements to establish the location of the fault in the system.
7. Correct all faults in the system.
8. Retest the system after repairs are completed.
9. Remove voltmeter and power analyzer meter from the system.

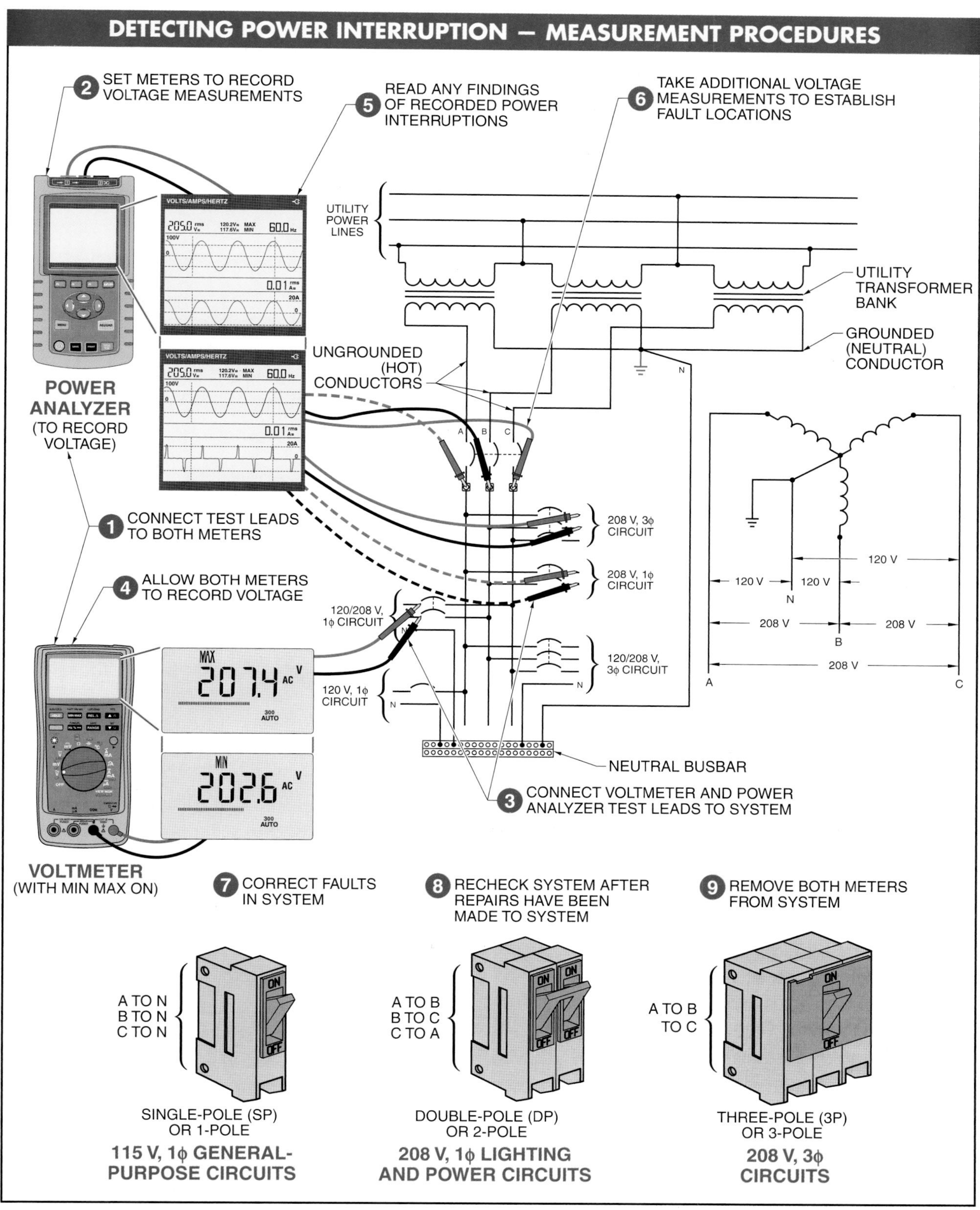

Figure 4-26. Power interruption problems cause problems on individual branch circuits and/or within larger sections of a power distribution system.

VOLTAGE CHANGES

Typically, the voltage in a power distribution system is within the acceptable range of +5% to −10%. But at times, voltage changes are caused by storms and electrical faults that raise or lower voltage enough to cause circuit and load problems. Voltage changes, such as voltage sags and swells, undervoltages, and overvoltages, are difficult to detect because voltage changes tend to come and go. For example, an overloaded circuit, transformer, or system may encounter an undervoltage problem (computers resetting, relays and motor starters dropping out) when loads are ON, but the voltage returns to normal levels when the loads are OFF. **See Figure 4-27.**

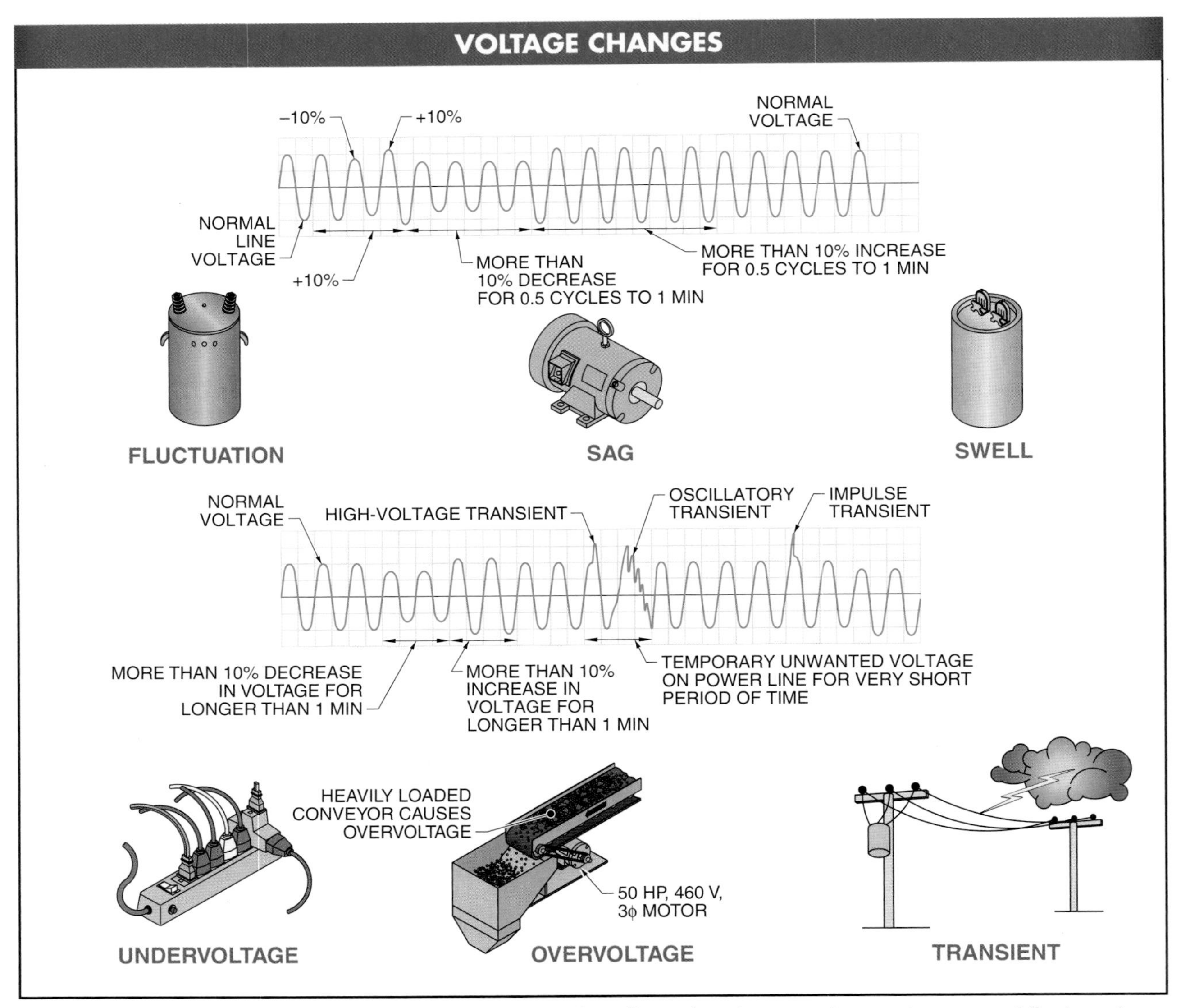

Figure 4-27. Voltage changes, such as voltage sags and swells, undervoltage, and overvoltage, are difficult to detect because voltage changes tend to come and go.

Voltage Fluctuations

A *voltage fluctuation* is an increase or decrease in the normal line voltage within the range of +5% to −10%. In the U.S. power distribution system, frequency (60 Hz) is constant and voltage is held within the normal range of +5% to −10%, while current constantly changes as loads are added and removed. Voltage fluctuations are commonly caused by overloaded transformers, unbalanced transformer loading, and/or high impedance caused by long circuit runs, undersized conductors, poor electrical connections, and/or loose connections in the system. Small voltage fluctuations typically do not affect equipment performance.

Voltage Sags

A *voltage sag* is a voltage drop of more than 10% (but not to 0 V) below the normal rated line voltage that lasts from 0.5 cycles up to 1 min. Voltage sags commonly occur when high-current loads are turned ON and the voltage on the power line drops below the normal voltage fluctuation (−10%) for a short period of time. Voltage sags commonly occur when large motors are switched ON or when temporary short circuits occur on utility power lines. When short circuits cause a breaker to open, a power interruption occurs. Voltage sags are often followed by voltage swells as voltage regulators overcompensate during a voltage sag.

Voltage varies within a circuit, and may be consistently, temporarily, or momentarily too high or too low. For example, voltage may be consistently too high if a circuit is near the source of a power distribution system. Voltage may be consistently too low if a circuit is near the end of a power distribution system, due to conductors being undersized, branch circuit runs being too long, large loads being turned ON, or the power supply transformer being undersized. Voltage may also be temporarily too low during certain time periods of a day (brownouts).

Every electrical or electronic load has a low voltage limit and a high voltage limit between which the load is designed to operate properly. Typically, most devices are listed for a given voltage with a voltage variation limit stated as plus or minus a percentage. When voltage dips below the lower voltage limit, electrical and electronic components are damaged, resulting in loss of memory, data loss, and/or equipment malfunction.

Voltmeters with a MIN MAX recording mode are used to measure voltage swells and voltage sags. To yield usable information about voltage changes, voltage measurements must be taken and recorded over time. Voltmeters with a MIN MAX recording mode are able to capture and display low-voltage (minimum voltage) conditions. However, measurements displayed from voltmeters with a MIN MAX recording mode do not indicate when or how long a low-voltage condition existed. Typically, a power analyzer meter with recording mode is the best meter to use for detecting voltage problems.

To acquire the most accurate voltage measurements, voltage must be measured and recorded at various times of the day and at various places within the facility. Measurements must also be routinely taken as part of a preventive maintenance program. When voltage fluctuations are found to be more than ±8%, a voltage regulator (stabilizer) must be added to the system. A *voltage regulator (stabilizer)* is a device that provides precise voltage control to protect equipment from voltage sags (voltage dips) and voltage swells (voltage surges).

Voltage Swells

A *voltage swell* is a voltage increase of more than 10% above the normal rated line voltage lasting from 0.5 cycles up to 1 min. Voltage swells commonly occur when large loads are turned OFF and voltage on the power line increases above the normal voltage fluctuation (+10%) for a short period of time. For example, a voltage swell commonly occurs in office areas of a plant when production lines with large loads are shut down.

Voltage swells are not as common as voltage sags. However, voltage swells are more destructive than voltage sags because voltage swells damage electrical equipment in very short periods of time. Even a very short high-voltage condition (voltage swell) causes permanent equipment or component damage.

Technical Tip

Voltage fluctuations are most common during summer months when air conditioning units increase the demand for power.

As with voltage sags, voltmeters with a MIN MAX recording mode are used to measure voltage swells because voltage must be measured over time to acquire the proper data. To acquire the most accurate voltage measurements, voltage must be measured and recorded at various times of the day and at various places within the facility.

Undervoltage

Undervoltage is a drop in voltage of more than 10% (but not to 0 V) below the normal rated line voltage for a period of time longer than 1 min. When voltage on the power lines drops below the normal voltage fluctuation (−10%) for long periods of time, an undervoltage condition exists. Undervoltage (low voltage) is more common than overvoltage (high voltage) on power distribution systems. Undervoltages are commonly caused by overloaded transformers, undersized conductors, conductor runs that are too long, too many loads on a circuit, or peak power usage periods (brownouts). A *brownout* is the reduction of the voltage level by a power company to conserve power during times of peak usage or excessive loading of the power distribution system.

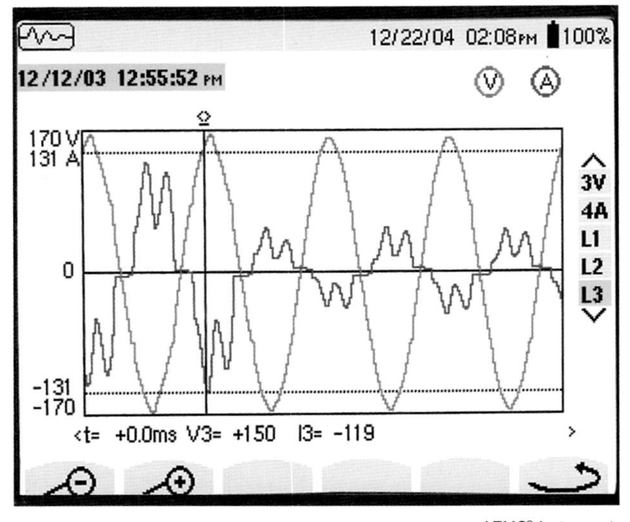

AEMC® Instruments

Hand-held oscilloscopes and power quality meters are used to visually indicate voltage fluctuations, voltage surges, undervoltages, and overvoltage conditions.

Overvoltage

Overvoltage is an increase of voltage of more than 10% above the normal rated line voltage for a period of time longer than 1 min. Depending on the cause of an overvoltage condition, voltage increases above the normal voltage fluctuation of +5% can occur for long periods of time. Overvoltages are caused when loads are near the beginning of a power distribution system, or when taps on a transformer are not wired correctly. A *tap* is one of many connection points along a transformer coil that are commonly provided at 2.5% increments.

Voltage Change Measurement Meters

The meters typically used to measure for power interruptions are also used to measure voltage variations. The best test instrument to use depends upon the application, the required results, and cost considerations. A voltage tester is used to test for high or low voltage conditions. A voltmeter with a MIN MAX recording function is used to record voltage variations over time, but cannot indicate how long or when a voltage change occurred. Recording meters are used to indicate exactly when a voltage variation took place and for what period of time.

Some recording meters are relatively simple and provide basic circuit data, while other recording meters can download their data to a desktop computer or laptop for more detailed circuit analysis. Basic power circuit testers record data and display data with lights for any low voltage or high voltage conditions. Power circuit testers with download capabilities record circuit changes and display the changes after the data is downloaded to a computer. **See Figure 4-28.**

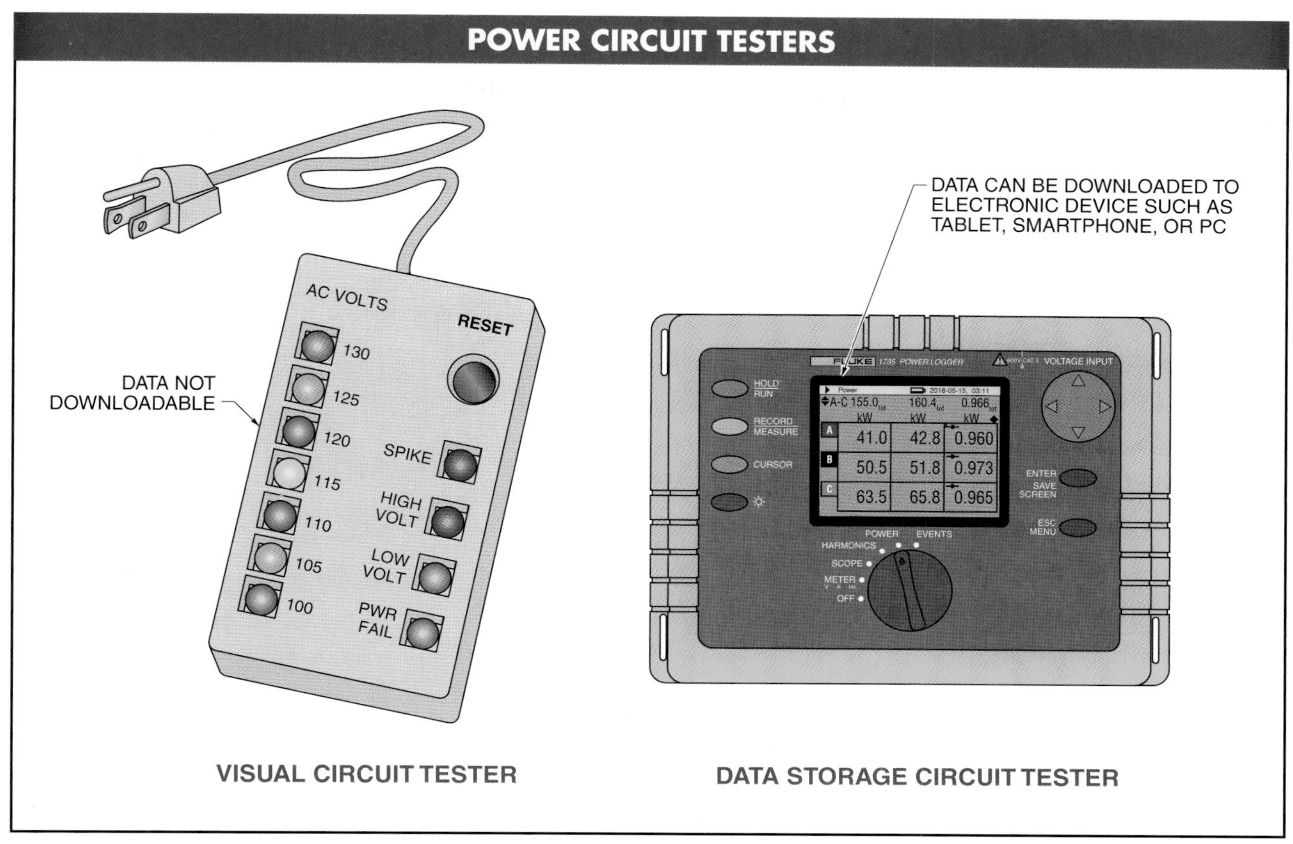

Figure 4-28. Recording meters are used to show exactly when voltage changes took place and for what period of time.

Power circuit testers can be connected to circuits for hours, days, weeks, or even months. When a problem does occur, power analyzers are checked for related data. A quick check of a power circuit tester connected to a system can determine whether a problem was in a facility or with the incoming power lines.

AC Voltage Measurement Procedures

Unlike power interruptions, voltage variations are a common part of all electrical systems. As loads are added and removed from a system, voltage varies in the system. As long as voltage variations remain in the acceptable range, no damage occurs to electrical equipment. However, when a circuit problem might be related to a low or high voltage condition, tests must be made for voltage variations. **See Figure 4-29.**

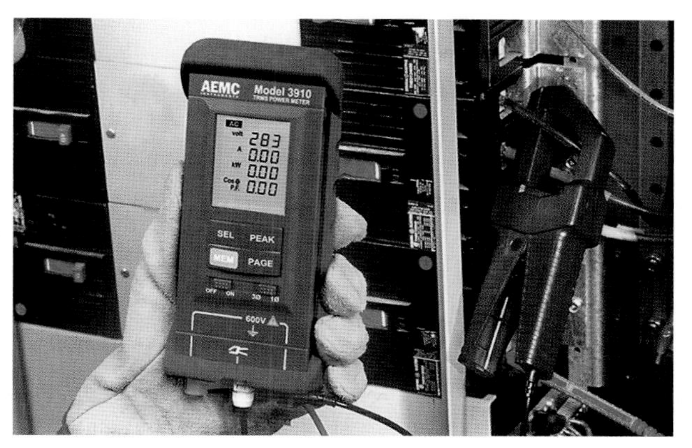

AEMC® Instruments

Recorders and handheld meters with recording functions are connected to circuits to capture low and high voltage conditions.

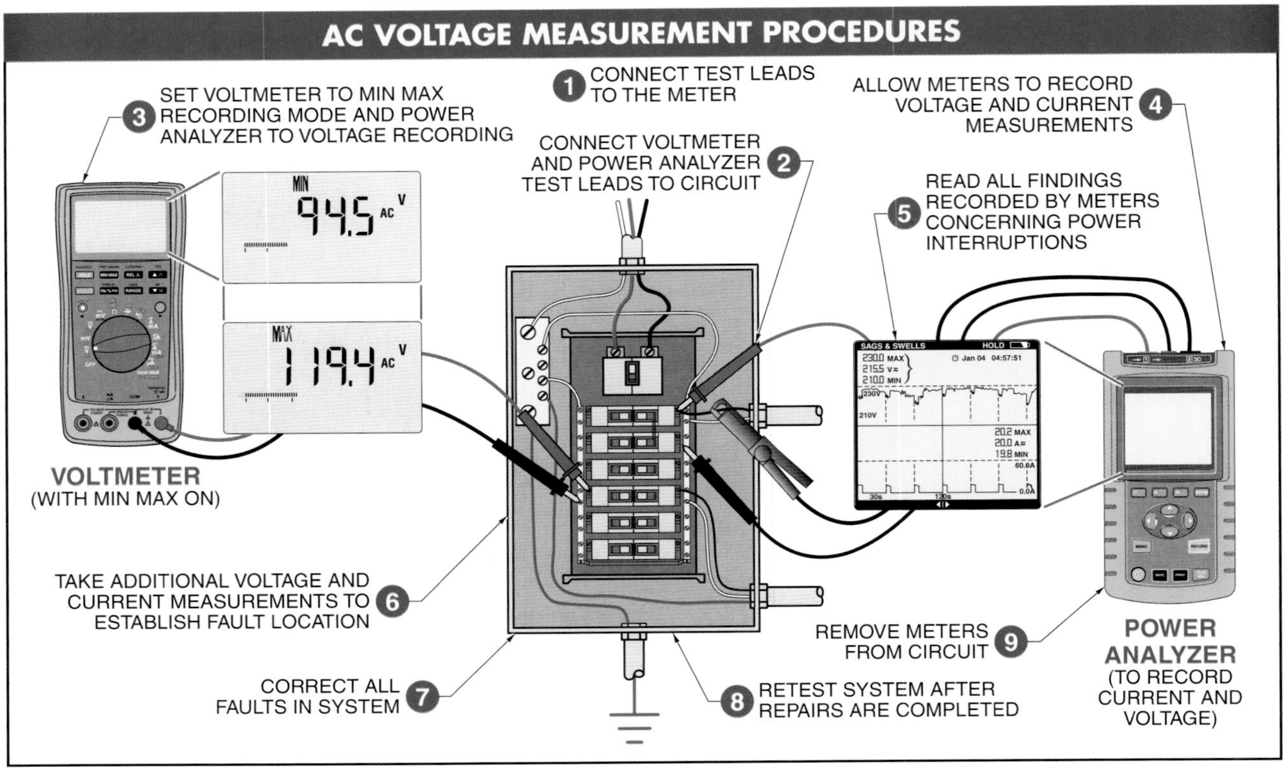

Figure 4-29. Voltage variations are a common part of all electrical circuits, and typically occur when loads are added to or removed from the system.

Before taking any voltage variation measurements using a digital multimeter and power analyzer meter, ensure the meters are designed to take measurements on the circuit being tested. Refer to the operating manual of the test instrument for all measuring precautions, limitations, and procedures. Always wear required personal protective equipment and follow all safety rules when taking the measurement. To test for voltage variations, apply the following procedures:

1. Connect the test leads of the voltmeter and power analyzer meter to the meter jacks as required.
2. Connect the voltage test leads of both meters to the circuit being tested.
3. Set the voltmeter to the MIN MAX recording mode, and the power analyzer to the voltage recording mode.
4. Allow the meters to record voltage measurements as loads are switched ON and OFF. Try to turn on all loads that may be on during peak usage (or run the test during peak usage times). Overloaded circuits are the number one cause of low voltage conditions.
5. Read and record any findings of measured (recorded) voltage changes.
6. Take additional voltage and current measurements as required to establish the location of the fault. Current measurements can be important because current measurements determine how much a circuit is being used (loaded).
7. Correct all faults. Balance loads (move some loads to less used circuits), adjust transformer taps to increase (or decrease) voltage levels, or increase transformer size.
8. Retest the circuit after repairs are completed.
9. Remove the voltmeter and power analyzer meter from the circuit.

TRANSIENTS

Research by the American Institute of Electrical Engineers (AIEE), service technicians, and equipment manufacturers clearly indicates that solid-state circuits and devices do not tolerate momentary voltage surges exceeding twice the normal operating voltage. Voltage surges are produced in all electrical systems from outside sources (lightning) and within the system by turning OFF loads that include coils (solenoids, magnetic motor starters, and motors). The voltage surge produced is called a transient voltage.

A *transient voltage (voltage spike)* is a temporary, undesirable voltage in an electrical circuit. Transient voltages range from a few volts to several thousand volts and last from a few microseconds up to a few milliseconds. *Oscillatory transient voltages* are transient voltages commonly caused by turning OFF high inductive loads and by switching OFF large utility power factor correction capacitors. Utility companies improve the power factor on the power lines by using power factor correction capacitors. An *impulse transient voltage* is a transient voltage commonly caused by a lightning strike that results in a short, unwanted voltage being placed on a power distribution system.

Transient voltages are caused by the sudden release of stored energy due to lightning strikes, unfiltered electrical equipment, contact bounce, arcing, and generators being switched ON and OFF. Transient voltages are produced from stored energy contained within a system. The size and duration of transient voltages depends on the value of inductance (L) and capacitance (C) of a system. Transient voltages range from a few volts to several thousand volts and occur at any point on the AC sine wave. Transient voltages on a 120 V power line reach several thousand volts or more. One high-voltage transient is all that is required to damage circuits or electrical equipment.

Technical Tip

Circuits are protected from transient voltages by permanently installed transient voltage surge suppressors (TVSSs). Installation and use requirements for TVSSs are defined in Article 285 of the NEC®.

Transient voltages differ from voltage swells and voltage sags by being larger in magnitude and shorter in duration, having a steep (short) rise time, and being erratic. Lightning is the most common source of transient voltages on utility power distribution systems. Lightning induces a surge that travels in two directions from the point of contact. In most cases, the transient voltage is dissipated by the utility company grounding and protection systems after the lightning has traveled 10 to 20 utility poles (or grounding/protection points). Damage to equipment is unlikely when equipment is located further than 20 grounding poles from the lightning strike. However, unprotected equipment is severely damaged when lightning strikes close to equipment that is in use, or when the system is not properly protected. All transient voltages are dampened (attenuated) as the transient travels through an electrical system.

High-level transient voltages caused by lightning strikes must be considered in all electrical applications because lightning may strike at any location. Low-level transient voltages are produced on the distribution system of a facility when loads are switched OFF. Transient voltages are produced continuously when using loads such as the following:

- solid-state power supplies
- magnetic motor starters
- fluorescent lighting using electronic ballasts
- variable speed motor drives
- DC motor drives
- solenoids
- soft-start motor starters
- electronically controlled welding equipment
- motors that include brushes (DC and universal)

Transient Voltage Measurement Procedures

Power analyzer meters are typically used to monitor transient voltages. The size, duration, and time of transient voltages can be displayed at a later time when using power analyzer meters. When transient voltages are identified as a problem within a facility, voltage surge suppressors (surge protection devices) must be used. A *surge protection device* is a device that limits the intensity of voltage surges that occur on the power lines of a power distribution system.

Transient voltage measurements must be taken any time equipment prematurely fails and a transient voltage is suspected. Transient voltage measurements are taken to verify that surge protection devices are working. A transient voltage measurement is different from a standard voltage measurement because a standard voltage measurement displays the RMS voltage value of a circuit, and transient voltage measurements display the peak voltage value of a circuit. For example, a standard voltmeter that displays 115 VAC is displaying the RMS voltage measurement of the circuit. The peak voltage of a 115 VAC RMS circuit is about 162 V (RMS voltage multiplied by 1.414 equals peak voltage). **See Figure 4-30.**

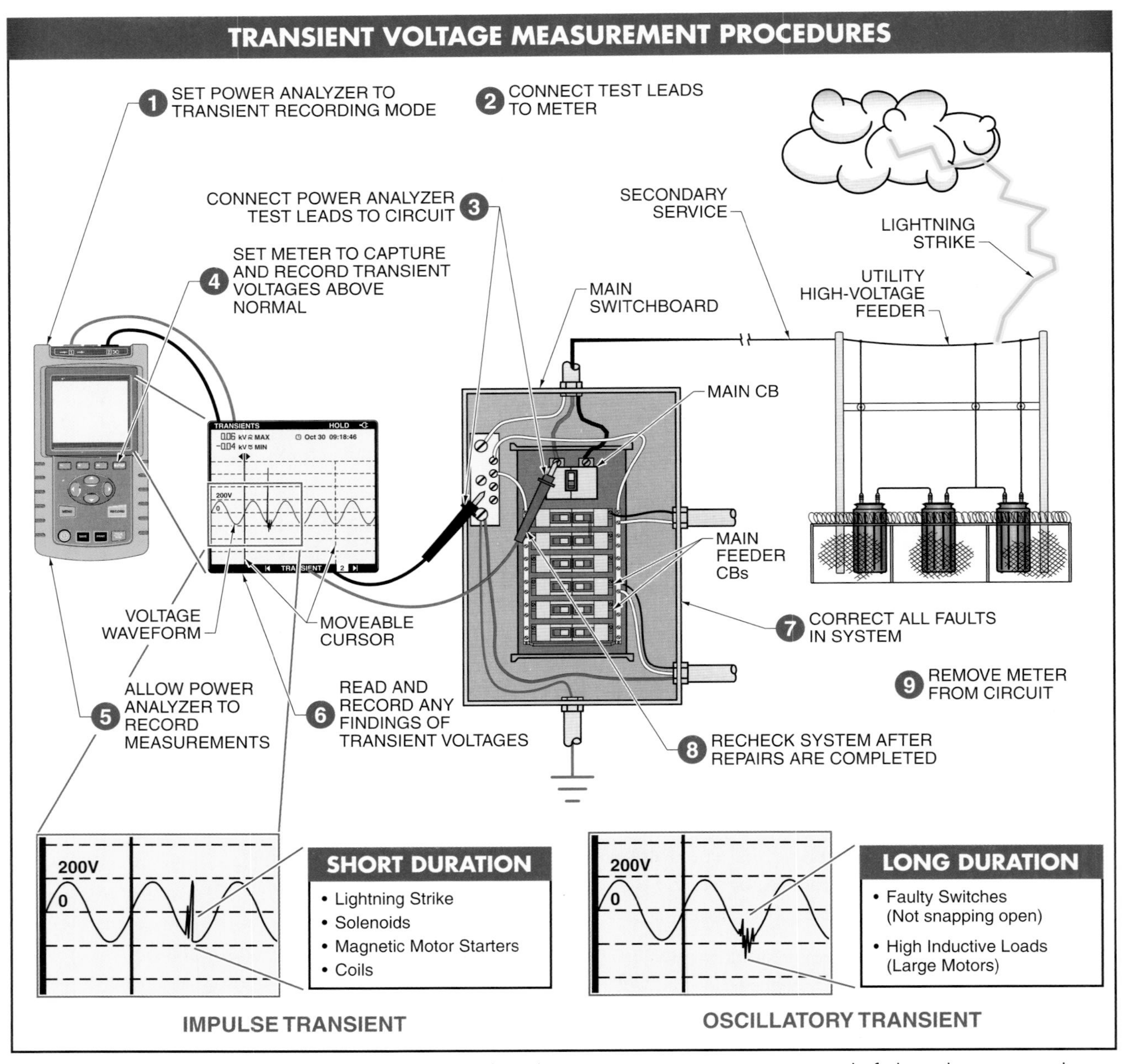

Figure 4-30. Transient voltage measurements must be taken any time equipment prematurely fails and transient voltages are suspected.

Before taking any transient voltage measurements using a power analyzer meter, ensure the meter is designed to take measurements on the system being tested. Refer to the operating manual of the test instrument for all measuring precautions, limitations, and procedures. Always wear required personal protective equipment and follow all safety rules when taking the measurement. To test for transient voltages, apply the following procedures:

1. Set the power analyzer meter to transient voltage recording mode.
2. Connect the test leads of the power analyzer meter to the meter jacks as required.
3. Connect the power analyzer meter voltage test leads to the system being tested. When using a power analyzer meter to record transient voltages, the test leads of the meter must be connected to the powered system before recording starts. The test leads must be connected first because a power analyzer meter will be looking for a voltage higher than the setting of the meter (50%, 100%, 150%, etc.). When the test leads are not connected before the meter starts recording, the meter will record 50% (or whatever the meter is set on) of nothing (0 V).
4. Set the transient voltage recording mode to record transient voltages at a set level (50%, 100%, 200%), or above normal voltage. Normal voltage is the voltage applied to the test leads before the meter is set to start recording transients. When a meter is connected to a standard 120 V circuit and is set to record transient voltages greater than 100%, the meter records any voltage over 340 V peak (120 V RMS multiplied by 1.414 equals 170 V peak, and the meter is set at 100%, so 170 V multiplied by 2 equals 340 V peak).
5. Allow the meter to record as loads are switched ON and OFF.
6. Read and record any findings of recorded transient voltages.
7. Correct for system transients by using surge suppressors to filter any transient voltages from the system.
8. Retest the system after repairs are completed.
9. Remove the power analyzer meter from the system.

Technical Tip

The total harmonic distortion (THD) at a typical electrical wall outlet in the United States is about 3%.

HARMONICS

A *harmonic* is a frequency that is an integer (whole number) multiple (second, third, fourth, fifth, etc.) of the fundamental frequency. The fundamental frequency on power distribution lines is 60 Hz (cycles per second) in North America and changes from the positive alternation to the negative alternation 60 times per second. For example, the second harmonic on a 60 Hz power distribution line is 120 Hz (60 Hz × 2). The second harmonic waveform (120 Hz) completes two cycles during one cycle of the fundamental waveform (60 Hz) over the same period of time. The third harmonic of the fundamental frequency is 180 Hz (60 Hz ×3). The third harmonic waveform (180 Hz) completes three cycles during one cycle of the fundamental waveform (60 Hz) over the same period of time.

Electronic circuits such as in energy-efficient lamps, variable speed motor drives, electronic ballasts used in lighting circuits, personal computers, printers, and medical test equipment that draw current in short pulses create harmonic distortion. Harmonics are a problem wherever there are a large number of personal computers and other nonlinear loads drawing current in short pulses. Equipment efficiency is improved when electronic equipment or circuits draw current in short pulses, but the short pulses cause harmonic distortion on the power lines.

Harmonic distortion is also created when diodes are used to rectify AC power and charge capacitors with the rectified DC voltage. The rectified DC voltage is used to charge large capacitors that draw current only during the peak of each half cycle of the AC sine wave to stay charged. During the rest of the sine wave, the capacitor draws no current. Harmonics cause "flat-topping" of the voltage waveform, lowering circuit peak voltage. In severe cases of flat-topping or when flat-topping and voltage sags occur, computers or other electronic equipment will continually reset due to insufficient peak voltage to charge capacitors.

Harmonics also cause motors to burn out, transformers to fail, circuit breakers to trip (nuisance tripping), and neutral conductors and other parts of the power distribution system to overheat. Severe overheating leads to electrical fires. When evaluating power quality, the incoming power, types (linear and nonlinear) and number of loads, and equipment used in the distribution system must all be tested.

Harmonic sequence is the phasor rotation with respect to the fundamental frequency (60 Hz). *Phasor rotation* is the order in which waveforms from each phase (phase A, phase B, and phase C) cross zero. Phasor rotation is simplified by using lines and arrows instead of waveforms to represent phase relationships. The phase sequence of a harmonic is important because the phase sequence determines the effect the harmonic has on the operation of loads and components such as conductors within a power distribution system. **See Figure 4-31.**

Positive sequence harmonics (fourth, seventh, tenth, etc.) have the same phase sequence as the fundamental harmonic (first), and cause additional heat in conductors, circuit breakers, and panels of a power distribution system. Negative sequence harmonics (second, fifth, eighth, etc.) have a phase sequence opposite the phase sequence of the fundamental (first) harmonic, which causes a rotating field in the opposite direction in motors. Similar to positive sequence harmonics, negative sequence harmonics cause additional heat in power distribution system components such as conductors, circuit breakers, and panel boards. Negative sequence harmonics also cause problems in induction motors because negative sequence harmonics cause part of the magnetic field of a motor to rotate in the reverse direction. The partial reverse rotation of the magnetic field is not enough to cause a motor to reverse direction, but the reverse rotation of the magnetic field does reduce forward torque of a motor and causes the motor to operate at higher than normal temperatures.

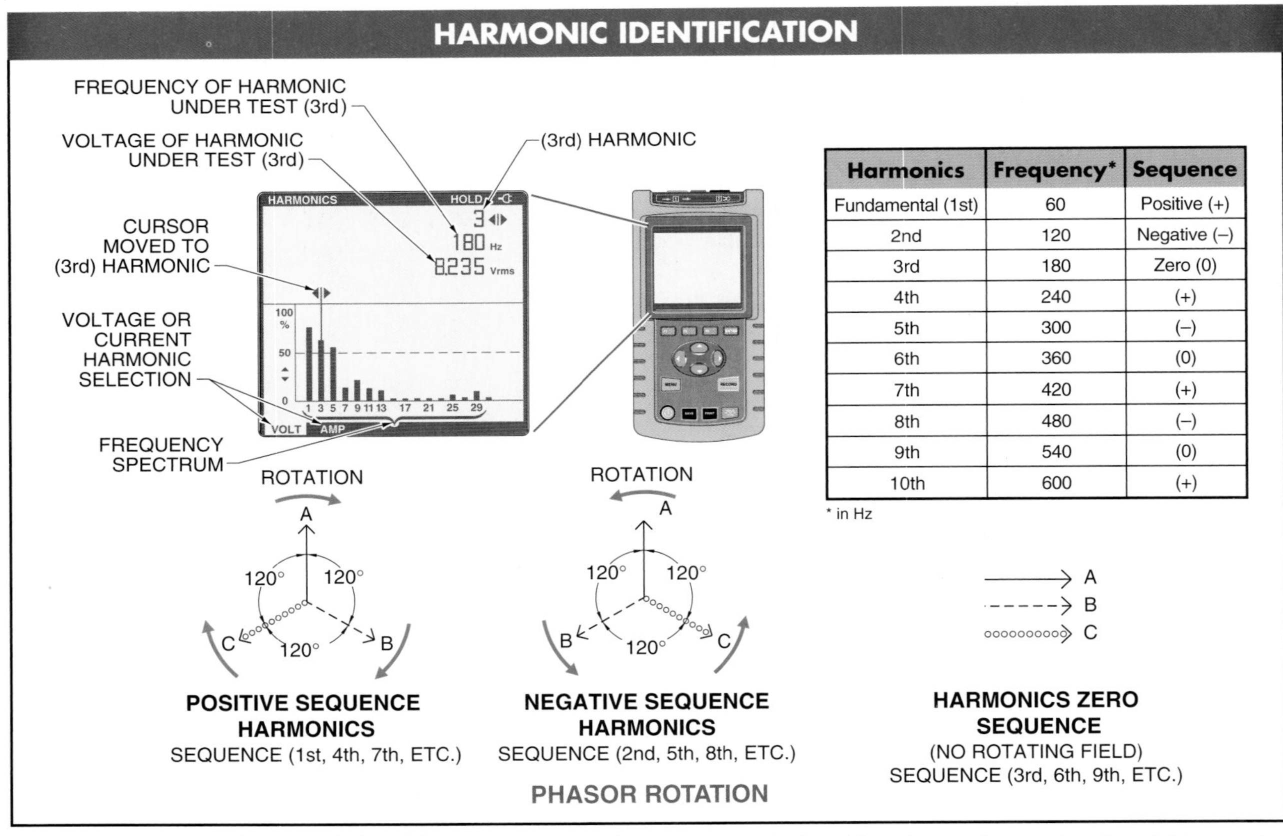

Harmonics	Frequency*	Sequence
Fundamental (1st)	60	Positive (+)
2nd	120	Negative (–)
3rd	180	Zero (0)
4th	240	(+)
5th	300	(–)
6th	360	(0)
7th	420	(+)
8th	480	(–)
9th	540	(0)
10th	600	(+)

* in Hz

Figure 4-31. Waveforms include odd harmonics, even harmonics, or both odd and even harmonics, in addition to the fundamental frequency.

Zero sequence harmonics (third, sixth, ninth, etc.) do not produce a rotating magnetic field in either direction. However, zero sequence harmonics do result in electrical equipment and system overheating. Zero sequence harmonics do not cancel, but add together in the neutral conductor of three-phase, 4-wire systems. A major system problem is created because there is no fuse or circuit breaker in the neutral conductor to limit current flow. Higher than normal currents in the neutral conductor are a fire risk.

Knowledge of the harmonics present on a power line is important when working on any power distribution system. A power analyzer meter can be used to measure the amount of voltage harmonics and current harmonics on a line. The intensity of each harmonic (second, third, fifth, etc.) present on the line and related information is indicated by data and the frequency spectrum on the graphic display of the meter.

Typically, even-numbered harmonics (second, fourth, sixth, eighth, etc.) tend to disappear or occur at levels that do not cause major problems. Likewise, higher harmonics such as 16th or 20th have smaller and smaller amplitudes, and are less important in affecting the overall operation of a power distribution system. However, odd-numbered harmonics are more likely to be present and do cause problems. For example, the third harmonic and odd multiples of the third harmonic (third, ninth, 15th, 21st, etc.), triplen harmonics (triplens), cause such problems as overloading of neutral conductors, telephone interference, and transformer overheating. **See Figure 4-32.**

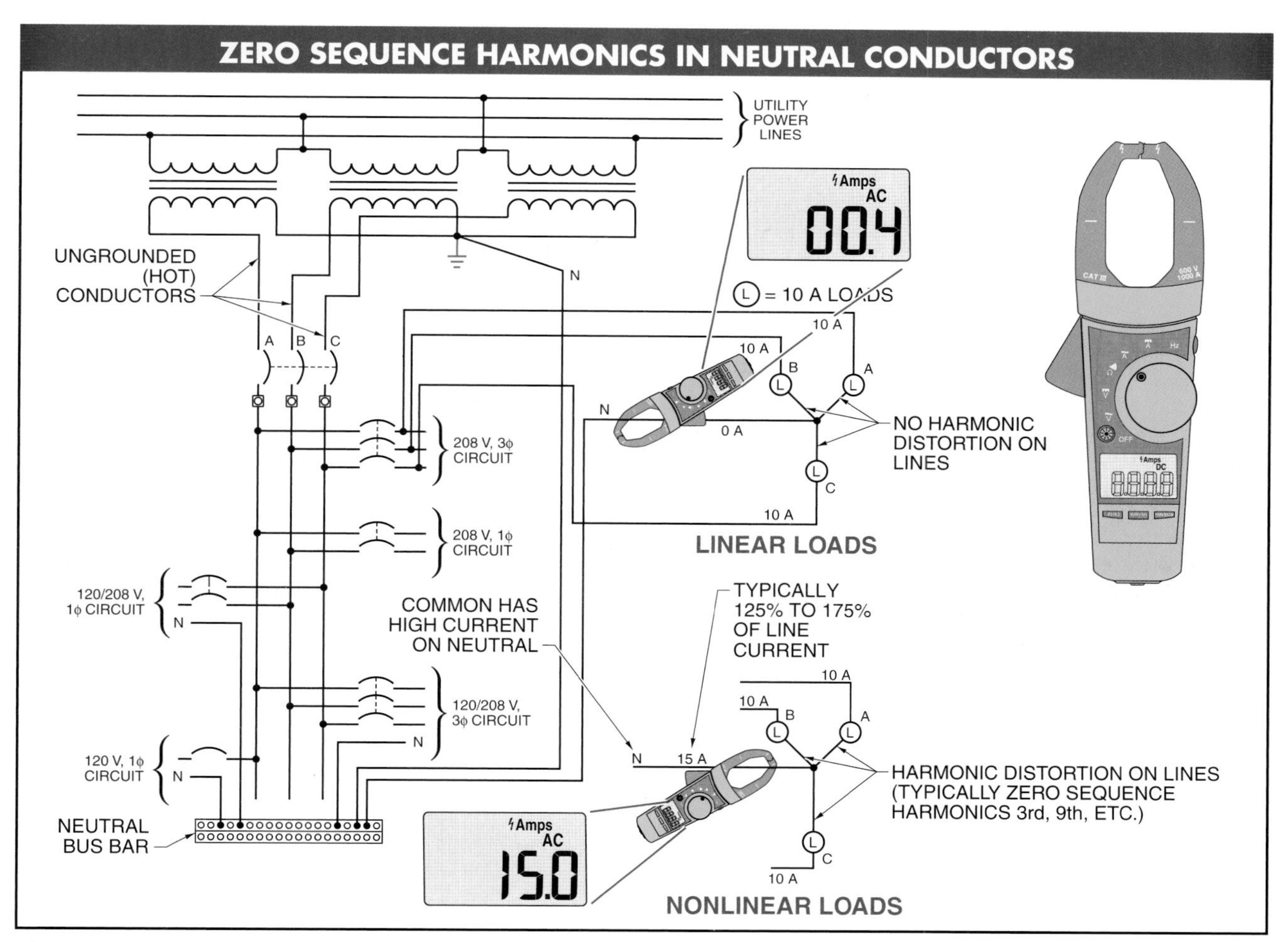

Figure 4-32. Zero sequence harmonics (third, sixth, ninth, etc.) do not produce a rotating field in either direction but add up on the neutral of a 4-wire system.

Total Harmonic Distortion

Total harmonic distortion (THD) is the amount of harmonics on a line compared with the fundamental frequency of 60 Hz. The THD considers all of the harmonic frequencies on a line. The greater the THD, the more distorted a pure 60 Hz sine waveform becomes. When troubleshooting a circuit for harmonics, measure the voltage THD and the current THD. The voltage THD must not be more than 5% and the current THD must not exceed 20% of the fundamental frequency.

Voltage THD is caused by harmonic currents on the line produced by nonlinear loads. High voltage distortion is a problem because voltage distortion becomes a carrier of harmonics to linear loads such as motors. Voltage harmonics cause problems (extra heat) in the power distribution system and to the loads connected to the system.

For an accurate measurement of the THD in a system, measure the THD at the transformer, not at the harmonic-generating load(s). Measuring for THD at the load(s) provides the highest THD reading because THD cancellation has not occurred along the system. When THD current is taken during full load, the THD is approximately equal to the total demand distortion (TDD). *Total demand distortion (TDD)* is the ratio of the current harmonics to the maximum load current. A TDD is a measurement taken when monitoring the system current harmonics over a period of time. A THD measurement is taken when testing or troubleshooting a system. The TDD is different from the THD because TDD is referenced to the maximum current measurement taken over time. The THD is a measurement of current on a power line only at the specific time of the measurement.

Harmonic Measurement Procedures

To test a circuit for harmonics, a power analyzer with a harmonic measurement function is required. Harmonic measurements are taken at the transformer, branch circuit, main power panel, or possibly the load. Harmonic distortion will be the greatest when measured at individual nonlinear loads and will be the least at the main panel because linear loads help equal out the total harmonics on a line. **See Figure 4-33.**

Before taking any harmonic measurements using a power analyzer meter, ensure the meter is designed to take measurements on the system being tested. Refer to the operating manual of the test instrument for all measuring precautions, limitations, and procedures. Always wear required personal protective equipment and follow all safety rules when taking the measurement. To measure harmonics, apply the following procedures:

1. Set the power analyzer to the harmonic measuring mode (and recording mode when required).
2. Connect the test leads and current clamp of the power analyzer meter to meter jacks as required.
3. Connect the voltage test leads and current clamp(s) of the meter to the system being tested.
4. Allow the meter to measure and record harmonics as loads are switched ON and OFF.
5. Read and record harmonic findings.
6. Correct any faults by adding harmonic filters to the system.
7. Retest the system after repairs are completed.
8. Remove the power analyzer meter from the system.

Troubleshooting Tip

TOTAL RESISTANCE—SERIES

Ohm's law states that the total resistance (R) of a circuit with series loads is the sum of the resistances of all the loads. To calculate total series resistance, apply the following formula:

$$R_T = R_1 + R_2 + R_3 + —\ldots$$

where

R_T = total resistance (in Ω)

R_1 = resistance 1 (in Ω)

R_2 = resistance 2 (in Ω)

R_3 = resistance 3 (in Ω)

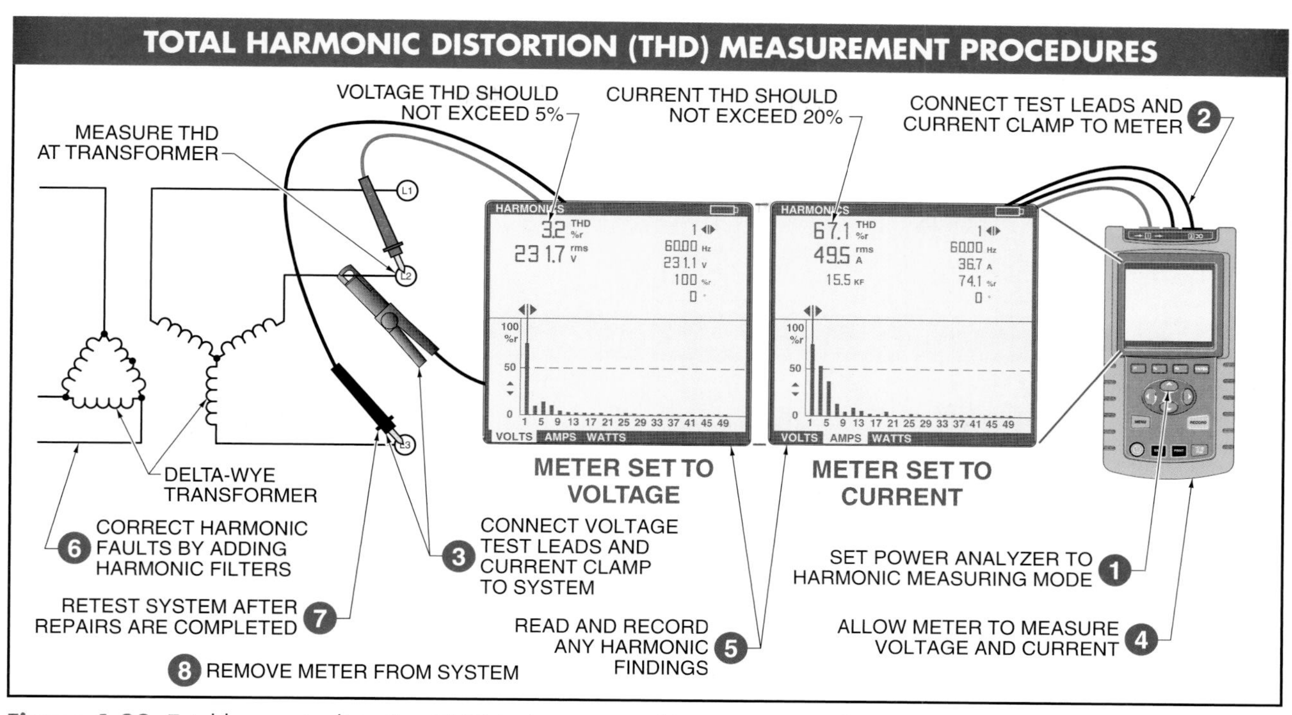

Figure 4-33. Total harmonic distortion (THD) is the amount of harmonics on a line (L1, L2, or L3) compared with the fundamental frequency (60 Hz).

POWER FACTOR

Power factor (PF) is the ratio of true power used in an AC circuit to apparent power delivered to the circuit. Power factor is commonly expressed as a percentage. True power equals apparent power only when the power factor is 100% or 1. When the power factor is less than 100% or 1, the circuit is less efficient and has a higher operating cost because not all current is performing work. To calculate power factor, apply the following formula:

$$PF = \frac{P_T}{P_A} \times 100$$

where

PF = power factor (in %)

P_T = true power (in W)

P_A = apparent power (in VA)

100 = constant

For example, the power factor of a small 1ϕ motor is typically very poor. What is the power factor of a ¼ HP, 1ϕ motor with 186.5 W of true power and 575 VA of apparent power?

$$PF = \frac{P_T}{P_A} \times 100$$

$$PF = \frac{186.5}{575} \times 100$$

$$PF = 0.3243 \times 100$$

$$PF = \mathbf{32.43\%}$$

The lower the power factor, the less efficient the circuit and the higher the overall operating cost. The overall operating cost is increased because every component in the system, such as transformers and conductors, must be sized for the higher current caused by a lower power factor. The power factor varies depending on types of loads.

Technical Tip

A common cause of a low power factor in industrial facilities is underloaded induction motors. The power factor of motors is much lower at partial loads than at full load. When replacing underloaded motors, the new motors should have smaller capacity. The factor can also be increased by installing synchronous motors or static capacitors across the line.

A power triangle identifies the relationship of true power, apparent power, reactive power, and power factor in a circuit. **See Figure 4-34.** A power triangle has a line (vector) that represents both magnitude and direction. Magnitude is indicated by the line length and direction is indicated by an angle of rotation from 0°. True power is represented by the horizontal vector line, apparent power is drawn lagging or leading true power by a line at angle theta (θ), and reactive power is represented by a vertical line that completes the triangle.

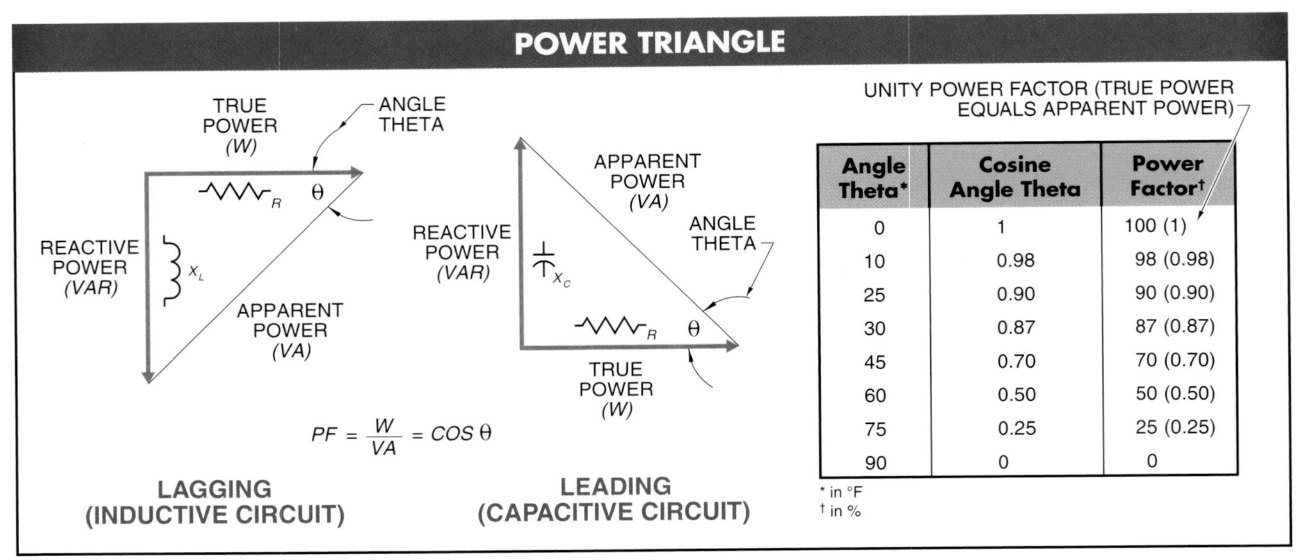

Angle Theta*	Cosine Angle Theta	Power Factor†
0	1	100 (1)
10	0.98	98 (0.98)
25	0.90	90 (0.90)
30	0.87	87 (0.87)
45	0.70	70 (0.70)
60	0.50	50 (0.50)
75	0.25	25 (0.25)
90	0	0

Figure 4-34. A power triangle is used to show the relationship of true power, apparent power, reactive power, and power factor in a circuit.

Power factor is lagging for inductive loads, leading for capacitive loads, and in phase for resistive loads. As circuit impedance increases, angle theta (θ) also increases, and as circuit impedance decreases, angle theta (θ) also decreases. Angle theta is used to find the power factor because the cosine of angle theta is equal to the circuit power factor. Power factor can be found by applying either of the following formulas:

$$PF = \frac{P_T}{P_A}$$

or $PF = \cos\theta$

where

PF = power factor (in %)

P_T = true power (in W)

P_A = apparent power (in VA)

$\cos\theta$ = cosine of angle theta

When the power factor is unity (cos θ is zero), true power is equal to apparent power. In most AC circuits, power factor is never equal to unity (1) because there is always some impedance on the power lines. The cos θ varies between 0 and 1, and power factor varies between 0 and 1 for most circuits.

A power triangle shows the mathematical relationship between the different types of power in a system by using basic trigonometry principles. In practice, power analyzer meters are used to take voltage and current measurements simultaneously. Power analyzer meters show the relationships among the different types of power and the power factor by displaying circuit measurements graphically. Power quality meters display true power (W or kW), apparent power (VA or kVA), reactive power (VAR or kVAR), and power factor (PF) of a circuit.

Poor Power Factor

Power factors of less than 1 increase the overall cost of operating a power distribution system. The lower the power factor, the higher the operating cost. Any time the power factor of a circuit drops to less than 85%, the circuit has a poor power factor. A poor power factor causes heat damage to insulation and other circuit components, reducing the amount of useful power available and requiring an increase in the sizes of conductors and equipment.

The lower the power factor, the higher the current required to supply power to the loads. For example, a 35 kW circuit load requires a 35 kVA transformer if the circuit power factor is 100%. However, if the circuit power factor is only 70%, a larger (50 kVA) transformer is required. **See Figure 4-35.**

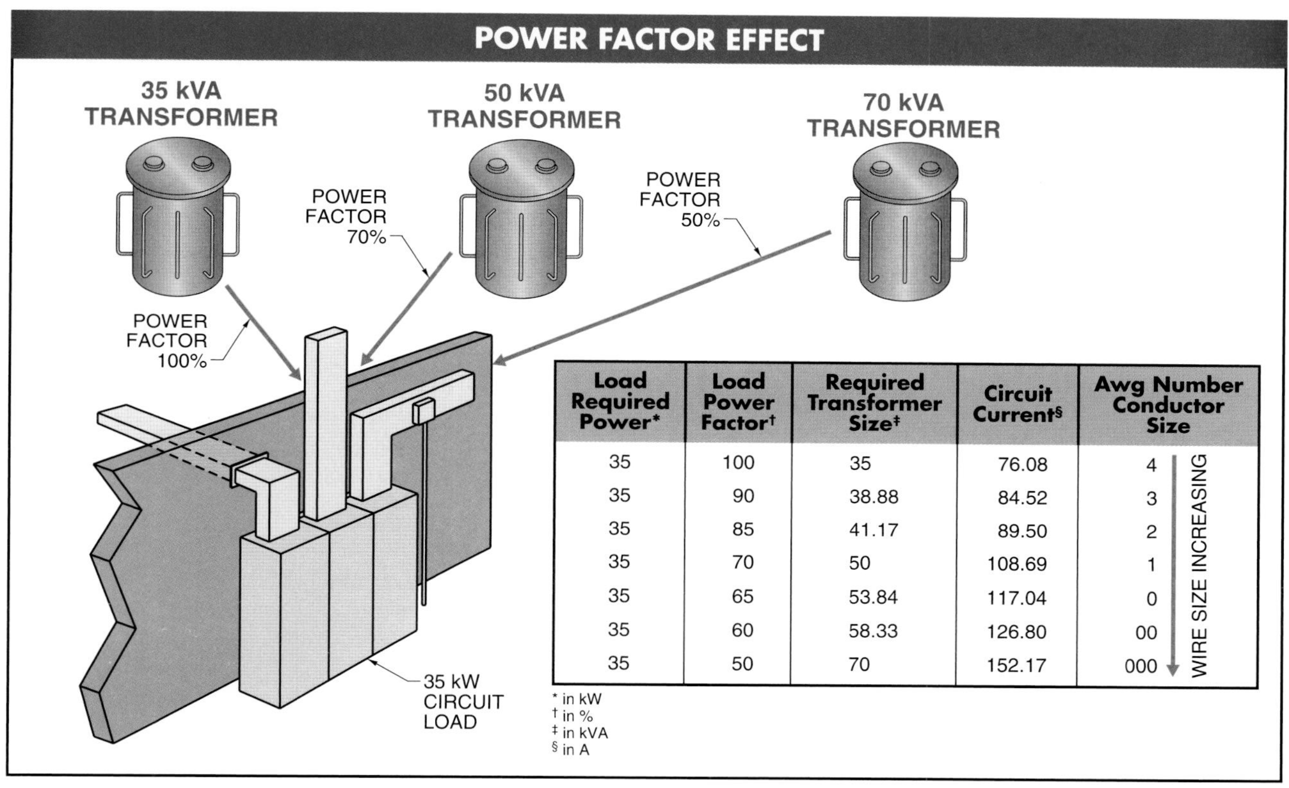

Load Required Power*	Load Power Factor†	Required Transformer Size‡	Circuit Current§	Awg Number Conductor Size
35	100	35	76.08	4
35	90	38.88	84.52	3
35	85	41.17	89.50	2
35	70	50	108.69	1
35	65	53.84	117.04	0
35	60	58.33	126.80	00
35	50	70	152.17	000

* in kW
† in %
‡ in kVA
§ in A

Figure 4-35. The lower the power factor, the higher the current required to supply the loads.

Poor power factor can be addressed with the addition of other components to a system. The efficiency of a motor is determined by the power out divided by the power in. Three-phase motors are more energy-efficient than single-phase or DC motors and are the most common motors used in commercial and industrial applications. Three-phase motors account for approximately 85% of all electric motors used and consume approximately 53% of all power produced. Efforts to improve energy efficiency and conservation of resources prompted the legislation of the Energy Policy Act (EPAct), which was passed by Congress and signed into law October 24, 1992. The Energy Policy Act took effect on October 24, 1997 and mandates efficiency standards for 3ϕ, 1 HP to 200 HP general-purpose motors.

Displacement Power Factor

Utility companies sell power based on the amount of true power (kW) a customer uses. However, utility companies supply apparent power (kVA) to customers. For example, assume a facility (company) is using 1 MW of true power per hour and has a power factor of 65%. The utility company must supply 1.539 MVA per hour (kVA = kW ÷ PF) of apparent power. If the power factor were corrected to 90%, the utility would only have to supply 1.111 MVA of apparent power to the company to do the same amount of work.

Because the power factor affects the cost of delivering power, utility companies typically charge larger customers with a poor power factor an additional charge for the kilovolt amps used. The additional charge is added to the electric bill as a power factor charge. Typically, an automatic increase occurs on the bill whenever power factor drops below 90% at a facility. The lower the power factor, the higher the additional power factor charge.

The two power factor measurements are total power factor (PF) and displacement power factor (DPF). *Total power factor (power factor* or *PF)* is the power factor equal to the total difference between true power and apparent power (VA) in a circuit. Power factor takes into account the power factor of the fundamental line frequency (60 Hz) and the power factor of any harmonic frequencies that may be on the power line.

Displacement power factor (DPF) is the power factor of the fundamental frequency (60 Hz) only. In any circuit that includes nonlinear loads, there are some harmonic frequencies and the total power factor is different from the displacement power factor. Any difference between the total power factor and the displacement power factor indicates that there are harmonics in the system.

Typically, utility company power factor charges are based on displacement power factor (DPF) values and not power factor (PF) values. A power analyzer meter measures and displays the PF and DPF of a circuit being tested. **See Figure 4-36.**

Companies find improving power factor is more economical than paying higher power costs. One way of correcting the power factor is to add capacitor banks to the power lines. Capacitor banks are designed to correct the power factor only at the fundamental frequency (60 Hz). When there is a large amount of harmonics on the power lines, sizing capacitors for the displacement power factor values will cause capacitor failure.

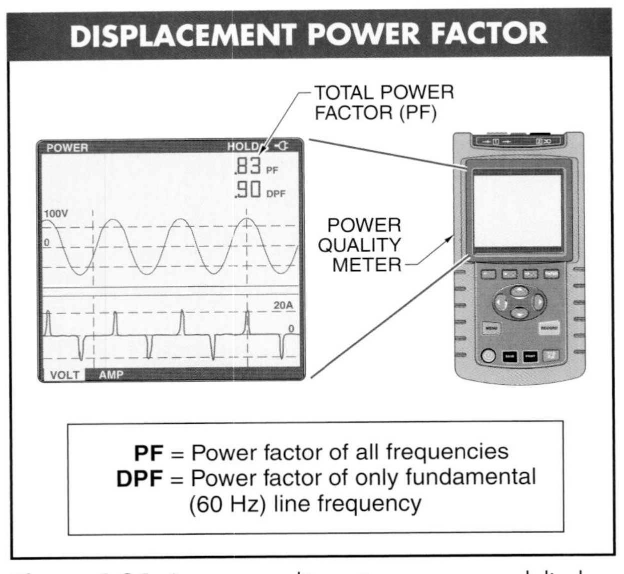

Figure 4-36. A power quality meter measures and displays power factor (PF) and displacement power factor (DPF).

Technical Tip

Typical power meter features include filtering, data logging, event triggering, and application software for use with a PC.

Power Factor Measurement Procedures

To test the power factor of circuits, a power quality analyzer meter with a power factor measurement function is required. Power factor measurements are taken at the transformer, branch circuit, main power panel, and possibly the load. **See Figure 4-37.**

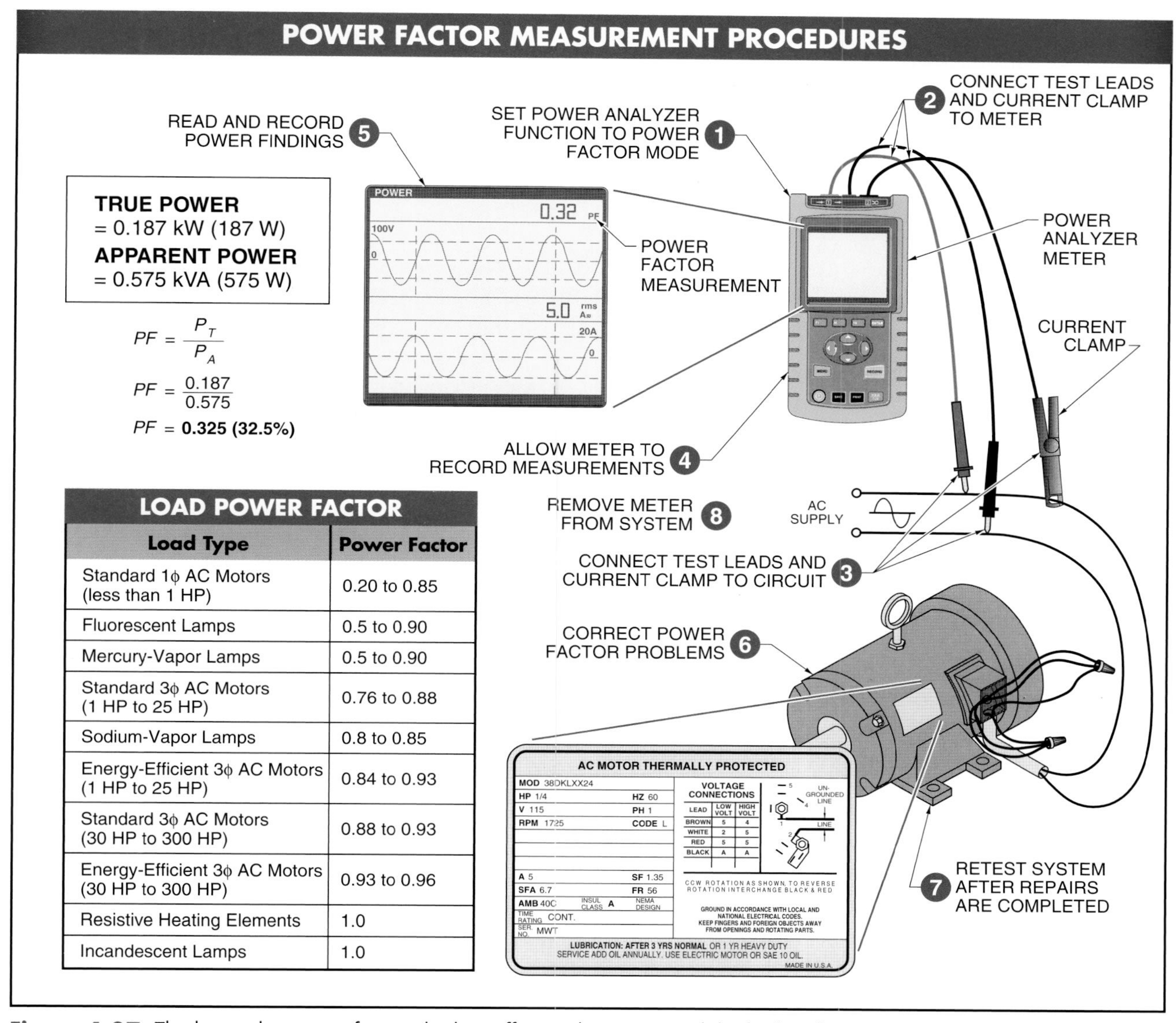

LOAD POWER FACTOR

Load Type	Power Factor
Standard 1φ AC Motors (less than 1 HP)	0.20 to 0.85
Fluorescent Lamps	0.5 to 0.90
Mercury-Vapor Lamps	0.5 to 0.90
Standard 3φ AC Motors (1 HP to 25 HP)	0.76 to 0.88
Sodium-Vapor Lamps	0.8 to 0.85
Energy-Efficient 3φ AC Motors (1 HP to 25 HP)	0.84 to 0.93
Standard 3φ AC Motors (30 HP to 300 HP)	0.88 to 0.93
Energy-Efficient 3φ AC Motors (30 HP to 300 HP)	0.93 to 0.96
Resistive Heating Elements	1.0
Incandescent Lamps	1.0

Figure 4-37. The lower the power factor, the less efficient the circuit and the higher the operating costs.

Before taking any power factor measurements using a power analyzer meter, ensure the meter is designed to take measurements on the system being tested. Refer to the operating manual of the test instrument for all measuring precautions, limitations, and procedures. Always wear required personal protective equipment and follow all safety rules when taking the measurement. To measure power factor, apply the following procedures:

1. Set the power analyzer meter to power factor measuring mode (and recording mode as required).
2. Connect the test leads and current clamp of the power analyzer meter to the meter jacks as required.
3. Connect the voltage test leads and current clamp(s) to the system being tested.
4. Allow the meter to measure and record voltage and current measurements (power).
5. Read and record the power findings.

6. Correct power factor problems by adding capacitor banks or other devices as required.
7. Retest the system after repairs are completed.
8. Remove the power analyzer meter from the system.

NOISE

Noise enters a power distribution system directly on the wires (L1, L2, and L3), on grounds, or through magnetic coupling of adjacent wires. Noise is typically produced on power lines from common mode noise and transverse mode noise. *Common mode noise* is noise produced between ground and hot lines, or between ground and neutral lines. *Transverse mode noise* is noise produced between hot and neutral lines. **See Figure 4-38.**

Arcing at motor brushes, ground faults, poor grounds, radio transmitters, and ignition systems, and the opening of electrical contacts also cause common mode noise. The opening of electrical contacts produces noise because an arc is created as the contacts are pulled apart. The higher the current in the circuit being opened, the larger and longer the arc. Transverse mode noise is also caused by welders, switched power supplies, and the firing of silicon-controlled rectifiers (SCRs) in electrical equipment.

Noise in a system can produce false signals in electronic circuits, leading to processing errors, incorrect data transfer, and printer errors. An oscilloscope is used to test for noise problems in electrical systems and electronic equipment. When noise is a problem, filter circuits and/or noise suppressors are added to circuits to reduce noise.

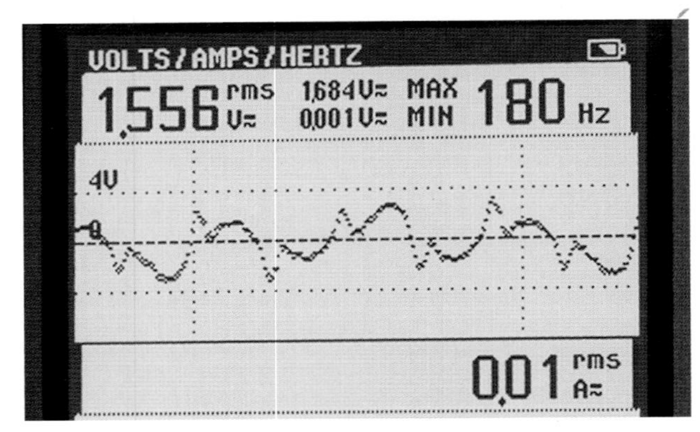

Fluke Corporation

Noise in an electrical system is indicated by using graphic display meters.

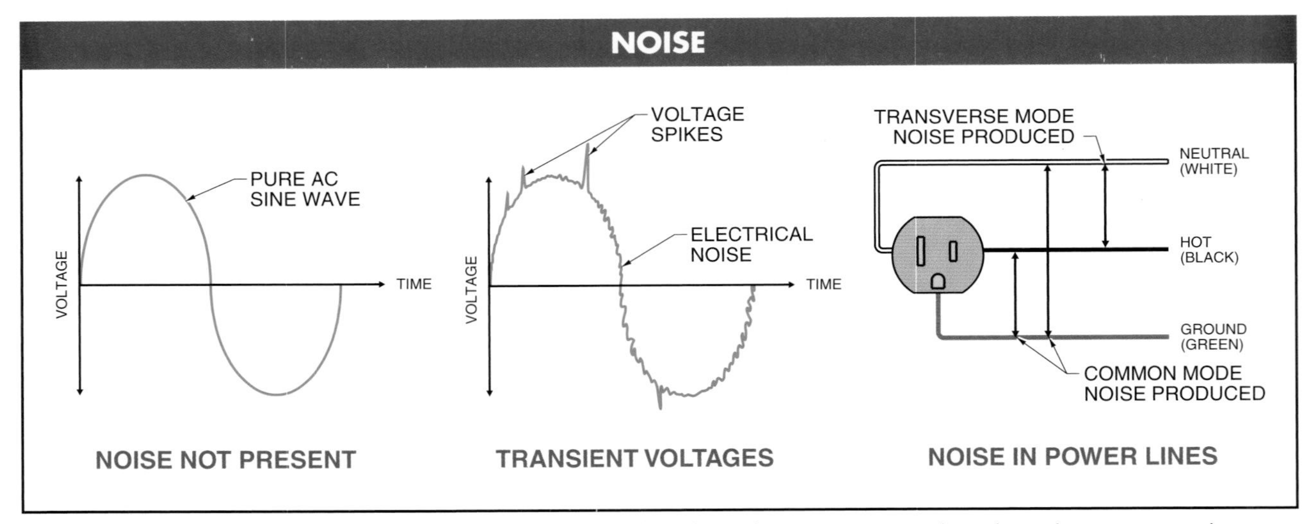

Figure 4-38. Noise can enter a power distribution system directly on the wires or grounds or through magnetic and capacitive coupling of adjacent wires.

POWER QUALITY TEST INSTRUMENTS

4

REVIEW

Name: ______________________________ Date: ____________________

________	**1.**	Voltage unbalance should not be more than what percent?
________	**2.**	What color is standard for grounding conductors?
________	**3.**	How many seconds can a momentary power interruption last?
________	**4.**	Which harmonic is 180 Hz?
T F	**5.**	Nonlinear loads are a common source of harmonic problems.
________	**6.**	Which test instrument is used to detect noise in an electrical signal?
________	**7.**	What colors are standard for neutral conductors?
________	**8.**	Is a programmable controller a linear or nonlinear load?
T F	**9.**	Some motors can continue to operate on only two out of three phases.
________	**10.**	Which test instrument is used to test phase unbalance?
________	**11.**	Current unbalance should not be more than what percentage?
T F	**12.**	Improper phase sequence reverses motor rotation.
________	**13.**	How many electrical degrees separate balanced phases?
________	**14.**	Is an incandescent lamp a linear or nonlinear load?
________	**15.**	What type of fluctuation is a voltage decrease of 15% for 32 sec?
T F	**16.**	High voltage unbalances can be created by small current unbalances.
T F	**17.**	Harmonics are produced by linear loads.
________	**18.**	Which test instrument is designed to measure temperature at a point without contact?
T F	**19.**	480 Hz is a triplen harmonic.
________	**20.**	What type of fluctuation is a voltage increase of 12% for 78 sec?

21. Explain the difference between linear and nonlinear loads.

22. What effects can harmonics have on loads?

23. How does low power factor affect the cost of operating a power distribution system?

24. Explain the first step when troubleshooting a power quality problem.

25. How can temperature measurements help troubleshoot power quality problems?

5 ELECTRONIC CIRCUIT TEST INSTRUMENTS

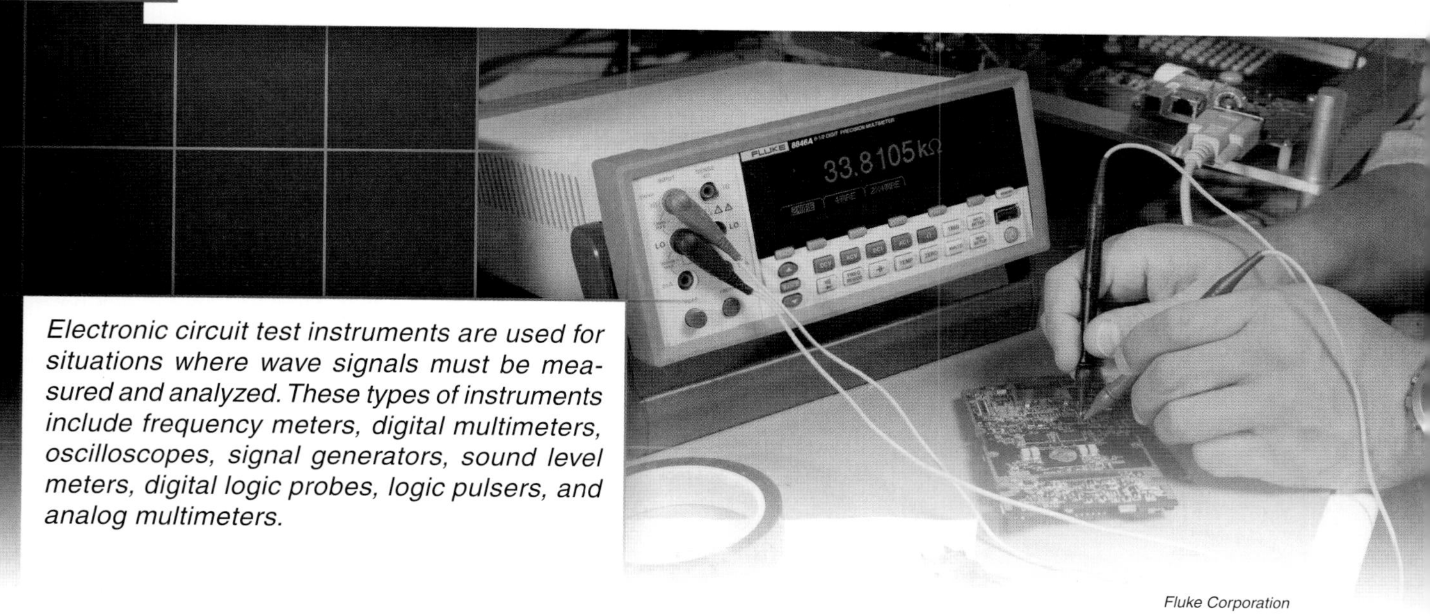

Electronic circuit test instruments are used for situations where wave signals must be measured and analyzed. These types of instruments include frequency meters, digital multimeters, oscilloscopes, signal generators, sound level meters, digital logic probes, logic pulsers, and analog multimeters.

Fluke Corporation

FREQUENCY

Frequency is the number of cycles per second of an AC sine wave. A *cycle* is one complete wave of alternating positive and alternating negative voltage or current. An *alternation* is one half of a cycle. A *period* is the time required to produce one complete cycle of a waveform. Frequency is measured in hertz (Hz). *Hertz (Hz)* is the international unit of frequency and is equal to one cycle per second. **See Figure 5-1.**

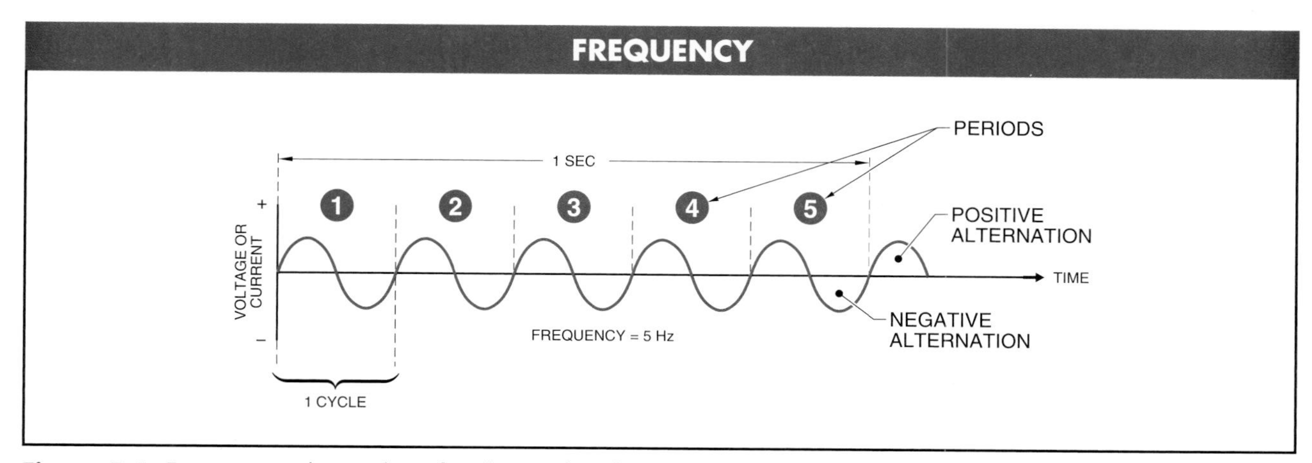

Figure 5-1. Frequency is the number of cycles produced per second and is measured in hertz (Hz).

The term frequency is typically used to describe electrical equipment operation, as in discussing power line frequency (which is normally 50 Hz or 60 Hz) and variable frequency drives (which normally use a 1 kHz to 20 kHz carrier frequency). The term frequency is also used to describe electronic equipment operation, as in the terms audio frequency (15 Hz to 20 kHz) and radio frequency (30 kHz to 300 kHz for low frequency, 300 kHz to 3 MHz for medium frequency, 3 MHz to 30 MHz for high frequency, and 30 MHz to 300 MHz for very high frequency).

Frequency Meters

A *frequency meter (frequency counter)* is a test instrument that is used to measure the frequency of an AC signal. A frequency meter can be a standalone meter whose primary function is to measure frequency, or a frequency-measuring function that is part of a digital multimeter. Frequency meters designed primarily for measuring the frequency of a signal have a wide frequency measuring range—from a few hertz to a gigahertz (GHz) or more. Digital multimeters that include a frequency measuring function can measure a wide range of frequencies but have a much lower measuring range, typically 5 Hz to 1 MHz. **See Figure 5-2.**

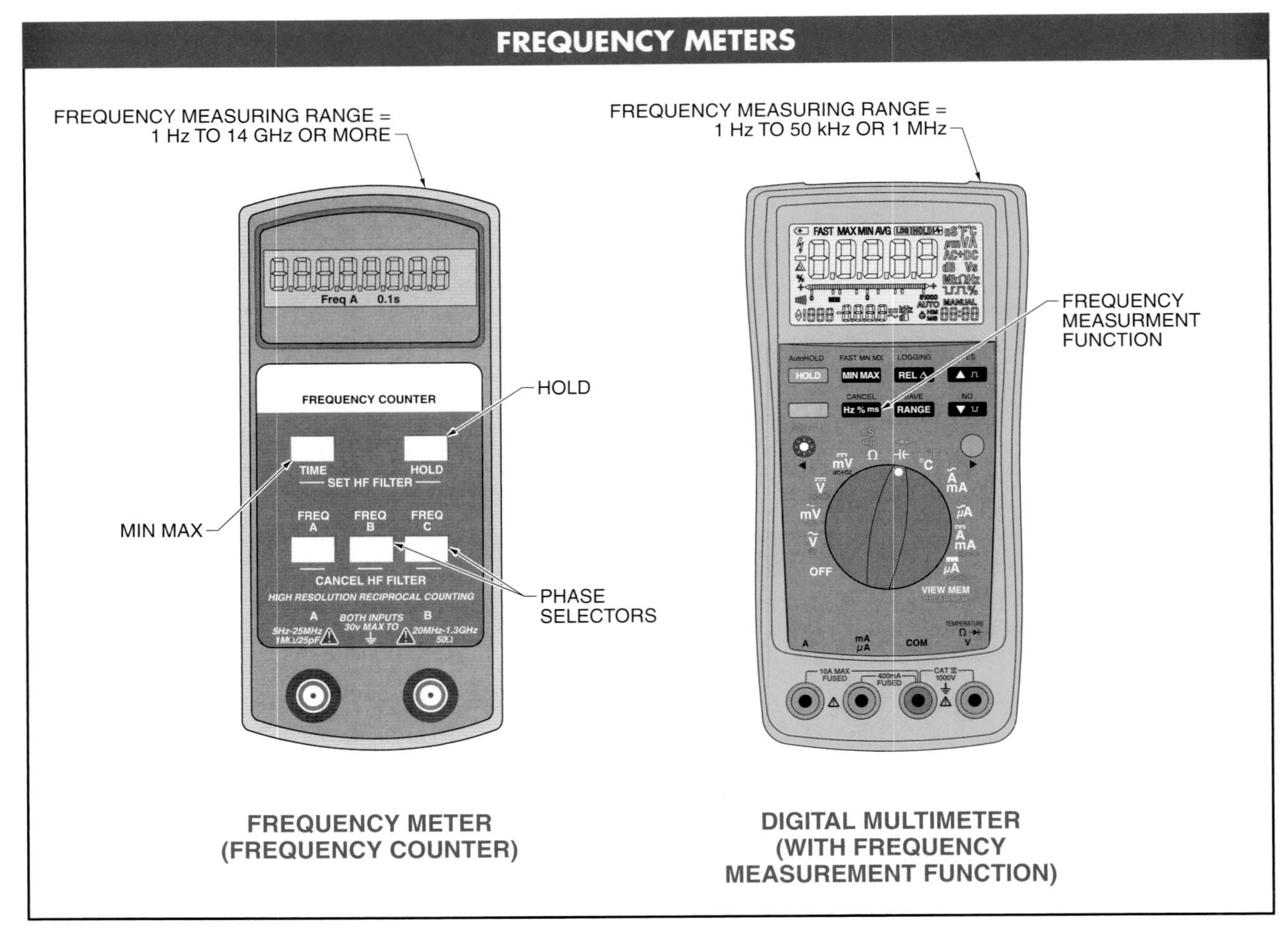

Figure 5-2. Frequency is measured with stand-alone frequency meters, or digital multimeters with a frequency measurement function.

Frequency Measurements

Electrical circuits and equipment are designed to operate at either a fixed or variable frequency. Equipment designed to operate at a fixed frequency performs abnormally when operated at a frequency other than the frequency specified by the manufacturer. For example, an AC motor designed to operate at 60 Hz rotates more slowly when the frequency is less than 60 Hz and rotates faster when the frequency is above 60 Hz. Any change in frequency to an AC motor causes a proportional change in motor speed. For example, a 5% reduction in frequency produces a 5% reduction in motor speed.

Frequency meters are used to measure frequency when troubleshooting electrical and electronic equipment. When frequency meters include a MIN MAX recording mode, frequency measurements are recorded over a specific time period. The MIN MAX recording mode is used to record frequency measurements the same way voltage, current, or resistance measurements are recorded. Frequency meters typically have a manual frequency range selection such as 1 Hz to 25 MHz or 20 MHz to 1 GHz. Some frequency meters automatically select the best frequency measurement range for the application. However, when the frequency of the measured voltage is outside the frequency measurement range, the meter cannot display an accurate measurement. Refer to the user's manual of a specific frequency meter for the frequency measurement range permitted by the meter.

Frequency Meter Frequency Measurement Procedures

Before taking any frequency measurements using a frequency meter, ensure the meter is designed to take measurements on the circuit being tested. **See Figure 5-3.** Refer to the operating manual of the test instrument for all measuring precautions, limitations, and procedures. To measure frequency using a frequency meter, apply the following procedures:

Safety Procedures

- Follow all electrical safety practices and procedures.
- Check and wear personal protective equipment (PPE) for the procedure being performed.
- Perform only authorized procedures.
- Follow all manufacturer recommendations and procedures.

1. Set the frequency meter to the frequency-measuring range required for the application.
2. Plug the test leads of the frequency meter into the meter jacks. The position of the test leads is arbitrary.
3. Connect the test leads to the circuit. The position of the test leads is arbitrary.
4. Read the displayed frequency. The abbreviation Hz, kHz, MHz, or GHz is displayed as part of the reading.
5. Remove the frequency meter from the circuit.

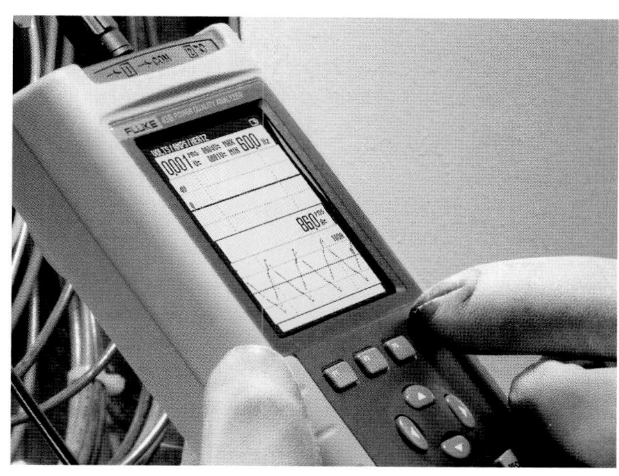

Fluke Corporation

Power quality analyzer meters display the frequency of electricity when connected to a circuit with a current clamp to measure current.

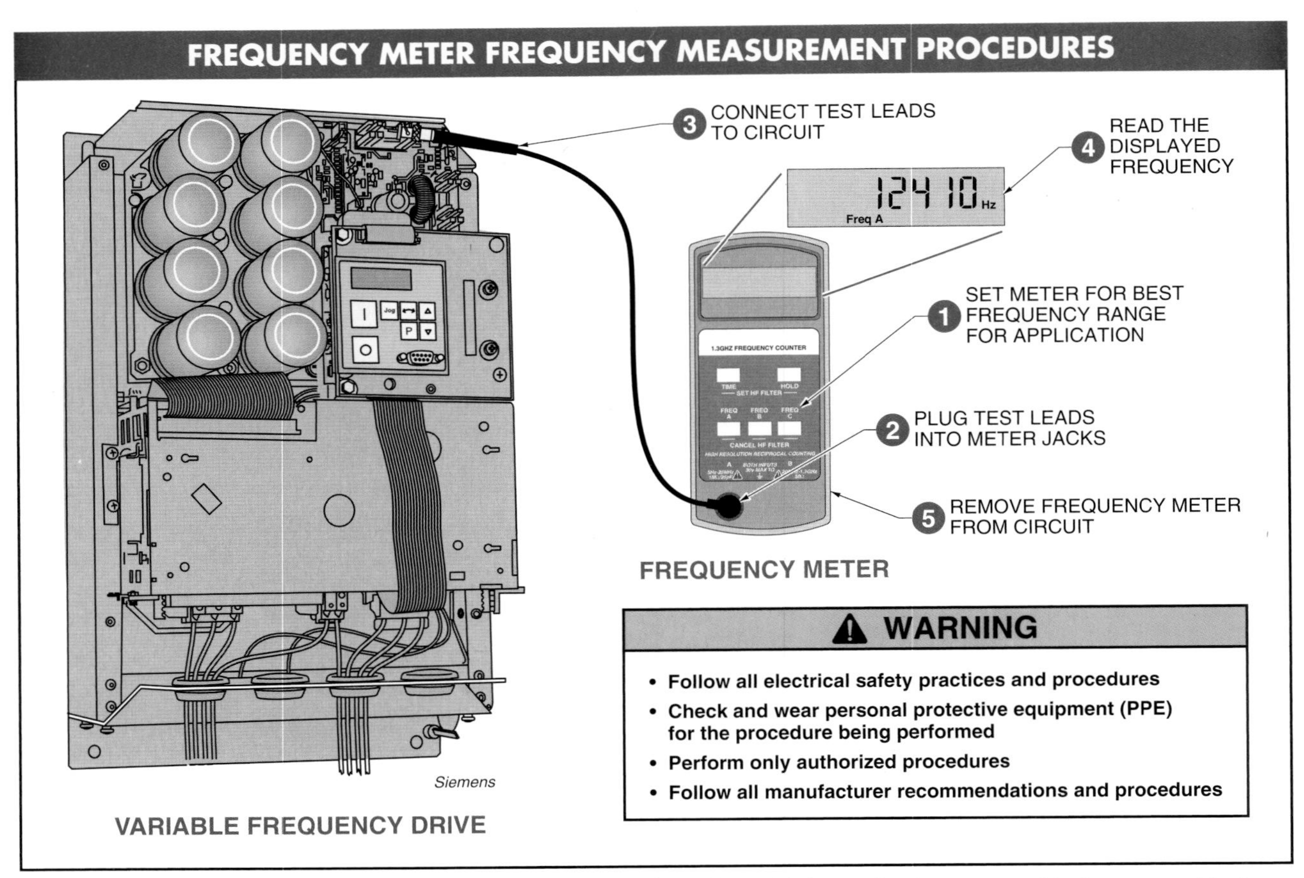

Figure 5-3. Measuring the frequency at various positions of an electrical device (such as a variable frequency drive) can identify defective device components.

Digital Multimeter Frequency Measurement Procedures

Before taking any frequency measurements using a digital multimeter, ensure the meter is designed to take measurements on the circuit being tested. **See Figure 5-4.** Refer to the operating manual of the test instrument for all measuring precautions, limitations, and procedures. Always wear required personal protective equipment and follow all safety rules when taking the measurement. To measure frequency using a digital multimeter with a frequency-measuring function, apply the following procedures:

1. Set the function switch to AC voltage. Set the range to the highest voltage setting when the voltage in the circuit is unknown. Multimeters typically power up in the automatic mode, which automatically selects the best measurement range based on the voltage and frequency present.
2. Plug the test leads of the frequency multimeter into the proper meter jacks.
3. Connect the test leads to the circuit.
4. Read the voltage displayed on the frequency multimeter.
5. With the meter still connected to the circuit, press the Hz button.
6. Read the frequency measurement displayed. The abbreviation Hz, kHz, MHz, or GHz is displayed as part of the reading.
7. Remove the frequency multimeter from the circuit.

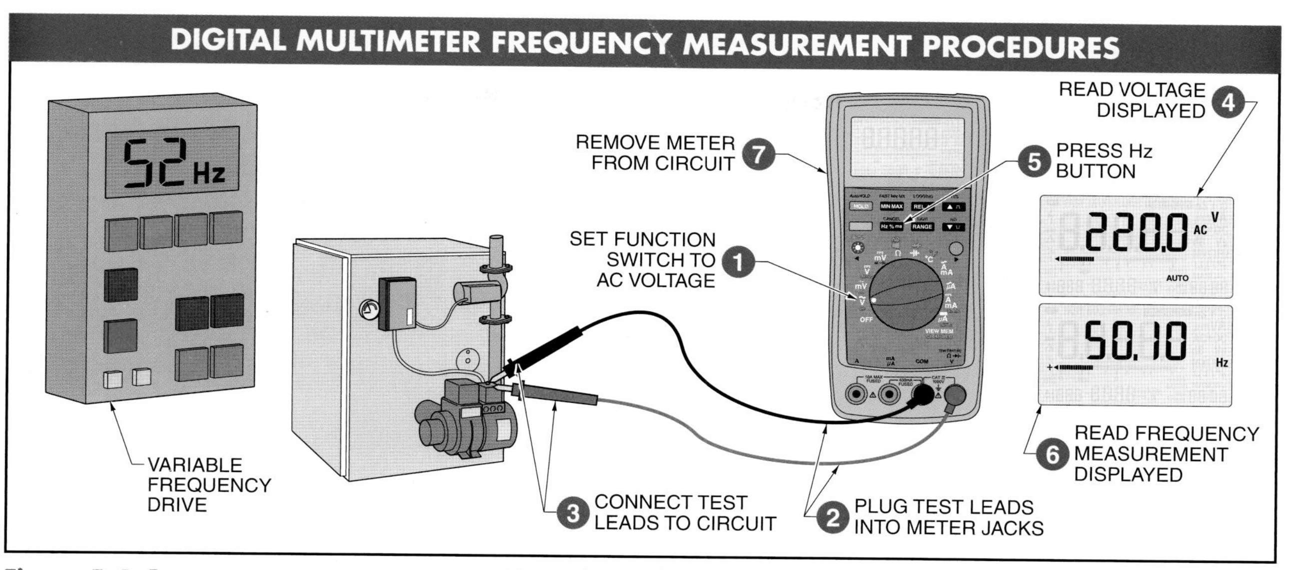

Figure 5-4. Frequency measurement using a DMM with a Hz button requires first measuring AC voltage.

In some circuits, there may be enough distortion on the power line to prevent an accurate frequency measurement. For example, AC variable frequency drives produce frequency distortions. To accurately check the frequency of a circuit with a variable frequency drive, use a meter that is designed to take frequency measurements on variable frequency drives. Meters designed to operate on circuits with solid-state devices include a bypass filter circuit to separate the power frequency from the carrier frequency and display only the power frequency.

OSCILLOSCOPES

Most electronic circuits are designed to take a signal (power, voice, picture, data, digital, etc.) and change the signal by amplifying, filtering, storing, displaying, or converting the signal from analog to digital or digital to analog. The testing of electronic circuits and equipment often requires measuring or observing electrical signals. Electrical signals are measured by using meters (voltmeters and ammeters), or are observed by using oscilloscopes. An *oscilloscope* is a test instrument that provides a visual display of voltages. An oscilloscope provides the waveform for the voltage in a circuit and also allows the voltage level, frequency, and phase to be measured. The two types of oscilloscopes are bench and handheld. Both types of oscilloscopes include the same basic features. **See Figure 5-5.**

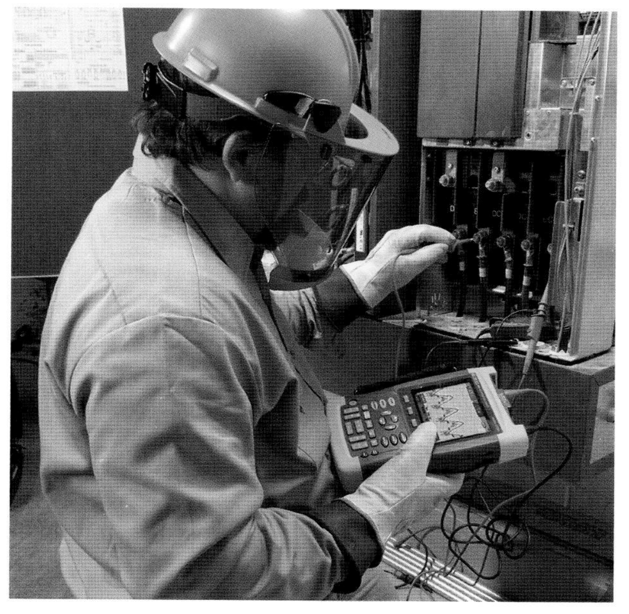

Fluke Corporation

Handheld oscilloscopes are used when signal measurements must be captured in the field.

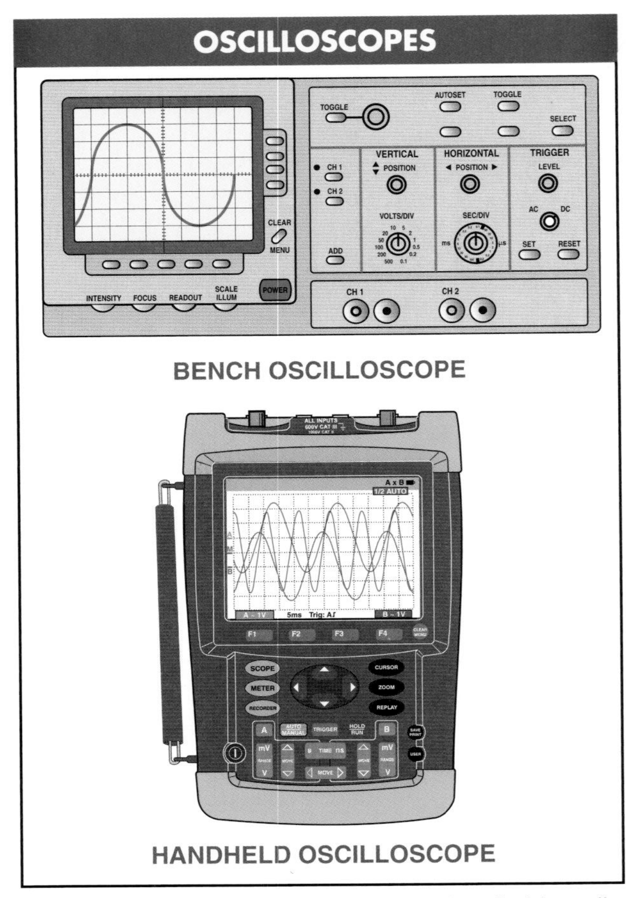

Figure 5-5. Bench oscilloscopes and handheld oscilloscopes include the same basic features and are referred to as oscilloscopes, scopes, or power quality meters.

A *bench oscilloscope* is a test instrument that displays the shape of a voltage waveform and is used mostly for bench testing electrical and electronic circuits. *Bench testing* is testing performed when equipment being tested is brought to a designated service area. Bench-type oscilloscopes are used to troubleshoot digital circuit boards, communication circuits, TVs, VCRs, DVDs, computers, and other types of electronic circuits and equipment.

Technical Tip

There are approximately 78 companies worldwide that manufacture oscilloscopes for testing purposes.

A *handheld oscilloscope* is a test instrument that displays the shape of a voltage waveform and is typically used for field testing. *Field testing* is testing performed when the test instrument is taken to the location of the equipment to be tested. Most handheld oscilloscopes are a combination oscilloscope and digital multimeter. Handheld oscilloscopes can also be referred to as portable oscilloscopes, Scopemeters®, PowerPads™, power quality meters, power analyzers, or power meters.

Technical Tip

Features to consider when choosing oscilloscopes include high-capacity hard discs, removable storage, nonvolatile memory, sample rate, and bandwidth.

Oscilloscope Features

All oscilloscopes display voltage wave shapes of electrical, electronic, and digital signals. However, how accurate a displayed wave shape or signal can be depends upon several oscilloscope features and specifications. The important specifications and features of an oscilloscope include sample rate, bandwidth, and whether the display has a dual trace.

Sample rate is the speed with which an oscilloscope takes a "picture" of an incoming signal. The higher the sample rate of an oscilloscope, the more accurate the trace of the signal being tested. Higher sample rates do increase the cost of an oscilloscope or any type of test instrument. Sample rate is specified as the number of wave captures in 1 sec and is listed in megasamples per second (MS/s) or gigasamples per second (GS/s). The more unstable (changing) the signal, the higher the sample rate of the test instrument must be to display an accurate trace. Typically, the specified sample rate of the test instrument is 3 to 5 times faster than the fastest cycle (frequency in Hz) to be captured. For example, to view a 10 MHz signal, a minimum 30 MS/s rated oscilloscope must be used.

Bandwidth is the width of a range of frequencies that have been specified as performance limits within which a meter can be used. There are two general bandwidth ratings, real-time and repetitive. *Real-time (single-shot) bandwidth* is the highest frequency an oscilloscope can capture in a single pass. The real-time bandwidth rating of an oscilloscope must be at least twice the frequency of the signal being tested. *Repetitive (equivalent-time) bandwidth* is the picture of the signal that the oscilloscope displays after multiple sweeps of the signal being tested. Bandwidth is typically rated in megahertz (MHz). The wider the bandwidth rating (20 MHz, 100 MHz, 200 MHz) of an oscilloscope, the greater the number of circuits the oscilloscope can be used to test.

Oscilloscope Display and Trace

An oscilloscope displays the voltage being tested on the screen of the unit. The screen of an oscilloscope contains scribed horizontal and vertical axes. The horizontal (x) axis represents time and the vertical (y) axis represents the amplitude of the voltage waveform. The scribed lines also divide the screen into equal divisions. The divisions are used to determine the voltage level and frequency of the displayed waveforms. **See Figure 5-6.**

A trace is established on the screen of an oscilloscope before the circuit being tested is connected to the oscilloscope. A *trace* is a reference point or line that is visually displayed on the face of the oscilloscope screen. A trace is normally positioned above or equally above and below the horizontal axis of the screen.

The starting point of a trace is located at the left side of the screen. *Sweep* is the movement of the displayed trace across the oscilloscope screen. The sweep of the oscilloscope trace is from left to right.

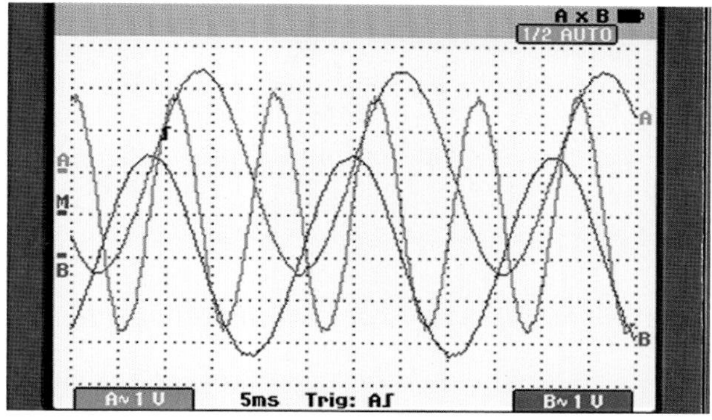

Fluke Corporation

Oscilloscope displays sometimes have color traces.

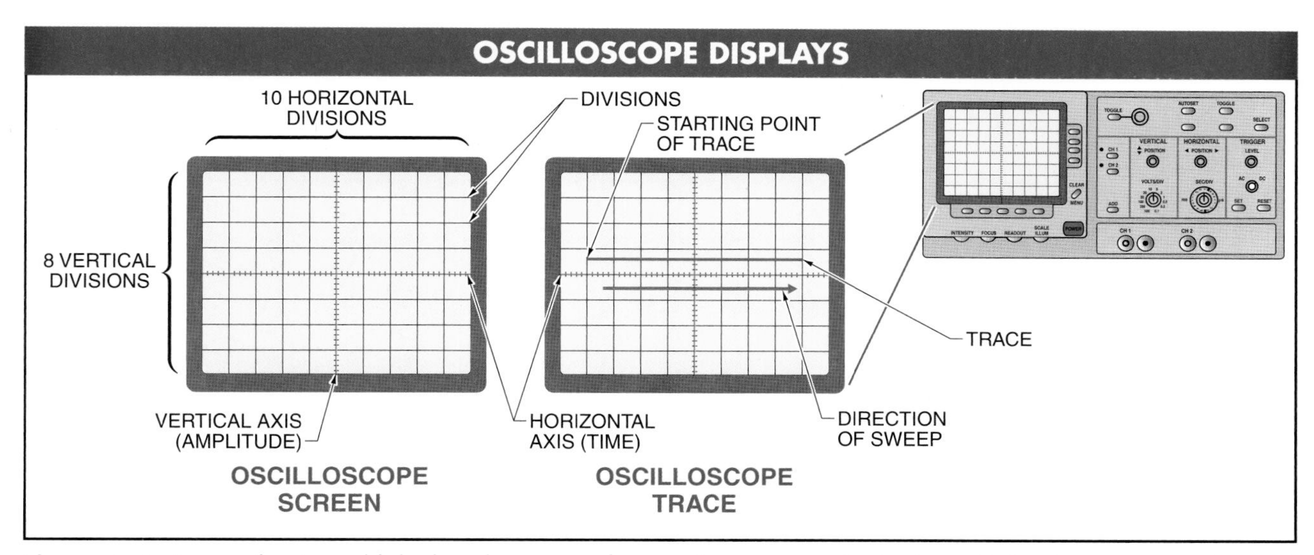

Figure 5-6. A trace that is established on the screen of an oscilloscope uses the horizontal (x) axis to represent time and the vertical (y) axis to represent the amplitude of the voltage waveform.

Dual-Trace Oscilloscopes

A *dual-trace oscilloscope* is an oscilloscope that displays two signal traces simultaneously. Dual-trace oscilloscopes include two separate inputs (channel 1 and channel 2) through which the two signals are sent into the oscilloscope for viewing and comparison. Dual-trace oscilloscopes are used for such applications as monitoring the input signal and output signal of a circuit simultaneously. By means of the dual trace, any signal gain, loss, distortion, or other changes to the signal are seen. **See Figure 5-7.**

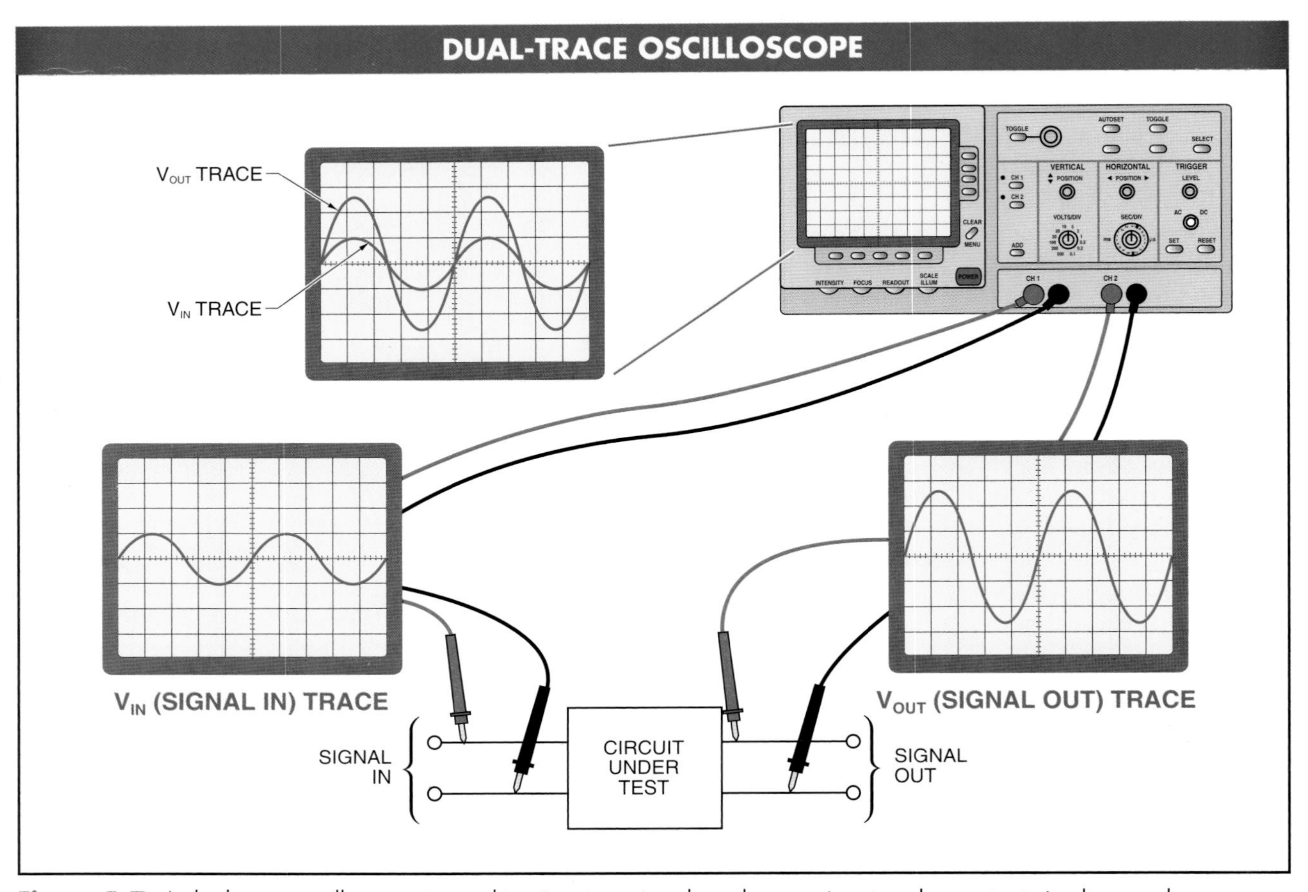

Figure 5-7. A dual-trace oscilloscope is used to view two signals such as an input and an output simultaneously.

Oscilloscope Adjustments

Manual controls on an oscilloscope are adjusted to view waveforms. Typical oscilloscope controls that are adjusted include intensity, focus, horizontal positioning, vertical positioning, volts per division, and time per division.

Intensity is the level of trace brightness. The intensity control of an oscilloscope sets the brightness level of the displayed voltage trace. The intensity level is kept as low as possible to keep the trace in focus. The focus control adjusts the sharpness of the displayed voltage trace.

The *horizontal control* adjusts the left-to-right position of the displayed voltage trace. The horizontal control sets the starting point of the trace. The vertical control adjusts the up-and-down position of the displayed voltage trace. **See Figure 5-8.**

The *volts/division (volts per division) control* selects the height of the displayed waveform. The setting determines the number of volts each vertical screen division represents. For example, when a waveform occupies four divisions and the volts/division control is set on 20 V, the peak-to-peak voltage (V_{p-p}) equals 80 V (4 × 20 = 80 V). Eighty volts peak-to-peak equals 40 V peak (V_{max}) (80 ÷ 2 = 40 V). **See Figure 5-9.** Forty volts peak equals 28.28 V_{rms} (40 × 0.707 = 28.28 V).

The *time/division (time per division) control* selects the width of the displayed waveform. The setting determines the length of time each cycle takes to move across the screen. For example, when the time/division control is set on 10 ms, each horizontal screen division equals 10 milliseconds (ms). When one cycle of a waveform equals four divisions, the displayed time equals 40 ms (4 × 10 = 40 ms). **See Figure 5-10.**

To determine the frequency of a displayed waveform, apply the following formula:

$$f = \frac{1}{T}$$

where

f = frequency (in hertz)

1 = constant

T = time period (in seconds)

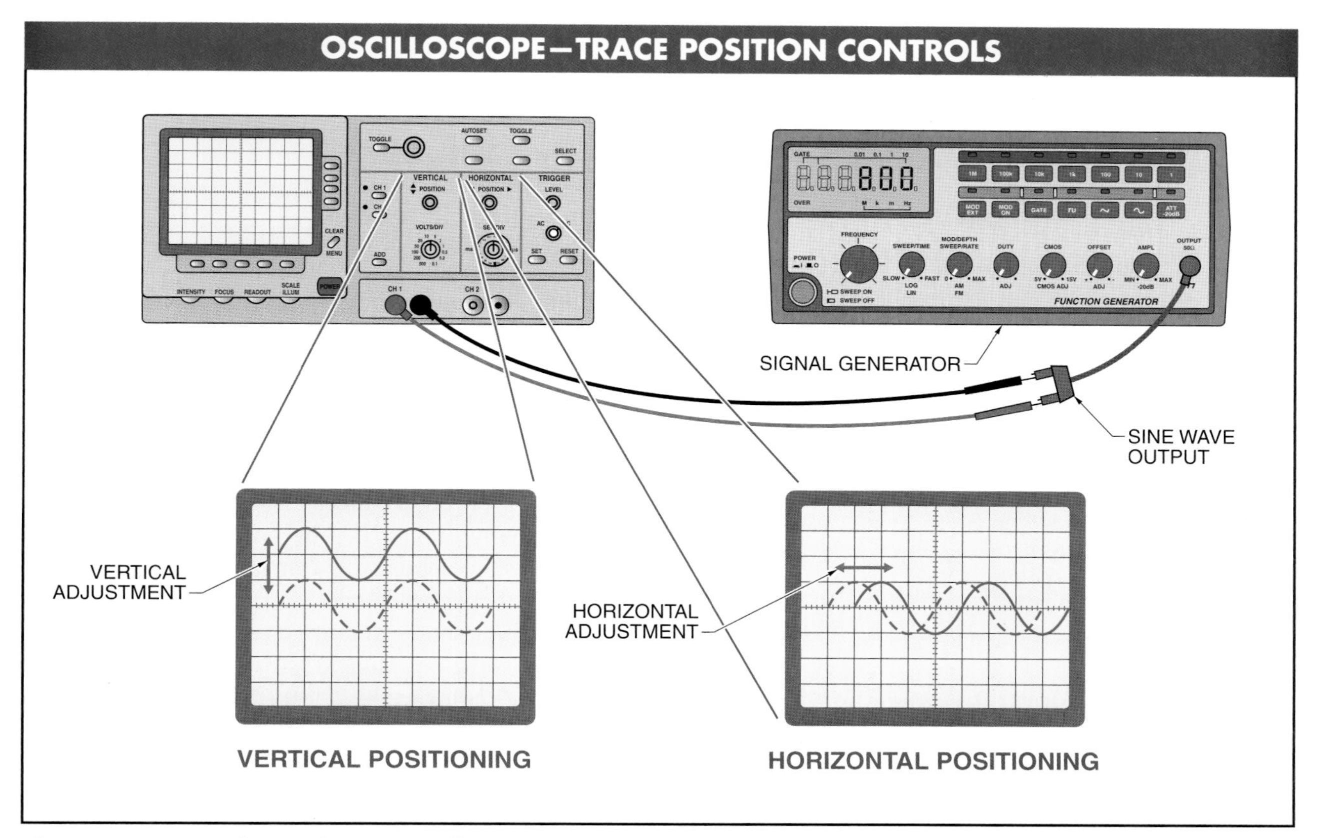

Figure 5-8. Manual controls on an oscilloscope such as intensity, focus, horizontal positioning, vertical positioning, volts per division, and time per division must be adjusted to view waveforms for usable information.

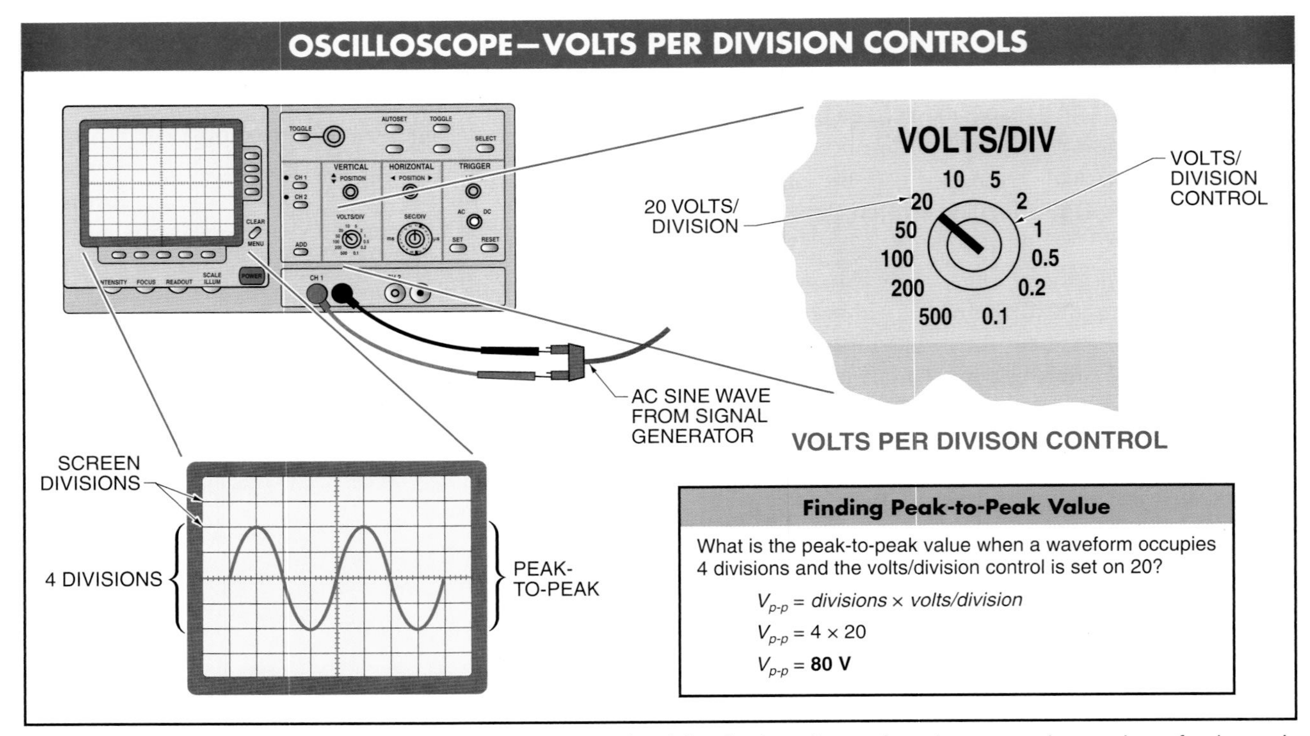

Figure 5-9. The volts per division control selects the height of the displayed waveform by setting the number of volts each horizontal screen division represents.

Example: Calculating Frequency

Find the frequency of a waveform when one cycle of a waveform occupies four divisions and the time/division control is set on 10 ms (10 ms = 0.01 sec).

1. Calculate time period.

   $T = 4 \times 0.01$

   $T = \mathbf{0.04\ sec}$

2. Calculate the frequency.

   $$f = \frac{1}{T}$$

   $$f = \frac{1}{0.04}$$

   $f = \mathbf{25\ Hz}$

Technical Tip

The accuracy of an oscilloscope is degraded at lower frequencies unless the device is capable of DC response.

Oscilloscope AC Voltage Measurement Procedures

An oscilloscope is connected in parallel with a circuit or component to measure voltage. An oscilloscope is connected to a circuit by a probe on the end of each test lead. A 1x probe (1 to 1) is used to connect the input of the oscilloscope to the circuit being tested when the test voltage is lower than the voltage limit of the scope.

A 10x probe (10 to 1) is used to divide the input voltage by 10. The voltage limit of an oscilloscope equals 10 times the normal rated voltage when a 10x probe is used. The level of measured voltage displayed on the oscilloscope screen must be multiplied by 10 to obtain the actual circuit voltage when a 10x probe is being used. For example, if the displayed oscilloscope voltage is 25 V while using a 10x probe, the actual circuit voltage is 250 V (25 × 10 = 250 V). **See Figure 5-11.**

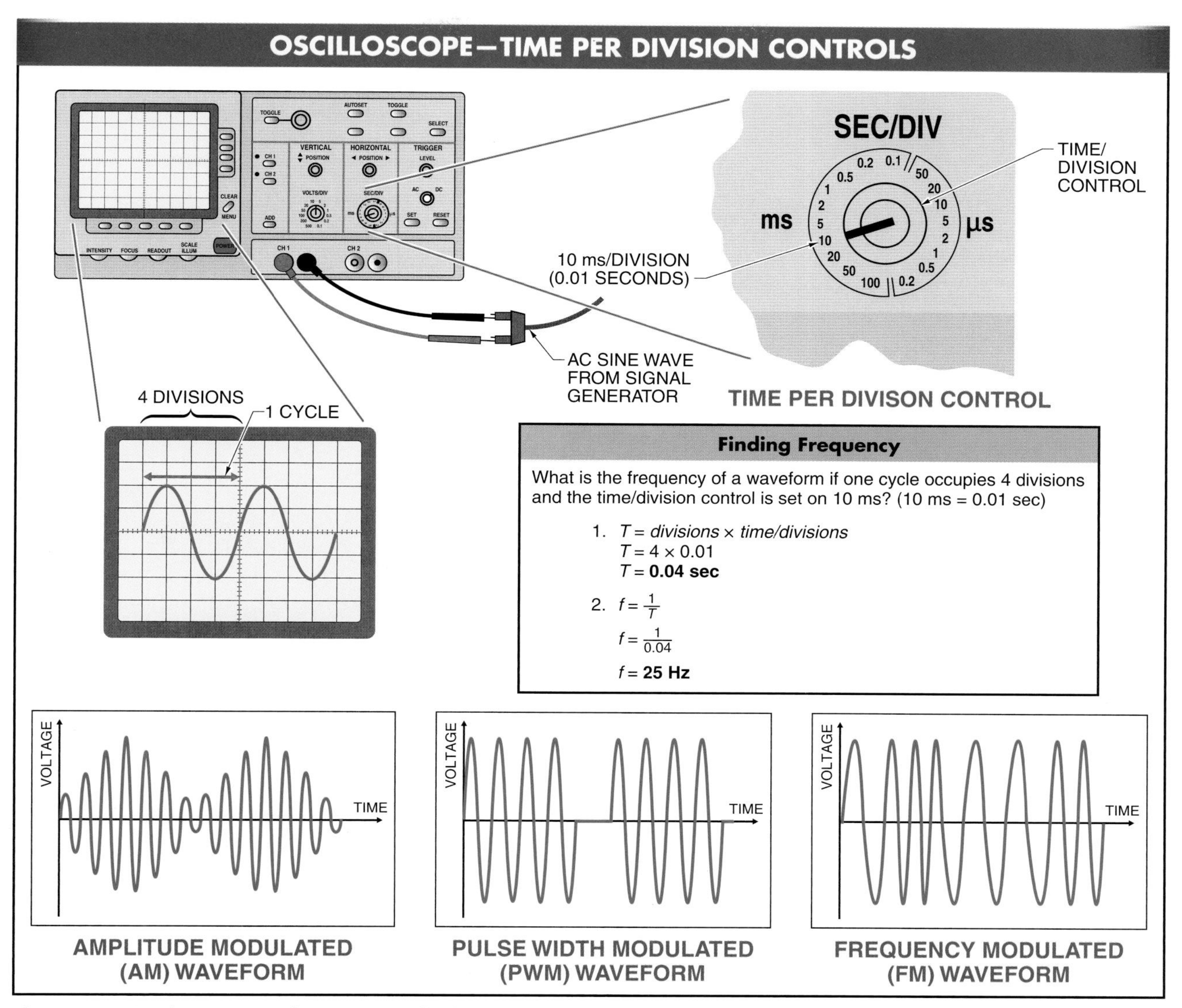

Figure 5-10. The time per division control selects the width of the displayed waveform by setting the length of time (per division) each cycle takes to move across the screen.

Troubleshooting Tip

TOTAL RESISTANCE—PARALLEL

Ohm's law states that the total resistance in a circuit containing parallel loads is less than the smallest resistance value of any load.

To calculate total resistance in a parallel circuit, apply the following formula:

$$R_T = \frac{1}{\frac{1}{R_1} + \frac{1}{R_2} + \ldots}$$

where

R_T = total resistance (in Ω)

R_1 = resistance 1 (in Ω)

R_2 = resistance 2 (in Ω)

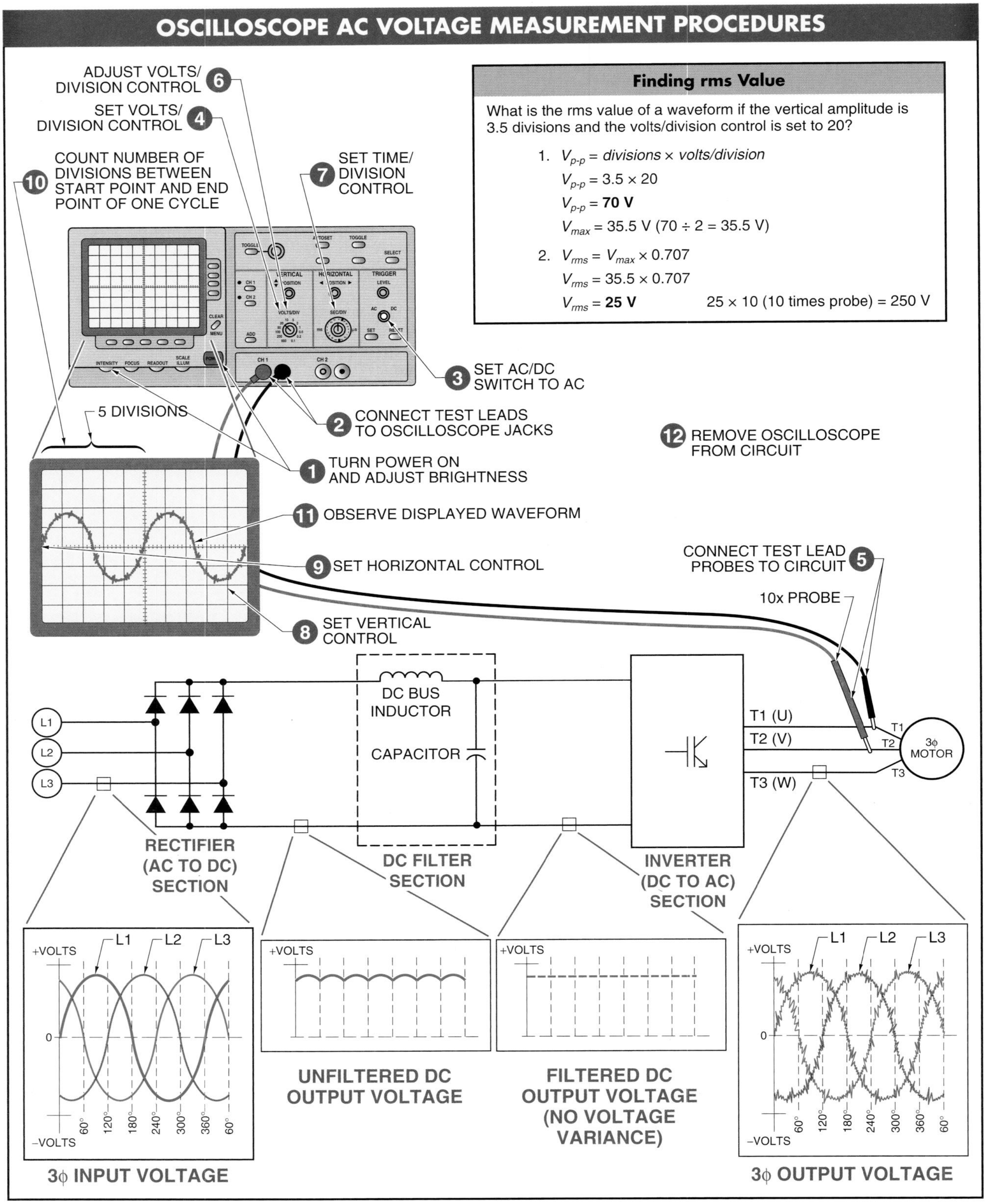

OSCILLOSCOPE AC VOLTAGE MEASUREMENT PROCEDURES

1. TURN POWER ON AND ADJUST BRIGHTNESS
2. CONNECT TEST LEADS TO OSCILLOSCOPE JACKS
3. SET AC/DC SWITCH TO AC
4. SET VOLTS/DIVISION CONTROL
5. CONNECT TEST LEAD PROBES TO CIRCUIT
6. ADJUST VOLTS/DIVISION CONTROL
7. SET TIME/DIVISION CONTROL
8. SET VERTICAL CONTROL
9. SET HORIZONTAL CONTROL
10. COUNT NUMBER OF DIVISIONS BETWEEN START POINT AND END POINT OF ONE CYCLE
11. OBSERVE DISPLAYED WAVEFORM
12. REMOVE OSCILLOSCOPE FROM CIRCUIT

5 DIVISIONS

10x PROBE

Finding rms Value

What is the rms value of a waveform if the vertical amplitude is 3.5 divisions and the volts/division control is set to 20?

1. $V_{p\text{-}p} = \textit{divisions} \times \textit{volts/division}$

   $V_{p\text{-}p} = 3.5 \times 20$

   $V_{p\text{-}p} = \mathbf{70\ V}$

   $V_{max} = 35.5$ V ($70 \div 2 = 35.5$ V)

2. $V_{rms} = V_{max} \times 0.707$

   $V_{rms} = 35.5 \times 0.707$

   $V_{rms} = \mathbf{25\ V}$ $\quad 25 \times 10$ (10 times probe) = 250 V

L1, L2, L3

RECTIFIER (AC TO DC) SECTION

DC BUS INDUCTOR

CAPACITOR

DC FILTER SECTION

INVERTER (DC TO AC) SECTION

T1 (U), T2 (V), T3 (W)

3φ MOTOR

+VOLTS, 0, –VOLTS, 60°, 120°, 180°, 240°, 300°, 360°, 60°

3φ INPUT VOLTAGE

UNFILTERED DC OUTPUT VOLTAGE

FILTERED DC OUTPUT VOLTAGE (NO VOLTAGE VARIANCE)

3φ OUTPUT VOLTAGE

Figure 5-11. Oscilloscopes sare connected in parallel with a circuit or component to measure voltage by 1x or 10x probes on the end of the voltage test lead.

Before using an oscilloscope to take AC voltage measurements, ensure the scope is designed to take measurements on the circuit being tested. Refer to the operations manual of the test instrument for all measuring precautions, limitations, and procedures. Always wear required personal protective equipment and follow all safety rules when taking the measurement. To use an oscilloscope to measure the AC voltage of a circuit, apply the following procedures:

1. Turn the power switch ON and adjust the trace brightness on the screen.
2. Connect the test leads with probes to oscilloscope jacks (channel 1) as required.
3. Set the AC/DC control switch to AC.
4. Set the volts/division control to display the voltage level being tested. Set the control to the highest value when the voltage level is unknown.
5. Connect the oscilloscope test lead probes to the circuit with the AC voltage being tested.
6. Adjust the volts/division control to display the full (peak-to-peak) waveform of the voltage being tested.
7. Set the time/division control to display several cycles of the voltage waveform being tested.
8. Set the vertical control to set the lower edge of the waveform on one of the lower lines.
9. Set the horizontal control so that the start of one cycle of the waveform begins at the vertical centerline on the oscilloscope screen.
10. Measure the vertical amplitude of the waveform by counting the number of divisions displayed (V_{p-p}).
11. Observe the displayed waveform for distortion or other problems.
12. Remove the oscilloscope from the circuit.

Technical Tip

Some digital scope specifications that should be considered when choosing an oscilloscope include sample rate, bandwidth, triggering, A/D resolution (degree of fineness on the screen), and data storage (RAM).

To calculate V_{rms}, first calculate V_{p-p}. V_{p-p} is calculated by multiplying the number of divisions by the volts/division setting. For example, when a waveform occupies four divisions and the volts/division setting is 10, V_{p-p} equals 40 V (4 × 10 = 40 V). V_{max} equals 20 V (40 ÷ 2 = 20 V).

V_{max} is multiplied by 0.707 to find V_{rms}. For example, when V_{max} is 20 V, V_{rms} equals 14.14 V (20 × 0.707 = 14.14 V). V_{rms} is the value of the voltage being tested as measured by an rms voltmeter.

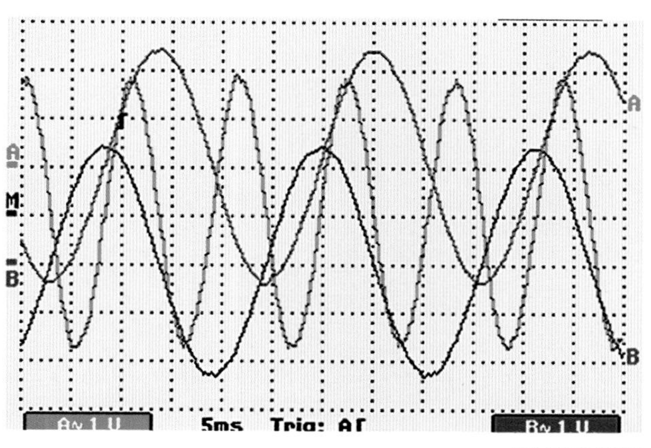

Fluke Corporation

Some oscilloscopes display all three phases of a system simultaneously for determining voltages and frequencies.

Oscilloscope or Handheld Oscilloscope Frequency Measurement Procedures

In AC applications with variable frequency drives, the ability to measure the frequency of the circuit is critical. Frequency meters are typically used in applications where high accuracy and direct numerical frequency readout are required. Oscilloscopes or handheld oscilloscopes provide very accurate readings for most frequency measurement applications. Oscilloscopes or handheld oscilloscopes also graphically show any distortion present in the circuit voltage. To measure frequency, oscilloscope or handheld oscilloscope probes are connected in parallel with the circuit or component being tested. **See Figure 5-12.**

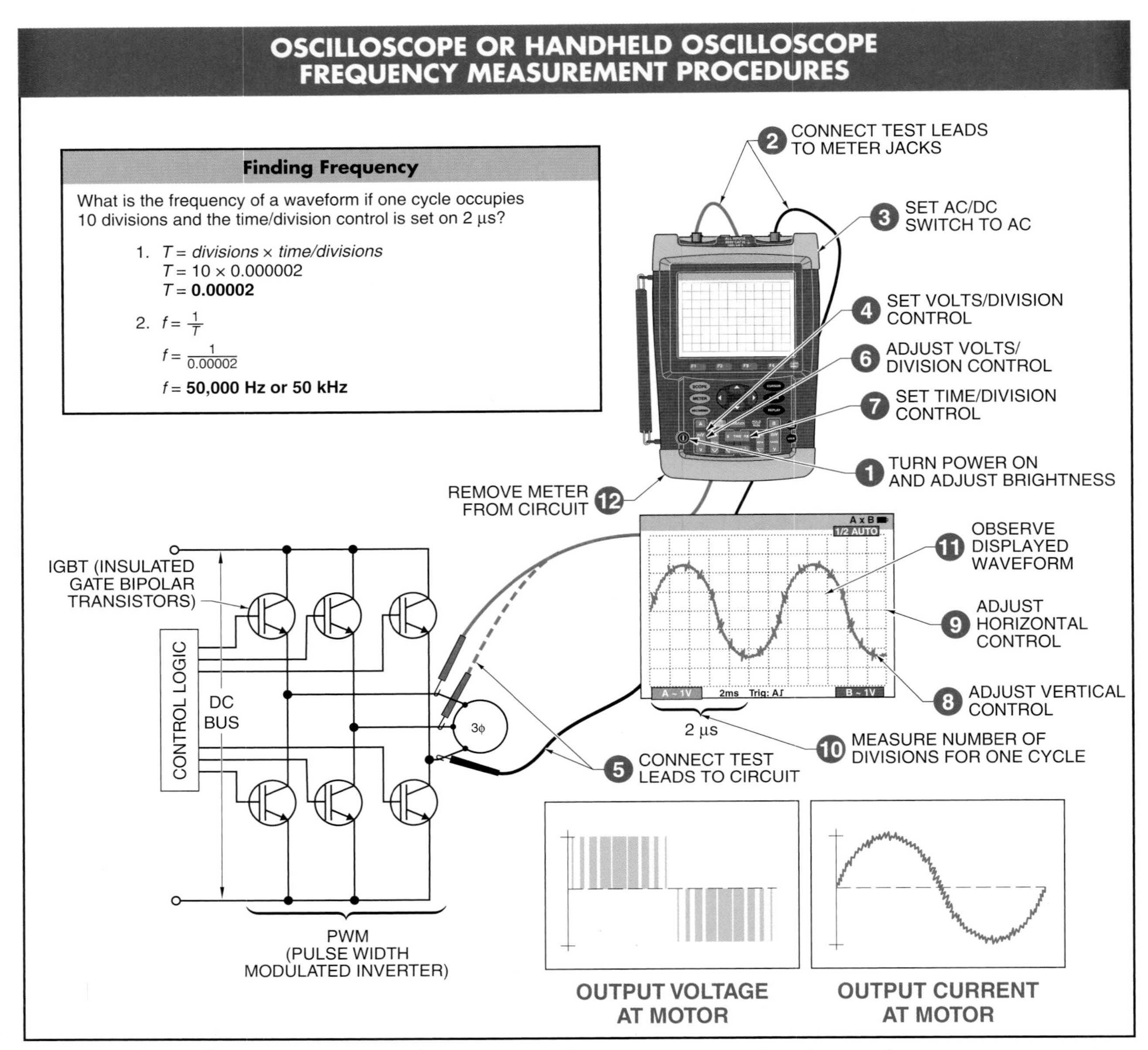

Figure 5-12. Oscilloscopes or power quality meters provide very accurate readings for most frequency measurement applications by displaying any distortion present in the circuit voltage being tested.

Before taking any frequency measurements using a handheld oscilloscope, ensure the meter is designed to take measurements on the circuit being tested. Refer to the operations manual of the test instrument for all measuring precautions, limitations, and procedures. Always wear required personal protective equipment and follow all safety rules when taking the measurement. To use a handheld oscilloscope to measure frequency, apply the following procedures:

1. Turn the handheld oscilloscope power switch ON and adjust the trace brightness.
2. Connect the test leads of the handheld oscilloscope to the meter jacks as required.

3. Set the AC/DC control to AC.
4. Set the volts/division control to display the voltage level being tested. Set the control to the highest value when the voltage level is unknown.
5. Connect the handheld oscilloscope probe to the AC voltage in the circuit being tested.
6. Adjust the volts/division control to display the vertical amplitude of the waveform being tested.
7. Set the time/division control to display approximately two cycles of the waveform being tested.
8. Set the vertical control so that the center of the waveform is on the centerline of the handheld oscilloscope screen.
9. Set the horizontal control so that the start of one cycle of the waveform begins at the vertical centerline on the handheld oscilloscope screen.
10. Measure the number of divisions between the start point and end point of one cycle of the waveform.
11. Observe the displayed waveform for distortion or other problems.
12. Remove the handheld oscilloscope from the circuit.

Troubleshooting Tip

HANDHELD OSCILLOSCOPES

Handheld oscilloscopes display signal frequencies by using internal calculations.

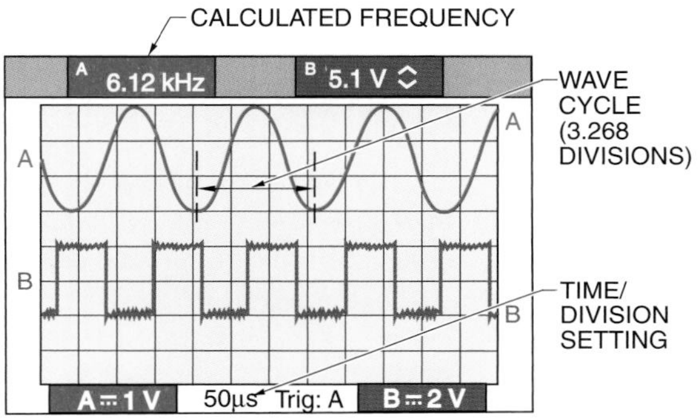

To determine frequency, multiply the number of measured divisions by the time/division setting. The resulting value is the time period for one cycle of the waveform. To determine the frequency of the waveform, divide the time period into 1.

Example: Calculating Frequency

Calculate the frequency when one cycle of a waveform occupies 10 divisions and the time/division setting is 2 μs (microseconds).

1. Calculate time period.

$T = 10 \times 0.000002$

$T = \mathbf{0.00002\ sec}$

2. Calculate frequency.

$$f = \frac{1}{T}$$

$$f = \frac{1}{0.00002}$$

$f = \mathbf{50{,}000\ Hz\ (50\ kHz)}$

When measuring the output frequency of a variable frequency drive, an oscilloscope or handheld oscilloscope can be used to look at the carrier frequency and the fundamental frequency. *Carrier frequency* is the frequency that controls the number of times the solid-state switches in the inverter section of a pulse width modulated (PWM) variable frequency drive turn ON and turn OFF per second. The higher the carrier frequency, the more individual pulses there are to reproduce the fundamental frequency. *Fundamental frequency* is the frequency of the voltage used to control motor speed. Carrier frequency pulses per fundamental frequency are found by applying the following formula:

$$P = \frac{F_{CARR}}{F_{FUND}}$$

where

P = pulses

F_{CARR} = carrier frequency (in Hz)

F_{FUND} = fundamental frequency (in Hz)

For example, what is the number of pulses per fundamental frequency when a carrier frequency of 1 kHz is used to produce a 60 Hz fundamental frequency?

$$P = \frac{F_{CARR}}{F_{FUND}}$$

$$P = \frac{1000}{60}$$

$$\mathbf{P = 16.66\ pulses}$$

A carrier frequency of 6 kHz used to produce a 60 Hz fundamental frequency has 100 individual carrier frequency pulses per fundamental cycle.

Fundamental frequency is the frequency of the voltage a motor uses, but the carrier frequency actually delivers the fundamental frequency voltage to the motor. The carrier frequency of most variable frequency drives can range from 1 kHz to about 20 kHz. The higher the carrier frequency, the closer the output sine wave is to a pure fundamental frequency sine wave. **See Figure 5-13.**

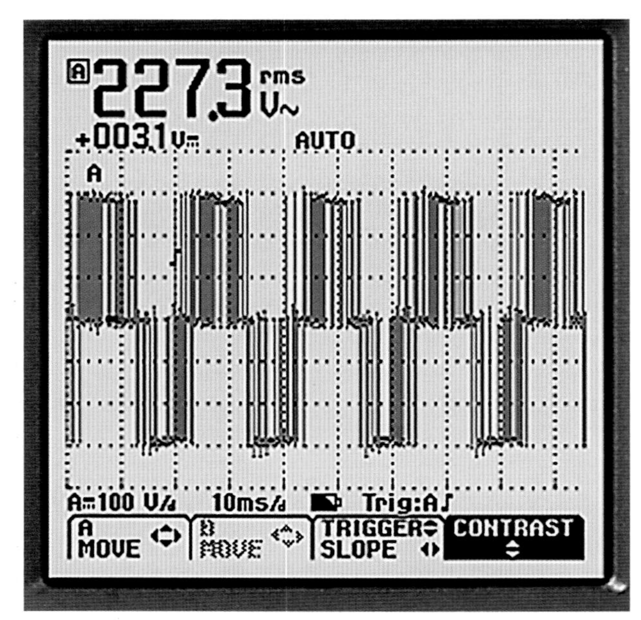

Fluke Corporation

Oscilloscopes are used to display the voltage output of variable frequency drives.

Increasing the frequency to a motor above the standard 60 Hz also increases the noise produced by the motor. Noise is noticeable in the 1 kHz to 2 kHz range because the noise is within the range of human hearing and is amplified by the motor. A motor connected to a variable frequency drive delivering a 60 Hz fundamental frequency with a carrier frequency of 2 kHz is about three times louder than the same motor connected directly to a pure 60 Hz sine wave through a magnetic motor starter. Motor noise is a problem in variable frequency drive applications such as HVAC systems in which the noise can carry throughout an entire building using the ductwork.

Most people can hear a tone from a motor when the tone is in the 1 kHz to 3 kHz range. The sound is heard as a high-pitched whine. A person can hear frequencies above 3 kHz, but the higher frequencies are not amplified by a motor as much as lower frequencies.

Fortunately, manufacturers have raised the carrier frequency of variable frequency drives beyond the range of human hearing to solve the noise problem. High carrier frequencies cause greater power losses (thermal losses) in variable frequency drives because of the solid-state switches in the inverter section of the drive. Variable frequency drives must be slightly derated or the size of the heat sinks increased to compensate for thermal increases. Derating a variable frequency drive decreases the power rating of the drive and increases the size of the heat sinks required, which adds additional cost to the drive.

The higher carrier frequencies of variable frequency drives are better (less heat is produced in the motor), but only up to a point (because of larger voltage spikes). A 6 kHz to 8 kHz carrier frequency simulates a pure sine wave better than a 1 kHz to 3 kHz carrier frequency and reduces heating in a motor because the more closely the voltage delivered to a motor simulates a pure sine wave, the cooler the motor operates. Even slightly reducing the temperature at which a motor operates increases the insulation life of the motor.

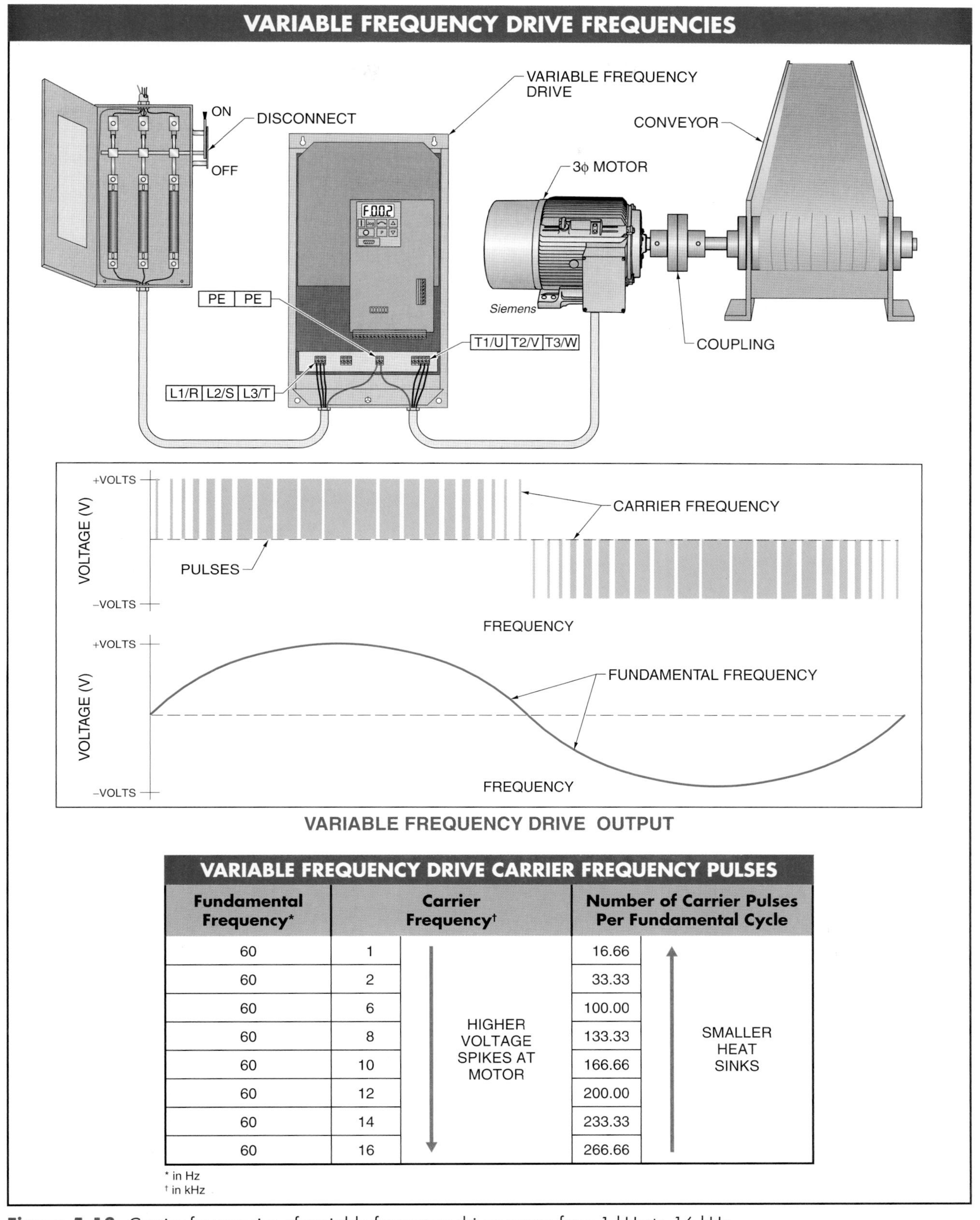

VARIABLE FREQUENCY DRIVE CARRIER FREQUENCY PULSES

Fundamental Frequency*	Carrier Frequency†		Number of Carrier Pulses Per Fundamental Cycle	
60	1	HIGHER VOLTAGE SPIKES AT MOTOR (↓)	16.66	SMALLER HEAT SINKS (↑)
60	2		33.33	
60	6		100.00	
60	8		133.33	
60	10		166.66	
60	12		200.00	
60	14		233.33	
60	16		266.66	

* in Hz
† in kHz

Figure 5-13. Carrier frequencies of variable frequency drives range from 1 kHz to 16 kHz.

Carrier frequency can be changed at a variable frequency drive to meet particular load requirements. The factory default value is usually the highest frequency, and changing to a lower frequency is accomplished through a parameter change, such as changing 12 kHz to 2.2 kHz. One effect of high carrier frequencies is that the fast switching of the inverter section of a variable frequency drive produces large voltage spikes that damage motor insulation. The voltage spikes become more of a problem as cable length between a variable frequency drive and motor increases.

Handheld Oscilloscope DC Voltage Measurement Procedures

When using a handheld oscilloscope to measure DC voltage, the displayed DC voltage is a flat line because pure DC does not vary in frequency. However, the DC voltage displayed on a handheld oscilloscope will show the level of DC voltage and any distortions to the DC voltage signal. Typically, the positive test lead probe is connected to a point in the circuit where the DC voltage is to be measured. The ground lead probe of the handheld oscilloscope is connected to the ground of the circuit. The voltage is positive when the trace moves above the centerline. The voltage is negative when the trace moves below the centerline. **See Figure 5-14.**

Before taking any DC voltage measurements using a handheld oscilloscope, ensure the meter is designed to take measurements on the circuit being tested. Refer to the operations manual of the test instruments for all measuring precautions, limitations, and procedures. Always wear required personal protective equipment and follow all safety rules when taking the measurement. To use a handheld oscilloscope to measure DC voltage, apply the following procedures:

1. Turn the handheld oscilloscope power switch ON and adjust the trace brightness.
2. Connect the test leads of the handheld oscilloscope to the meter jacks as required.
3. Set the AC/DC control to DC.
4. Set the volts/division control to display the voltage level being tested. Set the control to the highest value when the voltage level is unknown.
5. Connect the positive test lead probe of the handheld oscilloscope to a ground point in the circuit being tested.
6. Connect the ground test lead probe to a ground point of the circuit.
7. Set the vertical control so that the displayed voltage line (0 VDC) is in the center of the screen.
8. Remove the positive test lead probe from the ground point and connect the probe to the DC voltage being tested. The displayed voltage moves above or below the scope centerline, depending on the polarity of the DC voltage being tested.
9. Measure the vertical amplitude of the voltage from the centerline by counting the number of divisions from the centerline.
10. Observe the displayed waveform for distortion or other problems.
11. Remove the handheld oscilloscope from the circuit.

Multiply the number of displayed divisions by the volts/division setting to determine the DC voltage. For example, if a waveform is three divisions above the centerline and the volts/division control is set at 5 V, the voltage equals 15 VDC ($3 \times 5 = 15$ V).

Measuring DC voltages helps when troubleshooting DC circuits such as power supplies that rectify AC to DC. The shape of the waveform can be used to determine when any of the diodes in the rectifier circuit are shorted or open. **See Figure 5-15.**

Technical Tip

Oscilloscope specifications that indicate acceptable environmental conditions include operating temperature, maximum shock, and maximum vibration.

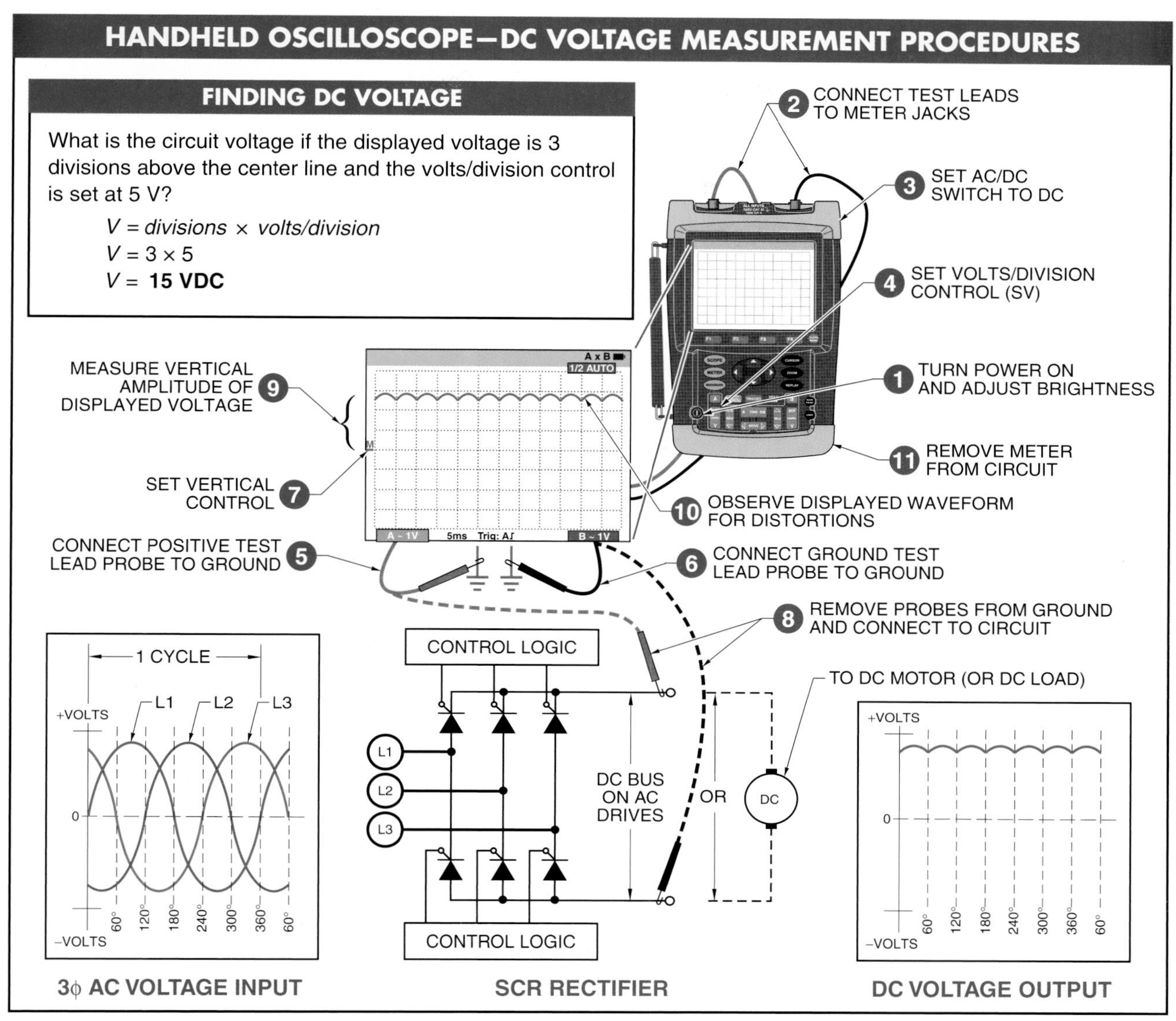

Figure 5-14. When using a power quality meter to measure DC voltage, the displayed DC voltage is a flat line because pure DC voltage does not vary in frequency, but power quality meters show the amount of DC voltage and any distortions of the DC voltage signal.

SIGNAL (FUNCTION) GENERATORS

Much of the work involved when troubleshooting electronics involves signal tracing. Signal tracing follows a signal through a circuit, device, or system to find where the signal disappears or becomes distorted. Signal tracing is accomplished using test instruments to either measure an electronic signal already within a system or circuit, or to measure an electronic signal intentionally injected into a system or circuit.

When troubleshooting or testing an operating circuit or system, test instruments are used to take measurements and observe a signal as the signal naturally appears at the measuring points. Based on test instrument measurements, circuits may be operating correctly, in need of repair, or in need of replacement.

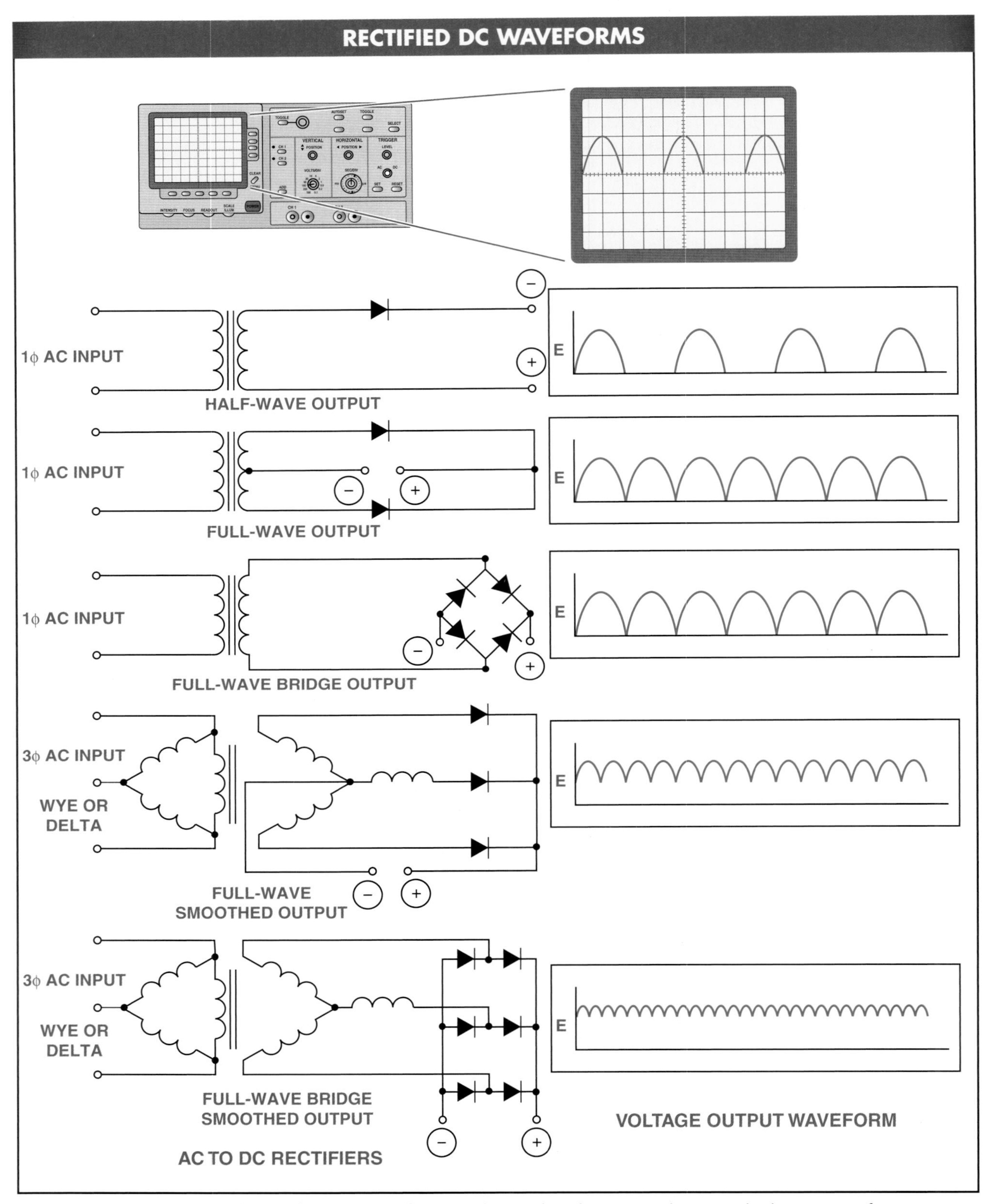

Figure 5-15. The shape of a rectified DC waveform can be used to determine when any diodes in a rectifier circuit are shorted or open.

A special electrical signal instead of a signal that would naturally be appearing within a circuit or system during normal operation is used to test electronic circuits and systems. Signal generators are typically used to produce test signals. A *signal (function) generator* is a test instrument that provides a known input signal to a component, circuit, or system for testing purposes. The injected signal is measured at various points as the signal travels through circuits to test when the signal disappears, or when it is overly attenuated (reduced in energy), distorted, or clipped. **See Figure 5-16.**

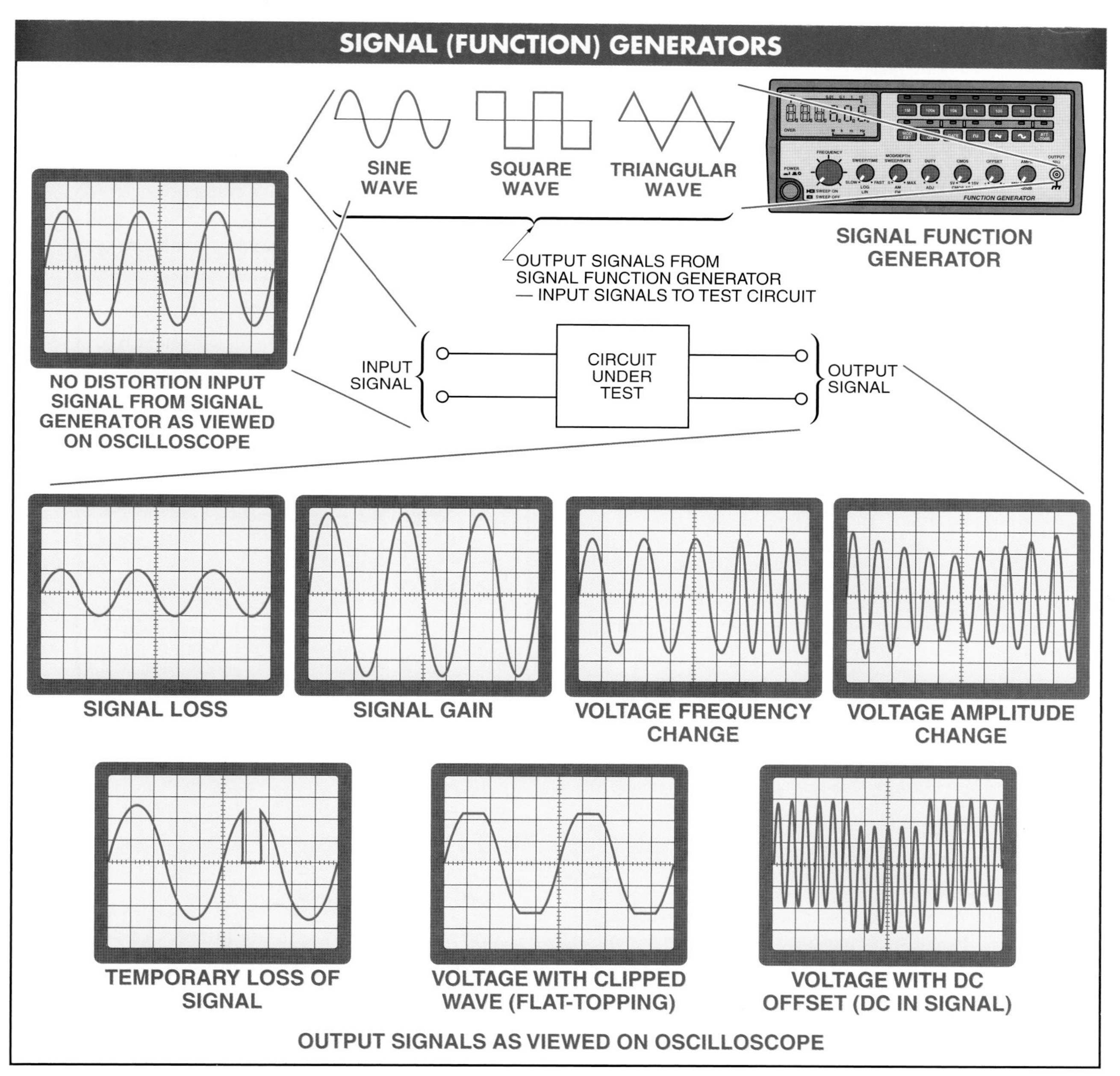

Figure 5-16. Signal (function) generators are test instruments that provide a known input signal to components, circuits, or systems for testing purposes.

Signal generators typically output sine waves, square waves, or triangular waves. Some also output other signals, such as digital pulse outputs for testing digital circuits. The output signal of a signal generator can be adjusted in amplitude and frequency to match the normal signals that would be sent through a circuit or system.

Signal Gain Measurements

Electronic circuits that are used for amplification are tested by injecting a known signal into the input point of the circuit and measuring the resulting output signal. When the output signal is greater than the input signal, the circuit has a gain. When the output signal equals the input signal, there is no gain; and when the output signal is less than the input signal, there is a loss of signal. *Gain* is a ratio of the amplitude of the output signal to the amplitude of the input signal.

Signal gain can be found by applying the following formula:

$$V_{gain} = \frac{V_{out}}{V_{in}}$$

where

V_{gain} = change in signal gain between the input and output signal of a circuit or system

V_{out} = measured signal output (in volts)

V_{in} = measured signal input (in volts)

Signal gain is only one measurement of a signal. In addition to testing for gain, a signal must also be tested to ensure the signal is not distorted or compromised (carrying noise). Oscilloscopes and handheld oscilloscopes are used to test the condition of a signal and display any distortions or other problems.

Decibel Gain

Voltage gain can also be expressed in dB. A *decibel (dB)* is an electrical unit used to express the ratio of the magnitudes of two electric values such as voltage or current. When an output signal is greater than the input signal, the gain (in dB) is a positive number (20 dB, 55 dB). When the output signal equals the input signal, the gain is 0 dB; and when the output signal is less than the input signal, the gain (in dB) is a negative number (–10 dB, –45 dB).

SOUND LEVEL (DECIBEL) METERS

Sound is energy that consists of pressure vibrations in the air. Pressure vibrations are produced by or originate from vibrating objects. Pressure vibrations travel outward in waves which cause recurring compression and rarefaction of air. *Compression* is an area of increased pressure in a sound wave produced when a vibrating object moves outward. *Rarefaction* is an area of reduced pressure in a sound wave produced when a vibrating object moves inward. **See Figure 5-17.**

Sound is heard when vibrations are picked up by the ears of a person.

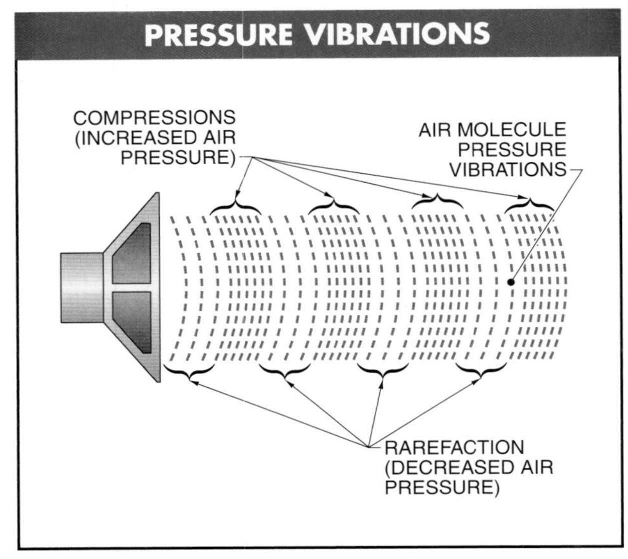

Figure 5-17. Sound is energy that consists of pressure vibrations in the air that are produced by or originate from vibrating objects.

Sound Frequency

Sound frequency (f) is the number of air pressure fluctuation cycles produced per second. Sound consists of various frequencies. Frequency is measured in hertz (Hz). Hertz is the international unit of frequency and is equal to one cycle per second.

The faster an object vibrates, the higher the frequency produced. The slower an object vibrates, the lower the frequency produced. As a frequency increases, wavelength decreases. *Wavelength* is the distance covered by one complete cycle of a sound wave as the wave passes through the air. Low frequencies produce deep, bass sounds that vibrate at several hundred vibrations per second. High frequencies produce high, shrill sounds that vibrate at several thousand vibrations per second. A sound with a 10 Hz frequency vibrates at 10 cycles per second. Every second, the air particles move up and down 10 times.

A person with good hearing can hear sounds with frequencies ranging from about 20 Hz to 20 kHz. As a person ages, the upper limit of hearing decreases. Alarms, horns, and bells typically operate at a specific frequency. Sirens typically operate at a frequency that varies slightly. Audio speakers operate at frequencies within the human hearing range. The *frequency spectrum* is the range of all possible frequencies. The *audio spectrum* is the part of the frequency spectrum that humans can hear (20 Hz to 20 kHz). **See Figure 5-18.**

Sound Intensity

Sound intensity (volume) is a measure of the amount of energy flowing in a sound wave. *Amplitude* is the distance that a vibrating object moves from a position of rest during vibration.

The larger the amplitude, the louder the sound produced. Sound waves that fluctuate in the air a small amount produce little sound. Eardrums do not have much vibration (movement) when a sound is considered soft. Sound waves that fluctuate in the air a large amount produce loud sounds. Eardrums have a lot of vibration (movement) when a sound is considered loud. Eardrums are damaged when sound waves are too large (the sound level is excessive). **See Figure 5-19.**

A *decibel (dB)* is an acoustical unit used to measure the intensity (volume) of sound. A decibel is $1/10$ of a bel (unit named after Alexander Graham Bell). A *bel* is the logarithm of an electric, acoustic, or other power ratio. Because the decibel scale is logarithmic, to raise the sound level 3 dB, the power level must be doubled. For any given sound, the loudness doubles for every increase of 10 dB or is reduced by $1/2$ for every decrease of 10 dB. For example, the sound level of normal conversation is one-half the sound level of a hair dryer. Sound waves lose intensity as the waves spread outward. Also, the loudness of a sound decreases as the distance the sound travels increases. **See Figure 5-20.**

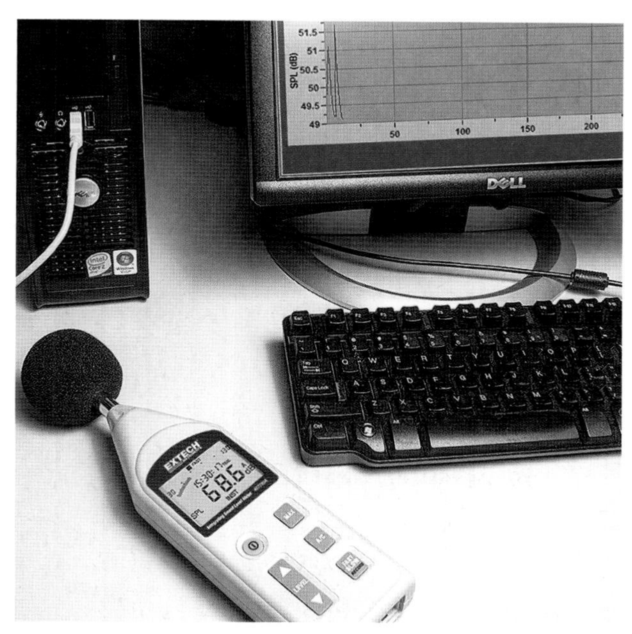

Extech by FLIR

Sound level meters detect sound waves and display the intensity of the sound waves in decibels.

The decibel level of a sound wave represents a ratio of the amount of sound to a referenced amount (normally 0 dB). Zero dB represents the absolute faintest sound that a normal human ear can possibly hear. As sound intensity increases, the decibel level also increases. Sound becomes painful when the amount of sound continues to increase. A sound intensity of 130 dB is considered the pain threshold for the human ear.

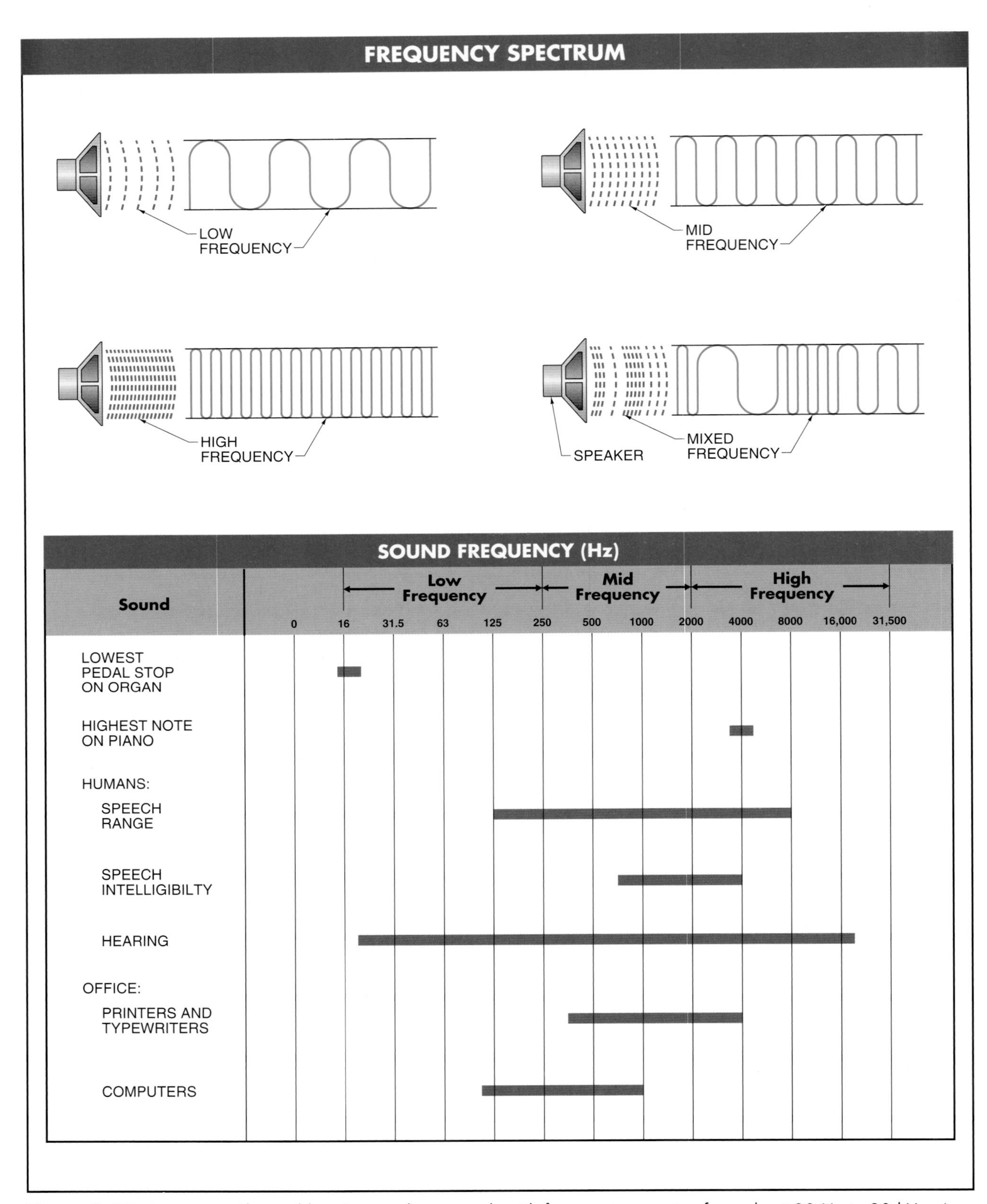

Figure 5-18. A person with good hearing can hear sounds with frequencies ranging from about 20 Hz to 20 kHz. As a person ages, the ability to hear in the upper limit of the frequency spectrum decreases.

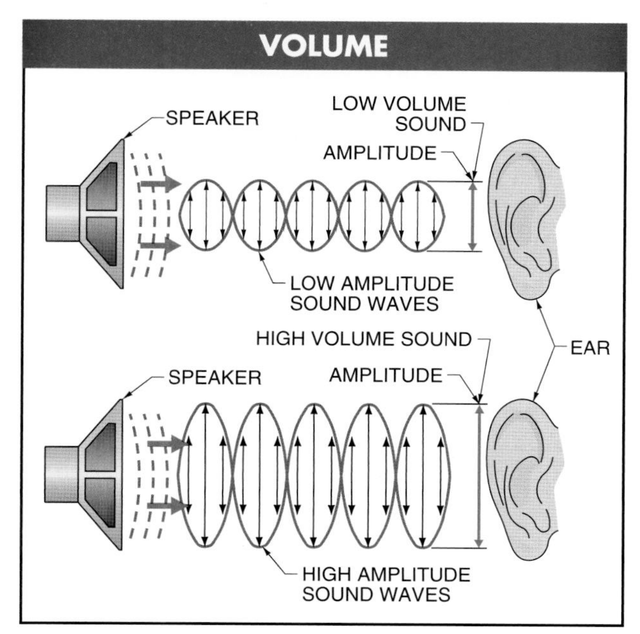

Figure 5-19. Sound intensity (volume) is a measure of the amount of energy flowing in a sound wave.

SOUND LEVELS

Decibel	Example	Loudness
180	Rocket engine	Deafening
160	Jet engine	
150	Explosion	
140	Loud rock music	
130	Air raid siren (pain threshold)	Very loud
120	Thunder	
110	Chain saw	
100	Subway	
90	Heavy truck traffic	
80	Vacuum cleaner	Loud
70	Busy street	
60	Hair dryer	Moderate
50	Normal conversation	
40	Running refrigerator	
30	Quiet conversation	Faint
20	Quiet living room	
10	Whisper	Very faint
0	Threshold of hearing	

Figure 5-20. A decibel (dB) is an acoustical unit used to measure the intensity (volume) of sound.

Technical Tip

Typical features for sound level meters include ratings for outdoor use and auxiliary output connections for use or measurement with other instruments. Some sound meters also include occupational vibration measurement capability.

The human ear is not equally sensitive over the entire frequency range. A 20 Hz sound at 80 dB is as loud as a 1 kHz sound at 20 dB to the human ear, because at 20 Hz the human ear is not very sensitive to sound. A typical human ear is the most sensitive to sounds between 3 kHz and 4 kHz. The human ear typically requires more sound pressure at low and very high frequencies for a person to perceive a sound to be as loud as a sound at 1 kHz. Sound sensitivity decreases with age. The hearing of older women drops to an upper range of about 10 kHz, and older men to about 8 kHz. Children and women tend to be more sensitive to sound than a majority of adult men.

Sound Level Meters

The level of sound (in dB) can be directly measured using a sound level meter. Sound level meters are used to measure sound levels in recording studios, factories, airports, and any other areas in which the exact amount of sound must be measured. **See Figure 5-21.**

 **CAUTION**

The sound levels that cause permanent hearing damage vary from individual to individual and with the length of exposure to the sound. According to OSHA, 30 min is all a human ear can handle at 110 dB. The human ear can handle 2 hr at 100 dB or 4 hr at 95 dB before the ear begins to be permanently damaged.

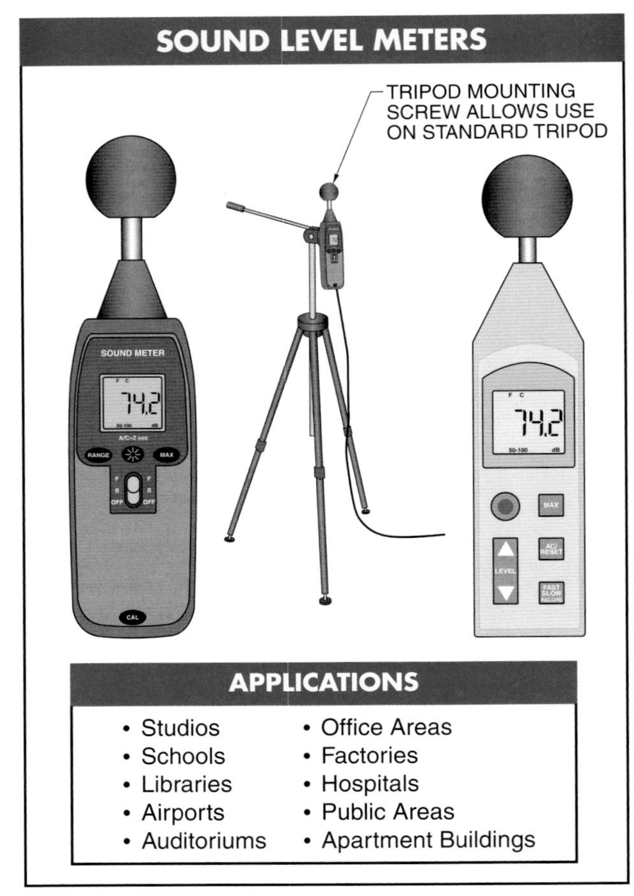

Figure 5-21. The level of sound (in dB) is directly measured using sound level meters in applications such as recording studios, factories, airports, and any other areas in which the exact amount of sound must be measured.

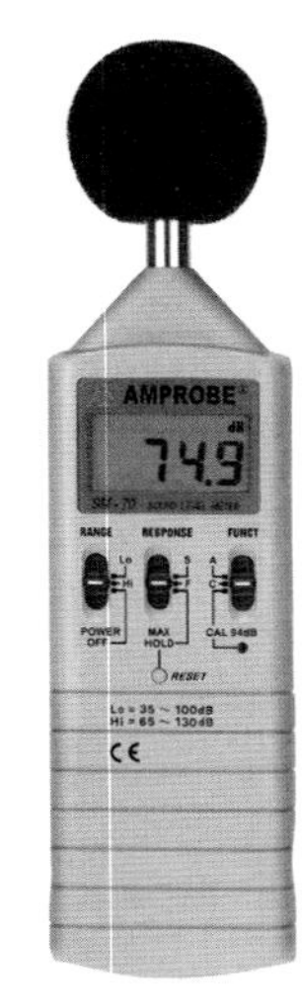

Amprobe/Advanced Test Products

Various sound level meters have different capabilities for measuring the level of sound frequencies across the sound spectrum.

DIGITAL LOGIC PROBES

Digital logic circuits make decisions for control circuits. A *digital signal* is a signal represented by one of two states. A digital signal is either a high (1) state or a low (0) state. A high signal state is typically 5 V, but can be from 2.4 V to 5 V. A low signal state is typically 0 V, but can be from 0 V to 0.8 V.

Digital logic gates are used to control electrical circuits. The AND, OR, and NOT logic gates are the three basic logic functions that make up most digital circuit logic. The NOT gate is used to invert the incoming signal to the gate. The NOR gate is a NOT, OR, or inverted OR gate. The NAND gate is a NOT, AND, or inverted AND gate. AND, OR, NOT, NOR, and NAND logic functions have the same meaning for digital logic, hard-wired electrical logic, and for relay logic. **See Figure 5-22.**

A *digital logic probe* is a special test instrument (DC voltmeter) that detects the presence or absence of a high or low signal. Displays on a digital logic probe include logic high, logic low, pulse light, memory, and TTL/CMOS. The high state light-emitting diode (LED) illuminates when the logic probe detects a high logic level (1). The low state LED illuminates when the logic probe detects a low logic level (0). **See Figure 5-23.**

The pulse LED flashes relatively slowly when the probe detects logic activity present in a circuit. Logic activity indicates that the circuit is changing between logic levels. The pulse light displays the changes between logic levels because the changes are typically too fast for the high and low LEDs to display.

The memory switch sets the logic probe to detect short pulses, usually lasting a few nanoseconds. Any change from the original logic level causes the memory LED to light and remain ON. The memory LED uses the pulse LED switch in the memory position. The memory switch is manually moved to the pulse position and back to the memory position to reset the logic probe.

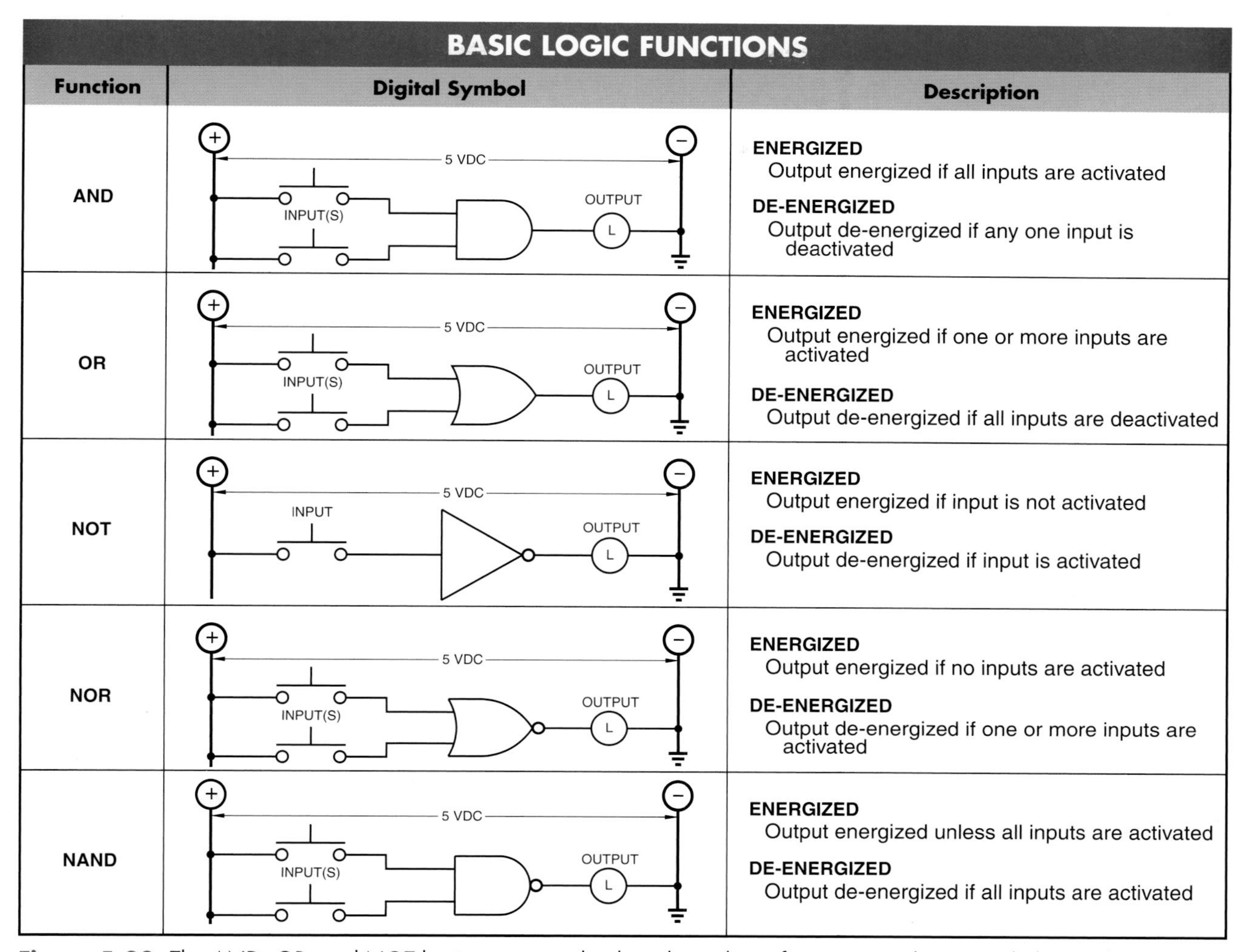

BASIC LOGIC FUNCTIONS

Function	Digital Symbol	Description
AND	+ − 5 VDC INPUT(S) OUTPUT L	**ENERGIZED** Output energized if all inputs are activated **DE-ENERGIZED** Output de-energized if any one input is deactivated
OR	+ − 5 VDC INPUT(S) OUTPUT L	**ENERGIZED** Output energized if one or more inputs are activated **DE-ENERGIZED** Output de-energized if all inputs are deactivated
NOT	+ − 5 VDC INPUT OUTPUT L	**ENERGIZED** Output energized if input is not activated **DE-ENERGIZED** Output de-energized if input is activated
NOR	+ − 5 VDC INPUT(S) OUTPUT L	**ENERGIZED** Output energized if no inputs are activated **DE-ENERGIZED** Output de-energized if one or more inputs are activated
NAND	+ − 5 VDC INPUT(S) OUTPUT L	**ENERGIZED** Output energized unless all inputs are activated **DE-ENERGIZED** Output de-energized if all inputs are activated

Figure 5-22. The AND, OR, and NOT logic gates are the three basic logic functions used to control electrical circuits.

The TTL/CMOS switch selects the logic family of integrated circuits (ICs) being tested. *Transistor-transistor logic (TTL) ICs* are a broad family of ICs that employ a two-transistor arrangement. The supply voltage for TTL ICs is 5.0 VDC, ± 0.25 V.

Complementary metal-oxide semiconductor (CMOS) ICs are a group of ICs that employ MOS transistors. CMOS ICs are designed to operate on a supply voltage ranging from 3 VDC to 18 VDC. Check circuit schematics for CMOS circuit voltages. The supply voltage for CMOS ICs should be greater than –5% of the rated voltage. CMOS ICs are noted for the ability to operate with exceptionally low power consumption.

Troubleshooting Tip

LOGIC PULSERS

Logic pulsers are used with logic probes to troubleshoot logic circuits. Logic pulsers inject a logic pulse or logic train into a circuit for a logic probe to detect.

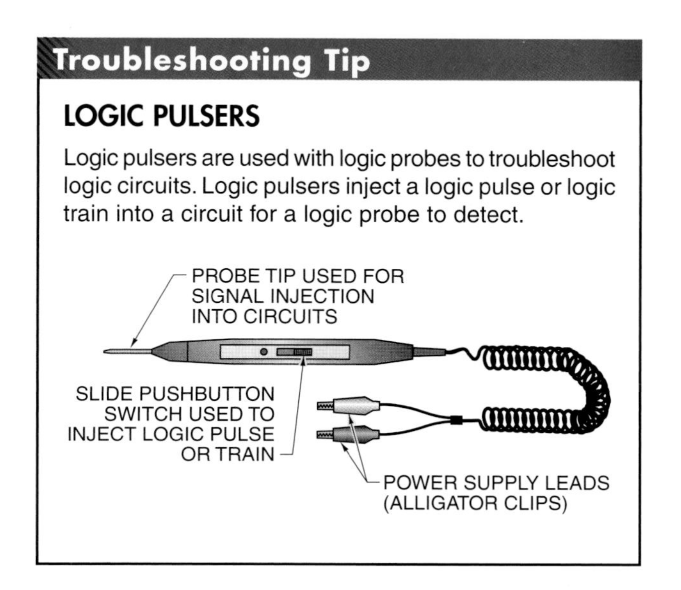

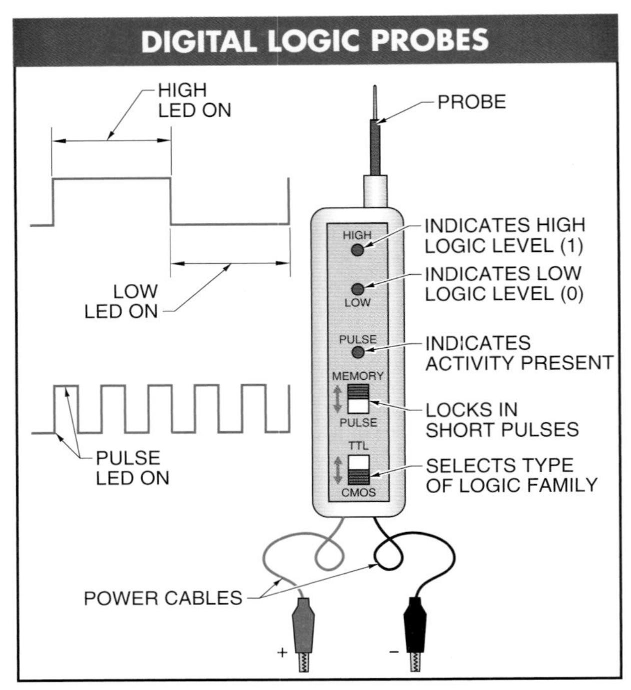

Figure 5-23. A digital logic probe is a special test instrument (a DC voltmeter) that detects the presence or absence of a high (1) or low (0) signal.

Technical Tip

A standard logic probe tests transistor-transistor logic, digital-transistor logic, resistor-transistor logic, horizontal-transistor logic, and complementary metal-oxide-semiconductor and metal-oxide-semiconductor digital circuitry.

Pull-up Resistors

A *floating input* is a digital input signal that is too high or too low at times. A floating input must never occur in a digital circuit. A *pull-up resistor* is an electronic component (resistor) that prevents a floating input condition. A pull-up resistor has one side connected to the digital power supply (at all times) and the other side connected to one side of the input to an IC. The connection produces a high-state signal on one side of the input to the IC. The other side of the IC input is connected to the circuit or ground. The high-state signal becomes a low-state signal when the input switch to the IC is closed. **See Figure 5-24.**

Digital Logic Probe Use

Digital circuits fail because the digital signal is lost somewhere between the circuit input and output stages. Finding the point where the signal is missing and repairing that area typically corrects the problem. Repairs typically involve replacing a component, section, or an entire PC board.

The power supply voltage must be tested with a voltmeter when a digital circuit or digital logic probe has intermittent problems. A digital logic probe may indicate a high signal when the supply voltage is too low for proper circuit or logic probe operation. **See Figure 5-25.**

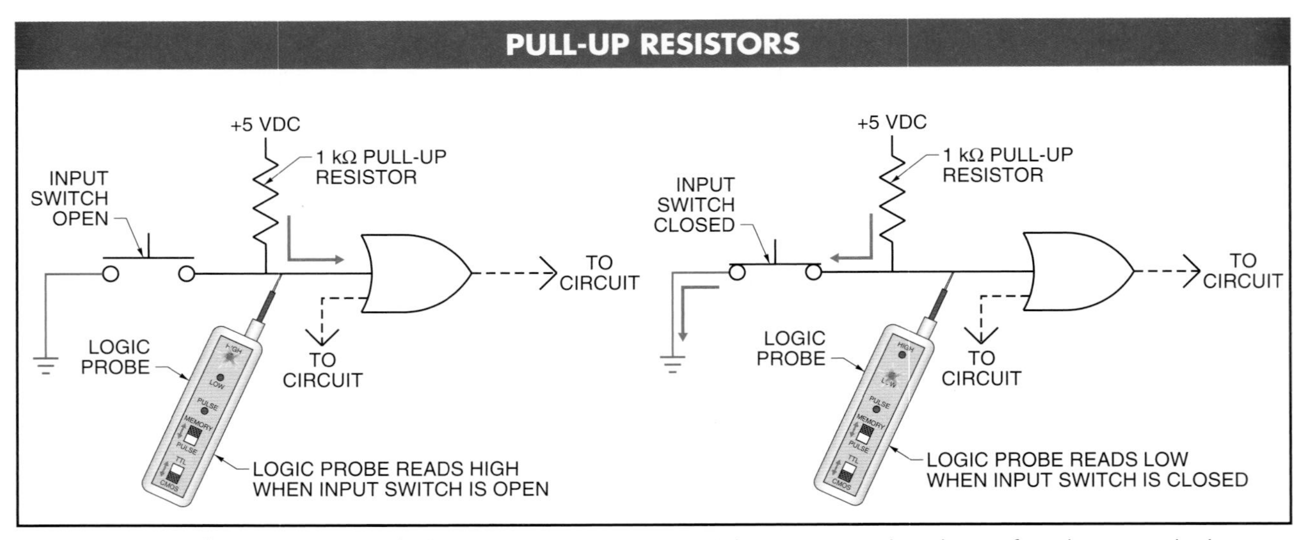

Figure 5-24. Pull-up resistors are electronic components (resistors) that prevent a digital input from being too high or too low (floating input).

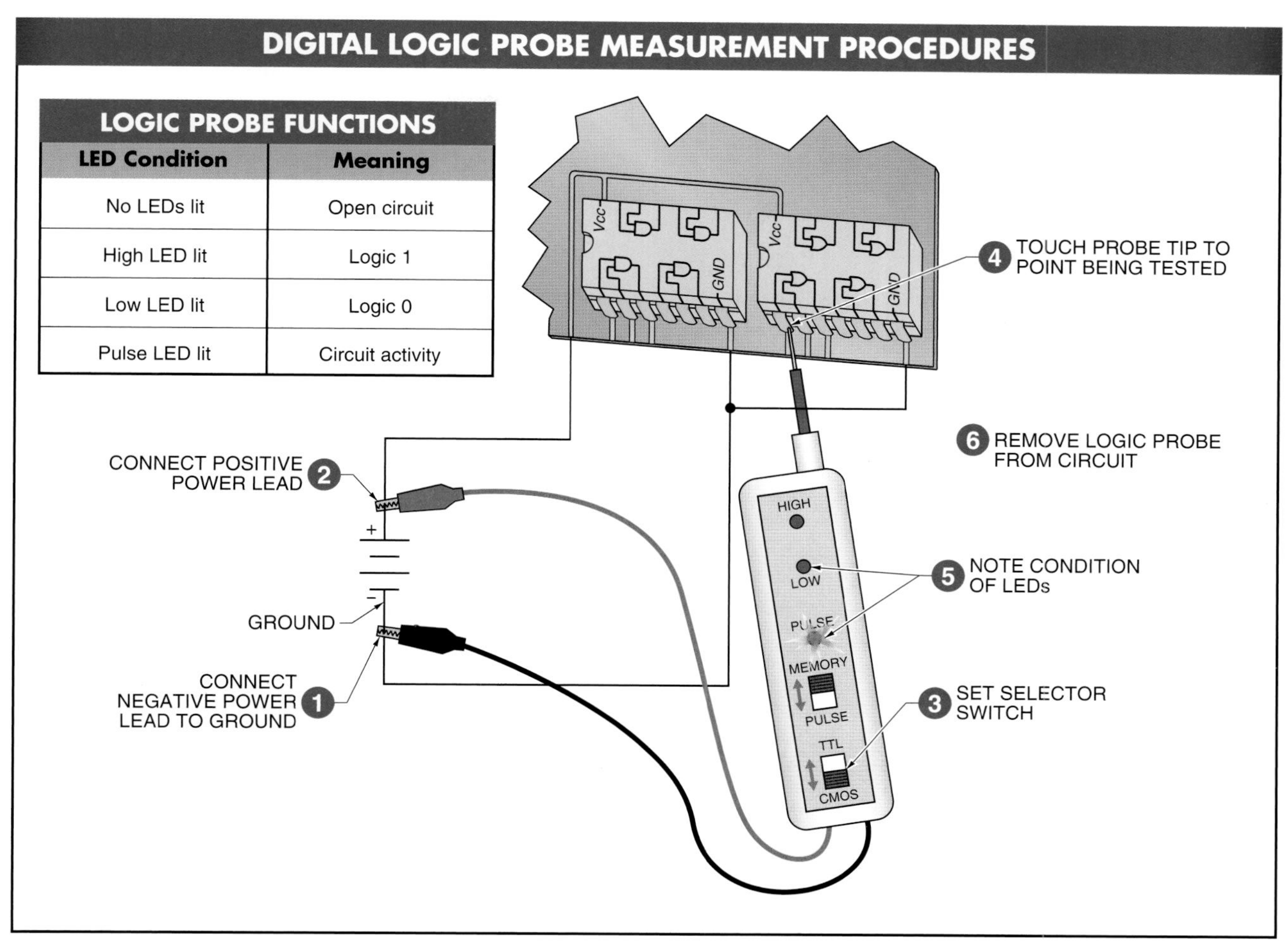

Figure 5-25. A digital logic probe can indicate a high signal when the supply voltage is too low for proper circuit or logic probe operation.

Before taking any digital measurements using a digital logic probe, ensure the probe is designed to take measurements on the circuit being tested. Refer to the operations manual of the test instrument for all measuring precautions, limitations, and procedures. Always wear required personal protective equipment and follow all safety rules when taking the measurement. To use a digital logic probe, apply the following procedures:

1. Connect the negative (black) power lead to the ground side of the digital power supply.
2. Connect the positive (red) power lead of the logic probe to the positive side of the digital power supply. The positive power supply is +5 VDC for TTL circuits.
3. Set the selector switch of the logic probe to the logic family (TTL or CMOS) being tested.
4. Touch the logic probe tip to the point in the digital circuit being tested. Start at the input side of the circuit and move to the output side of the circuit.
5. Observe the condition of the LEDs on the logic probe. Single-shot pulses are stored indefinitely by placing the switch in the memory position.
6. Remove the logic probe from the circuit.

Circuit Board Breaks

Broken traces (conducting paths) result when a PC board is subjected to mechanical stress. Soldering a bridge over the break repairs the break. A bridge is made by soldering a short piece of wire to both sides of the break. **See Figure 5-26.**

Technical Tip

Quality requirements met by suppliers of PC boards and services are CE, CSA, ISO 9001, ISO 9002, MIL-SPEC/STD, and UL®.

REPAIRING CIRCUIT BOARD BREAKS

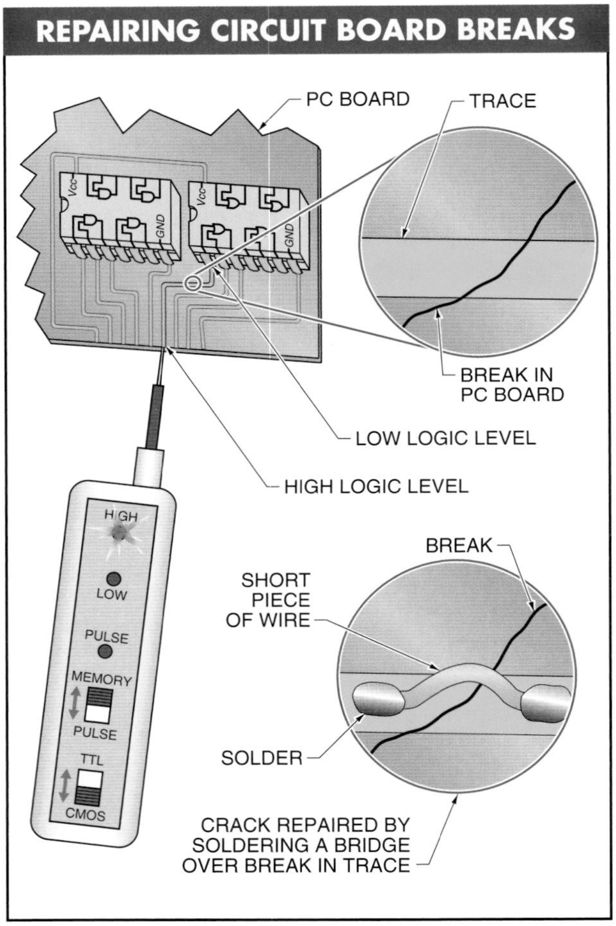

Figure 5-26. A broken trace (conducting path) results when a PC board is subjected to mechanical stress and is repaired by soldering a bridge (a short piece of wire) across the break.

TRANSISTOR TESTING

A *transistor* is a three-element device made of semiconductor material. A transistor is used in circuits as either a switch or an amplifier. Transistors are used as current control devices. The three elements of a transistor are the emitter (E), base (B), and collector (C). On a transistor symbol, the lead with the arrow is the emitter. A transistor is an NPN transistor when the arrow points away from the base. A transistor is a PNP transistor when the arrow points toward the emitter. **See Figure 5-27.**

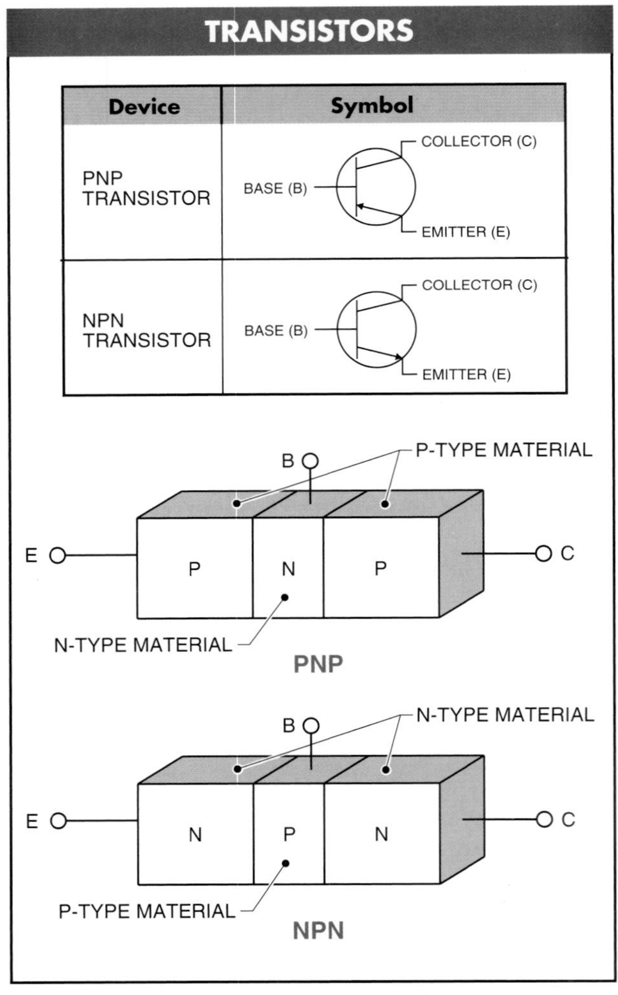

Figure 5-27. Transistors are used in electronic circuits as switches or signal amplifiers.

An *NPN transistor* is a transistor that has a thin layer of P-type material placed between two pieces of N-type material. A *PNP transistor* is a transistor that has a thin layer of N-type material placed between two pieces of P-type material. N-type material carries current in the form of electrons because the material is manufactured to have extra electrons. P-type material carries current in the form of positive charges (holes) because the material is manufactured to have a deficiency of electrons.

In a transistor, a very small current applied to one lead controls a large current flow through the other two leads. A transistor may be used to start or stop current flow (that is, to switch) or to increase current flow (to amplify). Any transistor may be used as a switch or an amplifier. Understanding how to identify the type of transistor (NPN or PNP) used as a switch and how to connect it into a circuit is important when installing or troubleshooting electronic devices, such as photoelectric switches. **See Figure 5-28.**

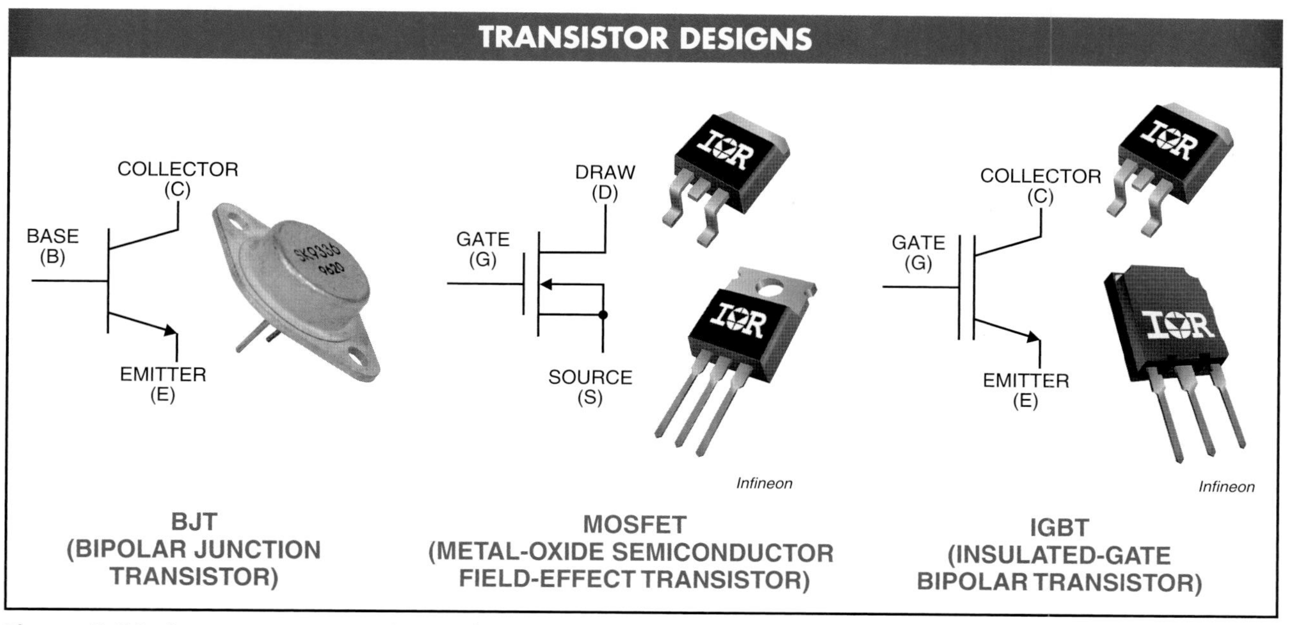

Figure 5-28. Some transistors are designed to operate more efficiently as switches while others are designed to operate more efficiently as amplifiers.

Transistors as Switches

Transistors were developed to replace mechanical switches. Transistors have no moving parts and can switch ON and OFF many times per second. Mechanical switches also have two conditions (ON and OFF or open and closed) but cannot be switched as fast as transistors. Both switch types have a very high resistance when open and a very low resistance when closed.

Transistors are operated as switches. For example, transistors are used to turn DC loads ON or OFF. **See Figure 5-29.** During transistor operation the resistance between the collector (C) and emitter (E) is determined by the current flow between the base (B) and emitter (E). When the collector-to-emitter resistance is high, similar to an open switch, no current flows between C and E. The DC load is not ON because there is no current flow.

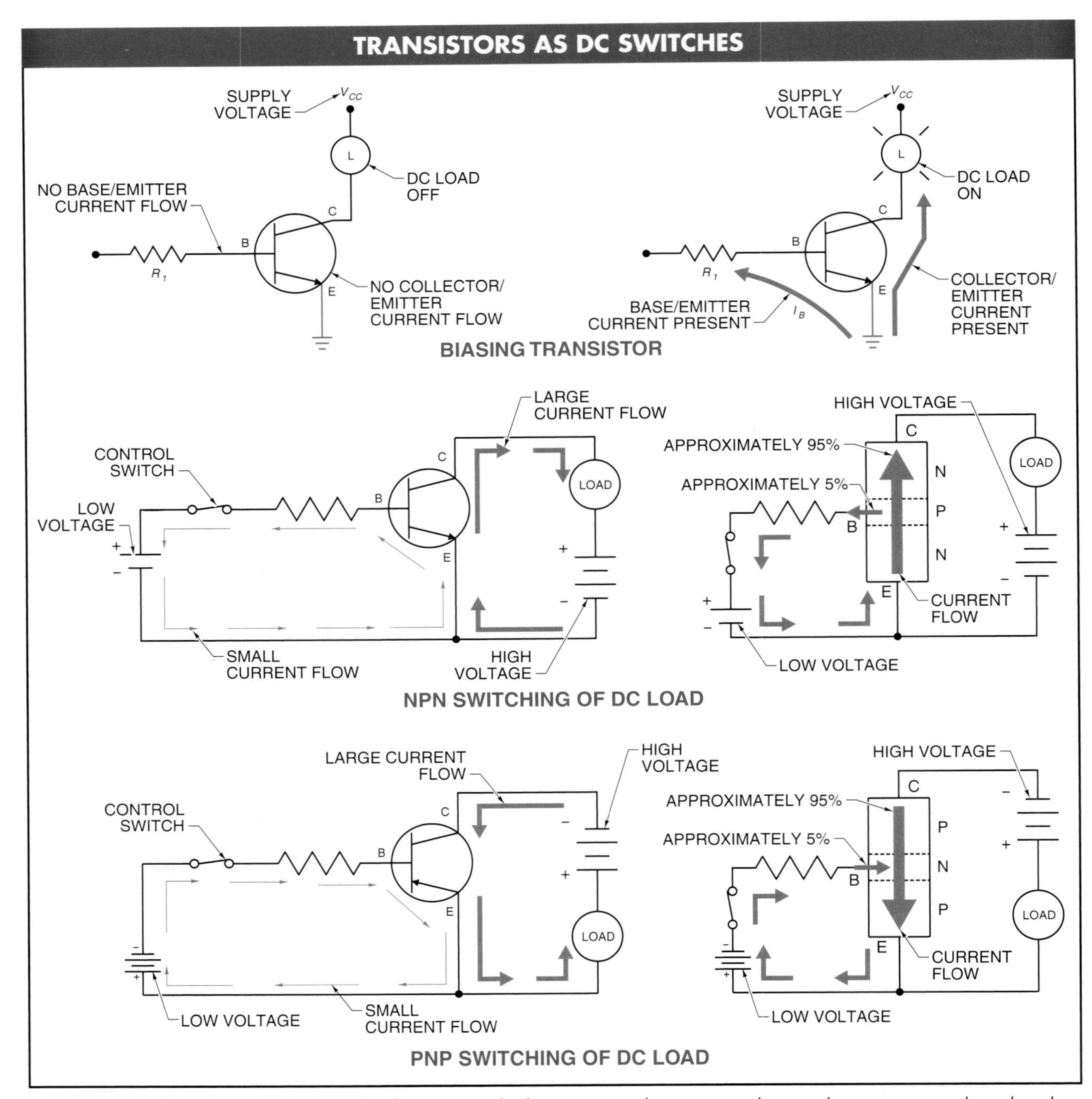

Figure 5-29. Transistors used as switches have a very high resistance when open and a very low resistance when closed.

When collector to emitter resistance is reduced to a very low value, similar to a closed switch, current flows between C and E, switching ON the DC load. A transistor switched ON is operating in the saturation region. The *saturation region* is the transistor operating condition where maximum current is flowing through the transistor (C to E). At saturation, the collector resistance is considered zero and the current flow is limited only by the resistance of the load.

The resistance of a DC load is the only current-limiting device in a circuit when a transistor reaches saturation. A transistor is operating in the cutoff region when the transistor is switched OFF. The *cutoff region* is the point at which a transistor is turned OFF and no current flows. At cutoff, all the voltage is across the open switch (transistor) and the collector-to-emitter voltage is equal to the supply voltage V_{CC}.

Transistors as Amplifiers

Transistors are used as AC amplification devices as well as DC switching devices. *Amplification* is the process of taking a small signal and increasing the size of the signal (gain). Transistor amplifiers are used to increase small signal currents and voltages so the signals can be used to do useful work. Amplification is accomplished by using a small signal to control the energy output from a large source such as a power supply. **See Figure 5-30.**

The primary objective when using a transistor as an amplifier is to produce gain. Gain is a ratio of the amplitude of the output signal to the amplitude of the input signal. Gain is a ratio of output to input and uses no unit of measure. Gain is used to describe current gain, voltage gain, and power gain. In each case, the output is compared to the input.

A single amplifier may not provide enough gain to increase the amplitude to where the output signal is useful. In such a case, two or more amplifiers are used to obtain the gain required. A *cascaded amplifier* is two or more amplifiers connected to obtain a required gain. For many amplifiers, gain is in the hundreds and even thousands.

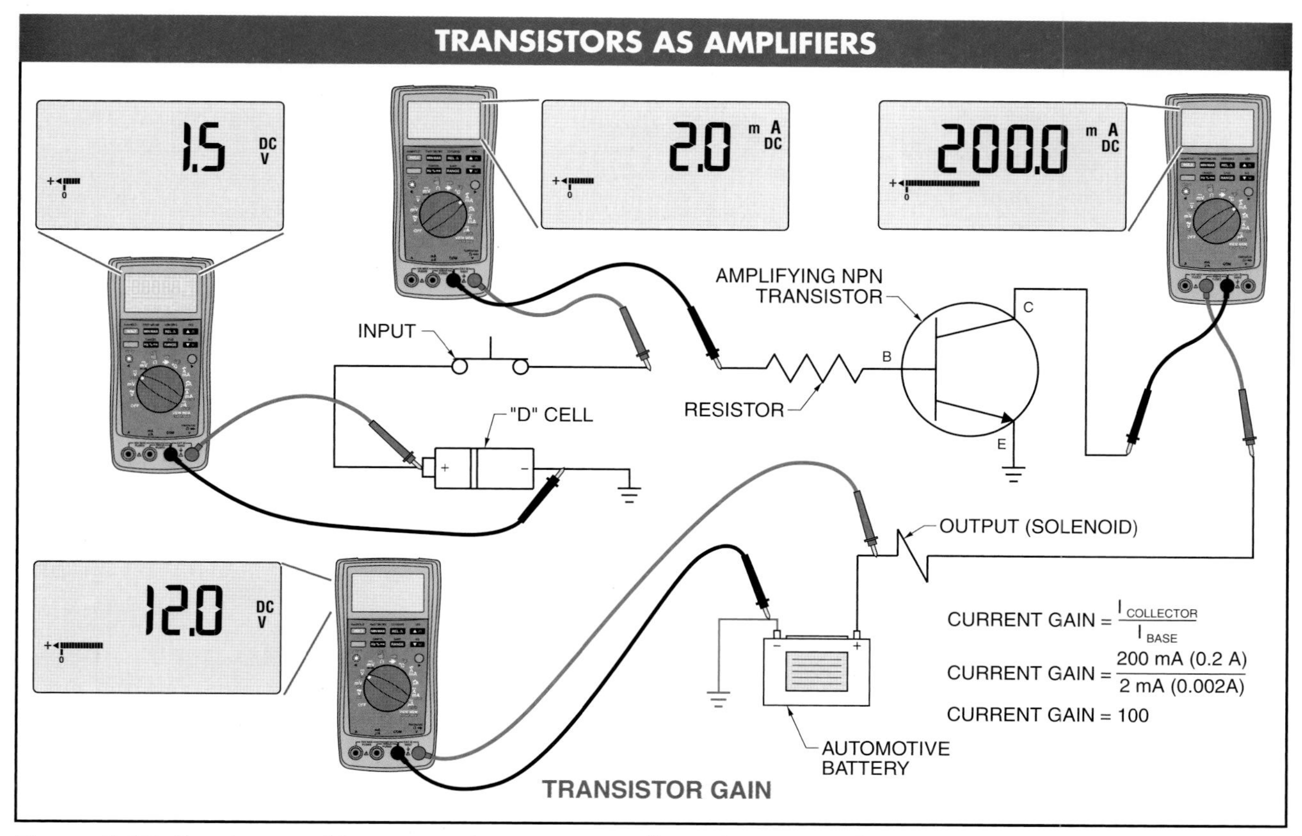

Figure 5-30. Transistor amplifiers are used to increase small signal currents and voltages so that the signals can be used to do useful work.

Transistor Testing Procedures

A transistor becomes defective because of excessive current or temperature. A transistor typically fails because of an open or shorted junction. The two junctions of a transistor are typically tested with an ohmmeter. **See Figure 5-31.**

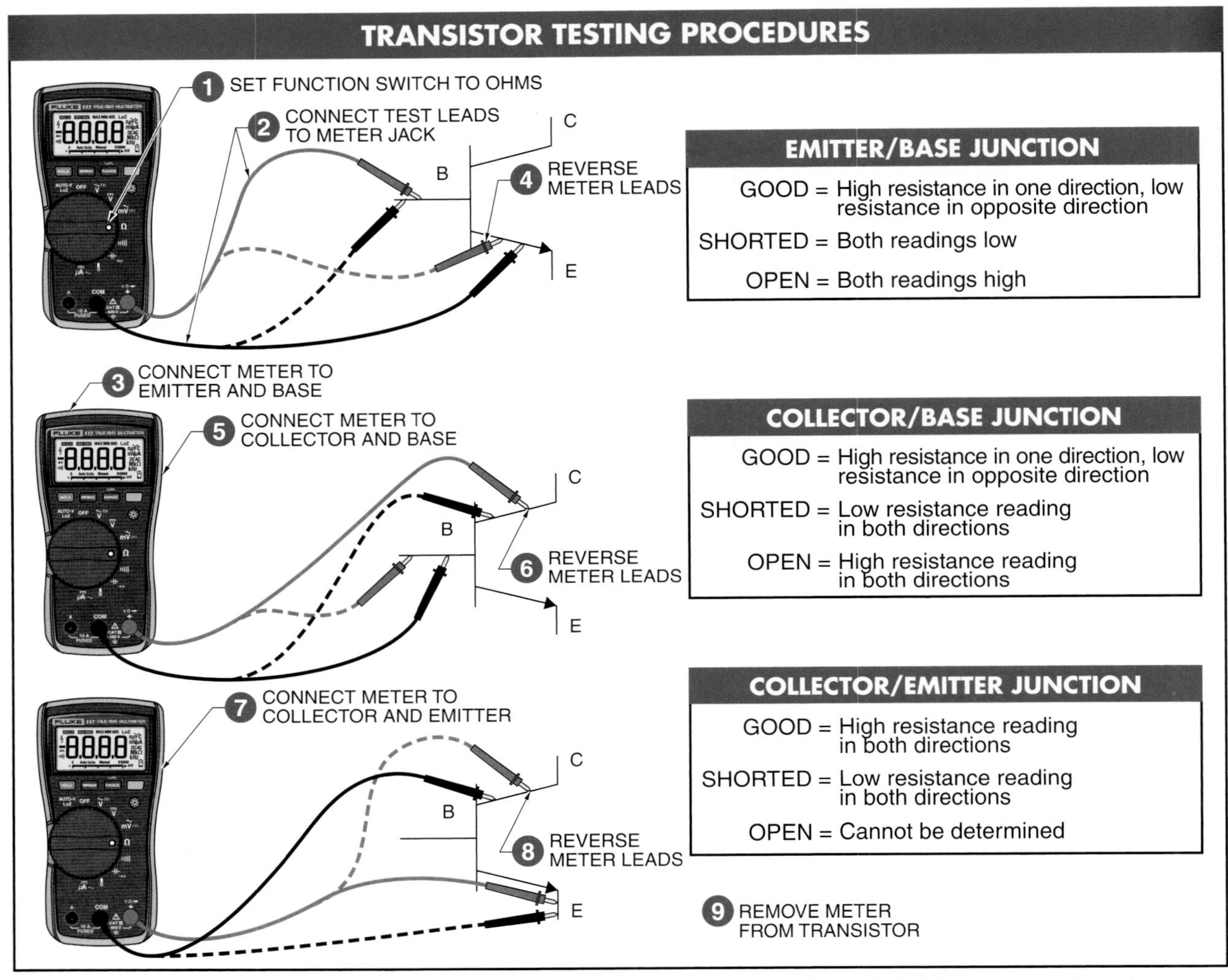

Figure 5-31. Transistors become defective due to excessive current or temperature and typically fail due to open or shorted junctions.

Troubleshooting Tip

TRANSISTOR TEST INSTRUMENTS

The in-circuit, field-service, and laboratory-standard are the three types of transistor testers used by electricians. In-circuit testers (high-impedance voltmeters) determine if a transistor is operating. Field-service testers (voltmeters and other test instruments) determine the precise operating conditions of a transistor. Laboratory-standard testers simulate an actual circuit to determine the operating characteristics of a transistor.

Before taking any resistance measurements using a digital multimeter, ensure the meter is designed to take measurements on the circuit being tested. Refer to the operations manual of the test instrument for all measuring precautions, limitations, and procedures. Always wear required personal protective equipment and follow all safety rules when taking the measurement. To test an NPN transistor for an open or shorted junction, apply the following procedures:

1. Set the function switch of the ohmmeter to ohms.
2. Connect the test leads of the ohmmeter to the meter jacks as required.
3. Connect the ohmmeter test leads to the emitter and base of the transistor. Measure the resistance.
4. Reverse the ohmmeter test leads and measure the resistance. The emitter to base junction is correct when the resistance is high in one direction and low in the opposite direction. The ratio of high to low resistance should be greater than 100:1. Typical resistance values are 1 kΩ (with the positive lead of the ohmmeter on the base) and 100 kΩ (with the positive lead of the ohmmeter on the emitter). The junction is shorted when both readings are low. The junction is open when both readings are high.
5. Connect the ohmmeter test leads to the collector and base of the transistor. Measure the resistance.
6. Reverse the ohmmeter test leads and measure the resistance. The collector to base junction is correct when the resistance is high in one direction and low in the opposite direction. The ratio of high to low resistance should be greater than 100:1. Typical resistance values are 1 kΩ (with the positive lead of the ohmmeter on the base) and 100 kΩ (with the positive lead of the ohmmeter on the collector).
7. Connect the ohmmeter test leads to the collector and emitter of the transistor. Measure the resistance.
8. Reverse the ohmmeter test leads and measure the resistance. The collector to emitter junction is correct when the resistance reading is high in both directions.
9. Remove the ohmmeter from the transistor.

The same test is used for an NPN transistor as is used for testing a PNP transistor. The difference in performing the test is that the ohmmeter test leads are reversed to obtain the same results.

THYRISTORS

A *thyristor* is a solid-state switching device that switches current ON by using a quick pulse of control current. A thyristor does not require that the control current remain ON once the thyristor is switched ON. The most common thyristors are the silicon controlled rectifier (SCR) and the triac.

A *silicon controlled rectifier (SCR)* is a thyristor that is triggered into conduction in only one direction, and is suited for DC current use. A *triac* is a thyristor that is triggered into conduction in either direction and is suited for AC current use. Thyristors have two terminals for load current and one terminal for control current. The three terminals of an SCR are the anode (A), cathode (K), and gate (G). The three terminals of a triac are the main terminal 1 (MT_1), main terminal 2 (MT_2), and gate (G). **See Figure 5-32.**

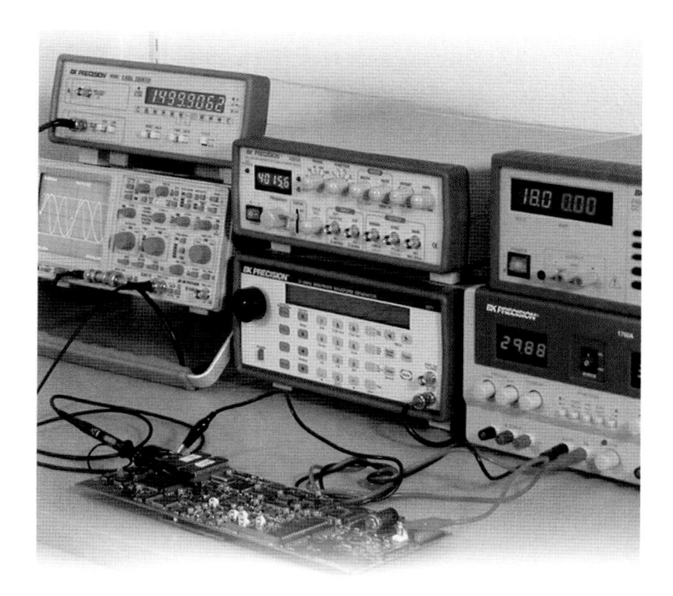

B&K Precision

Signal generators and various test instruments are used to test (field service) transistors for operating condition.

Thyristors are typically used to control large amounts of current. SCRs and triacs are used for most solid-state industrial switching circuits. SCRs and triacs control motor speed, heat output, brightness of lights, and other applications requiring solid-state switching for control.

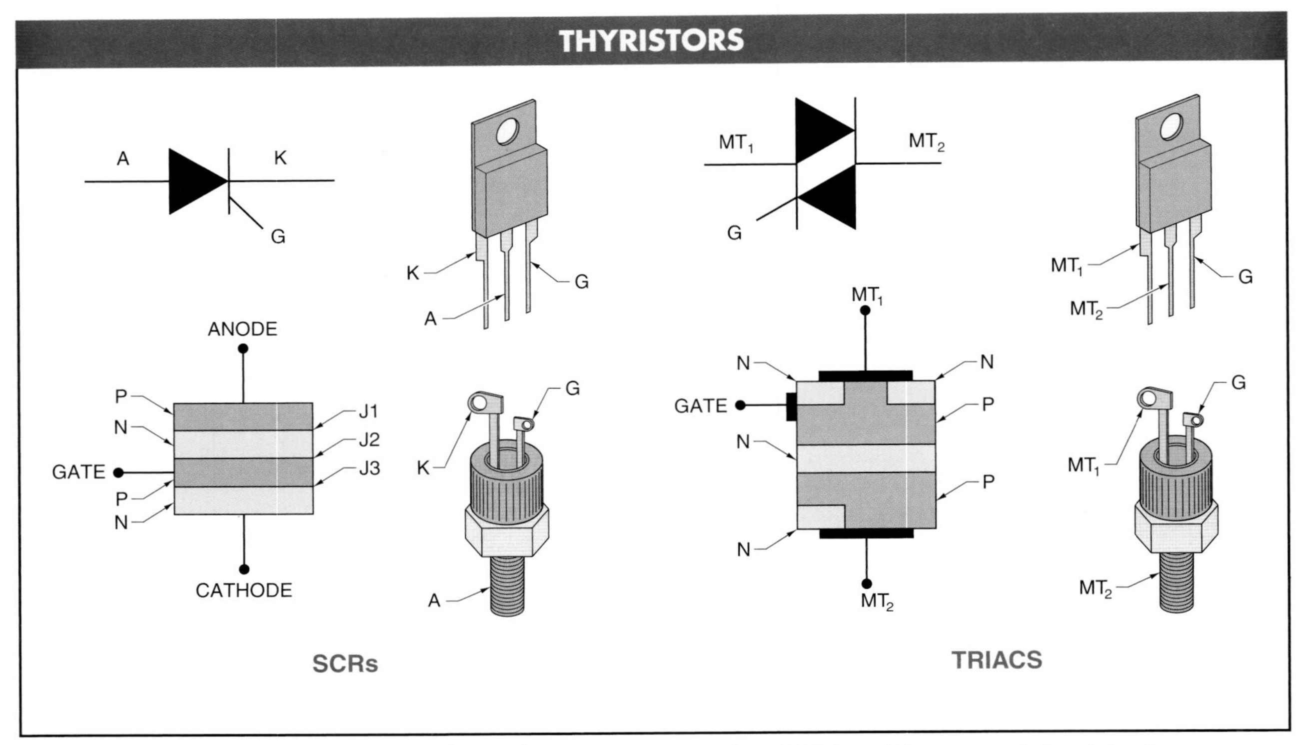

Figure 5-32. Thyristors operate as switches, either DC current switches (SCRs) or AC current switches (triacs).

Thyristor Triggering Methods

A thyristor is triggered into conduction by applying a pulse of control current to the gate. Once turned ON, a thyristor remains ON as long as there is a minimum level of holding current flowing through the load circuit. Reducing the current flowing through the load circuit below the holding current value turns a thyristor OFF. **See Figure 5-33.**

The correct method of turning a thyristor ON is to apply a proper signal to the gate of the thyristor (gate turn-on). Incorrect methods of turning on a thyristor are voltage breakover turn-on, static turn-on, and thermal turn-on.

Troubleshooting Tip

SILICON CONTROLLED RECTIFIERS

Silicon controlled rectifiers (SCRs) are four layer (PNPN) thyristors with an input control terminal (gate), an output terminal (anode), and a terminal common to both the input and output (cathode). SCRs are used where there is high current and voltage and are also used to control alternating currents. The change of sign of the current causes the device to automatically switch off. For example, an SCR can be used to bring a chemical mixture stored in a tank to a specific temperature and maintain that temperature. Disadvantages of SCRs are that, like diodes, they only conduct in one direction.

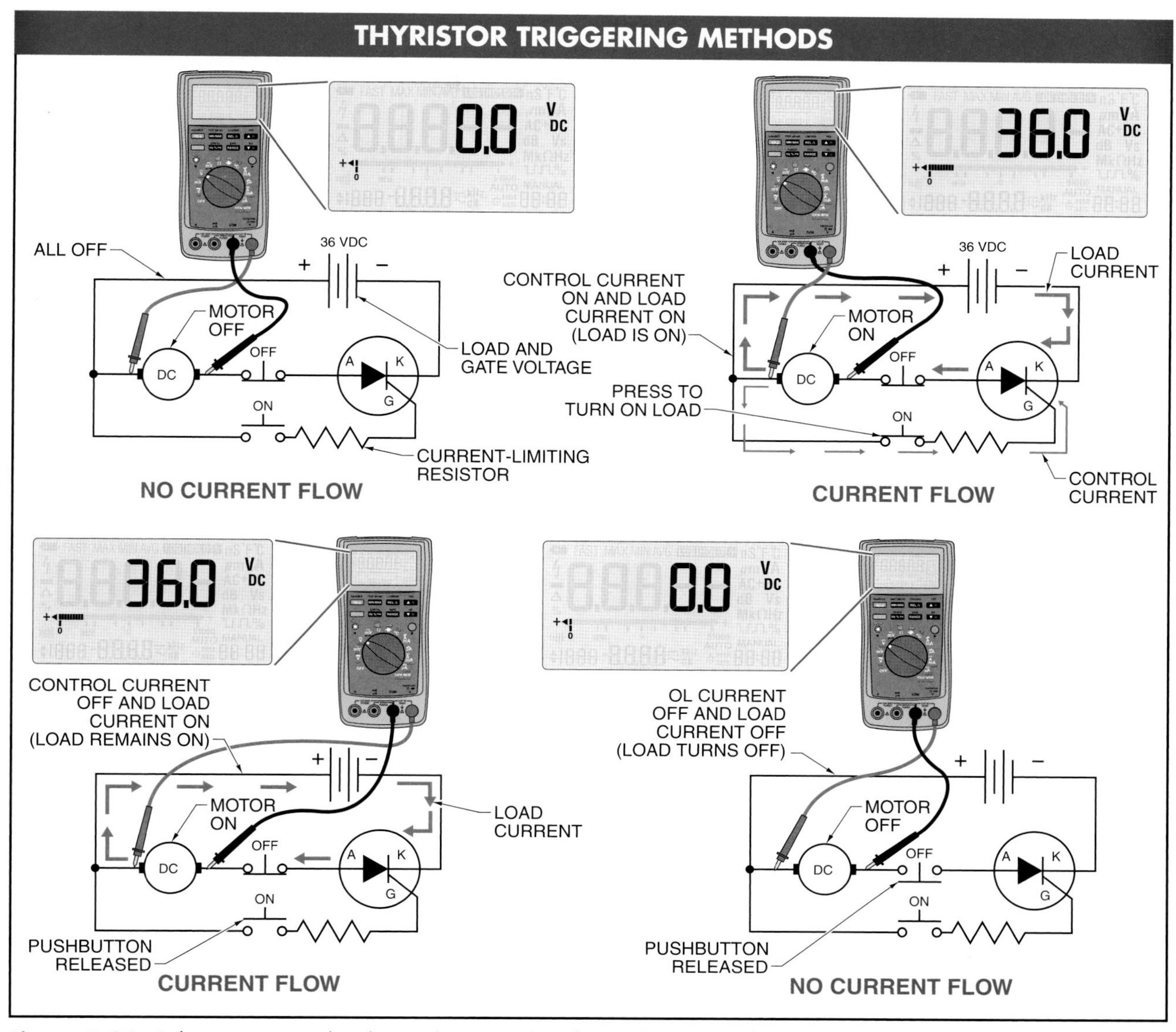

Figure 5-33. A thyristor is turned on by applying a pulse of control current to the gate. Once turned on, a thyristor remains on as long as there is a minimum level of holding current flowing through the load circuit.

Gate Turn-on

Gate turn-on is a method of turning on a thyristor that occurs when the proper signal is applied to the gate at the correct time. Gate turn-on is the only correct way to turn on a thyristor. For an SCR, the gate signal must be positive with respect to the cathode polarity for the thyristor to turn ON. **See Figure 5-34.**

Voltage Breakover Turn-on

Voltage breakover turn-on is a method of turning on a thyristor that occurs when the voltage across the thyristor terminals exceeds the maximum voltage rating of the device. Excessive voltage causes localized heating in a thyristor and damages the thyristor.

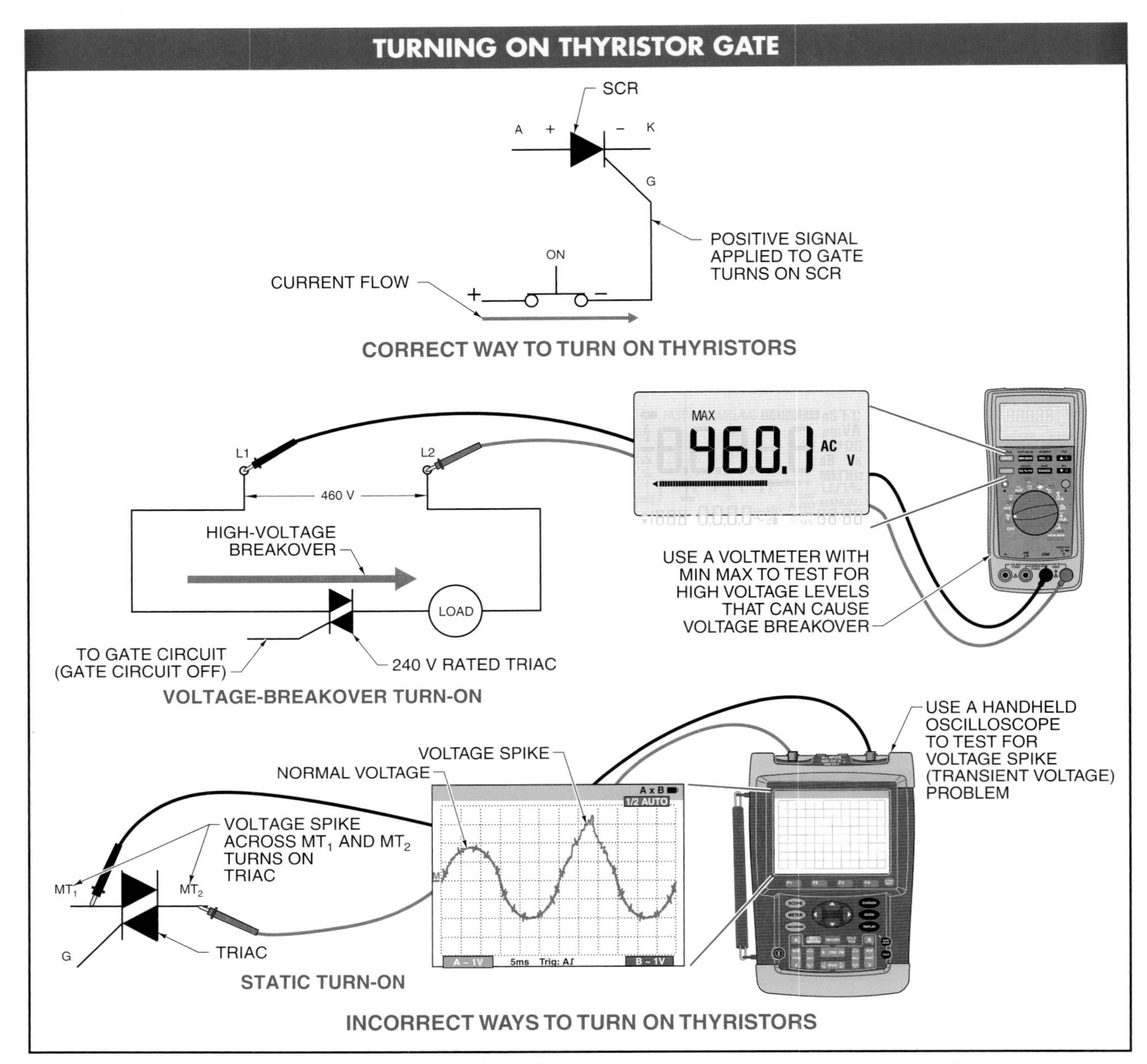

Figure 5-34. Thyristors can be turned on by methods that damage the thyristor such as voltage breakover turn-on, static turn-on, and heat turn-on.

Static Turn-on

Static turn-on is a method of turning on a thyristor that occurs when a fast-rising voltage is applied across the terminals of a triac. Manufacturers refer to the point of turn-on as the dv/dt rating. The dv/dt rating defines the level of voltage over a given time period that causes the device to turn ON. For example, a typical rating for a thyristor is 250 V/sec.

Static turn-on does not damage the thyristor provided the surge current is limited. A snubber circuit is typically added across the thyristor terminals to protect the thyristor when static turn-on is a problem.

Thermal Turn-on

All solid-state components are heat sensitive. *Thermal turn-on* is a method of turning on a thyristor that occurs when heat levels exceed the limit of the thyristor (typically 230°F or 110°C). When a solid-state thyristor is turned on by heat, the thyristor is typically destroyed. Using the correct heat sinks for a circuit eliminates thermal turn-on of thyristors.

Silicon Controlled Rectifier Testing Procedures

SCRs must be tested under operating conditions using an oscilloscope. An oscilloscope shows exactly how the SCR is or is not operating. Most high-power SCRs must be tested using a test circuit and an oscilloscope. Low-power and some high-power SCRs are tested using an analog multimeter. To ensure that the analog multimeter delivers enough output voltage to fire the SCR being tested (not all do), a known working SCR should be tested first. **See Figure 5-35.**

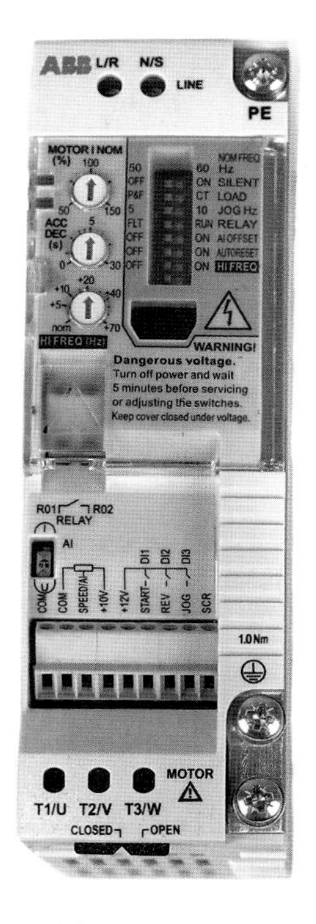

Heat sinks are used to dissipate heat from sensitive electronic equipment such as variable frequency drives.

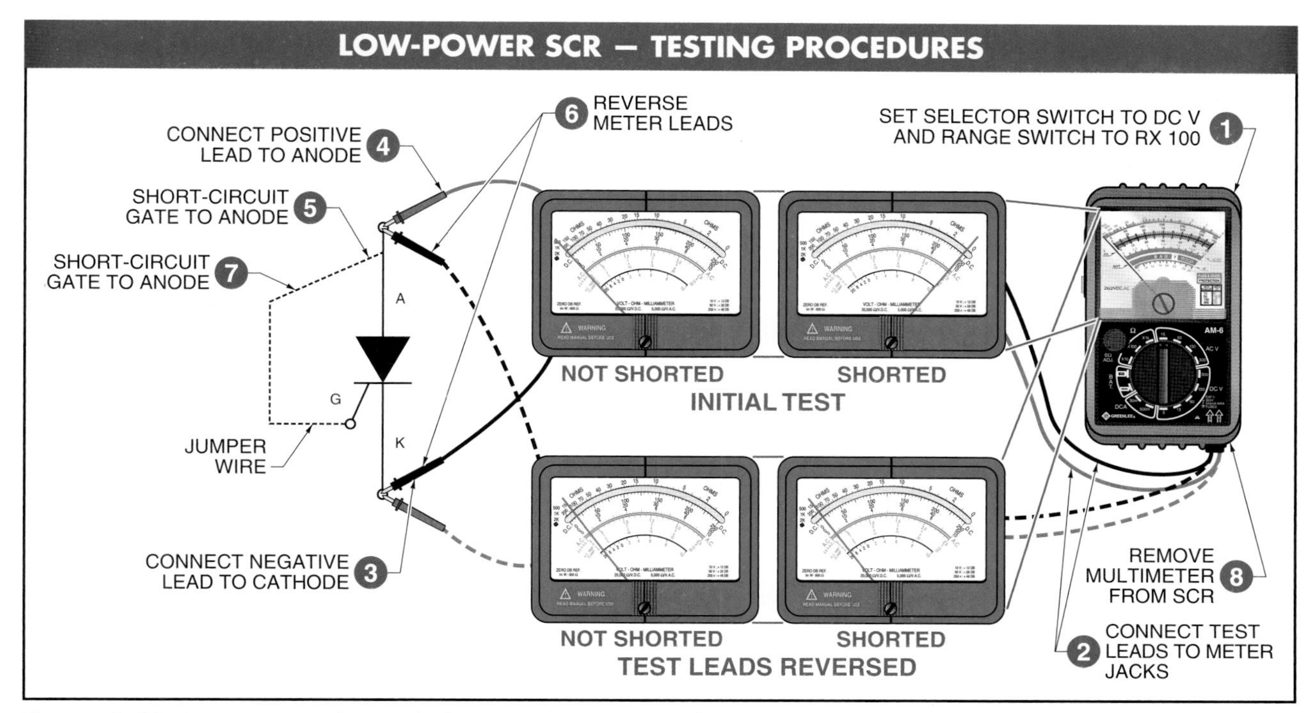

Figure 5-35. Low-power and some high-power SCRs are tested using an analog multimeter. A good SCR will have zero ohms and infinity ohms readings.

Before taking any resistance measurements using a multimeter, ensure the meter is designed to take measurements on the circuit being tested. Refer to the operations manual of the test instrument for all measuring precautions, limitations, and procedures. Always wear required personal protective equipment and follow all safety rules when taking the measurement. To test a low-power SCR using an ohmmeter, apply the following procedures:

1. Set the selector switch of the multimeter to DC V, and the range switch to the proper resistance range (X 100 is typical).
2. Connect the test leads of the multimeter to the meter jacks as required.
3. Connect the negative test lead of the multimeter to the cathode of the SCR.
4. Connect the positive test lead of the multimeter to the anode of the SCR. The ohmmeter should read infinity.
5. Short circuit the gate to the anode using a jumper wire. The multimeter should read almost 0 Ω. Remove the jumper wire. The low-resistance reading should remain.
6. Reverse the multimeter test leads so that the positive lead is on the cathode and the negative lead is on the anode. The multimeter should read almost infinity.
7. Short-circuit the gate to the anode with a jumper wire. The resistance displayed on the analog multimeter should remain high.
8. Remove the multimeter from the SCR.

Triac Testing Procedures

Triacs should be tested under operating conditions using an oscilloscope. An oscilloscope shows exactly how a triac is or is not operating. A DMM can be used to make a rough test with the triac out of a circuit. **See Figure 5-36.**

Before taking any resistance measurements using a digital multimeter, ensure the meter is designed to take measurements on the circuit being tested. Refer to the operations manual of the test instrument for all measuring precautions, limitations, and procedures. Always wear required personal protective equipment and follow all safety rules when taking the measurement. To test a triac using a DMM, apply the following procedures:

1. Set the DMM on the Ω scale.
2. Connect the negative lead to main terminal 1.
3. Connect the positive lead to main terminal 2. The DMM should read infinity.
4. Short-circuit the gate to main terminal 2 using a jumper wire. The DMM should read almost 0 Ω. The zero reading should remain when the lead is removed.
5. Reverse the DMM leads so that the positive lead is on main terminal 1 and the negative lead is on main terminal 2. The DMM should read infinity.
6. Short-circuit the gate of the triac to main terminal 2 using a jumper wire. The DMM should read almost 0 Ω. The zero reading should remain after the lead is removed.

Technical Tip

Triacs are used to control the speed of a motor. Since load current (or armature current) flows during both halves of the applied VAC, the motor rotates smoothly at all rotational speeds.

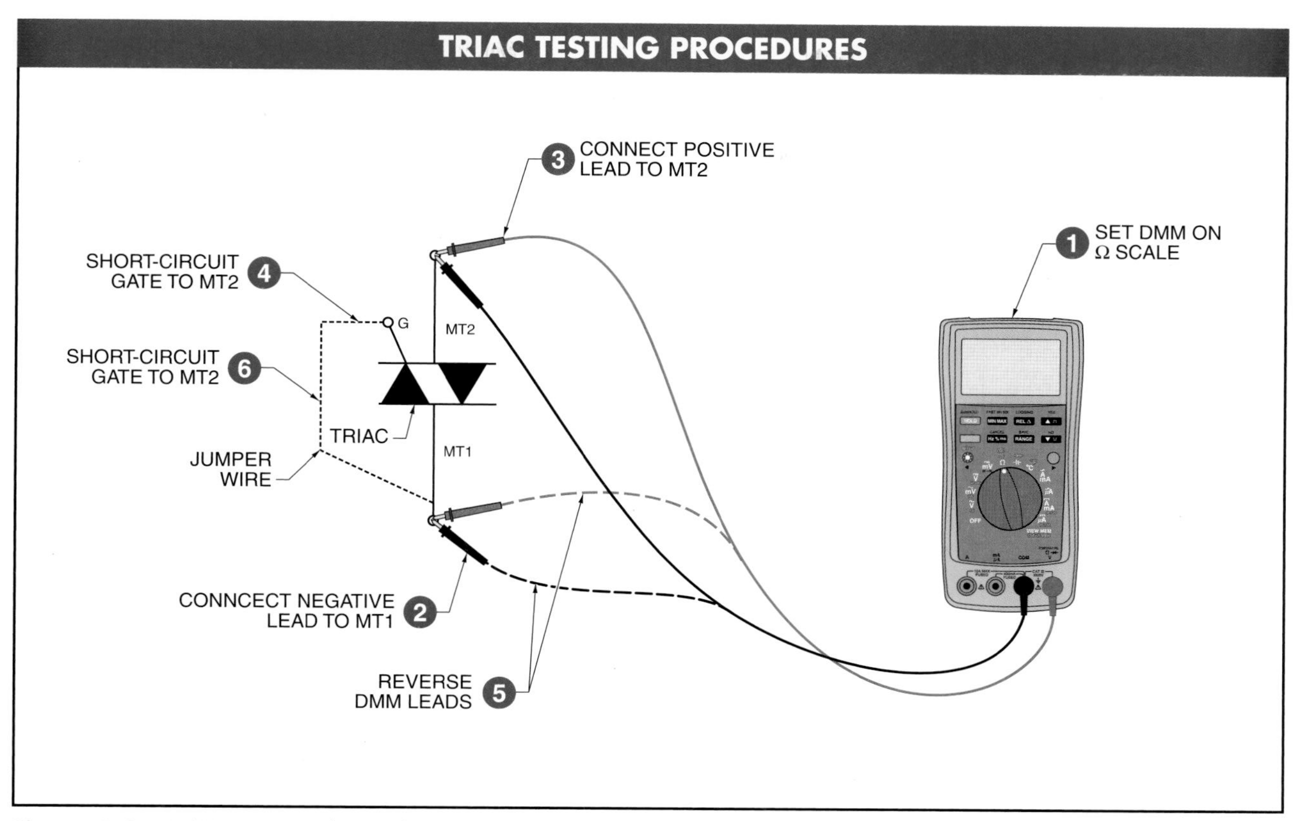

Figure 5-36. DMMs are used to make a rough test on triacs that are out of a circuit.

TROUBLESHOOTING MECHANICAL AND SOLID-STATE SWITCHES

Mechanical switches have been used since the first electrical circuits. Mechanical switches are used today, but more and more solid-state switches are being used to control the flow of current. Both types of switches cause problems and often require troubleshooting.

Mechanical Switches

Mechanical and solid-state switches are used to switch ON and OFF the flow of electricity in a circuit. A *mechanical switch* is any switch that uses contacts to start and stop the flow of current in a circuit. Mechanical switches have normally open, normally closed, or combination switching contacts. Mechanical switches may be manually, mechanically, or automatically operated.

Manually operated switches are used when a circuit is controlled by a person. Manually operated switches include pushbuttons, palmbuttons, foot switches, toggle switches, and keyboards. Mechanically operated switches are used when a circuit is controlled by the movement of an object. A limit switch is a typical mechanically operated switch. Automatically operated switches are used when a circuit is controlled by a change in the given conditions. Automatically operated switches include flow-, level-, pressure-, temperature-, humidity-, and gas-activated switches.

A suspected fault with a mechanical switch is tested using a voltmeter. A voltmeter is used to test the voltage into and out of a mechanical switch. **See Figure 5-37.**

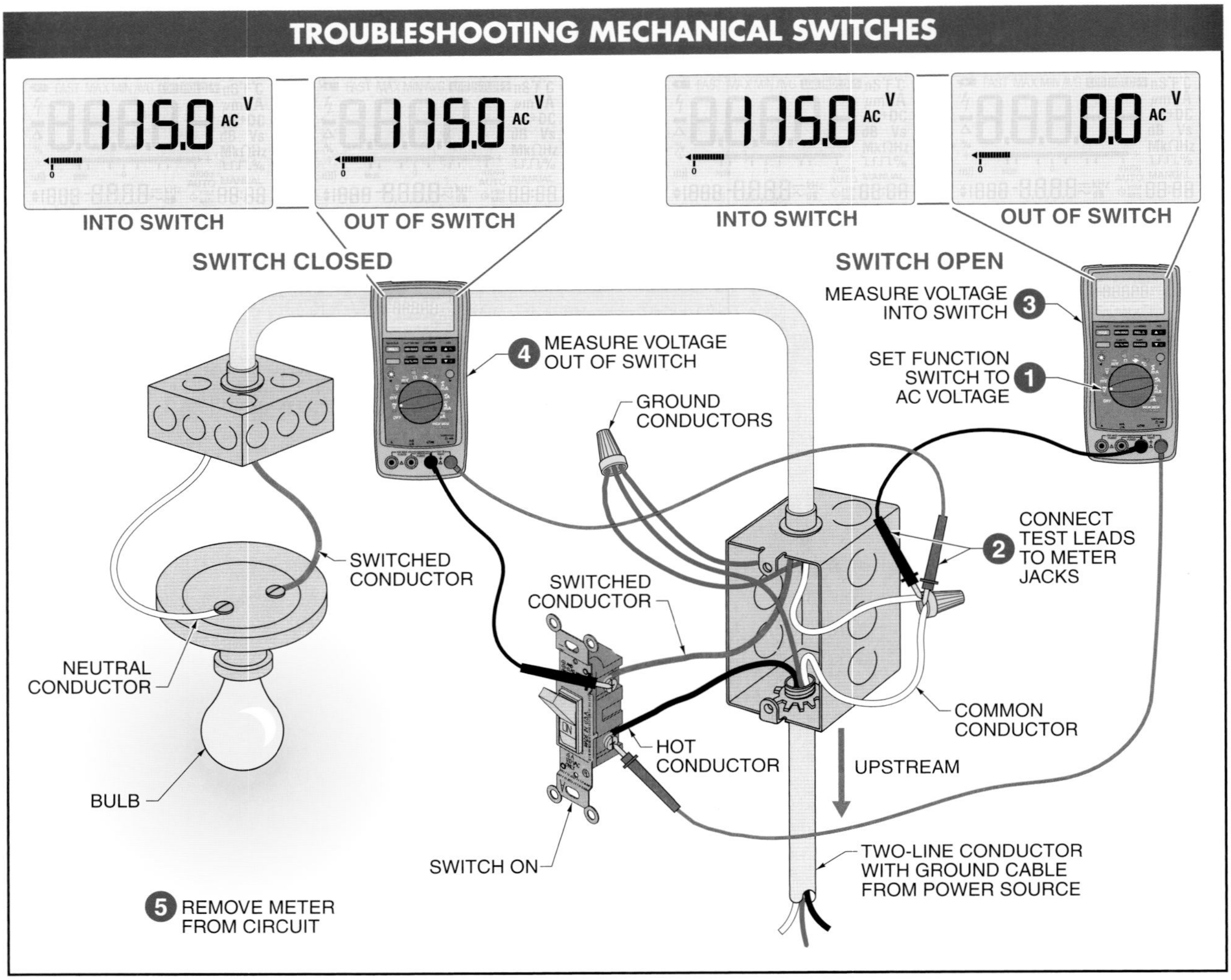

TROUBLESHOOTING MECHANICAL SWITCHES

Figure 5-37. Voltmeters are used to test a mechanical switch by testing the voltage into and out of the switch.

Before taking any AC voltage measurements using a digital multimeter, ensure the meter is designed to take measurements on the circuit being tested. Refer to the operations manual of the test instrument for all measuring precautions, limitations, and procedures. Always wear required personal protective equipment and follow all safety rules when taking a measurement. To test a mechanical switch, apply the following procedures:

1. Set the function switch of the voltmeter to AC voltage.
2. Connect the test leads of the voltmeter to the meter jacks as required.
3. Measure the voltage into the switch. Connect the voltmeter between the common and hot conductor of the switch. The negative voltmeter test lead can be connected to ground instead of the common when the common conductor is not available in the same box in which the switch is located. The problem is located upstream from the switch when there is no voltage present or the voltage is not at the correct level. The problem may be a blown fuse or open circuit. Voltage must be reestablished to the switch before the switch can be tested.

4. Measure the voltage out of the switch. Voltage must be present when the switch contacts are closed. There must not be voltage when the switch contacts are open. The switch has an open and must be replaced when there is no voltage reading in either switch position. The switch has a short and must be replaced when there is a voltage reading in both switch positions. Voltage drop of more than 100 mV indicates pitted or burned contacts.
5. Remove the voltmeter from the circuit.

WARNING

Always ensure power is OFF before replacing a switch in a circuit control. Use a voltmeter to ensure that power is OFF.

Solid-State Switches

A *solid-state switch* is a switch with no moving parts (contacts). A solid-state switch uses a triac, SCR, current sink (NPN) transistor, or current source (PNP) transistor to perform switching functions. Triacs are used for switching AC loads and SCRs are used for switching high-power DC loads. NPN and PNP transistors are used to switch low-power DC loads. Solid-state switches have outputs classified as normally open, normally closed, or combination switching. **See Figure 5-38.**

Testing Two-Wire Solid-State Switches

A *two-wire solid-state switch* has two terminals for connecting wires (not including a ground). A two-wire switch is connected in series with the controlled load. A two-wire solid-state switch is also called a load-powered switch because the switch draws operating current from the power source and allows the operating current to return to the power source through the load. The operating current (small amount that is not enough to power the load) flows through the load when the switch is not conducting (load OFF). Operating current is also called residual current or leakage current by some manufacturers and is measured with an ammeter when the load is OFF. **See Figure 5-39.**

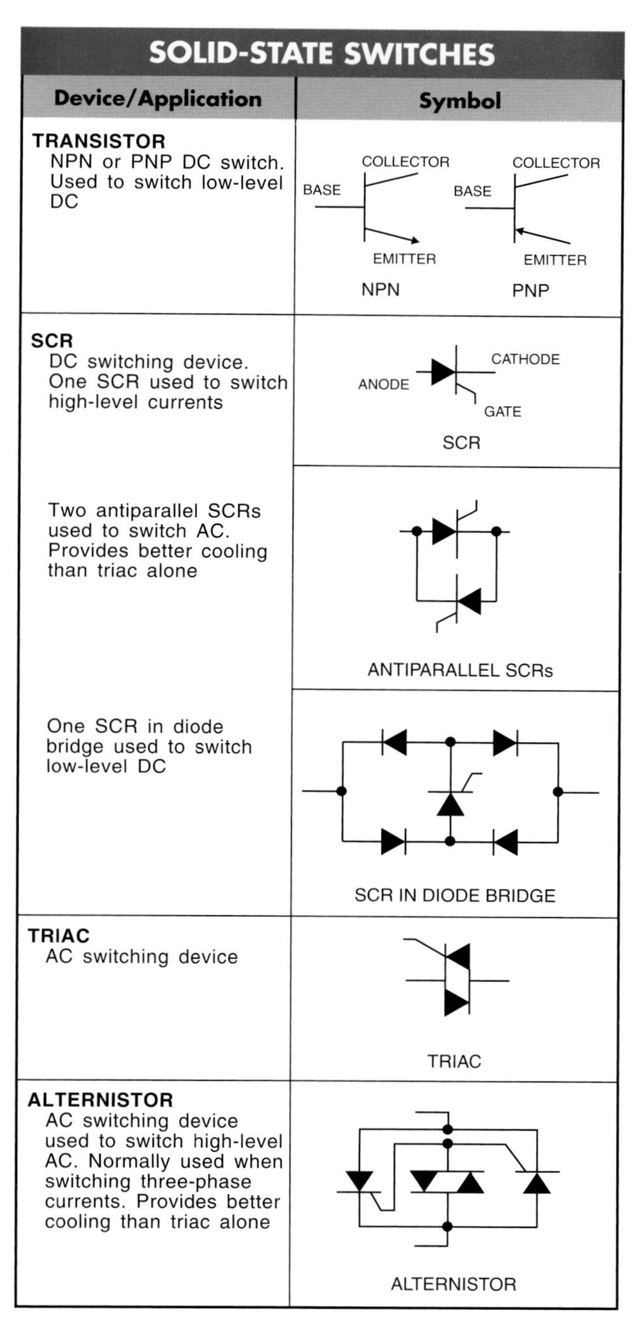

SOLID-STATE SWITCHES

Device/Application	Symbol
TRANSISTOR NPN or PNP DC switch. Used to switch low-level DC	COLLECTOR, BASE, EMITTER — NPN; COLLECTOR, BASE, EMITTER — PNP
SCR DC switching device. One SCR used to switch high-level currents	ANODE, CATHODE, GATE — SCR
Two antiparallel SCRs used to switch AC. Provides better cooling than triac alone	ANTIPARALLEL SCRs
One SCR in diode bridge used to switch low-level DC	SCR IN DIODE BRIDGE
TRIAC AC switching device	TRIAC
ALTERNISTOR AC switching device used to switch high-level AC. Normally used when switching three-phase currents. Provides better cooling than triac alone	ALTERNISTOR

Figure 5-38. Solid-state switches include transistors, silicon controlled rectifiers, triacs, and alternistors.

Technical Tip

The SCR differs from an ordinary semiconductor diode in that it will not pass significant current, even when forward biased, unless the anode voltage equals or exceeds the forward breakover voltage. When forward breakover voltage is reached, the SCR will switch ON and become highly conductive.

The total current in a circuit is a combination of the operating current and load current when the switch is conducting (load ON). A solid-state switching device must be rated high enough to carry the operating current and the load current required for the circuit. Total current is measured with an ammeter when the load is ON.

The current draw of a load must be sufficient to keep the solid-state switch operating when the switch is conducting (load ON). *Minimum holding current* is the minimum current that ensures proper operation of a solid-state switch. Minimum holding current values range from 2 mA to 20 mA for most solid-state switches.

Operating current and minimum holding current values are typically not a problem when a solid-state switch controls low-impedance loads such as motor starters, relays, and solenoids. Operating current and minimum holding current values are a problem when a switch controls high-impedance loads such as programmable controllers and other solid-state devices because the operating current is high enough to affect the load when the switch is not conducting. For example, a programmable controller could mistakenly use the operating current as an input signal.

Troubleshooting Tip

ADVANTAGES OF SOLID-STATE SWITCHES

- Long life, should never need replacement
- Diagnostic capability, red light indicates state
- Stable setpoint, should never need adjustment
- Seamless switching

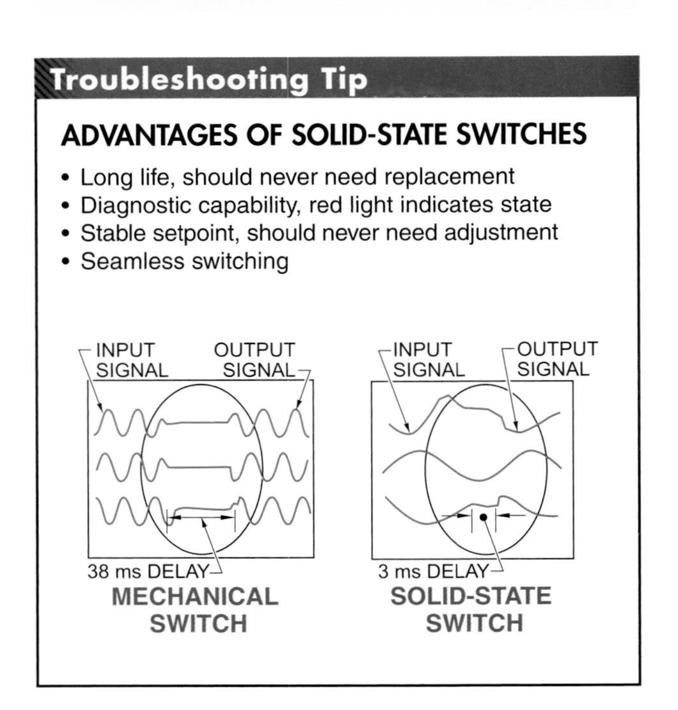

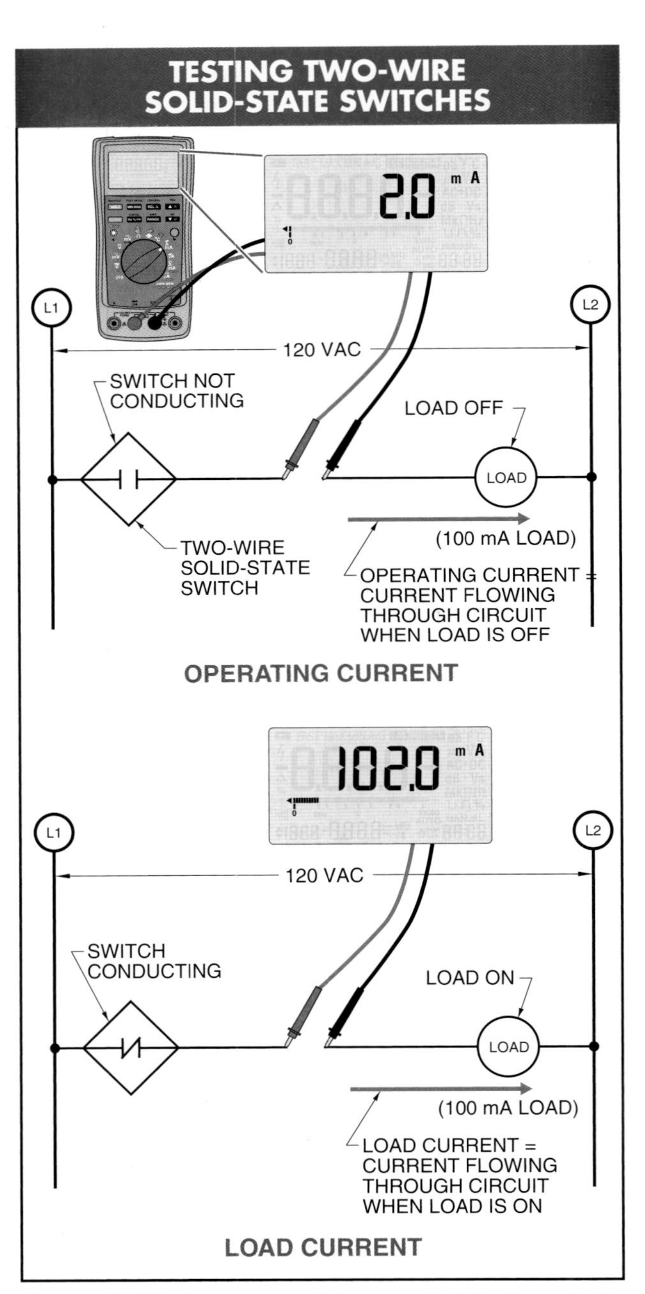

Figure 5-39. A two-wire solid-state switch is also called a load-powered switch because the switch draws operating current from the power source and allows the operating current to return to the power source through the load.

Two-wire solid-state switches connected in series affect the operation of a load because of the voltage drop across the switches. A two-wire solid-state switch drops about 3 V to 8 V. The total voltage drop across the switches equals the sum of the voltage drops across each switch. No more than three solid-state switches are allowed to be connected in series due to the voltage drop created by each switch. **See Figure 5-40.**

Two-wire solid-state switches connected in parallel affect the operation of a load because each switch has operating current that is flowing through the load. The load may turn ON when the operating current through the load becomes excessive. The total operating current equals the sum of the operating currents of each switch. No more than three solid-state switches should be connected in parallel.

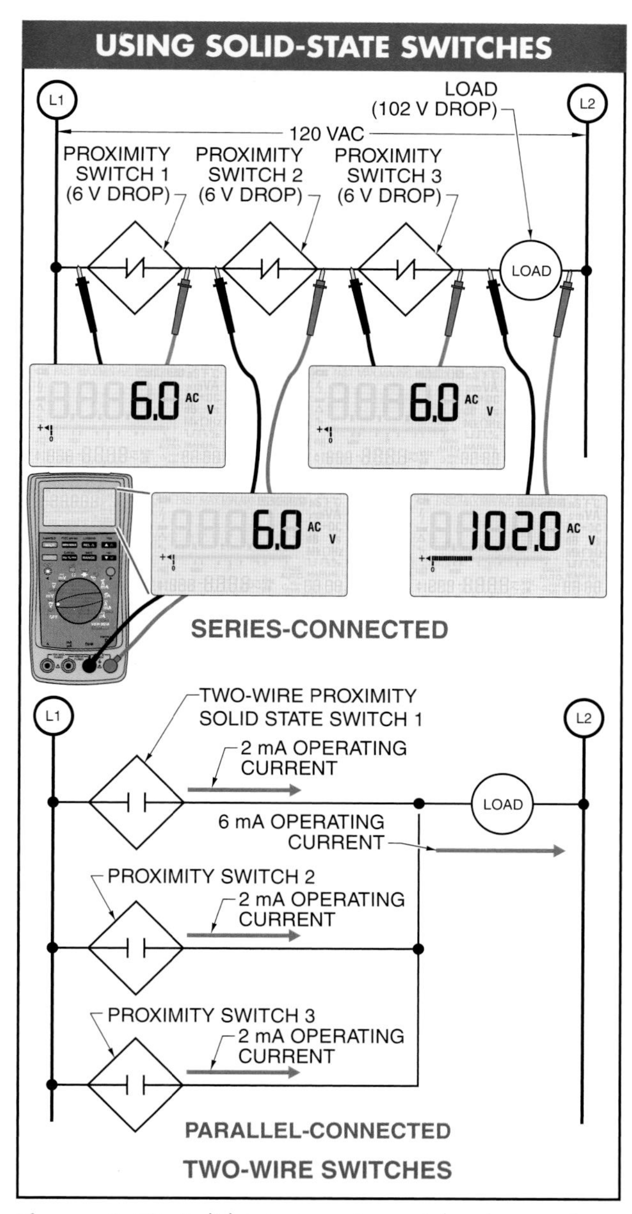

Figure 5-40. Solid-state two-wire switches have voltage drops of about 3 V to 8 V that must be accounted for when incorporating the switches for use.

A suspected fault in a two-wire solid-state switch is tested using a voltmeter. A voltmeter is used to test the voltage into the switch and out of the switch. **See Figure 5-41.**

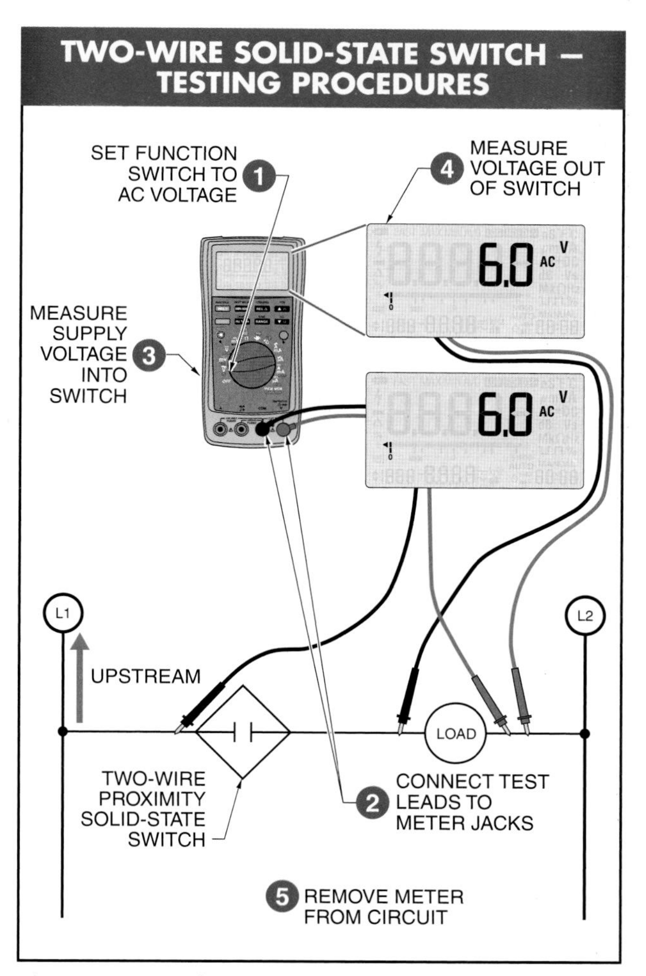

Figure 5-41. Voltmeters are used to test two-wire solid-state switches by testing the voltage into and out of the switches.

Before taking any AC voltage measurements using a digital multimeter, ensure the meter is designed to take measurements on the circuit being tested. Refer to the operations manual of the test instrument for all measuring precautions, limitations, and procedures. Always wear required personal protective equipment and follow all safety rules when taking a measurement. To test a two-wire solid-state switch, apply the following procedures:

1. Set the function switch of the voltage meter to measure AC voltage.
2. Connect the test leads of the voltage meter jacks as required.
3. Measure the supply voltage into the solid-state switch. The problem is located upstream from the switch when there is no voltage present. The problem may be a blown fuse or open circuit. Voltage to the solid-state switch must be reestablished before the switch is tested.
4. Measure the voltage out of the solid-state switch. The voltage should equal the supply voltage minus the rated voltage drop (3 V to 8 V) of the switch when the switch is conducting (load ON). Replace the switch when the voltage output is not correct.
5. Remove the voltmeter from the circuit.

Testing Three-Wire Solid-State Switches

A *three-wire solid-state switch* has three terminals for connecting wires (exclusive of ground). A three-wire solid-state switch is connected in series with the control load and is also connected to a power source. A three-wire solid-state switch draws operating current directly from a power source and does not allow the operating current to flow through the load.

A three-wire solid-state switch is also called a line-powered switch because the switch draws operating current from a power line. The two types of three-wire solid-state switches are the current source (PNP) switch and the current sink (NPN) switch. **See Figure 5-42.**

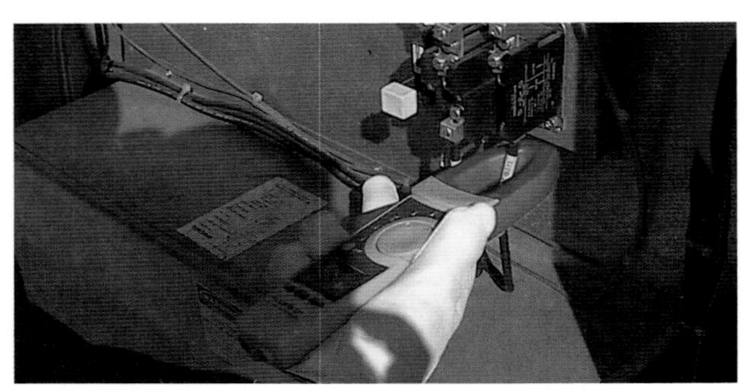

A clamp-on meter can be used to troubleshoot solid-state devices by testing for voltage, current, and resistance.

Always ensure power is OFF before replacing a control switch. Use a voltmeter to ensure that power is OFF.

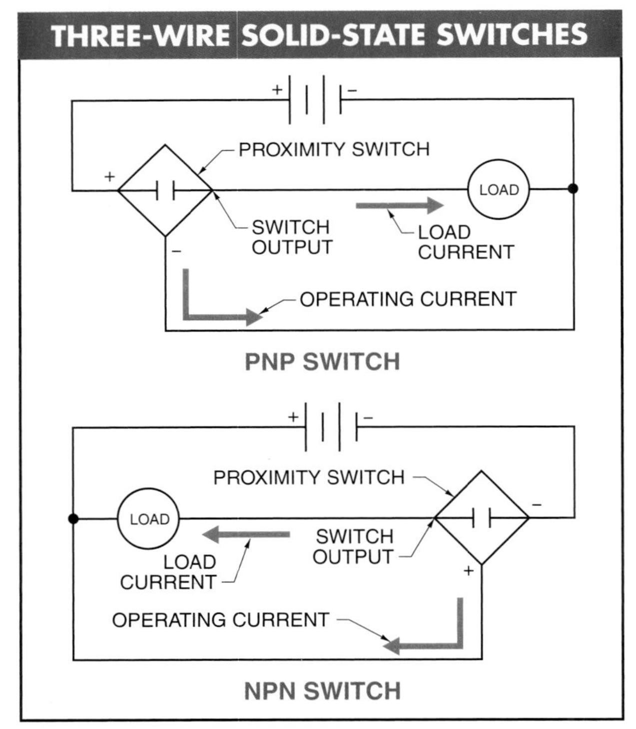

Figure 5-42. Three-wire solid-state switches (current source or current sink) are also called line-powered switches because the switch draws operating current from the power lines.

Three-wire solid-state switches connected in series affect the operation of the load because each switch downstream from the last switch must carry the load current and the operating current of each upstream switch. An ammeter is used to measure operating and total current values. The measured values must not exceed the maximum rating of the solid-state switch manufacturer. **See Figure 5-43.**

Three-wire solid-state switches connected in parallel affect the operation of the load because a nonconducting switch may be damaged by reverse polarity. A blocking diode must be added to each switch output to prevent reverse polarity on the switch.

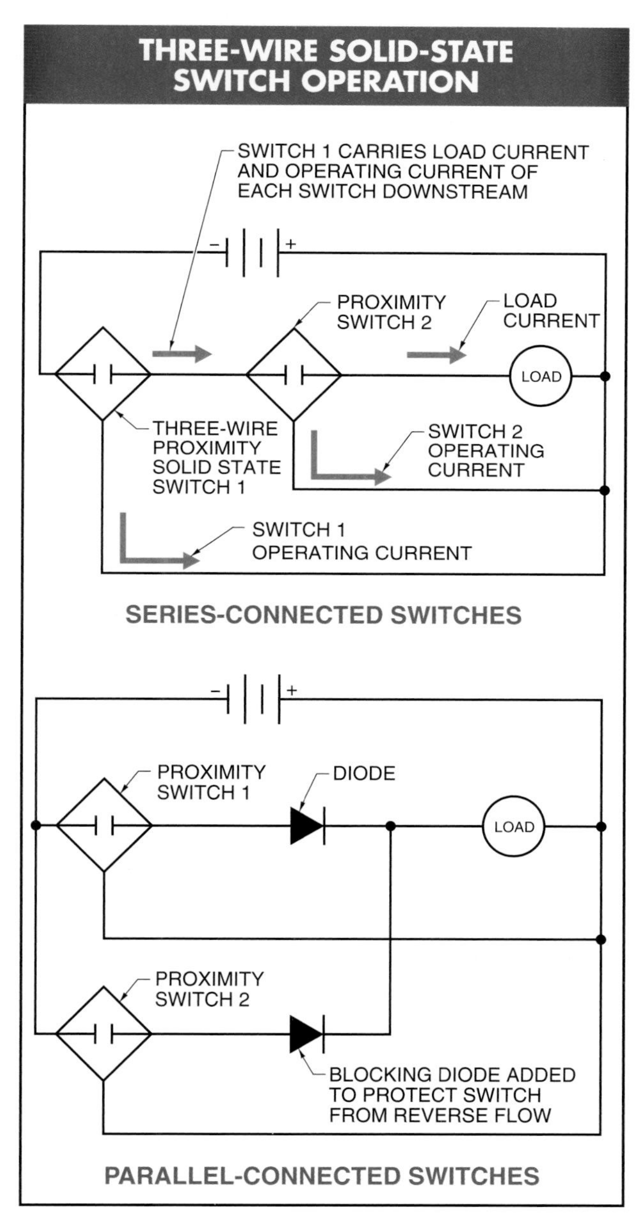

Figure 5-43. Three-wire solid-state switches connected in series affect the operation of the load because each switch must carry the load current and the operating current of each switch in the circuit.

Troubleshooting Tip

FINDING TOTAL RESISTANCE

Series Resistances

A series circuit has an 18 Ω resistance, a 44 Ω resistance, and a 31 Ω resistance. What is the total resistance of the circuit?

$$R_T = R_1 + R_2 + R_3$$

where

R_T = total resistance (in Ω)
R_1 = resistance 1 (in Ω)
R_2 = resistance 2 (in Ω)
R_3 = resistance 3 (in Ω)

$$R_T = R_1 + R_2 + R_3$$
$$R_T = 18 + 44 + 31$$
$$R_T = \mathbf{93}\ \Omega$$

Parallel Resistances

A parallel circuit has an 18 Ω resistance, a 44 Ω resistance, and a 31 Ω resistance. What is the total resistance of the circuit?

$$R_T = \frac{1}{\frac{1}{R_1} + \frac{1}{R_2} + \frac{1}{R_3}}$$

where

R_T = total resistance (in Ω)
R_1 = resistance 1 (in Ω)
R_2 = resistance 2 (in Ω)
R_3 = resistance 3 (in Ω)

$$R_T = \frac{1}{\frac{1}{R_1} + \frac{1}{R_2} + \frac{1}{R_3}}$$
$$R_T = \frac{1}{\frac{1}{18} + \frac{1}{44} + \frac{1}{31}}$$
$$R_T = \frac{1}{0.055 + 0.023 + 0.032}$$
$$R_T = \frac{1}{0.11}$$
$$R_T = \mathbf{9.09}\ \Omega$$

A suspected fault with a three-wire solid-state switch is tested using a voltmeter. A voltmeter is used to test the voltage into the switch and out of the switch. **See Figure 5-44.**

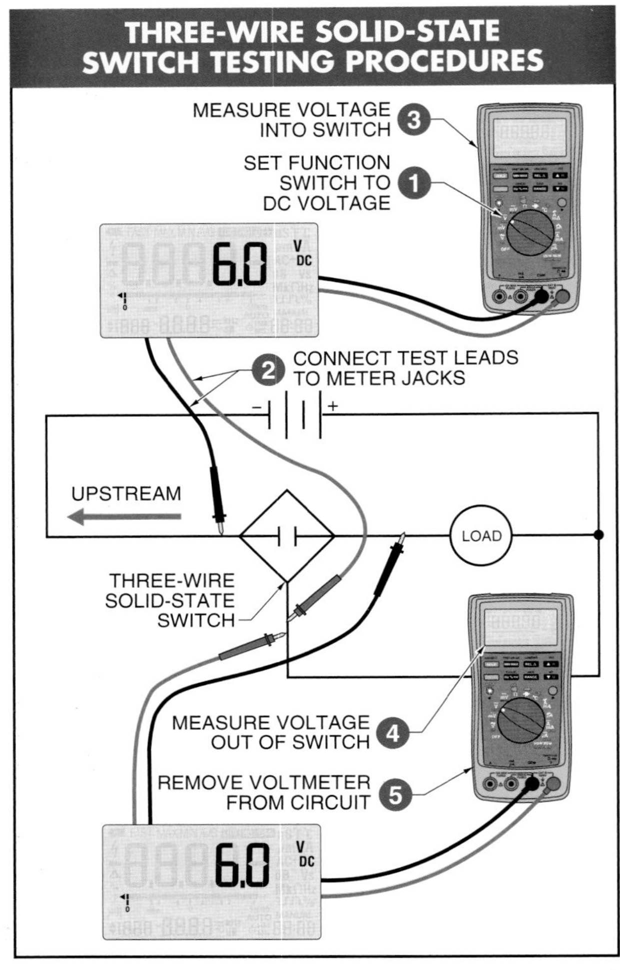

Figure 5-44. Voltmeters are used to test three-wire solid-state switches by testing the voltage into and out of the switches.

Before taking any AC voltage measurements using a digital multimeter, ensure the meter is designed to take measurements on the circuit being tested. Refer to the operations manual of the test instrument for all measuring precautions, limitations, and procedures. Always wear required personal protective equipment and follow all safety rules when taking a measurement. To test a three-wire solid-state switch, apply the following procedures:

1. Set the function switch of the voltmeter to DC voltage.
2. Connect the test leads of the voltmeter to the proper jacks of the meter.
3. Measure the voltage into the solid-state switch. The problem is located upstream from the switch when there is no voltage present or the voltage is not at the correct level. The problem may be a blown fuse or open circuit. Voltage to the solid-state switch must be reestablished before the switch is tested.
4. Measure the voltage out of the solid-state switch. The voltage should be equal to the supply voltage when the switch is conducting (load ON). Replace the solid-state switch when the voltage out of the switch is not correct.
5. Remove the voltmeter from the circuit.

WARNING

Make sure the voltage in the system under test does not exceed the voltage rating of the meter being used for the test.

ELECTRONIC CIRCUIT TEST INSTRUMENTS 5 REVIEW

Name: ______________________________ Date: ____________________

______ **1.** What unit is used to measure frequency?

T F **2.** Higher carrier frequencies produce fundamental frequencies closer to a pure sine wave.

______ **3.** What unit is used to measure the intensity of sound?

T F **4.** Rectifier circuits convert DC to AC.

______ **5.** Which electrical quantity is typically measured when testing mechanical switches?

T F **6.** Larger amplitude sound waves produce louder sounds.

______ **7.** Which type of digital logic gate inverts the incoming signal?

______ **8.** Which electrical quantity is typically measured when testing transistors?

______ **9.** Which type of digital logic gate de-energizes the output if one or more inputs are energized?

T F **10.** Gain is the ratio of output signal to input signal.

______ **11.** Does increasing frequency increase or decrease wavelength?

T F **12.** A manual switch is a switch with no moving contacts.

______ **13.** Does increasing oscilloscope sample rate increase or decrease the accuracy of the waveform trace?

______ **14.** If an oscilloscope measures 17 V with a 10x probe, what is the actual circuit voltage?

T F **15.** The decibel scale is linear.

T F **16.** Sound waves require a medium, such as air, to travel through.

______ **17.** What input condition is prevented by a pull-up resistor?

______ **18.** Which type of solid-state switch does not require the control current to remain ON for the switch to remain ON?

______ **19.** Which type of digital logic gate energizes the output only if all inputs are energized?

______ **20.** Digital logic probes measure a high or low state of what electrical quantity?

21. How can dual-trace oscilloscopes monitor signal gain?

22. What is the difference between fundamental frequency and carrier frequency?

23. Which three types of waves can be output by signal generators?

24. What are the two primary applications of transistors?

25. Explain at least one advantage and one disadvantage of high carrier frequencies in motor drives.

6 GROUNDING SYSTEMS AND EARTH GROUND TEST INSTRUMENTS

Ground test instruments are used for situations in which grounding electrode(s), grounding electrode system(s), and ground resistance are measured. These types of instruments include, but are not limited to, clamp-on ammeters, clamp-on leakage current meters, electrostatic locator meters, multimeters, three-terminal ground resistance meters, four-terminal ground resistance meters, and soil pH meters.

Fluke Corporation

GROUNDING AND BONDING

Proper grounding and bonding are required in any electrical system for installation and personnel safety. Safe wiring requires that a system be grounded where required by the NEC® and that all non-current-carrying conductive parts of equipment be properly grounded as required in the NEC®, OSHA requirements, equipment manufacturer requirements, and in accordance with all other applicable codes for a given installation. To promote accuracy in the use of grounding and bonding terminology, one must use the words in the context of how they are defined in the Code. Slang words and terms mean different things to people. Use of the defined terms results in accurate application of the NEC® rules that must be applied to installations and systems.

A *ground* is a low-resistance conducting connection between electrical circuits, equipment, and the earth. The term "ground" is simply defined in the NEC® as the earth. The term "grounded (grounding)" is defined in the NEC® as connected to the earth or to a conductive body that extends the earth connection. The term "ground fault" means that there is an unintentional conducting connection (usually accidental) to the earth or to a conducting body that extends the earth connection. The NEC® defines a ground fault as an unintentional, electrically conductive connection between an ungrounded conductor of an electrical circuit and the normally non-current-carrying conductors, metallic enclosures, metallic raceways, metallic equipment, or earth. Bonding is an essential function in electrical systems. The terms "bonded" and "bonding" are defined simply as connected together to establish continuity and conductivity. When non-current-carrying conductive parts of equipment are bonded together, they become one electrically.

An effective grounding electrode system requires that the grounding electrodes not only be installed correctly, but also be selected and designed, where possible, to last over the expected life of the electrical system. The NEC® does specify how long a grounding electrode system must last. The objective is to maintain good contact between the grounding electrodes and the earth. The grounding electrode system plays a minor role in facilitating overcurrent device operation. During a fault event, although the earth will be in the circuit, an extremely small amount of current is present over the grounding electrode conductors because of high levels of impedance between the earth and the source. The NEC® does not require a grounding electrode system to be tested when installed for a premises wiring system. Sometimes an engineering or design specification would require ground resistance testing to an established ground resistance level. This ground resistance value is constantly subject to change because of seasonal conditions, varying earth moisture levels, and the type of earth where the connection is made. To ensure that a grounding electrode system is connected effectively to the earth, there are several tests that can be made before, during, and after the grounding electrode system is installed. Grounding system testing includes taking voltage measurements (system voltage and static electricity voltages), current measurements (system currents and leakage currents), and resistance measurements (grounding system resistance, soil resistance, and soil pH measurements).

Soil resistance measurements are sometimes taken to determine an optimal location for the grounding electrode(s), grid, or all electrodes of the entire grounding electrode system. Soil resistance measurements are typically taken before a grounding electrode (grid or system) is installed. The resistance to earth of the grounding electrode is taken after the grounding electrode is installed to ensure that the grounding system meets the required maximum resistance specified by the designer, engineer, or owner.

The NEC® does not require grounding electrode resistance testing. The NEC® only requires that a single rod, pipe, or plate electrode have a resistance to ground (earth) that does not exceed 25 Ω. When the resistance to ground of a single rod, pipe, or plate electrode comes into question, a ground resistance test can be performed to demonstrate compliance. If the resistance to ground does exceed 25 Ω, an additional rod electrode can simply be installed no less than 6′ from the first one and at that point, the 25 Ω requirement no longer applies, and no ground resistance testing is required. Soil pH measurements are taken to determine if copper, stainless steel, or galvanized steel is the best material to use for a grounding electrode application. **See Figure 6-1.**

Test measurements and troubleshooting techniques used on grounding systems and components are somewhat different from measurements and troubleshooting techniques used with all other circuit conductors in electrical systems. The difference exists because typical electrical circuits have a clear designated path for current, a known voltage, and known electrical loads, but grounding electrode systems only have current present for a small duration of time when there is an event such as a ground fault experienced by the system.

Grounding electrode systems create a path to the earth from systems and equipment. They are not intended to be a fault-current clearing path. This is the function of the equipment grounding conductors and bonding conductors. During a fault event, current in the effective ground-fault current path (typically the equipment grounding conductor of a circuit) is heavy and can range from a couple of mA of unwanted static charge to induced voltages or fault currents of several thousand amps. Current to ground through a grounding electrode system is limited to lower values but is not constrained to follow any specific path available to earth as current would in normal current-carrying conductors of a circuit.

When measuring the resistance of electrical switches, components, insulation, and loads in electrical circuits, typically ohmmeters or megohmmeters are the test instrument of choice. Measuring the resistance of a grounding electrode (grid or system) buried in changing soil conditions and temperatures requires specialized measuring techniques using specialized test instruments. Persons using test instruments should be trained in their use and qualified to perform such testing. Many test instrument manufacturers offer training that can be provided to ensure that personnel are qualified to perform the testing desired.

GROUND TEST INSTRUMENTS AND METERS	
Test or Measurement	**Test Instrument**
Measuring circuit, system, or ground voltages	Standard voltmeter
Measuring circuit or system current	Standard clamp-on ammeter
Measuring ground leakage current	Clamp-on leakage current ammeter
Measuring static electrical charges	Electrostatic discharge meter
Measuring earth grounding system (ground rod/grid, ground connections, etc.) resistance without disconnecting any part of the grounding system	Clamp-on ground resistance meter
Measuring earth grounding system (ground rod/grid, ground connections, etc.) resistance before connecting the earth ground to the building grounding system	3-Terminal ground resistance meter
Measuring the earth's resistance (resistivity)	4-Terminal ground resistance meter
Measuring soil pH	pH meter

Figure 6-1. To install and troubleshoot grounding systems requires various types of test instruments.

Even when a grounding electrode system appears to be installed correctly, the system may still experience component, circuit, and system problems. Objectionable current over the grounding system in addition to higher resistance connections to ground can cause unwanted problems in audio systems, video systems, data distribution systems, computer systems, and other sensitive electronic equipment, however, the installation may still meet the minimum requirements of the NEC®. For this reason, manufacturers of sensitive electronic equipment often specify a required maximum ground resistance of 1 Ω or less for proper equipment operation. This is often included in the electrical engineering specifications for information technology (I.T.) rooms and associated sensitive electronic equipment.

For example, when multiple circuits share the same grounding electrode system, such as a theater stage that has electrical power cords, sound wires, and communication cables routed throughout the area, they can experience noise problems, such as speaker hum, unless the grounding electrode system has a resistance of 2 Ω or less. These problems can result from multiple causes. To solve speaker hum, illegal multiple ground connections are often established, but multiple grounds cause a serious "ground loop" problem. The term ground loop relates to circulating currents over multiple grounding paths that are present and moving in a circular direction. To simplify it, a ground loop is the circuit created by multiple earth grounding points containing a voltage potential between the ground points that is high enough to produce a circulating current in the grounding system. Ground loops can also cause electrical shock and operational problems for some electrical equipment.

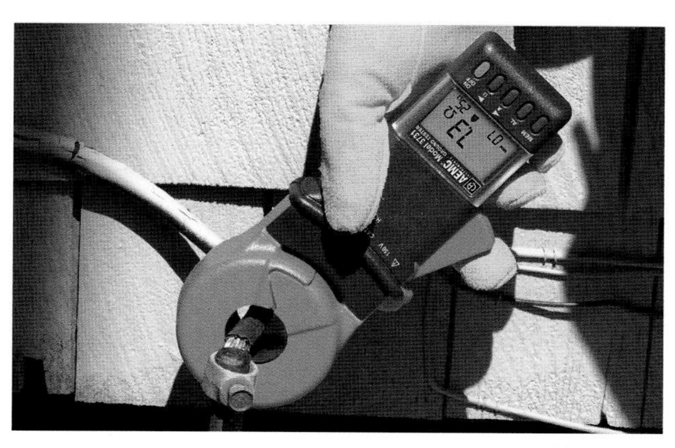

AEMC® Instruments

Per the NEC®, grounding systems must have less than 25 Ω of resistance to earth.

Grounding Electrode Systems

The function of a grounding electrode system is to provide a connection to the earth from electrical systems required to be grounded and to keep the electrical system and non-current-carrying parts of equipment at or as close to the same potential. The proper grounding of electrical tools, machines, equipment, and grounded distribution systems is an important factor in preventing dangerous conditions and results in electrical and electronic equipment operating correctly and safely.

An entire grounding system is not just a green (or bare) wire. It includes grounding electrode conductors, equipment grounding conductors, equipment bonding jumpers, and grounding electrodes, all working together in the system and each providing specific functionality. Grounding an electrical system to earth is accomplished by connecting one conductor of a system to a grounding electrode or grounding electrode system. The acceptable grounding electrodes recognized in the NEC® are provided in Section 250.52(A). **See Figure 6-2.**

Types of Grounding

Grounding and bonding systems can be broken down into three basic categories, which are building grounding (connections to earth through grounding electrodes), equipment grounding (connections of equipment to equipment grounding conductors), and electronic grounding (specific equipment grounding conductor installations that reduce electromagnetic interference in the grounding circuit) typically accomplished by the installation of isolated/insulated equipment grounding conductors and receptacles. Each method performs a different purpose and, when combined, the methods work together simultaneously to provide a safe and effective grounding system for personnel and equipment. **See Figure 6-3.**

Building Grounding Electrodes. A *building grounding electrode* is an electrode that is connected together with another electrode to form a grounding electrode system as required by NEC® Section 250.50. Each building or structure supplied with electric power must have a grounding electrode system. Often the grounding electrode system is inherent to the construction of a building or structure. Electrodes such as the metal frame of a building, concrete-encased electrodes, or an underground metal water pipe are examples of grounding electrodes that are inherent to building construction and must be included in the required grounding electrode system. A building grounding electrode system ensures that there is a connection to the earth from the service or system required to be grounded.

Building grounding also includes protecting the building, cooling tower, or outside structure from lightning strikes by providing the lightning or diversion path to ground through an installed lightning protection system. The ground terminals of a lightning protection system must also have an effective grounding connection to earth because of the high energy events created by lightning. A failure of any part of a lightning protection system when carrying a lighting strike significantly increases the chance of electrical flashover and can cause building damage and fire. Test equipment, such as a clamp-on ground resistance meter, and procedures used to measure the resistance of an electrical distribution system are also used to test the integrity of an earth grounding system.

Equipment Grounding Conductors. *Equipment grounding* is the connection of electrical equipment to ground to reduce the chance of electrical shock by grounding all non-current-carrying conductive parts. Another important reason for equipment grounding is to minimize the possibilities of electrical shock when a person comes in contact with electrical equipment or exposed conductive material. Equipment grounding conductors (EGCs) are defined in the NEC® as the conductive path that provides a ground-fault current path and connects normally non-current-carrying metal parts of equipment together and to the system grounded conductor, to the grounding electrode conductor, or both. The EGC performs grounding and bonding functions as well as serves as an effective ground-fault current path to facilitate overcurrent device operation in the event of a ground-fault event.

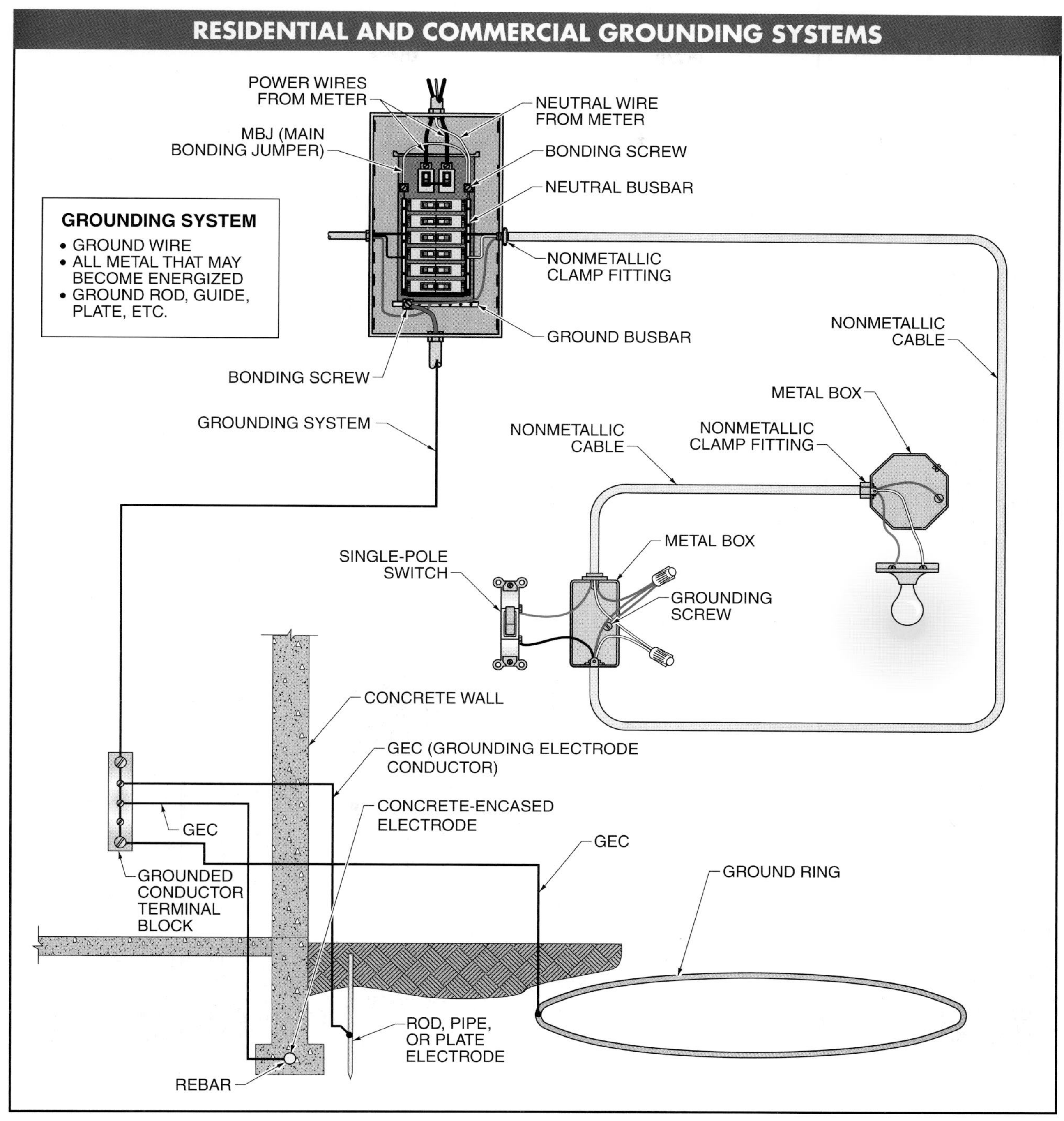

Figure 6-2. A grounding system provides a safe path for fault current to flow to earth ground when the grounding circuit is connected to a metal underground electrode or other grounding system.

Technical Tip

Article 250 of the NEC® covers bonding and grounding. The code includes specific requirements for systems, circuits, equipment to be grounded, circuit conductors, locations of grounding connections, types and sizes of bonding conductors and electrodes, methods of bonding and grounding, and conditions under which guards, isolation, or insulation may be substituted for grounding.

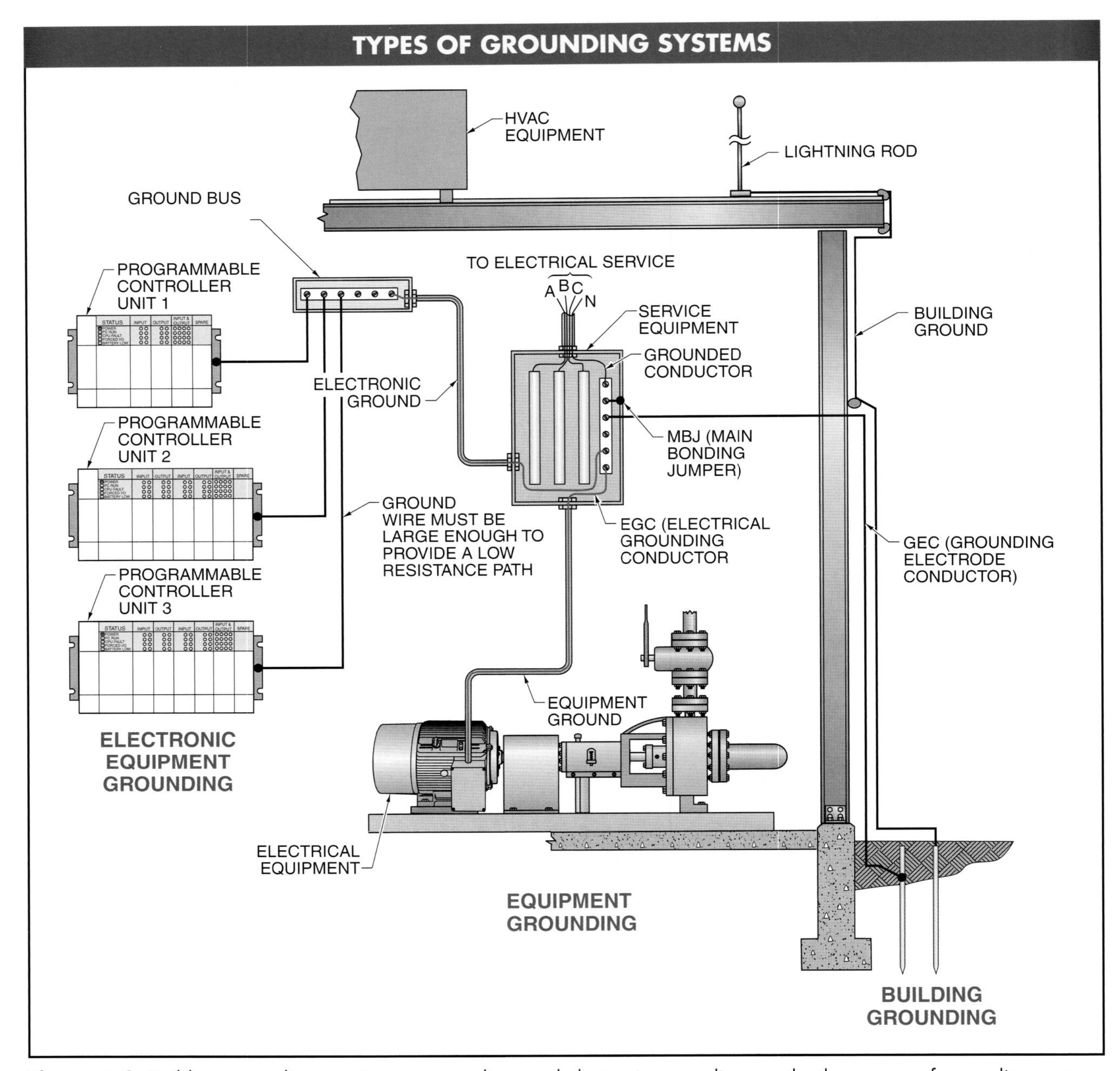

Figure 6-3. Building grounding, equipment grounding, and electronic grounding are the three types of grounding systems used to create a safe working environment for personnel and equipment.

Non-current-carrying metal parts that are connected to a grounding system include all metal boxes, raceways, enclosures, metal equipment parts, and any metal a person might touch that is around an electrical circuit. An equipment grounding circuit provides an effective path for unwanted fault current for the duration of time it takes an overcurrent device to clear. A fault current may develop because of insulation failure or a current-carrying wire making direct contact with a non-current-carrying metal part of a system.

In a properly grounded system, fault current must trip the overcurrent protection device (fuse or breaker). When a fuse opens or a breaker is tripped, the circuit is opened and current is no longer present in the circuit. Equipment grounding also helps to prevent electrical shocks from static electricity and static buildup in equipment. Static electricity can also cause fires and explosions when allowed to accumulate or if not properly handled in hazardous (classified) locations. NFPA 77, Recommended Practice on Static Electricity, contains in depth information about achieving protection against static electricity.

Electronic Equipment Grounding. *Electronic equipment grounding* is the connection of electronic equipment to ground to reduce the chance of electrical shock by grounding the equipment and all non-current-carrying exposed conductive parts. Electronic equipment grounding is used primarily to provide an effective and clean (isolated/insulated) EGC connection for electronic systems to enable better communication (less noise) with process control equipment and other systems. A properly installed isolated/insulated EGC often reduces static electrical charges, which allows signal integrity to be maintained for sensitive video, audio, data, medical, and security systems, programmable logic controllers (PLCs), computer numerical controls (CNC), and other electronic equipment. Signal reliability is difficult to maintain in electronic equipment where many signals are transmitting data at 5 V or less.

Electronic equipment grounding requirements must never conflict with minimum required building and equipment grounding requirements. Problems can occur because NEC®-minimum building and equipment grounding are primarily used for grounding to minimize the possibilities of electrical shock and fire. The primary use of building and equipment grounding conductors is to ensure an effective ground-fault current path to remove a fault current as fast as possible through the actuation of overcurrent protective devices.

Electronic equipment grounding enhances electronic noise reduction by using an isolated/insulated EGC to eliminate (dump) noise and other unwanted induced interference (signals).

Fluke Corporation

Equipment grounding systems are tested to verify proper operation. Proper grounding prevents shock to electricians when equipment is contacted.

The unwanted noise (objectionable current) that is removed to ground by electronic grounding systems is typically measured in milliamperes (mA) and continues to be present as long as the electronic equipment is connected. Manufacturers of electronic equipment and systems often specify a grounding electrode system with a resistance of 1 Ω, 3 Ω, 5 Ω, or less. This is by specification and not a requirement of the NEC®. The NEC® only provides the minimum installation requirements for electrical grounding and bonding systems.

A functional grounding and bonding system must always meet the minimum NEC® requirements for protection of the building, occupants, and electrical equipment. Specific or enhanced grounding schemes and systems that exceed the minimum requirements are typically specified for sensitive electronic equipment. All grounding requirements can be met by installing a grounding electrode system with the lowest possible resistance. Installing a grounding electrode system that will last is the objective and often a challenge. Test instruments are used to test new grounding system installations and also can be used for routine maintenance tests. Continued testing as part of a preventive maintenance program helps ensure that a grounding electrode system is operating correctly and safely.

MEASURING GROUND-FAULT CURRENTS

All electrical circuits are designed with a normal path for current, the circuit conductors. For example, the normal path of current for a standard 115 VAC circuit is from a hot conductor, through the load, and back through the neutral (grounded) conductor. Current should not be present in the EGC of an electrical circuit (24 V, 115 V, 230 V, 460 V) at any time. Current should only be present in the EGC when a fault occurs in the circuit. Typical faults that cause current through an EGC include ground faults, insulation breakdown, moisture, corrosion, damaged wires, and non-compliant neutral-to-ground connections on the load side of a service main bonding jumper or on the load side of a system bonding jumper in a separately derived system. NEC® Section 250.24(A)(5) provides this prohibition of load side neutral-to-ground connections. **See Figure 6-4.**

Clamp-On Leakage Current Ammeters

A short circuit creates a large amount of current from the ungrounded (hot) conductor through another ungrounded (hot) conductor or from one ungrounded (hot) conductor to the grounded (usually the neutral) conductor and thus rapidly causes the circuit fuses or breakers to trip. Fault currents must only flow through a ground conductor for a very short period of time (until the fuse or breaker opens the circuit).

Problems such as a higher resistance path between the ungrounded (hot) conductor and the EGC can cause current to be present in the EGC for as long as the circuit is in operation or until the combined load current and leakage current is enough to open the fuse or breaker. *Leakage current* is current that leaves the normal path of current flow (hot to neutral) and flows through other conductive paths. Leakage current can also flow through an EGC. The NEC® addresses leakage current in an EGC in Section 250.6, Objectionable Current.

Clamp-on ammeters are used to measure current in an energized conductor without opening the circuit. Standard clamp-on ammeters are designed to measure currents typically higher than 0.2 A (200 mA). Standard clamp-on ammeters are the best test instrument for measuring load or circuit currents. Clamp-on ammeters are used to measure current on any hot, switched, or grounded (often a neutral) conductor.

EGCs must not carry any current except during a ground-fault condition. When an EGC is carrying current, a problem exists and a safety concern is elevated. Although standard clamp-on ammeters are used to measure current on conductors, they are limited to measuring only ground-fault currents above the minimum-measurement limit of the meter. Because ground-fault currents of only a few milliamps are a problem and cause safety hazards (electrical shock, heat, or sparks), standard clamp-on ammeters must not be used as a test instrument for measuring leakage currents in grounding systems.

When leakage current or extremely small amounts of operating current must be measured with a clamp-on ammeter, a clamp-on leakage current ammeter must be used. A *clamp-on leakage current ammeter* is an ammeter that can measure currents as low as a few milliamps (mA). Most clamp-on leakage current ammeters include several measuring ranges (0 mA to 30 mA, 300 mA, 30 A, and 300 A) to make the ammeter more versatile. The difference between a standard clamp-on ammeter and a clamp-on leakage current ammeter is that a clamp-on leakage current ammeter can accurately measure currents of a few milliamps (using their lower measurement range). **See Figure 6-5.**

Technical Tip

A wireless DMM can be used to take readings from remote meters located throughout a system and display their measurements. Remote meters can transmit voltage, current, and temperature measurements to the central meter, allowing systems to be monitored at several locations.

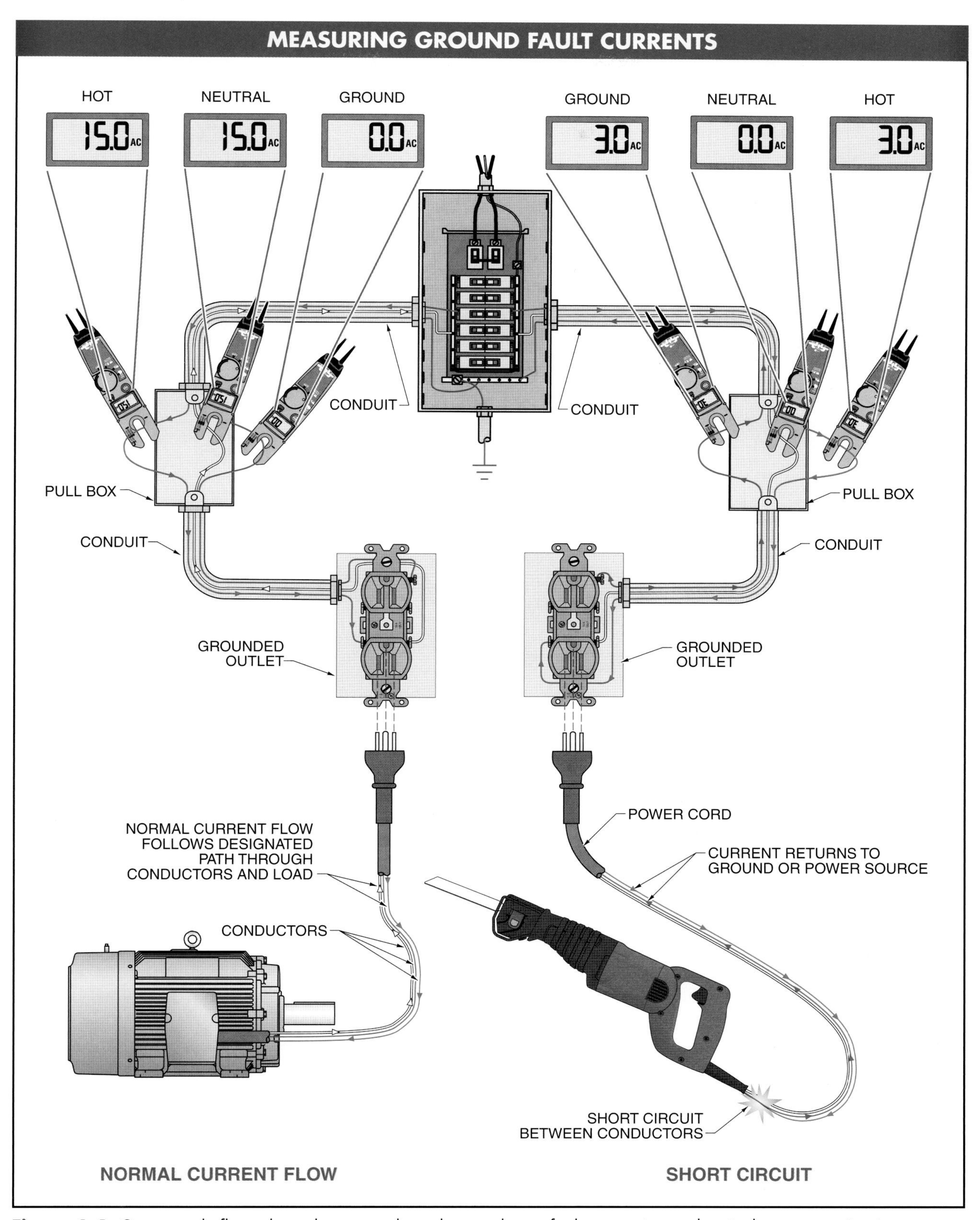

Figure 6-4. Current only flows through a ground conductor when a fault occurs in an electrical system or circuit.

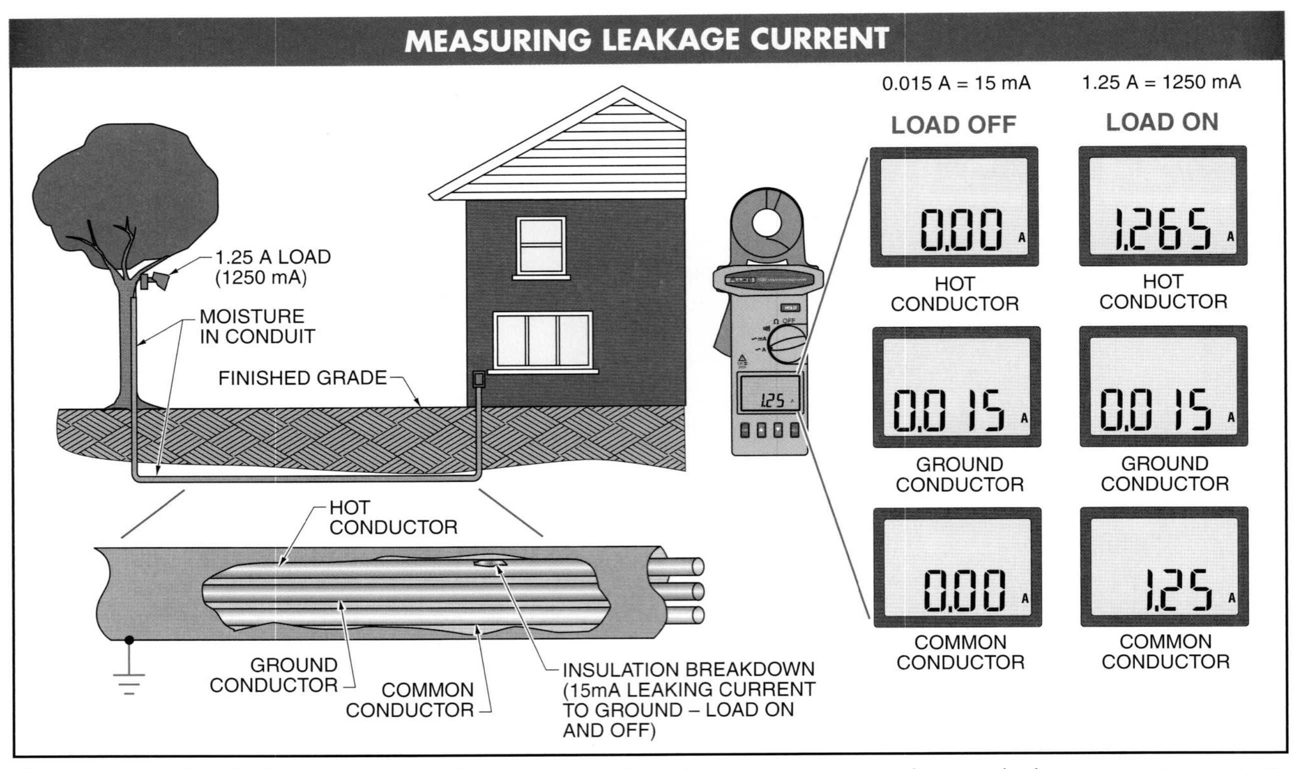

Figure 6-5. When leakage current must be measured with a clamp-on ammeter, a clamp-on leakage current ammeter is used that measures currents as low as a few milliamps (mA).

Clamp-On Leakage Current Ammeter Measurement Procedures

Clamp-on leakage current ammeters are typically used to measure leakage currents on grounding electrode conductors. Clamp-on leakage current ammeters are used to measure current on individual conductors to a load (motor or heating element), an individual circuit, a branch circuit, or the service grounding electrode conductor. For example, a clamp-on leakage current ammeter can be used to isolate a repeatedly tripping ground-fault circuit interrupter (GFCI) receptacle, breaker problem to a faulty GFCI, or an actual leakage current problem. When troubleshooting ground leakage problems, electricians should take measurements of the EGCs and grounding electrode system. When leakage current to ground is found, electricians must troubleshoot the electrical system to individual branch circuits and loads to identify and correct the problem. **See Figure 6-6.**

Safety Procedures

- Follow all electrical safety practices and procedures.
- Verify that the required test instrument training has been received by qualified persons in accordance with NFPA 70E.
- Check and wear personal protective equipment (PPE) adequate for the procedure being performed.
- Perform only authorized procedures.
- Follow all test instrument manufacturer recommendations and procedures.

Technical Tip

Most DMMs are accessorized to include an attachment that functions as a clamp-on ammeter.

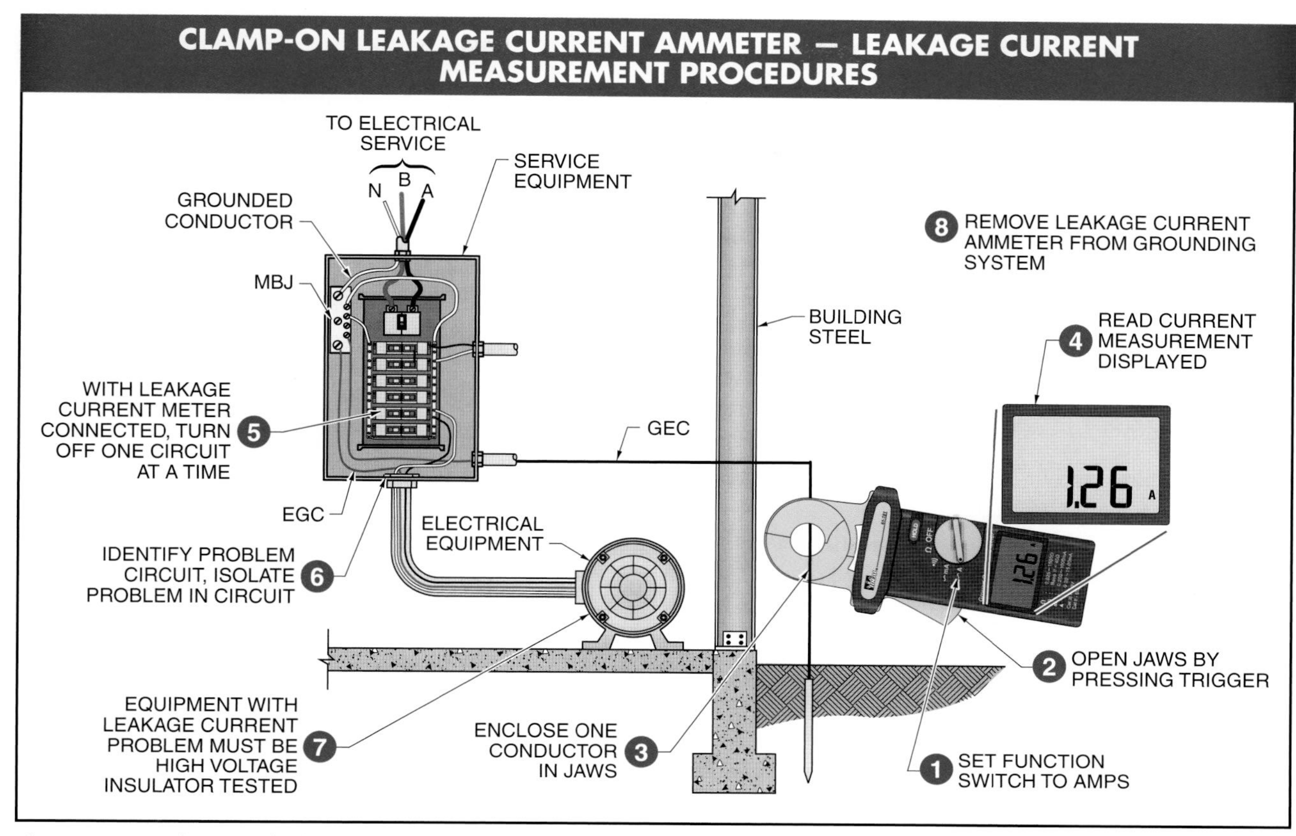

Figure 6-6. Clamp-on leakage current ammeters are used to measure leakage current from individual load conductors, individual circuits, branch circuits, or the main ground conductors.

Refer to the operating manual of the test instrument for all measuring precautions, limitations, and procedures. To test for leakage current on any circuit, apply the following procedure:

1. Set the function switch of the clamp-on ammeter to a measuring range higher than the expected circuit current. When the circuit current is unknown, start with the highest measuring range of the ammeter.
2. Open the jaws of the clamp-on ammeter by pressing against the trigger.
3. Enclose only one conductor or grounding electrode in the jaws. Ensure that the jaws are completely closed before taking any measurements.
4. Read the current measurement displayed. The greater the current in the grounding circuit, the greater the problem.

Then to isolate leakage current problems, continue with the procedure:

5. With the clamp-on leakage current ammeter connected to the main ground, turn off one circuit (circuit breaker) at a time and observe the ammeter reading. When the leakage-current measurement decreases significantly, the circuit with the problem has been identified.
6. Once the circuit with the problem is identified, isolate individual loads on the circuit by removing loads from the circuit and testing the individual grounds separately with the clamp-on leakage current ammeter.
7. Test any electrical equipment that indicates a leakage-current problem by performing a high-voltage insulation test using a megohmmeter.
8. Remove the clamp-on leakage current ammeter from the grounding system.

MEASURING STATIC ELECTRIC CHARGES

Proper grounding and bonding can also protect electrical and electronic equipment from false operation, failure, or damage from such problems as induced noise (low-level voltages). Grounding and shielding conductors can reduce unwanted noise. Reducing noise is important in low-voltage (less than 5 V), low-current (4 mA to 20 mA) signal circuits. Low-level signal circuits are seriously affected by electrical noise (magnetic fields) induced into control circuit wiring. Induced noise produces intermittent control circuit malfunctions.

Twisted wire pairs and coaxial cable help reduce noise induced by magnetic fields. Only one end of a shielded conductor should be grounded to prevent a ground loop problem, unless connection of the cable shield is required at both ends by the manufacturer or to meet an NEC® requirement. A grounded shield diverts unwanted induced interference to ground.

In addition to induced noise problems, a static electric charge can also cause false operation, failure, or damage to sensitive electrical and electronic equipment. Any moving objects, such as paper, conveyor belts, belt drives, and fabric, develop a very high static charge of several thousand volts (at low current levels) in a production facility. Static charges can produce annoying shocks and cause fires and explosions, especially in hazardous (classified) locations. A properly operating grounding electrode system reduces static charges.

Electrostatic Discharge Meters

Electrostatic discharge (ESD) is the movement of static electricity (electricity at rest) from the surface of one object to another object. Static electricity exists on plastic, fabric, paper, and other objects and is discharged when human skin or any other object at a different voltage potential comes in contact with the surface.

Electronic controls (SCRs, GTOs, BJTs, FETs, and IGBTs) that are part of an integrated electronic control circuit can be damaged by currents as small as 1 mA (0.001 A) and voltages of 10 V or more. On a dry day, an electrician can develop a static charge of several thousand volts, and touching an electronic circuit discharges the static charge into the circuit components. ESD damages or destroys semiconductors and other sensitive electronic components. The potential for damage caused by ESD must be understood when working with or around electronic devices such as computers, PLCs, and electric motor drives that include integrated circuits (ICs).

The two general categories of ICs are the transistor-transistor logic (TTL) family and the metal-oxide-semiconductor (MOS) family. MOS-based ICs are extremely susceptible to damage from ESD. Sensitive electronic components, such as MOS-based ICs, are shipped in static-shielded wrapping that guards against ESD damage. Electronic components are also susceptible to damage from improper handling before and after being installed in a circuit. Electronic components located on printed circuit (PC) boards are typically not replaced. The entire board is changed out. Due to the sensitivity of PC board components, care must be taken when replacing or working near PC boards.

To prevent ESD from damaging electronic components inside electronic equipment, take the following precautions:

- Discharge any static charge buildup by touching a conductive surface, such as grounded conduit, before touching electronic components or PC boards. Manufacturers recommend wearing a wrist strap when working with sensitive electronic components to prevent ESD. **See Figure 6-7.**
- Touch the insulated edge of PC boards and not the components when replacing PC boards inside electronic equipment.
- Ground the equipment being tested or serviced to ensure that all electronic circuits in the equipment are grounded.

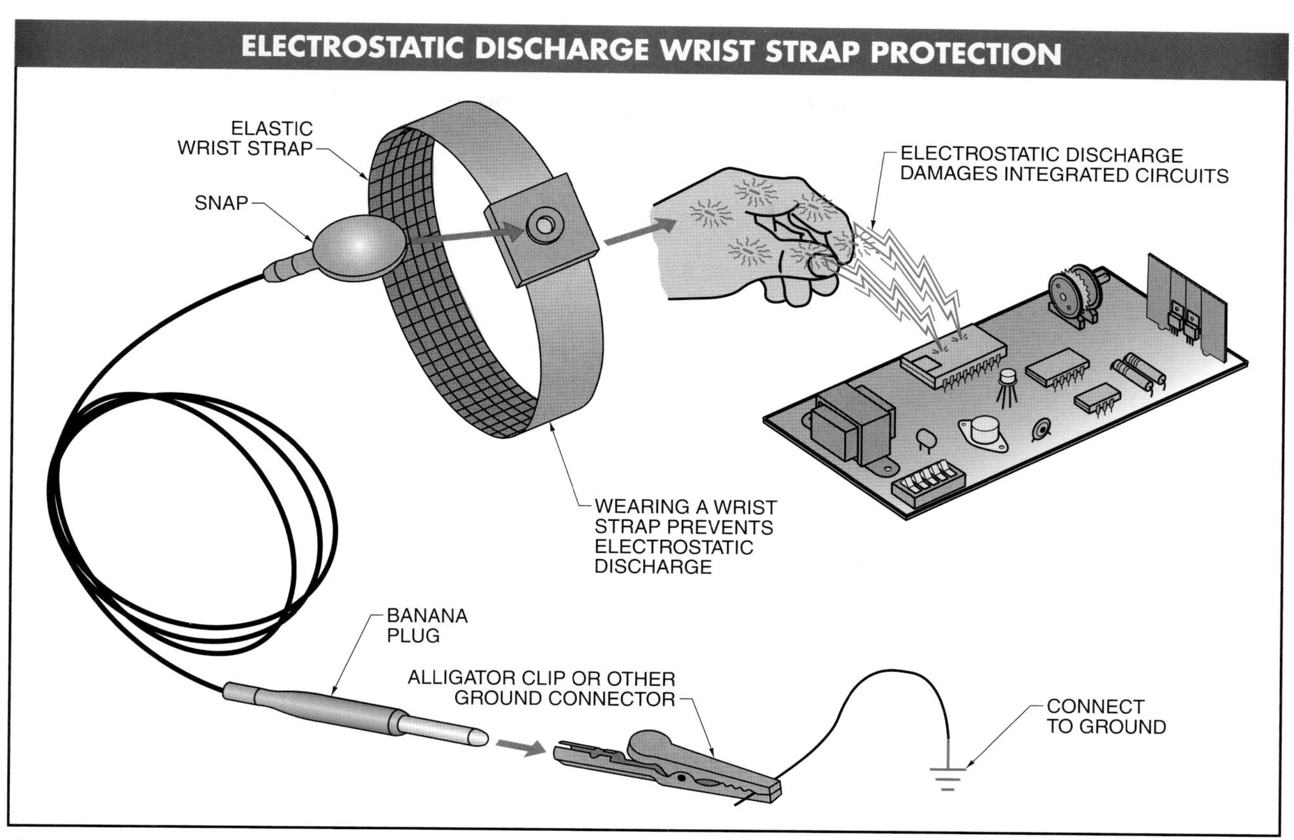

Figure 6-7. A wrist strap grounds an electrician when working with sensitive electronic equipment and components.

Electrostatic charges build up on almost any material and cause problems with electronic circuits and controls. When electrostatic charges are a problem, or can be a problem, an ESD meter is used to measure static charge on a surface or object. An ESD meter measures the amount of static charge contained in an object. Knowing the source and amount of static charge is the first step in eliminating the problem. Proper grounding is the typical solution to static charge problems, but other steps may be required to eliminate static charge problems completely.

Technical Tip

Electrical shock from touching an object is the result of the transfer of a static charge by contact. Lightning is an example of the transfer of a static charge by a spark. Static electricity has limited practical uses such as for electrostatic air filters and electrostatic spray painting.

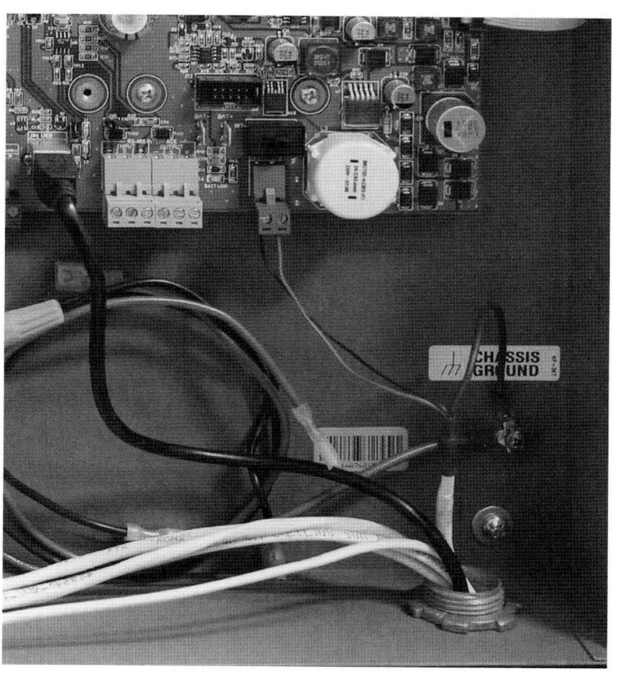

Electronic equipment must be grounded for protection from static electricity.

ESD Measurement Procedures

On conductive surfaces, static electric charges can move and do not typically cause a problem. However, on insulated surfaces, static electric charges build up and can be a problem. Voltages on a conductive surface are the same at all points of the surface and can be measured with a voltmeter. Static charge voltages on an insulated surface are not the same across the surface but are higher near the center of the surface and lower at the edges.

To measure the amount of static charge at various locations, an ESD meter is used. An ESD meter measures the voltage of a static charge and the polarity of the voltage without making contact with the surface. Avoiding contact with the surface having the static charge allows accurate measurements to be taken without discharging the static charge. ESD meters measure very high voltages because static charges can be tens of thousands if not hundreds of thousands of volts (at very low currents). **See Figure 6-8.**

Before taking any static charge measurements using an ESD meter, it should be ensured that the meter is designed to take measurements on the equipment being tested. The operating manual of the test instrument should be referred to for all measuring precautions, limitations, and procedures. The required PPE should be always be worn and all safety rules should be followed when taking the measurement. To measure an electrostatic charge, apply the following procedure:

1. Connect the ESD meter according to manufacturer recommendations. Typically this includes connecting the meter to earth ground.
2. Turn the ESD meter ON and zero the meter as recommended by the manufacturer.
3. Without touching the object, point the probe of the meter toward the object being tested. Some ESD meters provide a standard spacing arm that allows for constant distance measurements for each reading, typically 1″ or ½″ from the surface being tested.
4. Read the voltage of the static charge. It is normally read in kilovolts (kV).
5. When a static electric charge that is high enough to cause a problem is present, ground the charged area to earth ground.
6. Remove the ESD meter from the test area.

Troubleshooting Tip

ELECTROSTATIC METERS

To measure static electricity, a meter must be capable of measuring at least one of the following:

- amount of charge (voltage) on surfaces
- number of air ions (per cm^2 per second) that hit a surface
- voltage difference across the thickness of a material
- direct current (DC) electric field strength of air at the test point
- approximate conductivity of surfaces
- approximate attractive/repulsive force between charged surfaces

ELECTRICAL SERVICE GROUNDING

A grounding electrode is required at the main electrical service or at the source of a separately derived system. A *separately derived system (SDS)* is an electrical source, other than a service, that has no direct connection to the circuit conductors of any other electrical source other than those established by grounding and bonding connections. SDSs typically supply electrical power derived or taken from transformers, storage batteries, solar photovoltaic systems, or generators. Most SDSs are produced by the secondary side of a power distribution transformer.

An SDS is typically used to establish a new voltage level, lower the power source impedance, and isolate part of the power distribution system. Because an SDS does not have direct electrical connections to any other part of a supply distribution system (transformers magnetically couple), a new grounding electrode is required. A proper grounding electrode is required for safety and proper equipment operation and is established by making a connection between the SDS grounded conductor and a grounding electrode of the building or structure. See Section 250.30(A)(4) of the NEC® for SDS grounding electrode requirements.

Technical Tip

An ESD event may occur when an object that has a static electrical charge comes into contact with a conductive path.

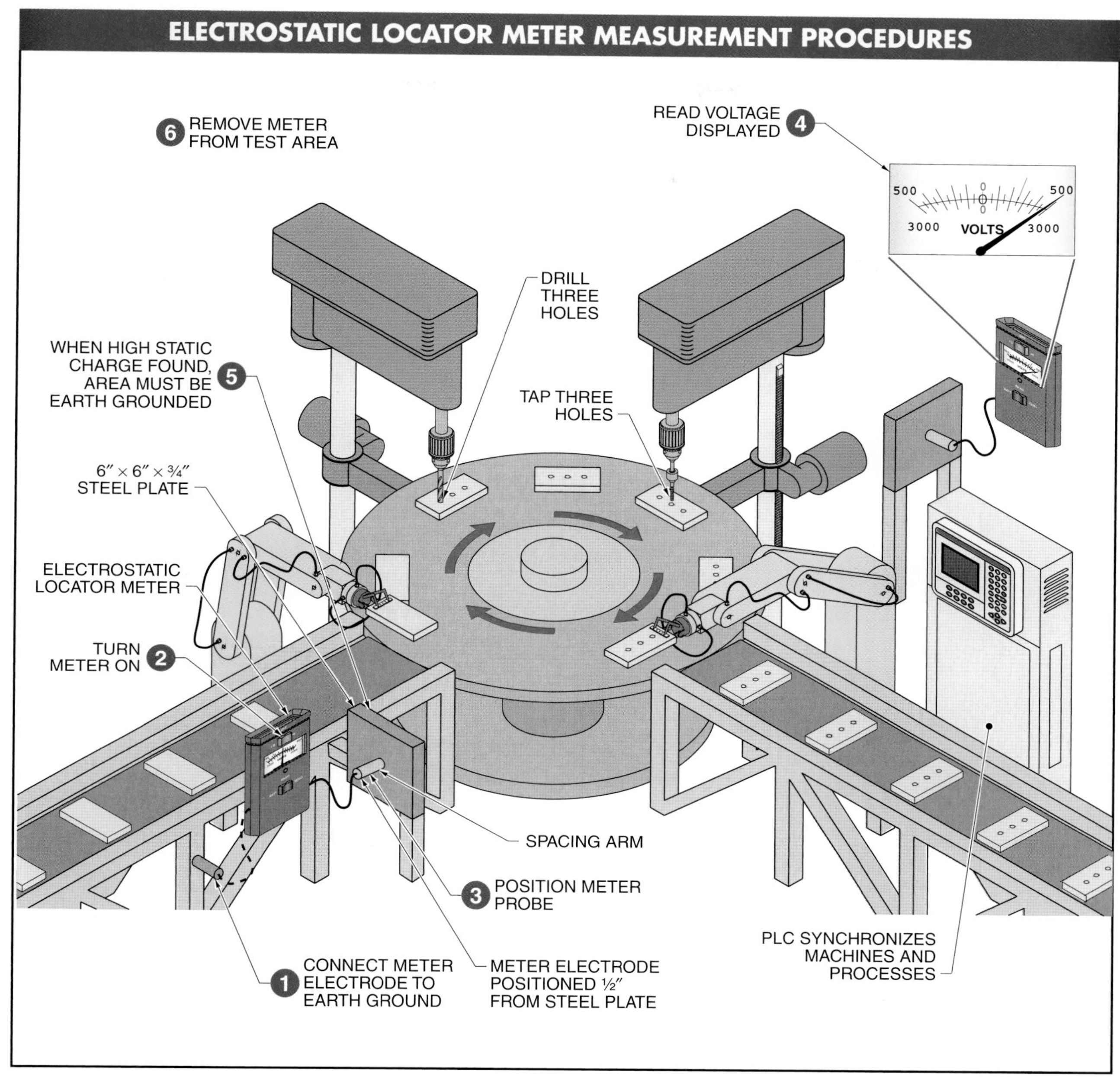

Figure 6-8. To measure the amount of the static charge at a location, an electrostatic discharge meter is used that can measure voltage in tens of thousands if not hundreds of thousands of volts.

Neutral-to-Ground Connections

A neutral-to-ground connection must be made at a grounded SDS or at the main service equipment only. The neutral-to-ground connection is made by connecting the neutral bus to the ground bus with a main bonding jumper. A *main bonding jumper (MBJ)* is a connection in a service panel that connects the equipment grounding conductor, the grounding electrode conductor, and the grounded conductor (neutral conductor). **See Figure 6-9.**

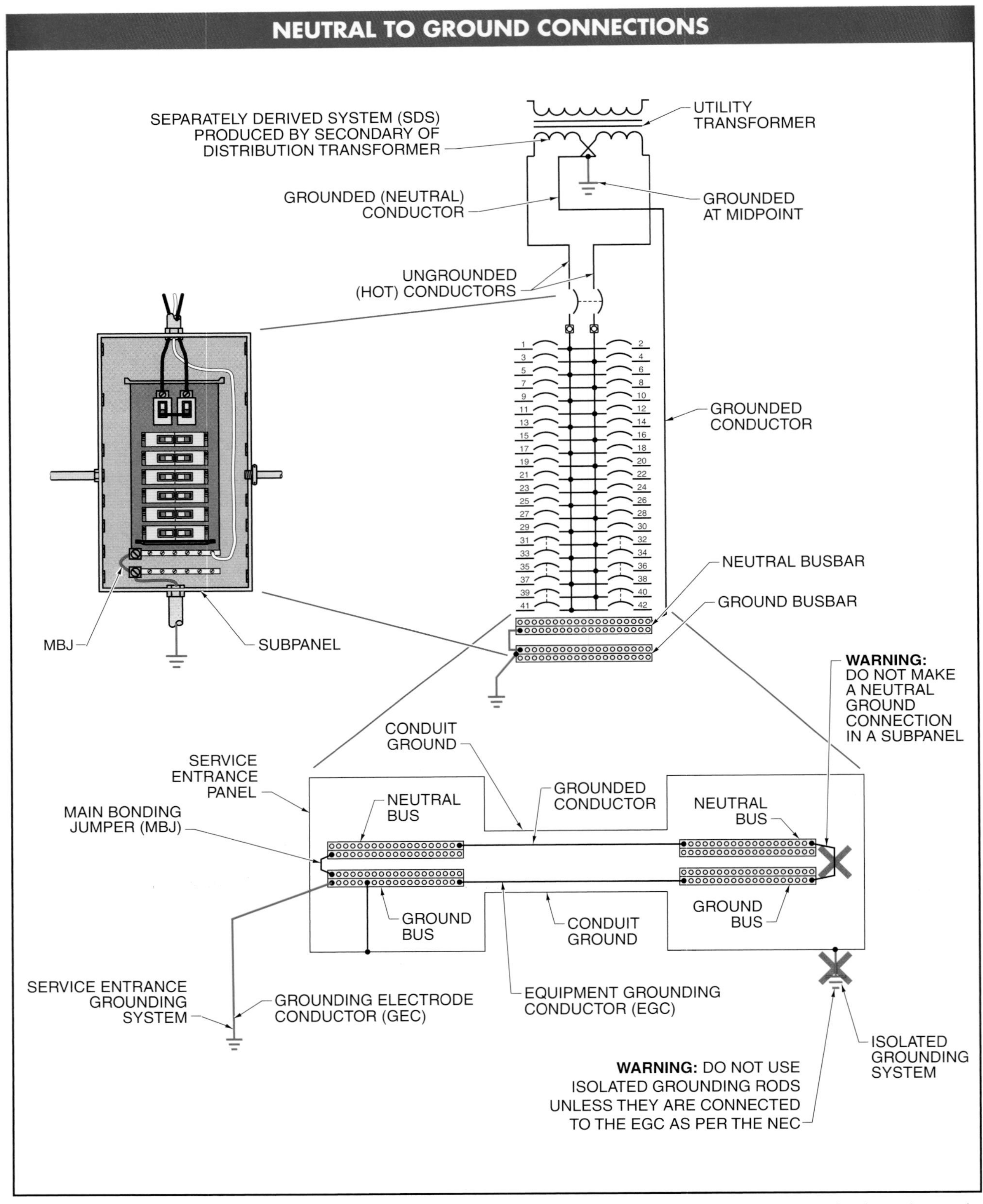

Figure 6-9. A neutral to ground connection must be made at the transformer or at the main service panel only and is made by connecting the neutral bus to the ground bus with a main bonding jumper (MBJ).

An *equipment grounding conductor (EGC)* is an electrical conductor that provides a low-impedance grounding path between electrical equipment and enclosures within a distribution system. A *grounding electrode conductor (GEC)* is a conductor that connects the grounded parts of a power distribution system (equipment grounding conductors, grounded conductors, and all metal parts) to the NEC®-compliant grounding electrode system. A *grounded conductor* is a conductor, such as a neutral conductor, that has been intentionally grounded.

Ground Loops

Neutral-to-ground connections must not be made in any panelboard, receptacles, or other equipment on the load side of the MBJ at the service or the system bonding jumper at an SDS. When a neutral-to-ground connection is made anywhere except in the main service panel, a parallel path for the normal return current from a load is created. The parallel path allows current through multiple metal parts of the system and other electrically conductive paths. This is often the cause of electrical noise in the grounding system. The NEC® prohibits neutral-to-ground connections that create ground loops because of safety concerns for persons and property.

In addition, power quality problems are created by such connections that are not compliant with the NEC®. In addition to not connecting the neutral-to-ground in subpanels, no additional isolated grounding electrodes can be established. An additional, separate, isolated grounding electrode creates two ground references that are typically at different voltage potentials. The two grounding electrodes result in current circulating and forming a ground loop between the two grounding electrodes in an attempt to equalize the difference in voltage potential.

A *ground loop* is a circuit that has more than one grounding point connected to earth ground, with a voltage potential difference between the grounding points high enough to produce a circulating current in the grounding system. Current circulation is caused as current flows from a higher voltage potential to a lower voltage potential. A voltage potential exists because there is a difference in impedance (total resistance, inductance, and capacitance) between the two grounding points. Section 250.54 of the NEC® permits the installation of an auxiliary grounding electrode, but it must be connected to the EGC of the circuit for which it is installed. **See Figure 6-10.**

Receptacle Ground Measurements

Power is delivered to loads at a convenient location by receptacles (outlets). Common receptacle types include the standard duplex, isolated ground, and GFCI receptacles. The standard duplex receptacle is the most common type used in general wiring. The isolated ground receptacle, identified by an orange triangle, is used to reduce electrical noise problems by running a separate ground conductor back to the electrical system ground. A GFCI receptacle detects ground faults and quickly disconnects power from the circuit with the ground fault.

When wiring a 120 V duplex receptacle, the neutral conductor is connected to the silver-colored screw, the ungrounded (hot) conductor is connected to the brass-colored screw, and the EGC is connected to the green screw. Voltage measurements taken at receptacles can be used to determine whether a neutral-to-ground connection exists that is not compliant with the NEC®.

Voltage measurements taken at a receptacle with a standard voltmeter are used to test a receptacle for proper wiring and to determine the approximate load on a branch circuit without exposing the circuit wires. A voltmeter with PEAK, MIN MAX, and other special measuring functions is used to test an electrical system for basic power quality problems by testing for peak voltage or minimum and maximum voltages that are recorded over time.

Technical Tip

To correct a ground loop problem, which is common with audio/visual systems, verify that the system is grounded to earth at one central point so that the circuit components share the same earth reference point. This can be accomplished by using a star arrangement of plug boards fed from a single AC receptacle.

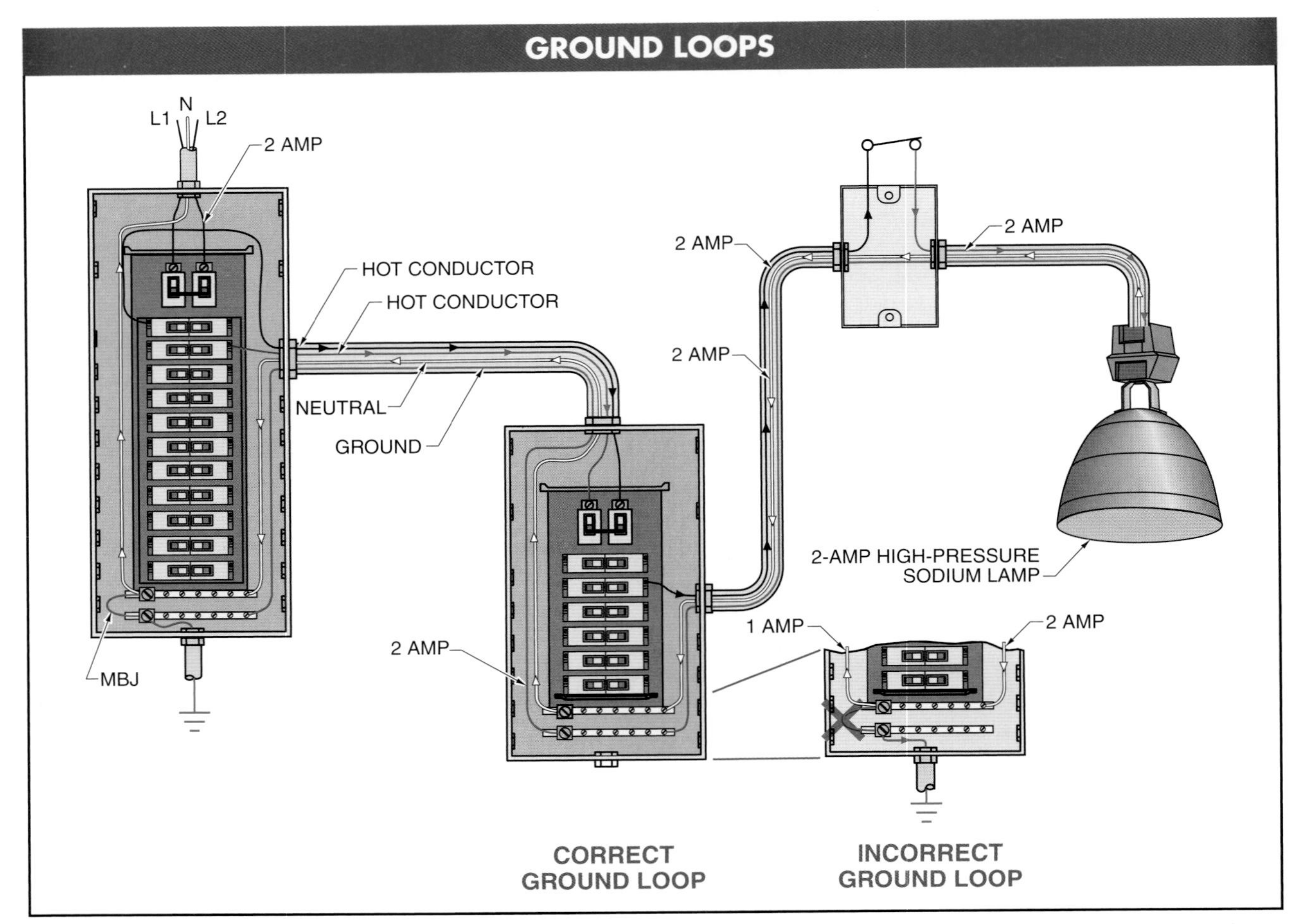

Figure 6-10. A ground loop is a circuit that has more than one grounding point connected to earth ground, with a voltage potential difference between the grounding points high enough to produce a circulating current in the grounding system.

Assured Grounding Programs. An assured grounding program can be used as an alternative to GFCI protection in limited situations. These programs must be in written format and enforced by a qualified person (or persons) for each specific jobsite. Assured grounding programs ensure that cord sets and receptacles, such as those used with two-prong and three-prong equipment that are not part of the permanent wiring, are properly grounded per NEC® 250.114, 250.138, 406.4(C), and 590.4(D).

A *two-prong Category II power cord device* is a device that has only two conductors extending from it, one hot and one neutral. Two-prong devices do not have a third (green wire grounding) prong on the power cord. Some two-prong devices are classified as double-insulated.

Double-insulated is a term used to describe an electrical product designed so that a single ground fault cannot cause a dangerous electrical shock through any exposed parts of the product that can be touched by an electrician. Double-insulated devices include not only the standard insulation used on conductors but also include extra insulating material between the energized parts of the device and the parts that can be touched.

To test an electrical device, leakage current is measured from the exposed metal parts to earth. When an electrical device such as a drill with plastic handles has no exposed metal, metal foil is applied to the exposed parts of the device and leakage current is measured between the metal foil and earth ground. The metal foil simulates a wet hand contacting the electrical device. **See Figure 6-11.**

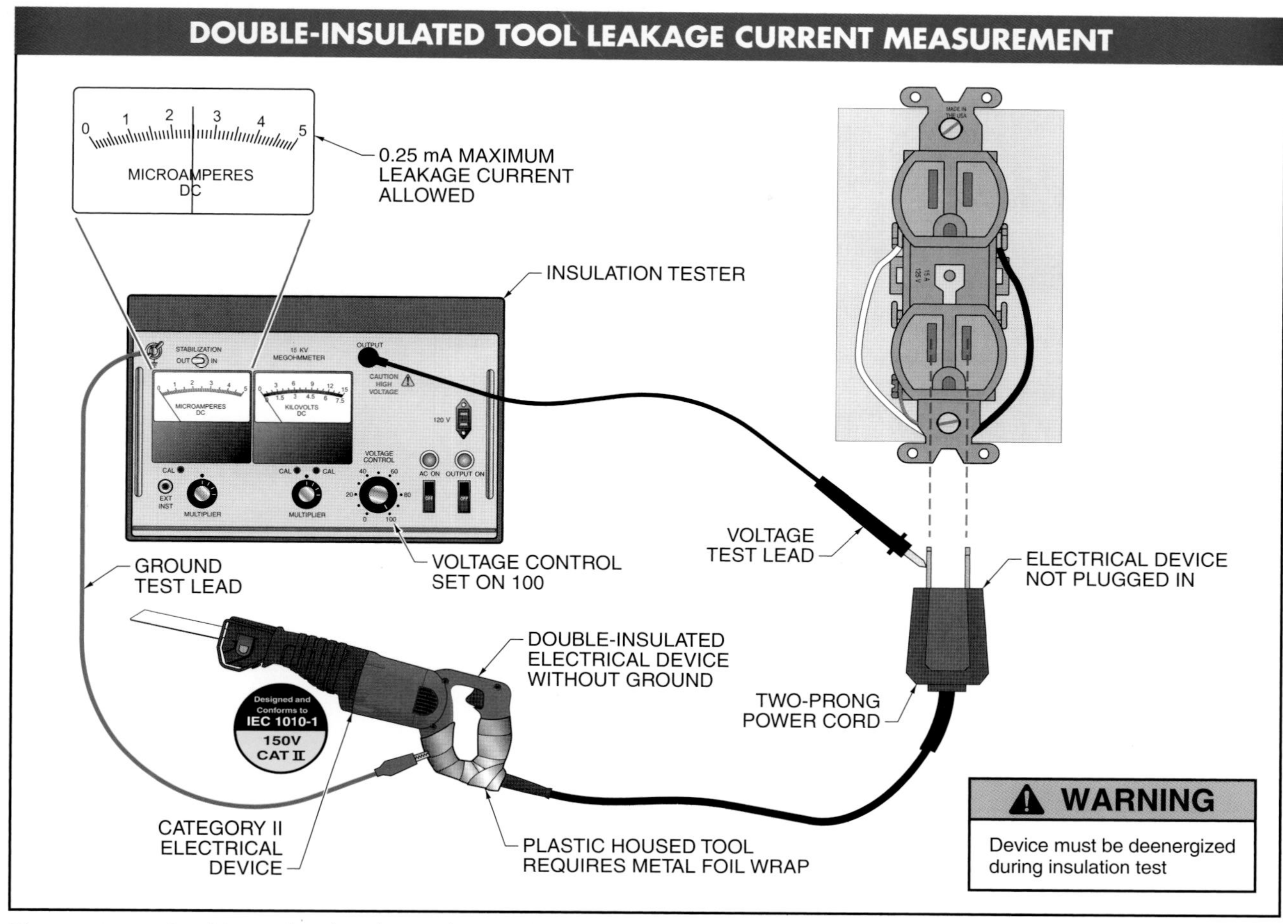

Figure 6-11. Double-insulated tools are designed so that a single ground fault cannot cause a dangerous electrical shock through any exposed parts of the tool.

During the test, the electrical device being tested is not plugged into a power source. The test instrument supplies the test voltage and measures the amount of leakage current. The specified maximum leakage current for a two-prong Category II device is typically 0.25 mA (0.00025 A). Equipment used in the medical field will have an even lower acceptable maximum leakage current limit.

When leakage current exceeds the specified limit, a three-prong power cord must be used. The ground (green) wire is added to carry the leakage current to ground by providing a low impedance (resistance) path from all non-current-carrying parts to earth ground.

A *three-prong Category I power cord device* is a device that has three conductors extending from it, one hot, one neutral, and one ground (green ground wire). Any leakage current flows through the ground (green) conductor back to ground during normal operation. The ground conductor prevents the exposed metal parts of the electrical device from becoming energized to the point of causing an electrical shock. **See Figure 6-12.**

The typical specified maximum leakage current for a three-prong Category I device is typically 0.75 mA (0.00075 A) for handheld electrical devices such as disc grinders, and 3.5 mA (0.0035 A) for electrical equipment such as floor buffers, small drill presses, and air compressors.

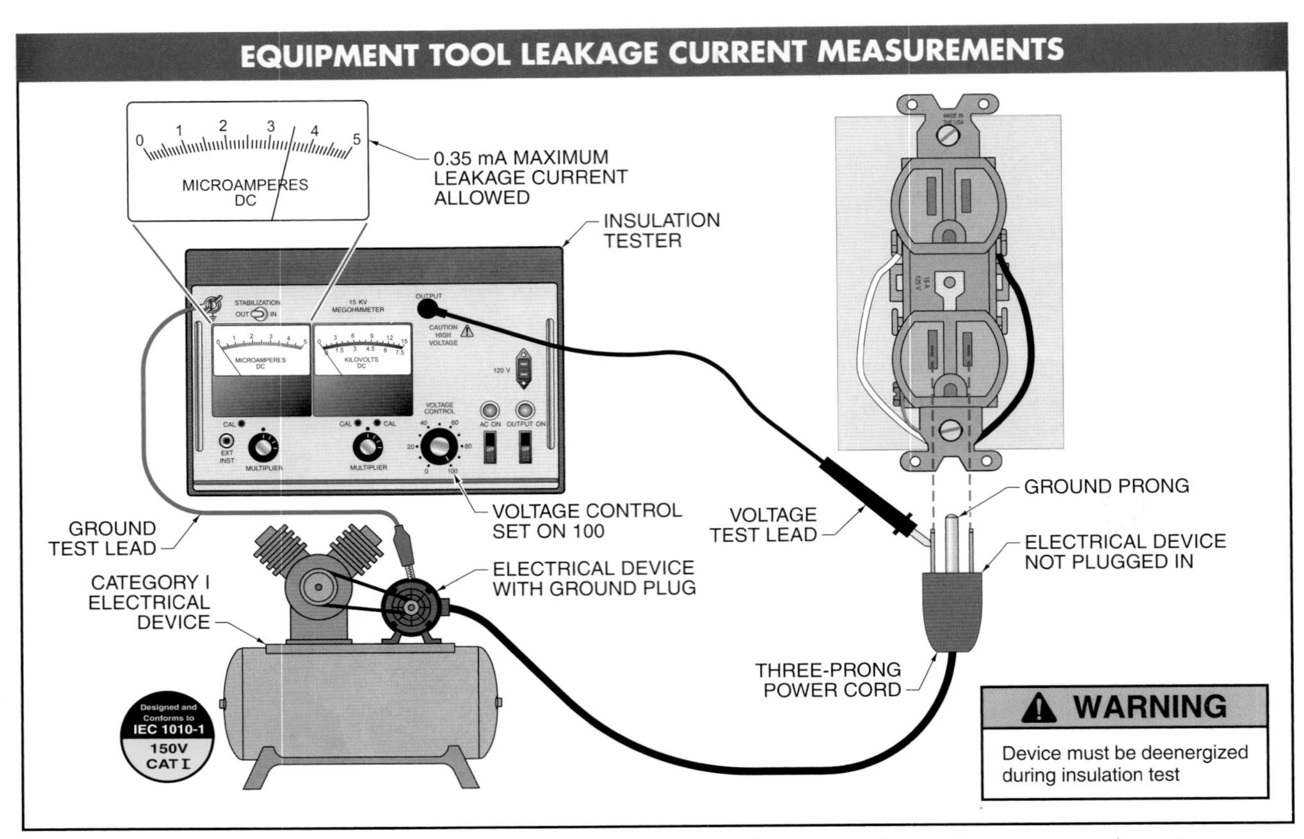

Figure 6-12. The specified maximum leakage current for Category I three-prong electrical equipment such as air compressors is 3.5 mA (0.0035 A).

GROUND CONTINUITY AND BOND TESTING

A grounding system is somewhat like seatbelts and air bags in automobiles; all are designed to protect individuals during an accident. In an electrical system, the ground removes (drains off) any leakage current from exposed non-current-carrying metal parts and directs any high fault current (short circuit) to ground.

In order to remove any leakage current from exposed non-current-carrying metal parts, the grounding system must connect all non-current-carrying metal parts to the ground prong of the power cord or to a permanent ground. In addition to being connected to ground, each part of a grounding system must have a low impedance (resistance) path to ground. To ensure that all non-current-carrying metal parts are connected to a grounding prong or permanent ground, a ground continuity test is performed. A *ground continuity test* is a test that verifies that a low impedance (resistance) path exists between all exposed conductive metal parts and the ground (green) conductor of the device. **See Figure 6-13.**

Even when the ground continuity test shows that all non-current-carrying parts are connected to the ground prong by a low impedance path, the grounding system may not be large enough to handle a fault current. A high fault current flowing through a small (undersized) conductor or path can blow apart (open) a grounding path before a fuse or circuit breaker removes power from the electrical device. When a grounding conductor or circuit is disabled, the ground is lost and an electrical shock, fire, or damage can take place. To ensure that a grounding system is properly sized, a ground bond test is performed. A *ground bond test* is a test that verifies that a grounding circuit (ground conductor and all grounding parts) of an electrical device is of sufficient size to carry a fault current to ground. The ground bond test passes a high current, typically 20 A to 30 A for typical appliances and tools, through the grounding circuit to determine if the circuit can carry a high fault current.

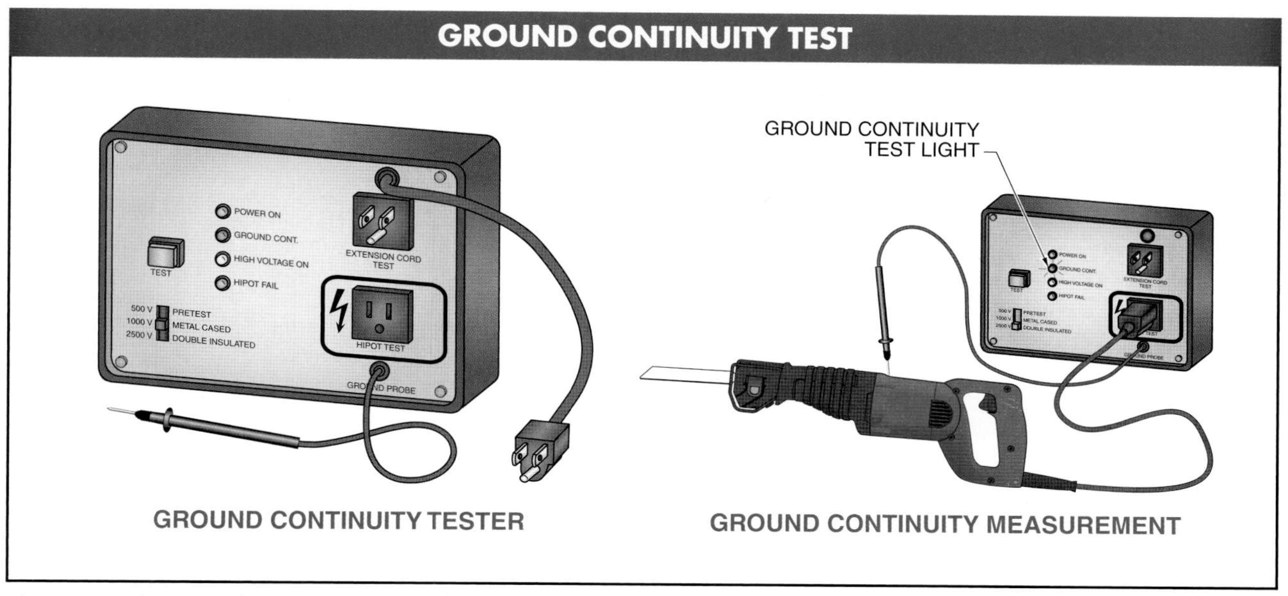

Figure 6-13. Ground continuity tests verify that all non-current-carrying metal parts of a tool are connected to a grounding prong or permanent ground.

Receptacle Grounding Measurement Procedures

Each receptacle in a system must be checked for proper wiring and possible problems. **See Figure 6-14.** Before taking any voltage measurements using a voltmeter, it must be ensured that the meter is designed to take measurements on the equipment being tested. The operating manual of the test instrument should be referenced for all measuring precautions, limitations, and procedures. The required PPE must be worn and all safety rules must be followed when taking the measurement. The following procedure is used to test receptacles (standard, isolated ground, or GFCI):

1. Measure the voltage between the ungrounded (hot) and grounded (neutral) conductor and between the ungrounded (hot) conductor and the EGC contact. When the receptacle is properly wired, there must be a measurement of approximately 115 V from the ungrounded conductor to the grounded (neutral) conductor and 115 V from the ungrounded conductor to the EGC. When voltage is present between the ungrounded (hot) conductor and neutral but not between the ungrounded (hot) conductor and the EGC, measure for voltage present between the grounded (neutral) and the equipment grounding contact. When approximately 115 V is measured from the grounded (neutral) to the EGC contact, the ungrounded conductor and the grounded conductor connections to the receptacle are reversed, which is a safety hazard that must be corrected.

   Note: A standard digital voltmeter should not trip a GFCI when measuring voltage from the ungrounded conductor to the EGC because standard digital voltmeters have an extremely high input impedance (resistance). When a standard digital voltmeter trips a GFCI during a test, use a meter with a high impedance rating.

2. Measure the voltage from the grounded (neutral) conductor to the EGC contact. There must be a low voltage on the grounded (neutral) conductor carrying current when a receptacle is supplying power to load on the branch circuit. The difference in voltage potential is caused by a voltage drop across the grounded (neutral) conductor because the conductor has resistance and is carrying current. The higher the power rating of the load(s), the greater the measured voltage on the neutral wire. A measurement of 0.5 V to 3 V indicates a non-NEC-compliant neutral-to-ground connection. The circuit must be diagnosed to locate it and make the required correction. A measurement of 5 V or higher indicates that a branch circuit is overloaded. A current measurement using a clamp-on ammeter must be taken at the circuit breaker inside the power panel.

3. Measure the rms voltage between the hot slot and the neutral slot of the receptacle with all loads turned ON and then with all loads turned OFF. If the voltage is low with loads ON (the maximum recommended voltage is 3% less than the voltage at the branch circuit breaker), the conductors may be too small or the branch circuit wiring may be too long.
4. Measure the peak voltage. The peak voltage is 1.414 times the measured rms voltage. When the peak voltage is low, the voltage is flat-topping. *Flat-topping* is a condition that occurs when a sine wave has lower peaks from current drawn only at the peaks of the voltage. Flat-topping is typically caused by nonlinear loads such as computers. When peak voltage is low, use a power quality meter to display the voltage waveform and test for flat-topping by examining the voltage waveform. Test for voltage harmonics and record voltage sags, swells, and transients over time to analyze the quality of power present for loads.

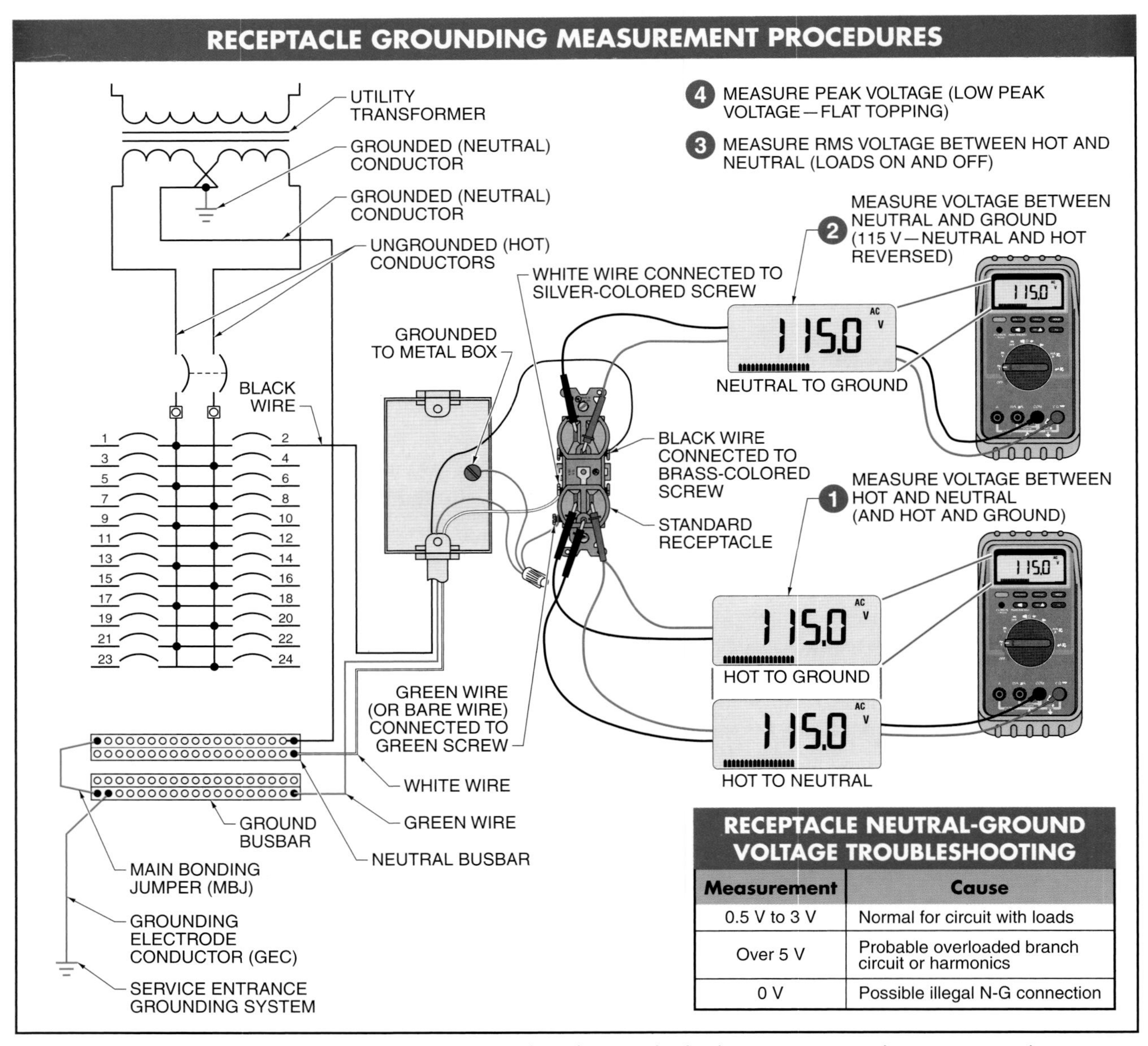

Measurement	Cause
0.5 V to 3 V	Normal for circuit with loads
Over 5 V	Probable overloaded branch circuit or harmonics
0 V	Possible illegal N-G connection

Figure 6-14. Voltage measurements taken at a receptacle with a standard voltmeter or receptacle tester are used to test a receptacle for proper wiring and to determine the approximate load on a branch circuit with or without exposing the circuit wires.

GROUNDING SYSTEM MEASUREMENTS

In a grounding electrode system, the grounding electrode (grid or system) provides the physical connection to earth and is the physical device that is used to dissipate fault events into the earth. Existing grounding electrodes and installed grounding electrodes are two types of grounding electrode systems.

An *existing grounding electrode* is a standard part of a facility that includes metal underground water pipe electrodes, metal frame electrodes of the building, and the reinforcing bars in concrete foundations (concrete-encased electrodes). An *installed grounding electrode* is a grounding electrode other than electrodes inherent to the building construction and must be installed. This includes ground rods, metallic plates, buried copper conductors, or other conductive structures buried in the ground installed for grounding purposes. Grounding rods are the most common type of installed grounding electrode. **See Figure 6-15.**

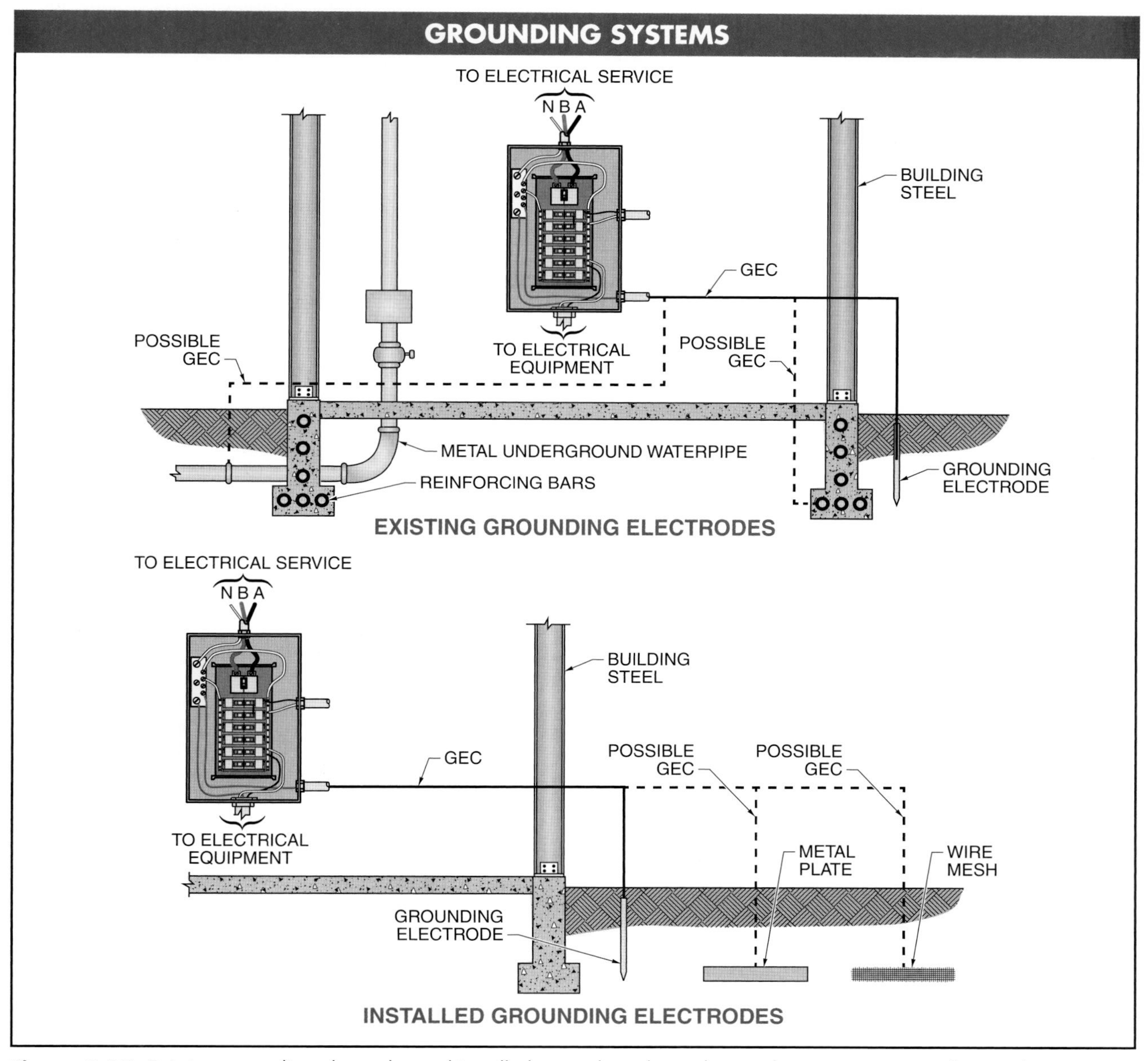

Figure 6-15. Existing grounding electrodes and installed grounding electrodes are the two main types of grounding systems.

There are two basic resistance tests for grounding electrode systems. The resistance of an installed grounding system is measured directly using clamp-on ground resistance meters. The resistance of a grounding system can also be measured indirectly by connecting a three-terminal or four-terminal ground resistance meter. Three-terminal and four-terminal ground resistance meters measure the resistance of the soil from the grounding system to electrodes driven at different locations.

Clamp-On Ground Resistance Meters

Clamp-on ground resistance meters directly measure the resistance of a grounding electrode or small grounding grid. Clamp-on ground resistance meters are used at any time, regardless of ground condition (frozen, wet, or dry). Clamp-on ground resistance meter measurements do not require any part of the ground system to be disconnected (opened) during the test.

Although a clamp-on ground resistance meter appears to be a standard clamp-on ammeter and has the same clamp-on feature, the two meters are totally different. A standard current clamp-on meter measures the strength of the magnetic field produced by current flow when placed around a conductor. The clamp-on ground resistance meter includes a transmitter part that sends out a signal through the conductor (typically the grounding electrode system) and a receiver part that receives the transmitted signal back from the transmitter and uses the received information to determine the resistance of the grounding electrode system.

Although the main function of a ground resistance meter is to measure the resistance of a grounding system, the meters include several other measuring functions that are related to taking ground resistance measurements. **See Figure 6-16.** The additional functions typically include the following:

- It measures higher circuit current like a standard clamp-on current meter. This function allows the current in the circuit to be tested before any resistance measurement is taken. A small amount of current is acceptable, but higher currents prevent the accurate measurement of ground resistance. Most ground resistance measurements require that the current flow be less than 5 A before an accurate resistance measurement can be taken.
- It measures leakage currents like a clamp-on leakage current ammeter. The lower current rating allows for the troubleshooting of leakage current problems.
- It includes data HOLD and other useful measurement functions typically included on multifunction ground resistance meters.

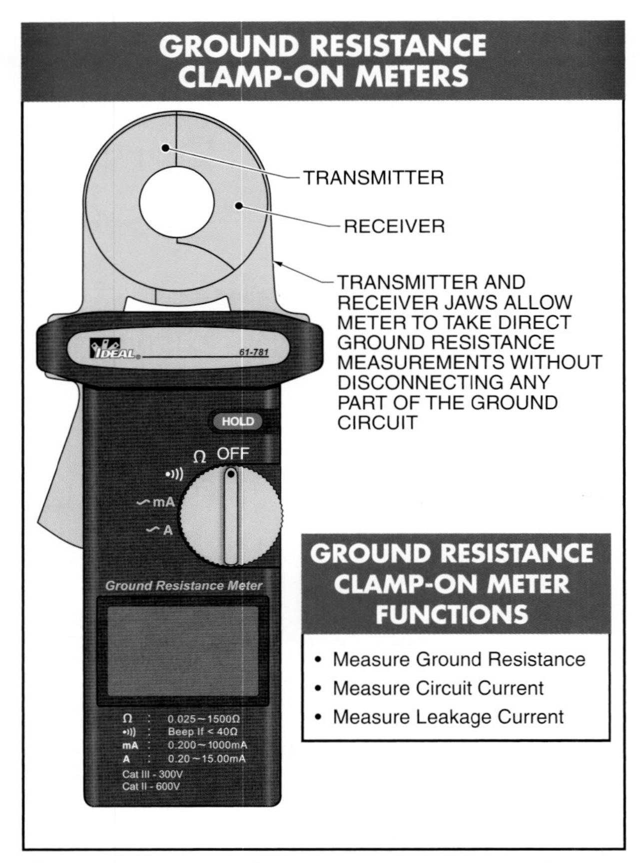

Figure 6-16. Ground resistance meters measure the resistance of a grounding system by transmitting a signal through a grounding conductor and using a receiver to determine (calculate) the resistance of the grounding conductor or system.

Technical Tip

Some test-instrument manufacturers now offer products with wireless capabilities, which allow technicians to take readings from a safe distance. The readings are transferred to a remote device such as a smartphone, tablet, or PC. Readings can also be transferred to a cloud-based storage system.

Clamp-on Ground Resistance Measurement Procedures

Before taking any current or resistance measurements using a clamp-on ground resistance meter, ensure the meter is designed to take measurements on the equipment being tested. Refer to the operating manual of the test instrument for all measuring precautions, limitations, and procedures. Always wear required personal protective equipment and follow all safety rules when taking the measurement. **See Figure 6-17.** The following procedures are used to take ground resistance measurements on grounding conductors, grounding electrodes, and other ground connections used with service entrances, transformers, utility grounds, transmission tower grounds, and communication ground systems:

1. Determine the best position for taking a ground resistance measurement. The clamp-on ground resistance meter can be placed around the grounding conductor leading to the grounding system or to the grounding electrode. The jaws of a clamp-on ground resistance meter must be completely closed around the grounding conductor or grounding electrode.
2. Set the function switch on the ground resistance meter to standard current measurement (or use a standard clamp-on ammeter). Measure the current in the grounding system at the location in which the grounding resistance meter will be connected. When the ammeter measures more than 1 A, there is a ground leakage current problem that must be corrected before proceeding with ground resistance measurements. Manufacturers of ground resistance meters specify the maximum current allowed in which a meter can take an accurate measurement (typically around 5 A).
3. Set the function switch of the ground resistance meter to Ω when the current through the grounding system is low. Place the jaws of the meter around the grounding conductor and measure the grounding system resistance. A grounding resistance meter will display the resistance of the grounding system.
4. Take any additional measurements that are available, or needed, using the ground resistance meter. For example, most ground resistance meters have a ground leakage scale that allows for measuring small grounding leakage currents.
5. Record all measurements taken.
6. Remove the clamp-on ground resistance meter from the grounding circuit.

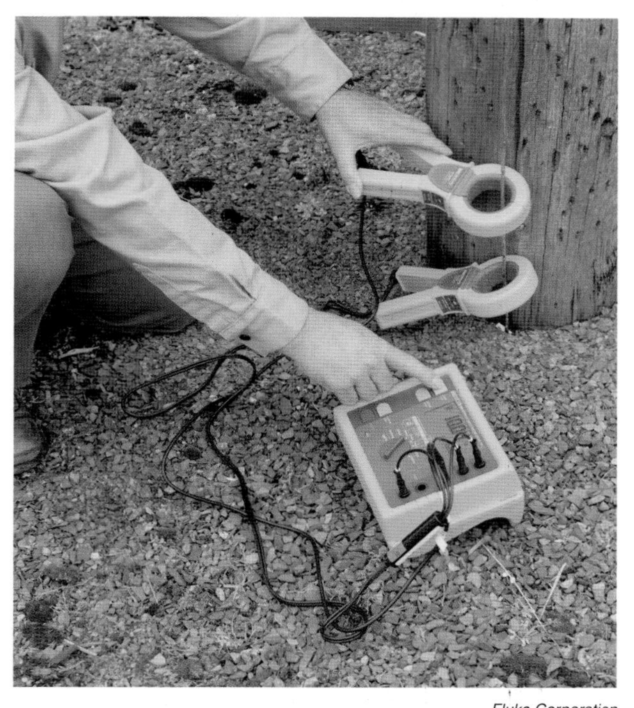

Fluke Corporation

Earth ground testers can be used to measure the integrity of a grounding system.

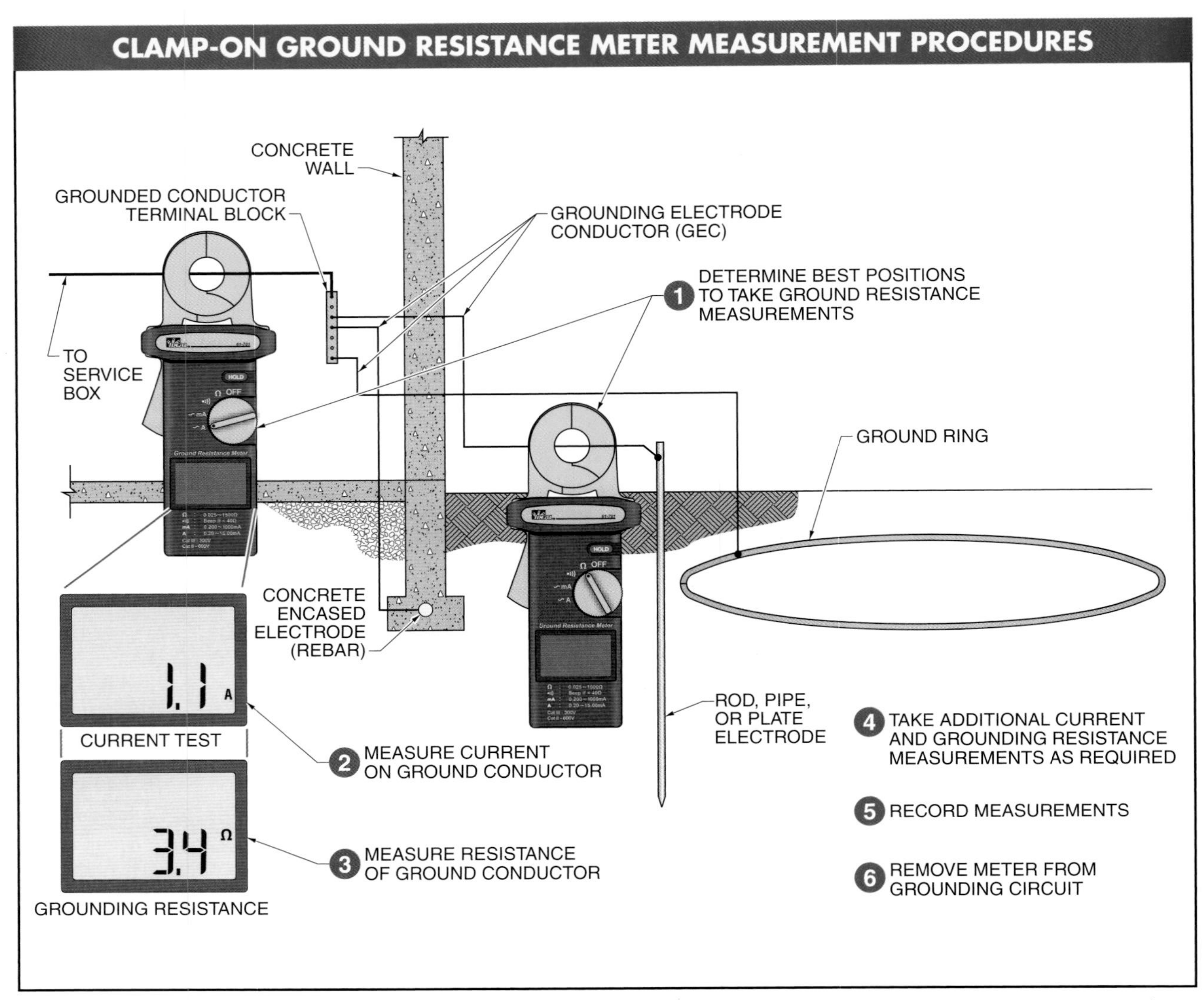

Figure 6-17. Ground resistance measurements are taken on grounding conductors used with service entrances, transformers, utility grounds, transmission tower grounds, and communication ground systems.

A high resistance reading indicates a problem. The higher the measured ground resistance, the poorer the grounding system will operate. High resistance indicates any of the following:

- Open, corroded, or damaged grounding system or connections. Open or loose connections and other damage can occur anyplace in the system (above or below the ground). Look for the obvious problems first, such as loose connections, corrosion, and damaged conductors.
- Insufficient grounding electrode, grid, or system size and incorrect type for soil conditions. Look for proper grounding practices, size, and number of electrodes typically used in the area.
- Changing soil resistance. Grounding systems are required to last a long time, but soil conditions can change over the years. Look for any construction changes that may have taken place since the system was installed. Another cause of high ground resistance is a falling water table that leaves the grounding system in drier soil compared to when the system was installed.

Measuring Grounding System Resistance

The principles of Ohm's law can be applied to measuring the grounding electrode resistance to ground after the grounding electrode (grid or system) is installed. Resistance measurements are taken to verify that the resistance of the grounding system is within required maximum resistance values. *Ohm's law* is the relationship between voltage, current, and resistance in an electrical circuit. Ohm's law states that current in a circuit is proportional to the voltage and inversely proportional to the resistance. **See Figure 6-18.**

When applying Ohm's law to determine earth resistance, one metal electrode, two probes, a power supply, an ammeter, and a voltmeter are required. Rod number 1 is the grounding electrode to be tested. Rods 2 and 3 are shorter rods (probes) driven into the earth about 8″ to 12″ apart for test purposes only.

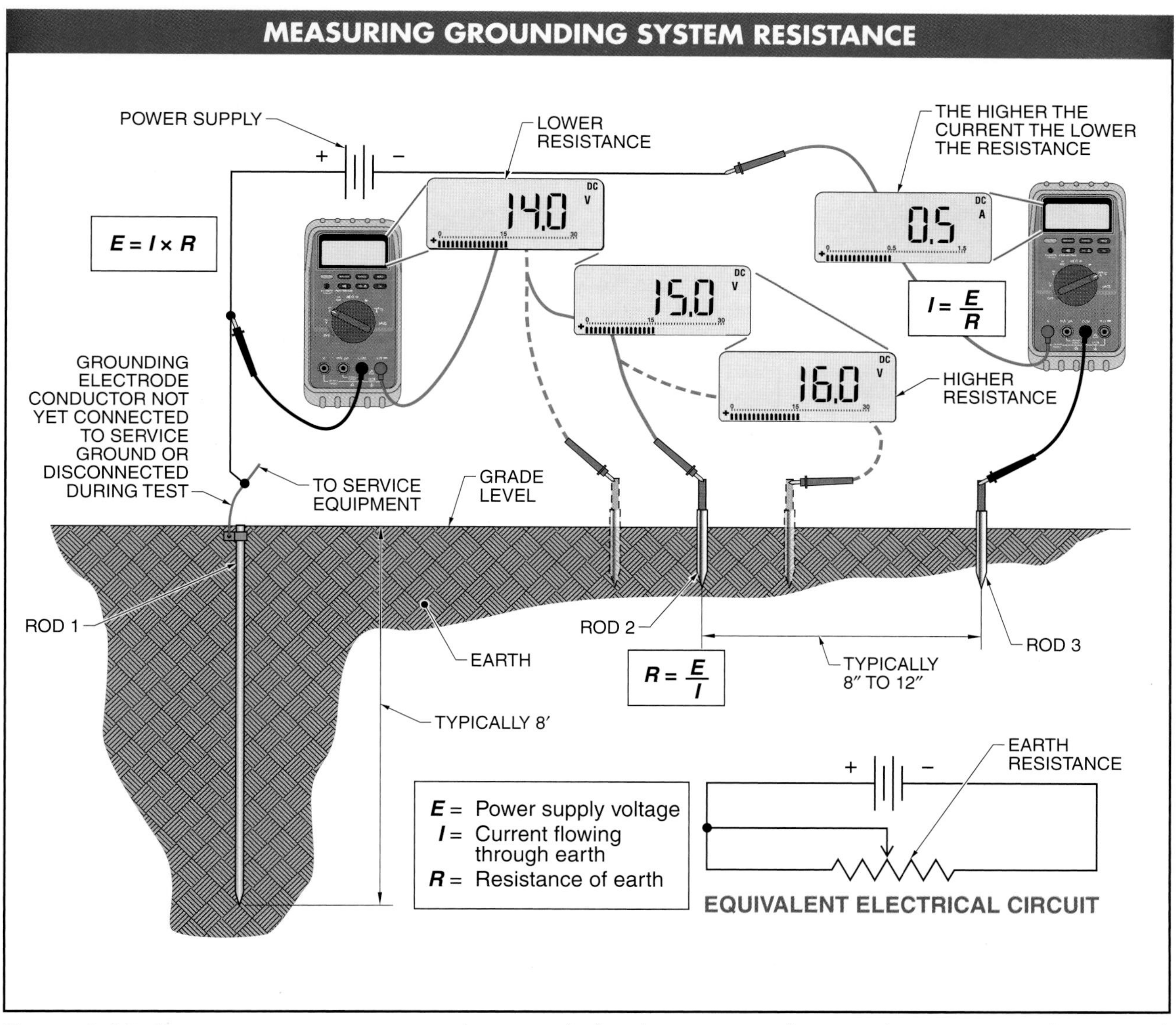

Figure 6-18. Resistance measurements are taken to verify that the resistance of a grounding system is within required maximum resistance values set by the NEC®.

The source voltage (from the power supply) is applied between the two outside rods (Rod 1 and Rod 3). An ammeter is connected into the circuit to measure the current draw on the power supply. The current drawn from the power supply is inversely proportional to the resistance of the circuit created (earth resistance). The lower the measured resistance, the higher the current capacity of the circuit, and the higher the measured resistance, the lower the current capacity of the circuit.

A voltmeter connected between Rods 1 and 2 measures the voltage potential difference of the earth between the two points. Rod 2 can be moved (in a straight line) to points between Rods 1 and 3. As Rod 2 is moved closer to Rod 3, the voltmeter reads a higher voltage. When Rod 2 touches Rod 3, the voltmeter will read the full voltage of the power supply. When Rod 2 is moved closer to Rod 1, the voltmeter indicates a lower voltage. When Rod 2 touches Rod 1, the voltmeter will read 0 V.

Ohm's law is used to calculate the resistance for each point. Ohm's law states that the resistance (R) in a circuit is equal to voltage (E) divided by current (I). To calculate resistance using Ohm's law, apply the following formula:

$$R = \frac{E}{I}$$

where

R = resistance (in Ω)

E = voltage (in V)

I = current (in A)

Ohm's law can be applied to find the resistance of the earth grounding system between Rod 1 and Rod 2 by applying the following formula:

$$R_{ground\text{-}system} = \frac{E_{1\text{-}2}}{I_{earth}}$$

where

$R_{ground\text{-}system}$ = resistance of the grounding system between Rod 1 and Rod 2 (in Ω)

$E_{1\text{-}2}$ = voltage measured between Rods 1 and 2 (in V)

I_{earth} = current flowing through the earth between Rods 1 and 3 (in A)

For example, what is the resistance of the grounding system between Rod 1 and Rod 2, when the ammeter reads 0.5 A and the voltmeter reads 15 V?

$$R_{ground\text{-}system} = \frac{E_{1\text{-}2}}{I_{earth}}$$

$$R_{ground\text{-}system} = \frac{15\text{ V}}{0.5\text{ A}}$$

$$R_{ground\text{-}system} = \mathbf{30\ \Omega}$$

The calculated resistance of a grounding system increases as Rod 2 is moved closer to Rod 3 and decreases as Rod 2 is moved closer to Rod 1. Using a series of measurements and Ohm's law, the resistances of the earth and grounding electrode at various distances from the grounding electrode (Rod 1) to Rod 2 can be calculated and plotted. **See Figure 6-19.**

For most measurements, a point is reached where the rate of increase in the resistance of the earth is low and remains relatively constant for a set distance. The area of measurement where the resistance remains relatively constant is referred to as the plateau area. Beyond the plateau area, the resistance of the earth from Rod 1 begins to increase again. The resistance of the earth rises sharply as Rod 2 becomes close to Rod 3.

Field tests show that the acceptable value of the grounding system resistance is typically obtained on the curve when Rod 2 is placed about 62% of the distance from the grounding electrode (Rod 1) to Rod 3. In the example shown, the distance is 62′ (0.62 × 100′).

Doubling the distance between Rods 1 and 3 to 200′ would result in placing Rod 2 at 124′ (0.62 × 200′) to obtain the typical earth resistance. Doubling the distance between Rod 1 and Rod 3 would in theory double the resistance between the rods and that would reduce the current by half. Thus, the 62% placement of Rod 2 typically works for most test distances. Thus, the 62% measurement is considered to be the actual ground resistance for most applications.

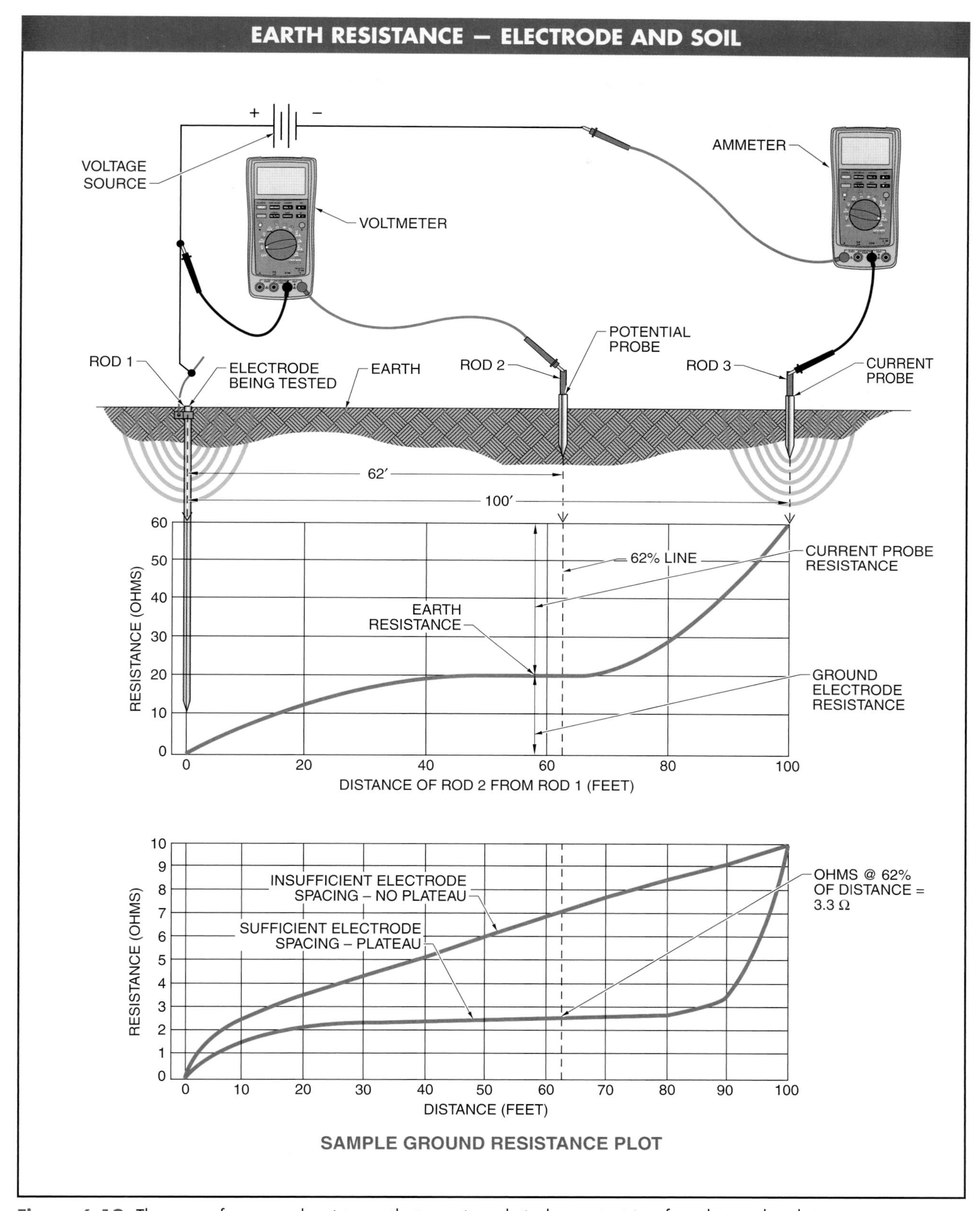

Figure 6-19. The area of measured resistance that remains relatively constant is referred to as the plateau area.

AEMC® Instruments

Rod 2 (R2) is placed to start earth resistance measurements at 62% of the Rod 1 (R1) to Rod 3 (R3) distance.

During an earth resistance test, Rod 3 must be placed far enough from Rod 1 (the grounding electrode being tested) so that the effective resistance areas to ground of Rod 1 and Rod 3 are not overlapping. The overlapping can be avoided by placing Rod 1 and Rod 3 far apart. Although the required distance between Rod 1 and Rod 3 varies with soil conditions, 100′ is typically an acceptable distance. When Rod 1 and Rod 3 are not placed far enough apart, the measurements when plotted will indicate an incorrect placement. **See Figure 6-20.**

Troubleshooting Tip

EARTH RESISTANCE TESTERS

The maximum resistance value R_{MAX} that can be measured in mode MΩ depends on the rated voltage selected for the test.

Rated Voltage Selected for Test*	R_{MAX} Equals Maximum Resistance Value†
50	99.9
100	199.9
250	499.9
500	999.9
1000	1999.9

* volts (DC)
† resistance (MΩ)

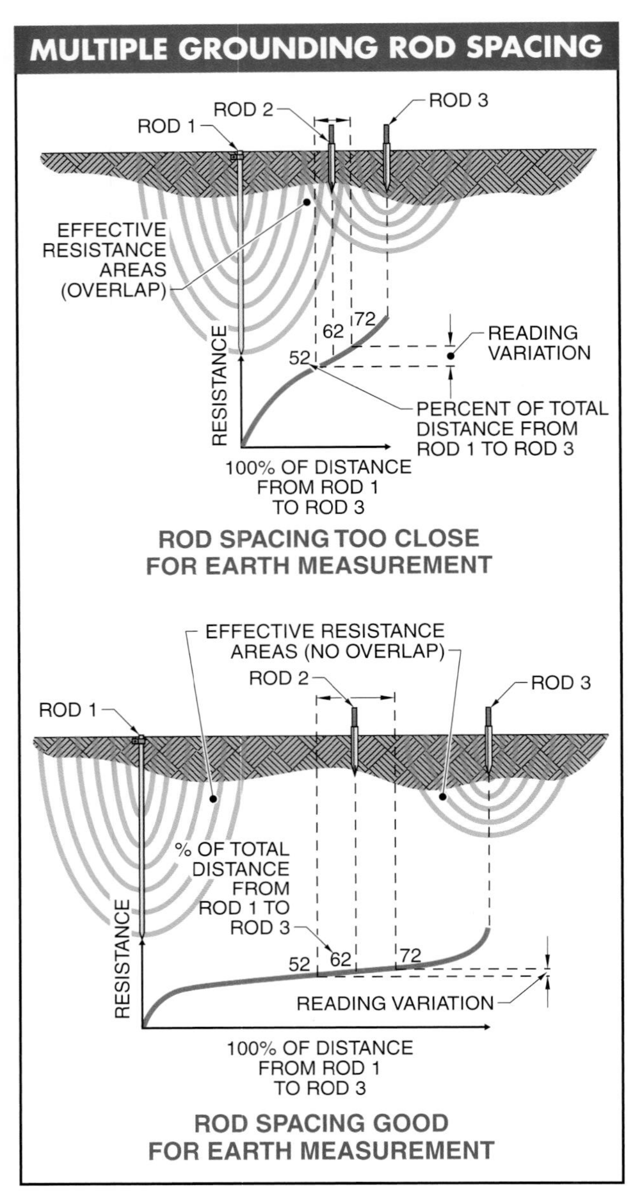

Figure 6-20. Grounding rod overlapping must be avoided by spacing Rod 1 and Rod 3 far enough apart (100′) so accurate measurements can be taken and plotted.

When Rod 1 and Rod 3 are not placed far enough apart, the measured resistance will continue to increase as Rod 2 is moved closer to Rod 3. There will be no leveling off (plateau) of resistance measurements, indicating that the distance between Rod 1 and Rod 3 must to be increased and the measurements retaken.

THREE-TERMINAL GROUND RESISTANCE MEASUREMENTS

When performing an earth resistance test, a separate voltmeter, ammeter, and power supply are used to measure the grounding system resistance between two points (Rod 1 and Rod 2). Although the measurements are taken using individual meters and a power supply, in practical use, a self-contained ground resistance meter that includes all the required components for measuring ground resistance is used. A ground resistance tester used for measuring earth resistance is typically called a three-terminal ground resistance tester. **See Figure 6-21.**

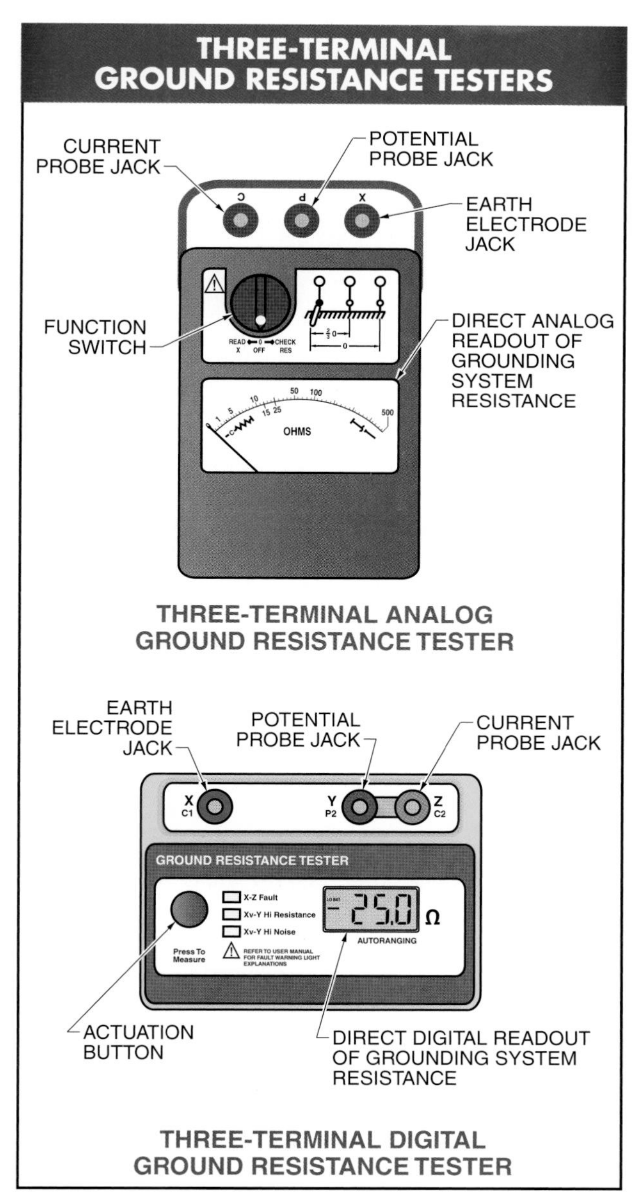

Figure 6-21. Three-terminal ground resistance testers include a voltmeter, ammeter, and power supply to measure the grounding system resistance between any two points.

A ground resistance tester includes a power supply, a voltmeter, an ammeter, a display for direct readout of resistance, and all the required components for measuring earth resistance or the resistance of a grounding system. Ground resistance testers output a voltage that is at a frequency other than 60 Hz to ensure an accurate measurement.

The measurement taken by a three-terminal ground resistance tester is often called the Fall-of-Potential method (or Three-Terminal method) because the voltmeter test leads connected to the moving rod (Rod 2) are actually measuring the voltage potential difference between the grounding electrode (grid or system) and Rod 2. The terminology used to identify Rod 1, Rod 2, and Rod 3 changes depending upon the manufacturer of the test instrument, but the measuring principles are the same. **See Figure 6-22.**

GROUND RESISTANCE TESTER—JACK TERMINOLOGY

Grounding Rod – Electrode	Meter Type Jack		
	1	2	3
Rod 1 – Rod X (Earth electrode)	X	C1 or P1	T1
Rod 2 – Rod Y (Potential probe)	P	P2	T2
Rod 3 – Rod Z (Current probe)	C	C2	T3

Figure 6-22. The terminology used to identify Rod 1, Rod 2, and Rod 3 changes depending upon the manufacturer of the test instrument, but the measuring principles are the same.

Practices for Spacing Grounding Electrodes

No set distance between the placement of Rod 1 and Rod 2 can be provided because the distance is dependent upon how deep the electrode is in the earth, the thickness and type of material of the electrode, and the type of soil the grounding electrode is placed in. But, some general practices can be applied for most applications. Use the following practices for the spacing of the test probes when taking a three-terminal earth resistance measurement:

1. When testing a single earth electrode, Rod 3 is placed 50′ from the grounding electrode (Rod 1), and Rod 2 is driven in the earth at a distance of 31′ from Rod 1 (0.62 × 50′ = 31′).
2. For a small grid of two earth electrodes, Rod 3 is placed 100′ from the grounding electrode (Rod 1), and Rod 2 is driven in the earth at a distance of 62′ from Rod 1 (0.62 × 100′ = 62′). A 100′ spacing between Rod 1 and Rod 2 is typically adequate for most grounding system resistance measurements (except for the very largest grounding systems).
3. For a large grid of electrodes consisting of several electrodes or plates that are connected, the distance between Rod 1 and Rod 3 must be increased to 200′ or more. Rod 2 is placed at 62% of the chosen distance.

Three-Terminal Ground Resistance Measurement Procedures

Before taking any earth and electrode resistance measurements using a three-terminal ground resistance tester, ensure the tester is designed to take measurements on the soil and electrode being tested. Refer to the operating manual of the test instrument for all measuring precautions, limitations, and procedures. Always wear required personal protective equipment and follow all safety rules when taking the measurement. **See Figure 6-23.**

When measuring the resistance of the soil for a grounding system using a three-terminal ground resistance tester, apply the following procedures:

1. Use manufacturer recommended measurement procedures for measuring earth resistance. Most ground resistance tester manufacturers also include a sheet for recording measurements. **See Appendix.**
2. Ensure the grounding electrode and system is not connected to the building ground.
3. Push or drive Rod 3 into the ground at about 100′ from the grounding electrode or grounding system being tested.
4. Push or drive Rod 2 into the ground at about 62′ (62% point) from the grounding electrode or grounding system being tested.
5. Connect the test leads of the ground resistance tester to the grounding electrode and test probes (Rod 2 and Rod 3).
6. Measure and record the earth resistance measurements.
7. Move Rod 2 ten feet to either side of the 62′ point (52′ and 72′ from Rod 1). When the three measurements are basically the same, the plateau area has been found and the 62′ reading is the reading of the grounding system resistance.
8. When the plateau for earth resistance is not found because Rod 3 is spaced too close to Rod 1, increase the distance between Rod 1 and Rod 3 and perform the earth resistance test again.
9. Remove the grounding resistance tester from the grounding system and earth.

Technical Tip

Per Article 250 of the NEC®, grounding electrode conductors should be attached to the grounding electrode by means of (1) a bolted clamp of cast bronze, brass, or cast malleable iron, (2) a pipe fitting or plug, and (3) a device that is screwed into the pipe fitting or plug. Soldered lugs, pipes, and fittings are not permitted.

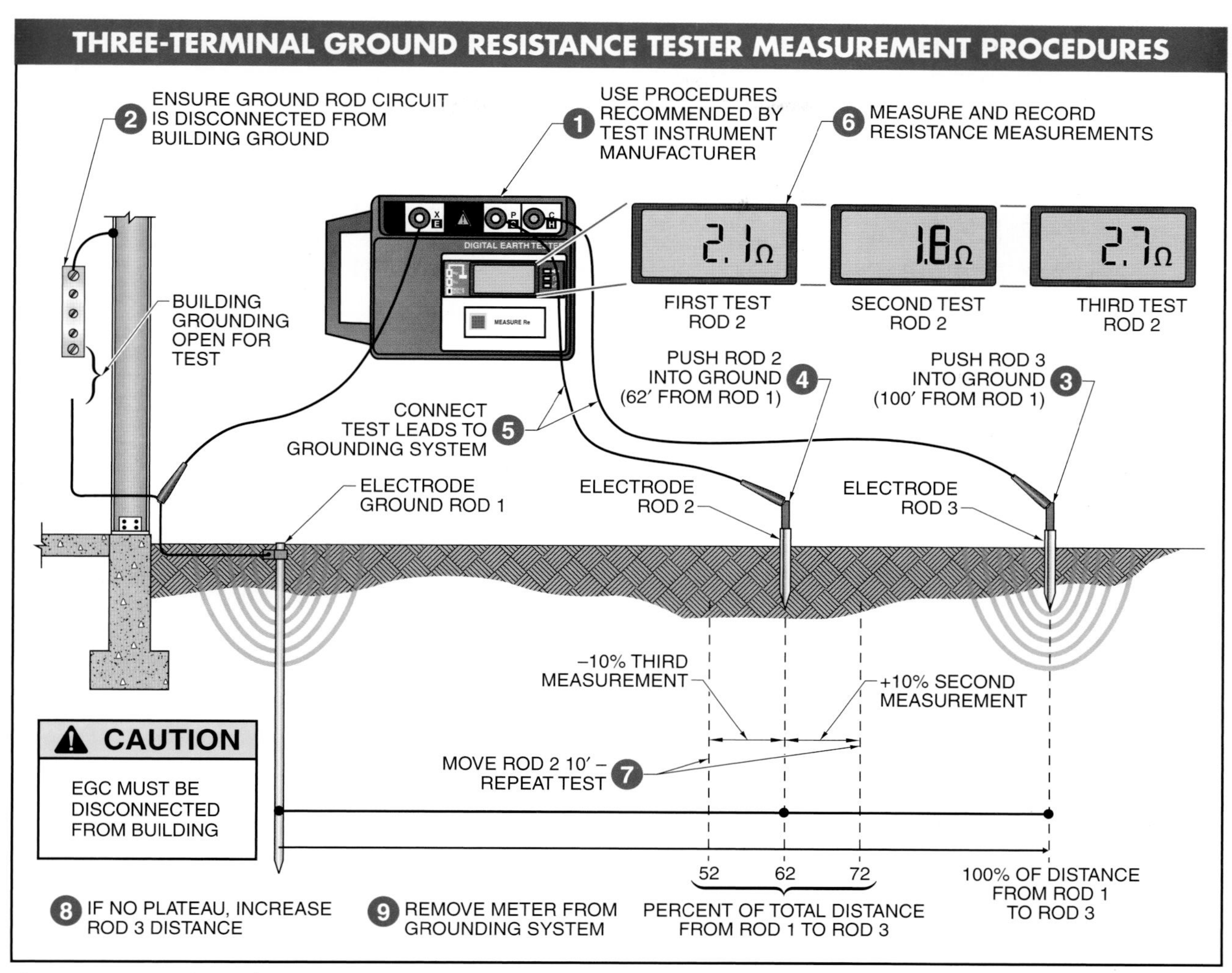

Figure 6-23. Before taking any earth and electrode resistance measurements, electricians must ensure that the grounding electrode and system are not connected to the building ground.

Grounding Electrode Resistance Factors

When the resistance of the earth grounding system is within the range specified by codes and standards, no further ground resistance modifications or measurements need to be taken. However, when the earth resistance is too high, modifications to the earth grounding system must be made. The three factors that affect the amount of resistance an earth grounding system has are resistance of the grounding electrode, contact resistance between the electrode and earth, and resistance of the earth surrounding the electrode.

Megger Group Limited

The measurements taken by a three-terminal ground resistance tester measure the voltage potential between the grounding electrode and Rod 2.

Grounding Electrode Resistance. The resistance of a grounding electrode (grid or system) is very low due to the metal composition (galvanized steel, stainless steel, or copper) of the electrode and the size and shape of the electrode. Increasing the number of grounding electrodes and/or the size (length and/or diameter) of an electrode lowers the total resistance (law of resistance in a parallel circuit). When adding additional grounding electrodes, the rules of parallel resistance do not apply exactly mathematically. For example, adding a second grounding electrode of the same size (resistance) will not reduce the resistance of the grounding circuit by half, as occurs with parallel connected resistors of the same value. **See Figure 6-24.**

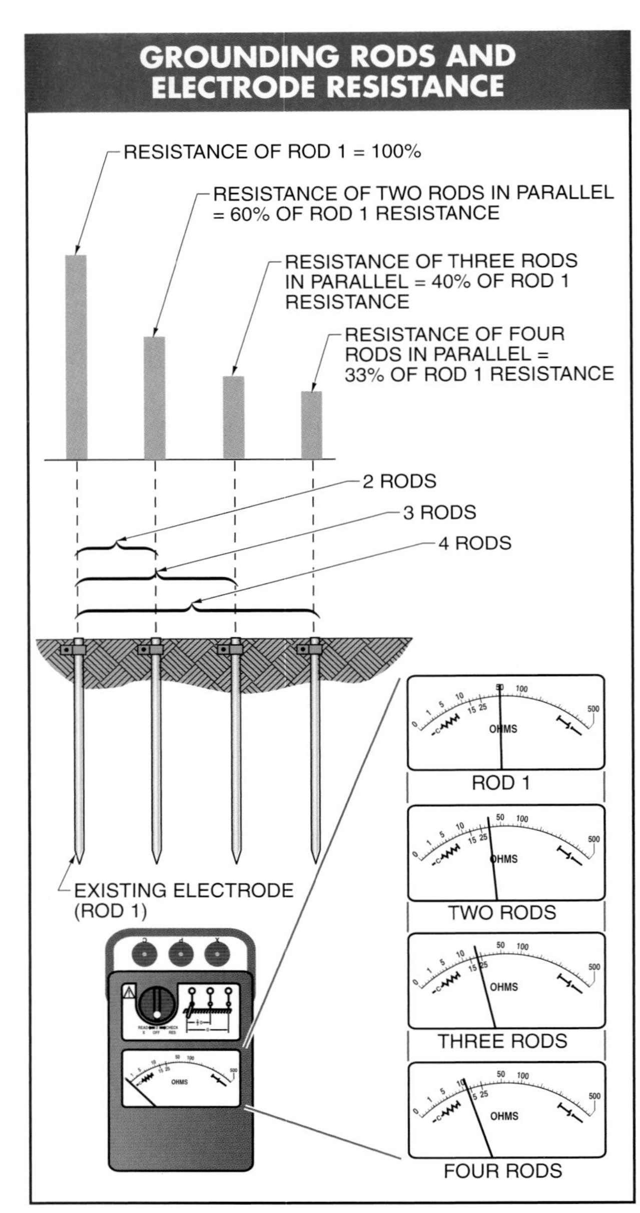

Figure 6-24. When adding additional grounding electrodes, the rules of parallel resistance do not exactly apply mathematically.

Contact Resistance between Grounding Electrode and Earth. A grounding electrode placed in the earth typically makes good contact with the earth during a normal installation. When the earth is packed solid against a bare metal electrode, very good electrical contact is made. Water can be used to compress the soil around an electrode and lower the resistance of the grounding system. Dry conditions and freezing temperatures expand the soil around grounding electrodes and increase the resistance of the system.

Resistance of Surrounding Earth. The resistance of the earth differs from location to location and thus is the major variable factor in determining the resistance of a grounding system. There are occasions when an acceptable level of earth resistance cannot be found for the installation of a grounding electrode. When an acceptable resistance cannot be found, three methods are used to reduce the resistance of the earth. To reduce the total resistance of a grounding system do the following:

- Increase the size of the grounding electrode. Use a longer and/or thicker electrode to increase contact area with the earth, lowering the resistance of the grounding system.
- Use multiple grounding electrodes. Spacing of multiple electrodes is important and all NEC® and other requirements and recommendations must be followed.
- Chemically treat the soil. Chemically treating the soil lowers the resistance of the grounding system but caution must be used because chemically treated soil typically increases the corrosion process of metal parts because salts—typically calcium chloride ($CaCO_3$) and sodium chloride (NaCl)—are used. Also, chemically treated soil may cause an environmental problem.

FOUR-TERMINAL GROUND RESISTANCE MEASUREMENTS

Grounding provides a low impedance (resistance) path to earth for high fault currents from circuit shorts and lightning, and a path for low-level currents such as radio frequency interference, static electricity, and other unwanted currents in an electrical or electronic system.

When a grounding system (electrode, grid, or loop) is placed in contact with the earth, the earth becomes part of the electrical system. The earth is used to carry fault currents away from the building and other areas of the electrical system where the fault current may cause damage such as electrical shock, equipment failure, or fire. The ability of the earth to carry fault currents is based on the resistance of the earth at the point of grounding electrode placement. The lower the resistance of the earth, the higher the current carrying capability of the grounding system. Soil resistance is measured by using four-terminal ground resistance testers. **See Figure 6-25.**

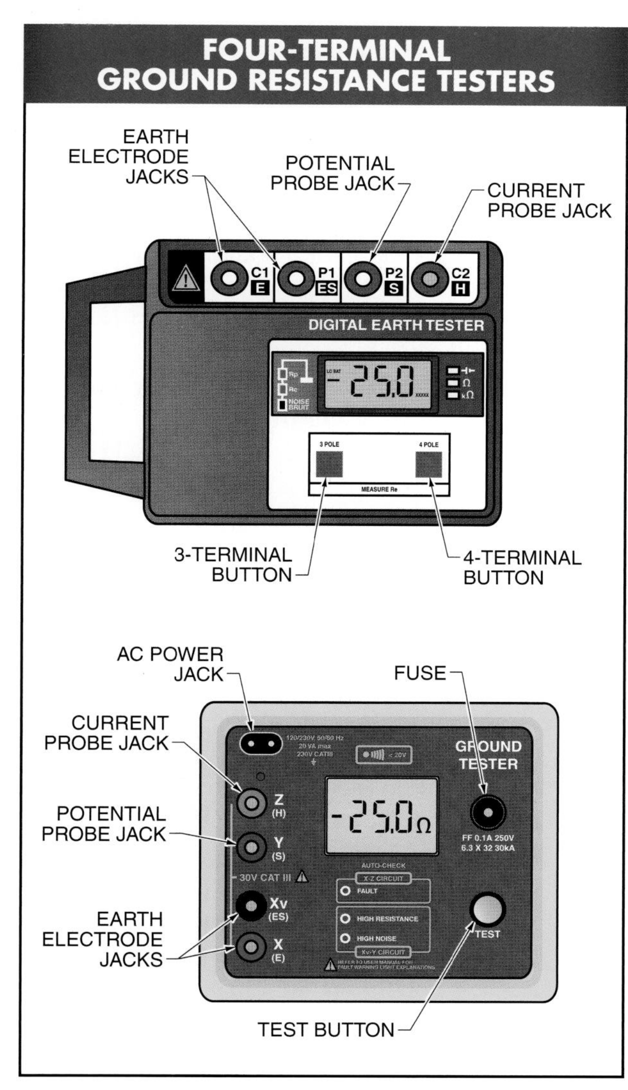

Figure 6-25. Soil resistance is measured by using four-terminal ground resistance testers.

Technical Tip

The total grounding system resistance decreases if additional ground rods are added in parallel, or if the area of the grounding system coming in contact with the earth is increased.

Soil Resistance (Resistivity)

The resistance (resistivity) of the earth (soil) varies with soil type, moisture content, temperature, and other factors. **See Figure 6-26.** The resistance of soil is measured to determine the following:

- The type of soil conditions (chemical and moisture composition) in an area; the condition of the soil to be used for grounding is useful from a geological standpoint as well as an electrical standpoint
- The corrosive effects the soil may have on a grounding system, and what material is best to use for the grounding system given the soil conditions
- The best design and type (electrode, grid, loop, or plate) of grounding system to be used

The earth (soil) is part of the electrical system and can carry current like any metal conductor. However, unlike low resistance metal conductors, soil has a much higher resistance than metal conductors. Besides using chemicals, soils are made to have a low resistance by making sure that the earth grounding system is in contact with a large area of soil. A large area of soil contact allows the laws of resistance for parallel circuits to be applied (the more resistances placed in parallel, the lower the total resistance).

SOIL RESISTANCE (RESISTIVITY)						
Material	Resistivity (Ω/cm)			Resistivity of ⅝ in. (16 mm) x 10 ft (3 m) Rod (Ω)		
	Avg	Min	Max	Avg	Min	Max
Fills, ashes, brine, cinders, waste, salt marsh	2370	590	7000	8	2	23
Clay, shale, gumbo, loam	4060	340	16,300	13	1.1	54
Soil, with added sand	15,800	1020	135,800	52	4	447
Gravel, sand, stones, with little clay or loam	94,000	59,000	456,000	311	195	1516

Temperature*		Resistivity*
°C	°F	Ω/cm
20	68	7200
10	50	9900
0	32 (water)	13,800
0	32 (ice)	30,000
–5	23	79,000
–15	14	330,000

* based on 15.2% moisture (sandy loam)

Figure 6-26. The resistance (resistivity) of the earth (soil) varies with soil type, moisture content, temperature, and other factors.

Troubleshooting Tip

CALCULATING EARTH RESISTANCE

What is the resistance of the earth between the grounding electrode and Rod 2 when 500 V at 250 A is applied?

$$R = \frac{E}{I}$$

where

R = resistance (in Ω)
E = voltage (in V)
I = current (in A)

$$R = \frac{E}{I}$$

$$R = \frac{500}{250}$$

$$R = \mathbf{2}\ \Omega$$

Four-Terminal Ground Resistance Measurement Procedures

A four-terminal ground resistance tester is used to measure ground resistance. When measuring ground resistance, four electrodes are driven into the ground. Unlike the three-terminal ground resistance testers, a four-terminal tester does not use the grounding electrode as one of the probes because the tester is used to measure the resistance of the earth, not the resistance of the earth and grounding electrode. **See Figure 6-27.** However, four-terminal ground resistance testers can be used as three-terminal grounding system resistance testers by connecting two of the four tester terminals together (typically C1 and P1).

Before taking any earth resistance measurements using a four-terminal ground resistance tester, ensure the tester is designed to take measurements on the soil being tested. Refer to the operating manual of the test instrument for all measuring precautions, limitations, and procedures. Always wear required personal protective equipment and follow all safety rules when taking the measurement. When measuring the resistance of the earth (soil) using a four-terminal ground resistance tester, apply the following procedures:

1. Use the recommended measurement procedures from the test instrument manufacturer for measuring earth resistance. Ensure the grounding electrode or system is not connected to the building ground.
2. Push or drive all four test instrument probes (Rods 1 through 4) into the ground at the depth and spacing recommended by the manufacturer. Studies show that the spacing of the probes determines the depth of the resistance measurement. In other words, when the distance between the probes is 6′, the resistance measurement displayed on the tester is the resistance of the earth to a depth of 6′.
3. Measure and record the resistance measurement of the earth (soil) at several different distances (4′, 6′, and 8′ apart). Any change in resistance measurements indicates that the soil conditions are changing over the measured surface or measured distance.
4. Remove the ground resistance tester from the grounding system.

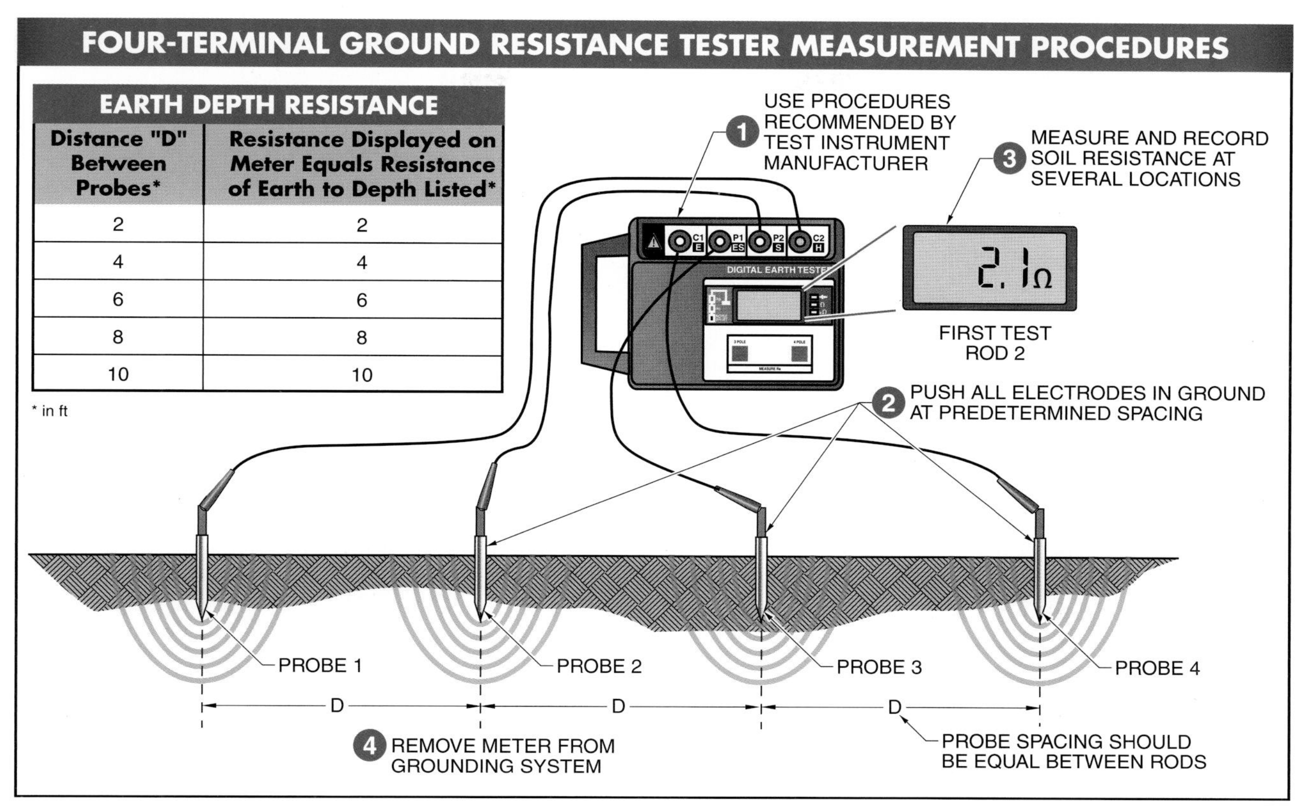

EARTH DEPTH RESISTANCE	
Distance "D" Between Probes*	Resistance Displayed on Meter Equals Resistance of Earth to Depth Listed*
2	2
4	4
6	6
8	8
10	10

* in ft

Figure 6-27. When measuring ground resistance, a four-terminal ground resistance tester does not use the grounding electrode as one of the probes because the tester is used to measure the resistance of soil only, not the resistance of the soil and grounding electrode.

Soil pH Measurement Procedures

The *pH* of a solution or substance is the degree of acidity or alkalinity of a solution or substance. A scale of 0 to 14 is used to measure pH, with 7 being the neutral point. Pure distilled water has a measured pH of 7. The lower the measured pH is below 7, the greater the acidity of the substance. For example, orange juice has a pH of 4.3, lemon juice has a pH of 2.3, and battery acid has a pH of 0.3. The higher the measured pH is above 7, the greater the alkalinity of the substance. For example, borax has a pH of 9.3, and milk of magnesia has a pH of 10.3. **See Figure 6-28.**

Before taking pH measurements using a pH meter, ensure the meter is designed to take measurements of the substance being tested. Refer to the operating manual of the test instrument for all measuring precautions, limitations, and procedures. Always wear required personal protective equipment and follow all safety rules when taking the measurement. To measure the pH of soil for a grounding electrode, apply the following procedures:

1. Gather a few samples of soil where the grounding electrode will be placed. Acquire samples that are 1′ or more down from the surface.
2. Turn the pH meter ON.
3. Take a pH reading at several locations in the soil.
4. Record the pH readings. Average the measurements.
5. Turn the pH meter OFF and remove the probe from the soil.
6. Clean the probe.

Note: It is important that soil acid does not damage the probe during storage.

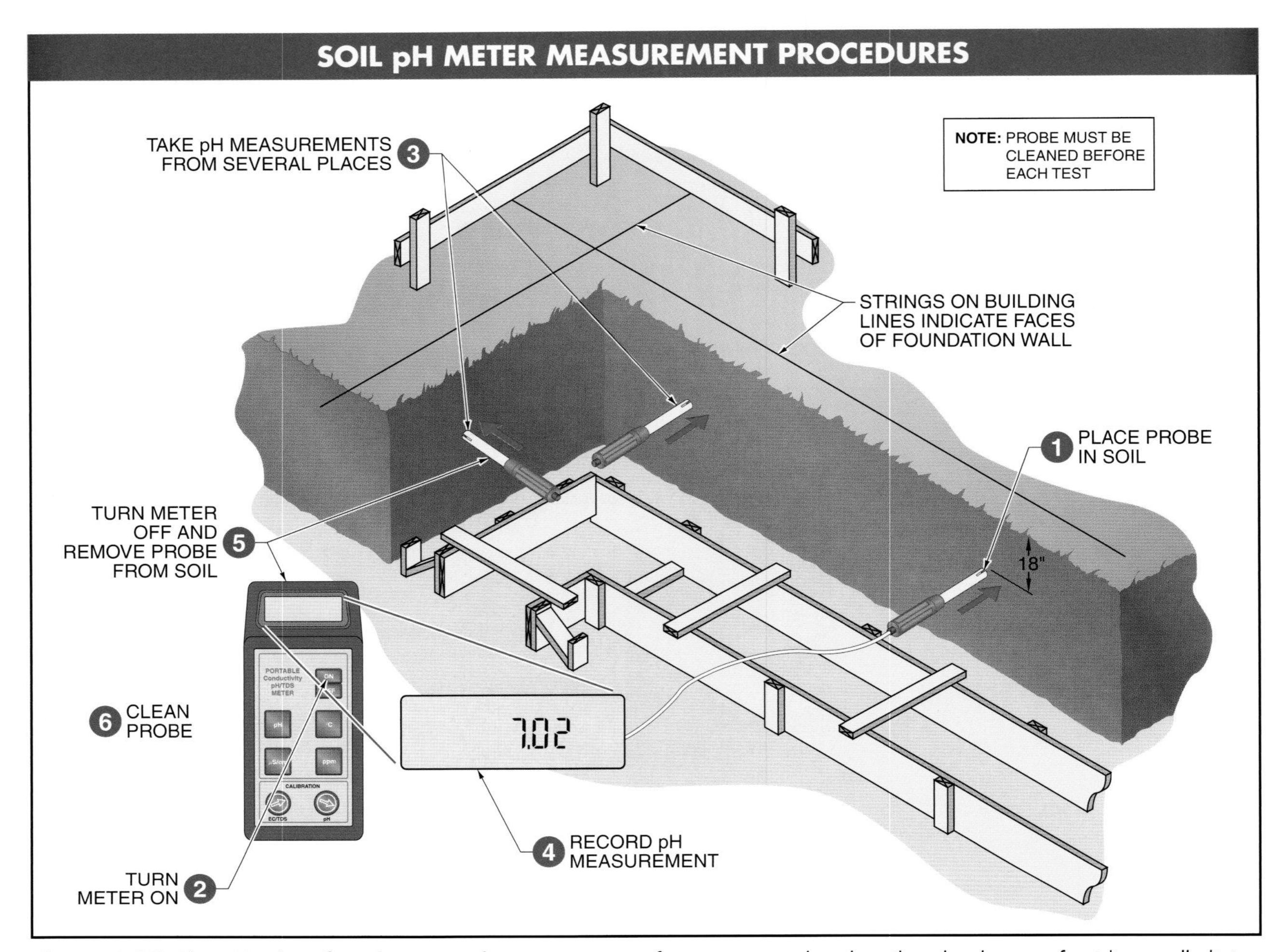

Figure 6-28. The pH value of a solution or substance is a unit of measurement that describes the degree of acidity or alkalinity.

When soil pH is high, corrective actions must be taken to reduce grounding electrode corrosion. Corrective actions include the following:

- Using solid copper, copper-clad (coated), or stainless-clad steel grounding electrodes. Copper-clad electrodes typically last three times longer than galvanized electrodes.
- Using a low resistance grounding cement. Grounding cement is specifically designed to be applied around a buried (entrenched) grounding electrode. The cement is used in areas that have a high soil resistivity to reduce electrode resistance to earth ground. The cement is also used in high pH soil conditions to reduce the corrosion effect on grounding electrodes.
- When possible, using a concrete-encased electrode, the metal frame of a building, or other approved grounding system.

GROUNDING SYSTEMS AND EARTH GROUND TEST INSTRUMENTS

Name: ______________________ Date: ______________

______ **1.** What problem is caused by multiple ground connections in a circuit?

______ **2.** Which type of receptacle disconnects from the power circuit when it detects ground faults?

______ **3.** What effect does high soil pH have on grounding electrodes?

______ **4.** Lightning rods are part of which category of grounding?

T F **5.** A grounded conductor is usually color-coded green.

______ **6.** What is the accepted relative distance between Rod 1 and Rod 2 for measuring earth resistance?

______ **7.** Which type of component is particularly sensitive to electrostatic discharge?

______ **8.** What is the conductor that connects electrical equipment to the grounding system called?

______ **9.** What is the maximum grounding electrode resistance required by the NEC®?

T F **10.** A grounding system is designed to provide a safe path for fault current to flow to ground.

T F **11.** Electronic grounding reduces the noise in electronic signals.

______ **12.** Which category of grounding includes connecting electrical components to the grounding system?

______ **13.** What type of soil measurement determines the best material for grounding electrodes?

T F **14.** Neutral to ground connections can be made at subpanels.

______ **15.** What type of clamp-on ammeter is capable of measuring current as low as a few milliamps?

______ **16.** What unit is used to measure ground resistance?

T F **17.** The grounding system for circuits that include sensitive electronic equipment must be significantly less than 25 Ω.

______ **18.** What is the name of the conductor that connects the grounded parts of the electrical system to the grounding electrode?

______ **19.** Which two buses are connected by the main bonding jumper?

______ **20.** What will a voltmeter measure between the hot and ground conductors of a properly grounded 115 VAC receptacle?

21. Why must grounding systems have low resistance?

22. How does an isolated ground receptacle reduce electrical noise?

23. What are examples of existing grounding electrodes?

24. Explain how the rod spacing in an earth resistance test affects the measurements.

25. What are examples of typical causes of fault currents?

7 MEDIUM VOLTAGE TEST INSTRUMENTS

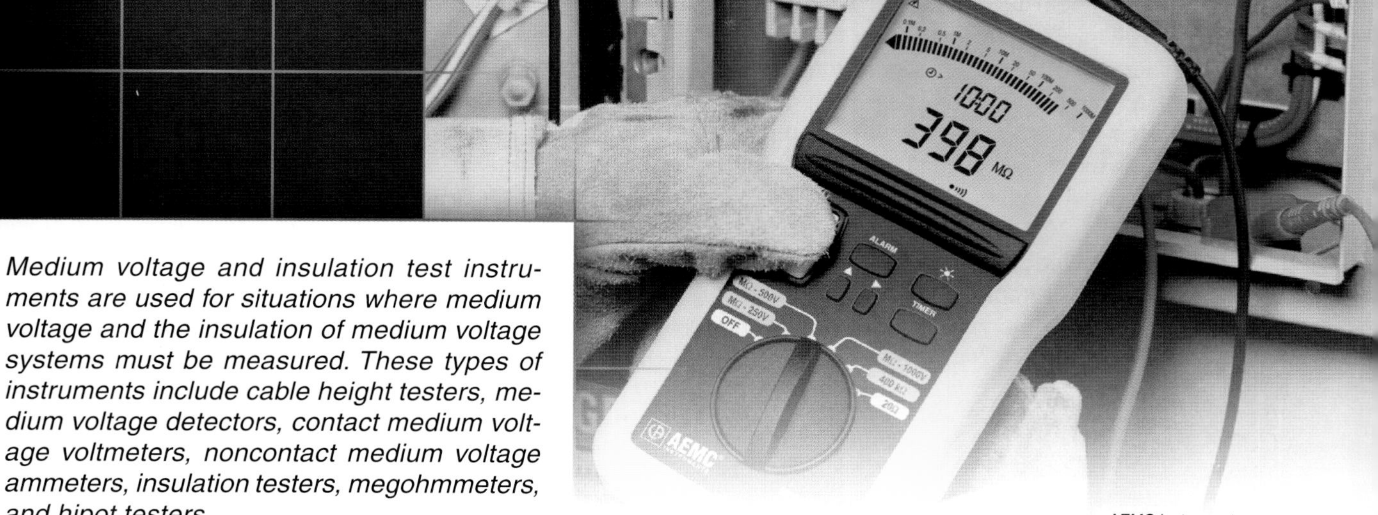

Medium voltage and insulation test instruments are used for situations where medium voltage and the insulation of medium voltage systems must be measured. These types of instruments include cable height testers, medium voltage detectors, contact medium voltage voltmeters, noncontact medium voltage ammeters, insulation testers, megohmmeters, and hipot testers.

AEMC Instruments

TESTING MEDIUM VOLTAGE SYSTEMS

Medium voltage systems require that greater precautions be used when taking measurements. Medium voltage systems also require the use of proper settings on test instruments with multiple voltage ranges and often require the use of specialized test instruments and PPE designed for working around and taking measurements on medium voltage systems.

The term "medium voltage" has various meanings depending upon the context in which it is used and the type of electrician (cable installers, residential/commercial journeymen, industrial journeymen, power distribution linemen) working on and around the system.

Various electrical trade organizations also define voltage levels differently. For example, ANSI/IEEE 1585-2002 (R2007) refers to medium AC voltage as 1 kV to 35 kV, while NECA/NEMA 600-2003 refers to medium AC voltage as 600 V to 69 kV AC, and ANSI C84.1-2016 refers to medium voltage as 2.4 kV to 69 kV. **See Figure 7-1.**

STANDARD NOMINAL THREE-PHASE SYSTEM VOLTAGES PER ANSI C84.1-2016

Voltage Class	Three-Wire*	Four-Wire
Low Voltage		208 Y/120
	240	240/120
	480	480 Y/277
	600	
Medium Voltage	2400	
	4160	4160 Y/2400
	4800	
	6900	
		8320 Y/4800
		12,000 Y/6930
		12,470 Y/7200
		13,200 Y/7620
	13,800	13,800 Y/7970
		20,780 Y/12,000
		22,860 Y/13,200
	23,000	
		24,940 Y/14,400
	34,500	34,500 Y/19,920
	46,000	
	69,000	
High Voltage	115,000	
	138,000	
	161,000	
	230,000	
Extra-High Voltage	345,000	
	500,000	
	765,000	
Ultra-High Voltage	1,100,000	

* in V

Figure 7-1. While NECA/NEMA 600-2003 refers to medium AC voltage as 600 V to 69 kV, ANSI C84.1-2016 refers to medium voltage as 2.4 kV to 69 kV.

Regardless of the definition of a voltage level by any organization, an understanding of what the system is, the required PPE, the test instruments designed for that voltage and system, and what the readings should be and their meanings is required. For example, a commercial journeyman working on a 480Y/277 V system will consider the 480 V the high voltage. A lineman working around a 34.5 kV and 115 kV system may define the 34.5 kV system as medium voltage and the 115 kV system as high voltage. Both systems must be considered high voltage because they are both extremely dangerous.

There are many reasons why medium voltage systems must be tested. These reasons include inadequate preparation, poor assembly techniques, installation instructions that were not followed, lack of training and experience, improper installation, and equipment that was installed in the improper environment. The most common reason for testing is to avoid the costs associated with outages and repairs, in addition to other reasons.

When medium voltage systems are tested, many standards can be referenced. However, in the United States, NFPA 70E, 2018 edition, is one of the most commonly referenced standards when testing medium voltage systems, specifically the following section:

- NFPA 70E 120.5(7) states that adequately rated portable test instruments must be used.
  - Exception No. 1 states that an adequately rated permanently mounted test device can be used.
  - Exception No. 2 states that on electrical systems over 1000 V, noncontact test instruments can be used.
  - Informational Note No. 1 references UL 61010-1 for systems 1000 V or less.
  - Informational Note No. 2 references IEC 61243-1, 61243-2, and 61243-3 for voltage detectors exceeding 1000 V.

Troubleshooting Tip

Unlike standard DMMs, medium voltage test instruments can produce a voltage that can be fatal. It is essential to follow all safety procedures and regulations when using medium voltage testers and equipment. Verify that the unit is properly grounded, all parts of the tester (leads, power cords, attachments, etc.) are in good condition, and that the terminal ratings of the unit are used when testing.

Conductor Insulation

All electrical conductors carrying power must be protected against contacting other conductors, metal parts, and people. Insulation around conductors protects the conductor from damage and isolates the electrical power contained within the conductor. However, not all energized parts of an electrical circuit are protected by insulation. When energized parts of an electrical circuit are exposed, such as where conductors are terminated at fuse or circuit breaker panels, distance (air space) is used as the insulator. Energized parts in a circuit that are separated by air space use the air space as the insulator. The greater the distance between energized electrical conductors or parts, the greater the resistance between the parts. The higher the voltage, the greater the air space that must be maintained to create a high enough resistance to prevent an arc.

Measuring Cable Heights

Overhead power lines (cables) carry extremely high voltages and are quite dangerous. Because of the high voltages in cables, the minimum height from the ground to where cables connect to buildings or structures is set by regulations. When first installed, cable distances are measured for conformity of standards and practices. But cable heights must also be remeasured occasionally because all cables sag over time and/or the height from the earth to the cable can change due to road resurfacing or landscaping. Changes in cable height can also indicate rotting wood poles, broken or damaged anchoring, or damaged cables.

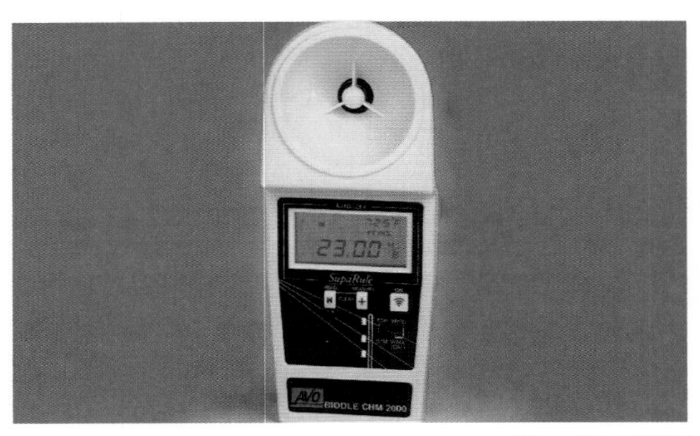

Megger Group Limited

Cable height meters are used to measure cable heights, cable sag, and the spacing between cables. Some cable height meters can also be used to measure horizontal distances of up to 300′.

Measuring the height of cables is performed by inspectors and utility (power and telecom), transportation, and building maintenance companies. For example, an inspector measures cable heights on a construction site to ensure compliance with OSHA standards for cable clearance. A *cable height meter* is a test instrument that is used to measure the height of a power cable in feet and inches or in meters. Cable height meters are typically able to read multiple cable heights simultaneously. Typically, the heights of the lowest and highest cables at the measuring point are recorded. Also, the spacing between the cables is measured and recorded. Cable height meters work by using ultrasonic waves and do not have to come in contact with any cable. To measure cable height using an ultrasonic cable tester, apply the following procedures. **See Figure 7-2.**

Safety Procedures

- Follow all electrical safety practices and procedures.
- Check and wear personal protective equipment (PPE) for the procedure being performed.
- Perform only authorized procedures.
- Follow all manufacturer recommendations and procedures.

1. Place the ultrasonic cable height tester on the ground at each location in which cable height measurements are to be made.
2. Record all required measurements.
3. Check measured values against code distances.

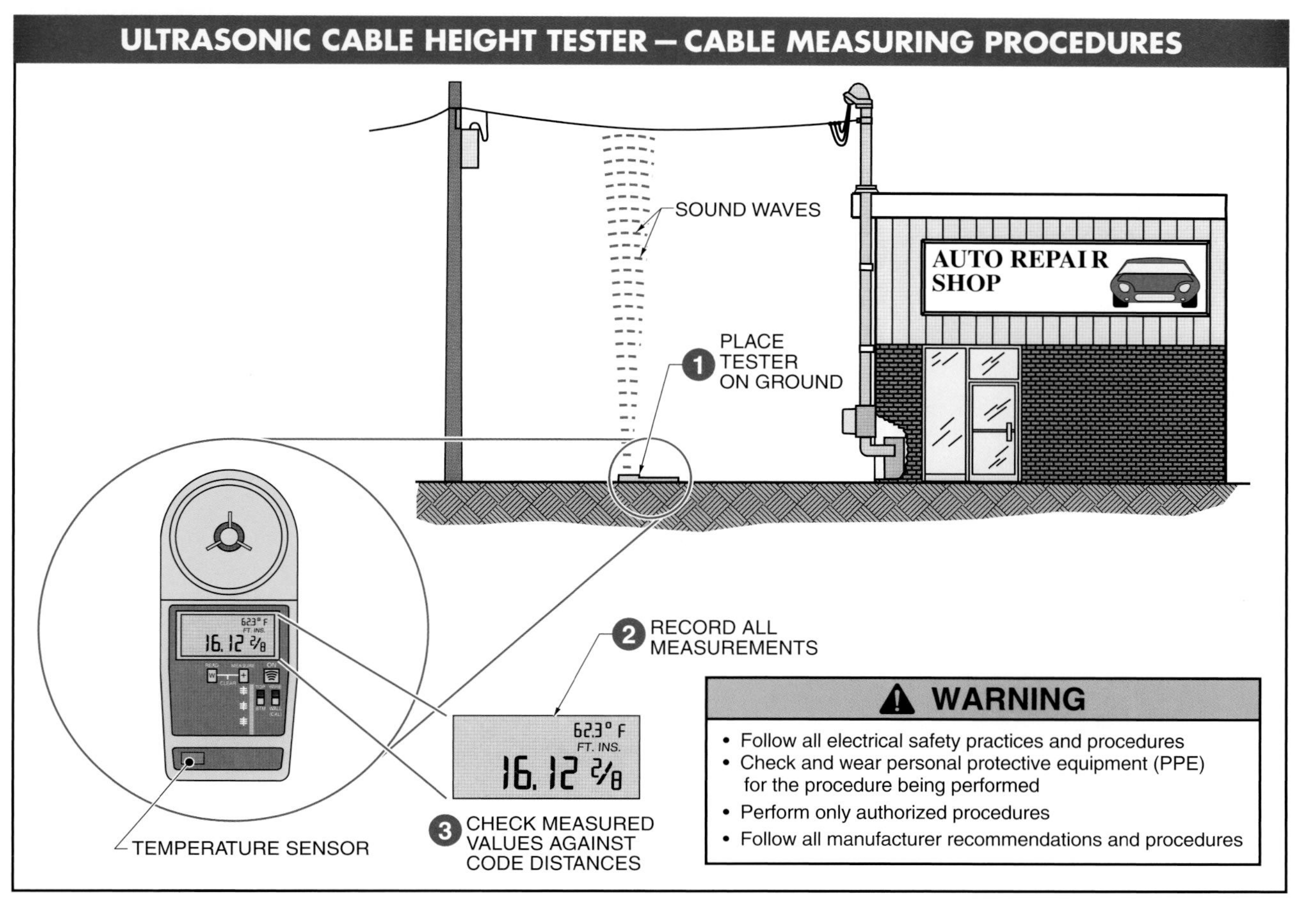

Figure 7-2. Ultrasonic cable height testers determine the height of overhead cables, the heights of the lowest and highest cables, and the spacing between cables.

Medium Voltage Detectors

Before taking a voltage measurement on any circuit with a medium voltage detector, a three-step procedure must be followed. *Note:* For safety information on high voltage detectors and related equipment, refer to OSHA 29 CFR 1910.269, which covers the operation and maintenance of electric power generation, control, transformation, transmission, and distribution lines and equipment, or to IEEE 510 – *Guide for Electrical Safety in High-Voltage Testing*. The three steps are as follows:

1. Check the medium voltage detector on a known (energized) voltage source or piezo verifier before taking a measurement on an unknown voltage source. Testing a medium voltage detector ensures the detector is operating properly.
2. Take the voltage measurement wearing all personal protective equipment (PPE) and following all applicable high voltage detector safety rules.
3. After taking the measurement on an unknown voltage source, retest a medium voltage detector on an energized circuit or with a piezo verifier for a known reading.

The three-step procedure prevents a malfunctioning medium voltage detector or blown fuse from allowing a false reading (or no reading) of an energized circuit. Having a test instrument that operates properly is important for any measurement, but absolutely required when taking medium voltage measurements.

To pretest a standard test instrument such as a multimeter, measuring the voltage on a known (energized) voltage source is easy because any standard receptacle (outlet) can be used. To pretest a medium voltage detector on a medium voltage system with hundreds, thousands, tens of thousands, or more volts is not practical or safe. Because of safety, medium voltage detectors are prechecked and postchecked using a test instrument that is specifically designed to safely test medium voltage detectors. **See Figure 7-3.**

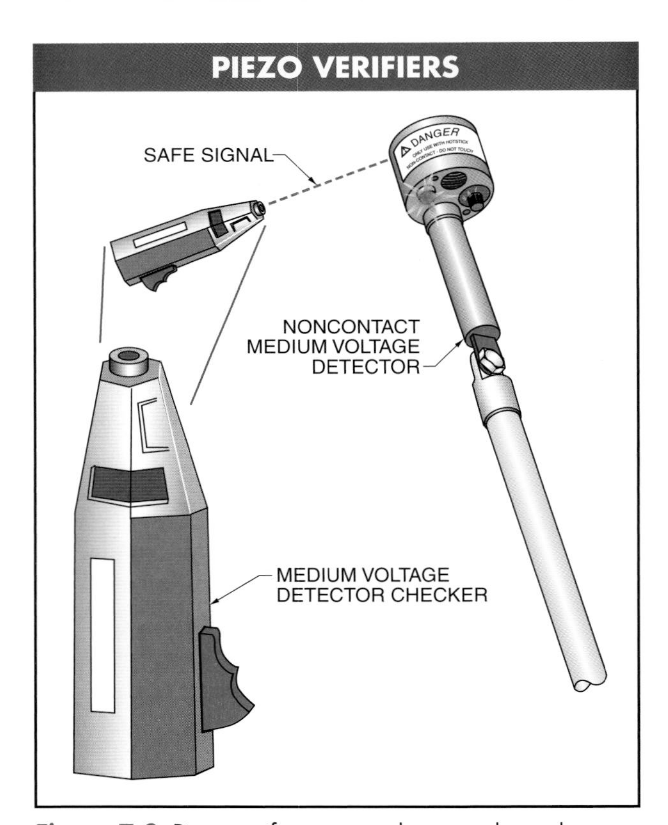

Figure 7-3. Piezo verifiers are used to test voltage detectors before and after use.

WARNING

Never use a medium voltage detector without first testing the detector with a piezo verifier that is specifically designed to pretest and posttest a medium voltage detector before and after the detector is used on actual circuits.

Technical Tip

Piezoelectricity is used for applications such as test instrumentation. Piezoelectricity is the ability of certain crystalline materials to produce a voltage when subjected to mechanical stress. This effect is also reversible as piezoelectric crystals, when exposed to an externally applied voltage, will change their shape by a small amount.

A *medium voltage detector checker (piezo verifier)* is a test instrument used to verify that a medium voltage detector is operating properly before any actual medium voltage measurements are taken. A piezo verifier sends out a safe signal that simulates a medium voltage for testing medium voltage detectors. When a medium voltage detector is operating properly, the simulated voltage signal is read and displayed as an actual measurement.

Noncontact Medium and Low Voltage Detectors

Voltage can be detected or measured. To detect voltage, a noncontact voltage detector is used that only indicates that voltage is present. A *noncontact voltage detector* is a test instrument that indicates the presence of voltage without displaying the actual amount of voltage present. Although a noncontact voltage detector can be used for troubleshooting, per NFPA 70E 120.5(7), Exception No. 2, "On electrical systems over 1000 volts, noncontact test instruments shall be permitted to be used to test each phase conductor."

Noncontact voltage detectors are available for low voltage detection (typically up to 600 V) and medium voltage detection. Medium voltage detectors detect voltages as high as thousands of volts.

Noncontact voltage detectors are used as:

- a first-line warning device that indicates the presence of voltage. After voltage is detected, a voltmeter that can read the actual circuit voltage must be used to measure the detected voltage.
- a basic test tool that is part of a preliminary troubleshooting procedure to detect when voltage is present at key points in a system. Key troubleshooting testing points include fuses and circuit breaker panels, pull boxes, and receptacles.

Noncontact voltage detectors must never be used as a device that is 100% reliable in indicating that there is no dangerous voltage present. When a noncontact voltage detector indicates that voltage is present, there is voltage and danger. When a noncontact voltage detector does not indicate that voltage is present, double-check the voltage source with another test instrument before physically coming in contact with the voltage source being tested. Also, use a device that indicates the level (115 V, 230 V, etc.) and/or type (AC or DC) of voltage present in a system.

Before performing tests using a noncontact voltage detector, ensure the detector is designed to perform a test on the system or wires being tested. **See Figure 7-4.** Refer to the operating manual of the test instrument for all measuring precautions, limitations, and procedures. Always wear required personal protective equipment and follow all safety rules when taking the measurement. To use a noncontact voltage detector to indicate the presence of voltage, apply the following procedures:

1. Pretest a noncontact voltage detector with a piezo verifier or on a known power source to ensure the noncontact voltage detector is operating properly.
2. Place the noncontact voltage detector near an area in which voltage might be present or an area scheduled for work.
3. When a noncontact voltage detector signals the presence of voltage, there is voltage present.
4. When a noncontact voltage detector does not signal the presence of voltage, do not assume there is no voltage; test for voltage with another test instrument to verify that no voltage is actually present.

WARNING

Best practice is to not assume that because a noncontact high voltage detector does not signal the presence of voltage, there is no voltage present. Noncontact voltage detectors are only the first line of defense in ensuring that power is not present. Always follow use of a noncontact voltage detector by testing for voltage with an actual medium voltage meter that measures and displays the actual voltage of a circuit. When using a medium voltage detector, always use a piezo verifier that is specifically designed to pretest and posttest a medium voltage detector.

Contact Medium Voltage Meters

All electrical circuits and systems are designed to operate at a set voltage level or voltage range. When the voltage level is not correct, there is a problem in the circuit or system. When power systems have too low or too high a voltage level, problems occur all along the electrical system. The voltage level in a circuit or system is important information for an electrician to know, no matter how low or how high the voltage is.

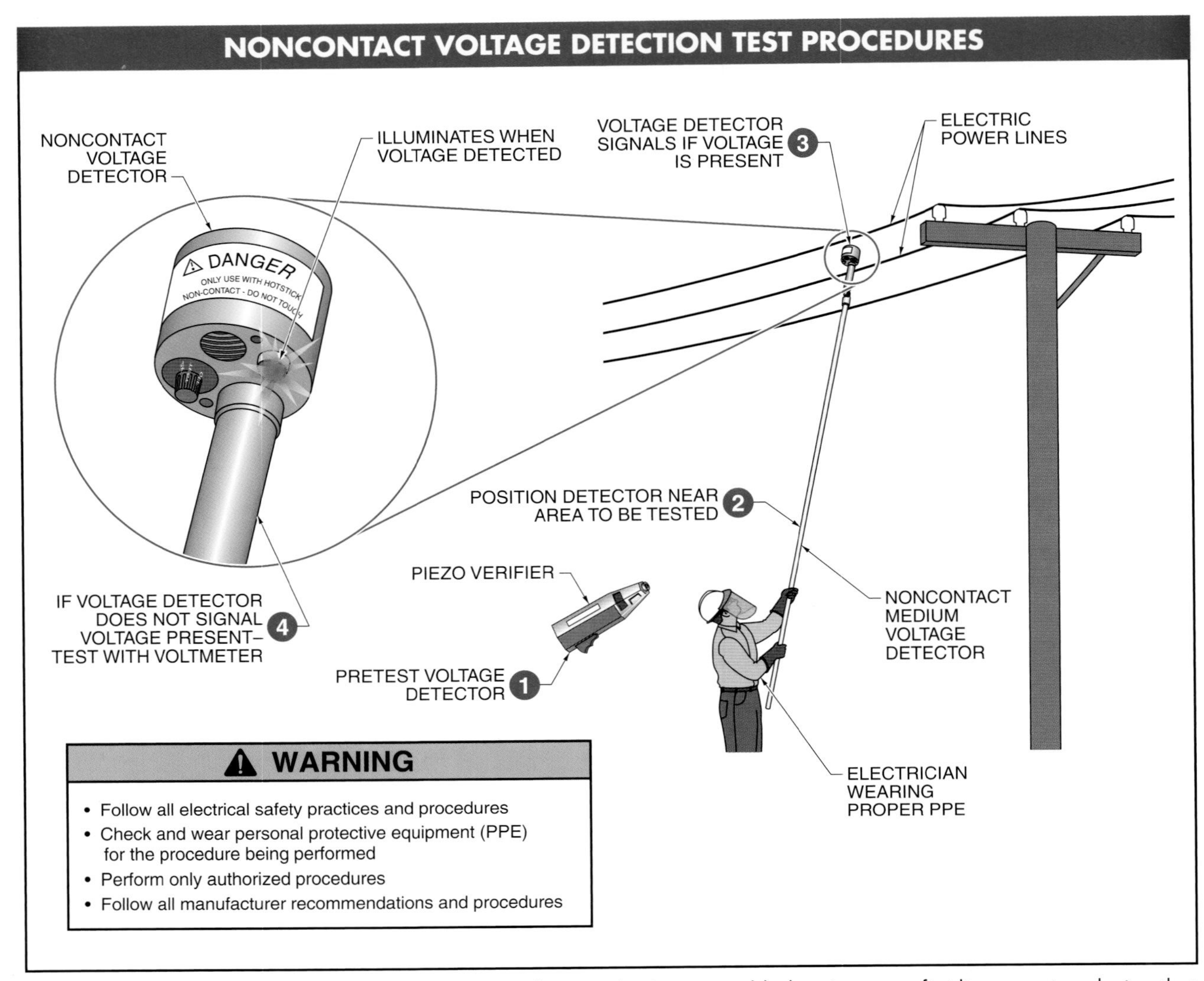

Figure 7-4. Noncontact voltage detectors are used primarily during troubleshooting as a first-line warning device that indicates the presence of voltage.

Medium voltage measurements are taken with contact medium voltage meters. A *contact medium voltage meter* is a voltmeter specifically designed to take voltage measurements on medium voltage cables. **See Figure 7-5.**

Contact medium voltage meters are used to do the following:

- verify that power has been removed before working on cables
- measure the actual voltage level on cables
- test fuses and other devices in electrical systems to measure the voltage level into and out of devices
- test grounding systems by taking voltage to ground measurements

Technical Tip

NEC® Article 450 can be referenced for medium voltage equipment installation and safety procedures.

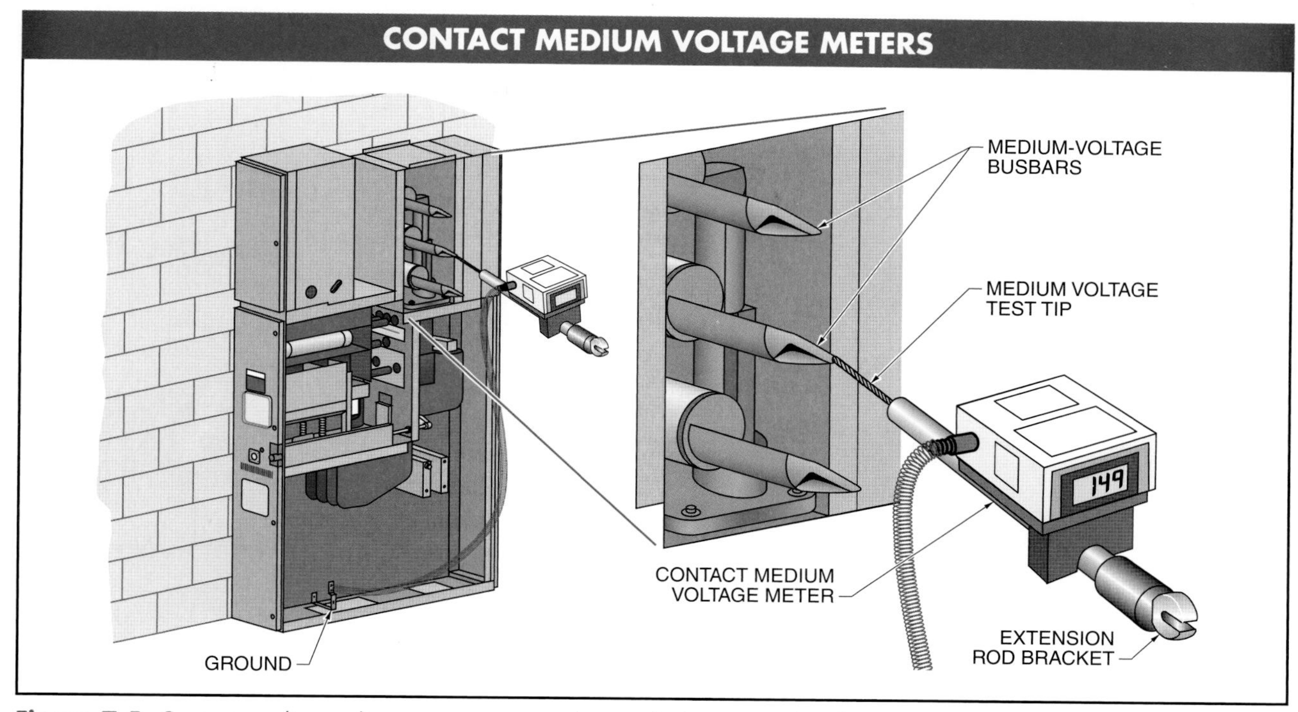

Figure 7-5. Contact medium voltage meters are used to verify that power has been removed before working on cables or to measure the actual voltage on cables.

Medium Voltage Test Probe

A *medium voltage test probe* is a DMM accessory used to increase the voltage measurement range above the DMM listed range. A medium voltage test probe has a very high resistance, which is added to the DMM input resistance circuit. A medium voltage test probe reduces the voltage to a DMM to approximately 1⁄1000 of the actual voltage present at the tips of the test probes. For example, 25,000 V (25 kV) at a test probe tip can be reduced with a medium voltage test probe to approximately 25 V at the DMM jacks. **See Figure 7-6.**

Medium voltage test probes designed to work with standard DMMs are different from contact medium voltage meters that are specifically designed to take medium voltage measurements on medium-power (medium current) cables. Medium voltage test probes used with DMMs are designed to take measurements on medium voltage, low current applications such as TV picture tubes and electronic ignition systems. Medium voltage test probes should not be used to take measurements in high current applications such as medium voltage power distribution systems, induction-type heaters, X-ray equipment, and broadcast transmitters.

Troubleshooting Tip

MEDIUM VOLTAGE, LOW CURRENT SAFETY PROCEDURES

When working with medium voltage, low current circuits with the power OFF, first discharge the power across large power-supply capacitors with a 2 W or greater resistor of 100 Ω to 500 Ω per volt of approximate value. Monitor the circuit while discharging and verify that there is no residual charge with a voltmeter or DMM set to measure voltage. For example, if working with medium voltage capacitors (used in power distribution systems to improve the systems' power factor, which increases the transmission capacity and reduces losses within the system), first discharge the capacitor contact. Use a 1 MΩ to 10 MΩ or 1 W or greater wattage on the end of an insulating stick or the probe of a medium voltage meter. Discharge to the metal frame connected to the outside of the capacitor.

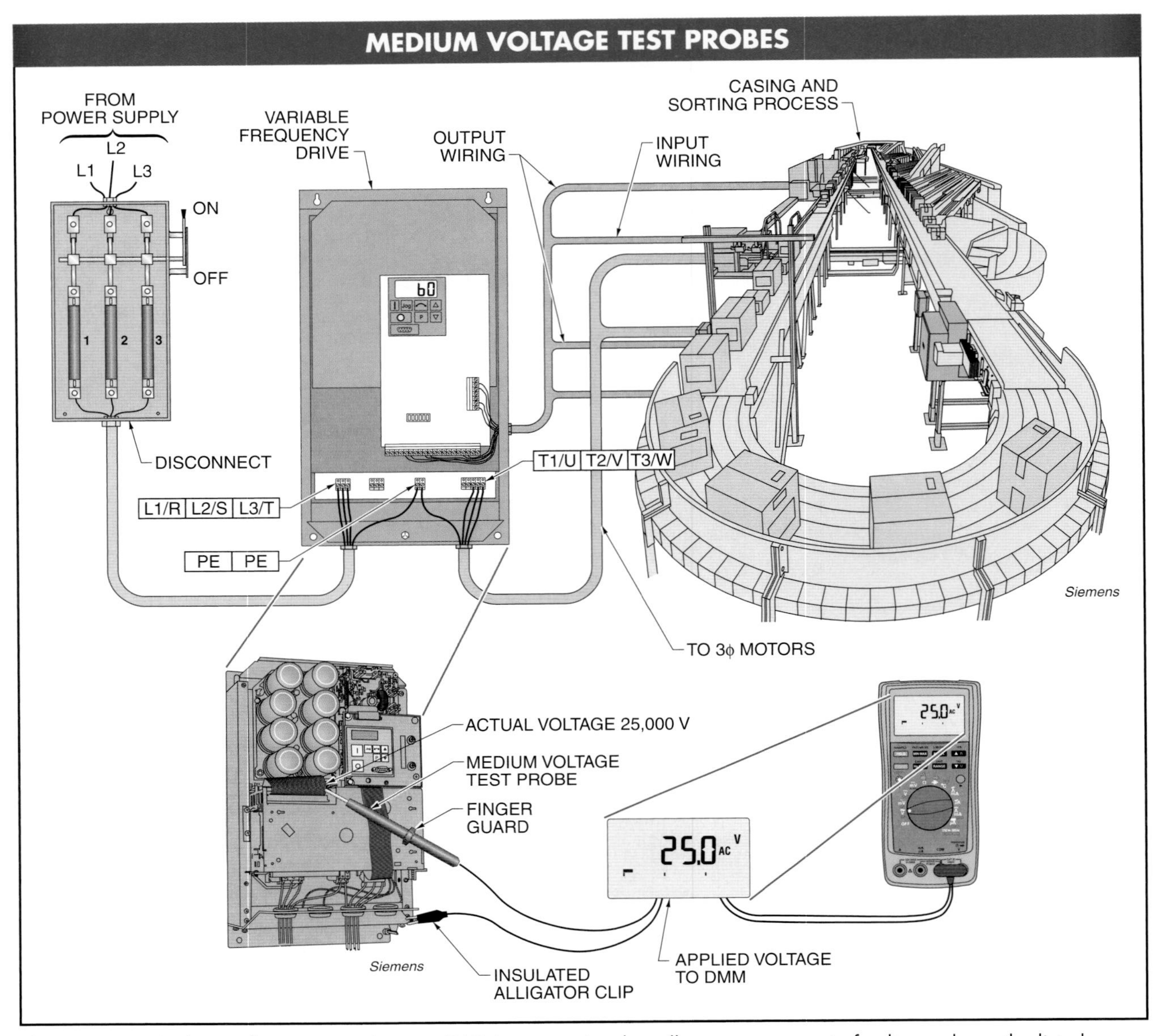

Figure 7-6. Medium voltage test probes are DMM accessories that allow measurement of voltage above the listed range of the test instrument.

Medium Voltage Phasing Testers

Three-phase circuits include three individual ungrounded (hot) cables. The three cables are identified as phase A (L1), B (L2), and C (L3). The cables must be maintained in the proper phase sequence in any three-phase distribution system. Accidental reversal of any two phases of a three-phase system can be disastrous. For example, motors depend upon proper phase sequencing to determine direction of rotation.

Phase sequencing can be tested on medium voltage cables by using a medium voltage phasing tester. A *medium voltage phasing tester* is a test instrument specifically designed to identify the three phases of a three-phase distribution system as L1, L2, and L3. Medium voltage phasing meters are typically included as a function of contact medium voltage meters. **See Figure 7-7.**

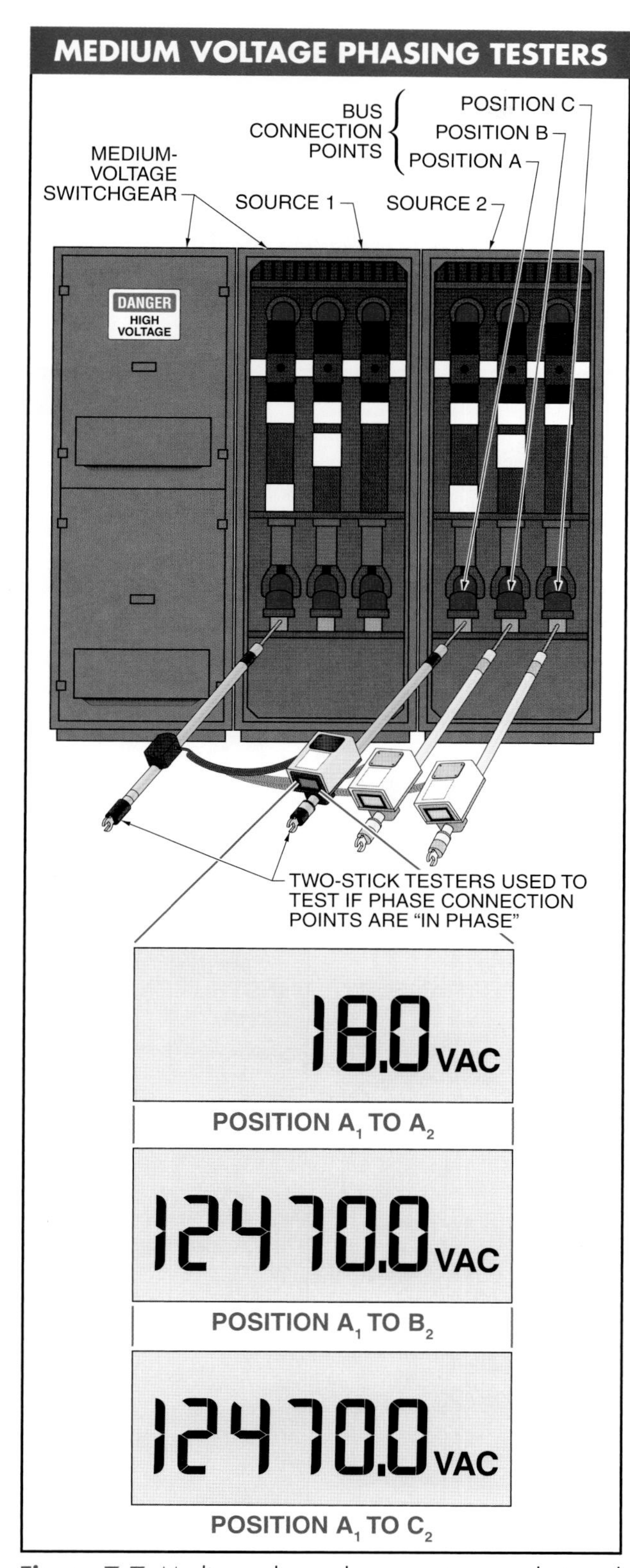

Figure 7-7. Medium voltage phasing testers are designed to identify the three phases of a three-phase distribution system as positions A, B, and C.

Noncontact Medium Voltage Ammeters

All electrical circuits and systems are designed to operate within a set current range. Once the high end of the current range is reached, the overcurrent protection devices (fuses and circuit breakers) of the circuit must remove power from the circuit. When the current level exceeds the conductor, component, or system current rating, there is a problem in the circuit or system. When troubleshooting a circuit, knowing the current level in the circuit is a must.

Medium current measurements are taken with noncontact medium voltage ammeters. A *noncontact medium voltage ammeter* is a test instrument specifically designed to take current measurements on medium voltage cables. Noncontact medium voltage ammeters are designed to be used as current clamps but have open ends (a U shape) to allow safer current measurements. **See Figure 7-8.**

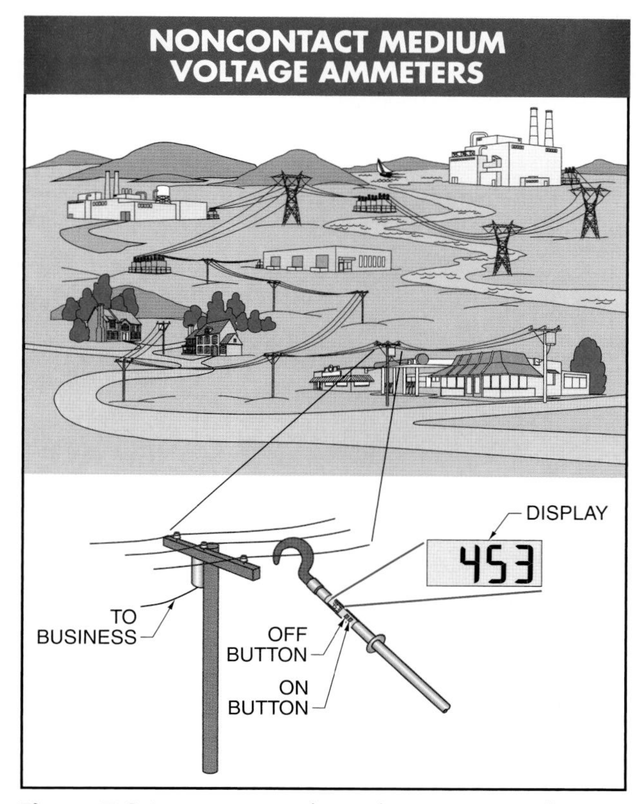

Figure 7-8. Noncontact medium voltage ammeters allow current measurements to be safely taken on medium voltage cables.

INSULATION RESISTANCE

No matter how large or how small a circuit, system, or electrical load is, electrical conductors (wires) are used to deliver the proper voltage, current, and power. Electrical conductors are always covered with some type of insulating material. The insulation prevents the current from flowing outside the designated path through the conductor.

No insulation can prevent all current flow through the insulation to ground or other conductors. In fact, some current (leakage current) flows through all insulation. Typically, leakage current is so small the current does not cause any problems and is ignored until the leakage reaches a point that the leakage starts causing electrical shocks, unwanted temperatures, or equipment damage.

The higher the resistance of insulation, the less leakage current there is flowing through the insulation. Insulation has the highest resistance when first placed into service on conductors and electrical devices. Ultimately, all insulation deteriorates over time and the resistance of the insulation decreases. Deterioration is caused by moisture, extreme temperatures, dust, dirt, oil, vibration, pollutants, and mechanical stress or damage. Even air spacing used as insulation between electrical parts will have a lower resistance over time because of dirt and moisture.

Insulation resistance typically becomes lower over time or drops suddenly. A sudden drop in insulation resistance is caused by damage such as cuts and breaks, or sudden environmental changes such as flooding, high temperature extremes, or contact with corrosive materials.

Knowing the condition of insulation is important no matter whether the insulation is the type used on conductors, or any other form of insulation used to separate energized electrical parts from each other. This includes such insulating devices such as the air spacing between electrical lugs and terminals, or any nonconducting materials (plastic and rubber) used to house electric appliances and tools so that an electrician is isolated from energized parts.

Knowledge of the actual condition of insulation can only be acquired by taking electrical measurements. Electrical measurements are taken between the conductors that carry current when energized and other parts of the system in which no current should ever flow under normal operating conditions. The two basic measurements that are taken to test insulation resistance are to measure the actual leakage current flowing through the insulation, and to measure the actual resistance of the insulation. **See Figure 7-9.**

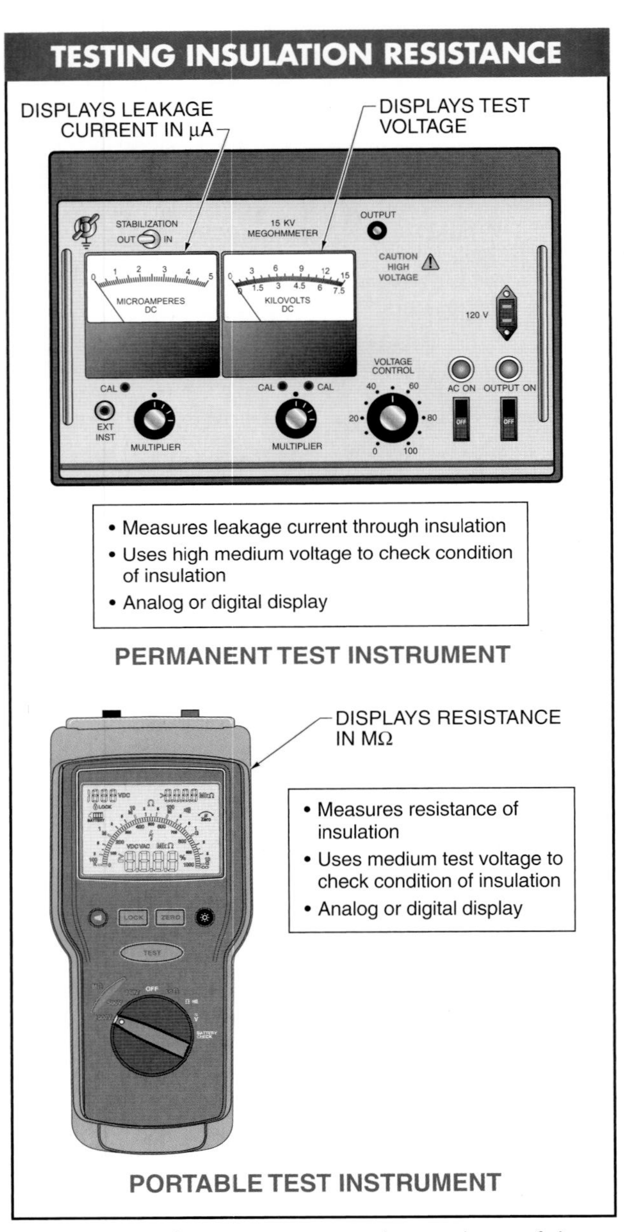

Figure 7-9. Electricians can test the condition of the insulation on conductors or other energized components by measuring leakage current or insulation resistance.

The best way to determine immediately if insulation is good and performing as required is to use a test instrument that applies a test voltage between the part of the circuit that will be carrying current and the parts of the circuit that are insulated from the current. The higher the measured flow of leakage current, the weaker (having less resistance) the insulating material is.

Leakage current measurements indicate the condition of insulation at the time of the test. However, the best way to keep track of the condition of insulation over time is to regularly take resistance measurements. Tracking insulation resistance can predict when insulation will fail and when an electrical device or circuit will fail. Preventive maintenance programs include tracking insulation resistance over time. Even when insulation resistance is not tracked over time, resistance measurements are taken to troubleshoot problems in a circuit.

Technical Tip

Insulation resistance values vary with temperature and the amount of moisture in the insulation. The temperature, humidity, and other similar factors should be recorded at the time the insulation resistance test is made.

General Resistance Rules

Ohm's law can be used to calculate the amount of current that will flow through a circuit. Ohm's law states that current (I) is equal to the voltage (E) divided by the resistance (R). Ohm's law also indicates that the higher the resistance of insulation, the lower the leakage current will be.

A general rule when measuring the resistance of insulation states that the absolute minimum total insulation resistance should be approximately 1 megohm (1 MΩ or 1,000,000 Ω) for each 1000 V (1 kV) of operating voltage. Using the general rule, leakage current through insulation must not be greater than 1 mA (0.001 A).

$$I = \frac{V}{R}$$

$$I = \frac{1000}{1,000,000}$$

$$\mathbf{I = 0.001\ A\ (1\ mA)}$$

The 1 MΩ per 1 kV rule is a minimum resistance rule. In practicality, insulation has a much higher resistance and is often required by codes and standards to be higher for most electrical devices. Resistance values of 10 to 100 times the minimum are preferred and are often required. **See Figure 7-10.**

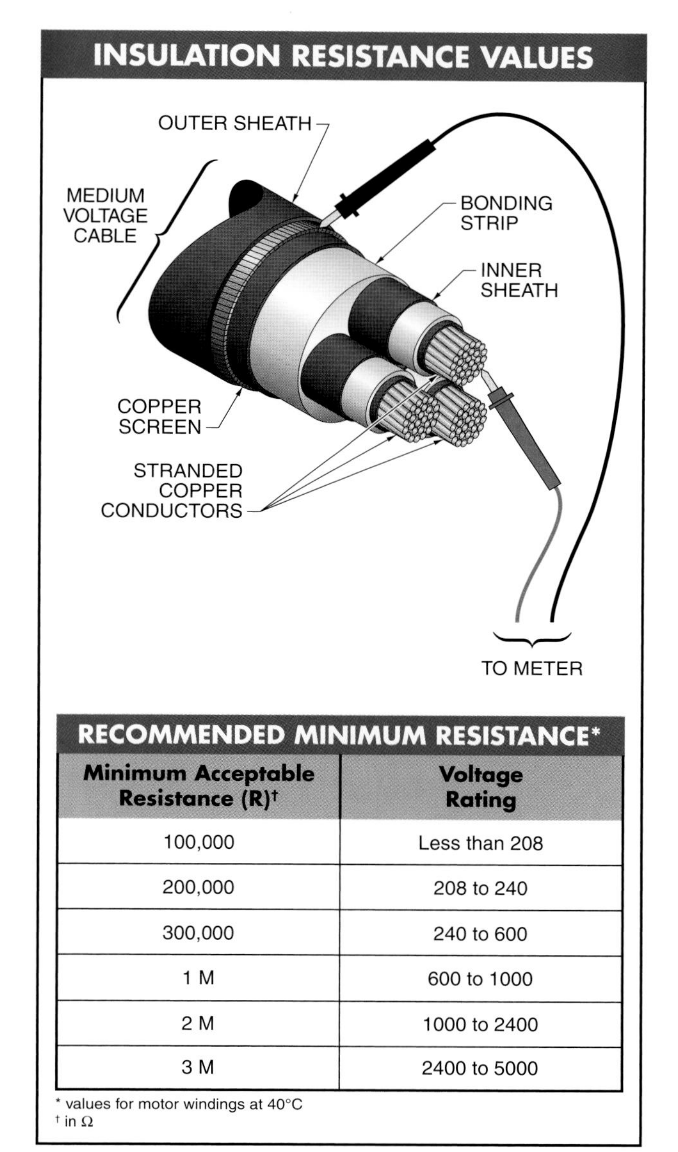

RECOMMENDED MINIMUM RESISTANCE*

Minimum Acceptable Resistance (R)†	Voltage Rating
100,000	Less than 208
200,000	208 to 240
300,000	240 to 600
1 M	600 to 1000
2 M	1000 to 2400
3 M	2400 to 5000

* values for motor windings at 40°C
† in Ω

Figure 7-10. A general rule for insulation is that the resistance of the insulation should be approximately 1 MΩ for each 1000 V of operating voltage.

Required Resistance

Although the general rule for required resistance can be applied when troubleshooting most motors and other electrical devices, there are times when higher resistance specifications must be applied. Most stated specifications list the maximum amount of acceptable leakage current and not the actual resistance. For example, a 3-wire handheld electrical appliance or tool must be insulated enough so as to allow no more than 0.75 mA (0.00075 A) of leakage current to flow through the exposed parts to ground. Per Ohm's law, the resistance required to limit current flow to 0.75 mA on a 120 V appliance would be 160,000 Ω, or 1,333,333 Ω when using 1000 V. Typical resistance values are one-third higher than the 1 MΩ per 1 kV rule. Medical equipment and electrical devices rated as "double-insulated" have much higher insulation ratings.

INSULATION TESTING

There are several reasons for testing the resistance of insulation on conductors, electrical parts, circuits, and components. Insulation resistance measurements are taken to verify quality control of manufactured electrical devices, ensure electrical devices meet codes and standards (safety compliance), determine a device's performance over time (preventive maintenance), and determine causes of failure (troubleshooting). Resistance testing is classified into the following kinds of tests:

- design test
- production test
- acceptance test
- verification test
- preventive maintenance test
- fault locating test

Design Tests

Manufacturers of electrical devices such as appliances and tools, and components such as wires and fuses want to manufacture safe and reliable products. Meeting minimum required codes and standards helps ensure products are built for safe operation. Exceeding minimum requirements helps establish the reputation of a company for producing reliable and quality products. Designing and building to the highest standards also helps reduce product liability issues that may arise from electrical shocks, fire, and injuries.

As a product is being developed, mechanical, functionality, and electrical design tests are performed. Electrical design tests are typically conducted in a laboratory to determine how electrical components will perform. Insulation resistance levels are tested prior to manufacturing any product.

Design tests are performed on newly designed components or on parts purchased from other companies that are being incorporated into the design of a product. Design tests test a component to failure. To test insulation, a medium voltage is applied to each component until the insulation of the components fails and conducts a higher than acceptable leakage current. **See Figure 7-11.**

Although a component is destroyed during the test, the electrical limits of the component are established. Knowing a component's limits and keeping records of test results is part of good design work. Design testing must not only be performed when first designing a product, but also whenever there is a product modification. Many component failures are traced back to the failure of substitute parts that were used because of cost cutting or from a change in suppliers.

Technical Tip

Design standards include specifications of materials, physical measurement processes, performance of products, and services rendered. The purpose of design standards is to obtain operational/manufacturing economies, increase interchangeability of products, and promote uniform product characteristics. Trade associations, national or international standards organizations, and individual manufacturers establish these standards.

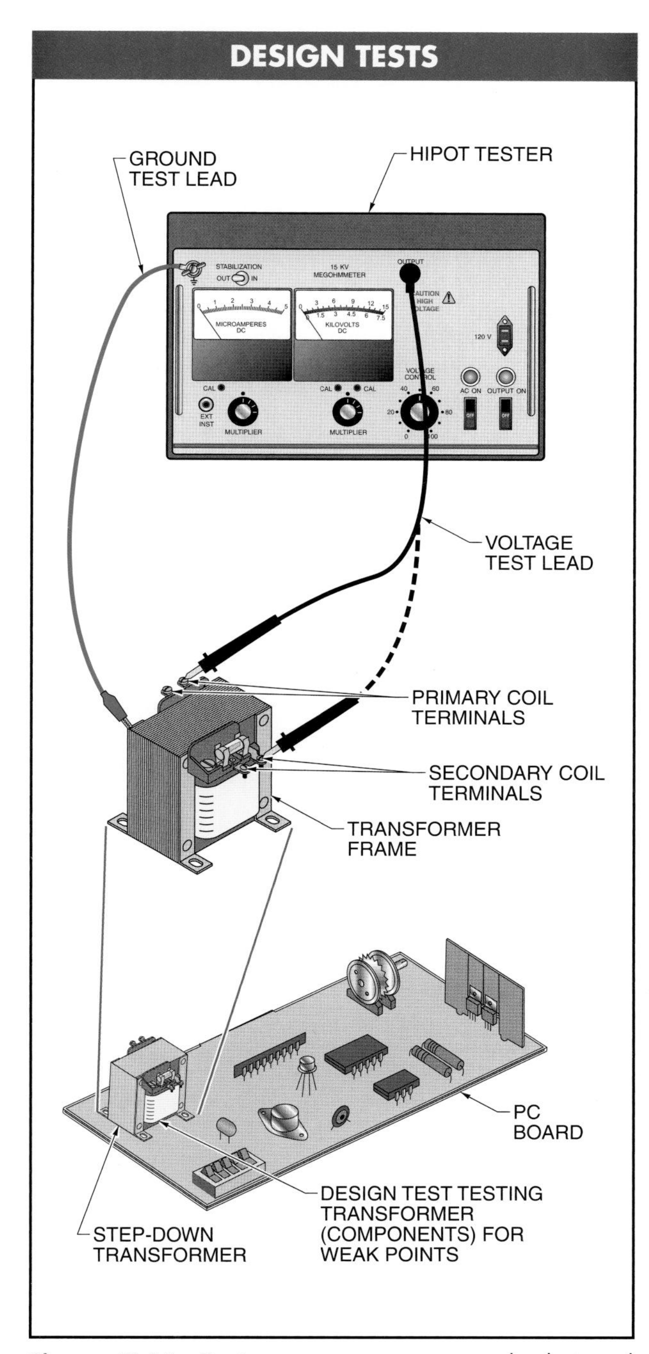

Figure 7-11. Design tests are run on newly designed components and test the component to failure.

Production Tests

Once a designed product is accepted, the product is placed into production. Production requirements are different from design requirements in that the production process must take into consideration building the product as quickly and economically as possible. To ensure what works in a lab also works to specifications after production, production tests on individual products must be made.

Production tests are performed to meet codes and standards and ensure quality control. Product defects typically start showing up during product testing. Product testing can be of the nondestructive or destructive type. **See Figure 7-12.**

Nondestructive testing applies tests to ensure minimum standards are being met. For nondestructive testing, the tests are typically performed at a level higher than the minimum requirements but not so high as to intentionally destroy the device. When a product meets the minimum standards, the product is placed back on the production line.

Destructive testing applies tests that are designed to push a product to failure. The failure limit of a component is determined and documented. Destructive testing establishes the weakest components and parts of a device. Some failures are associated with production methods, such as electrical components being placed too close together or damaged during production.

Independent testing laboratories such as Underwriters Laboratories® (UL) allow insignias to be placed on products. The UL requires that production tests be performed to ensure that the product continues to meet UL standards. An independent testing laboratory performs the initial test on a product before issuing an approval. Continued product approval requires documented proof of product safety and reliability.

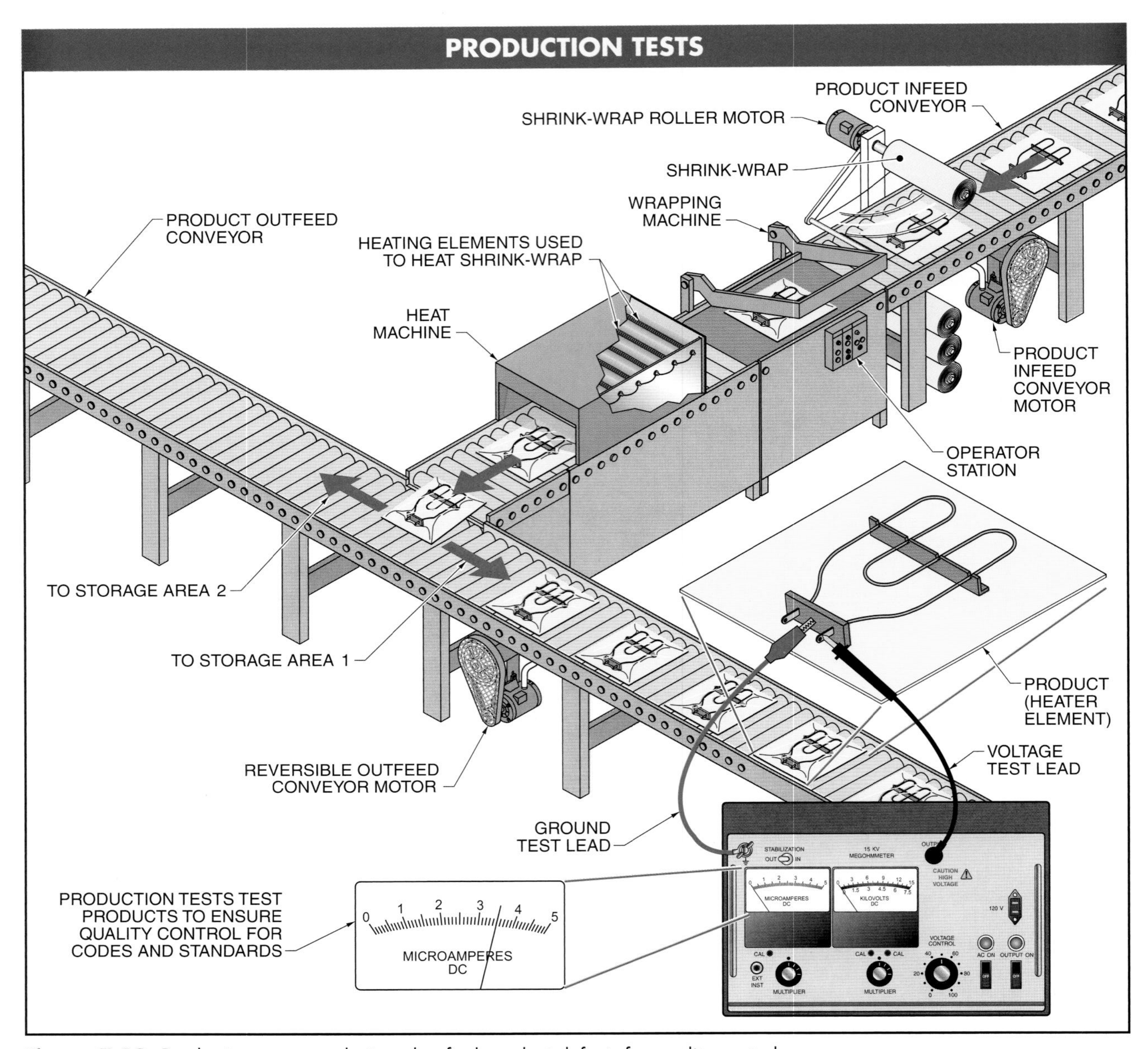

Figure 7-12. Production tests are designed to find product defects for quality control.

Acceptance Tests

After electrical equipment leaves the manufacturer the equipment is shipped and installed. During storage, shipping, and installation a part of a component or circuit can be damaged. Acceptance testing is performed immediately after installation but before the system is put in service. Acceptance testing is used to test for damaged equipment, installed cable damage, and proper installation (spacing, tightening) of electrical components. **See Figure 7-13.**

Design and production testing are performed by the manufacturer. The contractor installing the equipment and the owner of the equipment perform acceptance testing. The contractor performs acceptance testing to ensure the equipment is safe and installed correctly. The owner of the equipment performs acceptance testing to ensure that what is being installed and paid for is what was specified and purchased.

TESTING MV CIRCUIT BREAKER

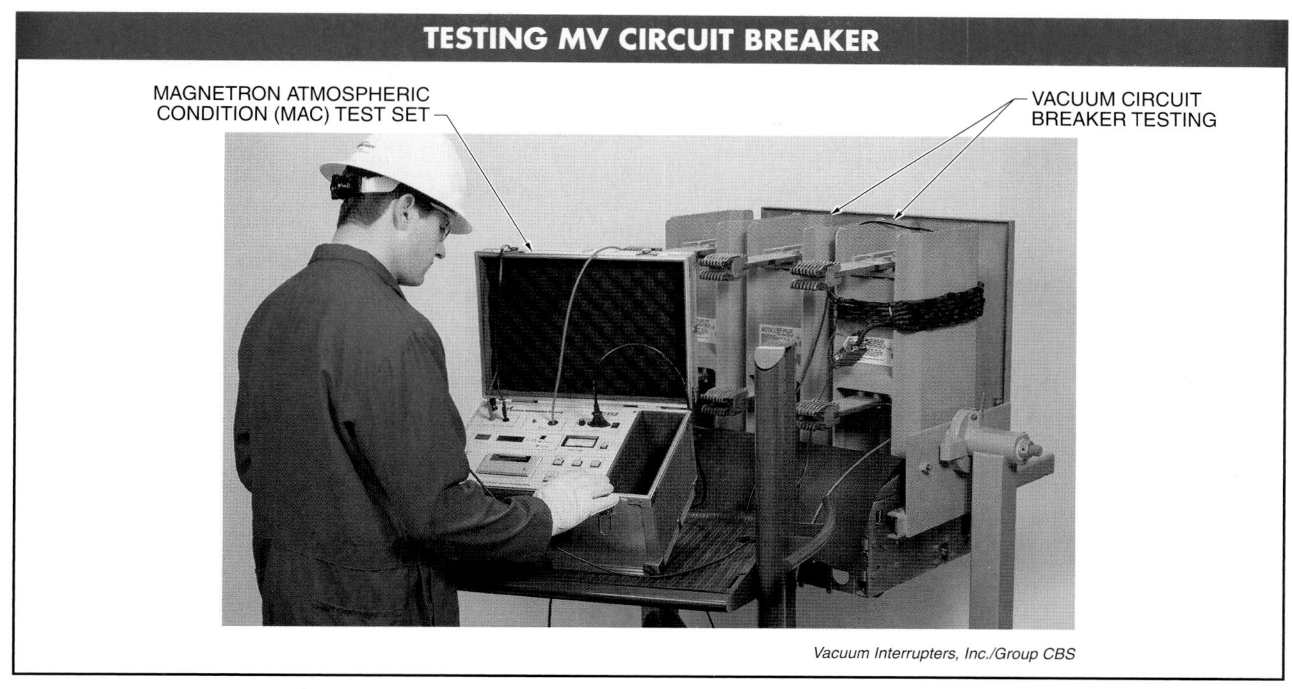

Vacuum Interrupters, Inc./Group CBS

Figure 7-13. Acceptance tests are performed immediately after equipment installation but before the system is put in service to locate damage caused by shipping, storage, and installation.

Technical Tip

Consulting engineering firms perform acceptance tests on such areas as electrical systems, concrete, and structural steel.

Greenlee Textron, Inc.

Cable length testers can measure the length of electrical and communication cables and can also indicate the distance to a fault in the cable (open or short), given access to only one end of a cable with two or more conductors.

Verification Tests

An acceptance test only indicates that the equipment of a system has been safely installed and not damaged. However, once a system is fully operational, potential problems may occur that were not considered. Potential problems can include the following:

- Additional loading of the system was not considered when the system was designed. In general, more loads than expected are added to an electrical system over time.
- Load types that produce unforeseen problems are added to the system. Load types include harmonic-producing loads as well as loads that produce high inrush currents and transients.
- Environmental conditions were not originally factored in, or have changed. Environmental conditions include increased temperature levels, dirt, dust, corrosive materials, and poor housekeeping.

To ensure equipment and circuits are safely and properly working after the equipment has been placed in operation, a verification test is performed. Verification tests are typically performed shortly after the equipment or circuit is placed in full operation and before the warranty or guarantee has elapsed.

Preventive Maintenance Tests

A good preventive maintenance program can detect and eliminate problems before problems create downtime. A main part of an electrical preventive maintenance program is the testing of equipment and conductor insulation. The condition of conductor insulation is a good indication of the condition of the equipment and electrical system in general.

Failing insulation must be corrected so that a system does not fail at an inopportune time. In general, as any system is operated over long periods of time, conductor insulation quality deteriorates at a predictable rate. By taking resistance measurements over time, conductor insulation failure (or expected life) can be predicted. **See Figure 7-14.**

Measurements over time are required because there are no two operating conditions that are the same. For example, a light fixture with an operating lifetime expected by the manufacturer to be 10 years may last less than one year, or more than 20 years. The worse the environmental conditions a light fixture must operate in, the shorter the operating life. The higher the ambient temperature a light fixture must operate in, the shorter the operating life.

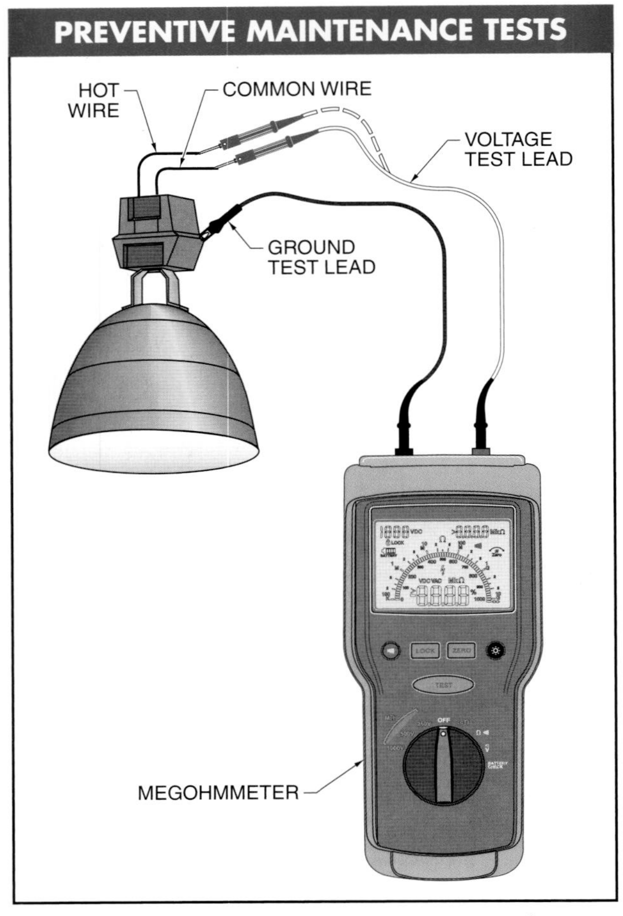

Figure 7-14. Preventive maintenance tests take resistance measurements of a device or conductor over time to predict failure.

Troubleshooting Tip

PREVENTIVE MAINTENANCE RECORDKEEPING

By maintaining accurate preventive maintenance records as reference guides, future problems with similar equipment can be alleviated.

Fault Locating Tests

Even when equipment is manufactured to high specifications, properly installed, correctly sized, and preventive maintenance is performed, equipment does fail. Failure typically occurs because one weak or damaged part of a circuit fails, or an entire system fails because the end of the system insulation life has been reached. When a component, equipment, circuit, or system fails, troubleshooting is performed to locate the fault. Troubleshooting requires knowledge of the equipment, circuit, and test instruments needed to perform the troubleshooting. Knowledge of insulation testers is essential because insulation testing involves testing a circuit or component using high voltage.

MEASURING INSULATION RESISTANCE

A standard ohmmeter such as on a digital or analog multimeter can be used to measure the resistance of insulation by taking a resistance measurement from one conductor to another conductor or to ground on a de-energized circuit. However, the resistance measurement may not be accurate for two basic reasons:

1. A standard ohmmeter measures resistance by applying a low voltage from the batteries of the meter through the component being tested. The resistance measurement taken is based on the meter battery voltage, typically 6 VDC or 9 VDC, which is applied to the component or conductor. When testing insulation resistance, the actual voltage that the device will operate at, or a higher voltage, must be used to measure the resistance. The closer the test voltage is to the actual voltage used in the circuit, the more accurate the resistance reading will be. For this reason a test instrument that measures resistance at a medium voltage must be used when testing the condition of insulation.
2. A standard ohmmeter is designed to take measurements that typically range from zero ohms to several million ohms. However, the resistance of good insulation between conductors and ground must be millions of ohms, if not tens of millions of ohms. To measure insulation resistance a test instrument that is designed to measure extremely high resistances must be used.

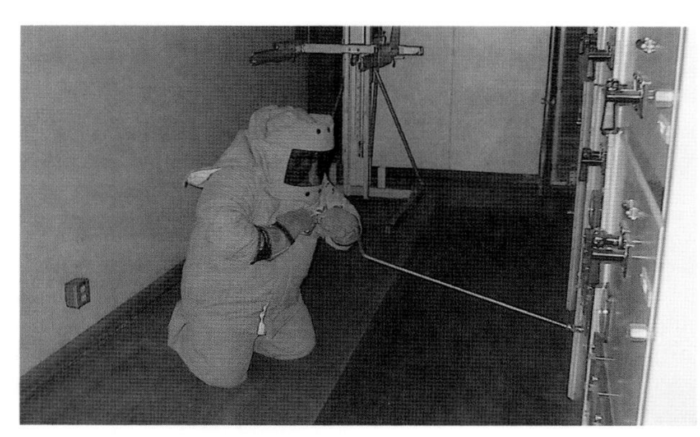

Proper PPE must always be worn when working with any type of medium voltage equipment.

Technical Tip

The properties of electrical insulation materials include dielectric strength, impulse strength, resistivity, dielectric loss, dielectric constant, tensile strength, chemical resistance, moisture resistance, and flammability (or the ability to self-extinguish).

There are two basic test instruments used to measure insulation resistance. The two test instruments are hipot (high potential or high [medium] voltage) testers and megohmmeters. **See Figure 7-15.** The major differences between a hipot tester and a megohmmeter are:

- A *hipot tester* is a test instrument that measures insulation resistance by measuring leakage current. Hipot testers apply a high (medium) test voltage between two different conductors or between a conductor and ground and measure the leakage current. An excessive amount of leakage current (in amps) indicates a lower resistance or breakdown in insulation. Hipot tests almost always involve applying a test voltage that is several times higher than the specified operating voltage of the cable or device being tested.
- A megohmmeter performs the same basic test as a hipot tester in that the meter also applies a medium test voltage to the circuit or component being tested, but displays the measurement as resistance in ohms (typically MΩ).
- A hipot tester indicates the condition of insulation by displaying the amount of leakage current through the insulation and a megohmmeter indicates the condition of insulation by displaying the amount of resistance that the insulation actually has when a medium voltage is applied to the conductors.

Hipot testers and megohmmeters are available in both digital and analog versions. Hipot testers are also called dielectric testers, production line testers, and hipot and current leakage testers. Megohmmeters and some hipot testers are also called insulation testers. Also, because both hipot testers and megohmmeters perform high voltage tests on insulation, some models are available that include both testers in one.

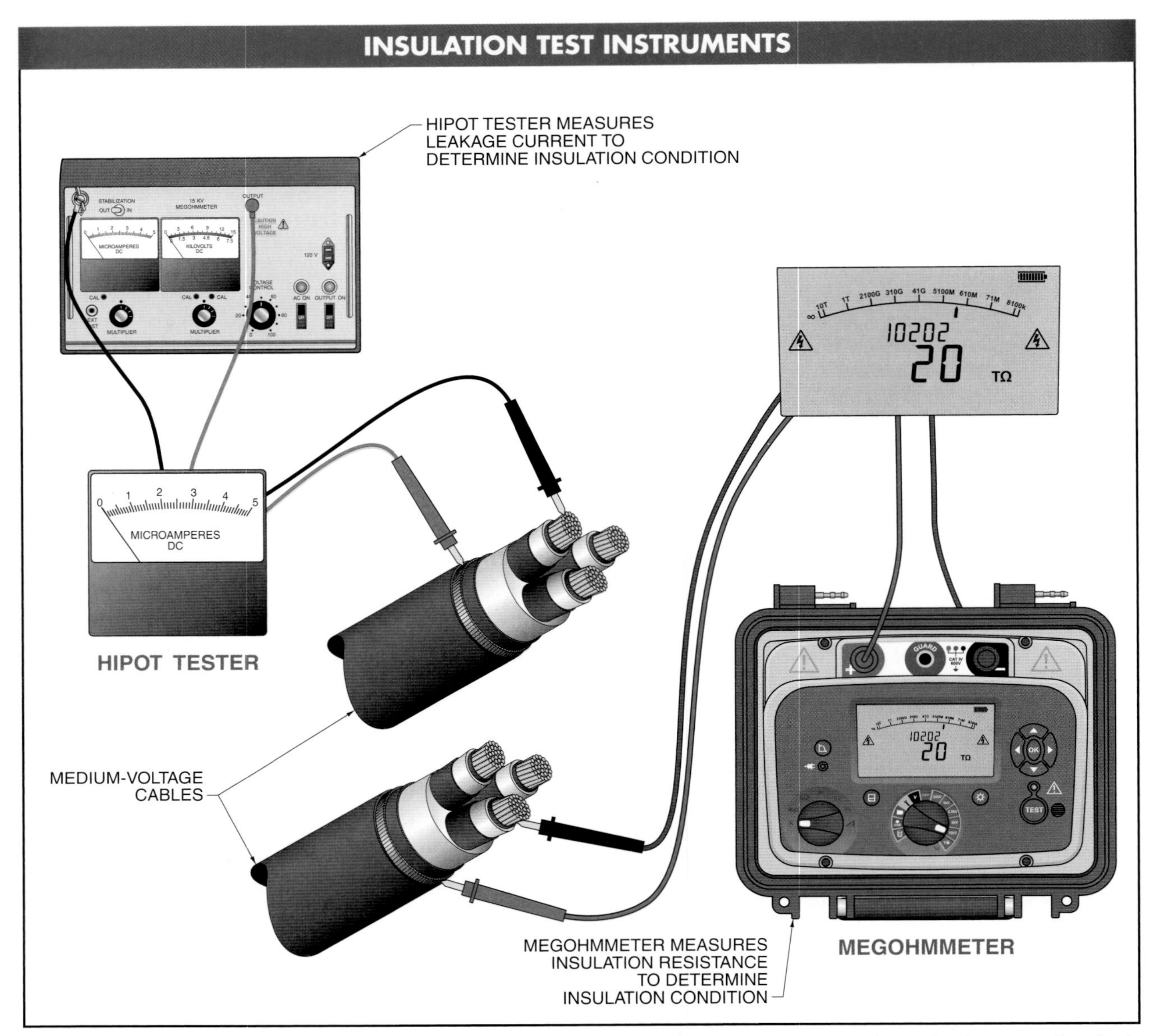

Figure 7-15. The two test instruments typically used to measure insulation resistance are hipot (high potential or high [medium] voltage) testers and megohmmeters.

Troubleshooting Tip

HIPOT TESTER SPECIFICATIONS

Specifications to consider when specifying hipot testers include AC and DC output voltage, AC and DC output current, resistance range, insulation limit, and test time. Output voltage can be either AC or DC. The type of test will determine what level of voltage will be supplied. Output voltage when hipot testing is usually very high. Output current can be either AC or DC. The type of test will determine what level of current will be supplied. In hipot testing the output current is usually very low. Testers can also measure the resistance range. The insulation limit specification determines the voltage range the insulator can withstand. Many tests performed for electrical safety have a test time requirement. This is the overall time of the test.

HIPOT INSULATION TESTING

Ohm's law states that a relationship exists between voltage, current, and resistance in an electrical circuit. Voltmeters are used to measure the voltage in a circuit. The measured voltage is accurate as long as the voltage being measured is within the measuring range of the meter. Ammeters are used to measure the current in a circuit. The measured current is accurate as long as the current being measured is within the measuring range of the meter. Ohmmeters used to measure resistance may show different values because a resistance measurement taken at one battery voltage level may not be the same resistance value measured when a different battery voltage level is used. Specific conditions require a more selective resistance testing method when trying to accurately take certain resistance measurements.

To solve the varying measurement problem, a tester that can accurately determine the true resistance of a circuit or insulation using normal operating voltage must be used. A hipot tester uses Ohm's law to determine the resistance of insulation by applying a voltage to the circuit or device being tested and measuring the current flowing through the insulation. The current flowing through the insulation is dependent upon the resistance of the insulation.

WARNING

Hipot test voltages can cause severe or fatal shocks.

ABB Motors and Mechanical Inc.

Hipot insulation testing is typically preferred on medium voltage cables and high horsepower motors.

Hipot Testing

A *hipot test* is a test performed on a product or circuit with a hipot tester to ensure that there is no chance of an electrical shock or that the component was not damaged during installation. The purpose of a hipot test on appliances and tools is to ensure that a person does not receive an electrical shock during normal operation of the device. The purpose of a hipot test on cable is to ensure the cable was not damaged during installation and will perform as required once the normal circuit power is applied.

A hipot test intentionally overstresses insulation by applying a much higher than normal voltage. The reason for the much higher voltage is to detect defects, damage, and other weaknesses that are present. When a weak or damaged point occurs in a circuit or cable, arcing will occur at the weak or damaged point.

There is a concern that a test performed with a hipot tester or megohmmeter at a medium voltage will damage the device or cable being tested by an arc that occurs at a weak point. In fact, the arc will damage, or further damage, the device or cable being tested when there is a weak or damaged point. The point of a hipot test is that when there is a problem the problem is found under controlled conditions rather that at some random time. There is never a good time for a fault to cause downtime or an electrical shock.

A medium test voltage should not, under normal circumstances, damage a device or cable in good working order because the device should have a very high margin of safety built into the insulation of the device. All electrical insulation must be able to withstand a much higher test voltage than the maximum listed operating voltage rating. For example, a conductor rated at 600 V is not going to break down at 601 V, or even at a voltage several times higher than the 600 V rating.

Some testing agencies require 100% hipot testing of every electrical product produced before the product can be sold. By doing so, no product that could cause an electrical shock from damage caused during production or by a faulty component will leave the factory. Even when not required by an outside testing or regulating agency, many manufacturers hipot test every electrical device produced.

Also, cables that are installed underground or above ground are typically tested to ensure there are no weak points before the actual power is connected and applied. When a device being tested passes a hipot test, the device is unlikely to cause an electrical shock or have a cable fail during the normal expected life of the device.

AC or DC Hipot Testing

A hipot test can be performed by applying AC or DC voltage to the product being tested. Choosing whether to use AC or DC voltage for the test depends upon the results desired, regulating and certifying agency requirements, manufacturer requirements or recommendations, the tester's preference and experiences, and what the customer wants. In general:

- When a product is designed to operate from an AC power source, an AC hipot test is used.
- When a product is designed to operate from a DC power source, a DC hipot test is used.
- When a product such as cable is designed to operate with either AC or DC voltage, both AC and DC hipot testing must be used.
- Regardless of whether a product is designed for AC or DC voltage, the requirements of any regulatory agencies involved with the product must be followed.
- Both AC and DC testing can damage a product if manufacturer guidelines for using the test instrument are not followed. Even when guidelines are followed, damage occurs when there is a fault or damage in the product being tested. Everyone involved—the customer, installing contractor, or regulating agency—must understand that a hipot test is designed to find faults and when a fault is found repairs must be made.

CAUTION

A hipot tester uses medium voltage for testing. Always follow the recommended procedures and safety rules of the manufacturer. After performing insulation tests with a hipot tester, discharge the device being tested using the discharge function built into the tester or specialized discharge jumpers.

DC Hipot Testing

When performing a DC hipot test, a DC voltage is applied to the product being tested. The voltage is applied in ever increasing steps, with time intervals between each step. Anytime DC voltage is applied between a conductor and insulator, there will be an electrical charge held between the conductor and insulator (principle of capacitance). The time intervals between voltage increases allow the charge to even out. The time intervals are required because, when a new charge is first applied to the conductor and insulator, a high current is initially drawn that soon stabilizes. Any current flowing after the current has stabilized is the leakage current through the insulation. On long cable runs, the reading will be delayed because of the effects of the longer charging time. **See Figure 7-16.**

CAUTION

After performing a DC voltage hipot test, the device being tested is charged and can cause an electrical shock. To prevent an electrical shock, the device being tested must be discharged. Hipot testers typically include an automatic discharge feature that discharges the test voltage to ground. A voltmeter must be used after discharging to verify there is no longer a charge on the device.

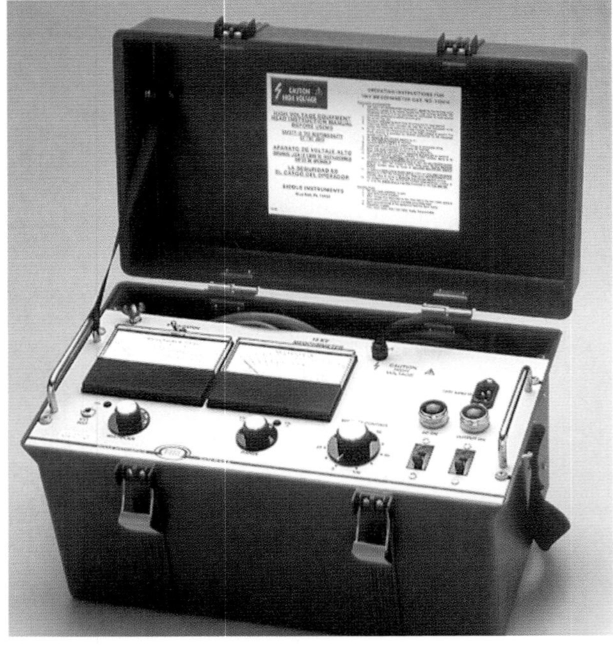

Megger Group Limited

Hipot testers are instruments that test for insulation integrity between the conductor and the insulation shielding.

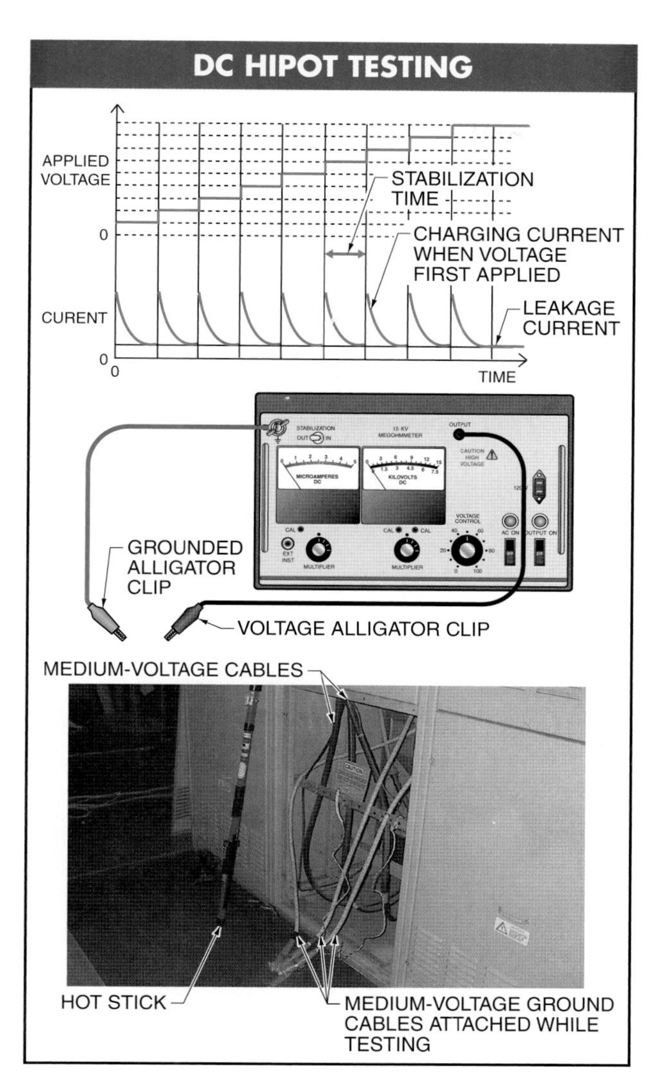

Figure 7-16. DC hipot testing requires that DC voltage be applied to a product in ever-increasing steps with time intervals between each step.

When the measured leakage current increases above the specified limits or starts increasing sharply, the insulation has reached the breakdown point (failure point). At the failure point the test can be stopped without damaging the device being tested. The actual applied voltage and time interval between voltage steps are determined by following the recommended test procedures of the hipot manufacturer for the tester being used. Hipot testers are available that can be programmed to automatically perform a test through each step and discharge any stored electrical charges.

Troubleshooting Tip

HIPOT TESTER FEATURES

Features found on hipot testers include built-in calibration, warning buzzer, front panel lockout, memory or storage capability, multiple test setup, PLC interface, rapid shutoff, remote control, selectable output frequency, and warning indicator lights. Displays available with hipot testers include analog meters, digital meters, and LED indicators. Interfaces include GPIB, RS232, printer ports, scanner ports, and printouts.

AC Hipot Testing

When performing an AC voltage hipot test, an AC voltage is applied to the product being tested. When applying AC voltage, the voltage must not be applied in steps for most equipment as with DC hipot testing. AC voltage hipot testing eliminates the longer test times required with DC testing in which the DC voltage is applied in increasing steps with time intervals between steps. Also, an AC hipot test does not require the discharging of stored energy after the test is finished.

However, there are disadvantages to AC hipot testing. AC hipot testing can be more damaging to insulation. Unlike DC hipot testing in which a point is reached at which the insulation breakdown can be detected and the test stopped before damage is done, AC hipot testing typically indicates only a pass or fail of the product. When a product fails, the damage has already occurred. Product failure is not necessarily bad because the weak point that failed would have failed anyway over time. The general thought is that there is no good time for any failure, but being able to predict a failure during testing allows for a controlled failure. The ability to predict failure is why hipot testing is performed before a product is placed in service or sold and during scheduled preventive maintenance shutdowns.

To prevent damage during an AC hipot test, some hipot testers actually apply AC voltage at a very low frequency (VLF). A low frequency AC voltage of less then 1 Hz is typically used instead of the standard 60 Hz AC power. The lower frequency causes less damage to cables and allows longer cables to be tested.

Hipot Test Voltages

A hipot test uses a high (medium) voltage to intentionally stress the insulating ability (dielectric) of insulation. A hipot test applies a medium voltage between conductors or between the device being tested and ground to verify that the insulation can provide an electrical barrier sufficient to prevent a problem. Test voltages are much higher than normal operating voltages. Typical test voltages used to test electrical appliances and tools are as follows:

- For AC operated circuits (Class 1) with a ground, the recommended test voltage is typically 1000 VAC plus twice the normal operating voltage. For a 120 VAC rated device the test voltage would be 1250 VAC. Also, for a 240 VAC rated device the test voltage would be 1500 VAC.
- When an AC voltage operated device is tested using DC voltage, the AC test voltage must be multiplied by 1.414 (rms to peak value) to acquire the proper DC voltage level for testing. For example, a 1250 VAC test voltage would be 1768 VDC.
- For AC voltage operated circuits without a ground (Class 2 – double-insulated), the required test voltage is typically higher. Test voltages of 2000 VAC or more are often specified for 120 VAC operated devices.

The test voltage is applied by a test instrument with the device being tested disconnected from any external power source. The test voltages are applied between the parts of the product which an electrician may come in contact with under normal operating conditions and the electrical conductors which would be applying power to the device.

MEGOHMMETER INSULATION TESTING

A hipot test is typically performed after a product is manufactured to ensure there are no manufacturing faults and after a product (typically cable) is installed to ensure there was no damage during installation. A hipot tester does not directly measure insulation resistance, but rather determines if insulation is good or bad by measuring the amount of leakage current through the insulation.

When the actual resistance value of insulation is to be measured, a test instrument that measures resistance must be used. Since insulation resistance is extremely high, the test instrument (megohmmeter) must be able to measure and display resistance values in the millions of ohms (MΩ).

Electrical Apparatus Service Association, Inc.

Megohmmeters can be used for testing defective motor winding insulation.

Megohmmeters

A *megohmmeter* is a high-resistance ohmmeter used to measure insulation deterioration on various wires by measuring high resistance values during medium voltage test conditions. A megohmmeter forces a medium DC voltage into a conductor or motor winding being tested and measures the leakage current through the insulation to calculate the resistance of the insulation. A megohmmeter measurement is displayed in megohms, which are calculated using Ohm's law.

Megohmmeters are manufactured in a variety of styles. Most megohmmeters have a function switch or selector switch to choose the appropriate test voltage. The megohmmeter display can be an analog display or a digital display. The power source of a megohmmeter can be battery power or 120 VAC. Some models have dual power sources such as batteries and 120 VAC. **See Figure 7-17.**

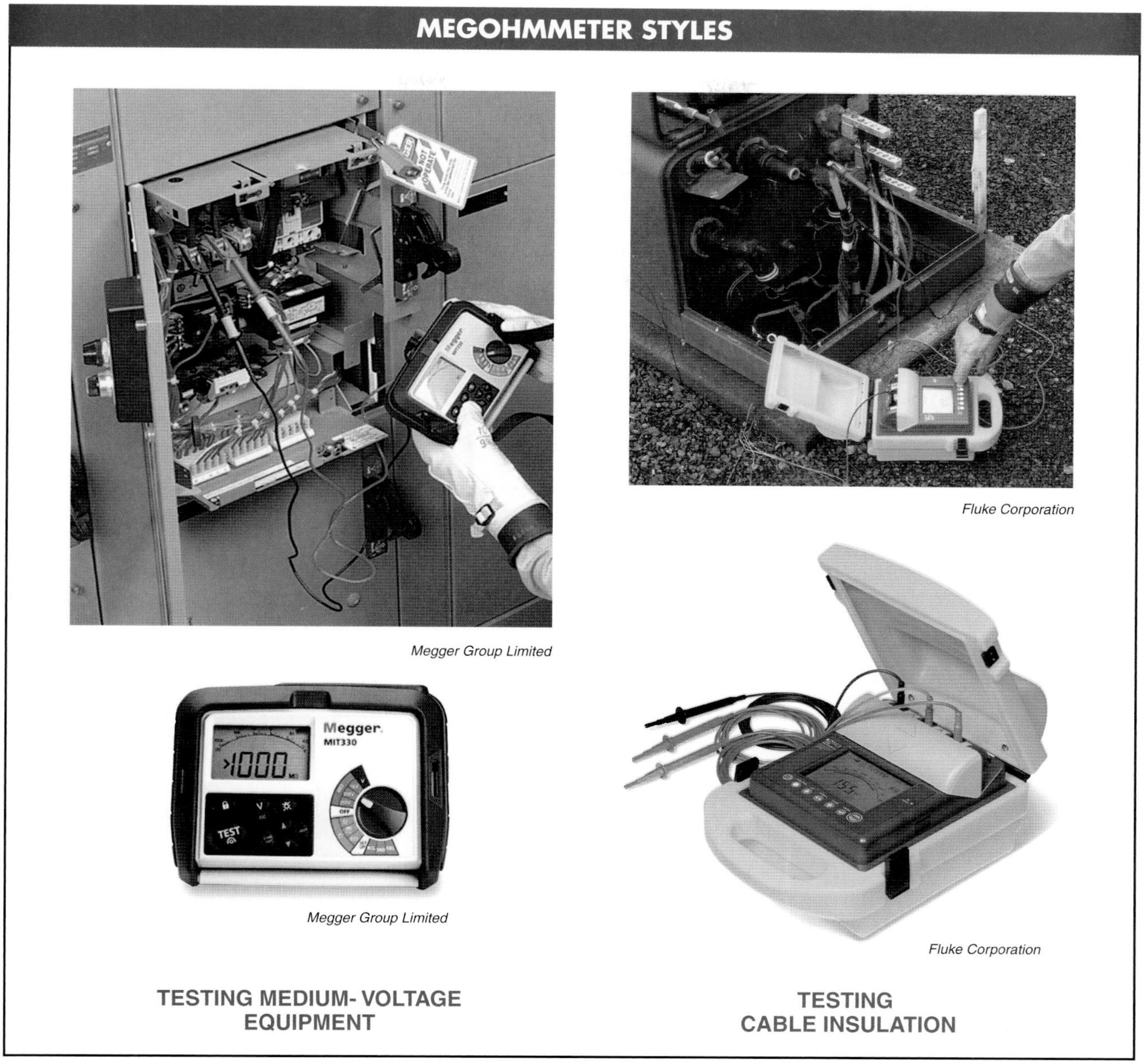
MEGOHMMETER STYLES
Megger Group Limited
Fluke Corporation
Megger
MIT330
>1000
TEST
OFF
Megger Group Limited
Fluke Corporation
TESTING MEDIUM- VOLTAGE EQUIPMENT
TESTING CABLE INSULATION

Figure 7-17. Megohmmeters are high-resistance-range meters powered by batteries or 120 VAC.

Troubleshooting Tip

READING ANALOG MULTIMETER SCALES

When reading an analog scale, add the primary, secondary, and subdivision readings. For example, with the dial switch set on "R x 100," the reading is 5000 Ω. When set on "R x 1," the reading is 50 Ω.

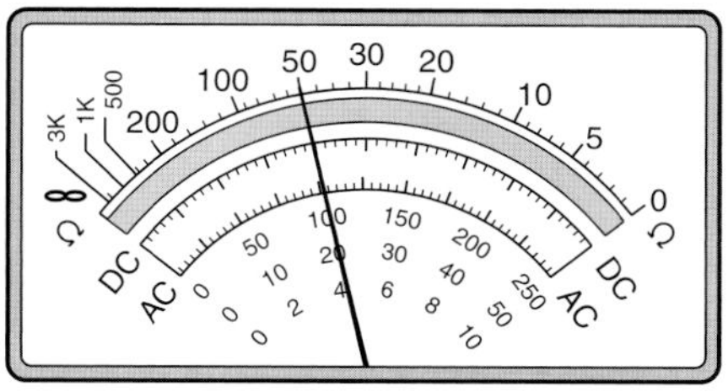
Ω ∞
3K
1K
500
200
100
50
30
20
10
5
0
Ω
DC
AC
0
50
10
0
2
0
100
20
4
150
30
6
200
40
8
250
50
10
DC
AC

Insulation in good working order has a high resistance. Insulation in poor working order has a low resistance. Insulation is damaged by moisture, oil, dirt, excessive heat, excessive cold, corrosive vapors, aging, and vibration. The ideal megohmmeter measurement is infinite resistance between a conductor or motor winding and ground. Infinite resistance is depicted by an infinity symbol (∞). Often a megohmmeter measurement is less than infinite.

Megohmmeter measurements typically follow the general rule of thumb of 1 MΩ of resistance for every 1000 V of insulation rating. When using a megohmmeter, the test voltage is typically rounded up to the nearest 1000 V. For example, wire used in 480 VAC or 240 VAC distribution systems has a rating of 600 V. For testing purposes consider the wire to have 1000 V insulation. The stator winding insulation for inverter duty motors has a rating of approximately 1500 V. For testing purposes consider the stator winding to have 2000 V rated insulation.

Several megohmmeter readings must be taken over a period of time because the resistance of good insulation varies with time. Megohmmeter readings are typically taken when an electrical device such as a motor is installed and at regular intervals thereafter. An electrical device is in need of service when a megohmmeter measurement is below the minimum acceptable value.

CAUTION

A megohmmeter uses medium voltage during testing (up to 5000 V). Avoid touching the test leads to any ground. Always follow the recommended procedures and safety rules of the manufacturer. After performing insulation tests with a megohmmeter, connect the device being tested to ground through a 5 kΩ, 5 W resistor if the megohmmeter does not include a discharge function.

Types of Megohmmeter Insulation Resistance Testing

There are three basic types of insulation tests used to determine the condition of insulation. The three insulation tests are the insulation spot test, the dielectric absorption test, and the insulation step voltage test.

Insulation Spot Testing. An insulation spot test is a short-term test in which a megohmmeter is connected to a test conductor over about a 60 sec time period.

Dielectric Absorption Testing. A dielectric absorption test is a longer term test in which a megohmmeter is connected to a test conductor at various time intervals that last about 10 minutes.

Insulation Step Voltage Testing. An insulation step voltage test is a test in which a megohmmeter is connected to a test conductor and applies an ever increasing voltage level (500 V, 1000 V, 1500 V) to measure the insulation resistance at various voltage levels. Some megohmmeters automatically record all resistance measurements taken during testing.

Insulation Spot-Test Measurement Procedures

An *insulation spot test* is a test that verifies the integrity of insulation on electrical devices such as stator windings, load conductors, cables, switchgear, heaters, and transformers. For example, an insulation spot test is taken when a motor is placed in service and every 6 months thereafter. An insulation spot test should also be taken after a motor has had maintenance or been rewound. **See Figure 7-18.**

During an insulation spot test, the test leads of the meter are connected across the conductor and insulation being tested. The test leads are also connected across any other conductor that may come in contact with the insulation being tested. The test voltage is applied for 60 sec to allow for the most accurate measurements.

Interpretation of the measured resistance values requires knowledge of previous resistance measurements, so recordkeeping is important. Megohmmeter manufacturers include charts for recording and plotting resistance measurements over time.

Technical Tip

When performing insulation tests that require working within a specific time frame, it is important to maintain the exact amount of time for each test. For example, if a spot test is performed for 65 sec, each recurring spot test should be performed and recorded for 65 sec.

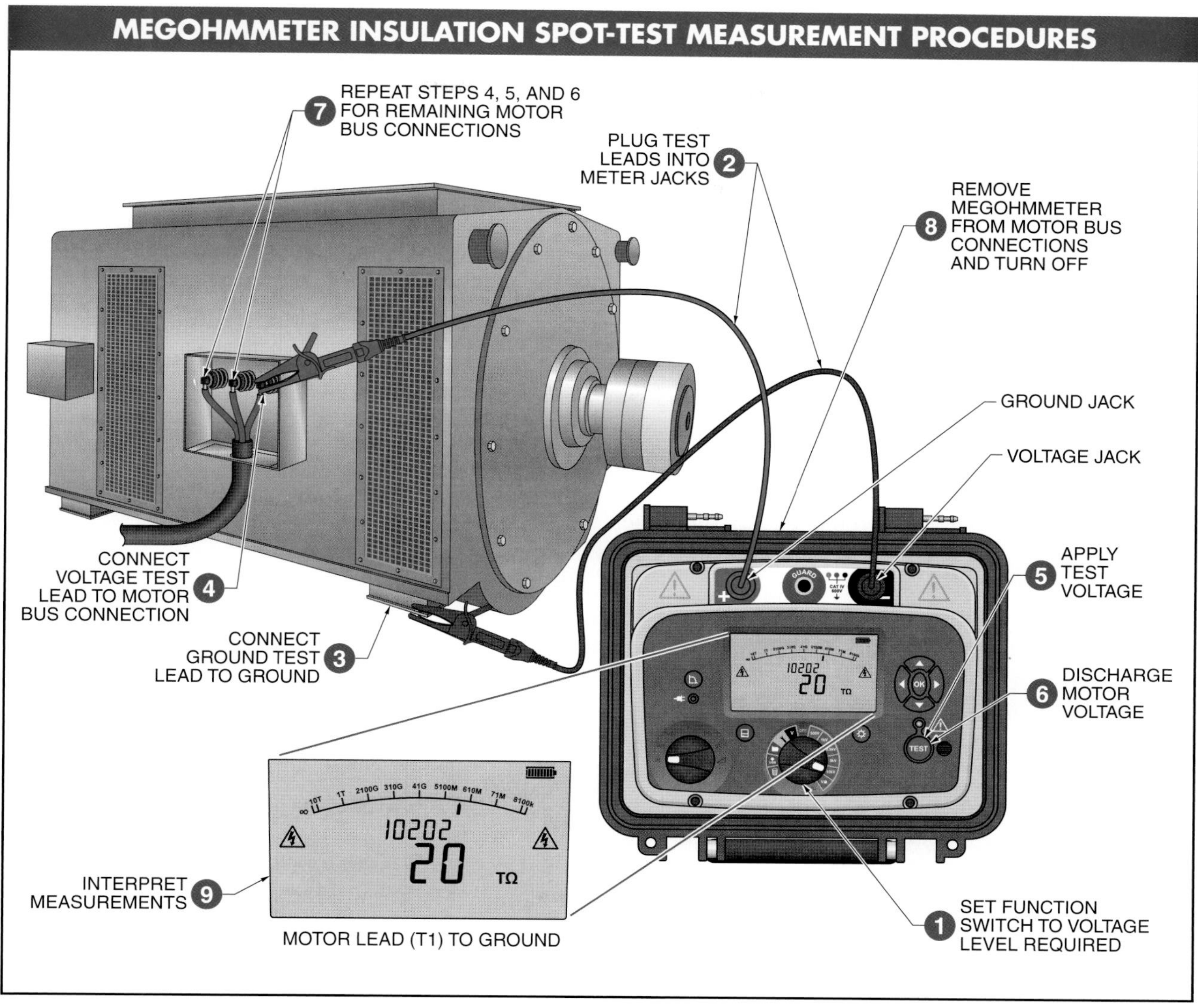

Figure 7-18. Insulation spot tests verify the integrity of insulation on electrical devices such as stator windings, load conductors, cables, switchgear, heaters, and transformers.

Before taking any insulation spot-test measurements using a megohmmeter, ensure the megohmmeter is designed to take measurements on the system or wires being tested. Refer to the operating manual of the test instrument for all measuring precautions, limitations, and procedures. Always wear required personal protective equipment and follow all safety rules when taking the measurement. To perform an insulation spot test using a megohmmeter, apply the following procedures:

1. Set the function switch of the megohmmeter to the proper test voltage level. The test voltage is typically set higher than the voltage rating of the insulation being tested to stress the insulation. The 10 kV setting is typically used for motors and conductors operating at 4160 VAC. When a megohmmeter does not have a 10 kV setting, use the voltage setting closest to but not greater than 10 kV.
2. Plug the test leads of the megohmmeter into the proper meter jacks.

3. Connect the black test lead of the megohmmeter to a grounded surface.
4. Connect the red test lead of the megohmmeter to one of the motor bus connections or an individual conductor.
5. Apply the test voltage for 60 sec. Record the megohmmeter reading. Record the lowest reading on an insulation spot-test graph when all readings are above the minimum acceptable reading. The lowest reading is used because a motor or a set of feeder conductors is only as good as the weakest point.
6. Discharge the circuit being tested.
7. Repeat steps 4, 5, and 6 for the remaining motor bus connections or individual conductors.
8. Remove the megohmmeter from the motor leads and turn OFF the meter to prevent battery drain.
9. Interpret the measurements taken.

Megohmmeter readings must be interpreted. **See Figure 7-19.** A motor installed outdoors and tested two days in a row can have two different readings depending on the weather (foggy conditions one day would result in low MΩ and sunny conditions the next day would result in high MΩ). In general, megohmmeter readings are the most useful when taken semiannually over a period of years. A sudden drop in a resistance measurement of a motor such as 100 MΩ to 2 MΩ over a six-month period is an indication of a problem, even when the measurement is above the accepted value. A large difference between resistance measurements of motor leads (L1 = 20 MΩ, L2 = 21 MΩ, and L3 = 1 MΩ) also is an indication of a problem.

The cause of low resistance readings must be determined. The cause can be moisture, dirt, or damaged insulation. Typically, low resistance readings require the motor or conductors to be repaired or replaced. The repaired or replaced items must be tested with a megohmmeter before being placed into service.

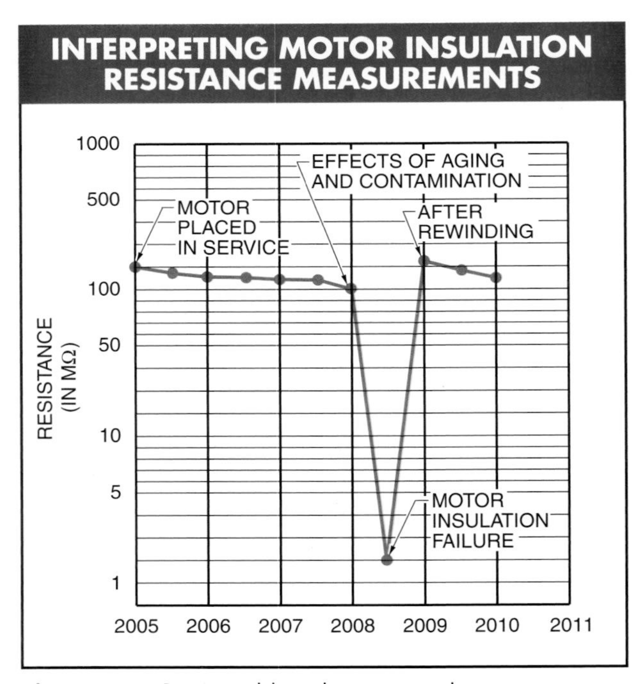

Figure 7-19. A sudden drop in insulation resistance measurements of a motor, as from 100 MΩ to 2 MΩ over a six-month period, is an indication of a problem.

Dielectric Absorption Test Measurement Procedures

A *dielectric absorption test* is a test that verifies the absorption characteristics of insulation in good working order. Insulation that has been contaminated will not pass a dielectric absorption test. The test is typically performed over a 10-minute time period. Typically, the resistance measurements are taken every 10 sec for the first minute and every minute thereafter.

The measured resistance values are plotted using graph paper provided by the manufacturer of the test instrument. The slope of the curve determines the condition of the insulation. Insulation in good working order will show a continuous increase in resistance over time. Insulation that is damaged with cracks, moisture, or contamination will have a relatively flat curve. **See Figure 7-20.**

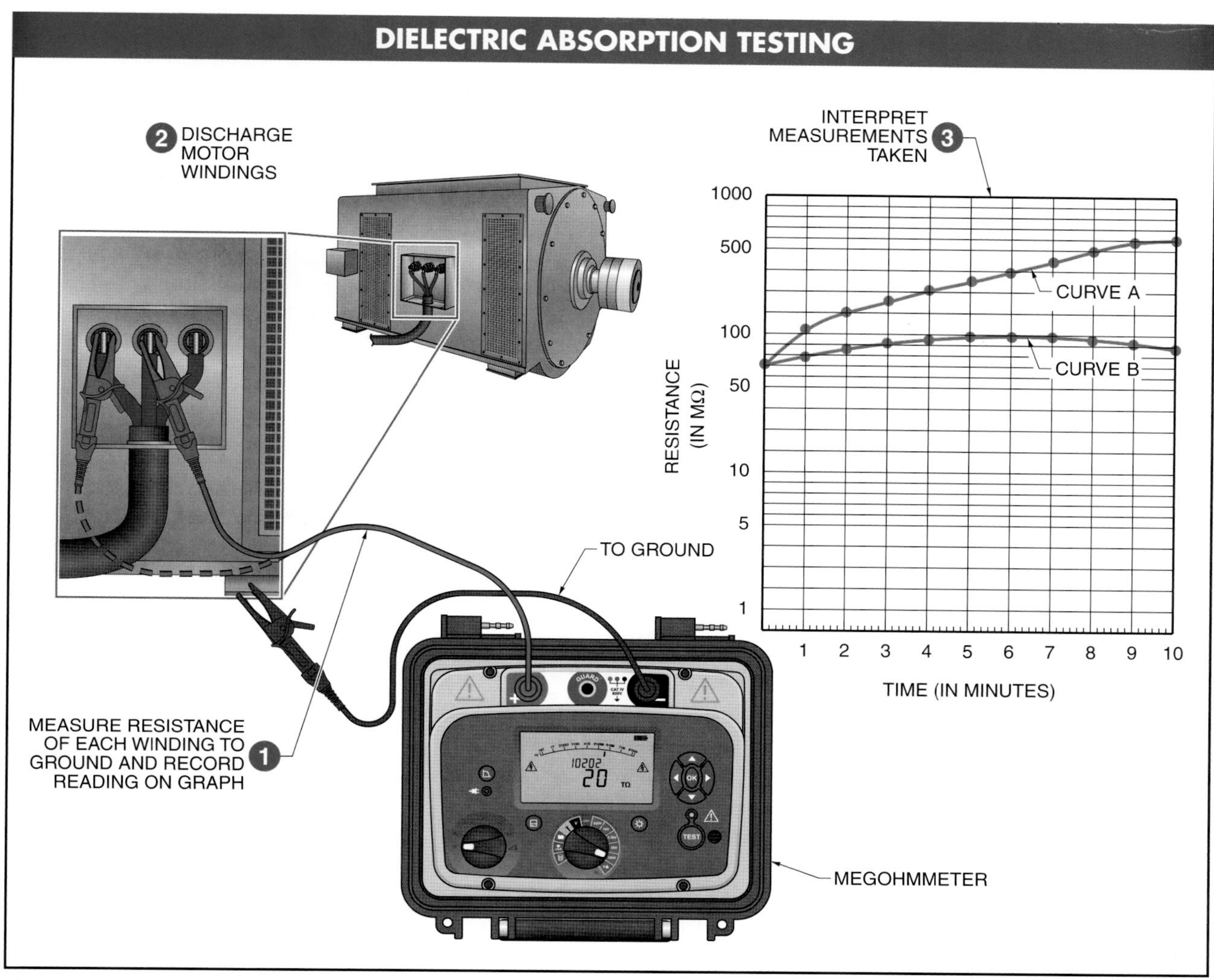

Figure 7-20. A dielectric absorption test verifies the absorption characteristics of insulation with measurements that are plotted on graph paper to provide a curve that is used to determine the condition of the insulation.

Before taking any dielectric absorption test measurements with a megohmmeter, ensure the megohmmeter is designed to take measurements on the system or wires being tested. Refer to the operating manual of the test instrument for all measuring precautions, limitations, and procedures. Always wear required personal protective equipment and follow all safety rules when taking the measurement. To perform a dielectric absorption test, apply the following procedures:

1. Connect a megohmmeter to measure the resistance of each winding lead to ground. Service the motor when a reading does not meet the minimum acceptable resistance values. Record the lowest meter reading on a dielectric absorption test graph. Record the readings at 10-second intervals for the first minute and every minute thereafter for 10 min.
2. Discharge the motor windings.
3. Interpret the measurements taken.

A polarization index is obtained by dividing the values of the 10-minute measurement by the value of the 1-minute measurement. The polarization index is an indication of the condition of the insulation of a motor or conductor. A low polarization index indicates excessive moisture or contamination. **See Figure 7-21.**

For example, when a 1-minute measurement of Class B insulation is 80 MΩ and the 10-minute measurement is 90 MΩ, the polarization index is 1.125 ($\frac{90\ M\Omega}{80\ M\Omega} = 1.125$). The insulation must contain excessive moisture or contamination.

MINIMUM ACCEPTABLE POLARIZATION INDEX VALUES	
Insulation	**Value**
Class A	1.5
Class B	2.0
Class F	2.0

Figure 7-21. A polarization index is obtained by dividing the values of the 10 min measurement by the value of the 1 min measurement, which provides an indication of the condition of the insulation of a motor or conductor.

Technical Tip

The 10 min to 1 min measurement of polarization index (PI) testing is covered by IEEE Standard 43-2000.

Insulation Step Voltage Test Measurement Procedures

An *insulation step voltage test* is a test that creates electrical stress on internal insulation cracks to reveal aging or damage not found during other motor or conductor insulation tests. An insulation step voltage test is similar to a hipot test, except the displayed measurement on the megohmmeter is in ohms, not amps. The insulation step voltage test is performed only after an insulation spot test has been completed. **See Figure 7-22.**

As with a hipot test, when insulation fails due to high voltage, damage occurs. Insulation failure is not necessarily bad because the weak point that failed would have typically failed over time anyway. The general thought is that there is no good time for any failure, but being able to predict a failure during testing allows for a controlled failure. Due to the ability to predict failure, insulation step voltage testing and hipot testing are typically performed before a product is placed in service or sold and during scheduled preventive maintenance shutdowns.

Before taking any insulation step voltage test measurements using a megohmmeter, ensure the megohmmeter is designed to take measurements on the system or wires being tested. Refer to the operating manual of the test instrument for all measuring precautions, limitations, and procedures. Always wear required personal protective equipment and follow all safety rules when taking the measurement. To perform an insulation step voltage test, apply the following procedures:

1. Set the function switch of the megohmmeter to proper voltage level and connect the test leads of the meter to measure the resistance of each winding lead to ground. Take a resistance measurement every 60 sec. Record the lowest reading.
2. Place the meter leads on the winding lead that has the lowest reading.
3. Increase the megohmmeter test voltage by increments of 500 V starting at 5000 V and ending at 10,000 V. Record measurements every 60 sec.
4. Discharge the motor windings.

Interpret the results of the test to determine the condition of the insulation. The resistance of insulation in good working order that is thoroughly dry remains approximately the same as the voltage levels increase. The resistance of deteriorated insulation decreases substantially as the voltage levels increase.

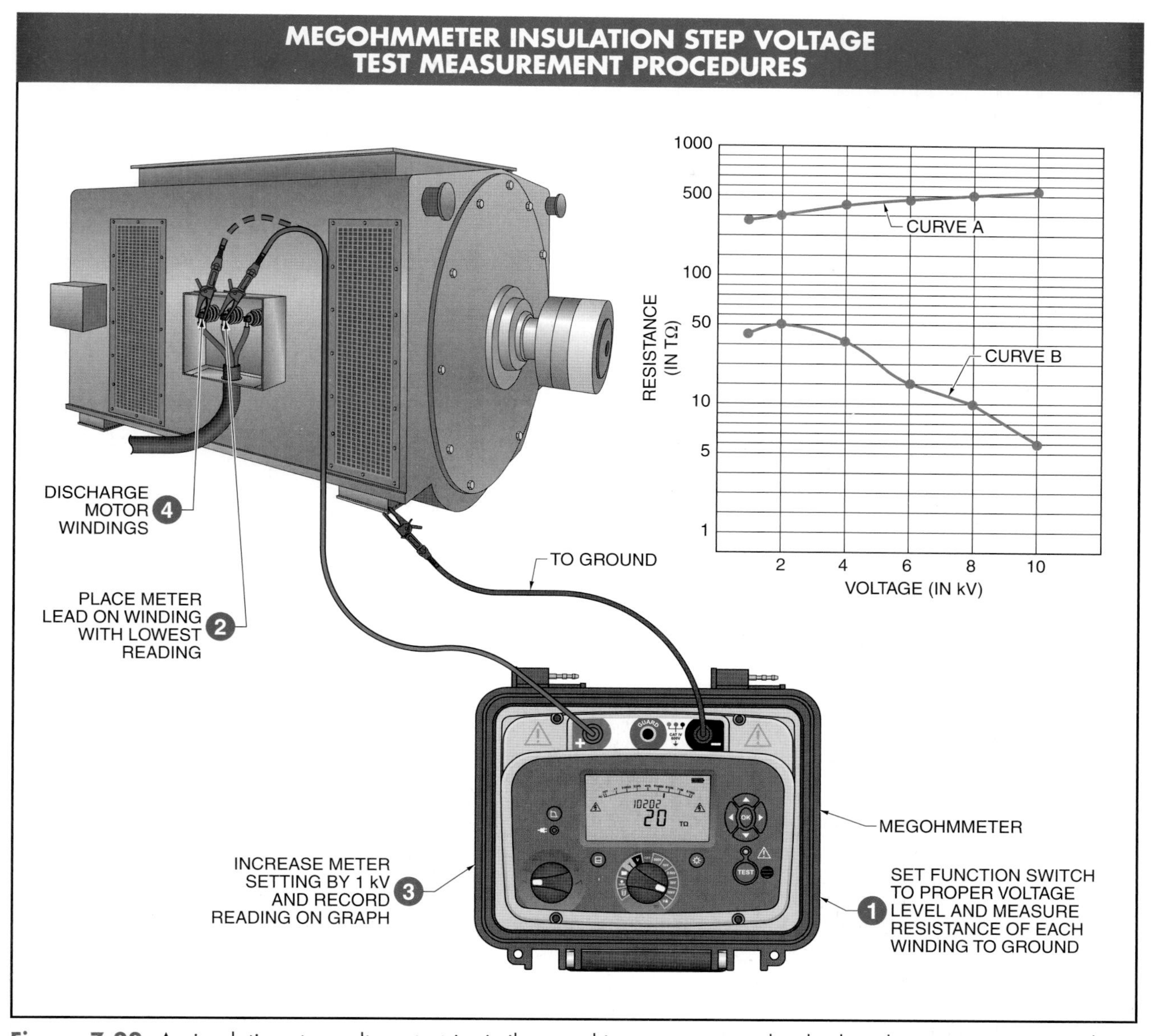

Figure 7-22. An insulation step voltage test is similar to a hipot test, except the displayed measurement is in ohms, not amps.

SAFETY CONSIDERATIONS

Appropriate PPE must be worn and safe work practices must be followed at all times when entering or approaching potential arc-flash boundaries to take measurements. NFPA 70E®, 2018, specifies arc-flash PPE categories in Table 130.7(C)(15)(a) and (b). Once the appropriate arc-flash PPE category is determined, Table 130.7(C)(15)(c) must be used to identify the required PPE for the task. In addition, Informative Annex H—Guidance on Selection of Protective Clothing and Other Personal Protective Equipment (PPE) contains Table H.2—Simplified Two-Category, Arc Rated Clothing System, which provides a simplified version of the arc-flash PPE categories.

NFPA 70E® Article 250, Personal Safety and Protective Equipment, covers the maintenance, inspection, and testing of protective equipment and protective tools. Additional requirements for working with equipment and systems of 600 V and higher are provided by OSHA:

- OSHA 1910.333(b)(2)(iv)(B) *Selection and Use of Work Practices*—"working on or near exposed deenergized parts."

  A qualified person shall use test equipment to test the circuit elements and electrical parts of equipment to which employees will be exposed and shall verify that the circuit elements and equipment parts are deenergized. The test shall also determine if any energized condition exists as a result of inadvertently induced voltage or unrelated voltage backfeed even though specific parts of the circuit have been deenergized and presumed to be safe. If the circuit to be tested is over 600 volts, nominal, the test equipment shall be checked for proper operation immediately after this test.

- OSHA 1910.333(c)(3) *Selection and Use of Work Practices*—"working on or near exposed energized parts." This section pertains to working near overhead power lines.

  If work is to be performed near overhead power lines, the lines shall be deenergized and grounded, or other protective measures shall be provided before work is started. If the lines are to be deenergized, arrangements shall be made with the person or organization that operates or controls the electric circuits involved to deenergize and ground them. If protective measures, such as guarding, isolating, or insulating, are provided, these precautions shall prevent employees from contacting such lines directly with any part of their body or indirectly through conductive materials, tools, or equipment.

MEDIUM VOLTAGE TEST INSTRUMENTS

Name: ______________________ Date: ______________________

______________ **1.** Which type of waves do cable height meters use to measure cable heights?

T F **2.** A circuit or component must be de-energized before taking resistance measurements.

______________ **3.** Which type of product testing is performed on newly designed components?

______________ **4.** Which type of insulation test is a short-term test that verifies the integrity of insulation?

______________ **5.** Which type of insulation test is similar to a hipot test except that the displayed measurement is in ohms instead of amps?

T F **6.** AC hipot tests are performed at 60 Hz.

T F **7.** A medium voltage test probe accessory displays a voltage on the attached multimeter that is approximately $\frac{1}{1000}$ of the actual voltage.

______________ **8.** Which type of product testing is performed periodically while the product is in service?

______________ **9.** Which type of test instrument is used to verify that a medium voltage detector is operating properly?

T F **10.** Insulation tests are taken over time (many years) to monitor any slow deterioration.

______________ **11.** Which electrical quantity does a hipot tester measure to determine insulation resistance?

______________ **12.** What is the general rule of thumb for insulation resistance to minimize leakage current?

T F **13.** It is normal for insulation resistance to lower gradually over time.

______________ **14.** Which type of product testing is performed after the product is installed but before it is put into service?

______________ **15.** What value calculated from resistance measurements indicates that insulation contains excessive moisture or contamination?

T F **16.** A medium voltage detector displays the level of voltage present on a power line.

______________ **17.** Which type of hipot test applies voltage in increasing steps?

______________ **18.** Which test instrument takes current measurements of medium voltage power lines?

______________ **19.** What will happen at a weak or damaged point in a conductor during a hipot test?

______________ **20.** Which type of test verifies a low impedance path to ground from all exposed metal parts and the grounding conductor?

21. Why is it necessary to test a medium voltage detector before and after use?

22. What are several causes of insulation deterioration or damage?

23. Why do some product testing methods require the product to be tested to failure?

24. Why might a standard ohmmeter not accurately measure insulation resistance?

25. How can preventive maintenance such as insulation testing reduce troubleshooting and downtime?

8 INSTRUMENTATION AND PROCESS CONTROL TEST INSTRUMENTS

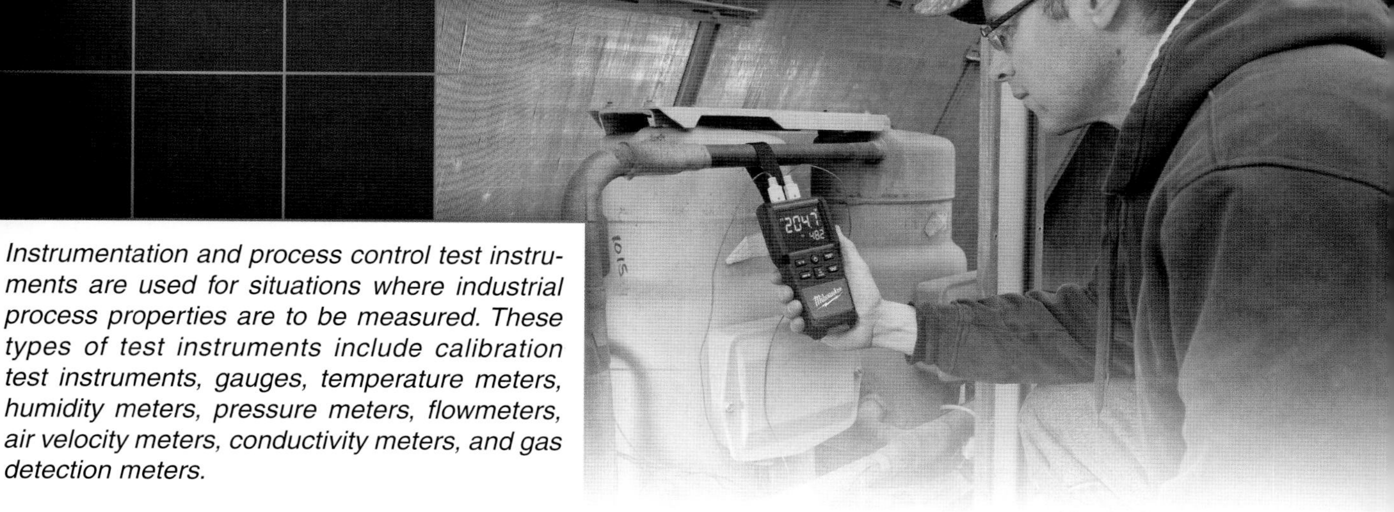

Instrumentation and process control test instruments are used for situations where industrial process properties are to be measured. These types of test instruments include calibration test instruments, gauges, temperature meters, humidity meters, pressure meters, flowmeters, air velocity meters, conductivity meters, and gas detection meters.

Milwaukee Tool Corporation

INSTRUMENTATION

Test instruments are designed to take and display measurements of electrical properties (voltage and current) and environmental (gases, wind speed, and light), physical (weight and thickness), and operating conditions (speed, pressure, and flow rate). Instruments are designed to be used as permanent devices within a facility for process control. Test instruments are designed to be used as portable devices and acquire the same information (signals) that instruments generate, but are typically used for troubleshooting.

Troubleshooting Tip

EXPLOSIONPROOF APPARATUS

An explosionproof apparatus is equipment that is enclosed in a case that is capable of withstanding any explosion that may occur within it without permitting the ignition of flammable gases or vapors on the outside of the enclosure.

Which test instrument is used depends upon the application, required accuracy of the measurement, operating environment, and desired results. The application determines the type of measurement taken, such as voltage, pressure, or whether a gas is present. The required accuracy of a measurement determines the model used (the higher the accuracy, the higher the cost). The operating environment determines the grade of test instruments that must be used—whether standard grade (the lowest cost), commercial grade, or industrial grade (the highest cost) test instruments. The operating environment determines whether an instrument must be designed to operate in a hazardous location or another type of special location. For example, explosionproof instruments are required in Class I, Division I locations. The desired results determine whether a test instrument with an analog display, digital display, or graphic display is used. The desired results also determine when to only display measurements, when to take and record (log) measurements, when to take only one type of measurement, or when to take various types of measurements.

Process Control and Instrumentation

Instrumentation is a large part of the process control industry. *Instrumentation* is the use of gauges and other measurement mechanisms to determine the value of electrical, environmental, physical, and operating conditions. A *process* is an operation or sequence of operations in which the substance being treated is changed. The change can be from one energy state to another state (cold to hot), a change in composition (wood pulp to paper), or a change in size or shape (rocks to gravel). Processes produce the products consumed each day such as food, fuel, coatings, pharmaceuticals, paper, and chemicals. A coatings manufacturing batch operation is an example of a process. **See Figure 8-1.**

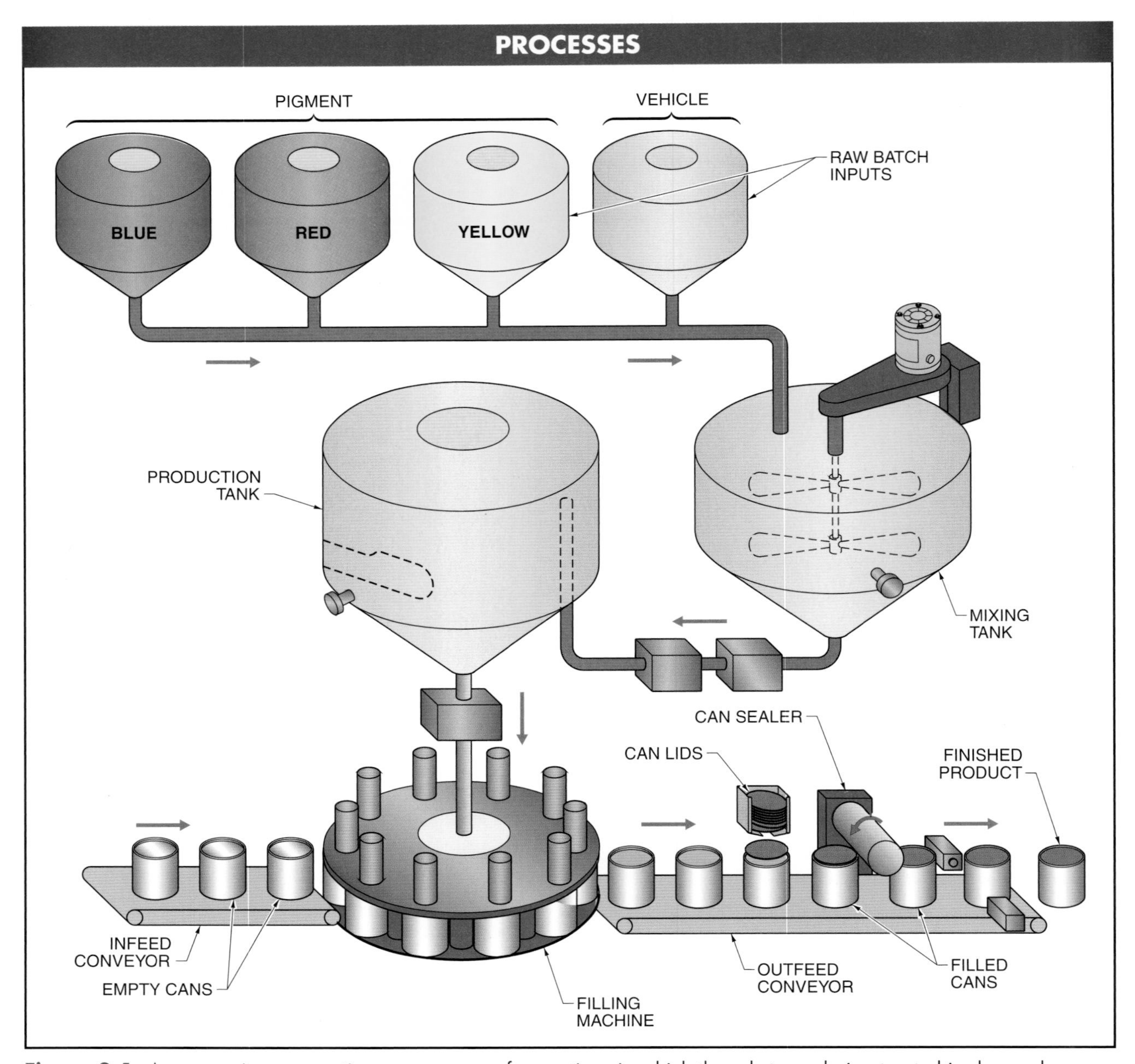

Figure 8-1. A process is an operation or sequence of operations in which the substance being treated is changed.

Processes require the use of instrumentation to measure variables such as temperature, humidity, pressure, vacuum, flow, conductivity, voltage, current, resistance, speed, time, and other process operating conditions. Without the ability to accurately measure process variables, standards, quality control, and documentation would be limited.

Instruments installed as part of a process of making a product are of a permanent type. Test instruments used as part of a troubleshooting system or when performing preventive maintenance tasks are of the portable type.

Portable and Permanent Test Instruments

A portable test instrument is a test instrument that takes one or more measurements and is designed to only be temporarily connected to a circuit or system when taking measurements. A permanent test instrument is a test instrument that takes one or more measurements and is designed to be permanently connected to a circuit or system when taking measurements. Portable test instruments are typically used by electricians when troubleshooting a circuit or system. Permanent test instruments are used by labs, quality control departments, and inspectors. Portable and permanent test instruments and meters measure the same types of variables. **See Figure 8-2.**

Portable test instruments are designed to be hand carried and typically include accessories such as various types of test leads and subject matter attachments. Test instruments must be kept in protective cases or carrying cases (with the operating manual) when not in use. Because portable test instruments are used in many different locations, the limits (CAT rating and maximum measurement limits) of the test instrument must be understood and applied to each application. Permanent test instruments are designed to be installed, calibrated, viewed, and recalibrated as required.

To ensure that the correct instrument is being used for a given application, the ratings of permanent and portable test instruments must exceed the application ratings. Test instrument ratings must exceed the application ratings because the test instrument is expected to be used over and over again without damage.

Milwaukee Tool Corporation

Portable test instruments allow technicians to easily take measurements and readings in the field.

Technical Tip

To maintain optimum accuracy, digital test instruments should be calibrated at least once a year by a qualified agency.

Calibration Test Instruments

All test instruments are designed to take and display a measured quantity. Test instruments are also designed to take measurements and display the measurements within certain accuracy limits. The accuracy of a measurement is always within the specified rating of the test instrument when the instrument is operated within the manufacturer specified limits.

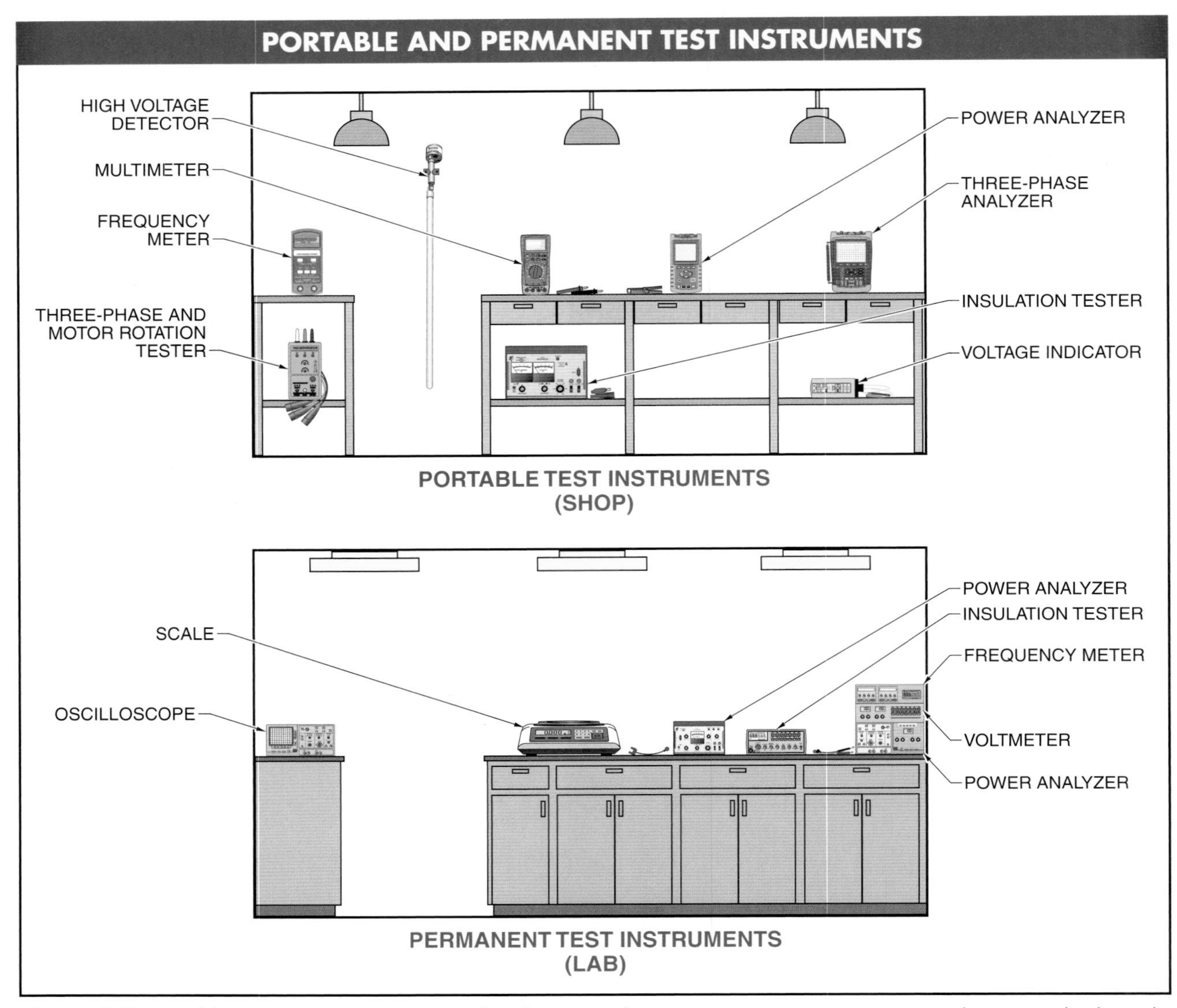

Figure 8-2. Portable test instruments are carried to the point of testing, permanent test instruments have samples brought to the lab for testing.

Although the accuracy of any measurement is important, the accuracy of test instruments must be maintained as close as possible to the actual value. To ensure accuracy, calibration test instruments are used to test (and retest after adjustments are made) other test instruments using circuit or process parameters. To ensure accuracy, calibration instruments must be properly understood, applied, and maintained.

Calibration test instruments measure variable quantities the same way as standard test instruments, except calibration test instruments have a much higher accuracy rating. Calibration test instruments also include special features that are used during testing and calibration of standard test instruments. For example, calibration test instruments typically can output voltage and current as well as measure voltage and current. Calibration test instruments are used to test and adjust variable frequency drives and PLCs to ensure proper operation. **See Figure 8-3.**

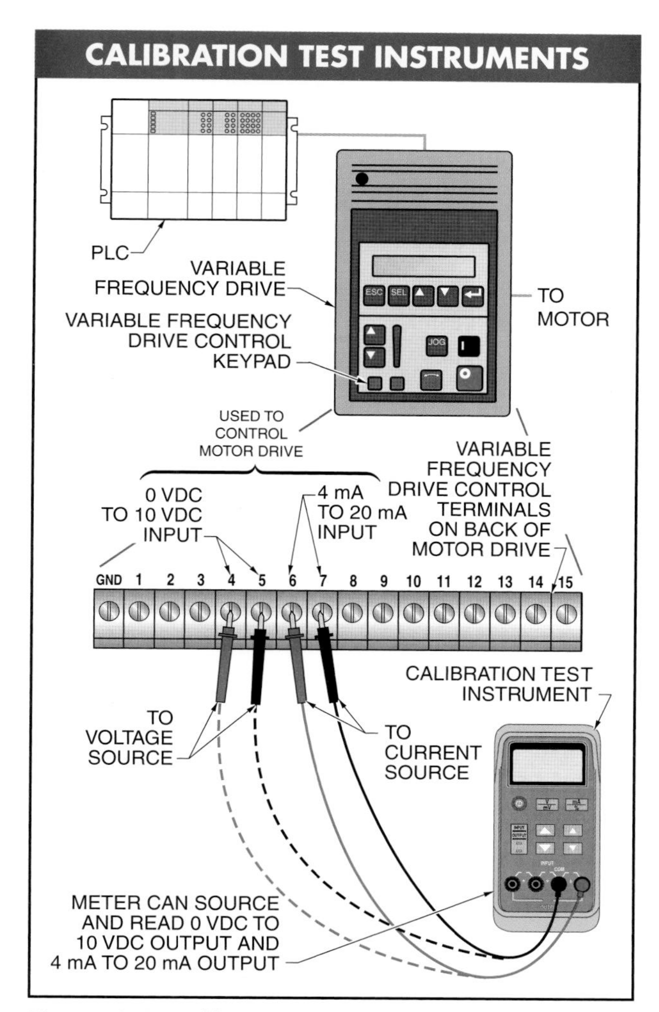

Figure 8-3. Calibration test instruments that can output a voltage and current as well as measure a voltage and current are used to check and adjust variable frequency drives and PLCs to ensure proper operation.

There are applications in which the displayed measurement must be as close as possible to the actual value. For example, a flowmeter can have an accuracy rating of ±2.5% or ±1.5%, depending upon the model. When a flowmeter is used to measure the flow rate of a product that is being sold or added to a process, the difference between models can produce a 1% error above the already possible 1.5% error on the best model. The 1% represents a 10 gal. per 1000 gal. error, which is not acceptable for the quality control or accounting departments of most companies.

Accuracy Specification

Measurements are typically not exact because the test instrument taking the measurement is not exact. All test instruments have an accuracy rating that determines how close the measurement of the instrument will be to the actual value. Manufacturers specify the accuracy of their instruments to a certain plus or minus percentage of the total range or scale of the instrument. Manufacturers typically specify control element accuracy as the worst-case accuracy over the entire range. There are various ways manufacturers list the accuracy of meters. **See Figure 8-4.** For example, when a pressure gauge has an indicated range of 0 psi to 60 psi, the accuracy of the meter reading can be listed as follows:

±0.5% of Full Face Value.

- ±0.5% × 60 psi = ±0.3 psi (0.005 × 60 = 0.3).
- At any point within the range of the meter (0 to 60 psi) the reading will be within ±0.3 psi in error. The lower the measured value, the greater the percentage of error.
- A 60 psi reading has an accuracy reading equal to ±0.5% (psi = ±0.3 psi ÷ 60 psi = ±0.005 = 0.5 %).
- A 30 psi reading has an accuracy reading equal to ±1% (psi = ±0.3 psi ÷ 30 psi = ±0.01 = 1%).
- A 10 psi reading has an accuracy reading equal to ±3% (psi = ±0.3 psi ÷ 10 psi = ±0.03 = 3%).

±0.5% of Reading.

- Percentage of reading provides a more accurate reading at all measurements within the test instrument rating.
- A 60 psi reading has an accuracy reading of = ±0.5%.
- A 30 psi reading has an accuracy reading of = ±0.5%.
- A 10 psi reading has an accuracy reading of = ±0.5%.

Technical Tip

When calibrating a test meter or instrument, the material or equipment of the reference standard should be as close as possible to the actual material or equipment and problem. Control samples used during the calibration should be measured for accuracy prior to every test.

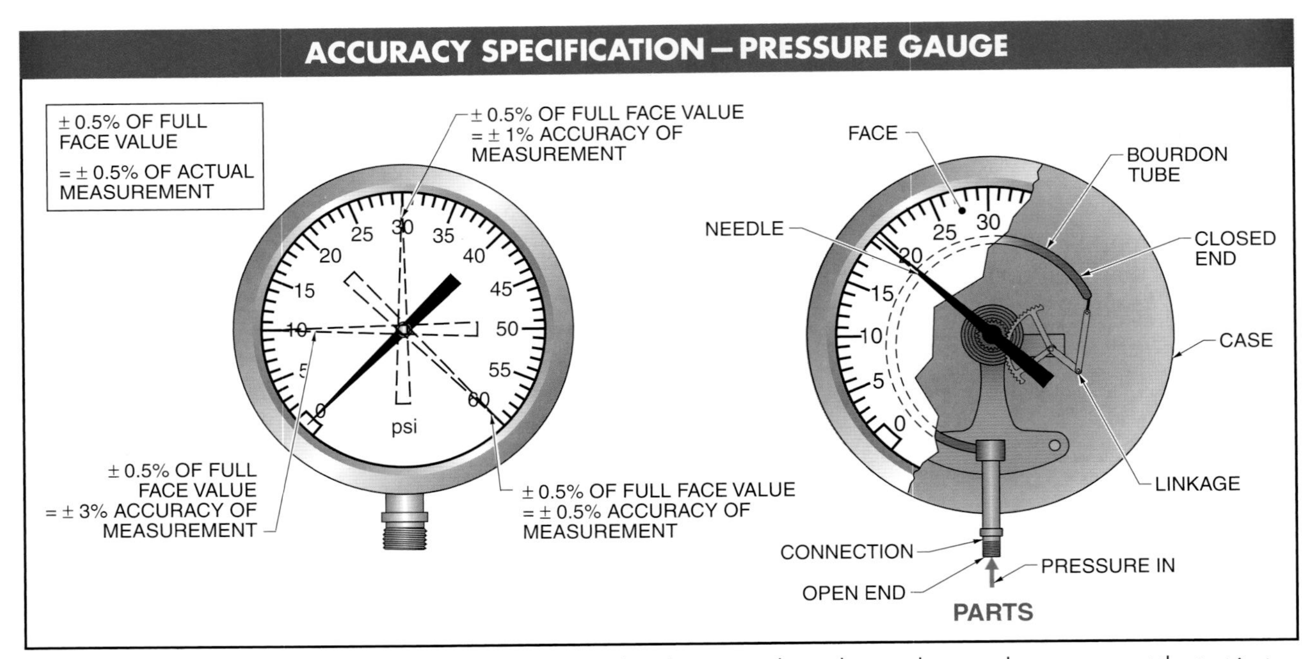

Figure 8-4. All test instruments have an accuracy rating that determines how close to the actual measurement the test instrument's measurement should be.

Single-Function and Multifunction Meters and Attachments

Single-function test instruments are designed to take a single measurement such as temperature only, humidity only, or resistance only. Single-function test instruments cost less and perform well in applications that require only a single measurement. Multifunction test instruments such as multimeters cost more, but offer choices as to the measurements taken.

In addition to single-function and multifunction test instruments, a standard multimeter (digital or analog) can be used with attachments (adapters) that allow the multimeter to measure almost any type of variable. **See Figure 8-5.** Most attachments output 1 mV DC per unit of measurement (per degree temperature, per rpm, or per dB). Most multimeters include a mV DC setting in addition to the standard DC setting to allow taking measurements with subject matter attachments. The mV DC setting allows for a more precise measurement because of the higher resolution. *Resolution* is the degree of precise measurement a test instrument is capable of taking.

TEMPERATURE TEST INSTRUMENTS

Temperature test instruments are used to measure the intensity of heat at a measuring point. *Heat* is thermal energy. Anything that transfers heat is a heat source. Electrical energy may be converted into thermal energy. Electricity is used to heat almost any gas, liquid, or solid. Electricity is used to produce heat in many residential, commercial, and industrial applications.

Heat produced from electricity, gas, coal, and other fuels is used in residential and commercial applications to heat rooms, heat water, cook food, and dry clothes. Most industrial processes call for the heating of liquids, solids, and gases. Heat is used to refine and process metals, refine crude oil, process foods, remove moisture, shape metals and plastics, produce steam, join metals and plastics, harden metals, and cure finishes such as paint, enamels, and varnishes.

Technical Tip

Heat transfer rate (Q_r) is the amount of heat (Q) supplied by a heating element per unit time, usually per minute.

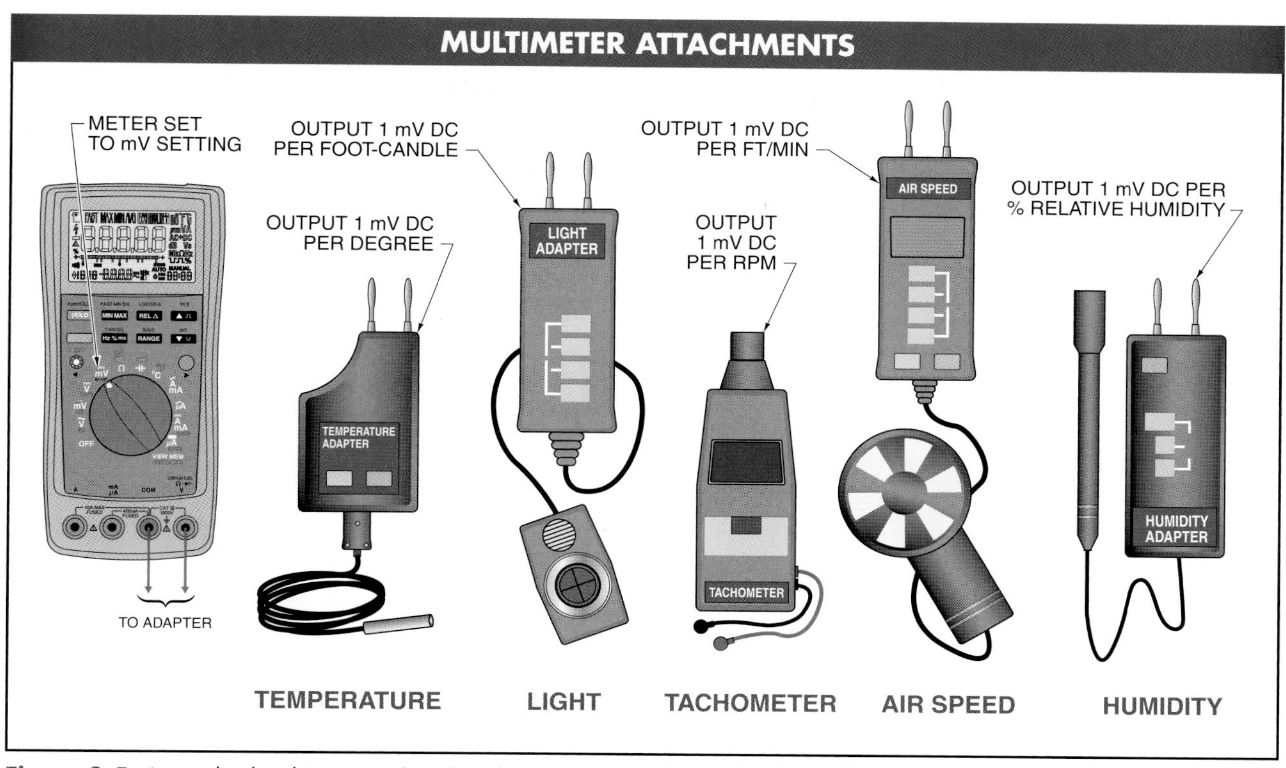

Figure 8-5. A standard multimeter, either digital or analog, can be used with attachments (adapters) that allow the multimeter to measure almost any type of variable.

Heat is measured in British thermal units (Btu) or calories (cal). The British thermal unit is a unit of measurement used to measure the amount of heat in the U.S. system of measurements. The calorie is a unit of measurement used to measure the amount of heat in the metric system of measurements. A *British thermal unit (Btu)* is the amount of heat required to raise 1 lb of water 1°F. A *calorie* is the amount of heat required to raise 1 gram (g) of water 1°C. One Btu is equivalent to 252 calories, or 0.252 kilocalories. **See Figure 8-6.**

Although the unit of heat energy is the Btu or calorie, electrical heating requires electrical energy to produce the heat. In the electrical field, electrical energy is measured in watts. Most electric heating elements are rated in watts. The amount of electrical energy used by a heating element is dependent upon the power rating of the heating element (in watts) and the amount of time (in hours) the heating element is ON.

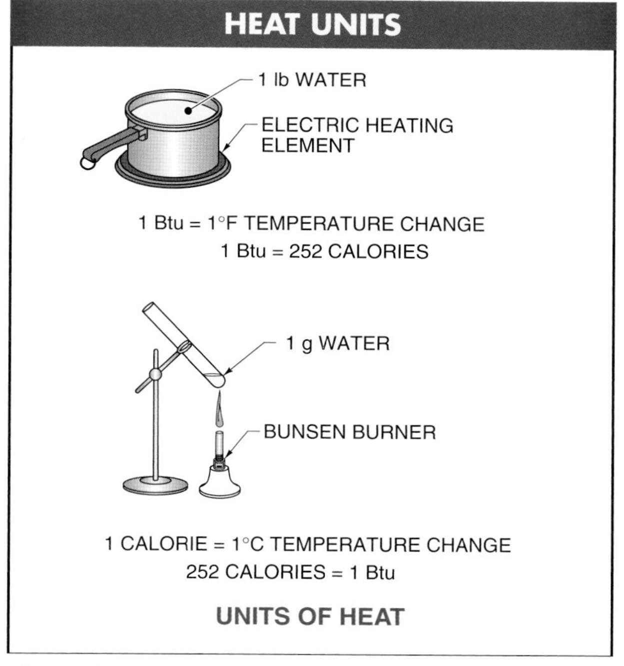

Figure 8-6. One Btu is equivalent to 252 calories, or 0.252 kilocalories. A calorie is the amount of heat required to raise 1 g of water 1°C.

Temperature

Temperature is the measurement of the intensity of heat. The amount of temperature increase depends upon the amount of heat produced, the mass of the object being heated, and the material of which the heated object is composed. The higher the amount of produced heat, the faster the temperature increases in the mass of the object being heated. The larger the mass of an object being heated, the slower the temperature rise of the object being heated. The better the conductivity of the material that is being heated, the faster the heat transfer. The rate at which a material conducts heat depends on the thermal conductivity rating of the material.

Thermal conductivity is the ability of a material to conduct heat in the form of thermal energy. The higher the thermal conductivity rating, the faster the material conducts (transfers) heat. The lower the thermal conductivity rating, the slower the material conducts (transfers) heat.

Gases and liquids, such as air and water, have poor thermal conductivity. Solids, such as aluminum and copper, have good thermal conductivity. The thermal conductivity number for a given material is based on the ability of the material to transfer heat. The number is based on the amount of heat transferred through 1 sq ft of surface area for a given thickness in Btu per hour, per 1°F difference through the material. Aluminum (thermal conductivity = 128) is a much better thermal conductor than steel (thermal conductivity = 26.2). Since aluminum is a much better conductor of heat, aluminum is frequently used for heat sinks. **See Figure 8-7.** A *heat sink* is a device that conducts and dissipates heat away from an electrical component.

Temperature Conversion

Temperature is typically measured in degrees Fahrenheit (°F) or degrees Celsius (°C). Because both Fahrenheit and Celsius systems are commonly used in the electrical field and on most temperature test instruments, converting one unit to the other may be required. **See Figure 8-8.**

To convert a Fahrenheit temperature reading to Celsius, subtract 32 from the Fahrenheit reading and divide by 1.8. To convert Fahrenheit to Celsius, apply the following formula:

$$°C = \frac{°F - 32}{1.8}$$

where

°C = degrees Celsius

°F = degrees Fahrenheit

32 = difference between scales

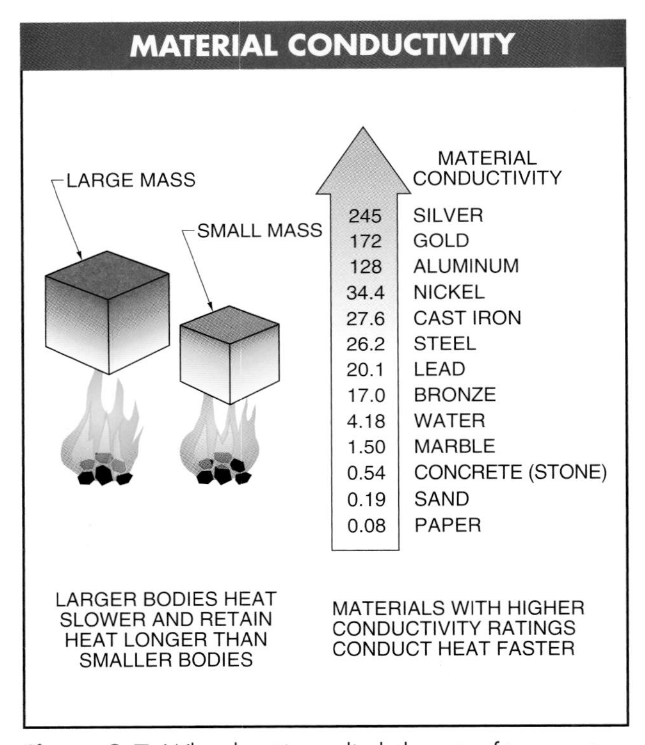

Figure 8-7. When heat is applied, the rate of temperature increase is affected by mass and conductivity.

Technical Tip

Specific heat capacity (c) of a substance is the quantity of heat required to increase the temperature of a unit mass of the substance by 1 degree. The specific heat capacity of a substance is a constant, the numerical value of which depends on the substance. For example, the specific heat capacity (c) of aluminum is 0.212, of ice is 0.430, and of water is 1.000.

1.8 = ratio between scales

For example, a digital thermometer indicates a reading of 158°F. Convert the Fahrenheit temperature to Celsius.

$$°C = \frac{°F - 32}{1.8}$$

$$°C = \frac{158 - 32}{1.8}$$

$$°C = \frac{126}{1.8}$$

$$°C = \mathbf{70}$$

To convert a Celsius temperature reading to Fahrenheit, multiply the Celsius reading by 1.8 and add 32. To convert Celsius to Fahrenheit, apply the formula:

$$°F = (1.8 \times °C) + 32$$

where

$°F$ = degrees Fahrenheit

1.8 = ratio between scales

$°C$ = degrees Celsius

32 = difference between scales

For example, a digital thermometer indicates a reading of 200°C. Convert the Celsius temperature to Fahrenheit.

$$°F = (1.8 \times °C) + 32$$

$$°F = (1.8 \times 200) + 32$$

$$°F = 360 + 32$$

$$°F = \mathbf{392°F}$$

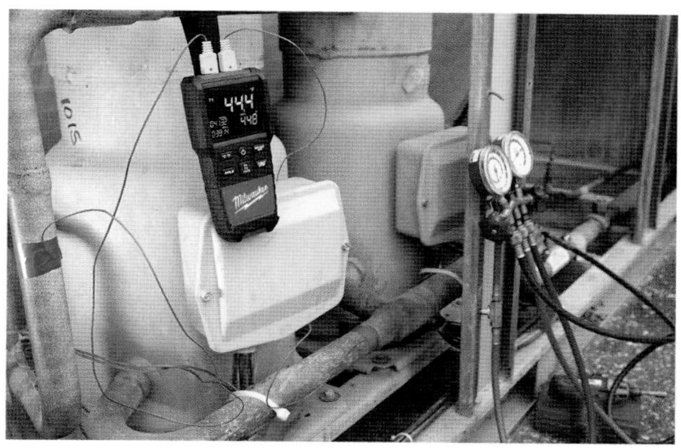

Milwaukee Tool Corporation

Interchangeable temperature probes used with temperature meters include the universal wire thermocouple and the pipe clamp type.

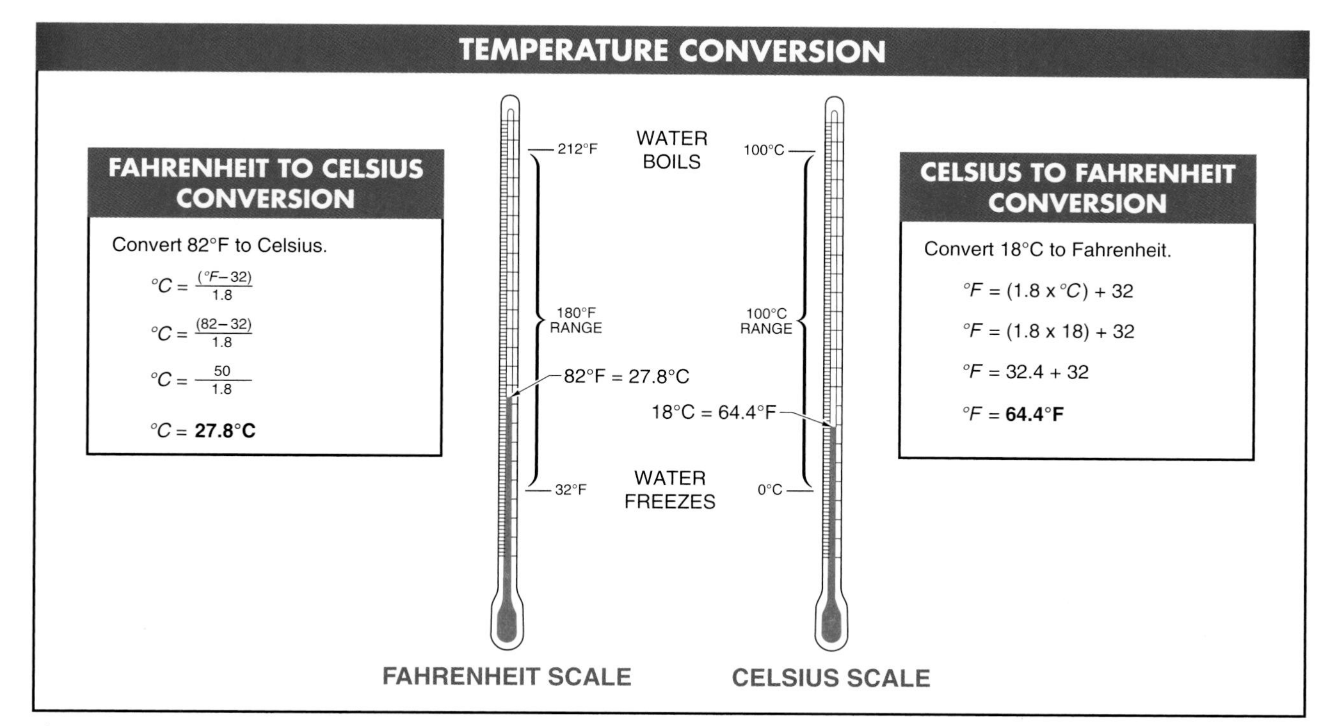

Figure 8-8. Temperature is commonly expressed in degrees Fahrenheit (°F) or degrees Celsius (°C).

Temperature Probes

Test instruments can measure temperature with a probe to take the temperature of the test point. A *temperature probe* is the part of a temperature test instrument that measures the temperature of liquids, gases, surfaces, and pipes. The temperature probe required depends on the material being measured, the temperature measurement range, and the accuracy required.

Contact Temperature Probes. A *contact temperature probe* measures temperature at a single point by direct contact with the area being measured. Contact temperature probes are used to measure the temperature of various solids, liquids, and gases, depending on the type of probe used. Contact temperature probes are connected directly to a temperature measuring test instrument, or the probes are connected to a temperature module between the probe and instrument. **See Figure 8-9.**

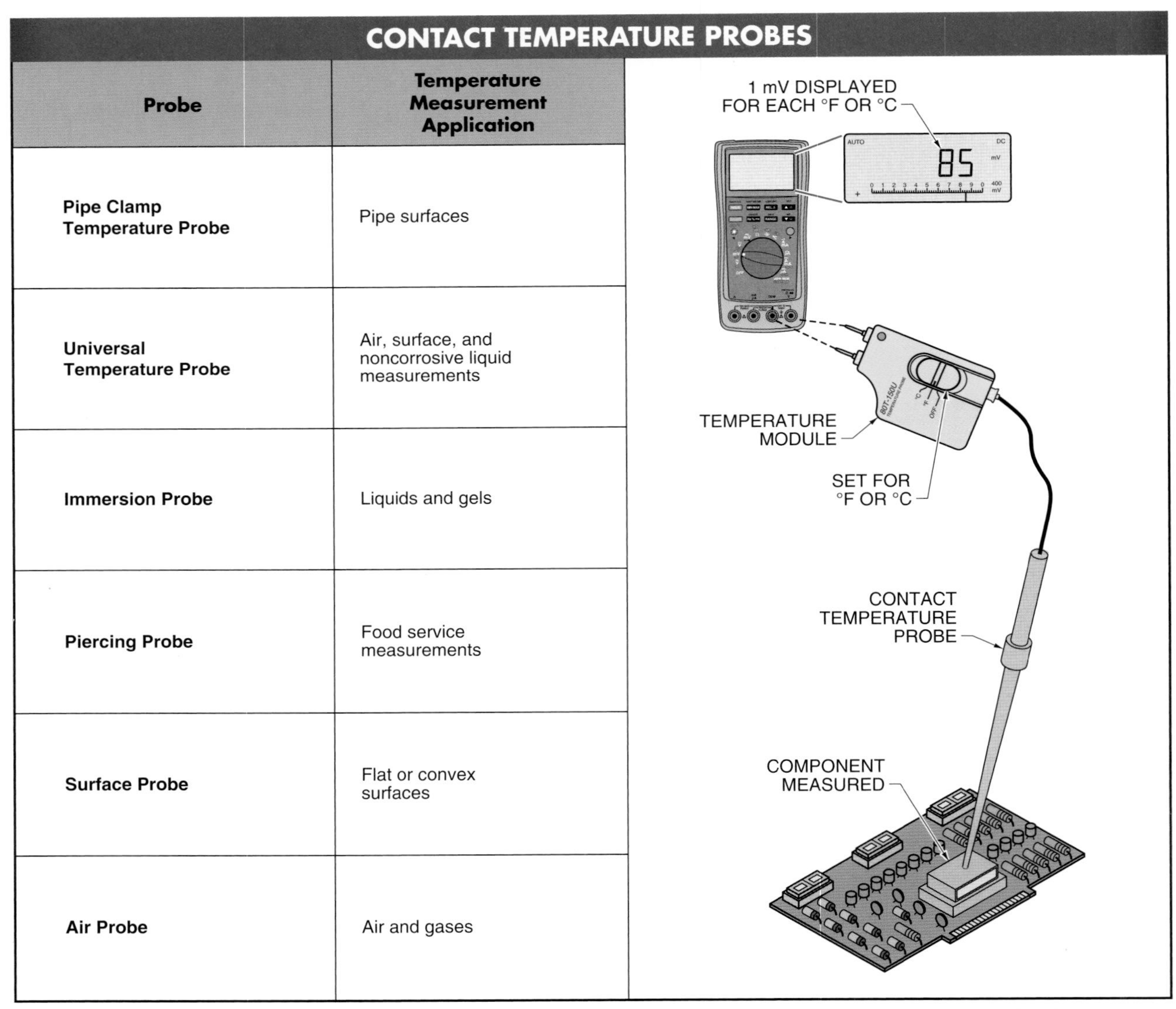

CONTACT TEMPERATURE PROBES

Probe	Temperature Measurement Application
Pipe Clamp Temperature Probe	Pipe surfaces
Universal Temperature Probe	Air, surface, and noncorrosive liquid measurements
Immersion Probe	Liquids and gels
Piercing Probe	Food service measurements
Surface Probe	Flat or convex surfaces
Air Probe	Air and gases

Figure 8-9. A contact temperature probe connected to a DMM can measure temperature at a single point by direct contact with the area measured.

The most common contact temperature probe uses a thermocouple for temperature measurement. A *thermocouple* is a device that produces electricity when two different metals that are joined together are heated. The voltage produced is proportional to the applied heat (measured temperature). The higher the temperature applied to a thermocouple, the higher the voltage produced by the thermocouple. The voltage produced is typically limited to a few millivolts. The voltage produced in mV DC is converted to a temperature measurement and displayed as degrees, typically in Fahrenheit or Celsius.

Noncontact Temperature Probes. A *noncontact temperature probe* is a device used for taking temperature measurements on energized circuits or on moving parts. An *infrared temperature probe* is a noncontact temperature probe that senses the infrared energy emitted by a material. All materials emit infrared energy in proportion to the temperature at the surface of the material. Infrared temperature probes are commonly used to take temperature measurements of electrical distribution systems, motors, bearings, switching circuits, and any other equipment where electrical heat buildup is critical. A noncontact temperature probe is an accessory for a DMM. **See Figure 8-10.**

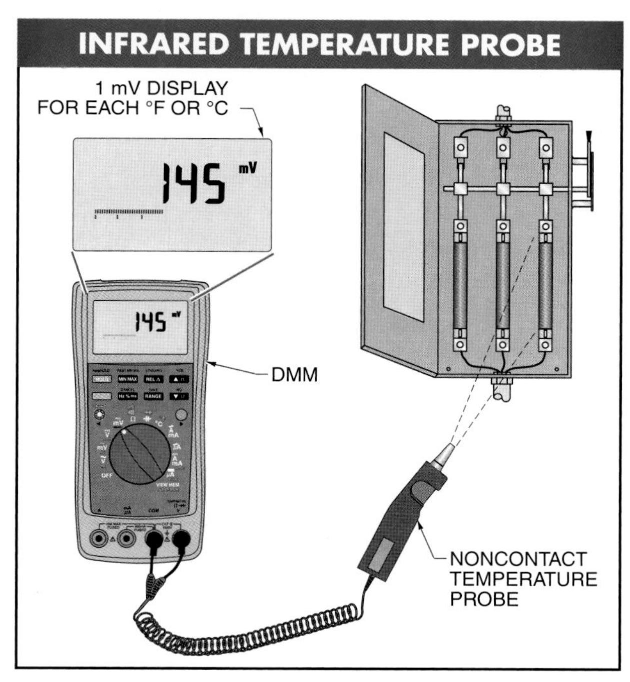

Figure 8-10. An infrared temperature probe connected to a DMM can be used to assess the condition of electrical connections and operating equipment.

CAUTION

When taking a temperature measurement, always use a temperature probe that is rated higher than the highest possible temperature to be measured.

Temperature Measurement Procedures

Temperature measurements can be taken using contact or noncontact temperature test instruments (noncontact thermometers). Considerations when taking a temperature measurement are identifying the best place in the circuit or system to take the measurement, and identifying the best type of temperature instrument and attachment to use for the application. **See Figure 8-11.**

To measure temperature using a temperature measuring test instrument, apply the following procedures:

Safety Procedures

- Follow all electrical safety practices and procedures
- Check and wear personal protective equipment (PPE) for the procedure being performed
- Perform only authorized procedures
- Follow all manufacturer recommendations and procedures

1. Turn the test instrument ON.
2. Ensure the test instrument measuring range exceeds the highest possible temperature measurement at the test point.
3. Set the test instrument to measure degrees Fahrenheit (°F) or degrees Celsius (°C).

4. Connect the test instrument to the test point, or point the noncontact thermometer in the direction of the equipment to be measured that is within the operating range recommended by the manufacturer.
5. Read and record the displayed value.
6. Remove the test instrument from the test point.
7. Repeat procedure for all additional measurements that are required.
8. Turn the temperature test instrument OFF.

Technical Tip

The electrical conductivity of a substance will change depending on the amount of humidity in the air.

HUMIDITY AND MOISTURE TEST INSTRUMENTS

Humidity is the amount of moisture (water vapor) in the air as a gas. Humidity varies according to the temperature and pressure of the air. The warmer the air is, the more water vapor the air can hold and the higher the humidity. When air cannot hold any more moisture, the air is saturated. Sensible humidity is typically expressed as relative humidity or dewpoint. *Relative humidity (% RH)* is the amount of moisture in the air compared to the amount of moisture the air can hold at saturation. Relative humidity is the most common method of expressing the amount of water vapor contained in air. For example, when air contains only half of the water vapor the air can hold, the relative humidity is 50%. Fog results when the relative humidity of air is 100%.

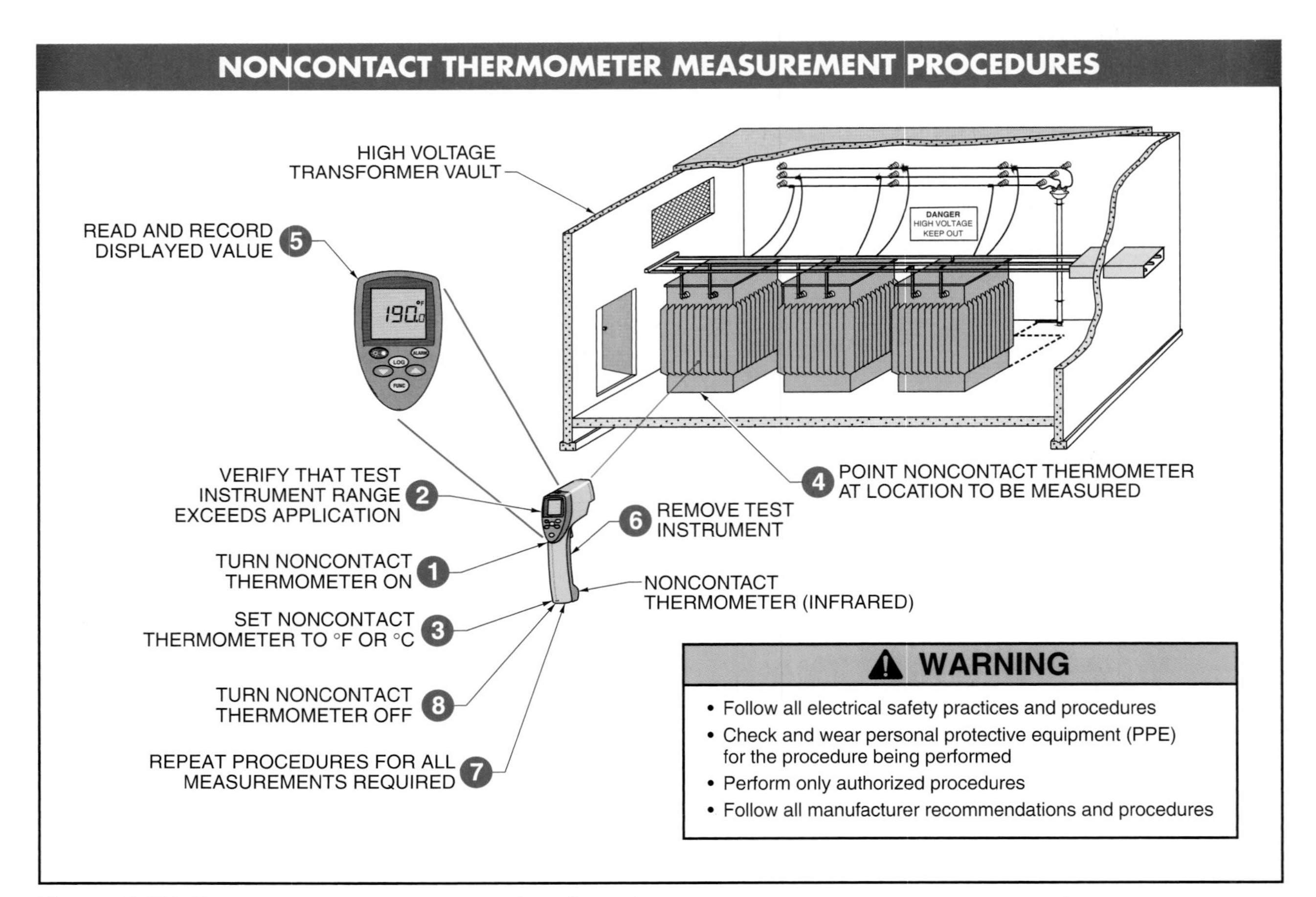

Figure 8-11. Temperature measurements can be taken using contact or noncontact temperature test instruments.

Moisture meters are used to measure the moisture content of building materials such as wood and insulation. Moisture meters are also used to measure the moisture content of industrial products such as corn or soybean germ.

Moisture Test Instrument Measurement Procedures

Measuring and controlling moisture is an important part of commercial and industrial applications such as in drying processes, test chamber regulation, heat-treating operations, pharmaceutical production, raw material storage, food processing, and HVAC systems. Moisture test instruments can sense a number of different factors, with relative humidity (degree of saturation) and absolute humidity (quantity of water vapor in a mixture of air and water) being the most common. Measuring and controlling moisture is an important part of the industrial process. The three main applications for moisture test instruments are measuring moisture in corn, in bulk solids or powders, and in fuels or liquids. **See Figure 8-12.** The electrical output from moisture test instruments can be analog current or voltage, frequency, or digital. Moisture test instruments are available in benchtop, handheld, or equipment-mounted models.

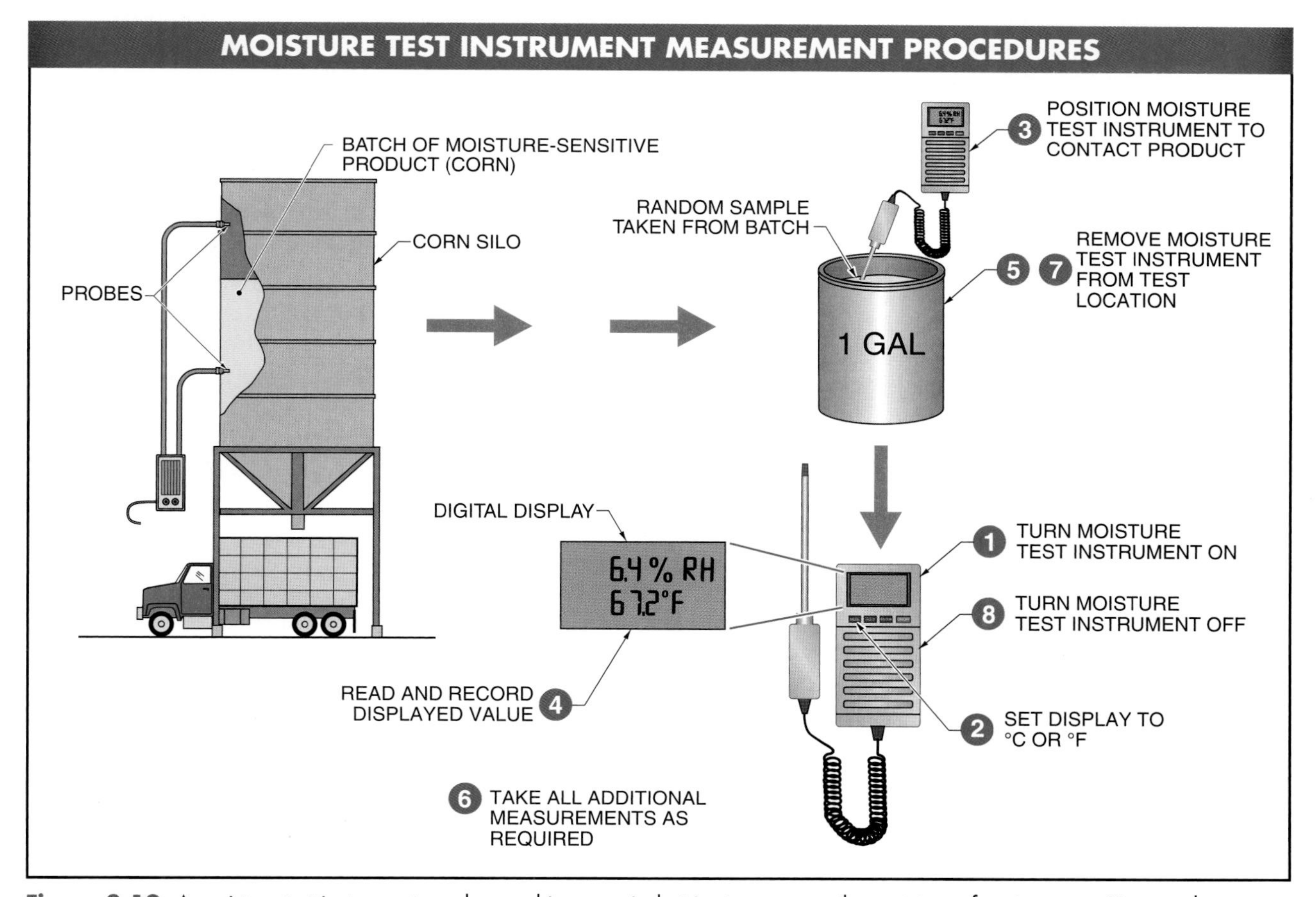

Figure 8-12. A moisture test instrument can be used in many industries to measure the moisture of moisture-sensitive products.

Before taking any moisture measurements using a moisture test instrument, ensure the instrument is designed to take measurements in the area being tested. Refer to the operating manual of the test instrument for all measuring precautions, limitations, and procedures. Always wear required personal protective equipment and follow all safety rules when taking the measurement. To measure moisture using a moisture test instrument, apply the following procedures:

1. Turn the moisture test instrument ON.
2. Most moisture test instruments will also measure the temperature of the air at the test point. Set the test instrument to display degrees Fahrenheit (°F) or degrees Celsius (°C).
3. Position the moisture test instrument for sampling the air or bring the test instrument in contact with the product being tested.
4. Read and record the displayed value.
5. Remove the test instrument from the test area or product.
6. Take all additional measurements that are required.
7. Remove the moisture test instrument from the test area or product being tested.
8. Turn the moisture test instrument OFF.

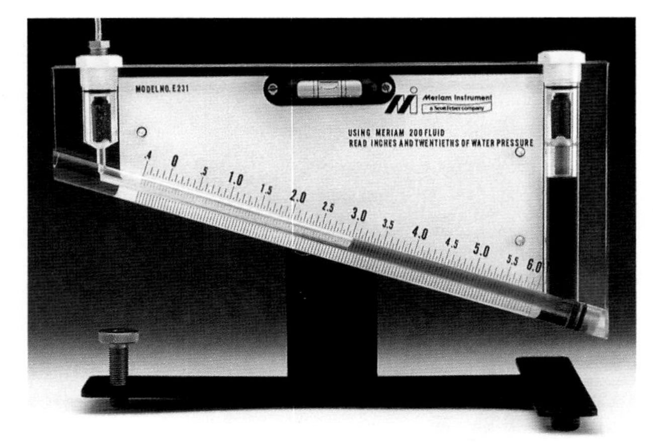

Meriam Process Technologies

Benchtop manometers are used to measure pressure in laboratory settings, quality control departments, or inspection areas.

Troubleshooting Tip

HUMIDITY RATIO

Humidity ratio (W) is the ratio of the mass (weight) of the moisture in a quantity of air to the mass of the air and moisture together. Humidity ratio indicates the actual amount of moisture found in the air. Humidity ratio is expressed in grains (gr) of moisture per pound of dry air (gr/lb) or in pounds of moisture per pound of dry air (lb/lb). A grain is a unit of measure that equals 1/7000 lb. For example, 1 lb of air may contain 78 gr or 0.0111 lb of moisture. Humidity represents latent heat. Latent heat is heat identified by a change of state and no temperature change. Therefore, latent heat cannot be measured with a thermometer.

PRESSURE TEST INSTRUMENTS

Pressure is force per unit of area. The exerted force always produces a deflection or change in the volume or dimension on the area to which it is applied. Pressure readings are taken by technicians and engineers to ensure safe and proper operation of equipment and facilities. Pressure is expressed in pounds per square inch (psi). Low pressures are expressed in inches of water column (in. wc). One psi equals 27.68 in. wc.

A difference exists between pressure and force. For example, a block of steel (10 sq in. footprint) weighing 200 lb that is on a table exerts only 20 psi (200 lb ÷ 10 sq in. = 20 psi). Also, a 200 lb force exerted over 10 sq in. of a fluid produces 20 psi in the fluid system. Force is measured with a scale (fixed, portable, or hanging) and pressure is measured with a gauge. **See Figure 8-13.**

The amount of pressure measured is displayed or stated as one of several types of pressure. When a pressure measurement is given and no unit is specified, then gauge pressure is presumed. **See Figure 8-14.** Pressure may be denoted as one of the following:

- Pounds per square inch = psi
- Gauge pressure = psig
- Absolute pressure = psia
- Differential pressure = psi or psid
- Inches of water = in. wc

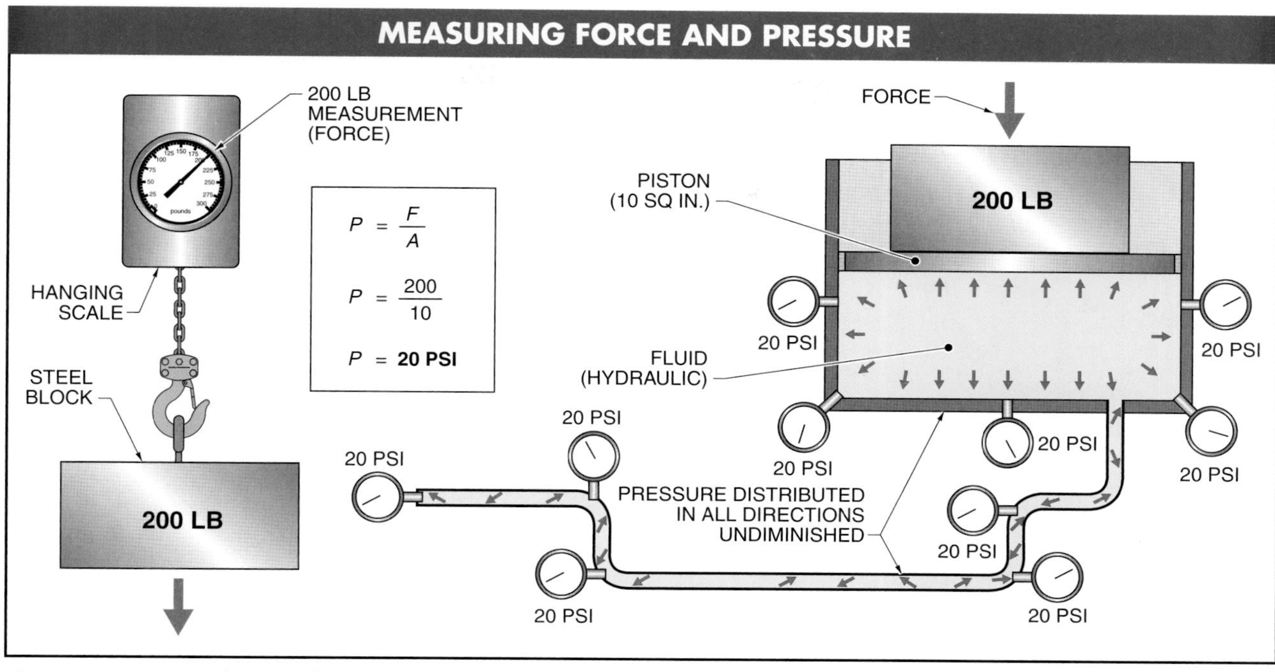

Figure 8-13. A scale is used to measure force while a pressure gauge is used to measure pressure.

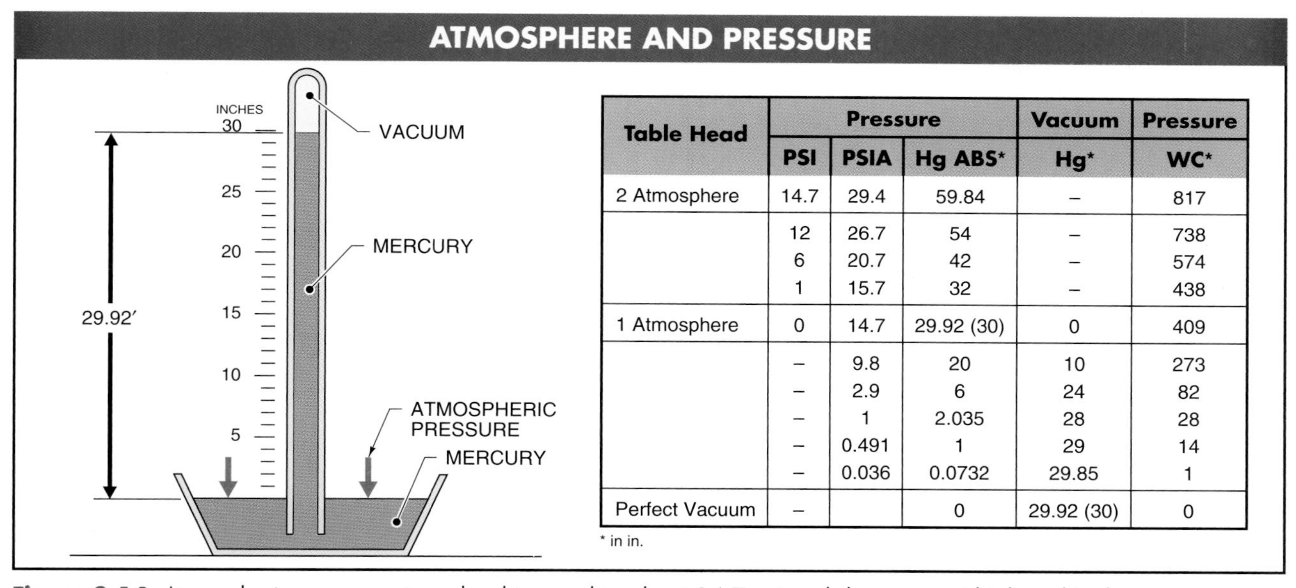

Table Head	Pressure			Vacuum	Pressure
	PSI	PSIA	Hg ABS*	Hg*	WC*
2 Atmosphere	14.7	29.4	59.84	–	817
	12	26.7	54	–	738
	6	20.7	42	–	574
	1	15.7	32	–	438
1 Atmosphere	0	14.7	29.92 (30)	0	409
	–	9.8	20	10	273
	–	2.9	6	24	82
	–	1	2.035	28	28
	–	0.491	1	29	14
	–	0.036	0.0732	29.85	1
Perfect Vacuum	–		0	29.92 (30)	0

* in in.

Figure 8-14. Atmospheric pressure at sea level is equal to about 14.7 psi and decreases at higher altitudes.

Gauge pressure is pressure relative to atmospheric pressure (barometric pressure). For example, when a tire gauge is used to measure the pressure in a tire, the tire pressure is measuring gauge pressure. Gauge pressure is the pressure above atmospheric pressure and vacuum is the pressure below atmospheric pressure. Normal atmospheric pressure is the pressure of the atmosphere at sea level, at 68°F, and with 36% humidity. Atmospheric pressure is 14.7 psia. **See Figure 8-15.**

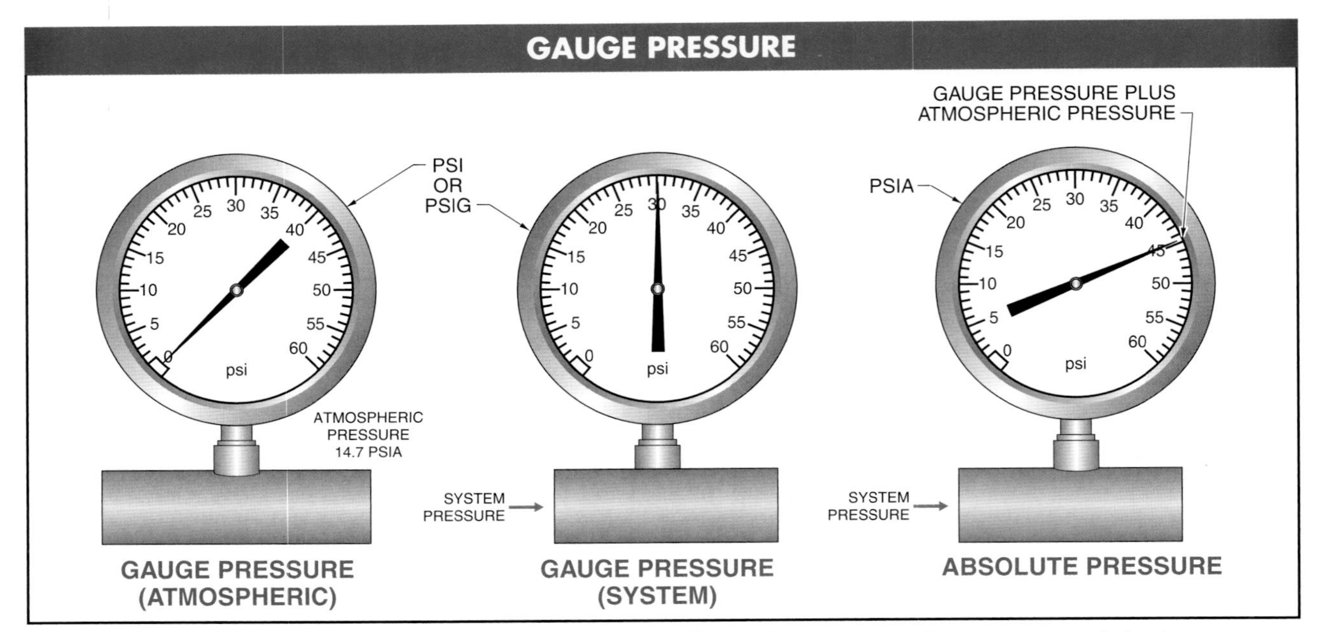

Figure 8-15. Gauge pressure shows the numerical value of the difference between atmospheric pressure and absolute pressure.

Absolute pressure is pressure measured from perfect vacuum, which is the zero point on the pressure scale. Absolute pressure measurements are not affected by changes in atmospheric pressure. Absolute pressure = gauge pressure + 14.7 psi. *Differential pressure* is the difference between two pressures where the reference pressure can be any pressure (not necessarily zero). *Vacuum* is an absolute pressure value expressed in inches of mercury, starting at atmospheric pressure and increasing in value (maximum is 29.92 in. Hg) as pressure drops to a perfect vacuum.

CAUTION

Fluids and gases under pressure are extremely dangerous. Ensure that there is no pressure in the system prior to connecting the pressure test instrument to the system that must be opened. The system must be completely purged, with all pressure removed.

Pressure Test Instrument Measurement Procedures

In many applications, measurement of system pressure or vacuum is required when troubleshooting or servicing the system. **See Figure 8-16.** Common pressure measuring instruments include differential pressure meters and manometers.

Determine the best place in the circuit/system to take the measurement. When a quick disconnect fitting for inserting a pressure test instrument is not included, the system must be opened to connect the pressure test instrument.

Before taking any pressure measurements using a pressure test instrument, ensure the instrument is designed to take measurements on the system being tested. Refer to the operating manual of the test instrument for all measuring precautions, limitations, and procedures. Always wear required personal protective equipment and follow all safety rules when taking the measurement. To measure pressure using a pressure test instrument, apply the following procedures:

1. Turn the pressure test instrument ON.
2. Ensure the test instrument measuring range exceeds the highest possible measurement in the system.
3. Connect the test instrument to the purged system.

4. Pressurize the system for pressure measurement.
5. Read and record the displayed pressure value on the pressure test instrument.
6. Repeat procedures for all additional measurements that may be required.
7. Purge the system of all pressure.
8. Remove the test instrument from the system.
9. Turn the pressure test instrument OFF.

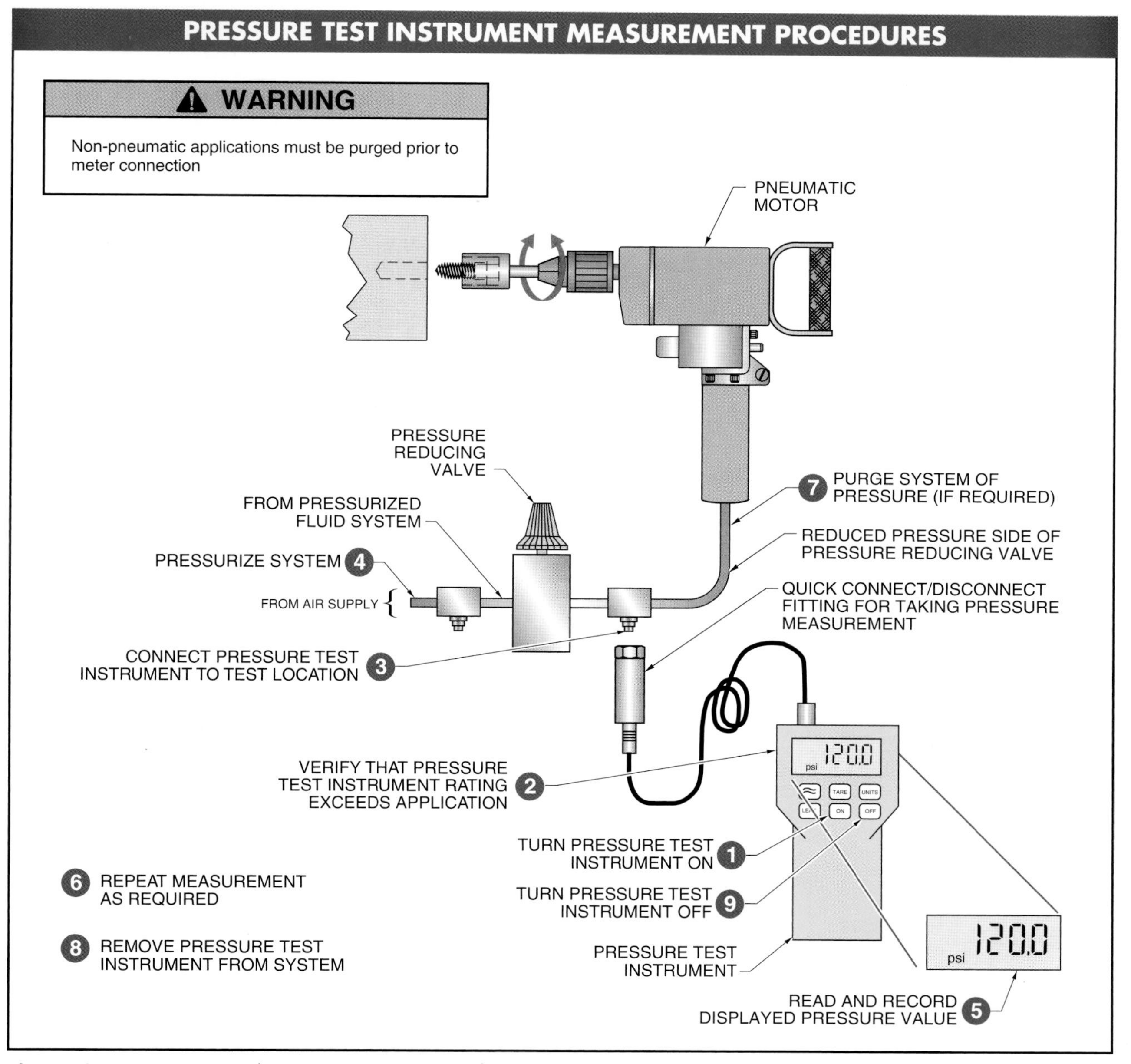

Figure 8-16. In many applications, measurement of system pressure or vacuum is required when troubleshooting or servicing the system.

FLOWMETERS

All processes require the flow of material. The amount of product flow must be measured and known at all phases of receiving, production, and delivery. A flowing material can be a gas, vapor, liquid, or solid particles. Applications that use flowmeters to measure flow include the following:

- boilers
- cooling lines
- air compressors
- fluid pumps
- food processing systems
- machine tools
- sprinkler systems
- water treatment systems
- heating processes
- refrigeration systems
- chemical processing and refining

A flow measurement is actually a flow rate measurement. Flow measurement units include the following:

- Gas = cubic feet per minute (cfm) or cubic feet per hour (cfh)
- Vapor (steam) = pounds per hour (pph)
- Liquid = gallons per minute (gpm), cubic feet per second (cfs), cubic feet per minute (cfm), barrels per hour (bph), or barrels per day (bpd)
- Solid (powders) = pounds per hour (pph) or tons per hour (tph)
- Slurry (liquid and solid mixture) = pounds per hour (pph) or tons per hour (tph)

Flow is measured directly by in-line fixed meters, or can be closely approximated by non-contact flowmeters. **See Figure 8-17.** In-line fixed-flowmeters are the most accurate, but may create problems by interfering with the flow of less viscous products such as powders and slurries.

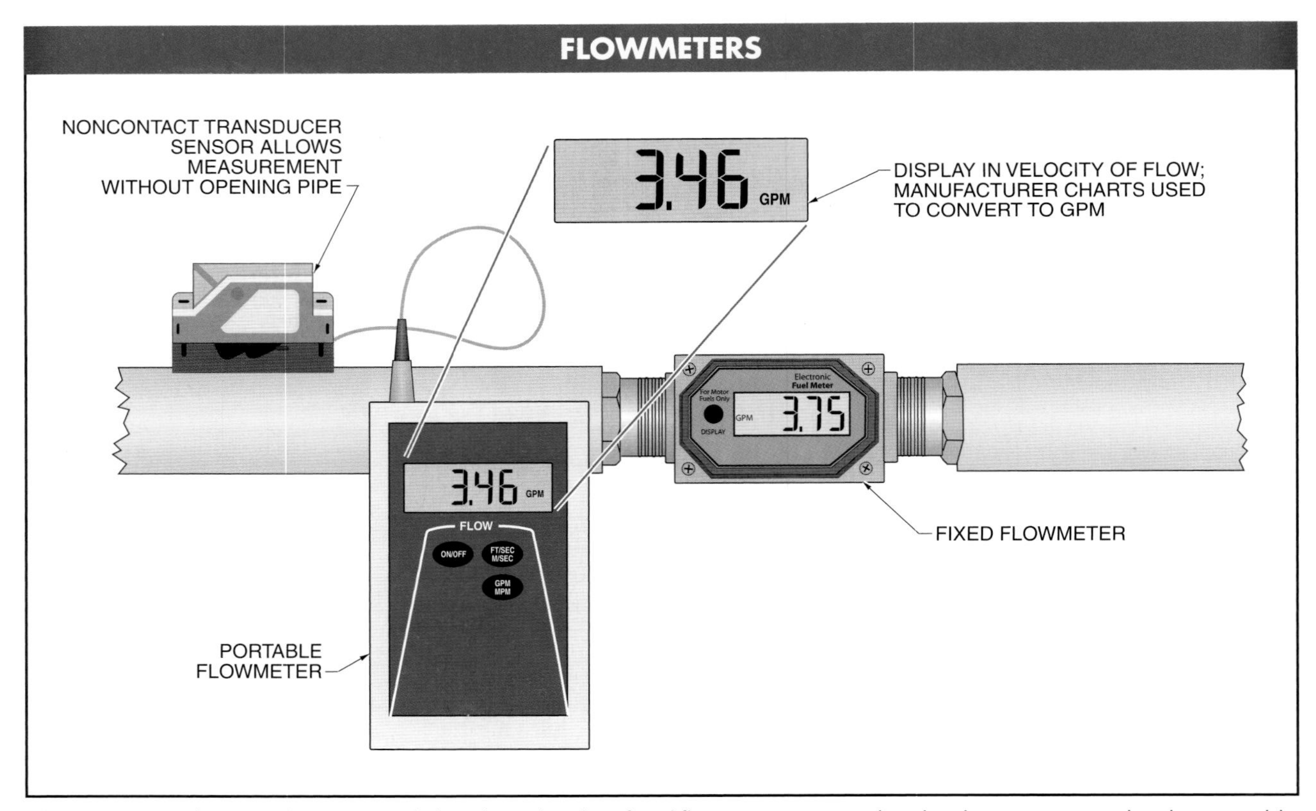

Figure 8-17. Flow can be measured directly with in-line fixed-flowmeters, or can be closely approximated with a portable noncontact flowmeter.

Noncontact flowmeters allow for a close approximation (often within ±2.0% or less, depending on product) measurement of flow. Noncontact flowmeters are ideal for testing or troubleshooting a system in which the indication of flow is more important than the exact value of the flow. Noncontact flowmeters operate by sending a signal into the pipe and measuring the reflections returning to the flow instrument. Thus, the readout is a velocity readout (ft/sec, etc.), but the velocity can easily be converted to a flow rate (gpm) by referring to manufacturers' charts, or the conversion is done automatically if the test instrument includes the required software. The most common industrial noncontact flowmeters utilize ultrasonic technology.

Flowmeter Measurement Procedures

Problems occur when the flow is stopped or slowed. Flow can be stopped by frozen pipes, clogged pipes, or improperly closed valves (manual or automatic). Flow measurements are taken using contact or noncontact flowmeters. **See Figure 8-18.**

Before taking any flow measurements using a flowmeter, ensure the instrument is designed to take measurements on the system being tested. Refer to the operating manual of the test instrument for all measuring precautions, limitations, and procedures. Always wear required personal protective equipment and follow all safety rules when taking the measurement. To measure flow using a flow test instrument, apply the following procedures:

1. Purge the system of pressure prior to connecting an in-line flow test instrument to the system that is to be opened.
2. Turn the flowmeter ON.
3. When a noncontact flowmeter is used, ensure the test instrument meets the requirements (allowable pipe material and size) set by the manufacturer for applications.
4. Connect the noncontact flowmeter to the transducer at the system test location.
5. Read and record the displayed flow value.
6. Repeat procedures for all additional tests that may be required.
7. Remove the flowmeter from the system.
8. Turn the flowmeter OFF.

GE Panametrics

Handheld flowmeters can be used to measure flow rate without opening the system.

AIR VELOCITY TEST INSTRUMENTS

The movement (flow) of air is important to various processes and applications. *Velocity* is the speed at which air, liquids, or solids travel through a system. The movement of air over heating elements and cooling elements is required in heating and air conditioning systems. When air has excessive velocity, air does not transfer the appropriate amount of heat from heat exchangers in the system. Air velocity test instruments are used to measure the speed of airflow. Air velocity test instruments display air velocity in:

- knots (kn)
- miles per hour (mph)
- feet per second (fps)
- feet per minute (fpm)
- meters per second (mps)
- kilometers per hour (km/h)

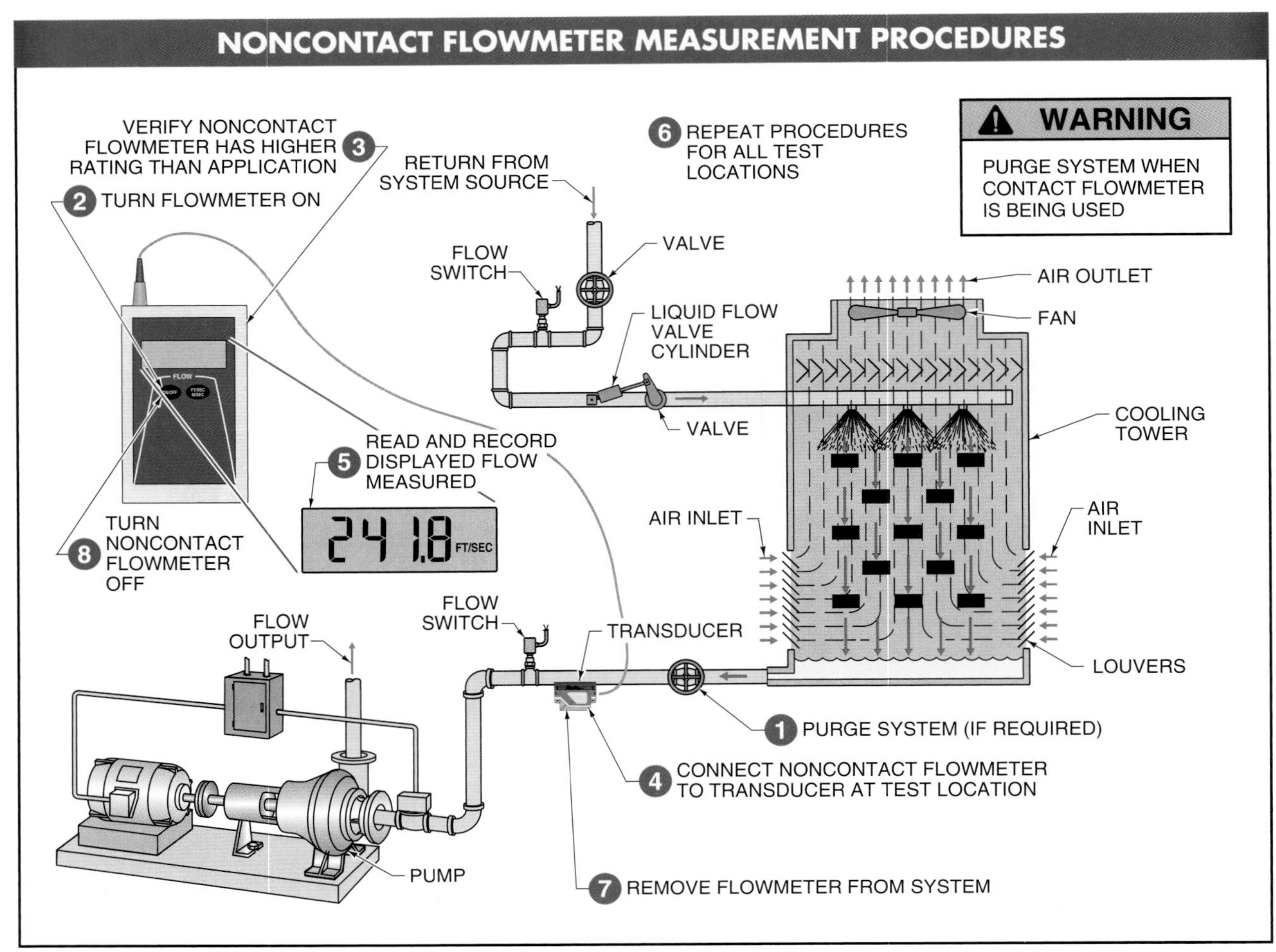

Figure 8-18. A noncontact flowmeter may be used along various locations of complex piping systems to check for problems that cause the flow to stop or slow down.

Air Velocity Test Instrument Measurement Procedures

Air velocity test instruments are used to test the speed of air coming out of ventilation ducts, exhaust systems, and other applications that require a measurement of air velocity. **See Figure 8-19.** A common use for air velocity test instruments is monitoring the flow of air when servicing heating, air conditioning, and ventilating equipment. Air velocity test instruments are also used for outdoor venues such as a sports stadium, aircraft control, monitoring weather conditions (anemometers), and other outdoor airflow applications. Some types are available with a temperature indicator in addition to a velocity indicator.

Before taking any air velocity measurements using an air velocity test instrument, ensure the instrument is designed to take measurements on the system being tested. Refer to the operating manual of the test instrument for all measuring precautions, limitations, and procedures. Always wear required personal protective equipment and follow all safety rules when taking the measurement. To measure air velocity using an air velocity test instrument, apply the following procedures:

1. Turn the test instrument ON.
2. Set the selector switch of the air velocity test instrument to display the desired units (fps, mph, or mps).

3. Position the air velocity test instrument at the test location.
4. Read and record the displayed value.
5. Repeat procedures for additional measurements that may be required.
6. Remove the air velocity test instrument from the test location.
7. Turn the air velocity test instrument OFF.

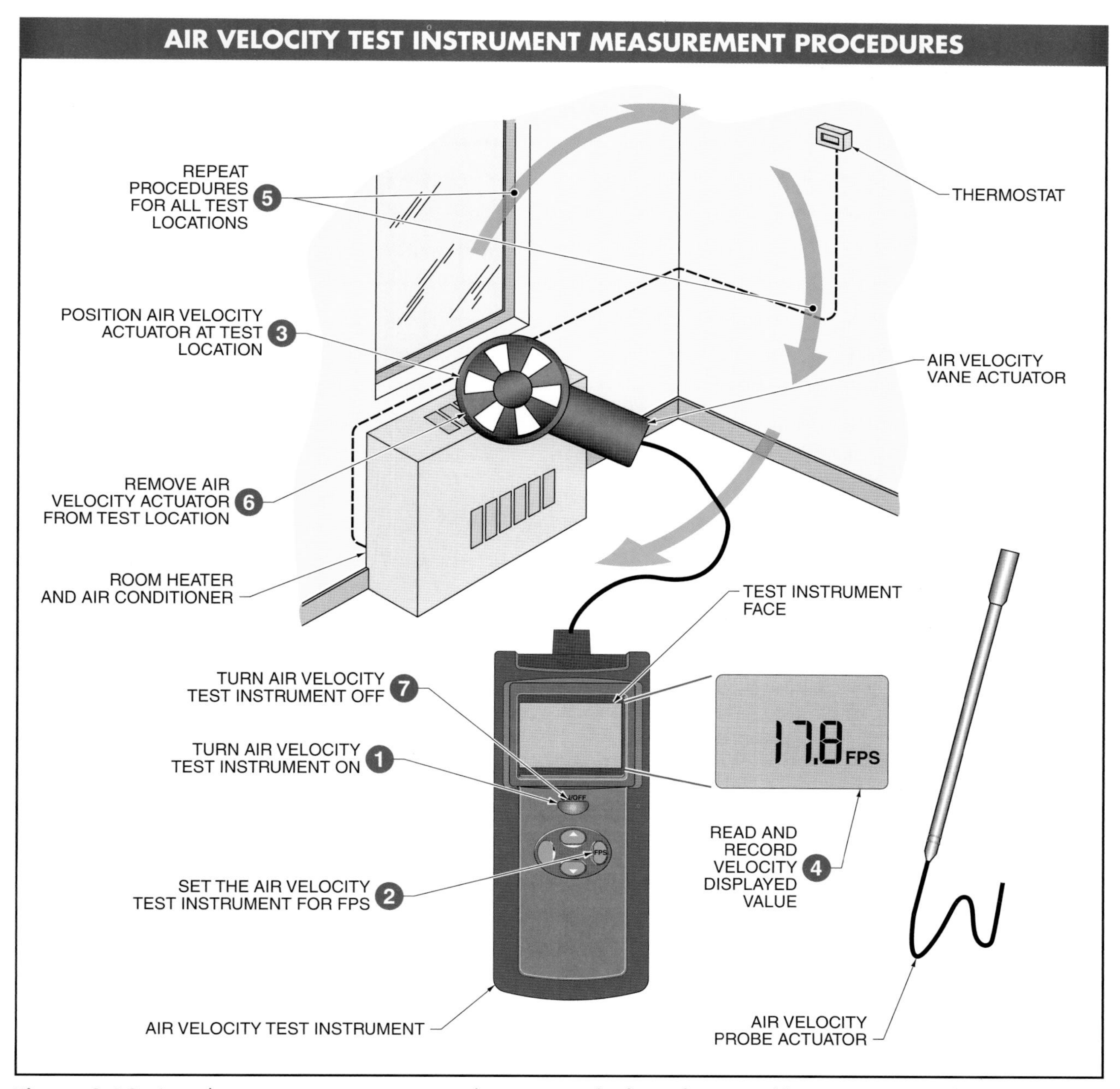

Figure 8-19. Air velocity test instruments are used to measure the flow of air out of heating units, ventilation ducts, and exhaust systems, and for other applications that require a measurement of airflow.

Troubleshooting Tip

CONDUCTOR APPLICATIONS

Wires and cables are used in building electrical systems to transmit electric power from the entry point of electric service (the utility entry point in the building) to the various outlets, fixtures, and utility devices. Electrical wires are designed for 600 V operations but are commonly used at voltages below that value such as 120 V, 240 V, or 480 V. Insulations that are typically used include thermoplastics, natural and synthetic rubber, and TPO (thermoplastic olefin) compounds. Rubber insulations are usually covered with an additional outer jacket such as braids or PVC, for better abrasion resistance. Electrical wires for use in buildings are separated by type according to application by the NEC®. Classification is by a letter that designates the insulation type and application characteristics. For example, type R indicates rubber or synthetic rubber insulation and TW indicates thermoplastic insulation for use in wet environments. The letter H indicates heat resistance. Other commercial insulation materials include silicone, polyethylene, polypropylene, or variations of these materials. Electrical conductors are available in duplex and multiple conductor assemblies with each individual conductor covered with an insulating material and with the entire assembly covered by an outer jacket. Bare wires and cables are used for outdoor applications such as power distribution and transmission lines. Insulators that are made from porcelain, or glass in older systems, support bare wire conductors along power transmission lines.

CONDUCTIVITY TEST INSTRUMENTS

Conductivity (G) is the ability of a substance or material to conduct electric current. Conductivity is the opposite of resistance. All substances conduct electric current to some degree. Insulating materials such as glass, rubber, and plastic have extremely low conductivity and metals such as silver, copper, and aluminum have very high conductivity. The amount of conductivity a material has can be measured using a conductivity meter.

The measurement of conductivity is required for industrial applications in which the quality of water or the composition of a liquid solution must be known. Many applications exist where conductivity of a solution or object must be known, for example in hospitals; breweries; hydroelectric plants; boiler feedwater systems; and food processing, petroleum processing, agriculture, chemical processing, mining, marine, wastewater treatment, manufacturing, and power generation operations.

The unit of conductance is the siemens (S), which is the metric equivalent of the mho (1 mho = 1 siemens). Conductivity is typically measured between opposite sides of a 1 cm cube of the material being tested. The 1 cm cube measurement indicates the conductivity in units of siemens/centimeter (S/cm). The unit S/cm is too large for most conductance tests. For this reason, the unit of conductivity on most conductivity meters is mS/cm (millisiemens) or µS/cm (microsiemens). **See Figure 8-20.** For the best conductance test results, use a conductivity test instrument that has automatic temperature compensation.

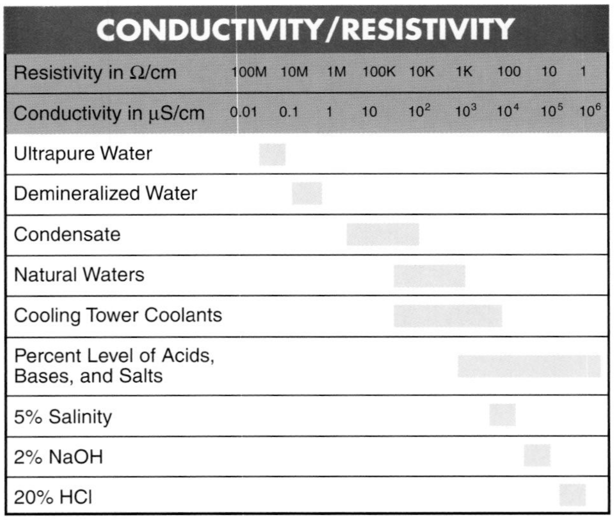

CONDUCTIVITY/RESISTIVITY

Resistivity in Ω/cm	100M	10M	1M	100K	10K	1K	100	10	1
Conductivity in µS/cm	0.01	0.1	1	10	10^2	10^3	10^4	10^5	10^6
Ultrapure Water									
Demineralized Water									
Condensate									
Natural Waters									
Cooling Tower Coolants									
Percent Level of Acids, Bases, and Salts									
5% Salinity									
2% NaOH									
20% HCl									

Figure 8-20. The measurement of conductivity is required for industrial applications in which the quality of water or the composition of a liquid solution needs to be known.

Conductivity Test Instrument Measurement Procedures

Before taking any conductivity measurements using a conductivity test instrument, ensure the instrument is designed to take measurements on the object being tested. Refer to the operating manual of the test instrument for all measuring precautions, limitations, and procedures. Always wear required personal protective equipment and follow all safety rules when taking the measurement. **See Figure 8-21.** To measure the amount of conductivity using a conductivity test instrument, apply the following procedures:

1. Turn the conductivity test instrument ON.
2. Set the selector switch of the conductivity test instrument to the highest range (or expected range).
3. Mix a solution (when required) and/or take readings at various locations. Fill a nonconductive container (made of glass, rubber, or plastic) with a sample of the solution to be tested.
4. Place the measuring probe in the solution being tested or on opposite sides of a 1 cm cube. Keep the probe in the solution or on the cube until the reading has stabilized.
5. Read and record the displayed value. *Note:* To ensure proper readings, calibrate the conductivity test instrument on a regular basis, as specified by the manufacturer.
6. Remove the conductivity test instrument from the solution or cube.
7. Repeat procedures for all additional measurements that may be required to properly calculate an average.
8. Turn the conductivity test instrument OFF.

CAUTION

To ensure an accurate conductivity measurement, the solution within the radius of the meter probe location must be representative of the solution as a whole.

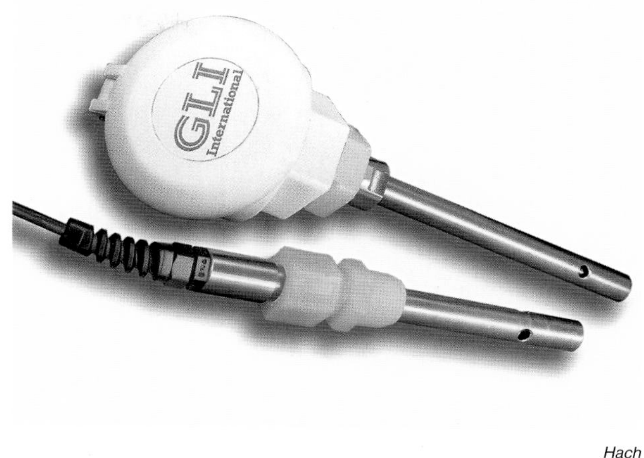

Hach

A conductivity tester probe is immersed in a liquid solution to measure the liquid's ability to conduct electrical current.

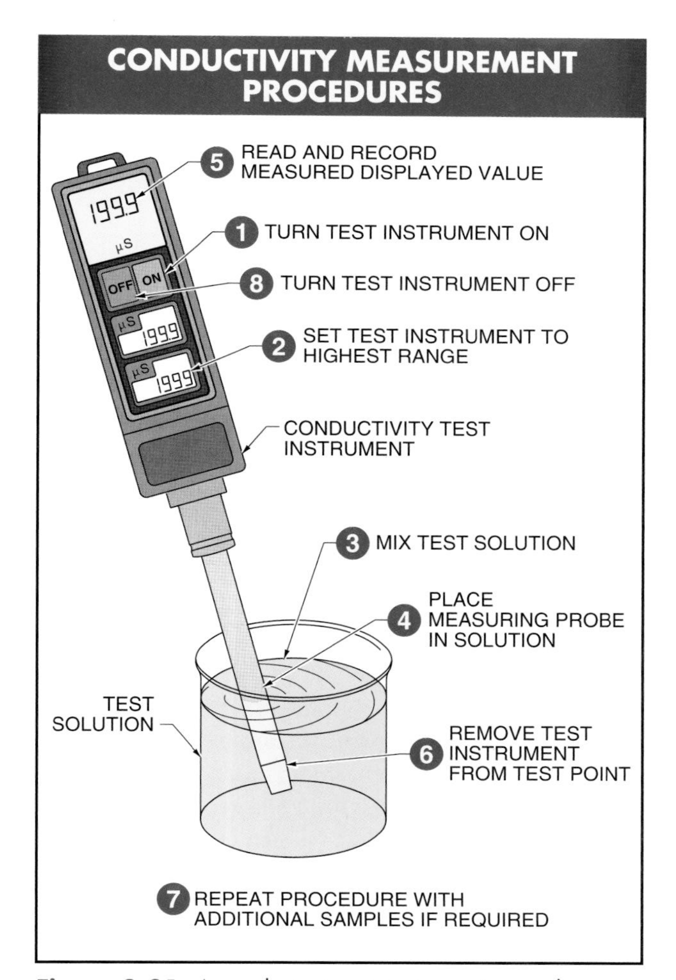

Figure 8-21. A conductivity test instrument is used to measure the conductivity of liquid solutions.

Troubleshooting Tip

REFRIGERANT RECOVERY

All refrigerants must be recovered before disposing of any type of equipment that has refrigerant in its system. Recovery equipment typically has desiccant packages to trap moisture, and an in-line particulate filter of 15 micron size to trap solids. Long hoses between the air conditioning or refrigeration unit and the recovery machine must be avoided to prevent excess pressure drops. Long hoses also cause an increase in refrigerant emissions if refrigerants escape to the atmosphere, because a longer hose has more volume. To facilitate refrigerant recovery, the EPA requires that service apertures or process stubs be installed on all appliances containing Class I or Class II refrigerants.

GAS AND REFRIGERANT TEST INSTRUMENTS

Process gases, refrigerant gases, and other gases are kept under pressure in storage tanks when not being used in a system. With any system under pressure, leaks can develop. Large leaks are quickly detected by smell or loss of process requirements (cooling). Small leaks can be much harder to detect. Gas and refrigerant leak detection test instruments are used to detect large and small leaks. **See Figure 8-22.** A *leak detector* is a device that is used to detect refrigerant or other gas leaks in pressurized air conditioning, refrigeration, or process systems. Leak detection is performed by using leak detection methods or one of many different types of leak detectors available such as soap bubbles, or electronic, fluorescent, ultrasonic, fixed, or halide torch leak detectors.

Soap bubbles typically from a spray bottle are used to pinpoint leaks in a system by covering the piping with a soapy solution. As the refrigerant or other gas leaks, bubbles are formed at the point of the leak. Soap bubbles are used as a safe way to detect leaks from a system. Soap bubbles can only be used in a system that is fully pressurized.

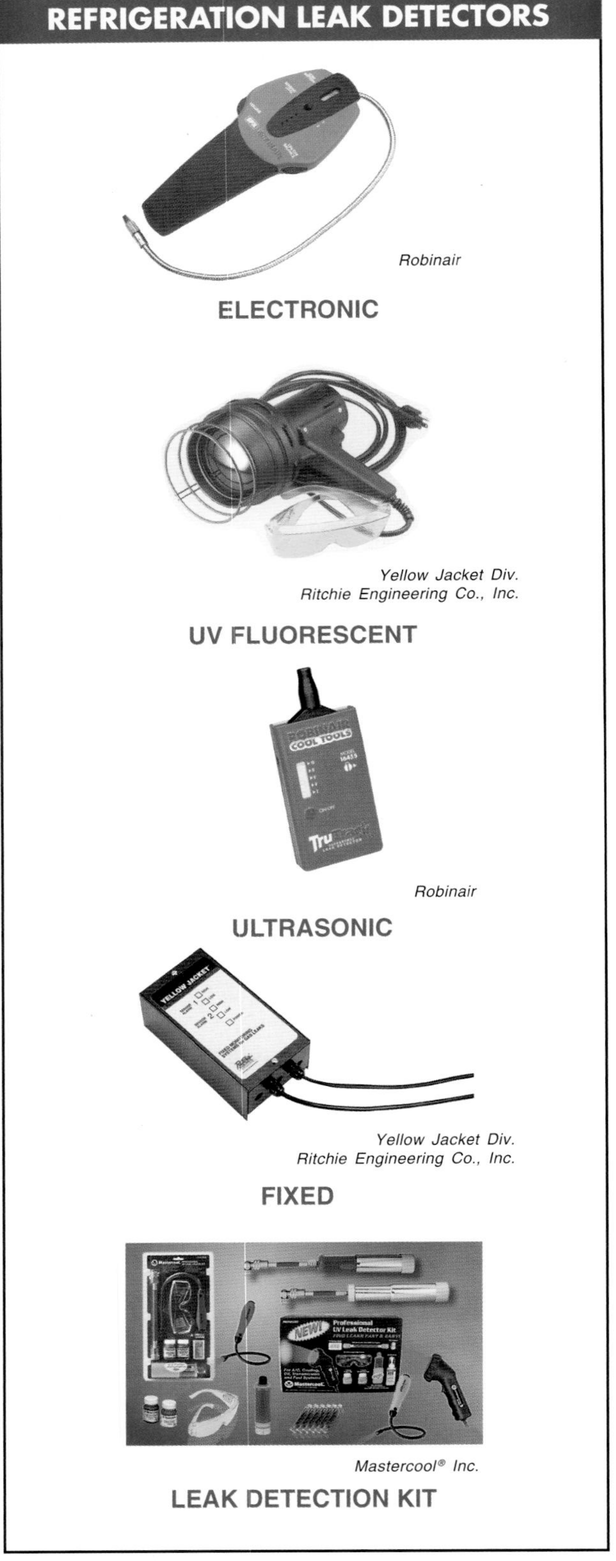

Figure 8-22. A leak detection device detects refrigerant leaks in air conditioning or refrigeration systems.

An *electronic leak detector* is a leak detector that detects the presence of gases. Handheld electronic leak detectors are considered the industry standard for detecting the location of refrigerant leaks. Electronic and ultrasonic leak detectors detect the general area of small refrigerant leaks. Electronic leak detectors work well for detecting small leaks, but large leaks can cause false positives. A *fluorescent leak detector* is a leak detector that uses a UV light to detect fluorescent dye that is added to a system. Fluorescent leak detection works well as long as the system can be allowed to run after the dye has been injected into the system. An *ultrasonic leak detector* is a leak detector that listens for the sounds created by a leak. Ultrasonic leak detectors can pinpoint the location of a leak but cannot determine the size of a leak or the type of refrigerant or gas that is leaking. A *fixed leak detector* is a stationary leak detector system with sensors and controllers to detect one specific type of refrigerant. A *halide torch leak detector* is a leak detector that uses a torch flame that changes color depending on the type of gas pulled across a copper element. Halide torch leak detectors can determine the type and amount of gas that is leaking. Halide torch leak detectors are the least common type of leak detector used because of safety concerns created by the open flame.

Nitrogen (actually nitrogen with a trace of R-22 refrigerant) can also be used to pressurize a system to determine if there are any leaks in the system. Nitrogen must not be added to a fully charged system and released to the atmosphere. Electricians must always verify the maximum test pressure allowed in a system by checking the pressure rating on the nameplate of the equipment being tested.

Standing pressure and standing vacuum are pressure tests used to determine the presence of a leak and possibly the size of the leak. Neither standing pressure nor standing vacuum can be used to determine the location of a leak. Leaks in a system that is not pressurized are found by charging the system with nitrogen and a small amount of refrigerant. The system is charged to a specified pressure and then tested using an electronic leak detector.

Troubleshooting Tip

EPA EVACUATION REQUIREMENTS

Since July 13, 1993, technicians have been required to evacuate refrigeration and air conditioning equipment to establish vacuum levels when opening equipment. An EPA-approved equipment testing organization must certify recovery and recycling equipment. Technicians who add refrigerant to equipment to top off the system are not required to evacuate the system.

Gas and Refrigerant Handling

Gases and refrigerants are dangerous when allowed to leak out of sealed systems and mix with air. When gases and refrigerants are properly contained in cylinders or other containers, hazards are reduced. All gases and refrigerants, regardless of quantity, must be handled with extreme care.

Gas and refrigerant containers are dangerous when exposed to open flames or high temperatures. Most gases and refrigerants boil at very low temperatures, and when heated, the compounds change chemically, generating toxic fumes. Always refer to the manufacturer-recommended safety procedures when handling gases and refrigerants used in air conditioning, refrigeration, and process systems. Gas and refrigerant safety rules include the following:

- Store gas and refrigerant containers in a clean, dry area out of direct sunlight. Never heat gas and refrigerant containers above 125°F.
- Comply with fire regulations concerning storage quantities, types of approved containers, and proper labeling.
- Never allow a gas or refrigerant to come in contact with skin. Gases and refrigerants can poison through skin pores, dry out skin, and cause frostbite. Always use gloves and face protection.
- Never allow gas or refrigerant vapors to build up in a low or confined area. Some gases and fluorocarbon refrigerants are heavier than air and can cause suffocation, heart irregularities, or unconsciousness due to lack of oxygen when exposure exceeds acceptable levels.

- Read the label on the gas or refrigerant cylinder to identify contents and verify color coding.
- Never use oxygen or compressed air to pressurize air conditioning, refrigeration, or process systems to check for leaks. When mixed with compressor oil, oxygen or compressed air can cause an explosion. Air conditioning or refrigerant systems with an R-134a refrigerant charge must be leak-checked with pressurized nitrogen. **See Figure 8-23.**
- Immediately clean up any type of spill.

The ideal gas or refrigerant is environmentally friendly, nonflammable, nontoxic, and able to perform as intended in the air conditioning, refrigeration, or process system. Gases and refrigerants are not completely safe, but can be used safely.

ANSI/ASHRAE 34 classifies refrigerants into safety groups. A1 refrigerants are the safest, and B3 refrigerants are the most toxic and flammable. The letters A and B indicate level of toxicity (B being higher), and numbers 1, 2, and 3 indicate level of flammability (3 being the highest). R-11, R-12, R-22, R-500, R-502, and R-134a are classified as A1 refrigerants and R-123 is classified as a B1 refrigerant.

An oxygen-deprivation sensor is required to detect low oxygen levels in work areas. Typically, oxygen alarms will alarm at 19.5% or less by volume. Monitoring of rooms for the right amount of oxygen is required for all areas with refrigerants. A self-contained breathing apparatus (SCBA) must be worn when a large leak has occurred. When SCBA is not available, electricians must ventilate the area or vacate the area immediately.

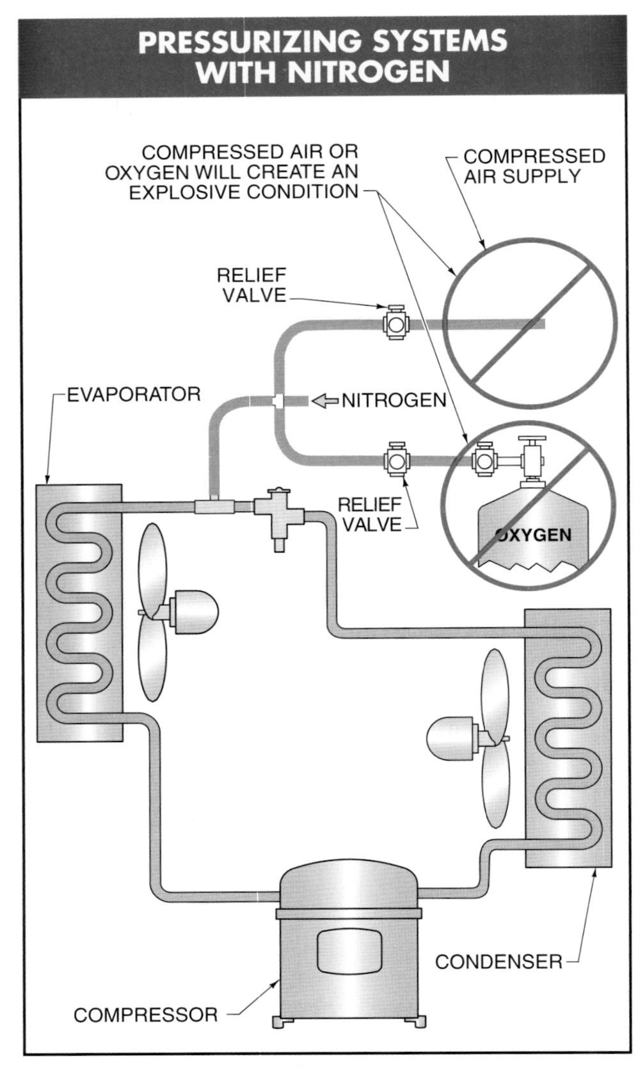

Figure 8-23. Compressed air or oxygen used to pressurize a system creates an explosive condition.

Gas and Refrigerant Leak Test Instrument Measurement Procedures

To obtain any information on any refrigerant, a material safety data sheet (MSDS) can be obtained from the manufacturer. When working with any refrigerant, the technician must review the MSDS for that refrigerant.

Before performing any leak tests using a gas or refrigerant leak test instrument, ensure the instrument is designed to take measurements on the system being tested. Refer to the operating manual of the test instrument for all measuring precautions, limitations, and procedures. **See Figure 8-24.** Always wear required personal protective equipment and follow all safety rules when taking the measurement. To measure gas or refrigerant leaks using a gas or refrigerant leak test instrument, apply the following procedures:

1. Turn the leak test instrument ON.
2. When the expected leak is small, set the selector switch to the lowest (most sensitive) range.
3. Position the leak detector probe of the test instrument at the location of the suspected leak.
4. Read and record the displayed value or indicator lights.
5. Repeat procedures for all additional measurements required to locate the leak.
6. Remove the leak test instrument from the testing location.
7. Turn the leak test instrument OFF.

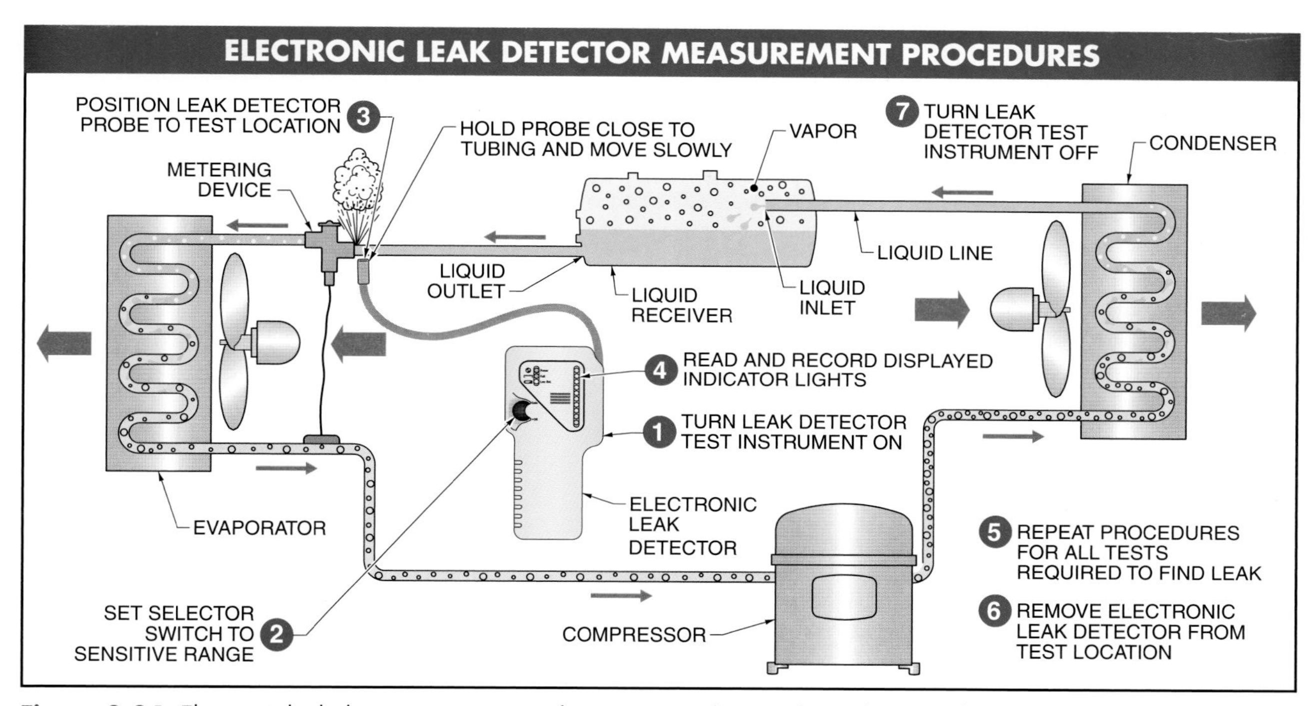

Figure 8-24. Electronic leak detectors are extremely sensitive and can indicate the exact location of a leak.

THICKNESS TEST INSTRUMENTS

The thickness of any solid object can be measured if full access to all sides of the object is possible. However, thickness measurements are difficult to take on objects that are installed and in operation. Thickness measurements on objects such as pipes are used to indicate size and condition of the pipe. Thickness test instruments are used to measure the thickness of pipe walls, steel, plastic, cast iron, and solid materials of any shape. Various inspectors use thickness test instruments as part of preventive maintenance programs, quality control (QC) programs, research and development (R&D), and for troubleshooting manufacturing processes.

Thickness Test Instrument Measurement Procedures

Thickness test instruments measure thickness by transmitting energy waves through the test material (object being tested) and interpreting the return signals as measurements. **See Figure 8-25.**

Before taking any thickness measurements using a thickness test instrument, ensure the instrument is designed to take measurements on the object being tested. Refer to the operating manual of the test instrument for all measuring precautions, limitations, and procedures. Always wear required personal protective equipment and follow all safety rules when taking the measurement. To measure the thickness of an object using a thickness test instrument, apply the following procedures:

1. Turn the thickness test instrument ON.
2. Connect the thickness test instrument to the object.
3. Read and record the displayed value.
4. Repeat procedure for all additional measurements that may be required.
5. Remove the thickness test instrument from the object.
6. Turn the thickness test instrument OFF.

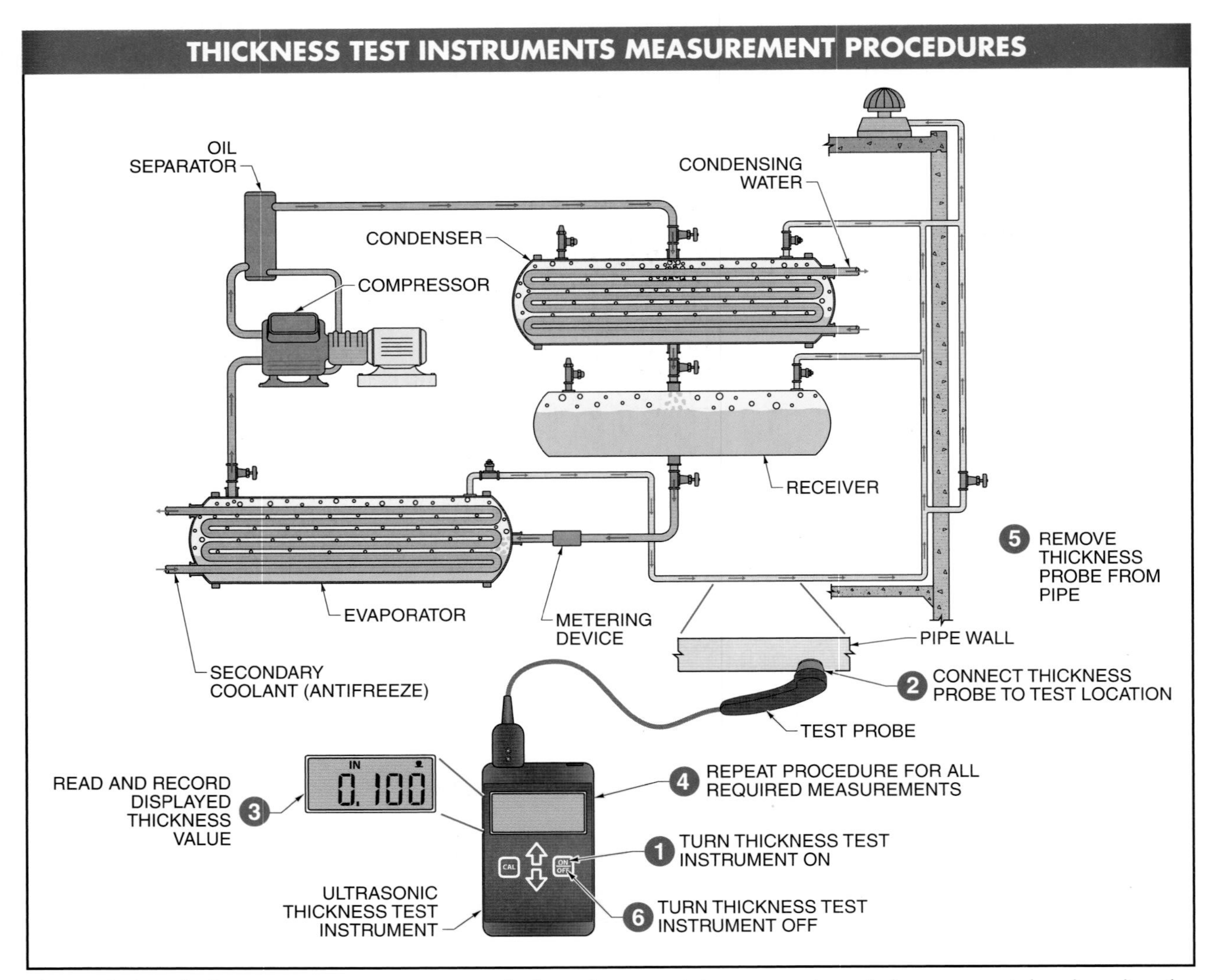

Figure 8-25. Thickness test instruments measure thickness by transmitting energy through the test material and reading the measured return signals.

INSTRUMENTATION AND PROCESS CONTROL TEST INSTRUMENTS

Name: ______________________________ Date: ____________________

______ **1.** What unit is used to measure conductivity?

T F **2.** Resolution is the percentage of accuracy in an instrument.

T F **3.** Conductivity is the opposite of capacitance.

______ **4.** What is the most common unit for measuring pressure?

______ **5.** What is the accuracy (in psi) of a gauge that reads 80 psi and lists its accuracy as ±0.5% of reading?

______ **6.** Which heat energy unit is defined as the amount of heat needed to raise the temperature of 1 lb of water by 1°F?

T F **7.** Pressure is force per unit of area.

______ **8.** Which test instrument would most likely be used when determining if a corroded pipe wall is still structurally sound?

______ **9.** Which two characteristics of a gas affect its relative humidity?

______ **10.** What type of test instrument is used to check the accuracy of standard test instruments?

______ **11.** Which has a higher thermal conductivity, water or gold?

T F **12.** The amount of uncertainty in a reading remains constant for any reading on an instrument that lists its accuracy as a percentage of full face value.

______ **13.** What is gauge pressure at 25.3 psia?

T F **14.** Moisture test instruments measure the conductivity of the air.

______ **15.** On a pressure gauge that measures up to 120 psi, does a reading of 100 psi ±0.5% of reading have a higher degree of accuracy than a reading of 50 psi ±0.5% of full face value?

T F **16.** Relative humidity is the amount of moisture in the air compared to the amount of moisture required to saturate the air.

______ **17.** What type of energy does a noncontact temperature probe measure?

______ **18.** Which type of gas leak detector uses a dye that is added to a refrigerant system?

______ **19.** Which indicates a larger change in heat energy, one Celsius degree or one Fahrenheit degree?

______ **20.** Which type of gas leak detector listens for the sounds created by a leak?

21. What types of changes can occur to substances in processes?

22. Give at least six examples of process variables that are measured with instrumentation.

23. Explain the difference between absolute pressure and gauge pressure, including how each measures vacuum.

24. How can a soapy solution be used to detect refrigerant or other gas leaks?

25. What are the major advantages and disadvantages of in-line and noncontact flowmeters?

9 SPECIAL MAINTENANCE TEST INSTRUMENTS

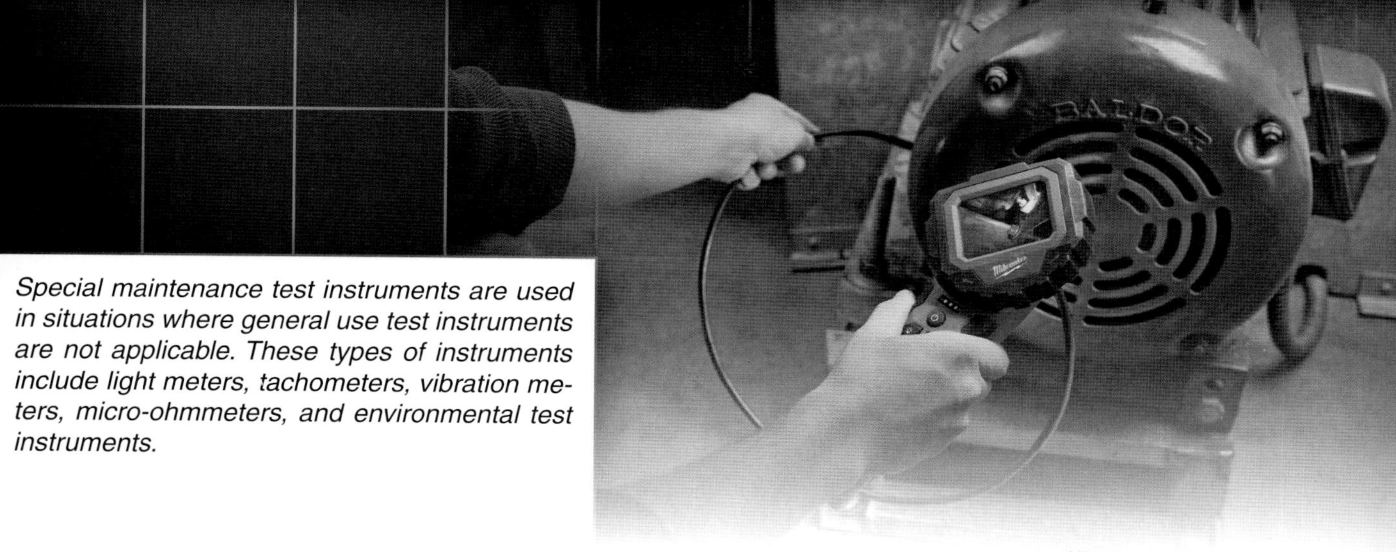

Special maintenance test instruments are used in situations where general use test instruments are not applicable. These types of instruments include light meters, tachometers, vibration meters, micro-ohmmeters, and environmental test instruments.

Milwaukee Tool Corporation

LIGHT METERS

A *light meter* is a test instrument that measures the amount of light in footcandles (fc) or lumens (lm). Facility engineers and electricians use light meters to evaluate and document lighting levels to ensure proper lighting that conforms to standards for working, safety, and security areas, plus lamp performance over time. The amount of light illuminating a surface is dependent upon the number of lumens produced by the light source, and the distance from the light source to the surface. The lower the lumen output, or the greater the distance from the light source, the lower the amount of light on a surface.

The light level required on a surface varies widely for each lighting application. For example, an operating table in a hospital requires much more light than the lobby of the hospital. **See Figure 9-1.**

Light

Light is the portion of the electromagnetic spectrum that produces radiant energy. The electromagnetic spectrum ranges from cosmic rays that have extremely short wavelengths to electric power frequencies that have extremely long wavelengths. **See Figure 9-2.** Light can be in the form of visible light or invisible light. *Visible light* is the portion of the electromagnetic spectrum that the human eye can perceive.

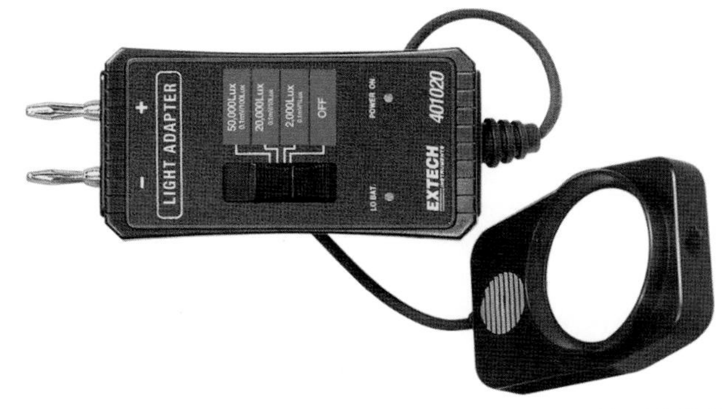

Extech by FLIR

Light-measuring adapter accessories can be added to some digital multimeters to make light meters.

RECOMMENDED LIGHT LEVELS

INTERIOR LIGHTING

AREA	LIGHT LEVEL*
Assembly	
Rough, easy seeing	30
Medium	100
Fine	500
Auditorium	
Exhibitions	30
Banks	
Lobby, general	50
Writing areas	70
Teller station	150
Canning	
Cutting, sorting	100
Clothing Manufacturing	
Pattern making	50
Shops	100
Garages (Auto)	
Parking	10
Repair	50
Hospital/Medical	
Lobby	30
Dental chair	1000
Operating table	2500
Machine Shop	
Rough bench	50
Medium bench	100
Materials handling	
Picking stock	30
Packing, labeling	50
Offices	
Regular office work	100
Accounting	150
Detailed work	200
Printing	
Proofreading	150
Color inspecting	200
Schools	
Auditoriums	20
Classrooms	60-100
Indoor gyms	30-40
Stores	
Stockroom	30
Service area	100
Warehousing, storage	
Inactive	5
Active	30

* in footcandles

EXTERIOR LIGHTING

AREA	LIGHT LEVEL*
Building	
Light surface	15
Dark surface	50
Loading/unloading area	20
Parking areas	
Industrial	2
Shopping	5
Storage yards (Active)	20
Street	
Local	0.9
Expressway	1.4
Car lots	
Front line	100-500
Remaining area	20-75

* in footcandles

SPORTS LIGHTING (PROFESSIONAL)

AREA	LIGHT LEVEL*
Baseball	
Outfield	100
Infield	150
Boxing	
Professional	200
Championship	500
Football	100
Hockey	50
Racing	
Auto, horse	20
Dog	30
Skating	
Rink	5
Soccer	100
Tennis Courts	
Recreational	15
Tournament	30
Volleyball	
Recreational	10
Tournament	20

* in footcandles

FOOTCANDLE - A UNIT OF MEASURE OF THE INTENSITY OF LIGHT FALLING ON A SURFACE, EQUAL TO ONE LUMEN PER SQUARE FOOT

Figure 9-1. Recommended light levels in footcandles are set to allow safe performance of tasks in an area, on a bench, or at a machine.

Technical Tip

Light is usually measured by illuminance or by lumens. Illuminance is the amount of light falling on a given surface and uses foot-candles (fc) as a unit of measure. The light outdoors on a sunny day is about 1200 fc, compared to 350 fc on a cloudy day. The lumen is a unit of measurement of the amount of energy given off by a lamp. A 60 W lamp would be around 800 lumens while a 250 W halogen lamp would be about 4000 lumens.

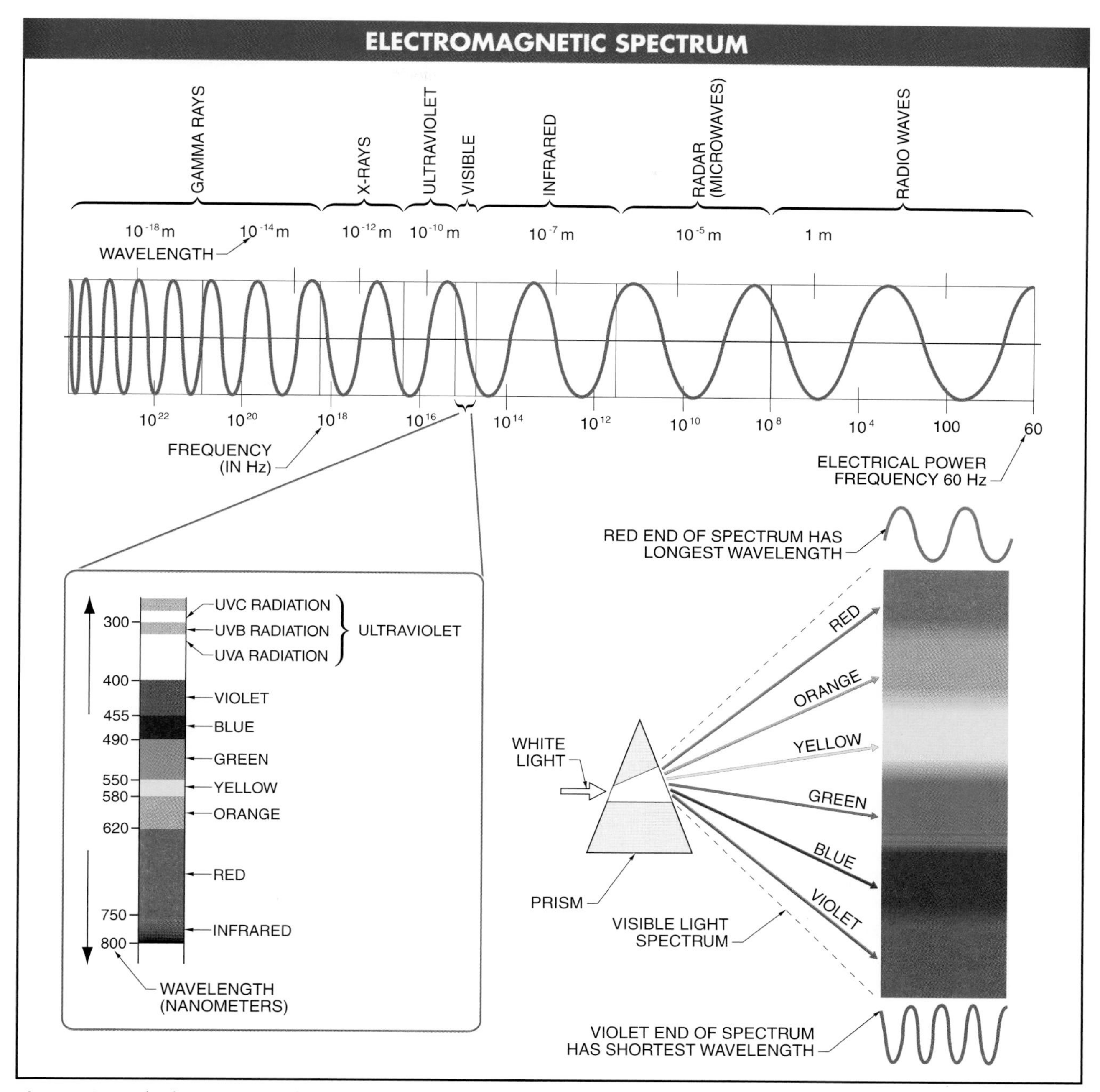

Figure 9-2. The human eye only perceives visible light, which is a small portion of the electromagnetic spectrum.

Visible light includes the part of the electromagnetic spectrum that ranges from violet to red light. Visible light is produced by standard incandescent lamps, fluorescent lamps, HID (high-intensity discharge) lamps, and LED lamps. *Invisible light* is the portion of the electromagnetic spectrum on either side of the visible light spectrum. Invisible light includes ultraviolet and infrared light. Special purpose lamps, such as germicidal lamps, black lamps, sun lamps, and infrared heating lamps, are designed to produce light energy that is outside the visible light spectrum. **See Figure 9-3.**

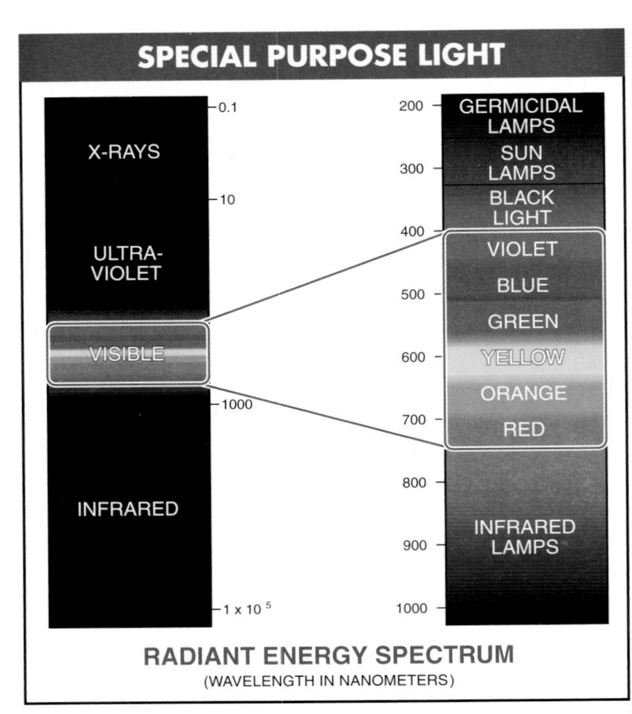

Figure 9-3. Light outside the visible light spectrum is used to perform special tasks such as heating.

The color of light is determined by the wavelength of the light. Visible light having short wavelengths produces the color violet. Visible light having long wavelengths produces the color red. Wavelengths between violet and red produce blue, green, yellow, and orange. When a light source such as the sun produces energy over the entire visible spectrum in approximately equal quantities, the combination of colored light produces white light. When a light source such as a low-pressure sodium lamp produces energy that mostly lies in a narrow band of the spectrum, such as yellow-orange, a nonwhite light is produced. **See Figure 9-4.**

Because the energy produced by low-pressure sodium lamps is in the yellow-orange area of the light spectrum, the color distortion of light-colored objects viewed under low-pressure sodium lamps can be extreme. However, since low-pressure sodium lamps produce more light per watt of power than most other types of lamp, low-pressure sodium lamps are used in applications such as street lighting where color distortion can be tolerated.

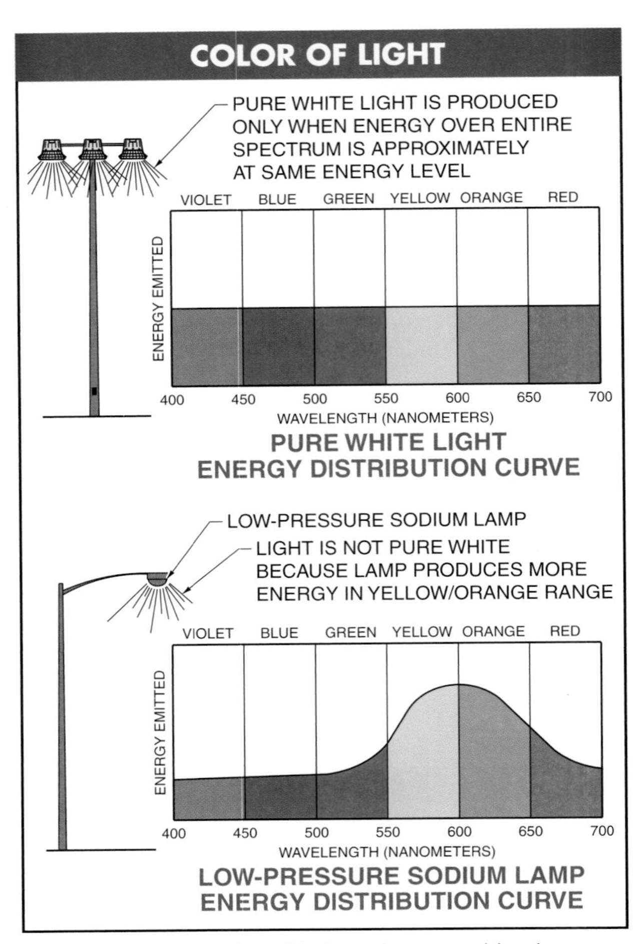

Figure 9-4. The color of light is determined by the wavelength of the light. An equal mixture of all wavelengths creates white light.

A *light-emitting diode (LED)* is a diode that emits light when forward current is applied. In an LED, light is produced when current is passed through the diode. LEDs are commonly used as visual indicators because they have an extremely long life. LED lighting products produce light approximately 90% more efficiently than incandescent light bulbs. Thermal management, or assurance that the component will not overheat, is generally the single most important factor in the successful performance of an LED over its lifetime. The higher the temperature at which the LEDs are operated, the more quickly the light will degrade and the shorter the useful life of the LED. While LED lighting is more efficient, versatile, and lasts longer than other types of lighting products, it is still susceptible to overheating.

LED lighting is the most modern and efficient light source available on the market. LEDs produce about 90% more light than an incandescent bulb. They also have the longest life of any light source and are available in different colors and shapes. They are available in extremely small sizes, in standard bulb sizes, as fluorescent bulb replacements, and in large sizes for street and bay-type lighting. LEDs do not "burn out" (fail) in the manner other bulb types do. However, over time, heat diminishes the light output. LEDs have internal heat sinks to pull the heat away from the bulb. For this reason, thermal (heat) management is a part of LED installation to ensure a long, reliable life.

Lamps

A *lamp* is an electrical system output device that converts electrical energy into light. The amount of light a lamp produces is expressed in lumens. A *lumen (lm)* is a unit used to measure the total amount of light produced by a light source. For example, a standard 40 W incandescent lamp produces about 480 lm, and a standard 40 W fluorescent lamp produces about 3100 lm. Manufacturers rate lamps (light bulbs) by the total amount of light (lumens) produced by the lamps. Since the number of lumens is the total amount of light, lumens are comparable to the amount of current (amperes) in an electrical circuit or the amount of flow (gpm) in a hydraulic system. *Efficacy* is the ratio of lumens generated by a lamp to the watts consumed by the lamp. Efficacy is also referred to as lumens per watt.

The light produced by a light source causes illumination. *Illumination* is the effect that occurs when light falls on a surface. The unit of measure of illumination is the footcandle or lux. The *footcandle (fc)* is a unit of illumination equal to 1 lm per sq ft. The *lux* is the unit of illumination equal to 1 lm per sq m. One lux is equal to 1 lumen per square meter. One footcandle equals 10.76 lux and 1 lux equals 0.0929 footcandles.

Light Meter Measurement Procedures

The light falling on a surface may be direct or indirect. *Direct light* is light that travels directly from a lamp to the surface being illuminated. *Indirect light* is light that is reflected from one or more objects to the surface being illuminated. For indoor applications, both direct and indirect light are considered and measured. For outdoor applications, only direct light is considered and measured. **See Figure 9-5.**

To measure the amount of light with a light meter, apply the following procedures:

Safety Procedures

- Follow all electrical safety practices and procedures.
- Check and wear personal protective equipment (PPE) for the procedure being performed.
- Perform only authorized procedures.
- Follow all manufacturer recommendations and procedures.

1. Turn the lamps ON in an area where the amount of light is being measured.
2. Turn the light meter ON.
3. Set the function switch of the light meter to footcandles or lux mode.
4. Set the selector switch of the light meter to the expected range, for example 40 fc or 400 fc.
5. Set the light meter to record light measurements with MIN MAX and/or PEAK modes.
6. Take several light level measurements at various locations and at various angles.
7. Record all light level measurements.
8. Record any support information, such as when measurements were taken and if shades were open or closed.
9. Turn the light meter OFF.

Technical Tip

While high-power LED lamps are more efficient than other types of lamps, they still produce heat. Heat in an LED lamp is dissipated by incorporating a metal heat sink block into the design of the lamp.

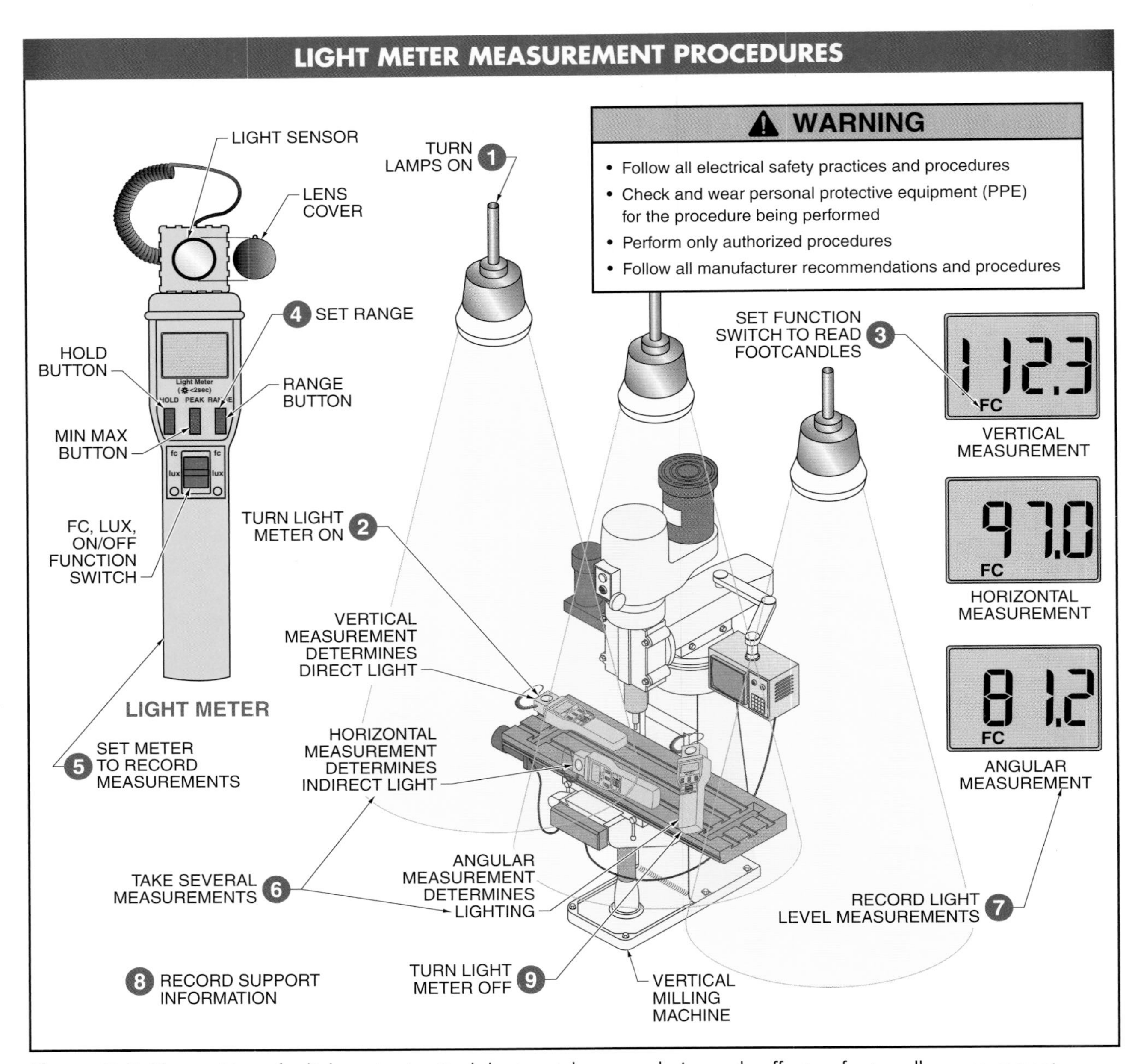

Figure 9-5. The position of a light meter (vertical, horizontal, or angular) greatly affects a footcandle measurement.

TACHOMETERS

A *tachometer* is a test instrument that measures the speed of a moving object. Speed measurements are taken using contact tachometers, laser tachometers, photo tachometers, or strobe tachometers. The type of tachometer used for an application depends upon the required output measurement (rpm, ft/min, m/min) and expected results (accuracy of measurement). Rotating speeds are displayed in revolutions per minute (rpm) and linear speeds are displayed in feet per minute (ft/min, fpm) or meters per minute (m/min, mpm). *Rotational speed* is the distance traveled per unit of time in a circular direction. *Linear speed* is the distance traveled per unit of time in a straight line.

Typically, contact tachometers are the least expensive and simplest to use. However, not all tachometers need to come in physical contact with the moving object, which can cause safety problems while taking the measurements. Noncontact tachometers such as photo tachometers and laser tachometers cost about the same (photo tachometers are slightly lower in price) and do not require direct contact with the moving object being tested. Photo tachometers have been used for many years but are being replaced with laser tachometers as the price of laser tachometers continues to drop. Models of tachometers are available that include a contact tachometer and a laser (or photo) tachometer in one unit. **See Figure 9-6.**

Strobe tachometers are the most expensive, but include the ability to time moving parts by use of a flashing light (strobe light) in addition to taking rpm measurements. Strobe tachometers are also used to test for vibration problems and alignment of parts such as cams and fan blades, and are used to test liquid spray patterns of paint and glue.

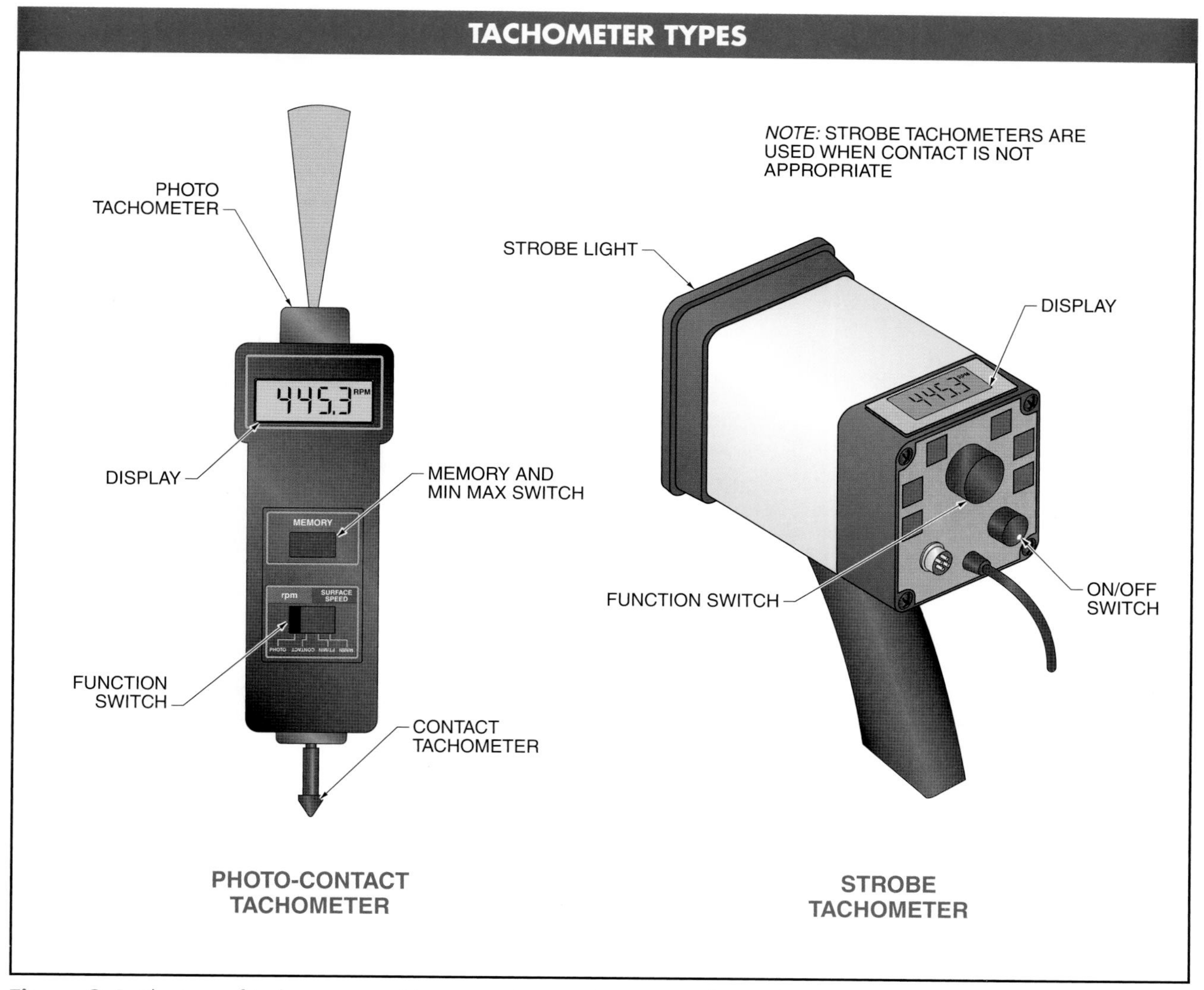

Figure 9-6. The type of tachometer used for an application depends on the application (contact or noncontact), the desired measurement output (rpm, ft/min, or m/min), and expected results (accuracy).

Contact Tachometers

A *contact tachometer* is a test instrument that measures the rotational speed of an object through direct contact of the tachometer tip with the object being measured. Contact tachometers are used to measure the speed of rotating objects such as motor shafts, gears, belts, and pulleys, and linear speeds of moving conveyors and press webs, for example. Contact tachometers measure speeds from 0.1 rpm to about 25,000 rpm.

Contact Tachometer Measurement Procedures

Use caution when working around moving objects. Use a photo, laser, or strobe tachometer when there is danger of contacting moving objects. **See Figure 9-7.**

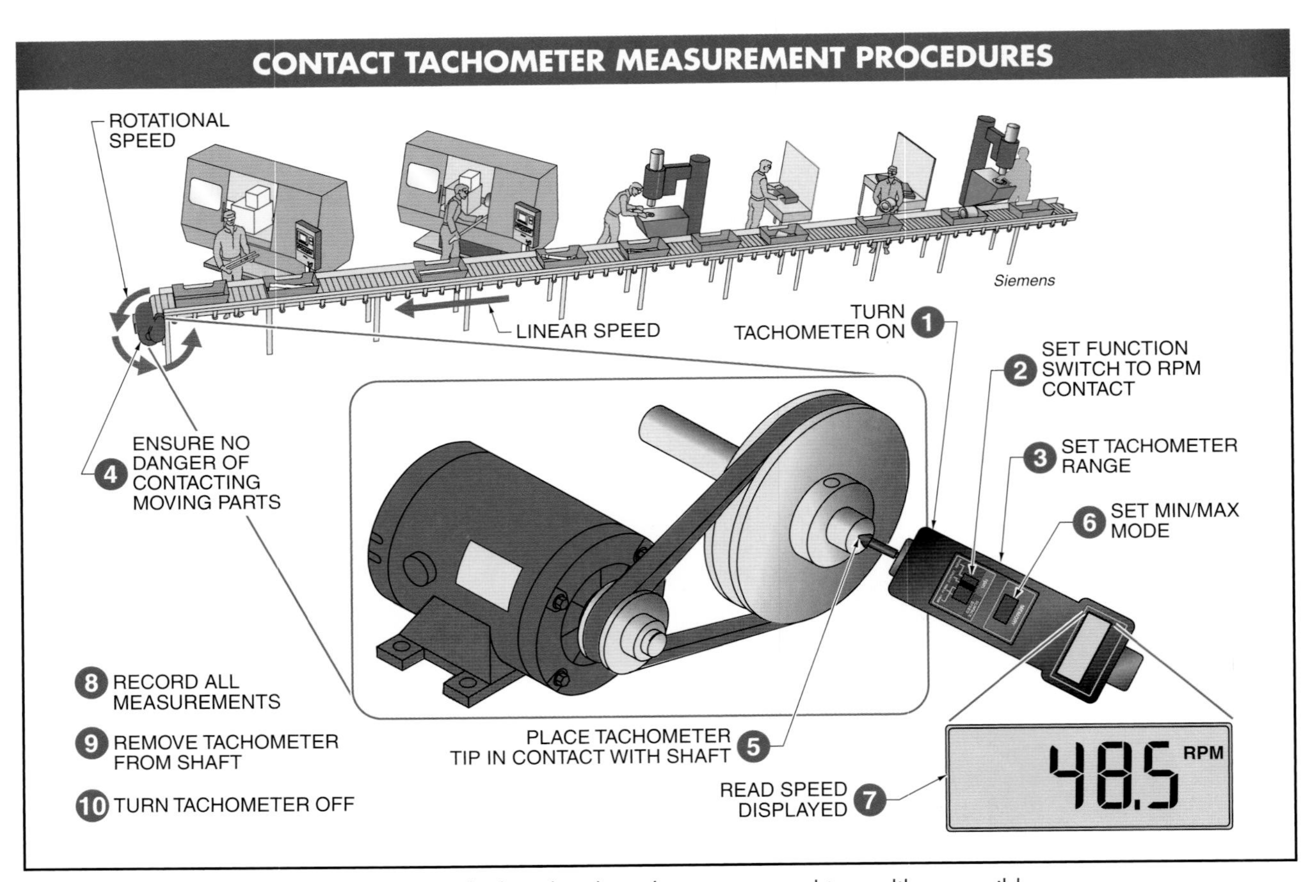

Figure 9-7. Contact tachometers are used when the object being measured is readily accessible.

Troubleshooting Tip

TACHOMETER PARAMETERS

The most important parameters to consider when specifying tachometers are operating speed range and accuracy. Operating speed range is the range of rotary speed measurement the tachometer can monitor. The accuracy is measured in rpm. The sensor technology used in tachometers can be contact type, photoelectric, inductive, or Hall effect. Hall effect tachometers use Hall effect technology to determine rotational speed.

Before taking any measurements using a contact tachometer, ensure the tachometer is designed to take measurements on the object being tested. Refer to the operating manual of the test instrument for all measuring precautions, limitations, and procedures. Always wear required personal protective equipment and follow all safety rules when taking the measurement. To measure the speed of an object with a contact tachometer, apply the following procedures:

1. When a contact tachometer has a digital display, turn the tachometer ON.
2. Set the function switch of that tachometer to the required rpm, ft/min, or m/min mode.
3. Set the selector switch of the tachometer to the required range—for example, 0 rpm to 100 rpm, 100 rpm to 1000 rpm, or 1000 rpm to 10,000 rpm.
4. Ensure there is no danger of any part of the human body coming in contact with moving parts.
5. Place the tip of the tachometer in direct contact with the rotating object.
6. Set any special recording modes (MIN MAX, LAST) that may be included on the tachometer.
7. Read the speed displayed on the meter.
8. Record all measurements.
9. Remove the tachometer from the rotating object.
10. Turn the tachometer OFF.

Extech by FLIR

Contact tachometers, which measure motor speed by directly contacting a rotating shaft with the tachometer's probe, usually have a narrower measurement range than laser or photo tachometers.

Photo Tachometers

A *photo tachometer* is a test instrument that uses light beams to measure the speed of an object without directly contacting the object. A photo tachometer measures speed by focusing a light beam on a reflective area or a piece of reflective tape and counting the number of reflections per minute (internal calculation).

A photo tachometer is used in applications in which the rotating object cannot be reached or touched. A photo tachometer cannot measure linear speeds. Photo tachometers measure speeds from 1 rpm to about 100,000 rpm.

Photo Tachometer Measurement Procedures

Exercise caution when working around moving objects. Turn OFF and lockout/tagout disconnect switches of motors and rotating machinery as required. **See Figure 9-8.**

Before taking any measurements using a photo tachometer, ensure the tachometer is designed to take measurements on the rotating object being tested. Refer to the operating manual of the test instrument for all measuring precautions, limitations, and procedures. Always wear required personal protective equipment and follow all safety rules when taking the measurement. To measure speed (rpm) of an object with a photo tachometer, apply the following procedures:

1. When the object being measured does not already have a reflective surface that can be used to take the rpm measurement, turn OFF and lockout/tagout the device being measured.
2. Place reflective tape on the object to be measured.
3. Turn the rotating device being measured ON.
4. Turn the tachometer ON.
5. Set the function switch of the tachometer to the required rpm, ft/min, or m/min mode.

6. Set the selector switch of the tachometer to the required range—for example, 0 rpm to 100 rpm, 100 rpm to 1000 rpm, or 1000 rpm to 10,000 rpm.
7. Ensure there is no danger of any part of the human body coming in contact with rotating objects.
8. Point the light beam of the photo tachometer at the reflective tape.
9. Set any special recording modes (MIN MAX, LAST) that may be included on the tachometer.
10. Read the speed displayed on the meter.
11. Record all measurements.
12. Turn the tachometer OFF.

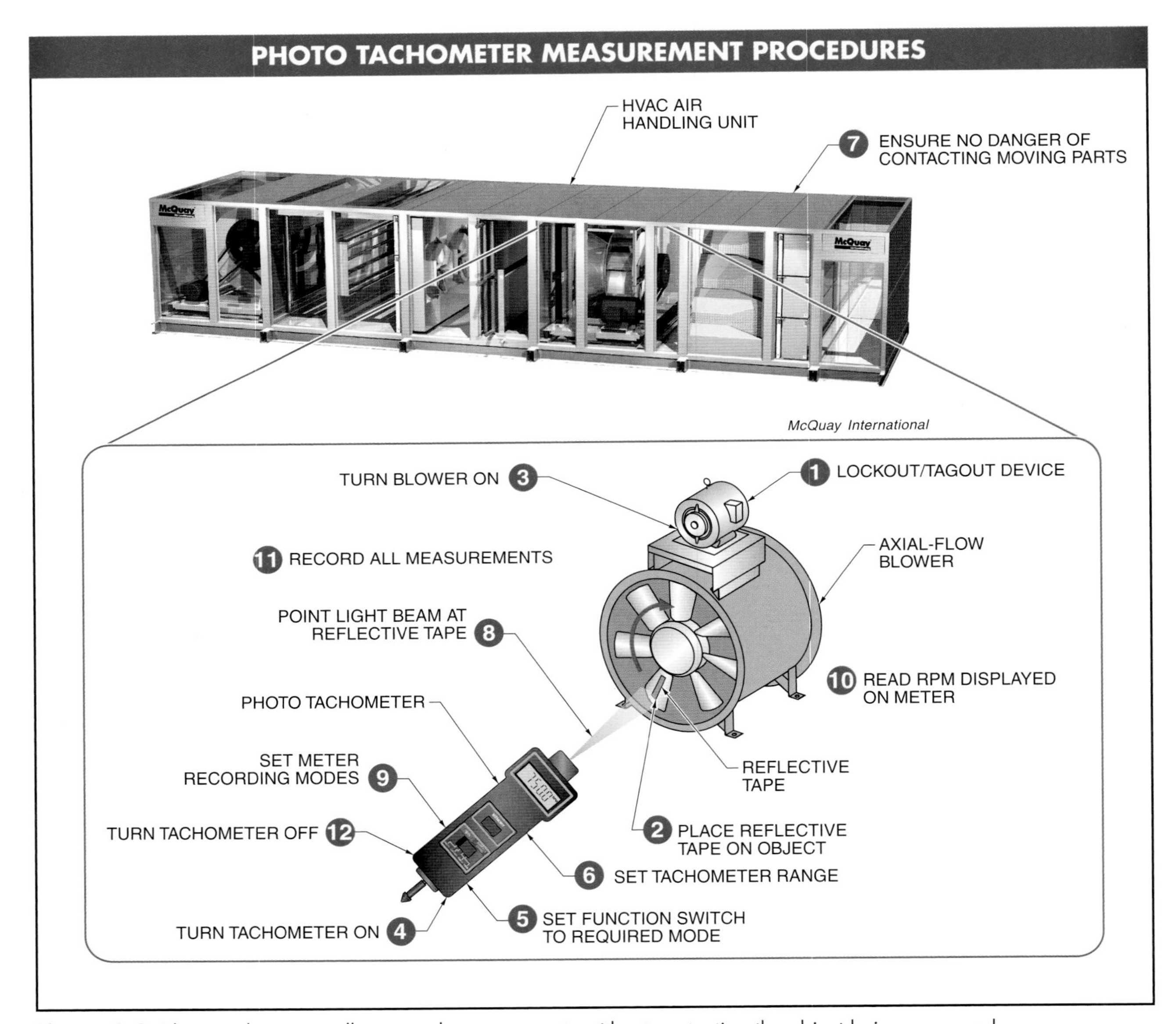

Figure 9-8. Photo tachometers allow speed measurements without contacting the object being measured.

Laser Tachometers

A *laser tachometer* is a test instrument that uses a laser light to measure the speed of an object without directly contacting the object. A laser tachometer measures speed by focusing a laser light beam on a reflective area or a piece of reflective tape. Laser tachometers work well in areas of high ambient light because red laser light is easier to see than the white light emitted by photo tachometers.

Laser tachometers allow speed measurements in hard-to-reach or dangerous areas. Because a laser light beam has a concentrated beam of light, laser tachometers are better than photo tachometers when taking measurements in confined areas. Similar to photo tachometers, laser tachometers measure speeds from 1 rpm to about 100,000 rpm.

Laser Tachometer Measurement Procedures

Exercise caution when working around moving objects. Turn OFF and lockout/tagout all disconnect switches of motors and rotating machinery as required. **See Figure 9-9.**

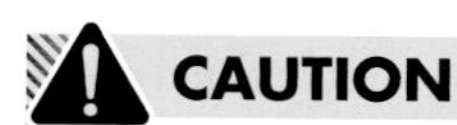

Never look directly into the laser light beam. Lasers can permanently damage eyes. Avoid areas of reflected laser beams. A laser reflection is as hazardous as a direct laser beam, and sources of reflection should be removed from the laser work area.

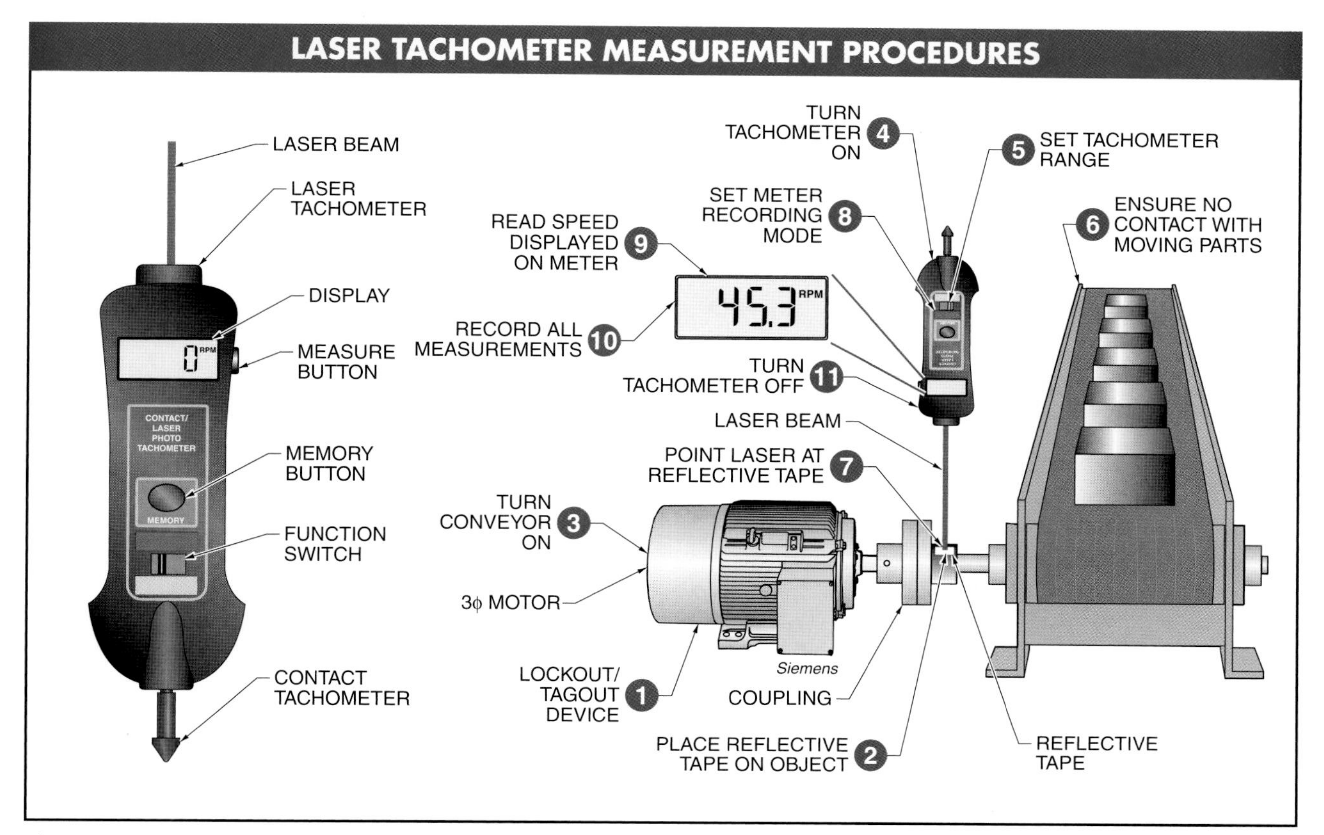

Figure 9-9. Laser tachometers allow speed measurements from hard-to-reach surfaces without contacting the object.

Before taking any measurements using a laser tachometer, ensure the tachometer is designed to take measurements on the object being tested. Refer to the operating manual of the test instrument for all measuring precautions, limitations, and procedures. Always wear required personal protective equipment and follow all safety rules when taking the measurement. To measure the speed of an object with a laser tachometer, apply the following procedures:

1. When the object being measured does not already have a reflective surface that can be used to take an rpm measurement, turn OFF (lockout/tagout) the device being measured.
2. Place reflective tape on the object being measured.
3. Turn the device being measured ON.
4. Turn the tachometer ON.
5. Set the tachometer to the required range—for example, 0 rpm to 100 rpm, 100 rpm to 1000 rpm, or 1000 rpm to 10,000 rpm.
6. Ensure there is no danger of any part of the human body coming in contact with moving objects.
7. Point the laser beam of the laser tachometer at the reflective tape.
8. Set any special recording modes (MIN MAX or LAST) that may be included on the tachometer.
9. Read the speed displayed on the meter.
10. Record all measurements.
11. Turn the tachometer OFF.

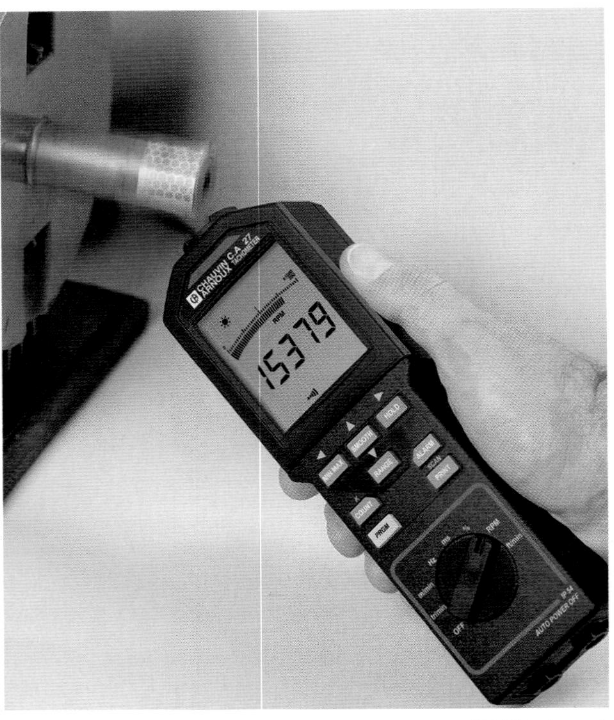

AEMC® Instruments

Reflective tape on a motor shaft bounces laser light back to a laser tachometer, which uses the pulses to calculate shaft rotational speed.

Strobe Tachometers

A *strobe tachometer* is a test instrument that uses a flashing light to measure the speed of a moving object. A strobe tachometer measures speed by synchronizing the flash rate of a light with the speed of the moving object.

Strobe tachometers can take speed measurements through glass, which eliminates the possibility of direct exposure to hazardous areas. Strobe tachometers are also used for analysis of motion and vibration. Strobe tachometers measure speeds from 20 rpm to about 100,000 rpm, or 20 ft/min to about 12,500 ft/min.

Strobe Tachometer Measurement Procedures

Exercise caution when working around moving objects. Turn OFF and lockout/tagout disconnect switches of motors and rotating machinery as required. **See Figure 9-10.**

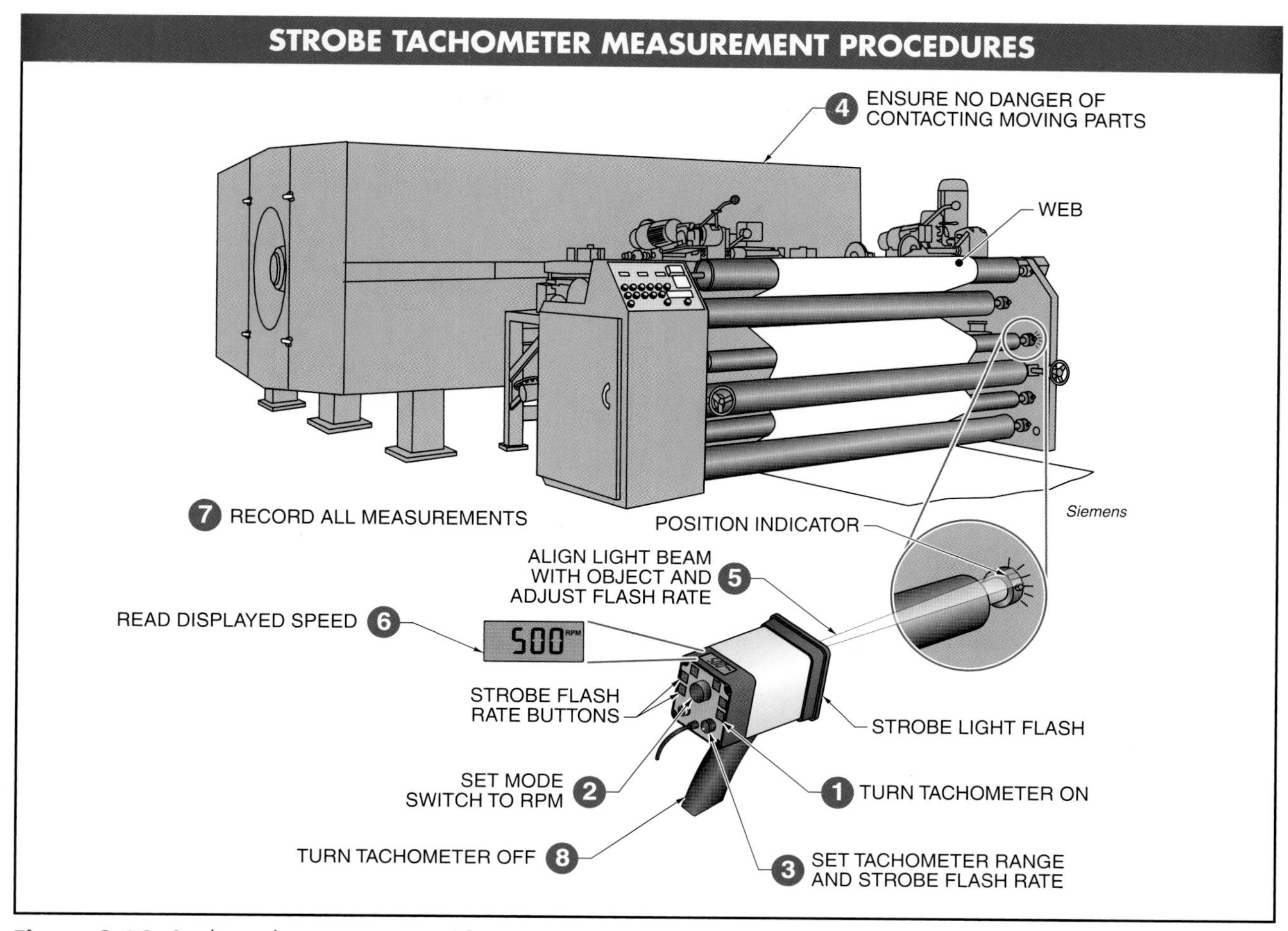

Figure 9-10. Strobe tachometers are used for speed measurements, but can also be used for timing and vibration analysis.

Before taking any measurements using a strobe tachometer, ensure the tachometer is designed to take measurements on the object being tested. Refer to the operating manual of the test instrument for all measuring precautions, limitations, and procedures. Always wear required personal protective equipment and follow all safety rules when taking the measurement. To measure the speed of an object with a strobe tachometer, apply the following procedures:

1. Turn the strobe tachometer ON.
2. Set the mode switch of the tachometer to the required rpm, ft/min, or m/min units.
3. Set the range switch of the tachometer to either the 0.1 ft/min to 40 ft/min range or to the 0.2 ft/min to 5000 ft/min range. Adjust strobe flash rate.
4. Ensure there is no danger of any part of the human body coming in contact with moving objects.
5. Align the visible light beam with the object being measured and reduce flash rate to freeze image.
6. Read the speed displayed on the meter.
7. Record all measurements.
8. Turn the tachometer OFF.

VISUAL INSPECTION TEST INSTRUMENTS

Electricians and maintenance personnel are required to work on systems and in areas where everything cannot be easily seen. Problems that cannot be seen, such as blocked pipes and ducts, leaks behind walls, and obstacles that may be in the way when trying to install devices in panels, can be accounted for with visual inspection.

Visual inspection test instruments are used to look at areas that are typically out of sight. Visual inspection test instruments can be as simple as a light and flexible fiber cable. Visual inspection test instruments include display screens (with or without cameras) and extension cables that allow viewing 100′ or more away. **See Figure 9-11.**

Visual Inspection Test Instrument Measurement Procedures

To view an area out of normal sight using a visual inspection instrument, ensure the inspection instrument is designed to view the area in question. Refer to the operating manual of the visual inspection instrument for all measuring precautions, limitations, and procedures. Always wear required personal protective equipment and follow all safety rules when viewing an area. To view areas that are out of normal sight with a visual inspection instrument, apply the following procedures:

Safety Procedures

- Lockout/tagout all sources of electrical power.

1. Ensure all power is OFF in the area to be viewed.
2. Turn the visual inspection test instrument ON.
3. Test the operation of the visual inspection test instrument on a known area.
4. Make any adjustments to light or other settings for proper viewing of display.
5. Carefully insert the viewing cable into the area being inspected by observing what is shown on the camera display.
6. Record and/or take pictures of any problems or obstacles that are viewed. Record distances and types of obstacles.
7. Remove viewing cable from electrical equipment.
8. Turn the visual inspection test instrument OFF.

Troubleshooting Tip

USING A DIGITAL CONDUCTIVITY METER FOR VISUAL INSPECTION

A digital conductivity meter uses eddy current techniques for taking nondestructive (visual) measurements and readings. Eddy current techniques are extremely sensitive in detecting surface and near-surface defects such as cracks, folds, and carbide content. This technology is limited to metallic components—the material must be electrically conductive—but it provides an instantaneous measurement that makes it well suited to high-speed, in-line inspection on a production line. A digital conductivity (eddy current) meter is useful for sorting a wide variety of nonferrous metals and alloys and for measuring changes in physical properties affecting conductivity. Such changes include hardening, annealing, heat-treating, corrosion, aging, and variations in alloys.

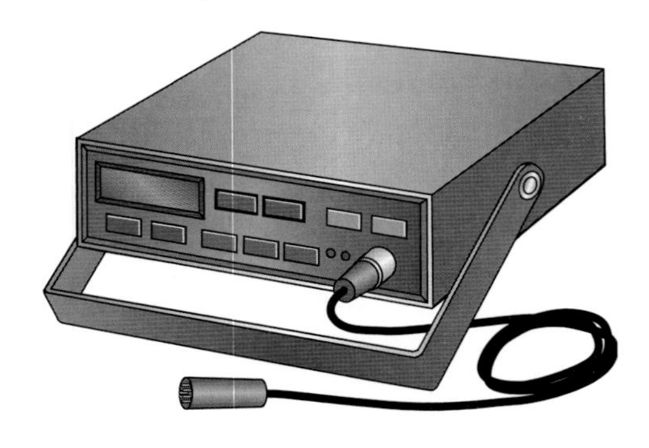

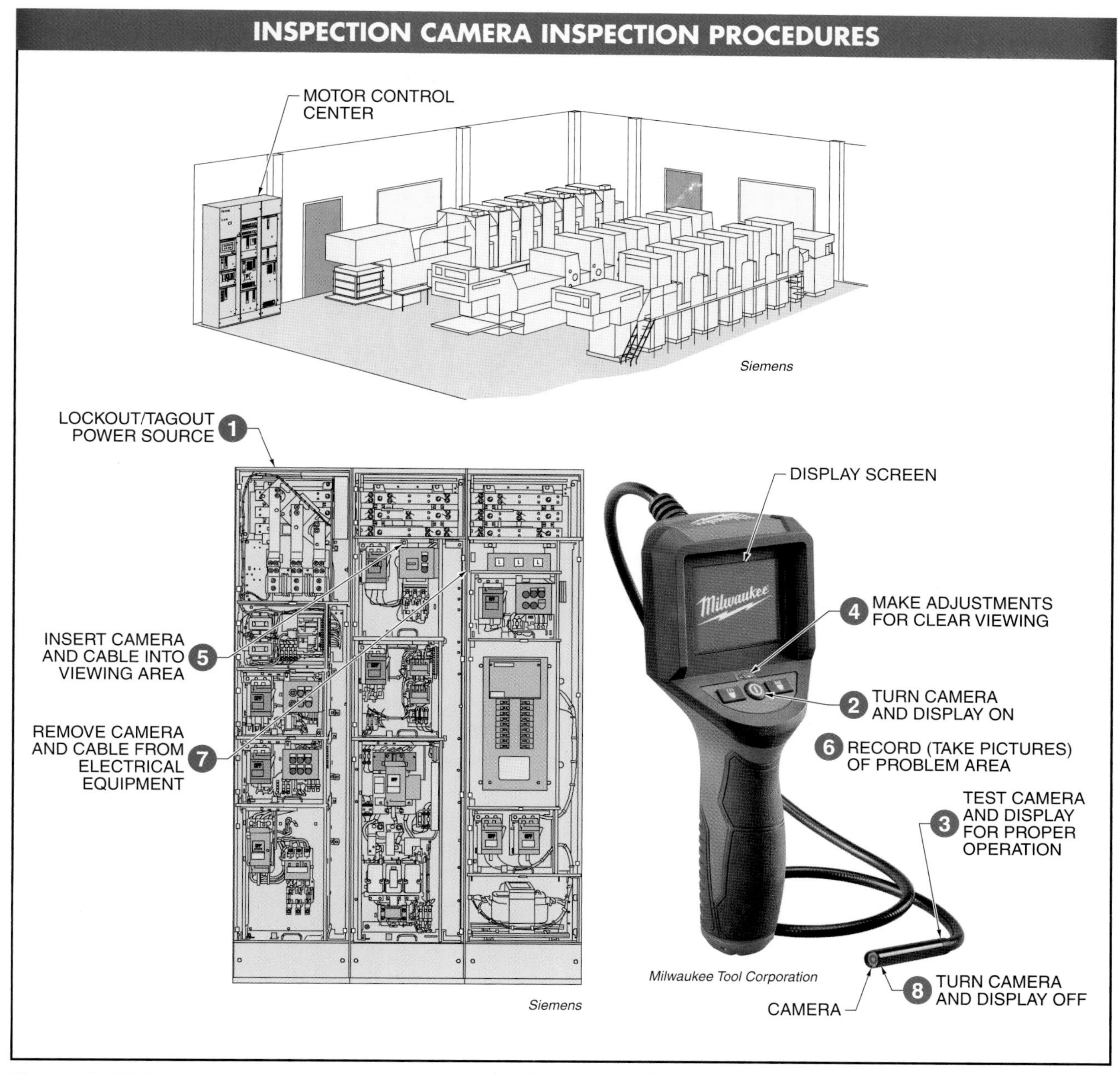

Figure 9-11. Visual inspection test instruments provide electricians with a method to view areas behind panels and inside ducts that typically are out of normal sight.

MICRO-OHMMETERS

Resistance (R) is the opposition to the flow of electrons in a circuit. Most resistance measurements are acceptable when taken with multimeters, but when a high resistance measurement is required, megohmmeters are used. However, when very accurate low resistance measurements are required, micro-ohmmeters are used. A *micro-ohmmeter* is a test instrument that accurately measures resistance using microhms (μΩ). **See Figure 9-12.** Some micro-ohmmeter models measure resistance using 0.1 microhms (0.0000001 Ω).

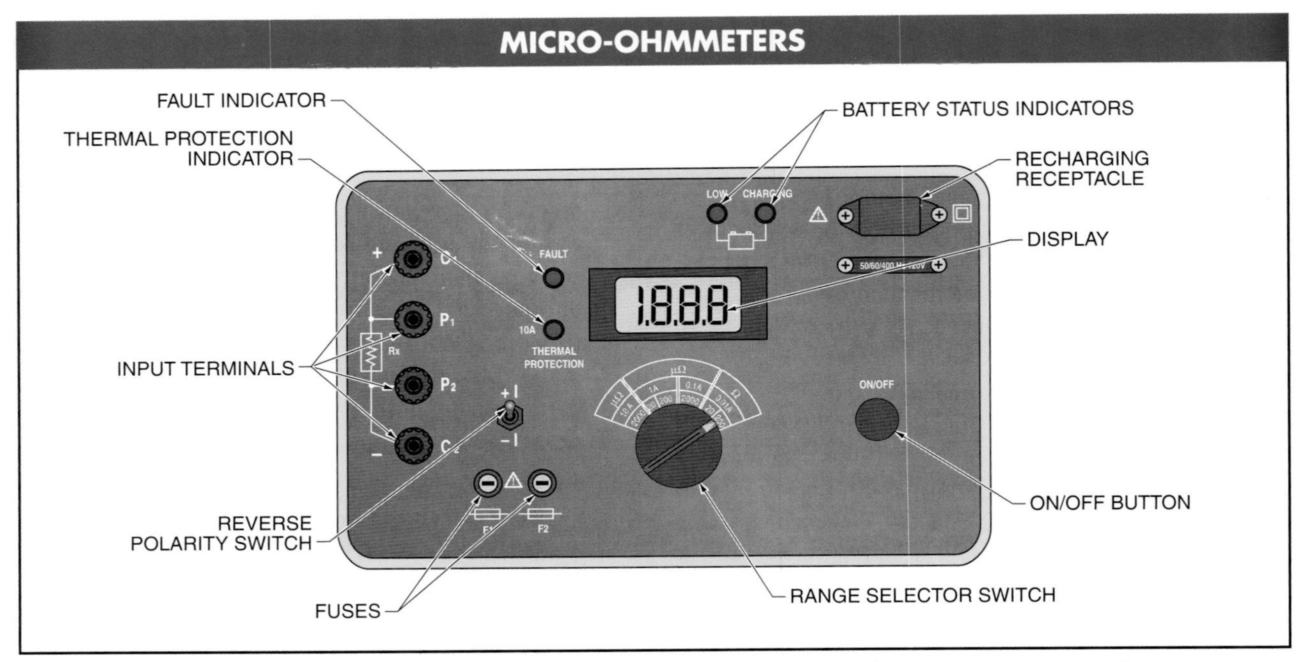

Figure 9-12. Micro-ohmmeters are highly sensitive ohmmeters that measure very low resistances in microhms.

Micro-ohmmeters are used to measure the resistance of electrical devices that typically have an extremely low resistance during proper circuit operation. Due to extremely low resistance operation, micro-ohmmeters are used to test and troubleshoot:

- contacts of switches and breakers
- busbars and cable joint connections
- grounding connections
- motor and transformer windings
- fuses
- wires and cables

Micro-ohmmeter Measurement Procedures

Micro-ohmmeters are also used to check nonelectrical connections such as aircraft frame bonds and welded joints for proper low resistance. In both electrical and mechanical applications, the higher the resistance of a connection, the more likely it is that a problem exists with the connection. In any electrical system, the higher the resistance of a connection, the greater the voltage drop and power drop (heat) across the connection. **See Figure 9-13.**

Before taking any measurements using a micro-ohmmeter, ensure the micro-ohmmeter is designed to take measurements on the object being tested. Refer to the operating manual of the test instrument for all measuring precautions, limitations, and procedures. Always wear required personal protective equipment and follow all safety rules when taking the measurement. To take low resistance measurements of an object using a micro-ohmmeter, apply the following procedures:

Safety Procedures

- Lockout/tagout all sources of electrical power.

1. Ensure all circuit power is OFF.
2. Turn the micro-ohmmeter ON.
3. Set the selector switch of the micro-ohmmeter to the μΩ, mΩ, or Ω range.

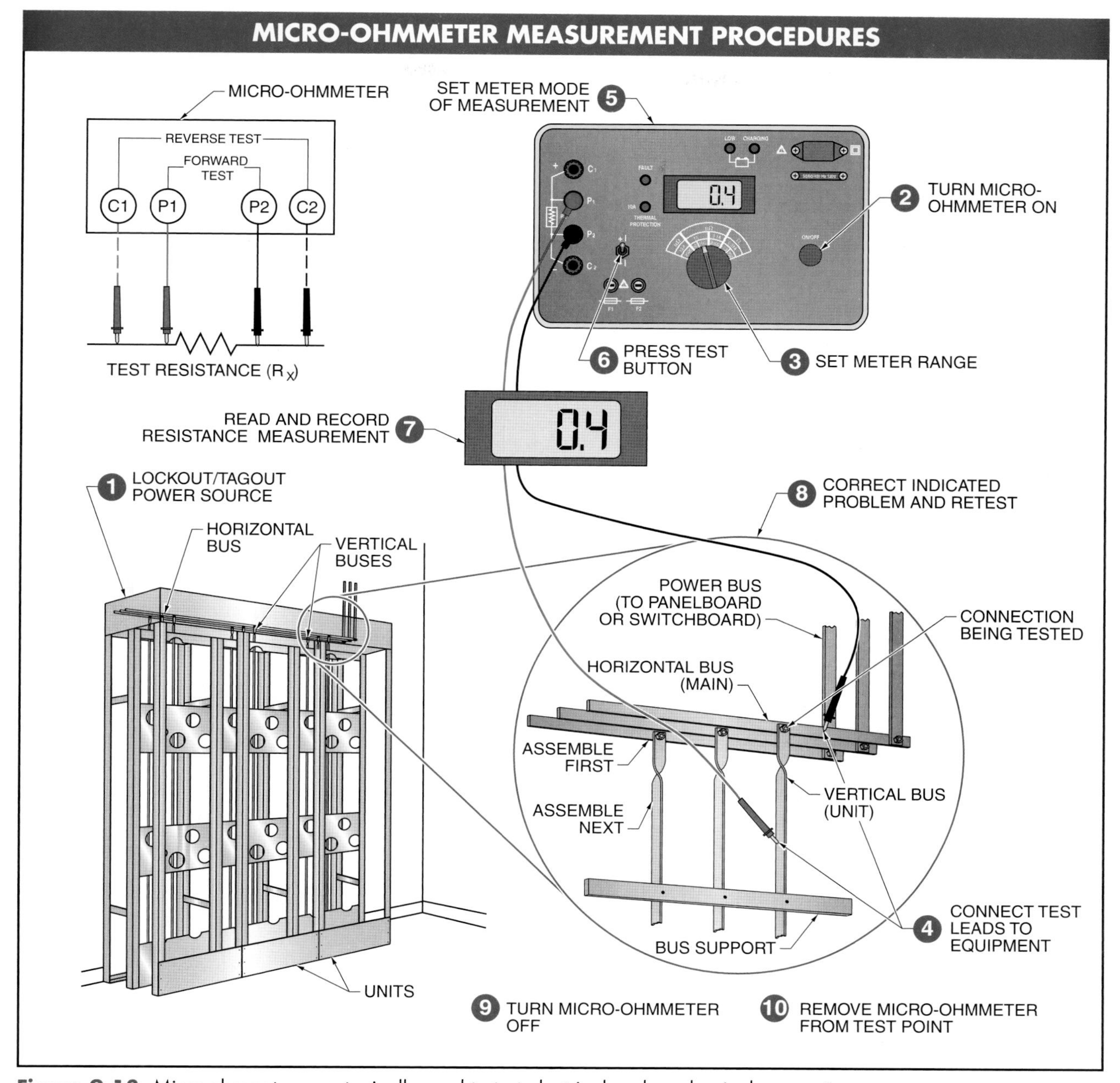

Figure 9-13. Micro-ohmmeters are typically used to test electrical and mechanical connections.

4. Connect the test leads of the micro-ohmmeter across the connection (circuit) being measured. Some micro-ohmmeters have four test leads that must be connected. Micro-ohmmeters with four leads use a forward test current and a reverse test current to provide a measurement that cancels out the resistance of the test leads.
5. Set the function switch of the micro-ohmmeter to normal or inductive mode. Normal mode is used for most pure resistance measurements and inductive mode for motor and transformer winding measurements.
6. Press the test button or switch.

7. Read and record resistance measurement.
8. When the resistance reading is higher than normal (or specifications), correct the problem and retest the device.
9. Remove the micro-ohmmeter from the point of testing (circuit).
10. Turn the micro-ohmmeter OFF.

MSA Safety, Inc.

Before working in an area where gas may be present, workers must have the proper portable gas detectors that measure the atmosphere in an area for lack of oxygen, explosive vapors, toxic gases, and other hazardous conditions.

ENVIRONMENTAL TEST INSTRUMENTS

For certain tasks, electricians are required to work in areas of confined space. A *confined space* is an area that has limited openings for entry and/or unfavorable natural ventilation that could cause a dangerous air contamination condition. The dangers in a confined space include elevated heat and cold levels, noise levels, toxic gas levels, combustible gas levels, and dust levels. When an electrician is required to work in a confined space (or any area of concern), test instruments that can measure the environmental conditions in a confined space must be used.

 **CAUTION**

Do not enter an area of known potential hazard unless equipped with proper personal protective equipment and working with a partner who can monitor safety conditions at all times, and unless all personnel have been trained and authorized to be in the hazardous area.

Gas Detectors

Gas detectors are available that measure in parts per million (ppm) and display (with LCD and alarms) the levels of gases. Gases that are typically measured include the following:

- carbon monoxide (CO)
- oxygen (O_2)
- hydrogen sulfide (H_2S)
- ammonia (NH_3)
- sulfur dioxide (SO_2)
- methane (CH_4)
- chlorine (Cl_2)
- nitrogen dioxide (NO_2)
- hydrogen cyanide (HCN)

Gas detectors are available as single gas or multiple gas detectors. Single gas detectors typically detect the more common gases that cause problems, such as carbon monoxide (CO). Today, gas detectors that measure three, four, or five gases are becoming more popular and can be found for use with uncommon gases. **See Figure 9-14.**

 **CAUTION**

Gas detectors and meters are used to provide an indication of potential problems, but must never be relied on for 100% accuracy in indicating a problem. A second gas detector or meter must be used by a partner (never work alone in a confined or hazardous area) to verify meter measurements. Hazardous areas can also have multiple fixed gas detectors for measurement verification.

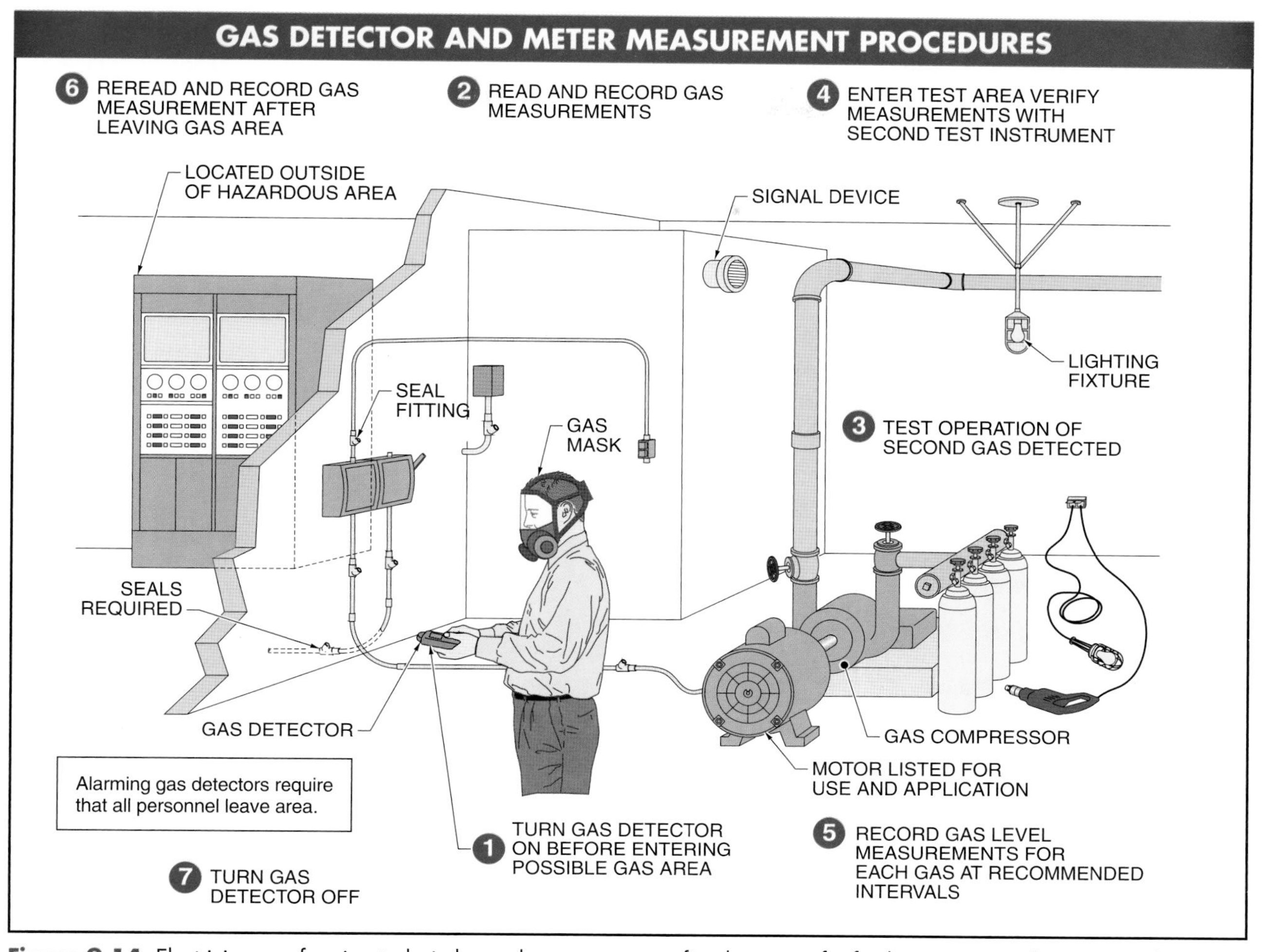

Figure 9-14. Electricians performing tasks in hazardous areas or confined spaces of a facility are required to use gas detectors.

Gas Detector Measurement Procedures

Before taking any measurements using a gas detector, ensure the detector is designed to take measurements of the gas being tested. Refer to the operating manual of the test instrument for all measuring precautions, limitations, and procedures. Always wear required personal protective equipment and follow all safety rules when taking the measurement. To measure gases in an area using a gas detector, apply the following procedures:

Safety Procedures

- Never work alone in a confined space or hazardous area.
- When a gas detector alarms, all personnel must leave the area immediately.

1. Before entering the area of potential problem, turn the gas detector ON.
2. Read and record the measurements displayed.

3. Have another electrician (partner) test the operation of a second gas detector.
4. Enter the area being tested for gas. Read and record the displayed value of each gas. Verify measurements with partner and second test instrument.
5. Record (with confined space inventory) the measured level of each gas as required.
6. After leaving the confined area, reread and record the measured levels of each gas detected.
7. Turn the gas detector OFF.

Technical Tip

Control radioactive hazards by limiting exposure time, using protective barriers, and maintaining a safe working distance.

Radiation Detectors

Electricians are sometimes required to work in areas that contain radioactive materials. Radioactive materials are found in facilities such as nuclear power plants, hospitals, research labs, and nondestructive inspection areas. All electricians in areas containing radioactive materials are required to monitor radiation levels at all times. Other professions such as police departments, fire departments, and other emergency response agencies must also have and know how to use radiation detectors.

Electromagnetic radiation, cosmic radiation, and nuclear radiation are the types of radiation that can be found in facilities. Nuclear radiation is radiation created by radioactive atoms. High levels of nuclear radiation are known to cause radiation sickness or death in humans. Radiation detectors are used to measure the levels of radiation, record radiation levels over time, and warn of danger when levels exceed a specific threshold.

CAUTION

Radiation detectors are used to provide an indication of potential problems, but must never be relied on for 100% accuracy in indicating a problem. A second radiation detector must be used by a partner (never work alone in a confined or hazardous area) to verify meter measurements. Hazardous areas can have multiple fixed radiation detectors for measurement verification.

Radiation Detector Measurement Procedures

Before taking any measurements using a radiation detector, ensure the detector is designed to take measurements in the area being tested. **See Figure 9-15.** Refer to the operating manual of the test instrument for all measuring precautions, limitations, and procedures. Always wear required personal protective equipment and follow all safety rules when taking the measurement. To measure radiation levels of an area using a radiation detector, apply the following procedures:

Safety Procedures

- When a radiation detector alarms, all personnel must leave the area.

1. Before entering the area of potential problem, turn the radiation detector ON.
2. Read and record radiation measurements displayed.
3. Have partner test the operation of a second radiation detector.
4. Enter the area where work is being performed. Read and record radiation measurements displayed. Check measurements with measurements of partner.
5. Read and record the measurement level of radiation at the required or recommended time intervals.

6. After leaving the area, reread and record the measured levels of radiation.
7. Turn the radiation detector OFF.

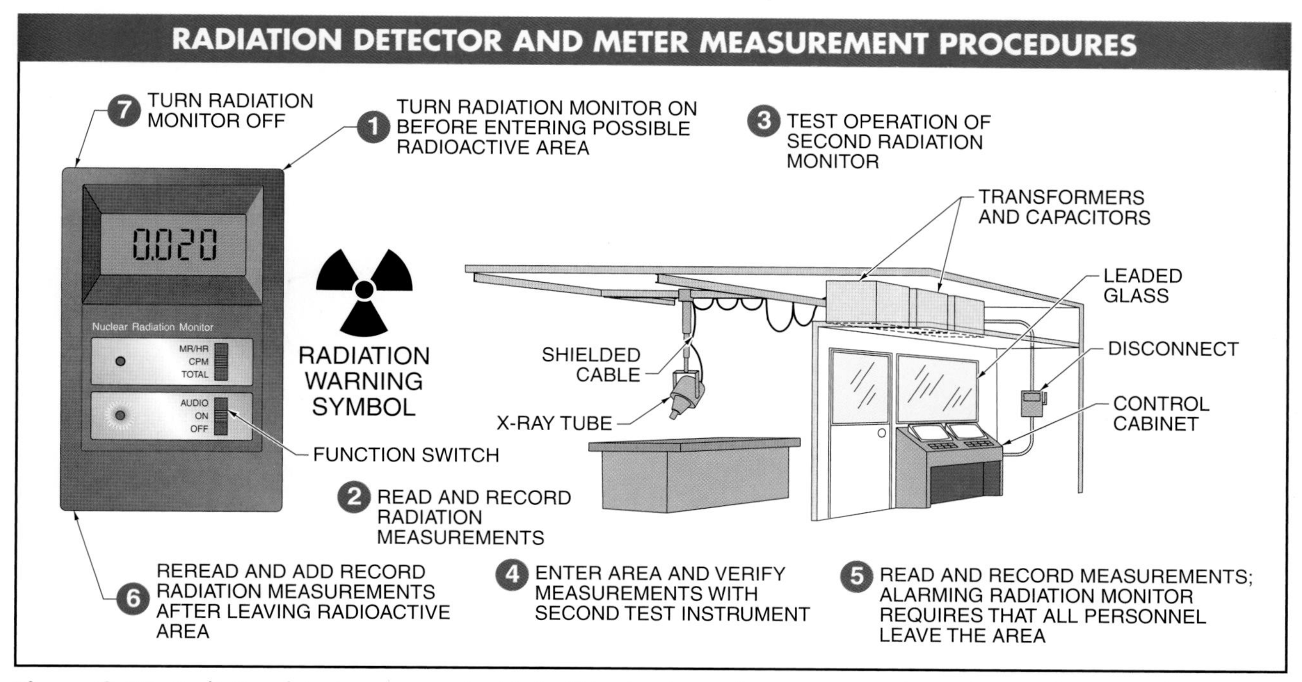

Figure 9-15. Radiation detectors are required in facilities having radioactive materials or when performing tasks on instrumentation having radioactive materials.

CIRCUIT TRACERS

A *circuit tracer* is a two-piece test instrument that includes a transmitter that is plugged into a receptacle, or connected with alligator clips to conductors, and a receiver that provides an audible and visual indication of the circuit to which the transmitter is connected. Circuit tracers are test instruments designed to locate hidden powered conductors, nonpowered conductors, and noncurrent-carrying conductors and metal pipes. They are used to locate and trace the paths of conductors in walls, ceilings, floors, and underground outdoor cables and pipes. They are also used to find broken or open cables and conductors as well as short circuits in branch circuit runs. Circuit tracers are also used to identify a specific fuse or breaker that is protecting a circuit as well as to troubleshoot miswired or shared neutral conductors that may interfere with AFCI and combination AFCI breakers. Circuit tracers are rated for specific voltage ranges.

When a circuit tracer is used, a transmitter is connected to one end of the circuit under test and a receiver is used to trace the circuit by scanning the area under test. The transmitter emits a signal at a fixed frequency that the receiver is tuned to receive, sense, and locate. As the receiver gets close to the transmitted signal, the strength of the signal displayed on the receiver and/or the receiver's audible signal increases. For example, a transmitter can be connected to a receptacle and the receiver used to locate the circuit path, the panel to which the circuit is connected, and the circuit breaker belonging to that receptacle. **See Figure 9-16.**

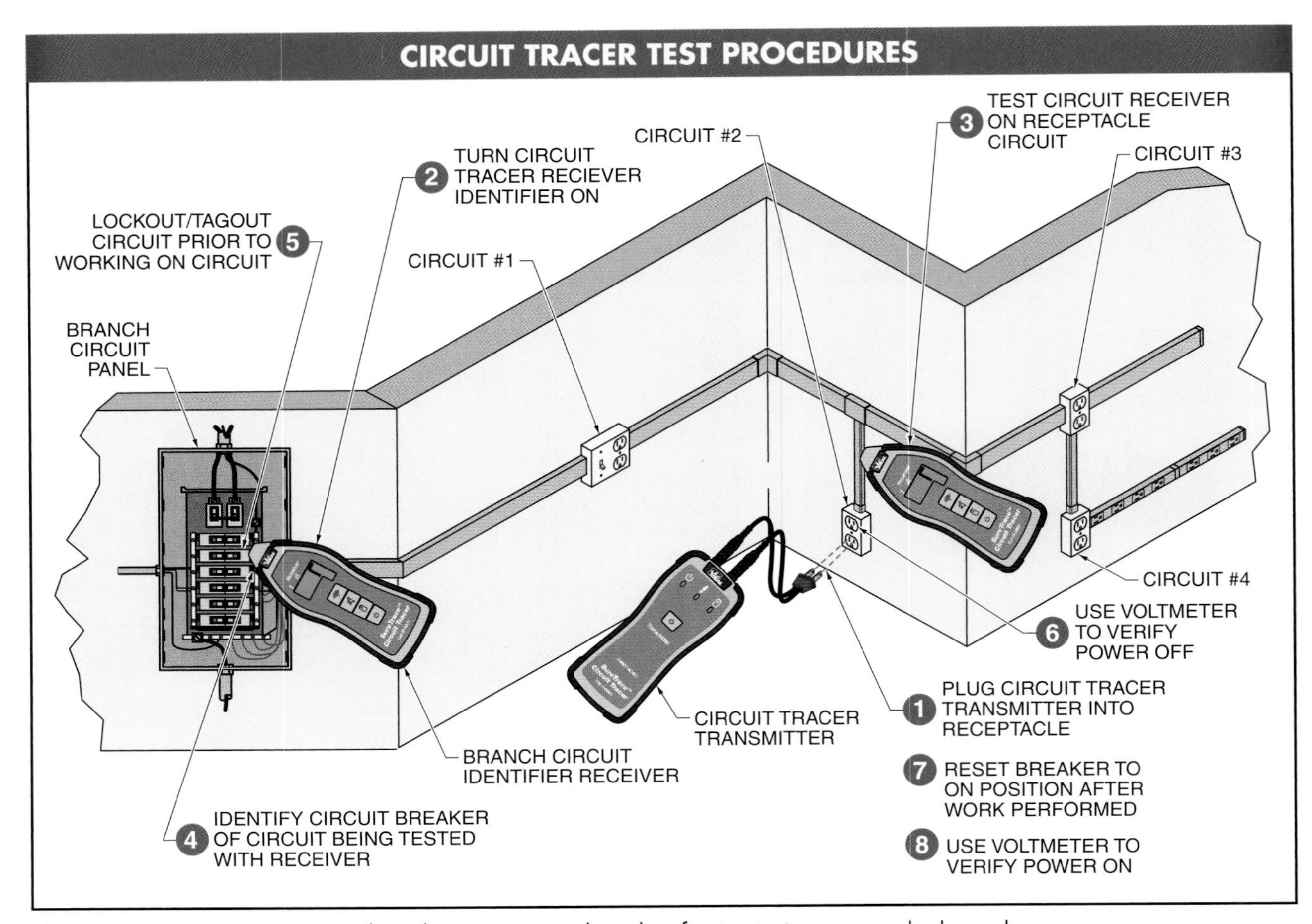

Figure 9-16. Circuit tracers reduce the time required to identify circuits in an unmarked panel.

Before work can be performed on any electrical circuit that does not require power, all power must be removed from the circuit. Removing power on most branch circuits is accomplished by turning OFF the circuit breakers and using lockout/tagout procedures. Sometimes, the circuit breaker that controls the circuit to which power is to be removed is not clearly marked or identifiable. Turning OFF an incorrect breaker might unnecessarily require any loads (systems with latching circuits, computers, patient monitoring, etc.) to be reset. To identify the circuit breaker for a given circuit, circuit tracers are used.

Before performing any tests using a circuit tracer, ensure that the circuit tracer is rated for the voltage carried on the circuit being tested. Refer to the operating manual of the test instrument for all measuring precautions, limitations, and procedures. Always wear the required PPE and follow all safety rules when taking the measurement. To identify the circuit connected to a specific circuit breaker using a circuit tracer, apply the following procedure:

1. Plug the circuit tracer transmitter into the receptacle (outlet) that is to be identified and turn it ON. *Note:* Some OEMs require the use of a plug adapter when connecting to a receptacle.
2. Turn ON the circuit tracer receiver. Verify that the receiver is set to the lowest sensitivity mode or the sensitivity mode recommended by the OEM to properly identify breakers in a panel.
3. To verify that the receiver is operational, test the receiver of the circuit tracer at the same receptacle to which the transmitter is connected.

4. Use the receiver to identify and mark any part of the circuit, or the circuit breaker that controls the circuit being tested, by scanning all of the circuit breakers in a consistent manner looking for the highest discernable reading.
5. When the circuit being tested requires power to be turned OFF, open the circuit breaker and apply a lockout/tagout on the breaker of the circuit being tested.
6. Use a voltmeter to verify that power is OFF before working on the identified circuit.
7. After work has been performed on the circuit, reset the circuit breaker to the ON position.
8. Use a voltmeter to verify that power is restored to the branch circuit.

Miswired Neutral Conductor Identification Procedures

AFCI and combination circuit breakers are sensitive to shared neutral conductors. They will often trip the moment a load is applied if they are wired to an incorrectly spliced neutral between two separate branch circuits. This situation can be easily identified and corrected by using a circuit tracer and applying the following procedure:

1. Identify the AFCI breaker(s) that are tripping.
2. De-energize the panel and isolate the neutral conductors for the breaker(s) and their associated branch circuits.
3. Verify continuity between the affected neutral conductors with a DMM set to measure continuity.
4. Attach the transmitter alligator clips to the neutral conductor(s) under test.
5. Use the receiver to scan each receptacle and switch box for the strongest signal. Once located, remove the faceplate to expose the receptacle, separate the misspliced neutral conductors, and then reconnect them properly.

IDEAL Industries, Inc.

Circuit tracers can be used to accurately locate buried conductors and cables.

Circuit Breaker Identification Procedures

Sometimes circuit breakers are not labeled or are labeled incorrectly. When a circuit breaker is not labeled or is mislabeled, the circuit that it controls can be identified with a circuit tracer by applying the following procedure:

1. Plug the transmitter into the receptacle in question (connect alligator clips as required).
2. At the panel, set the receiver in the lowest sensitivity mode and scan both rows of circuit breakers while holding the receiver square and at right angles to the circuit breakers. Also scan the circuit breakers in the area of the panel where the branch circuit conductors have been connected.
3. Check for the single highest reading. Turn OFF the circuit breaker and verify that the receiver is indicating a loss of signal.

Dead Short Circuit Identification Procedures

A *dead short circuit (dead short)* is an electrical circuit that has a resistance of zero. A dead short can result from the "hot" electric feeder making contact with a metal object, such as a metal outlet box or a ground or neutral wire. The main indicator of a dead short is a tripped breaker. To identify a dead short with a circuit tracer, apply the following procedure:

1. Use a DMM to perform a continuity test between all of the exposed conductors to identify which conductors are short-circuited together.
2. Attach the test leads with alligator clips to the conductors in question.
3. Set the receiver to the highest sensitivity setting and scan along cable runs and outlets, checking for those areas where the signal strength changes. The signal may suddenly fall in intensity or drop to almost zero. If that happens, reduce the sensitivity setting on the receiver and continue scanning to pinpoint the most likely problem area.

Hidden Conductor Locating Procedures

Sometimes, conductors are not visible because they are located behind walls, ceilings, and floors. A circuit tracer can be used to locate conductors that are not visible. To locate hidden conductors using a circuit tracer, apply the following procedure:

1. Plug the transmitter into a receptacle or attach the test lead alligator clips to the hot and neutral conductors.
2. Set the receiver to the highest sensitivity mode.
3. Hold and rotate the receiver in the area where conductors may be located to find the highest numerical reading; this is necessary to follow signal strength while tracing. Signal variation occurs due to bends, twists, and conductors that run deeper or shallower along their path; this may require constant adjustment of the receiver's angle to properly trace a circuit.
4. Adjust the signal strength on the receiver as required. If the reading is too high, reduce the strength. If the reading is too weak, utilize the remote return path method for the transmitter, and then repeat step 3.
5. Continue tracing while following the highest reading until the end of the cable is found.

Underground Conductor Location Procedures

Conductors and pipes that are located underground can present a hazard and must be located prior to performing any digging or excavation work. A circuit tracer can be used to safely and accurately locate underground conductors and pipes. To locate underground conductors and pipes with a circuit tracer, apply the following procedure:

1. Connect only the test lead alligator clips to the hot and neutral conductors. **CAUTION:** Use all appropriate safety procedures and lockout/tagout as required. Plug leads into the transmitter.
2. Turn the transmitter and receiver ON.
3. Walk about 10′ in the direction towards where the line is buried; while holding the receiver level with respect to the ground and pointing in one consistent direction, walk at about 90° to the pathway of the buried conductor while searching for the highest reading on the receiver. Be sure to walk across the line of the conductor noting an increase of readings to a maximum value and then falling off when passing over the conductor. Mark the location of the highest reading.
4. Walk another 10′ further and repeat the step 3, again looking for the centerline and highest reading. Mark the spot of highest reading.
5. Repeat steps 1 through 4 for the length of the run.

Remote Return Paths

Sometimes, signals can be weak or not registered on a tester. A *remote return path* is a separate pathway located at a distance away from an adjacent pathway in an electrical circuit. Remote return paths are used because in some instances, due to the proximity of two conductors (such as when conductors are pressed closely together in a conduit), the signal can actually cancel itself out. Providing separate pathways at some distance from each other via a remote return path can alleviate this problem. It is not unlike the cancellation effect experienced when measuring current. For example, a clamp meter cannot be clamped around the hot and neutral conductors simultaneously to measure current, as the field strength cancels itself out when return current flows back and 180° out-of-phase with the inbound hot conductor. **See Figure 9-17.**

The strongest tracing signal is generated on a closed loop system where the signal can exit one terminal on the transmitter and return to the adjacent terminal through the conductor pathway. This could be as simple as placing an incandescent lightbulb on the circuit to allow the signal to travel through the filament and return to the transmitter.

In some circumstances, the signal generated by the transmitter is weakened by the conductors or pathways that are available to use. Creating a remote return path using a conductor placed away from the conductor to be traced can often greatly increase the signal strength. With remote return paths, rather than having the signal "go out" on the hot conductor and "return" on the neutral conductor, the signal is sent out on one conductor and uses a suitable alternate conductor looped outside the wall or conduit to bring the signal back to the transmitter. This large loop can often generate a powerful signal for situations where a weakened signal is difficult to locate.

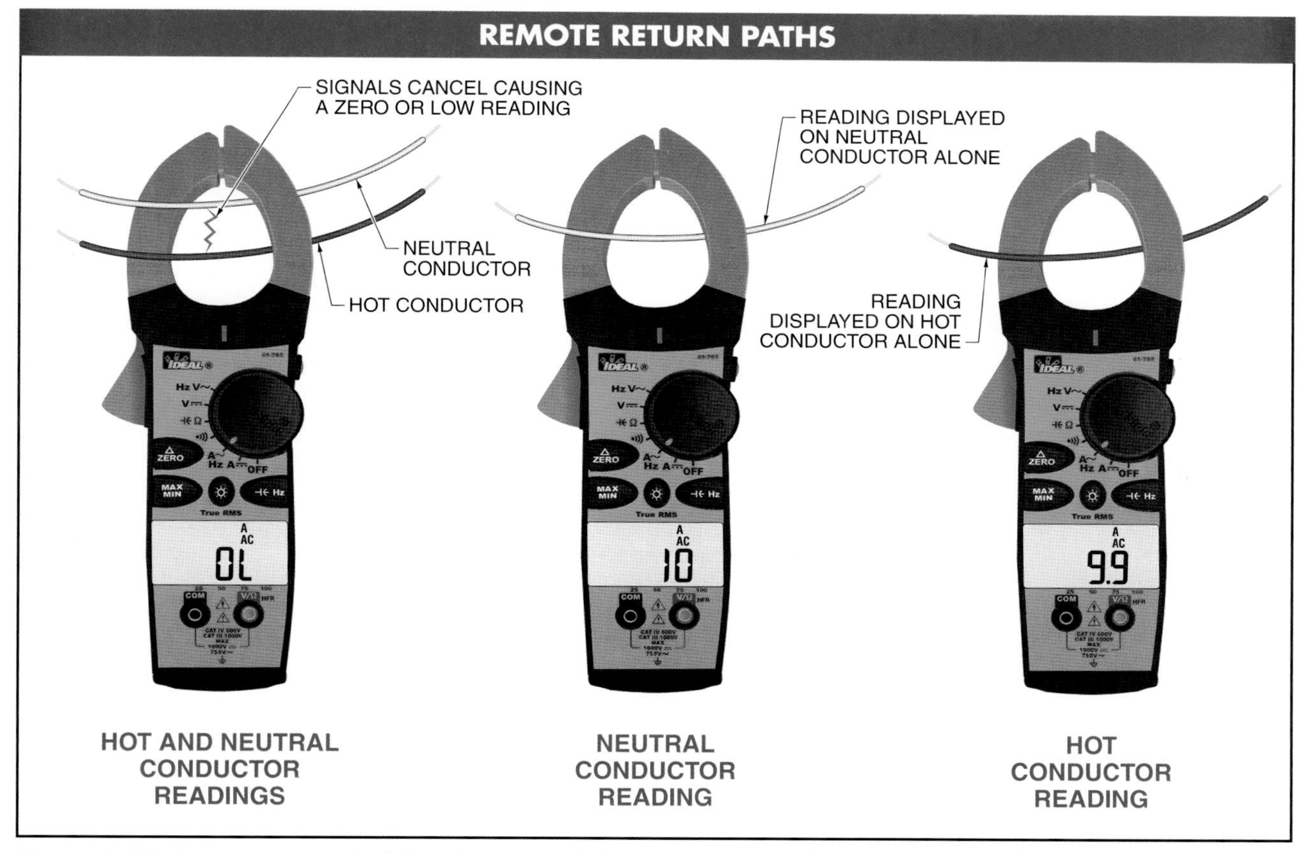

Figure 9-17. Remote return paths follow the principal of current readings in hot and neutral conductors canceling out when return current flows back and 180° out-of-phase with the inbound hot conductor.

VIBRATION METERS

Vibration is a continuous periodic change in displacement with respect to a fixed reference. Vibrations cause mechanical and electrical problems by causing parts and connections to loosen and/or fail. Vibrations also create undesirable noise problems.

Vibration measurements are taken when the transducer of a vibration meter is connected to the object being tested. Transducers convert vibrations into electrical signals which are used to drive the direct digital readout (in in./sec or mm/sec) of a typical vibration meter. To acquire detailed analyses of vibration problems, vibration analyzers with vibration software are used. Vibration analyzers include graphic displays for detailed analysis of vibration measurements and problems. **See Figure 9-18.**

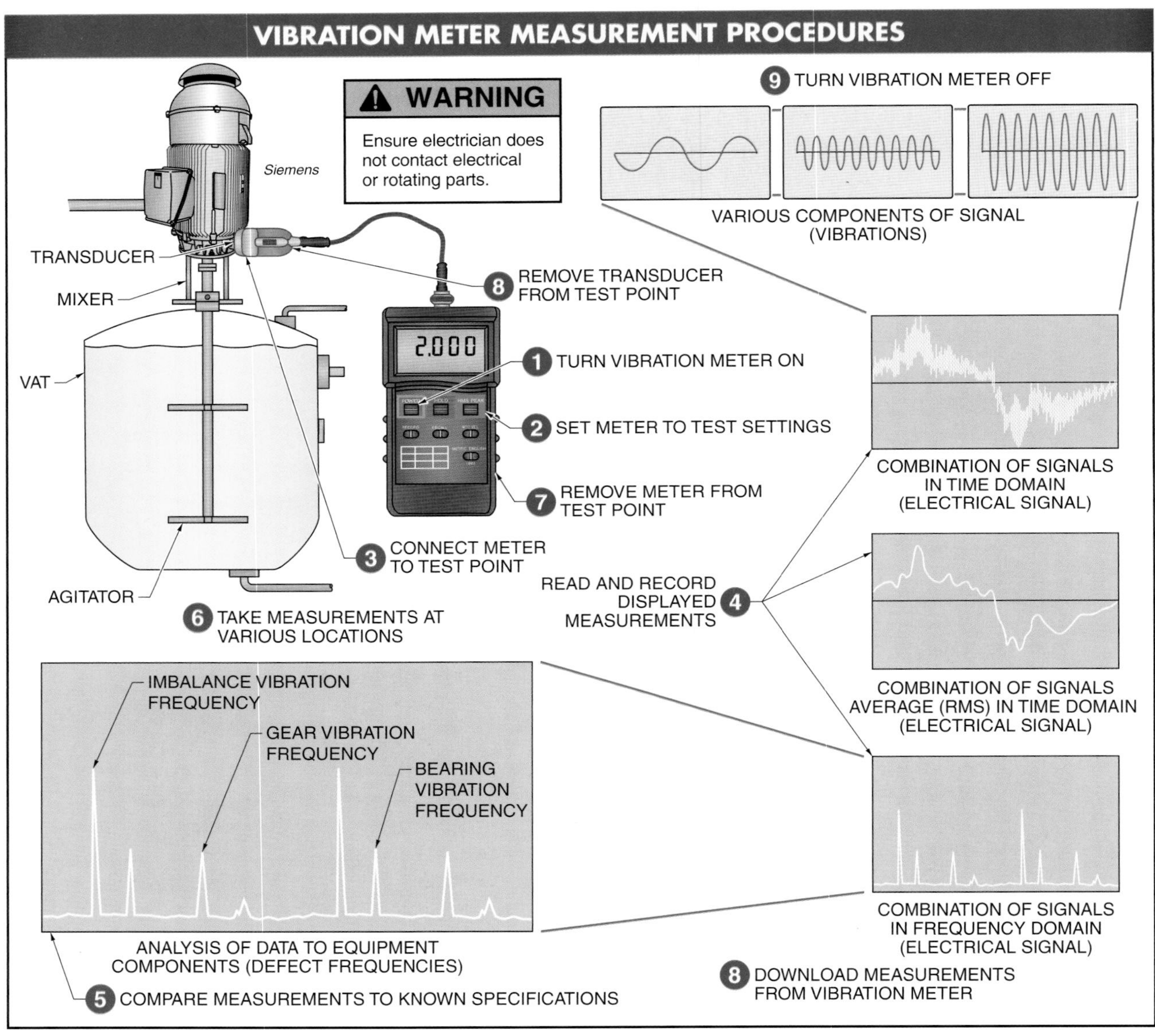

Figure 9-18. Vibration measurements are taken when the transducer, which converts vibrations to electrical signals, of a vibration meter is connected to the object being tested.

Before taking any measurements using a vibration meter, ensure the meter is designed to take measurements on the object being tested. Refer to the operating manual of the test instrument for all measuring precautions, limitations, and procedures. Always wear required personal protective equipment and follow all safety rules when taking the measurement. To measure the amount of vibration using a vibration meter, apply the following procedures:

Safety Procedures

- Ensure that there is no danger from electrical parts or mechanical parts at the point at which the vibration meter is to be connected.

1. Turn the vibration meter ON.
2. Set the function switch of the vibration meter to the required rpm or in./sec mode.
3. Connect the transducer of the vibration meter to the point of testing.
4. Read and record the measurements displayed.
5. Compare the displayed measurement to known specifications, manufacturer recommendations, and/or the meter manufacturer charts. For example, some manufacturers include charts that indicate readings as GOOD, SATISFACTORY, UNSATISFACTORY, or UNACCEPTABLE.
6. Take several measurements at various locations so comparisons can be made and problems isolated.
7. Remove the transducer from the point of testing.
8. Download any measurements when the vibration meter allows for downloading of information.
9. Turn the vibration meter OFF.

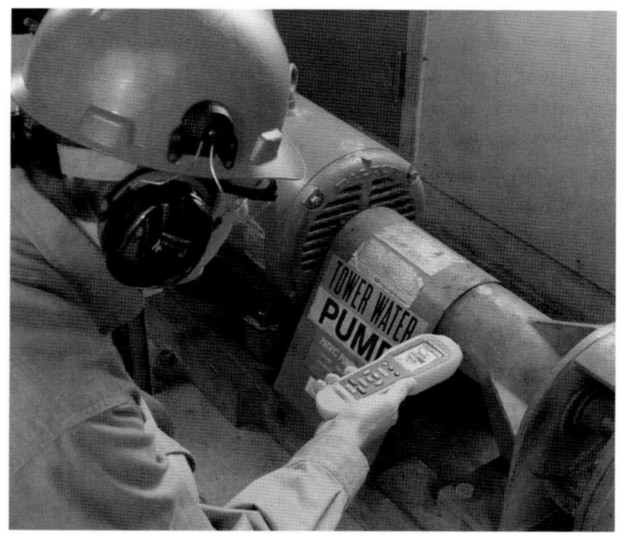

Fluke Corporation.

Hand held vibration meters are used for comprehensive machine condition monitoring.

LASER MEASURING METERS

Standard tape measures (powered return steel tape) have traditionally been used to take measurements. Lasers are typically red beams of light that are used to aim an ultrasonic signal to take a measurement without having to connect anything between the measuring points. Laser measuring meters measure distance by sending out an ultrasonic signal which bounces off solid objects and returns to the receiver. The time taken to travel the distance is used to calculate and display the distance. Laser measuring meters are typically 99.5% accurate. Some laser measuring meters also calculate and display area and volume based on distance measurements recorded. **See Figure 9-19.**

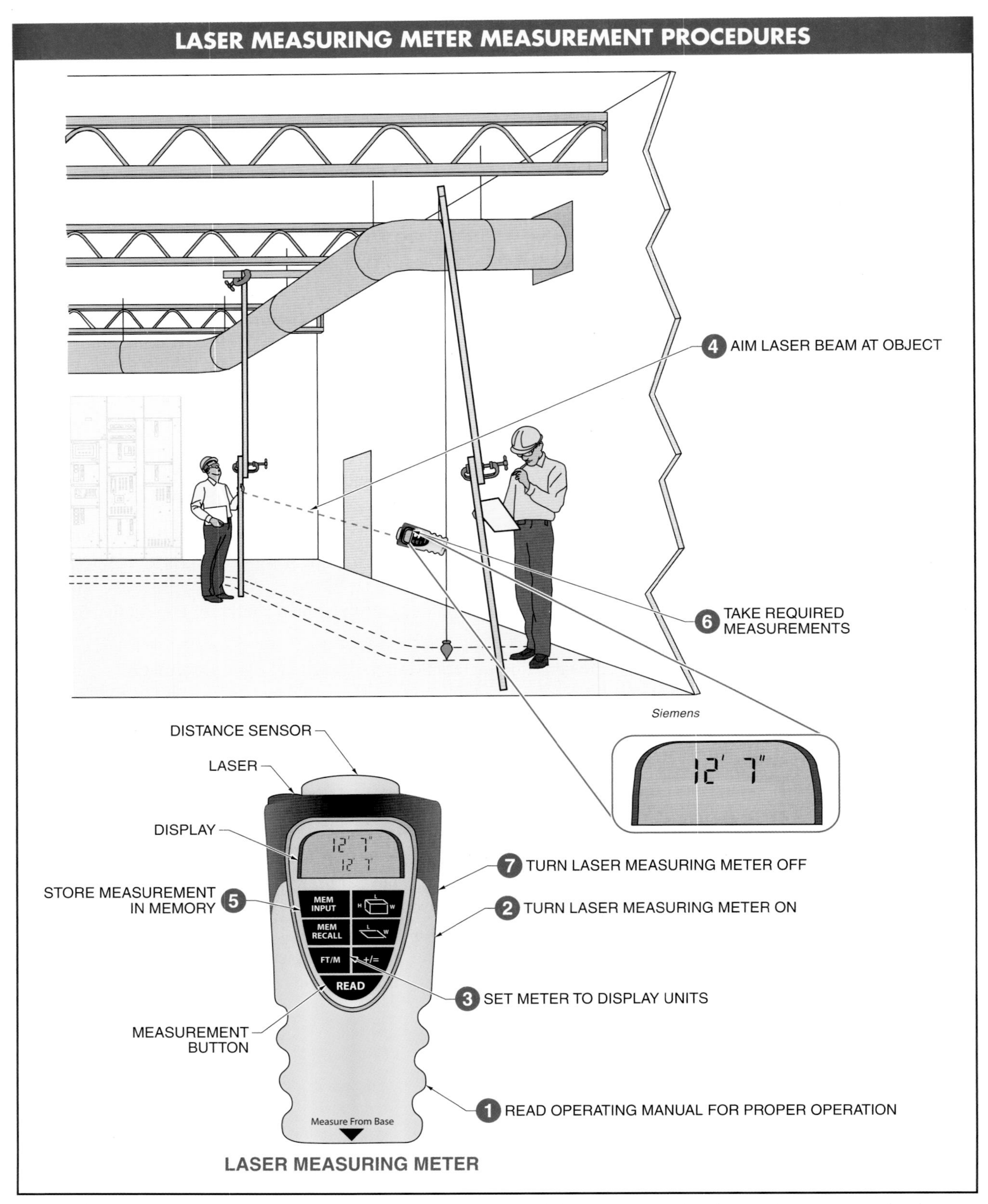

Figure 9-19. Laser measuring meters measure distance by determining the time it takes for an ultrasonic signal to bounce off a solid object.

Before taking any measurements using a laser measurement meter, ensure the meter is designed to take measurements in the area being measured. Refer to the operating manual of the test instrument for all measuring precautions, limitations, and procedures. Always wear required personal protective equipment and follow all safety rules when taking the measurement. To measure distance using a laser measuring meter, apply the following procedures:

1. Before using a laser measuring meter, read the operating manual. Reading the manual may not be necessary for taking distance measurements, but reading the manual is important for understanding how to store measurements and use the stored measurements to display area and volume calculations.
2. Turn the laser measuring meter ON.
3. Set the function switch of the meter to display the required feet or meter mode.
4. Aim the laser beam at an object located at the distance to be measured.
5. Store the measurement in the memory of the laser measuring meter (or write it down).
6. Take all required measurements.
7. Turn the laser measuring meter OFF.

Technical Tip

Modern test equipment includes Bluetooth-enabled wireless tools that send measurements from electrical test instruments directly to "the cloud" or directly from the test instrument to a smartphone. An app can then be used to send the measurement from a smartphone to secure cloud storage. Measurements can be shared with several different technicians or separated and distributed from a specific piece of equipment.

WIRELESS REMOTE MEASUREMENT AND DISPLAY METERS

Test instruments that include remote transmission of measured values allow individual meters to send recorded values to any other meter, smartphone/tablet, or computer that can display transmitted values and data. Wireless meters use standard wireless communication methods, such as Bluetooth. Wireless meters can take measurements like nonwireless meters at the point of measurement by an individual; but once connected, they also allow measurements to be taken without a person and viewed from a safe location. Real-time measurements can be viewed anytime as well as previously recorded measurements, images, and data.

Wireless meters can measure, record, transmit, and remotely display values such as AC/DC voltage, AC/DC current, resistance, temperature, and power. Wireless meters can also measure and transmit data such as thermal images, vibration data, power quality measurements, and insulation information. The most commonly used wireless meter is a DMM, but the following wireless models are also available:

- Clamp meters
- Temperature meters
- Infrared cameras
- Insulation testers
- Vibration meters
- Power quality meters
- Power and energy loggers
- Portable oscilloscopes
- Instrumentation loop calibrators

All measured, recorded, and transmitted values can be displayed on a computer or smart device such as a phone, tablet, or television by using the manufacturer's downloadable software app. All measurements can also be sent to a cloud computing service ("the cloud") or another data storage device for unlimited access. By using a computer or smartphone, individuals located at the point of measurement or any remote location can view measured values, data, and videos through direct communication with the meter. **See Figure 9-20.**

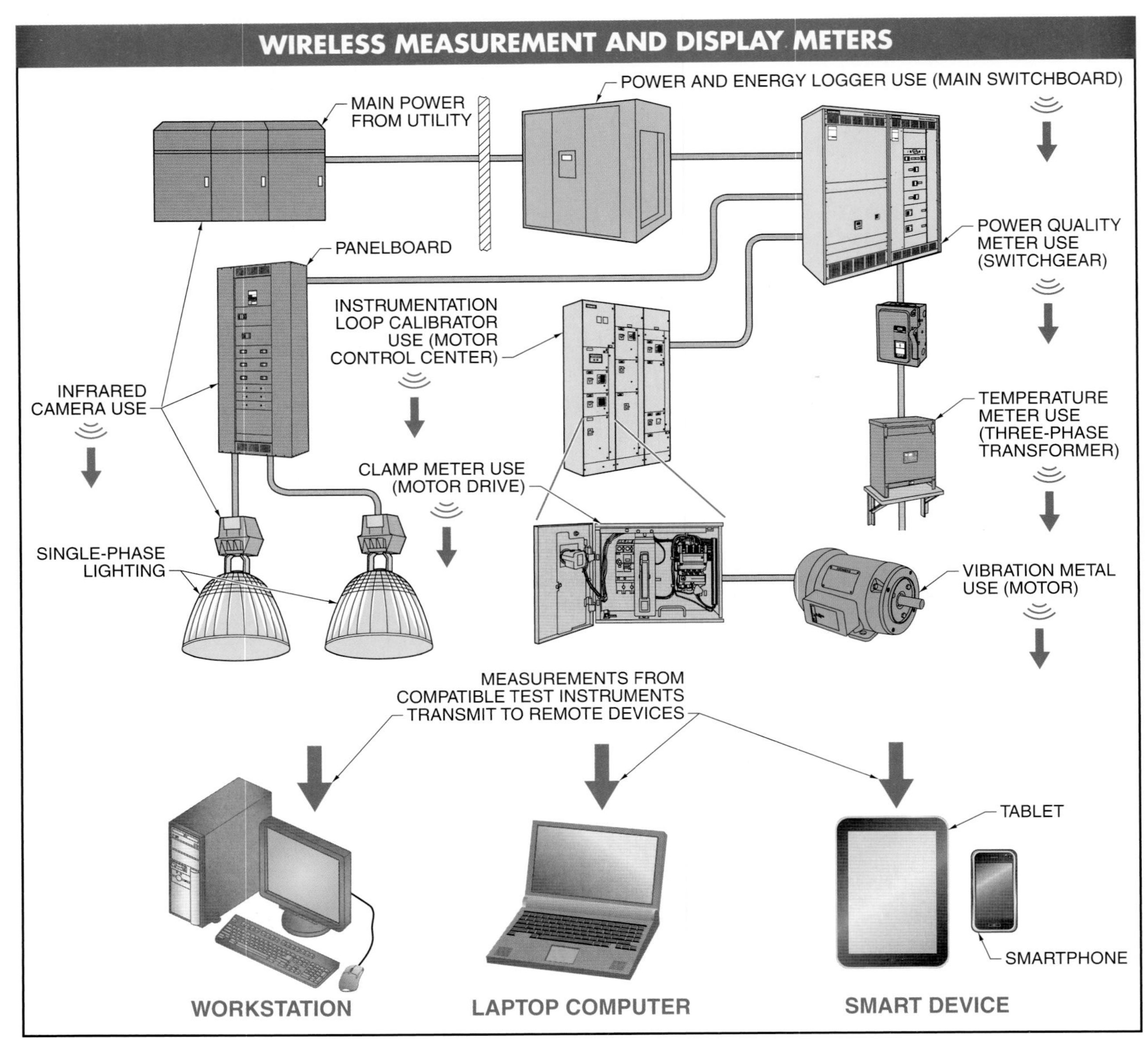

Figure 9-20. Test instruments that include remote transmission of measured values allow individual meters to send recorded values to any other meter, smartphone/tablet, or computer that can display transmitted values and data.

SPECIAL MAINTENANCE TEST INSTRUMENTS

9 REVIEW

Name: ______________________________ Date: ____________________

______ **1.** Which unit is used to measure the amount of light emitted by a lamp?

______ **2.** Which type of tachometer uses a flashing light to measure speed?

T F **3.** Recommended light levels in buildings vary by industry, venue, and event.

______ **4.** Which type of test instrument is used to measure rotational speed?

______ **5.** Which types of tachometers use a reflective area or piece of reflective tape on the rotating object to count revolutions?

______ **6.** Where is the transmitter unit of a branch circuit identifier plugged in?

T F **7.** All power must be turned OFF to a component before testing it with a micro-ohmmeter.

______ **8.** Which type of test instrument accurately measures very small resistances?

T F **9.** Rotational speed can be measured in feet per minute (fpm).

______ **10.** Which unit of illumination is equal to one lumen per square foot?

______ **11.** How does a gas detector indicate when the levels of a gas become unsafe?

______ **12.** Which part of a vibration meter is attached to the object being measured?

T F **13.** Laser and photo tachometers usually have wider measurement ranges than contact tachometers.

______ **14.** What is the illumination (in fc) of 50 lumens on 5 sq ft of area?

______ **15.** Which unit of illumination is equal to 1 lumen per square meter?

T F **16.** Micro-ohmmeters can be used to check mechanical connections such as welds and fastened joints.

T F **17.** Motors must be locked out and tagged out before measurements are taken with a tachometer.

______ **18.** What is the illumination of 280 fc converted to lux?

______ **19.** Is the wavelength of infrared energy longer or shorter than visible light?

______ **20.** Which color are the shortest wavelengths of visible light?

21. Why are noncontact tachometers safer than contact tachometers?

22. What are some of the potential dangers of a confined space?

23. How does the wavelength of energy from light sources affect the appearance of colors?

24. How is a branch circuit identifier used to identify the correct circuit breaker for a particular receptacle?

25. What is the difference between how light is measured for indoor and outdoor applications?

10 TROUBLESHOOTING

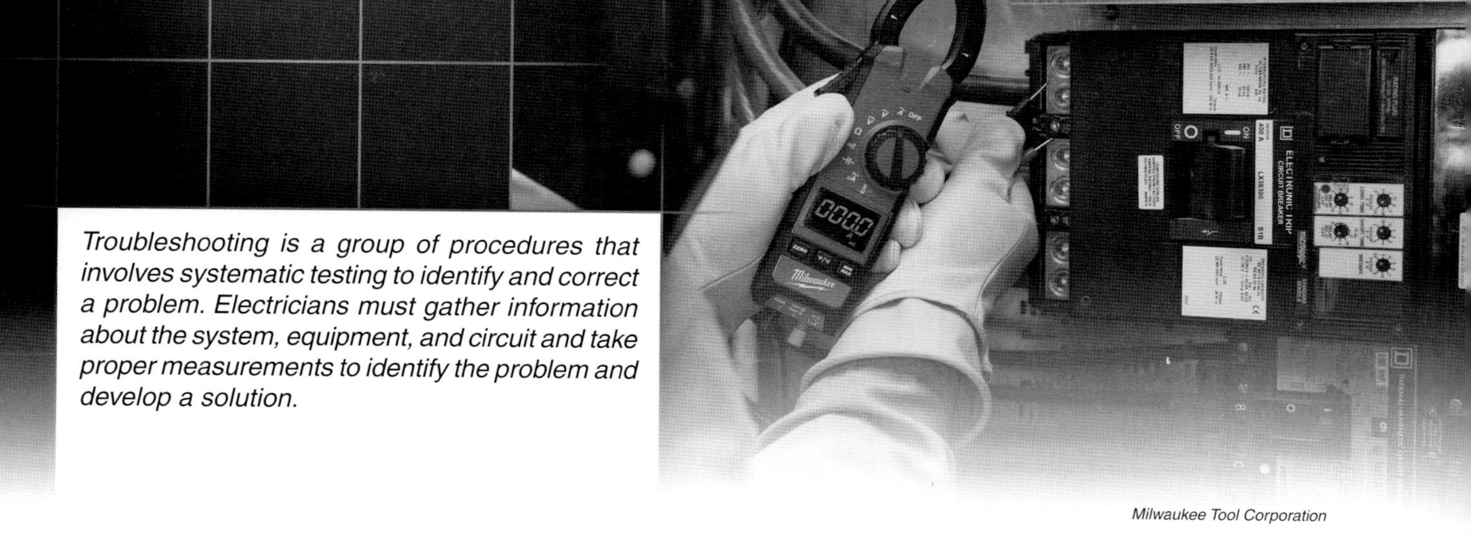

Troubleshooting is a group of procedures that involves systematic testing to identify and correct a problem. Electricians must gather information about the system, equipment, and circuit and take proper measurements to identify the problem and develop a solution.

Milwaukee Tool Corporation

TROUBLESHOOTING

Troubleshooting combines the logical application of knowledge of electrical circuitry and theory with troubleshooting techniques. The combination of knowledge, theory, and techniques forms a complete troubleshooting system that can be used to keep equipment operating at peak efficiency with minimal downtime or lost production. By utilizing a troubleshooting system, electrical and electronic equipment can be divided into various groups that use specific test instruments to test for deficiencies and malfunctions. Troubleshooting systems also aid in the repair of equipment in a timely and efficient manner.

IDEAL Industries, Inc.

Troubleshooting involves using test instruments to identify faulty or malfunctioning equipment.

TROUBLESHOOTING LEVELS

Troubleshooting is a logical step-by-step process used to find a problem in an electrical power system or process as quickly and easily as possible. A *system* is a combination of equipment, components, and/or modules that are connected to perform work or meet a specific need. A *process* is an operation or sequence of operations in which the substance being treated is changed. A *malfunction* is the failure of a system, equipment, or component to operate properly.

To locate and correct a malfunction quickly, troubleshooting is performed at different levels. The levels used when troubleshooting electrical and electronic systems are the system, equipment or unit, board or module, component, and chip levels. **See Figure 10-1.**

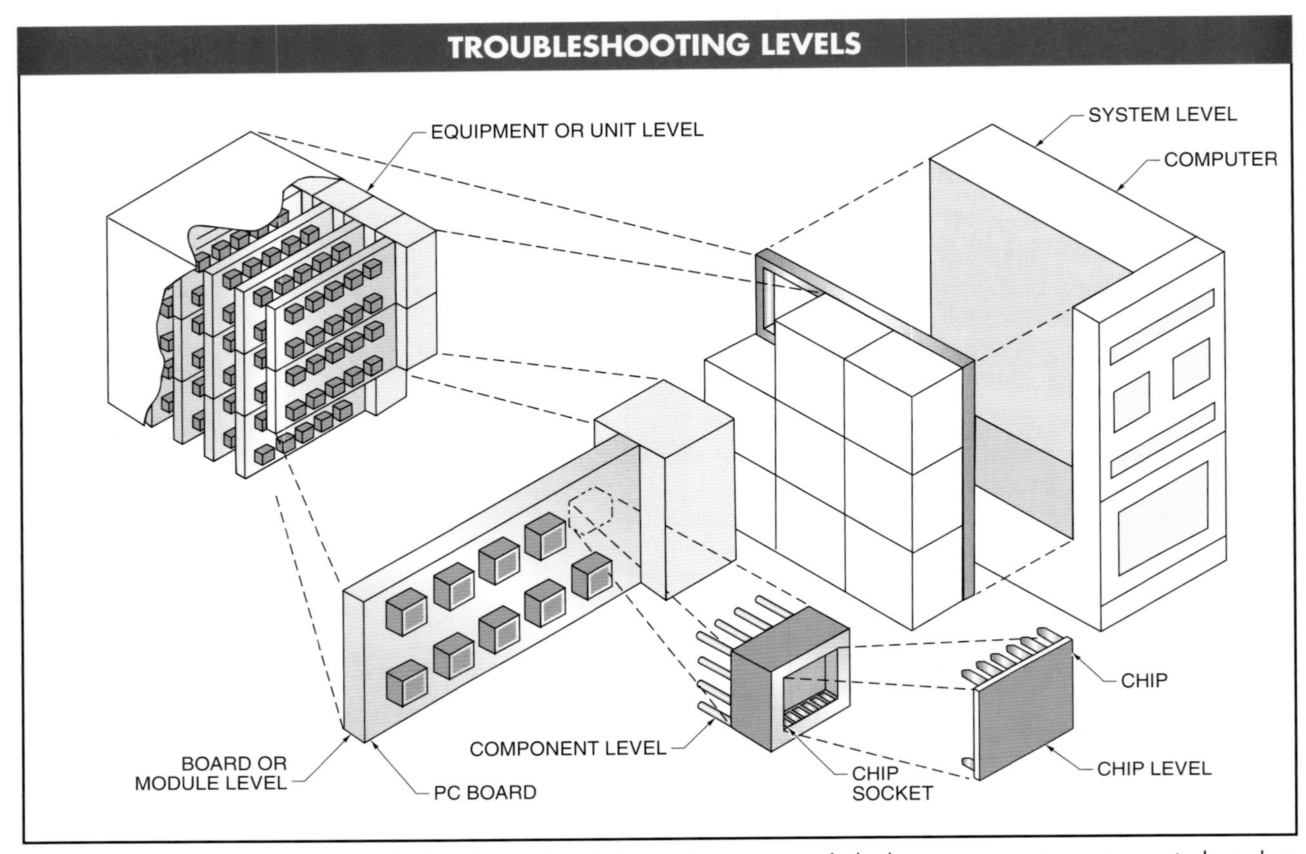

Figure 10-1. Troubleshooting levels for electrical and electronic systems include the system, equipment or unit, board or module, component, and chip.

System Level Troubleshooting

A *system* is a combination of interconnected individual units, modules, and components. In a system, the combination of units, modules, and components works together to produce the desired results.

A system may include main or handheld programming terminals; fixed, remote, or modular controllers; operator terminals; message displays; and interfaces. **See Figure 10-2.**

Programming terminals are typically located at a central location. Handheld programming terminals for local programming changes and troubleshooting are connected to any point in a system. The controllers are located at individual machines or processes throughout a facility.

Troubleshooting at the system level requires knowledge of all hardware operation, software, and interfaces in the system. *Hardware* is the physical components in a system. Hardware malfunctions occur when two or more pieces of equipment are not properly sending or receiving data.

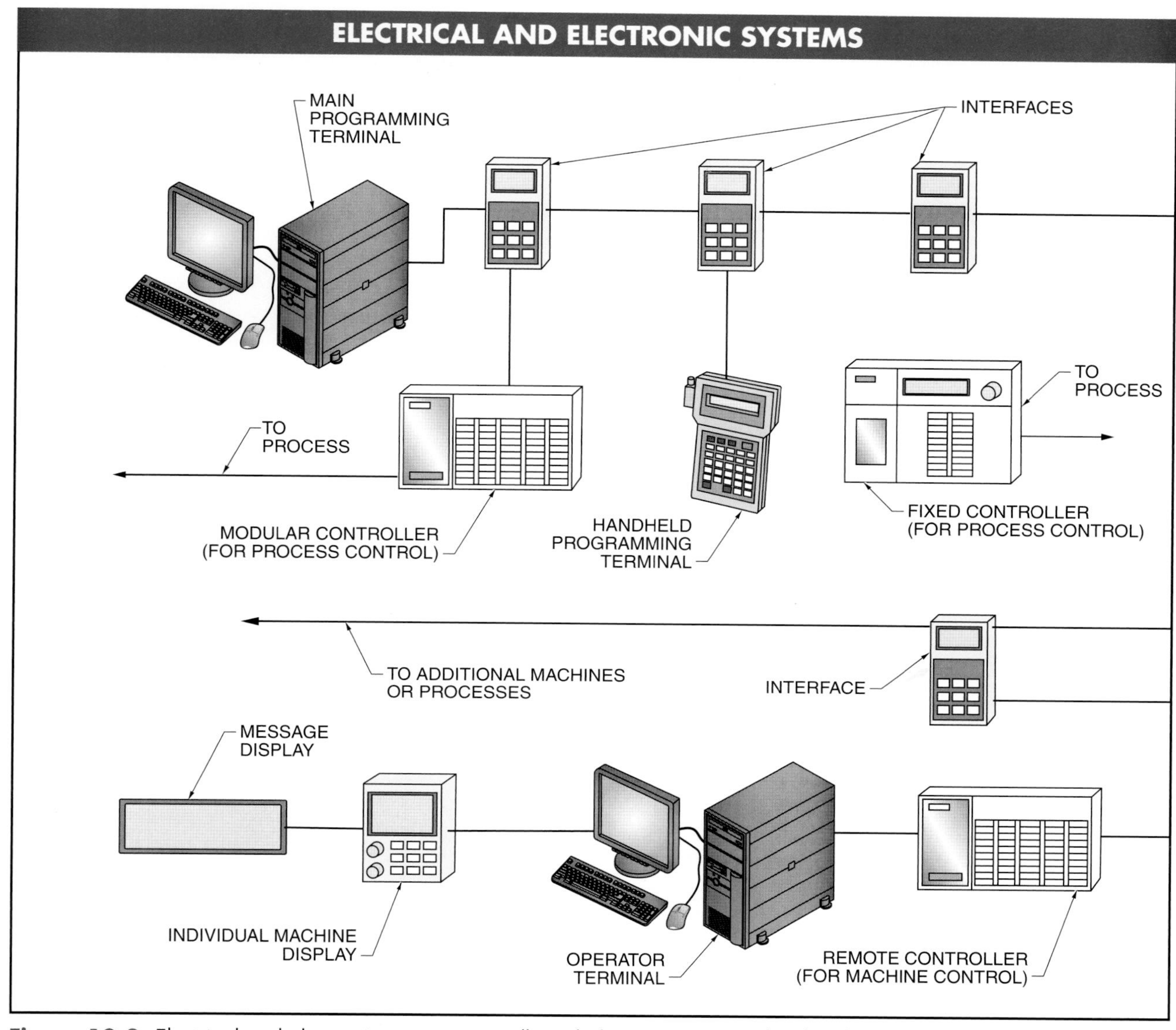

Figure 10-2. Electrical and electronic systems typically include a main terminal, a handheld programming terminal, fixed controllers, modular controllers, remote controllers, operator terminals, message displays, and interfaces.

Software consists of the programs and procedures that allow hardware to operate. Software malfunctions typically involve problems with lines of code (programming), drivers, and instruction sets.

An *interface* is an electronic device that allows communication between various components that are being used together in a system. Interfaces use USB, serial, and parallel ports. Interface problems occur when two or more pieces of equipment are working properly but are not able to communicate with each other.

A *systems analyst* is an individual who troubleshoots at the system level. A systems analyst understands the function and operation of each interface, connection, cable, program, piece of equipment, and language used. *System analysis* is the breakdown of system requirements and components performed when designing, implementing, maintaining, or troubleshooting a system.

System troubleshooting is performed on-site or off-site. *On-site troubleshooting* is troubleshooting at the location where the hardware is installed. *Off-site troubleshooting* is troubleshooting from a location other than where the hardware is installed. Off-site troubleshooting is performed when an on-site computer sends information about a malfunction to an off-site computer. An off-site computer is programmed to analyze the malfunctions of a system. When the source of a malfunction is determined, the off-site computer sends the information to the on-site computer. **See Figure 10-3.**

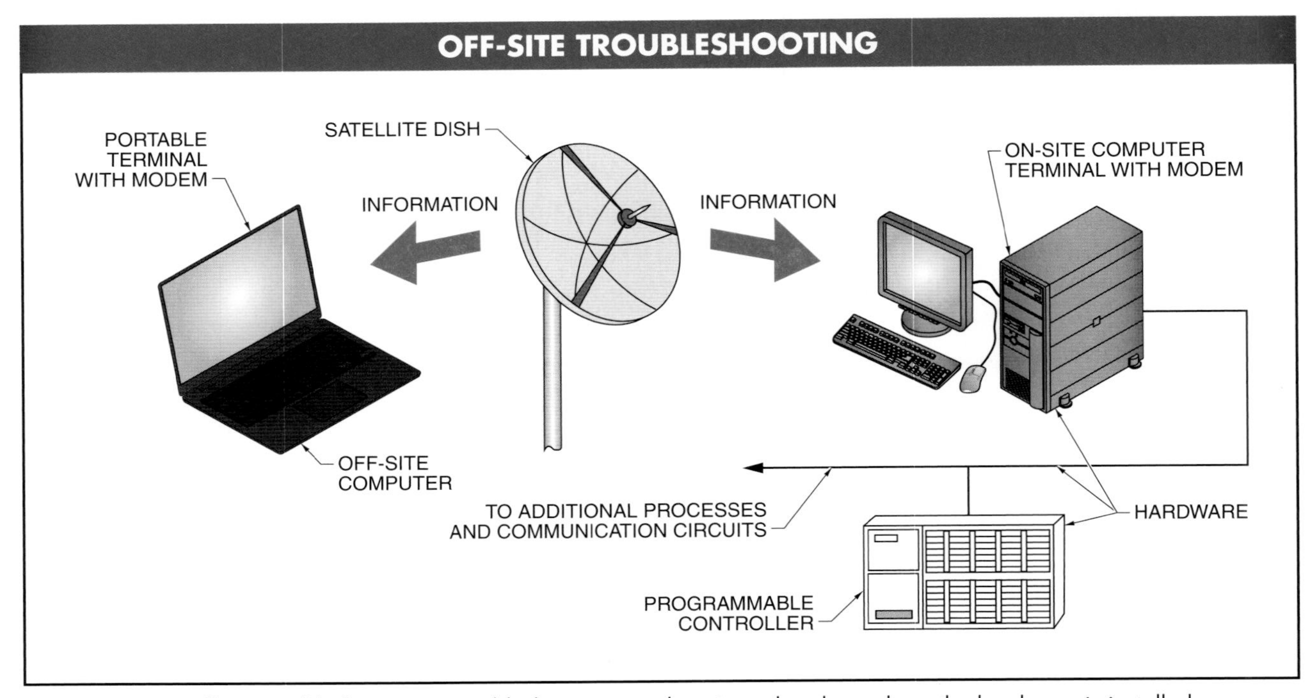

Figure 10-3. Off-site troubleshooting is troubleshooting at a location other than where the hardware is installed.

Equipment or Unit Level Troubleshooting

A *unit* is an individual component that performs a specific task by itself. Systems are made when units are connected together. For example, a television, Blu-ray player, DVR, and AM-FM receiver each are separate units. When connected, an entertainment system is created.

Troubleshooting at the equipment or unit level requires identification of the equipment or component that is malfunctioning. To identify a malfunctioning piece of equipment, substitution of components or testing of components is performed.

Substitution is the replacement of a malfunctioning piece of equipment or component. Substitution is performed when a piece of equipment or component is small and easily accessible, and when a replacement component is available.

Testing is performed when equipment or components are hard-wired or large in size, or when no replacement component is available. A *hard-wired unit* is a unit that has conductors connected to terminal screws, as with a programmable controller that has hundreds of hard-wired connections. Testing requires knowledge of test equipment usage and the ability to correctly interpret test instrument results.

For example, when a computer cannot operate a machine connected to a programmable controller, a malfunction exists at the programming terminal, interface, connecting cables, or programmable controller. A handheld programming terminal is typically plugged into the programmable controller to troubleshoot at the equipment or unit level without disconnecting any equipment. **See Figure 10-4.**

A handheld programming terminal monitors information from and sends information to the programmable controller. Monitoring indicates whether information is received from the main computer. When no information is received, a malfunction exists between the programmable controller and the main computer. When information is received, a malfunction exists between the programmable controller and the machine. When a malfunctioning piece of equipment is identified, the malfunctioning piece of equipment is replaced or serviced at the board or module level.

HANDHELD PROGRAMMING TERMINAL

MAIN PROGRAMMING TERMINAL
INTERFACE
CONNECTING CABLES
PROGRAMMABLE CONTROLLER
HANDHELD PROGRAMMING TERMINAL
TO MACHINE

Figure 10-4. A handheld programming terminal is plugged into a programmable controller to troubleshoot at the equipment or unit level without disconnecting any equipment.

Board or Module Level Troubleshooting

A *board* is a group of electronic components that are connected together on a printed circuit (PC) board and that perform a set task. A *module* is a group of electronic and electrical components that is housed in an enclosure and that performs a set task.

Troubleshooting at the board or module level requires identifying the malfunctioning board or module for replacement. When troubleshooting electronic equipment at the board or module level, the entire board or module (timer, counter, limit switch, photoelectric control, or motor starter) that contains the malfunction is replaced. For example, a plug-in module on a programmable controller has input terminals for pushbuttons or limit switches, and output terminals for lights, motor starters, and solenoids. **See Figure 10-5.** When a problem develops within a module, the module is typically removed and replaced. The replaced module is typically sent back to the manufacturer for servicing.

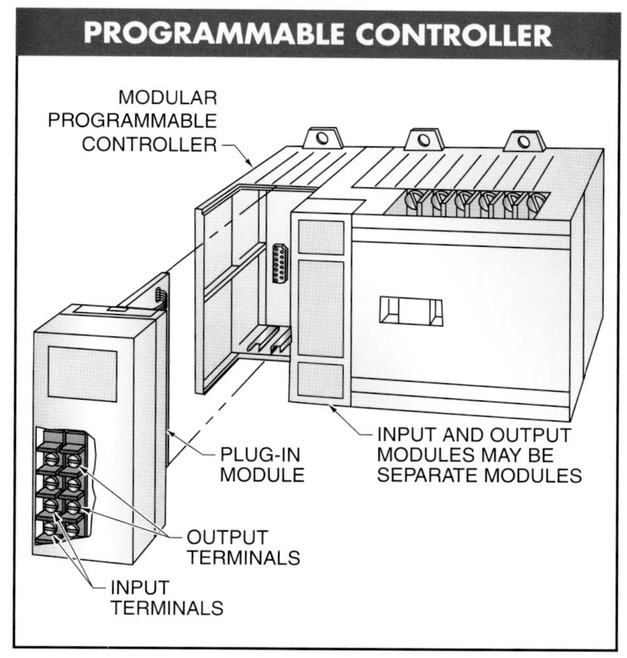

Figure 10-5. Plug-in modules such as input and output terminals of a programmable controller connect inputs such as pushbuttons or limit switches, and outputs such as lights, motor starters, and solenoids, to the programmable controller.

When replacing a PC board, voltage and current inputs and outputs of the board are measured. For example, the input signal to a PC board is 5 VDC and the output signal from the PC board is 5 VDC when pushbuttons 1, 2, and 3 are closed. The 5 VDC output signal can serve as the input signal to a solid-state relay which has a 3 VDC to 32 VDC input rating. When a PC board does not have the correct 5 VDC output voltage, the electrician must follow the circuit to locate the problem on the board. Actual readings vary from circuit to circuit and board to board, depending on the manufacturer and devices used. The voltage drop across the output of a solid-state relay is 1.5 VAC to 6 VAC because of the voltage drop across any solid-state switch. Because of the voltage drop across the relay output switch, the voltage at the motor is 117 VAC. **See Figure 10-6.**

Technical Tip

When replacing a PC board, all static electricity should be removed by using a static electricity grounding strap and a surface that is free from static electricity. Avoid touching PC board components.

B&K Precision

A precision probe attachment can be used with a digital multimeter to troubleshoot individual circuit board components.

Component Level Troubleshooting

A *component* is an individual device used on a board or module. Components include springs, chip sockets, resistors, diodes, contacts, transistors, and capacitors. Troubleshooting at the component level requires finding the exact component that is malfunctioning. Examples of malfunctioning components include bad transistors or capacitors inside a timer or photoelectric control, bad contacts inside a magnetic motor starter, or bad terminals inside a chip socket.

Troubleshooting at the component level is time consuming. Typically, it is more economical to replace an entire board or module than to troubleshoot at the component level and replace a component. A digital multimeter (DMM) set to measure voltage can be used to test components or a group of components on a PC board. **See Figure 10-7.**

Chip-Level Troubleshooting

A *chip* is an integrated circuit that is packaged in an insulating polymer case and is typically used to create digital circuits. The insulating polymer case has metal pins that protrude from the chip for connection to other chips for specific functions. Chips are always replaced rather than repaired.

TROUBLESHOOTING METHODS

A skilled electrician following a troubleshooting plan will find a malfunction quickly and efficiently. When a malfunctioning component is found, the component is replaced or repaired. Preventive maintenance is then performed to prevent future problems.

Preventive maintenance is the work performed to keep machines, assembly lines, production operations, and plant operations running with little or no downtime. Preventive maintenance programs allow equipment to be maintained in good operating condition with little downtime or troubleshooting required.

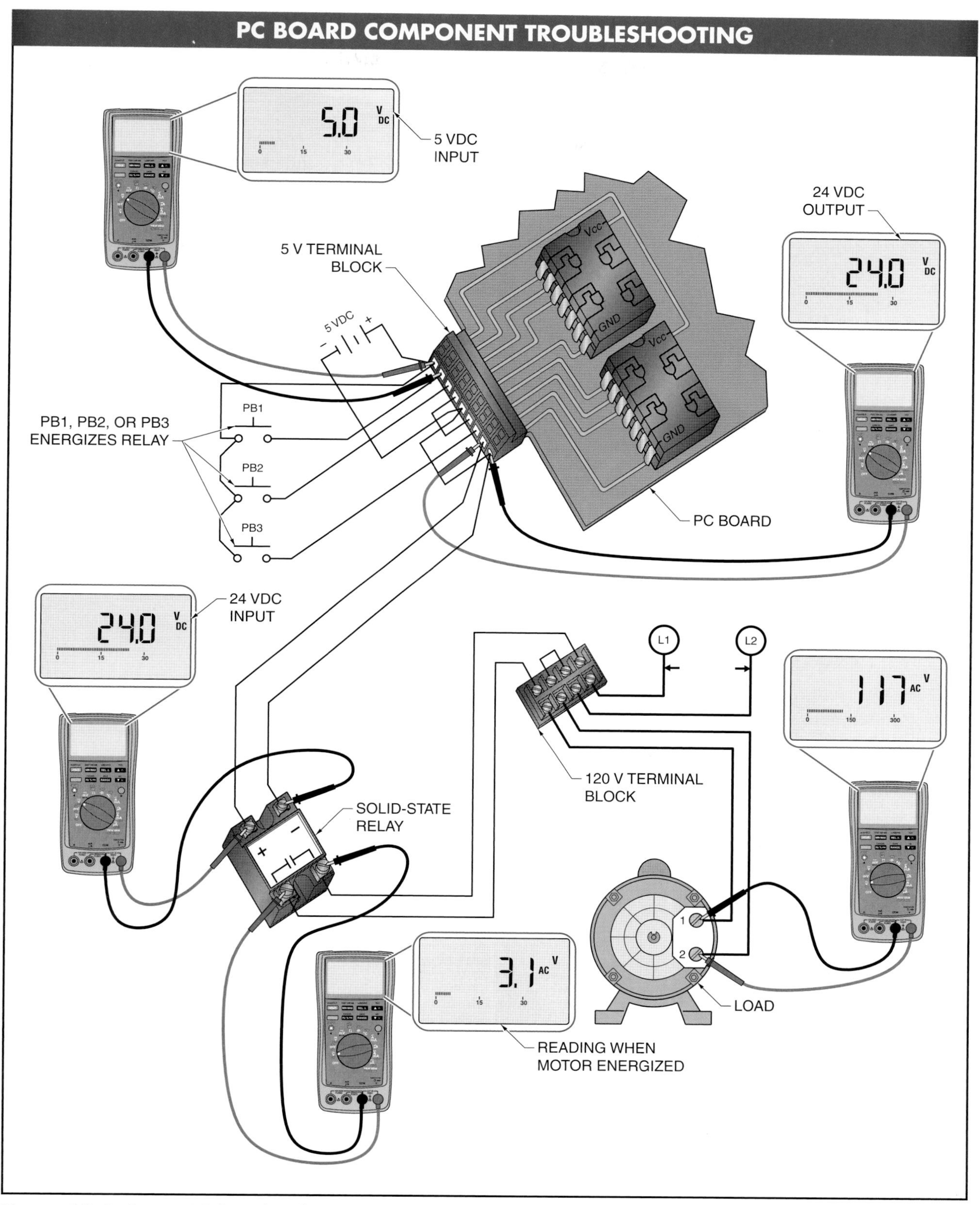

Figure 10-6. During PC board replacement, voltage inputs into the board and voltage outputs out of the board are measured.

Troubleshooting methods used in preventive maintenance programs include troubleshooting by knowledge and experience, troubleshooting by plant procedures, troubleshooting by manufacturer procedures, or a combination of all three methods.

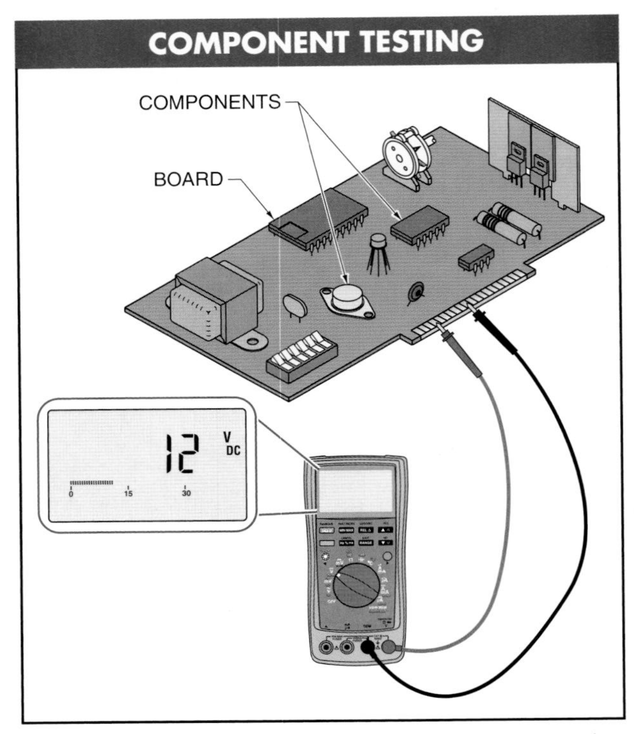

Figure 10-7. A DMM set to measure voltage is used to test components, or groups of components, on a PC board.

Troubleshooting by Knowledge and Experience

Troubleshooting by knowledge and experience is a method of finding a malfunction in a system or process by applying information acquired from past malfunctions. In certain circumstances, troubleshooting by knowledge and experience is only partially effective because the primary malfunction is not corrected. For example, a fuse may blow or a circuit breaker may trip on a machine, stopping production. Records show that changing the fuse or resetting the breaker allows production to continue, but the reason why the fuse blew or the circuit breaker tripped is not known.

Troubleshooting by Plant Procedures

Most plants (facilities) have procedures that must be followed when troubleshooting equipment. Facility procedures are typically developed by supervisors or operators and used to ensure safe and efficient troubleshooting of equipment by plant personnel. Facility procedures are specific to the system or process in use by a company. **See Figure 10-8.**

⚠ WARNING

Facility Procedures/Defective Module Replacement

1. Inform the machine operator that the power will be turned OFF.
2. Turn OFF, lock out, and tag the disconnect switch feeding power to the machine.
3. Using a DMM set to measure voltage, test to ensure that no voltage is present at the power terminals of the programmable controller.
4. Remove the conductor connected to each terminal screw. Mark each conductor with the same number as the terminal screw.
5. Pull the module locking lever out and down until it is perpendicular to the face of the module.
6. Slide the module out and away from the receptacle.
7. Insert the replacement module into the receptacle and lock it in place using the locking lever.
8. Reconnect the conductors to the terminal screws.
9. Inform the machine operator that the power will be turned ON.
10. Place all machine selector switches in the manual position.
11. Remove the lock and tag from the disconnect switch. Make sure all employees are clear of the machine.
12. Turn power ON.
13. Restart the machine.
14. Cycle the machine through one operation using the manual switches.
15. If the machine operates properly, place the selector switches in the automatic position.
16. Cycle the machine automatically 10 times.
17. Inform supervisor on duty if machine is not operating properly.
18. Fill out a company repair report if the machine is operating properly. Place one copy of the report in the maintenance folder inside the machine cabinet and return one copy to supervisor on duty.
19. Remain with the operator until the machine is back in operation.
20. Clear the area of tools and any debris resulting from the maintenance call.

Figure 10-8. Plant procedures are specific to the system or process used by an individual company.

Troubleshooting by Manufacturer Procedures

Troubleshooting by manufacturer procedures is a method of finding malfunctioning equipment by using the procedures and recommendations provided by the manufacturer. Manufacturer procedures differ from facility procedures in that manufacturer procedures are typically shorter and more general than plant procedures. **See Figure 10-9.**

Plant or manufacturer troubleshooting is typically in the form of flow charts. A *flow chart* is a diagram that indicates a logical sequence of steps for a given set of conditions. Flow charts help an electrician to follow a logical path when troubleshooting a malfunction. Flow charts use symbols and interconnecting lines to provide next-step directions.

Symbols used on flow charts include ellipses, rectangles, diamonds, and arrows. An ellipse indicates the beginning and end of a flow chart or section of a flow chart. A rectangle contains a set of instructions. A diamond contains a question stated so as to bring out a yes or no answer. The yes or no answer determines the direction to follow through the flow chart. An arrow indicates the direction to follow based on an answer. **See Figure 10-10.**

Troubleshooting Tip

SAFE TROUBLESHOOTING PRACTICES

Common electrical troubleshooting tasks involve questioning operators, testing the voltage level, current balance and loading, harmonics, grounding, loose connections and terminals, and bad branch circuit breakers. Before taking any measurements, electricians must become familiarized with the equipment that will be used. Read the instrument manufacturer's user manual and understand the WARNING and CAUTION sections. Do not use the instrument/meter for a purpose other than what it is intended for. If the equipment is used in a manner not specified by the manufacturer, the protections for the equipment itself and for the user may be impaired or the warranty voided.

⚠ WARNING

Manufacturer Procedures/Programmable Controller Troubleshooting & Replacement

1. If there is no indication of power (status lights OFF) on the programmable controller, measure the voltage at the incoming power terminal on the power supply module.
2. If correct voltage is present, replace power supply on programmable controller.
3. Remove power from programmable controller.
4. Disconnect the power lines from the power supply terminals.
5. Disconnect the processor power cable from the power supply output terminal.
6. Remove the four mounting screws on power supply to free power supply from main panel.
7. Grasp the power supply firmly and pull out.
8. Press the replacement power supply into the main panel.
9. Replace and tighten the four mounting screws.
10. Connect the processor power cable.
11. Connect the power lines.
12. Turn power ON. maintenance call.

Figure 10-9. Manufacturer procedures vary from plant procedures in that manufacturer procedures are shorter and more general than plant procedures.

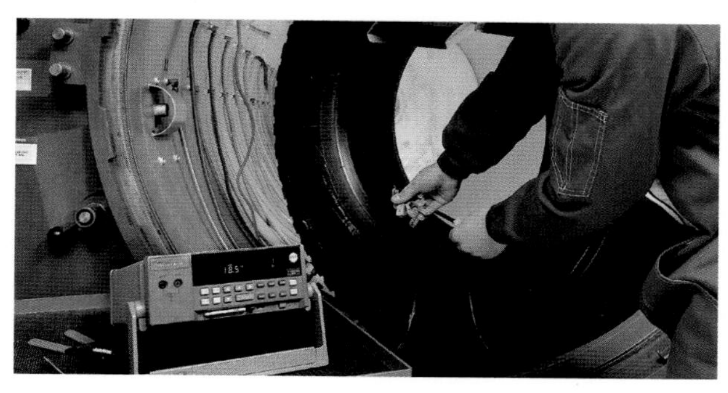

Fluke Corporation

Sometimes testing finished products reveals a problem in the manufacturing process that requires troubleshooting the equipment.

TROUBLESHOOTING PROCEDURES

A *troubleshooting procedure* is a logical step-by-step process used to find a malfunction in a system or process as quickly and easily as possible. When troubleshooting a system or component, a six-step troubleshooting procedure is followed.

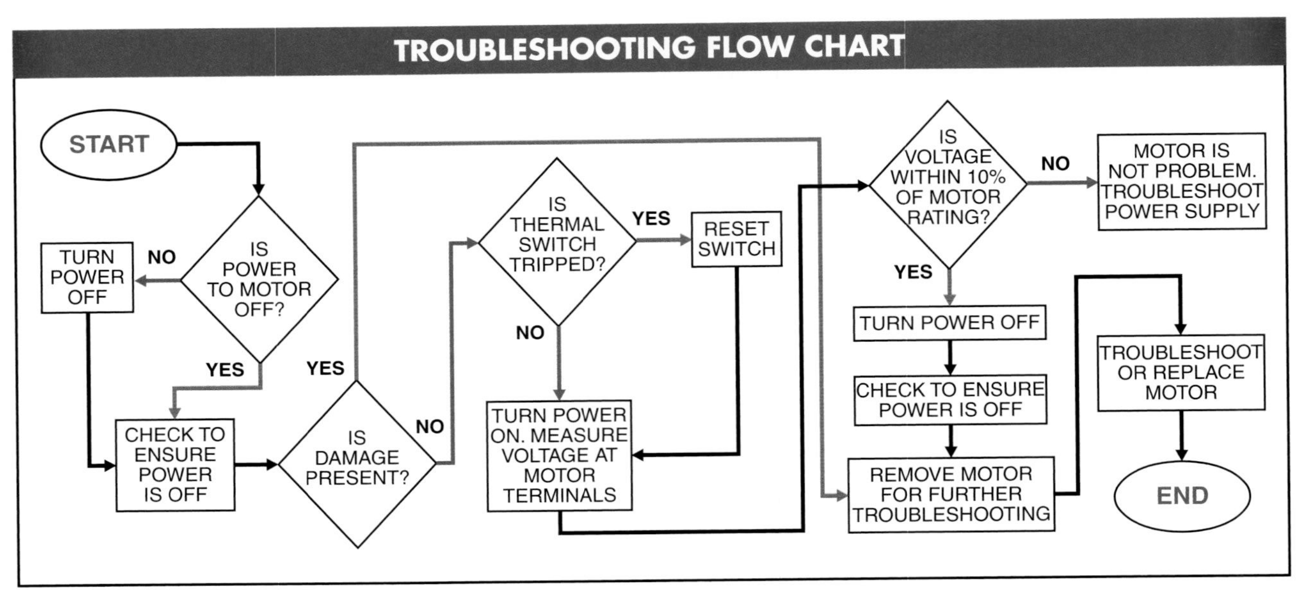

Figure 10-10. A flow chart is a diagram that shows a logical sequence of steps for a given set of conditions.

Six-Step Troubleshooting Procedures

A six-step troubleshooting procedure is used in the troubleshooting process to maintain and repair electrical systems and subsystems. Flow charts are used to properly perform the six-step troubleshooting procedure. **See Figure 10-11.** The six-step troubleshooting procedure includes the following:

1. Symptom recognition
2. Symptom elaboration
3. Hypothesizing probable faulty functions
4. Locating the faulty function
5. Locating the faulty circuit
6. Failure analysis

Symptom Recognition. Symptom recognition is the first step in the six-step troubleshooting process. *Symptom recognition* is the action of recognizing a malfunction in electronic equipment and/or systems. Observe a normally operating system or process during startup, slow operation, fast operation, and shutdown. Determine what part of a system is not working properly and also the condition of the system outputs (energized and de-energized). Visually inspect for problems such as jammed, burned, or broken material. Test fuses, circuit breakers, and overload resets. Tap into the knowledge of the operator on the machine and check for recent modifications and unauthorized changes.

Additional questions to consider during the first step (symptom recognition) of the troubleshooting process include, What effect does a long downtime have on the process or product? Is it more economical to repair or replace? Can the defective part be replaced with a spare or be fixed in the maintenance shop (manufacturer shop) where time is not important to production? Symptom recognition questions must be answered prior to moving forward with the analysis and repair of malfunctioning equipment.

Symptom Elaboration. Symptom elaboration is the second step in the six-step troubleshooting process. *Symptom elaboration* is the process of obtaining a more detailed description of the machine malfunction. Information concerning the system and electronic components is gathered through the use of operational controls and built-in indicating instruments to aid in the evaluation of a possible malfunction. All information, regardless of how minor the information may seem, is listed when gathered to aid in the solution of a malfunction.

Technical Tip

When following the six-step troubleshooting procedure, visual inspection and knowledge of normal equipment operation are important because they are common to all six steps.

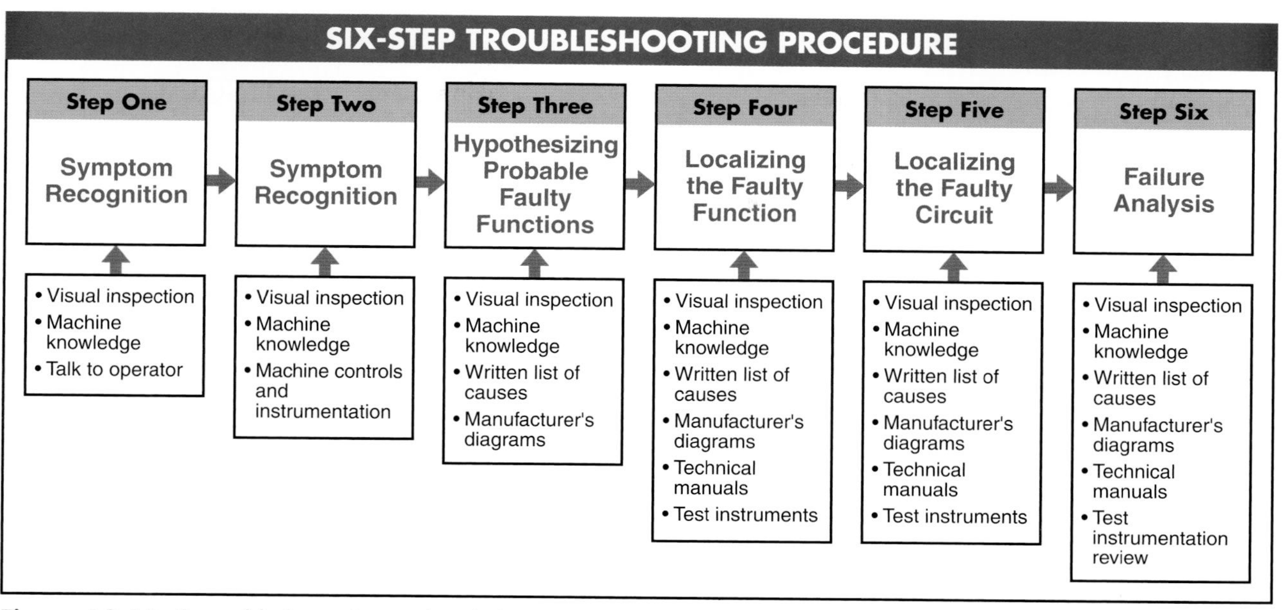

Figure 10-11. To troubleshoot electrical and electronic systems, a six-step troubleshooting procedure is performed.

Hypothesizing Probable Faulty Functions. Hypothesizing probable faulty functions is the third step in the six-step troubleshooting process. *Hypothesizing probable faulty functions* is the theoretical analysis of what could be causing a machine to malfunction. Once the probable malfunctions have been listed, the third step is to gather and refer to as much information as possible from the manufacturer, including technical service manuals, machine prints, and spare parts lists. A bill of materials, when available, should also be included in this process. By gathering machine information and consulting with local equipment dealers and manufacturer representatives, the probable cause of a malfunction can be narrowed down to a specific area (or areas in multifunction machines). Most manufacturer technical service manuals have a corrective maintenance section that lists probable causes and corrective actions. When all information has been reviewed, a decision must be made as to the most likely cause of the malfunction.

Locating the Faulty Function. Locating the faulty function is the fourth step in the six-step troubleshooting process. *Locating the faulty function* is the process where the determination is made as to which section of a multifunction machine is malfunctioning or what part of a single-function machine is actually at fault. Locating a fault involves the use of visual inspections from step 1, the lists developed in step 2, and the technical manuals, prints, and materials lists gathered in step 3. Testing devices are used in step 4 of the troubleshooting process, but not down to the circuit level. All knowledge gathered up to this point should be utilized to locate the source of the malfunction to a particular section of the equipment.

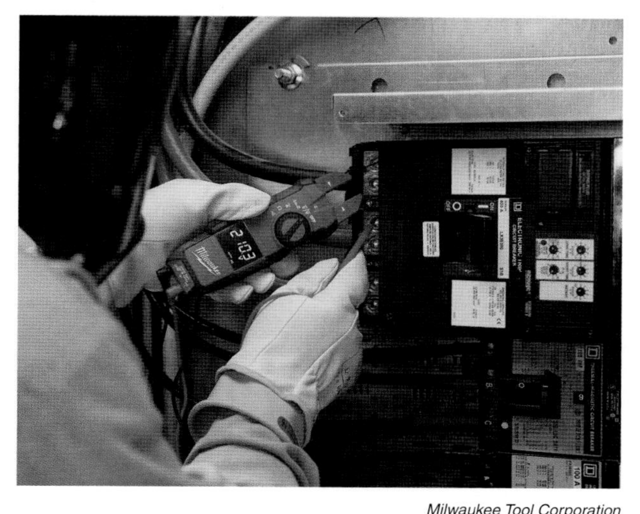

Milwaukee Tool Corporation

Troubleshooting usually requires the use of test instruments such as ammeters to locate a fault within a system.

Locating the Faulty Circuit. Locating the faulty circuit is the fifth step in the six-step troubleshooting process. *Locating the faulty circuit* is a troubleshooting process that requires that a circuit be tested along various points for the presence or absence of a signal, depending on the type of circuit. Extensive use of test instrumentation is required to test circuits for a signal. The signal must be traced from a source to the point where the signal is lost. For example, a DMM set to measure voltage is used to test a circuit that has a suspected short. By following the flow path of circuit voltage, the location of the actual short circuit is easily identified. When the source of a fault has been located to a specific part of a circuit, the sixth and final step in the troubleshooting process, failure analysis, is performed.

Failure Analysis. Failure analysis is the sixth and final step in the six-step troubleshooting process. *Failure analysis* is the testing of a circuit using test instruments to identify the actual faulty component in the circuit. The identification and replacement of the faulty component does not complete step 6 of the troubleshooting process. The cause of the component failure must also be determined. For example, a motor connected to a pump creates product flow through a piping system. **See Figure 10-12.** The pump develops a higher than normal pressure when the pipe that carries the product is partially blocked. To create flow against higher pressures, the motor of the pump draws additional current. The additional current blows a fuse, stopping the motor. The component that failed (the fuse in this case) may not be where the problem originated, but a secondary malfunction that is the result of a primary malfunction of another component in the system. The primary malfunction is the scale buildup in the pipe. The secondary malfunction is that the blown fuse stopped the motor. Changing the fuse only solves the problem until the fuse blows again.

Technical Tip

Although the recommended practice is to install a green wire for the grounding conductor, conduit is sometimes used exclusively for this purpose. When checking ground bonds, it is important to check the tightness of all conduit connections.

Troubleshooting Tip

SIX-STEP TROUBLESHOOTING PROCEDURE APPLICATION

The six-step troubleshooting procedure can be applied to industrial electrical systems in such areas as transformers and tap changers, switchgear, motors, generators, uninterruptible power supplies (UPS), cables, cable joints and terminations, batteries, standby power systems, and power factor.

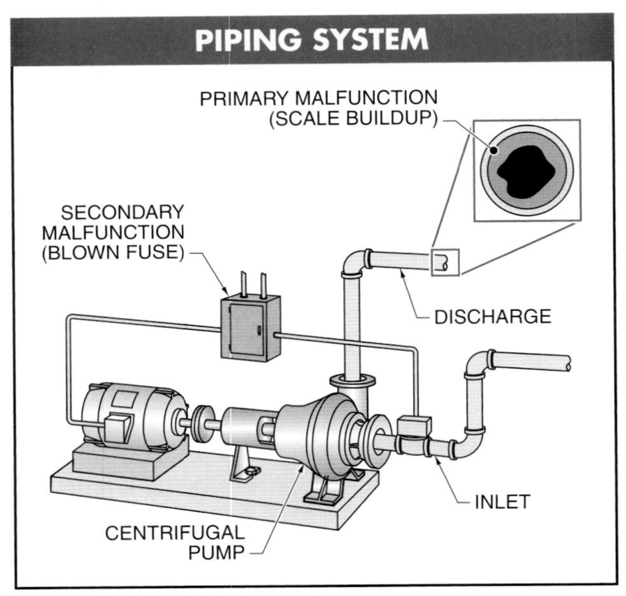

Figure 10-12. Secondary malfunctions occur repeatedly when the primary malfunction is not serviced.

When the symptoms of a malfunction are directly related to a faulty component, electricians can assume that the malfunction of the equipment can be corrected by replacing or repairing the faulty component. When symptoms are similar to specific malfunctions created by other components, the information gathered in steps 3 and 4, as well as the knowledge of the machine operator and electrician, must be utilized to determine when other components are the cause of a fault. Various test instruments are used to determine when there is a probable fault with other sections of the equipment.

Document Findings

When the six-step troubleshooting procedure has been successfully completed and the equipment is returned to service, the electrician must create a detailed document (maintenance or machine log) of what tasks were performed to solve the problem. A machine log must include the following:

- Initial symptoms or equipment malfunction
- A list of steps taken to determine the problem
- What tests were performed
- What test instruments were used, along with the readings taken
- What was done to correct the problem
- A list of components replaced or repaired
- A list of components tested when a malfunction is not corrected
- Length of machine downtime (date and time of initial shutdown, date and time of return to service)
- Names of personnel involved in the troubleshooting process and repair of equipment
- Description of outside services used, if applicable
- Suggestions to help prevent the malfunction from recurring; examples include replacing parts at a certain time, adding additional fuses, or changing the circuit design or operation

Machine log information is stored either electronically or manually for future reference when troubleshooting a machine or similar equipment. A *repair and service record form* is an example of documentation used for the troubleshooting and repair of a particular piece of equipment. **See Figure 10-13.** All components used in the repair of the malfunctioning equipment must be reordered at this time.

EQUIPMENT RELIABILITY

When troubleshooting or servicing a piece of equipment, the age of the equipment must be considered. Most equipment follows a typical life expectancy curve. **See Figure 10-14.**

Break-in Period

A *break-in period* is the time just after the installation of a new piece of equipment in which defects resulting from defective parts, poor manufacturing quality, contamination, or environmental stress appear. The highest failure rate for most equipment occurs during the break-in period. Troubleshooting a malfunctioning piece of equipment during the break-in period includes the following:

- Determining if the equipment is correctly installed. The equipment may be connected to an improper voltage level, improperly grounded, or installed in an environment for which the equipment was not designed.
- Looking for signs of improper operator use.

Useful Life

Useful life is the period of time after the break-in period when most equipment operates properly. Equipment can fail during the useful life period of a life expectancy curve. Troubleshooting a malfunctioning piece of equipment during the useful life period includes the following:

- Checking for signs of improper maintenance, such as dirty or clogged filters. Most equipment requires periodic maintenance. Periodic maintenance typically includes cleaning and lubricating.
- Determining when equipment is properly protected from improper environmental conditions, such as higher-than-normal operating temperatures, high humidity, salt, corrosive chemicals, and dirt.
- Locating signs of damage caused by misuse or accidents. Accidents often occur due to careless operator use or improper environmental conditions. Improper environmental conditions include voltage surges and flooding.

REPAIR AND SERVICE RECORD FORMS

MOTOR REPAIR AND SERVICE RECORD

Motor File #: 004632 Serial #:

Date Installed: 1/9 Motor Location: Equipment Room

MFR: Dayton Type: D Frame: 145T

HP: 10 Volts: 230 Amps: 15

RPM: 1725 Filter Sizes:

Date	Operation	Mechanic
2/7	Checked Current Drain - - OK	
3/14	Checked Belts - Realigned Belts	
5/16	Cleaned Motor	

SEMIANNUAL MOTOR MAINTENANCE CHECKLIST

Motor File #: 004632 Serial #:

Date Installed: 1/9 Motor Location: Equipment Room

MFR: Dayton Type: D Frame: 145T

HP: 10 Volts: 230 Amps: 15

RPM: 1725 Date Serviced: 7/11

Step	Operation	Mechanic
1	Turn OFF and lock out all power to the motor and its control circuit.	RNS
2	Clean motor exterior and all ventilation ducts.	|
3	Check motor's wire raceway.	|
4	Check and lubricate bearings as needed.	|
5	Check drive mechanism.	|
6	Check brushes and commutator.	|
7	Check slip rings.	|
8	Check motor terminations.	|
9	Check capacitors.	|
10	Check all mounting bolts.	|
11	Check and record line-to-line resistance.	|
12	Check and record megohmmeter resistance from L1 to ground.	|
13	Check motor controls.	|
14	Reconnect motor and control circuit power supplies.	|
15	Check line-to-line voltage for balance and level.	|
16	Check line current draw against nameplate rating.	|
17	Check and record inboard and outboard bearing temperatures.	|

Figure 10-13. A machine log (Repair and Service Record form) is an example of documentation used when troubleshooting and repairing a specific piece of equipment.

Wear-out Period

Wear-out period is the period of time after the useful life period of a piece of equipment ends and typical equipment failures occur. Troubleshooting a piece of equipment in the wear-out period includes the following:

- Determining when a malfunction must be repaired immediately or when the equipment will last long enough that it can be repaired during scheduled downtime.
- Determining when equipment is safe. Deterioration occurs in older equipment. When conductor insulation has deteriorated or when safety guards are not present, the equipment will not be safe, even after a malfunction is repaired.

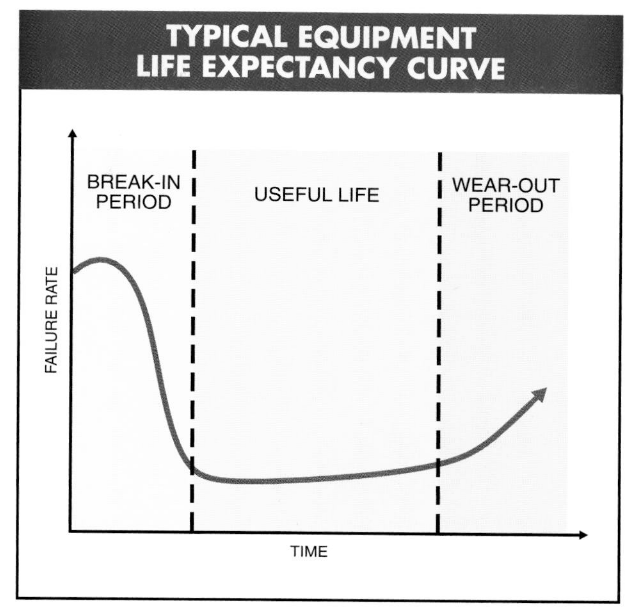

Figure 10-14. Most equipment follows a typical life expectancy curve.

TROUBLESHOOTER RESPONSIBILITIES

All industries have systems and/or processes that must operate properly in order to produce product. To keep a system and/or processes operating properly, trained personnel are required to troubleshoot and repair any malfunction that occurs to a system or process. Personnel who troubleshoot and repair systems or processes are part of the maintenance/service department of a company. The responsibilities of the maintenance/service personnel include the following:

- Installing new equipment according to up to date prints and codes. Prints may be paper or electronic. Paper prints are the most common and are typically provided with each component or machine when purchased. Paper prints include diagrams that indicate power connections, assembly or connection procedures, and replacements parts. All prints use standard symbols and abbreviations to indicate circuit operation. **See Figure 10-15.** Electronic prints that include the circuit schematic and troubleshooting procedures are also available from most manufacturers.

  Electronic prints are prints captured and stored electronically. They are used for large or complicated equipment. Electronic prints are often complete sets of prints that include setup, operating, troubleshooting, and preventive maintenance procedures. These prints are typically sold as an extra package and typically show a logical start-up procedure for a new system or process.
- Follow a logical start-up procedure for a new system or process. Perform general checks of fluid levels, guard positions, and belt tightness, and check for loose parts prior to turning power ON. Turn ON and test one part of a machine or system at a time. Test all manual operations first. Determine if all safety equipment and features are working properly. Anticipate component failure. Operate a machine long enough to allow pressure to build, belts to slip, and oil to heat up. The majority of component failure on new equipment occurs during the first few hours or days of operation. Problems can be mechanical that produce an electrical malfunction (limit switch).

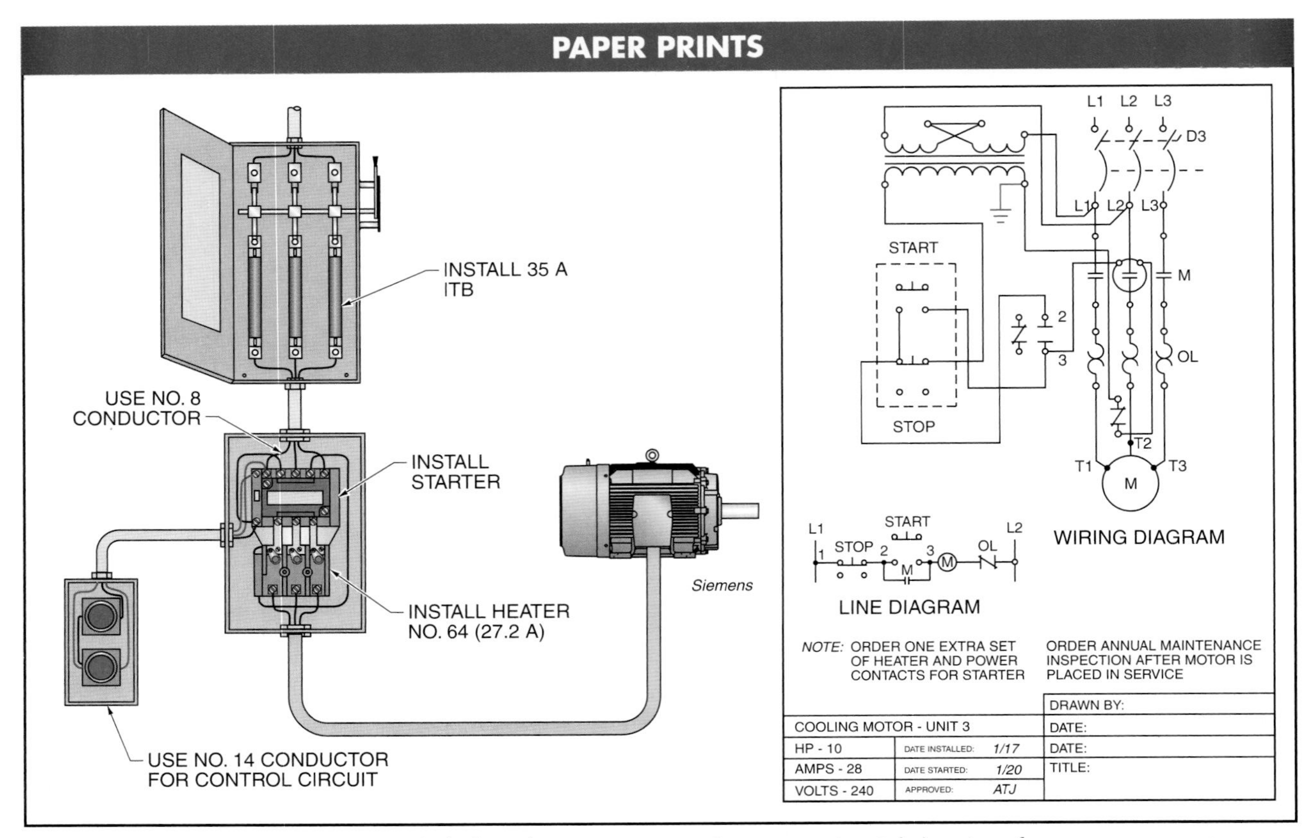

Figure 10-15. Paper prints may include line diagrams, wiring diagrams, pictorial drawings for power connections, assembly or connection procedures, and replacement part charts.

- Apply a preventive maintenance program. Preventive maintenance is action taken to maintain equipment in good operating condition to ensure uninterrupted operation. Preventive maintenance includes tightening belts, adding lubricant, cleaning, changing filters, calibrating, general inspection, and setting control limits. Preventive maintenance is typically performed according to a set schedule. Monthly, semiannual, and annual preventive maintenance schedules are common.
- Prevent malfunctions. Every process or operation has potential problems that cause malfunctions. Preventing malfunctions requires close attention to the entire process, including ensuring that the correct size and type of parts are fed into a machine, that the machine is operated properly, and that improvements, modifications, or design changes are made.
- Identify malfunctions that occur. A skilled troubleshooter identifies the cause of a malfunction before the malfunction is repeated. Note all malfunctions that occur. The true cause of a malfunction may not be clear. For example, a part that jams in a machine may be caused by a part that is not within specifications, a machine that is not within specifications, operator misuse, improper lubrication, or a control malfunction. Most malfunctions are identified after the malfunction has appeared several times.
- Repair equipment in an efficient manner. Anticipate and have the required tools and test instruments for a troubleshooting call. Troubleshoot from simple to the complex. Ask the opinions of operators and other personnel that are familiar with the system. Determine if problems can be fixed temporarily to allow production to continue until the end of the operation. Maintain good records of maintenance, equipment condition, and downtime. Provide justification for overtime, new tools and equipment, and additional personnel.

- Handle emergencies. The maintenance department may handle emergencies that occur in a facility until additional help arrives. Emergencies that may require trained maintenance personnel include general emergencies, trade-related emergencies, and operational emergencies.

  General emergencies include fires, floods, accidents, and power outages. Personnel must understand facility evacuation procedures, emergency telephone numbers, basic first aid procedures, and the meaning of various alarms.

  Trade-related emergencies include electrical shock, burns, and exposure to specific hazardous conditions. Maintenance personnel must understand power disconnection procedures, shock treatment, and CPR.

  Operational emergencies result from hazardous materials, fire hazards, and safety problems. Personnel must understand specific chemical, fire, and physical safety procedures.

It is important to remain informed about the latest test instruments, procedures, and technologies. Trade and technical magazines provide information on new equipment, techniques, and technologies. Manufacturers and local suppliers offer training seminars concerning various product lines and updated prints. The latest equipment, procedures, and technologies are often available through local trade schools and company training courses.

TROUBLESHOOTING 10

REVIEW

Name: ______________________________ Date: ____________________

______________ **1.** What are the physical components in a system called?

T F **2.** It is never economical to replace an entire board rather than an individual component.

______________ **3.** Which troubleshooting level would include troubleshooting a water pump in a boiler system?

______________ **4.** Which troubleshooting method involves using documentation provided by equipment manufacturers?

______________ **5.** Which step in the troubleshooting process involves using test instruments extensively to diagnose faulty equipment?

______________ **6.** Which troubleshooting level would include troubleshooting an individual winding in an electric motor?

T F **7.** Personnel are expected to document their procedures and findings after troubleshooting equipment or systems.

______________ **8.** What part of a system is composed of programs and procedures for operating hardware?

T F **9.** Manuals, wiring diagrams, and other equipment documents are important resources for troubleshooting.

T F **10.** A module is an integrated circuit that is used to create digital circuits.

______________ **11.** Which troubleshooting method involves using procedures developed by the facility?

______________ **12.** Which troubleshooting level would include troubleshooting a PLC?

T F **13.** Most equipment failures occur during the break-in period.

______________ **14.** Which troubleshooting level would include troubleshooting an LCD message display circuit?

______________ **15.** What device allows communication between various components used together in a system?

______________ **16.** Which type of maintenance is performed to keep equipment and process running with little or no downtime?

______________ **17.** Which type of troubleshooting is performed away from the facility with the malfunction?

_______________ **18.** Which step in the troubleshooting process involves interviewing the machine operator about malfunctions and modifications?

T F **19.** The system recognition step in the troubleshooting process involves researching all the available documentation on a piece of equipment.

_______________ **20.** Which step in the troubleshooting process involves determining the cause of the failure?

21. How are equipment failures distributed during typical equipment life expectancy?

22. What are troubleshooter responsibilities in regards to preventing malfunctions on a machine?

23. How could documenting a troubleshooting process help prevent future failures?

24. How are flow charts similar to the six-step troubleshooting procedure?

25. How are the five troubleshooting levels used to diagnose a failure?

APPLICATIONS APPENDIX

INTRODUCTION TO TEST INSTRUMENTS

Name: ______________________ Date: ______________________

Application 1-1 — Electrical Prefixes

Prefixes are used with electrical and electronic components and test instruments to simplify large and small numbers. The prefixes most commonly used in the electrical field are mega- (M), kilo- (k), milli- (m), and micro- (μ). Some test instruments also use other, less common prefixes for certain measurements, such as pico- (p) for capacitor measurements and giga- (G) for high resistance measurements. For example, one picofarad (1 pF) is equal to 0.000000000001 F and one gigohm (1 GΩ) is equal to 1,000,000,000 Ω. See Metric Prefixes.

METRIC PREFIXES

Multiples and Submultiples	Prefixes	Symbols	Meaning
1,000,000,000,000 = 10^{12}	tera	T	trillion
1,000,000,000 = 10^{9}	giga	G	billion
1,000,000 = 10^{6}	mega	M	million
1000 = 10^{3}	kilo	k	thousand
100 = 10^{2}	hecto	h	hundred
10 = 10^{1}	deka	da	ten
Base Unit 1 = 10^{0}	—	—	—
.1 = 10^{-1}	deci	d	tenth
.01 = 10^{-2}	centi	c	hundredth
.001 = 10^{-3}	milli	m	thousandth
.000001 = 10^{-6}	micro	μ	millionth
.000000001 = 10^{-9}	nano	n	billionth
.000000000001 = 10^{-12}	pico	p	trillicnth

Very small or very large numerical values can be awkward to use or convey, but can be made more manageable by changing the measuring unit. A base unit is any unit that does not include a prefix, such as 0.015 A, 1500 V, and 2,300,000 W. Values can be simplified by converting the numerical value and adding a prefix to the unit symbol. For example, 0.015 A = 15 mA, 1500 V = 1.5 kV, and 2,300,000 W = 2.3 MW. See Metric Conversions.

METRIC CONVERSIONS												
Initial Units	Final Units											
	giga	mega	kilo	hecto	deka	base unit	deci	centi	milli	micro	nano	pico
giga		3R	6R	7R	8R	9R	10R	11R	12R	15R	18R	21R
mega	3L		3R	4R	5R	6R	7R	8R	9R	12R	15R	18R
kilo	6L	3L		1R	2R	3R	4R	5R	6R	9R	12R	15R
hecto	7L	4L	1L		1R	2R	3R	4R	5R	8R	11R	14R
deka	8L	5L	2L	1L		1R	2R	3R	4R	7R	10R	13R
base unit	9L	6L	3L	2L	1L		1R	2R	3R	6R	9R	12R
deci	10L	7L	4L	3L	2L	1L		1R	2R	5R	8R	11R
centi	11L	8L	5L	4L	3L	2L	1L		1R	4R	7R	10R
milli	12L	9L	6L	5L	4L	3L	2L	1L		3R	6R	9R
micro	15L	12L	9L	8L	7L	6L	5L	4L	3L		3R	6R
nano	18L	15L	12L	11L	10L	9L	8L	7L	6L	3L		3R
pico	21L	18L	15L	14L	13L	12L	11L	10L	9L	6L	3L	

A conversion table can be used to quickly change between a base unit and equivalent units that use prefixes. For each prefix combination, the table lists the direction and number of places the decimal point is moved to change the numerical value. To convert a large numerical value into a smaller number with large units, the decimal point is moved to the left and the corresponding prefix (k, M, G, etc.) is added. To convert a small numerical value into a larger number with small units, the decimal point is moved to the right and the corresponding prefix (m, μ, p, etc.) is added.

For example, to change 470,000 W to kW, the decimal point is moved three places to the left and the prefix "k" added (470,000 W = 470 kW). Likewise, to change 0.024 A to mA, the decimal point is moved three places to the right and the prefix "m" added (0.024 A = 24 mA).

Convert each of the measurements.

1. ______________ 0.420 mA = ___ A

2. ______________ 0.420 mA = ___ μA

3. ______________ 1.230 kΩ = ___ Ω

4. ______________ 1.230 kΩ = ___ MΩ

5. ______________ 0.954 MHz = ___ kHz

6. ______________ 0.954 MHz = ___ Hz

7. ______________ 351 ms = ___ s

8. ______________ 351 ms = ___ μs

9. ______________ 45 μF = ___ F

10. ______________ 45 μF = ___ pF

11. ______________ 0.640 V = ___ mV

12. ______________ 0.640 V = ___ μV

13. ______________ 21 μA = ___ A

14. ______________ 21 μA = ___ mA

15. ______________ 572 MΩ = ___ GΩ

16. ______________ 572 MΩ = ___ kΩ

Application 1-2—Electrical Quantities

Test instruments are used to measure electrical quantities, such as the voltage in a circuit. When voltage is a variable in electrical formulas, such as Ohm's law ($E = I \times R$) or the power formula ($P = E \times I$), it is represented by the capital letter "E," for electromotive force. However, when using a test instrument to measure voltage, the unit symbol for voltage (V) is used on the meter and meter display, not the letter "E."

For each electrical quantity being measured by a test instrument, it is important to know the unit of measurement and the abbreviations used to represent the electrical quantity on both the test instrument and in an electrical formula. See Common Electrical Quantities.

COMMON ELECTRICAL QUANTITIES

Variable	Name	Unit of Measure and Abbreviation
E	voltage	volt—V
I	current	ampere—A
R	resistance	ohm—Ω
P	power	watt—W
P	power (apparent)	volt-amp—VA
C	capacitance	farad—F
L	inductance	henry—H
Z	impedance	ohm—Ω
G	conductance	siemens—S
f	frequency	hertz—Hz
T	period	second—s

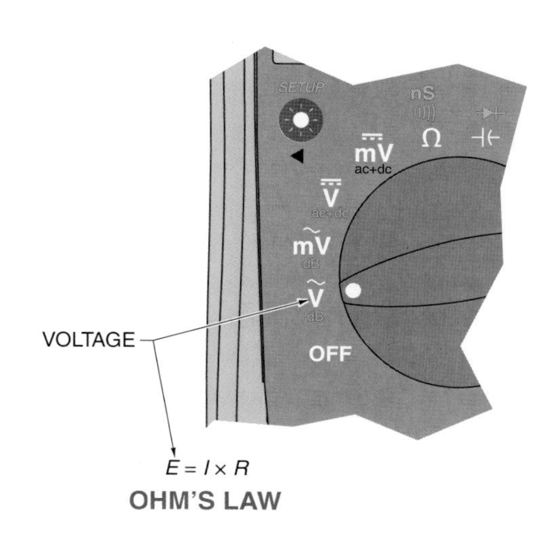

For each of the symbols on the test instruments, state the name of the electrical quantity being measured, the unit of measurement for the electrical quantity being measured, and the equivalent abbreviation for the electrical quantity being measured that is used as a variable in electrical formulas. For example, a meter function switch set to "V" will measure "voltage" in units of "volts," which is represented by the abbreviation "E."

1. _______________ What is the name of the electrical quantity the test instrument is set to measure?

2. _______________ What is the base unit of measurement for the electrical quantity to be measured?

3. _______________ What is the abbreviation for the electrical quantity being measured, as it is used in an electrical formula (as a variable)?

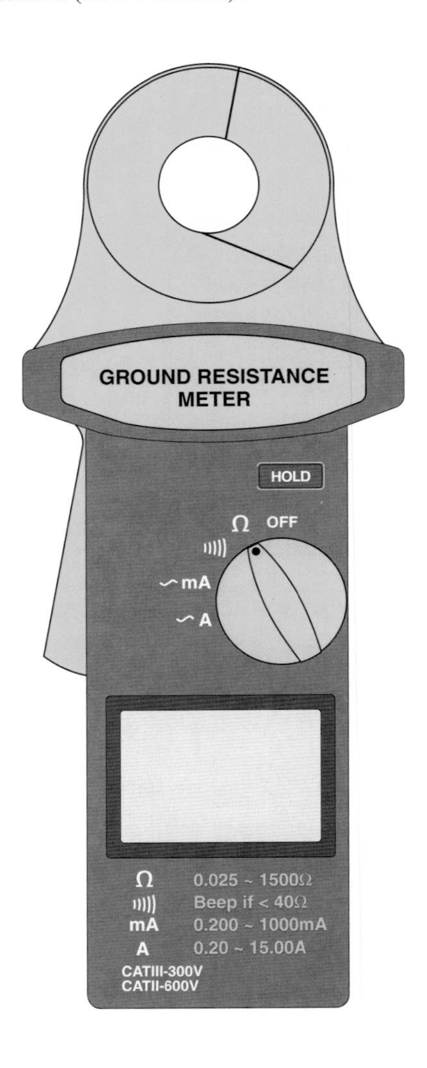

4. ______________ What is the name of the electrical quantity the test instrument is set to measure?

5. ______________ What is the base unit of measurement for the electrical quantity to be measured?

6. ______________ What is the abbreviation for the electrical quantity being measured, as it is used in an electrical formula (as a variable)?

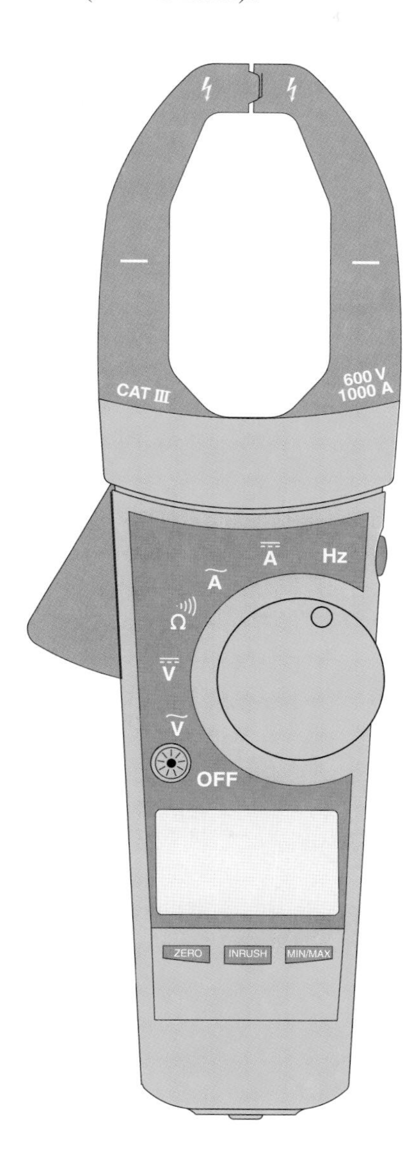

7. ______________ What is the name of the electrical quantity the test instrument is set to measure?

8. ______________ What is the base unit of measurement for the electrical quantity to be measured?

9. ______________ What is the abbreviation for the electrical quantity being measured, as it is used in an electrical formula (as a variable)?

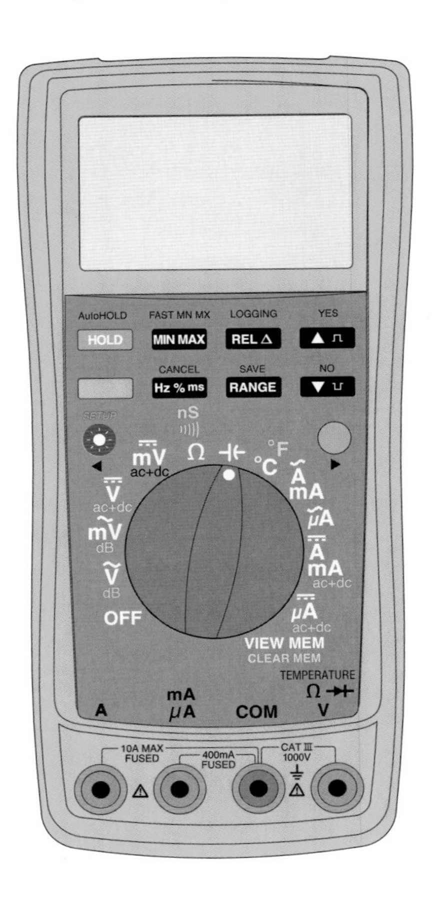

Application 1-3 — Test Instrument Settings and Connections

Test instruments can take safe and accurate measurements only if they are properly set to the electrical measurement to be measured, the meter test leads are connected to the correct meter jacks, and the meter is properly connected to the circuit to be measured. One of the most common mistakes made when using a test instrument is the meter's function switch setting not matching the connection of the meter's test leads—for example, setting a meter to measure voltage (V) but connecting the test leads to the current (A) jacks. Another common mistake is setting the meter to measure VAC and connecting it to a VDC part of a circuit (or setting the meter to VDC and connecting it to a VAC part of a circuit).

1. Set (draw in) the function switch on the multimeter to use the continuity test mode to test the normally open (NO) contacts on a pushbutton before it is wired into a circuit.
2. Connect the two test leads to the correct jacks on the multimeter.

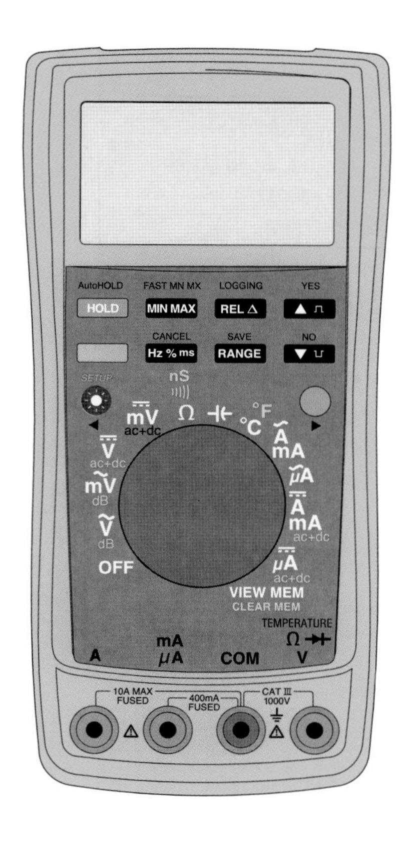

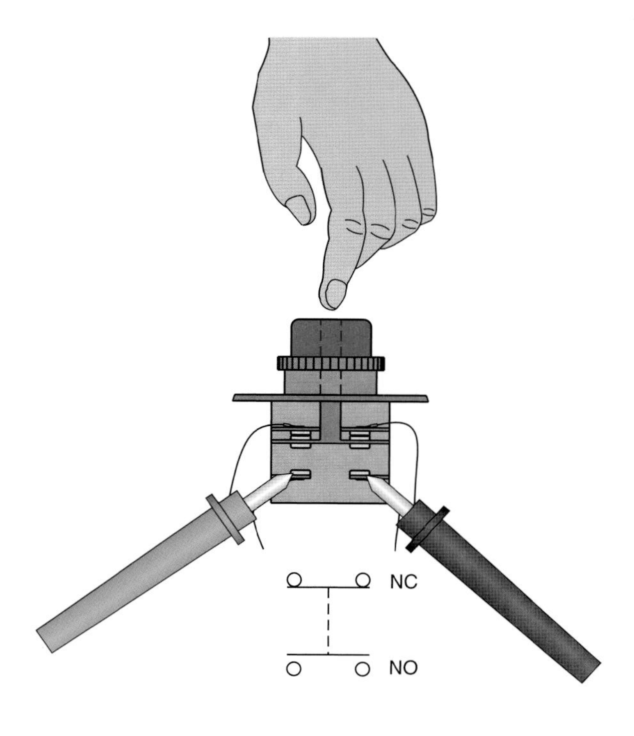

3. Set (draw in) the function switch on the multimeter to take voltage measurements on the step-down transformer in a door chime circuit.

4. Connect the two test leads to the correct jacks on the multimeter.

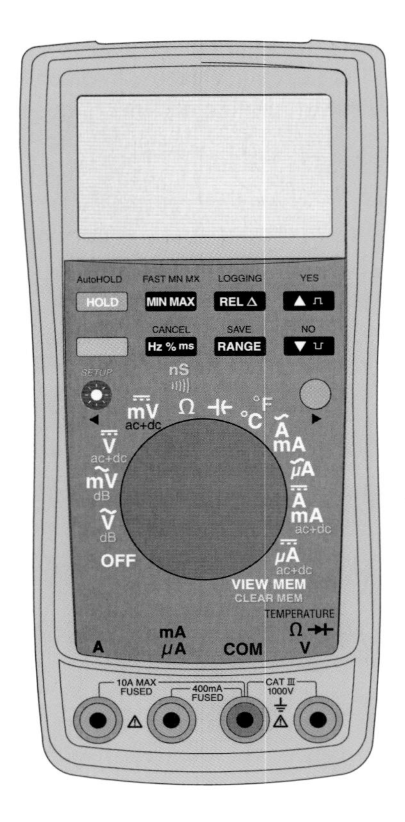

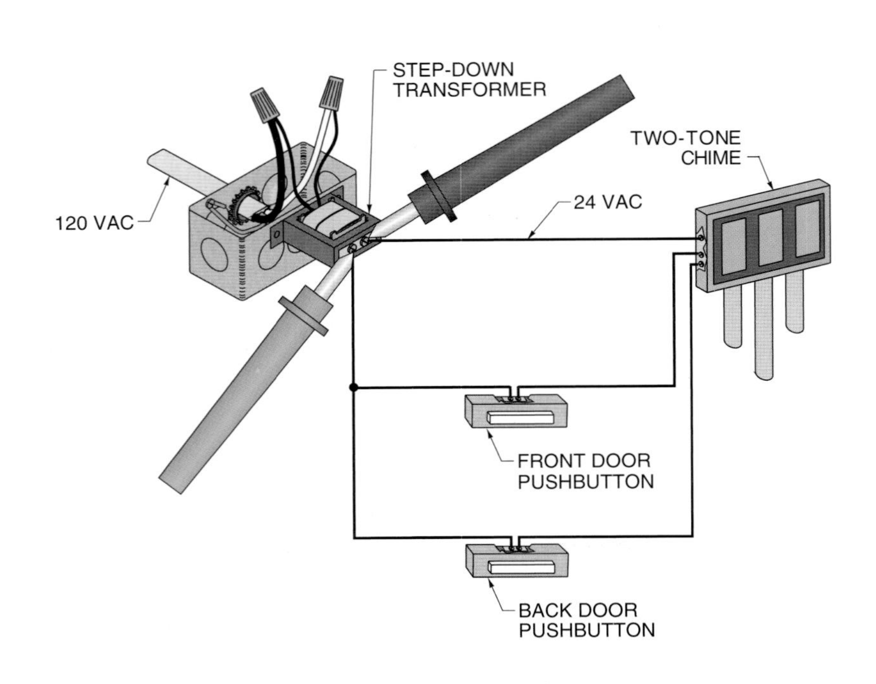

5. Set (draw in) the function switch on Multimeter 1 to take in-line current measurements to determine the current draw of the motor starter coil.

6. Connect the two test leads connected to the motor starter coil circuit to the correct jacks on Multimeter 1.

7. Set (draw in) the function switch on Multimeter 2 to take voltage measurements on the circuit connecting the solid-state relay to the PLC computer.

8. Connect the two test leads connected to the SSR to the correct jacks on Multimeter 2.

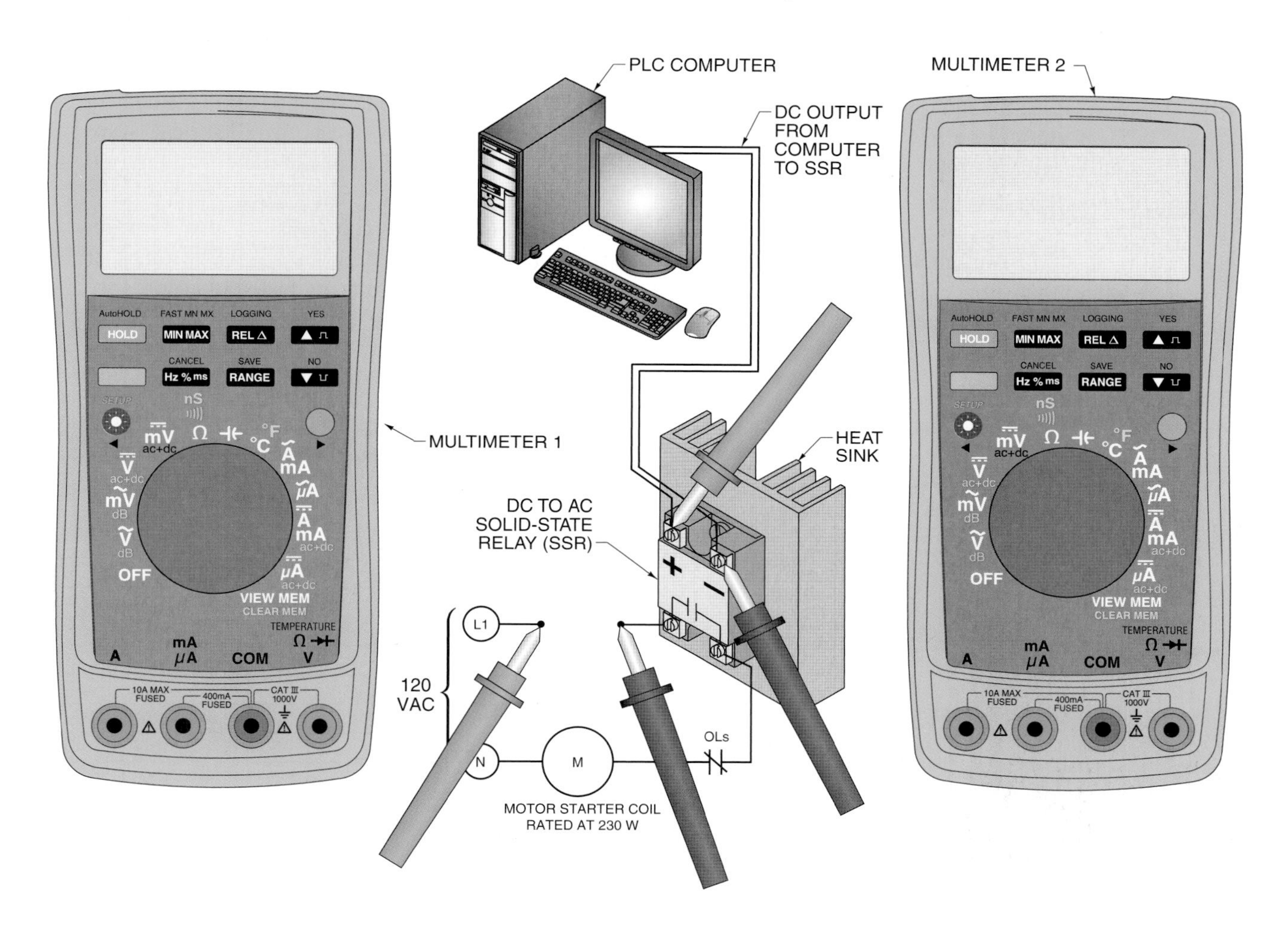

Application 1-4—Electrical Glove Sizes

Per NFPA 70E, personal protective and safety requirements must be followed when working around energized electrical circuits. Approved rubber electrical gloves must be worn anytime electrical measurements are taken on energized circuits in which there is a chance of an electrical shock (usually 50 V and above).

The three types of gloves an electrician may wear include insulated (rubber) gloves, cotton liners, and leather outer gloves. An insulated (rubber) glove (required) provides a high enough resistance to prevent electricity from entering the hand. A cotton liner (optional) is worn inside the insulated glove to add comfort and aid hand movement. A leather outer glove must be worn to protect the insulated rubber glove.

An individual's glove size must be known in order to select a glove that fits best and provides optimum comfort and protection. To determine glove size, the hand is held flat with the fingers together and thumb extended. See Glove Sizing. The circumference around the knuckles is measured and rounded up to the nearest ½″ (7, 7½, 8, etc.). The glove size is obtained by adding ½″ to the rounded measurement. See Electrical Glove Sizes. A string may be used if a flexible tape measure is not available. The length of the string is measured with a standard tape measure or ruler.

Dielectric rubber gloves are typically seamless with a curved hand design for ease of use with hot and live electrical equipment. Gloves are available in solid black or in two-tone options (yellow/black or orange/black) that show color when the glove is worn through or damaged. Gloves are tested in accordance with ASTM D120 specifications.

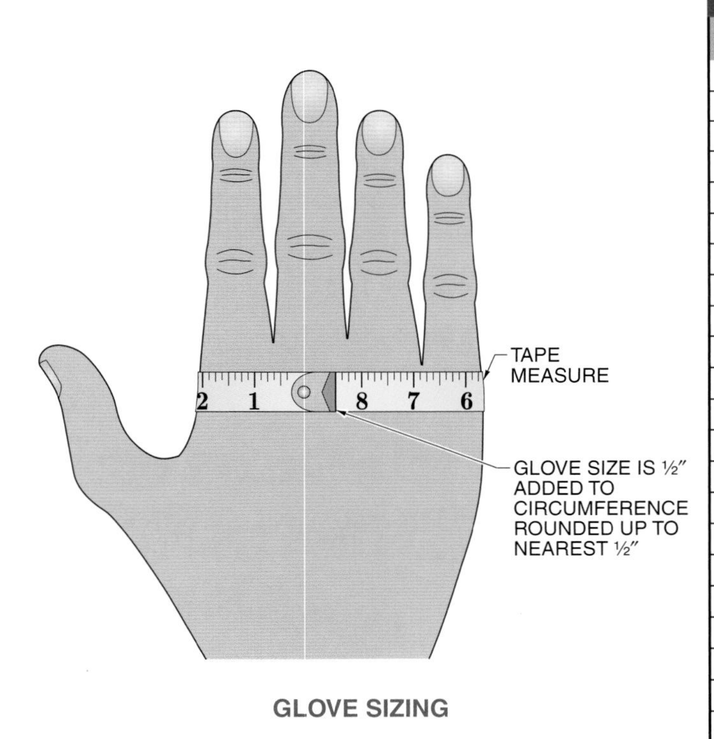

GLOVE SIZING

ELECTRICAL GLOVE SIZES

Model Number	Size	Protection*
7-1000	7	1000
7-20k	7	20,000
7.5-1000	7.5	1000
7.5-20k	7.5	20,000
8-1000	8	1000
8-20k	8	20,000
8.5-1000	8.5	1000
8.5-20k	8.5	20,000
9-1000	9	1000
9-20k	9	20,000
9.5-1000	9.5	1000
9.5-20k	9.5	20,000
10-1000	10	1000
10-20k	10	20,000
10.5-1000	10.5	1000
10.5-20k	10.5	20,000
11-1000	11	1000
11-20k	11	20,000
11.5-1000	11.5	1000
11.5-20k	11.5	20,000
12-1000	12	1000
12-20k	12	20,000

* in V

Determine the proper glove size and model number.

1. __________ What is the proper glove size for a hand with a circumference of 6¼″?

2. __________ Which glove model is required for a hand with a circumference of 6¼″ when working around 120 V or less?

3. __________ What is the proper glove size for a hand with a circumference of 10⅓″?

4. __________ Which glove model is required for a hand with a circumference of 10⅓″ when working around 480 V or less?

5. __________ What is the proper glove size for a hand with a circumference of 8¾″?

6. __________ Which glove model is required for a hand with a circumference of 8¾″ when working around 1200 V or less?

7. __________ What is the proper glove size for a hand with a circumference of 9⅛″?

8. __________ Which glove model is required for a hand with a circumference of 9⅛″ when working around 1.5 kV or less?

Application 1-5 — Temperature

Electrical circuits are used to transfer, switch, control, or convert electrical energy into some other form of energy, such as light, sound, or mechanical motion. These processes also produce heat. Electric heat lamps and electric heating elements produce heat intentionally. Unwanted heat is produced in an electrical circuit any time current flowing through a circuit encounters resistance, such as when current flows through conductors (wire) and encounters a bad (loose, corroded, etc.) connection or a faulty switch (having worn contacts). The higher the resistance of the undersized conductor, bad connection, or faulty switch, the greater the amount of heat produced at that point.

Test instruments can be used to measure the amount of heat at different locations within an electrical system. By taking temperature measurements throughout an electrical circuit, existing problems can be found (troubleshooting), and potential future problems can be found (preventive maintenance).

Temperature is usually measured in degrees Fahrenheit (°F) or degrees Celsius (°C). Some temperature measurement instruments can be set to display a temperature measurement in both degrees Fahrenheit and degrees Celsius, and others can only display the temperature measurement in either degrees Fahrenheit or degrees Celsius, but not both. It is important to know how to convert between degrees Fahrenheit and degrees Celsius when taking and recording temperature measurements. See Temperature Conversion.

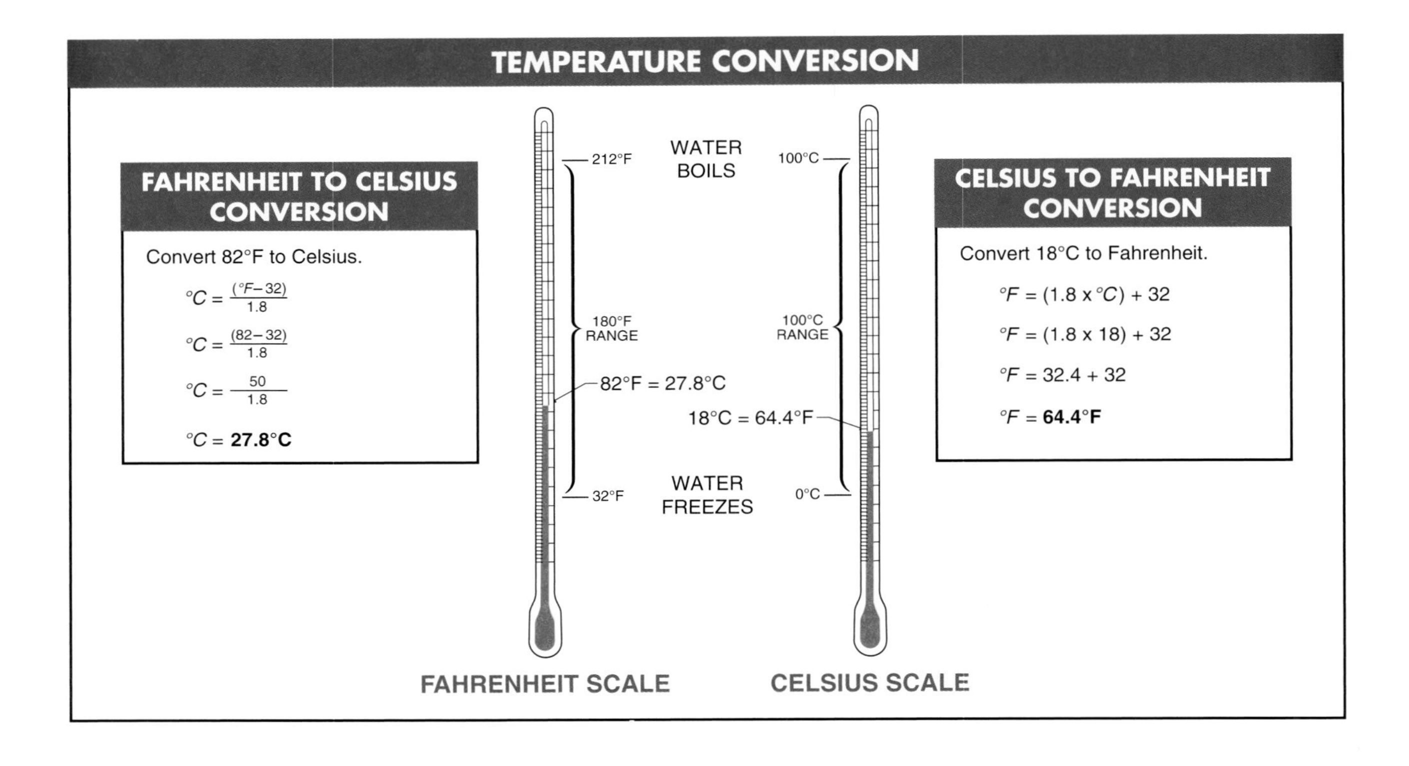

Convert each of the temperature measurements.

1. ______________ What is the electrical connection temperature in degrees Celsius?

2. ______________ What is the ambient temperature in degrees Celsius?

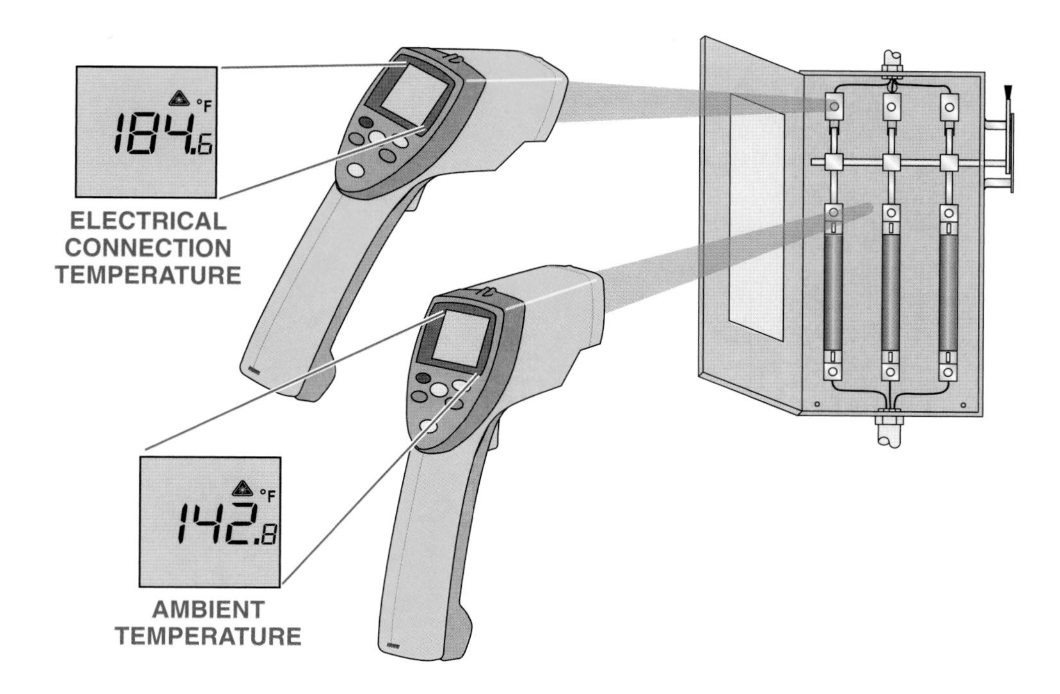

3. ______________ What is the electrical connection temperature in degrees Fahrenheit?

4. ______________ What is the ambient temperature in degrees Fahrenheit?

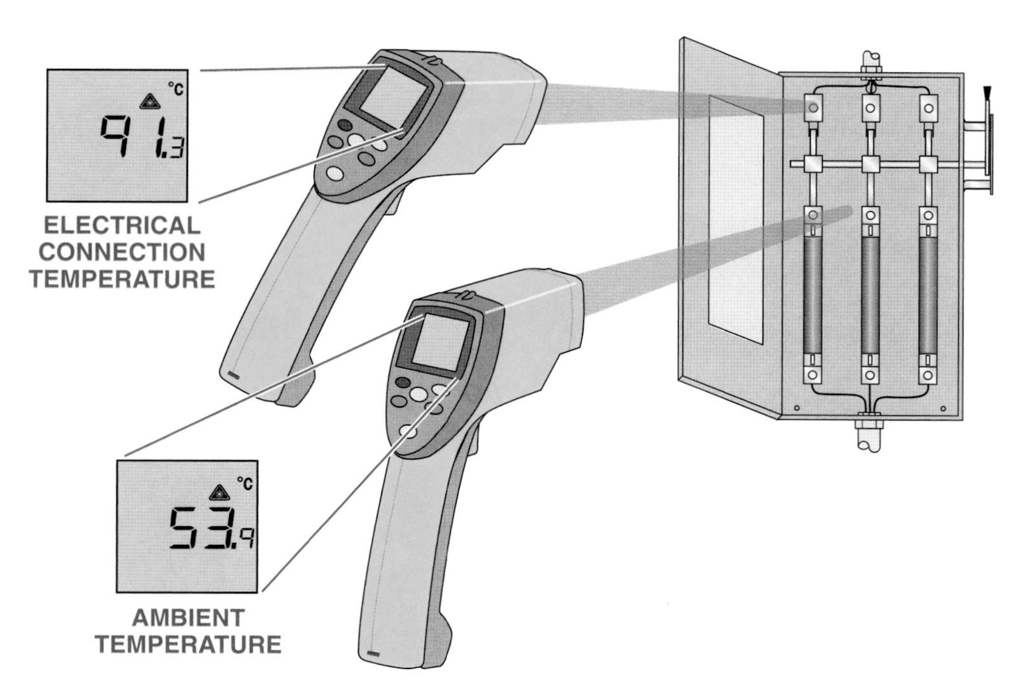

Application 1-6—MIN MAX Recording Time

Test instruments take instantaneous measurements. For example, a voltmeter connected to a receptacle (outlet) will display the voltage at the receptacle at that moment. Instantaneous measurements are useful for most situations, but sometimes measurements over a period of time are required to diagnose some problems. For example, the voltage at receptacles will fluctuate within an acceptable level (+5% to –10%), even if the receptacle is properly wired and operating normally. However, if there is a power quality or wiring problem, large voltage variations can occur and cause problems in loads connected to the receptacle, such as causing computers to reboot. In order to find a voltage variation problem, measurements must be taken over a period of time.

Any multimeter with a MIN MAX recording function can be used to take voltage (or other electrical quantity) measurements over time. The meter saves any new minimum or maximum values that last for at least a certain period of time. This measurement duration depends on the meter's make and model and is specified in milliseconds on the meter or in the manual. Shorter durations indicate that the meter is capable of detecting shorter fluctuations.

Peak voltage measurements are taken over time to detect and measure transients on a power line. A transient voltage (voltage spike) is a temporary, undesirable voltage in an electrical circuit. Transient voltages range from a few volts to several thousand volts and last from a few microseconds up to a few milliseconds. In order to capture and record transients (peak voltages), the meter must have a much faster minimum measurement duration than is required for taking general MIN MAX recordings. Meters with PEAK MIN MAX functions have the capability of changing the duration to a smaller value for measuring transients.

1. ______________ How long (in sec) must a new maximum or minimum voltage last before this multimeter records a new measured value?

2. ______________ Could the MIN MAX function on this multimeter record a momentary power interruption on a standard 60 Hz power line that lasted for three complete cycles?

3. ______________ For this multimeter, for how many complete cycles would a momentary power interruption have to exist before the MIN MAX function recorded the voltage change?

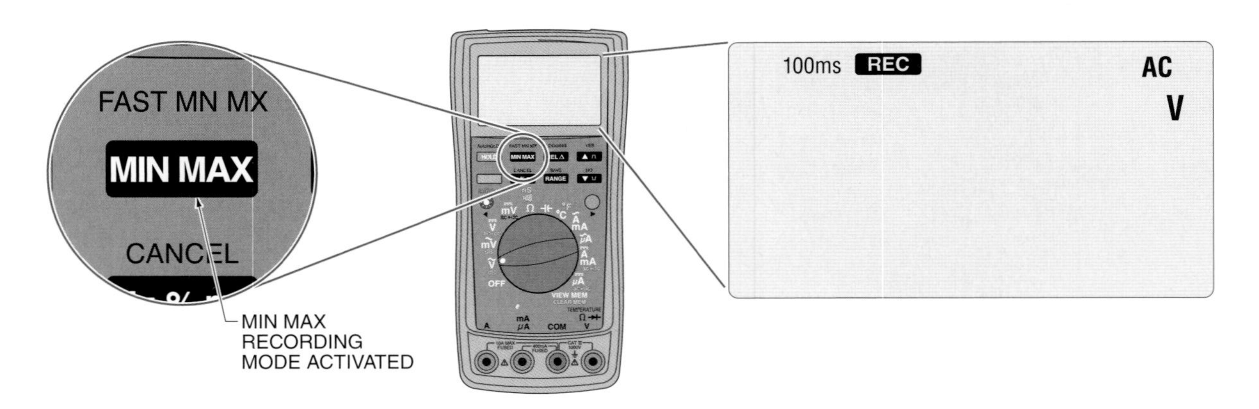

4. _______________ How long (in sec) must a new maximum or minimum voltage last before this multimeter records a new measured value?

5. _______________ If there are 60 complete cycles in one second (60 Hz), for how many milliseconds does one cycle (1 Hz) last?

6. _______________ Could the PEAK MIN MAX function on this multimeter record a momentary power interruption on a standard 60 Hz power line that lasted for one-half cycle?

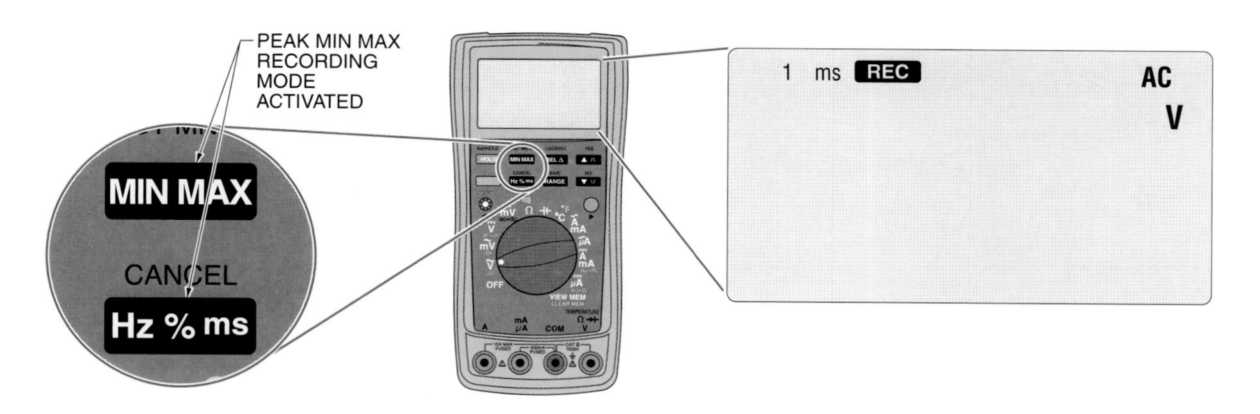

Application 1-7—CAT Ratings

IEC Standard 1010 classifies the applications in which test instruments may be used into four overvoltage installation categories (CAT I–CAT IV). They categorize the magnitude of transient voltage a test instrument must withstand when used on a power distribution system. Test instruments are designed and marked for the maximum category in which they can safely be used. Applications require test instruments with a CAT rating the same or higher than the application category. See IEC 1010 Overvoltage Installation Categories.

IEC 1010 MEASUREMENT CATEGORIES

Category	In Brief	Examples
CAT I	Electronic	• Protected electronic equipment • Equipment connected to (source) circuits in which measures are taken to limit transient overvoltage to an appropriately low level • Any high-voltage, low-energy source derived from a high-winding-resistance transformer, such as the high-voltage section of a copier
CAT II	1ϕ receptacle-connected loads	• Appliances, portable tools, and other household and similar loads • Outlets and long branch circuits • Outlets at more than 30′ (10 m) from CAT III source • Outlets at more than 60′ (20 m) from CAT IV source
CAT III	3ϕ distribution, including 1ϕ commercial lighting	• Equipment in fixed installations, such as switchgear and polyphase motors • Bus and feeder in industrial plants • Feeders and short branch circuits and distribution panel devices • Lighting systems in larger buildings • Appliance outlets with short connections to service entrance
CAT IV	3ϕ at utility connection, any outdoor conductors	• Refers to the origin of installation, where low-voltage connection is made to utility power • Electric meters, primary overcurrent protection equipment • Outside and service entrance, service drop from pole to building, run between meter and panel • Overhead line to detached building

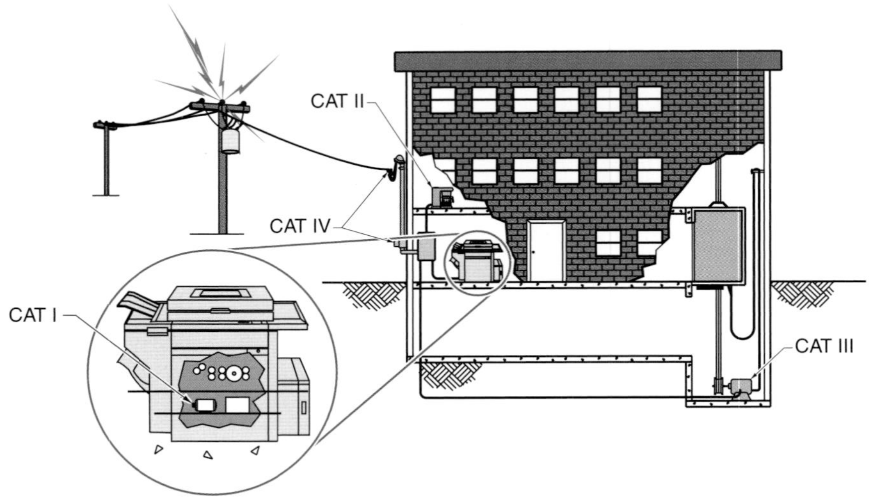

1. ______________ What is the minimum CAT rating required for a test instrument taking measurements at an electric meter to test for proper voltage on service entrance conductors?

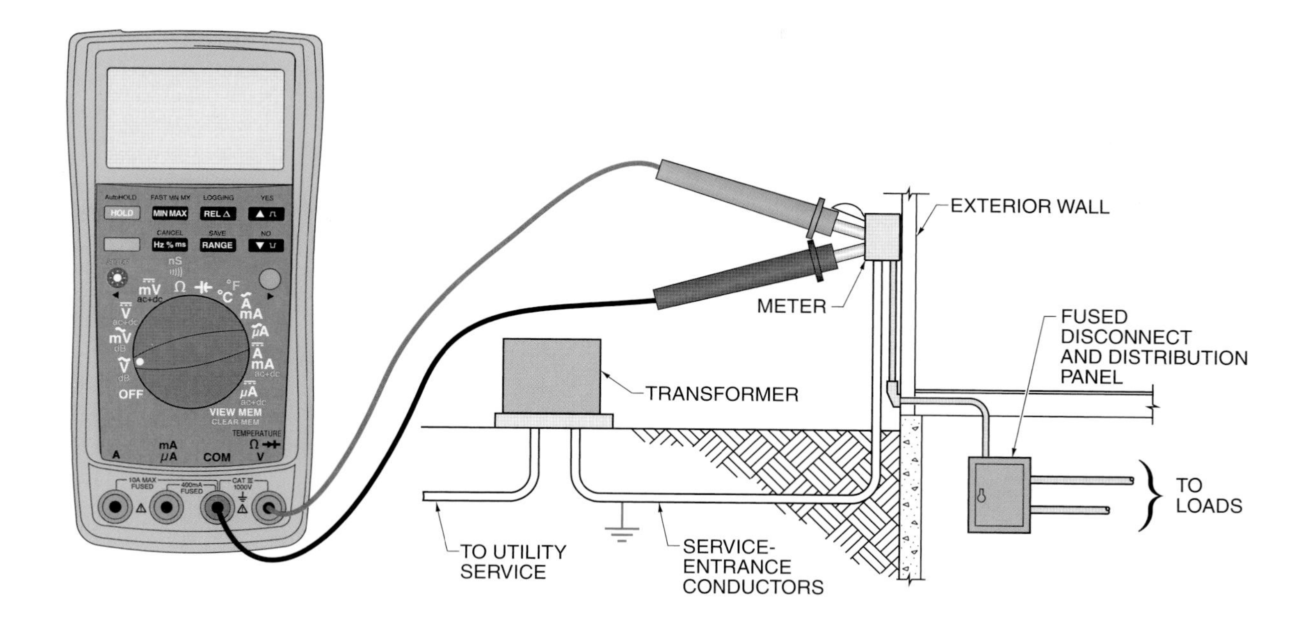

2. ______________ What is the minimum CAT rating required for a test instrument taking measurements on a soft drink beverage gun to test electrical connections?

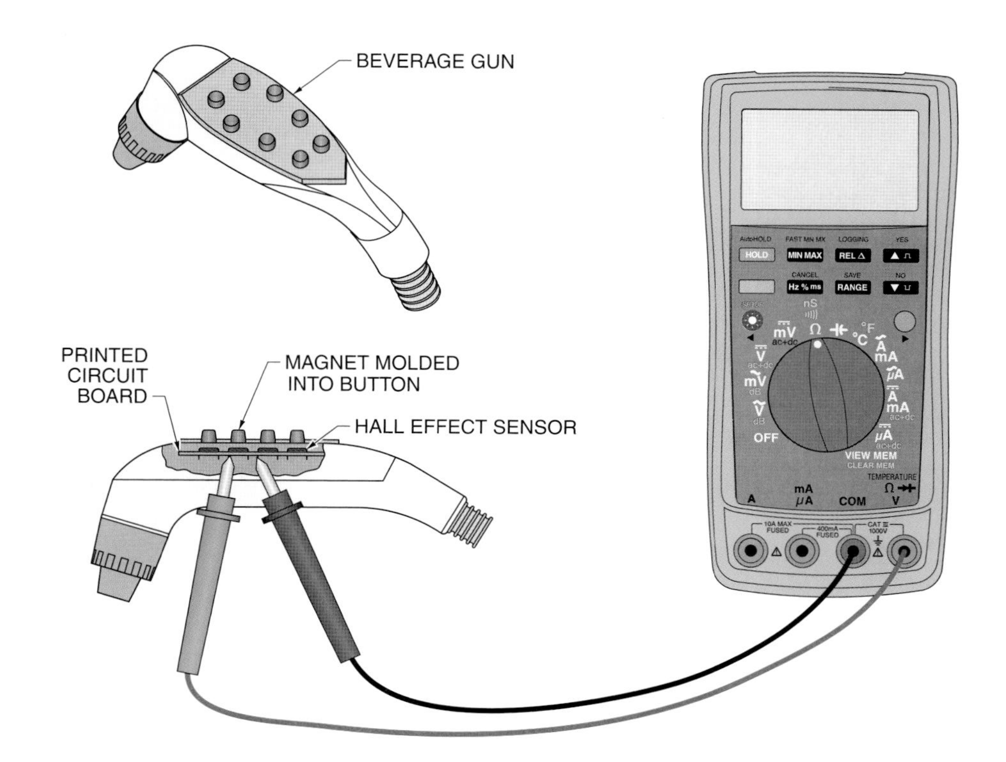

3. _______________ What is the minimum CAT rating required for a test instrument taking measurements at a receptacle to make sure it is wired correctly?

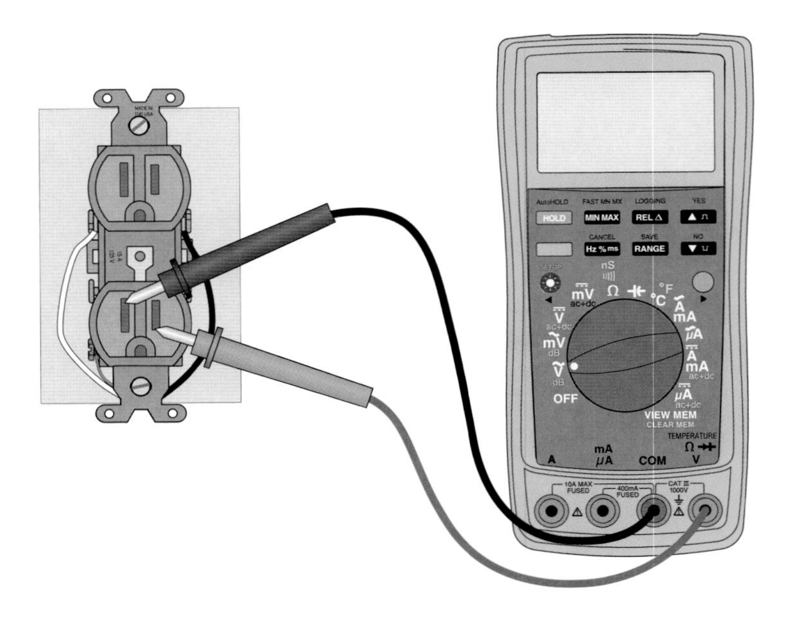

4. _______________ What is the minimum CAT rating required for a test instrument taking measurements on photoelectric switches used to detect truck positions at a loading dock?

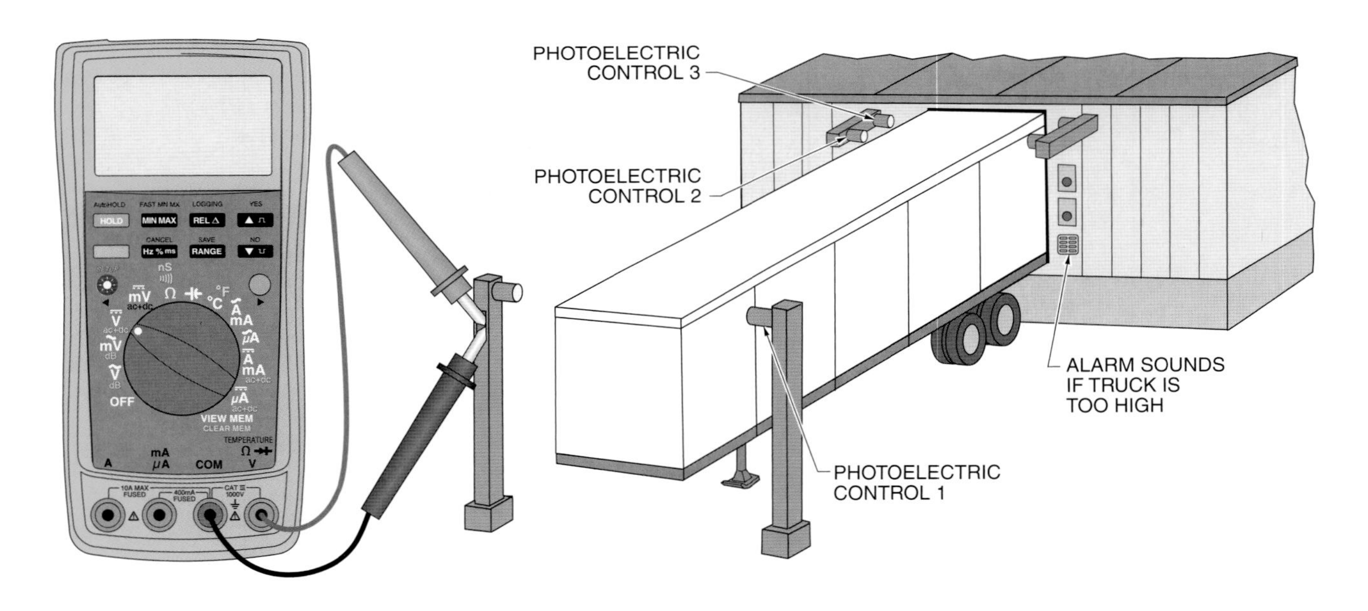

Application 1-8 — Lockout/Tagout

Lockout is the process of removing the source of electrical power and installing a lock that prevents the power from being turned on. Tagout is the process of placing a danger tag on the source of power, which indicates that the power may not be turned on until the danger tag is removed. Lockout/tagout procedures help prevent an electrical shock when working on electrical equipment and are required whenever work can be done without power. However, lockouts and tagouts cannot be used when the work requires electrical power to be on in order to take electrical measurements with test instruments.

Determine if a lockout/tagout device is required and appropriate for each service call. Write "YES" if lockout/tagout is required and write "NO" if lockout/tagout is not required and cannot be used for the service call.

1. ____________ Is lockout/tagout required for a service call where cartridge heating elements are changed to a larger size to produce more heat on the press platens?

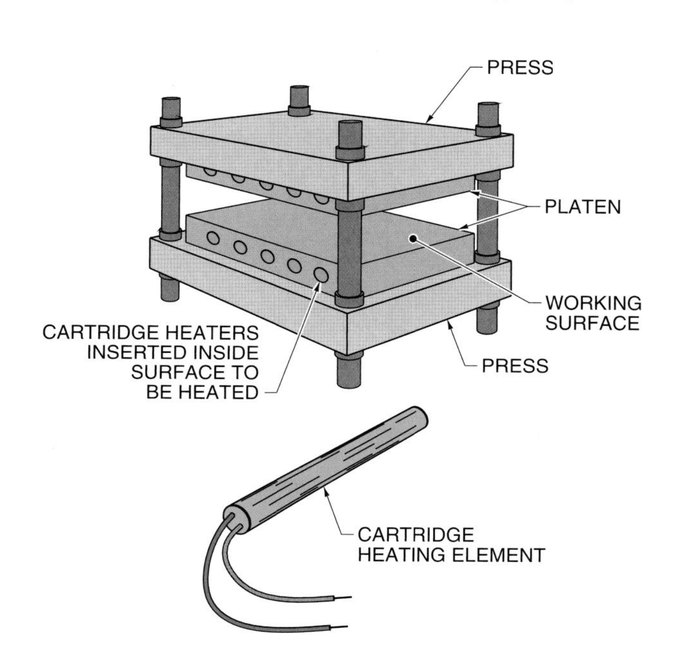

2. ______________ Is lockout/tagout required for a service call where voltage is measured at the heating element to make sure the SCRs are properly controlling the voltage level at the heating element?

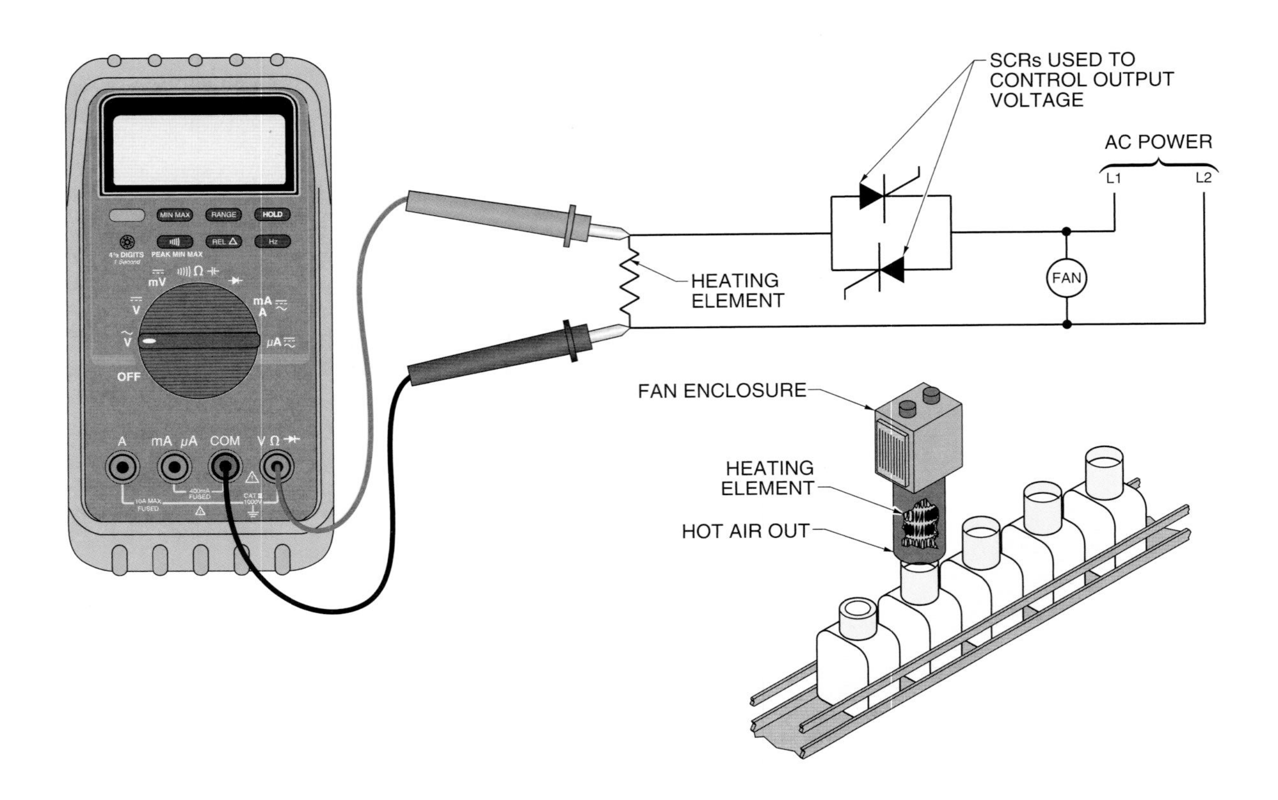

3. ______________ Is lockout/tagout required for a service call where the condition (tightness and wear) of the motor/compressor belt is checked?

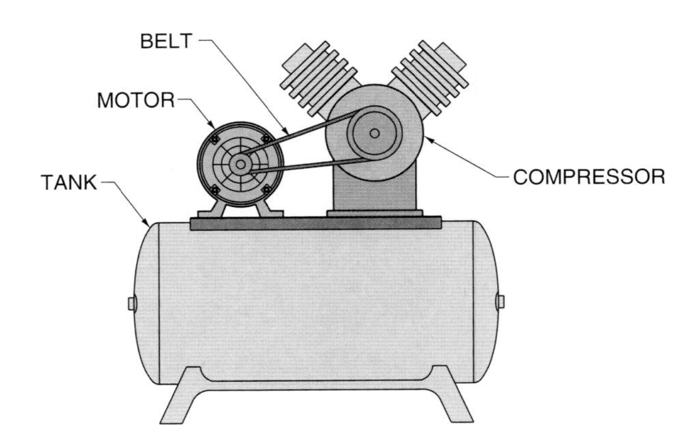

4. ______________ Is lockout/tagout required for a service call where a clamp-on ammeter is used to measure the current draw of the conveyor motor at different product load conditions?

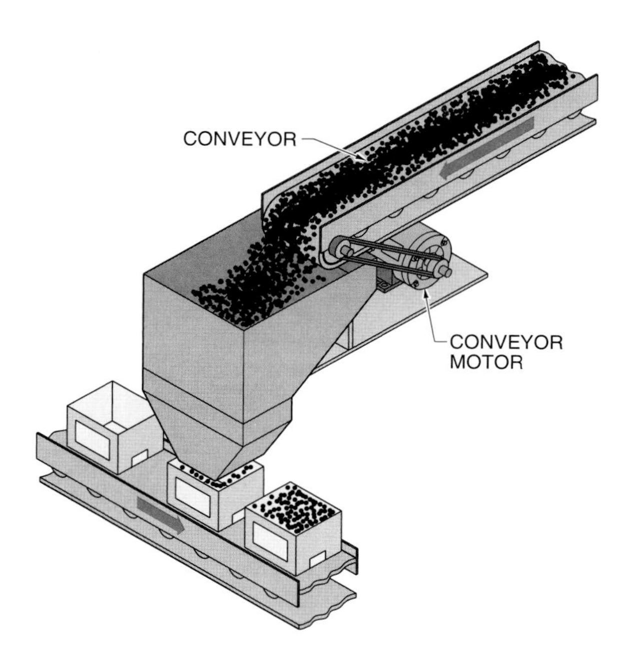

Application 1-9—Conductor Sizing

A conductor is a low-resistance material that carries electricity to different parts of a circuit. Copper is the most common conductor material used for inside wiring because it has a very low resistance. Low resistance means less heat and less voltage drop across the conductor.

Aluminum is also used for conductors, but must be sized larger than copper conductors to compensate for aluminum's higher material resistance. However, since it is much lighter than copper (even when sized up), aluminum is used for outside power distribution lines where weight is a major factor. Copper-clad aluminum conductors have copper bonded to aluminum cores, which balances the lower resistance of copper with the lower weight of aluminum.

Conductor Sizes

Conductors are sized based on the American Wire Gauge (AWG) numbering system. See Conductor Sizes. Smaller AWG numbers indicate larger conductors and more current-carrying capacity. For example, a No. 12 AWG copper conductor is larger in diameter than a No. 14 copper conductor and may carry more current. Conductors that are No. 8 and smaller may be either solid or stranded. Conductors that are larger than No. 8 are stranded. See Stranded and Solid Conductors.

CONDUCTOR SIZES

AWG	Diameter*	Area	AWG	Diameter*	Area
00 (2/0)	0.3648	●	8	0.1285	●
0 (1/0)	0.3249	●	10	0.1019	●
1	0.2893	●	12	0.0808	●
2	0.2576	●	14	0.0641	●
3	0.2294	●	16	0.0508	●
4	0.2043	●	18	0.0403	•
6	0.1620	●	20	0.0320	•

* in in.

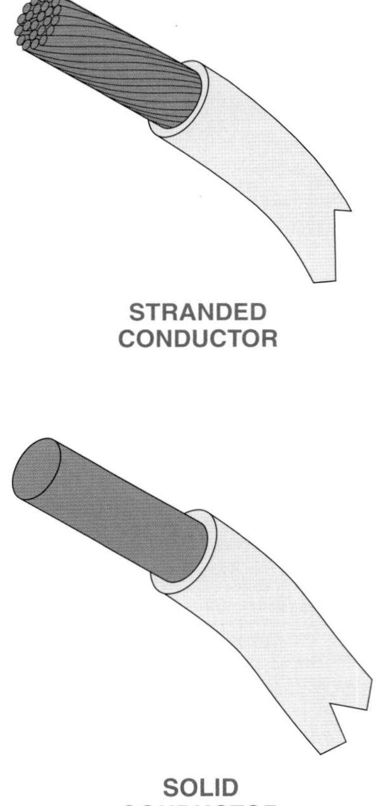

Conductors that are No. 18 and No. 16 are normally used for the power cords of small appliances, plug-in lamps, and speakers. Conductors that are No. 14 and No. 12 are used for wiring most lighting circuits and supplying power to standard receptacle outlets. Conductors that are No. 10 through No. 4 are used for wiring residential and commercial appliances, such as electric ranges, water heaters, furnaces, and air conditioners. They are also used for supplying power to subpanels and large motors. Conductors that are No. 3 and larger are used for supplying power to main service panels.

Ampacity

Ampacity is a rating based on the amount of continuous current that insulated conductors can safely carry without damaging the insulation. Ampacity depends on the conductor material and size, the type of insulation, the ambient temperature at the conductor location, and the number of current-carrying conductors run together. NEC® tables list the ampacities of various conductors and can be used to determine the minimum conductor size or type for a particular installation. See Ampacities of Insulated Conductors.

AMPACITIES OF INSULATED CONDUCTORS*

Type of Insulation	Types TW, UF	Types RHW, THHW, THW, THWN, XHHW, USE, ZW	Types TBS, SA, SIS, FEP, FEPB, MI, RHH, RHW-2, THHN, THW-2, THWN-2, USE-2, XHH, XHHW-2, ZW-2	Types TW, UF	Types RHW, THHW, THW, THWN, XHHW, USE	Types TBS, SA, SIS, THHN, THW-2, THWN-2, RHH, RHW-2, USE-2, XHH, XHHW-2, ZW-2
AWG	**COPPER**			**ALUMINUM OR COPPER-CLAD ALUMINUM**		
18	—	—	14	—	—	—
16	—	—	18	—	—	—
14	20	20	25	—	—	—
12	25	25	30	20	20	25
10	30	35	40	25	30	35
8	40	50	55	30	40	45
6	55	65	75	40	50	60
4	70	85	95	55	65	75
3	85	100	110	65	75	85
2	95	115	130	75	90	100
1	110	130	150	85	100	115
1/0	125	150	170	100	120	135
2/0	145	175	195	115	135	150
3/0	165	200	225	130	155	175
4/0	195	230	260	150	180	205
Ambient Temperature†	**Correction Factor for Ambient Temperature**					
21–25	1.08	1.05	1.04	1.08	1.05	1.04
26–30	1.00	1.00	1.00	1.00	1.00	1.00
31–35	0.91	0.94	0.96	0.91	0.94	0.96
36–40	0.82	0.88	0.91	0.82	0.88	0.91
41–45	0.71	0.82	0.87	0.71	0.82	0.87
46–50	0.58	0.75	0.82	0.58	0.75	0.82
51–55	0.41	0.67	0.76	0.41	0.67	0.76
56–60	—	0.58	0.71	—	0.58	0.71
61–70	—	0.33	0.58	—	0.33	0.58
71–80	—	—	0.41	—	—	0.41

* Based on ambient temperature of 30°C (86°F) and not more than three current-carrying conductors in a raceway, cable, or earth (directly buried).
† in °C

Conductors that carry more current create more heat, requiring insulation material that can withstand higher temperatures. Various insulation materials have different temperature breakdown ratings. The most common types of insulation material are designated TW, RHW, and SA.

Correction Factors

Additional conditions must be considered when determining conductor ampacity. If the ambient temperature at the installation location is higher than 30°C (86°F), the ampacity must be derated using a correction factor. The ambient temperature used to determine the correction factor should be the highest expected temperature for that location, which may be high near sources of heat, or for warm climates during summer months.

For example, an installation calls for a copper conductor with TW insulation that must carry 40 A of current. Using the table, the size of the conductor must be at least No. 8 AWG. However, the conductor will be installed near a bakery oven and the ambient temperature may reach as high as 120°F (49°C). The correction factor for this type of conductor at this temperature is 0.58. Therefore, the No. 8 conductor is derated to 23.2 A (40 A × 0.58 = 23.2 A). For a similar conductor to carry 40 A in this location, it must be a larger size. The minimum size for this installation is found by derating the ampacities of other conductors to find one with at least 40 A of corrected ampacity. In this case, at least a No. 4 TW conductor is required (70 A × 0.58 = 40.6 A). Another option is to use a conductor with insulation rated for higher temperatures, such as SA. A No. 8 copper conductor with SA insulation could be used instead (55 A × 0.82 = 45.1 A).

Another condition that must be accounted for is the number of current-carrying conductors in a raceway, cable, or underground because the heat created by conductors that are close together reduces their ampacity. When four or more current-carrying conductors are run together, a correction factor is applied to derate their ampacities. See Correction Factor for More than Three Current-Carrying Conductors. This correction is applied after ampacity is adjusted for ambient temperature. Non-current-carrying conductors, such as grounding conductors, are not counted when determining this correction factor.

CORRECTION FACTOR FOR MORE THAN THREE CURRENT-CARRYING CONDUCTORS

Number of Current-Carrying Conductors	Correction Factor
4 – 6	0.80
7 – 9	0.70
10 – 20	0.50
21 – 30	0.45
31 – 40	0.40
Over 40	0.35

For example, if four copper TW current-carrying conductors are called for in the bakery installation, a correction factor of 0.80 must also be applied to the ampacity rating. The No. 4 conductor is now derated to 32.5 A, so it is no longer adequate. Therefore, No. 2 is the minimum size for this type of conductor for this installation (95 A × 0.58 = 55.1 A; 55.1 A × 0.80 = 44.1 A).

In each of the following conductor installation applications, the ambient temperature is measured during the highest expected normal high temperature time. Using the tables and the information provided, determine the minimum AWG conductor size for the installation application.

As with many electrical installations, the current value is not always directly given. Ohm's law and the power formula can be used to find an unknown electrical quantity when two other electrical quantities are given or measured.

1. ______________ What is the minimum conductor size for this installation?

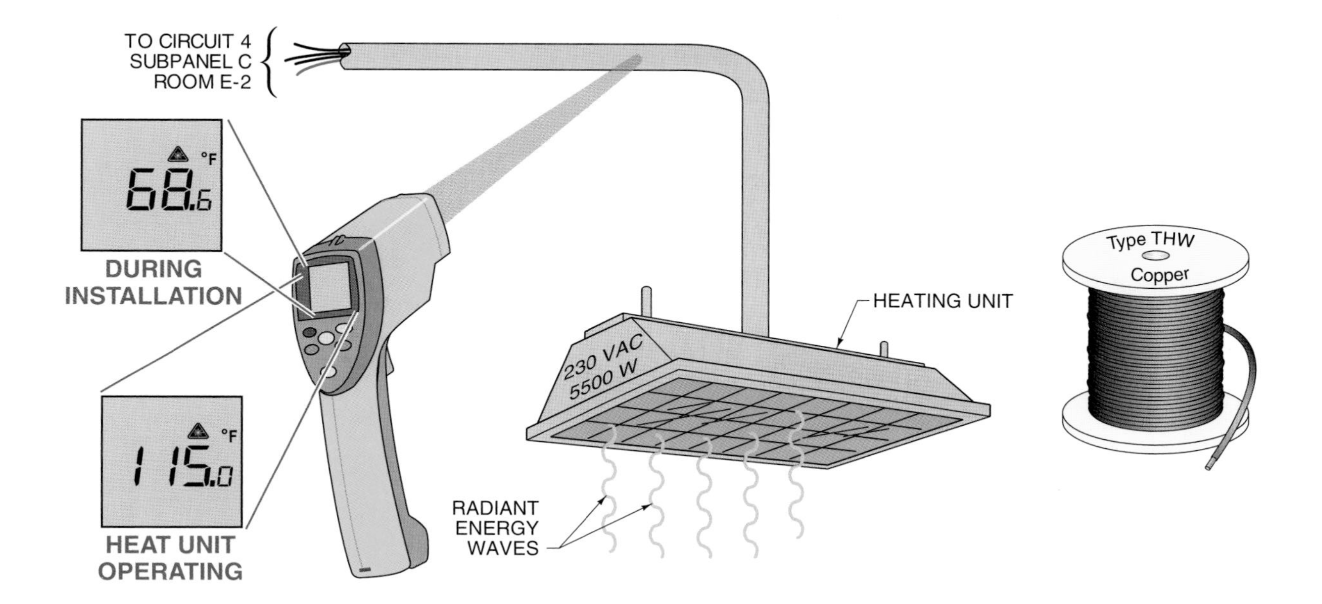

2. ______________ What is the minimum conductor size for the branch circuit in this installation?

3. ______________ What is the minimum conductor size for the power supply in this installation?

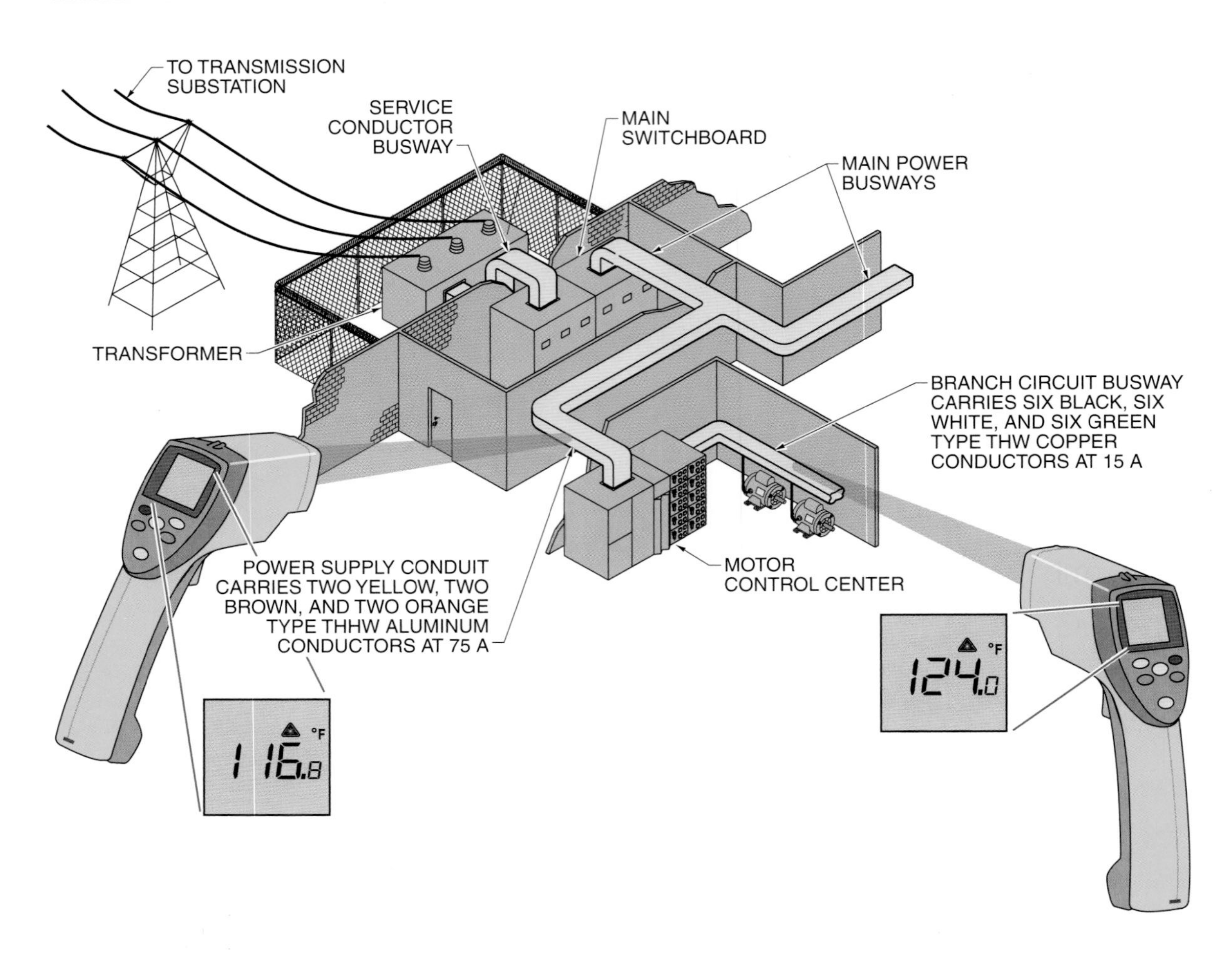

4. ______________ What is the minimum conductor size for this installation?

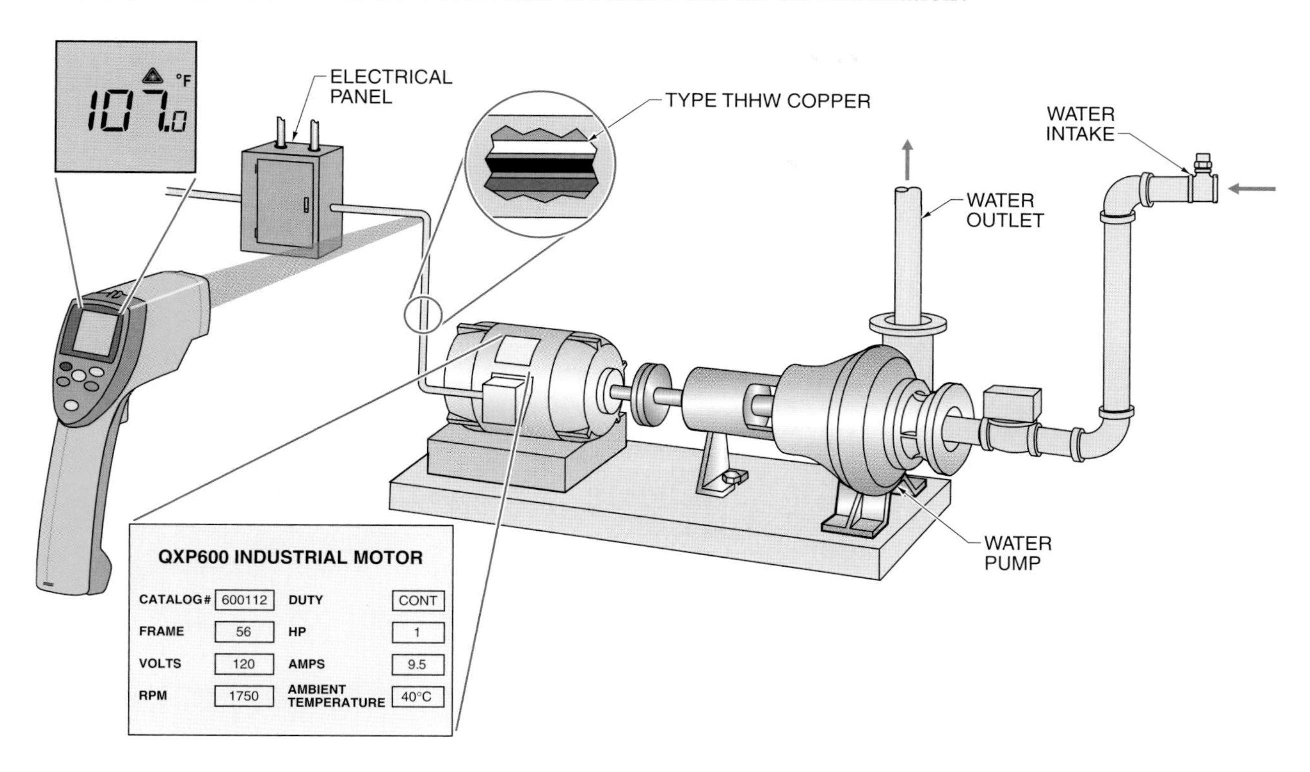

5. ______________ What is the minimum conductor size for this installation?

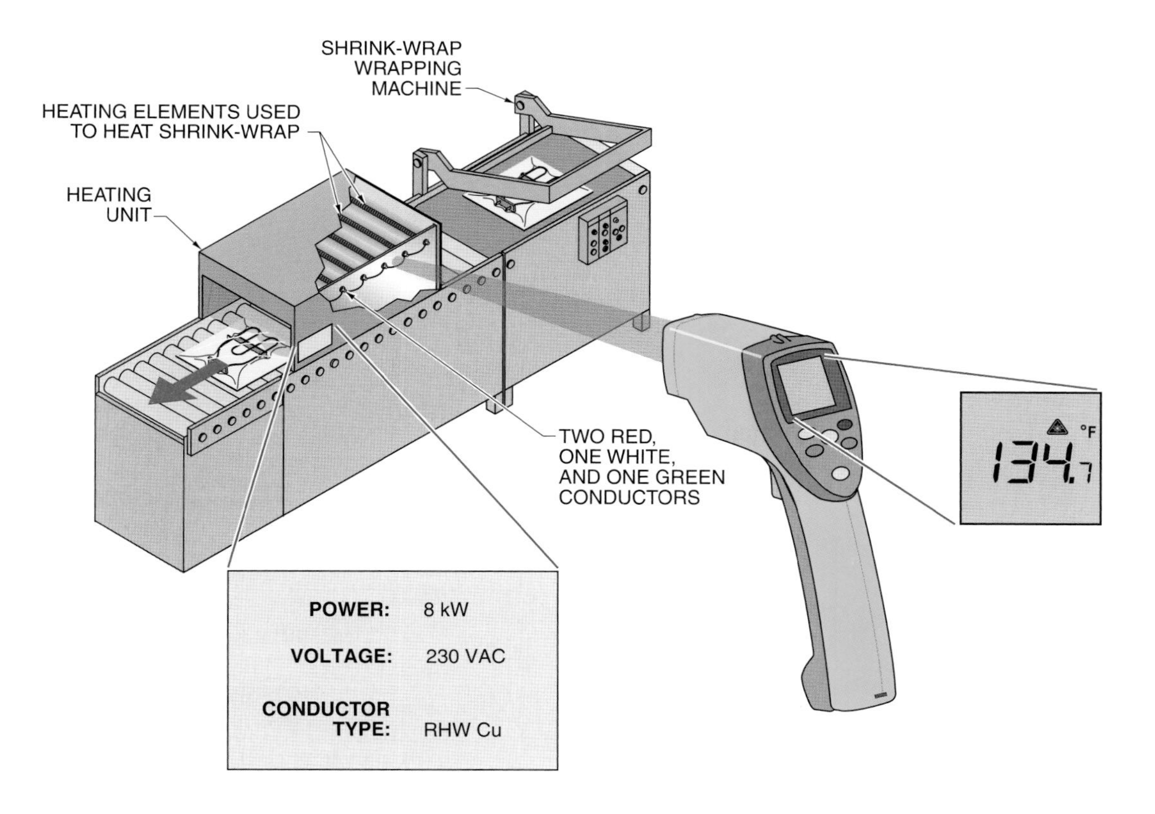

Application 1-10—Branch Circuit Voltage Drop Measurement

In addition to the ambient temperature and the number of current-carrying conductors, another factor limits the amount of current any given conductor can safely carry. The length of the conductor from the service panel (circuit breaker protecting the branch circuit) to the end of the run (furthest point) and back must be limited so that no more than 3% of the supplied voltage is dropped across the conductor. The amount of voltage drop depends on the current carried by the conductor and the resistance of the conductor. Resistances vary with conductor type and size. See Conductor Resistances.

CONDUCTOR RESISTANCES*

AWG	COPPER		ALUMINUM	
	Solid	Stranded	Solid	Stranded
18	7.77	7.95	12.80	13.10
16	4.89	4.99	8.05	8.21
14	3.07	3.14	5.06	5.17
12	1.93	1.98	3.18	3.25
10	1.21	1.24	2.00	2.04
8	0.764	0.778	1.26	1.28
6	—	0.491	—	0.808
4	—	0.308	—	0.508
3	—	0.245	—	0.403
2	—	0.194	—	0.319
1	—	0.154	—	0.253

* in ohms per 1000 ft at 75°C (167°F)

Calculating Voltage Drop

To determine the voltage drop across a conductor run, first calculate the total resistance of the conductor by applying the following formula:

$$R_R = R_C \times L$$

where

R_R = resistance of conductor for length of run (in Ω)

R_C = resistance of conductor per unit length (in Ω/ft)

L = length of run (in ft)

Then, using Ohm's law, the voltage drop across the conductor can be calculated by applying the following formula:

$$E_D = I \times R_R$$

where

E_D = voltage drop across conductor (in V)

I = current in conductor (in A)

R_R = resistance of conductor for length of run (in Ω)

For example, what is the voltage drop across a 300 ft run (150 ft each way) of No. 6 AWG stranded copper conductor carrying 55 A? From the table, a No. 6 copper conductor has a resistance of 0.491 Ω per 1000 ft at 75°F. This is equivalent to 0.000491 Ω/ft (0.491 Ω ÷ 1000 ft = 0.000491 Ω/ft).

$$R_R = R_C \times L$$

$$R_R = 0.000491\ \Omega/\text{ft} \times 300\ \text{ft}$$

$$R_R = \mathbf{0.147\ \Omega}$$

$$E_D = I \times R_R$$

$$E_D = 55\ \text{A} \times 0.147\ \Omega$$

$$E_D = \mathbf{8.09\ V}$$

The percentage voltage drop can be calculated by applying the following formula:

$$\%E_D = \frac{E_D}{E} \times 100$$

where

$\%E_D$ = percentage voltage drop (in %)

E_D = voltage drop across conductor (in V)

E = voltage supplied at service panel (in V)

100 = conversion factor

For example, what is the percentage voltage drop of the same No. 6 copper conductor run with a supplied voltage of 230 V?

$$\%E_D = \frac{E_D}{E} \times 100$$

$$\%E_D = \frac{8.09\ \text{V}}{230\ \text{V}} \times 100$$

$$\%E_D = \mathbf{3.5\%}$$

A voltage drop of 3.5% is not acceptable under NEC® guidelines. To reduce the voltage drop, the conductor must be larger, the run shorter, the supplied voltage greater, or the current load less. Usually, the circuit voltage, current, and run length cannot be changed, so sizing up the conductor to reduce the resistance is the best choice. A No. 4 copper conductor in the same circuit would drop only 5.08 V (2.2%) and a No. 3 would drop 4.04 V (1.8%).

Measuring Voltage Drop

Calculating the voltage drop across a conductor to determine the correct conductor size should be done in the planning stage of an electrical project. However, conductor length is not always known accurately and the circuit may be modified later by adding or changing loads. Also, even if the conductor is properly sized for a circuit, any loose or corroded connections will increase the total resistance between the power panel and load. Therefore, the most accurate way to determine the voltage drop on a branch circuit is by taking voltage measurements.

Voltage is measured under no-load and full-load conditions to determine conductor voltage drop. To measure voltage drop in a branch circuit, apply the following procedure:

1. Turn all loads connected to the branch circuit OFF.
2. Measure the voltage at the furthest outlet (or load that is OFF) on the branch circuit to obtain the circuit's no-load voltage.
3. Turn all loads connected to the branch circuit ON.
4. Measure the voltage at the furthest outlet (or load that is ON) on the branch circuit to obtain the circuit's full-load voltage.

The percent of voltage drop in the branch circuit is calculated by applying the following formula:

$$\%V_D = \frac{V_{NL} - V_{FL}}{V_{NL}} \times 100$$

where

$\%V_D$ = percentage voltage drop (in V)

V_{NL} = no-load voltage (in V)

V_{FL} = full-load voltage (in V)

100 = conversion factor

For example, a branch circuit delivering power to a furnace has a measured full-load voltage of 112.5 V (furnace ON) and a measured no-load voltage of 115.8 V (furnace OFF). What is the percentage voltage drop?

$$\%V_D = \frac{V_{NL} - V_{FL}}{V_{NL}} \times 100$$

$$\%V_D = \frac{115.8 - 112.5}{115.8} \times 100$$

$$\%V_D = \frac{3.3}{115.8} \times 100$$

$$\%V_D = \mathbf{2.8\%}$$

The measured voltage drop across the conductor is less than 3%, so the conductor is properly sized for this circuit.

Determine the percent of voltage drop for each application from both the design specifications and using the measured values.

1. ______________ What is the percent of voltage drop, as calculated from the design specifications?

2. ______________ Is the calculated voltage drop within the acceptable limit?

3. ______________ What is the percent of voltage drop, as determined from the measured values?

4. ______________ Is the measured voltage drop within the acceptable limit?

5. What might account for the significant difference between the calculated and actual voltage drops?

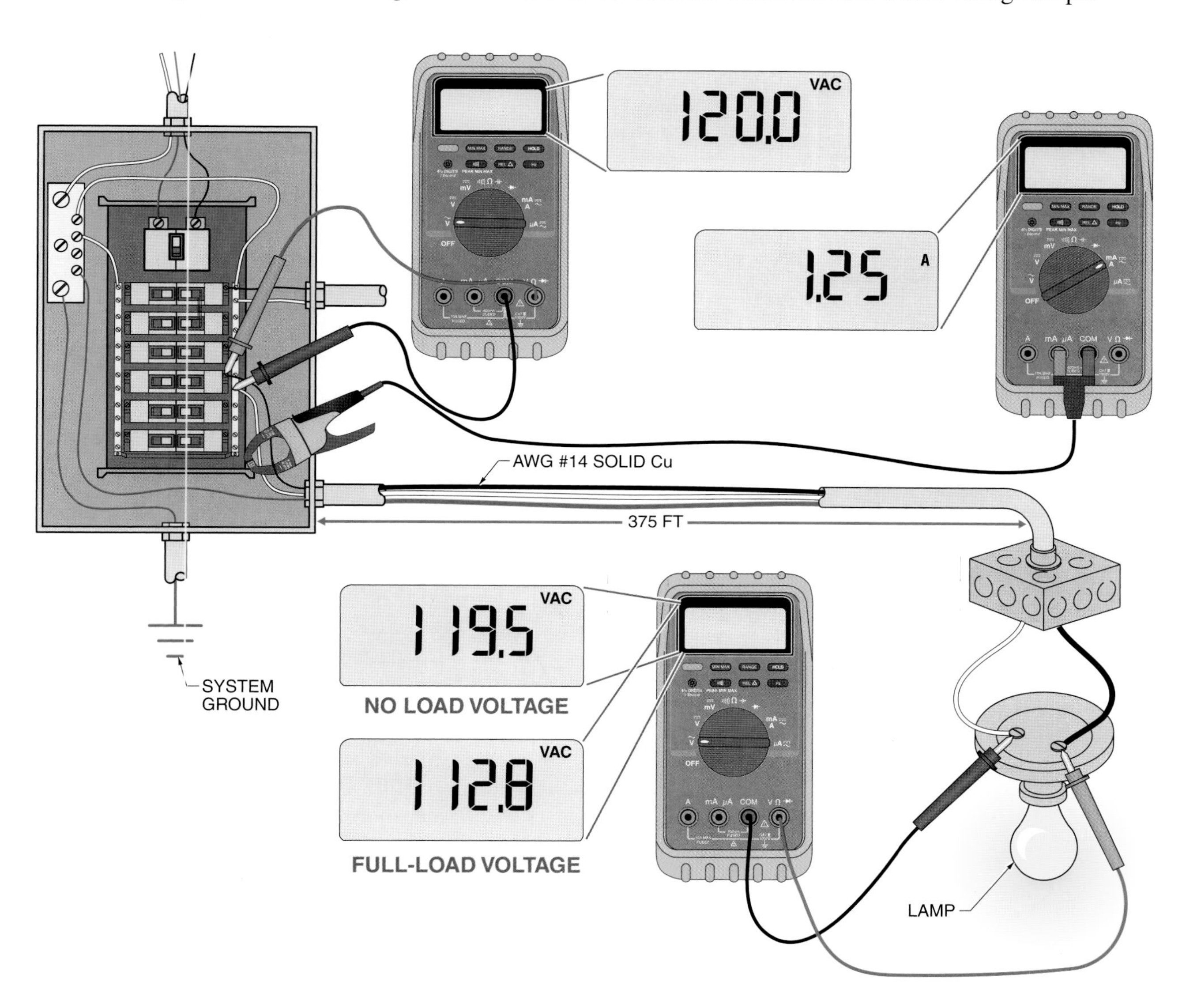

6. _______________ What is the percent of voltage drop, as calculated from the design specifications?

7. _______________ Is the calculated voltage drop within the acceptable limit?

8. _______________ What is the percent of voltage drop, as determined from the measured values?

9. _______________ Is the measured voltage drop within the acceptable limit?

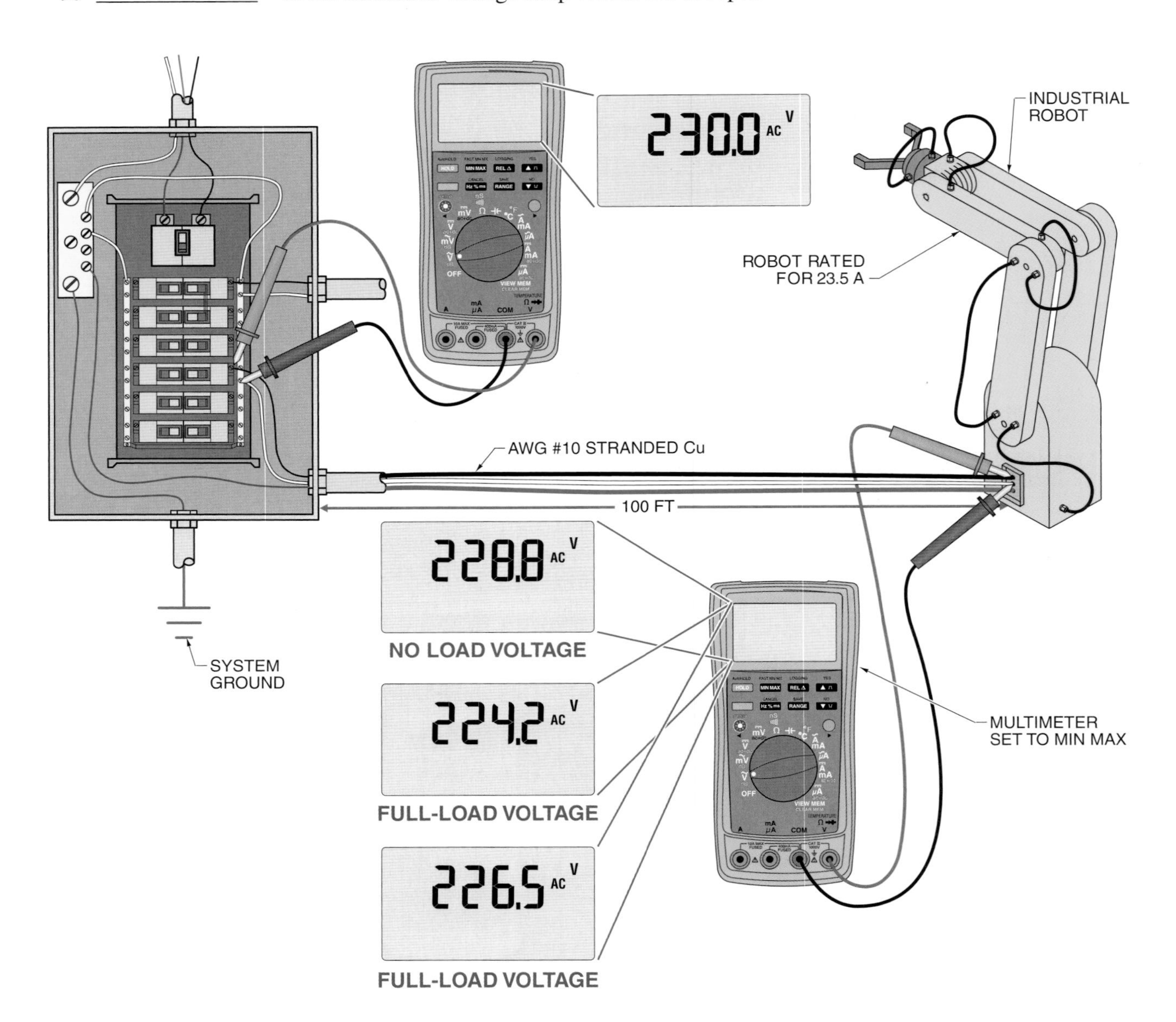

10. ______________ What is the percent of voltage drop, as calculated from the design specifications?

11. ______________ Is the calculated voltage drop within the acceptable limit?

12. ______________ What is the percent of voltage drop, as determined from the measured values?

13. ______________ Is the measured voltage drop within the acceptable limit?

14. What might account for the significant voltage drop in this circuit?

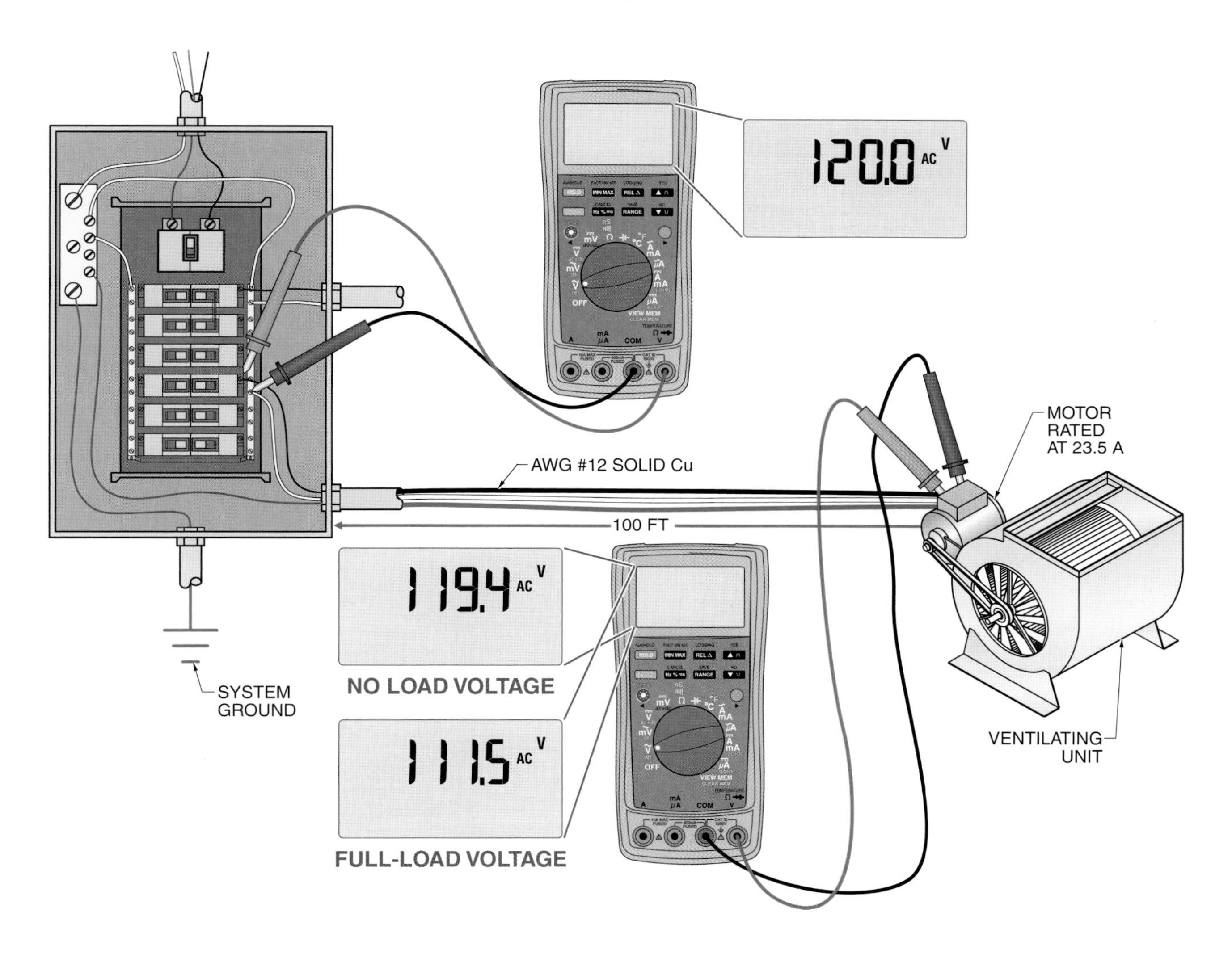

15. _______________ What is the percent of voltage drop, as calculated from the design specifications?

16. _______________ Is the calculated voltage drop within the acceptable limit?

17. _______________ What is the percent of voltage drop, as determined from the measured values?

18. _______________ Is the measured voltage drop within the acceptable limit?

19. _______________ What would be the calculated percent of voltage drop if the conductors were AWG #12 stranded aluminum instead?

20. _______________ Would the calculated voltage drop with the aluminum conductor be within the acceptable limit?

21. _______________ If stranded aluminum wire were required, what size would be needed to keep the voltage drop below the limit?

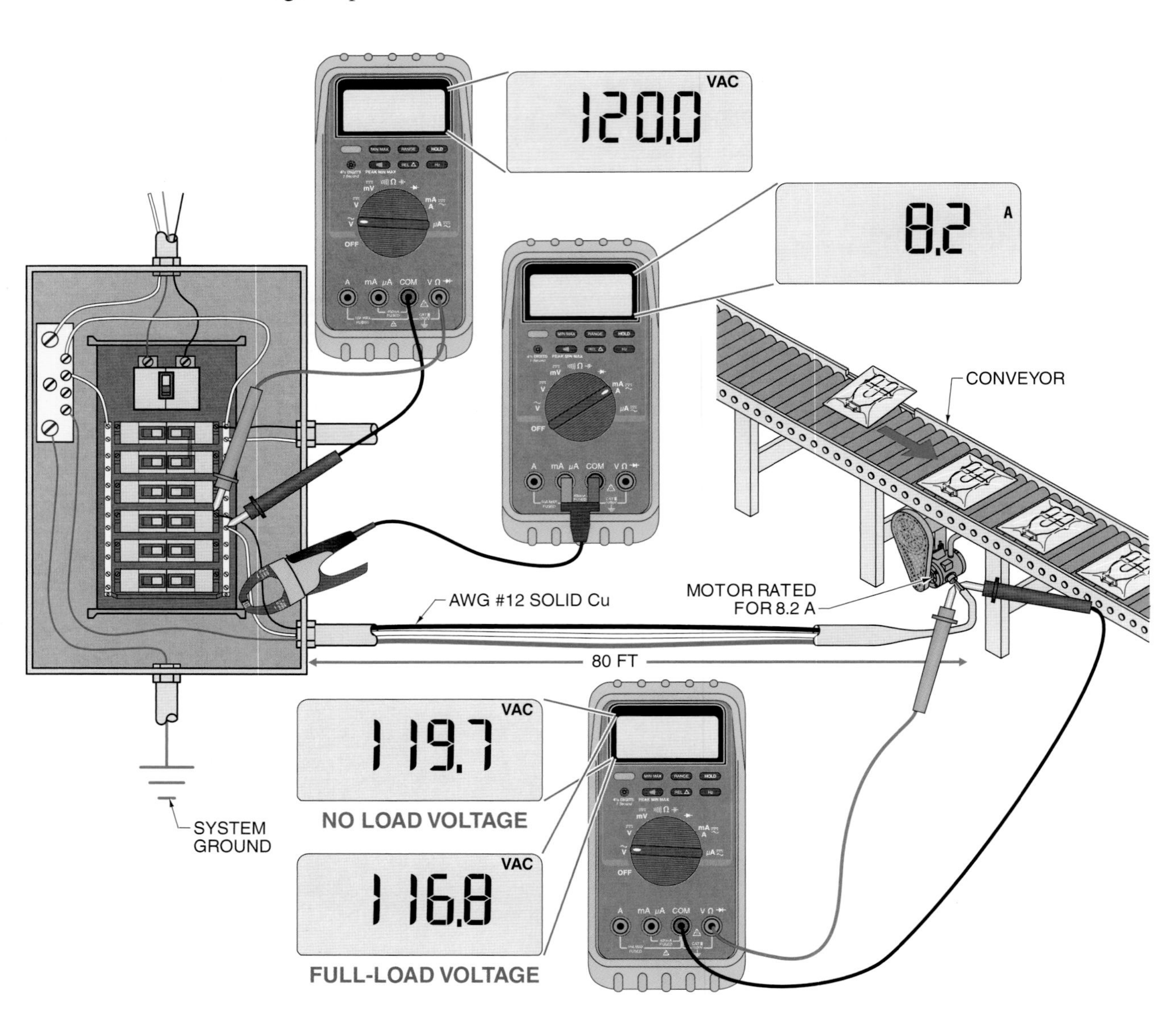

Application 1-11 — Fluorescent Lamp Output

The light output of fluorescent lamps is affected by both power quality and ambient temperature. A fluorescent lamp uses a transformer called a ballast to limit current flow and supply a high starting voltage for the lamp. A supply voltage higher or lower than the rating of the ballast affects the life of the lamp and ballast, and the lamp's light output (in lumens). The voltage supplied to a fluorescent lamp should be within ±7% of the lamp rating. See Fluorescent Lamp Fixture and Fluorescent Lamp Voltage Characteristics.

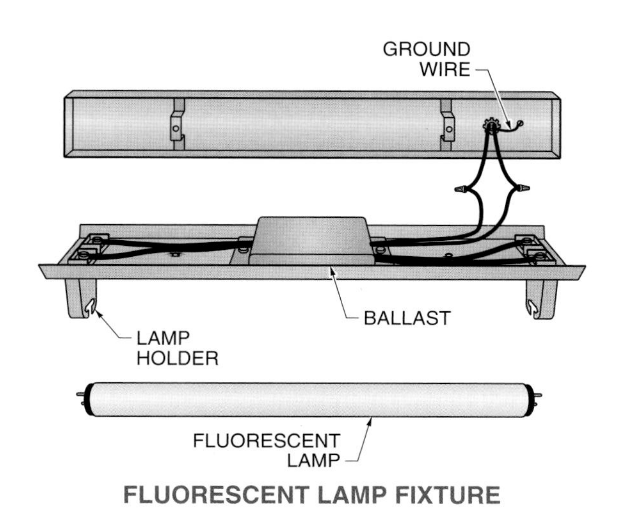

FLUORESCENT LAMP FIXTURE

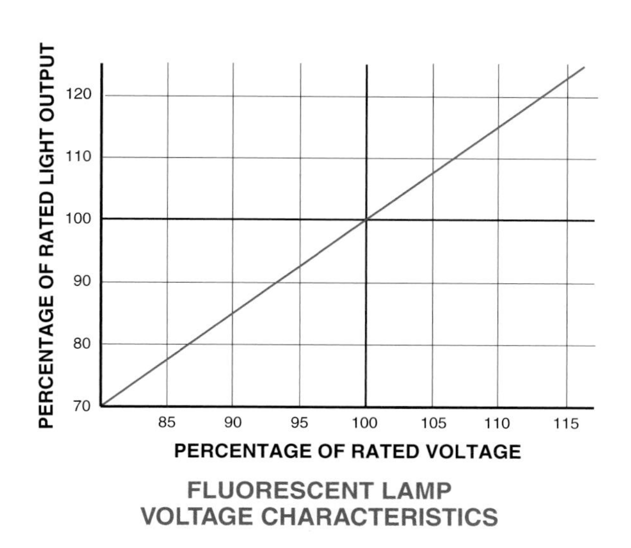

FLUORESCENT LAMP VOLTAGE CHARACTERISTICS

Standard indoor fluorescent lamps are designed to deliver a peak light output at approximately 75°F ambient temperature, considered to be room temperature for most fluorescent lamp installations. Moderate changes in ambient temperature (50°F to 105°F) have little effect (less than 10%) on the light output. Temperatures lower than 50°F or higher than 105°F have a greater effect on light output. See Fluorescent Lamp Temperature Characteristics.

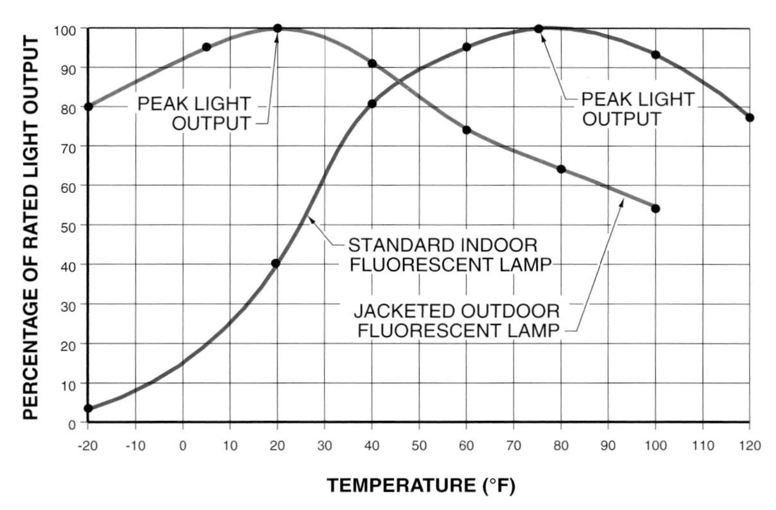

FLUORESCENT LAMP TEMPERATURE CHARACTERISTICS

Fluorescent lamps in cold conditions do not start well and put out less light. Cold lamps require more time to start because the cathode at the end of the lamp must heat up to release electrons into the mercury vapor. Then, even after the lamp is ON, it delivers less light because low temperature affects the mercury vapor pressure inside the lamp.

Outdoor-rated florescent lamps include an outer glass jacket. The jacket helps retain heat, which effectively shifts the peak light output to a lower ambient temperature point. Jacketed lamps are recommended for outdoor and cold indoor applications such as freezer warehouses, subways, and tunnels.

When accounting for both the ballast voltage and the ambient temperature to determine the actual light output, the two percentage values are multiplied. For example, if a low voltage reduces light output to 80% of the rated value and a cold environment reduces light output to 60% of the rated value, the effect of both together results in an actual light output of 48% of the rated value (80% × 60% = 48%).

Use the voltage measurement to determine the percentage of rated light output based on voltage and based on ambient temperature. Then determine the actual percentage of rated light output based on both voltage and temperature.

1. ______________ What is the percentage of rated light output based on the ballast voltage?

2. ______________ What is the percentage of rated light output based on the ambient temperature?

3. ______________ What is the actual percentage of rated light output based on both the ballast voltage and ambient temperature?

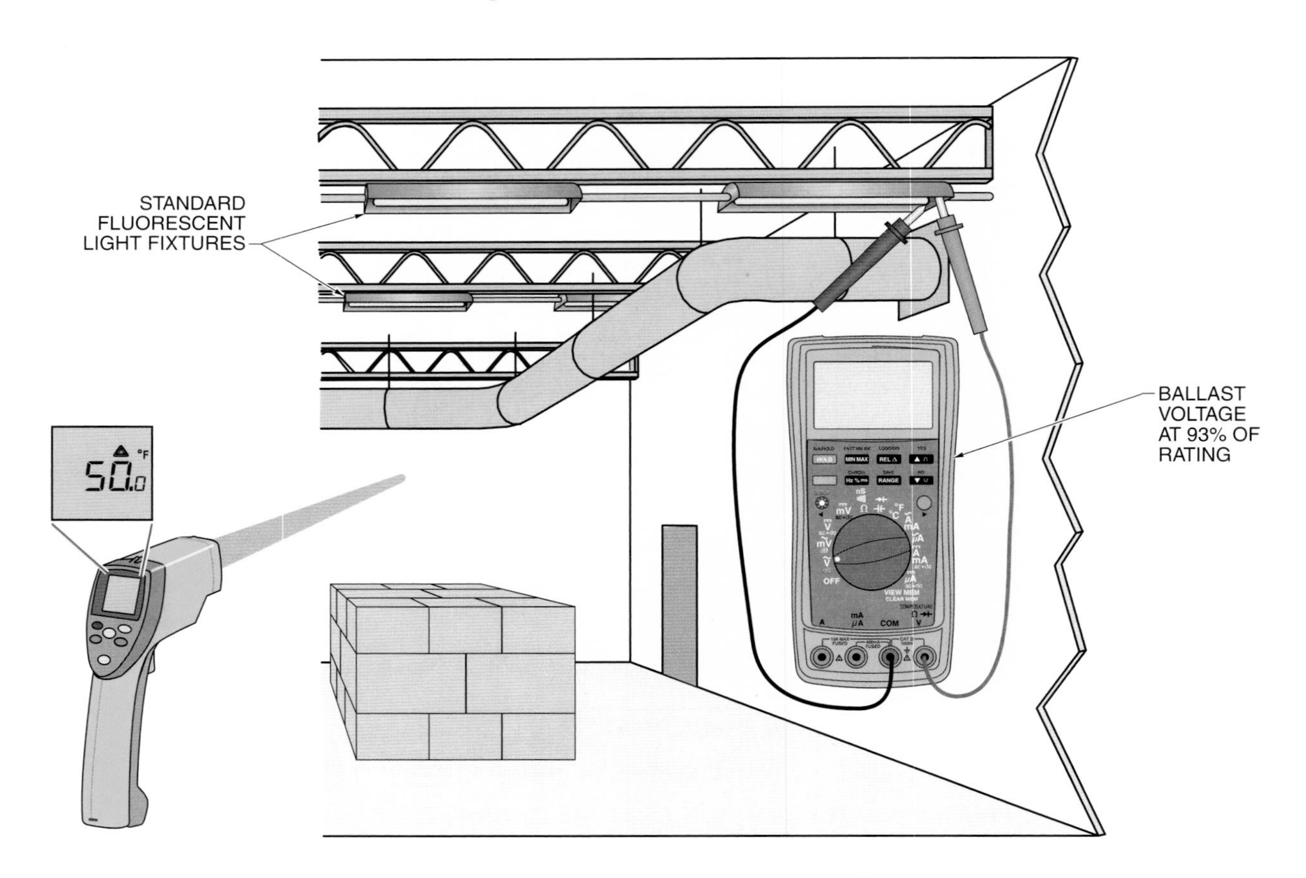

4. _______________ What is the percentage of rated light output based on the ballast voltage?

5. _______________ What is the percentage of rated light output based on the ambient temperature?

6. _______________ What is the actual percentage of rated light output based on both the ballast voltage and ambient temperature?

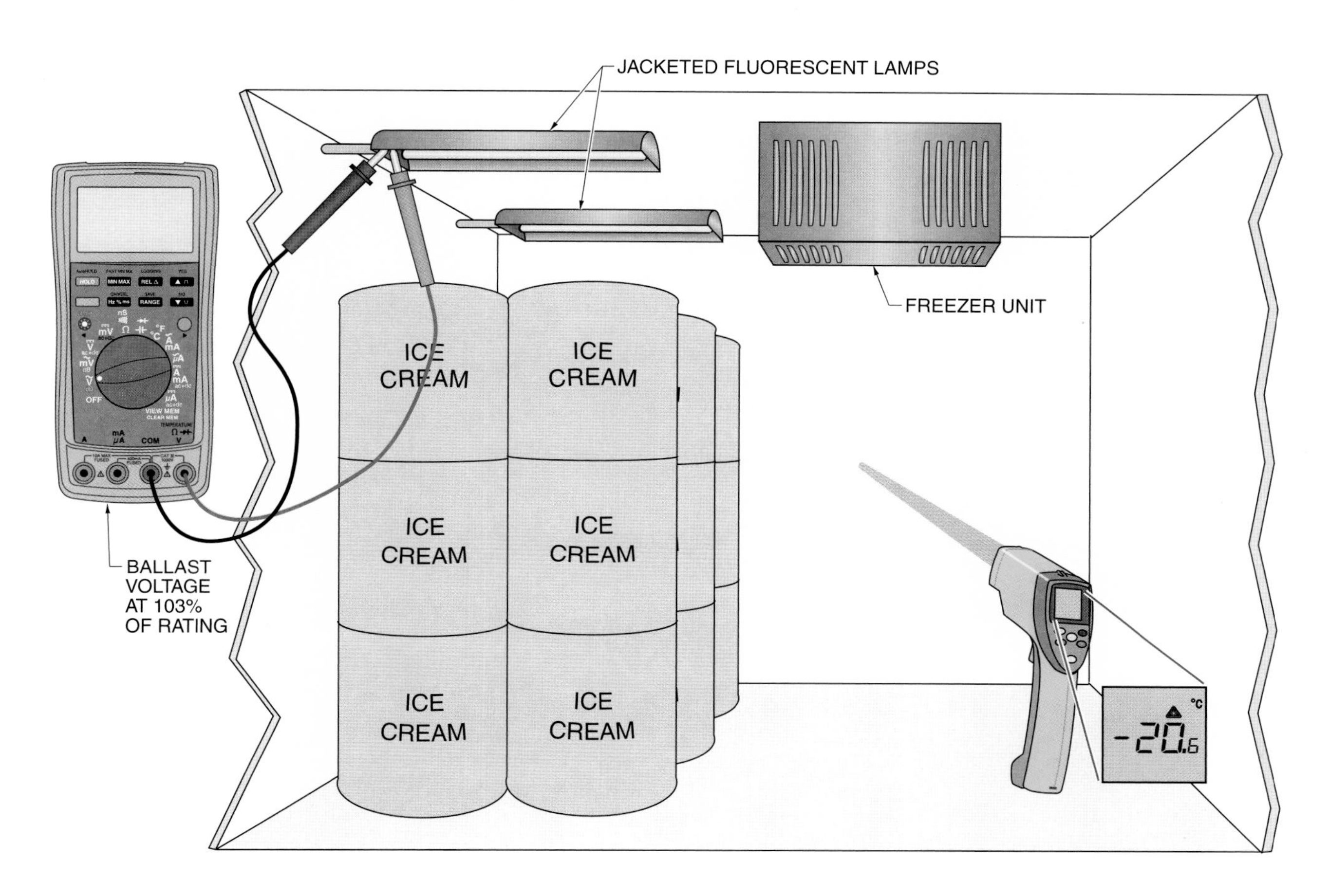

GENERAL USE TEST INSTRUMENTS

2 APPLICATIONS

Name: ______________________ Date: ______________

Application 2-1 —Testing Receptacles

Test instruments are used to test electrical components and circuits to ensure they are properly installed and working correctly. Even a correctly installed component may not be working properly if there is a component malfunction or the component is not correctly sized for the load or system. After electrical components and circuits are working properly, faults can develop over time. To find the fault, test instruments can be used to troubleshoot the individual components and circuit.

Using test instruments to test a new circuit or troubleshoot a faulty circuit or component requires the knowledge of how to properly set and connect the test instrument into the circuit or to the component and then knowing what the measured values actually mean.

In each of the following applications, a test instrument is used to test a 120 VAC/15 A rated receptacle (outlet). The acceptable voltage range for receptacles is –10% to +5%. See Receptacles.

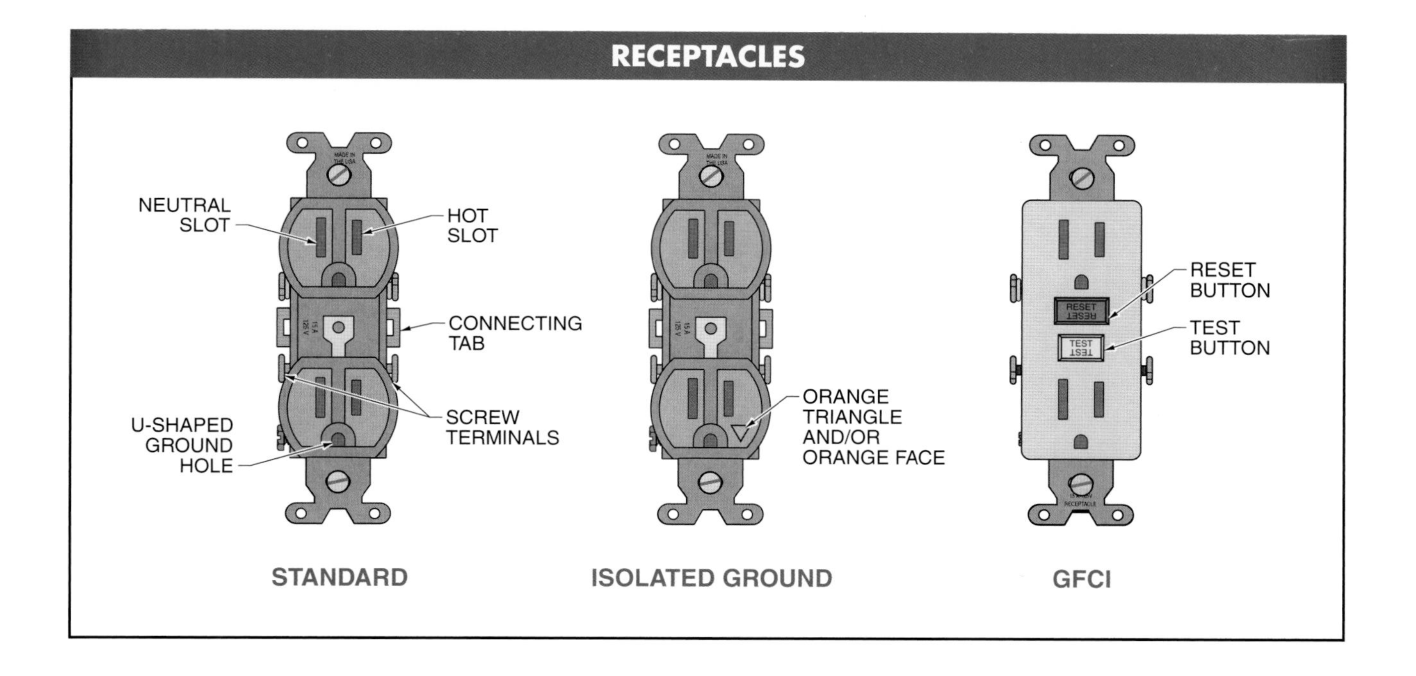

Testing Receptacles with a Test Light

1. Connect the black test lead of Test Light 1 to the neutral side of the receptacle and the red test lead to the hot side of the receptacle.

2. ______________ If the receptacle and circuit are working properly, should Test Light 1 be ON or OFF?

3. Connect the black test lead of Test Light 2 to the grounded side of the receptacle and the red test lead to the hot side of the receptacle.

4. ______________ If the receptacle and circuit are working properly, should Test Light 2 be ON or OFF?

5. Connect the black test lead of Test Light 3 to the grounded side of the receptacle and the red test lead to the neutral side of the receptacle.

6. ______________ If the receptacle and circuit are working properly, should Test Light 3 be ON or OFF?

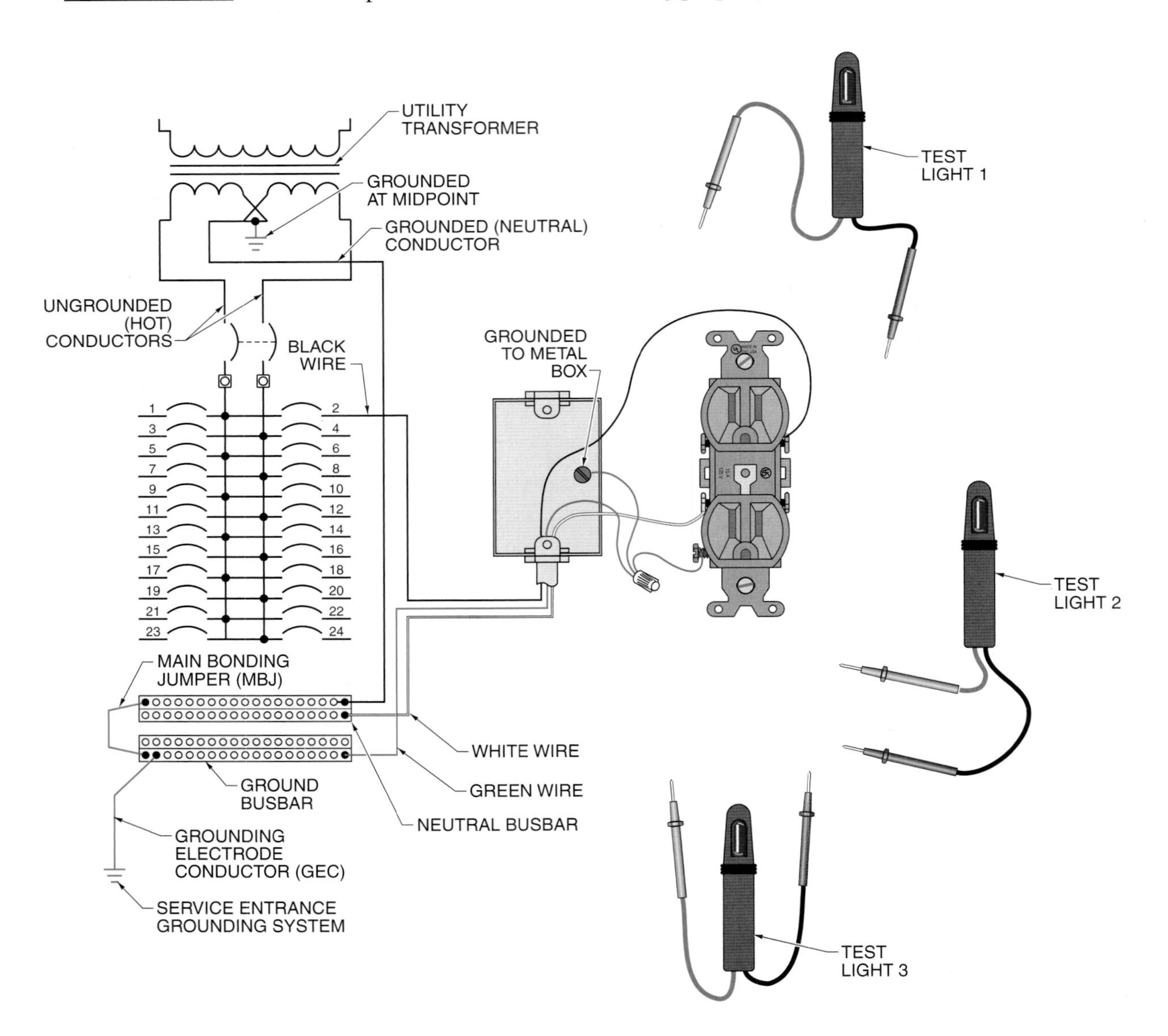

Testing Receptacles with a Voltage Indicator

7. ____________ Voltage Indicator 1 is held near one of the slots of Receptacle A. If the receptacle and circuit are working properly, should Voltage Indicator 1 tip be glowing or not glowing?

8. ____________ Voltage Indicator 2 is held near one of the slots of Receptacle B. If Voltage Indicator 2 tip is glowing, is the receptacle wired correctly?

9. If Receptacle B is not wired correctly, what is the problem with the circuit?

10. ____________ Voltage Indicator 3 is held near one of the slots of Receptacle B. If the receptacle and circuit are working properly, should Voltage Indicator 3 tip be glowing or not glowing?

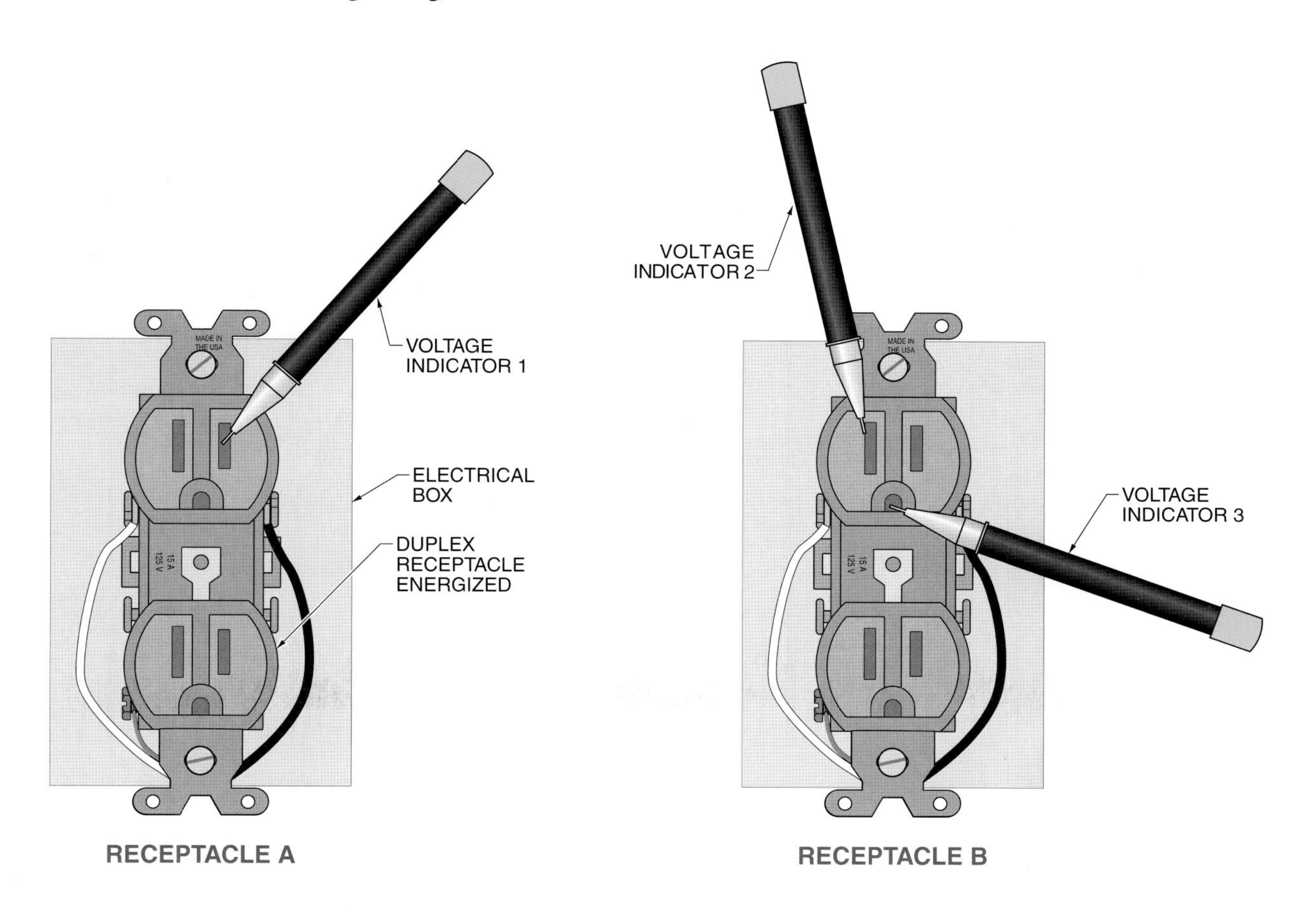

Testing Receptacles with a Voltage Tester

11. Connect the black test lead of Voltage Tester 1 to the neutral side of the GFCI receptacle and the red test lead to the hot side of the receptacle.

12. ______________ If the GFCI receptacle and circuit are working properly, what should Voltage Tester 1 be displaying?

13. Connect the black test lead of Voltage Tester 2 to the grounded side of the GFCI receptacle and the red test lead to the hot side of the receptacle.

14. ______________ If the GFCI receptacle and circuit are working properly, what should happen when Voltage Tester 2 is connected to the circuit?

15. ______________ If the GFCI receptacle and circuit are working properly, what should Voltage Tester 3 be displaying?

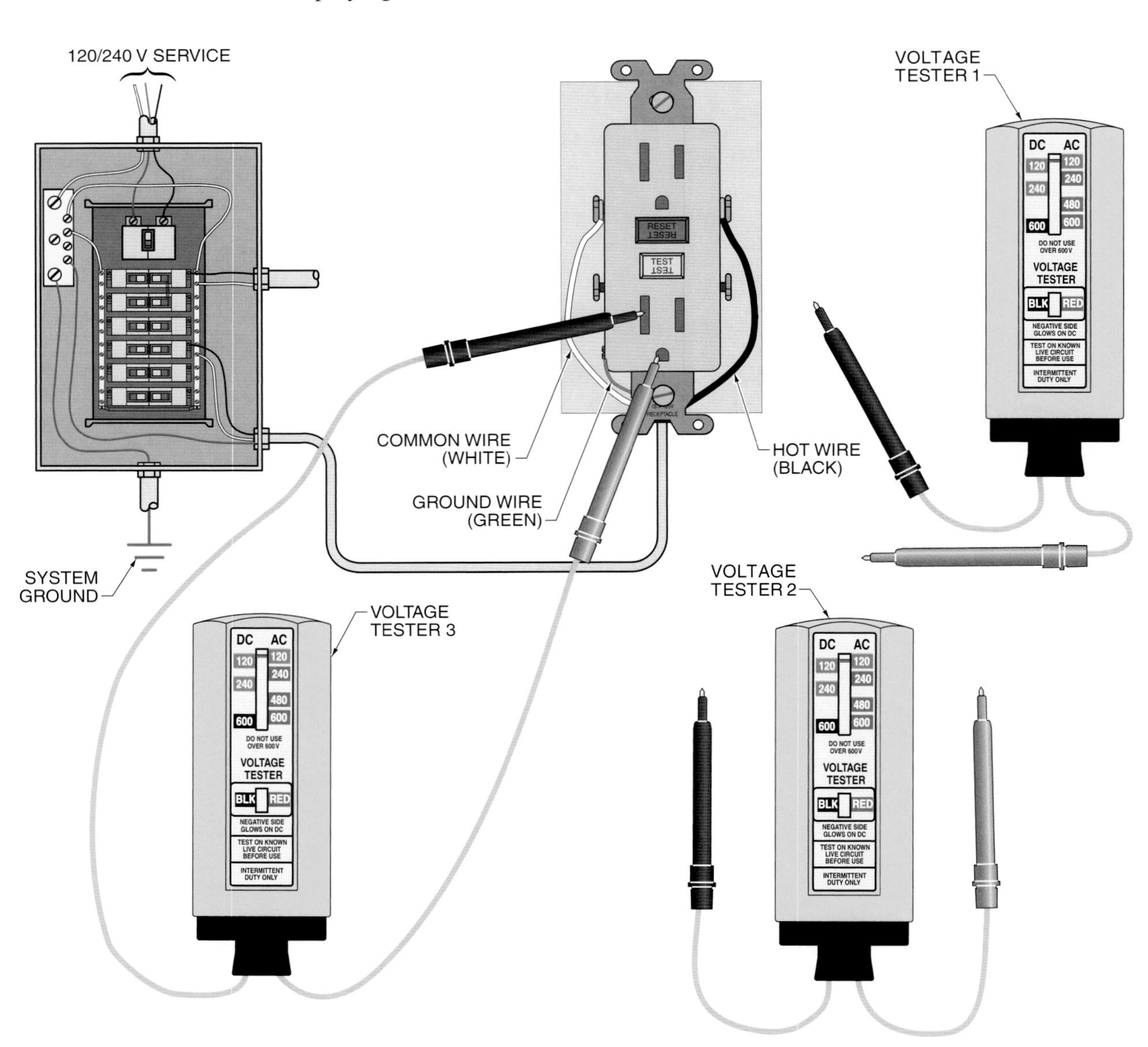

Testing Receptacles with a Basic Multimeter

16. Set (draw in) the multimeter function switch to take voltage measurements at the receptacle.

17. Connect the two test leads to the correct multimeter jacks.

18. Connect the black test lead to the neutral side of the receptacle and the red test lead to the hot side of the receptacle.

19. ______________ If the receptacle and circuit are working properly, what is the minimum voltage measurement that should be measured for the circuit to be in the standard acceptable voltage range?

20. ______________ If the receptacle and circuit are working properly, what is the maximum voltage measurement that should be measured for the circuit to be in the standard acceptable voltage range?

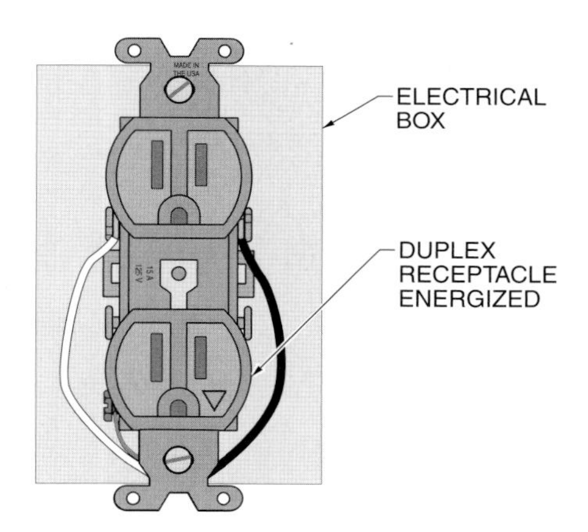

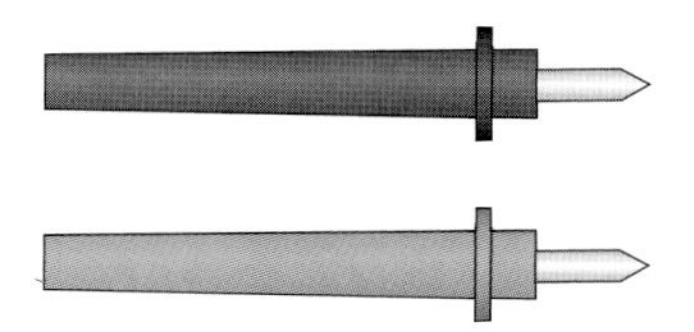

Testing Receptacles with an Analog Multimeter

21. Set (draw in) the analog multimeter function and range switches to take voltage measurements at the receptacle.

22. Connect the black test lead to neutral and the red lead to hot. What is the expected reading?

23. Connect the black test lead to the ground side of the receptacle and the red test lead to the hot side of the receptacle.

24. Draw the position that the pointer of the analog multimeter should be at if the circuit is wired correctly.

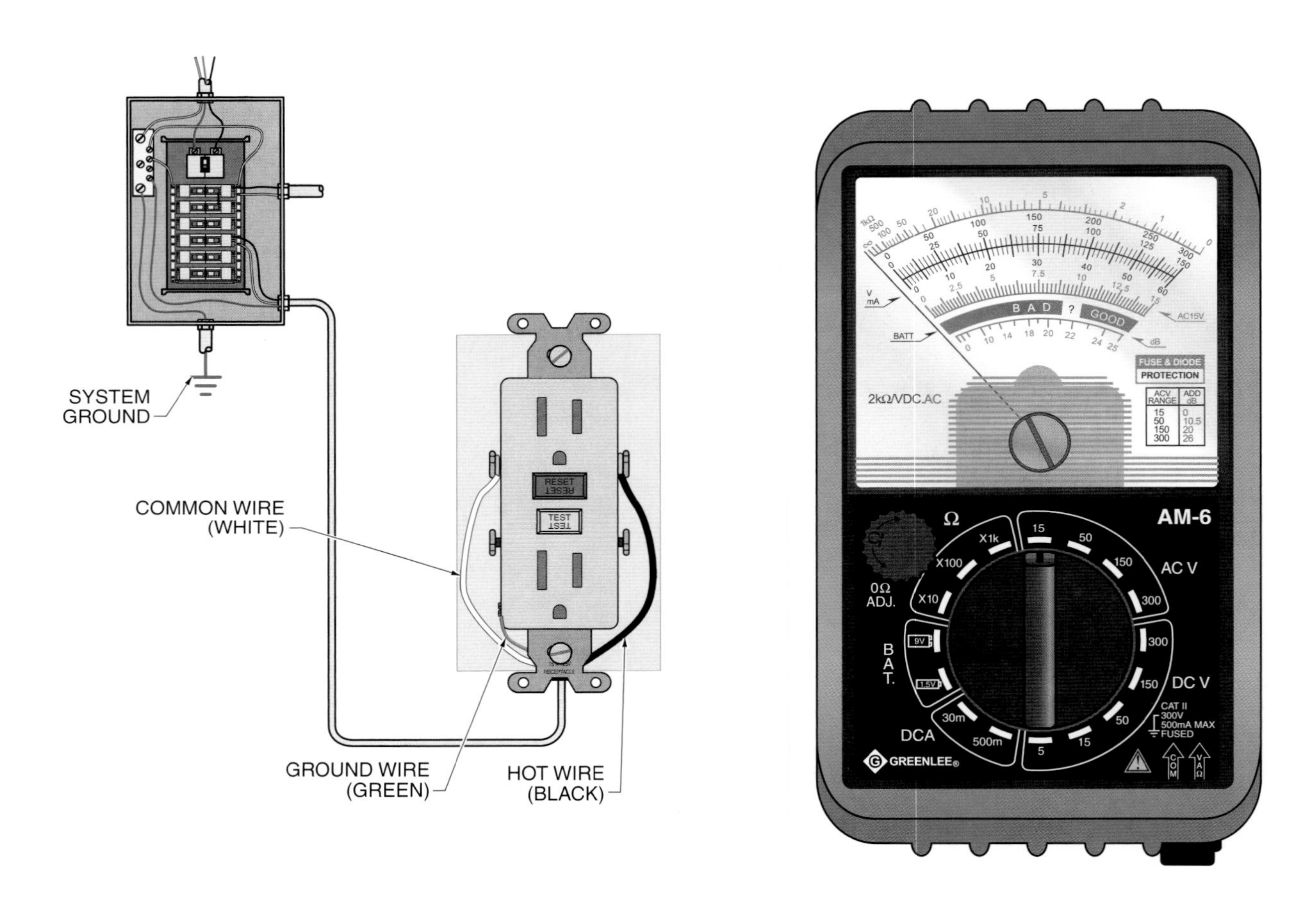

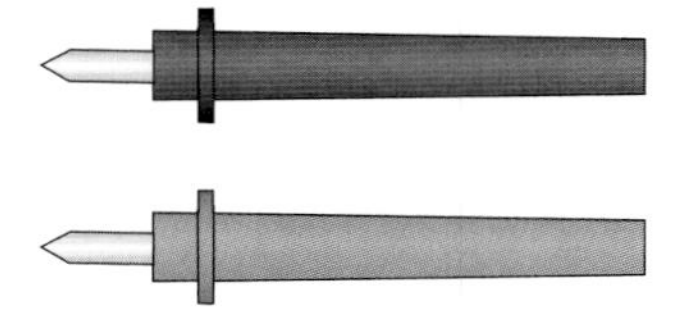

Testing Receptacles with an Advanced Multimeter

25. Set (draw in) the multimeter function switch to take voltage measurements at the receptacle.

26. Connect the two test leads to the correct jacks on the multimeter.

27. Connect the black test lead to the neutral side of the receptacle and the red test lead to the hot side of the receptacle.

28. If measurements are to be taken over time to see if the voltage in the circuit is fluctuating, circle the multimeter function button that would be used.

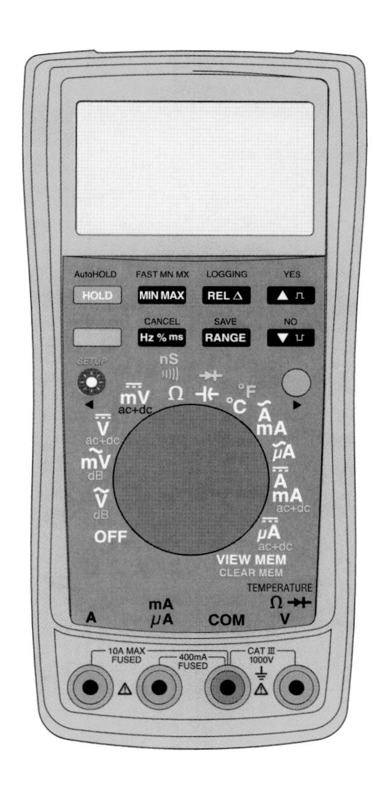

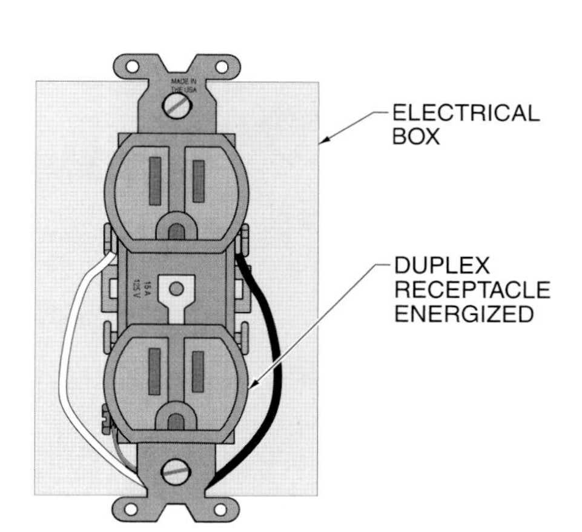

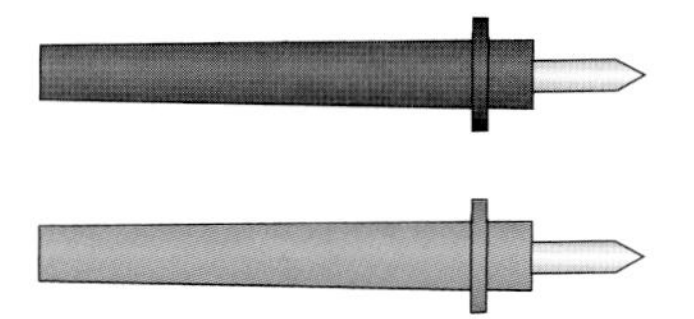

Application 2-2—Test Instrument Abbreviations and Symbols

All test instruments use abbreviations and symbols to simplify electrical terms. Understanding the abbreviations and symbols used with test instruments is required in order to properly set and use test instruments. Identify each abbreviation and symbol used on each test instrument.

Ground Resistance Meter

1. ______________ Symbol A = ___.

2. ______________ Symbol B = ___.

3. ______________ Symbol C = ___.

4. ______________ Symbol D = ___.

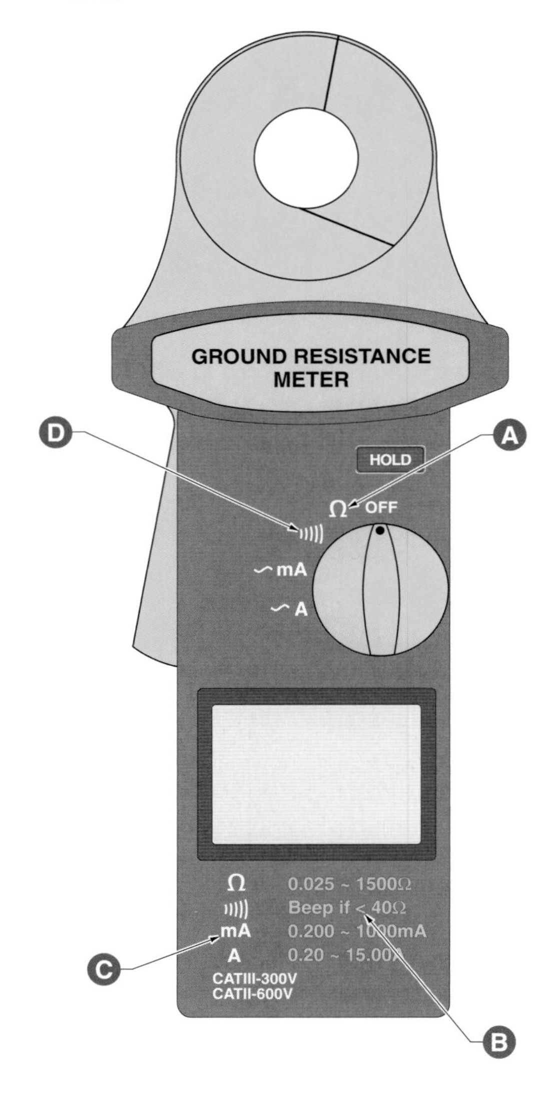

Analog Multimeters

5. ______________ Symbol A = ___.

6. ______________ Symbol B = ___.

7. ______________ Symbol C = ___.

8. ______________ Symbol D = ___.

9. ______________ Symbol E = ___.

10. ______________ Symbol F = ___.

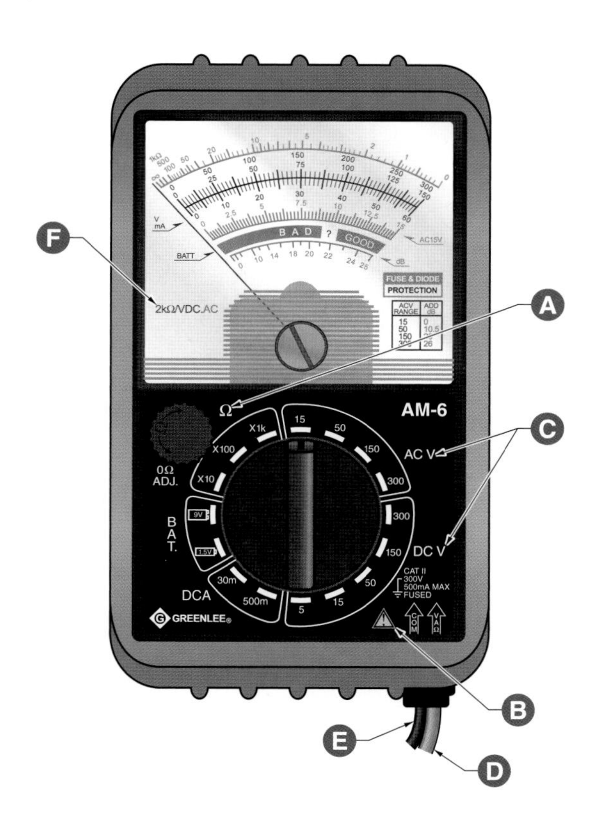

Digital Multimeters

11. ______________ Symbol A = ___.

12. ______________ Symbol B = ___.

13. ______________ Symbol C = ___.

14. ______________ Symbol D = ___.

15. ______________ Symbol E = ___.

16. ______________ Symbol F = ___.

17. ______________ Symbol G = ___.

18. ______________ Symbol H = ___.

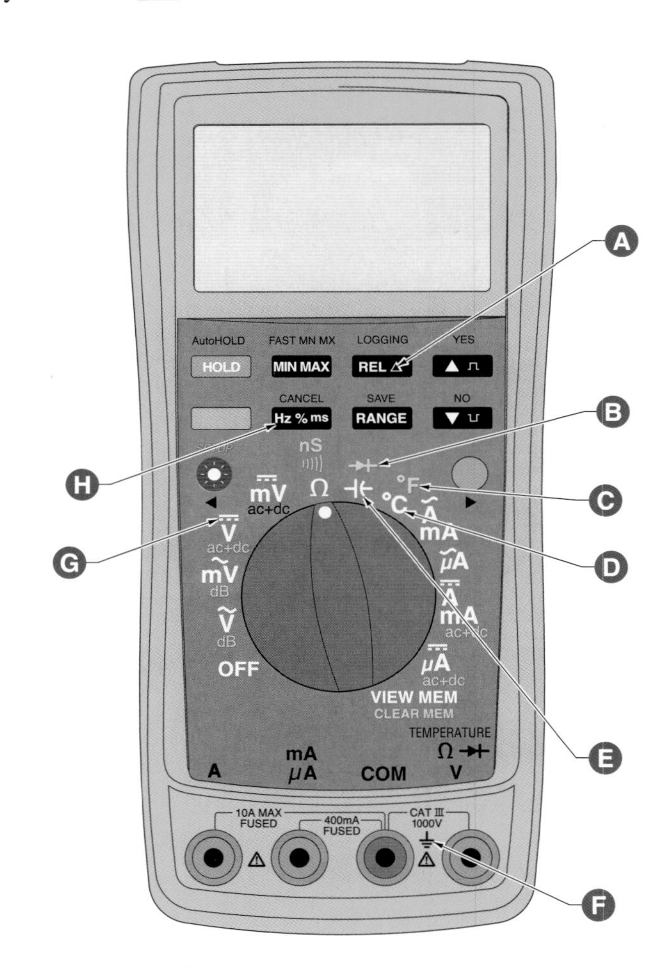

Application 2-3 — Troubleshooting Switches

Test instruments are used to take measurements to verify that a circuit or component is working properly. Before a test instrument is connected into a circuit, the test instrument must be set properly. When connecting a test instrument into a circuit, approximate test instrument readings should be anticipated if the test instrument readings are going to be used to help determine circuit problems. Test instrument readings that are not understood cannot aid in understanding the circuit or finding a problem.

1. Set (draw in) the function switch on Multimeter 1 to measure voltage at the switch.

2. Set (draw in) the function switch on Multimeter 2 to measure voltage at the lamp.

3. ______________ If the circuit is working properly, what should Multimeter 1 read when the two-way switch is in the open (OFF) position?

4. ______________ If the circuit is working properly, what should Multimeter 2 read when the two-way switch is in the open (OFF) position?

5. ______________ If the circuit is working properly, what should Multimeter 1 read when the two-way switch is in the closed (ON) position?

6. ______________ If the circuit is working properly, what should Multimeter 2 read when the two-way switch is in the closed (ON) position?

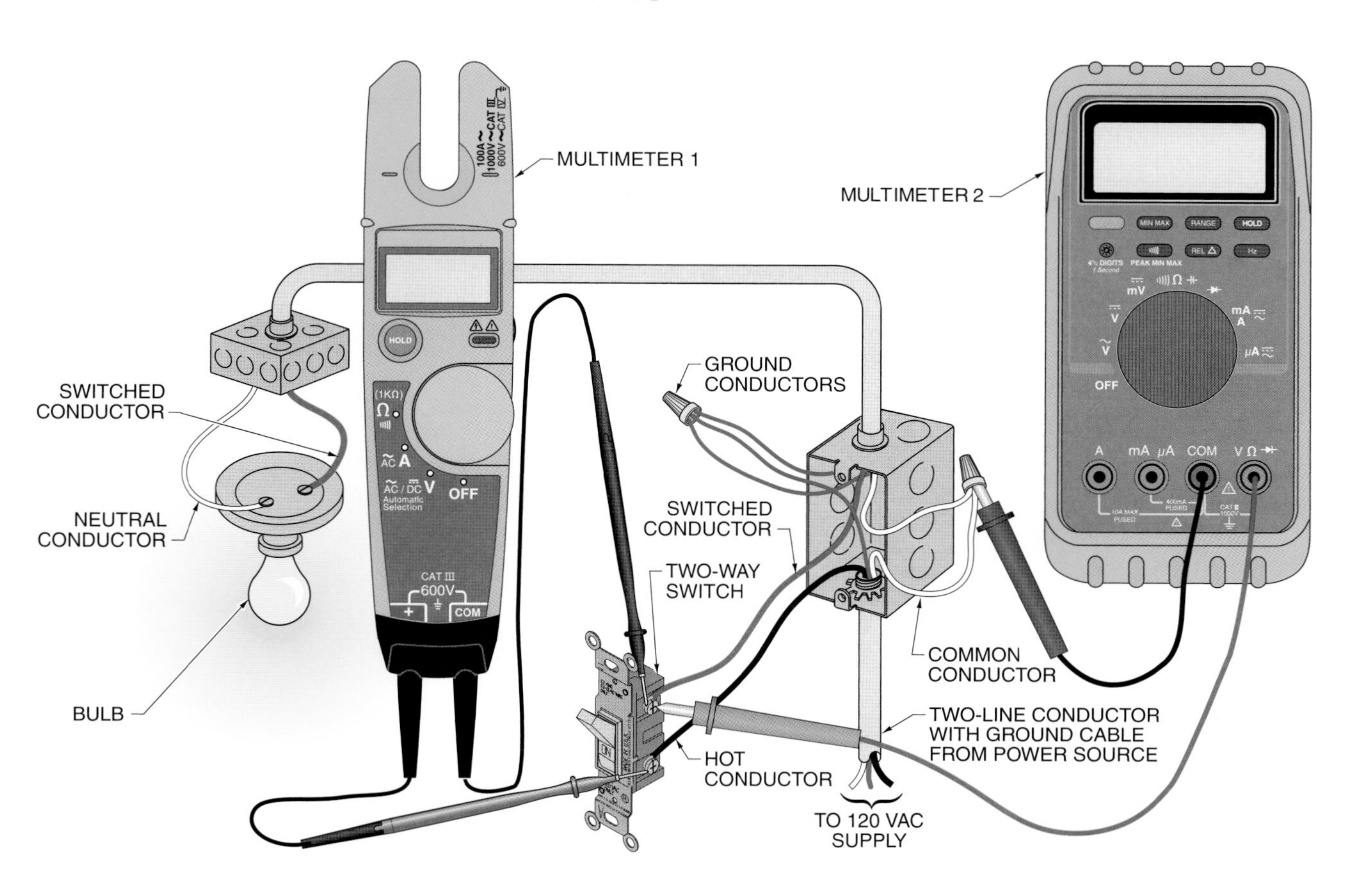

Application 2-4—Understanding Current Measurements

Current measurements are taken when testing or troubleshooting a circuit. Current measurements are used to determine how much load is on a circuit. Current measurements should be taken over time or at different operating times because current can vary as a circuit's load condition changes (loads turned ON and OFF, and/or motor loads change).

The incoming power supply and motor are configured for 220 VAC.

1. Set (draw in) the function switch on Multimeters 1, 2, and 3 to measure the current of the motor.
2. ______________ What is the expected current measurement if the motor is fully loaded to the nameplate horsepower rating?
3. ______________ What is the expected current measurement if the motor is fully loaded to the nameplate horsepower rating and operating at its service factor rating?
4. ______________ What is the expected current measurement if the motor is operating at 80% of its full load rating?
5. ______________ If the incoming power supply and motor are wired for 440 VAC, what is the expected current measurement if the motor is fully loaded to the nameplate horsepower rating?

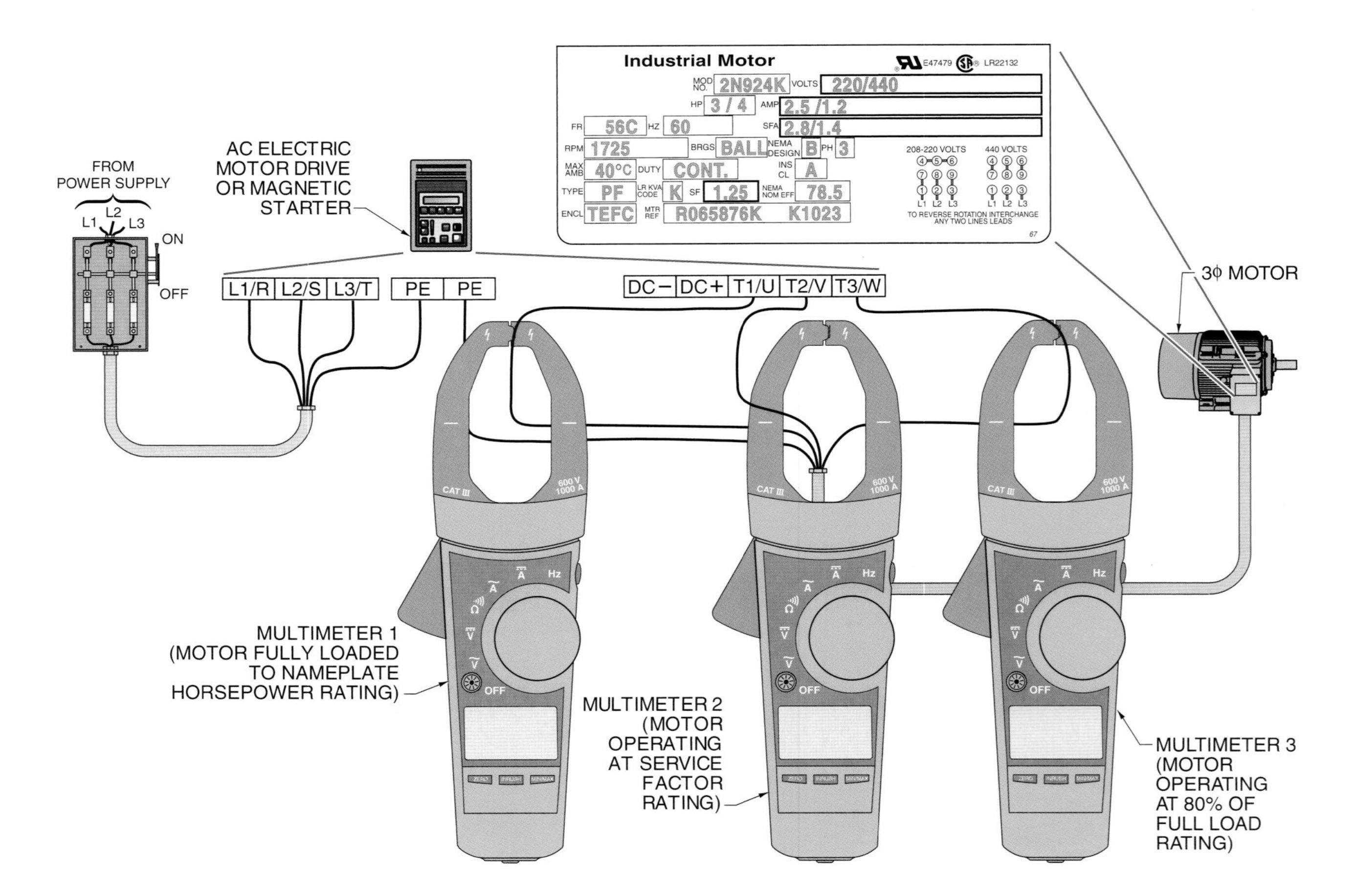

Application 2-5 — Motor Winding Resistance

All motors have motor windings (conductors) that develop magnetic fields that produce rotation when connected to power. All motor winding conductors have resistance. The more horsepower a motor produces, the larger the size of winding conductor required. The larger the size of the conductor, the smaller the AWG number. The larger the conductor, the less the resistance. Taking resistance measurements and applying the laws of resistance in series and parallel circuits can help verify connections to determine if a motor is wired correctly before any power is applied to the motor.

Resistance in Series Circuits

The total resistance in a circuit containing series-connected components (motor windings) equals the sum of the resistances of all components. The resistance in the circuit increases if components (motor windings) are added in series and decreases if components are removed. To calculate total resistance in a series circuit, apply the following formula:

$$R_T = R_1 + R_2 + R_3 + \ldots$$

where

R_T = total circuit resistance (in Ω)

R_1 = resistance 1 (in Ω)

R_2 = resistance 2 (in Ω)

R_3 = resistance 3 (in Ω)

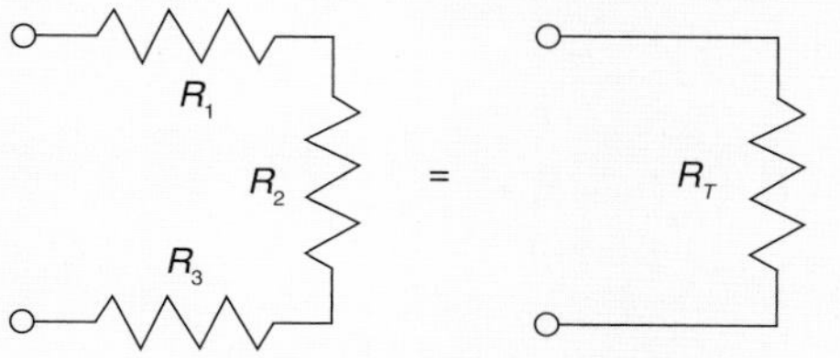

R_1 R_2 R_3 = R_T

For example, what is the total resistance in a circuit that has 2 Ω, 4 Ω, and 6 Ω resistors connected in series?

$$R_T = R_1 + R_2 + R_3$$

$$R_T = 2 + 4 + 6$$

$$R_T = \mathbf{12\ \Omega}$$

Resistance in Parallel Circuits

The total resistance in a circuit containing parallel-connected components (motor windings) is less than the smallest resistance value. The total resistance decreases if loads are added in parallel and increases if loads are removed. To calculate total resistance in a parallel circuit containing two resistors, apply the following formula:

$$R_T = \frac{R_1 \times R_2}{R_1 + R_2}$$

where

R_T = total circuit resistance (in Ω)

R_1 = resistance 1 (in Ω)

R_2 = resistance 2 (in Ω)

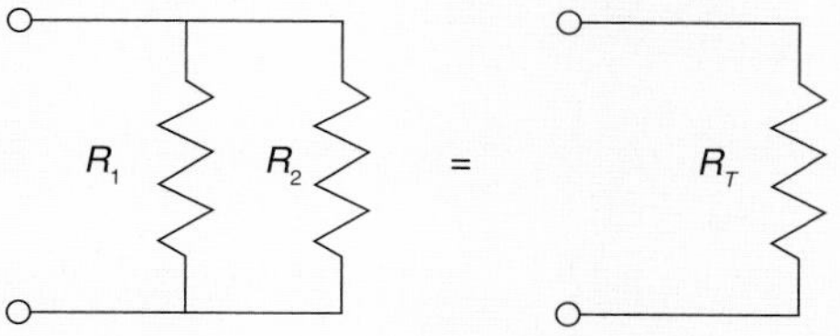

R_1 R_2 = R_T

For example, what is the total resistance in a circuit containing resistors of 16 Ω and 24 Ω connected in parallel?

$$R_T = \frac{R_1 \times R_2}{R_1 + R_2}$$

$$R_T = \frac{16 \times 24}{16 + 24}$$

$$R_T = \frac{384}{40}$$

$$R_T = \mathbf{9.6\ \Omega}$$

The total resistance of more than two resistors with different values, connected in parallel is calculated by applying the following formula:

$$R_T = \frac{1}{\frac{1}{R_1} + \frac{1}{R_2} + \frac{1}{R_3} + \cdots}$$

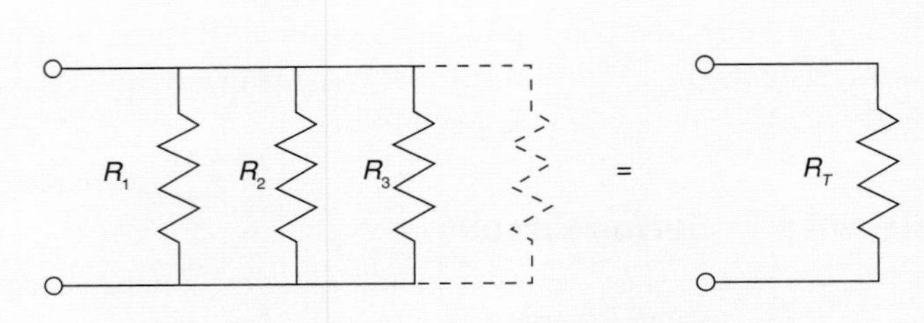

where
R_T = total circuit resistance (in Ω)
R_1 = resistance 1 (in Ω)
R_2 = resistance 2 (in Ω)
R_3 = resistance 3 (in Ω)

For example, what is the total resistance in a circuit containing resistors of 16 Ω, 24 Ω, and 48 Ω connected in parallel?

$$R_T = \frac{1}{\frac{1}{R_1} + \frac{1}{R_2} + \frac{1}{R_3}}$$

$$R_T = \frac{1}{\frac{1}{16} + \frac{1}{24} + \frac{1}{48}}$$

$$R_T = \frac{1}{0.0625 + 0.0417 + 0.0208}$$

$$R_T = \frac{1}{0.125}$$

$$R_T = \mathbf{8\ \Omega}$$

Testing Single-Voltage, Wye-Connected, Three-Phase Motor Wiring

The motor winding resistance for Motor Coil T1 to internal connection point is 4 Ω before the motor leads are connected (as given by the motor manufacturer data). After the motor leads are connected to the motor side of the motor starter, resistance measurements are taken before power is applied to verify that the motor is connected correctly.

1. ______________ What is the resistance between T1 and T2?

2. ______________ What is the resistance between T2 and T3?

3. ______________ What is the resistance between T3 and T1?

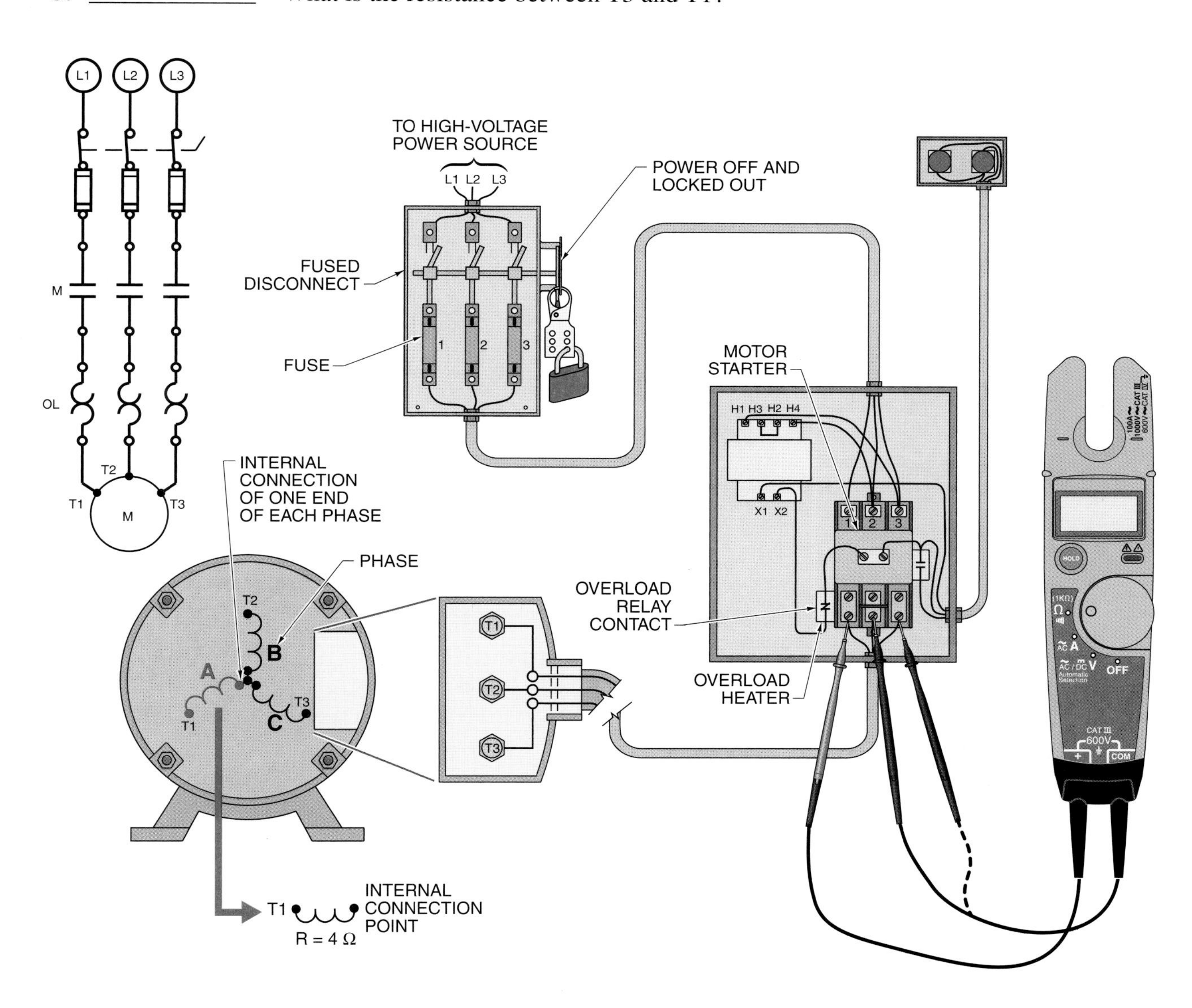

Testing Single-Voltage, Delta-Connected, Three-Phase Motor Wiring

The motor winding resistance for Motor Coil A is 6 Ω before the motor internal leads are connected (as given by the motor manufacturer data). After the motor leads are connected to the motor side of the motor starter, resistance measurements are taken before power is applied to verify that the motor is connected correctly.

4. ____________ What is the resistance between T1 and T2?

5. ____________ What is the resistance between T2 and T3?

6. ____________ What is the resistance between T3 and T1?

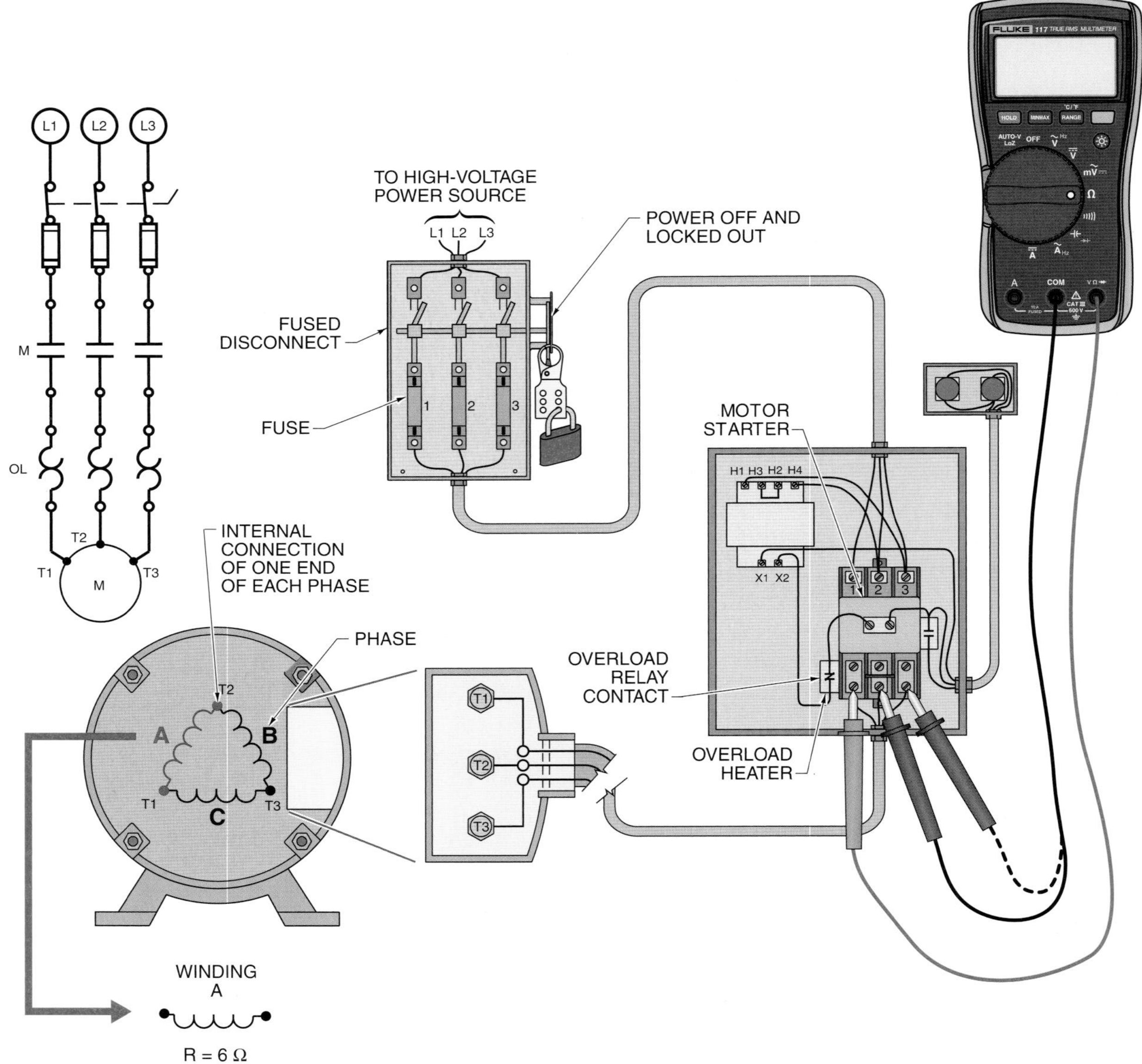

Testing Dual-Voltage, Wye-Connected, Three-Phase Motor Wiring

The motor winding resistance for Motor Coil T1 to T4 is 12 Ω before the motor leads are connected. After the motor leads are connected to the motor side of the motor starter, resistance measurements are taken before power is applied to verify that the motor is connected correctly.

7. ____________ What is the resistance between T1 and T2?

8. ____________ What is the resistance between T2 and T3?

9. ____________ What is the resistance between T3 and T1?

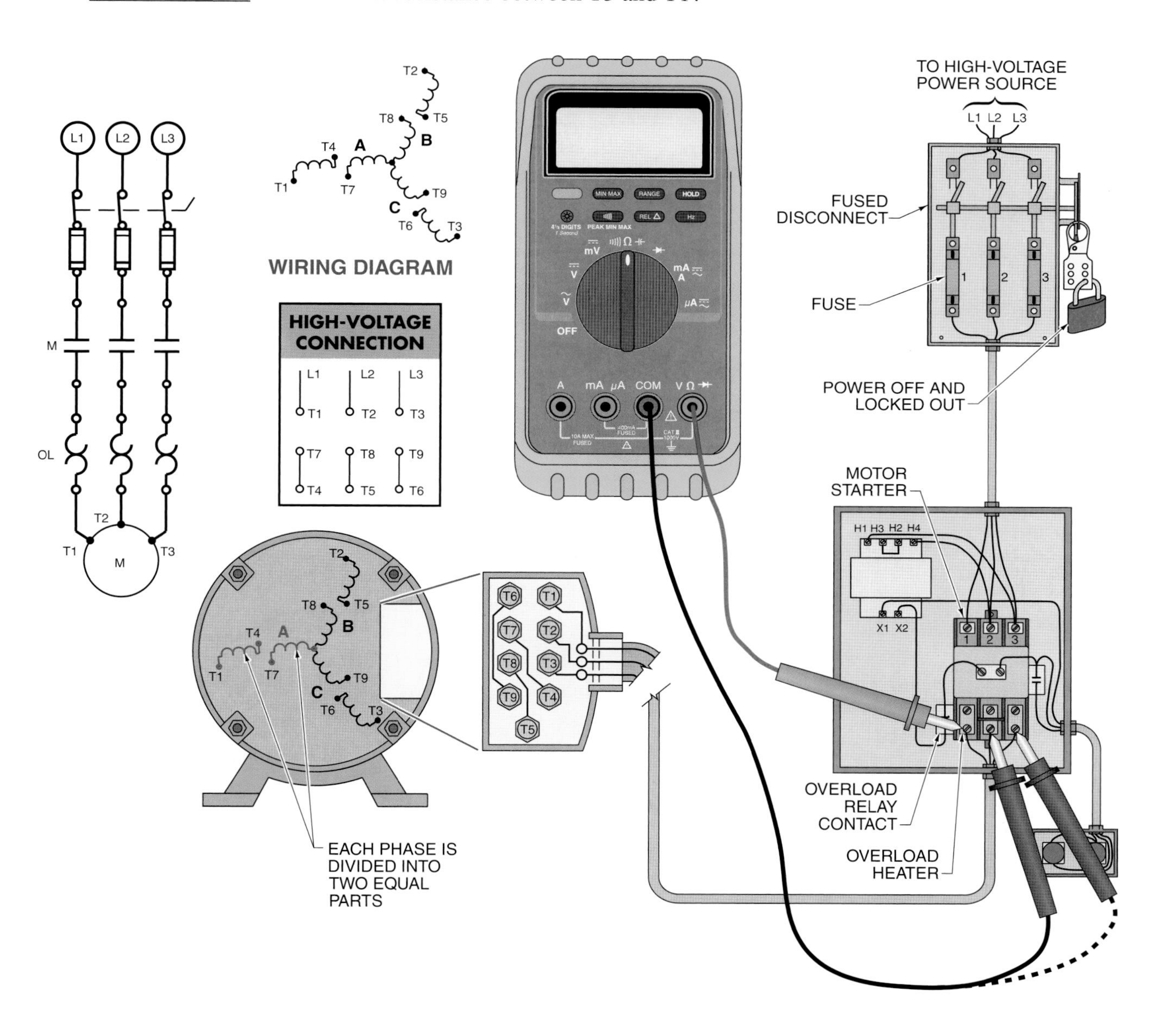

Testing Dual-Voltage, Delta-Connected, Three-Phase Motor Wiring

The motor winding resistance for Motor Coil T1 to T4 is 10 Ω before the motor leads are connected. After the motor leads are connected to the motor side of the motor starter, resistance measurements are taken before power is applied to verify that the motor is connected correctly.

10. ______________ What is the resistance between T1 and T2?

11. ______________ What is the resistance between T2 and T3?

12. ______________ What is the resistance between T3 and T1?

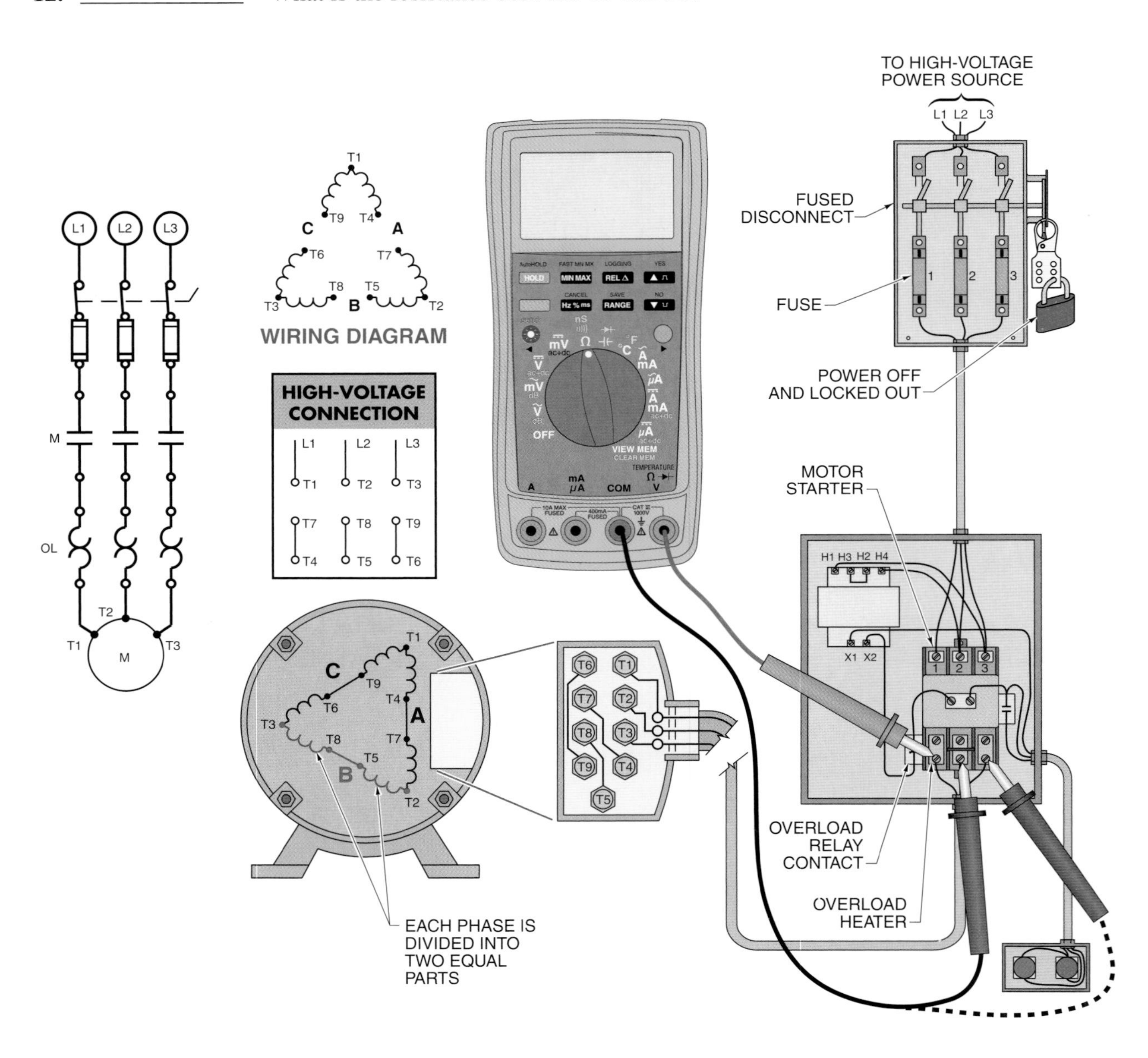

Application 2-6 — Selecting Appropriate Test Instruments

Some test instruments are designed for one specific purpose while others are designed for a range of tasks. Every test instrument has advantages and disadvantages that determine its usefulness in a given situation. Selecting the appropriate test instrument for a particular measurement or troubleshooting application requires an understanding of which electrical quantity needs to be measured (volts, amps, ohms, etc.), how the measured quantity is to be displayed (maximum value, numerical value, waveform, etc.), and the limits of the test instrument (CAT rating, current limit, etc.).

In general, the least complicated piece of test equipment that can perform the required task is recommended. For example, a test light, receptacle tester, voltage tester, circuit analyzer, and multimeter can all be used to check a receptacle (outlet). If only an indication of whether the receptacle is powered (live) is needed, the test light will work. If the receptacle must be tested for power and correct wiring, the receptacle tester will work. However, if more advanced measurements or tests (voltage levels, voltage recorded over time, peak voltage, etc.) are required at the receptacle, then a voltage tester, circuit analyzer, or multimeter must be used.

All limits on voltage and current measurements (whether the current jacks are fused or unfused, the condition of the test leads, etc.) must be considered as well as factors such as the CAT ratings of the application and the test instrument.

Select a test instrument from the following that best meets the measurement requirement for each situation. Also, give the reason for the selection. The simplest test instrument that can perform the required task should be selected, even if more than one test instrument could be used.

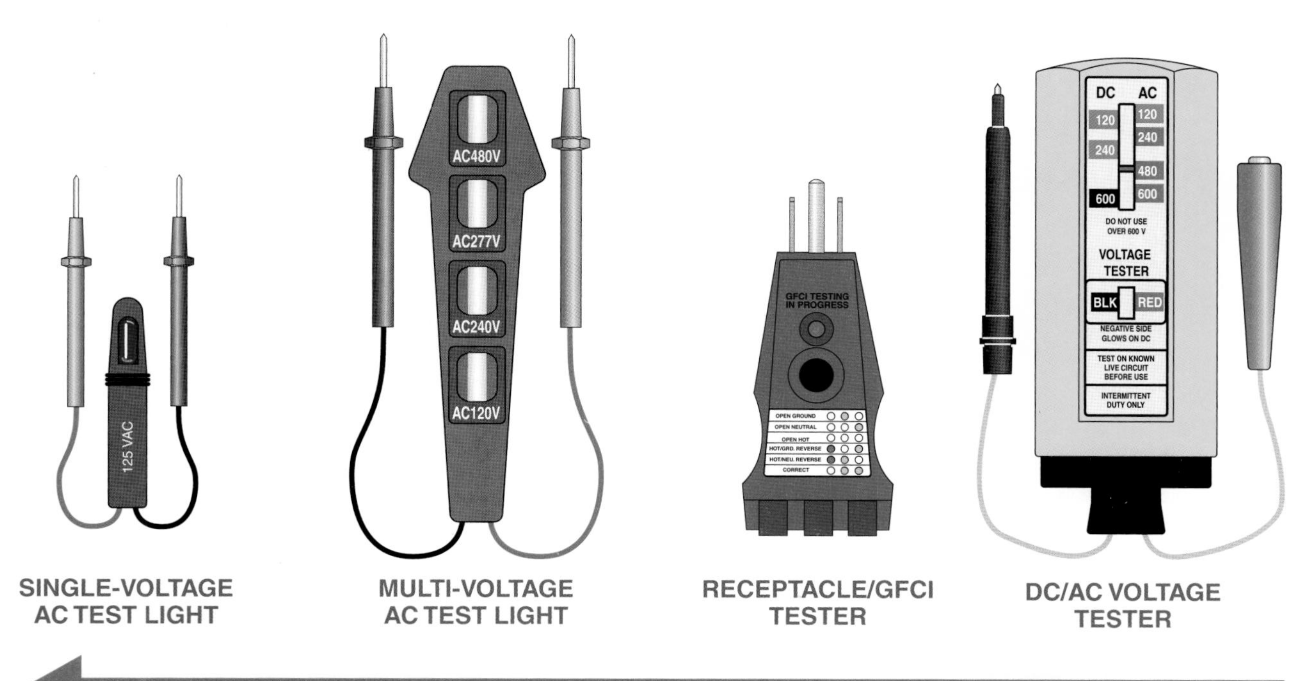

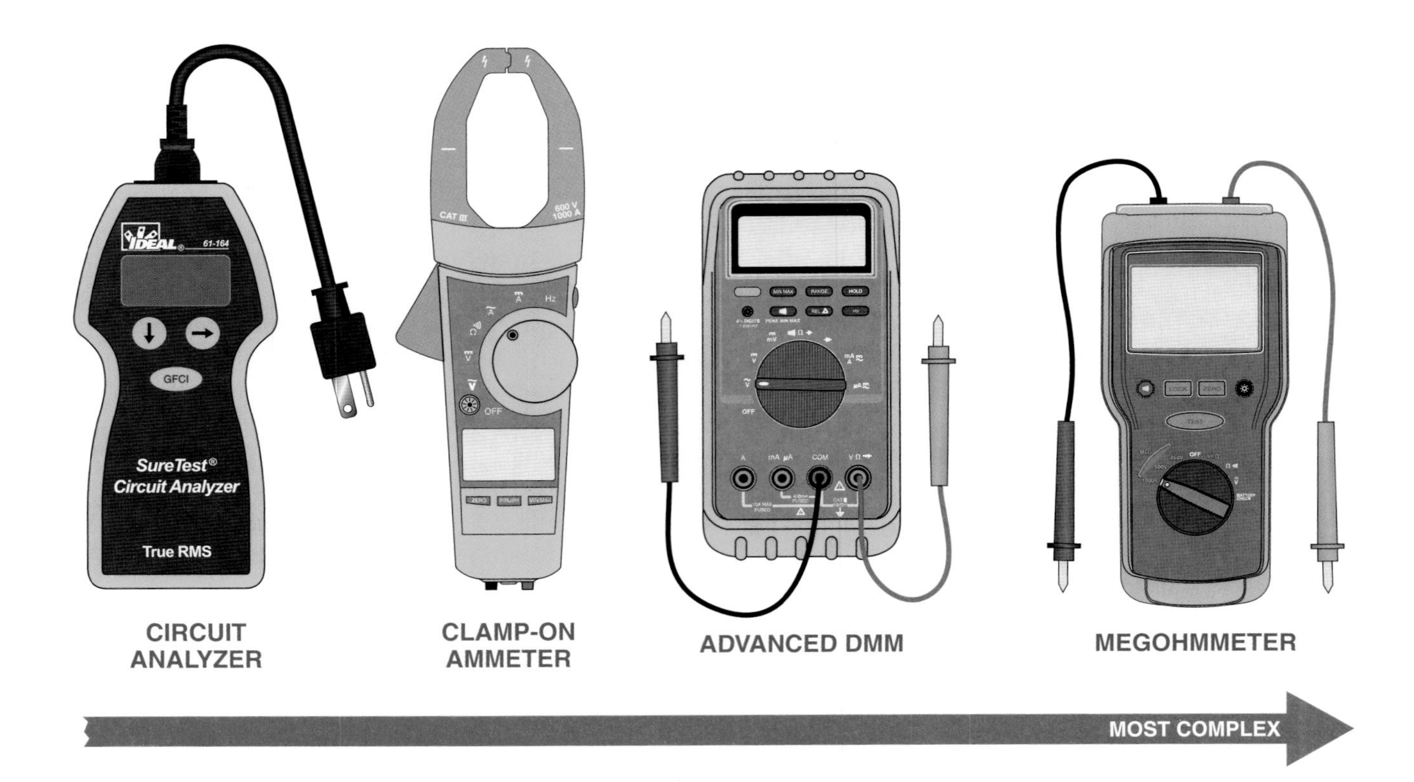

A service call requests that several receptacle circuits in an office be checked at the branch circuit power panel at a high usage time to see if any of the circuits are near their current limit (rating of the circuit breaker).

1. ______________ Which is the simplest test instrument that meets the service call requirements?

2. Why is this test instrument the best choice?

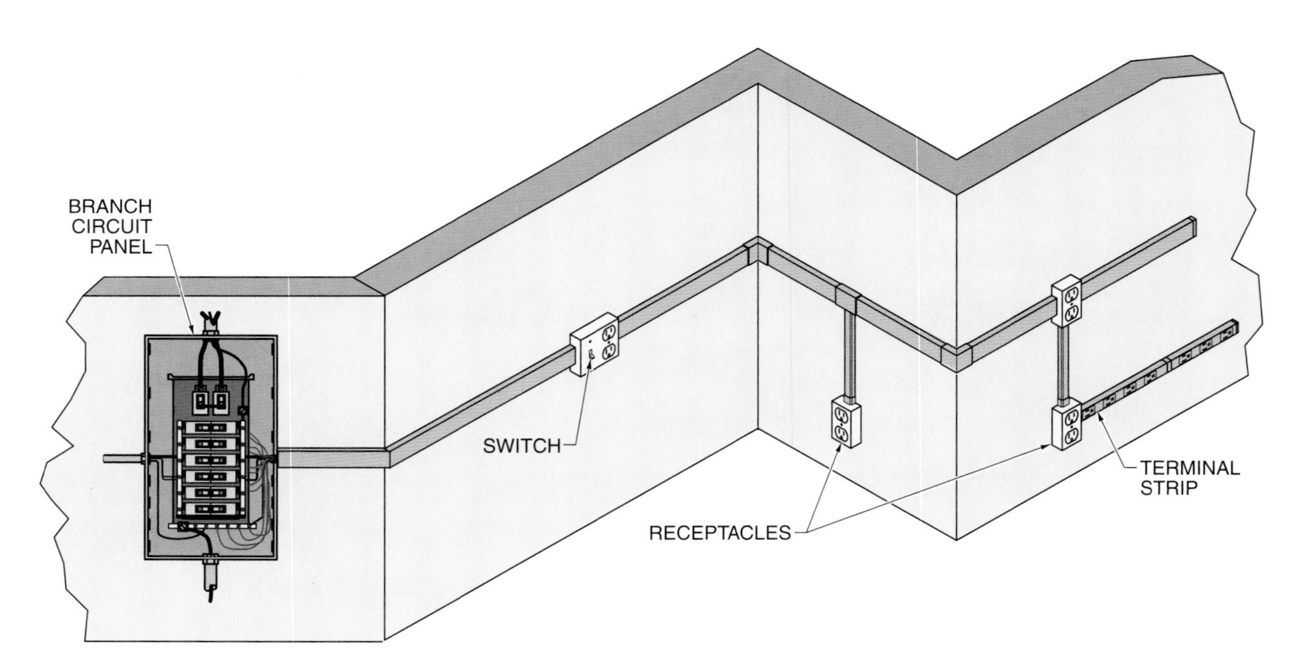

A service call requests that all the receptacles in the office (including the terminal strip) be checked to make sure they are powered and grounded, and that any GFCI receptacles be tested before office computers are plugged into them.

3. _____________ Which is the simplest test instrument that meets the service call requirements?

4. Why is this test instrument the best choice?

When the office equipment is connected, the receptacles should be retested to compare the rms voltage to the peak voltage. (The peak voltage should be 1.414 times the rms voltage.) The line impedance should also be tested to make sure the branch circuit conductors are not undersized and the circuit runs are not too long.

5. _____________ Which is the simplest test instrument that meets the service call requirements?

6. Why is this test instrument the best choice?

A service call states that some computers are automatically rebooting during high-usage times when printers and fax machines on the same circuit are operating. The voltage should be measured and recorded for one complete working day (10 hr) to determine the lowest voltage during that time.

7. _____________ Which is the simplest test instrument that meets the service call requirements?

8. Why is this test instrument the best choice?

A service call requests that a switch be tested to determine if it controls the receptacle next to the switch. It may control both sides, only one side, or not control the receptacle at all.

9. _____________ Which is the simplest test instrument that meets the service call requirements?

10. Why is this test instrument the best choice?

VOICE-DATA-VIDEO (VDV) TEST INSTRUMENTS

3 APPLICATIONS

Name: ______________________________ Date: ____________________

Application 3-1 — Cable Insulation Testing

Conductors are coated with insulation to prevent contact with other conductors, metal parts, and people. Unwanted contact results in electrical shorts that can cause communications failure, equipment damage, and hazardous situations. Insulation prevents electrical shorts because it is a high-resistance material that resists electron flow. Insulation resistance is reduced by physical damage from cuts, crimps, or stress, and deterioration from age, heat, UV light, or moisture. Since cables are usually inside hidden cable runs, damaged insulation is not always easily found. The resistance of communication cable insulation is measured by testing the conductors in a run alongside each other and ground using a megohmmeter. See Insulation Resistance Testing.

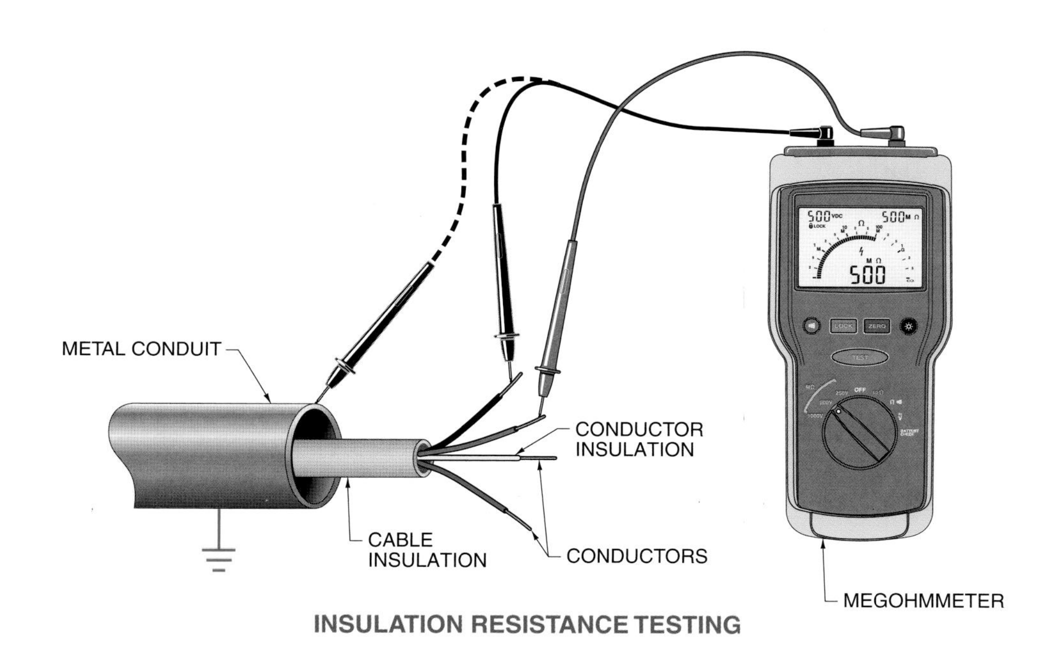

INSULATION RESISTANCE TESTING

The resistance of cable insulation should be tested when cables are installed, as part of a predictive maintenance schedule (every couple of years), and when there is a problem in the system that has not been solved. Older systems should be checked more frequently because aging accelerates the deterioration of insulation.

All insulation resistance tests are made with power OFF to cables and cables disconnected at both ends from any devices. For low voltage cables (rated for 1000 V or less), the megohmmeter applies 1000 VDC for 1 min to measure resistance. The resistance measured from conductor to conductor and conductor to ground must be a minimum of several megohms and will probably measure hundreds of megohms on good insulation.

The true measure of the condition of insulation is obtained by taking measurements over time, such as annually. Small changes in resistance value over time are acceptable. For example, a change from 300 MΩ to 275 MΩ is only about an 8.3% change and does not indicate a problem. However, a change from 300 MΩ to 200 MΩ is a 33% change and indicates a problem. See Insulation Resistance Changes.

INSULATION RESISTANCE CHANGES*

% Difference from Baseline \ Baseline Measurement	20	40	60	80	100	250	500	750	1000	1500	2000	Differenct Interpretation
10	18	36	54	72	90	225	450	675	900	1350	1800	Acceptable operating range for most equipment
20	16	32	48	64	80	200	400	600	800	1200	1600	
30	14	28	42	56	70	175	350	525	700	1050	1400	Requires additional testing and inspection to verify proper operation; check for environmental contamination
40	12	24	36	48	60	150	300	450	600	900	1200	
50	10	20	30	40	50	125	250	375	500	750	1000	
60	8	16	24	32	40	100	200	300	400	600	800	Indicates a potential problem; perform tests on system until problem is located and corrected
70	6	12	18	24	30	75	150	225	300	450	600	
80	4	8	12	16	20	50	100	150	200	300	400	
90	2	4	6	8	10	25	50	75	100	150	200	
100	0	0	0	0	0	0	0	0	0	0	0	

* in MΩ

Note: Baseline measurements are used as comparison readings for field measurements. A baseline measurement is typically the manufacturer's factory specifications or the last recorded reading from a preventive maintenance procedure.

1. ______________ What is the percentage change between the first and second reading?

2. ______________ Is this percentage change acceptable?

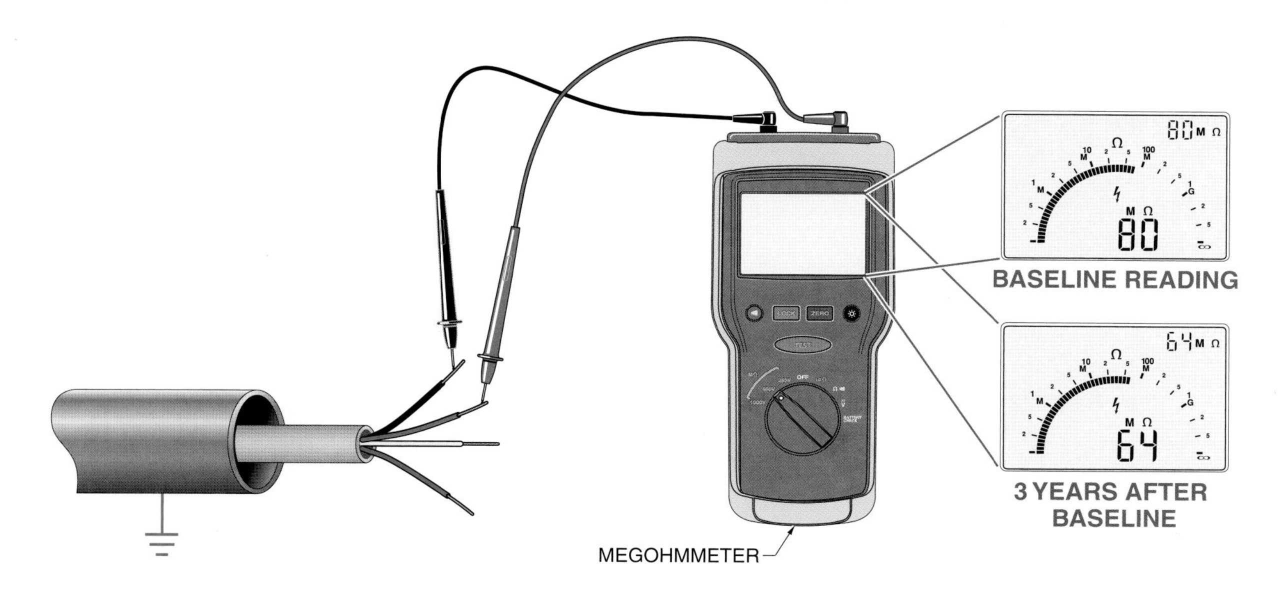

3. ______________ What is the percentage change between the first and second reading?

4. ______________ Is this percentage change acceptable?

5. ______________ What is the percentage change between the second and third reading?

6. ______________ Is this percentage change acceptable?

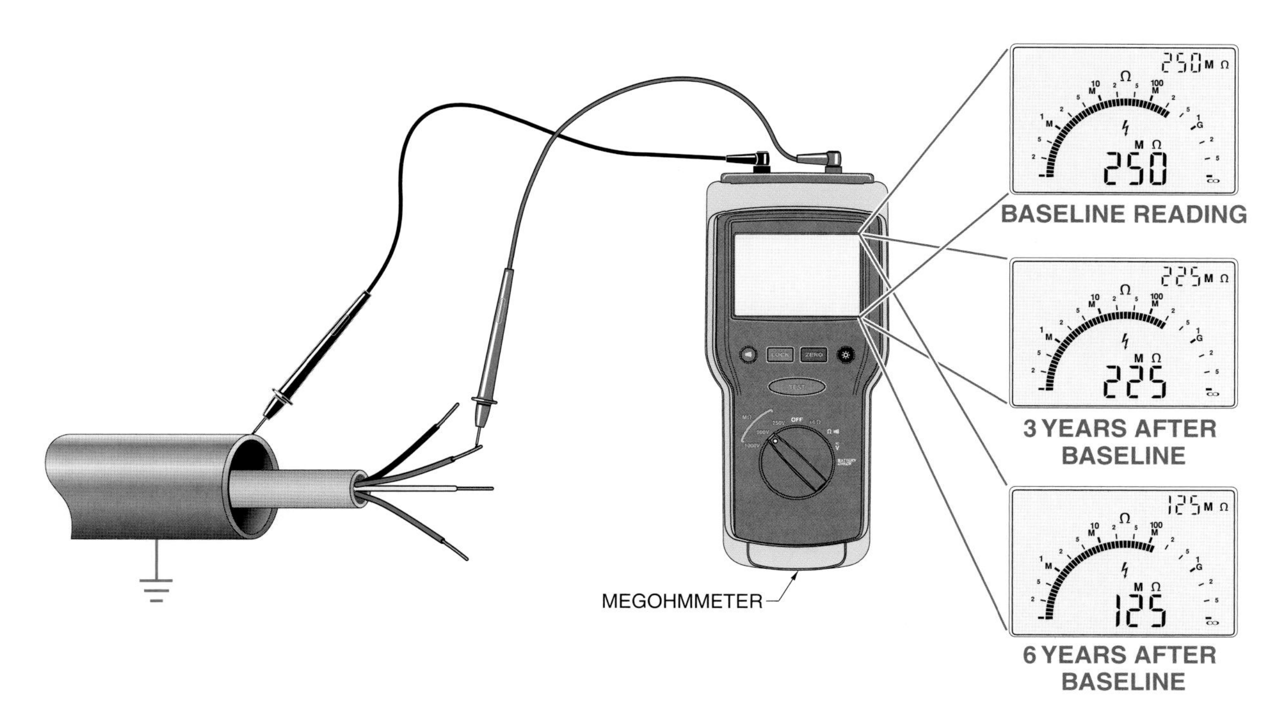

Application 3-2—Wire Mapping

When experiencing a communications problem, one of the first tests to perform is a wire map test, which checks the order and arrangement of wires at two connection points. Wire map testers determine if a cable contains a fault or wiring error, or is wired correctly. See Correct Wiring. A fault is a short or open and is usually caused by either physical damage to the wire or incomplete terminations at the connectors. Wiring errors include three types of incorrect wire arrangements. A reversed pair has the two wires of a pair reversed at one end. A set of crossed pairs has the two pairs reversed at one end. A set of split pairs forms two pairs at each end by mixing one wire from each of two twisted pairs. See Wiring Errors.

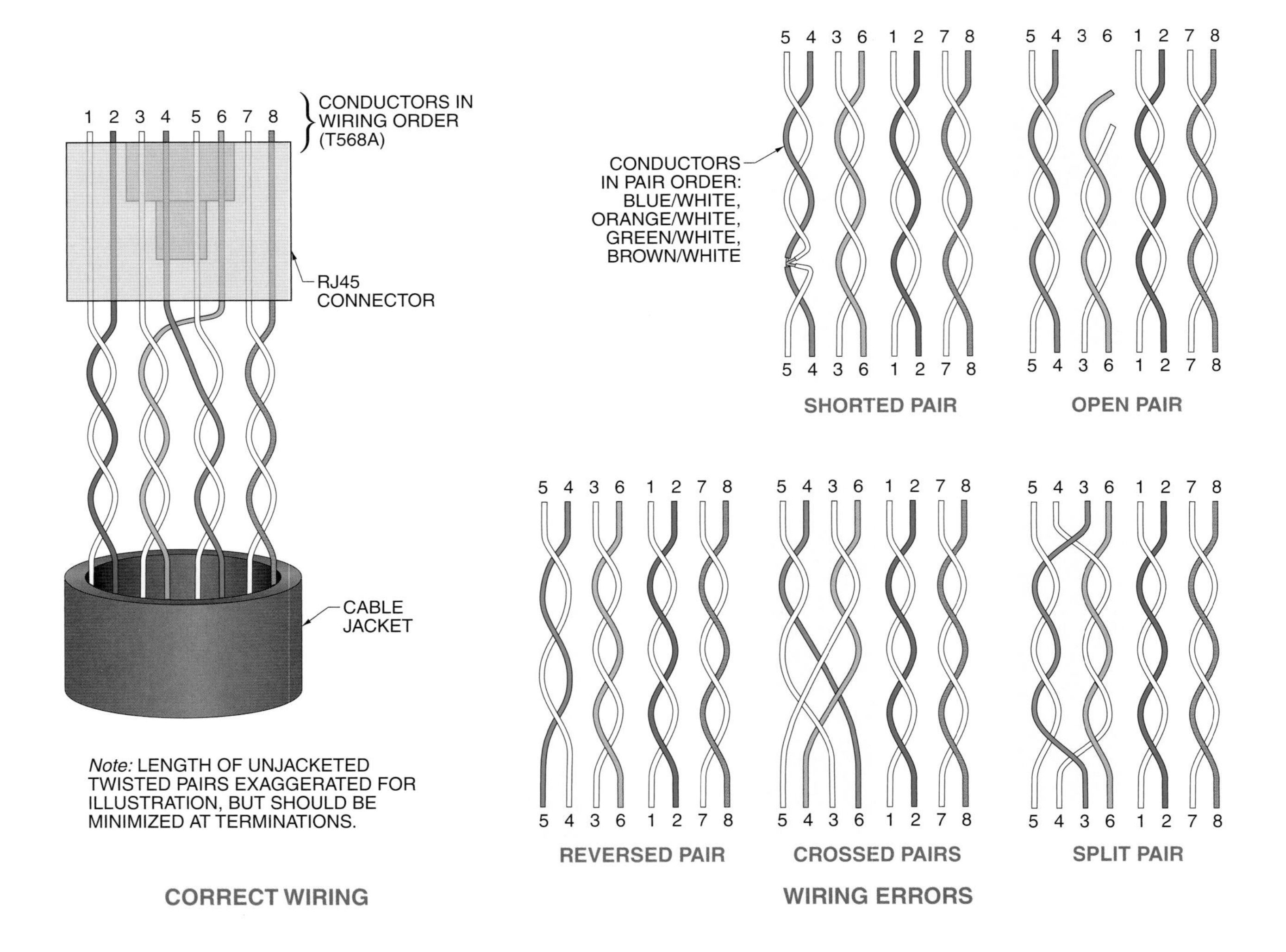

Wire map tests can be performed by individual testers or by more advanced cable analyzers that include wire mapping as one of the several tests they perform. The tester may require a separate unit to be placed at the opposite end of the cable from the tester to help map the wires at both points. When interpreting the results on the meter display, be sure to note the wire numbers. Some meters may display the pairs in different orders.

1. ______________ What type of error does Wire Map Test 1 show?

2. ______________ What type of error does Wire Map Test 2 show?

3. How could the error in Wire Map Test 2 be corrected?

4. ______________ What type of error does Wire Map Test 3 show?

5. How could the error in Wire Map Test 3 be corrected?

6. ______________ What type of error does Wire Map Test 4 show?

7. ______________ What type of error does Wire Map Test 5 show?

8. How could the error in Wire Map Test 5 be corrected?

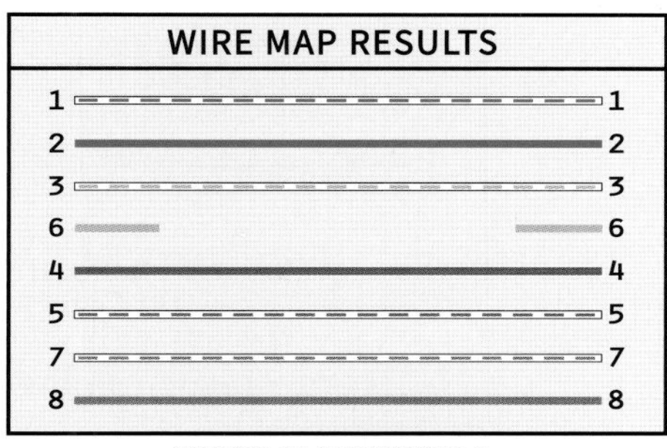

WIRE MAP TEST 1

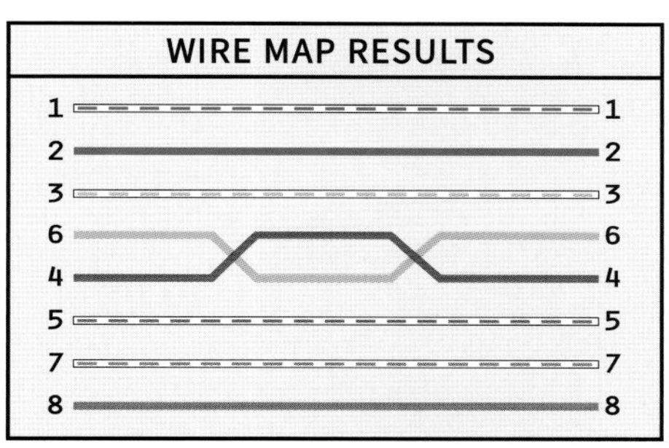

WIRE MAP TEST 2

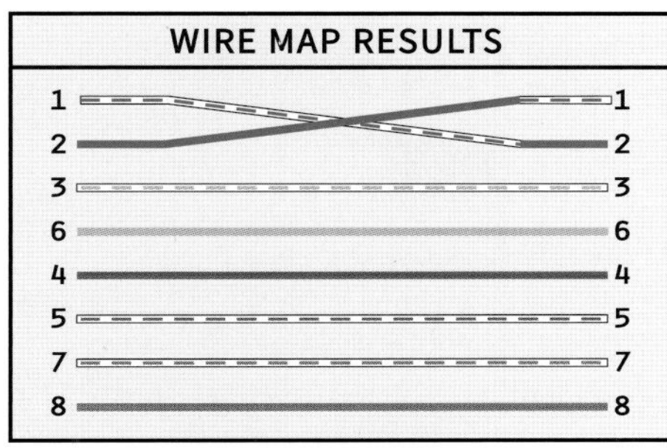

WIRE MAP TEST 3

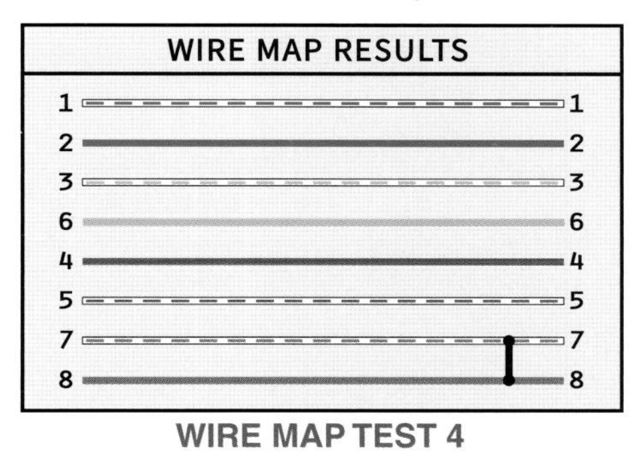

WIRE MAP TEST 4

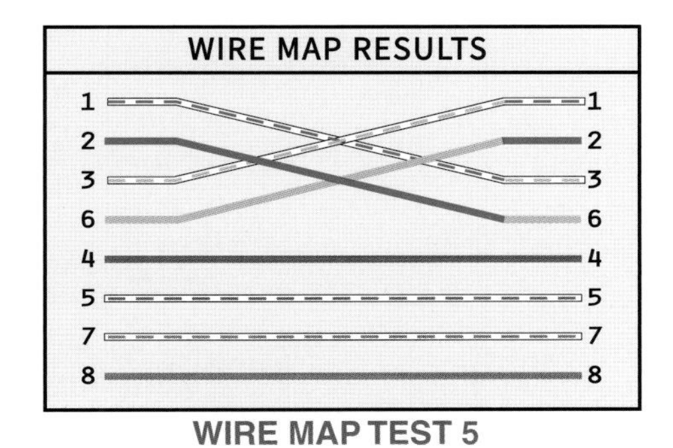

WIRE MAP TEST 5

Application 3-3—Time Domain Reflectometer Waveforms

A time domain reflectometer (TDR) test measures conductor length and locates faults (shorts or opens) and poor connections on conductors. The meter sends out a pulse and measures the time it takes for a reflection of the pulse to return. Reflections are caused by the end of the conductor, faults, and poor connections. By comparing the reflection pulse's travel time, shape, and polarity to the original transmitted pulse, the meter can determine the distance to and cause of the reflection.

A properly terminated conductor with no faults has uniform impedance and absorbs the pulse without reflecting it. A fault in a conductor produces an abrupt change in the conductor's impedance that causes reflections of the pulse to return to the TDR. The amplitude of the reflection is proportional to the change of impedance. An open in the conductor represents an abrupt increase in the conductor's impedance, which sends a reflection of the pulse back to the meter with the same polarity. An unterminated conductor will also reflect the pulse from the end of the conductor, which is like an open, even if the conductor has no faults. See Open Cable. A short in the conductor represents an abrupt decrease in the conductor's impedance, which sends a reflection of the pulse back to the meter with the opposite polarity. See Shorted Cable.

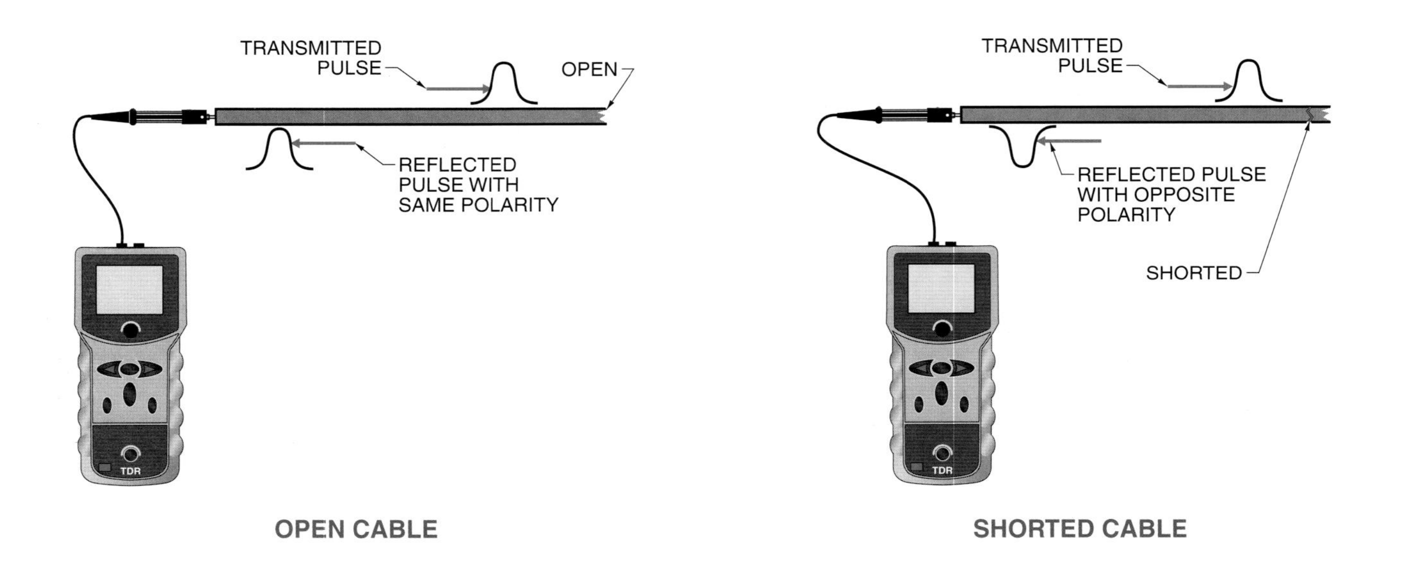

Most TDRs interpret the signals internally and only display the results, which usually include the type of fault (if any) and the distance estimate to the fault. However, some TDRs will display the actual waveform of the pulses. The initial pulse appears as a positive waveform on the display of impedance change versus time. Any reflections appear as subsequent waves of variable amplitudes and either polarity. Partial and complete opens appear as reflections of the same polarity. See TDR Waveform of Opens. Partial and complete shorts appear as reflections of the opposite polarity See TDR Waveform of Shorts.

There are many other types of faults or wiring components that can be detected by analyzing a TDR waveform, including water infiltration, splices, taps, and radio frequency (RF) interference. Bad connections are relatively common faults and appear as a high impedance reflection followed closely by a low impedance reflection, forming an "S" shape. See TDR Waveform of Bad Connection.

The physical location of a fault can be calculated from its reflection's travel time (using the time scale). Many TDRs will calculate distances internally and show a result for the location of the display curser, or just substitute the time scale for a distance scale (typically in meters). The location can also be approximated when the waveform shows the end of the cable. For example, if a waveform shows a partial short about halfway between the initial pulse and a complete open (the end of an unterminated cable), then the partial short is located about halfway down the conductor.

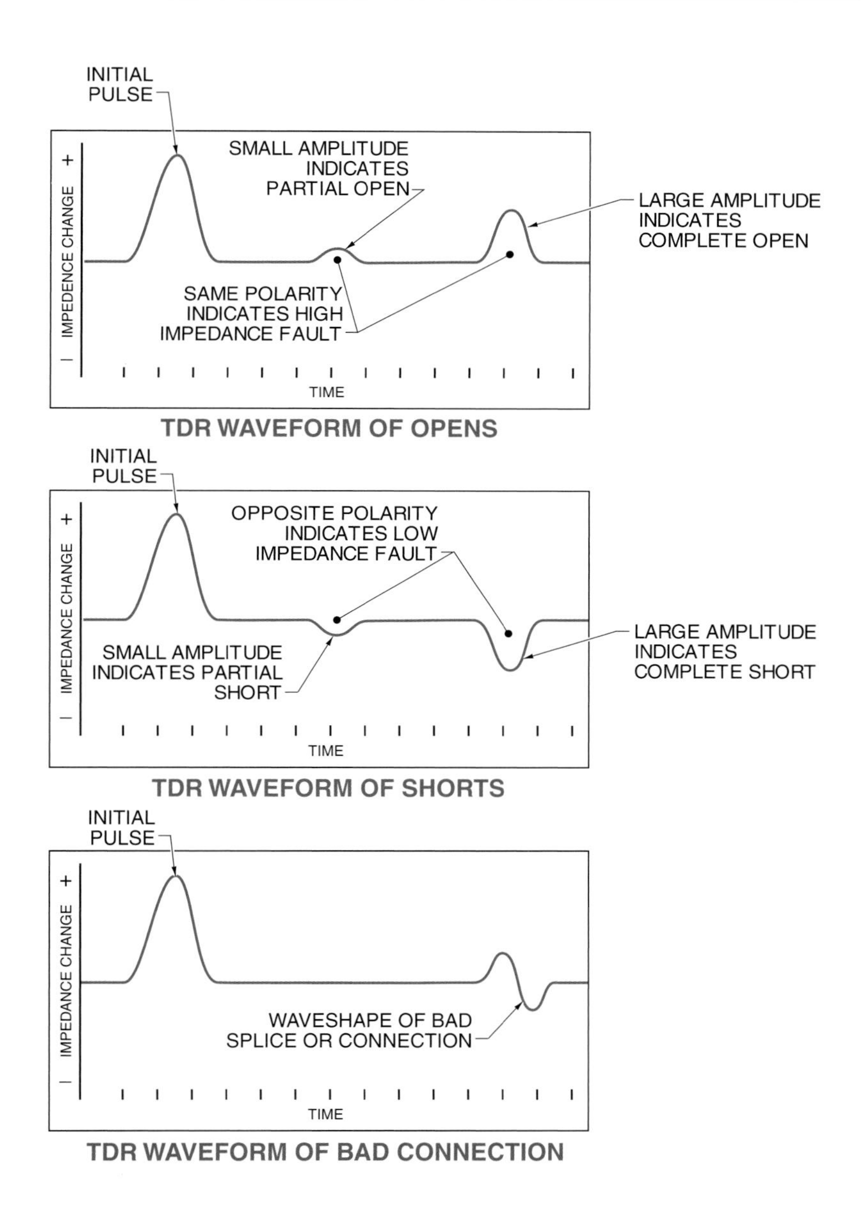

A waveform TDR tests four conductors in a communications cable installation and displays the results as waveforms.

1. ______________ What type of fault (if any) is present on Conductor 1?

2. ______________ Is Conductor 1 properly terminated at the far end?

3. ______________ What type of fault (if any) is present on Conductor 2?

4. ______________ Is Conductor 2 properly terminated at the far end?

5. ______________ What type of fault (if any) is present on Conductor 3?

6. ______________ Is Conductor 3 properly terminated at the far end?

7. ______________ What type of fault (if any) is present on Conductor 4?

8. ______________ Is Conductor 4 properly terminated at the far end?

9. ______________ Approximately how far down Conductor 4 is the fault?

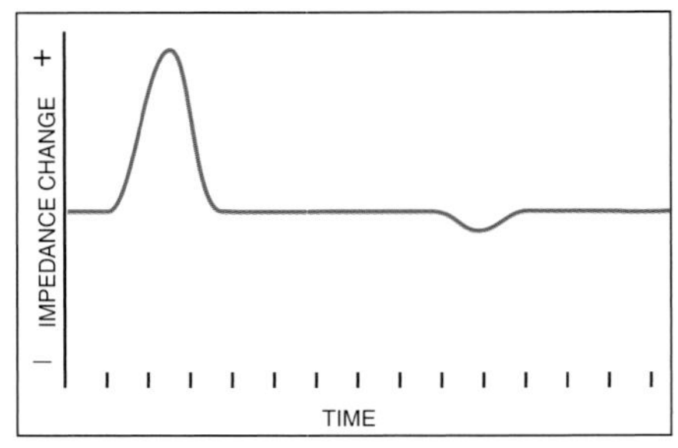

TDR WAVEFORM OF CONDUCTOR 1

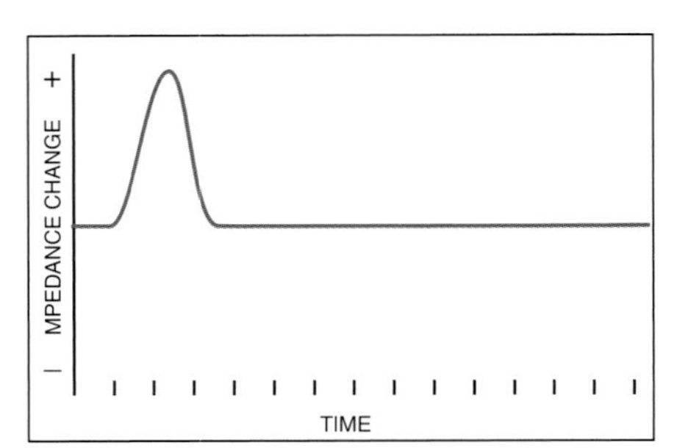

TDR WAVEFORM OF CONDUCTOR 2

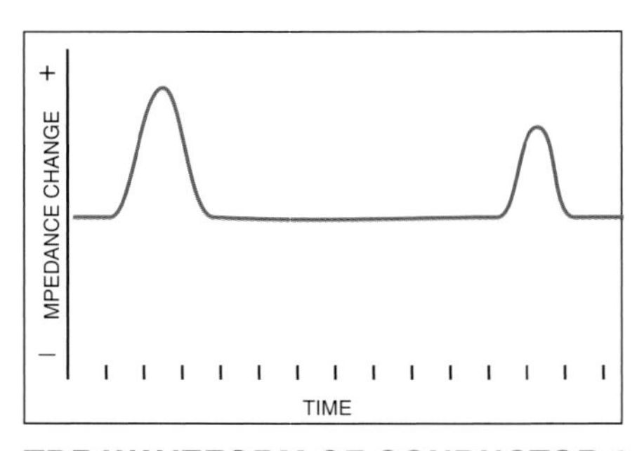

TDR WAVEFORM OF CONDUCTOR 3

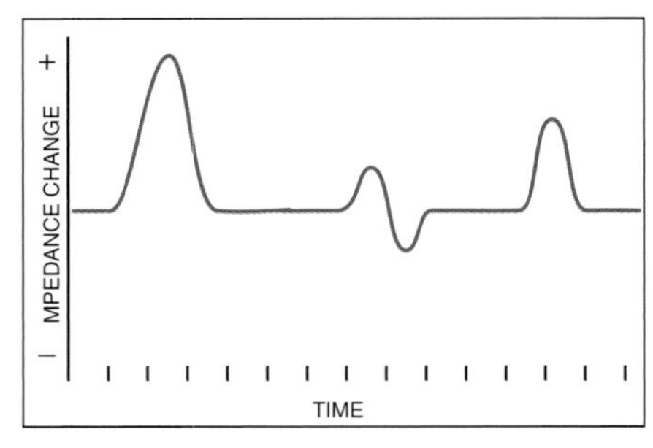

TDR WAVEFORM OF CONDUCTOR 4

Application 3-4 — Attenuation

A transmitted signal becomes weaker as it travels down a conductor. If the signal becomes too weak for the system hardware to interpret the signal, the transmitted data is lost. In copper cabling, signal loss occurs because of cable resistance, impedance mismatch, insulation breakdown, crosstalk, and other problems. In fiber optic cable, signal loss is caused by problems including dirty connections, scratches on fiber ends, or misaligned connections. An attenuation test is a test that measures the attenuation (power loss) of cable. Two meters are used to measure a conductor's attenuation. The remote unit is connected to one end of the conductor to transmit a fixed signal and the meter is connected to the other end of the conductor to measure the strength of the transmitted signal.

Attenuation is measured in decibels (dB). For fiber optic cable, the decibel is a logarithmic ratio of output power to input power. For every 3 dB of loss, the output power is approximately one half the input power, and for every 10 dB of loss, the output power is approximately one tenth of the input power.

Decibels are useful because they can be easily added together, as opposed to ratios, which must be multiplied. For example, one length of fiber optic cable attenuates a signal to one half of the input power (–3 dB). Another length attenuates a signal to one tenth of the input power (–10 dB). If a signal travels the length of both cables, only 5% of the input power is left at the end ($\frac{1}{2} \times \frac{1}{10} = \frac{1}{20} = 5\%$). If the two cables are connected together, the total loss from the cables is 95%, which instead can be easily expressed as –13 dB by adding the decibel values together (–3 dB + [–10 dB] = –13 dB). See Compounding Attenuation.

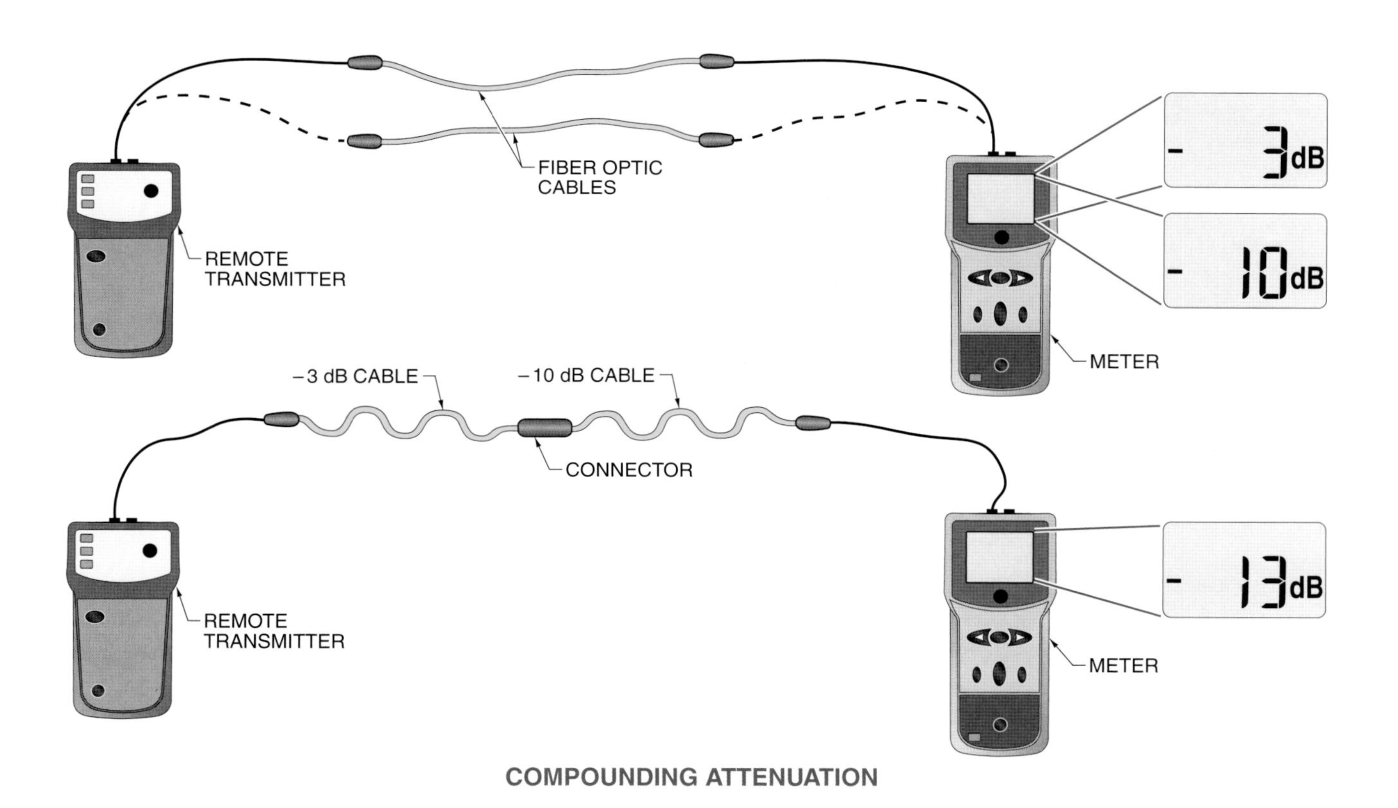

COMPOUNDING ATTENUATION

Incoming data lines can be easily connected to various wall jacks throughout a building by making connections on a patch panel with very short cables. In this scenario, assume the patch cables contribute no additional signal losses.

1. _______________ What percentage of power is lost from Wall Jack 1 to the patch panel?
2. _______________ What percentage of power is lost from Wall Jack 2 to the patch panel?
3. _______________ What percentage of power is lost from Wall Jack 3 to the patch panel?
4. _______________ If Wall Jack 1 is patched to the building network, what is the total loss (in dB)?
5. _______________ If Wall Jack 1 is patched to the building network, what is the approximate total loss (in %)?
6. _______________ If Wall Jack 2 is patched to the building network, what is the total loss (in dB)?
7. _______________ If Wall Jack 2 is patched to the building network, what is the approximate total loss (in %)?
8. _______________ If Wall Jack 3 is patched to the building network, what is the total loss (in dB)?
9. _______________ If Wall Jack 3 is patched to the building network, what is the approximate total loss (in %)?

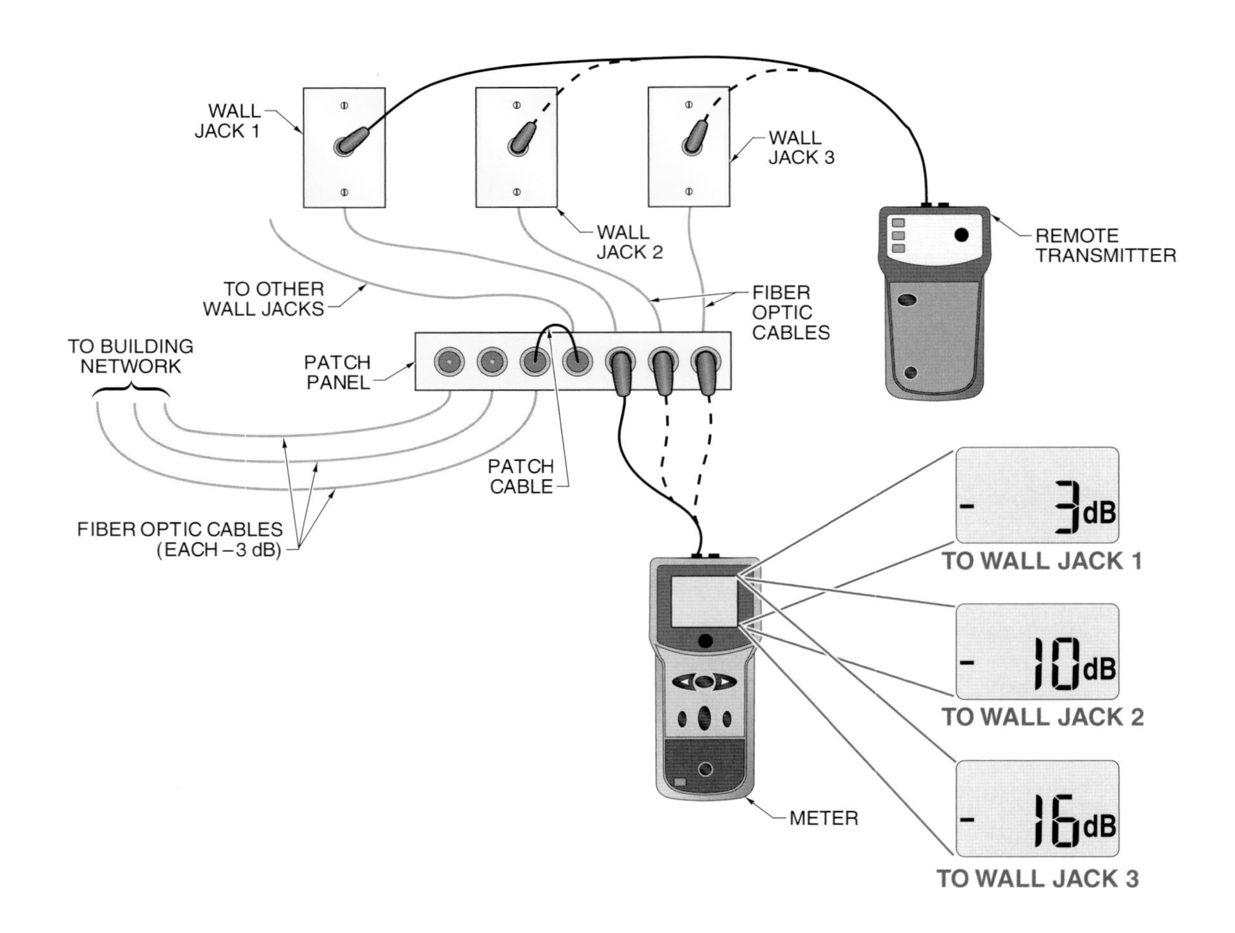

Application 3-5 — Fiber Optic Video Microscope Views

Fiber optic video microscopes are used to view the endfaces (terminations) of fiber optic cables. To make a proper fiber optic connection, the fiber must be cut cleanly and the endface polished smooth. A good termination will appear as a clear image in a fiber optic video microscope, with a light gray core surrounded by darker cladding. See Fiber Video Optic Microscope.

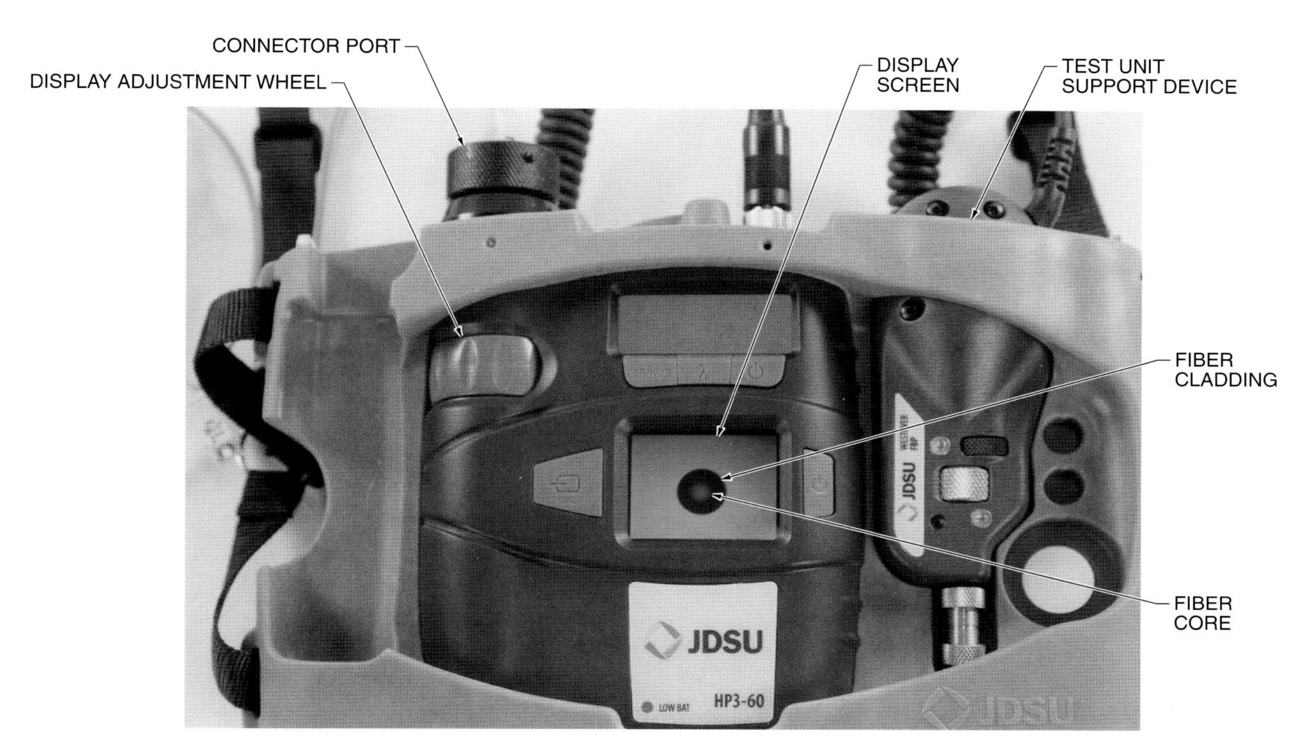

FIBER OPTIC VIDEO MICROSCOPE

Images with dirt, cracks, scratches, or fuzziness (roughness) indicate problems with the termination. Some problems, such as light scratches or roughness on the end surface of the fiber, can be resolved by cleaning or re-polishing. Others, such as cracks and shattered ends, require the fiber to be cut and terminated again.

Fiber optic video microscopes can only be used to view the end of a fiber optic cable. Problems such as chipped, broken, or cracked fibers can also occur anywhere along the fiber cable run, but these require an optical time domain reflectometer (OTDR) to diagnose.

From the following list, choose the problem that is represented in each fiber optic video microscope image. Each problem is represented only once.

A. The fiber optic cable is broken or completely blocked.
B. The fiber optic endface is dirty and needs cleaning or repolishing.
C. The fiber optic endface is scratched and needs repolishing or re-termination.
D. The fiber optic cable is cracked across the core and should be replaced.
E. The fiber optic cable is shattered within the cladding and should be re-terminated.
F. The fiber optic endface has a rough surface or has excess epoxy and needs repolishing.

1. ______________ Which problem is represented in Microscope Image 1?

2. ______________ Which problem is represented in Microscope Image 2?

3. ______________ Which problem is represented in Microscope Image 3?

4. ______________ Which problem is represented in Microscope Image 4?

5. ______________ Which problem is represented in Microscope Image 5?

6. ______________ Which problem is represented in Microscope Image 6?

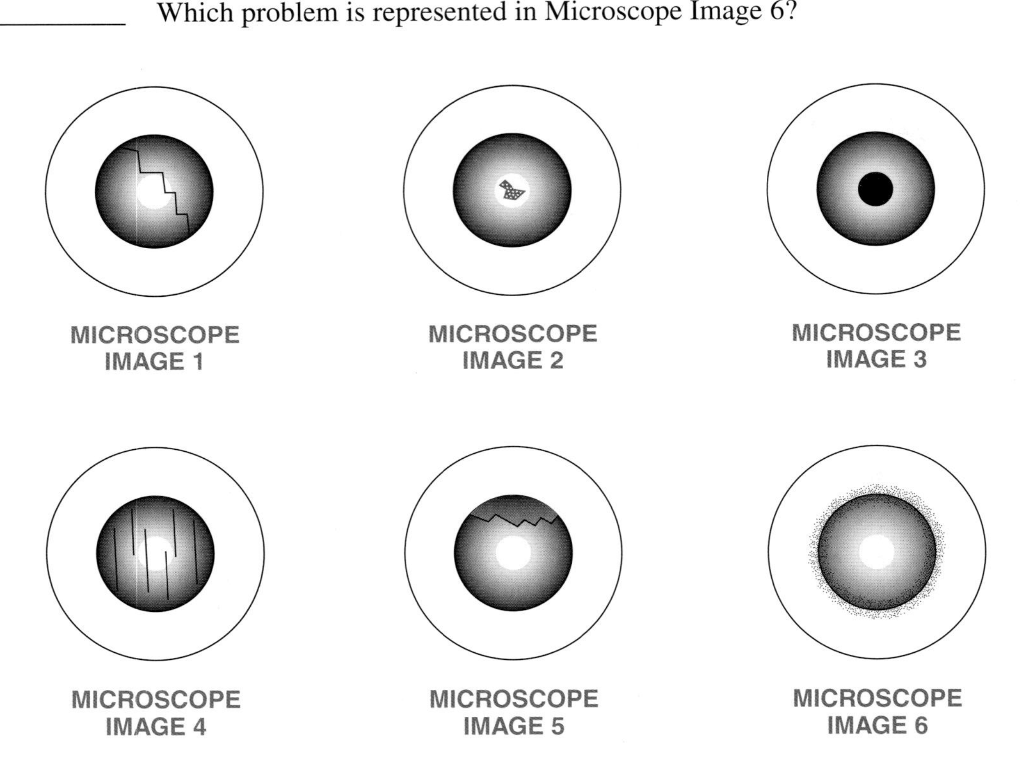

POWER QUALITY TEST INSTRUMENTS

Name: ______________________________ Date: ____________________

Application 4-1 — Transformer Output

Three single-phase transformers can be connected in a wye or delta configuration to develop three-phase voltage. A wye configuration is a three-phase transformer connection that has one end of each transformer coil connected together. The remaining end of each coil is connected to the three incoming power lines (primary side) or used to supply power to the loads (secondary side). A delta configuration is a transformer connection that has each of the three transformer coils connected end-to-end to form a closed loop. Each connection point is connected to an incoming power line (primary side) or used to supply power to the loads (secondary side). The voltage output and type available for the loads is determined by whether the transformer is connected in a wye or delta configuration. See Transformer Configurations.

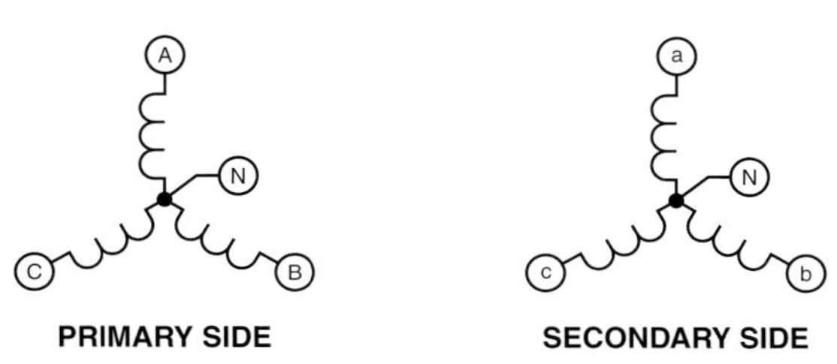

Wye Configuration

Delta Configuration

TRANSFORMER CONFIGURATIONS

The primary side of a three-phase transformer bank may be either a wye or delta configuration, depending on what the utility company has determined to be the best configuration for transmitting and balancing their system. The secondary side of the transformer bank may be either a wye or delta configuration, depending on what is best for the loads. See Wye-to-Wye Step-Down Transformer Bank and Delta-to-Delta Step-Down Transformer Bank.

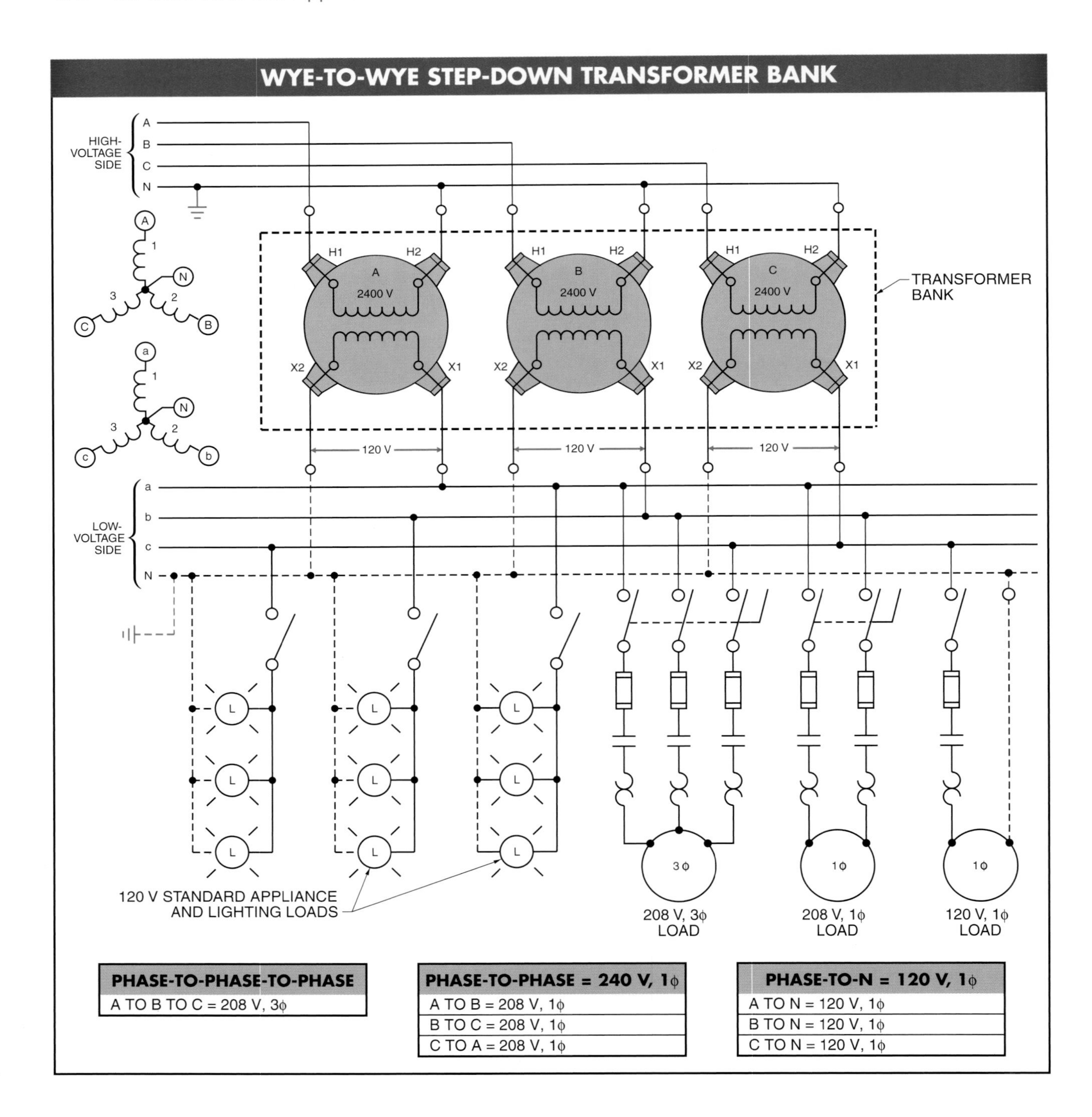

A wye-connected secondary is commonly used in schools, stores, and offices and provides three different types of service:

- 208 V, three-phase service for 3ϕ motor loads and three-phase heaters
- 208 V, single-phase service for 1ϕ motor loads, single-phase heaters, and single-phase lamps
- 120 V, single-phase service for single-phase lighting and small appliance loads

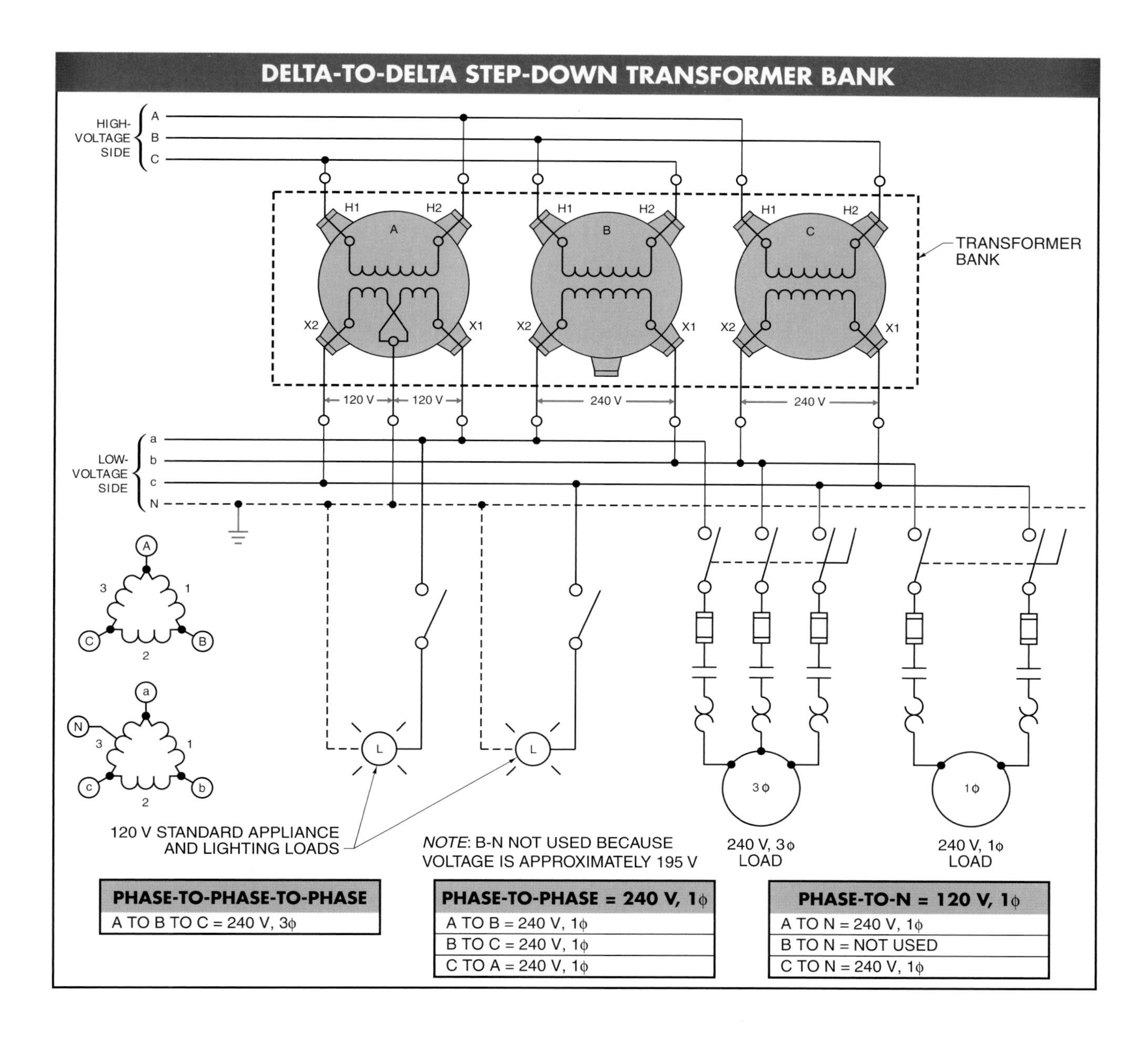

A delta-connected secondary provides three different types of service:

- 240 V, three-phase service for 3ϕ motor loads and three-phase heaters
- 240 V, single-phase service for 1ϕ motor loads, single-phase heaters, and single-phase lamps
- 120 V, single-phase service for single-phase lighting and small appliance loads

In a basic transformer, if the primary voltage changes (up or down), then the secondary voltage will also change (up or down). However, most power distribution transformers are adjustable because they must maintain the correct voltage output to loads on the secondary side. Taps on the primary side change the ratio of the transformer coils so that the secondary output remains constant even if the primary voltage changes. For example, a common transformer type changes 480 V (primary side) to 120 V (secondary side). The ratio of the transformer is 4:1 (480 V:120 V = 4:1). If the 480 V service sags down to 456 V and the ratio remains constant, then the secondary falls to 114 V (456 V ÷ 4 = 114 V). Instead, the primary side tap is changed so that the ratio becomes 3.8:1, which produces 120 V (456 V ÷ 3.8 = 120 V). Tap adjustments may be manual or automatic. Tap positions are often labeled with the primary side voltage that produces the desired secondary output at that position. See Transformer Taps.

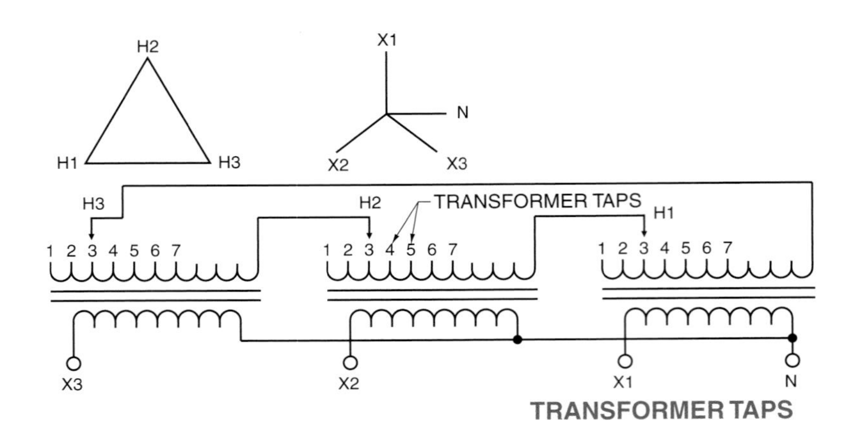

TAP CONNECTIONS		
Primary Voltage	**Tap**	**Secondary Voltage**
503	1	
493	2	208 V (LINE-TO-LINE)
480	3	
466	4	
456	5	120 V (LINE-TO-NEUTRAL)
443	6	
433	7	

TRANSFORMER TAPS

Answer the following questions based on the transformer nameplate data. Assume the system loads have a perfect power factor.

1. ____________ Is the primary side of this transformer designed for a wye or delta power supply connection?

2. ____________ Is the secondary side of this transformer designed for a wye or delta load connection?

3. ____________ If this transformer is used to supply 120 V loads, what is the normal primary to secondary voltage ratio?

4. ____________ How much power (in VA) can this transformer bank deliver without being overloaded?

5. ____________ If the loads are balanced (spread out evenly over the three power lines/individual transformers), how much power (in VA) can each of the three individual transformers deliver without being overloaded?

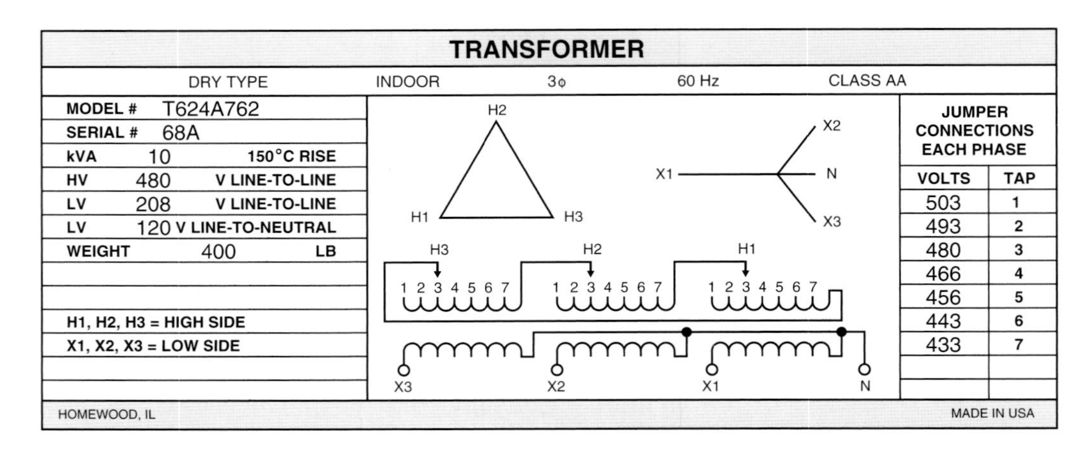

Answer the following questions based on the transformer nameplate data and the circuit.

6. Set (draw in) the function switch position on Multimeter 1 for taking voltage measurements at the transformer.

7. Connect Multimeter 1 to measure the voltage for any load connected to L2 and L3.

8. Set (draw in) the function switch position on Multimeter 2 for taking voltage measurements at the transformer.

9. Connect Multimeter 2 to measure the voltage for any load connected to L1 and neutral.

10. ______________ At the current tap position, what is the primary to secondary voltage ratio of the transformer if the secondary voltage is 120 V?

11. ______________ If Multimeter 2 displays 124 V, what is the primary side voltage?

12. ______________ If Multimeter 2 displays 124 V, which tap position should the primary tap be moved to?

13. ______________ If Multimeter 2 displays 114 V instead, what is the primary side voltage?

14. ______________ If Multimeter 2 displays 114 V, which tap position should the primary tap be moved to?

15. ______________ If the loads connected to the system have a perfect power factor (100% or 1), what is the maximum current that Multimeter 3 should read?

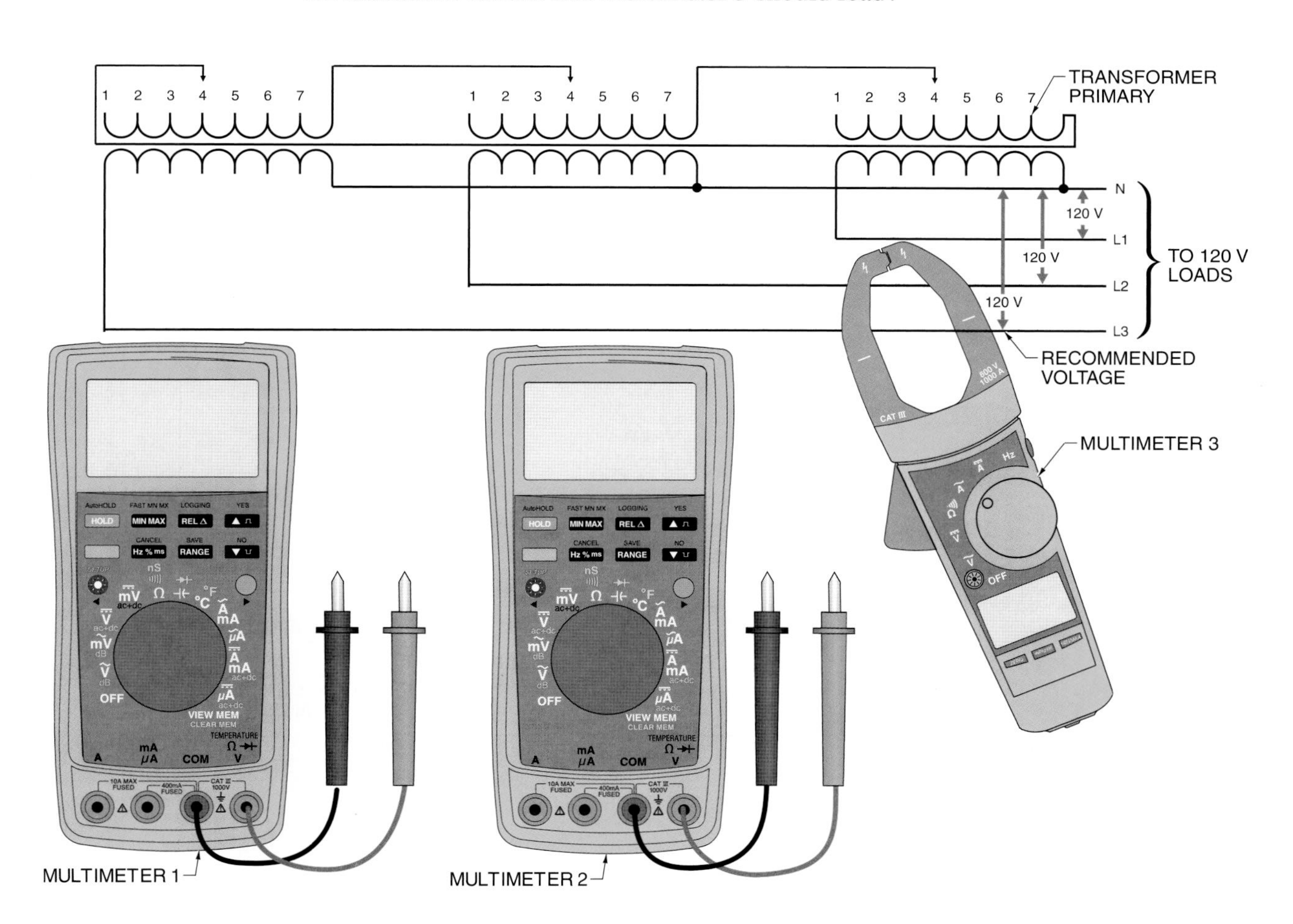

Application 4-2—Building Power Distribution

Electric power companies generate and distribute high-voltage, three-phase power. Residential, commercial, and most industrial customers use low-voltage, three-phase (208 V to 480 V) and single-phase (120 V to 277 V) power. Some industrial customers use three-phase voltages higher than 480 V.

Transformers connected in wye or delta configurations step down the high voltage from the transmission lines to the voltage level required by the consumer. Three transformers (except in an open-delta system, which uses only two) are interconnected to deliver single-phase and three-phase voltages to commercial and industrial customers. One transformer is normally used for delivering single-phase voltage to residential customers.

The type, size, and voltage level of a power distribution system are determined by the customer's power needs and the utility company's distribution system. For example, power produced for a large commercial application such as a college, business, or hotel is delivered to the building as high voltage from the utility company. An in-plant substation reduces the high voltage from the utility company for use in building circuits.

A substation is an assembly of equipment installed for switching, changing, or regulating the voltage of electricity. A substation can be a large outdoor utility distribution center or an in-plant distribution center. An in-plant substation contains transformers, switchboards, switchgear, transfer switches, and secondary switches that distribute low voltage levels (208 V to 480 V) to feeder panels, other secondary transformers, busway systems, and branch-circuit panels. See Building Power Distribution.

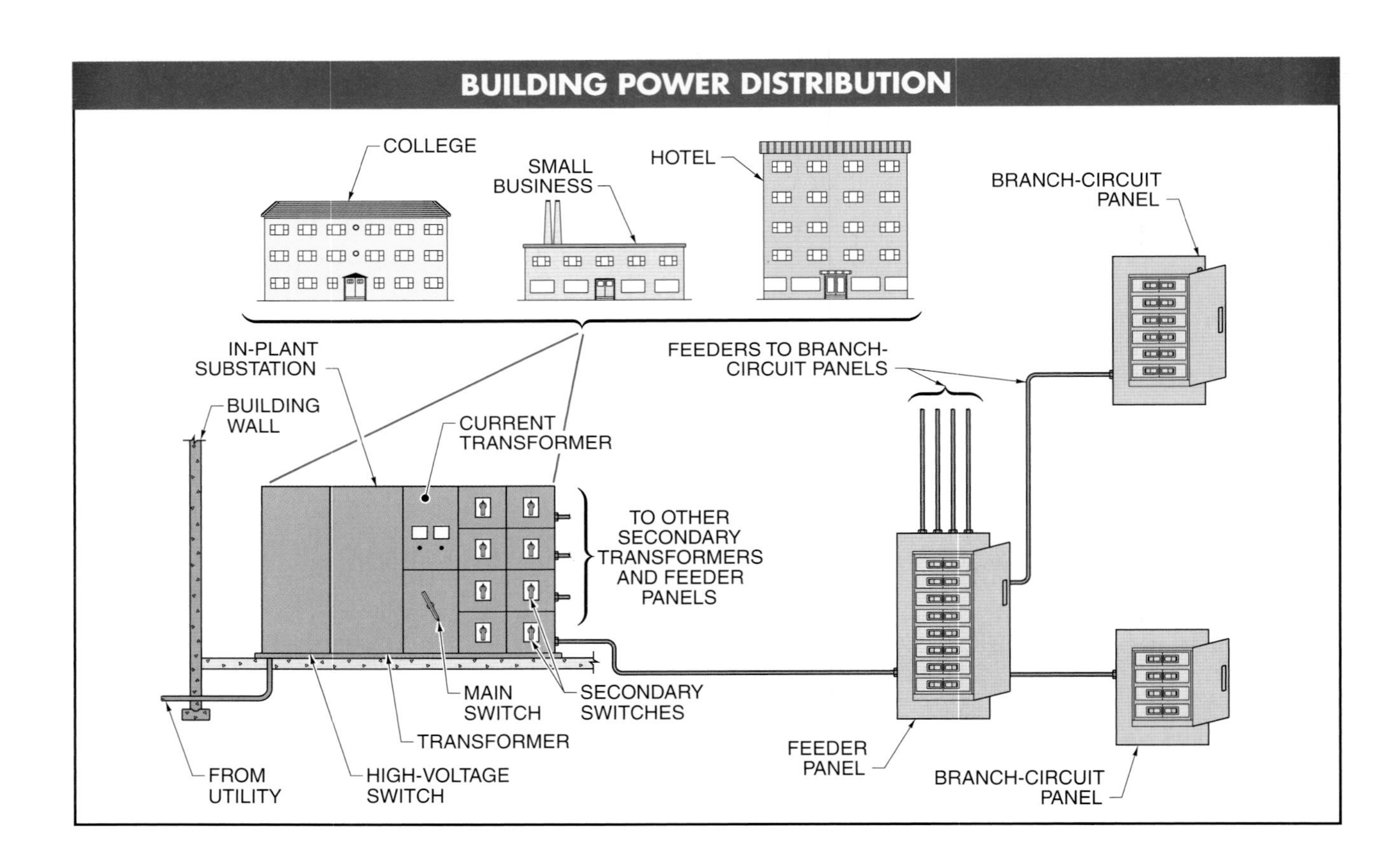

Power quality measurement and troubleshooting activities vary with the type of distribution system—whether it is a residential, commercial, or industrial application. When troubleshooting power quality problems, measurements such as voltage, current, power, or harmonics are taken throughout the building distribution system. An understanding of the different types of power distribution systems, power quality problems, and test instruments is required when troubleshooting a problem. A Power Quality Troubleshooting Checklist can be used to help diagnose problems or identify potential problem areas. See Appendix.

Conductor Color Coding

The type of system (wye or delta) and voltage levels must be known when taking electrical measurements or troubleshooting an electrical circuit. Standard conductor color codes help determine the type of electrical circuit. Conductor color coding is also helpful when troubleshooting, so it is even used sometimes in applications that do not require every conductor to be color-coded. Always use standard colors if possible.

Conductors are covered with an insulating material that is available in different colors. The advantage of using different colors on conductors is that the color shows the function of each conductor. Some colors have a standard meaning. For example, the color green always indicates a conductor used for grounding. Other colors may have more than one meaning depending on the circuit. For example, a red conductor may be used to indicate a hot wire in a 230 V circuit or switched wire in a 115 V circuit.

Green, or green with a yellow stripe, is the standard color for a grounding conductor. Green is used to indicate a grounding conductor regardless of the voltage level (such as 115 V, 230 V, or 460 V) or circuit (single-phase or three-phase). A grounding conductor is a conductor that normally carries current only during a fault (short circuit). Grounding conductors may also be bare in some applications.

The colors white or natural gray are used for the neutral conductor. A neutral conductor is a current-carrying conductor that is intentionally grounded. Neutral conductors carry current from a load (such as a lamp, heating element, or motor) back to the power source. Neutral conductors are connected directly to loads and never connected through fuses, circuit breakers, or switches.

An ungrounded (hot) conductor is a current-carrying conductor that is connected to loads through fuses, circuit breakers, and switches. Ungrounded conductors can be any color other than white, natural gray, green, or green with a yellow stripe. Black, red, blue, orange, and yellow are usually used for ungrounded conductors, with black being the most common. The exact color used to indicate different hot conductors may vary. For example, the colors used to identify A (L1), B (L2), and C (L3) in a three-phase system depend on the configuration of the system. The exception to this is listed in NEC® Article 110.15, which states that in a 4-wire delta-connected secondary system, the higher voltage phase must be colored orange (or clearly marked) because it is too high for low-voltage (115 V) single-phase power and too low for high-voltage (230 V) single-phase power. See Circuit Conductor Color Coding.

Receptacles (outlets) have different configurations to prevent the connection of equipment to the wrong source of power (voltage type and level). The receptacle must be wired to the correct transformer/service type in order for the receptacle to deliver the proper voltage.

CIRCUIT CONDUCTOR COLOR CODING

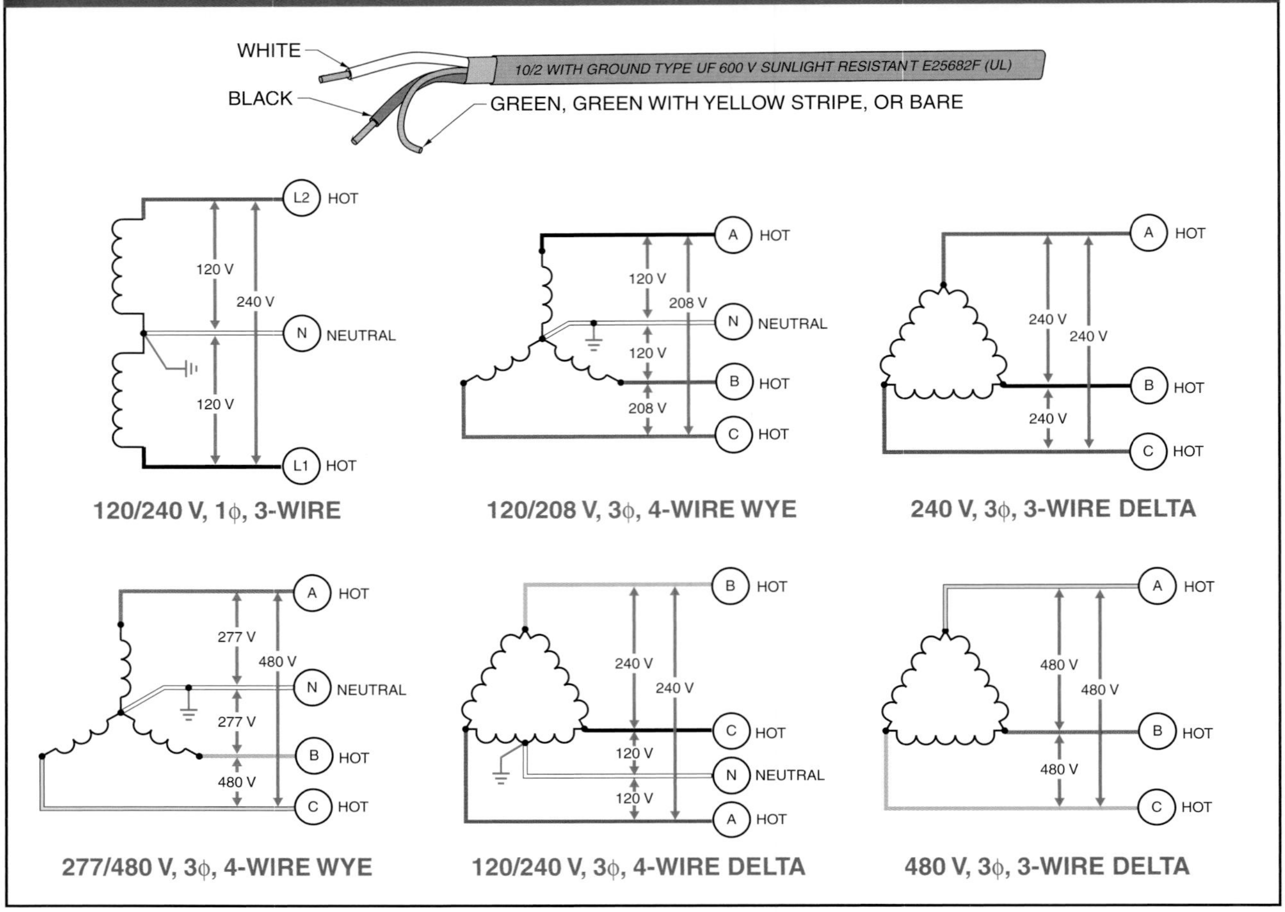

WHITE
BLACK
10/2 WITH GROUND TYPE UF 600 V SUNLIGHT RESISTANT E25682F (UL)
GREEN, GREEN WITH YELLOW STRIPE, OR BARE

L2 HOT
120 V
240 V
N NEUTRAL
120 V
L1 HOT

120/240 V, 1ϕ, 3-WIRE

A HOT
120 V
208 V
N NEUTRAL
120 V
B HOT
208 V
C HOT

120/208 V, 3ϕ, 4-WIRE WYE

A HOT
240 V
240 V
B HOT
240 V
C HOT

240 V, 3ϕ, 3-WIRE DELTA

A HOT
277 V
480 V
N NEUTRAL
277 V
B HOT
480 V
C HOT

277/480 V, 3ϕ, 4-WIRE WYE

B HOT
240 V
240 V
C HOT
120 V
N NEUTRAL
120 V
A HOT

120/240 V, 3ϕ, 4-WIRE DELTA

A HOT
480 V
480 V
B HOT
480 V
C HOT

480 V, 3ϕ, 3-WIRE DELTA

1. What are the voltage, phase type, and number of wires for this service?

2. ______________ What are the two possible colors of Conductor 1?

3. ______________ What is the color of Conductor 2?

4. ______________ What is the color of Conductor 3?

5. ______________ What are the two possible colors of Conductor 4?

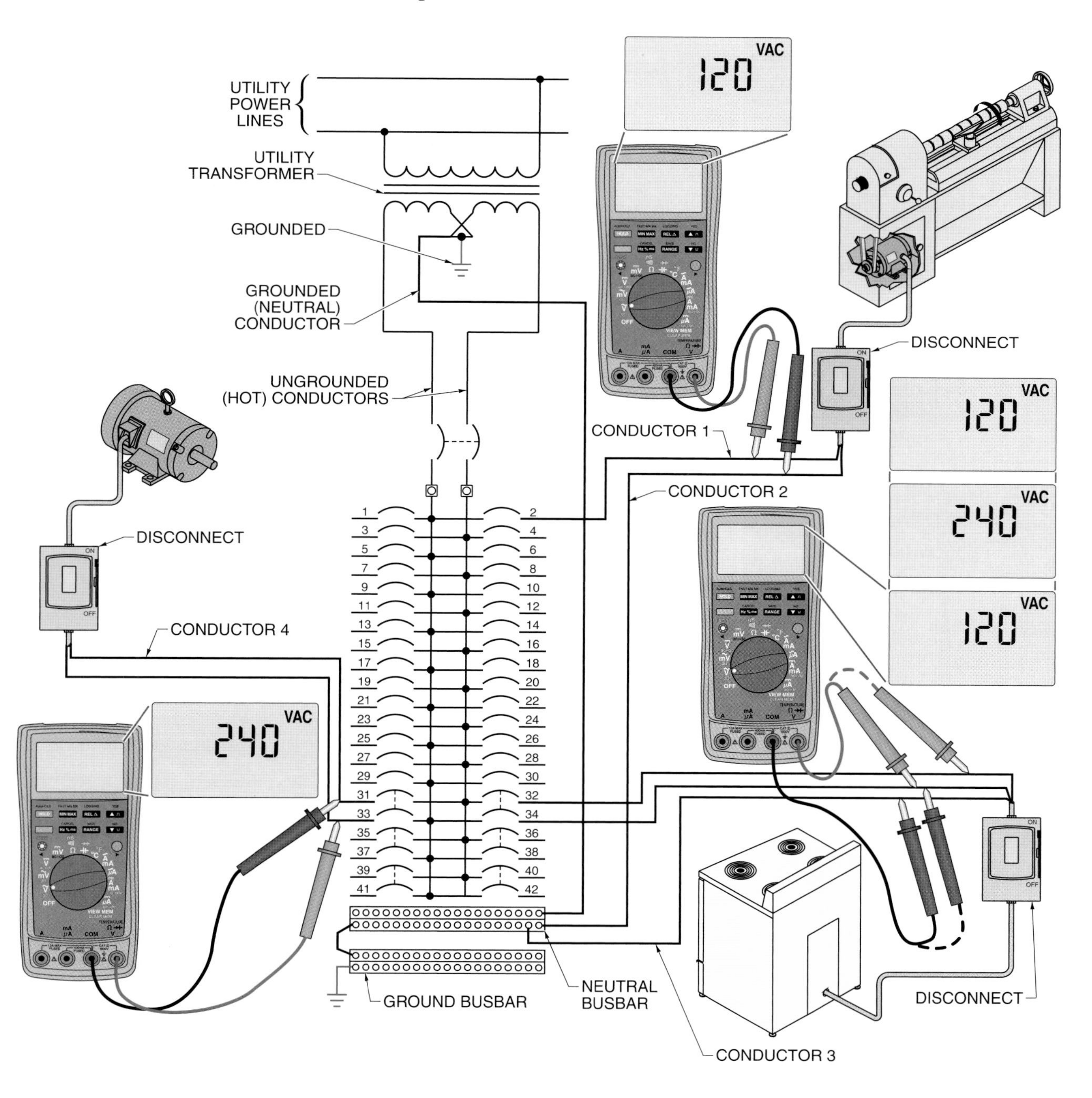

6. What are the voltage, phase type, and number of wires for this service?

7. ______________ What is the color of Conductor 1?

8. ______________ What is the color of Conductor 2?

9. ______________ What is the color of Conductor 3?

10. ______________ What is the color of Conductor 4?

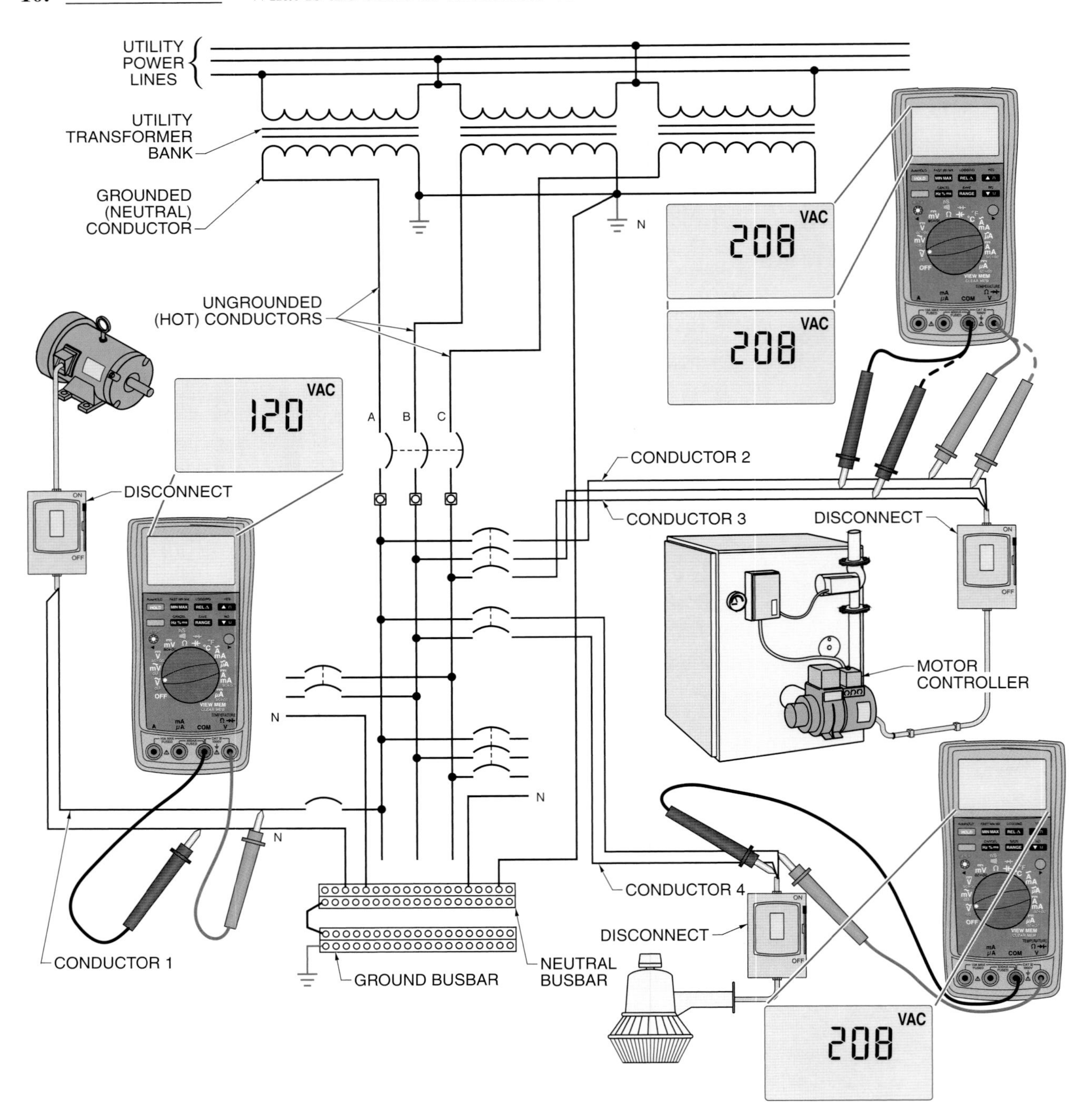

11. What are the voltage, phase type, and number of wires for this service?

12. ______________ What is the color of Conductor 1?

13. ______________ What is the color of Conductor 2?

14. ______________ What is the color of Conductor 3?

15. ______________ What is the color of Conductor 4?

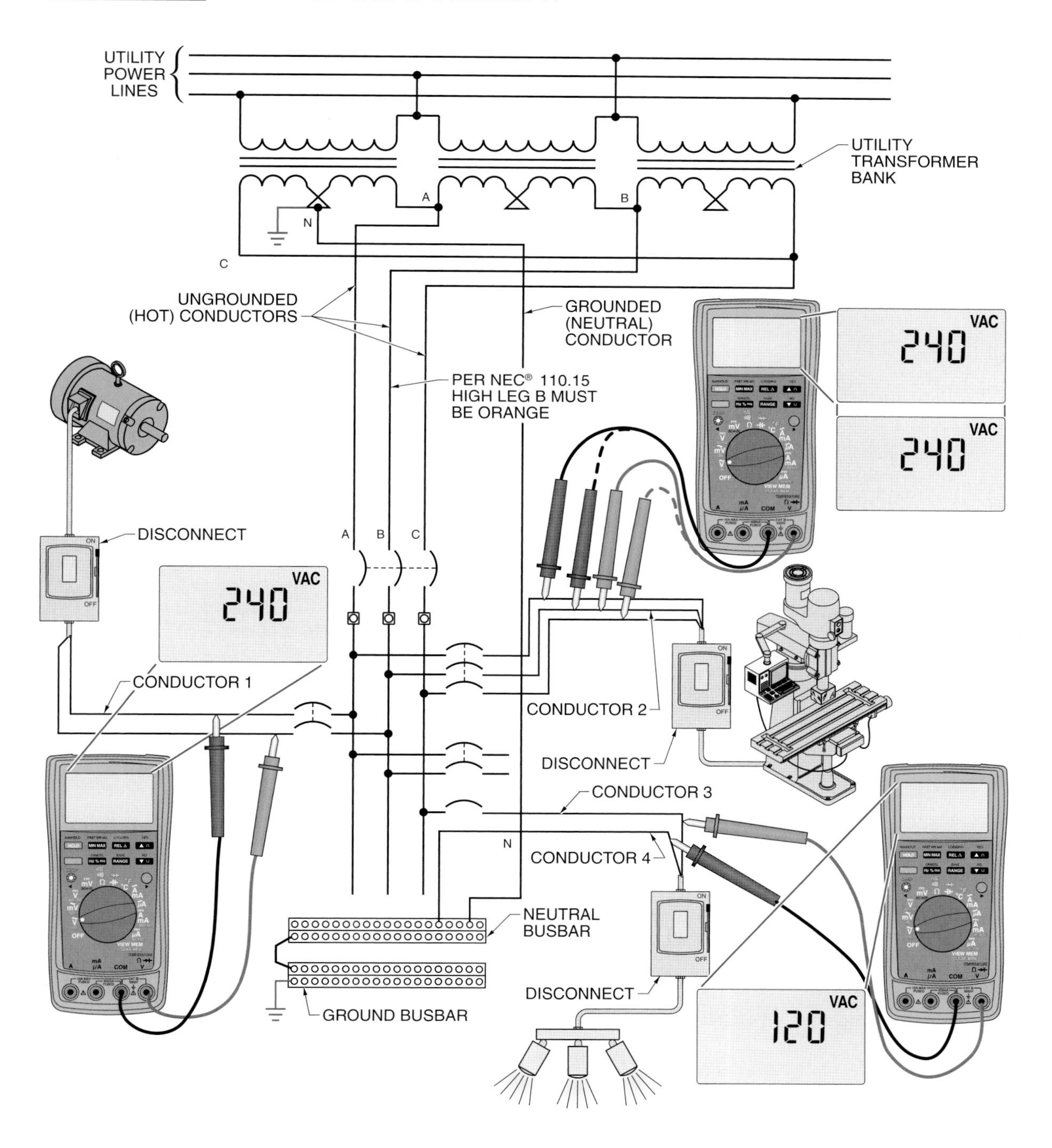

16. What are the voltage, phase type, and number of wires for this service?

17. _______________ What is the color of Conductor 1?

18. _______________ What is the color of Conductor 2?

19. _______________ What is the color of Conductor 3?

20. _______________ What is the color of Conductor 4?

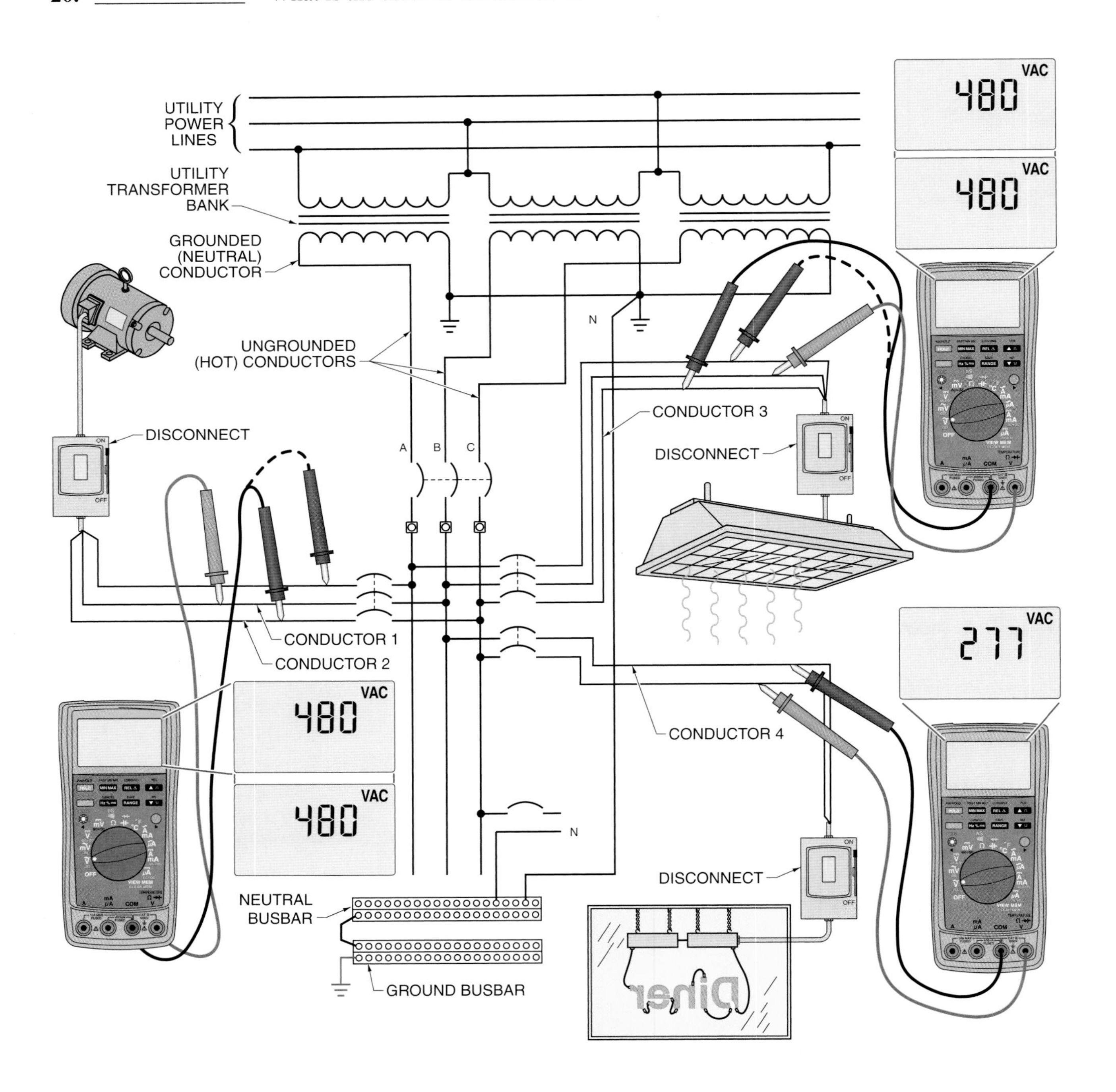

21. ______________ What is the voltage and phase type for the commercial receptacle?

22. ______________ What is the voltage and phase type for the residential receptacle?

23. ______________ What is the voltage and phase type for the industrial receptacle?

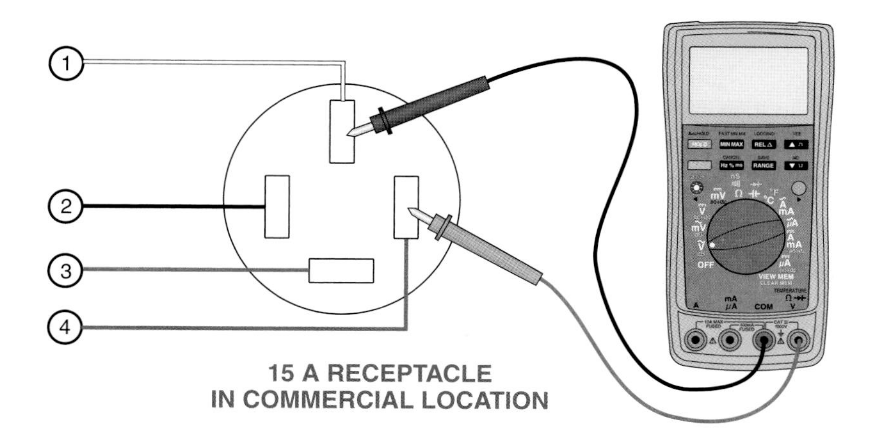

15 A RECEPTACLE IN COMMERCIAL LOCATION

VOLTAGE MEASUREMENTS
1 TO 2 = 120 V
1 TO 4 = 120 V
1 TO 3 = 120 V
2 TO 3 = 208 V
2 TO 4 = 208 V
4 TO 3 = 208 V

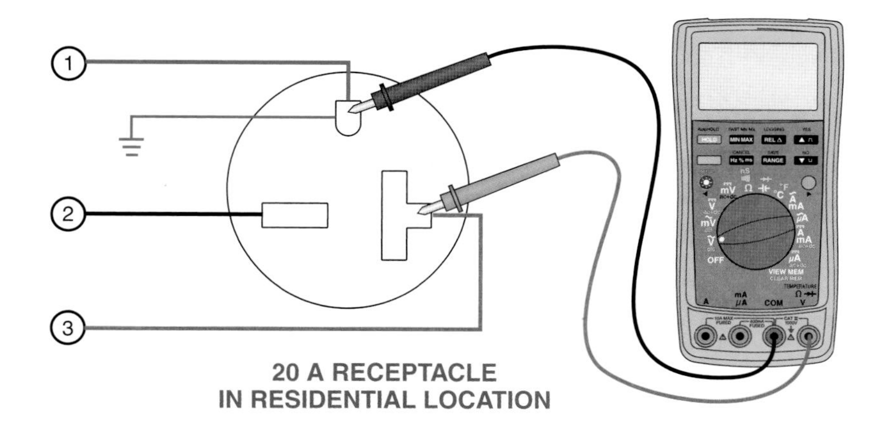

20 A RECEPTACLE IN RESIDENTIAL LOCATION

VOLTAGE MEASUREMENTS
1 TO 2 = 120 V
1 TO 3 = 120 V
2 TO 3 = 120 V

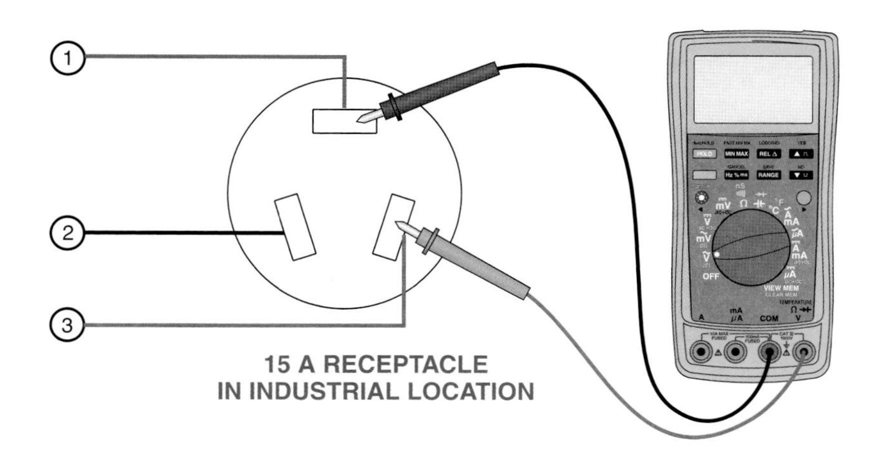

15 A RECEPTACLE IN INDUSTRIAL LOCATION

VOLTAGE MEASUREMENTS
1 TO 2 = 240 V
1 TO 3 = 240 V
2 TO 3 = 240 V

Application 4-3—Transformer Overloading

Transformers are used to deliver power to a set number of loads. As loads in a system are switched ON and OFF, the power delivered by the transformer changes. For example, at night, the power output required from a transformer in an office building may be low. During business hours, the power output required from the transformer may be high. In order for a transformer to deliver enough power at all times, the transformer must be able to deliver enough power during peak load demands. Peak load is the maximum output requirement of a transformer. See Transformer Load Cycle.

A transformer is overloaded when it is required to deliver more power than the listed rating. A transformer is usually not damaged when overloaded for a short time because it takes some time for the extra load to raise the transformer's internal temperature to damaging levels. A transformer may be damaged if overloaded for a long time because of improper cooling or power quality problems that cause additional heat buildup. Nonlinear loads cause additional heat from harmonics, which increase at higher frequencies. Nonlinear loads also produce additional heat from the skin effect caused by the higher frequency harmonics.

Transformers that deliver power to many nonlinear loads can be overloaded because of the increased harmonics. Older transformers that delivered enough power when most loads were linear may not be able to deliver enough power when nonlinear loads are added. This occurs even if the total current draw is the same or less than it was when the transformer was delivering power to only linear loads. This is a common problem for transformers delivering power to office buildings, schools, and libraries because large numbers of computers have been added in recent years.

Transformer manufacturers list the length of time a transformer may safely be overloaded at a given peak level. For example, a transformer that is overloaded 15 times its rated current has a permissible overload time of 5.5 sec. The rated permissible overload is based on 60 Hz linear loads that decrease transformer heating as the loads are removed. Additional heating occurs when different load types are connected to a transformer. See Transformer Overloading.

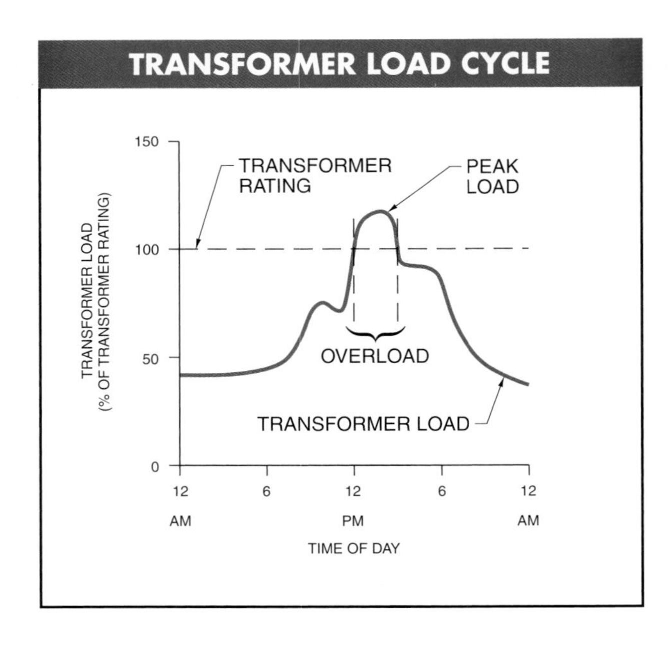

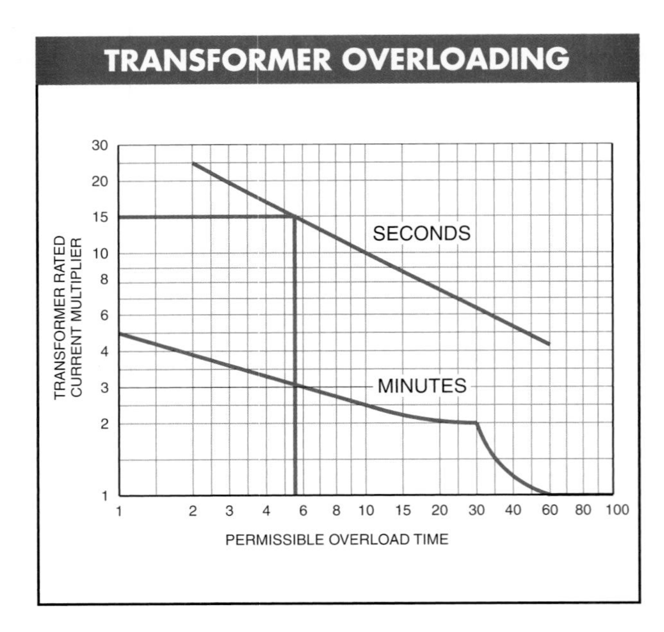

A K-rated transformer is a transformer designed to handle the extra heating effects caused by harmonic currents. A K factor represents the amount of harmonics produced by nonlinear loads connected to power lines, and can be measured with a power quality meter. Larger K factors indicate a greater amount of harmonics present. K-rated transformers have a K-rating such as K-4, K-13, or K-20. Higher K-ratings indicate greater heat dissipation capabilities. See K-Rated Transformers.

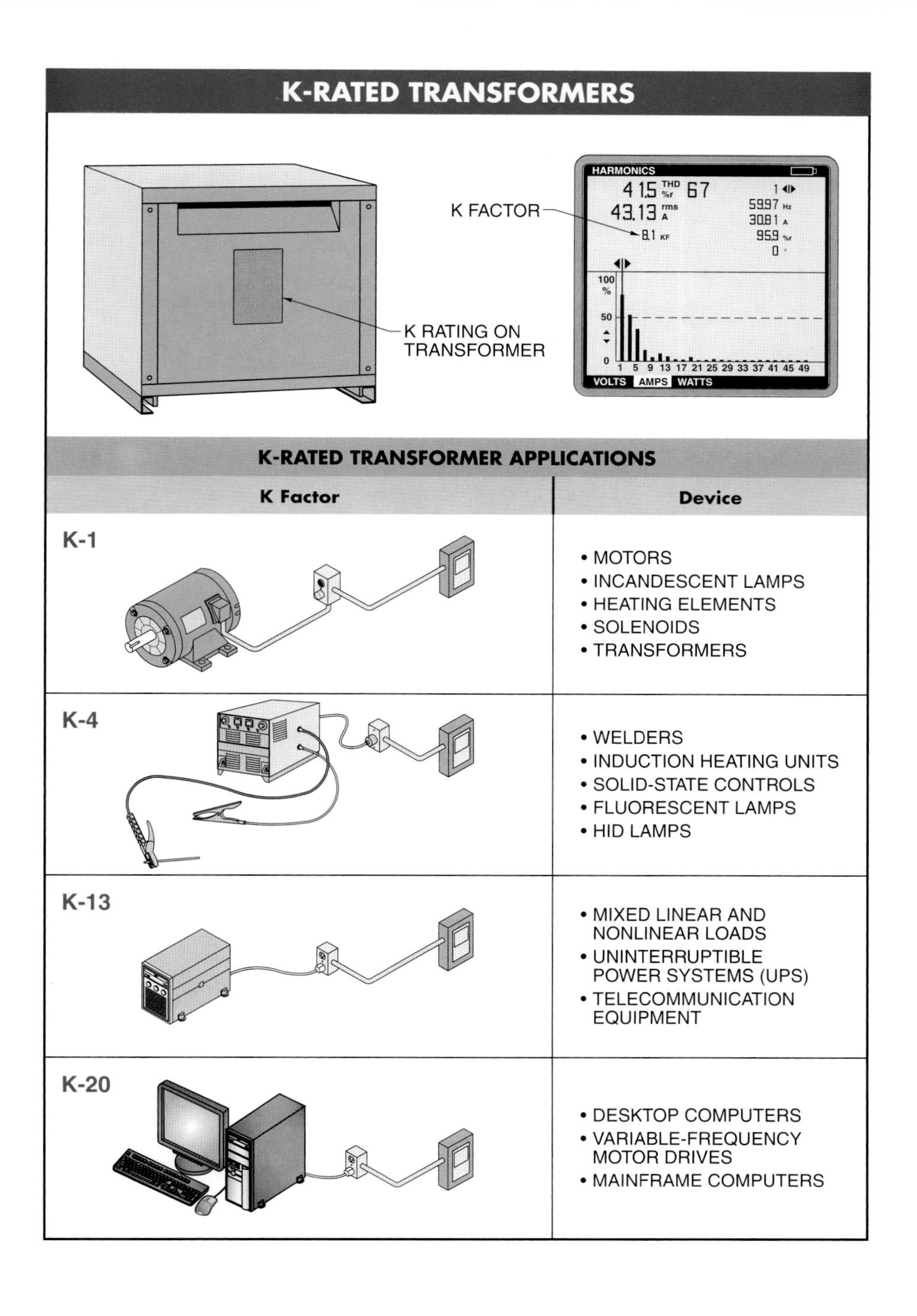

Linear loads without any harmonics, such as motors, incandescent lamps, and heating elements, have a K factor of 1 (K-1). Nonlinear loads with some harmonics, such as welding machines, induction heating units, and solid-state controls, have a K factor of 4 (K-4). Branch circuits that include a mix of linear and nonlinear loads, such as uninterruptible power systems and telecommunication equipment, have a K factor of 13 (K-13). Circuits that include mostly nonlinear loads, such as desktop computers and variable-frequency motor drives, have a K factor of 20 (K-20). The heating effect caused by the harmonics of a K-4 load is four times that of K-1. The heating effect of a K-20 load is 20 times that of K-1.

K-rated transformers have a larger neutral conductor and special windings to reduce the effects of harmonics and the extra heat they produce. If a transformer does not have a K-rating listed, it is assumed to be a standard transformer with a K-rating of K-1. To determine the required transformer K-rating, a K factor measurement is taken with a power quality meter when the system is fully loaded. The K factor measurement will probably not equal the exact K-rating of available transformers, so the K factor should be rounded up to the next available rating to ensure proper transformer operation under full load.

1. ______________ What is the lowest recorded voltage measurement?

2. ______________ What is the highest recorded voltage measurement?

3. ______________ What is the lowest recorded current measurement?

4. ______________ What is the highest recorded current measurement?

5. What caused the voltage to dip to the lowest recorded value?

6. What indicates that the transformer is under the greatest load?

7. ______________ If the transformer is overloaded to twice its rating, how long can it be safely overloaded?

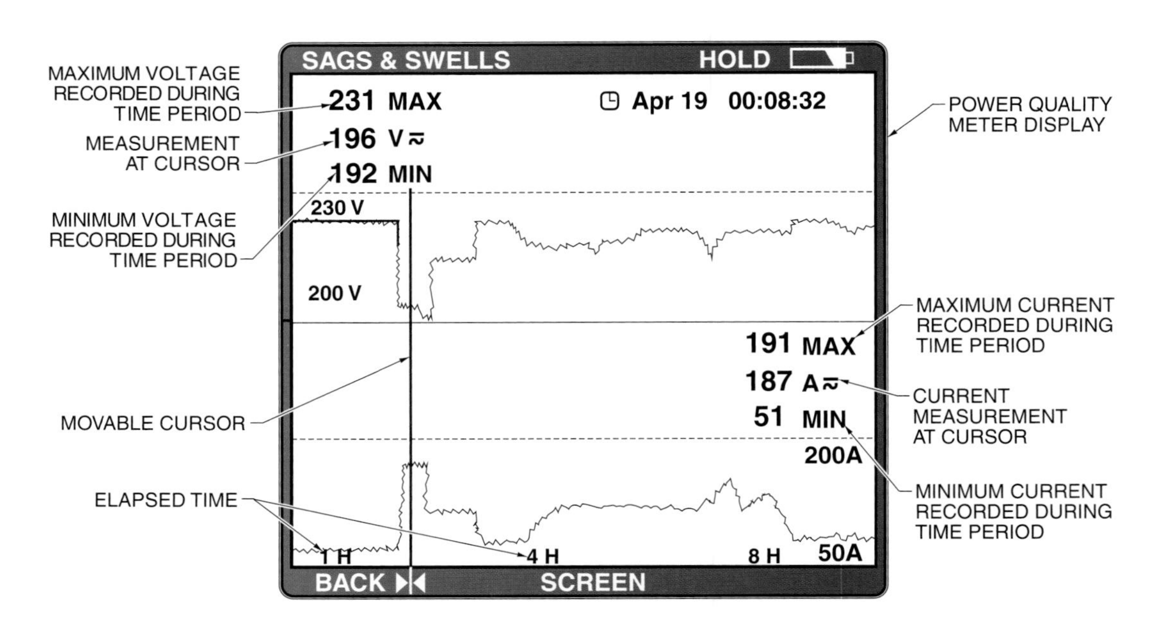

8. ______________ What is the voltage at the cursor?

9. ______________ What is the current at the cursor?

10. ______________ If the circuit loads are rated to operate at 230 VAC, what is the maximum percent voltage drop during the recording time?

11. ______________ If the service panel is rated at 200 A, what is the maximum percent of current drawn?

12. ______________ If the maximum current was recorded at 12 PM, does this coincide with a typical transformer load cycle?

13. ______________ If the loads are mostly computers, printers, and copy machines, which K rating should this transformer have?

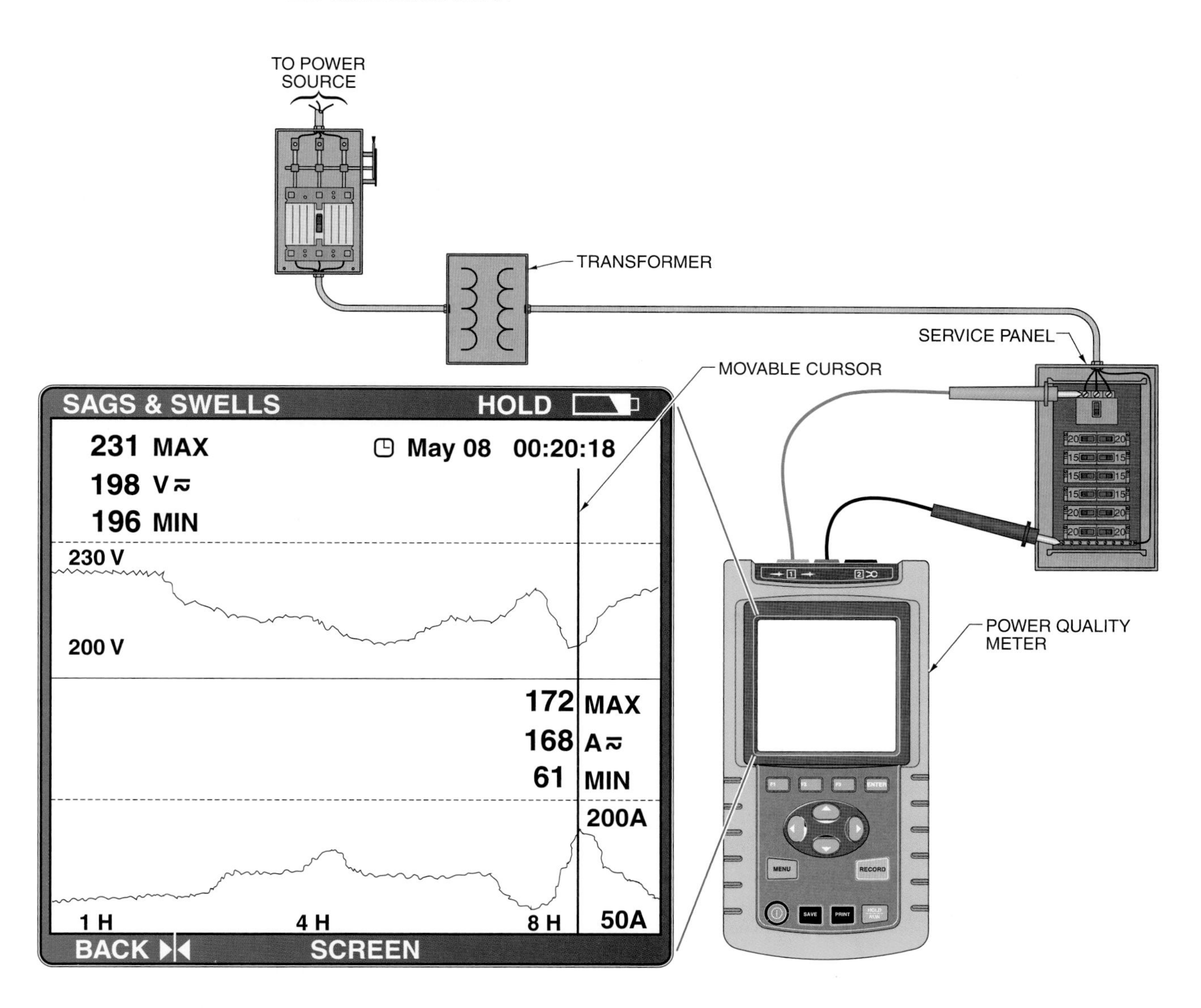

14. ______________ Is a high current draw causing the voltage to dip?

15. ______________ Is the voltage dip caused by the loads connected to the transformer or a problem up-stream from the transformer?

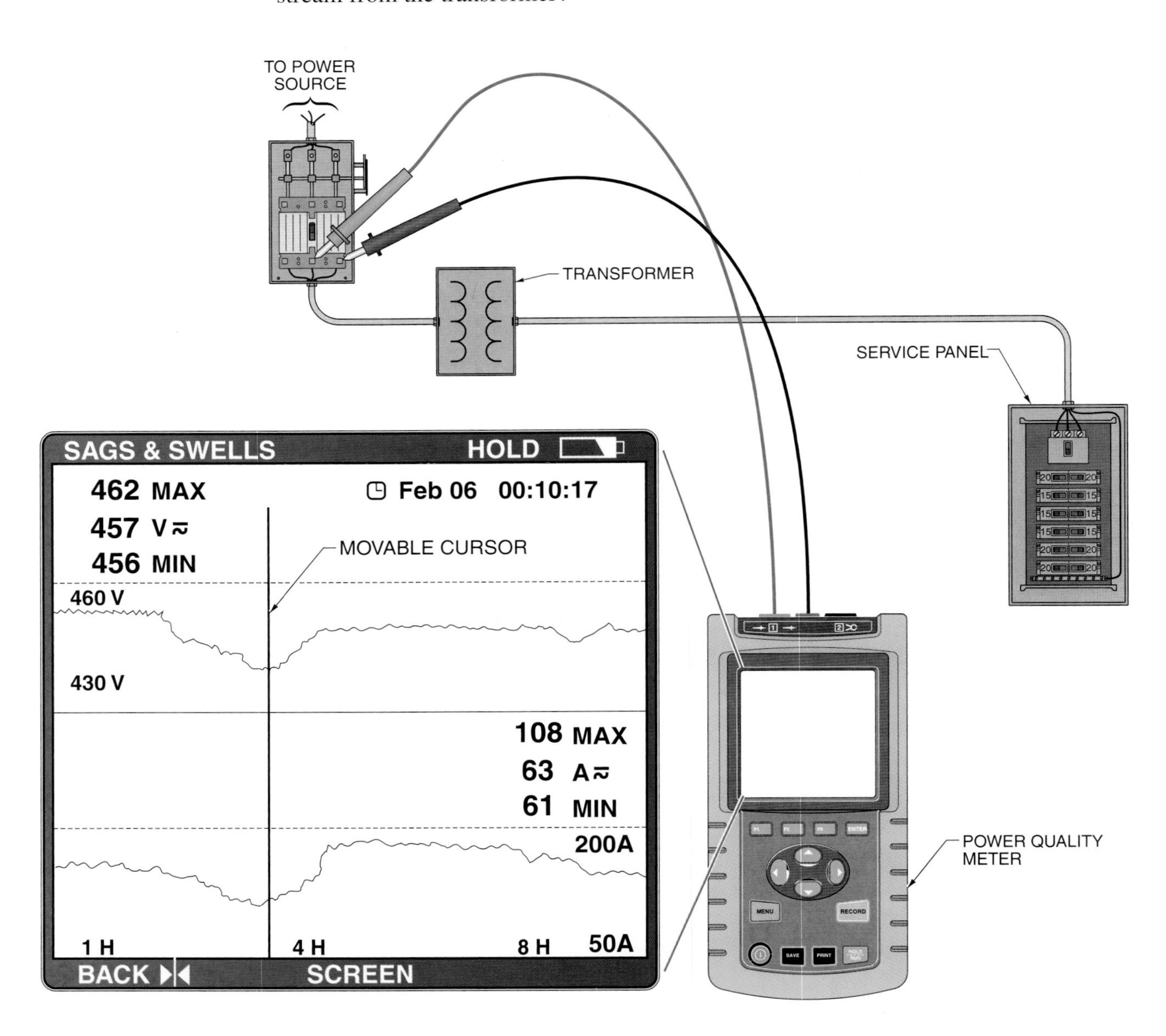

Application 4-4 — Overheated Neutral

The electrical service to most residences is 120/240 V, single-phase. The lower voltage line (120 V) is used for general-purpose receptacles and lighting, and is obtained by connecting loads between the neutral conductor and either hot wire (L1 or L2). The higher voltage line (240 V) is used for heating, cooling, and cooking, and is obtained by connecting loads between the two hot wires (L1 and L2).

The current is 180° out of phase between each hot conductor. If they are equal in magnitude, they cancel each other in the neutral wire. For example, if a 10 A linear load is connected to L1 and neutral and a 10 A linear load is connected to L2 and neutral, there is no current flowing in the neutral (10 A – 10 A = 0 A). See Same Current Draw.

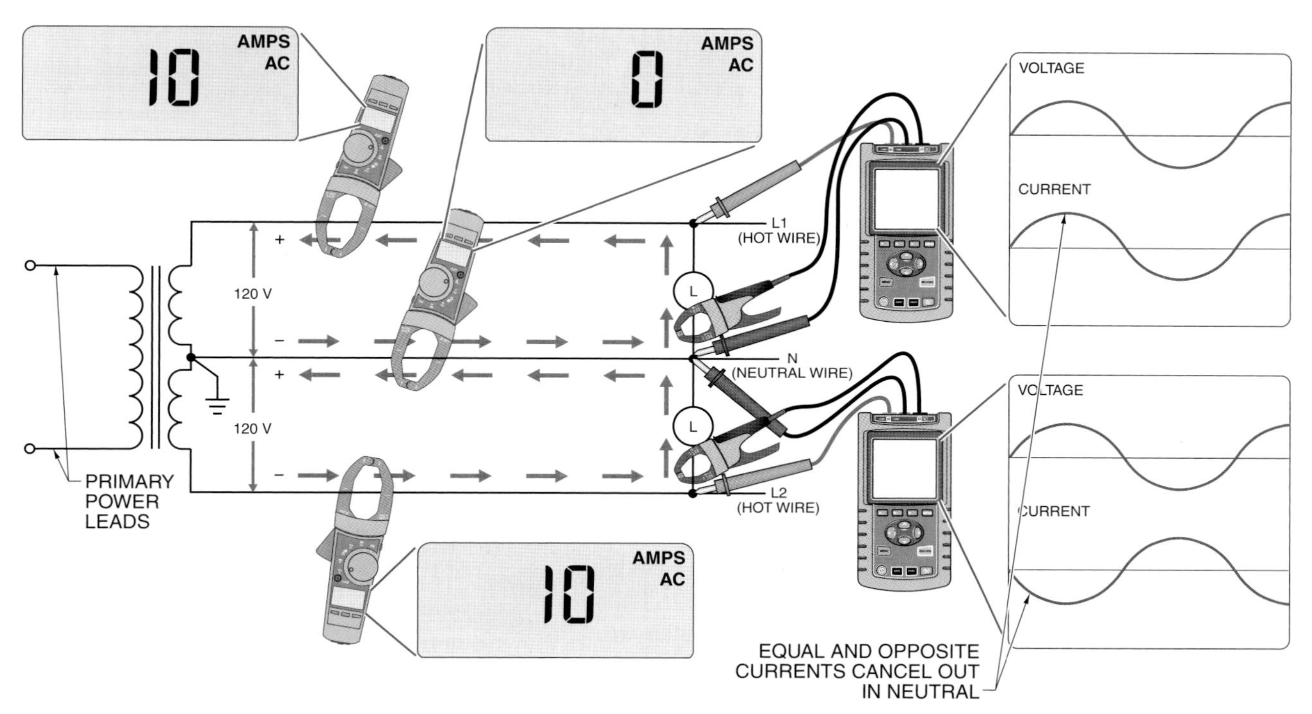

SAME CURRENT DRAW

If linear loads connected between L1 and neutral and L2 and neutral do not have the same current draw, the difference (unbalance) in current draw flows through the neutral. For example, if a 20 A linear load is connected to L1 and neutral and a 5 A linear load is connected to L2 and neutral, there is 15 A of current in the neutral (20 A – 5 A = 15 A). Because the neutral carries only the unbalance current, current in the neutral never exceeds the highest current level in either hot line (L1 or L2) when linear loads are connected to the transformer. See Different Current Draw.

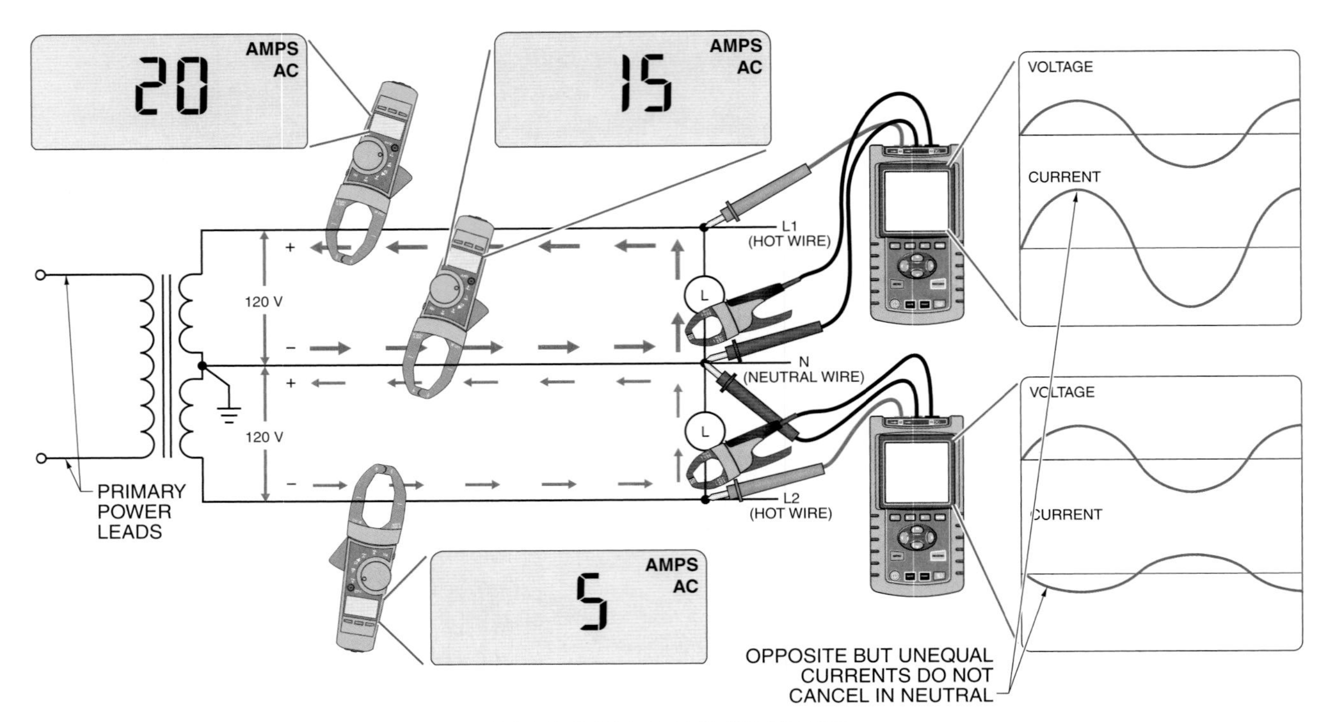

DIFFERENT CURRENT DRAW

Although an ammeter can measure the current cancellation in the neutral, the ammeter cannot show that the cancellation takes place due to the 180° phase shift. A power quality meter can be used to measure the current in the conductors and show the phase shift.

The electrical service to most commercial buildings is a wye-connected, three-phase system. This system is used because it can provide a large amount of balanced single-phase power in addition to three-phase power. In a three-phase, four-wire system supplying power to linear loads, the fundamental 60 Hz currents cancel in the neutral conductor, so one common neutral is usually used.

However, when single-phase nonlinear loads are connected to a three-phase, 4-wire system, neutral current can exceed an individual phase current. The neutral current can be between 125% and 225% of the highest phase current. The nonlinear loads produce triplen harmonics, which are odd-numbered multiples of the third harmonic (such as 3rd, 9th, 15th, and so on). Triplen harmonics do not cancel, but add together in the neutral conductor. See Harmonics from Nonlinear Loads.

The third harmonic current is usually responsible for most of the neutral current because it is typically the harmonic with the highest current value. High neutral current is dangerous because it causes overheating in the neutral conductor. Because there is no circuit breaker to limit current, overheating of the neutral can become a fire hazard. Excessive current in the neutral conductor can also cause higher than normal voltage drops between the neutral and ground at 120 V outlets.

When nonlinear loads are connected to power lines, any harmonics present cause the 60 Hz fundamental frequency to become distorted. This distortion causes inaccurate measurements in test instruments that are not rated "true-rms." To prevent inaccurate measurements, only a true-rms multimeter or power quality meter should be used when taking voltage or current measurements.

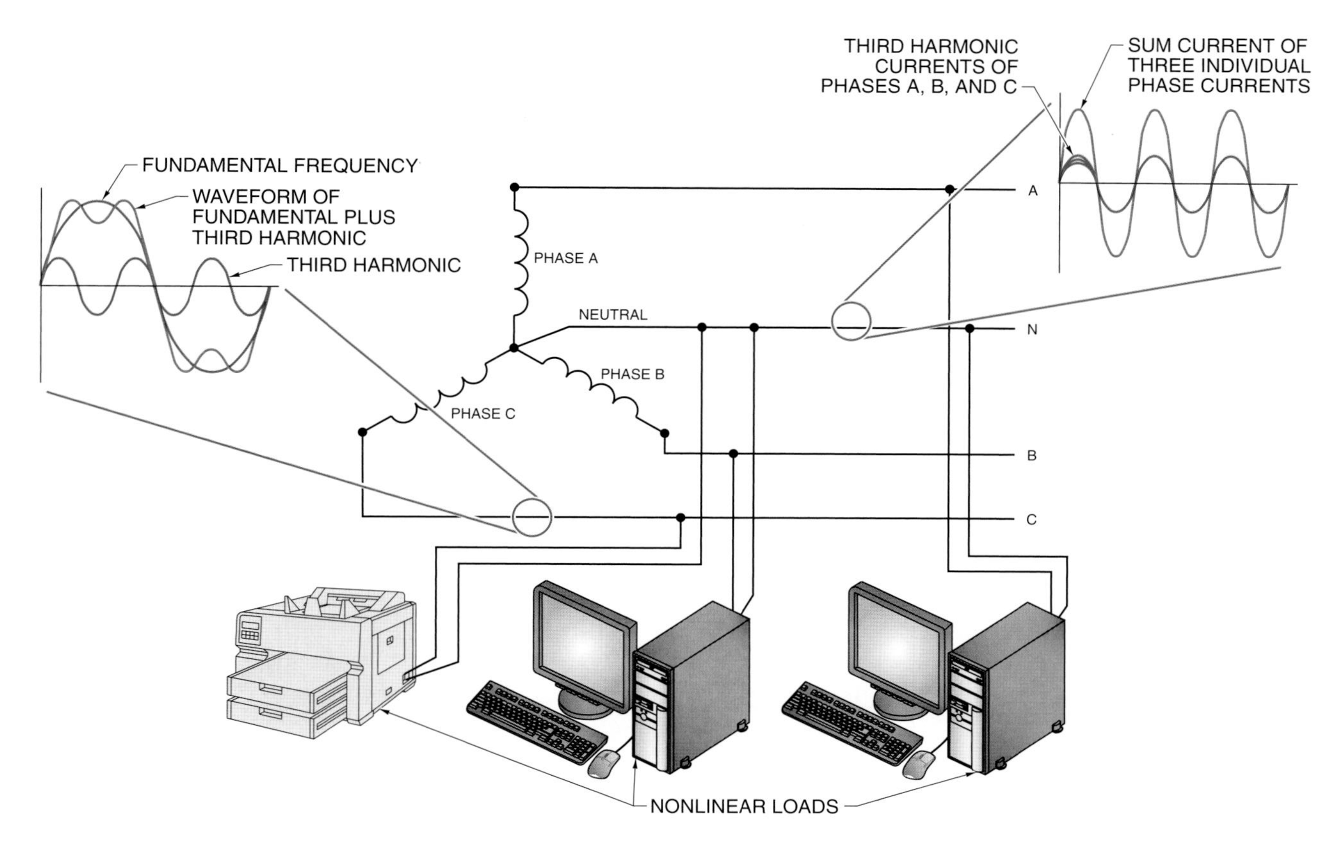

HARMONICS FROM NONLINEAR LOADS

A power quality meter can also display the percent of harmonic current on the power lines. The power quality meter usually displays the odd-number harmonics (1, 3, 5, and so on) since it is the odd-numbered harmonics that cause system and load problems, such as overheating. See Harmonics Display.

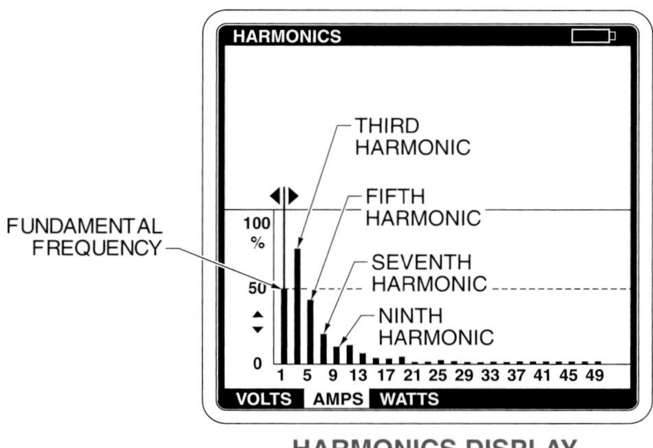

HARMONICS DISPLAY

1. ______________ Which measurement shows that the loads connected to the system are all linear loads?

2. ______________ Which measurement shows that the loads connected to the system are more linear than nonlinear?

3. ______________ Which measurement shows that many of the loads must be nonlinear?

4. ______________ What frequency would the multimeter display for Measurement 1?

5. ______________ What frequency would the multimeter display for Measurement 2?

6. ______________ What frequency would the multimeter display for Measurement 3?

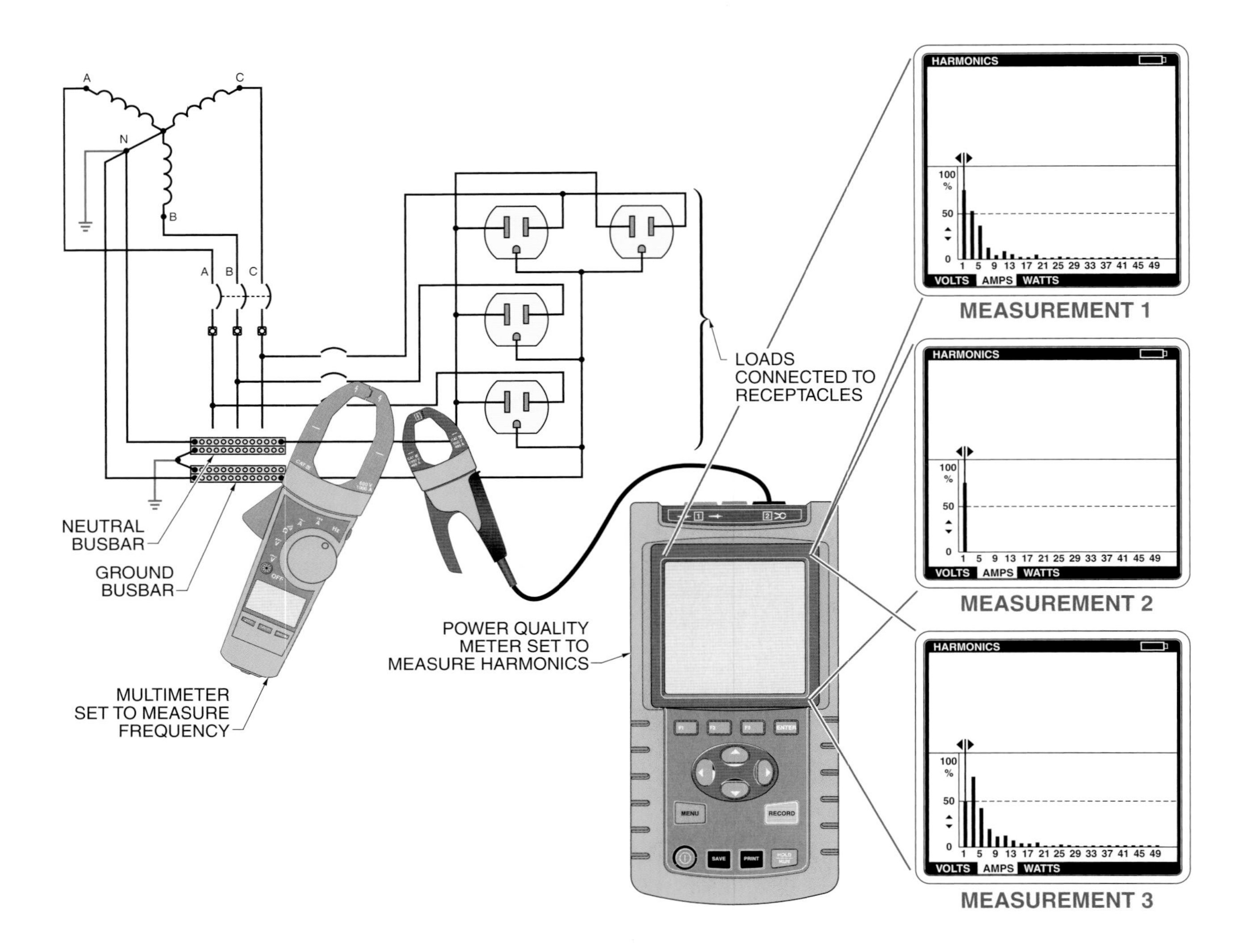

Application 4-5 — Voltage Changes

It is normal for AC voltage to vary slightly, and fluctuations within +5% to –10% of the voltage rating of loads are usually not a problem. However, more serious voltage changes, such as power interruptions, voltage sags, voltage swells, undervoltages, and overvoltages, are a problem. Voltmeters with a recording mode (MIN MAX) can be used to identify voltage changes within the meter's recording specifications. For example, a meter with a 100 ms (0.1 sec) capture time can record any voltage changes that occur for at least 6 cycles (0.1 sec × 60 cycles per sec = 6 cycles). This means the meter can capture most voltage changes, but not transient voltages, which typically only last a fraction of a cycle. See Voltage Changes.

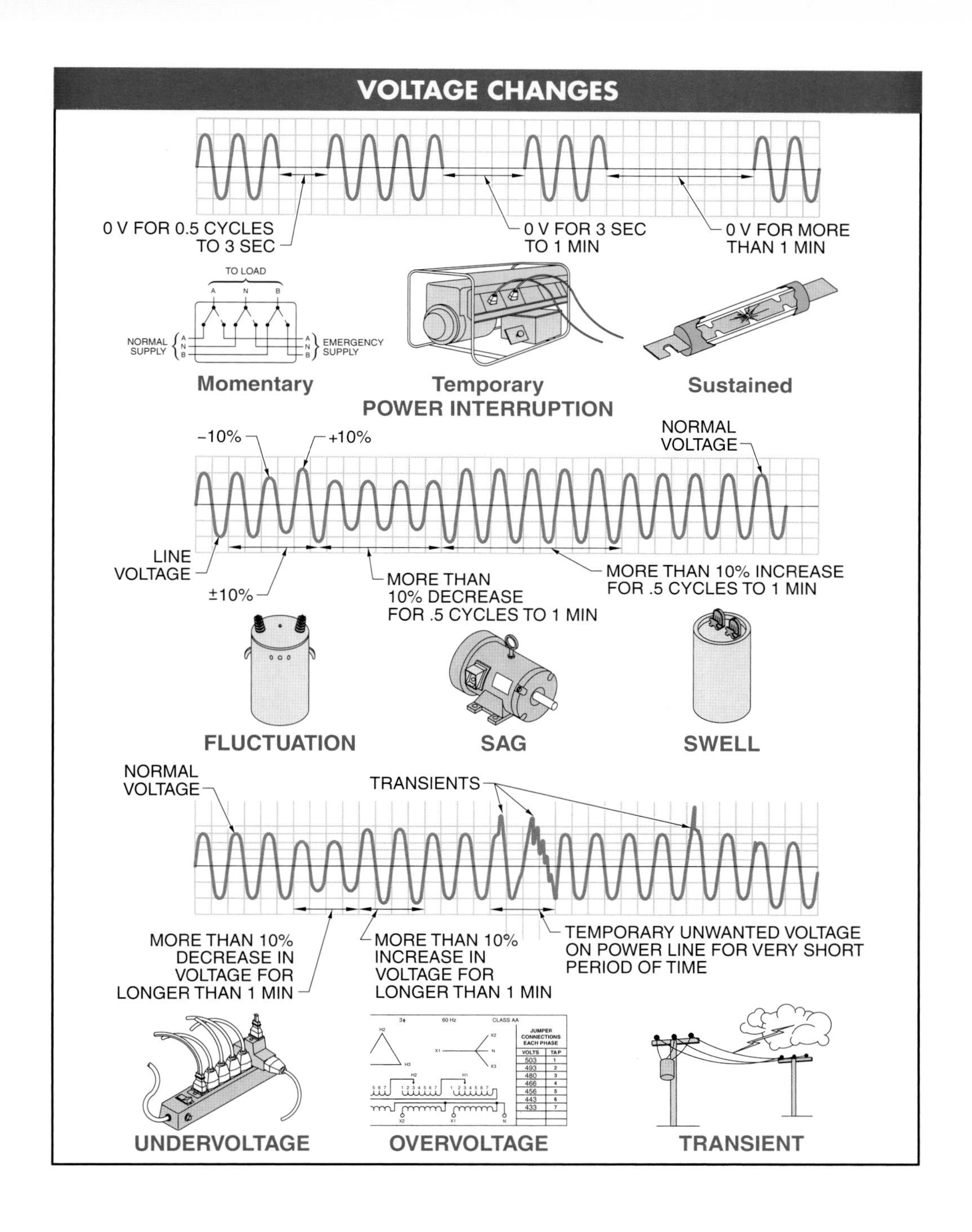

It is reported that computers connected to Circuit 4 have been automatically rebooting. To diagnose the problem, voltage measurements are taken over time at different points in the power distribution system. Answer the following questions based on the voltage measurement results.

1. Why are voltage measurements taken at the receptacles on Circuit 4?

2. ______________ Is there a problem at Circuit 4 that could cause computers to automatically reset?

3. Why are voltage measurements taken at the main circuit breaker in Panelboard 2?

4. ______________ Is there a problem at Circuit 4 only?

5. Why are voltage measurements taken at the main circuit breaker in Panelboard 1?

6. ______________ Is there a problem throughout the entire in-plant power distribution system?

7. Why are voltage measurements taken at the main power feed into the switchboard?

8. Where is the most likely problem and a place where additional voltage measurements need to be taken?

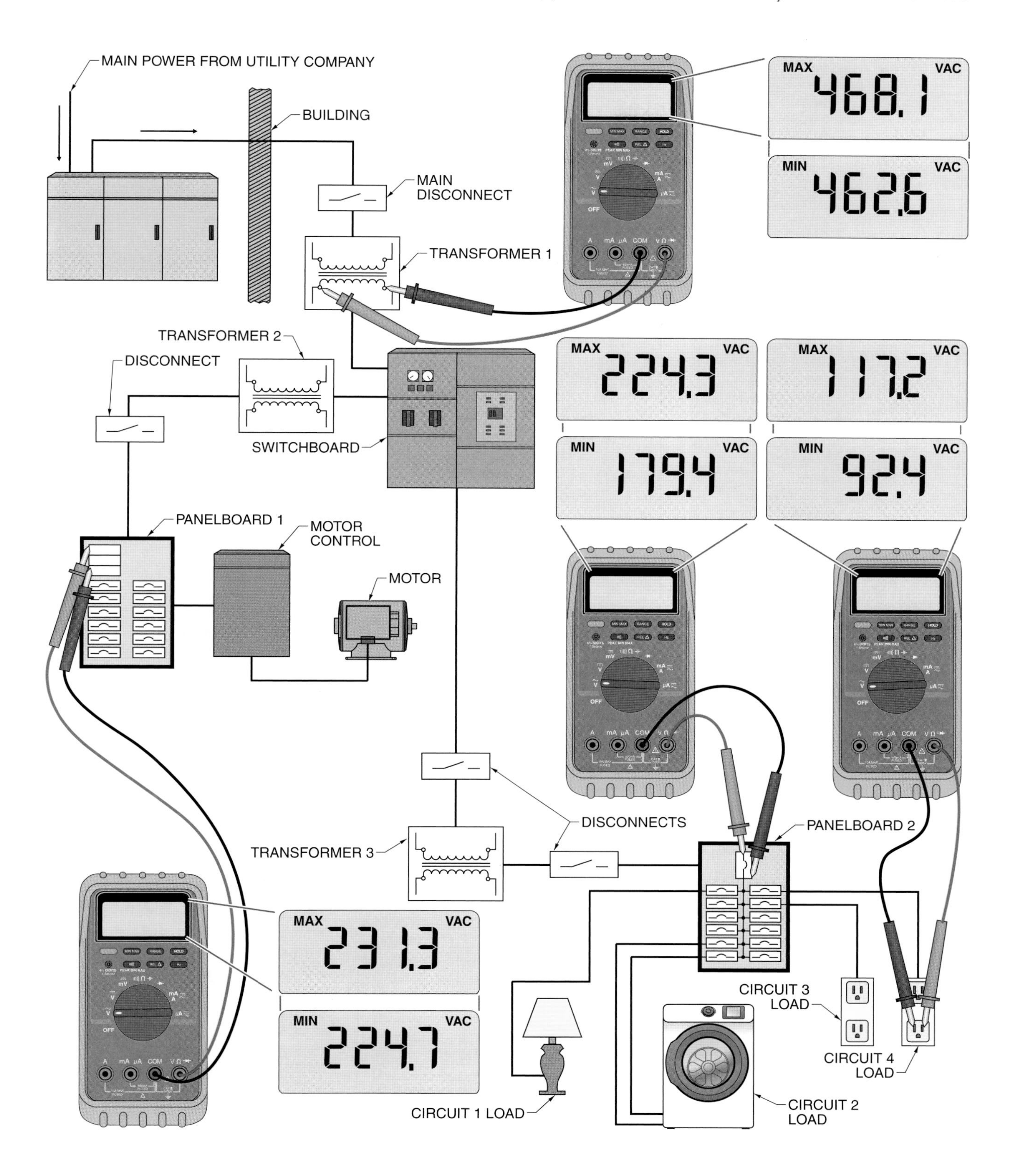
MAIN POWER FROM UTILITY COMPANY
BUILDING
MAIN DISCONNECT
TRANSFORMER 1
MAX 468.1 VAC
MIN 462.6 VAC
TRANSFORMER 2
DISCONNECT
SWITCHBOARD
MAX 224.3 VAC
MIN 179.4 VAC
MAX 117.2 VAC
MIN 92.4 VAC
PANELBOARD 1
MOTOR CONTROL
MOTOR
DISCONNECTS
TRANSFORMER 3
PANELBOARD 2
MAX 231.3 VAC
MIN 224.7 VAC
CIRCUIT 3 LOAD
CIRCUIT 4 LOAD
CIRCUIT 1 LOAD
CIRCUIT 2 LOAD

Application 4-6—High Crest Factor

Normal AC voltage and current values generate a waveform in the shape of the sine function. Certain loads may distort the sinusoidal (sine function) waveform. Crest factor is the measure of the amount of distortion in nonsinusoidal waveforms.

AC voltage values are stated and measured as peak, peak-to-peak, average, or rms values. The peak voltage value (V_{max}) of a sine wave is the maximum value of either the positive or negative alternation. The positive and negative alternation values are equal in a sine wave.

The peak-to-peak voltage value (V_{p-p}) is the value measured from the maximum positive alternation to the maximum negative alternation. Some voltmeters are designed to measure peak-to-peak voltage. See Peak-to-Peak Voltage.

The average voltage value (V_{avg}) of a sine wave is the mathematical mean of all instantaneous voltage values in the sine wave. The average voltage value is equal to 0.637 times the peak value of a standard sine wave. See Average Voltage.

The root-mean-square voltage value (V_{rms}), or effective value, of a sine wave is the voltage value that produces the same amount of heat in a pure resistive circuit as DC of the same value. The rms value is equal to 0.707 times the peak value in a sine wave. See rms Voltage.

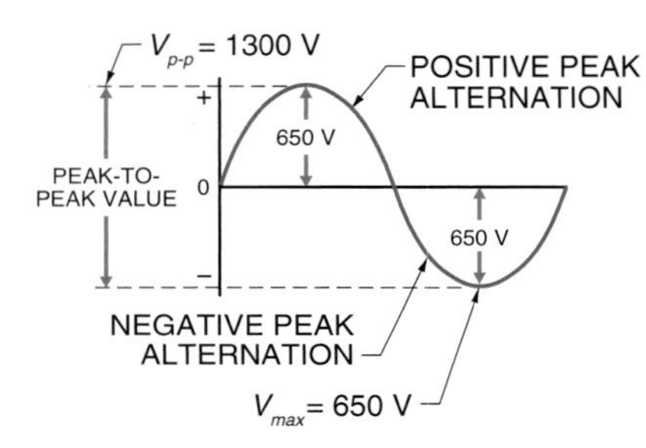

PEAK-TO-PEAK VOLTAGE

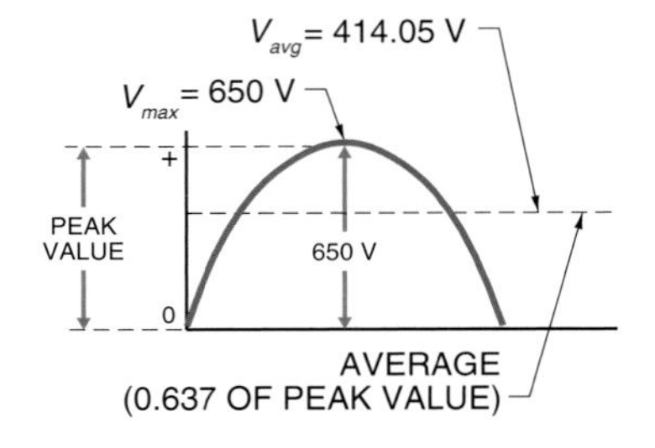

AVERAGE VOLTAGE

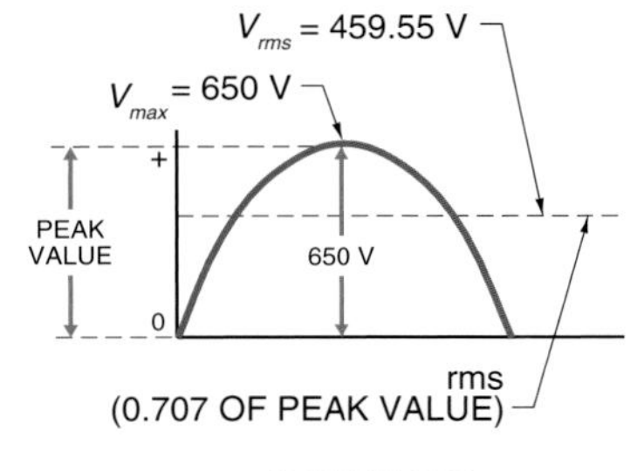

rms VOLTAGE

Some voltmeters and ammeters, especially older models, measure the peak voltage or current value and then display an equivalent rms value that is calculated by multiplying the peak value by 0.707. For example, a voltmeter connected to a 120 VAC receptacle may measure the circuit's peak voltage (170 VAC) and display the equivalent rms value (170 VAC × 0.707 = 120 VAC). This value is accurate if the waveform is a perfectly sinusoidal waveform.

Linear loads do not distort the sine wave. For example, when a magnetic motor starter is used to control a motor, the motor and motor starter load will not distort the AC voltage sine wave or AC current sine wave. See Linear Load Crest Factors.

A nonsinusoidal waveform has a distorted appearance when compared with a pure sine waveform. Nonsinusoidal waveforms are present in equipment such as variable-speed motor drives, light dimmers, computers, and most circuits that use solid-state electronics. These types of loads are called nonlinear loads.

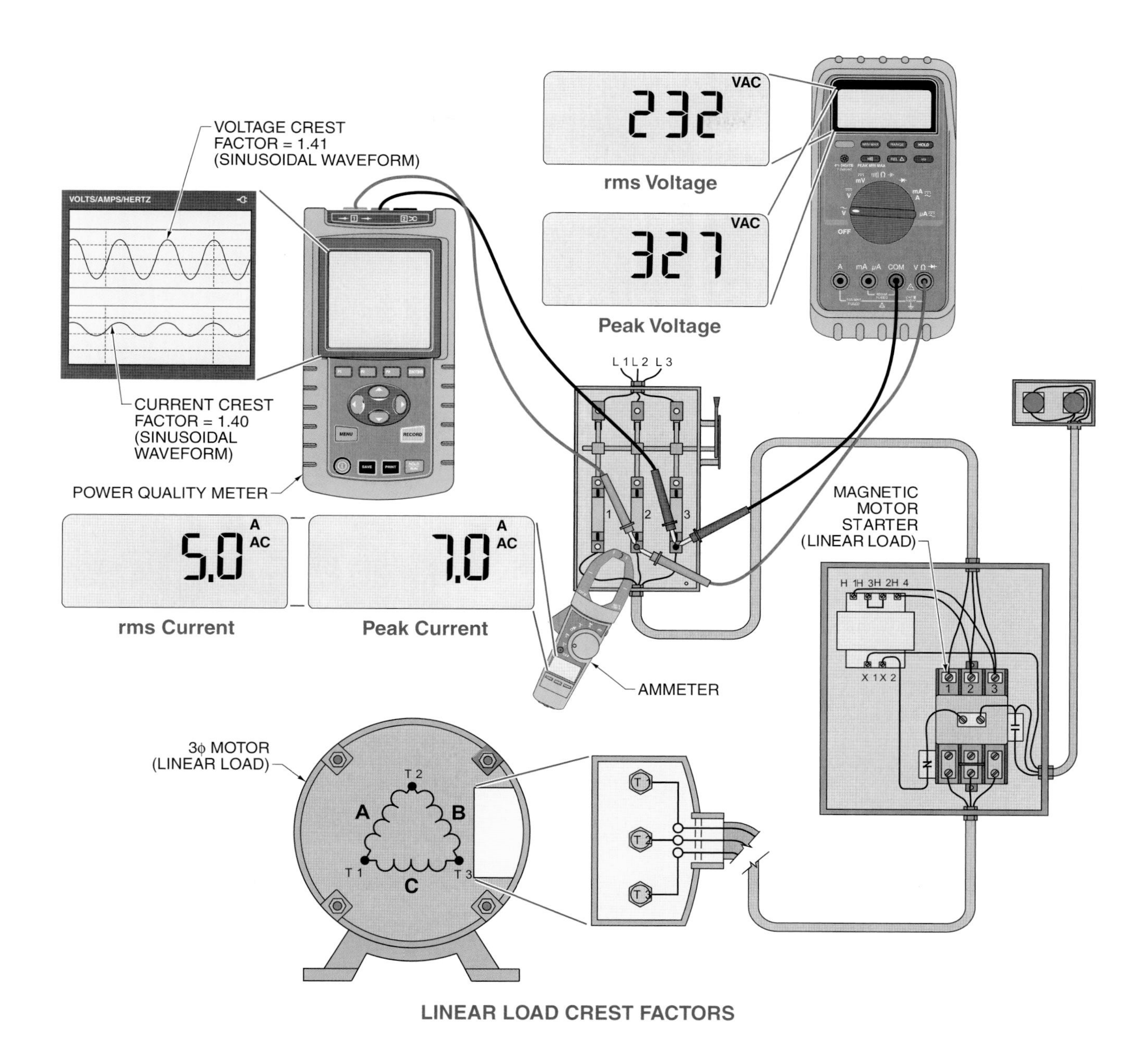

LINEAR LOAD CREST FACTORS

If the AC waveform is distorted, a calculated rms value is not correct. Voltage measurement errors can range from about 1% to more than 50%. This is a problem as systems include more nonlinear loads. Meters that are rated "true-rms" measure rms values directly, so they are the most accurate meters for any waveform shape, including nonsinusoidal. Only true-rms test instruments should be used in any circuit that includes a nonlinear load. For example, when a solid-state motor starter is used to control a motor, the motor starter load will distort the AC current waveform. See Nonlinear Load Crest Factors.

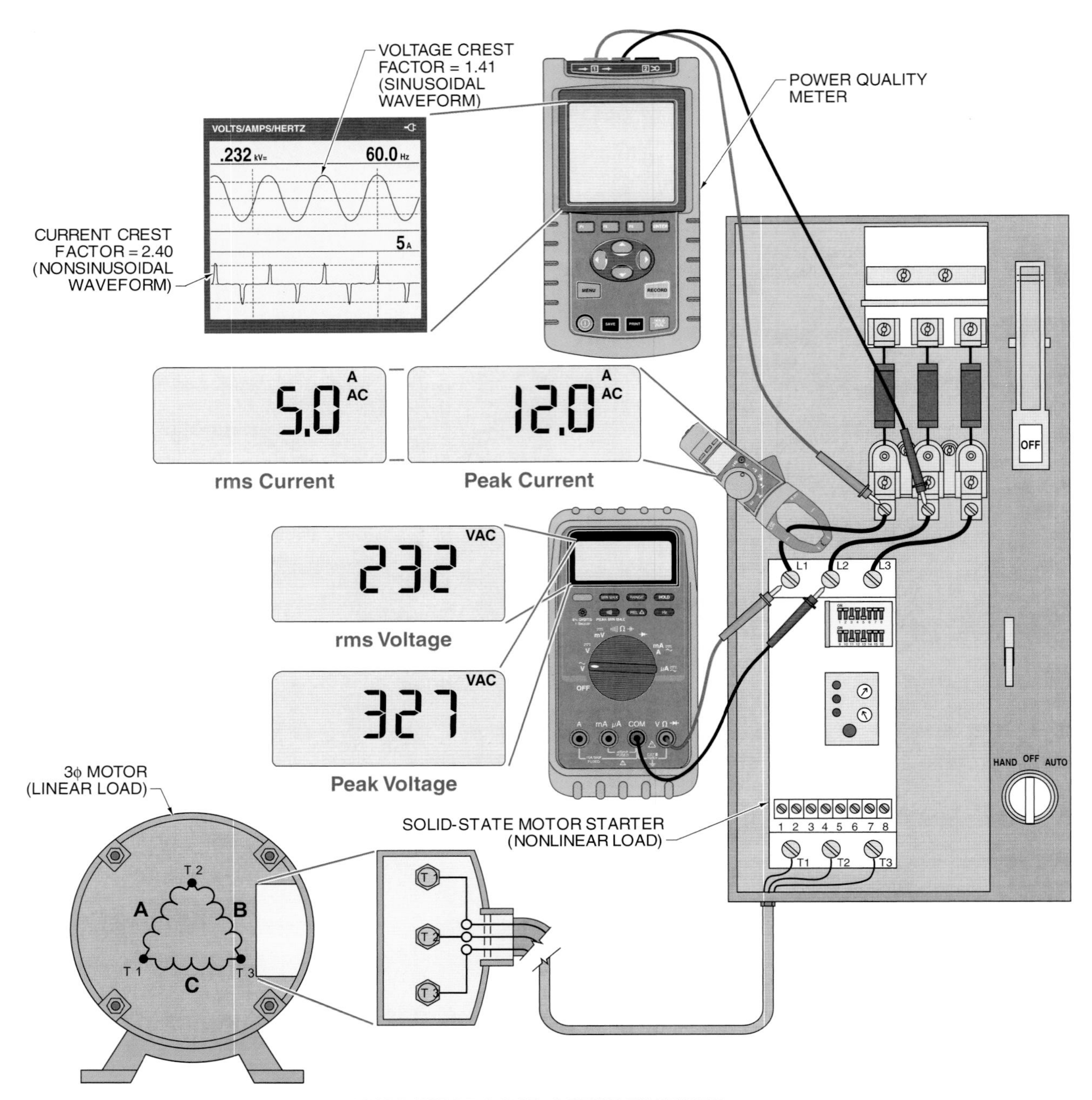

NONLINEAR LOAD CREST FACTORS

Crest factor is the ratio of the peak voltage value to the rms voltage value (or the peak current value to the rms current value). A pure sinusoidal waveform has a crest factor of 1.41. Distortion of the waveform changes the crest factor. A greater change in crest factor from 1.41 indicates a more distorted waveform, which reduces the accuracy of calculated rms values. See High Crest Factors.

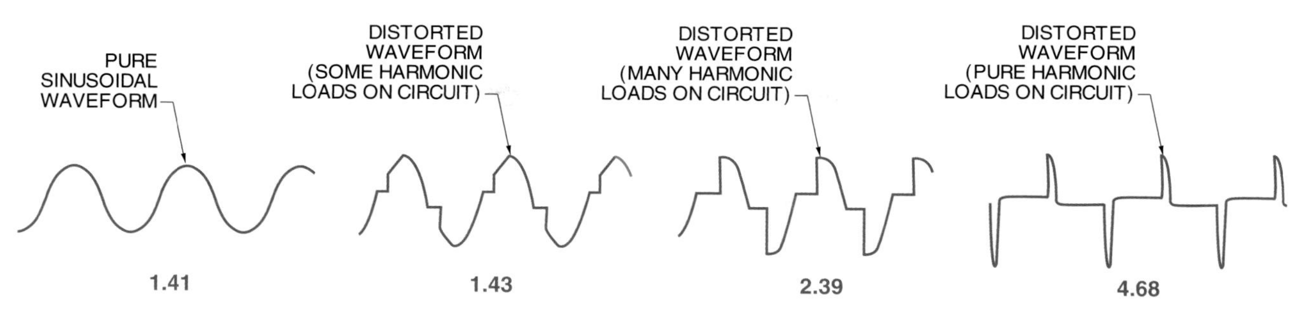

HIGH CREST FACTORS

Crest factor is calculated by applying the following formula:

$$CF = \frac{V_p}{V_{rms}}$$

where

CF = crest factor

V_p = peak voltage (in V)

V_{rms} = rms voltage (in V)

For example, what is the crest factor of a voltage waveform with peak voltage of 294 V and true rms voltage of 208 V?

$$CF = \frac{V_p}{V_{rms}}$$

$$CF = \frac{294}{208}$$

$$CF = \mathbf{1.41}$$

The crest factor is 1.41, so the voltage waveform must be sinusoidal. However, the same power circuit has a peak current of 12.7 A and true rms current of 5.3 A, resulting in a crest factor of 2.4 (12.7 A ÷ 5.3 A = 2.4). This indicates that many nonlinear loads in this circuit are distorting the current waveform.

1. ______________ What is the crest factor of the voltage?

2. ______________ What is the crest factor of the current?

3. Draw the expected voltage waveform for the circuit in the oscilloscope screen.

4. Draw the expected current waveform for the circuit in the oscilloscope screen.

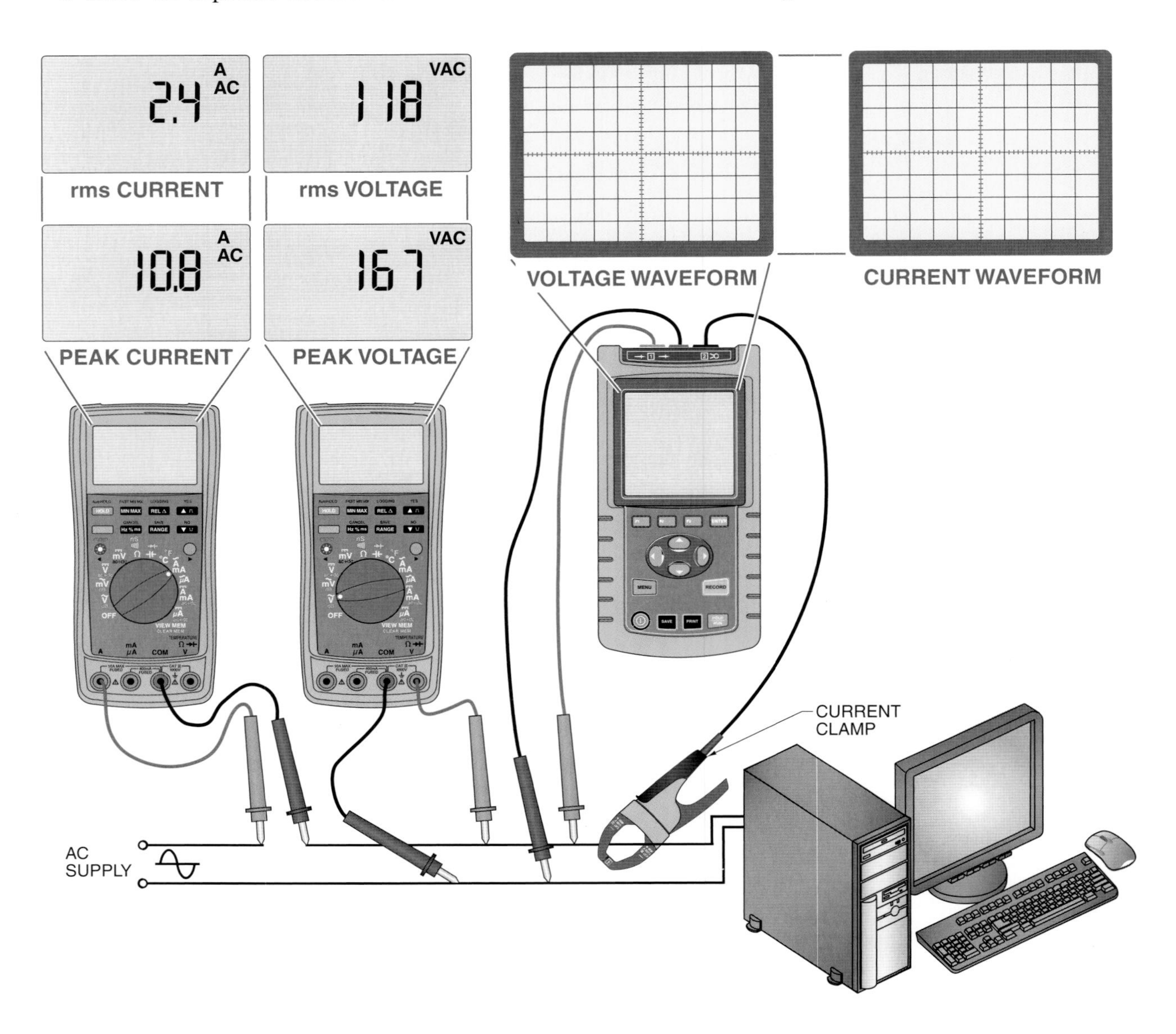

5. _______________ What is the crest factor of the voltage?

6. _______________ What is the crest factor of the current?

7. Draw the expected voltage waveform for the circuit in the oscilloscope screen.

8. Draw the expected current waveform for the circuit in the oscilloscope screen.

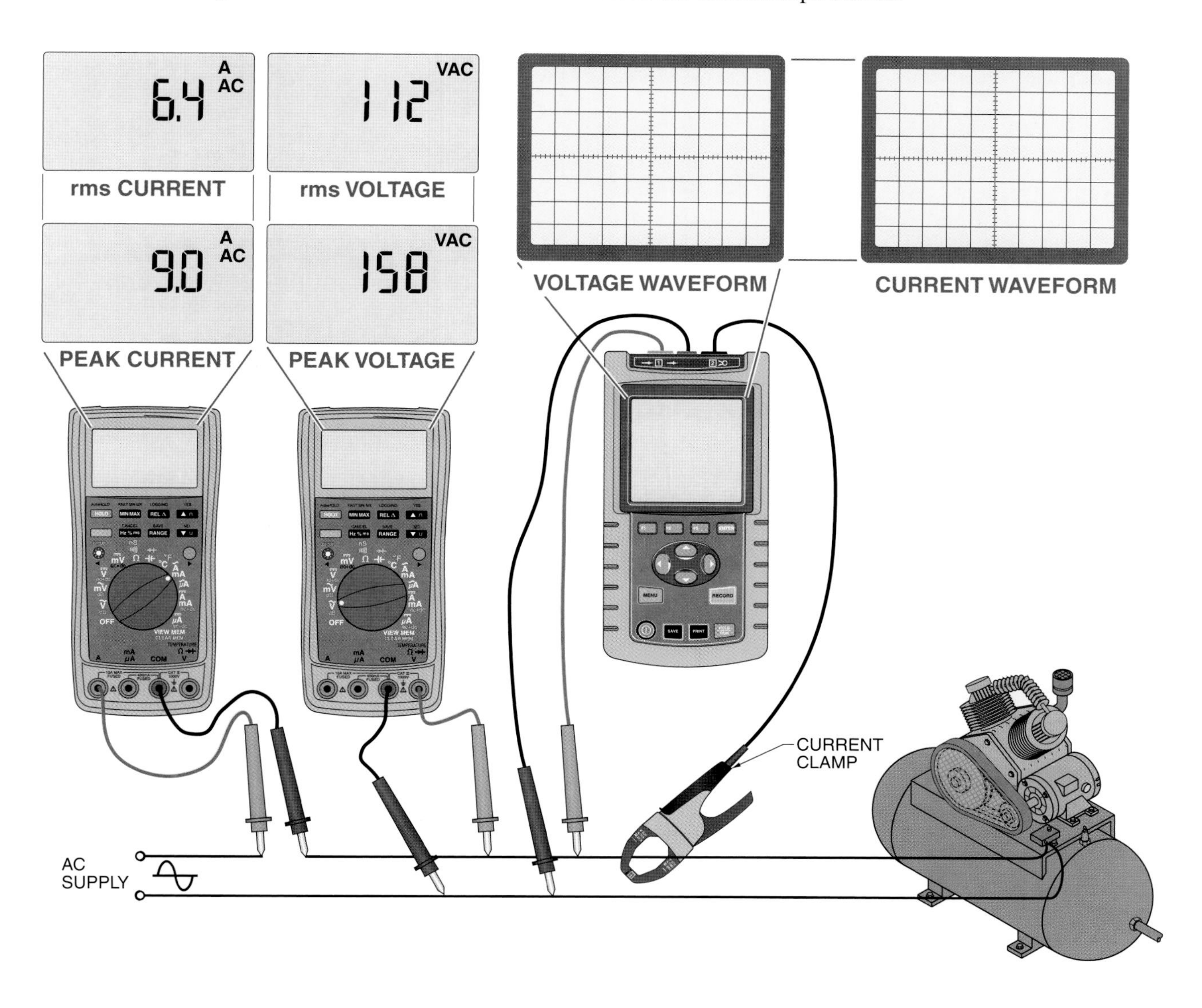

9. ______________ What is the crest factor of the voltage?

10. ______________ What is the crest factor of the current?

11. Draw the expected voltage waveform for the circuit in the oscilloscope screen.

12. Draw the expected current waveform for the circuit in the oscilloscope screen.

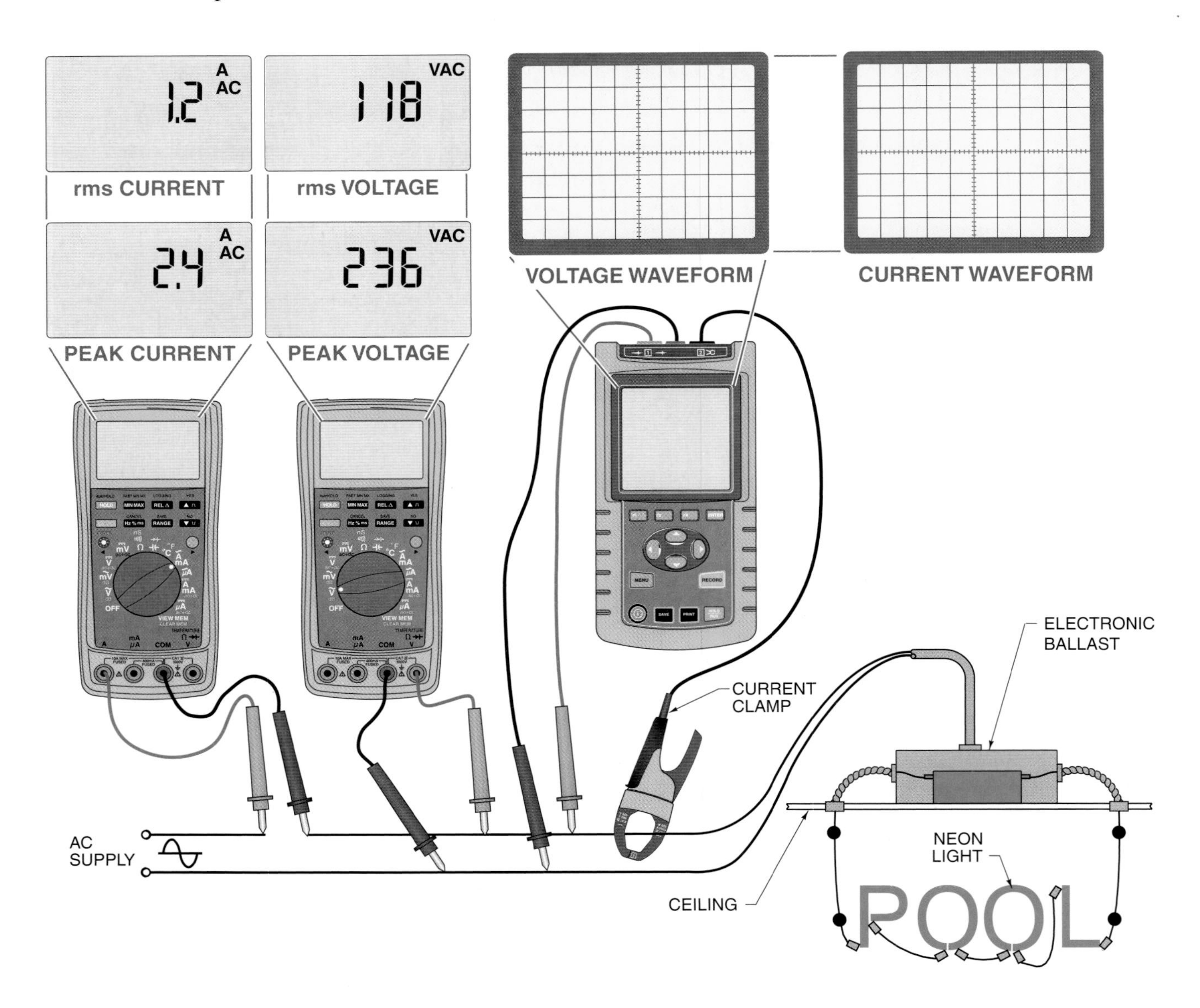

13. ______________ What is the crest factor of the voltage?

14. ______________ What is the crest factor of the current?

15. Draw the expected voltage waveform for the circuit in the oscilloscope screen.

16. Draw the expected current waveform for the circuit in the oscilloscope screen.

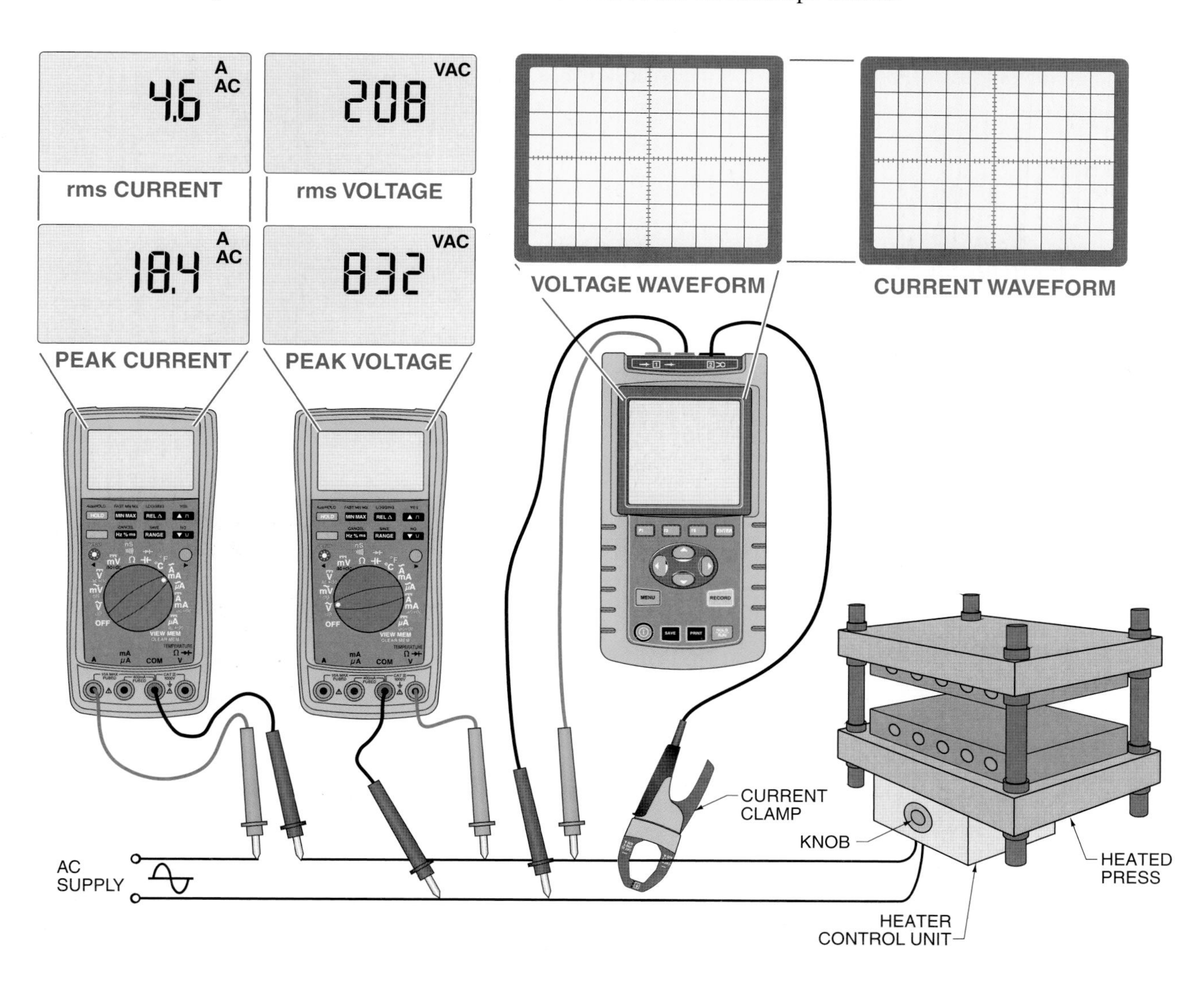

Application 4-7—Low Crest Factor

Nonlinear loads causing higher than normal current crest factors (greater than 1.41) can also produce lower than normal voltage crest factors (less than 1.41). When the high current is drawn at the voltage peak, it causes the voltage peak to be pulled down. This pulling down of voltage peaks is called "flat-topping." See Flat-Topped Voltage Waveform and Distorted Current Waveform.

Severe flat-topping can cause the normal voltage sine wave to appear like a square wave. This shape results in the peak and rms values being closer together. When crest factor is calculated, the value is less than 1.41. (The minimum possible crest factor is 1.0, which describes a perfect square wave.)

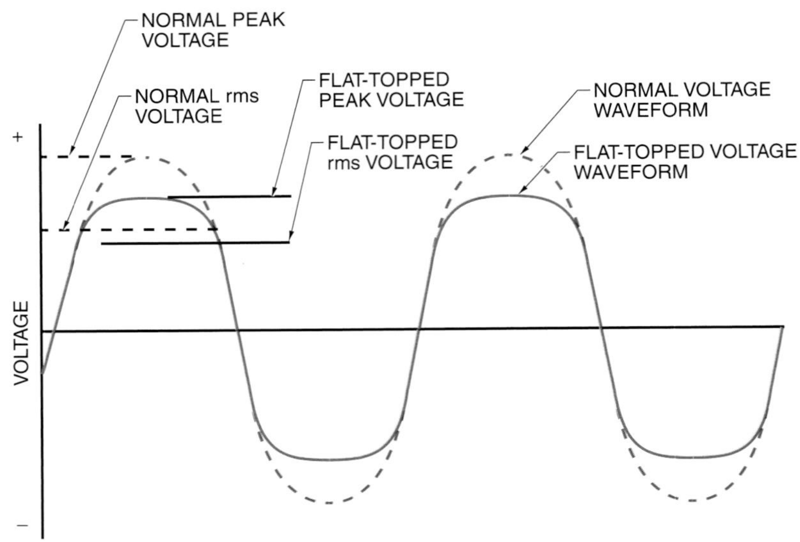

FLAT-TOPPED VOLTAGE WAVEFORM

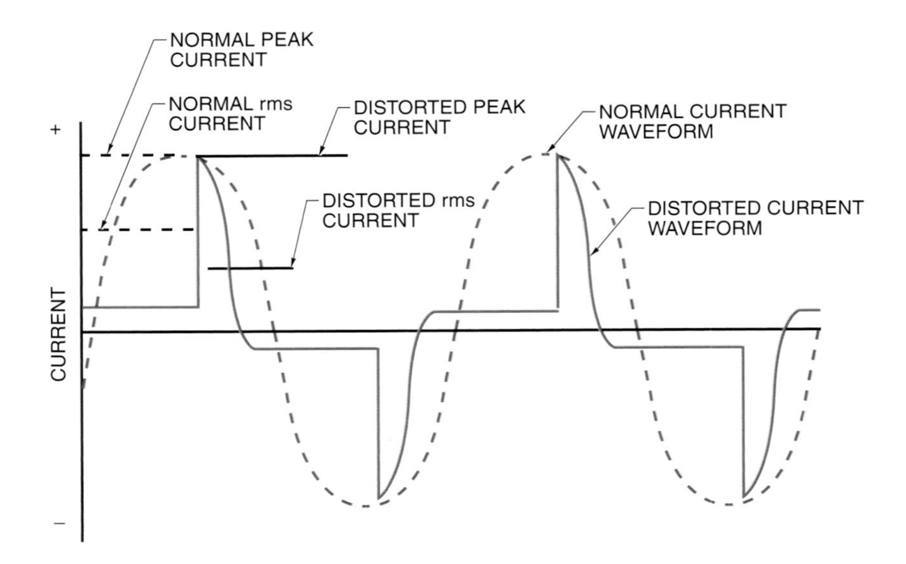

DISTORTED CURRENT WAVEFORM

A voltage flat-topping problem can be detected by calculating crest factor from peak and rms voltages. By calculating current crest factor as well, the problem can be attributed to distortion in the current waveform rather than some other problem.

1. ______________ What is the voltage crest factor?

2. ______________ What is the current crest factor?

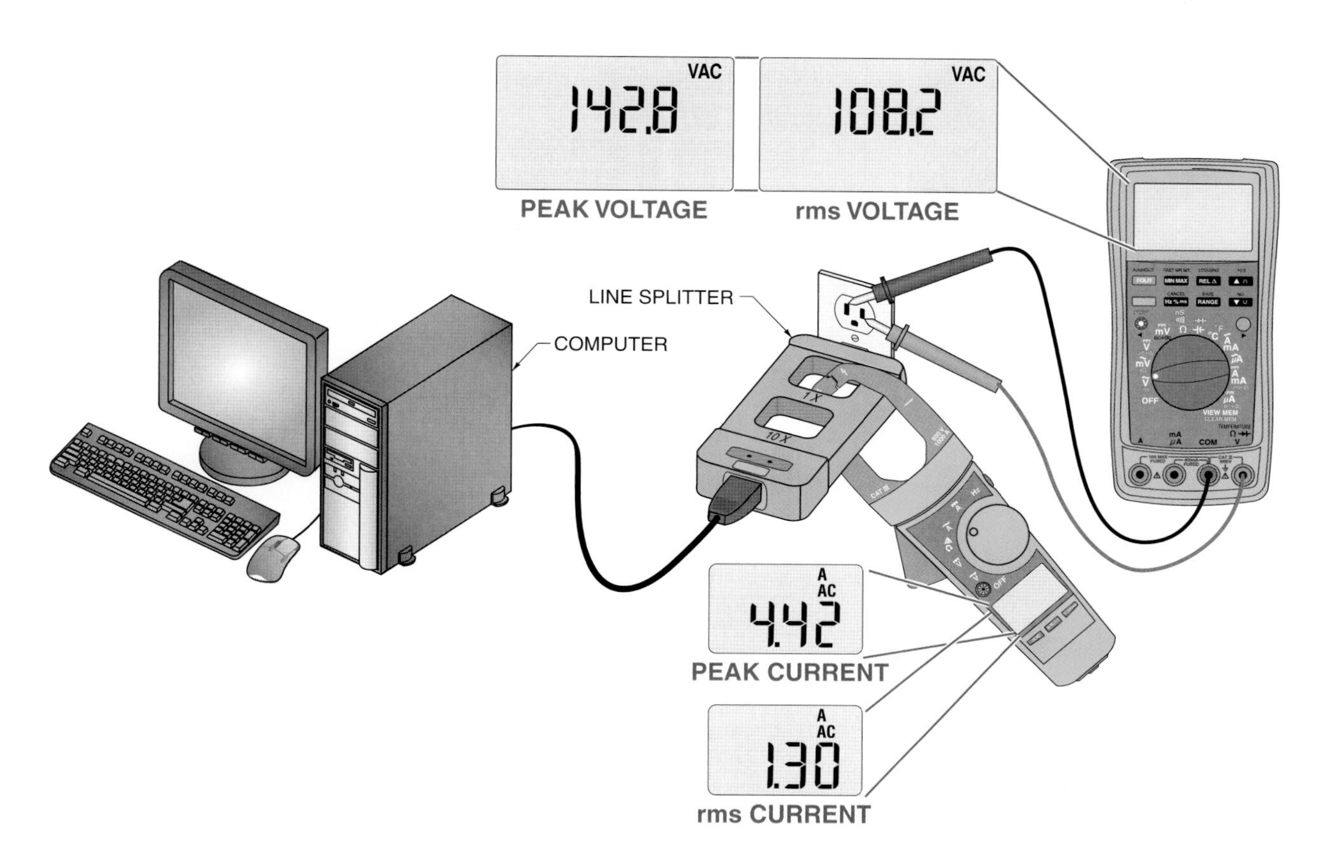

3. _______________ What is the voltage crest factor?

4. _______________ What is the current crest factor?

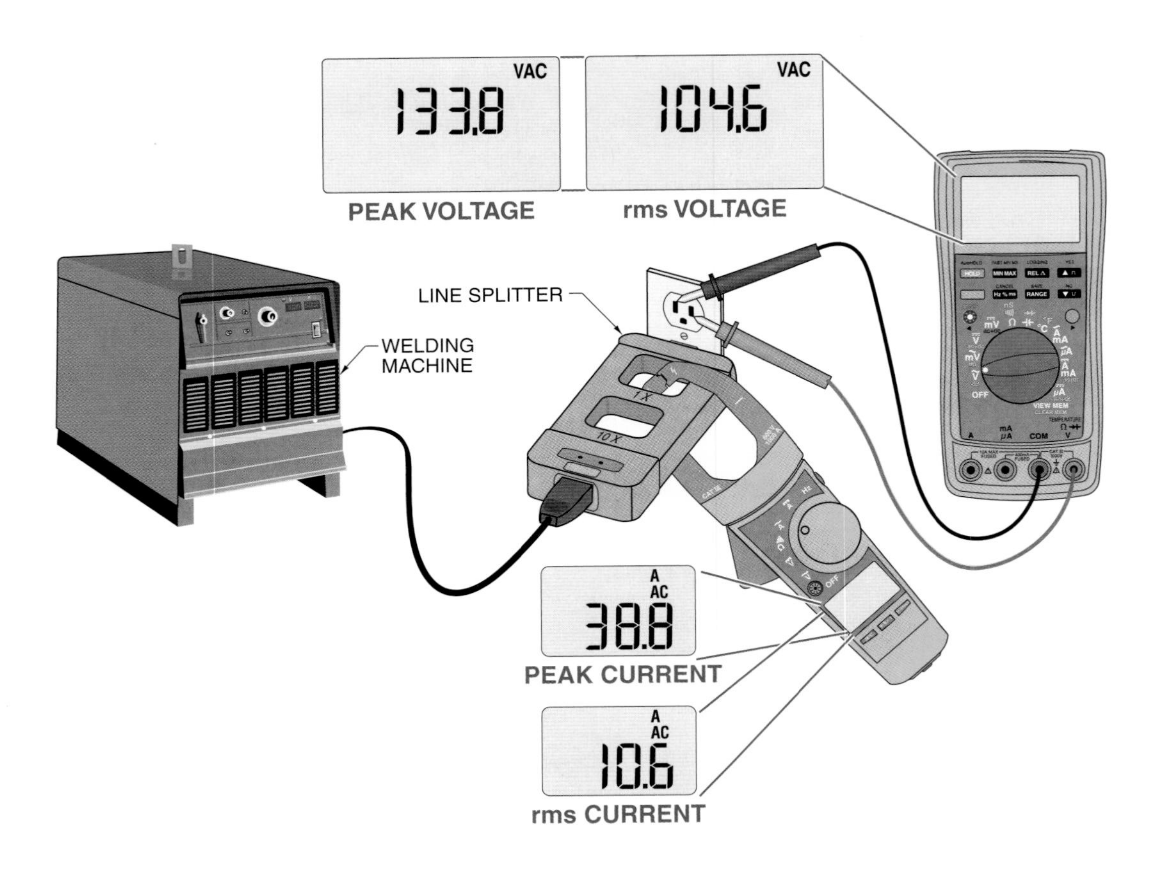

5. ______________ What is the voltage crest factor?

6. ______________ What is the current crest factor?

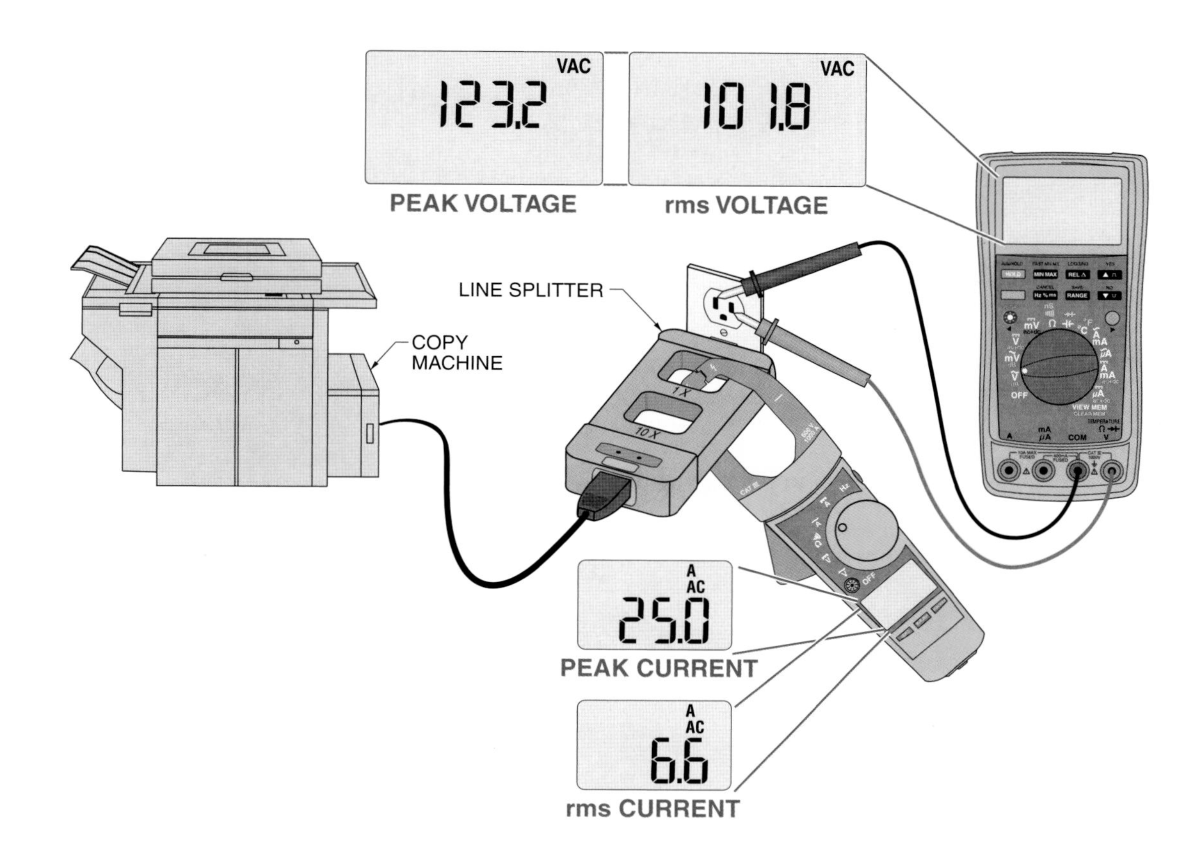

Application 4-8 — Voltage and Current Unbalance

Multimeters can be used to measure voltage and current unbalance in a three-phase power system by taking readings at each line. Three separate measurements must be taken and recorded. Unbalance is determined by calculating the percent difference between the largest deviation and the average value.

A three-phase power quality meter can also be used to measure the voltage and current unbalance. This is the quickest method because the meter is connected to each power line with voltage leads and current clamps and takes readings simultaneously. The meter also makes the calculations required to determine unbalance and displays the result directly.

1. _______________ What is the percentage of voltage unbalance at the motor?

2. _______________ Is the voltage unbalance within acceptable limits?

3. _______________ What is the percentage of current unbalance at the motor?

4. _______________ Is the current unbalance within acceptable limits?

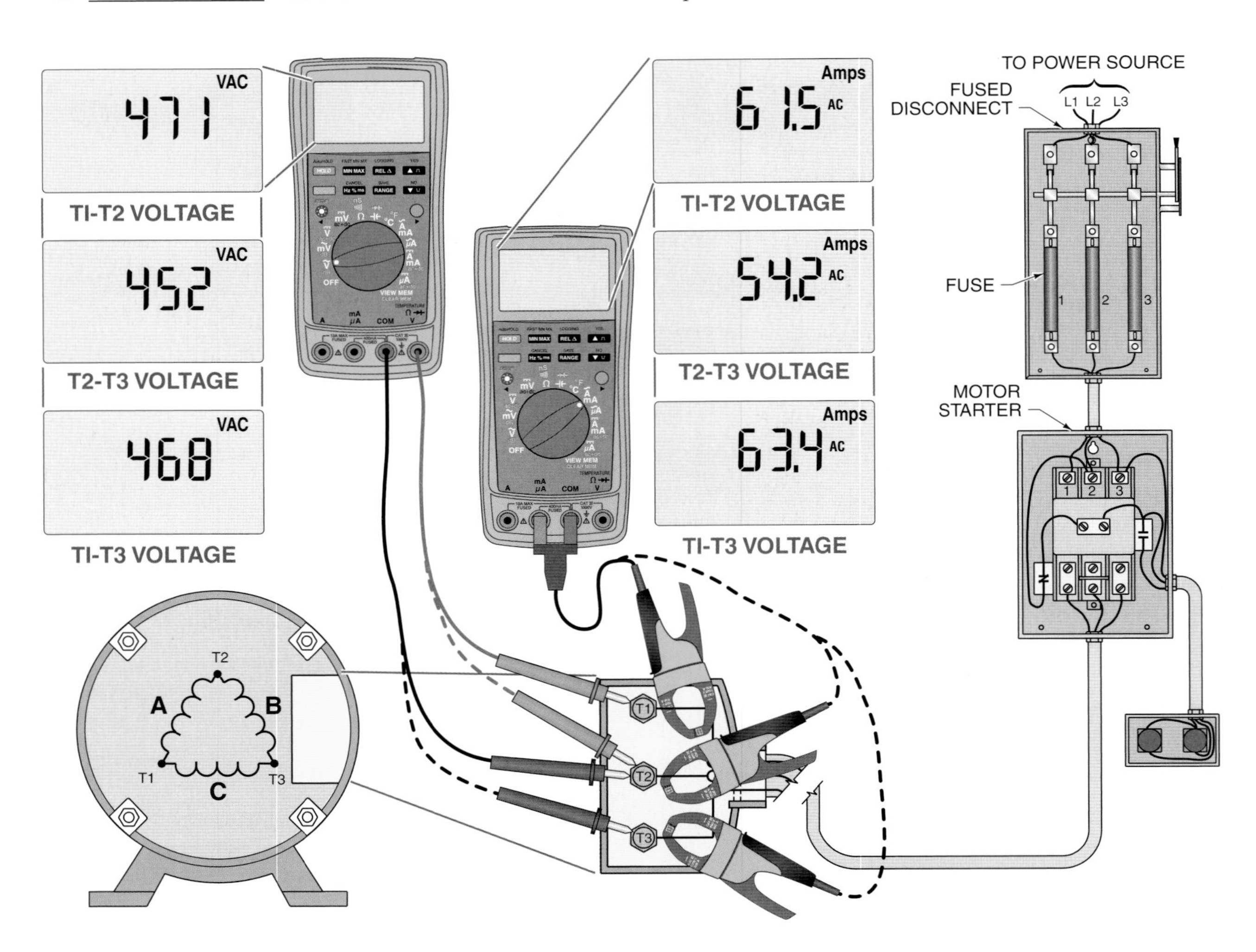

5. ______________ What is the percentage of voltage unbalance at the motor?

6. ______________ What is the percentage of current unbalance at the motor?

7. ______________ Based on the motor's nameplate information, are the current measurements within acceptable limits?

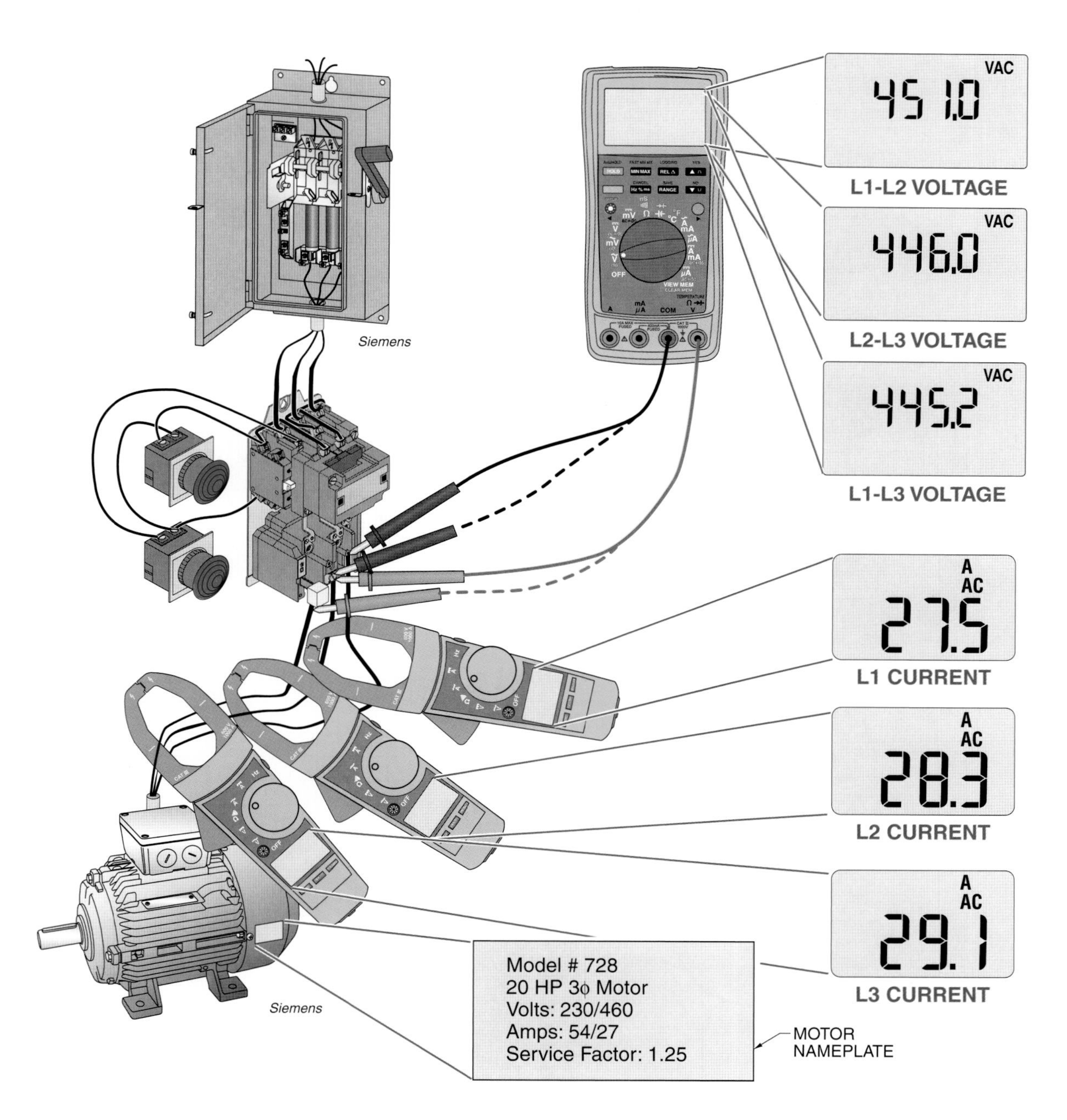

8. ______________ What is the percentage of voltage unbalance at the motor?

9. ______________ What is the percentage of current unbalance at the motor?

10. ______________ Are the current measurements within an acceptable range for a 20 HP motor with a nameplate current that could not be read? (The motor might be in a remote location, or the nameplate damaged or missing.)

11. ______________ Do the measurements indicate that the voltage waveform is sinusoidal or distorted?

12. ______________ Do the measurements indicate that the current waveform is sinusoidal or distorted?

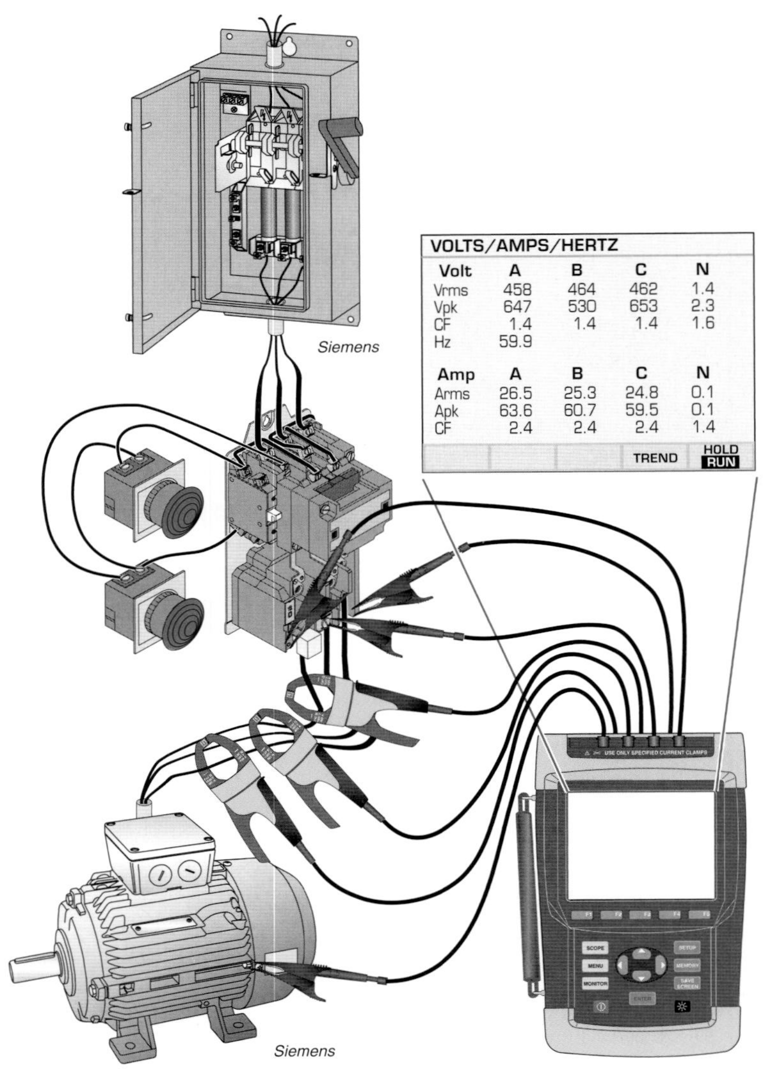

FULL-LOAD CURRENTS – 3φ, AC INDUCTION MOTORS

Motor Rating*	Current†			
	208 V	230 V	460 V	575 V
¼	1.11	0.96	0.48	0.38
⅓	1.34	1.18	0.59	0.47
½	2.2	2.0	1.0	0.8
¾	3.1	2.8	1.4	1.1
1	4.0	3.6	1.8	1.4
1½	5.7	5.2	2.6	2.1
2	7.5	6.8	3.4	2.7
3	10.6	9.6	4.8	3.9
5	16.7	15.2	7.6	6.1
7½	24.0	22.0	11.0	9.0
10	31.0	28.0	14.0	11.0
15	46.0	42.0	21.0	17.0
20	59	54	27	22
25	75	68	34	27
30	88	80	40	32
40	114	104	52	41
50	143	130	65	52
60	169	154	77	62
75	211	192	96	77
100	273	248	124	99
125	343	312	156	125
150	396	360	180	144
200	—	480	240	192
250	—	602	301	242
300	—	—	362	288
350	—	—	413	337
400	—	—	477	382
500	—	—	590	472

* in HP
† in A

Application 4-9 — Circuit Power in Resistive Loads

Ohm's law and the power formula can be applied to circuits in which electrical resistance (such as heating elements or incandescent lamps) is the only significant opposition to the flow of current. Theoretical power can be calculated by using the voltage rating of the load and a measured resistance. See Calculating Power.

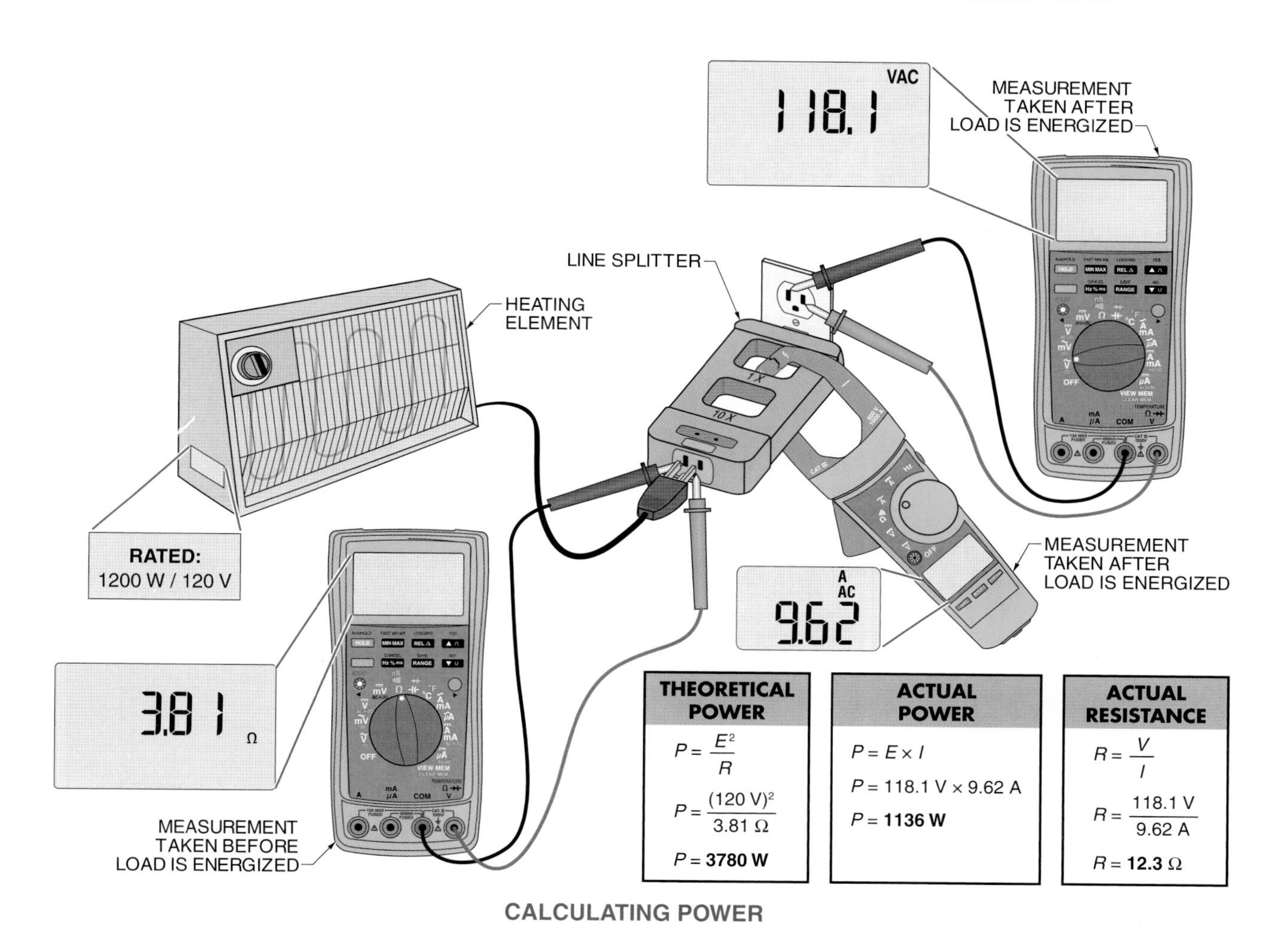

CALCULATING POWER

Actual power output is easily calculated by multiplying voltage by current, which are both measured when the load is energized. Also, resistance calculated from these values is more accurate than directly measured resistance. This is because some loads change resistance significantly when power is applied. For example, a lamp has a much higher resistance when energized than when de-energized. This higher resistance reduces the lamp's actual power output from the lamp's theoretical power output. However, resistance measurements cannot be taken when a circuit is powered, so actual resistance in energized loads must be derived from other measurements.

1. ______________ What is the theoretical power output of the load?

2. ______________ What is the actual power output of the load?

3. ______________ What is the actual resistance of the load?

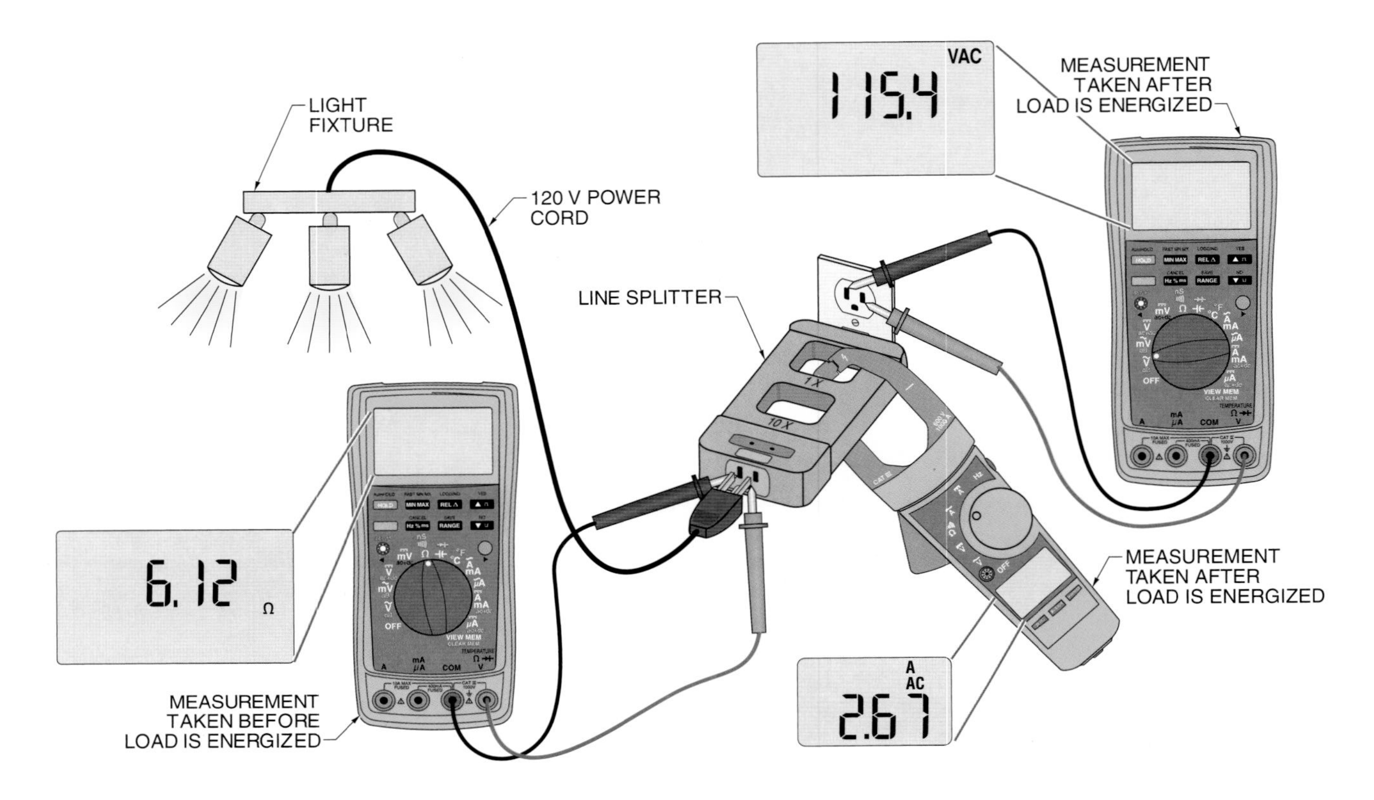

4. ______________ What is the theoretical power output of the load?

5. ______________ What is the actual power output of the load?

6. ______________ What is the actual resistance of the load?

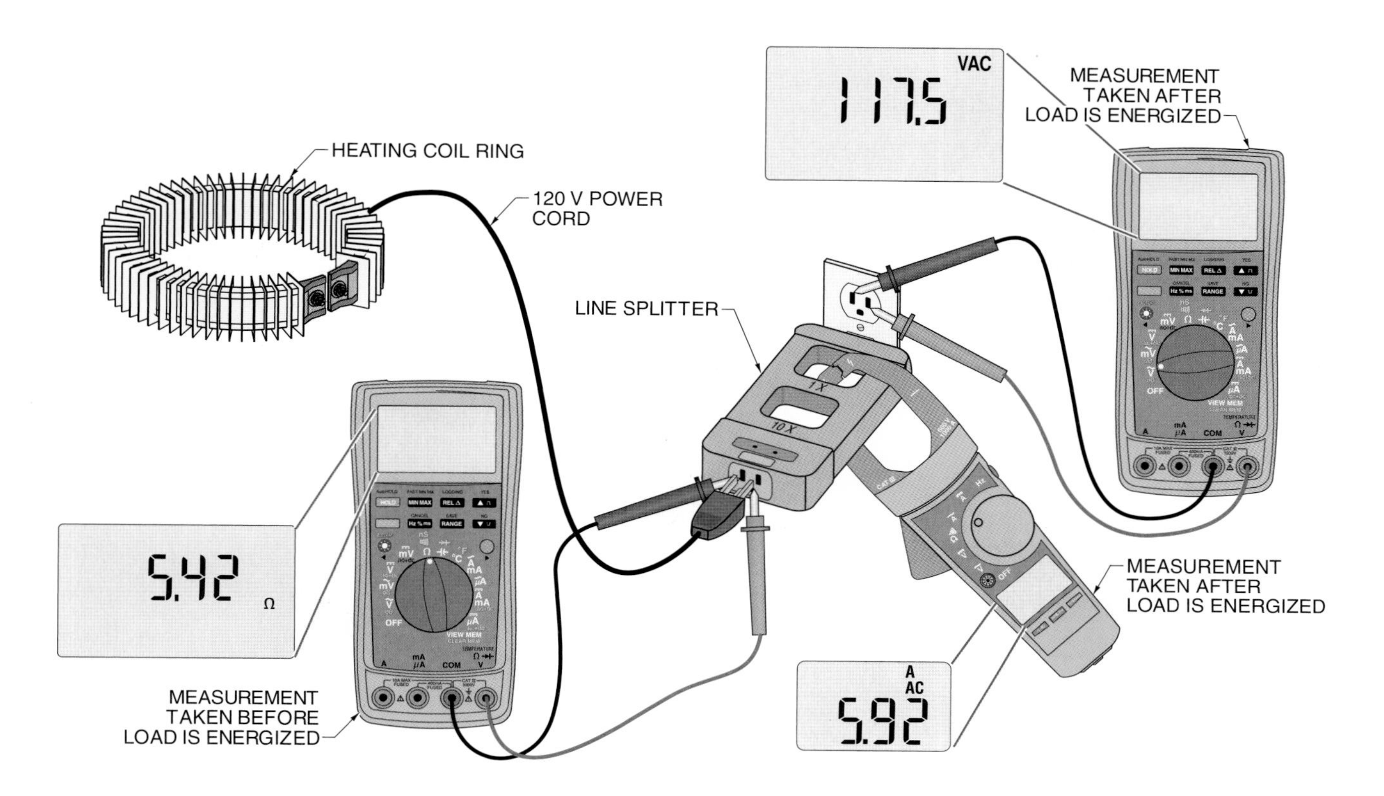

7. ______________ What is the theoretical power output of the load?

8. ______________ What is the actual power output of the load?

9. ______________ What is the actual resistance of the load?

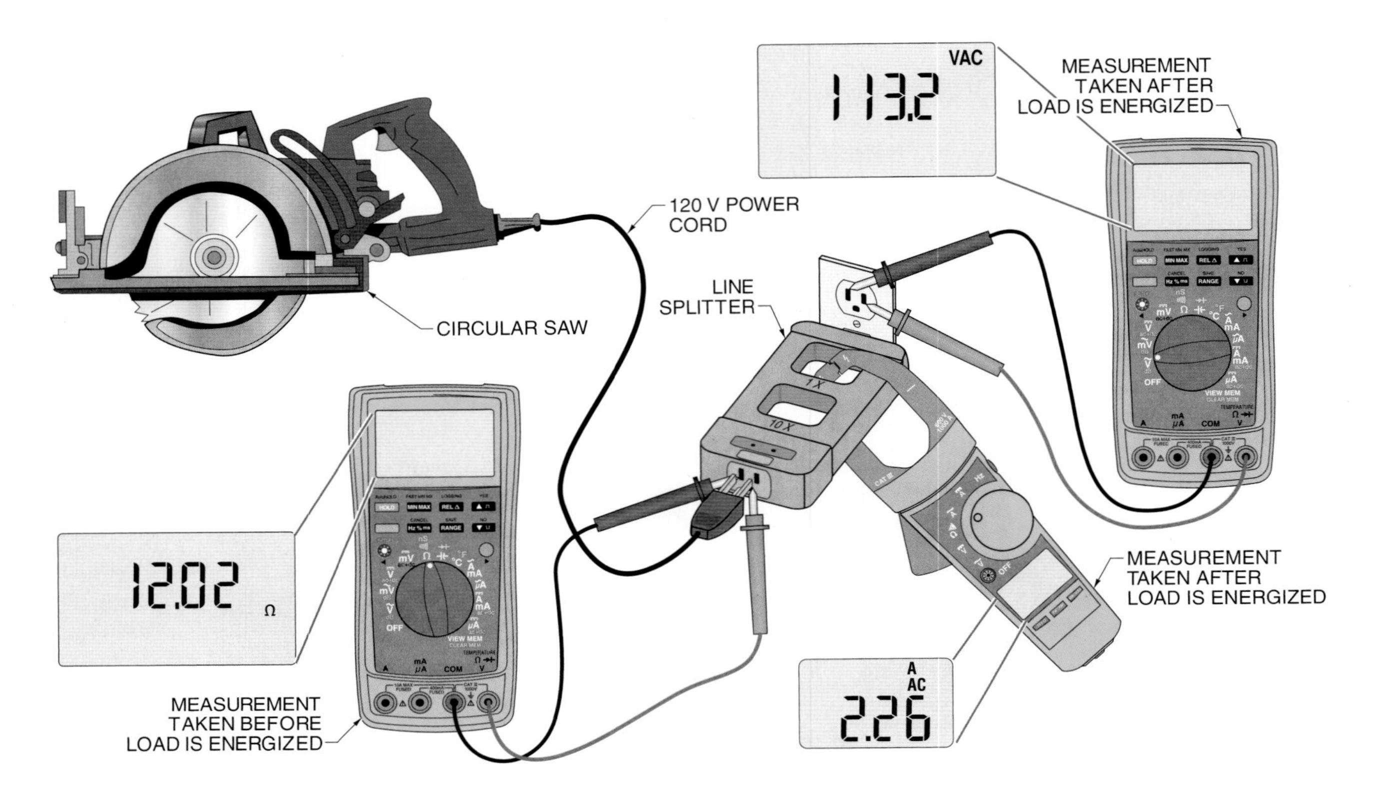

ELECTRONIC CIRCUIT TEST INSTRUMENTS 5

APPLICATIONS

Name: ______________________________ Date: ____________________

Application 5-1 — Line Splitters

Taking current measurements requires either opening the circuit for an in-line current measurement (usually 10 A or less), or using a clamp-on current ammeter or attachment. It is usually easier to take clamp-on meter measurements than in-line measurements because they do not require opening or modifying the circuit in any way. Another advantage of using a clamp-on ammeter or attachment is that they can measure several hundreds (or thousands) of amps without the potential safety hazard of opening the circuit.

There are two disadvantages to using a clamp-on ammeter for current measurements. First, clamp-on ammeters are generally not designed to take small current measurements (typically less than 1 A). The second is that clamp-on ammeters must be enclosed around only one conductor. This can be difficult when trying to measure the current of loads such as computers, power tools, portable heaters, or any other load that uses a two- or three-conductor power cord.

A line splitter is a measurement accessory that is designed to make measuring current easier for loads that are low-current devices or for when it is difficult to find a single-conductor measuring point. Most line splitters include a 1x slot and a 10x slot where the current clamp can easily be placed. When placed in the 1x slot, the ammeter displays the actual circuit current and is used for measurements 1 A or higher. When placed in the 10x slot, the ammeter displays 10 times the actual circuit current and is used for measurements less than 1 A. By multiplying the current by 10, the clamp-on meter can take a more accurate measurement. When using the 10x slot, the meter reading must be divided by 10 in order to get the actual current measurement. See Line Splitter.

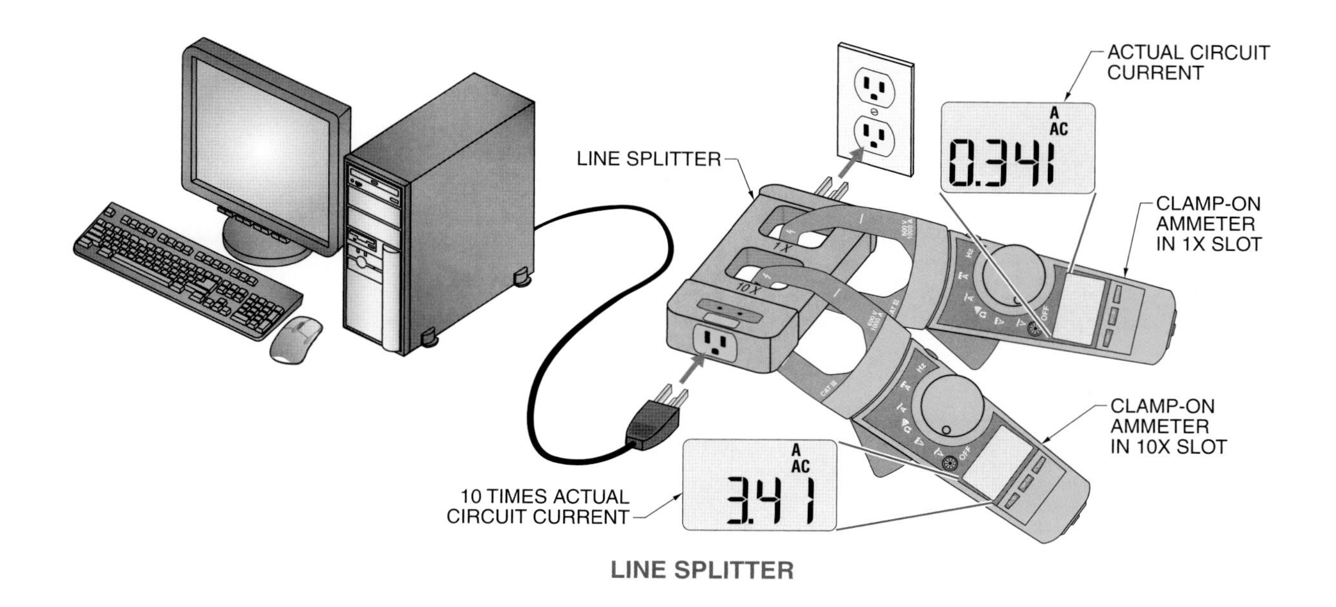

LINE SPLITTER

When using a clamp-on attachment accessory on a multimeter, careful attention must be paid to the output type. Some clamp-on attachments output 1 mA/A and others output 1 mV/A. For example, an actual current of 2.45 A will read as 2.45 mA on some clamp-on attachments and 2.45 mV on others. The clamp's output type, mA or mV, will also determine the correct setting and connection of the multimeter.

1. Connect the clamp attachment to the proper meter jacks on the multimeter.

2. Set (draw in) the function switch on the multimeter to the correct position.

3. ______________ If the meter reads 84.35, what is the actual current draw (in A) of the load under test?

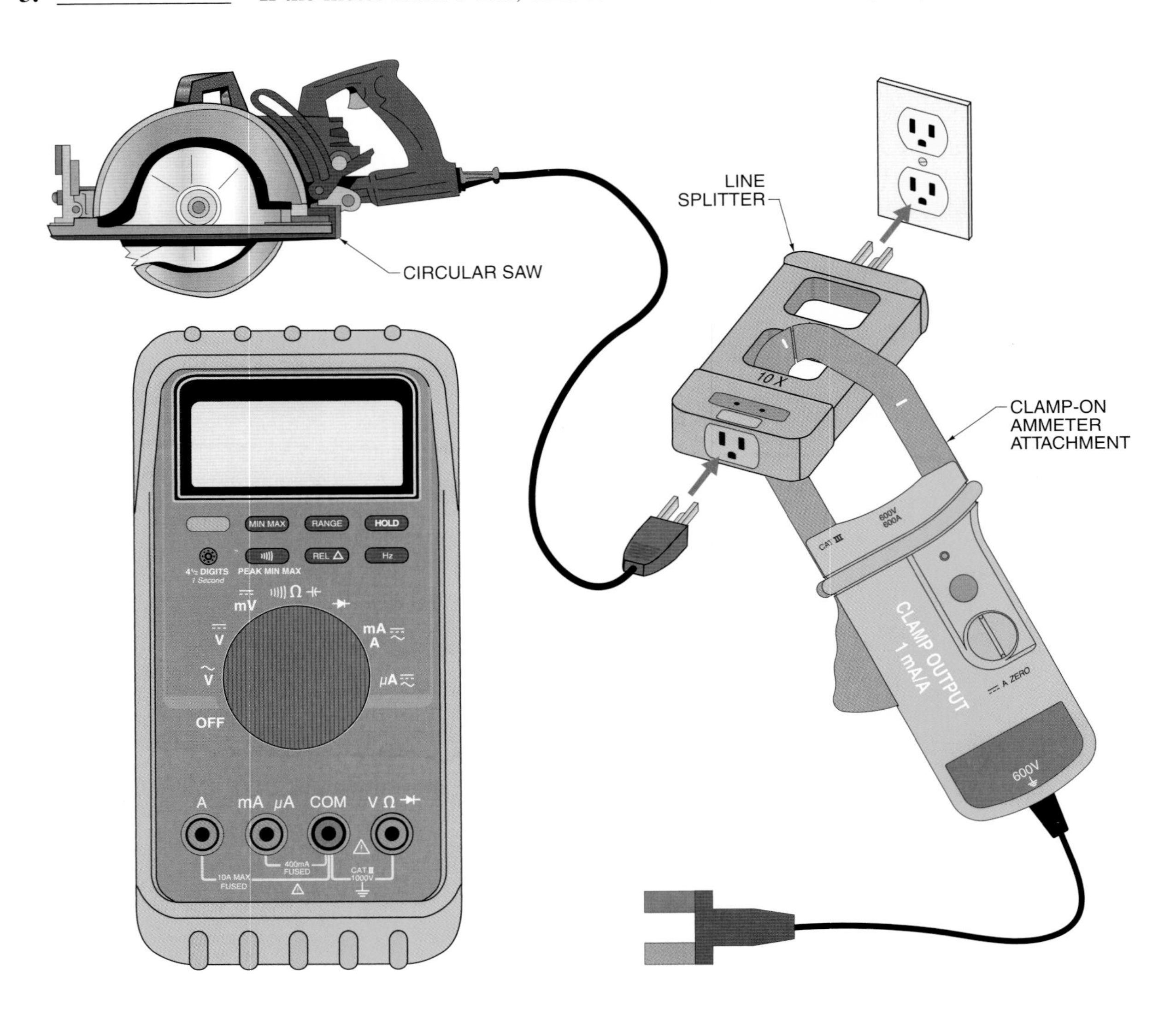

4. Connect the clamp attachment to the proper meter jacks on the multimeter.

5. Set (draw in) the function switch on the multimeter to the correct position.

6. ______________ If the meter reads 0.097, what is the actual current draw (in A) of the load under test?

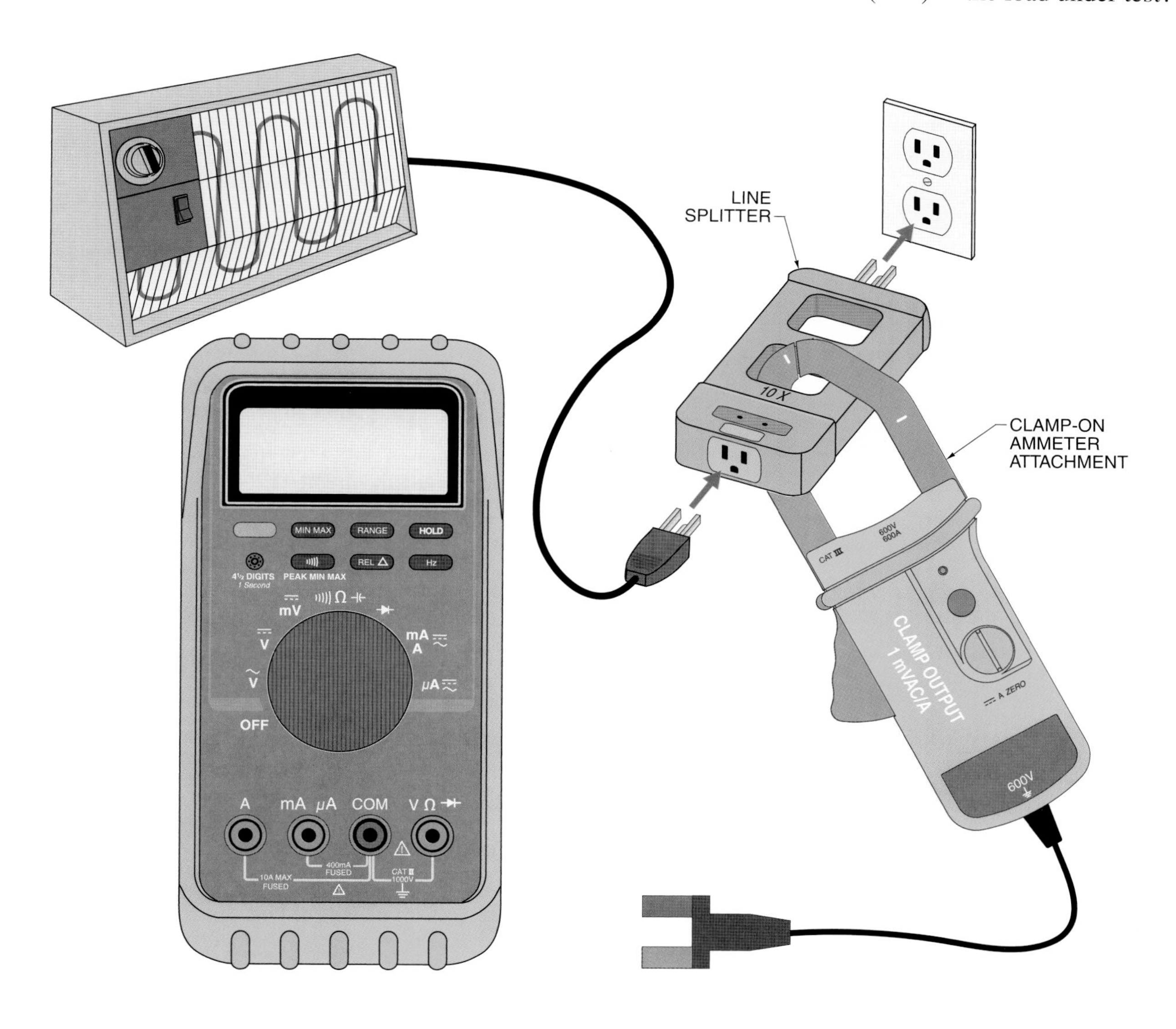

Application 5-2—Troubleshooting Hidden Diodes

When AC powers DC equipment, diodes are used to convert the AC into DC. The most familiar examples of this are the power converters used to recharge cell phones or laptop computers from 120 VAC receptacles. The power converters are plugged into a 120 VAC receptacle and deliver 9 VDC, 12 VDC, or other DC voltages. These are common, and a troubleshooter usually knows when to set a meter for an AC voltage measurement or a DC voltage measurement. A troubleshooting problem exists when the troubleshooter assumes a circuit is AC that is actually DC, or the troubleshooter assumes a circuit is DC that is actually AC. In such cases, the incorrect setting of the meter (set on AC when it should be on DC) will cause measurement errors and incorrect diagnoses about the actual problem.

Several rules can help eliminate measurement error when troubleshooting an unknown voltage type, or when measurements do not seem right.

Voltage Measurement Rules

- Most meters can be set to measure AC voltage and be connected to a DC voltage without damaging the meter, circuit, or circuit components. However, the measured results will not be correct.
- Most meters can be set to measure DC voltage and be connected to an AC voltage without damaging the meter, circuit, or circuit components. However, the measured results will not be correct.
- If the voltage type (AC or DC) is unknown, take a measurement with the meter set to measure DC voltage. Reverse the meter test leads and take the measurement again. If the voltage is DC at the test point, the two measured values will be the same but one will have a negative (–) reading (for example, 12 VDC and –12 VDC).

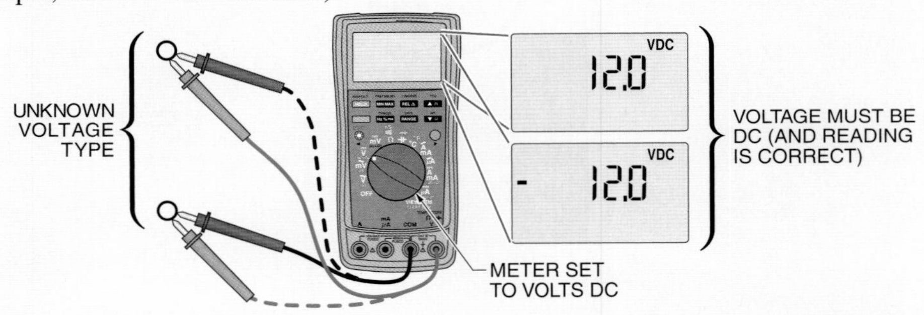

- If the two measured values are both negative and fluctuating (like ghost voltages), the voltage is not DC. Set the meter to measure AC and retake the measurements. If the voltage at the test point is AC, both readings will be the same (for example, 117.5 VAC and 117.5 VAC) and neither one will be negative.

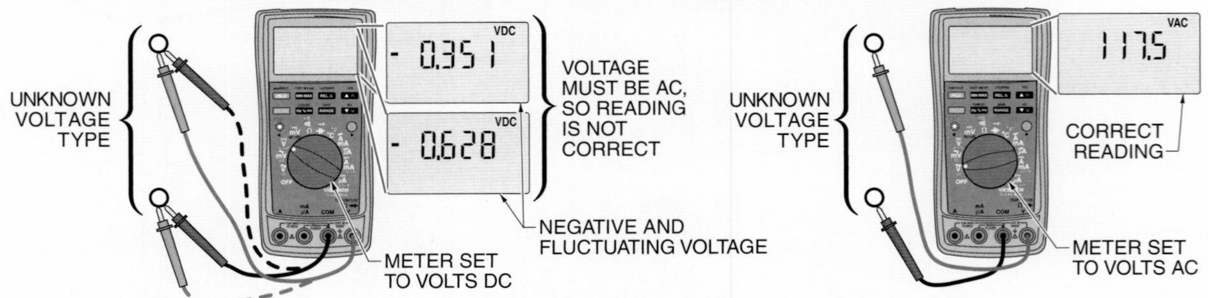

- If the measurements do not clearly indicate whether the voltage at the test point is DC or AC, a graphic display voltmeter (scope type) must be used to observe the voltage waveform. The viewed waveform may show that the voltage includes both AC and DC elements.

Troubleshooting a circuit that includes a solenoid is a good example of test measurements that may not seem correct, even if there is no problem. AC solenoids are rated either intermittent- or continuous-duty. An intermittent solenoid uses AC through a coil to develop a magnetic field and move the solenoid plunger. The AC voltage will heat the coil because it is constantly varying in magnitude and direction. However, the heat buildup will not be a problem if the solenoid is only energized for a short time (usually less than 60 seconds). See Intermittent-Duty Solenoid.

If the solenoid is to be energized for a longer time period, then either the coil must be physically larger to dissipate the heat or the solenoid must use DC instead. Intermittent-duty-rated solenoids are converted to continuous-duty by placing a diode in series with the coil. The diode converts the AC voltage into half-wave DC voltage. Half-wave DC voltage allows for a cooler operating solenoid because the voltage never reverses direction and is only applied during the half-cycle in which the diode allows current to flow through the coil. Since there are still 60 half-cycles of voltage per second (from the 60 Hz AC supply), the coil's magnetic field is still strong enough to hold the solenoid plunger in place. See Continuous-Duty Solenoid.

Since the solenoid's nameplate lists the solenoid voltage as AC (for example 120 VAC) and the diode is usually hidden under the cover of the solenoid, it can be difficult to determine that the AC-rated device is actually operating on DC.

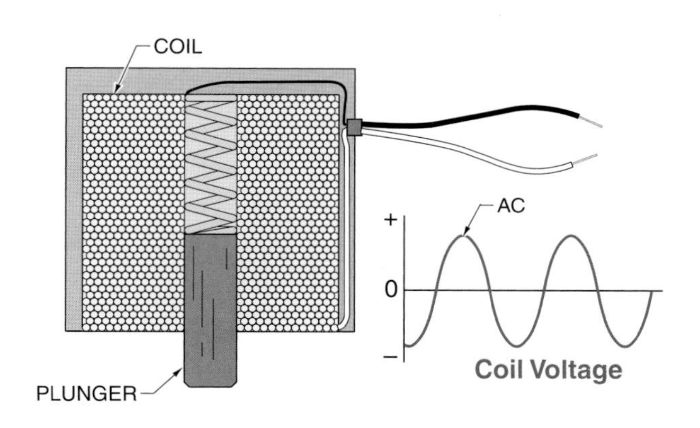

INTERMITTENT-DUTY SOLENOID

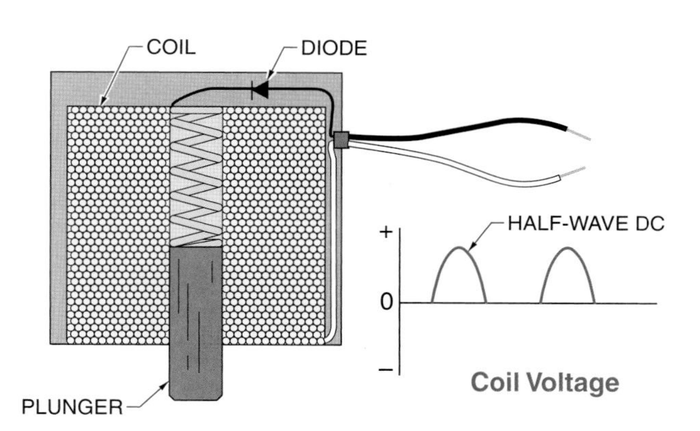

CONTINUOUS-DUTY SOLENOID

An electrician troubleshoots a switch and solenoid in a product handling unit that is operating erratically. Since the solenoid is rated at 120 V/30 W, the multimeters used to measure voltage in the circuit are set to measure VAC.

1. ______________ What voltage should Multimeter 2 display when the pushbutton is open?

2. ______________ What voltage should Multimeter 2 display when the pushbutton is closed?

3. ______________ What voltage should Multimeter 1 display when the pushbutton is closed?

4. ______________ Is the voltage between Test Points 2 and 3 AC or DC?

When the pushbutton is open, Multimeter 1 displays unexpected readings that fluctuate between very small voltage values. The electrician suspects that there may be a diode in the solenoid.

5. ______________ Is the voltage between Test Points 1 and 2 AC or DC?

6. ______________ Which measurement function should be set on Multimeter 1 to measure the correct type of voltage?

7. ______________ Will Multimeter 1 read a constant voltage while the pushbutton is open?

8. ______________ If the diode shorted out, which type of voltage (AC or DC) would be applied to the solenoid when the pushbutton closed?

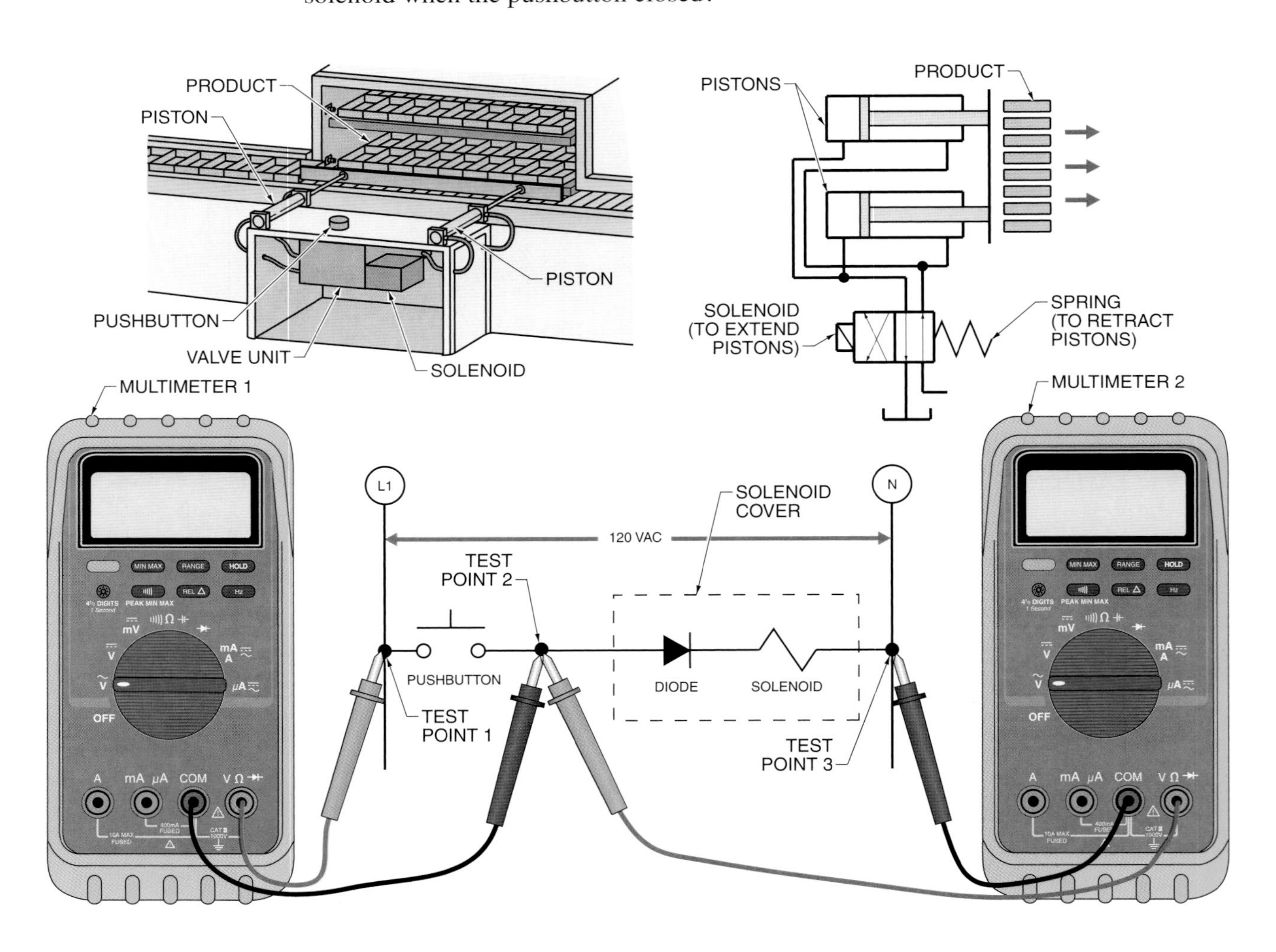

9. Draw the waveform pattern that would be viewed on Oscilloscope 1 when the pushbutton is open.

10. Draw the waveform pattern that would be viewed on Oscilloscope 2 when the pushbutton is open.

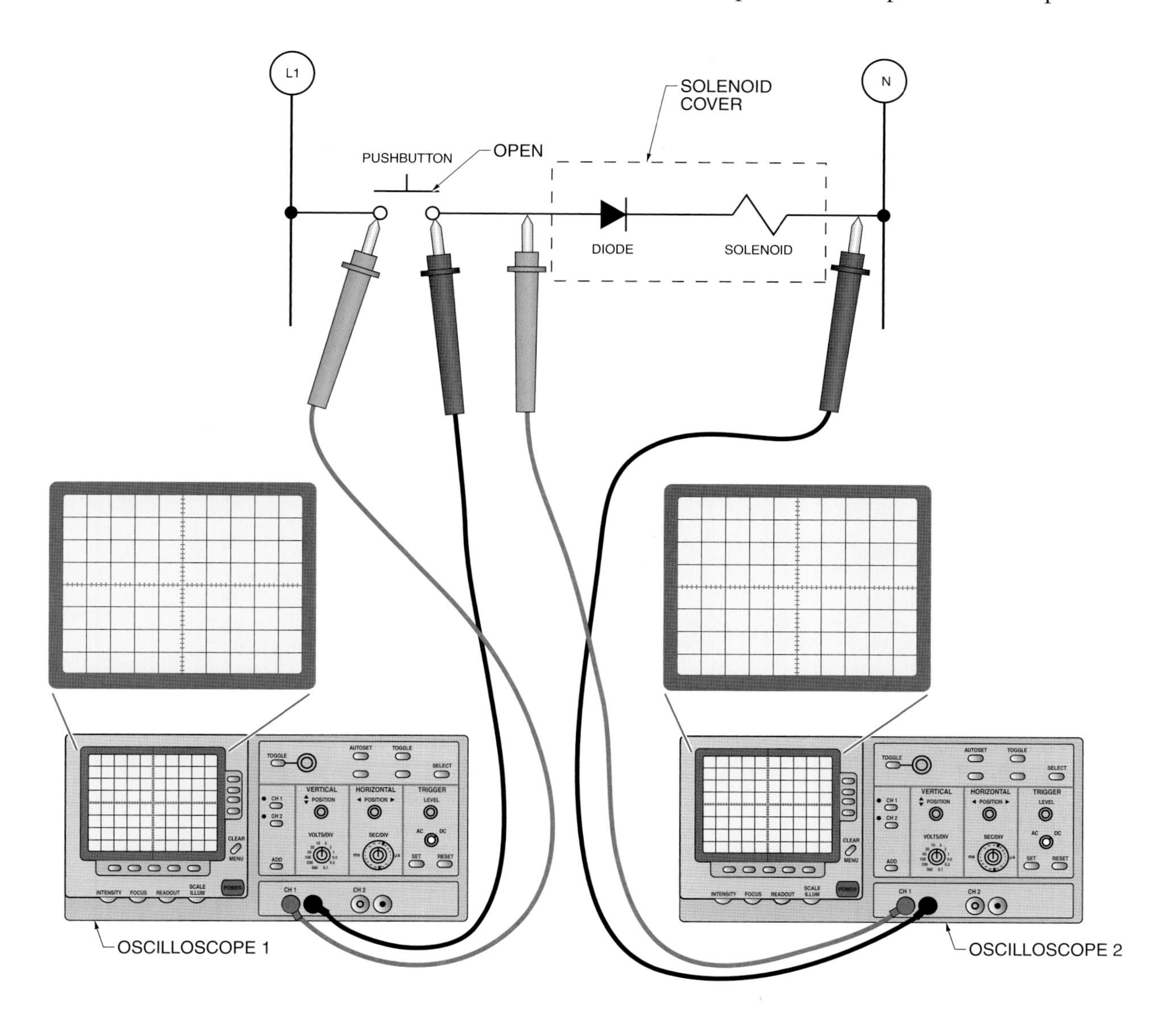

11. Draw the waveform pattern that would be viewed on Oscilloscope 3 when the pushbutton is closed.

12. Draw the waveform pattern that would be viewed on Oscilloscope 4 when the pushbutton is closed.

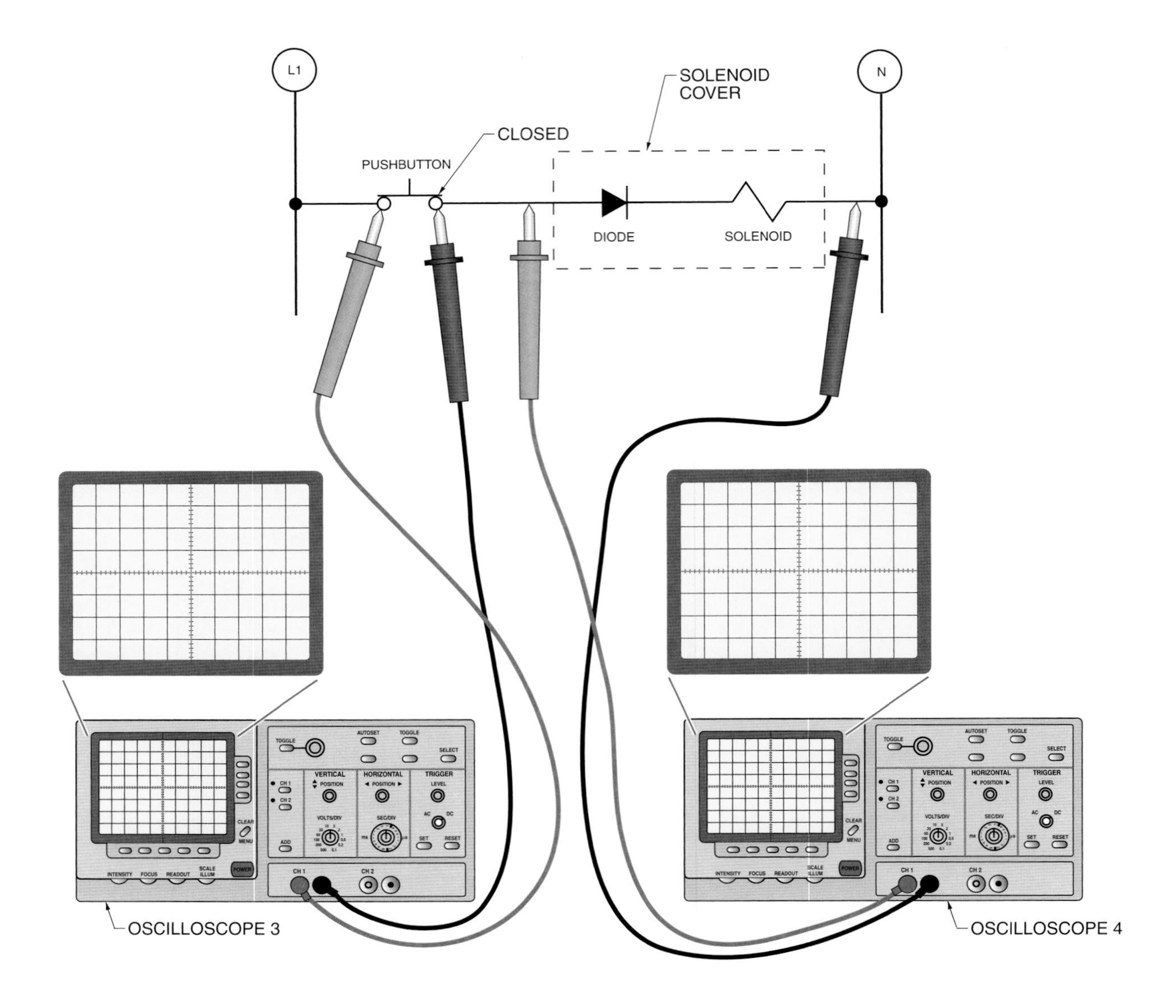

Application 5-3—Automotive Charging System Testing

At the heart of the automobile's electrical and electronic systems is a power supply that must deliver the required power (voltage and current) at all times. The battery is the initial power source to start the automobile and supplies enough power to keep some electronic systems operational when the automobile is not running. However, once the automobile is running, it is the alternator that delivers enough power to operate all the electrical devices and recharge the battery. The alternator is a 3ϕ generator that includes a rectifier (diode) circuit to convert the generated AC into DC. A voltage regulator is used to maintain the proper DC voltage level. See Battery Charging Circuit.

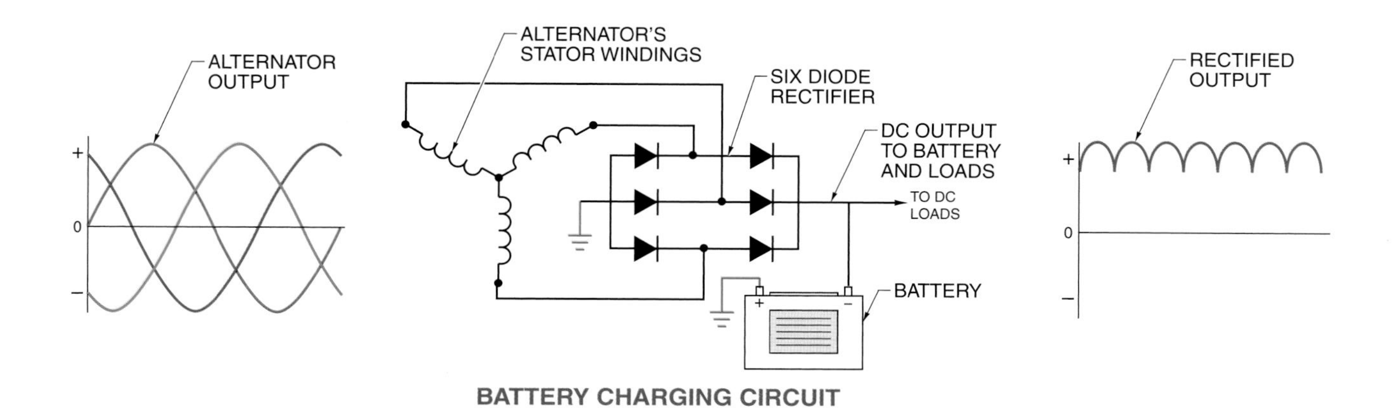

BATTERY CHARGING CIRCUIT

When problems occur in a residential, commercial, or industrial AC electrical system, the power source (power panels, transformers, etc.) is a good place to start troubleshooting. Making sure that the supply power is at the correct voltage and current levels, properly grounded, and not overloaded helps determine potential problems that may be showing up downstream of the power supply. Likewise, a good troubleshooting starting point in automobile electrical circuits is to make sure the power source (battery or alternator) is delivering enough power and at the correct levels at all times. Testing the automobile's charging system will determine the condition of the power supply.

With the engine turned OFF, a meter will display the no-load battery voltage, which should range from 12.4 V to 12.6 V. To test the charging system, the engine is started and brought up to approximately 1500 rpm. A meter set to relative mode will display a positive or negative voltage change, which should range from 0.5 VDC to 2 VDC. If the measured value is higher than 2 VDC, the system is overcharging the battery. If the measurement is a negative value (lower than the no-load voltage) a negative sign appears in front of the measurement and the charging system is not charging the battery.

To fully test the charging system, increase the engine rpm to approximately 2000 rpm and turn ON all major accessories (air conditioning, lights, etc.). The meter display will show the full-load voltage that is above or below the no-load voltage. The full-load voltage should be at least 0.5 V.

1. Set (draw in) the multimeter function switch to measure the voltage at the battery.
2. Connect the multimeter test leads to the proper battery terminal posts.
3. Circle the multimeter function button that zeroes out the no-load voltage measurement.
4. _______________ If the no-load voltage measurement was 12.42 V and the meter displays 0.72 V during the 1500 rpm test, what is the actual output of the charging system?
5. _______________ If the meter displays –0.43 during the 1500 rpm test, what is the actual output of the charging system?
6. _______________ If the voltage measurement decreases to –1.32 V during the 2000 rpm test, what is the actual output of the charging system?
7. What do the negative values indicate?

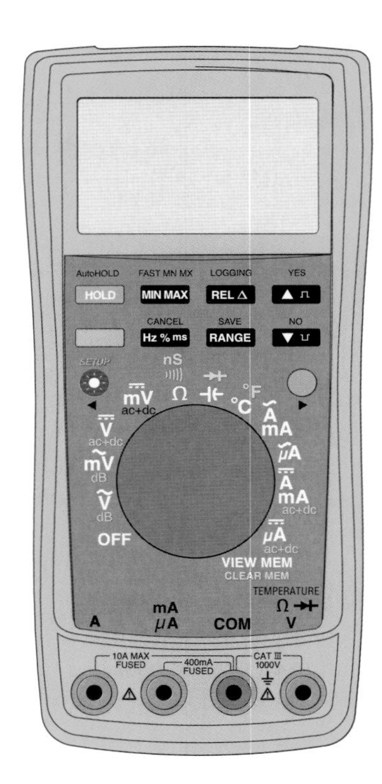

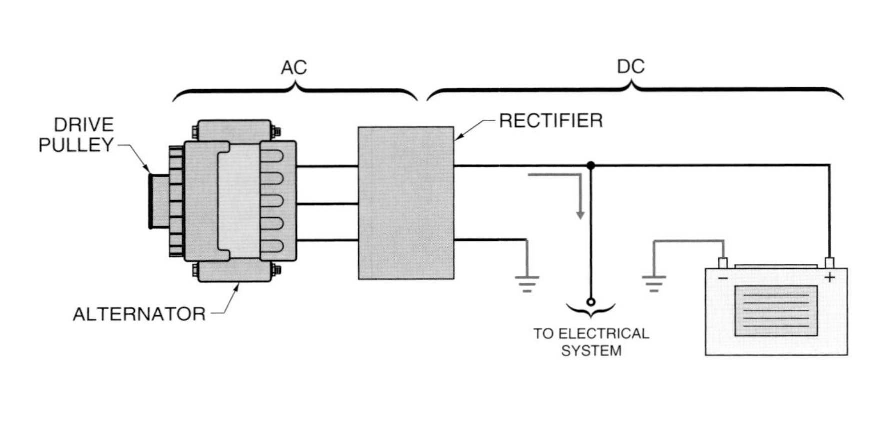

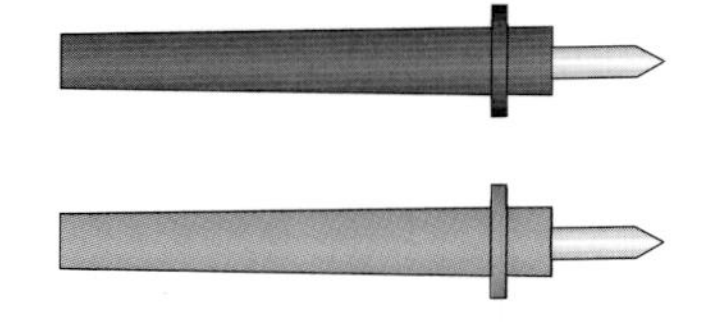

Application 5-4—Digital Logic Probes

Digital logic gates (AND, OR, NOT, etc.) are combined to develop various logic functions that can solve almost any control application. Digital logic probes are used to test and troubleshoot digital circuits. In most cases, a digital circuit is replaced (not serviced) if the digital circuit is identified as defective. However, it is important to understand a digital logic system and know how to locate a fault instead of relying on changing boards to solve a problem.

A coin changer circuit is an example of a digital logic circuit. AND and OR gates can be combined to produce the logic required in a basic coin changer. One set of inputs is activated as coins are inserted into a vending machine. Another set of inputs is activated when the cost of the product is determined. If the cost of the product selected is less than the coin inserted, the digital circuit makes the decisions required to release the correct change. For example, if a quarter is inserted and a 10¢ product is selected, the coin changer should give 15¢ in change. To give 15¢ in change, the coin changer must energize CR1 (1 nickel) and CR2 (1 dime). See Coin Changer Circuit.

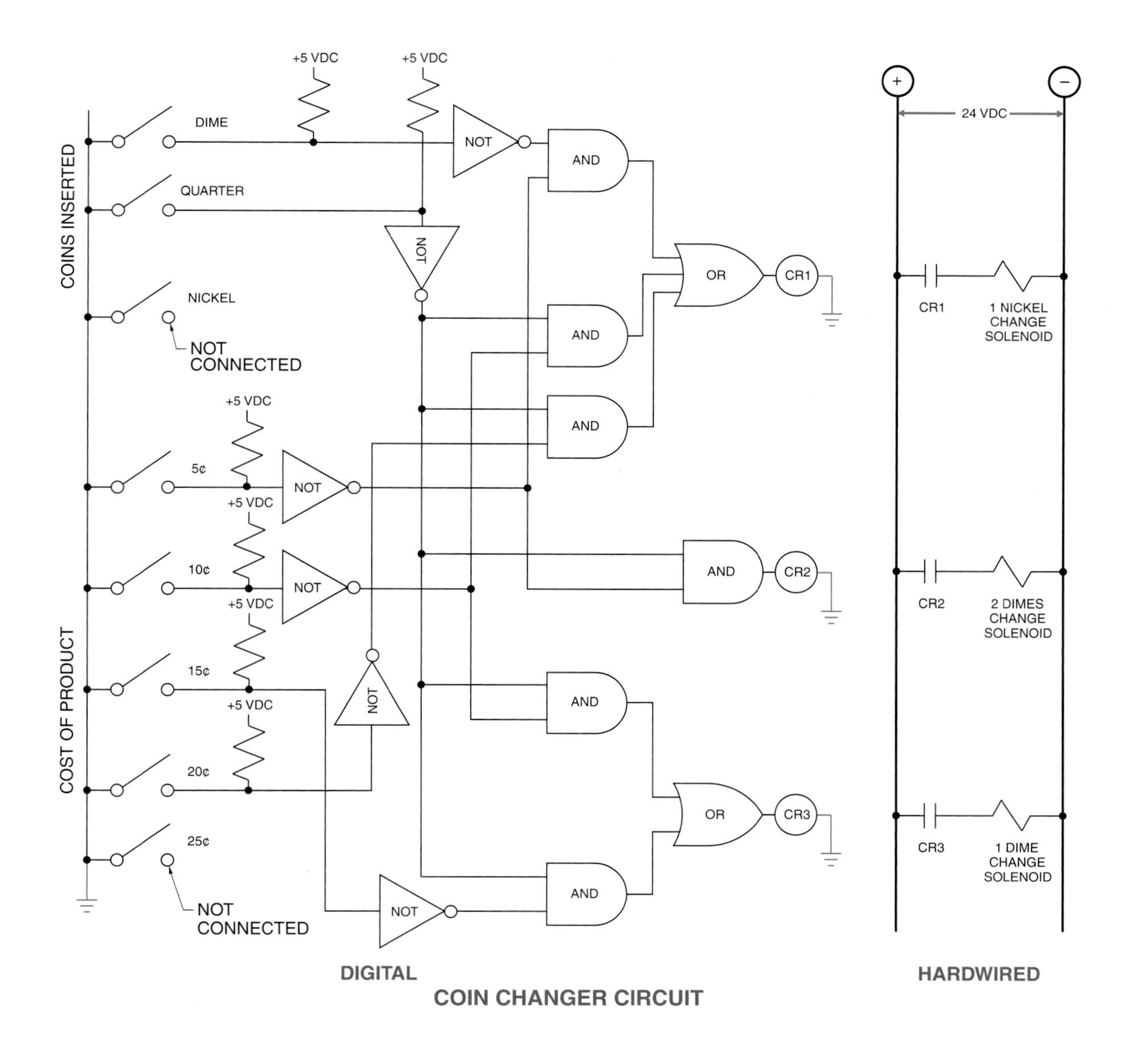

COIN CHANGER CIRCUIT

To simplify the example of a coin changer circuit, it is assumed that only one coin is inserted for the purchase of the product, and the cost of the product selected is 5¢, 10¢, 15¢, 20¢, or 25¢. The NICKEL coin insert switch is intentionally not connected in this simplified circuit because there would not be any change given if only a nickel is used. Likewise, the 25¢ coin product selection switch is also intentionally not connected because there would not be any change given if a 25¢ product is selected and the highest value coin that can be inserted is a quarter.

In a digital circuit, no logic gate can be left floating (not connected to a high or low at all times). To prevent a gate from floating, pull-up resistors are used. The pull-up resistor connects the gate to a HIGH (+5 VDC) as long as the switch is open. When the switch is closed, the gate is connected to ground and thus goes LOW (0 VDC). The pull-up resistor also limits the amount of current flowing through the circuit when the switch is connected to ground. A NOT gate (also called inverter) is used to change a HIGH to a LOW anytime the switch is open. When the switch is closed and there is a LOW at the inverter input, the inverter delivers a HIGH to the coin changer circuit. The inverter receives 5 VDC from the power supply connected to the digital chip (but not shown on the digital print).

Voltage readings are taken before any coins are inserted into the Coin Changer.

1. ______________ What voltage should Multimeter 1 display?

2. ______________ What voltage should Multimeter 2 display?

A dime is inserted into the Coin Changer and the 5¢ product switch is selected.

3. ______________ Should Logic Probe 1 indicate a HIGH or LOW?

4. ______________ Should Logic Probe 2 indicate a HIGH or LOW?

5. ______________ Should Logic Probe 3 indicate a HIGH or LOW?

6. ______________ Should Logic Probe 4 indicate a HIGH or LOW?

7. ______________ Should Logic Probe 5 indicate a HIGH or LOW?

8. ______________ Should Logic Probe 6 indicate a HIGH or LOW?

9. ______________ Should Logic Probe 7 indicate a HIGH or LOW?

10. ______________ Should Logic Probe 8 indicate a HIGH or LOW?

A quarter is inserted into the Coin Changer and the 5¢ product switch is selected.

11. ______________ Should Logic Probe 1 indicate a HIGH or LOW?

12. ______________ Should Logic Probe 2 indicate a HIGH or LOW?

13. ______________ Should Logic Probe 3 indicate a HIGH or LOW?

14. ______________ Should Logic Probe 4 indicate a HIGH or LOW?

15. ______________ Should Logic Probe 5 indicate a HIGH or LOW?

16. ______________ Should Logic Probe 6 indicate a HIGH or LOW?

17. ______________ Should Logic Probe 7 indicate a HIGH or LOW?

18. ______________ Should Logic Probe 8 indicate a HIGH or LOW?

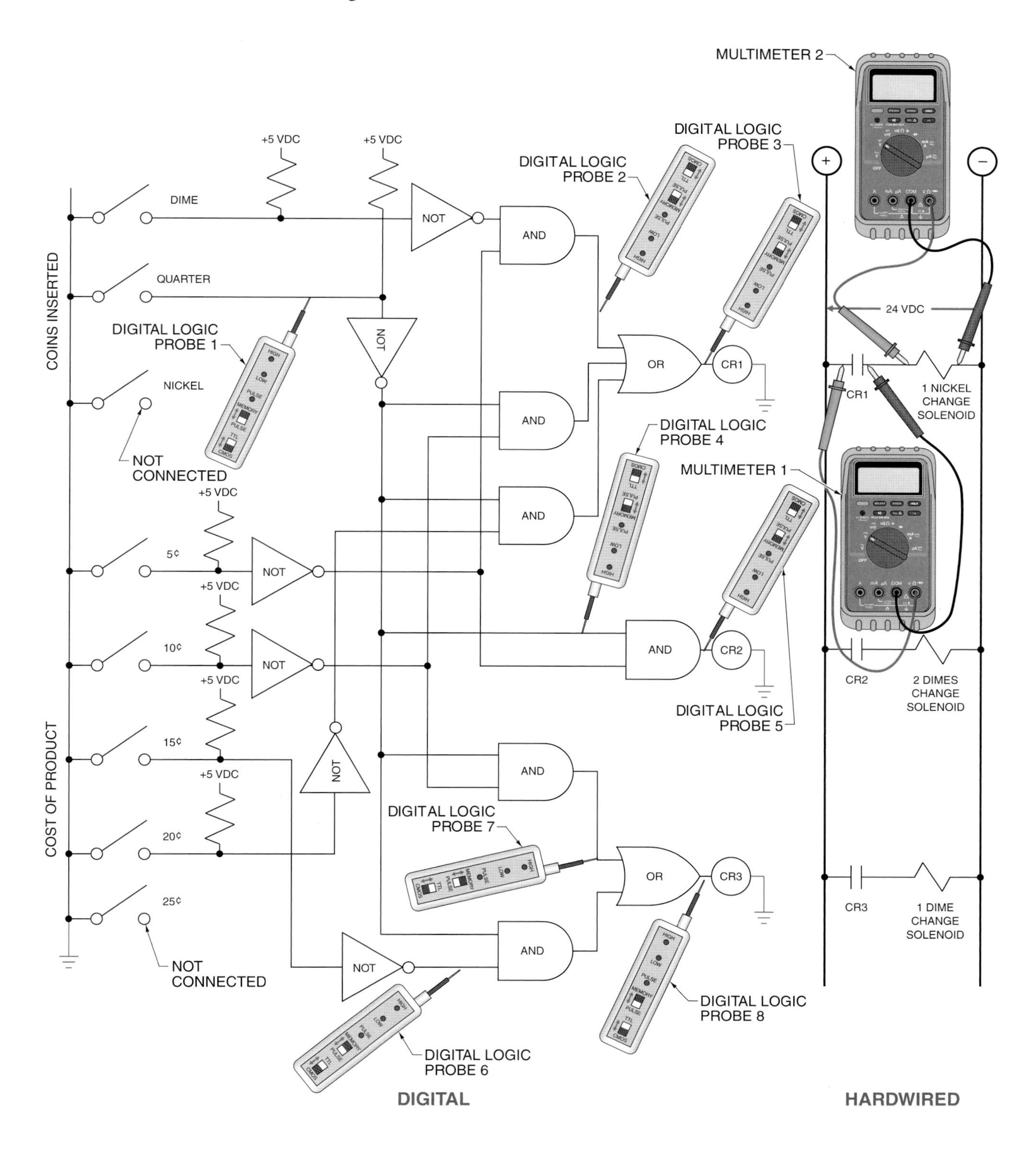

Application 5-5 — Troubleshooting Battery Chargers

A battery charger recharges DC devices by producing a high current flow through a weak battery and a trickle current flow through a fully charged battery. The battery charger automatically switches from high current flow to trickle current flow when the battery is fully charged. See Battery Charger and Battery Charger Circuit.

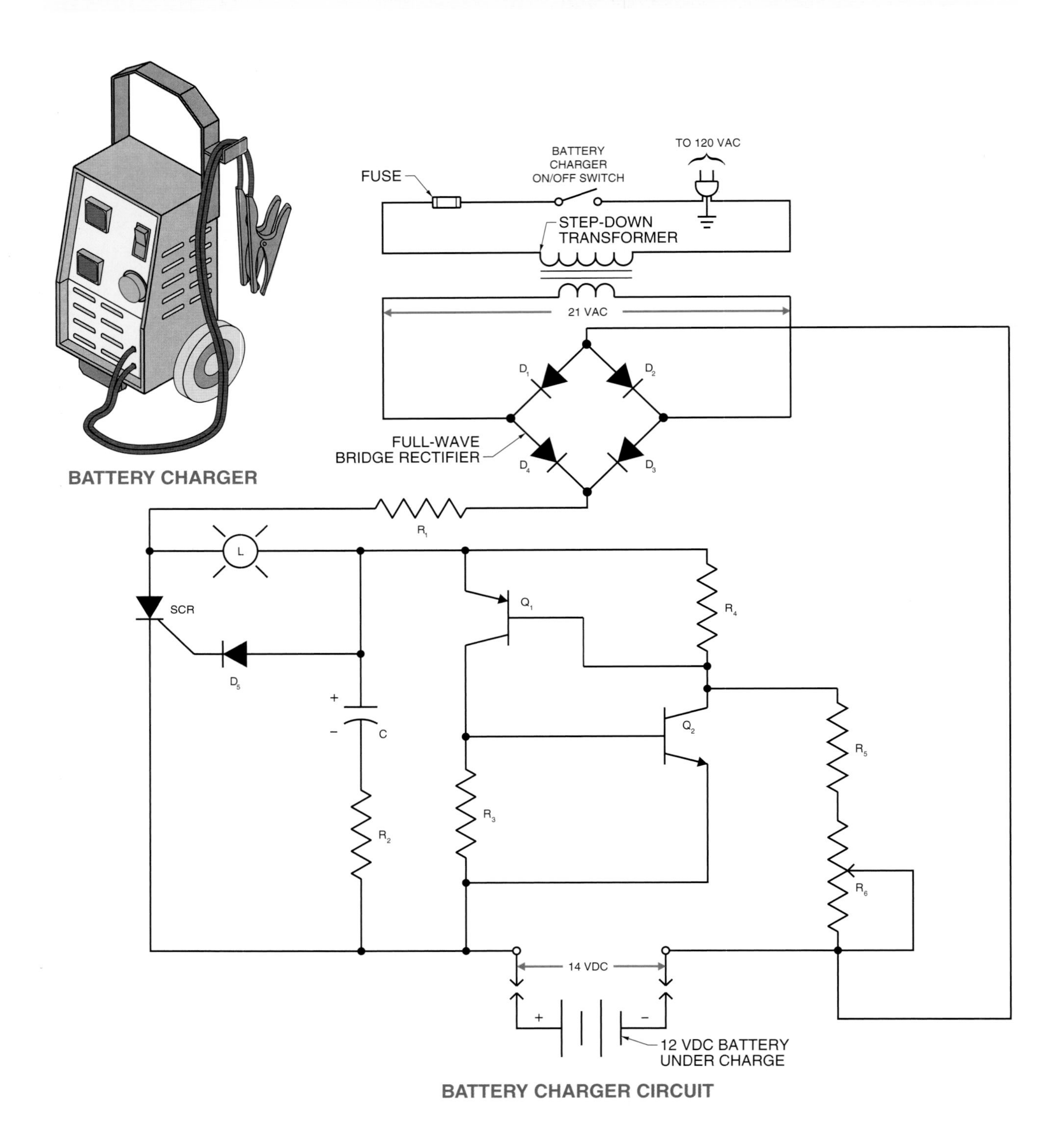

BATTERY CHARGER

BATTERY CHARGER CIRCUIT

In a battery charger designed to recharge 12 VDC batteries, 120 VAC is applied to a step-down transformer when the battery charger ON/OFF switch is closed. A fuse installed in the primary side of the transformer protects the battery charger from excessive current flow. The step-down transformer reduces the supply voltage to approximately 21 VAC. A full-wave bridge rectifier circuit changes (rectifies) the AC into full-wave DC.

A typical battery charger circuit consists of an SCR (silicon controlled rectifier) circuit and a transistor circuit. The SCR circuit controls the high current flow. The transistor circuit controls the trickle current flow. The SCR and transistor circuits are connected in parallel with each other and in series with the battery under charge.

A weak battery connected to the battery charger provides a low resistance current path. The current flows through resistor R_1, the lamp, capacitor C, resistor R_2, and the battery. The capacitor charges and provides enough positive charge on the anode of diode D_5 so that the diode conducts and allows current to flow. The current flowing through the diode triggers the gate of the SCR, which allows high current DC to flow. Current flows through the SCR as long as the weak battery under charge is drawing a high current. The transistor circuit acts like an open switch when the SCR circuit is ON.

A fully charged battery produces a higher voltage in the transistor circuit. This triggers the transistor into conduction, which allows the capacitor to discharge through the transistor circuit. The discharging of the capacitor lowers the voltage at D_5 to the point where it does not allow current to flow to the gate of the SCR. Then the only path for current flow is through the transistor circuit.

The transistor circuit allows only a trickle current to flow because the circuit has a higher resistance than the SCR circuit. The voltage drop across the lamp is great enough to allow the lamp to turn ON when the transistor circuit is conducting. The lamp remains ON as long as the battery under charge is connected into the circuit. The point at which the charging circuit switches to a trickle current flow is set with R_6.

R_1 is placed in series with the charging circuit to limit the current flow if a dead battery is placed on the battery charger. The limited current prevents damage to the battery charger's circuit.

A weak battery is connected to the battery charger and the charger is switched ON.

1. Draw the waveform that should appear on Oscilloscope 1.
2. Draw the waveform that should appear on Oscilloscope 2.
3. Draw the waveform that should appear on Oscilloscope 3.
4. Draw the waveform that should appear on Oscilloscope 4.

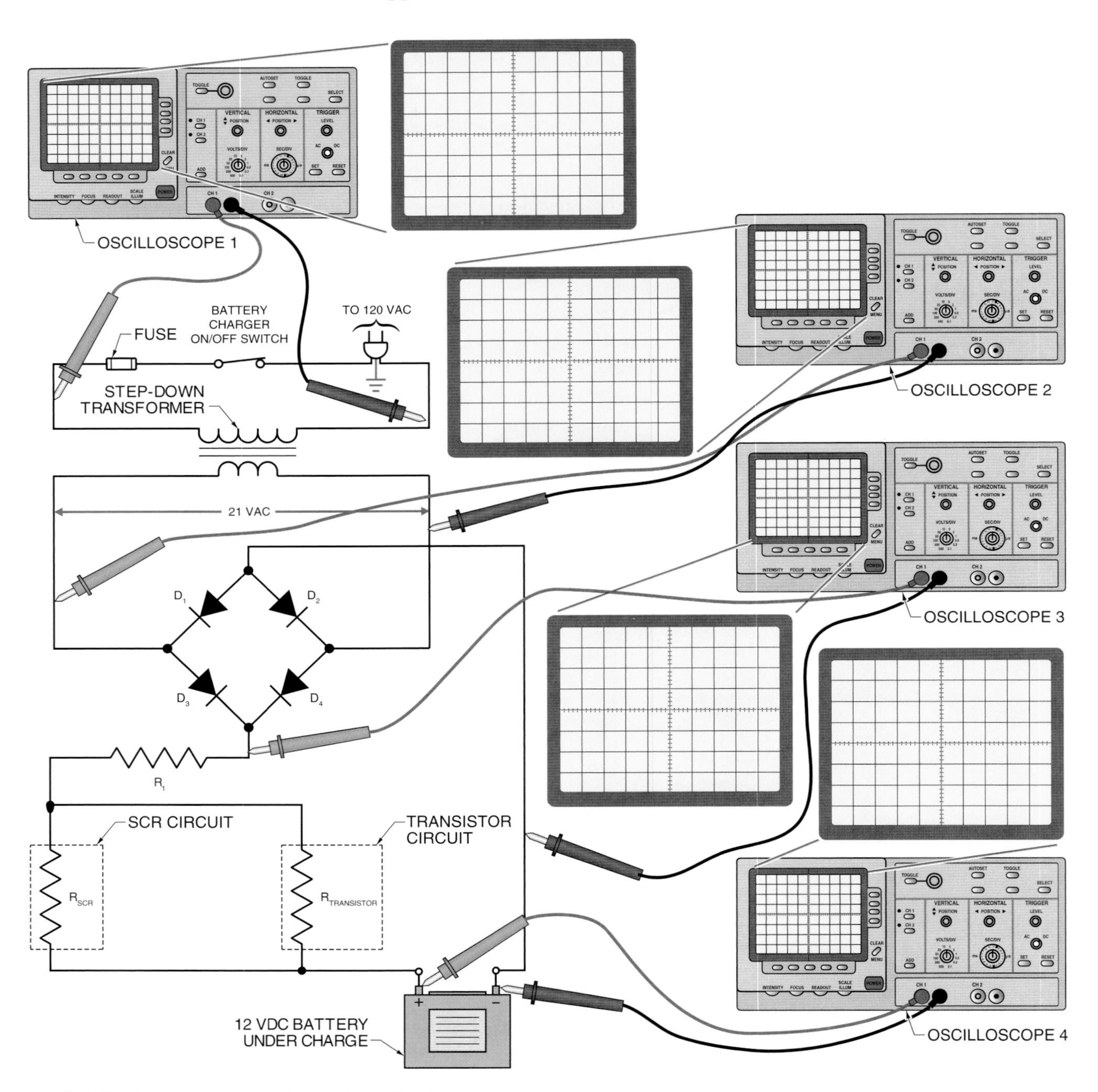

SIMPLIFIED BATTERY CHARGER CIRCUIT

Application 5-6—Troubleshooting Photoelectric or Proximity Switches

Solid-state photoelectric and proximity switches are designed to last a long time if properly used. Photoelectric and proximity switches do fail if they experience high transient voltages, draw higher than rated current, or operate in high ambient temperatures. Solid-state photoelectric and proximity switches can be tested using a voltmeter.

When DC photoelectric or proximity switches are required to operate an AC load, a solid-state relay (SSR) is used as the interface between the DC circuit and the AC circuit. The switch is used to operate the SSR input and the SSR output controls the AC load.

Photoelectric and proximity switches are available with either a PNP transistor output (current source, positive switching) or an NPN transistor output (current sink, negative switching). See PNP Transistor Switch and NPN Transistor Switch. The wiring diagram on the switch is used to identify the type of switching used. To determine the type, the placement of the load (L) on the diagram indicates whether the load is connected directly to the positive of the DC power supply at all times (NPN-type switch), or the load is connected directly to the negative of the DC power supply at all times (PNP-type switch). Photoelectric and proximity switches are available with a normally open (NO) output, normally closed (NC) output, or both an NO and NC output in the same switch.

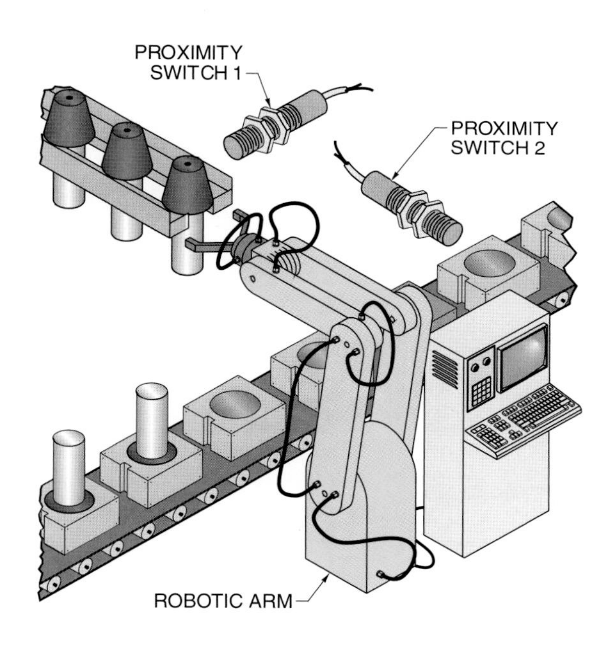

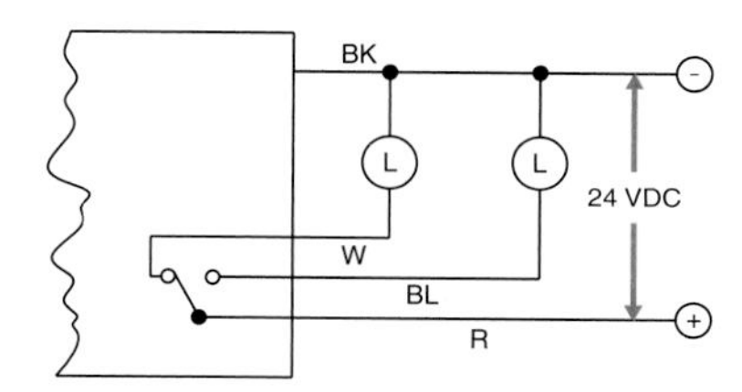

PNP TRANSISTOR SWITCH

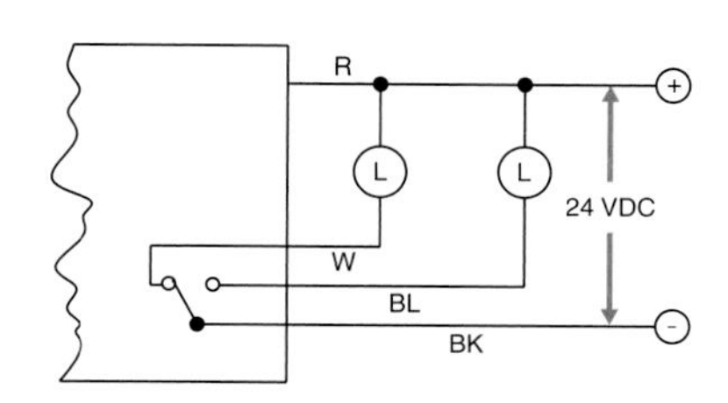

NPN TRANSISTOR SWITCH

Proximity Switch 1 is not activated (no target in front of it).

1. ______________ Is the proximity switch an NPN or PNP-type switch?

2. ______________ What is the expected voltage reading for Multimeter 1?

3. ______________ What is the expected voltage reading for Multimeter 2?

4. ______________ What is the expected voltage reading for Multimeter 3?

Proximity Switch 1 is activated (a target in front of it).

5. ______________ What is the expected voltage reading for Multimeter 1?

6. ______________ What is the expected voltage reading for Multimeter 2?

7. ______________ What is the expected voltage reading for Multimeter 3?

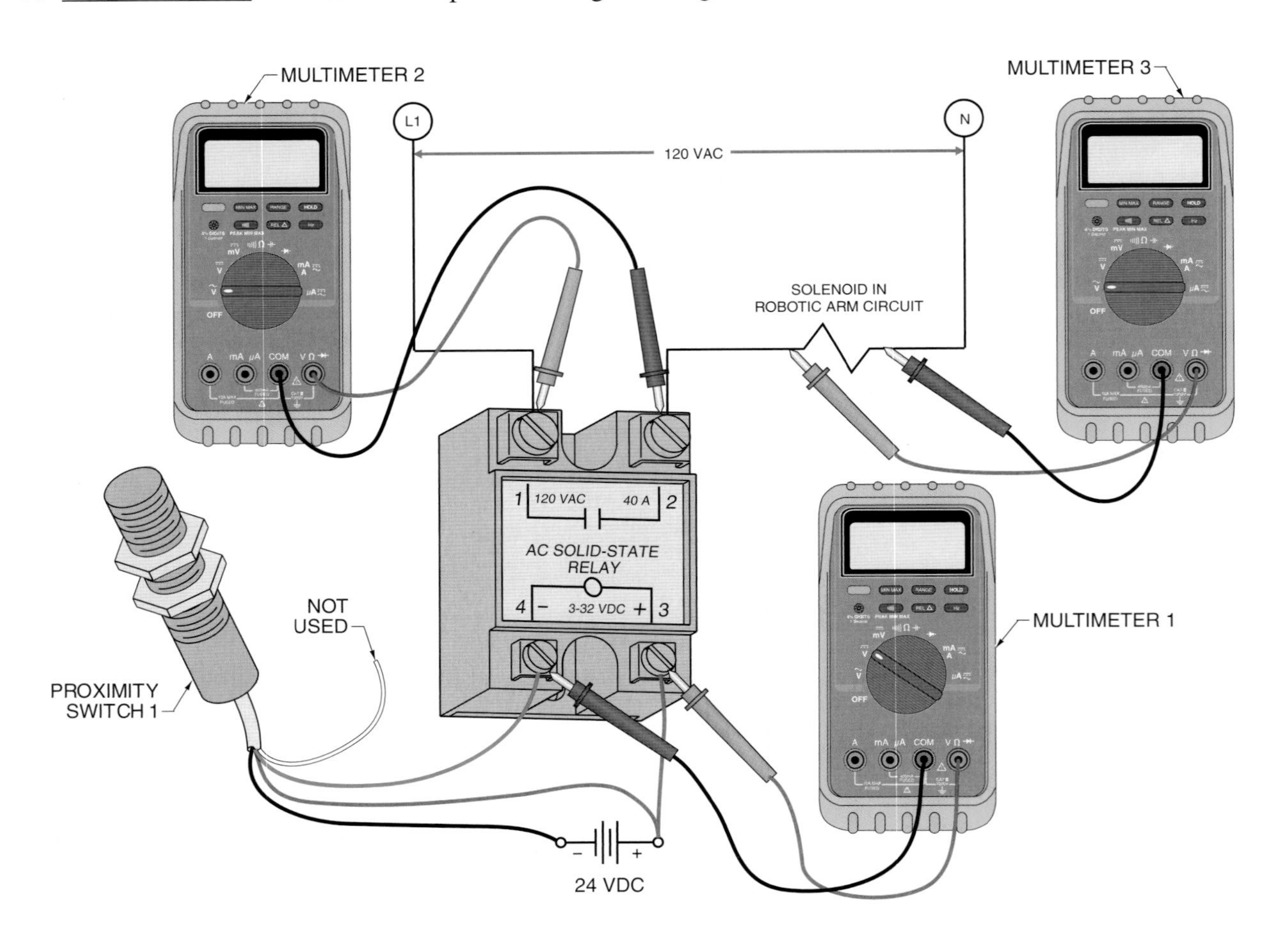

Proximity Switch 2 is not activated (no target in front of it).

8. ______________ Is the proximity switch an NPN or PNP-type switch?

9. ______________ What is the expected voltage reading for Multimeter 1?

10. ______________ What is the expected voltage reading for Multimeter 2?

11. ______________ What is the expected voltage reading for Multimeter 3?

Proximity Switch 2 is activated (a target in front of it).

12. ______________ What is the expected voltage reading for Multimeter 1?

13. ______________ What is the expected voltage reading for Multimeter 2?

14. ______________ What is the expected voltage reading for Multimeter 3?

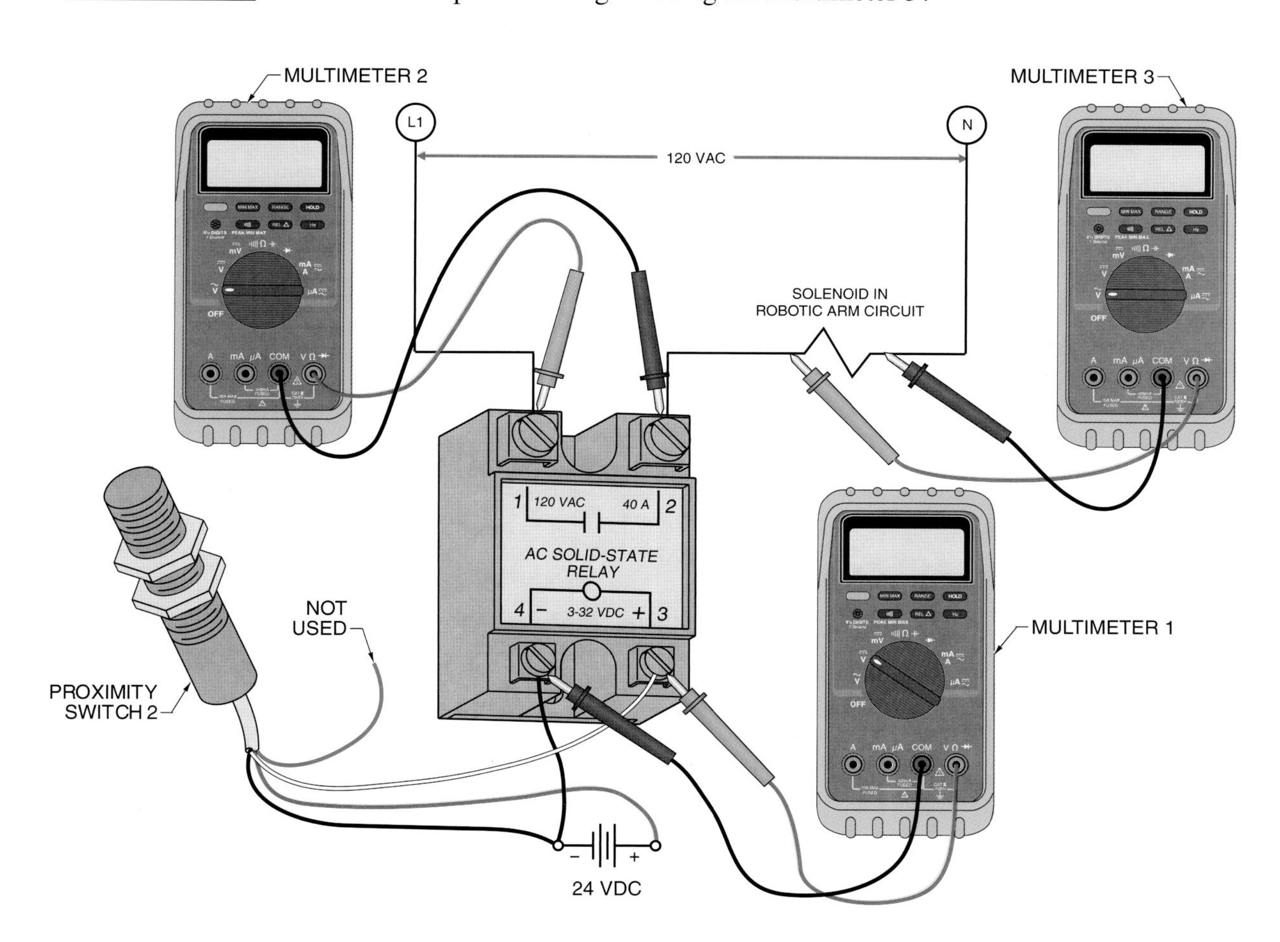

Application 5-7—Photovoltaic Output

A photovoltaic cell (solar cell) is a device that converts solar energy to electrical energy. Photovoltaic cells produce a voltage when exposed to light. Photovoltaic cells are being used to directly or indirectly power an ever greater number of electronic circuits. Photovoltaic cells can be used to directly power electronic devices such as handheld calculators. However, most photovoltaic cells are used as part of a battery-powered system in which the photovoltaic cells power the electronic circuits when light is present and charge batteries to supply power when light is not available. Applications of photovoltaic cell and battery-powered devices include school crossing or warning signs, portable traffic lights (in construction zones), and remote weather stations.

Photovoltaic cells are rated by the amount of energy they convert. Most manufacturers rate the output in terms of volts (V) and milliamps (mA). Photovoltaic cells follow the same laws as batteries when connected in series and parallel. A circuit of cells connected in series produces a voltage output equal to the sum of the individual cells' voltage outputs, but the current stays the same. For example, 11 cells are connected in series. If each cell produces 100 mA at 0.5 V, then the output of the circuit is 100 mA at 5.5 V. See Photovoltaic Cells in Series.

A circuit of cells connected in parallel produces a current output equal to the sum of the individual cells' current outputs, but the voltage stays the same. For example, if the same 11 cells are connected in parallel, the circuit's output is 1100 mA at 0.5 V. See Photovoltaic Cells in Parallel. By combining series and parallel circuits, any desired voltage and current combination can be designed.

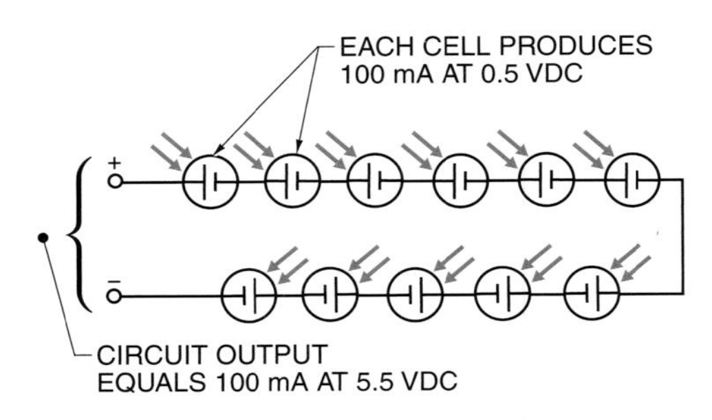

PHOTOVOLTAIC CELLS IN SERIES

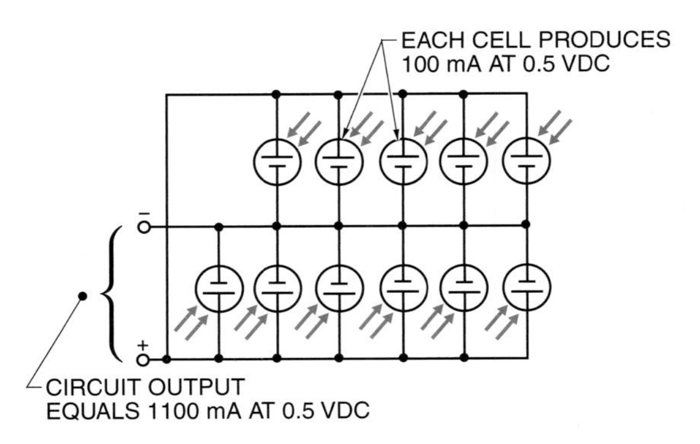

PHOTOVOLTAIC CELLS IN PARALLEL

Each solar cell is rated for a maximum of 1 VDC at 40 mA.

1. ______________ What is the total maximum voltage output (in V) of Circuit 1?

2. ______________ What is the total maximum current output (in mA) of Circuit 1?

3. ______________ What is the total maximum power output (in W) of Circuit 1?

4. ______________ What is the total maximum voltage output (in V) of Circuit 2?

5. ______________ What is the total maximum current output (in mA) of Circuit 2?

6. ______________ What is the total maximum power output (in W) of Circuit 2?

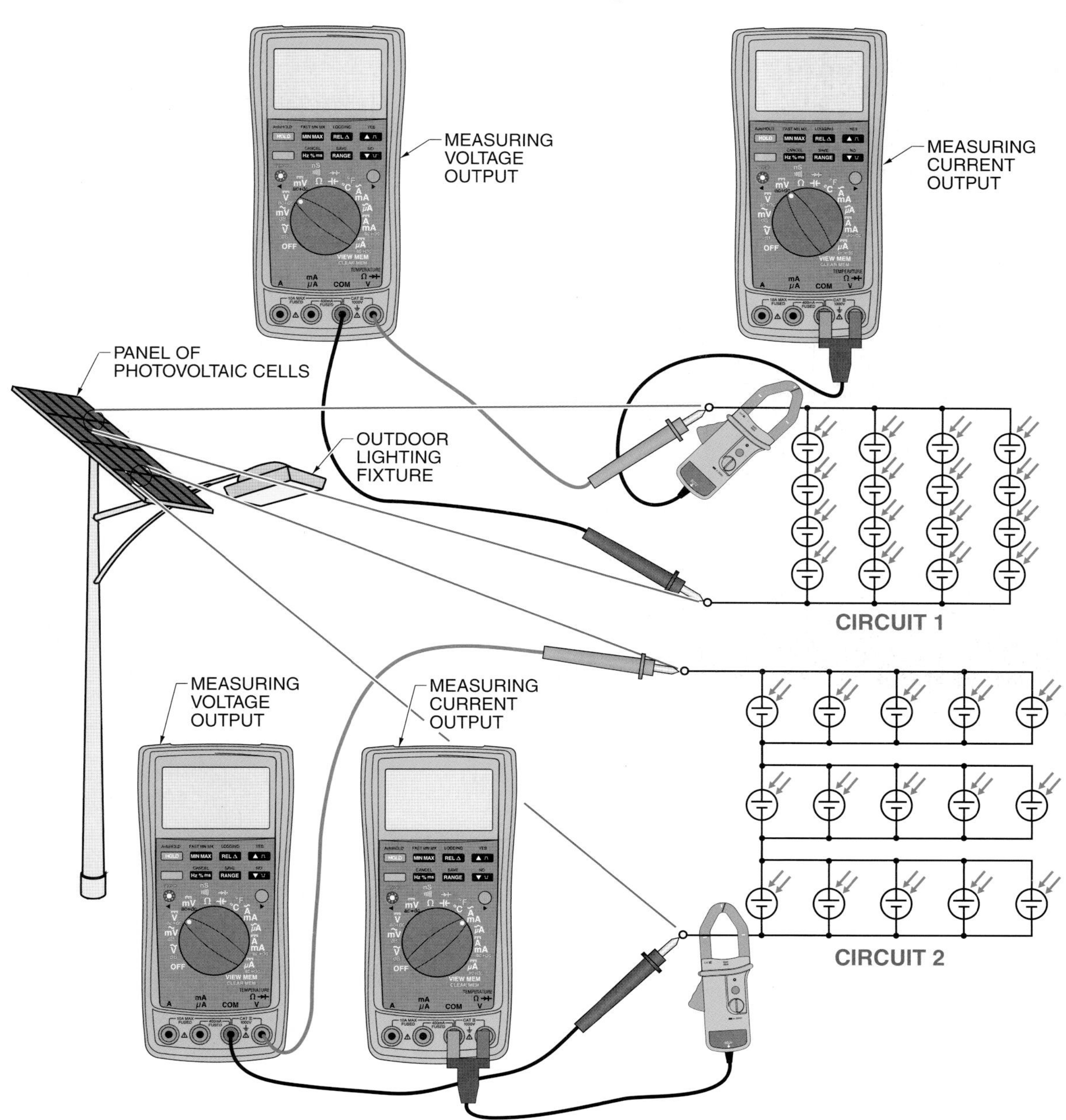

Application 5-8—Troubleshooting Motor Drives

An electric motor drive is an electronic device that controls the direction, speed, torque, and other operating functions of an electric motor, in addition to providing motor protection and monitoring functions. Smaller motor drives (less than 5 HP) are usually not serviced but instead replaced when faulty. Larger motor drives (5 HP and higher) are usually serviced. Individual parts of the drive, such as power supply diodes, capacitors, and large output power transistors can be replaced. Whole sections of a drive, such as the rectifier section, output power transistor section, or control circuit section can also be replaced, eliminating the need to service down to the individual component level.

Before repairing or replacing an electric motor drive, the drive should first be identified as the problem. Test instruments are used to take measurements in the drive system to identify the circuit's problem.

Testing the Power Supply

All electric motor drives have an input voltage rating, which does not need to be the same as the drive's output (and motor) voltage rating. For example, drives less than 5 HP can have voltage input ratings of 120 VAC, 1ϕ; 240 VAC, 1ϕ; or 240 VAC, 3ϕ; and deliver a 240 VAC, 3ϕ output to drive 3ϕ motors. The actual input voltage should be within +5% to –10% of the drive's input voltage rating. A voltmeter is used to measure the voltage at the drive's input. See Testing Input Voltage.

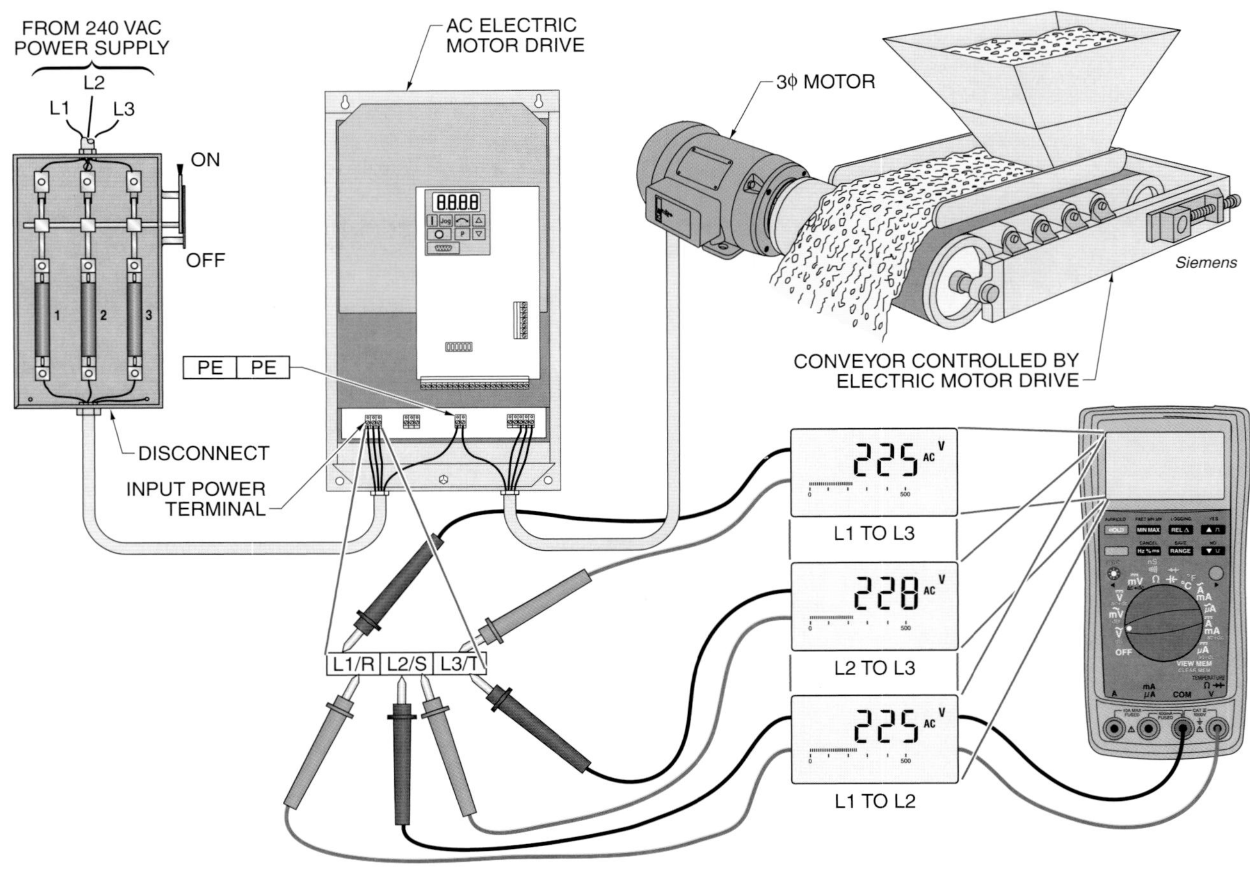

TESTING INPUT VOLTAGE

If the drive does not have the correct voltage, or has no voltage, at its input terminals, test the fuses or circuit breakers protecting the drive. Begin by measuring the voltage coming into the fuses. If the voltage is not correct, troubleshoot upstream from the fuses. If the voltage is within an acceptable limit (+5% to –10%), test the fuses. Fuses are tested one at a time by first measuring the voltage into a fuse and then moving the meter lead from the top of the fuse to the bottom of the same fuse to test the voltage out of the fuse. If voltage is the same out of the fuse as into the fuse, the fuse is good. If there is voltage going into the fuse, but no voltage coming out, the fuse is bad. See Testing Power Supply Voltage.

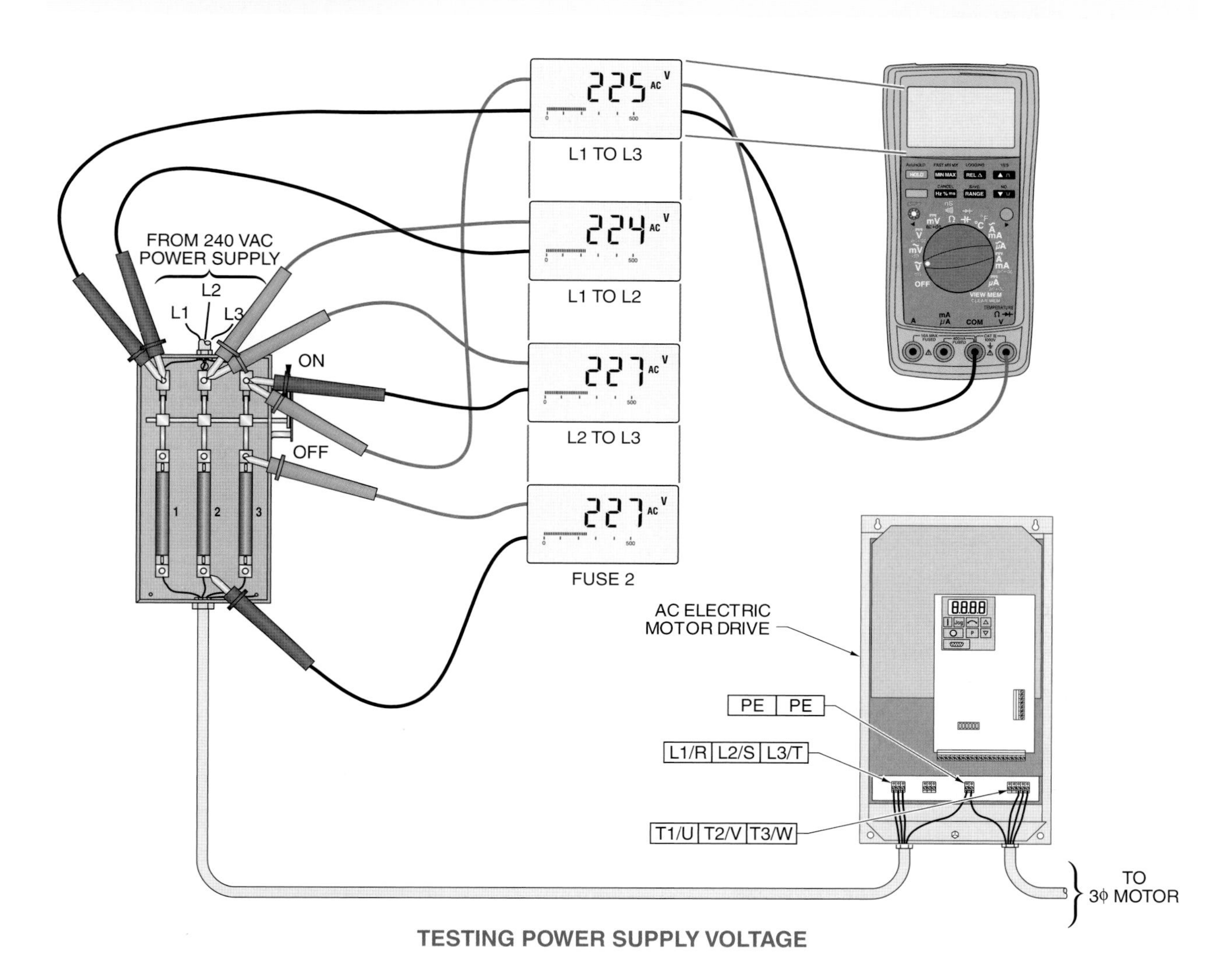

TESTING POWER SUPPLY VOLTAGE

Testing the DC Bus Section

An electric motor drive rectifies the incoming AC supply voltage to DC voltage. The DC bus (link) section of a drive is the rectified (and filtered) DC voltage. The drive inverts the DC back into AC at a controlled (voltage level and frequency) 3ϕ voltage. The DC bus voltage should be 1.4 times the drive's rated output (motor) voltage. Changes in the incoming power supply voltage will proportionally affect the DC bus voltage. For example, if the incoming power supply voltage is 5% less, then the DC bus voltage will be 5% less.

Testing Output Current

It is important to measure the voltage into and out of a drive to make sure the voltage is within an acceptable range. However, measuring current will give a more accurate picture of how much the system is being loaded. If the measured current is equal to the nameplate current rating of the motor, the motor is fully loaded. If the measured current is less than the nameplate current rating of the motor, the motor is underloaded. If the measured current is greater than the nameplate current rating of the motor, the motor is overloaded and/or faulty. The current rating of the drive and the power supply must be equal to (or, preferably, greater than) the highest measured current out of the drive. See Testing Drive Current.

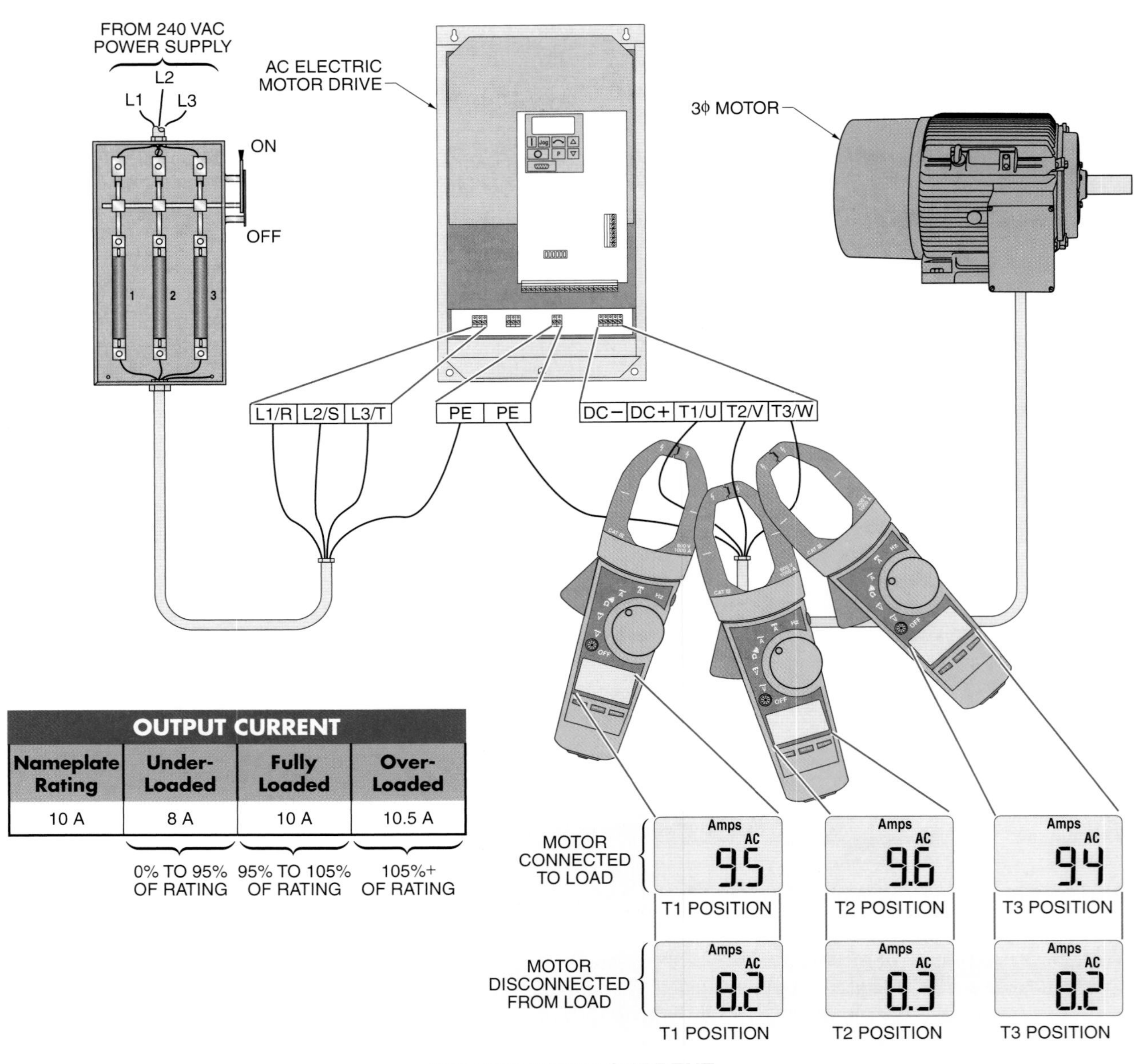

OUTPUT CURRENT			
Nameplate Rating	Under-Loaded	Fully Loaded	Over-Loaded
10 A	8 A	10 A	10.5 A

TESTING DRIVE CURRENT

To determine if there is a motor problem or if the motor is overloaded, disconnect the load from the motor and take the current measurements again. The current measurements of an unloaded motor should be less than the motor's nameplate rated current.

The input voltage to a motor drive should be measured when the motor connected to the drive is OFF (no load measurement) and when the motor is fully loaded (full load measurement) to ensure the drive is not overloaded and the power supply is delivering enough power.

1. ______________ What is the expected voltage reading from L1 to L3?

2. ______________ What is the expected voltage reading from L1 to L2?

3. ______________ What is the expected voltage reading from L2 to L3?

4. ______________ What is the expected voltage reading from L1 to ground?

5. ______________ What is the expected voltage reading from L2 to ground?

6. ______________ What is the expected voltage reading from L3 to ground?

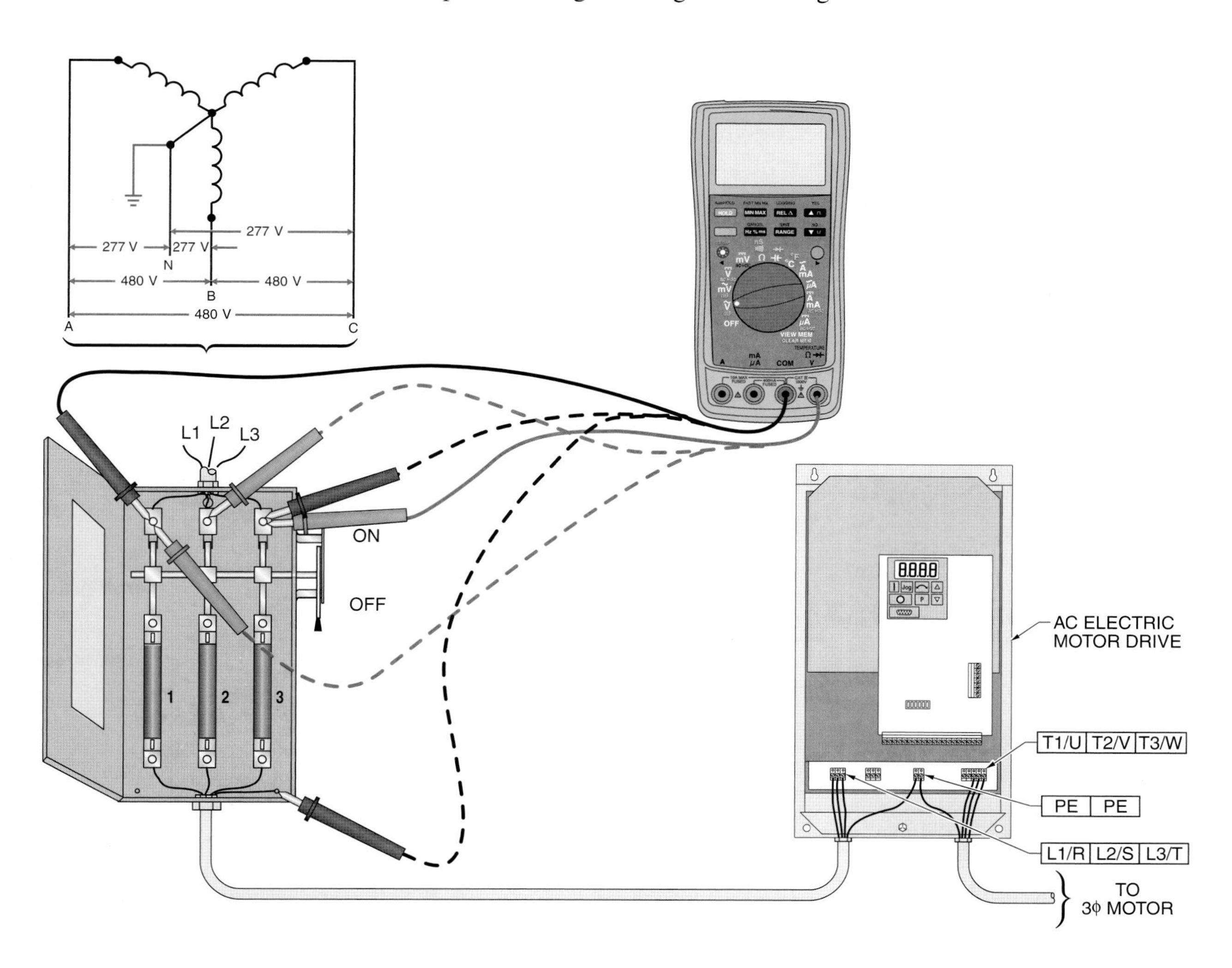

7. _______________ What is the percent voltage drop in the drive/motor system?

8. _______________ Is this voltage drop in the acceptable range?

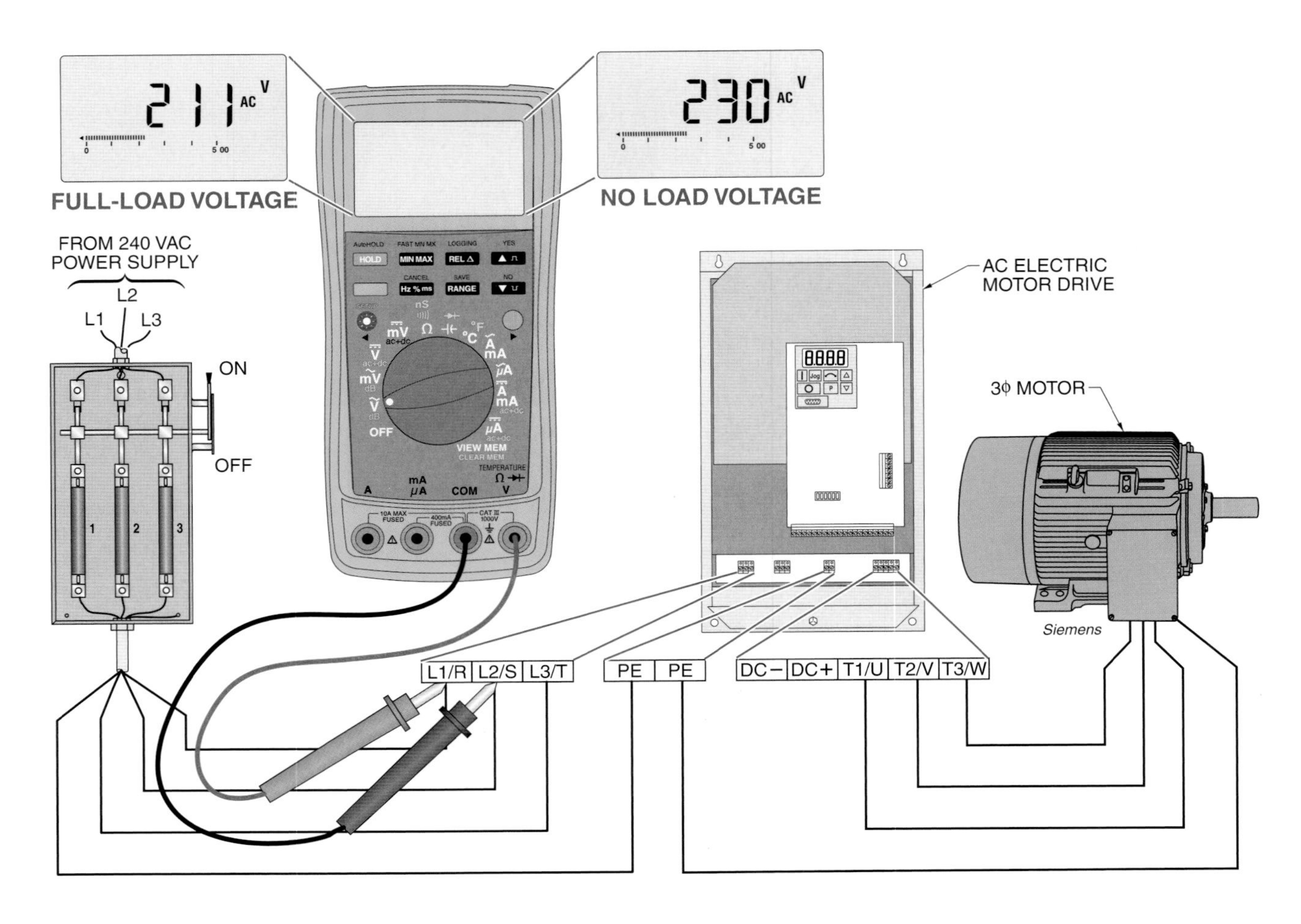

9. _______________ If the incoming phase-to-phase voltage is 208 VAC, what is the DC bus voltage?

10. _______________ If the incoming phase-to-phase voltage is 220 VAC, what is the DC bus voltage?

11. _______________ If the incoming phase-to-phase voltage is 230 VAC, what is the DC bus voltage?

12. _______________ If the incoming phase-to-phase voltage is 460 VAC, what is the DC bus voltage?

13. _______________ If the incoming phase-to-phase voltage is 480 VAC, what is the DC bus voltage?

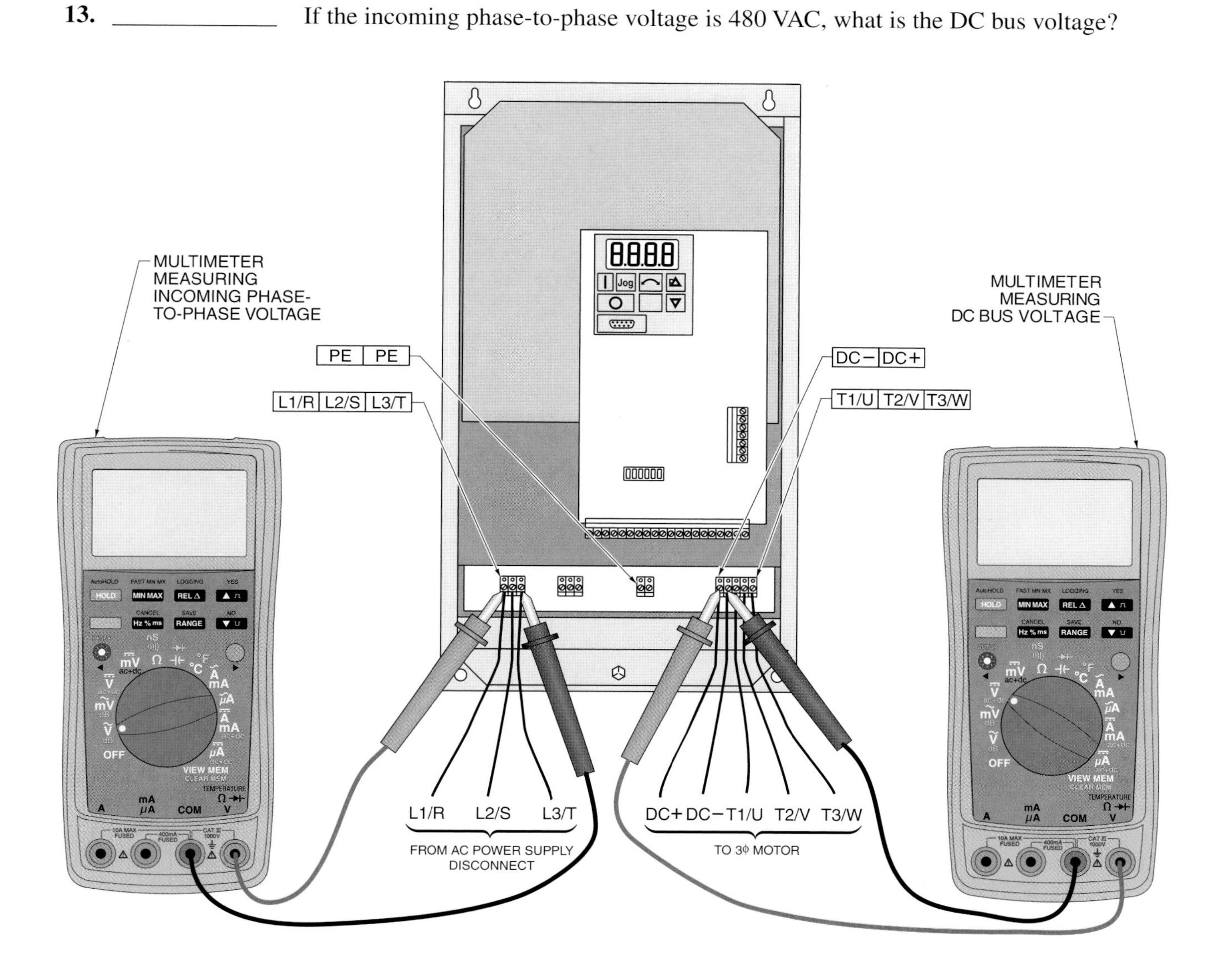

14. ______________ What is the expected full-load current reading for Ammeter 1?

15. ______________ What is the expected full-load current reading for Ammeter 2?

16. ______________ What is the expected full-load current reading for Ammeter 3?

17. ______________ What is the expected full-load current reading for Ammeter 4?

18. ______________ What is the expected full-load current reading for Ammeter 5?

19. ______________ What is the expected full-load current reading for Ammeter 6?

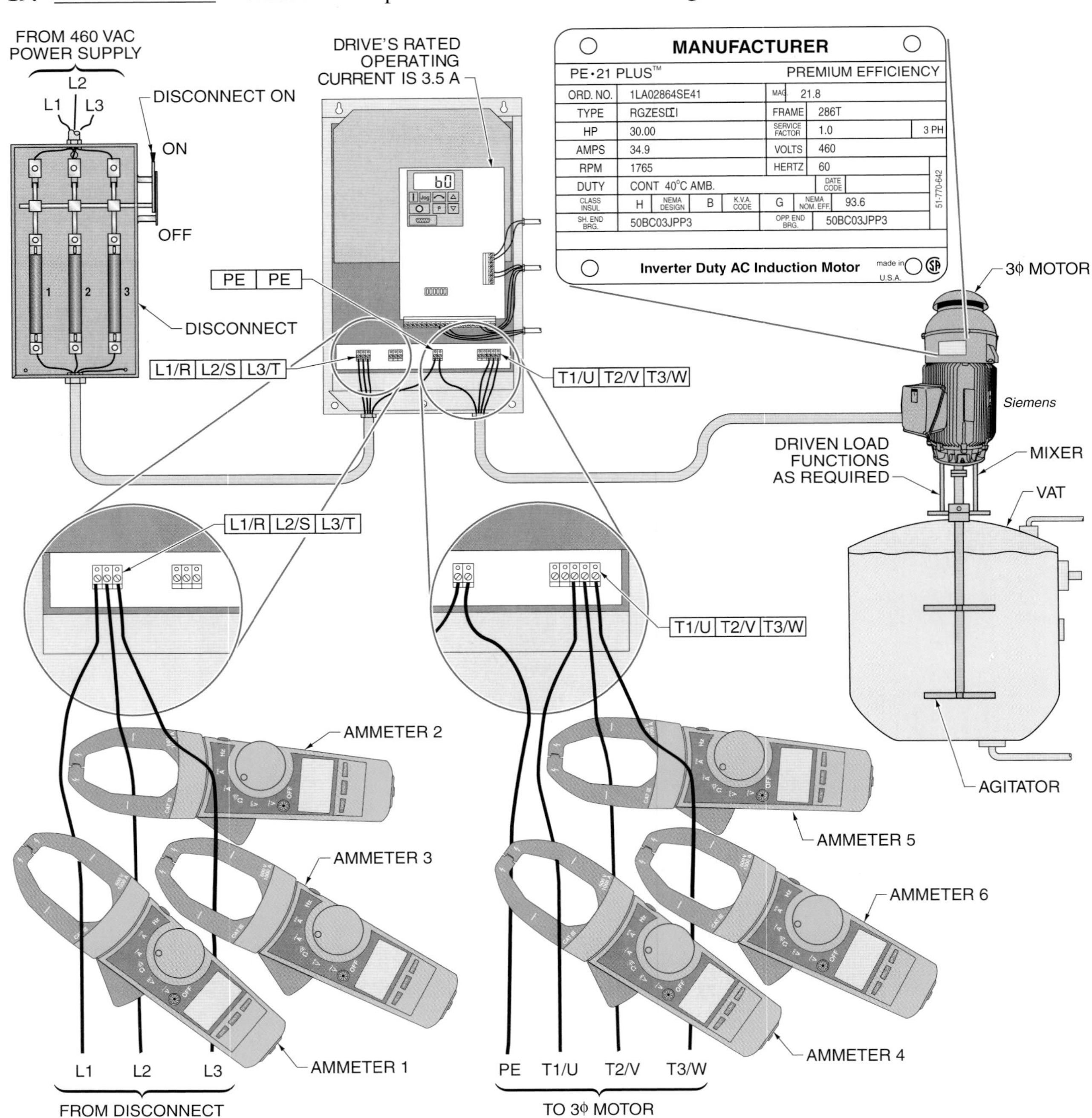

Application 5-9—Troubleshooting PLC Inputs and Outputs

A programmable logic controller (PLC) is a solid-state control device that is programmed and reprogrammed to automatically control machines, security systems, lighting systems, and industrial processes. A PLC contains a power supply, input and output modules, processor, and programming terminal. See Programmable Logic Controller.

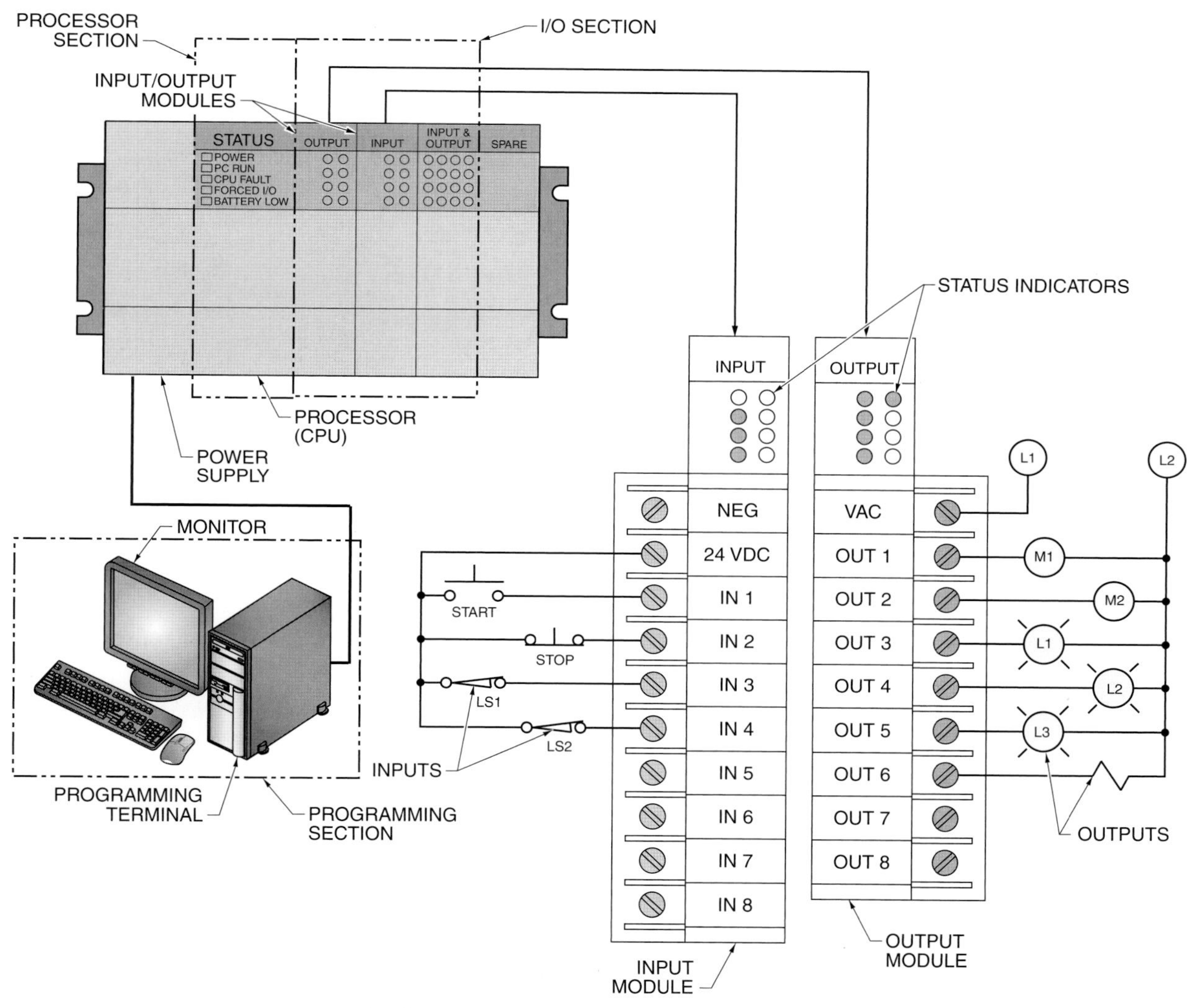

PROGRAMMABLE LOGIC CONTROLLER

The power supply provides necessary voltage levels required for the internal operation of the PLC. The power supply also provides power for the input and output modules. The input and output sections function as the eyes, ears, and hands of the PLC. The input section receives information from pushbuttons, temperature switches, pressure switches, photoelectric and proximity switches, and other sensors. The output section delivers the output voltage required to control alarms, lights, solenoids, starters, and other loads. It is the input and output sections in which most PLC troubleshooting that is not software-related takes place.

The processor section is the brain of the PLC. The processor section is the section of a PLC that organizes all control activity by receiving inputs, performing logical decisions according to the program, and controlling the outputs. The programming section of a PLC is the section that allows input into the PLC program through a keyboard.

1. Set (draw in) the function switch on Multimeter 1 to test the PLC input section.
2. Connect the test leads of Multimeter 1 to the meter and to the PLC input section to test Input 1.
3. ____________ What is the expected voltage reading for Multimeter 1 when the switch on Input 1 is open?
4. ____________ What is the expected voltage reading for Multimeter 1 when the switch on Input 1 is closed?
5. Set (draw in) the function switch on Multimeter 2 to test the PLC input section.
6. Connect the test leads of Multimeter 2 to the meter and to the PLC input section to test the voltage output of the PLC input module.
7. ____________ What is the expected voltage reading for Multimeter 2 if there are no problems?

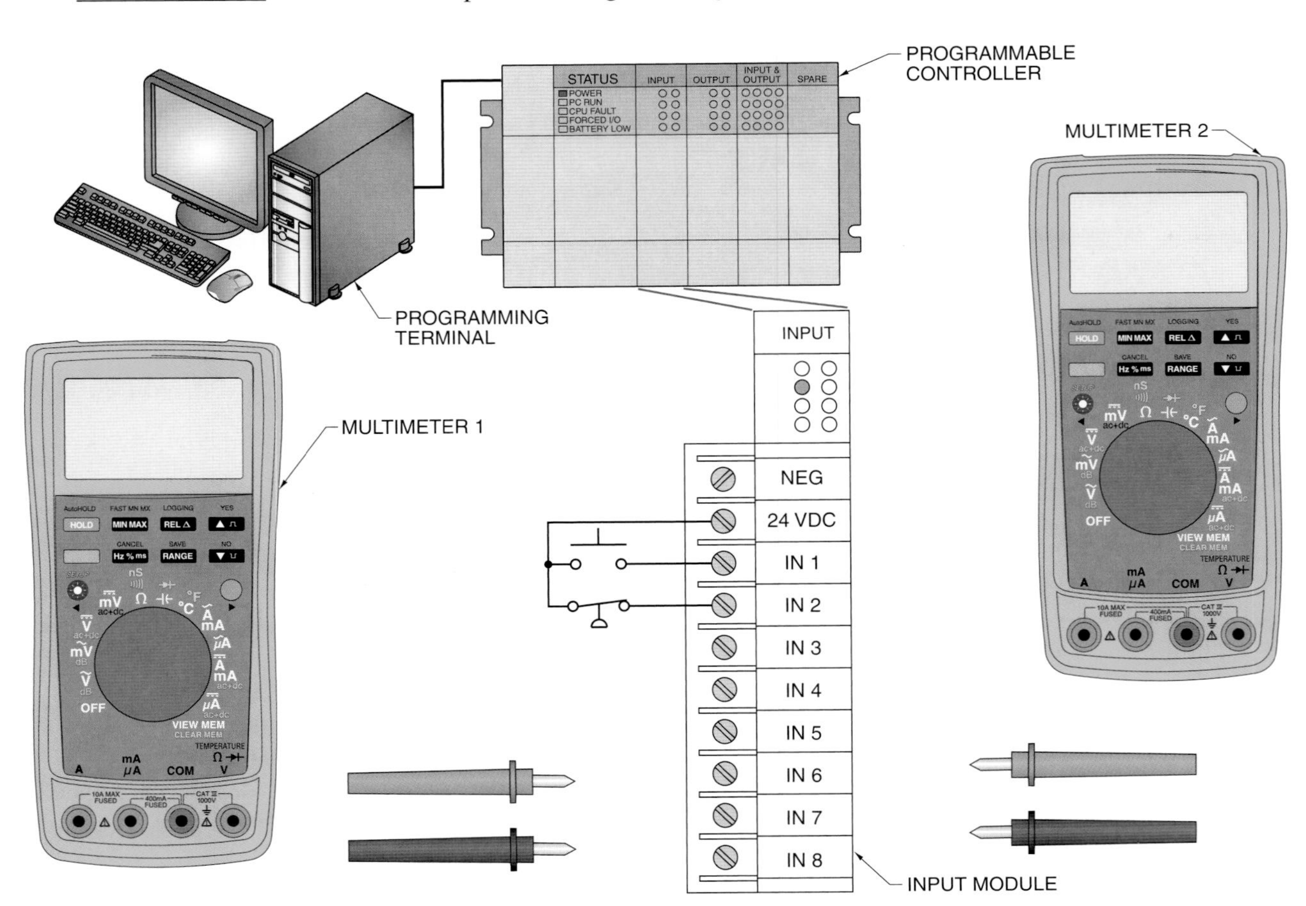

8. Set (draw in) the function switch on Multimeter 3 to test the PLC output section.

9. Connect the test leads of Multimeter 3 to the meter and to the PLC output section to test Output 1.

10. ______________ What is the expected voltage reading for Multimeter 3 when the output is energized?

11. Set (draw in) the function switch on Multimeter 4 to test the PLC output section.

12. Connect the test leads of Multimeter 4 to the meter and to the supply voltage for the output loads.

13. ______________ What is the expected voltage reading for Multimeter 4 if there are no problems?

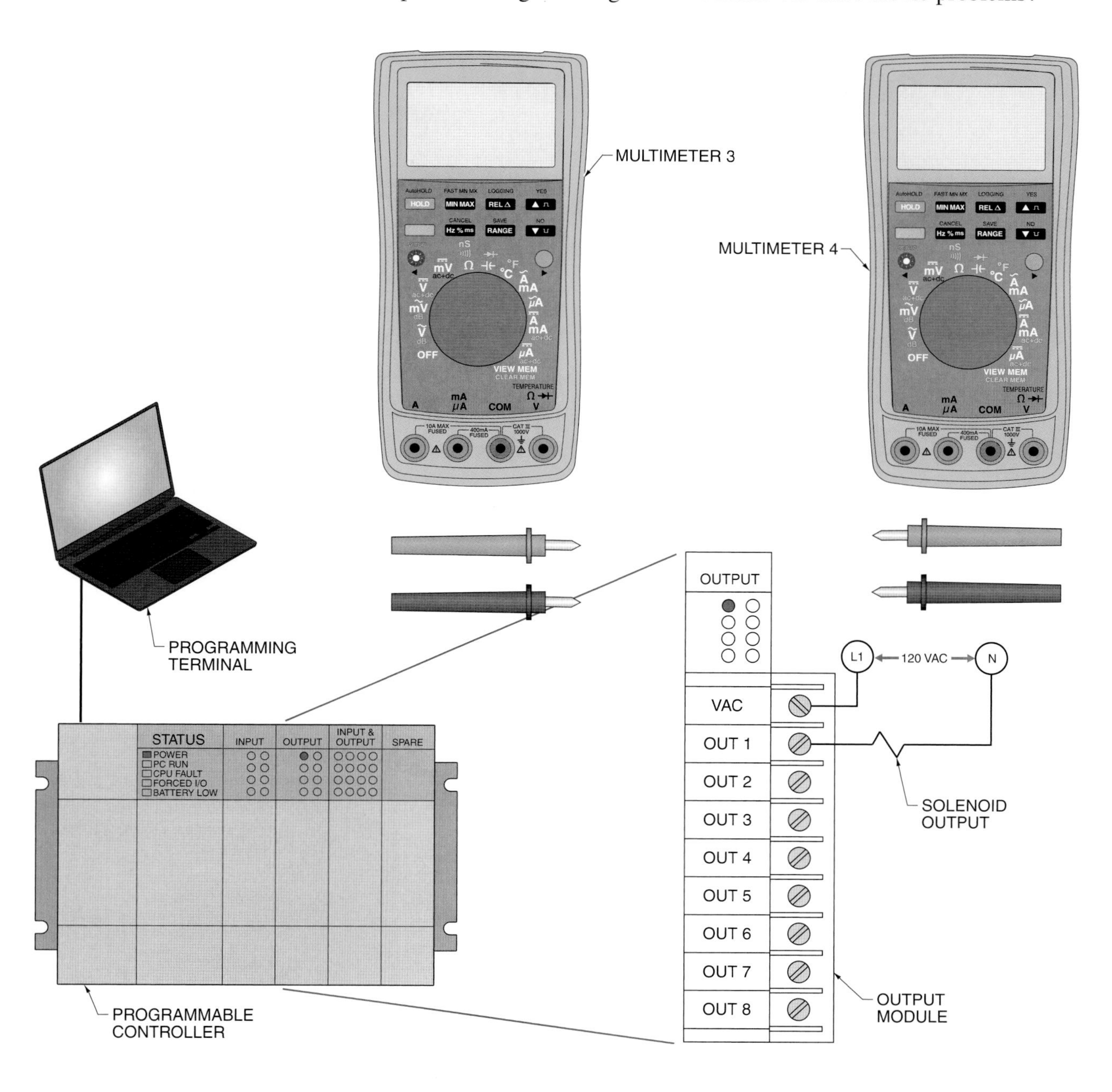

14. Set (draw in) the function switch on Multimeter 5 to test the PLC input section.

15. Connect the test leads of Multimeter 5 to the meter and to the PLC input section to test the proximity switch output.

16. Set (draw in) the function switch on Multimeter 6 to test the PLC input section.

17. Connect the test leads of Multimeter 6 to the meter and to the PLC input section to test the voltage supplied to the proximity switch.

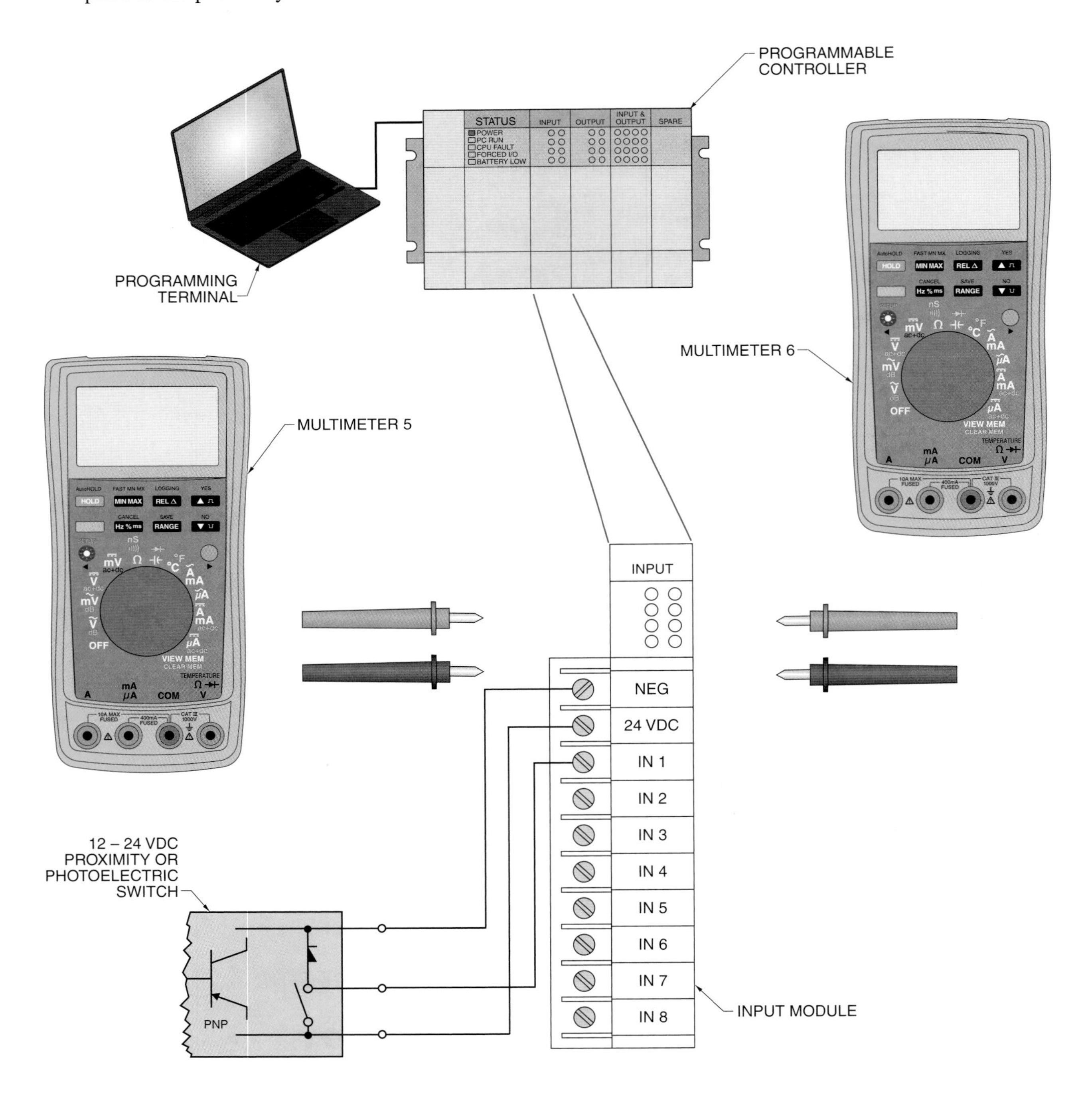

GROUNDING SYSTEMS AND EARTH GROUND TEST INSTRUMENTS

Name: ______________________________ Date: ____________________

Application 6-1 — System Grounding

Whenever working on or around an electrical circuit or system, the system should be checked to ensure it is grounded. Grounding electrical systems, circuits, and equipment makes them safer by helping to prevent electrical shocks and fires. See Grounding System.

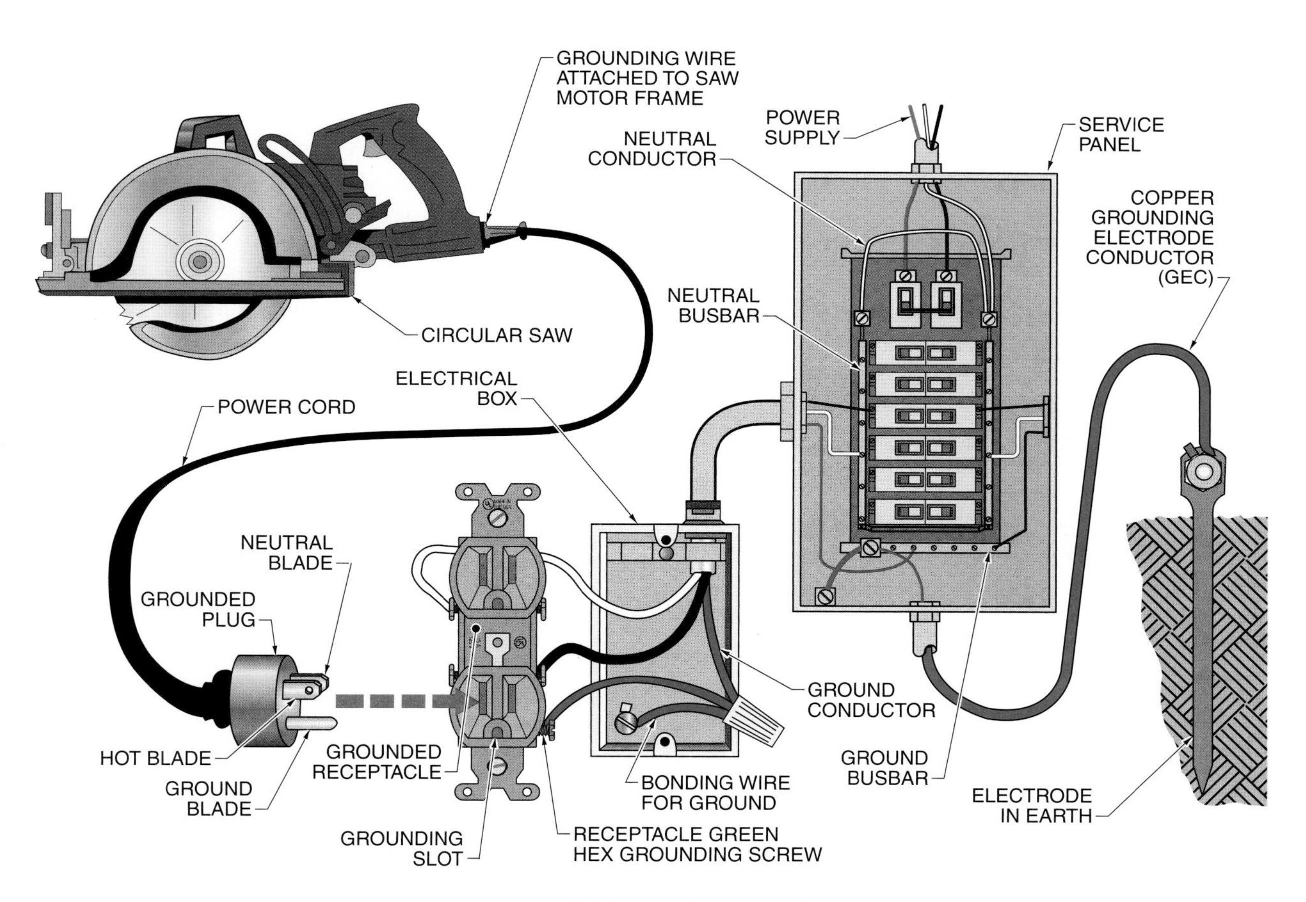

GROUNDING SYSTEM

If an electrical system, circuit, or piece of equipment is grounded, a voltage is present between a hot conductor (ungrounded energized conductor) and ground. A hot conductor is designed to carry current to loads in an electrical circuit and has a voltage potential between it and earth ground equal to the circuit's voltage, such as 120 VAC, 208 VAC, or 480 VAC. Hot conductors are usually colored black, red, blue, orange, brown, or yellow, though black is the most common. Hot conductors should always be properly fused to protect the circuit's conductors and loads from overloads.

A grounded conductor is designed to carry current away from loads in electrical circuits and has been intentionally grounded at a designated location in the electrical system. The grounded conductor is typically called the neutral conductor and is usually colored white or natural gray. Unlike the hot conductor, the grounded (neutral) conductor is not fused.

Grounding connects all exposed non-current carrying metal parts of a system, circuit, or equipment to the earth. Unlike the hot conductor and grounded conductor, the grounding conductor does not carry any circuit or load currents unless there is a problem. The grounding conductor is used to provide a low-resistance path for any fault current to flow to earth ground. Green and green with a yellow stripe are the standard insulation colors for grounding conductors, though grounding conductors can also be bare (no insulation). Grounding conductors, like grounded conductors, are not fused.

In order to test the grounding of a system, circuit, or piece of equipment, an understanding of the different types of services is required. Testing grounding requires knowing the different types of services and how to take voltage measurements with test instruments.

120/240 V, 1φ, 3-Wire Service

A 120/240 V, 1φ, 3-wire service is commonly used to supply power to residential or light commercial buildings. This service provides 120 V, 1φ; 240 V, 1φ; and 120/240 V, 1φ circuits. In residential applications, this service is commonly used for lighting and small appliance use. In commercial applications, this service is commonly used for office equipment, commercial refrigerators, hotel hot tubs and saunas, motors less than 5 HP, cooking equipment, and security equipment. When using many high-power devices, a large power panel or additional power panels may be used. See 120/240 V, 1φ, 3-Wire Service.

120/208 V, 3φ, 4-Wire Service

A 120/208 V, 3φ, 4-wire service is the most common service used for commercial buildings such as offices and schools. It is used to supply customers that require a large amount of 120 V, 1φ power; 208 V, 1φ power; or low-voltage 208 V, 3φ power. This service includes three ungrounded (hot) lines and one grounded (neutral) line. Each hot line has 120 V to ground when connected to the neutral line.

The 120 V circuits are balanced to equally distribute the power from the three hot lines by alternately connecting the 120 V circuits to the power panel so that the phases (A to N, B to N, C to N) are divided among individual load circuits. Likewise, 208 V, 1φ loads, such as 208 V lamps and heating appliances, should also be balanced between phases (A to B, B to C, C to A). Three-phase loads, such as heating elements designed for 3φ power, can be connected to phases A, B, and C. See 120/208 V, 3φ, 4-Wire Service.

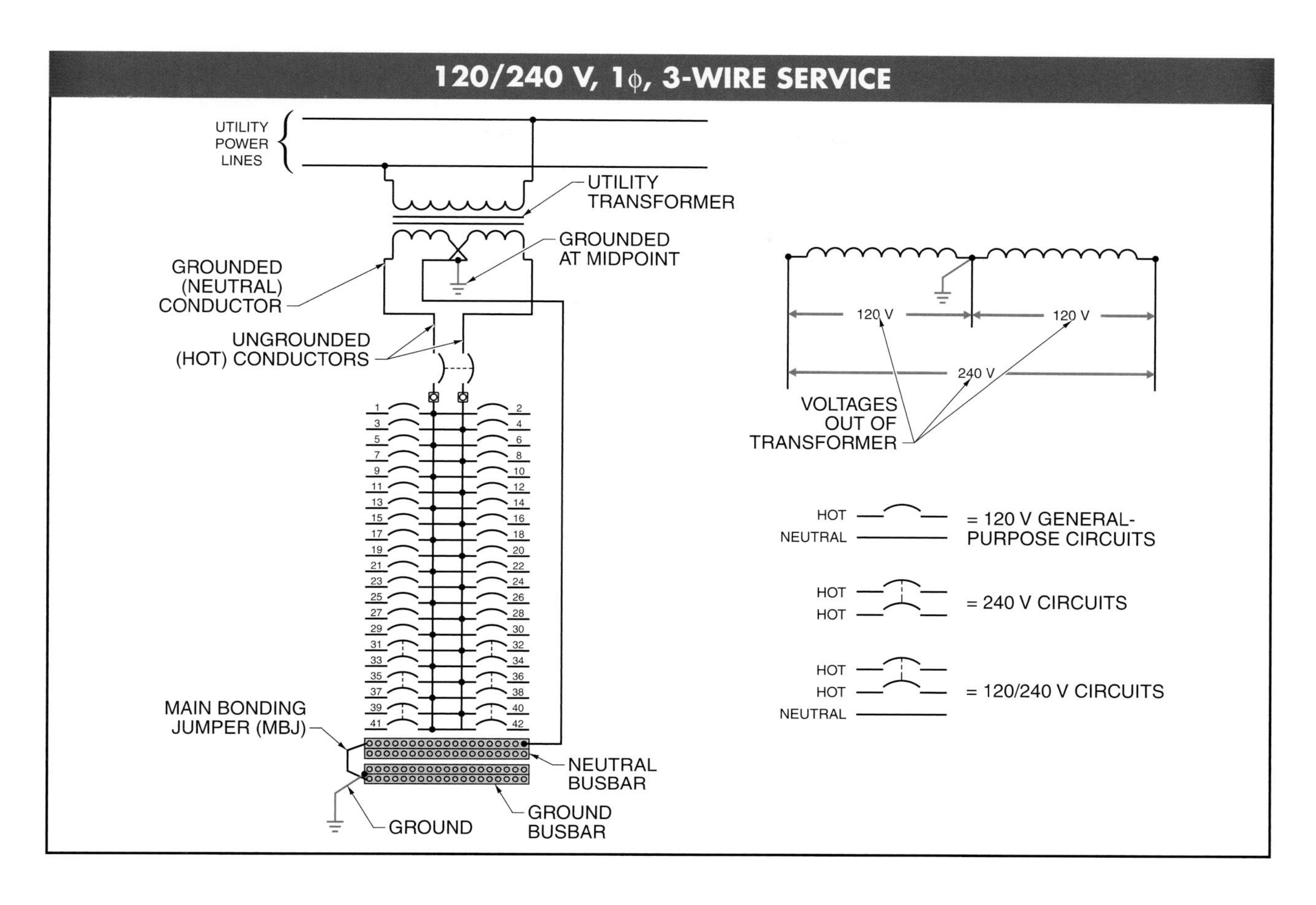
120/240 V, 1ϕ, 3-WIRE SERVICE
UTILITY POWER LINES
UTILITY TRANSFORMER
GROUNDED AT MIDPOINT
GROUNDED (NEUTRAL) CONDUCTOR
UNGROUNDED (HOT) CONDUCTORS
MAIN BONDING JUMPER (MBJ)
NEUTRAL BUSBAR
GROUND BUSBAR
GROUND
120 V
120 V
240 V
VOLTAGES OUT OF TRANSFORMER
HOT
NEUTRAL
= 120 V GENERAL-PURPOSE CIRCUITS
HOT
HOT
= 240 V CIRCUITS
HOT
HOT
NEUTRAL
= 120/240 V CIRCUITS

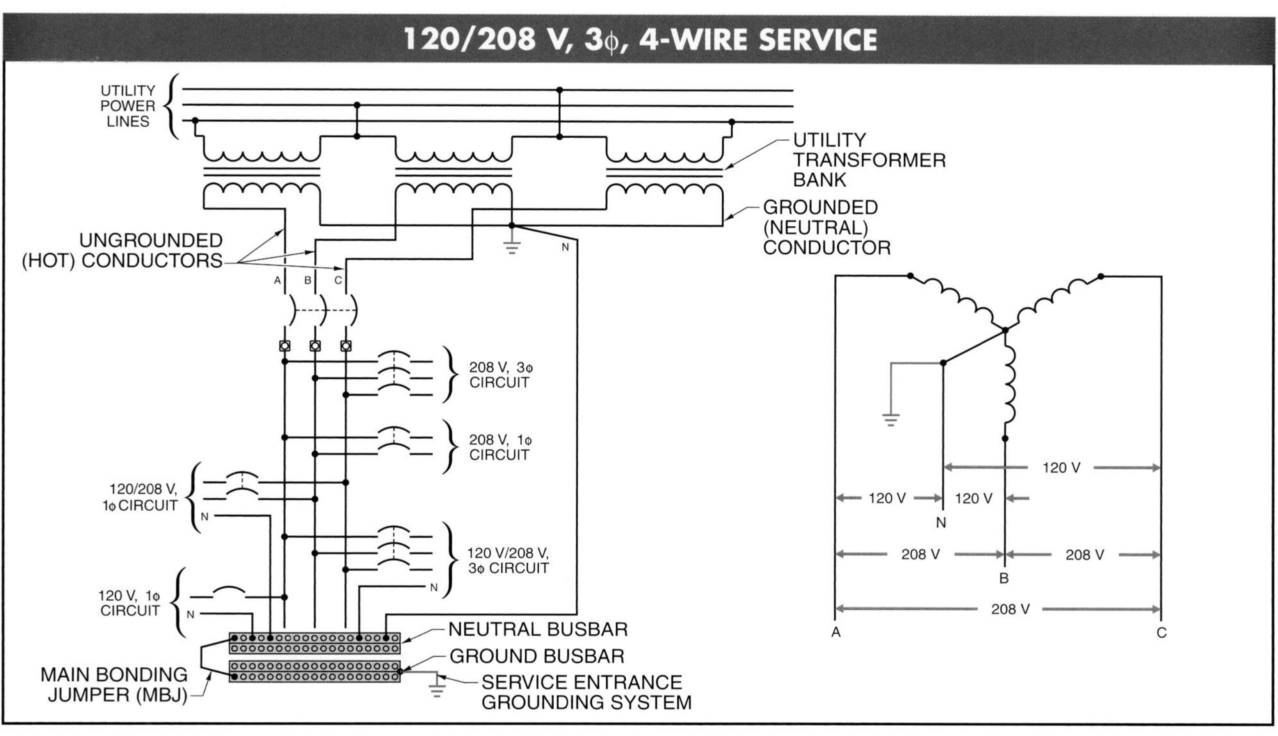
120/208 V, 3ϕ, 4-WIRE SERVICE
UTILITY POWER LINES
UTILITY TRANSFORMER BANK
GROUNDED (NEUTRAL) CONDUCTOR
UNGROUNDED (HOT) CONDUCTORS
A
B
C
N
208 V, 3ϕ CIRCUIT
208 V, 1ϕ CIRCUIT
120/208 V, 1ϕ CIRCUIT
120 V/208 V, 3ϕ CIRCUIT
120 V, 1ϕ CIRCUIT
NEUTRAL BUSBAR
GROUND BUSBAR
MAIN BONDING JUMPER (MBJ)
SERVICE ENTRANCE GROUNDING SYSTEM
120 V
120 V
120 V
208 V
208 V
208 V

120/240 V, 3ϕ, 4-Wire Service

A 120/240 V, 3ϕ, 4-wire service is common in commercial and industrial applications. It is used to supply customers that require a large amount of three-phase power with some 120 V and 240 V, 1ϕ power. Single-phase power is delivered by one of the three transformers and three-phase power is delivered by using all three transformers. The 120 V, 1ϕ power is provided by center tapping one of the transformers. Because only one transformer delivers all of the 120 V power, this service is used in applications that require mostly three-phase power or 240 V, 1ϕ power. In many applications, the total amount of 120 V power used is small when compared to the total amount of three-phase power used. Each transformer may be center tapped if large amounts of 120 V power are required. See 120/240 V, 3ϕ, 4-Wire Service.

Phase B shall be colored orange (or clearly marked) per the NEC® when the switchboard or panelboard is fed from a 120/240 V, 3ϕ, 4-wire, delta-connected service. There is approximately 195 V between phase B and the neutral. This is considered an unreliable source of power because 195 V is too high for standard 115 V or 120 V loads and too low for standard 230 V or 240 V loads.

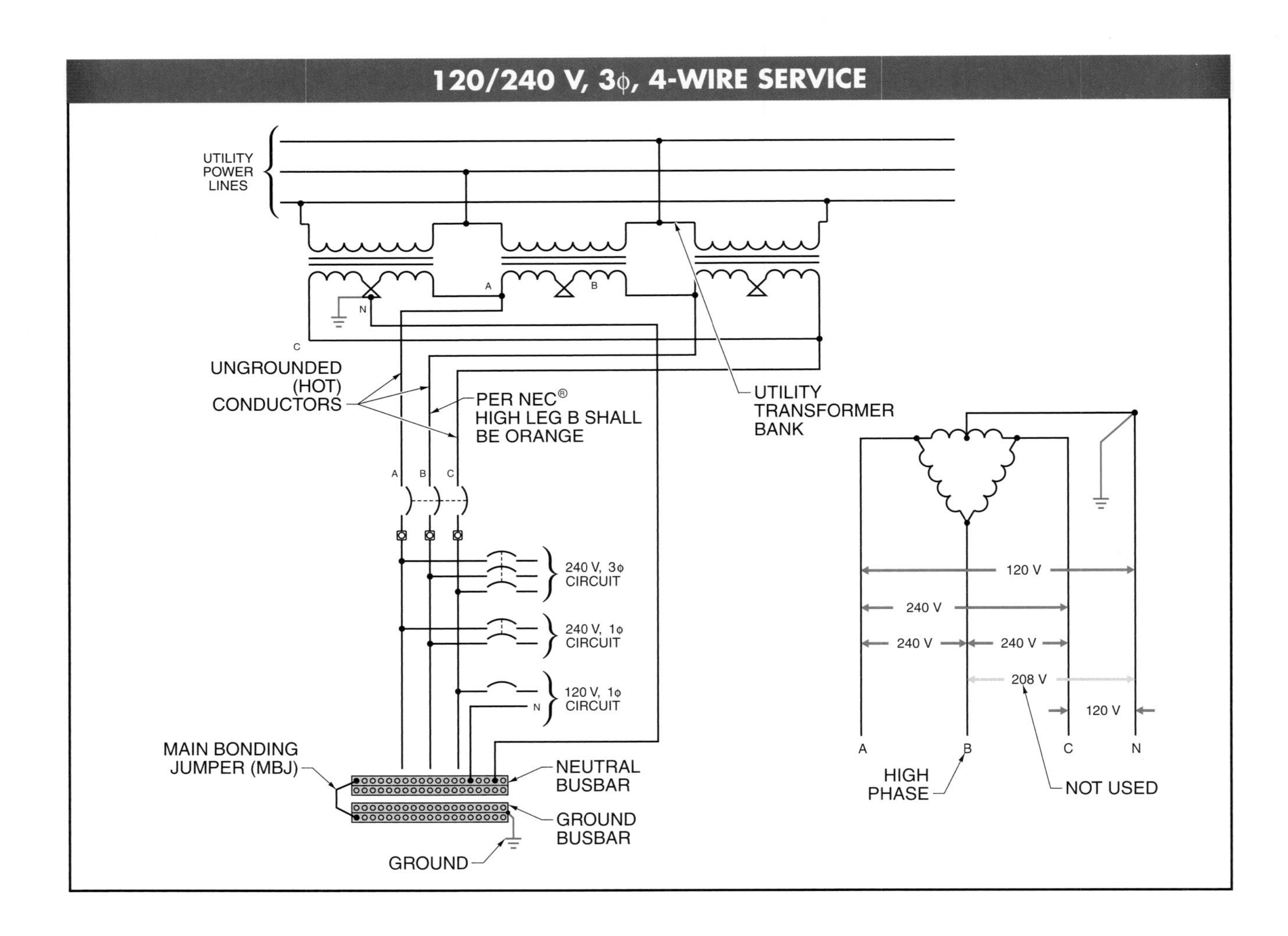

277/480 V, 3ϕ, 4-Wire Service

A 277/480 V, 3ϕ, 4-wire service is the same as a 120/208 V, 3ϕ, 4-wire service, except the voltage levels are higher for industrial applications. This service includes three ungrounded (hot) lines and one grounded (neutral) line. Each hot line has 277 V to ground when connected to the neutral or 480 V when connected between any two hot (A to B, B to C, or C to A) lines. This service provides 277 V, 1ϕ or 480 V, 1ϕ power, but not 120 V, 1ϕ power. Additional transformers can be connected to the 277/480 V, 3ϕ, 4-wire service to reduce the voltage to 120 V, 1ϕ. See 277/480 V, 3ϕ, 4-Wire Service.

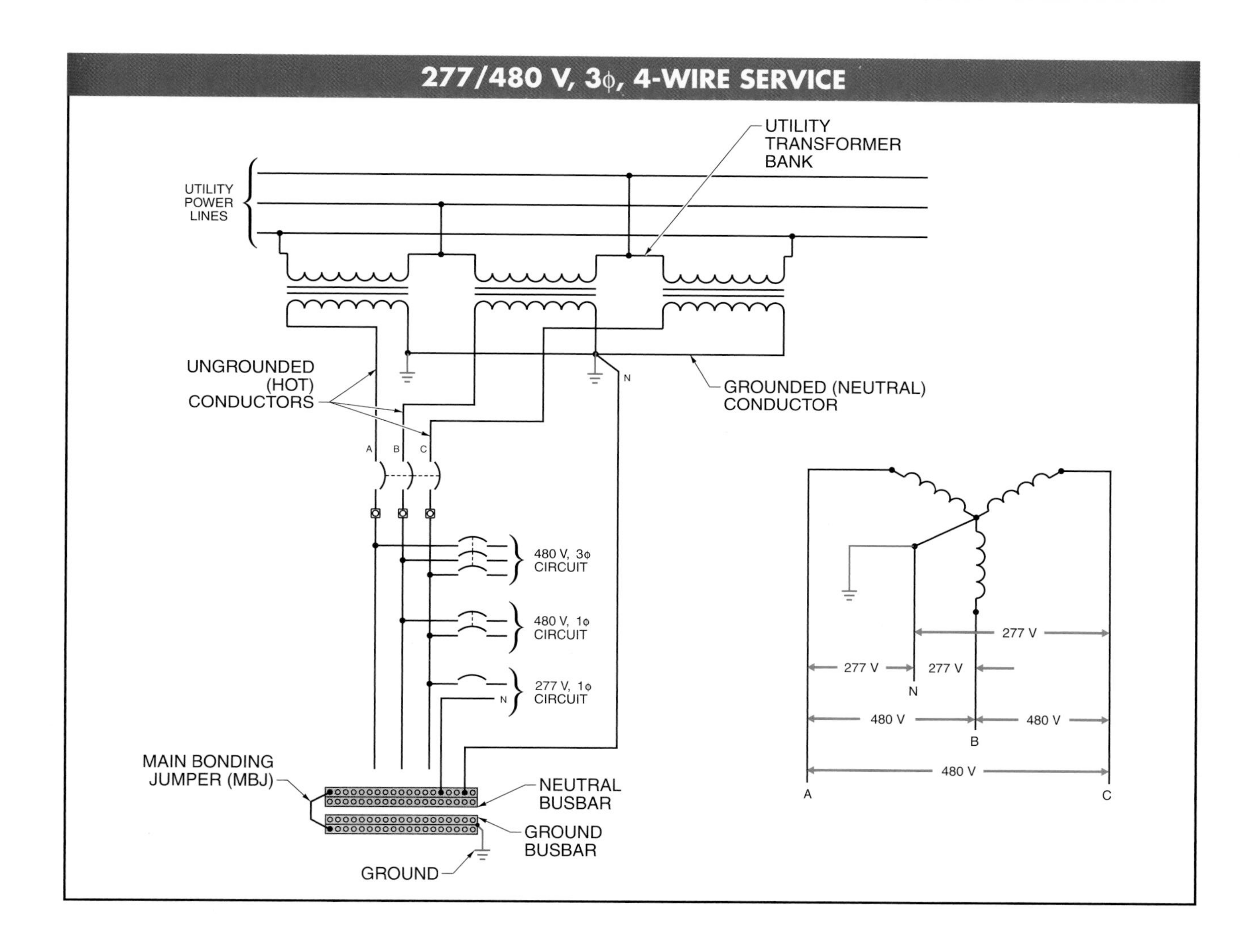

Determine the type of electrical service (voltage levels, phase type, and number of wires) delivered to each service panel and predict the readings of Multimeter 2 and Multimeter 3 if the system is properly grounded.

1. What type of service is delivered to the service panel?

2. ______________ What should Multimeter 2 read?

3. ______________ What should Multimeter 3 read?

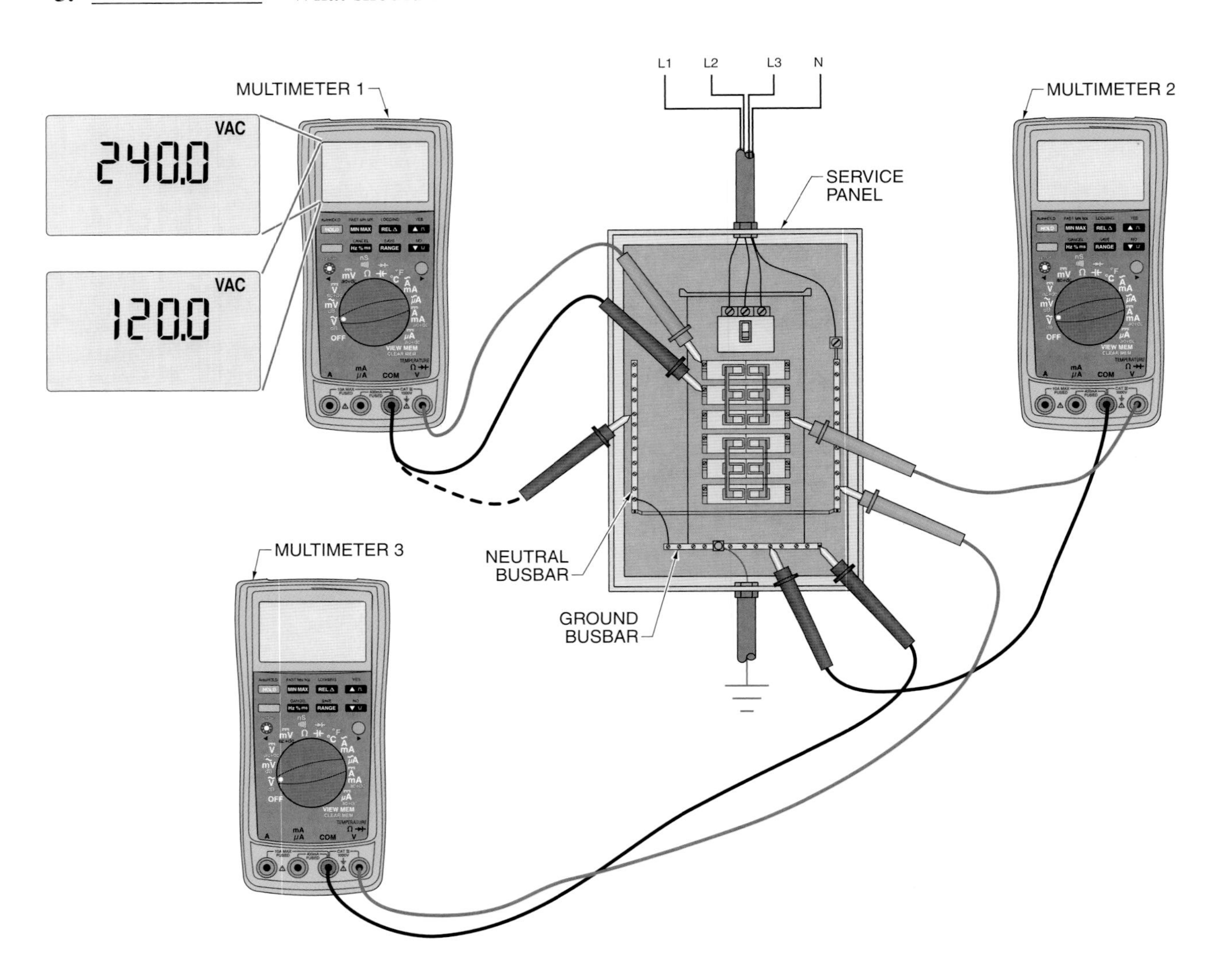

4. What type of service is delivered to the service panel?

5. ______________ What should Multimeter 2 read?

6. ______________ What should Multimeter 3 read?

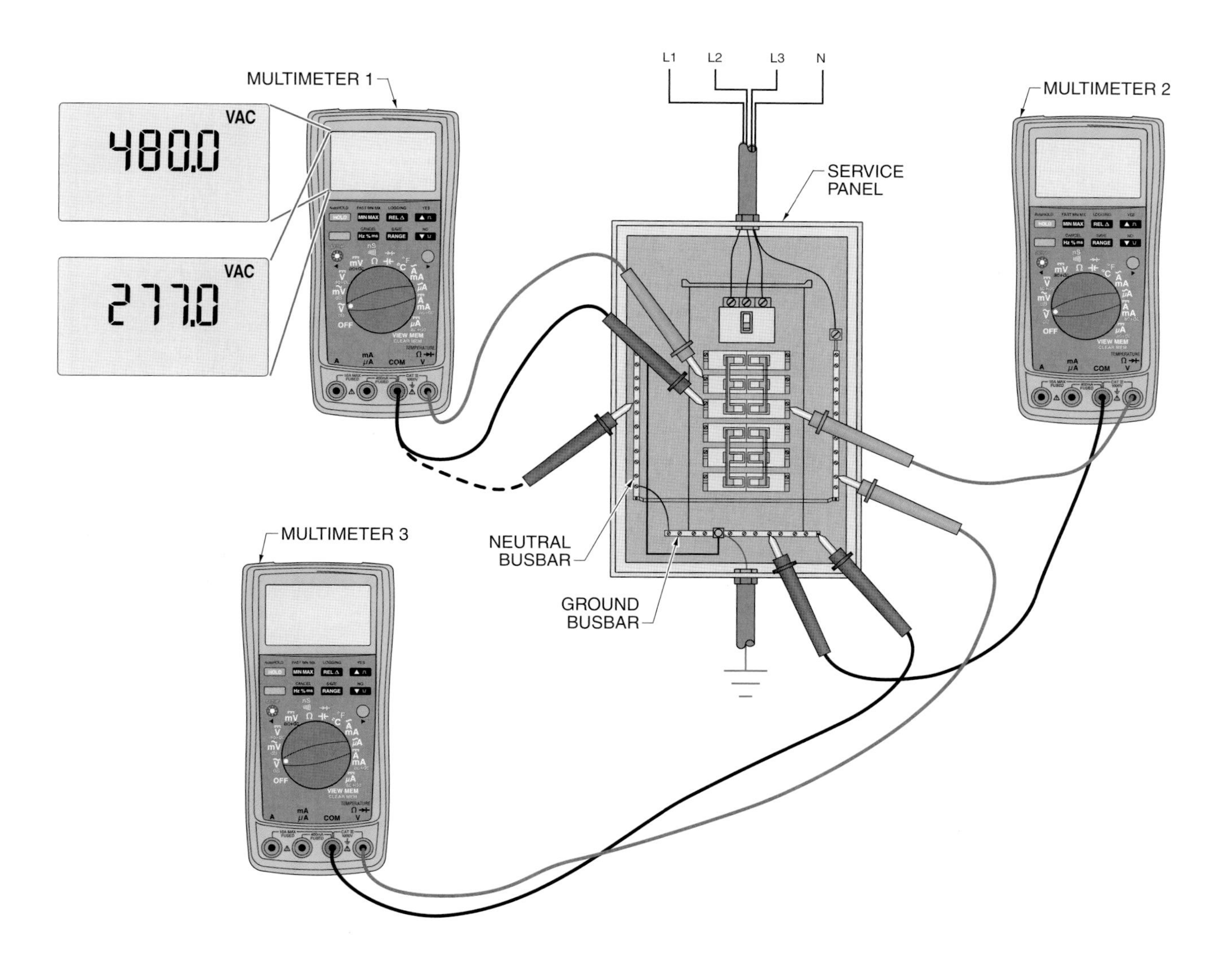

7. What type of service is delivered to the service panel?

8. ______________ What should Multimeter 2 read?

9. ______________ What should Multimeter 3 read?

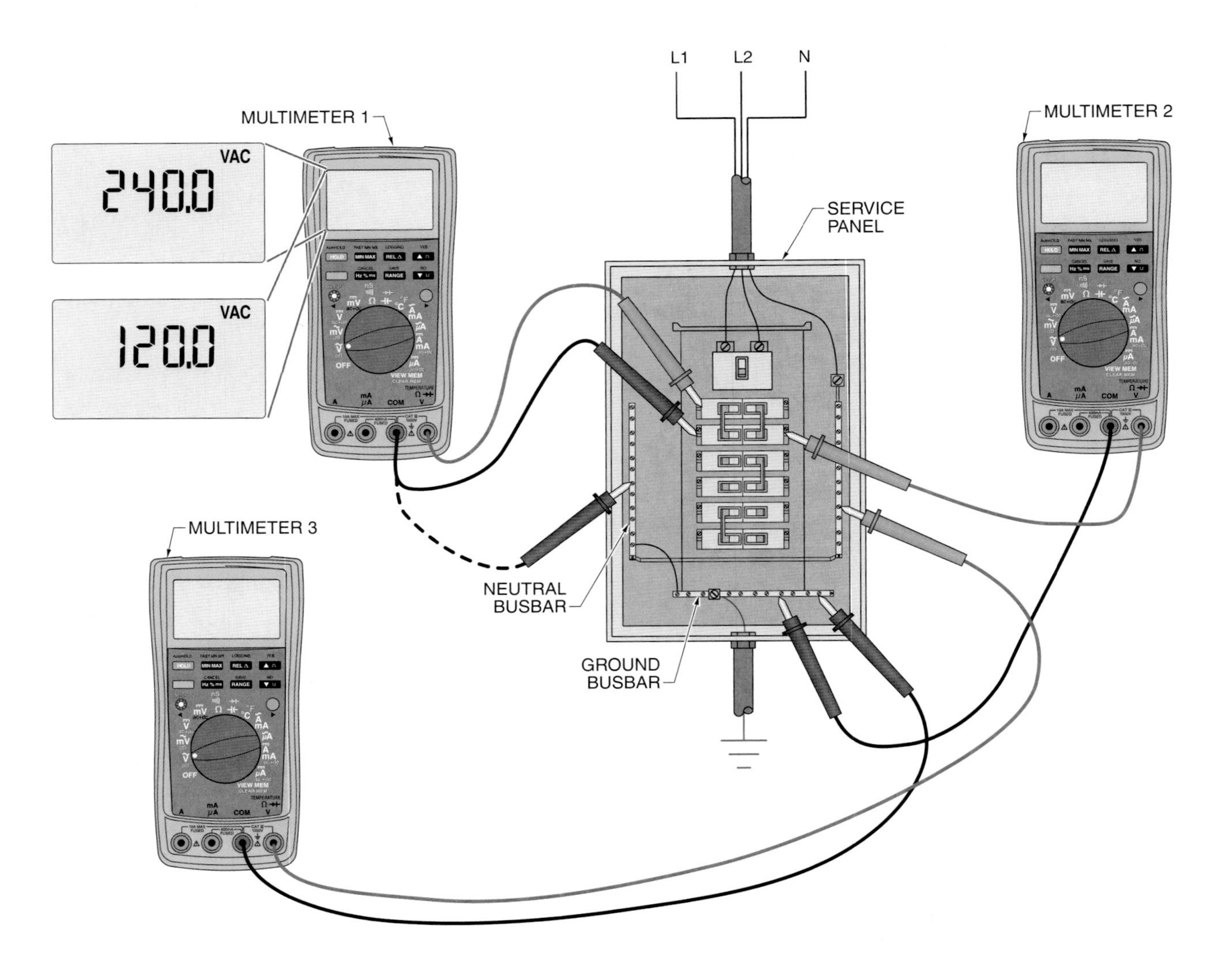

10. What type of service is delivered to the service panel?

11. ______________ What should Multimeter 2 read?

12. ______________ What should Multimeter 3 read?

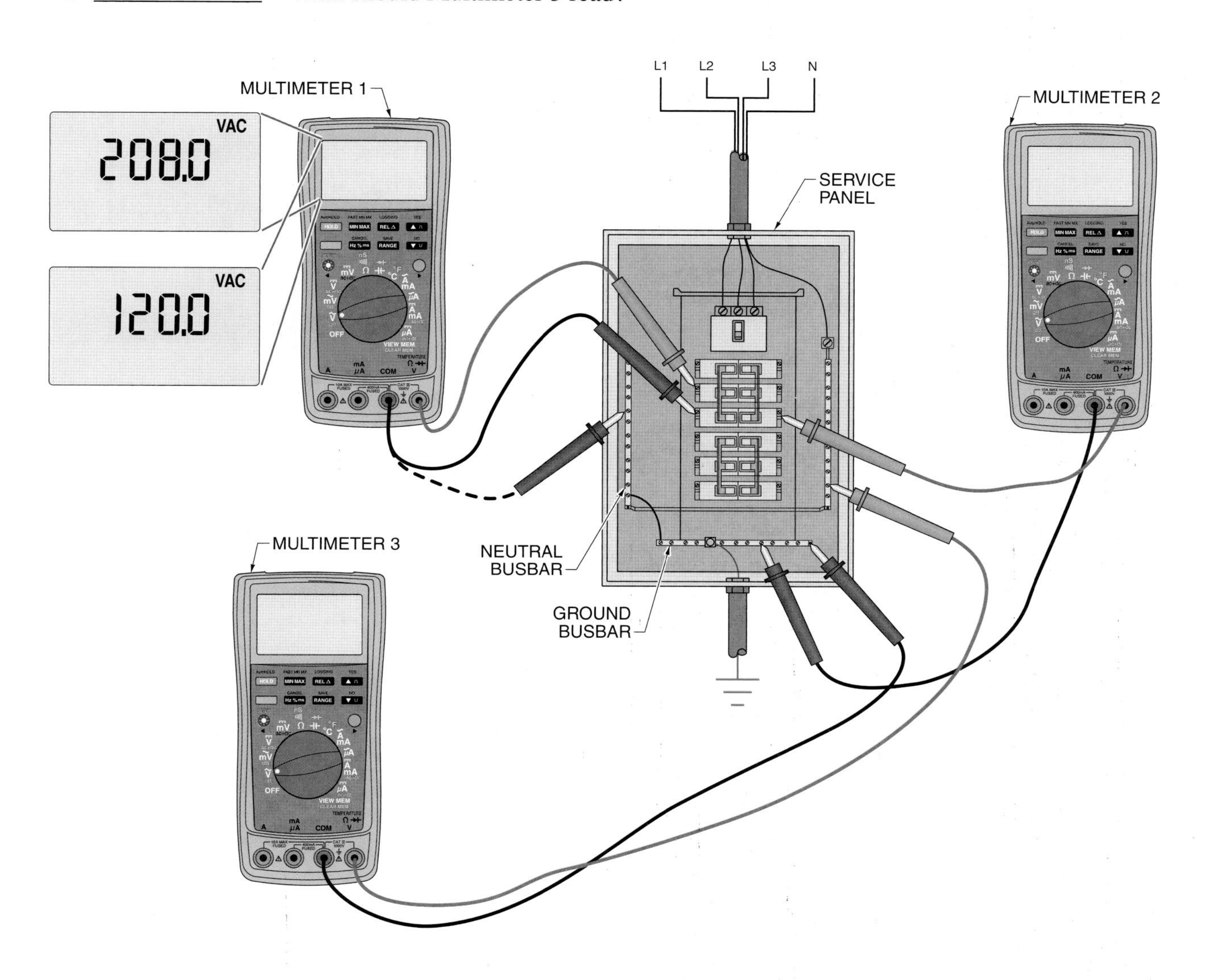

Application 6-2—Grounding Electrode Resistance

All electrical systems must be grounded by connecting the grounding system to the earth with a grounding electrode, the metal frame of the building, concrete-encased electrodes, a grounding ring, or an underground metal water pipe as per NEC® requirements. Ideally, a solid steel grounding electrode with a minimum diameter of 5⁄8″ and length of 8′ is driven vertically into noncorrosive soil that has good conductivity (is moist year around). The top of the electrode should be flush with or below ground level, unless protected against physical damage and at least 8′ is in contact with the soil.

In reality, less than optimal soil and ground conditions mean that alternative grounding installations may be needed. Therefore, the NEC® allows some modifications to grounding electrode installation, though the electrode size cannot be reduced and the grounding system resistance must be less than 25 Ω at all times (year-round soil conditions). See Grounding Electrode Installation Requirements.

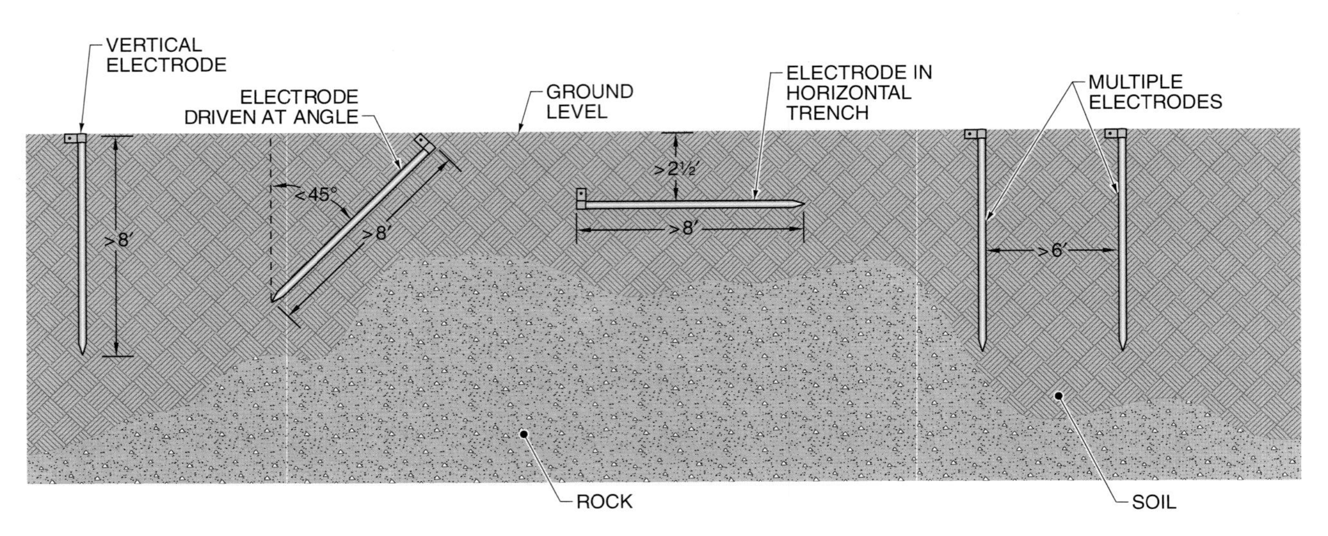

GROUNDING ELECTRODE INSTALLATION REQUIREMENTS

If rocky conditions prevent vertical installation, the grounding electrode can be driven at an angle not to exceed 45° from vertical. Alternatively, the grounding electrode may be buried in a trench that is at least 2½′ deep.

If one electrode exceeds the 25 Ω limit of resistance to ground, additional electrodes can be added to the system to lower the total resistance because of the law of resistance in parallel. However, the rules of calculating resistance in parallel do not apply to multiple grounding rods. The resistance is lowered by set percentages as each additional rod with the same individual resistance is added. The second rod lowers the total resistance to 60% of the first rod. The third rod lowers the total resistance to 40% of the first rod. The fourth rod lowers the total resistance to 33% of the first rod. The multiple electrodes should be at least 6′ apart and connected together at the top.

1. Set (draw in) the function switch position of the ground resistance meter to measure the worst-case leakage current.

2. Set (draw in) the function switch position of the ground resistance meter to measure leakage current of less than 1 A.

3. Set (draw in) the function switch position of the ground resistance meter to measure the grounding system resistance.

4. ______________ Is the measured ground resistance acceptable?

5. ______________ If a second ground electrode is driven into the ground to lower the resistance, what is the minimum distance that must be maintained between the two grounding electrodes?

6. ______________ If the second grounding electrode provides a resistance path equal to the first grounding electrode's resistance, what is the total ground resistance?

7. ______________ Is the total resistance of the two grounding electrodes acceptable?

8. ______________ If a third grounding electrode provides a resistance path equal to the first grounding electrode's resistance, what is the total ground resistance?

9. ______________ Is the total resistance of the three grounding electrodes acceptable?

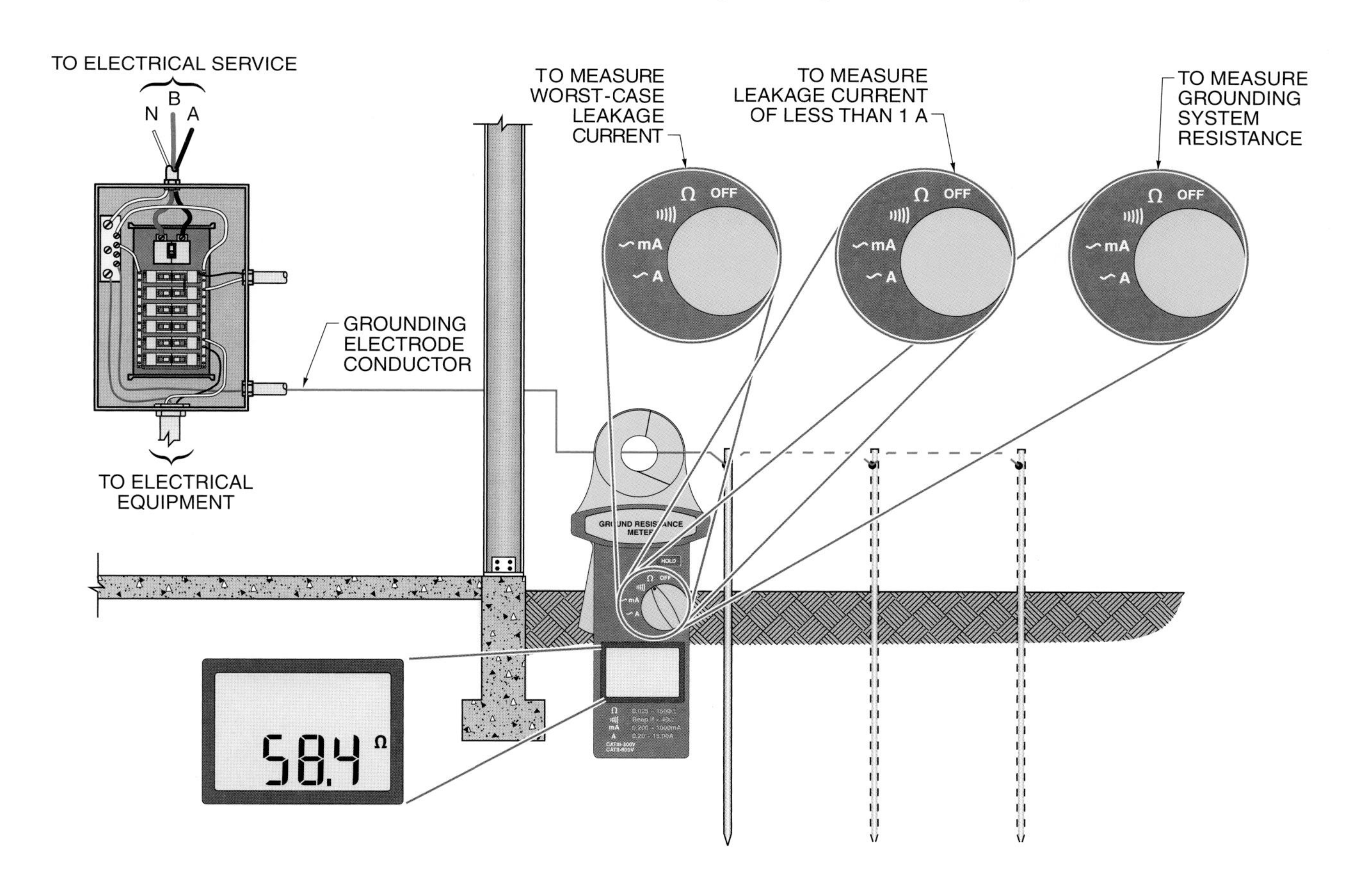

Application 6-3—Ground Faults

Proper grounding requires that a low-resistance path to earth ground be maintained from all non-current carrying metal throughout an electrical system. The low-resistance ground path can be compromised if the grounding electrode is not properly installed, a ground connection such as a conduit fitting or ground wire splice is no longer providing a low-resistance path to ground, or the ground path has opened.

Losing the ground path by itself does not cause an electrical shock, but does create an electrical hazard in that an electrical shock is now likely to occur. An electrical shock occurs when a person contacts an energized (hot) nongrounded conductor, and either a grounded conductor (neutral) or the ground (non-current carrying metal or ground itself).

A voltmeter can test for proper grounding by connecting between circuit components that should be grounded (all non-current carrying metal and the ground wire) and a known energized (hot) part of the circuit. If the portion of the system under test is properly grounded, the voltmeter will read the system voltage (typically around 115 VAC on standard residential circuits). If there is a partial ground connection (loose or corroded connection), the reading may be any voltage between 0 V and the full system voltage. Lower voltage readings indicate poorer ground connections. See Testing for Ground Faults.

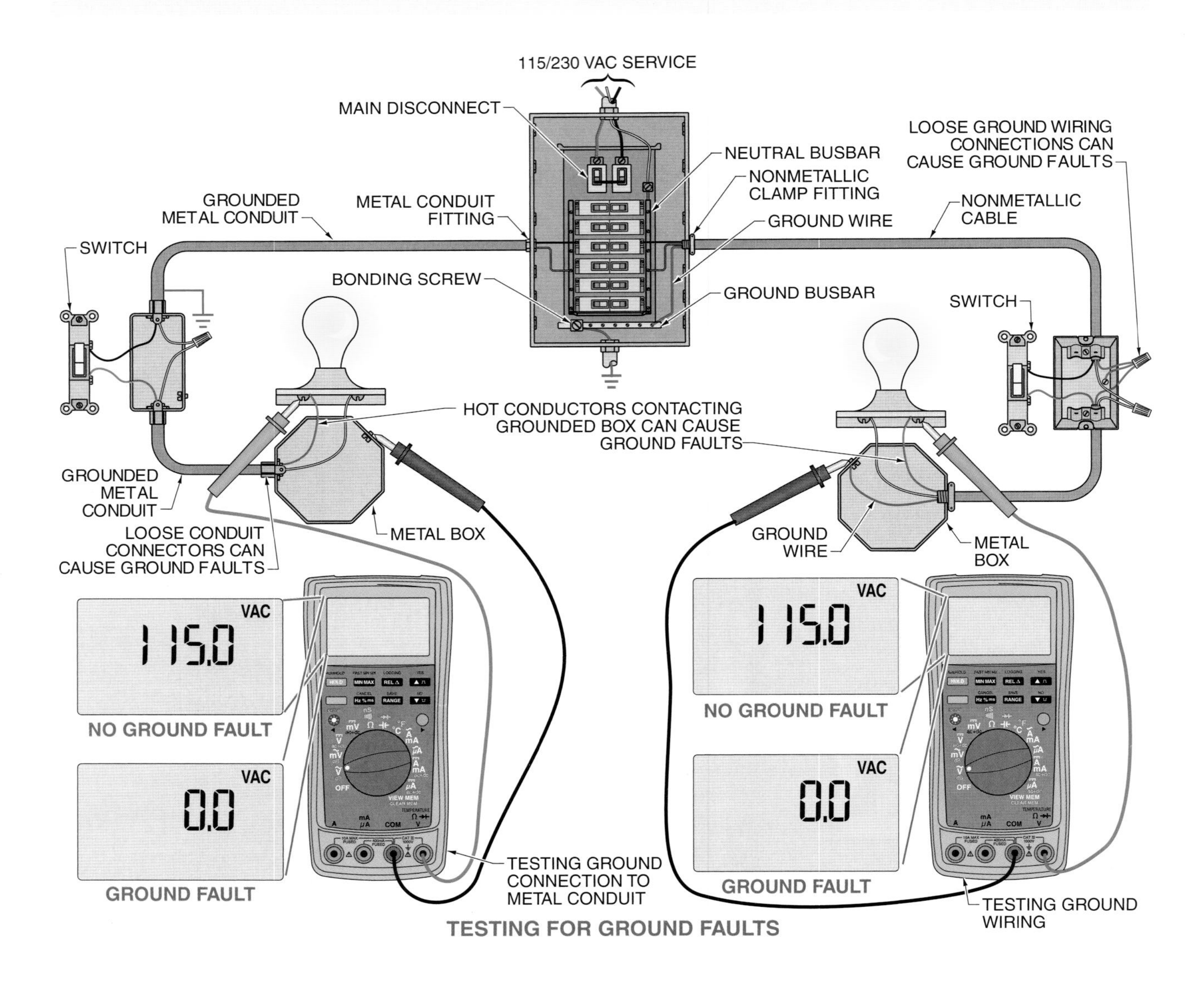

TESTING FOR GROUND FAULTS

1. ______________ If the system is properly grounded, what will Multimeter 1 read?

2. ______________ If the system is properly grounded, what will Multimeter 2 read?

3. ______________ If the system is properly grounded, what will Multimeter 3 read?

A person receives an electrical shock when touching the metal box holding Lamp 1. Testing the system again, Multimeter 1 reads 0 VAC, Multimeter 2 reads 0 VAC, and Multimeter 3 reads 115 VAC.

4. Is the fault most likely located between Lamp 1 and Lamp 2, between Lamp 2 and the switches, or between the switches and the service panel?

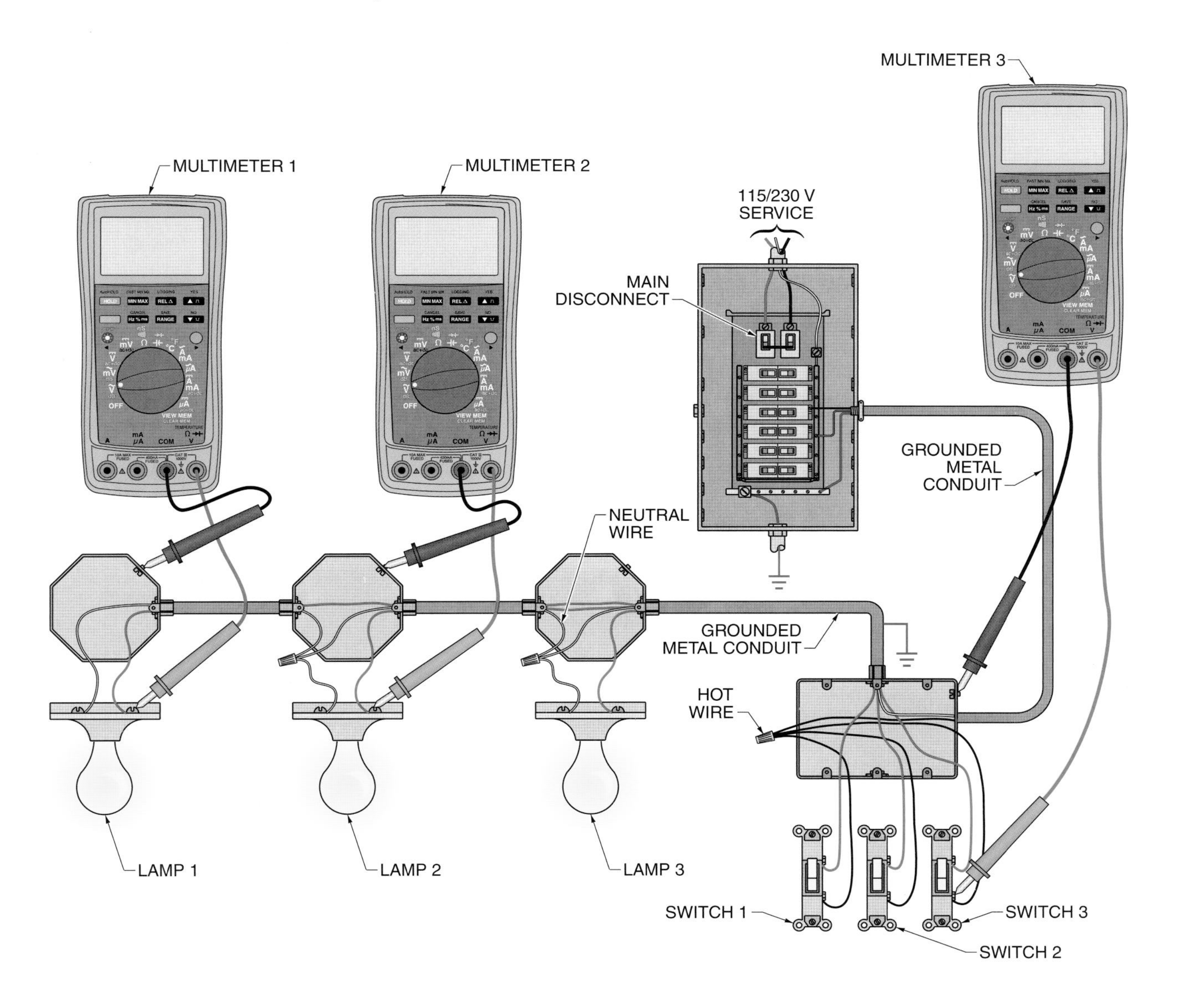

5. ______________ If the system is properly grounded, what will Multimeter 1 read?

6. ______________ If the system is properly grounded, what will Multimeter 2 read?

7. ______________ If the system is properly grounded, what will Multimeter 3 read?

8. ______________ If the system is properly grounded, what will Multimeter 4 read?

A person receives an electrical shock when touching the brass switch wall plate of Switch 1. Testing the system again, Multimeter 1 reads 115 VAC, Multimeter 2 reads 115 VAC, Multimeter 3 reads 115 VAC, and Multimeter 4 reads 52 VAC.

9. The most likely location of the fault is between which two components?

10. What would cause the 52 VAC reading on Multimeter 4?

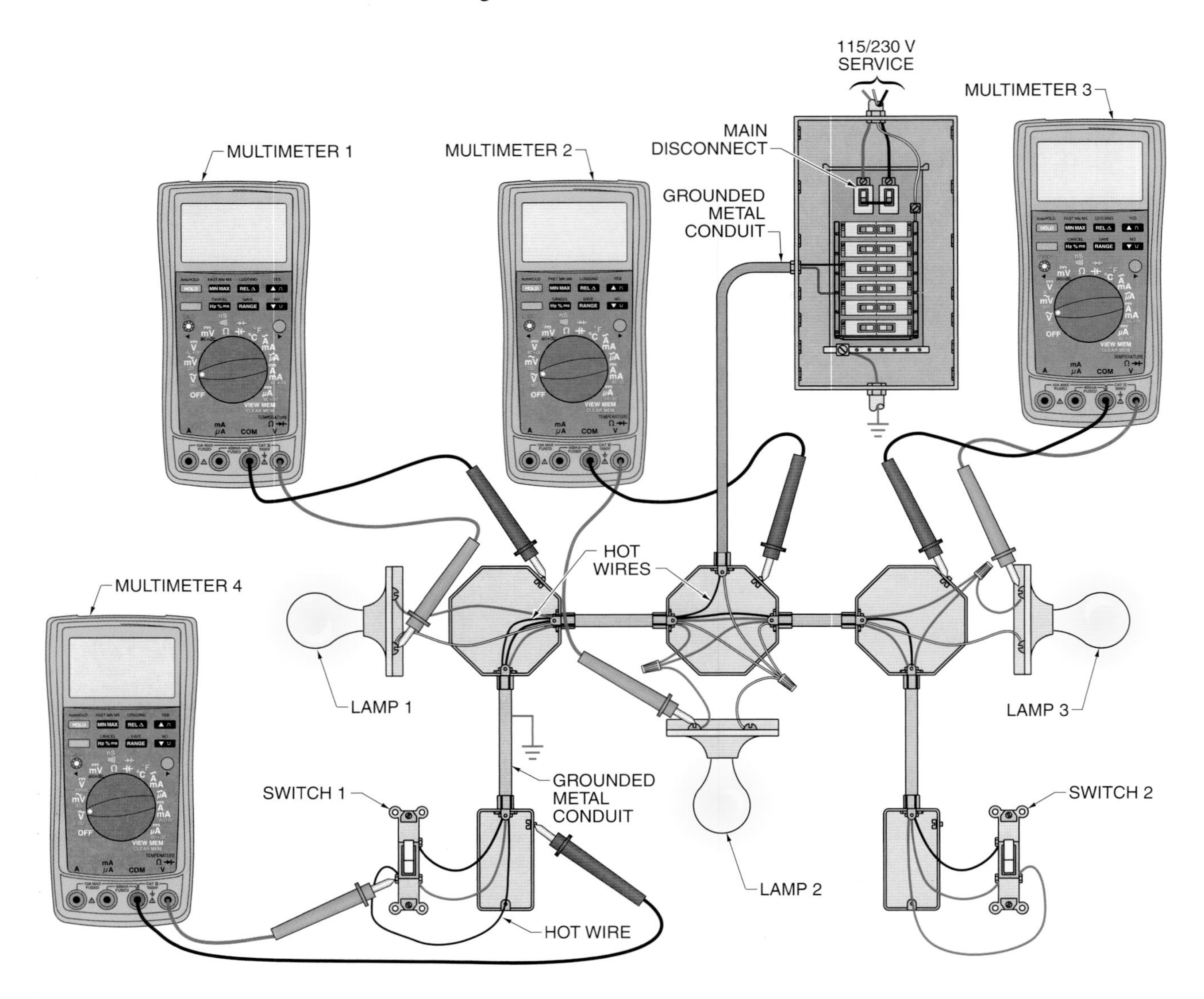

Application 6-4 — Ground Currents

In some circuits, the service entrance grounding conductor may have a small amount of current flowing to ground. Electronic devices such as computers, medical equipment, and sound equipment are highly susceptible to noise introduced into their signals from electromagnetic fields and other sources. The grounding conductor on electronic devices not only prevents electrical shocks but also drains the unwanted noise to ground. This excess current is very small, on the order of a few micro- or nanoamperes, so it probably cannot be measured by a standard clamp-on ammeter. An in-line ammeter set to microamperes or a leakage current ammeter can measure very small ground currents.

Currents flowing from the grounding conductors of multiple electronic devices add together at the service panel and flow to the earth-grounding electrode. Larger electrical systems, especially with more electronic devices, will produce higher total ground currents. For example, a school with many computers will have a higher ground current than a single-family residence. When a measured ground current is higher than expected on a grounding system, additional current measurements should be taken to determine the source of the current. See Measuring Ground Current.

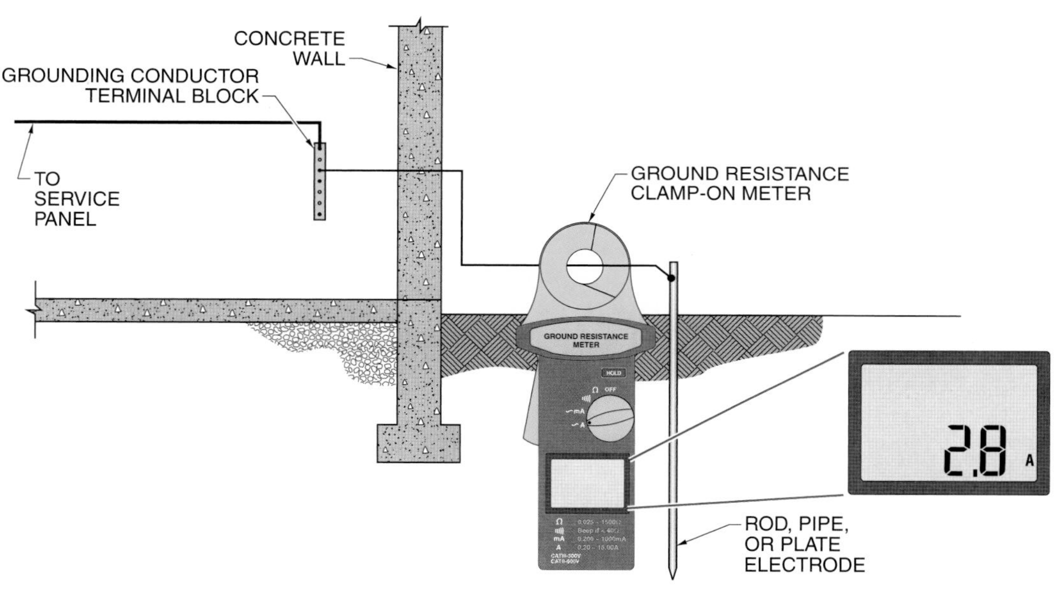

MEASURING GROUND CURRENT

A ground resistance clamp-on meter has measured a higher than expected current (2.8 A) on the outside building grounding system. For each of the given set of clamp-on current measurements taken inside the building, determine the most likely source of the problem from the following list.

A. Load 1
B. Load 2
C. Load 3
D. Other loads connected to the Main Panel
E. Other loads connected to Subpanel 1
F. Other loads connected to Subpanel 2

When taking ground current measurements inside the building, Ammeter 1 reads 3 mA, Ammeter 2 reads 0 mA, Ammeter 3 reads 0 mA, Ammeter 4 reads 2.75 A, and Ammeter 5 reads 1 mA.

1. ______________ Where is the most likely source of the ground current?

When taking measurements inside the building, Ammeter 1 reads 3 mA, Ammeter 2 reads 0 mA, Ammeter 3 reads 2.75 A, Ammeter 4 reads 4 mA, and Ammeter 5 reads 1 mA.

2. ______________ Where is the most likely source of the ground current?

When taking measurements inside the building, Ammeter 1 reads 3 mA, Ammeter 2 reads 0 mA, Ammeter 3 reads 0 mA, Ammeter 4 reads 4 mA, and Ammeter 5 reads 1 mA.

3. ______________ Where is the most likely source of the ground current?

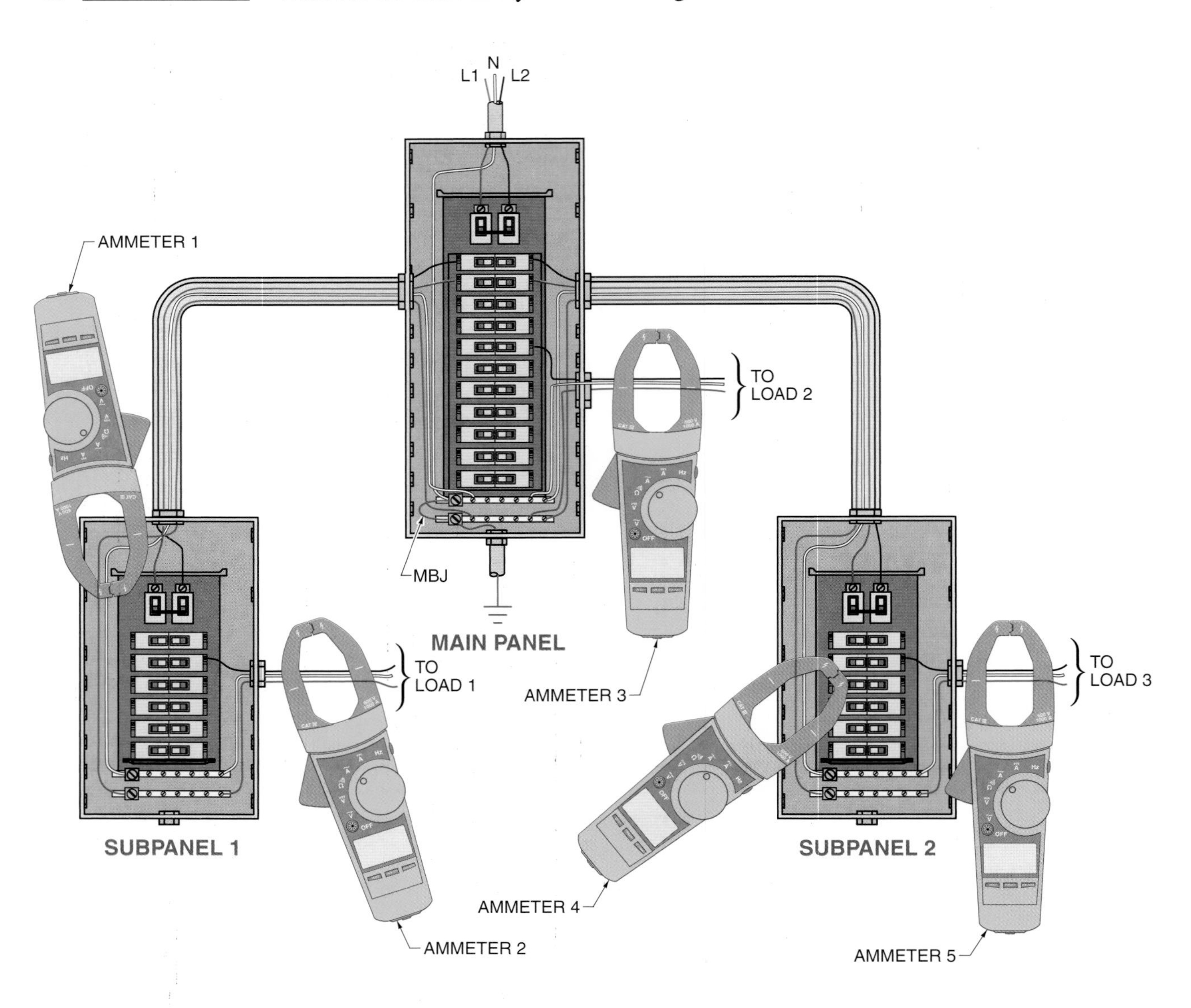

MEDIUM VOLTAGE TEST INSTRUMENTS

Name: ______________________________ Date: ____________________

Application 7-1 —Medium Voltage Dangers

It is generally understood that working around medium voltage is dangerous and that contacting medium-voltage lines can be fatal. However, several things about electric shocks are misunderstood. Although it is true that medium voltage can kill, it is also true that several thousand volts of static electricity may cause only a harmless shock. It is also possible to receive a fatal electric shock from standard 120 VAC residential circuits and appliances.

Actually, it is the combination of voltage and current that creates the potential for serious electric shock. First, the voltage must be high enough to cause current to flow and, once current flows, the current must be high enough to cause an electric shock.

Voltage causes current to flow once the resistance is low enough. A high resistance between two conducting materials prevents any current flow between them, even when the voltage is high. However, when the resistance decreases, current will start to flow. With high enough voltage, as little as 8 mA of current can cause a painful electric shock. See Effects of Electric Current.

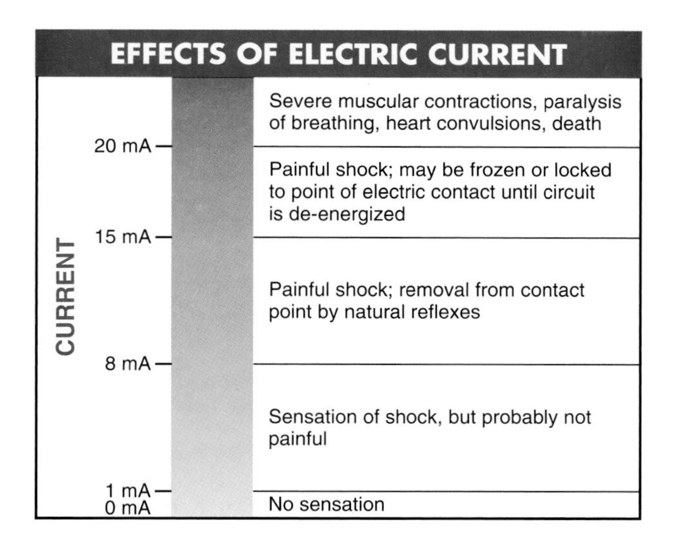

Increasing resistance limits or prevents current flow. Insulating materials such as electrical gloves, rubber-soled shoes, rubber insulating mats, and double-insulated tools all raise the resistance of a potential current path through the body, lowering the possible current flow if a hot (ungrounded) conductor is contacted.

Any voltage above 50 V must be considered dangerous, and all safety rules must be followed regarding personal protective equipment (PPE) and proper tools, equipment, and test instruments. Voltages below 50 V are also considered potentially hazardous when resistance is low enough to cause current to flow. For example, a few volts (6 V, 12 V, 24 V) is usually not enough to cause current flow through a person that has dry hands. This is why a person does not receive an electric shock when touching conductors with small voltages such as small batteries, exposed speaker wires, or computer peripheral cables. However, a person can receive an electric shock if their resistance is lowered enough. For example, people have received electric shocks in showers in which there is a few volts potential between the faucet and drain ground. Thus, measurements should always be taken to determine voltage and current levels.

Determine the amount of current that flows through the body when a person accidentally touches the ungrounded (hot) conductor at a 120 VAC receptacle.

1. ______________ How much current (in mA) flows through a body with 50,000 Ω of resistance?

2. ______________ How much current (in mA) flows through a body with 40,000 Ω of resistance?

3. ______________ How much current (in mA) flows through a body with 30,000 Ω of resistance?

4. ______________ How much current (in mA) flows through a body with 20,000 Ω of resistance?

5. ______________ How much current (in mA) flows through a body with 15,000 Ω of resistance?

6. ______________ How much current (in mA) flows through a body with 10,000 Ω of resistance?

7. ______________ How much current (in mA) flows through a body with 5000 Ω of resistance?

8. ______________ How much current (in mA) flows through a body with 3000 Ω of resistance?

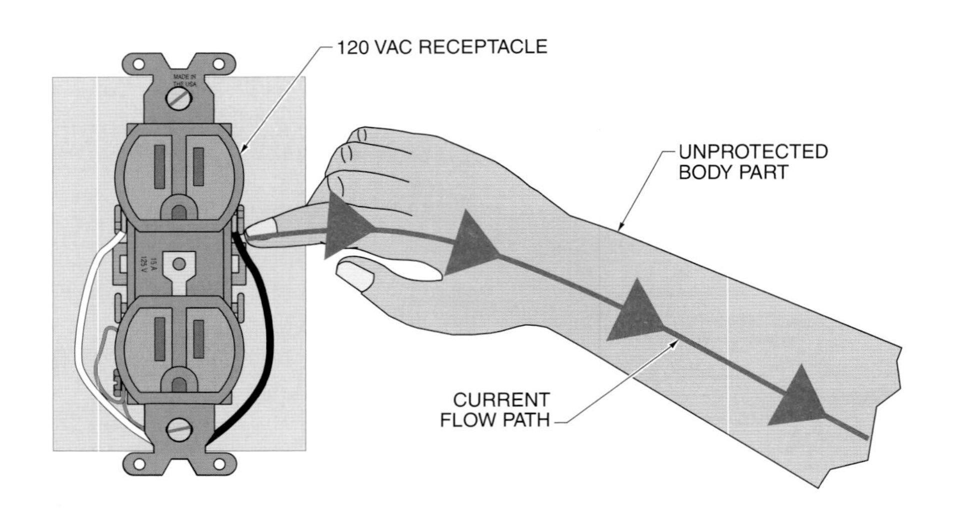

9. Plot the current values against resistance on the graph.

10. ______________ At approximately what resistance will this person begin to feel a painful shock?

11. ______________ At approximately what resistance may this person be frozen to the point of contact?

12. ______________ At approximately what resistance is this person in danger of serious harm or death?

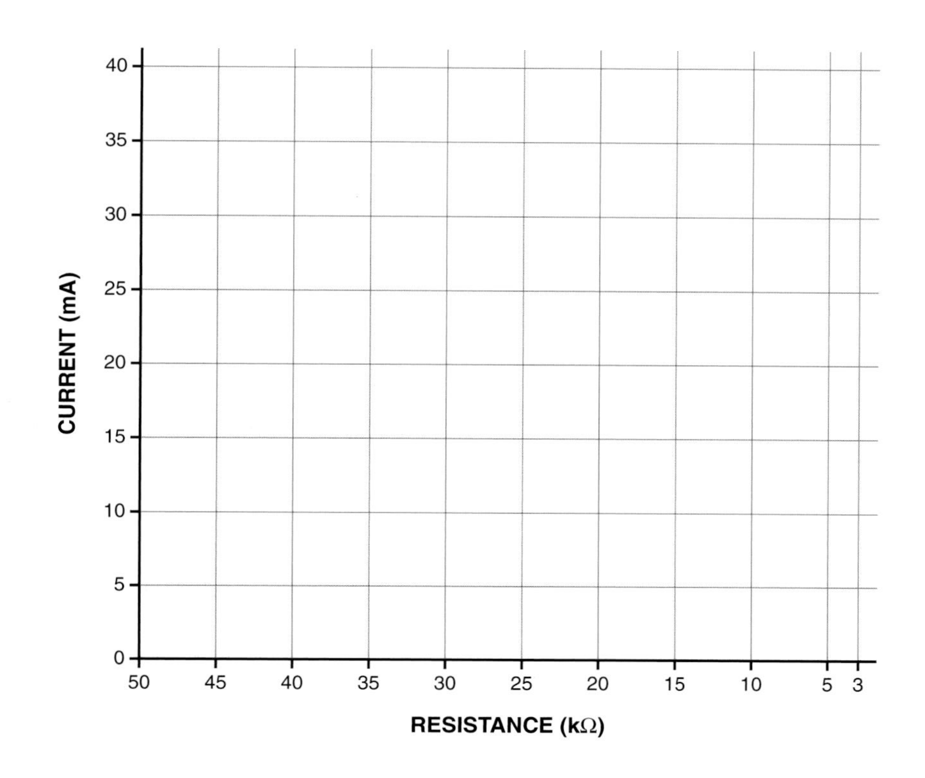

Application 7-2—Cable Height Tester

Electric power lines are run overhead on wood, steel, or reinforced concrete poles; underground in duct lines or conduits; or buried directly in the ground. All electric power lines are dangerous and can electrocute anyone in contact with them. However, accidental contact is more likely with overhead power lines because they are exposed and protected primarily by distance. Higher power line voltages and greater potential for pedestrian or vehicular traffic require that greater minimum clearances be maintained.

The NEC®, OHSA, and other regulating agencies cover minimum clearances (distance) required between electrical conductors and buildings, roofs, driveways, roads, the earth, or anything else on the ground. NEC® 230.24 covers minimum overhead clearance distances of service drop conductors not over 600 V.

Clearance from Ground

The minimum clearance for service drop conductors over the ground (final grade) varies with the voltage of the conductors and the type of traffic underneath. See Clearance from Ground.

- For 150 V or less over pedestrian only traffic (no vehicles), the clearance shall be at least 10′.
- For 300 V or less over pedestrian and car traffic (no trucks), the clearance shall be at least 12′.
- For greater than 300 V but not more than 600 V, over pedestrian and car traffic (no trucks), the clearance shall be at least 15′.
- For not more than 600 V over traffic including trucks, the clearance shall be at least 18′.

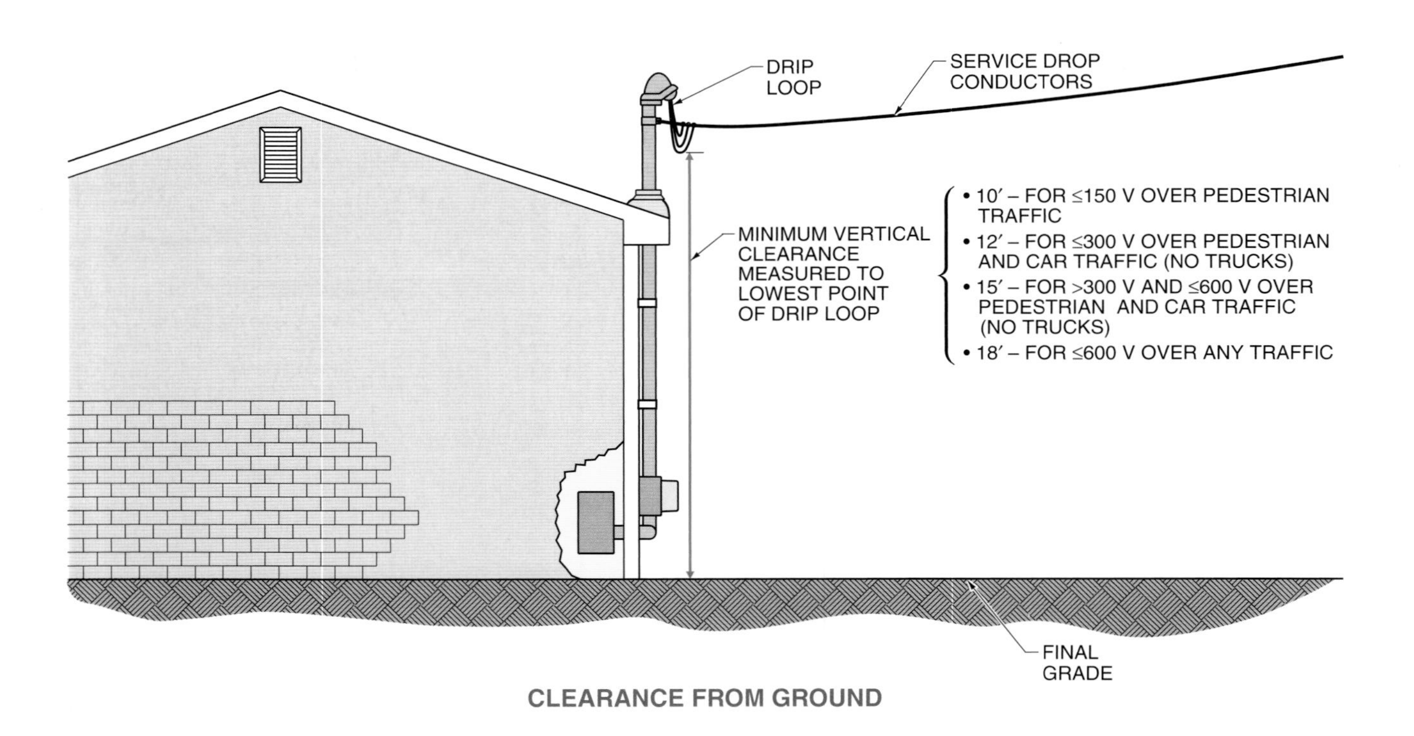

CLEARANCE FROM GROUND

Clearance over Roofs

The minimum clearance for service drop conductors over roofs varies with the voltage of the conductors, the roof type, and conductor termination. See Clearance over Roofs.

- For not more than 600 V over a roof, the clearance shall be 8′. That clearance shall be maintained in all directions for a minimum distance of 3′ from the edge of the roof.
- For not more than 600 V over a roof subject to pedestrian or vehicular traffic, the clearance shall be 18′.
- For not more than 300 V over a rooftop slope of 4″ in 12″ (about 18°) or greater, the clearance shall be 3′.
- For not more than 300 V when the conductors will be terminated at a through-the-roof raceway after not more than 6′ of the conductors pass over the roof, the clearance shall be 18″.

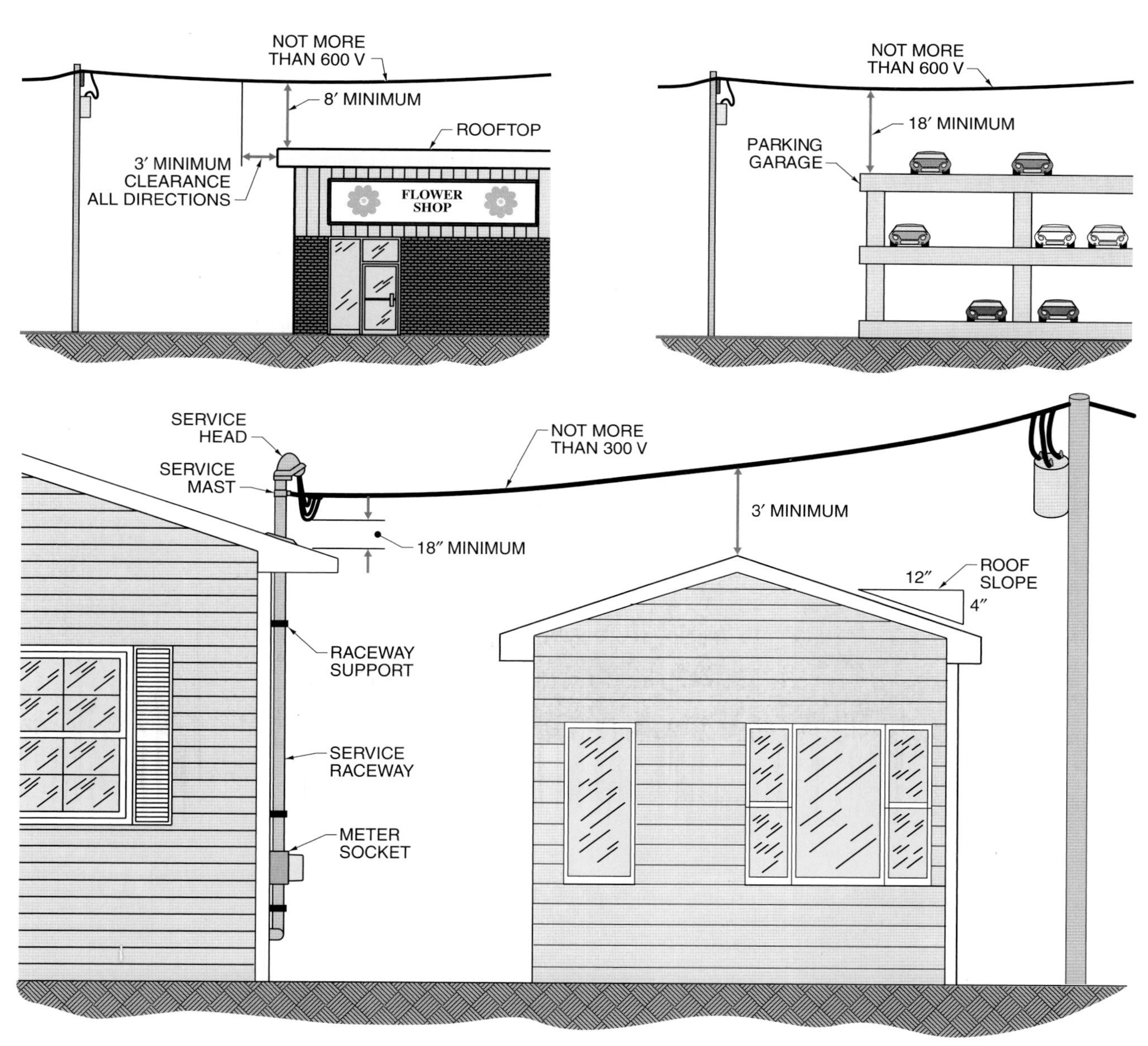

CLEARANCE OVER ROOFS

Even power lines that are properly installed can become dangerous during ice storms, tree fallings, structural additions, and landscape and earth grading changes because the power lines may become closer to the ground or buildings.

Also, power lines can sag because their high current load increases the temperature of the conductors. Sags can be from several feet for shorter spans (150′ to 300′) up to 20′ on long spans (1000′ to 1500′). Power lines usually sag the most on hot days, when electricity demand is at its peak.

For these reasons, the actual distance between a power line and the nearest structure must be measured with a cable height tester if there is any uncertainty about it meeting requirements.

The electric utility installs electrical service to an auto repair shop and measures the height of the installed cable over the surface of the flat roof and over the parking lot. The parking lot traffic consists of pedestrians and small vehicles.

1. ______________ Does this installation meet the minimum height clearances for this application?

2. ______________ If the shop does a large amount of arc welding, should the cable height be tested again because of the high electrical demand?

3. ______________ If the roof is resurfaced by adding 2″ of material, does the installation still meet the minimum height clearance?

4. ______________ If the shop begins receiving truck deliveries in their parking lot, does the installation still meet the minimum height clearance?

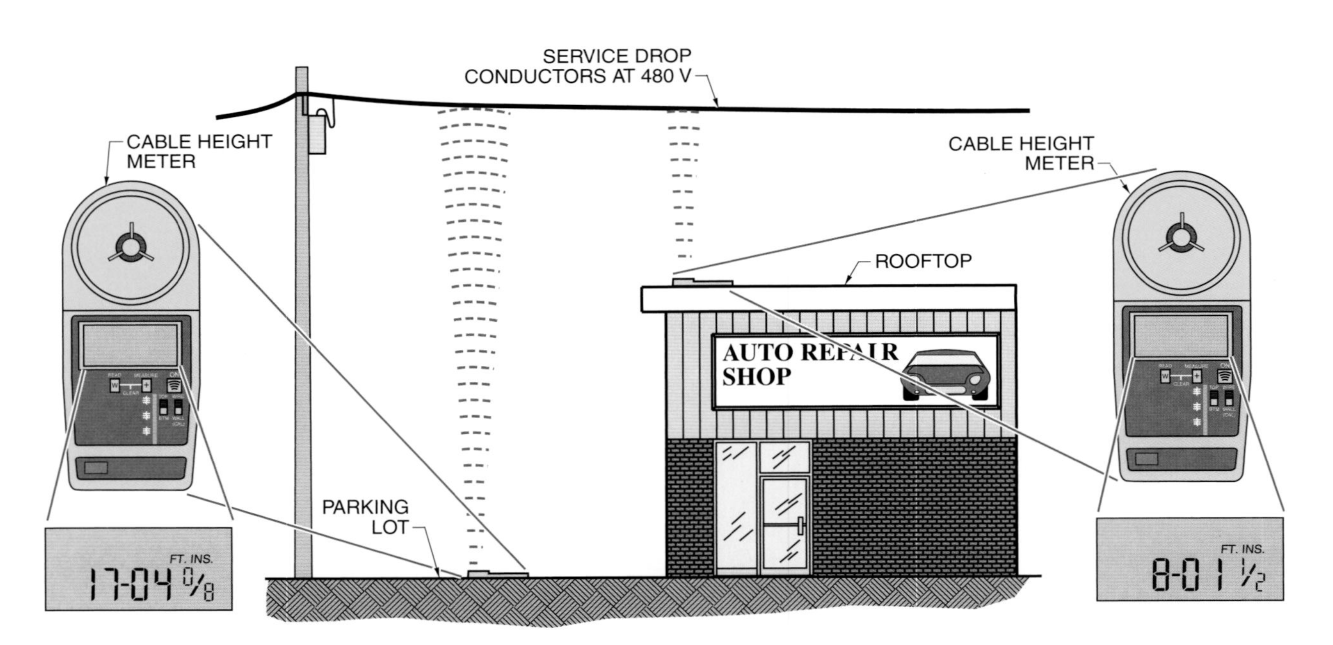

A homeowner wants to build a house addition and a detached garage in the backyard, both underneath the service drop conductors supplying the home.

5. ______________ Does the existing installation meet the minimum height clearances for this application?

6. ______________ If the garage is a flat-roof design, what is the maximum allowable garage height for that location?

7. ______________ If the garage is a sloped-roof design with a 5″ in 12″ slope, what is the maximum allowable garage height for that location?

8. ______________ If the addition to the home is on the side with the service mast, will the service mast need to be reconfigured to meet clearance requirements?

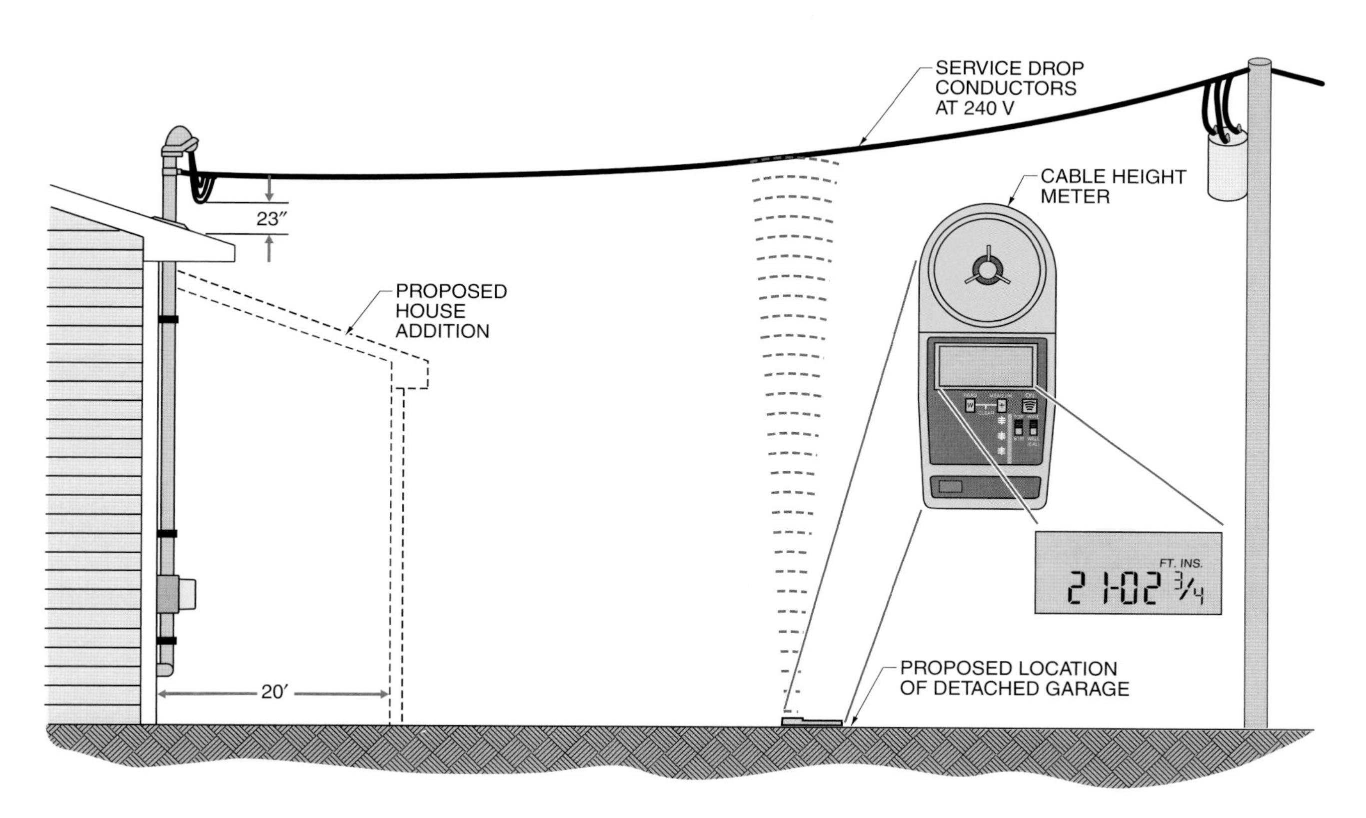

Application 7-3—Motor Winding Leakage Current

Excessive moisture, chemicals, and dirt cause wire insulation to break down. Insulation breakdown reduces resistance, which allows more leakage current between the windings, or between the windings and the frame of the motor. The excessive leakage current damages the windings and may cause hazardous short circuits.

A megohmmeter can measure the resistance between pairs of motor windings, and between motor windings and the frame of the motor (ground). The resistance measurements are taken with the motor disconnected or the power off and locked out. The amount of leakage current can be calculated by dividing the motor's rated voltage by the resistance measurements. Leakage currents up to 1 mA per 1000 V are considered acceptable.

1. ______________ What was the leakage current (in mA) between the winding and the frame when the motor was put into service?

2. ______________ What is the leakage current (in mA) between the winding and the frame one year later?

3. ______________ Has the insulation deteriorated unacceptably in one year?

4. ______________ If the motor was reconfigured for 230 V, what would the leakage current (in mA) be?

5. ______________ Would the leakage current in the 230 V configuration be acceptable?

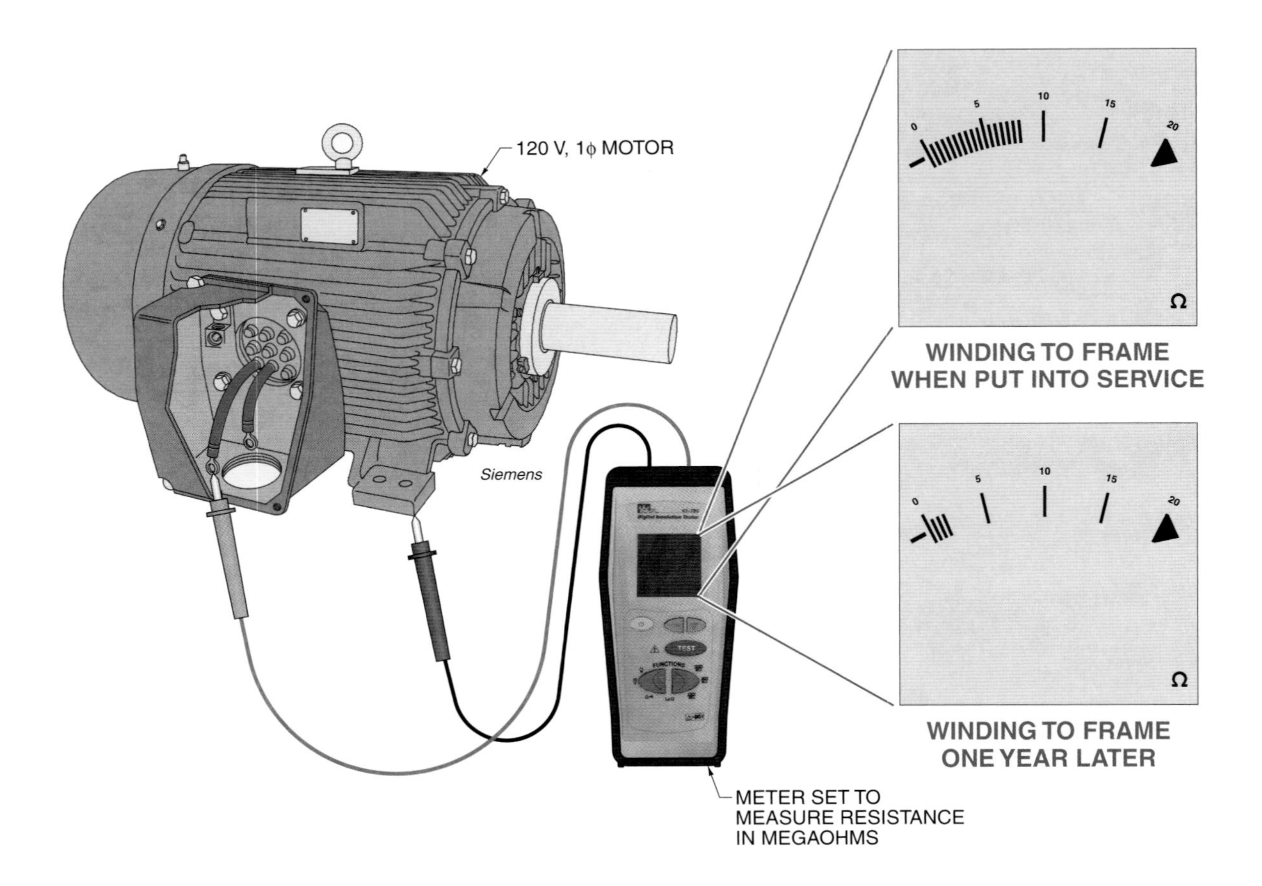

6. ______________ What was the leakage current (in mA) between Winding 1 and the frame when the motor was put into service?

7. ______________ What was the leakage current (in mA) between Winding 2 and the frame when the motor was put into service?

8. ______________ What was the leakage current (in mA) between Winding 3 and the frame when the motor was put into service?

9. ______________ What is the leakage current (in mA) between Winding 1 and the frame one year later?

10. ______________ What is the leakage current (in mA) between Winding 2 and the frame one year later?

11. ______________ What is the leakage current (in mA) between Winding 3 and the frame one year later?

12. ______________ Which winding's insulation has deteriorated unacceptably in one year?

13. ______________ Which winding's insulation must be monitored for further deterioration?

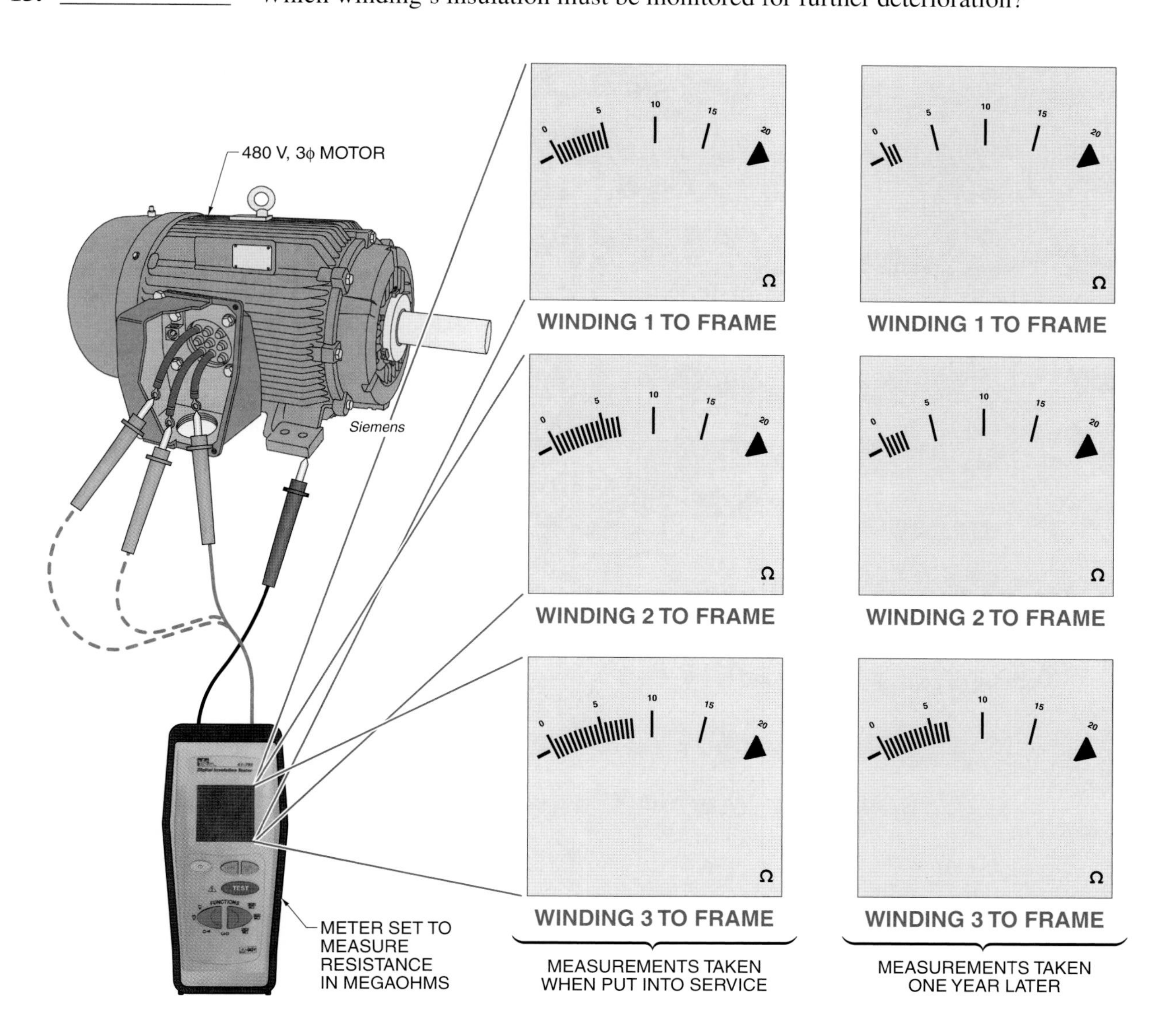

Application 7-4—Insulation Spot Testing

An insulation spot test uses a megohmmeter to measure the resistance of insulation on conductors inside motors. Measurements are taken between the windings and between each winding and ground. The insulation spot tests should be performed when a motor is placed into service and every six months afterward. The measurements are recorded on a chart or plotted on a graph for keeping track of the condition of the motor's insulation over time.

Three common types of DC motors include DC series, DC shunt, and DC compound motors. Each is insulation spot-tested in a similar way, but the internal configuration of the windings must be known in order to test at the proper terminal. A DC series motor has the field windings connected in series with the armature. See DC Series Motor. A DC shunt motor has the field windings connected in shunt (parallel) with the armature. See DC Shunt Motor. A DC compound motor has the field windings connected in both series and shunt with the armature. See DC Compound Motor.

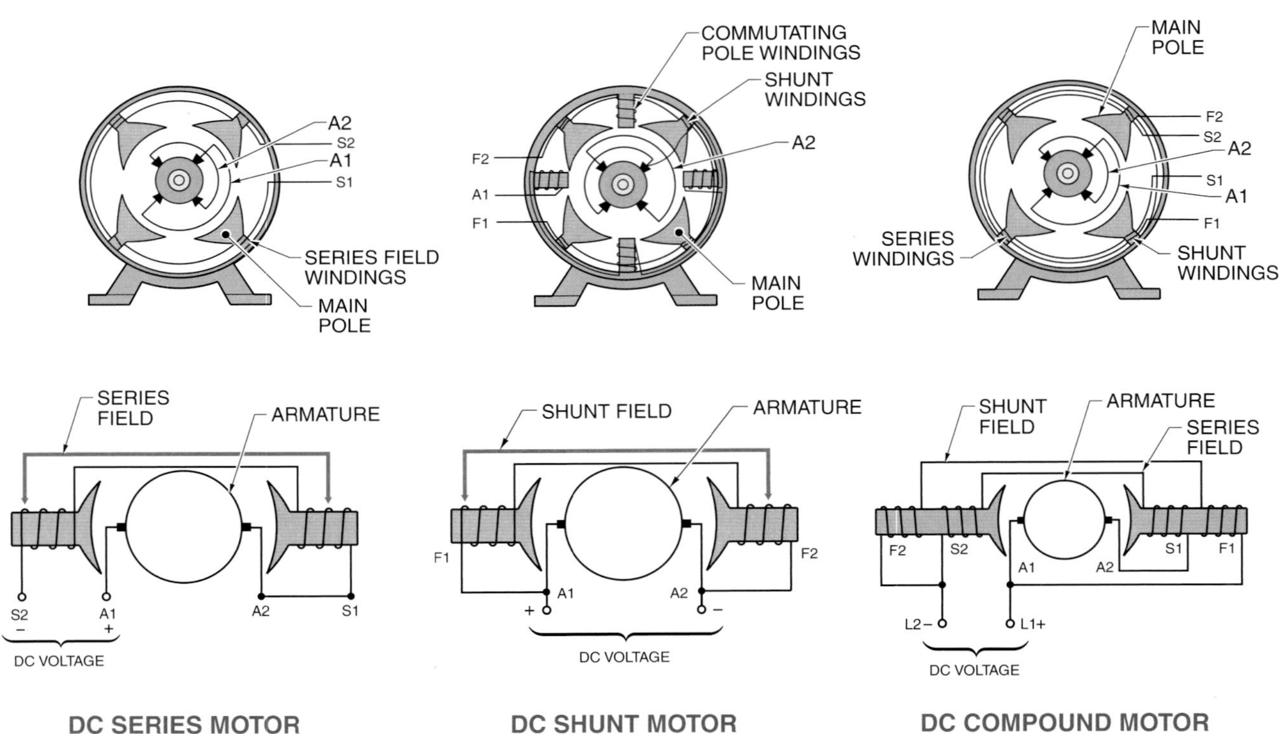

DC SERIES MOTOR

DC SHUNT MOTOR

DC COMPOUND MOTOR

1. Connect the megohmmeter leads for taking resistance measurements between S1 and ground (the motor frame).
2. Plot the insulation spot test measurements on the insulation spot test graph.
3. What is the problem with the winding insulation?

4. ______________ During which time period did the problem begin?

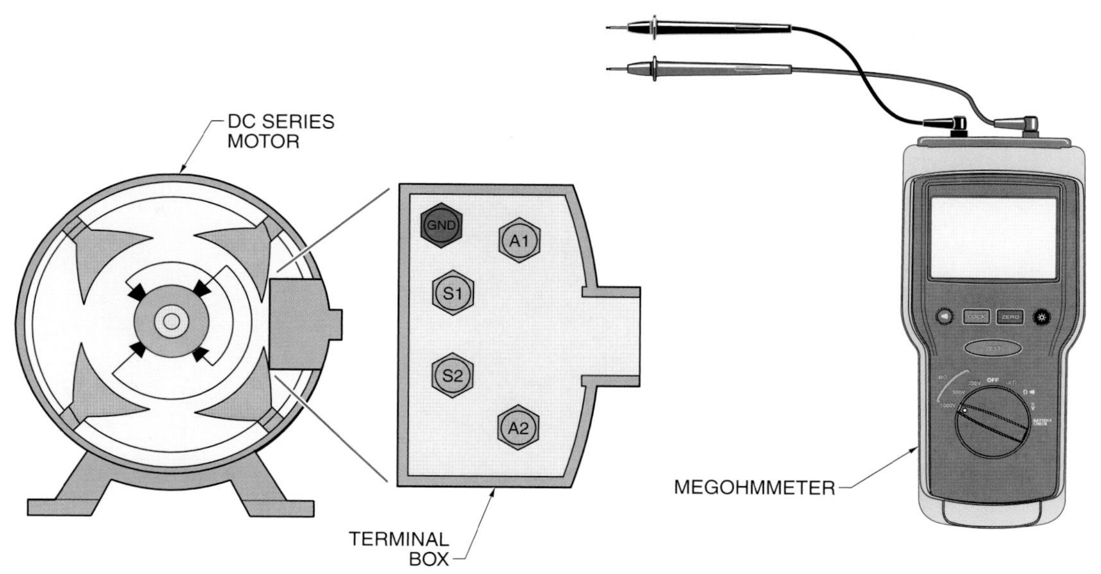

INSULATION SPOT TEST	
Test Date	**Resistance***
Jan, Year 1	400
Jul, Year 1	400
Jan, Year 2	300
Jul, Year 2	275
Jan, Year 3	250
Jul, Year 3	90
Jan, Year 4	20
Jul, Year 4	1
Jan, Year 5	0.5
Jul, Year 5	—
Jan, Year 6	—

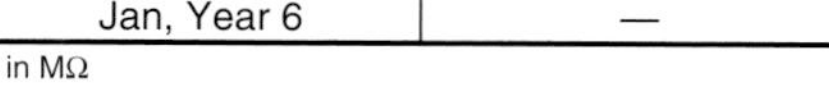

* in MΩ

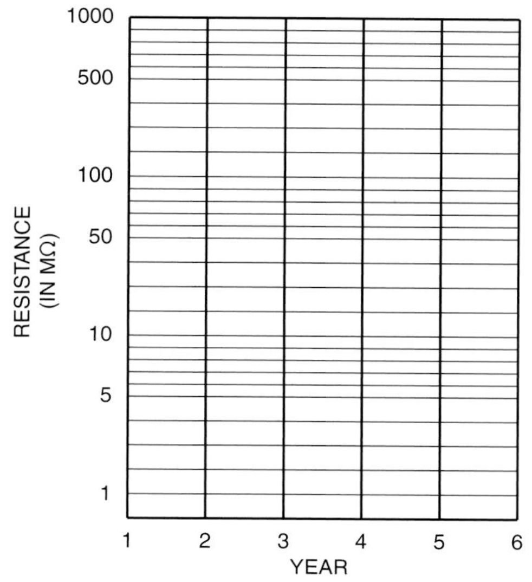

INSULATION SPOT TEST GRAPH

5. Connect the megohmmeter leads for taking resistance measurements between A1 and ground (the motor frame).

6. Plot the insulation spot test measurements on the insulation spot test graph.

7. _____________ Does the motor require service?

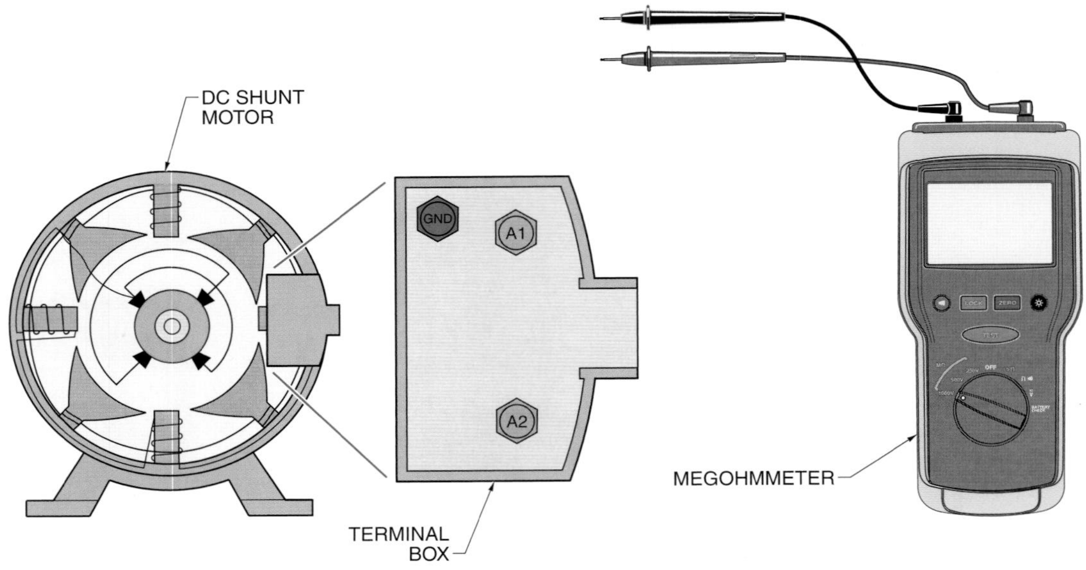

INSULATION SPOT TEST	
Test Date	**Resistance***
Jan, Year 1	500
Jul, Year 1	475
Jan, Year 2	450
Jul, Year 2	450
Jan, Year 3	425
Jul, Year 3	400
Jan, Year 4	400
Jul, Year 4	375
Jan, Year 5	375
Jul, Year 5	350
Jan, Year 6	350

* in MΩ

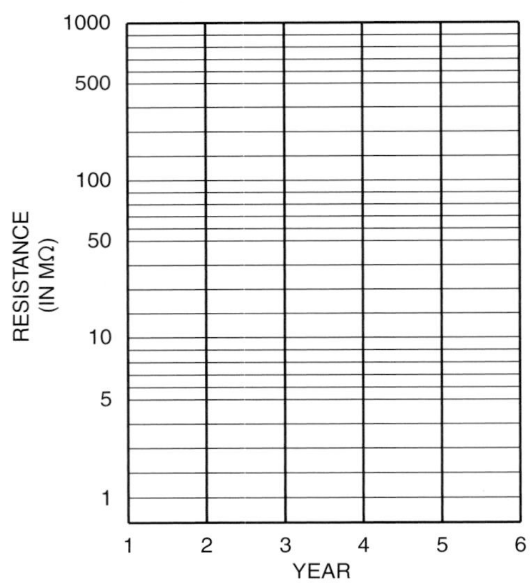

INSULATION SPOT TEST GRAPH

8. Connect the megohmmeter leads for taking resistance measurements between the first series winding terminal and the first shunt winding terminal.

9. Plot the insulation spot test measurements on the insulation spot test graph.

10. ______________ During which period was the motor serviced?

11. ______________ Did the servicing include replacing motor windings?

12. ______________ Is the winding insulation in the refurbished motor performing adequately?

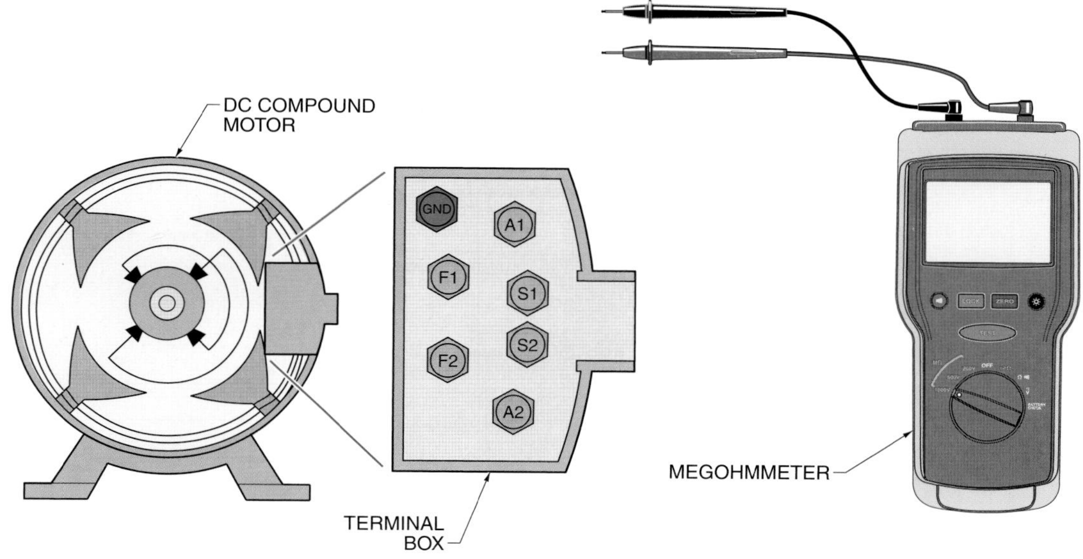

INSULATION SPOT TEST	
Test Date	**Resistance***
Jan, Year 1	400
Jul, Year 1	375
Jan, Year 2	300
Jul, Year 2	80
Jan, Year 3	10
Jul, Year 3	400
Jan, Year 4	350
Jul, Year 4	350
Jan, Year 5	325
Jul, Year 5	325
Jan, Year 6	300

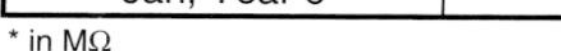

* in MΩ

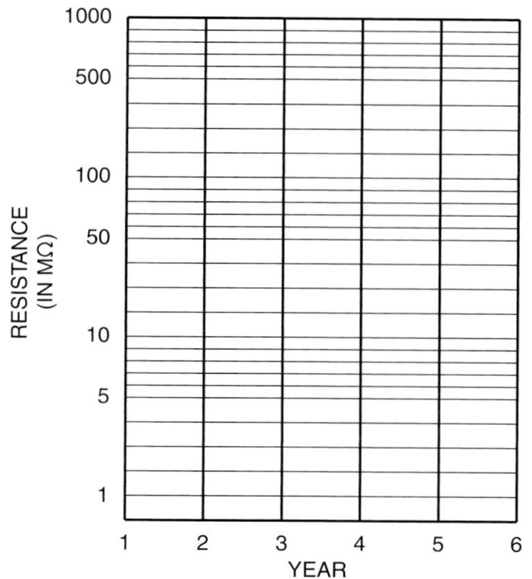

INSULATION SPOT TEST GRAPH

Application 7-5—Dielectric Absorption Testing

Insulation resistance also changes slightly over short periods (minutes) as voltage is applied to conductors. A dielectric absorption test measures the absorption characteristics of insulation by measuring resistance over a 10 min period. Measurements are recorded every 10 sec for the first minute and every minute thereafter. The readings can also be plotted on a graph to illustrate the changing resistance over time. The resistance of good insulation will increase continuously, appearing as an upward-sloping curve on the graph. The resistance of damaged or contaminated insulation will remain relatively constant, appearing as a flat curve. See Dielectric Absorption Test Graph.

Polarization index is a method of quantifying the minimum acceptable slope of the dielectric absorption test curve. The polarization index is calculated by dividing the values of the 10 min measurement by the 1 min measurement. A polarization index value that meets or exceeds the minimum value for the type of insulation tested indicates acceptable insulation. Insulation types are identified by class, such as Class A and Class F, and have different minimum polarization values, though most are between 1.5 and 2.0. See Polarization Index.

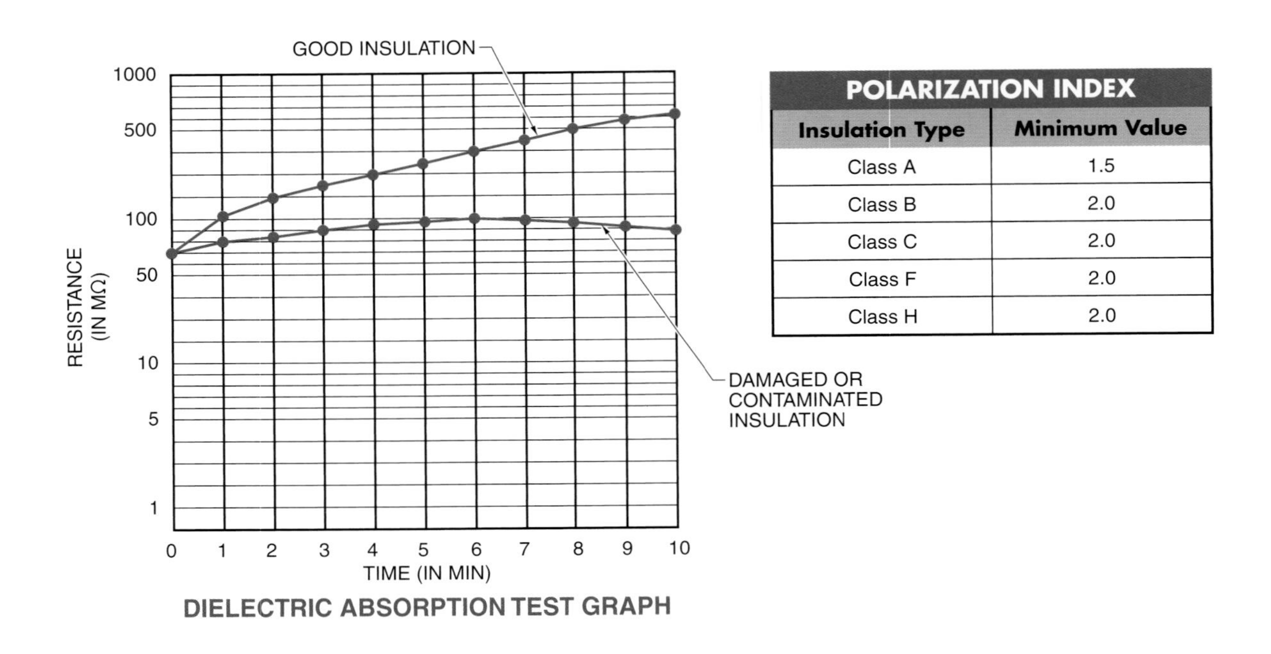

DIELECTRIC ABSORPTION TEST GRAPH

POLARIZATION INDEX	
Insulation Type	**Minimum Value**
Class A	1.5
Class B	2.0
Class C	2.0
Class F	2.0
Class H	2.0

It is usually not necessary to perform a dielectric absorption test and polarization index calculation on the insulation of every conductor in a motor or other load. Only one conductor with poor insulation will require servicing the load. A quicker insulation spot test on every conductor will identify the weakest insulation, which may then be further tested with a dielectric absorption test.

Megohmmeter measurements are taken on a 3ϕ motor from winding to winding and from each winding to ground. A dielectric absorption test is performed on the conductor with the lowest resistance measurement.

1. Plot the insulation test measurements on the dielectric absorption test graph.
2. ______________ What is the polarization index value?
3. ______________ Is the insulation resistance acceptable for Class B insulation?

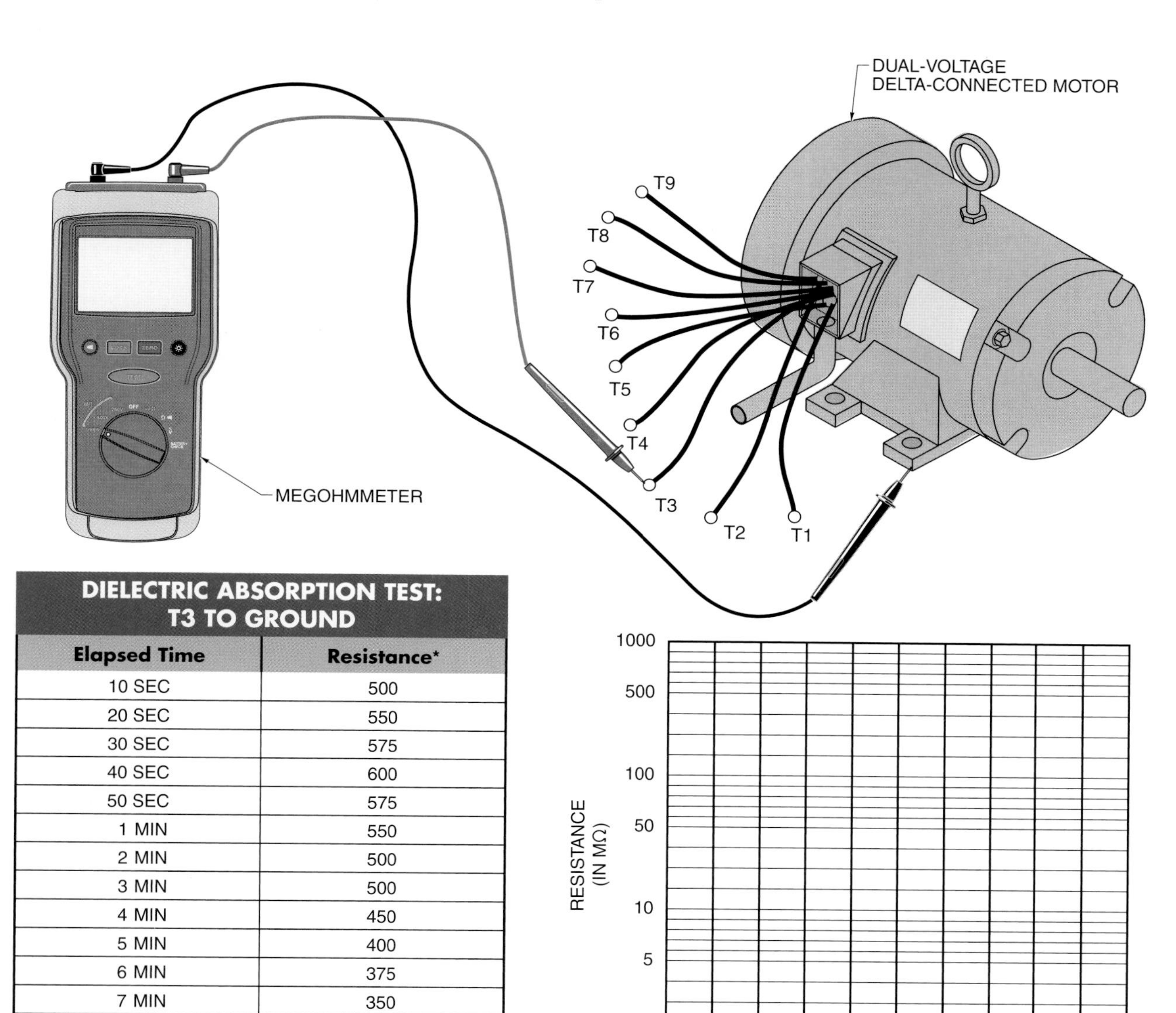

DIELECTRIC ABSORPTION TEST: T3 TO GROUND

Elapsed Time	Resistance*
10 SEC	500
20 SEC	550
30 SEC	575
40 SEC	600
50 SEC	575
1 MIN	550
2 MIN	500
3 MIN	500
4 MIN	450
5 MIN	400
6 MIN	375
7 MIN	350
8 MIN	300
9 MIN	275
10 MIN	250

* in MΩ

Megohmmeter measurements are taken on a heating element between ground and each wire leading to the heating element. A dielectric absorption test is performed on the conductor with the lowest resistance measurement.

4. Plot the insulation test measurements on the dielectric absorption test graph.

5. ______________ What is the polarization index value?

6. ______________ Is the insulation resistance acceptable for Class A insulation?

7. ______________ Is the insulation resistance acceptable for Class F insulation?

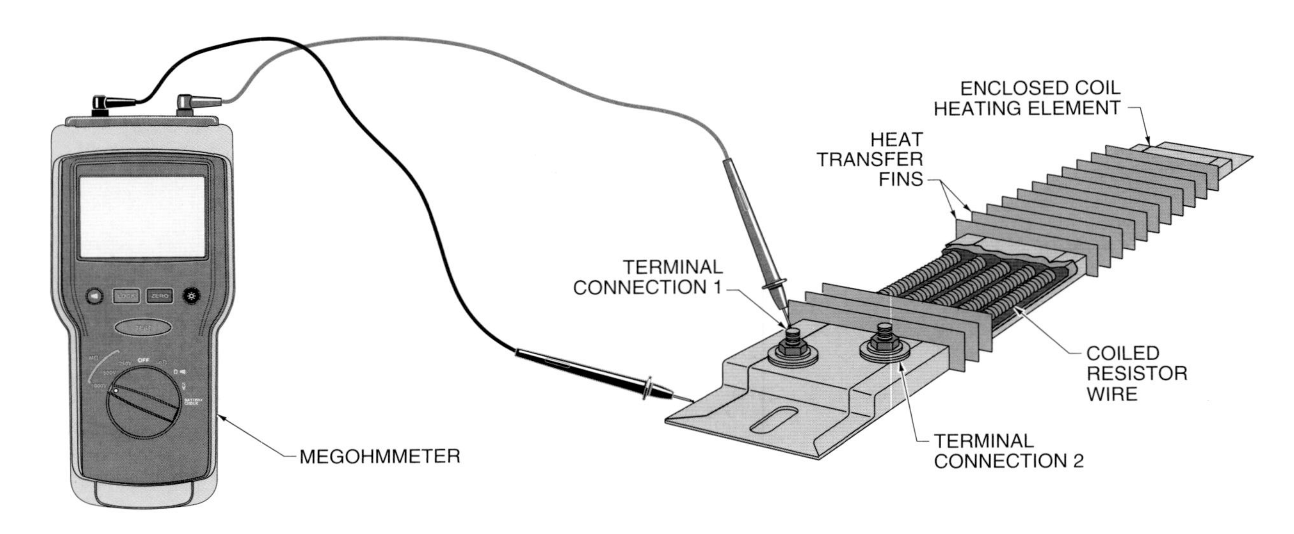

DIELECTRIC ABSORPTION TEST: TERMINAL 1 TO GROUND	
Elapsed Time	**Resistance***
10 SEC	100
20 SEC	150
30 SEC	175
40 SEC	175
50 SEC	200
1 MIN	200
2 MIN	225
3 MIN	275
4 MIN	300
5 MIN	325
6 MIN	325
7 MIN	350
8 MIN	350
9 MIN	325
10 MIN	350

* in MΩ

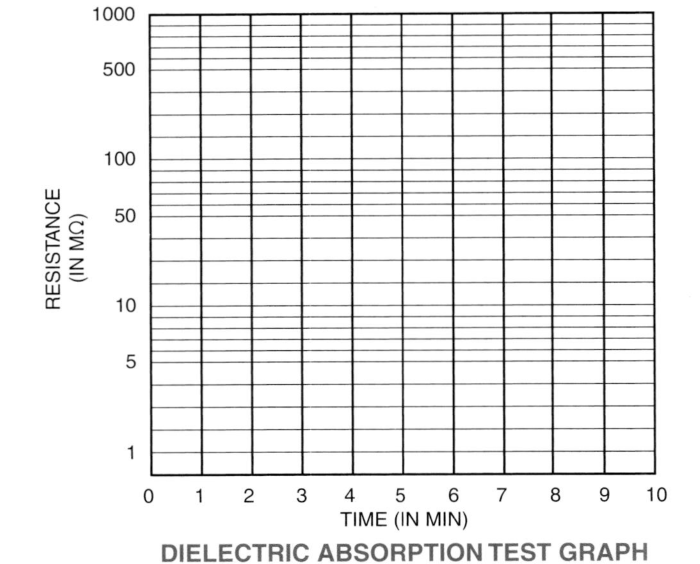

DIELECTRIC ABSORPTION TEST GRAPH

Application 7-6 — Insulation Step Voltage Testing

An insulation step voltage test reveals damaged or deteriorated areas on insulation by intentionally stressing the insulation to the point where weakened insulation fails. This test is only destructive when the insulation is already weakened to the point where it should be serviced anyway. The test should not damage insulation that is in good condition. This test is similar to hipot testing except that the displayed measurement is in ohms rather than amps.

During an insulation step voltage test, the resistance is measured as the voltage is increased incrementally from 1000 V to 5000 V. The resistance of good insulation remains relatively constant throughout the test. The resistance of damaged or deteriorated insulation decreases substantially as the voltage increases. See Insulation Step Voltage Test.

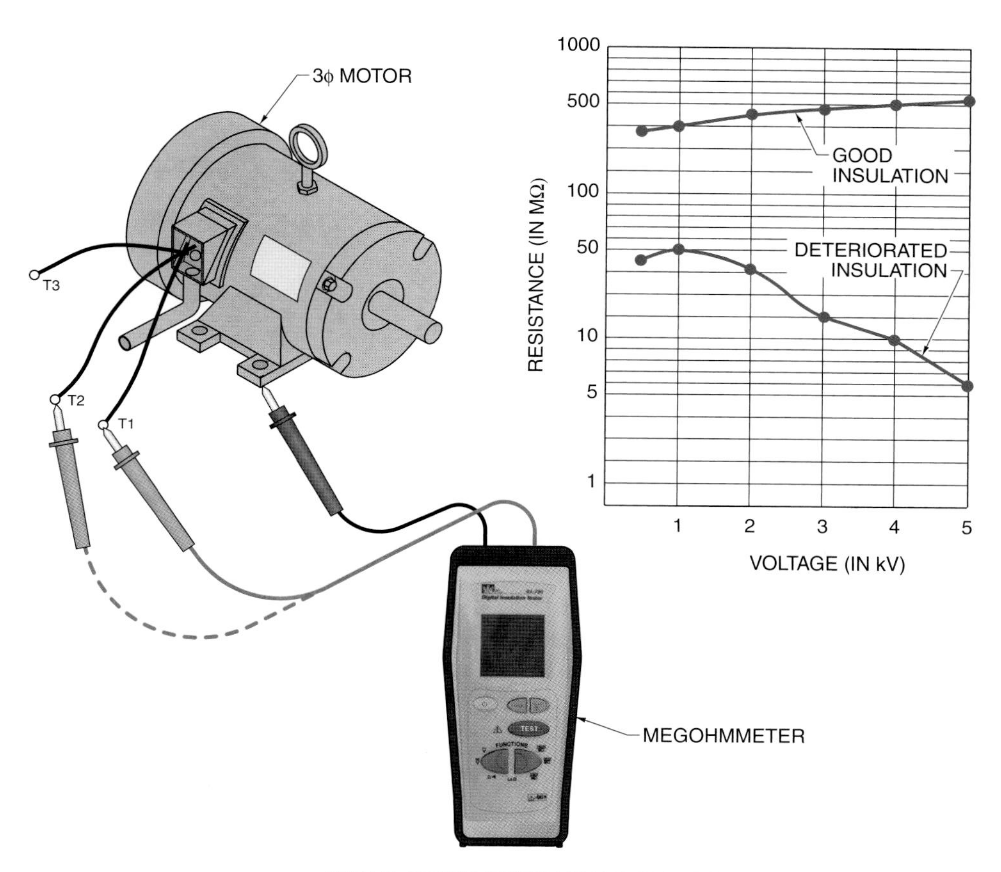

INSULATION STEP VOLTAGE TEST

Megohmmeter measurements are taken on a 3ϕ motor from winding to winding and from each winding to ground. An insulation step voltage test is performed on the conductors with the lowest resistance measurement.

1. Plot the insulation test measurements of T2 on the insulation step voltage test graph.
2. ______________ Is the insulation resistance acceptable?
3. Plot the insulation test measurements of T3 on the insulation step voltage test graph.
4. ______________ Is the insulation resistance acceptable?
5. ______________ What can be done to repair the motor?

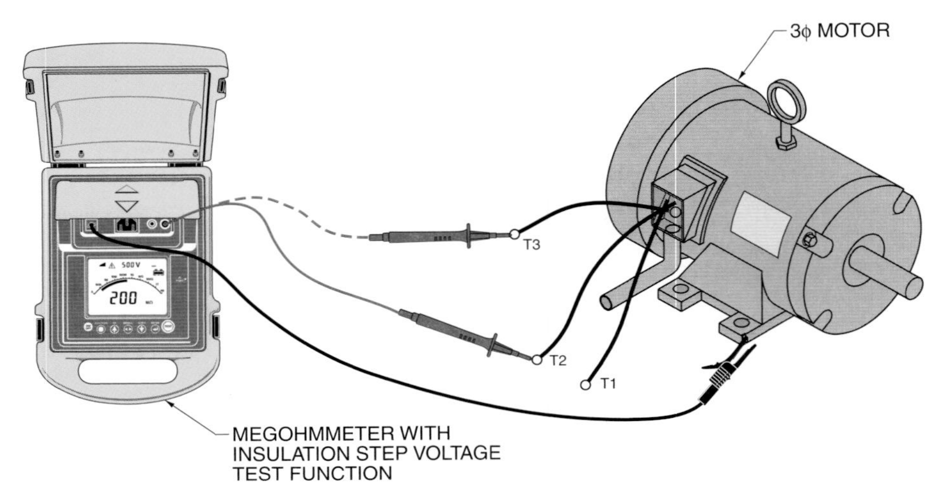

INSULATION STEP VOLTAGE TEST		
Voltage*	Resistance of T2 to Ground†	Resistance of T3 to Ground†
500	200	175
1000	200	200
1500	225	200
2000	225	150
2500	220	95
3000	220	80
3500	210	50
4000	200	40
4500	195	25
5000	190	10

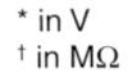
* in V
† in MΩ

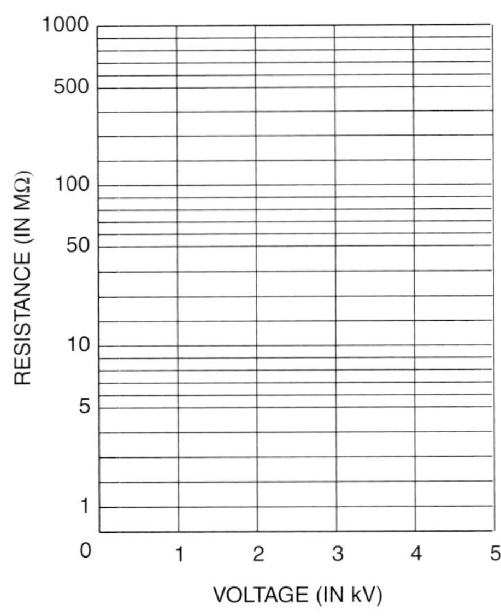

INSULATION STEP VOLTAGE TEST GRAPH

INSTRUMENTATION AND PROCESS CONTROL TEST INSTRUMENTS

8 APPLICATIONS

Name: ______________________ Date: ______________

Application 8-1 —Temperature Measurement Attachment

Test instruments are usually used to measure electrical quantities such as voltage, current, and resistance. Specialized instruments are often used to measure physical quantities such as speed, weight, and flow, and operational conditions such as temperature, humidity, and light. However, standard multimeters can also be used to take specialized measurements when used with sensor attachments that plug into the multimeter jacks. Attachments may be designed for digital or analog multimeters.

The readings are typically in mV or mA per unit being measured by the attachment, such as degrees, psi, or m/s. Careful attention must be paid to the output of each attachment so that the multimeter is set, connected, and read correctly. For example, an air velocity measurement attachment may output 1 mA/fps. The multimeter must be set to measure mA and the attachment must be connected to the mA and common jacks. The resulting measurement will display as mA, but is equivalent to fps (feet per second) of air velocity. See Multimeter Air Velocity Measurement Attachment.

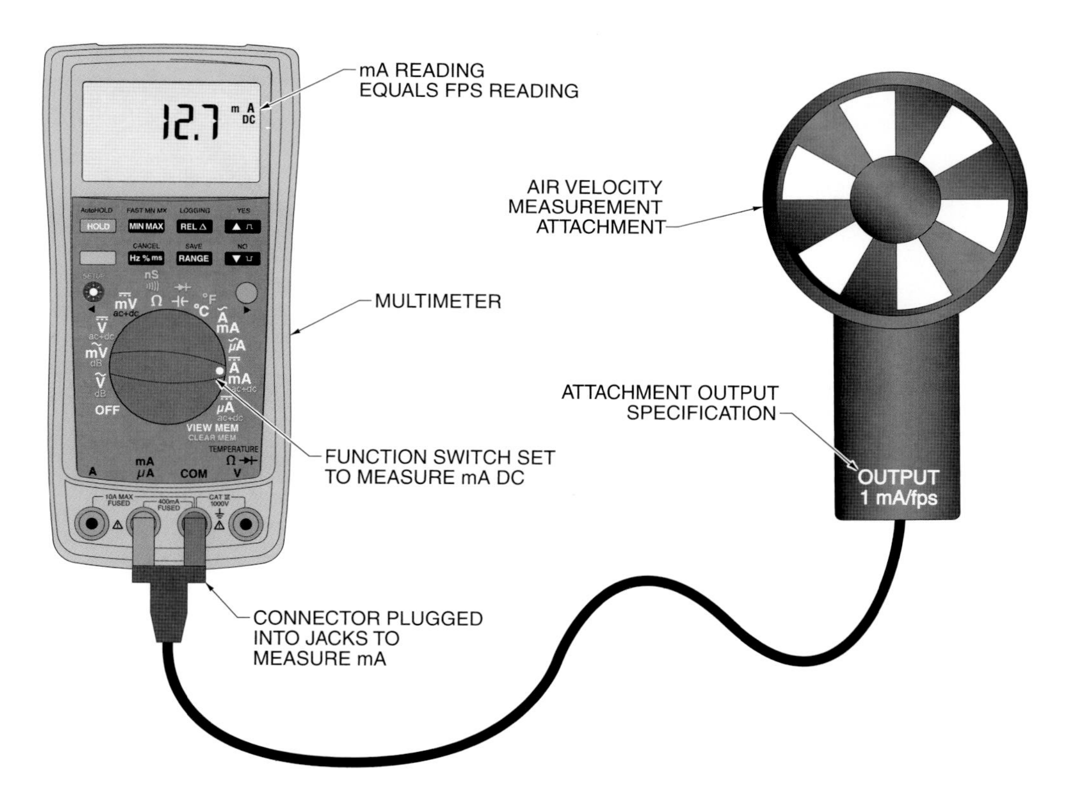

MULTIMETER AIR VELOCITY MEASUREMENT ATTACHMENT

A temperature measurement attachment is added to a DMM for taking temperature readings.

1. Set (draw in) the multimeter selector switch to match the attachment's output.

2. Connect the attachment test leads to the DMM to match the attachment's output.

Application 8-2—Thermocouples

A thermocouple is a temperature sensor that produces a voltage output nearly proportional to the temperature. A thermocouple consists of two dissimilar metals joined at one end. Current flows through the thermocouple when the welded joint is heated. A small voltage output, typically only a few millivolts DC, is measured at the open junction. Higher voltage indicates a higher temperature at the heated end. See Thermocouples.

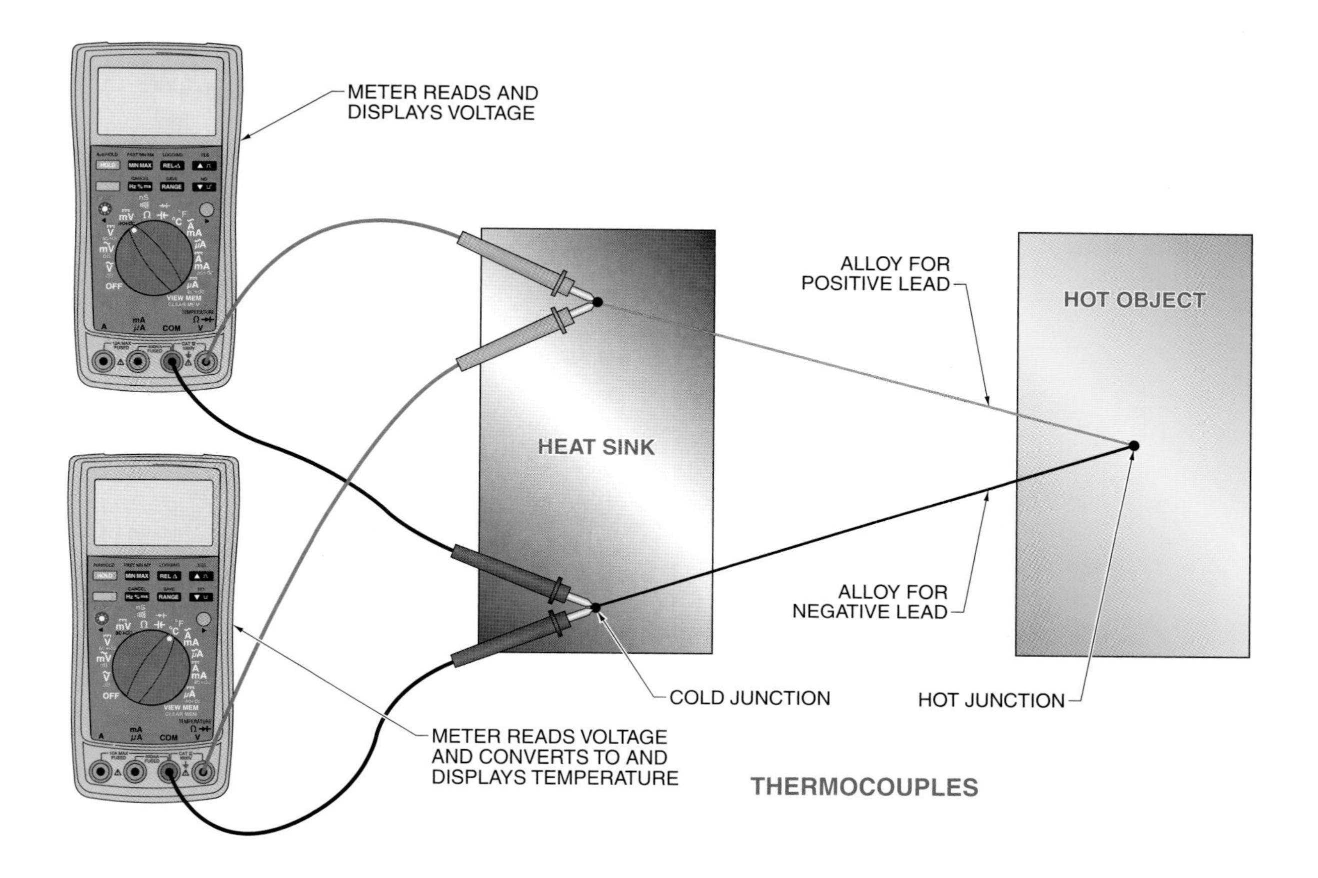

THERMOCOUPLES

The heated end of a thermocouple is known as the hot junction. The other end is known as the cold junction. For the thermocouple's output voltage to be converted into a true temperature measurement, some compensation for temperature differences between the two junctions must be made. A temperature compensator ensures that the correct thermocouple output is delivered to the measuring point, regardless of any changing ambient temperatures. The temperature compensator is placed between the thermocouple and the measuring controller or meter.

Thermocouples are the most common type of temperature sensor used in industrial heating applications. One of the reasons thermocouples are so common is that they have the highest temperature-measuring capability of any sensor type. Thermocouples can measure temperatures up to 3000°F. They are also low in cost, rugged, small, have a fast response time, and are available in a wide range of sizes.

Thermocouples come in a variety of assemblies, are suited to a variety of temperature ranges, and are available for a variety of operating conditions. Standard color-coded leads are used to indicate the thermocouple type and which lead is positive and which lead is negative. See Thermocouple Color Codes.

THERMOCOUPLE COLOR CODES

ANSI CODE	Alloy Combination		Thermocouple Grade		Extension Grade	
	Positive Lead	Negative Lead	Color Code*	Maximum Temperature Range	Color Code*	Maximum Temperature Range
J	IRON Fe (magnetic)	COPPER-NICKEL Cu-Ni (CONSTANTAN)	+ −	−210°C to 1200°C −346°F to 2193°F	+ −	0°C to 200°C 32°F to 392°F
K	NICKEL-CHROMIUM Ni-Cr (CHROMEL)	NICKEL-ALUMINUM Ni-Al (magnetic) (ALUMEL)	+ −	−270°C to 1372°C −454°F to 2501°F	+ −	0°C to 200°C 32°F to 212°F
T	COPPER Cu	COPPER-NICKEL Cu-Ni (CONSTANTAN)	+ −	−270°C to 400°C −454°F to 752°F	+ −	−60°C to 100°C −76°F to 212°F
E	NICKEL-CHROMIUM Ni-Cr (CHROMEL)	COPPER-NICKEL Cu-Ni (CONSTANTAN)	+ −	−270°C to 1000°C −454°F to 1832°F	+ −	0°C to 200°C 32°F to 392°F
N	NICKEL-CHROMIUM-SILICON Ni-Cr-Si	NICKEL-SILICON-MAGNESIUM Ni-Si-Mg	+ −	−270°C to 1300°C −454°F to 1832°F	+ −	0°C to 200°C 32°F to 392°F
R	PLATINUM-13% RHODIUM Pt-13% Rh	PLATINUM Pt	NOT ESTABLISHED	−50°C to 1768°C −58°F to 3214°F	+ −	0°C to 150°C 32°F to 392°F
S	PLATINUM-13% RHODIUM Pt-10% Rh	PLATINUM Pt	NOT ESTABLISHED	−50°C to 1768°C −58°F to 3214°F	+ −	0°C to 150°C 32°F to 300°F
B	PLATINUM-30% RHODIUM Pt-30% Rh	PLATINUM-6% RHODIUM Pt-6% Rh	NOT ESTABLISHED	50°C to 1820°C 122°F to 3308°F	+ −	0°C to 100°C 32°F to 212°F

* U.S. and Canadian color codes

The various types of thermocouples are different because of the pairs of metals that are used. Specific alloys may be better suited to certain operating environments, such as immersion in corrosive liquids. The different alloy combinations also produce thermocouples for different temperature ranges. Some thermocouples may have larger temperature ranges and others may have shorter ranges but are more accurate. A more steeply sloped plot on a temperature-voltage graph means that the thermocouple has a greater voltage change for each degree change in temperature. This indicates that the thermocouple allows more precise temperature measurement. See Thermocouple Temperature-Voltage Relationships.

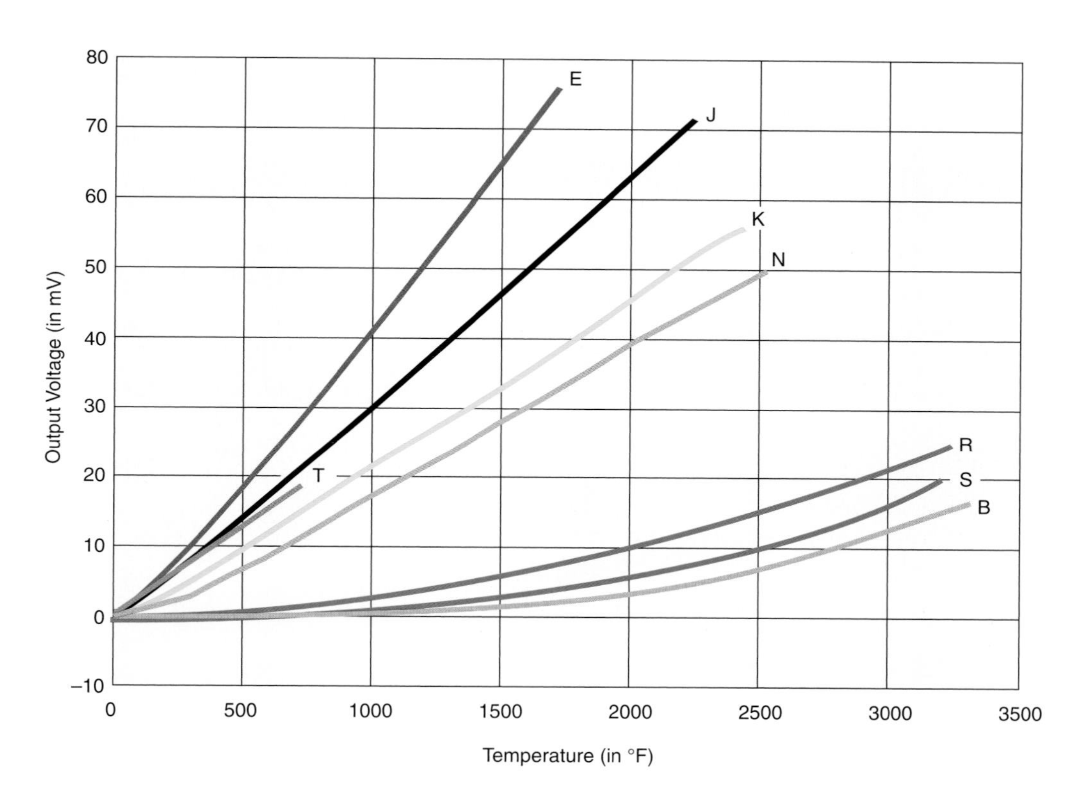

THERMOCOUPLE TEMPERATURE-VOLTAGE RELATIONSHIPS

A type J thermocouple monitors the temperature of a process fluid. A multimeter is used to read the voltage output directly.

1. Set (draw in) the function switch of the multimeter to test the output of the thermocouple.

2. Connect the multimeter test leads to the correct jacks on the multimeter.

3. Connect the multimeter test leads to test the output of the thermocouple at the temperature compensator.

4. ______________ If the multimeter reads 14 mV, what is the approximate temperature of the fluid?

5. ______________ If the multimeter reads 27 mV, what is the approximate temperature of the fluid?

6. ______________ If the multimeter reads 30 mV, what is the approximate temperature of the fluid?

The thermocouple is changed to a type E, which should be more precise in the temperature range of the process fluid.

7. ______________ If the fluid is 500°F, what should the multimeter approximately read?

8. ______________ If the fluid is 1000°F, what should the multimeter approximately read?

9. ______________ If the multimeter reads 30 mV, what is the approximate temperature of the fluid?

10. ______________ If the multimeter reads 24 mV, what is the approximate temperature of the fluid?

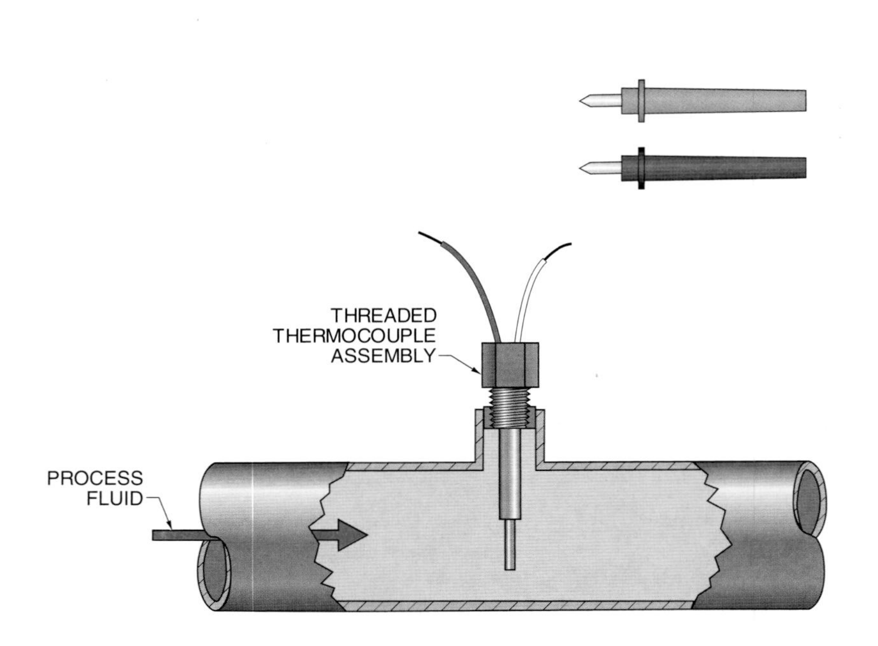

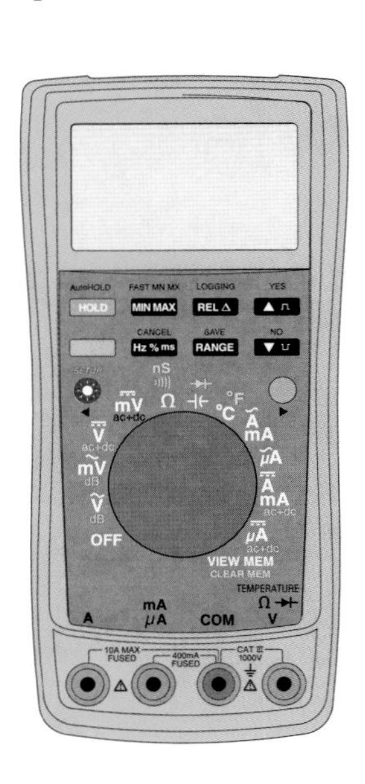

Application 8-3—Process Control Signals

Standard control signals are used to represent possible analog values of a process variable. Control signals may be electrical, using variable voltage or variable current, or pneumatic, using variable air pressure. For example, if a temperature in a process tank varies from 70°F to 170°F, a temperature sensor can be used to send a signal to a controller that reads the temperature and converts it to a standard signal between 4 mA DC and 20 mA DC. A controller with a variable current output would be calibrated (adjusted) to produce a 4 mA signal at 70°F and a 20 mA signal at 170°F. Thus, when the temperature controller reads 120°F (halfway between 70°F and 170°F), the controller would output a 12 mA signal (halfway between 4 mA and 20 mA) and any actuators controlled by the signal would be halfway between their two extreme states (such as halfway between fully open and fully closed).

A system regulates the temperature of a process fluid between 50°F and 150°F by cooling or heating the fluid as needed. The cooling and heating units each have a valve that opens an amount proportional to the amount of cooling or heating needed. A thermocouple measures the fluid's temperature and a temperature controller outputs the corresponding control signal that operates the valves. The multimeter measures the temperature controller's current output.

1. Set (draw in) the function switch on the multimeter for measuring the control signal.

2. Connect the multimeter test leads to the correct jacks on the multimeter.

3. ______________ If the fluid temperature is 100°F, what should the multimeter display?

4. ______________ If the fluid temperature is 100°F, what is the condition of the cooling valve?

5. ______________ If the fluid temperature is 100°F, what is the condition of the steam valve?

6. ______________ If the fluid temperature is 75°F, what should the multimeter display?

7. ______________ If the fluid temperature is 75°F, what is the condition of the cooling valve?

8. ______________ If the fluid temperature is 75°F, what is the condition of the steam valve?

9. ______________ If the fluid temperature is 125°F, what should the multimeter display?

10. ______________ If the fluid temperature is 125°F, what is the condition of the cooling valve?

11. ______________ If the fluid temperature is 125°F, what is the condition of the steam valve?

12. ______________ If the fluid temperature is 50°F, what should the multimeter display?

13. ______________ If the fluid temperature is 50°F, what is the condition of the cooling valve?

14. ______________ If the fluid temperature is 50°F, what is the condition of the steam valve?

15. ______________ What is the ideal temperature that this system is trying to maintain?

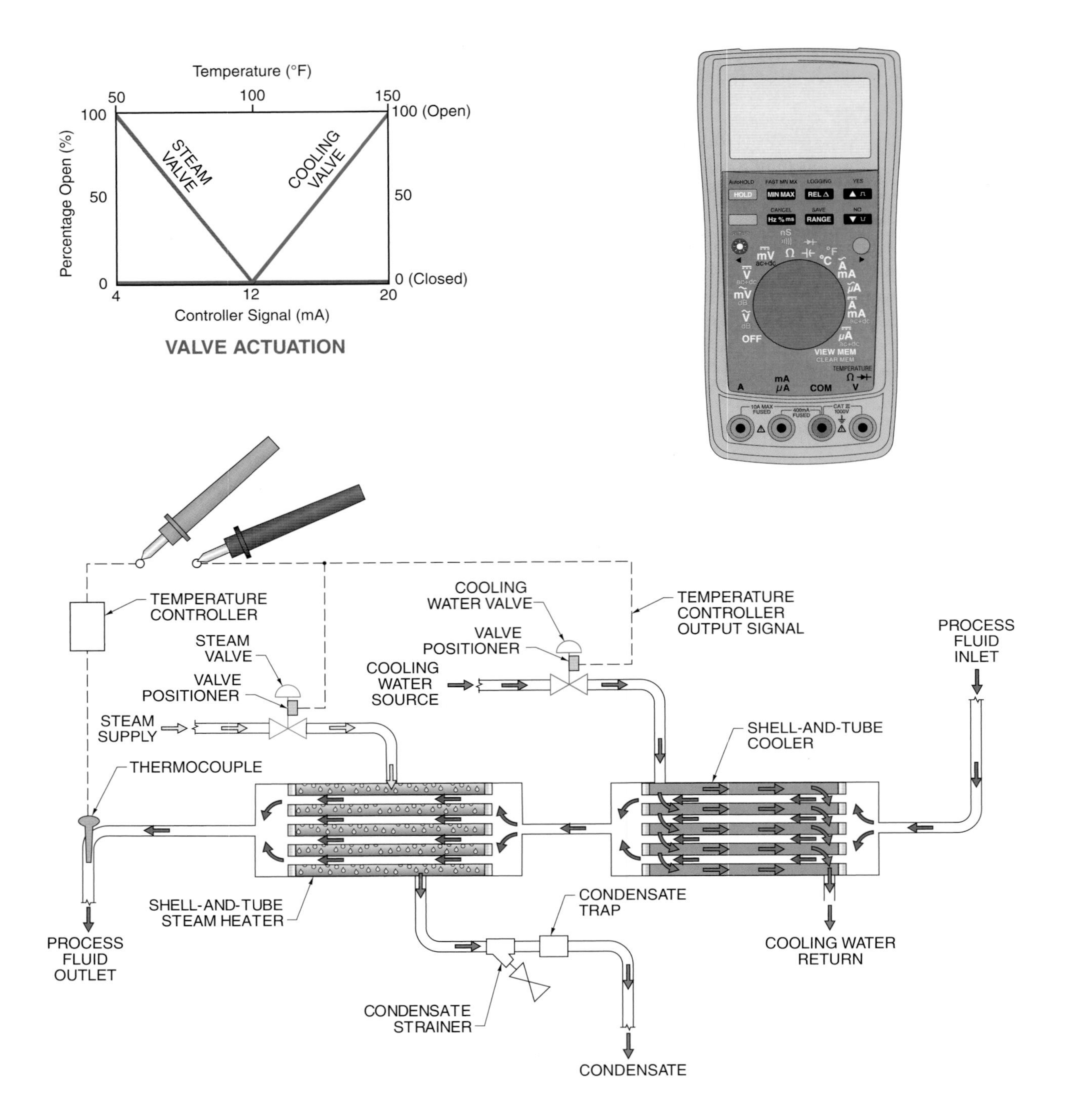

Temperature (°F)
50
100
150
100
100 (Open)
STEAM VALVE
COOLING VALVE
Percentage Open (%)
50
50
0
0 (Closed)
4
12
20
Controller Signal (mA)
VALVE ACTUATION
TEMPERATURE CONTROLLER
STEAM VALVE
VALVE POSITIONER
STEAM SUPPLY
THERMOCOUPLE
COOLING WATER VALVE
VALVE POSITIONER
COOLING WATER SOURCE
TEMPERATURE CONTROLLER OUTPUT SIGNAL
PROCESS FLUID INLET
SHELL-AND-TUBE COOLER
SHELL-AND-TUBE STEAM HEATER
PROCESS FLUID OUTLET
CONDENSATE TRAP
CONDENSATE STRAINER
CONDENSATE
COOLING WATER RETURN

Application 8-4 — Motor Drive Control Signals

Electric motor drives are ideal for controlling pump and fan motors, because the speed of the motor can be varied to control the flow rate. The motor is connected to the drive, which is connected to the incoming supply voltage. Control signals into the drive can be digital signals to turn the motor ON or OFF, or analog signals to control motor speed. See Motor Drive Control Signal Inputs and Motor Drive Input Examples.

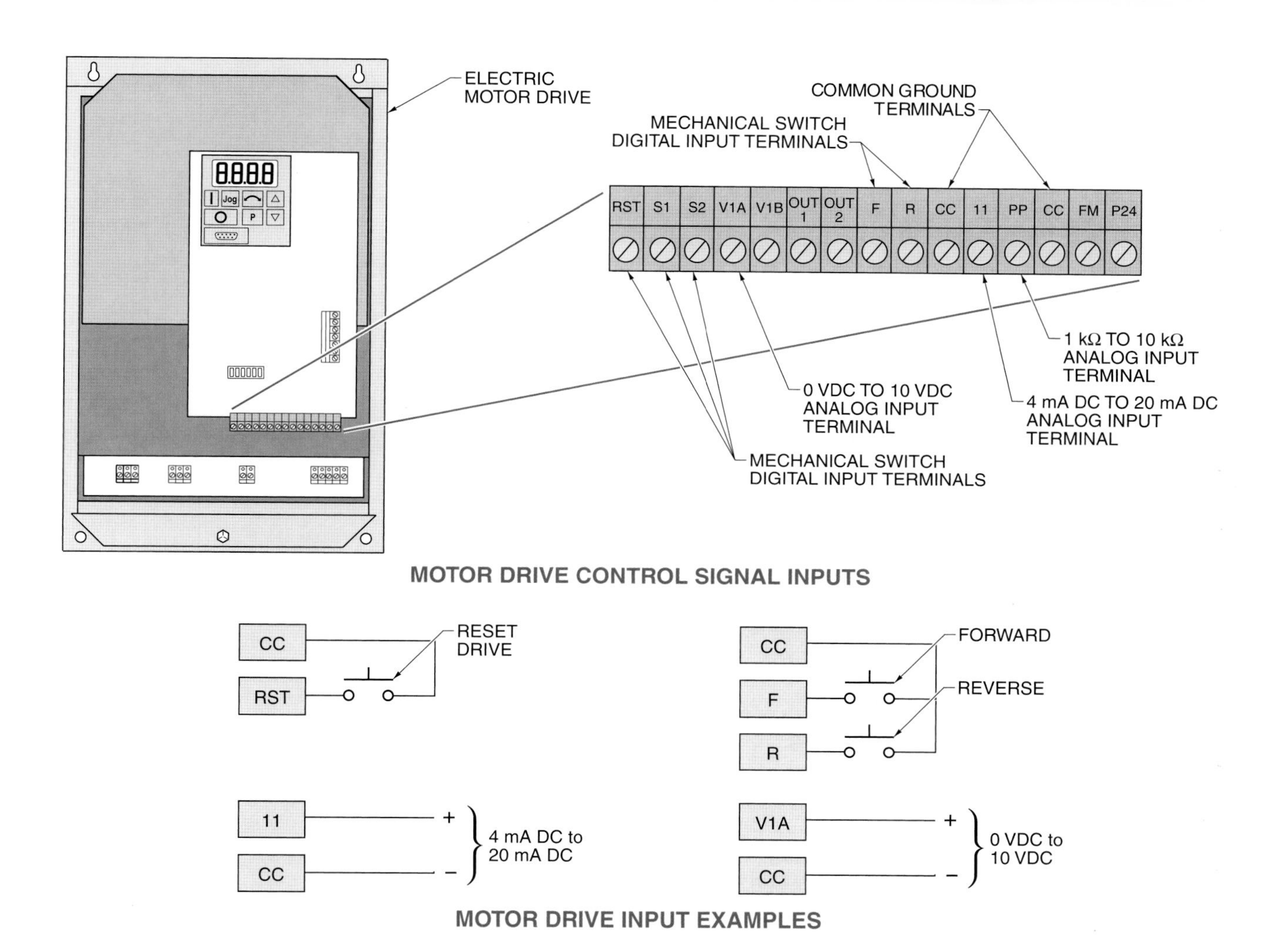

MOTOR DRIVE CONTROL SIGNAL INPUTS

MOTOR DRIVE INPUT EXAMPLES

The digital signals can be from mechanical switches, such as pushbuttons, pressure switches, or temperature switches, or from the output contacts of relays or PLCs. Relays, especially solid-state relays, are used to input control signals from electronic circuits.

The analog signals control the speed of the motor as a percent of total motor speed. For example, an analog signal of 0 VDC to 10 VDC would increment the motor speed by 10% of motor maximum speed for every 1 VDC of input signal. The maximum speed of a motor is the nameplate-rated speed (in rpm) at the nameplate-rated frequency (in Hz). Changing the frequency changes the speed of the motor.

An electric motor drive can be set to output a maximum frequency lower or higher than the motor's nameplate-rated frequency. Also, an analog signal input of 0 VDC to 10 VDC into a drive would set the frequency at 10% of the drive's (not necessarily the motor's) maximum frequency for every 1 VDC input. Therefore, the analog signal is incrementing the drive frequency up to a set maximum, which may or may not match the motor's maximum frequency. The drive output frequency can be calculated by applying the following formula:

$$f_{OUT} = \frac{S - S_{MIN}}{S_{MAX} - S_{MIN}} \times f_D$$

where

f_{OUT} = frequency output of drive (in Hz)

S = control signal (in V or mA)

S_{MAX} = maximum control signal (in V or mA)

S_{MIN} = minimum control signal (in V or mA)

f_D = maximum frequency setting of drive (in Hz)

For example, a 6 VDC signal (out of 0 VDC to 10 VDC) into a motor drive set for a maximum frequency of 30 Hz results in a frequency output of 18 Hz.

$$f_{OUT} = \frac{S - S_{MIN}}{S_{MAX} - S_{MIN}} \times f_D$$

$$f_{OUT} = \frac{6 - 0}{10 - 0} \times 30$$

$$f_{OUT} = 0.6 \times 30$$

$$f_{OUT} = \mathbf{18\ Hz}$$

The output frequency of the motor drive controls the speed of the motor. The speed of the motor is proportional to the drive frequency and can be calculated by applying the following formula:

$$n = \frac{f_{OUT}}{f_M} \times n_M$$

where

n = motor speed (in rpm)

f_{OUT} = frequency output of drive (in Hz)

f_M = nameplate-rated frequency of motor (in Hz)

n_M = nameplate-rated speed of motor (in rpm)

For example, a motor drive outputs 18 Hz to a motor with a nameplate rated speed of 3550 rpm at 60 Hz. The motor speed is 1065 rpm.

$$n = \frac{f_{OUT}}{f_M} \times n_M$$

$$n = \frac{18}{60} \times 3550$$

$$n = 0.3 \times 3550$$

$$n = \mathbf{1065\ rpm}$$

A motor with a nameplate-rated speed of 1740 rpm at 60 Hz is controlled by a motor drive. The motor drive is set for a maximum frequency output of 60 Hz and receives control signals of 0 VDC to 10 VDC.

1. Set (draw in) the function switch for measuring the analog control signal into the drive.

2. Connect the multimeter test leads to the correct jacks for measuring the control signal at the drive.

3. Connect the test leads to measure the control signal circuit of the drive.

4. ______________ What is the output frequency when the multimeter reads 4 VDC?

5. ______________ What is the motor speed when the multimeter reads 4 VDC?

The motor drive is reset to a maximum output frequency of 66 Hz.

6. ______________ What is the output frequency when the multimeter reads 4 VDC?

7. ______________ What is the motor speed when the multimeter reads 4 VDC?

8. ______________ What is the output frequency when the multimeter reads 6 VDC?

9. ______________ What is the motor speed when the multimeter reads 6 VDC?

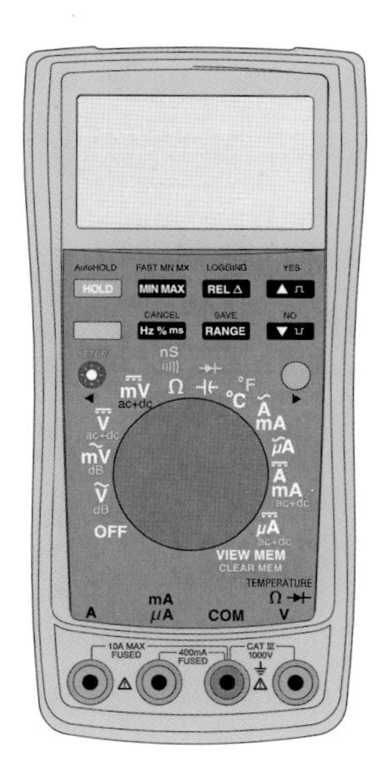

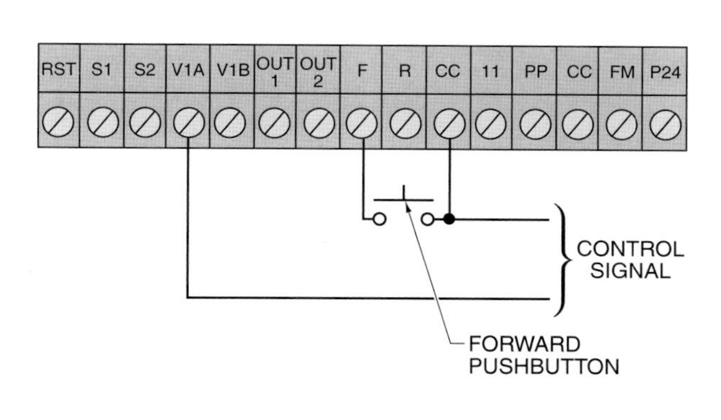

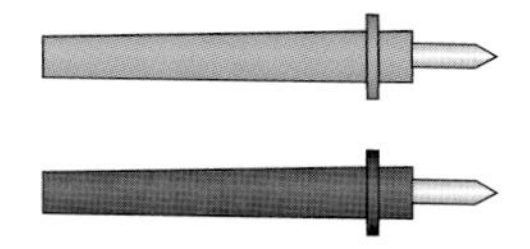

A motor with a nameplate-rated speed of 1740 rpm at 60 Hz is controlled by a motor drive. The motor drive is set for a maximum frequency output of 60 Hz and receives control signals of 4 mA DC to 20 mA DC.

10. Set (draw in) the function switch for measuring the analog control signal into the drive.

11. Connect the multimeter test leads to the correct jacks for measuring the control signal at the drive.

12. Connect the test leads to measure the control signal circuit of the drive.

13. ______________ What is the output frequency when the multimeter reads 6 mA?

14. ______________ What is the motor speed when the multimeter reads 6 mA?

15. ______________ What is the output frequency when the multimeter reads 18 mA?

16. ______________ What is the motor speed when the multimeter reads 18 mA?

The motor drive is reset to a maximum output frequency of 50 Hz.

17. ______________ What is the output frequency when the multimeter reads 10 mA?

18. ______________ What is the motor speed when the multimeter reads 10 mA?

19. ______________ What is the output frequency when the multimeter reads 15 mA?

20. ______________ What is the motor speed when the multimeter reads 15 mA?

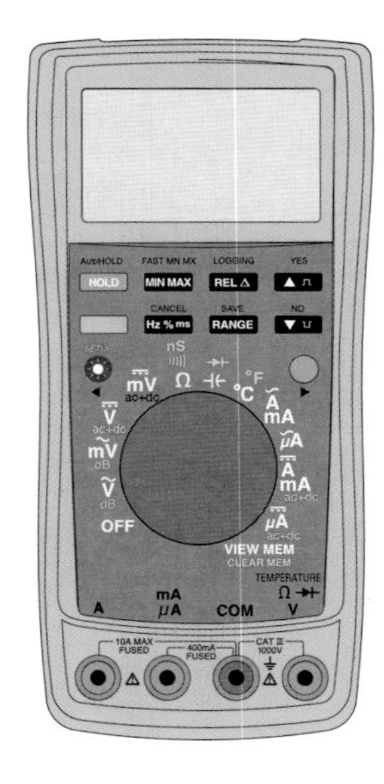

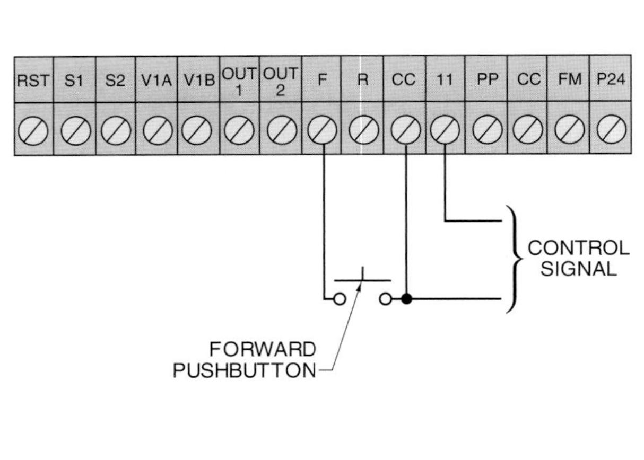

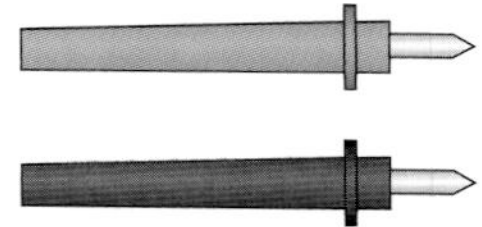

SPECIAL MAINTENANCE TEST INSTRUMENTS

9 APPLICATIONS

Name: ______________________ Date: ______________

Application 9-1 — Micro-Ohmmeter Measurements

Often, circuit devices such as contacts, switches, and splices are assumed to have a resistance so low that their effect on the circuit is negligible. See Neglecting Contact Resistance. However, very low resistance electrical connections, and even the conductors, will cause a voltage drop in a current-carrying circuit. The voltage drop results in power loss in the form of excess heat. If the circuit is very sensitive to voltage changes or excess heat, the small voltage drops can affect circuit operation.

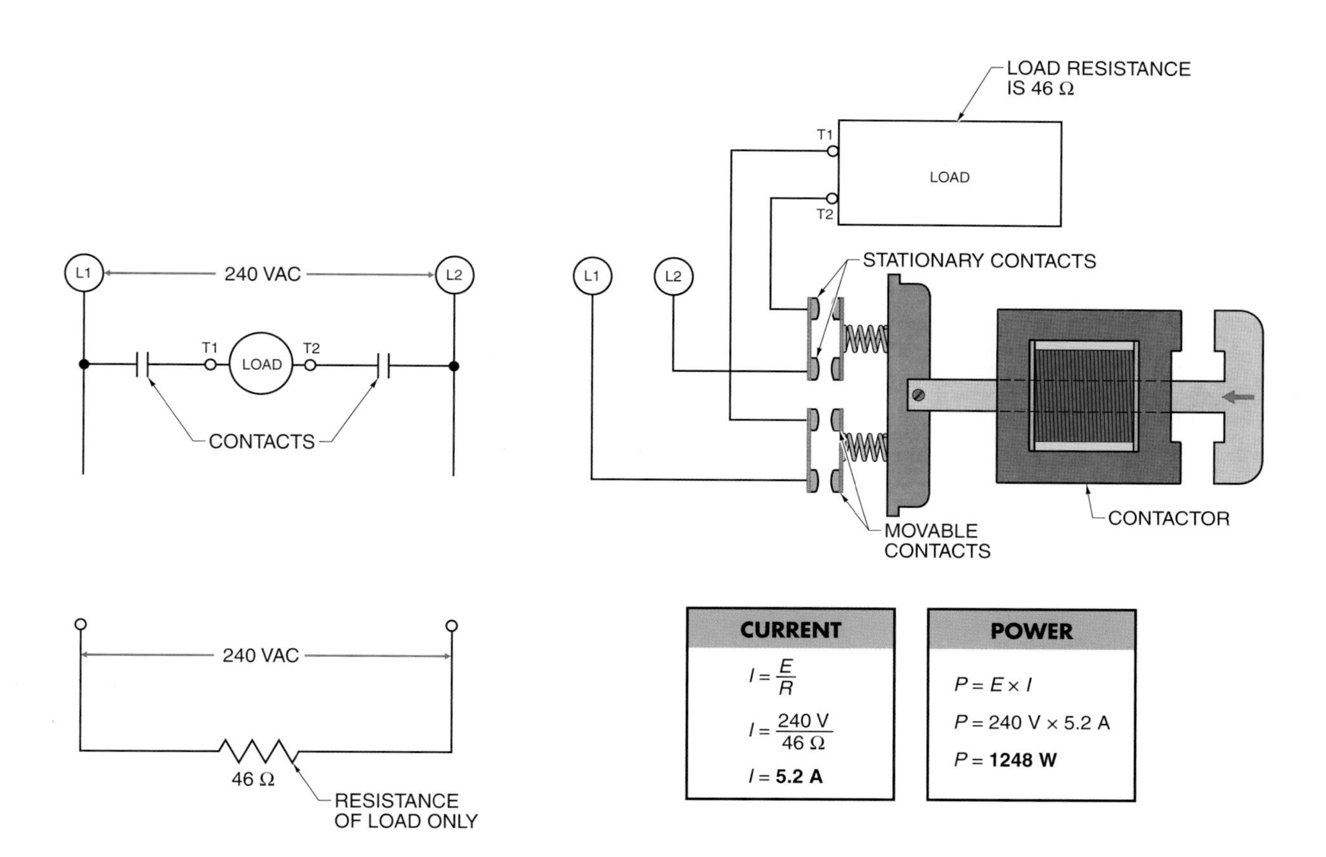

NEGLECTING CONTACT RESISTANCE

Micro-ohmmeters are used to measure very low resistances of less than 1 Ω, and some measure as low as 0.000001 Ω. Measuring the additional resistances of connections such as conductors, contacts, switches, and splices builds a more complete model of an actual circuit. Total circuit resistance, voltage drops, and power loss can be calculated with Ohm's law, the power formula, and the laws of series and parallel circuits. See Including Contact Resistance.

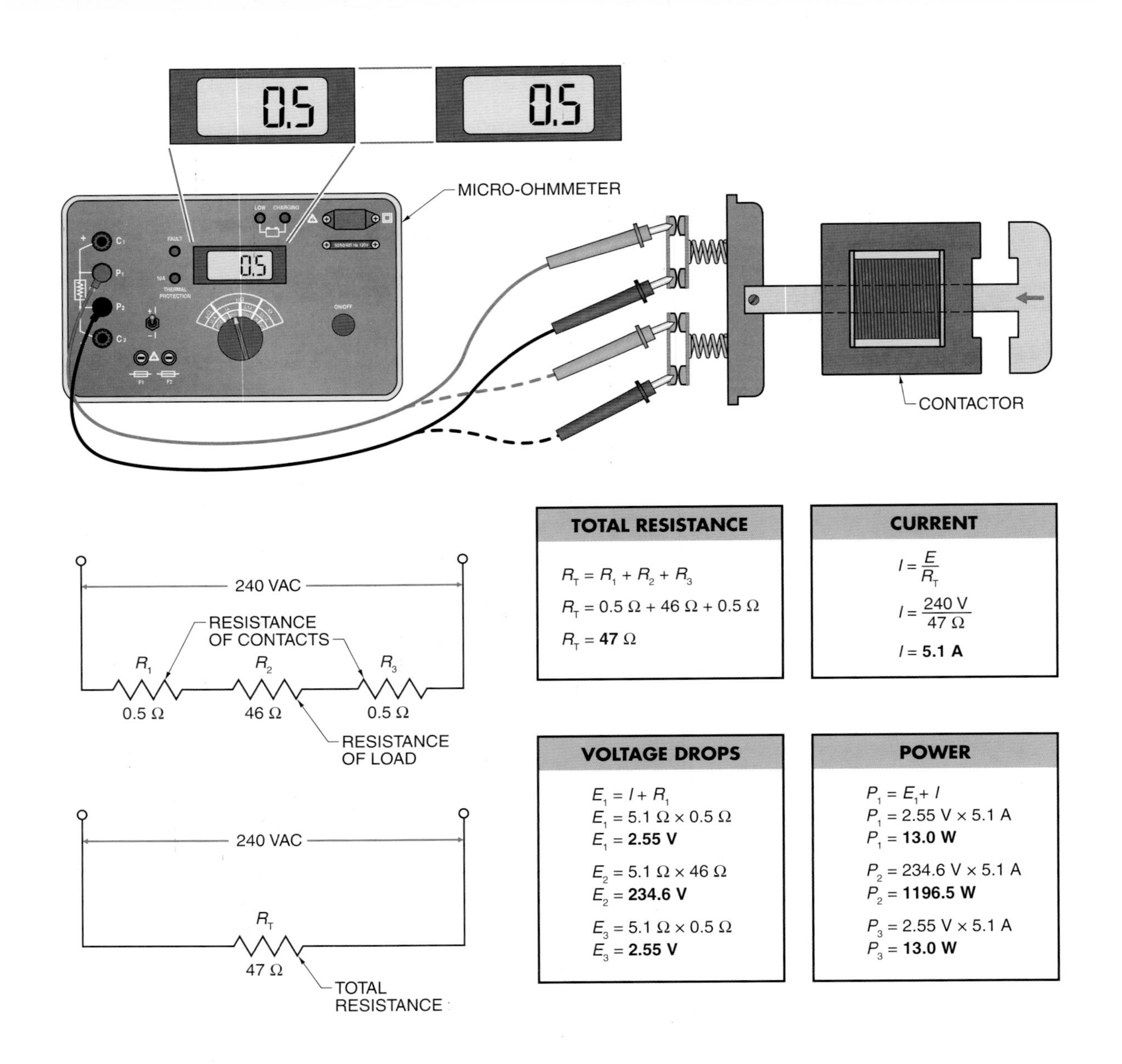

INCLUDING CONTACT RESISTANCE

A 240 VAC circuit has a contactor switching a load that has 158 Ω of resistance. The micro-ohmmeter measures 100 mΩ for each set of contacts.

1. ______________ What is the total circuit resistance?

2. ______________ What is the current (in A) through the circuit?

3. ______________ What is the voltage drop (in V) across each contact?

4. ______________ How much power (in mW) is produced at each contact?

5. ______________ How much power (in mW) is produced at the load?

6. ______________ If the resistance of the contacts is neglected, how much power is produced at the load?

A 240 VAC circuit has a contactor switching a load that has 67 Ω of resistance. The micro-ohmmeter measures 50 mΩ for each set of contacts.

7. ______________ What is the total circuit resistance?

8. ______________ What is the current (in A) through the circuit?

9. ______________ What is the voltage drop (in V) across each contact?

10. ______________ How much power (in mW) is produced at each contact?

11. ______________ How much power (in mW) is produced at the load?

12. ______________ If the resistance of the contacts is neglected, how much power is produced at the load?

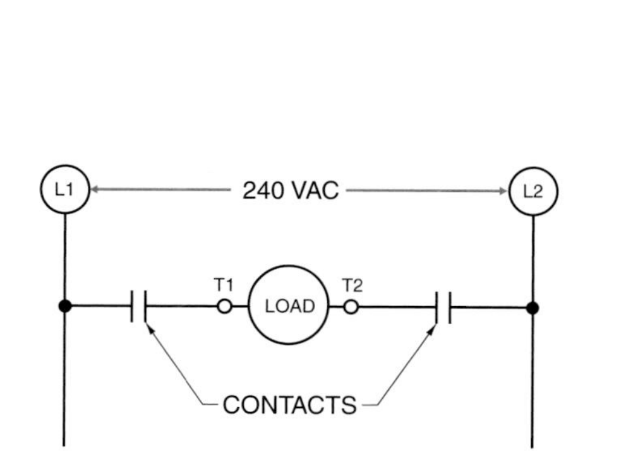

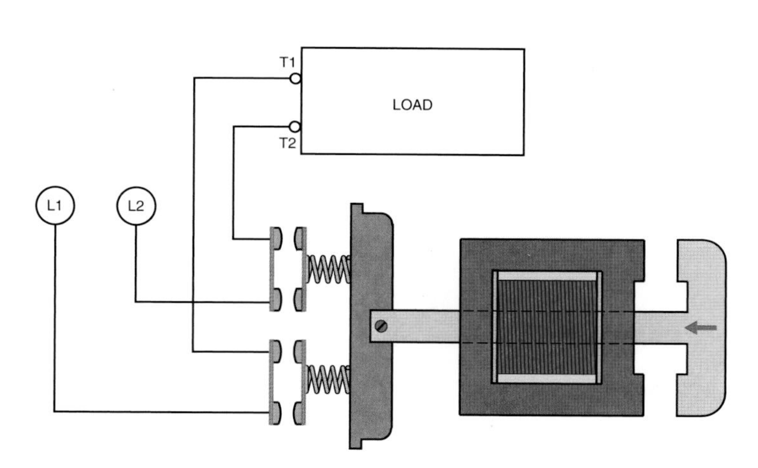

Application 9-2—Horsepower-Torque-Speed Relationship

The speed of a motor is measured using a contact tachometer, photo tachometer, or laser tachometer. Torque is the rotational force at the shaft of the motor. A motor under greater load requires more torque to turn the shaft. Horsepower is the combination of speed and torque together.

Horsepower, torque, and speed are related such that when any two are known, the third quantity can be determined. For example, if speed and horsepower are known, torque can be determined. If speed and torque are known, horsepower can be determined, and if torque and horsepower are known, speed can be determined.

Conversion charts can give quick approximations of the unknown value when the other two are known. Connecting points on any two scales defines a line which approximately indicates the third value where the line intersects the third scale. See Horsepower-Torque-Speed Conversion.

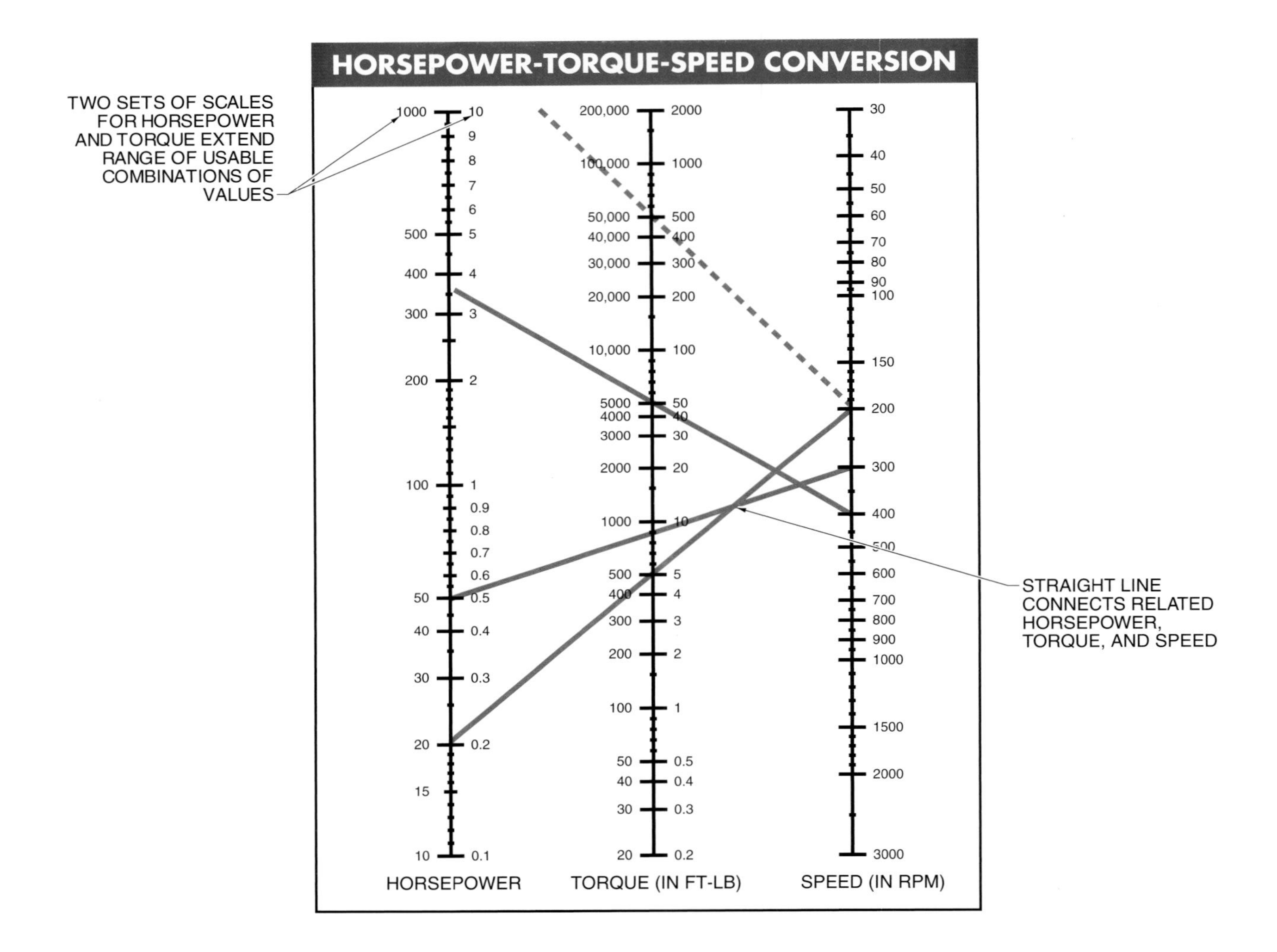

The horsepower and torque scales cover wider ranges by including two sets of numbers. The numbers on the left side of the scales are 100 times larger than the numbers on the right side. When using the conversion chart, the same sides of the horsepower and torque scales must be used together. The number values and the line connecting them determine which side to use.

For example, the line connecting 200 rpm with 500 ft-lb (on the right side of the torque scale) extends off the conversion chart and horsepower cannot be determined. However, connecting 200 rpm with 500 ft-lb on the left side of the torque scale yields a usable line. The result is approximately 20 HP. The left side of the horsepower scale must be used because the left side of the torque scale was used.

Also, the conversion chart shows that 50 ft-lb of torque at 400 rpm produces approximately 3.6 HP and that a ½ HP motor turning at 300 rpm produces approximately 9 ft-lb of torque.

Conversion charts are useful for approximating values such as torque, but calculating the torque gives a more accurate value. Torque is found by applying the following formula:

$$T = \frac{HP}{n} \times 5252$$

where

T = torque (in ft-lb)

HP = power (in HP)

n = rotational speed (in rpm)

For example, what is the full-load torque of a 60 HP, 240 V, 3ϕ motor turning at 1725 rpm?

$$T = \frac{HP}{n} \times 5252$$

$$T = \frac{60}{1725} \times 5252$$

$$T = 0.0348 \times 5252$$

$$T = \mathbf{182.7\ ft\text{-}lb}$$

A laser tachometer measures 120 rpm on a 50 HP conveyor motor.

1. ______________ What is the approximate torque (using the conversion chart)?
2. ______________ What is the calculated torque?
3. ______________ If the motor slows down to 100 rpm, what is the calculated torque?
4. ______________ If the motor speeds up to 200 rpm, what is the calculated torque?

A laser tachometer measures 1700 rpm on a 2 HP conveyor motor.

5. ______________ What is the approximate torque (using the conversion chart)?
6. ______________ What is the calculated torque?
7. ______________ If the motor speeds up to 2000 rpm, what is the calculated torque?
8. ______________ If heavier boxes are placed on the conveyor, what might happen to the motor speed?

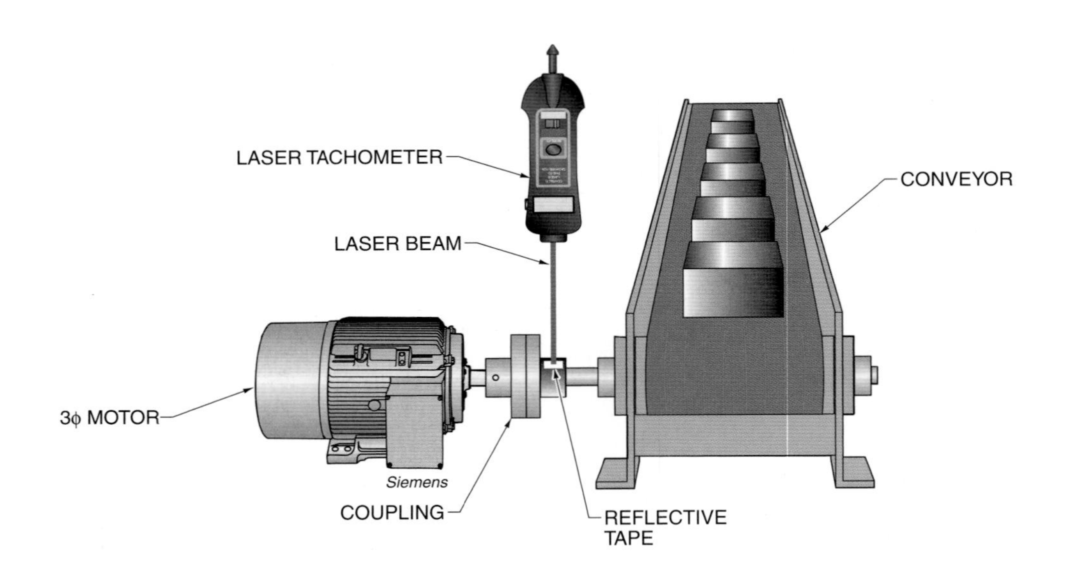

Application 9-3 – Carbon Monoxide Testing

A carbon monoxide (CO) probe is a multimeter attachment that measures the level of CO in an atmosphere. Carbon monoxide is a colorless and odorless toxic gas that can cause serious health problems or death by obstructing the transfer of oxygen in the bloodstream. Common sources of carbon monoxide emissions include malfunctioning or improperly installed furnaces or fireplaces, obstructed or undersized chimneys or flue exhausts, poorly maintained gas, oil, or kerosene appliances, and improperly ventilated combustion engines such as automobile engines, portable generators, and lawnmowers. See Carbon Monoxide (CO) Emission Sources.

CARBON MONOXIDE (CO) EMISSION SOURCES

Appliance	Fuel	Typical Problems
Gas furnaces; room heaters	Oil, natural gas, or LPG (liquefied petroleum gas)	1. Cracked heat exchanger 2. Not enough air to burn fuel properly 3. Defective/blocked flue 4. Poorly adjusted burner 5. Building not properly pressurized
Central heating furnaces	Coal or kerosene	1. Cracked heat exchanger 2. Not enough air to burn fuel properly 3. Defective grate
Room heaters; central heaters	Kerosene	1. Improper adjustment 2. Wrong fuel 3. Wrong wick or wick height 4. Not enough air to burn fuel 5. System not properly vented
Water heaters	Natural gas or LPG	1. Not enough air to burn fuel properly 2. Poorly adjusted burner 3. Misuse as a room heater 4. System not properly vented
Ranges; ovens	Natural gas or LPG	1. Not enough air to burn fuel properly 2. Poorly adjusted burner 3. Misuse as a room heater 4. System not properly vented
Stoves; fireplaces	Gas, wood, coal	1. Not enough air to burn fuel properly 2. Defective/blocked flue 3. Green or treated wood 4. Cracked heat exchanger 5. Cracked firebox

Symptoms of carbon monoxide poisoning include headaches, dizziness, sleepiness, and general weakness. High levels of carbon monoxide can be deadly. Personnel should be evacuated and carbon monoxide levels checked if anyone reports these symptoms. For maximum safety, carbon monoxide measurements should be taken whenever working on or around devices capable of producing carbon monoxide. Carbon monoxide levels are also measured when evaluating a confined space.

A carbon monoxide probe multimeter attachment measures the level of carbon monoxide in parts per million (ppm). Permissible carbon monoxide exposure levels are set by governing agencies. See Carbon Monoxide (CO) Level Standards.

CARBON MONOXIDE (CO) LEVEL STANDARDS	
0 to 1 PPM	Normal background levels
9 PPM	ASHRAE* limit for living areas
50 PPM	OSHA† enclosed space 8-hour average level
100 PPM	OSHA† exposure limit
200 PPM	Mild headache, fatigue, nausea, and dizziness
800 PPM	Dizziness, nausea, and convulsions; death within 2 to 3 hours

* American Society of Heating, Refrigeration, and Air-Conditioning Engineers, Inc.
† Occupational Safety and Health Administration

A service call requires that a warehouse be checked for carbon monoxide levels. The warehouse staff uses a variety of electric, propane-powered, and gas-powered vehicles to transport products.

1. Connect the carbon monoxide probe attachment to the correct multimeter jacks.

2. Set (draw in) the correct function switch position for taking carbon monoxide level measurements.

3. ______________ If the multimeter measures levels between 10 mV and 85 mV at various warehouse locations, do the carbon monoxide levels meet the OSHA exposure limit for an 8 hr shift?

4. ______________ If the multimeter measures levels between 10 mV and 85 mV at various warehouse locations, do the carbon monoxide levels meet the OSHA exposure limit for a 10 hr shift?

5. ______________ If the multimeter measures levels between 10 mV and 125 mV, and an average level of 30 mV, at various warehouse locations, are the carbon monoxide levels above a dangerous level at any location within the warehouse?

6. ______________ If the multimeter reads 25 mV and the relative mode is activated, what will the multimeter read?

7. ______________ If the multimeter reads +30 mV REL, what is the actual level of carbon monoxide at that location?

8. ____________ If the multimeter reads –8 mV REL, what is the actual level of carbon monoxide at that location?

9. ____________ If the multimeter reads +52 mV MAX REL, what is the maximum level of carbon monoxide in the warehouse?

10. ____________ If the multimeter reads –12 mV MIN REL, what is the minimum level of carbon monoxide in the warehouse?

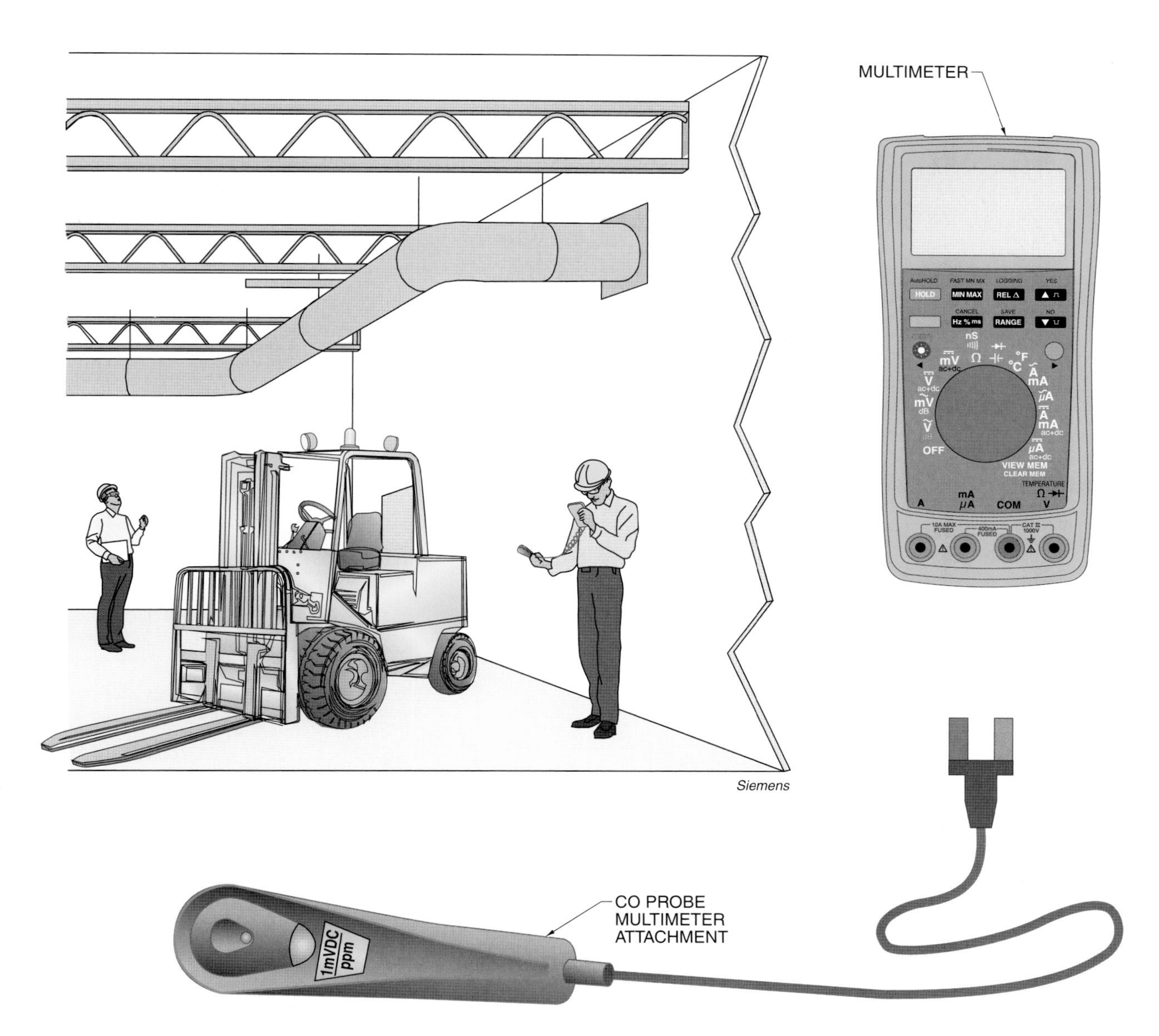

Application 9-4—Light Level Testing

The proper amount of light ensures that living and working conditions are safe and secure, and that environments are favorable for playing sports, displaying merchandise, or other activities. It is important to achieve good lighting in the right amount for the application. Electricians should know how to measure light levels and make recommendations for lighting types and levels in new installations, or additional lighting in existing systems.

Lamp (light bulb) manufacturers produce many different lamp types and sizes for various applications. Lamps are rated for their power consumption (in watts), light output (in lumens or lumens per watt), life expectancy (in hours), and features such as light color and energy efficiency. See Lamp Types.

LAMP TYPES

Type	Shape	Power Consumption*	Light Output†	Average Rated Life‡	Features
Mercury vapor		40 to 1000	50 to 60	24,000 +	Long life; low price
Metal-halide		35 to 1500	80 to 125	6000 to 20,000	Good color rendering; long life; high efficiency
High-pressure sodium		35 to 1000	80 to 150	15,000 to 24,000	Extremely long life; good lumen maintenance; high efficiency
Low-pressure sodium		18 to 180	190 to 200	18,000	Very long life; extremely high efficiency; poor color rendering
Fluorescent		4 to 215	55 to 100	6000 to 36,000	Wide choice of light colors; excellent color rendering; good uniformity; long life; very efficient
Incandescent		3 to 1500	15 to 25	500 to 8000	Easy to install; many shapes; low cost; instant start
Halogen		45 to 1500	18 to 22	2000 to 6000	Compact; high output; white light; easy to install
LED		<1 to 30	12 to 22	25,000	Extremely long life; plastic housing; available in various shapes and colors

* in W
† in lm/W
‡ in hr

Light meters are used to measure the amount of light at any location. The amount of light is usually specified and measured in footcandles (fc) or lumens (lm). Charts list recommended light levels for various applications such as offices, classrooms, and sporting events. Most light level charts and specifications use footcandle units. See Recommended Light Levels.

RECOMMENDED LIGHT LEVELS

Interior Lighting				Exterior Lighting	
Area	**Light Level***	**Area**	**Light Level***	**Area**	**Light Level***
Assembly		Machine shop		Building	
Rough, easy seeing	30	Rough bench	50	Light surface	15
Medium	100	Medium bench	100	Dark surface	50
Fine	500	Materials handling		Loading/unloading area	20
Auditorium		Picking stock	30	Parking areas	
Exhibitions	30	Packing, labeling	50	Industrial	2
Banks		Offices		Shopping	5
Lobby, general	50	Regular office work	100	Storage yards (active)	20
Writing areas	70	Accounting	150	Street	
Teller station	150	Detailed work	200	Local	0.9
Canning		Printing		Expressway	1.4
Cutting, sorting	100	Proofreading	150	Car lots	
Clothing manufacturing		Color inspecting	200	Front line	100 to 500
Patternmaking	50	Schools		Remaining area	20 to 75
Shops	100	Auditoriums	20		
Garages (auto)		Classrooms	60 to 100		
Parking	10	Indoor gyms	30 to 40		
Repair	50	Stores			
Hospital/Medical		Stockroom	30		
Lobby	30	Service area	100		
Dental chair	1000	Warehousing, storage			
Operating table	2500	Inactive	5		
		Active	30		

* in fc

Necessary light levels are also affected by the age of the people in that environment. Older people require more light to see as well as younger people. Since recommended light level charts typically assume an average age of 30, they may need to be adjusted if the average age of the people in an environment is other than 30. For an average age of 20, the values can be reduced to 75% of the recommended light levels because younger eyes require less light. However, the recommended light levels are higher for people over the age of 30. The recommended light levels are doubled for an average age of 40, tripled for an average age of 50, and multiplied by four for an average age of 60.

1. Connect the light meter attachment to the correct jacks on the multimeter.
2. Set (draw in) the multimeter function switch for taking light level measurements.

The multimeter and light meter attachment are used to take light level readings in a warehouse parking lot. The readings are 5.6 mV, 6.8 mV, and 6.2 mV for three locations in the parking lot.

3. ______________ What is the average light level (in fc)?
4. ______________ Is the light level adequate for the application?
5. ______________ Is the light meter attachment measuring direct light, indirect light, or both?
6. ______________ If an area of the parking lot is used for loading and unloading materials, is the light level adequate for that application?

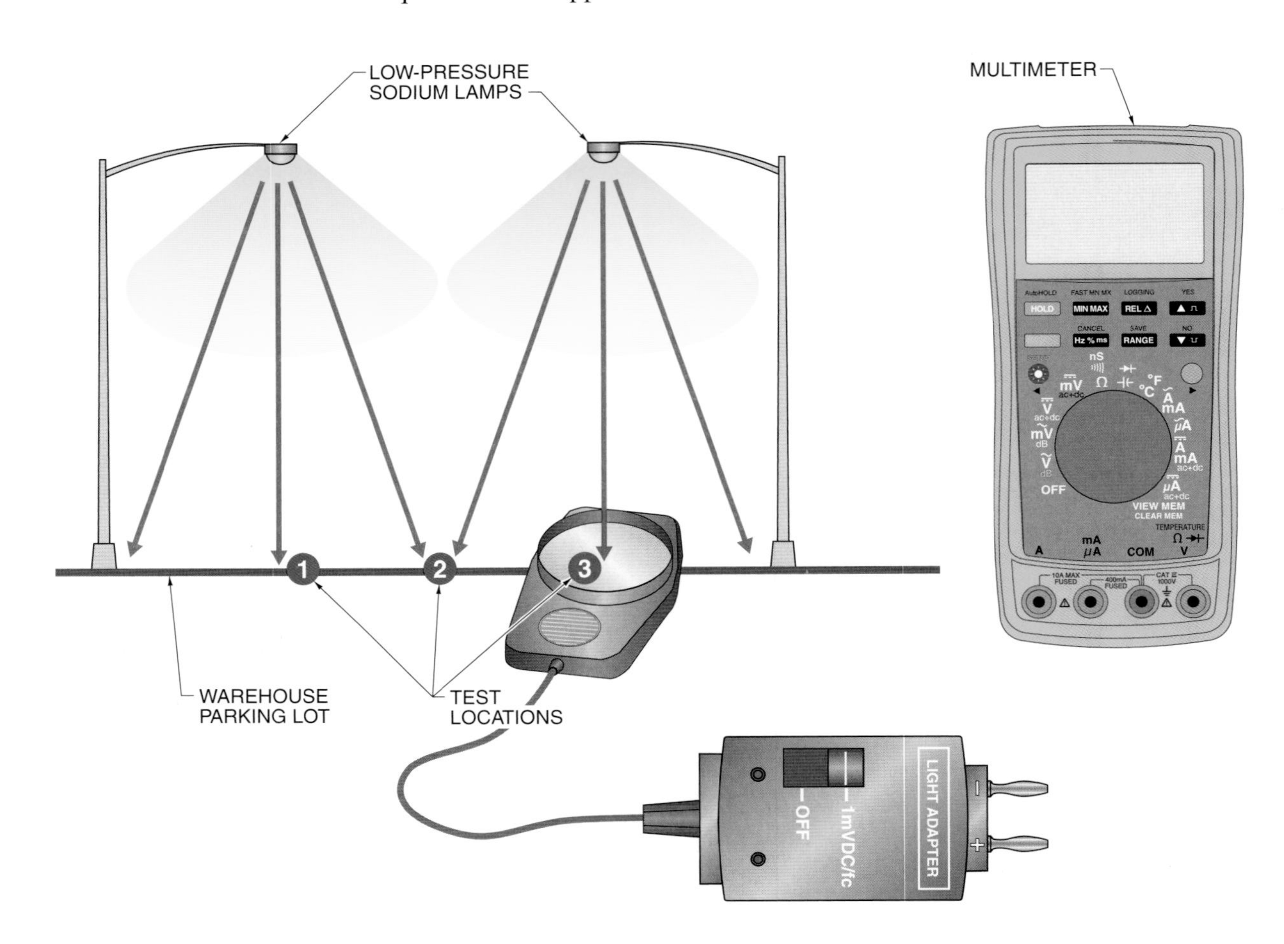

7. Connect the light meter attachment to the correct jacks on the multimeter.

8. Set (draw in) the multimeter function switch for taking light level measurements.

The multimeter and light meter attachment are used to take light level readings inside an active warehouse. The readings are 32.5 mV, 48.6 mV, 50.2 mV, 47.9 mV, 33.3 mV for five locations in the building.

9. ______________ What is the average light level (in fc)?

10. ______________ Is the light level adequate for the application?

11. ______________ If the average age of the warehouse staff is 40, is the light level adequate?

12. ______________ Is the light meter attachment measuring direct light, indirect light, or both?

13. Why does the test location in the center of the room have the highest reading?

14. ______________ If Lamp 2 burned out, which location would have the highest light level?

15. ______________ Which location would have the lowest light level?

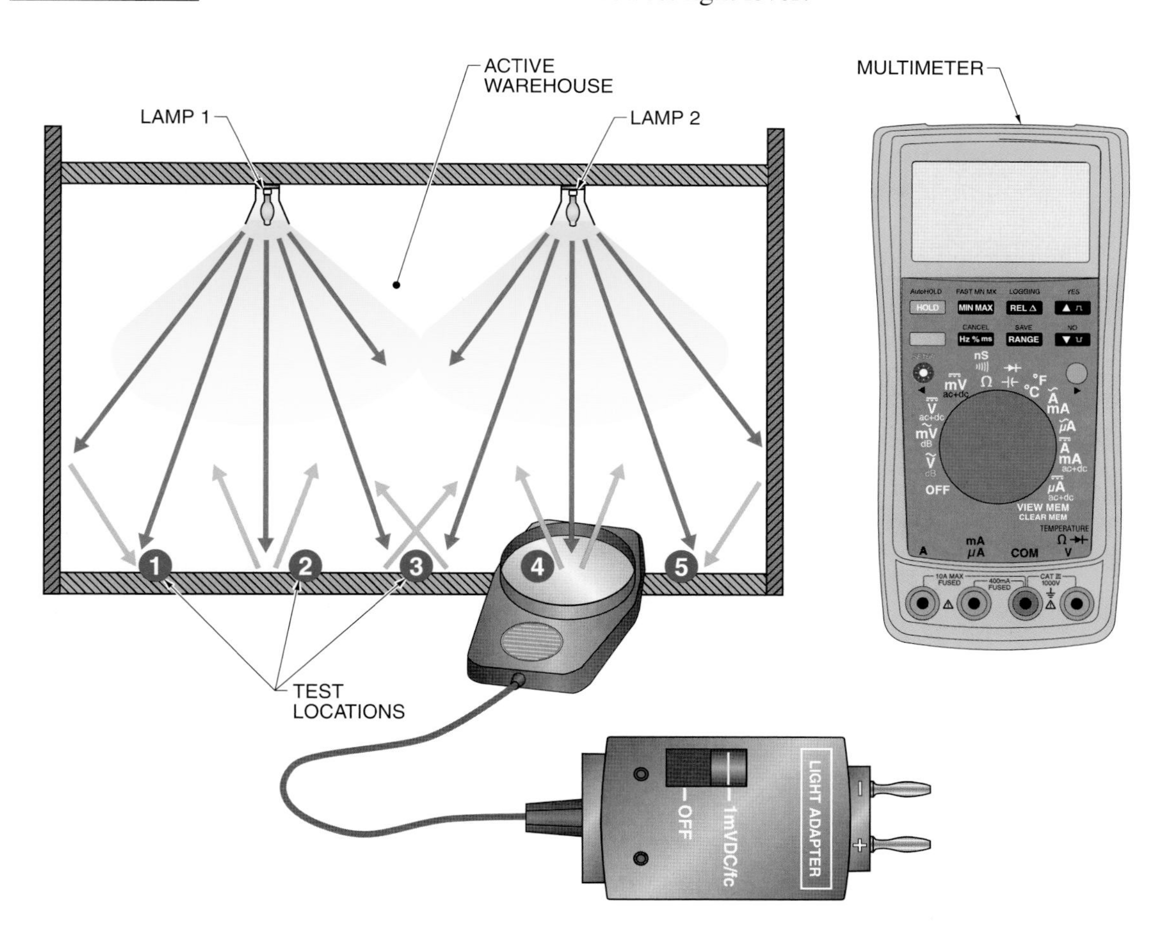

TROUBLESHOOTING 10
APPLICATIONS

Name: ______________________________ Date: ____________________

Application 10-1 — Troubleshooting Two-Way Switching Circuits

A two-way switch is a single-pole, single-throw (SPST) switch used to control a load (such as a lamp) from one location. When in the ON position, the switch makes a connection between two conductors and allows current to flow from one conductor to the other. When in the OFF position, the switch breaks the connection between the two conductors and current cannot flow. Two-way switches have ON and OFF positions clearly marked on them. See Two-Way Switching Diagram.

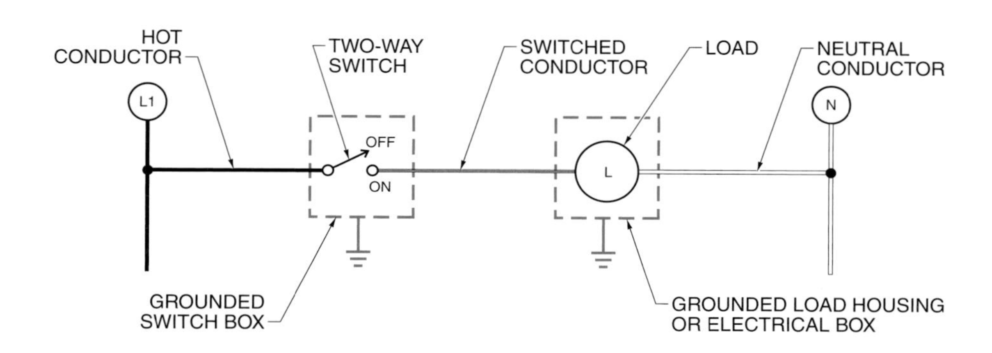

TWO-WAY SWITCHING DIAGRAM

Switches are wired on hot conductors, which are before the load. The conductor between the power supply and the switch remains the "hot" conductor, and the conductor between the switch and the load is called the "switched" conductor. The switched conductor is usually red.

The following tests can be applied to troubleshoot switching circuits. See Switching Circuit Troubleshooting.

1. Test for proper grounding. Connect a voltmeter between the metal box (or the green/bare ground wire in a plastic box) and the hot conductor (ungrounded conductor) going into the switch. There should be system voltage at all times, regardless of the switch position. This test should be done at both the switch box and the load's box. If the system is not properly grounded, ground the system before conducting further tests.
2. Test for correct system voltage. Connect the voltmeter between the hot conductor and the neutral conductor. There should be system voltage at all times, regardless of the switch position.

3. Test the switch for proper operation. Connect the voltmeter between the neutral conductor and switched conductor. There should be no voltage when the switch is in the OFF position and system voltage when the switch is in the ON position.
4. Test the load. Connect the voltmeter between the load terminals. There should be no voltage when the switch is in the OFF position and system voltage when the switch is in the ON position.

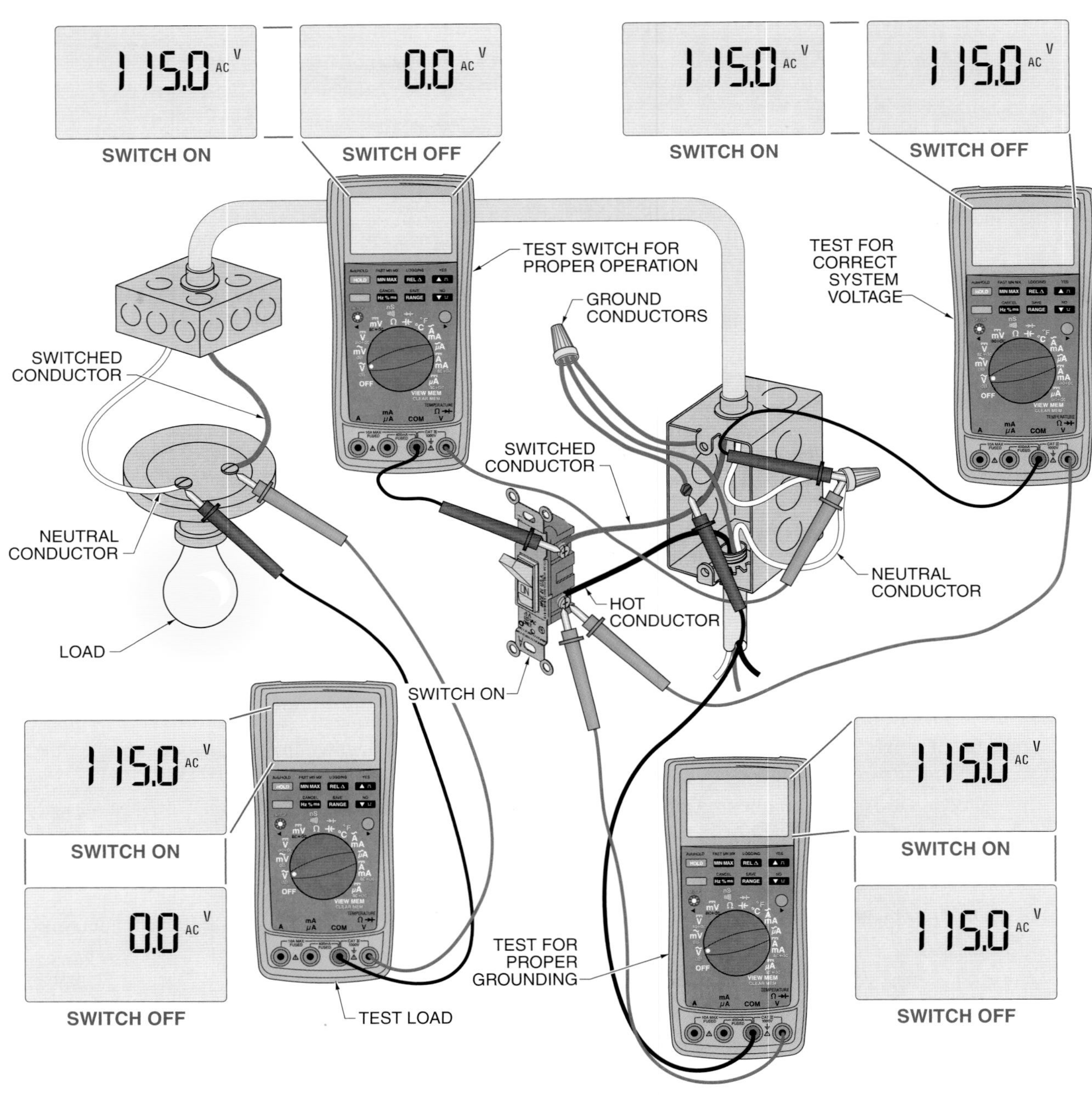

SWITCHING CIRCUIT TROUBLESHOOTING

1. Connect the test leads of Multimeter 1 to test for proper grounding at the box for Switch 1.
2. ______________ What should Multimeter 1 read when the switch is in the ON position?
3. ______________ What should Multimeter 1 read when the switch is in the OFF position?
4. Connect the test leads of Multimeter 2 to test for correct system voltage out of Switch 1.
5. ______________ What should Multimeter 2 read when the switch is in the ON position?
6. ______________ What should Multimeter 2 read when the switch is in the OFF position?
7. Connect the test leads of Multimeter 4 to test Switch 2 for proper operation.
8. ______________ What should Multimeter 4 read when the switch is in the ON position?
9. ______________ What should Multimeter 4 read when the switch is in the OFF position?
10. Connect the test leads of Multimeter 3 to test Lamp 2.
11. ______________ What should Multimeter 3 read when the switch is in the ON position?
12. ______________ What should Multimeter 3 read when the switch is in the OFF position?

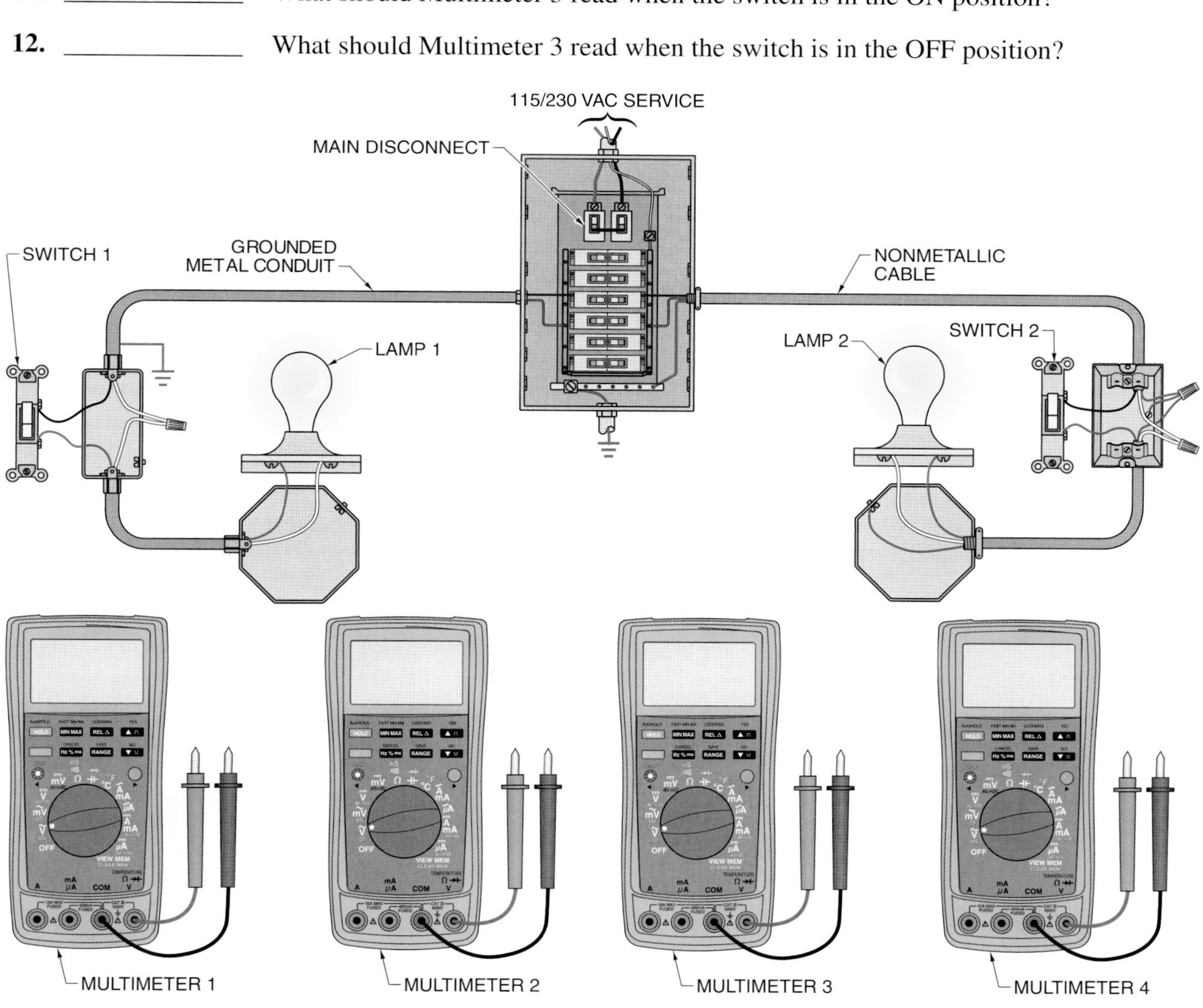

Application 10-2—Troubleshooting Three-Way Switching Circuits

The same tests performed on a two-way switching circuit can be used to troubleshoot a three-way switching circuit. A three-way switch is a single-pole, double-throw (SPDT) switch. Two three-way switches are used to control a load from two different locations. A three-way switch does not have designated ON and OFF positions because the two switches affect each other's operation. The position of one switch that allows the flow of current (turning the load ON) depends on the position of the other switch.

A three-way switch has one common terminal and two traveler terminals. The common terminal is connected to the hot conductor or the switched conductor, depending on the switch's location in the circuit. The traveler terminals are connected to traveler conductors, which connect the two switches. See Three-Way Switching Diagram.

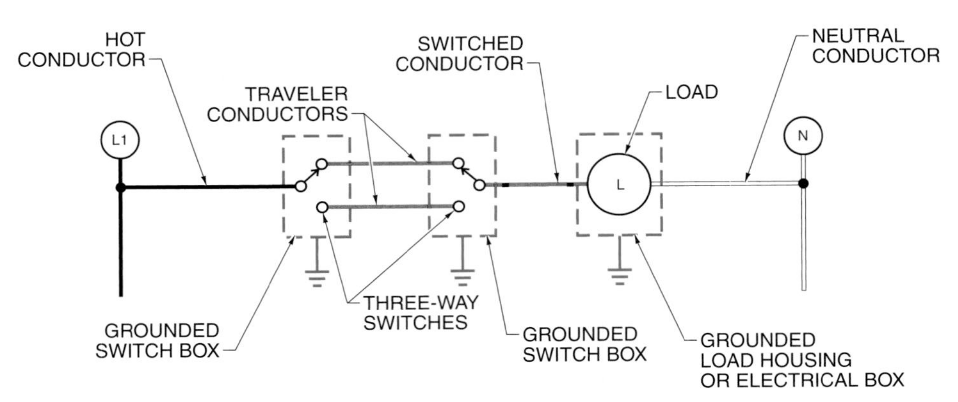

THREE-WAY SWITCHING DIAGRAM

A three-way switch operates like a pair of two-way switches. When the three-way switch is in one position, contact is made with one of the traveler terminals, similar to how a two-way switch makes contact in the ON position. When the three-way switch is in the other position, it makes contact with the other traveler terminal. In either position, one of the travelers is always connected to the common terminal.

When two three-way switches are used together in a circuit, either one can be used to switch the load ON. One switch connects one of the travelers to the circuit. The other switch either completes the circuit (turning the load ON) by connecting its common to the same traveler, or the circuit remains open (the load OFF) by connecting its common to the unconnected traveler.

A three-way switch circuit includes more conductors than a two-way switch, and each conductor must be identified when troubleshooting. The traveler conductors are color-coded red. The switched conductor in a three-way switching circuit is usually red or black and may be marked with colored tape to distinguish it from other conductors. The hot conductor is black and the neutral conductor is white. See Three-Way Switching Circuit.

Power can enter the conduit system at either one of the switches or at the load, but the switches must always be connected before the load. For example, if power enters at the load, the hot conductor is run through the conduit to one of the switches first. This is because it is the hot conductor that must be switched, not the neutral conductor. The travelers and the switched conductor then carry the current back to the load.

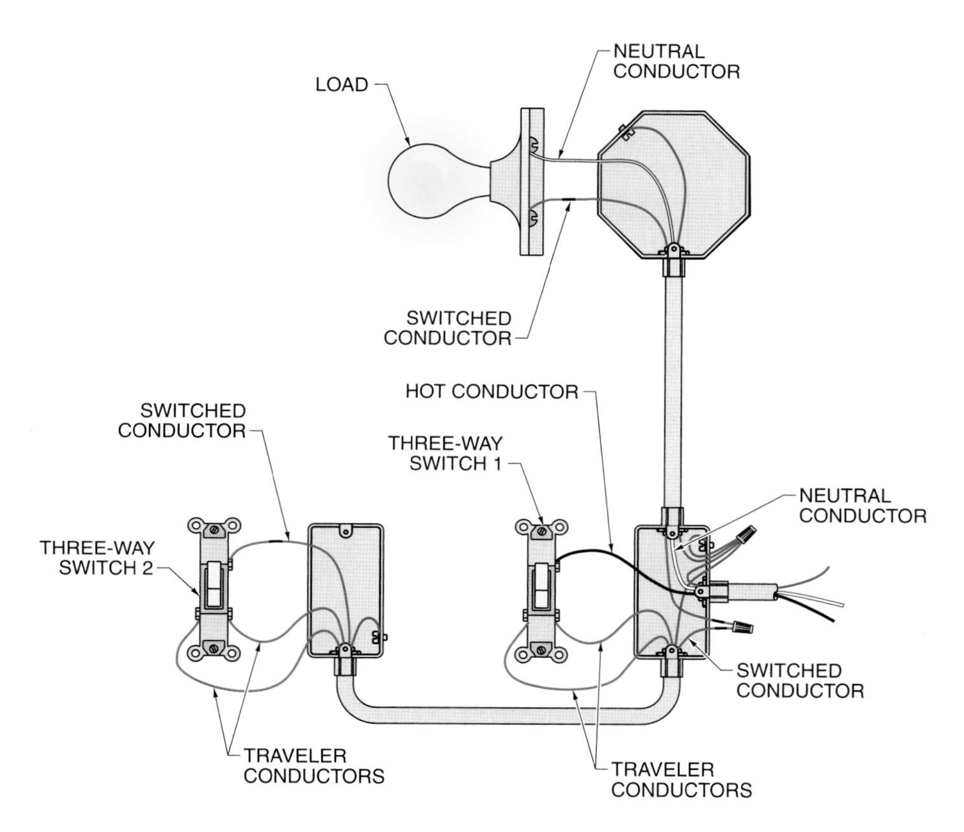

THREE-WAY SWITCHING CIRCUIT

1. ______________ What should Multimeter 1 read?
2. ______________ Which test is Multimeter 1 performing?
3. ______________ If Multimeter 2 reads 115 VAC, what should it read if Switch 2 changes positions?
4. ______________ If Multimeter 2 reads 0 VAC, what should it read if Switch 1 changes positions?
5. ______________ If Multimeter 3 reads 115 VAC for both positions of Switch 1, is there a problem in Switch 1 or Switch 2?

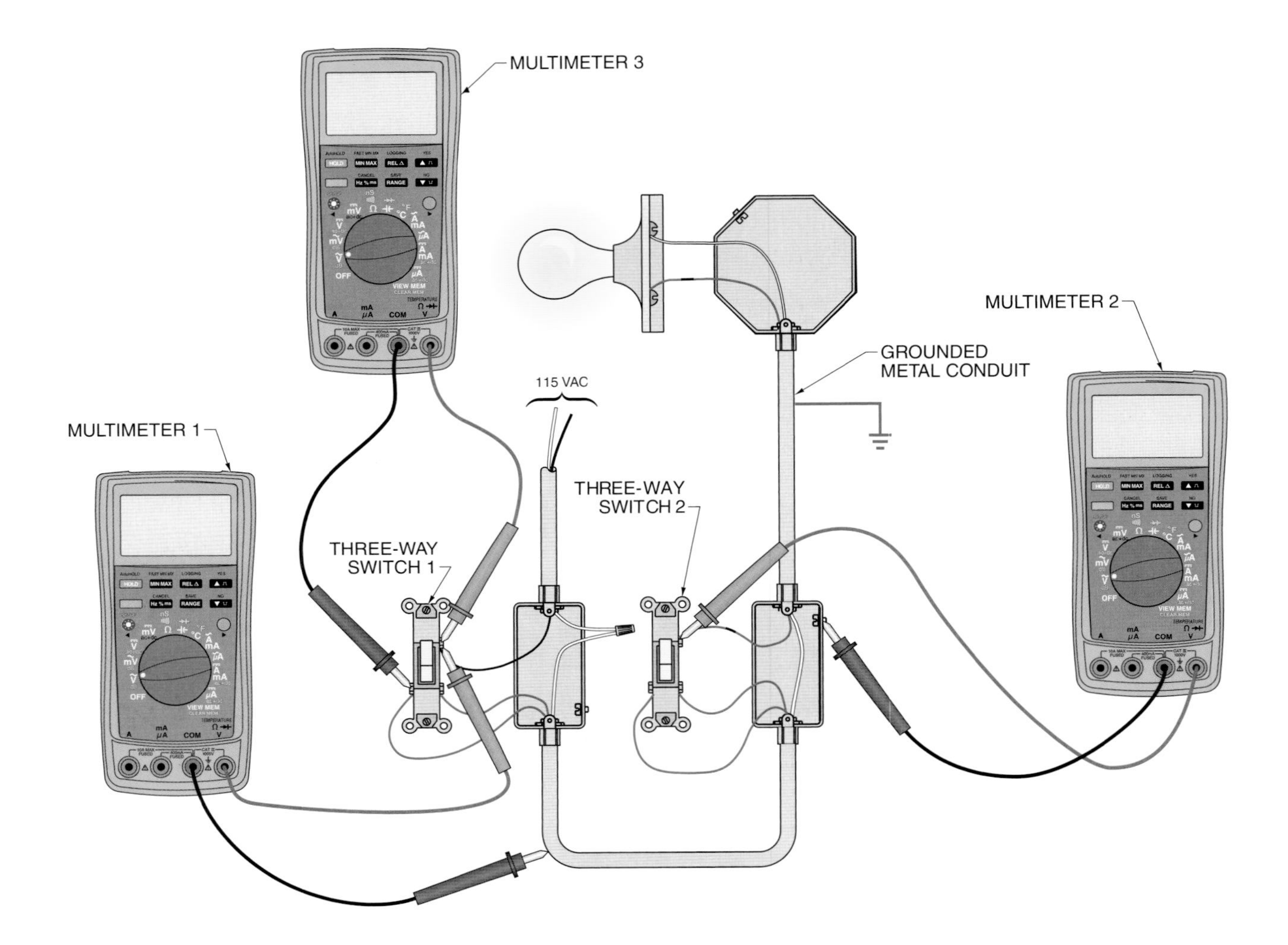

6. ______________ If Multimeter 1 reads 115 VAC, what should it read if Switch 1 changes positions?

7. ______________ If Multimeter 1 reads 0 VAC, what should it read if Switch 2 changes positions?

8. ______________ If Multimeter 1 reads 115 VAC for both positions of Switch 1, is there a problem in Switch 1 or Switch 2?

9. ______________ If Multimeter 1 reads 115 VAC for both positions of Switch 2, is there a problem in Switch 1 or Switch 2?

10. ______________ What should Multimeter 2 read?

11. ______________ If Multimeter 2 reads 0 V, what is the most likely problem?

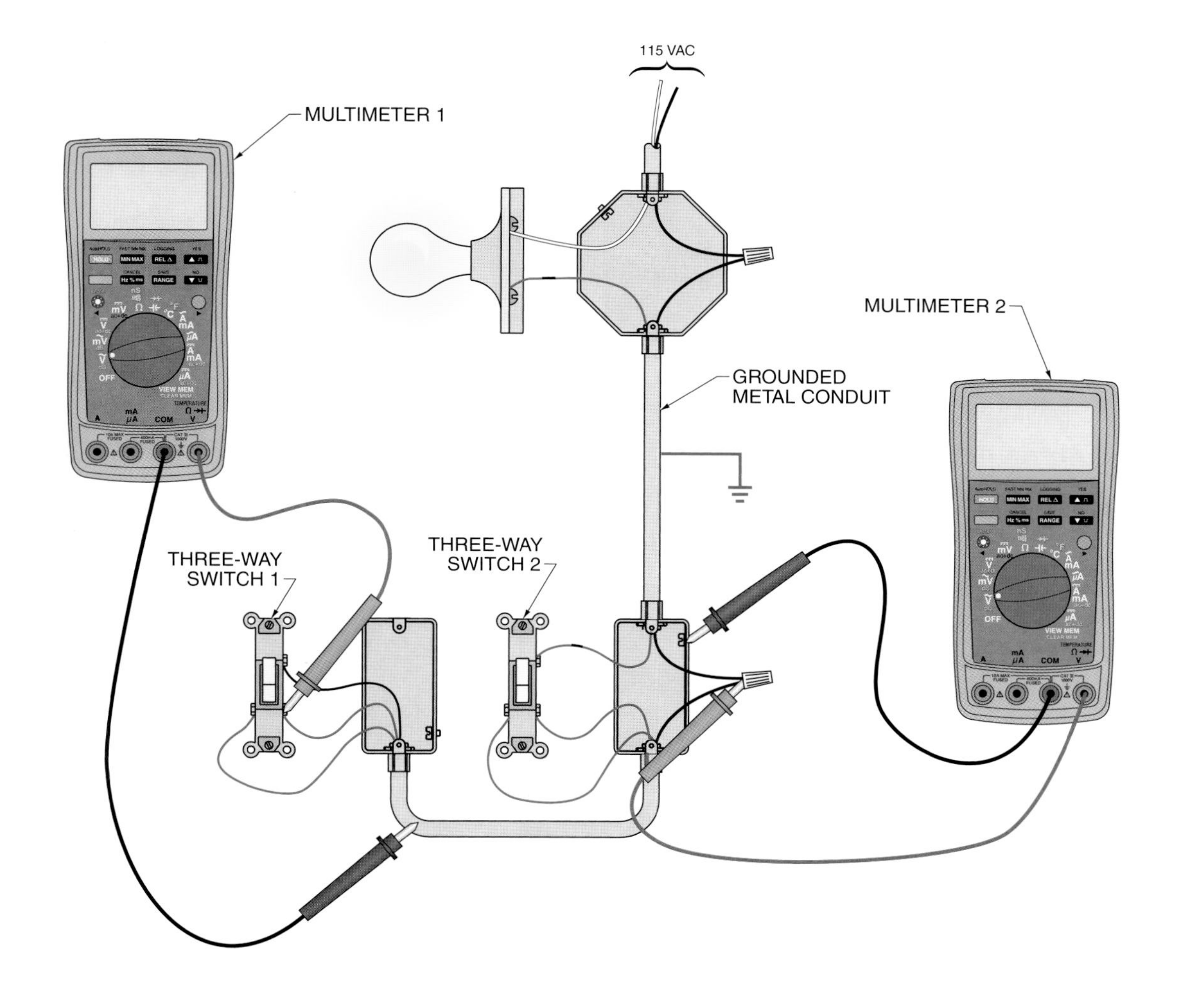

12. ______________ If Multimeter 1 reads 115 VAC, what should it read if Switch 1 changes positions?

13. ______________ If Multimeter 1 reads 115 VAC, what should it read if Switch 2 changes positions?

14. ______________ What should Multimeter 2 read?

15. ______________ Will changing the position of Switch 2 change the reading of Multimeter 2?

16. ______________ If Multimeter 1 reads 0 VAC for both positions of Switch 1 and Switch 2, could the problem be in the load?

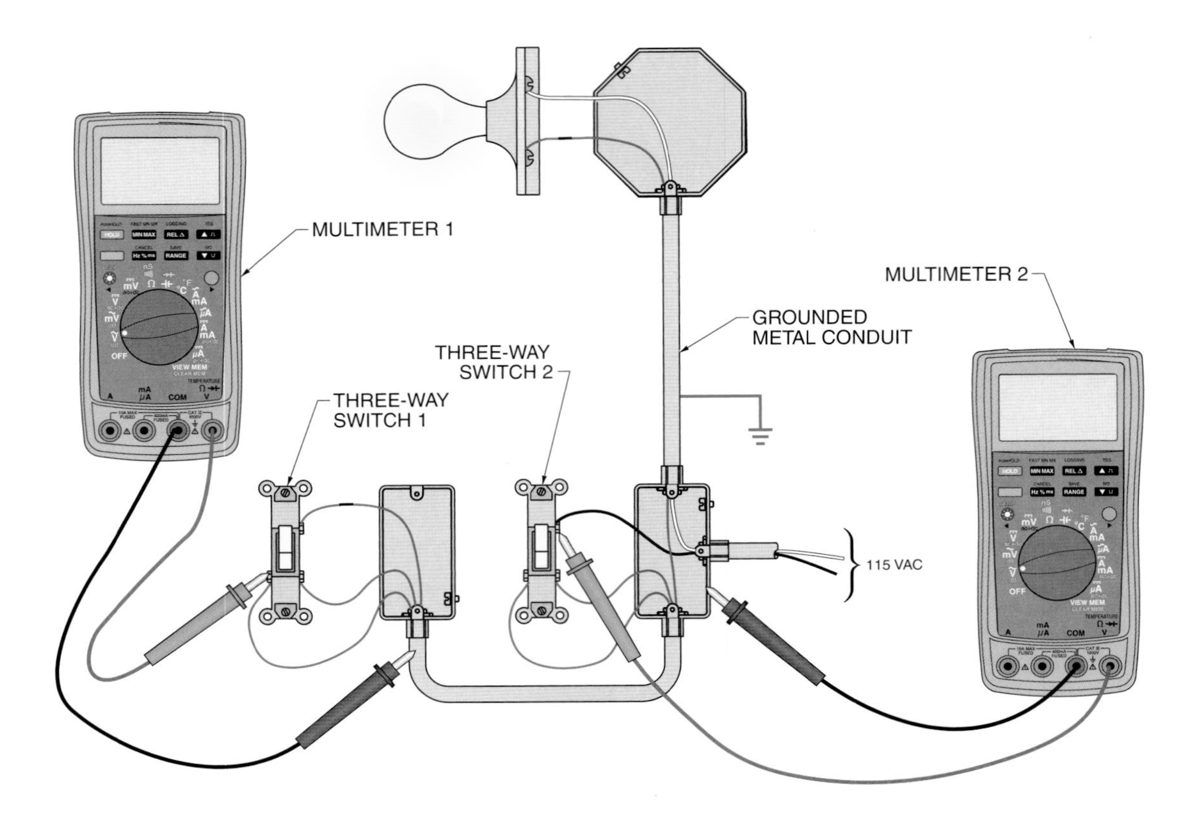

Application 10-3 — Troubleshooting Hardwired Circuits

Troubleshooting is a systematic process used to find a problem in an electrical circuit. The fact that an electrical circuit can be wired several different ways can complicate the troubleshooting process. A circuit can be hardwired, wired using terminal strips, or wired to a PLC. Each circuit functions exactly the same way, and the wiring type is not obvious to operators, supervisors, or even maintenance personnel until the electrical panels are opened. Each different wiring method requires a different troubleshooting method.

For example, a simple start-stop circuit can be wired by any of the three methods. Although a hardwired control circuit is the most common, it is the most difficult to troubleshoot. See Hardwired Circuit. A control circuit wired to a terminal strip is much easier to troubleshoot because most test instrument measurements can be taken at the terminal strip. See Terminal Strip Wired Circuit. Wiring a simple control circuit using a PLC may seem excessive at first, but with current demands for security and automation, this type of wiring is becoming more common. For example, a PLC can be programmed to automatically start (or stop) the motor for a process, keep track of the number of times the motor is energized, and automatically transmit information about routine maintenance requirements. See PLC-Wired Circuit.

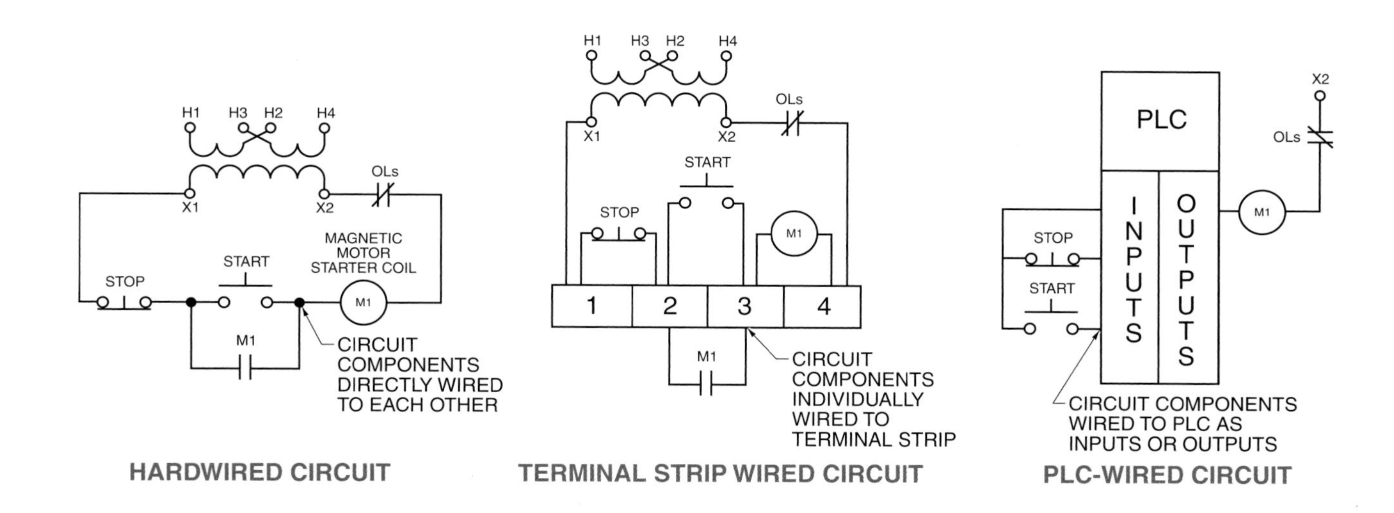

Direct hardwiring is the oldest and most straightforward wiring method used. In direct hardwiring, the power circuit (higher voltage) and control circuit (lower voltage) are wired point-to-point. In point-to-point wiring, each component in a circuit is connected directly to the next component specified on the wiring and line diagram. See Hardwired Reversing Circuit.

For example, the transformer X1 terminal is connected directly to the fuse, the fuse is connected directly to the stop pushbutton, the stop pushbutton is connected directly to the reverse pushbutton, the reverse pushbutton is connected directly to the forward pushbutton, and so on until the final connection from the overload (OL) contact is made back to the X2 terminal.

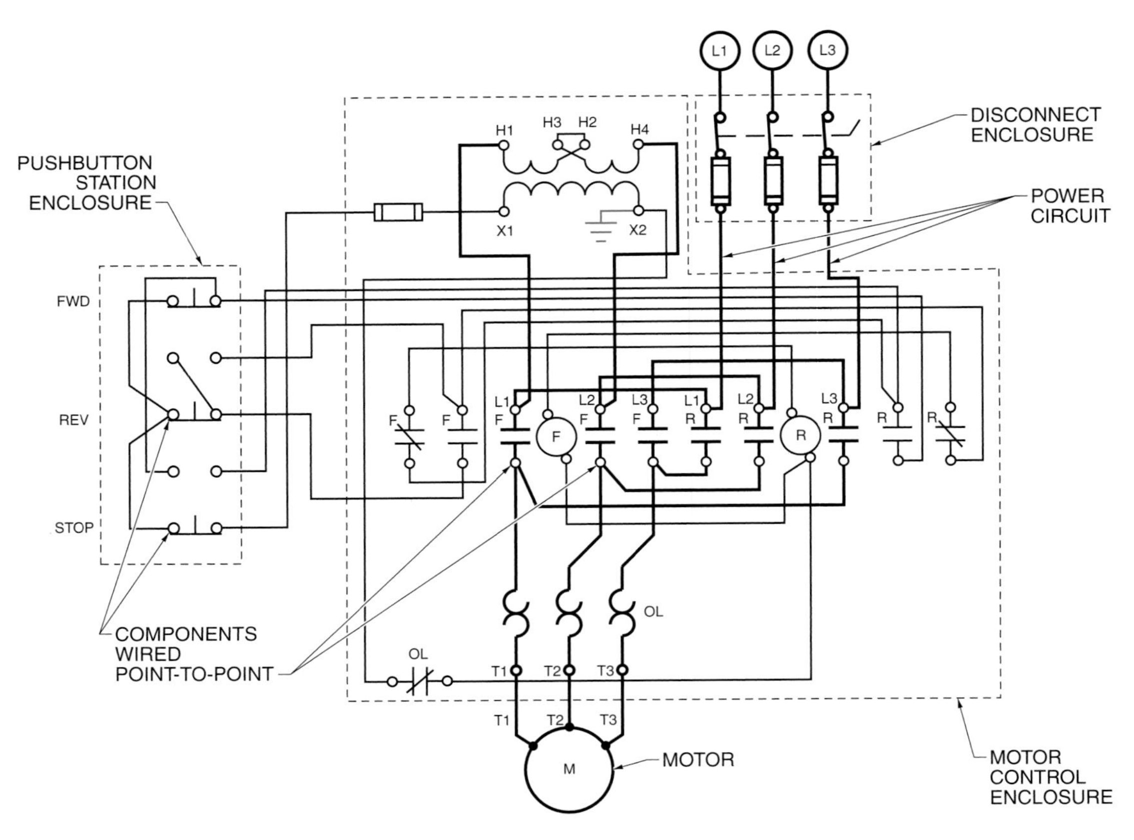

HARDWIRED REVERSING CIRCUIT

The disadvantage of a direct hardwired circuit is that circuit troubleshooting and modification can be time consuming. Gaining access to test points and around conductors with test instruments can be difficult in hardwired circuits. The original wiring may have been intended to be permanent, and test points could be protected by insulation or placed where there may not be enough room to use test leads or current clamps.

When working on hardwired circuits, conductors may need to be manually traced throughout a hardwired circuit, which is time consuming. Wiring diagrams help identify connections and conductor runs, but may not be complete, especially if modifications have been made.

Hardwired circuits are also difficult to modify. Any additional components must be inserted into the circuit without adversely affecting the circuit operation. Determining the necessary connections can be complicated, and the actual wiring may be difficult if conduits and boxes are crowded, if there is not enough room under terminal screws, or if permanent connections (such as soldered connections) must be undone. Modifications also complicate troubleshooting later if they were not documented on the original wiring diagrams.

Some circuit modifications, such as adding indicator lamps, may not cause a problem because they only require adding new conductors, and existing conductors do not need to be moved or removed. Other circuit modifications, such as adding switches, can be more difficult. For example, if forward and reverse limit switches are to be added to a circuit, some of the wiring must first be removed from the circuit, then new wiring for the limit switches must be added. See Modified Hardwired Reversing Circuit.

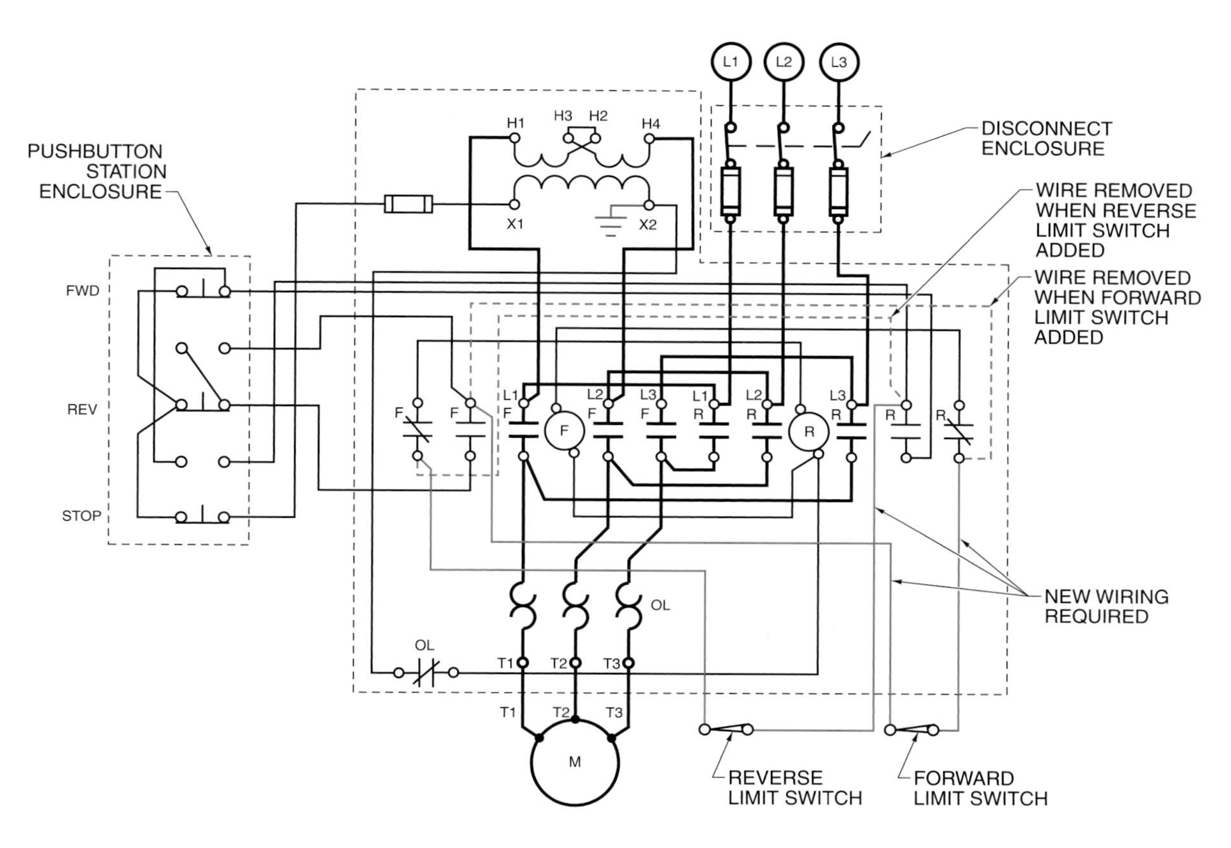

MODIFIED HARDWIRED REVERSING CIRCUIT

The three loads in this circuit are the motor, the forward coil, and the reverse coil. Multimeter 1 takes measurements in the disconnect enclosure after the fuses. Multimeter 3 measures 12.5 A on L1, 13.8 A on L2, and 11.9 A on L3.

1. ______________ If Multimeter 1 is connected after Fuse 1 and the motor is running in the forward direction, which loads will contribute to the measured current?

2. ______________ If Multimeter 1 is connected after Fuse 2 and the motor is running in the forward direction, which loads will contribute to the measured current?

3. ______________ If Multimeter 1 is connected after Fuse 3 and the motor is running in the forward direction, which loads will contribute to the measured current?

4. ______________ If Multimeter 2 reads 115 VAC when the reverse pushbutton is pressed, but the coil does not engage, which component (other than the coil itself) is most likely the problem?

5. ______________ What is the percentage of current unbalance at the motor?

6. ______________ Is the current unbalance acceptable?

7. ______________ What should Multimeter 4 read before any pushbutton is pressed?

8. ______________ What should Multimeter 4 read when the forward starter coil is on?

9. ______________ Will Multimeter 5 read the current to the motor?

10. ______________ What prevents both the forward and reverse coils from being energized at the same time?

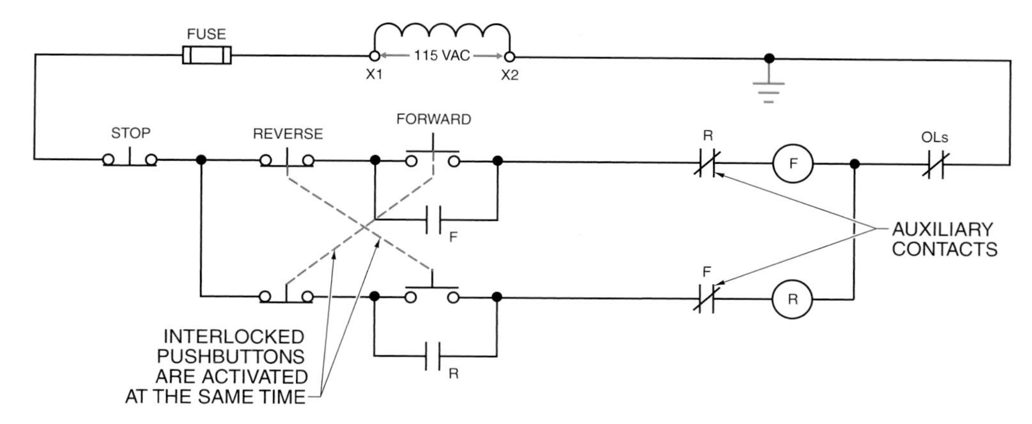

CONTROL CIRCUIT LINE DIAGRAM

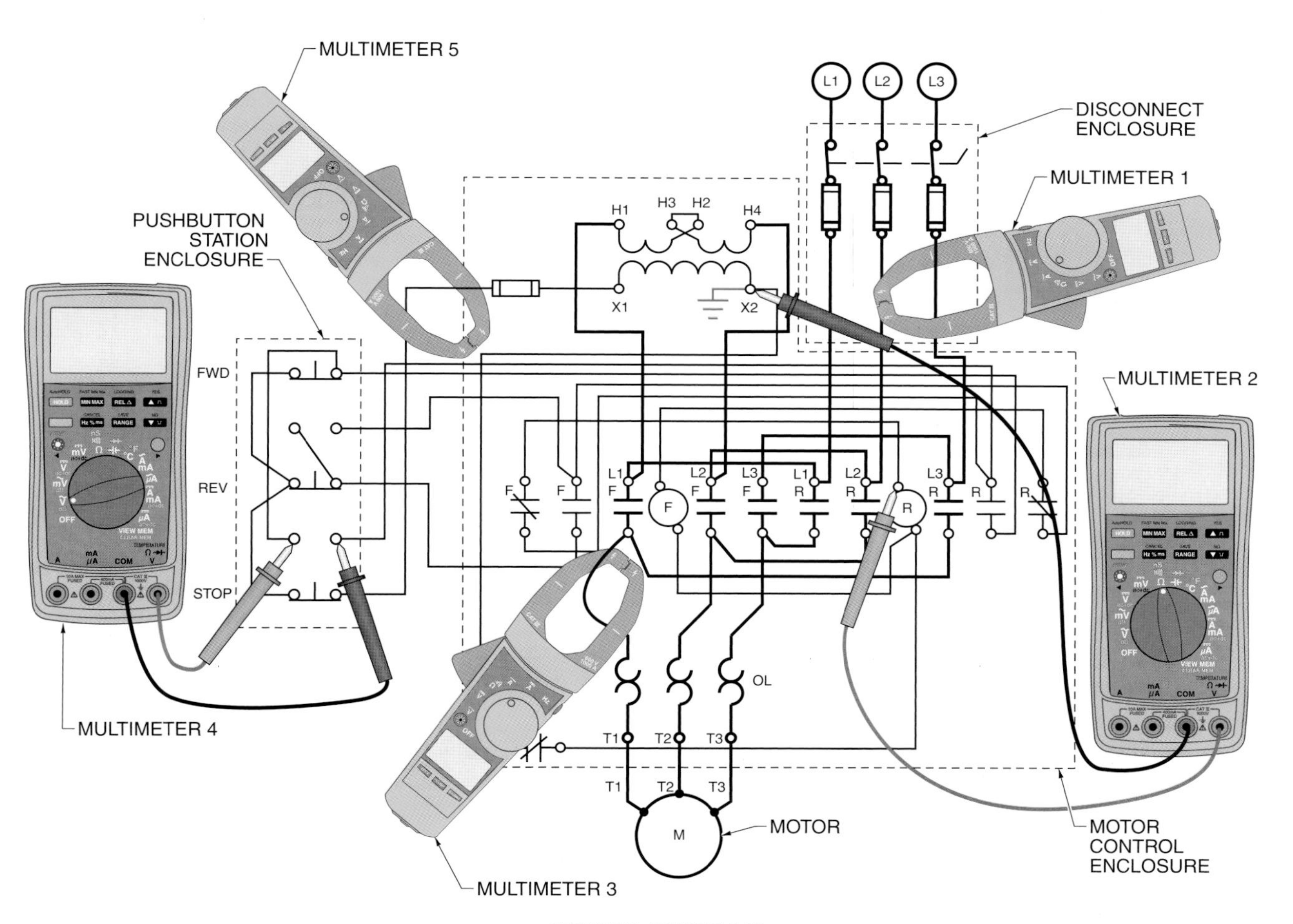

WIRING DIAGRAM

Application 10-4—Troubleshooting Terminal Strip Wired Circuits

Terminal strip wiring makes circuit troubleshooting quicker and easier because test instrument measurements can be taken directly at the terminal strip. The terminal strip is the connection point for all the components and devices in the circuit. On the line diagram, each component is assigned reference numbers to identify each connection. The wire reference numbers identify the terminals on the terminal strip where the connections are made. For example, the STOP pushbutton is connected between Terminals 1 and 3. The reverse pushbutton has four connections. The normally closed (NC) connections are made between Terminals 3 and 4 and the normally open (NO) connections are made between Terminals 7 and 8. See Line Diagram with Terminal Strip Numbers.

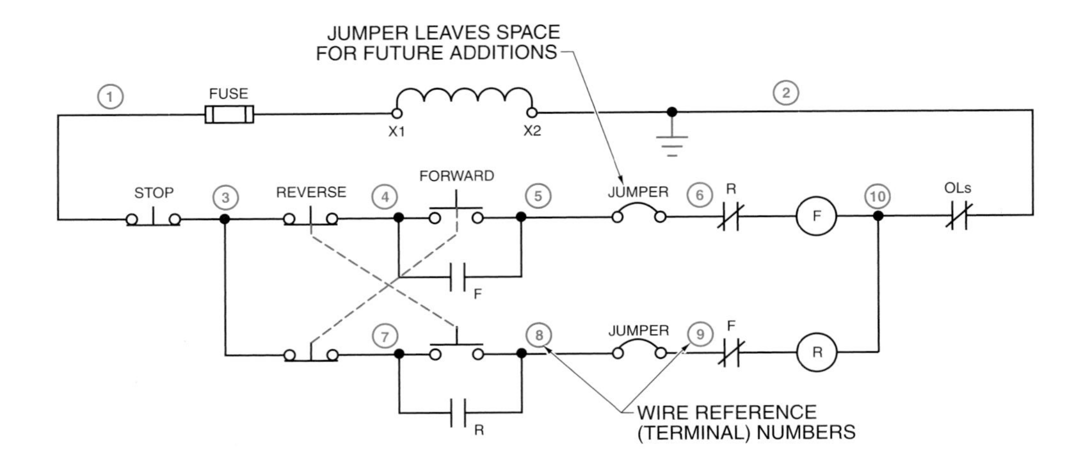

LINE DIAGRAM WITH TERMINAL STRIP NUMBERS

The terminal strip will typically use screw terminals to connect conductors. If many connections are to be made at a certain number, two or more terminals can be assigned to that number to accommodate the conductors. The like-numbered terminals are then connected with jumpers to make them electrically continuous. Jumpers can also be used between two terminal spots with different numbers when future modifications are anticipated. Later, the jumper can be removed and a component added between the two numbers.

Wire reference numbers are assigned from the top left to the bottom right. The power supply connections are usually assigned numbers "1" and "2". When troubleshooting a circuit with a terminal strip, a voltmeter is used to measure the voltage on the power supply terminals to verify the circuit is powered. If the voltage is correct, the black test lead is left on one terminal and the red test lead is moved to different terminals until the problem is located.

Using terminal strips also makes circuit modifications easier because connections are simple to add or undo. Standard options to the circuit, such as indicator lights or additional switches, may be specified with terminal strip numbers already assigned as a wiring guide. See Terminal Strip Wiring Diagram.

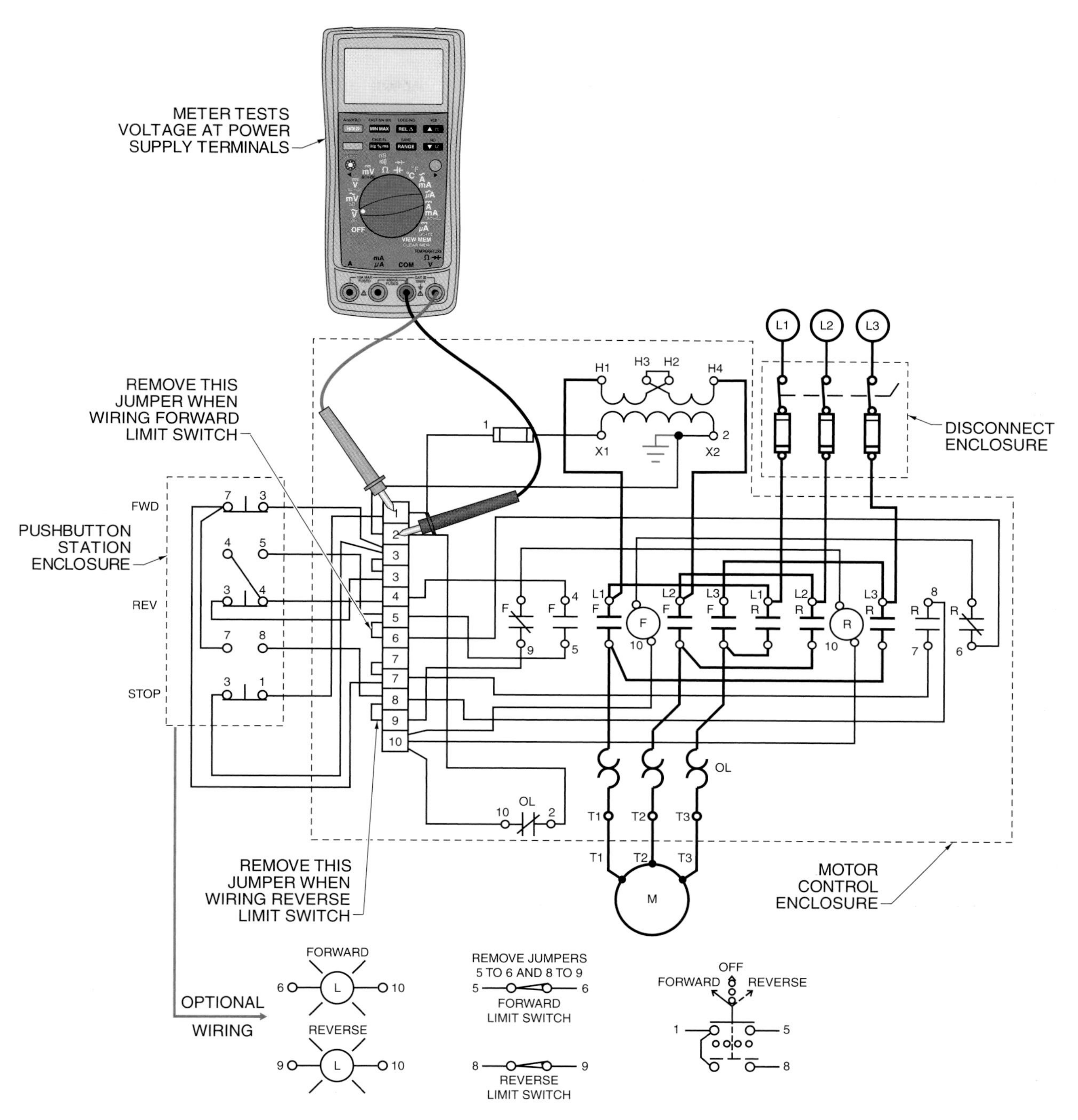

TERMINAL STRIP WIRING DIAGRAM

1. ______________ To which terminals is the STOP pushbutton connected?
2. ______________ To which terminals is the UP pushbutton connected?
3. ______________ To which terminals is the Bottom LS connected?
4. ______________ How many components connect to Terminal 6?
5. ______________ Which components are connected to Terminal 7?
6. ______________ What should Multimeter 1 read when the motor is ON?
7. ______________ What should Multimeter 2 read when the motor is OFF?
8. ______________ What should Multimeter 3 read when the motor is OFF?
9. ______________ What should Multimeter 4 read when the motor is OFF?

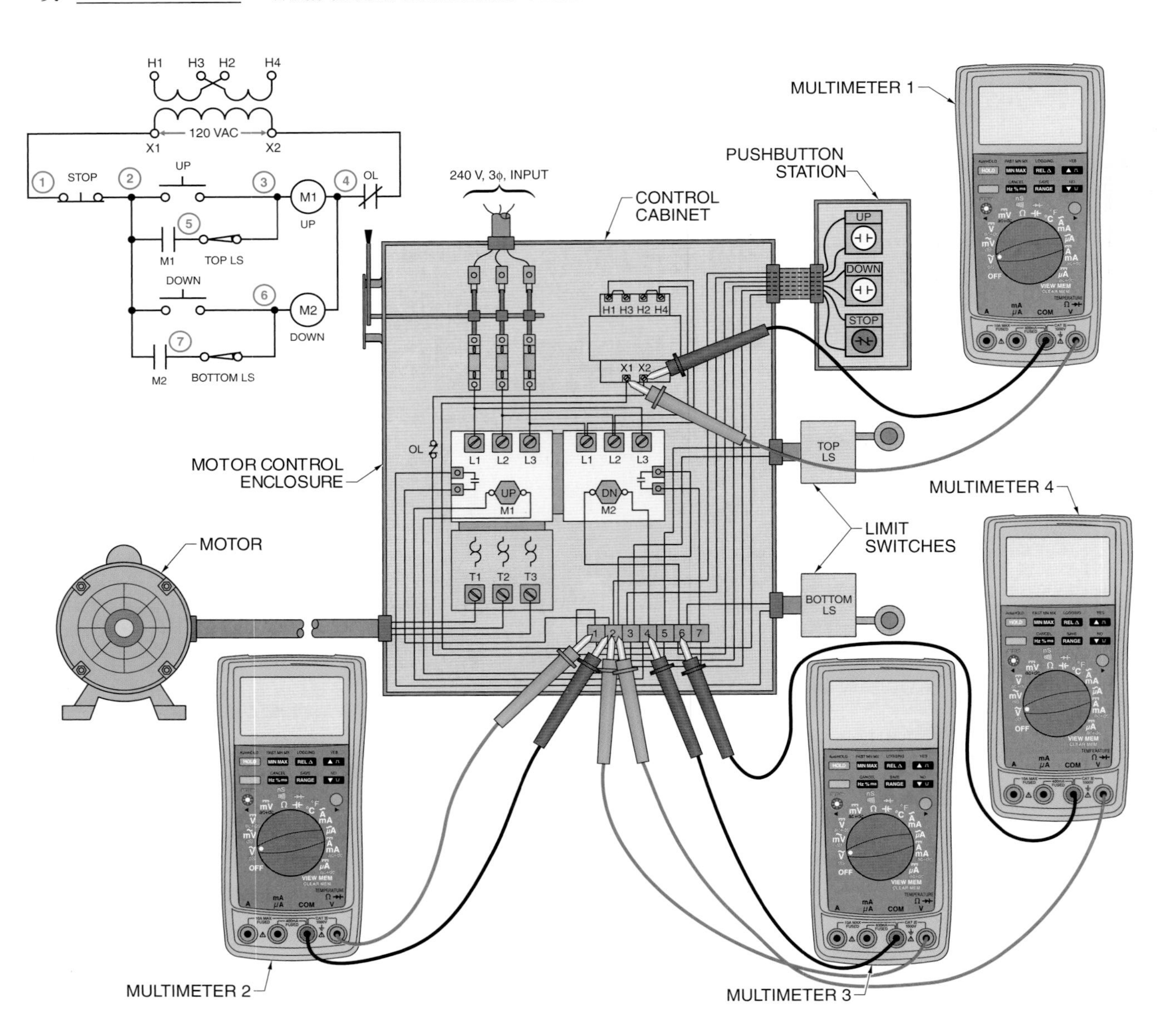

10. ______________ What should Multimeter 1 read?

11. ______________ If the control circuit fuse is blown (open), what will Multimeter 1 read?

12. ______________ What should Multimeter 2 read before any buttons are pressed?

13. ______________ What is the problem if Multimeter 3 reads 0 VAC and the fuse is good?

14. ______________ What should Multimeter 4 read before any buttons are pressed?

15. ______________ If the reverse pushbutton is pressed, what should Multimeter 4 read?

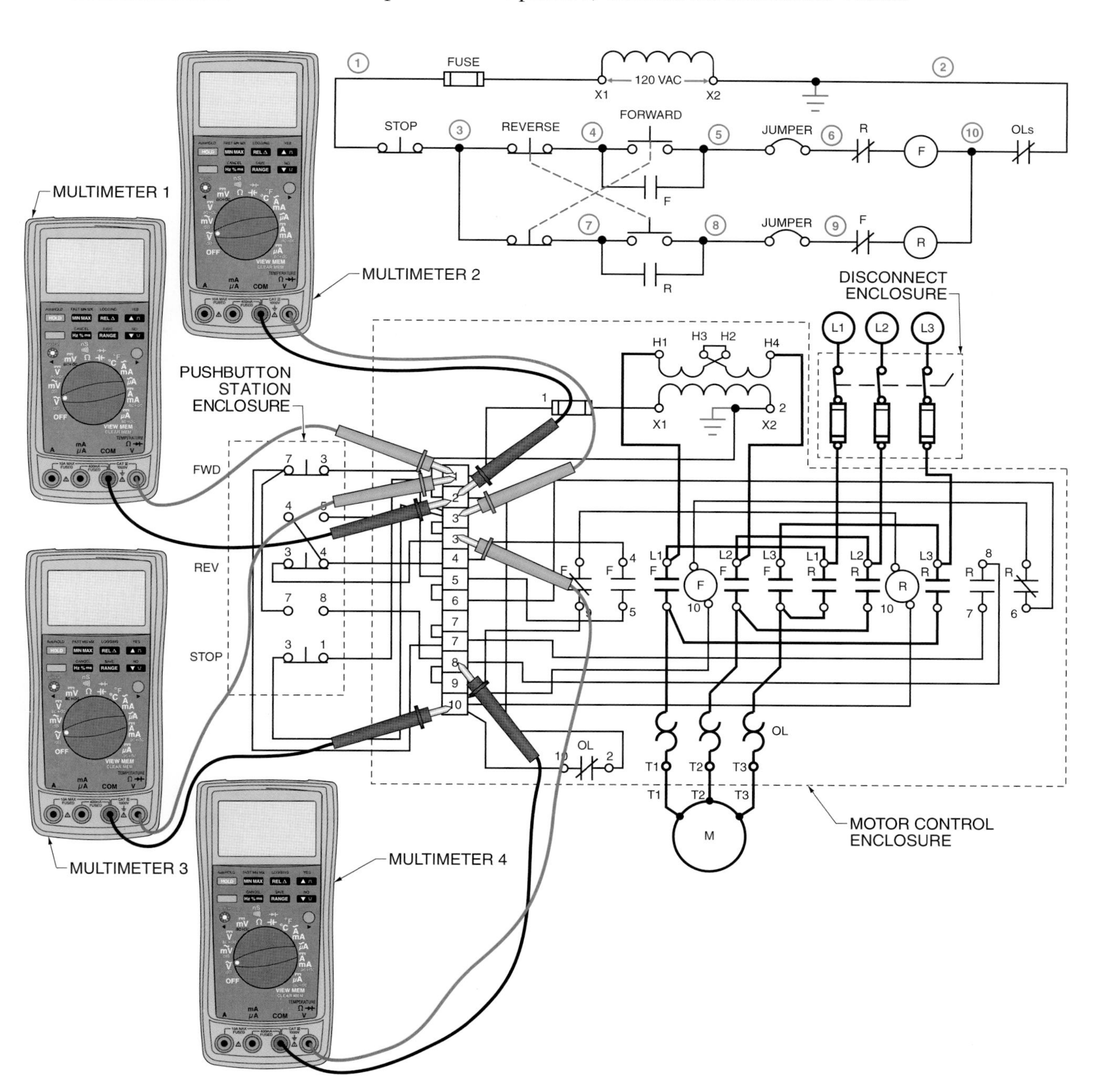

Application 10-5—Troubleshooting PLC-Wired Circuits

PLCs provide greater flexibility and monitoring capability for control circuits than hardwired or terminal strip circuits. A PLC can monitor all control functions, make decisions based on the programmed logic, and activate loads with digital (ON or OFF) or analog (variable) signals.

A PLC replaces much of a control circuit's wiring. The wiring logic becomes software logic instead, which is much easier to change. The inputs (pushbuttons, limit switches, and overload contacts) are wired directly to the PLC input module and the outputs (motor starter coils and indicator lamps) are wired directly to the PLC output module. The power circuit does not change. See PLC Control Circuit.

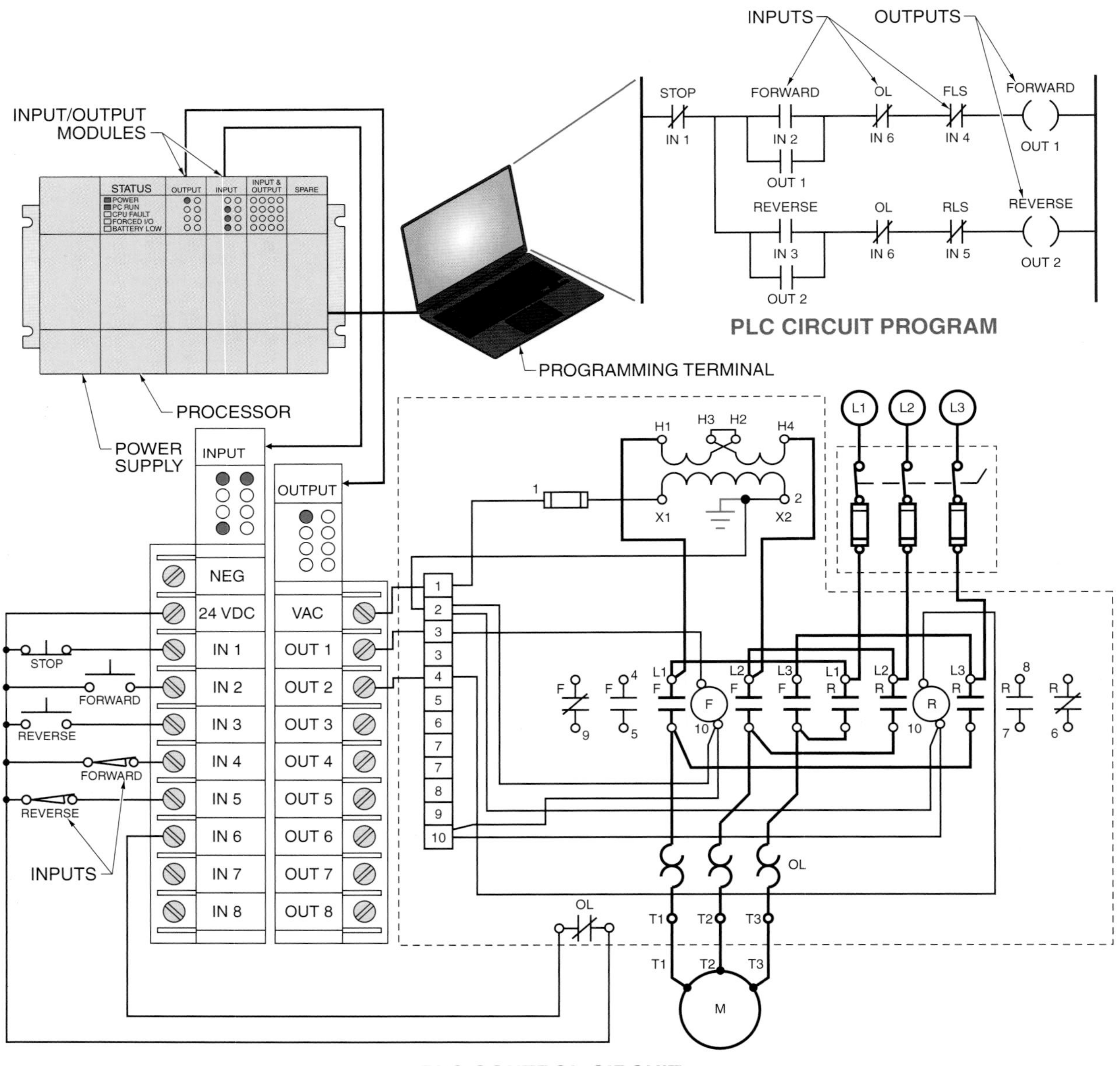

PLC CONTROL CIRCUIT

The circuit operation (logic) is programmed using PLC software and transferred to the PLC. The PLC program monitors and displays the condition (ON or OFF) of the circuit inputs and outputs. If changes in the control circuit are required, they can be reprogrammed and downloaded without changing the circuit wiring.

When programming inputs for a PLC, the actual input type (normally open or normally closed) and the way the input is programmed must be considered. When programming a PLC, a hardwired input can be programmed either normally closed or normally open, even if it is wired only one way. Often, hardwired normally closed inputs are programmed as normally open inputs. This is because PLCs will only energize outputs preceeded by programmed normally closed inputs when the input is activated (open), which is contrary to the wiring logic.

PLCs have several features that are helpful when troubleshooting. Indicator lights on the input and output modules display the current state of the inputs and outputs. If the indicator lights show an error in the control logic (an output is being activated when it should not be), the problem is probably in the PLC program. If the indicator lights show correct logic, but the components are acting unpredictably, the problem is probably in the wiring or component.

Voltage and current can be measured at the input and output modules, which are similar to terminal strips. Also, loads can be forced (activated against the program logic) ON or OFF if necessary. This can be useful for checking the operation of an output without waiting for the PLC to activate the load.

The three loads in this circuit are the motor, the forward coil, and the reverse coil.

1. ____________ Which input corresponds to the reverse pushbutton?

2. ____________ Which input corresponds to the overloads?

3. ____________ Which output corresponds to the forward coil?

4. ____________ Which input controls Output 2?

5. ____________ Which input affects both outputs?

6. ____________ What type and level of voltage is used in the input module?

7. ____________ What should Multimeter 1 read when Input 2 is activated?

8. ____________ What should Multimeter 2 read before the forward limit switch is activated?

9. ____________ Which output will be activated if Input 2 is activated?

10. ____________ Which output will be deactivated if Input 5 is activated?

11. ____________ If Input 2 is activated, what should Multimeter 3 read?

12. ____________ If Input 5 is activated, what should Multimeter 3 read?

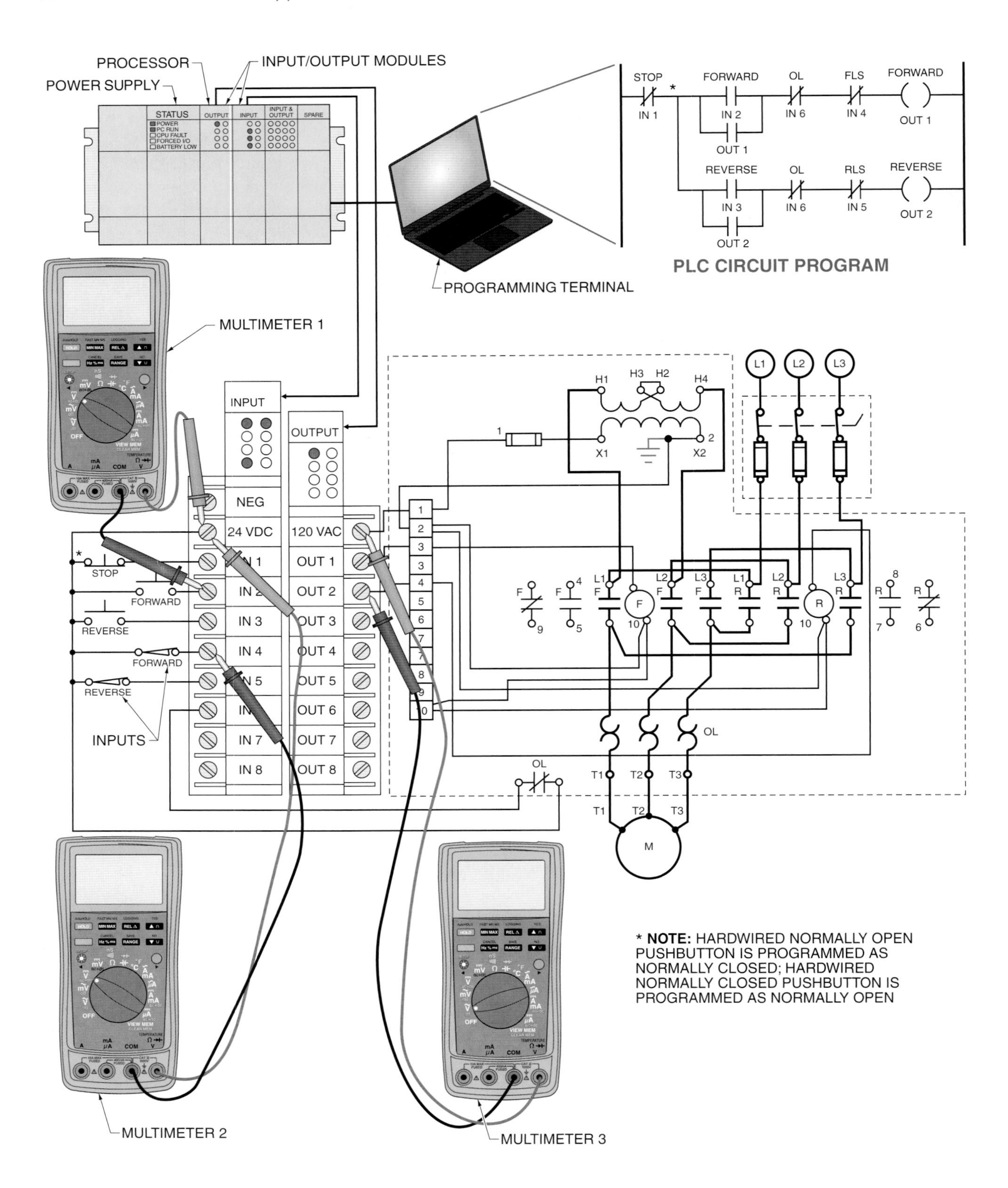
PROCESSOR
POWER SUPPLY
INPUT/OUTPUT MODULES
STATUS
OUTPUT
INPUT
INPUT & OUTPUT
SPARE
POWER
PC RUN
CPU FAULT
FORCED I/O
BATTERY LOW
PROGRAMMING TERMINAL
STOP
FORWARD
OL
FLS
FORWARD
IN 1
IN 2
IN 6
IN 4
OUT 1
OUT 1
REVERSE
OL
RLS
REVERSE
IN 3
IN 6
IN 5
OUT 2
OUT 2
PLC CIRCUIT PROGRAM
MULTIMETER 1
INPUT
OUTPUT
NEG
24 VDC
120 VAC
IN 1
IN 2
IN 3
IN 4
IN 5
IN 7
IN 8
OUT 1
OUT 2
OUT 3
OUT 4
OUT 5
OUT 6
OUT 7
OUT 8
STOP
FORWARD
REVERSE
FORWARD
REVERSE
INPUTS
H1
H3
H2
H4
X1
X2
L1
L2
L3
OL
T1
T2
T3
M
MULTIMETER 2
MULTIMETER 3
* NOTE: HARDWIRED NORMALLY OPEN PUSHBUTTON IS PROGRAMMED AS NORMALLY CLOSED; HARDWIRED NORMALLY CLOSED PUSHBUTTON IS PROGRAMMED AS NORMALLY OPEN

GENERAL APPENDIX

TEST INSTRUMENT TERMINOLOGY...		
Term	**Symbol**	**Definition**
AC		Continually changing current that reverses direction at regular intervals. Standard U.S. frequency is 60 Hz
AC COUPLING		Device that passes an AC signal and blocks a DC signal. Used to measure AC signals that are riding on a DC signal
AC/DC		Indicates ability to read or operate on alternating and direct current
ACCURACY ANALOG METER		Largest allowable error (in percent of full scale) made under normal operating conditions. The reading of a meter set on the 250 V range with an accuracy rating of ±2% could vary ±5 V. Analog meters have greater accuracy when readings are taken on the upper half of the scale
ACCURACY DIGITAL METER	 100.0 AC V	Largest allowable error (in percent of reading) made under normal operating conditions. A reading of 100.0 V on a meter with an accuracy of ±2% is between 98.0 V and 102.0 V. Accuracy may also include a specified number of digits (counts) that are added to the basic accuracy rating. For example, an accuracy of ±2% (±2 digits) means that a display reading of 100.0 V on the meter is between 97.8 V and 102.2 V
ALLIGATOR CLIP	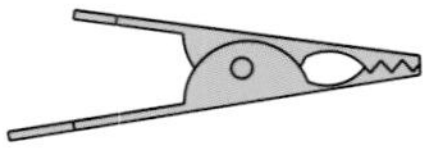	Long-jawed, spring-loaded clamp connected to the end of a test lead. Used to make temporary electrical connections
AMBIENT TEMPERATURE		Temperature of air surrounding a meter or equipment to which the meter is connected
AMMETER		Meter that measures electric current
AMMETER SHUNT		Low-resistance conductor that is connected in parallel with the terminals of an ammeter to extend the range of current values measured by the ammeter
AMPLITUDE	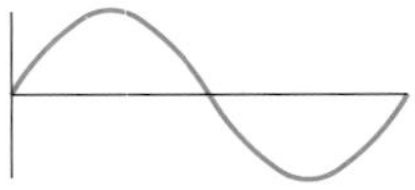	Continually changing current that reverses direction at regular intervals. Standard U.S. frequency is 60 Hz
ATTENUATION		Highest value reached by a quantity under test
AUDIBLE		Sound that can be heard
AUTORANGING		Function that automatically selects a meter's range based on signals received
AVERAGE VALUE	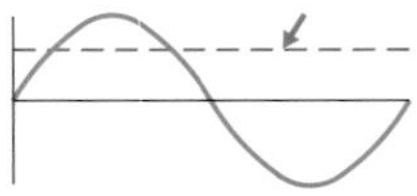	Value equal to 0.637 times the amplitude of a measured value

...TEST INSTRUMENT TERMINOLOGY...		
Term	**Symbol**	**Definition**
BACKLIGHT		Light that brightens the meter display
BANANA JACK		Meter jack that accepts a banana plug
BANANA PLUG		Long, thick terminal connection on one end of a test lead used to make a connection to a meter
BATTERY SAVE		Feature that enables a meter to shut down when battery level is too low or no key is pressed within a set time
BNC		Coaxial-type input connector used on some meters
CAPTURE		Function that records and displays measured values
CELCIUS	°C	Temperature measured on a scale for which the freezing point of water is 0° and the boiling point is 100°
CLOSED CIRCUIT		Circuit in which two or more points allow a predesigned current to flow
COUNTS		Unit of measure of meter resolution. A 1999 count meter cannot display a measurement of 1/10 of a volt when measuring 200 V or more. A 3200 count meter can display a measurement of 1/10 of a volt up to 320
DC		Current that constantly flows in one direction
DECIBEL (dB)		Measurement that indicates voltage or power comparison in logarithmic scale
DIGITS		Indication of the resolution of a meter. A 3½ digit meter can display three full digits and one half digit. The three full digits display a number from 0 to 9. The half digit displays a 1 or is left blank. A 3½ digit meter displays readings up to 1999 counts of resolution. A 4½ digit meter displays readings up to 19,999 counts of resolution
DIODE		Semiconductor that allows current to flow in only one direction
DISCHARGE		Removal of an electric charge
DUAL TRACE		Feature that allows two separate waveforms to be displayed simultaneously
EARTH GROUND		Reference point that is directly connected to ground

...TEST INSTRUMENT TERMINOLOGY...		
Term	**Symbol**	**Definition**
EFFECTIVE VALUE		Value equal to 0.707 of the amplitude of a measured quantity
FAHRENHEIT	°F	Temperature measured on a scale for which the freezing point of water is 32° and the boiling point is 212°
FREEZE		Function that holds a waveform (or measurement) for closer examination
FREQUENCY		Number of complete cycles occurring per second
FUNCTION SWITCH		Switch that selects the function (AC voltage, DC voltage, etc.) that a meter is to measure
GLITCH		Momentary spike in a waveform
GLITCH DETECT		Function that increases the meter sampling rate to maximize the detection of the glitch(es)
GROUND		Common connection to a point in a circuit whose potential is taken as zero
HARD COPY		Function that allows a printed copy of the displayed measurement
HOLD BUTTON	HOLD H	Button that allows a meter to capture and hold a stable measurement
LIQUID CRYSTAL DISPLAY (LCD)		Display that uses liquid crystals to display waveforms, measurements, and text on its screen
MEASURING RANGE		Minimum and maximum quantity that a meter can safely and accurately measure
NOISE		Unwanted extraneous electrical signals
OPEN CIRCUIT		Circuit in which two (or more) points do not provide a path for current flow
OVERFLOW		Condition of a meter that occurs when a quantity to be measured is greater than the quantity the meter can display
OVERLOAD	OL	Condition of a meter that occurs when a quantity to be measured is greater than the quantity the meter can safely handle for the meter range setting
PEAK		Highest value reached when measuring
PEAK-TO-PEAK		Highest and lowest voltage value of a waveform

...TEST INSTRUMENT TERMINOLOGY		
Term	**Symbol**	**Definition**
POLARITY		Orientation of the positive (+) and negative (–) side of direct current or voltage
PROBE		Pointed metal tip of a test lead used to make contact with the circuit under test
PULSE		Waveform that increases from a constant value, then decreases to its original value
PULSE TRAIN		Repetitive series of pulses
RANGE		Quantities between two points or levels
RECALL		Function that allows stored information (or measurements) to be displayed
RESOLUTION		Sensitivity of a meter. A meter may have a resolution of 1 V or 1 mV
RISING SLOPE		Part of a waveform displaying a rise in voltage
ROOT-MEAN-SQUARE		Value equal to 0.707 of the amplitude of a measured value
SAMPLE		Momentary reading taken from an input signal
SAMPLING RATE		Number of readings taken from a signal every second
SHORT CIRCUIT		Two or more points in a circuit that allow an unplanned current flow
TERMINAL	A mA µA COM VΩ 10A MAX FUSED 400mA MAX FUSED 1000V MAX	Point to which meter test leads are connected
TERMINAL VOLTAGE		Voltage level that meter terminals can safely handle
TRACE		Displayed waveform that shows the voltage variations of the input signal as a function of time
TRIGGER		Device which determines the starting point of a measurement
WAVEFORM		Pattern defined by an electrical signal
ZOOM		Function that allows a waveform (or part of waveform) to be magnified

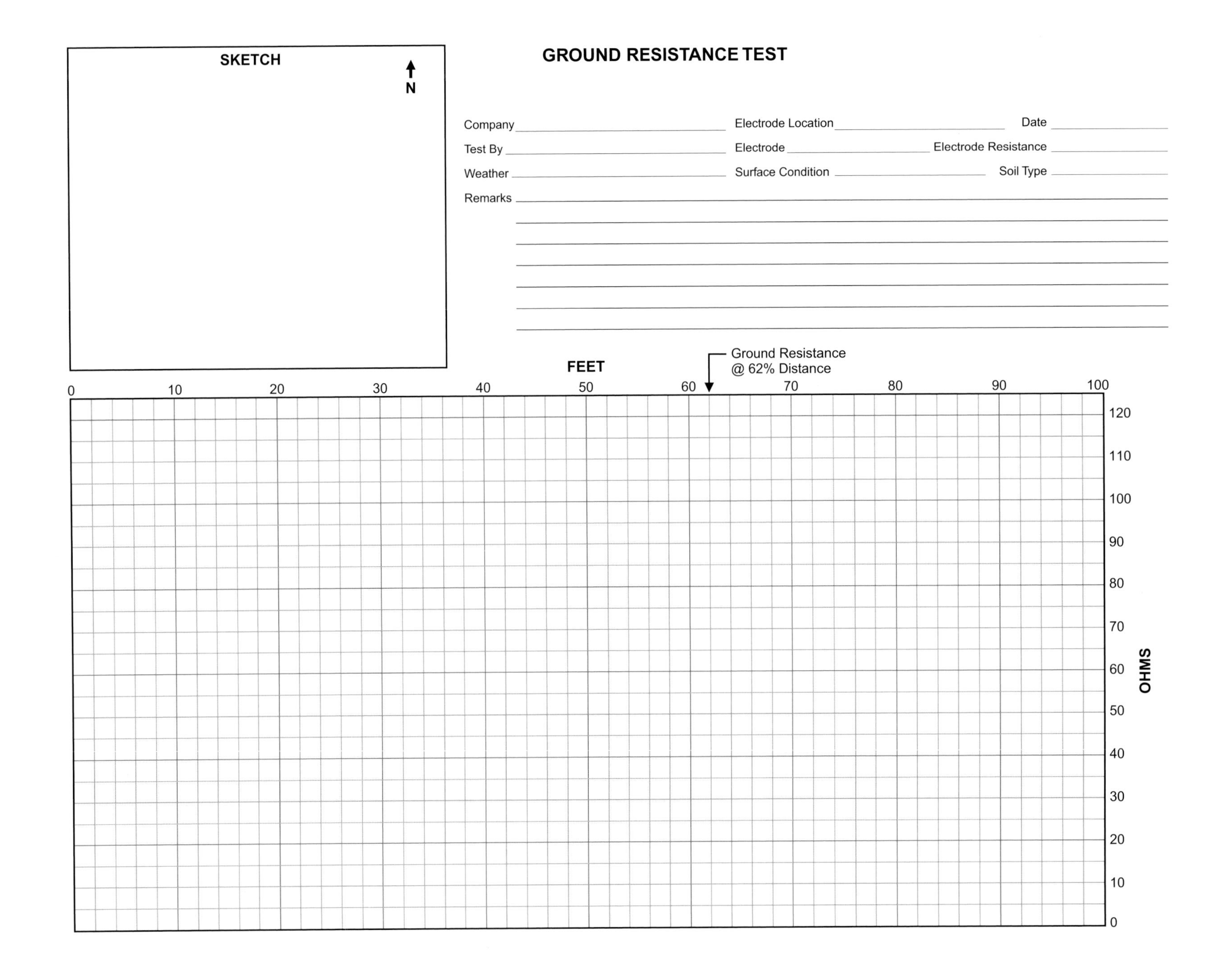

GROUND RESISTANCE TEST

Company ____________ Electrode Location ____________ Date ____________

Test By ____________ Electrode ____________ Electrode Resistance ____________

Weather ____________ Surface Condition ____________ Soil Type ____________

Remarks ____________

EXPECTED RESULTS		
Measurement/Test	**Expected Results**	**Possible Problem**
Line Voltage (rms)	110 V – 127 V	Low voltage = circuit overloaded, undersized conductor, high-resistance splice, too long of circuit run
Peak Voltage	1.41 times rms Voltage	Harmonics producing loads on circuit
Voltage Drop	< 5%	More than 5% = circuit overloaded, undersized conductor, high-resistance splice, too long of circuit run
Ground-Neutral Voltage	> 0.5 V – < 5 V	0 V = ground to neutral connection > 5 V = circuit overloaded
Ground Impedence	< 1 Ω to protect personnel < 0.25 to protect electronic equipment	Circuit overloaded, undersized conductor, high-resistance splice, too long of circuit run
GFCI Test	< 7 mA and at a time < 200 ms	Defective GFCI, improper wiring
Arc Fault Test	Breaker trips	Defective AFCI breaker, improper wiring

METRIC PREFIXES			
Multiples and Submultiples	**Prefixes**	**Symbols**	**Meaning**
$1,000,000,000,000 = 10^{12}$	tera	T	trillion
$1,000,000,000 = 10^{9}$	giga	G	billion
$1,000,000 = 10^{6}$	mega	M	million
$1000 = 10^{3}$	kilo	k	thousand
$100 = 10^{2}$	hecto	h	hundred
$10 = 10^{1}$	deka	da	ten
Base Unit $1 = 10^{0}$	—	—	—
$0.1 = 10^{-1}$	deci	d	tenth
$0.01 = 10^{-2}$	centi	c	hundredth
$0.001 = 10^{-3}$	milli	m	thousandth
$0.000001 = 10^{-6}$	micro	μ	millionth
$0.000000001 = 10^{-9}$	nano	n	billionth
$0.000000000001 = 10^{-12}$	pico	p	trillionth

METRIC CONVERSIONS												
Initial Units	**Final Units**											
	giga	**mega**	**kilo**	**hecto**	**deka**	**base unit**	**deci**	**centi**	**milli**	**micro**	**nano**	**pico**
giga		3R	6R	7R	8R	9R	10R	11R	12R	15R	18R	21R
mega	3L		3R	4R	5R	6R	7R	8R	9R	12R	15R	18R
kilo	6L	3L		1R	2R	3R	4R	5R	6R	9R	12R	15R
hecto	7L	4L	1L		1R	2R	3R	4R	5R	8R	11R	14R
deka	8L	5L	2L	1L		1R	2R	3R	4R	7R	10R	13R
base unit	9L	6L	3L	2L	1L		1R	2R	3R	6R	9R	12R
deci	10L	7L	4L	3L	2L	1L		1R	2R	5R	8R	11R
centi	11L	8L	5L	4L	3L	2L	1L		1R	4R	7R	10R
milli	12L	9L	6L	5L	4L	3L	2L	1L		3R	6R	9R
micro	15L	12L	9L	8L	7L	6L	5L	4L	3L		3R	6R
nano	18L	15L	12L	11L	10L	9L	8L	7L	6L	3L		3R
pico	21L	18L	15L	14L	13L	12L	11L	10L	9L	6L	3L	

COMMON PREFIXES		
Symbol	**Prefix**	**Equivalent**
G	giga	1,000,000,000
M	mega	1,000,000
k	kilo	1000
base unit	—	1
m	milli	0.001
μ	micro	0.000001
n	nano	0.000000001
p	pico	0.000000000001

ELECTRICAL/ELECTRONIC ABBREVIATIONS/ACRONYMS

Abbr/ Acronym	Meaning	Abbr/ Acronym	Meaning	Abbr/ Acronym	Meaning
A	Ammeter; Ampere; Anode; Armature	FU	Fuse	PNP	Positive-Negative-Positive
AC	Alternating Current	FWD	Forward	POS	Positive
AC/DC	Alternating Current; Direct Current	G	Gate; Giga; Green; Conductance	POT.	Potentiometer
A/D	Analog to Digital	GEN	Generator	P-P	Peak-to-Peak
AF	Audio Frequency	GRD	Ground	PRI	Primary Switch
AFC	Automatic Frequency Control	GY	Gray	PS	Pressure Switch
Ag	Silver	H	Henry; High Side of Transformer; Magnetic Flux	PSI	Pounds per Square Inch
ALM	Alarm			PUT	Pull-Up Torque
AM	Ammeter; Amplitude Modulation	HF	High Frequency	Q	Transistor
AM/FM	Amplitude Modulation; Frequency Modulation	HP	Horsepower	R	Radius; Red; Resistance; Reverse
		Hz	Hertz	RAM	Random-Access Memory
ARM.	Armature	I	Current	RC	Resistance-Capacitiance
Au	Gold	IC	Integrated Circuit	RCL	Resistance-Inductance-Capacitance
AU	Automatic	INT	Intermediate; Interrupt	REC	Rectifier
AVC	Automatic Volume Control	INTLK	Interlock	RES	Resistor
AWG	American Wire Gauge	IOL	Instantaneous Overload	REV	Reverse
BAT.	Battery (electric)	IR	Infrared	RF	Radio Frequency
BCD	Binary Coded Decimal	ITB	Inverse Time Breaker	RH	Rheostat
BJT	Bipolar Junction Transistor	ITCB	Instantaneous Trip Circuit Breaker	rms	Root Mean Square
BK	Black	JB	Junction Box	ROM	Read-Only Memory
BL	Blue	JFET	Junction Field-Effect Transistor	rpm	Revolutions per Minute
BR	Brake Relay; Brown	K	Kilo; Cathode	RPS	Revolutions per Second
C	Celsius; Capacitiance; Capacitor	L	Line; Load; Coil; Inductance	S	Series; Slow; South; Switch
CAP.	Capacitor	LB-FT	Pounds per Foot	SCR	Silicon Controlled Rectifier
CB	Circuit Breaker; Citizen's Band	LB-IN.	Pounds per Inch	SEC	Secondary
CC	Common-Collector Configuration	LC	Inductance-Capacitance	SF	Service Factor
CCW	Counterclockwise	LCD	Liquid Crystal Display	1 PH; 1ϕ	Single-Phase
CE	Common-Emitter Configuration	LCR	Inductance-Capacitance-Resistance	SOC	Socket
CEMF	Counter Electromotive Force	LED	Light-Emitting Diode	SOL	Solenoid
CKT	Circuit	LRC	Locked Rotor Current	SP	Single-Pole
CONT	Continuous; Control	LS	Limit Switch	SPDT	Single-Pole, Double-Throw
CPS	Cycles per Second	LT	Lamp	SPST	Single-Pole, Single-Throw
CPU	Central Processing Unit	M	Motor; Motor Starter; Motor Starter Contacts	SS	Selector Switch
CR	Control Relay			SSW	Safety Switch
CRM	Control Relay Master	MAX.	Maximum	SW	Switch
CT	Current Transformer	MB	Magnetic Brake	T	Tera; Terminal; Torque; Transformer
CW	Clockwise	MCS	Motor Circuit Switch	TB	Terminal Board
D	Diameter; Diode; Down	MEM	Memory	3 PH; 3ϕ	Three-Phase
D/A	Digital to Analog	MED	Medium	TD	Time Delay
DB	Dynamic Braking Contactor; Relay	MIN	Minimum	TDF	Time Delay Fuse
DC	Direct Current	MN	Manual	TEMP	Temperature
DIO	Diode	MOS	Metal-Oxide Semiconductor	THS	Thermostat Switch
DISC.	Disconnect Switch	MOSFET	Metal-Oxide Semiconductor Field-Effect Transistor	TR	Time Delay Relay
DMM	Digital Multimeter			TTL	Transistor-Transistor Logic
DP	Double-Pole	MTR	Motor	U	Up
DPDT	Double-Pole, Double-Throw	N; NEG	North; Negative	UCL	Unclamp
DPST	Double-Pole, Single-Throw	NC	Normally Closed	UHF	Ultrahigh Frequency
DS	Drum Switch	NEUT	Neutral	UJT	Unijunction Transistor
DT	Double-Throw	NO	Normally Open	UV	Ultraviolet; Undervoltage
DVM	Digital Voltmeter	NPN	Negative-Positive-Negative	V	Violet; Volt
EMF	Electromotive Force	NTDF	Nontime-Delay Fuse	VA	Volt Amp
F	Fahrenheit; Fast; Field; Forward; Fuse	O	Orange	VAC	Volts Alternating Current
FET	Field-Effect Transistor	OCPD	Overcurrent Protection Device	VDC	Volts Direct Current
FF	Flip-Flop	OHM	Ohmmeter	VHF	Very High Frequency
FLC	Full-Load Current	OL	Overload Relay	VLF	Very Low Frequency
FLS	Flow Switch	OZ/IN.	Ounces per Inch	VOM	Volt-Ohm-Milliammeter
FLT	Full-Load Torque	P	Peak; Positive; Power; Power Consumed	W	Watt; White
FM	Fequency Modulation	PB	Pushbutton	w/	With
FREQ	Frequency	PCB	Printed Circuit Board	X	Low Side of Transformer
FS	Float Switch	PH;	Phase	Y	Yellow
FTS	Foot Switch	PLS	Plugging Switch	Z	Impedance

THREE-PHASE VOLTAGE VALUES
For 208 V × 1.732, use 360
For 230 V × 1.732, use 398
For 240 V × 1.732, use 416
For 440 V × 1.732, use 762
For 460 V × 1.732, use 797
For 480 V × 1.732, use 831
For 2400 V × 1.732, use 4157
For 4160 V × 1.732, use 7205

POWER FORMULA ABBREVIATIONS AND SYMBOLS	
P = Watts	V = Volts
I = Amps	VA = Volt Amps
A = Amps	ϕ = Phase
R = Ohms	√‾ = Square Root
E = Volts	

OHM'S LAW AND POWER FORMULA

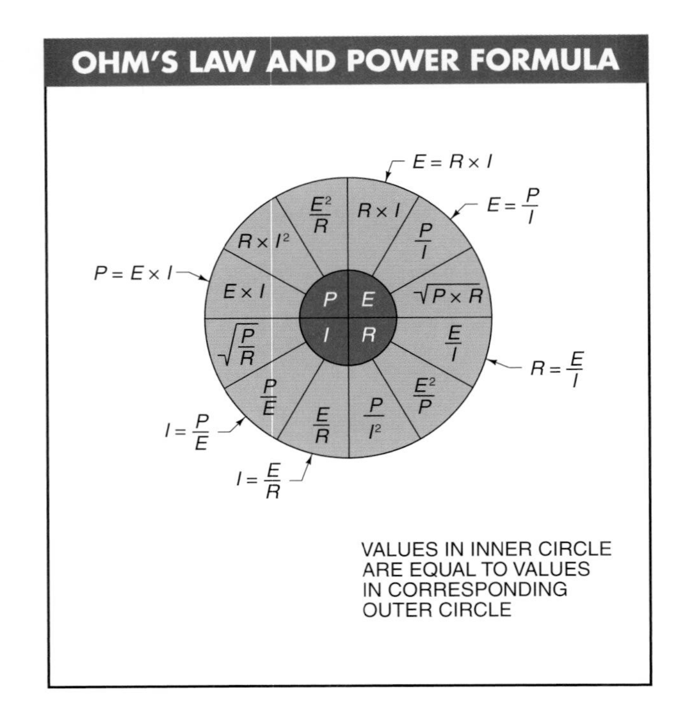

POWER FORMULAS – 1ϕ, 3ϕ					
Phase	To Find	Use Formula	Example		
			Given	Find	Solution
1ϕ	I	$I = \frac{VA}{V}$	32,000 VA, 240V	I	$I = \frac{VA}{V}$ $I = \frac{32,000\ VA}{240\ V}$ $I =$ **133 A**
1ϕ	VA	$VA = I \times V$	100 A, 240 V	VA	$VA = I \times V$ $VA = 100\ A \times 240\ V$ $VA =$ **24,000 VA**
1ϕ	V	$V = \frac{VA}{I}$	42,000 VA, 350 A	V	$V = \frac{VA}{I}$ $V = \frac{42,000\ VA}{350\ A}$ $V =$ **120 V**
3ϕ	I	$I = \frac{VA}{V \times \sqrt{3}}$	72,000 VA, 208 V	I	$I = \frac{VA}{V \times \sqrt{3}}$ $I = \frac{72,000\ VA}{360\ V}$ $I =$ **200 A**
3ϕ	VA	$VA = I \times V \times \sqrt{3}$	2 A, 240V	VA	$VA = I \times V \times \sqrt{3}$ $VA = 2 \times 416$ $VA =$ **832 VA**

NEMA ENCLOSURE CLASSIFICATION

Type	Use	Service Conditions	Tests	Comments
1	Indoor	No unusual	Rod entry, rust resistance	
3	Outdoor	Windblown dust, rain, sleet, and ice on enclosure	Rain, external icing, dust, and rust resistance	Does not provide protection against internal condensation or internal icing
3R	Outdoor	Falling rain and ice on enclosure	Rod entry, rain, external icing, and rust resistance	Does not provide protection against dust, internal condensation, or internal icing
4	Indoor/outdoor	Windblown dust and rain, splashing water, hose-directed water, and ice on enclosure	Hosedown, external icing, and rust resistance	Does not provide protection against internal condensation or internal icing
4X	Indoor/outdoor	Corrosion, windblown dust and rain, splashing water, hose-directed water, and ice on enclosure	Hosedown, external icing, and corrosion resistance	Does not provide protection against internal condensation or internal icing
6	Indoor/outdoor	Occasional temporary submersion at a limited depth		
6P	Indoor/outdoor	Prolonged submersion at a limited depth		
7	Indoor locations classified as Class I, Groups A, B, C, or D, as defined in the NEC®	Withstand and contain an internal explosion of specified gases, contain an explosion of specified gases, contain an explosion sufficiently so an explosive gas-air mixture in the atmosphere is not ignited	Explosion, hydrostatic, and temperature	Enclosed heat-generating devices shall not cause external surfaces to reach temperatures capable of igniting explosive gas-air mixtures in the atmosphere
9	Indoor locations classified as Class II, Groups E or G, as defined in the NEC®	Dust	Dust penetration, temperature, and gasket aging	Enclosed heat-generating devices shall not cause external surfaces to reach temperatures capable of igniting explosive gas-air mixtures in the atmosphere
12	Indoor	Dust, falling dirt, and dripping noncorrosive liquids	Drip, dust, and rust resistance	Does not provide protection against internal condensation
13	Indoor	Dust, spraying water, oil, and noncorrosive coolant	Oil explosion and rust resistance	Does not provide protection against internal condensation

Type
1
4
4X
7
9
12

IEC ENCLOSURE CLASSIFICATION

IEC Publication 529 describes standard degrees of protection that enclosures of a product must provide when properly installed. The degree of protection is indicated by two letters, IP, and two numerals. International Standard IEC 529 contains descriptions and associated test requirements to define the degree of protection that each numeral specifies. The following table indicates the general degrees of protection. For complete test requirements refer to IEC 529.

First Numeral*†	Second Numeral*†
Protection of persons against access to hazardous parts and protection against penetration of solid foreign objects.	Protection against liquids‡ under test conditions specified in IEC 529.
0 Not protected	**0** Not protected
1 Protection against objects greater than 50 mm in diameter (hands)	**1** Protection against vertically falling drops of water (condensation)
2 Protection against objects greater than 12.5 mm in diameter (fingers)	**2** Protection against falling water with enclosure tilted 15°
3 Protection against objects greater than 2.5 mm in diameter (tools, wires)	**3** Protection against spraying of falling water with enclosure tilted 60°
4 Protection against objects greater than 1.0 mm in diameter (tools, small wires)	**4** Protection against splashing water
5 Protection against dust (dust may enter during test but must not interfere with equipment operation or impair safety)	**5** Protection against low-pressure water jets
6 Dusttight (no dust observable inside enclosure at end of test)	**6** Protection against powerful water jets
	7 Protection against temporary submersion
	8 Protection against continuous submersion

Example: IP41 describes an enclosure that is designed to protect against the entry of tools or objects greater than 1 mm in diameter, and to protect against vertically dripping water under specified test conditions.

* All first and second numerals up to and including numeral 6 imply compliance with the requirements of all preceding numerals in their respective series. Second numerals 7 and 8 do not imply suitability for exposure to water jets unless dual coded; e.g., IP_5/IP_7

† The IEC permits use of certain supplementary letters with the characteristic numerals. If such letters are used, refer to IEC 529 for an explanation.

‡ The IEC test requirements for degrees of protection against liquid ingress refer only to water

HARMONICS CLASSIFICATION

Harmonics	Frequency*	Sequence
Fundamental (1st)	60	Positive (+)
2nd	120	Negative (–)
3rd	180	Zero (0)
4th	240	(+)
5th	300	(–)
6th	360	(0)
7th	420	(+)
8th	480	(–)
9th	540	(0)
10th	600	(+)

* in Hz

TRANSFORMER DERATINGS

Maximum Ambient Temperature*	Maximum Transformer Loading**
40	100
45	96
50	92
55	88
60	81
65	80
70	76

* in °C
** in %

CONDUCTOR COLOR CODING COMBINATIONS

Voltage*	Circuit	Conductor Colors
120	1ϕ, 2-wire with ground	One black (hot wire), one white (neutral wire), and one green (ground wire)
120/240	1ϕ, 3-wire with ground	One black (one hot wire), one red (other hot wire), one white (neutral wire), and one green (ground wire)
120/208	3ϕ, 4-wire wye with ground	One black (phase 1 hot wire), one red (phase 2 hot wire), one blue (phase 3 hot wire), one white (neutral wire), and one green (ground wire)
240	3ϕ, 3-wire delta with ground	One black (phase 1 hot wire), one red (phase 2 hot wire), one blue (phase 3 hot wire), and one green (ground wire)
120/240	3ϕ, 4-wire delta with ground	One black (first low phase hot wire), one red (second low phase hot wire), one orange (high phase leg wire), one white (neutral wire), and one green (ground wire)
277/480	3ϕ, 4-wire wye with ground	One brown (phase 1 hot wire), one orange (phase 2 hot wire), one yellow (phase 3 hot wire), one white (neutral wire), and one green (ground wire)
480	3ϕ, 3-wire delta with ground	One brown (phase 1 hot wire), one orange (phase 2 hot wire), one yellow (phase 3 hot wire), and one green (ground wire)

SELECTED DMM SYMBOLS

Symbol	Meaning	Symbol	Meaning	Symbol	Meaning
~	AC		See service manual	○	Switch position OFF (power)
⎓	DC	⧈	Double insulation	|	Switch position ON (power)
	AC or DC		Fuse	⊙	Manual Range mode
+	Positive		Battery		Warning: Dangerous or high voltage that could result in personal injury
–	Negative	H	Hold	⚠	Caution: Hazard that could result in equipment damage or personal injury
⏚	Ground	)))))	Audio beeper	1000 V MAX	Terminals must not be connected to a circuit with higher than listed voltage
±	Plus or minus	⊣⊢	Capacitor	△	Relative mode – displayed value is difference between present measurement and previous stored measurement
▶|	Diode	%	Percent	Ω	Ohms resistance
▶|)))))	Diode Test	▷	Move right		Meter display light
<	Less than	◁	Move left		
>	Greater than	⊘	No (do not use)		
△	Increase setting				
▽	Decrease setting				

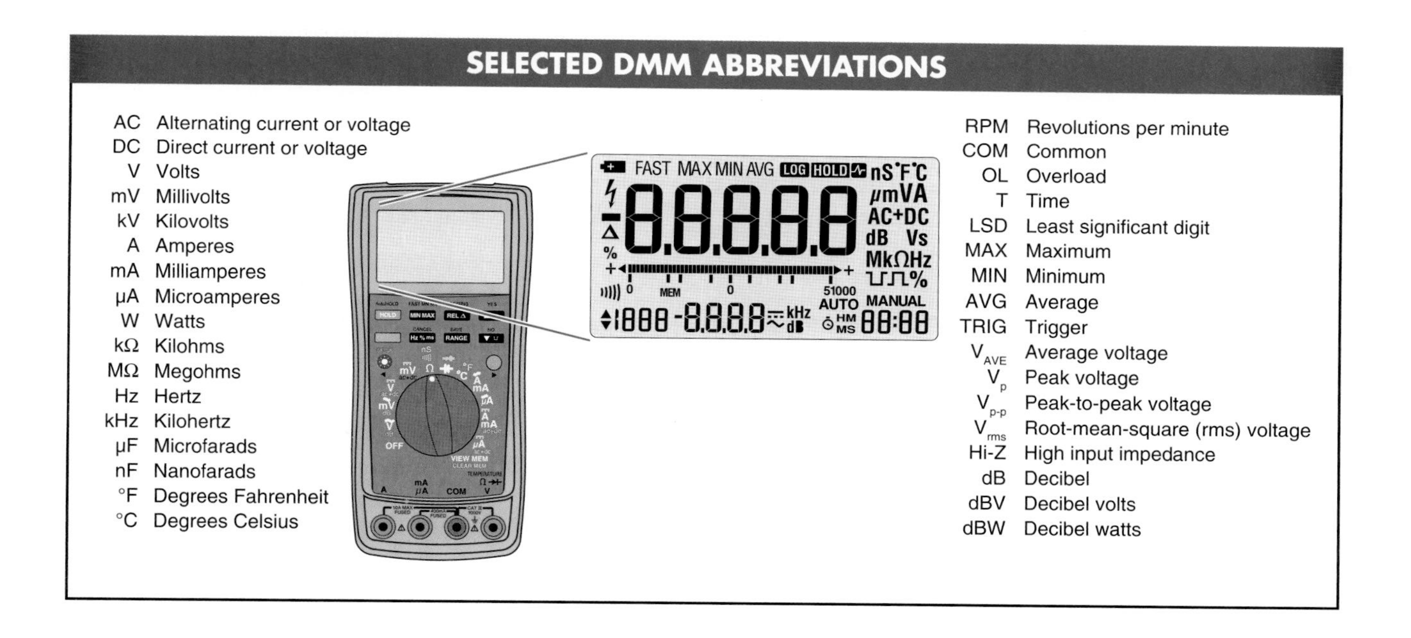

SELECTED DMM ABBREVIATIONS

Abbreviation	Meaning
AC	Alternating current or voltage
DC	Direct current or voltage
V	Volts
mV	Millivolts
kV	Kilovolts
A	Amperes
mA	Milliamperes
μA	Microamperes
W	Watts
kΩ	Kilohms
MΩ	Megohms
Hz	Hertz
kHz	Kilohertz
μF	Microfarads
nF	Nanofarads
°F	Degrees Fahrenheit
°C	Degrees Celsius
RPM	Revolutions per minute
COM	Common
OL	Overload
T	Time
LSD	Least significant digit
MAX	Maximum
MIN	Minimum
AVG	Average
TRIG	Trigger
V_{AVE}	Average voltage
V_p	Peak voltage
V_{p-p}	Peak-to-peak voltage
V_{rms}	Root-mean-square (rms) voltage
Hi-Z	High input impedance
dB	Decibel
dBV	Decibel volts
dBW	Decibel watts

TROUBLESHOOTING POWER QUALITY PROBLEMS

Power Quality Problem	Possible Effects on Load or System	UPS	Line Voltage Regulator	Transient (Surge) Suppressor	Harmonic Filters	Line Filters	Power Line Conditioner	K-Rated Transformer	Isolation Transformer	Zig-Zag Transformer	Proper Wiring and Grounding
Sustained Power Interruption (Decrease to 0 V for longer than 1 min)	Downtime, lost data, loss of heating/cooling, security/safety problems	●									
Temporary Power Interruption (Decrease to 0 V for 3 sec to 1 min)	Short shutdowns, lost data, need to restart/reset equipment and processes	●									
Momentary Power Interruption (Decrease to 0 V for .5 cycles to 3 sec)	Data errors, lost data, may need to restart/reset equipment and processes,lamps flicker, motor circuits trip OFF	●	●				●				
Sags (Decrease of 10% rated voltage or more, but not 0 V, lasting .5 cycles to 1 min)	Lost data, lamps flicker or HID lamp restarts, false alarms		●				●				●
Swells (increase of 10% rated voltage or more lasting .5 cycles to 1 min)	Accelerate equipment failure, lost data		●				●				
Undervoltage (Decrease of 10% rated voltage or more, but not 0 V, lasting longer than 1 min)	Lost data, lamps flicker or HID lamp restarts, false alarms		●				●				●
Overvoltage (Increase of 10% rated voltage or more, lasting longer than 1 min)	Insulation life expectancy reduced, fire hazard from overheated equipment, equipment damage, fast bulb burnouts		●				●				
Transients (Temporary, unwanted voltage on power lines)	Lost data, data errors, premature equipment failure			●			●		●		●
Harmonics (Multiple of the fundamental frequency of 60 Hz)	Fire hazard from overheated neutrals, transformers, and connectors. Lost data, equipment damage, data errors, false CB tripping, capacitor fuses blowing (from resonance)				●		●	●	●	●	●
Noise (Unwanted low level signal on lines)	Incorrect signals sent to input modules (PLCs, instrumentation, etc.), erratic equipment behavior, lost data, data errors					●	●		●		●

POWERS OF 10

1×10^4	=	10,000	=	$10 \times 10 \times 10 \times 10$	Read ten to the fourth power
1×10^3	=	1000	=	$10 \times 10 \times 10$	Read ten to the third power or ten cubed
1×10^2	=	100	=	10×10	Read ten to the second power or ten squared
1×10^1	=	10	=	10	Read ten to the first power
1×10^0	=	1	=	1	Read ten to the zero power
1×10^{-1}	=	0.1	=	1/10	Read ten to the minus first power
1×10^{-2}	=	0.01	=	$1/(10 \times 10)$ or 1/100	Read ten to the minus second power
1×10^{-3}	=	0.001	=	$1/(10 \times 10 \times 10)$ or 1/1000	Read ten to the minus third power
1×10^{-4}	=	0.0001	=	$1/(10 \times 10 \times 10 \times 10)$ or 1/10,000	Read ten to the minus fourth power

UNITS OF ENERGY

Energy	Btu	ft lb	J	kcal	kWh
British thermal unit	1	777.9	1.056	0.252	2.930×10^{-4}
Foot-pound	1.285×10^{-3}	1	1.356	3.240×10^{-4}	3.766×10^{-7}
Joule	9.481×10^{-4}	0.7376	1	2.390×10^{-4}	2.778×10^{-7}
Kilocalorie	3.968	3.086	4.184	1	1.163×10^{-3}
Kilowatt-hour	3.413	2.655×10^6	3.240×10^{-4}	860	1

UNITS OF POWER

Power	W	ft lbs/s	HP	kW
Watt	1	0.7376	0.341×10^{-3}	0.001
Foot-pound/sec	1.356	1	0.818×10^{-3}	1.356×10^{-3}
Horsepower	745.7	550	1	0.7457
Kilowatt	1000	736.6	1.341	1

STANDARD SIZES OF FUSES AND CBs

NEC® 240.6 (a) lists standard ampere ratings of fuses and fixed-trip CBs as follows:

15, 20, 25, 30, 35, 40, 45, 50, 60, 70, 80, 90, 100, 110, 125, 150, 175, 200, 225, 250, 300, 350, 400, 450, 500,600, 700, 800,1000, 1200, 1600, 2000, 2500, 3000, 4000, 5000, 6000

VOLTAGE CONVERSIONS

To Convert	To	Multiply By
rms	Average	0.9
rms	Peak	1.414
Average	rms	1.111
Average	Peak	1.567
Peak	rms	0.707
Peak	Average	0.637
Peak	Peak-to-peak	2

CAPACITORS

Connected in Series		Connected in Parallel	Connected in Series/Parallel
Two Capacitors	**Three or More Capacitors**		
$C_T = \frac{C_1 \times C_2}{C_1 + C_2}$ where C_T = total capacitance (in µF) C_1 = capacitance of capacitor 1 (in µF) C_2 = capacitance of capacitor 2 (in µF)	$\frac{1}{C_T} = \frac{1}{C_1} + \frac{1}{C_2} + \ldots$	$C_T = C_1 + C_2 + \ldots$	1. Calculate the capacitance of the parallel branch. 2. Calculate the capacitance of the series combinaltion. $C_T = \frac{C_1 \times C_2}{C_1 + C_2}$

SINE WAVES

Frequency	Period	Peak-to-Peak Value
$f = \frac{1}{T}$ where f = frequency (in hertz) 1 = constant T = period (in seconds)	$T = \frac{1}{f}$ where T = period (in seconds) 1 = constant f = frequency (in hertz)	$V_{p-p} = 2 \times V_{max}$ where 2 = constant V_{p-p} = peak-to-peak value V_{max} = peak value

Average Value	rms Value
$V_{avg} = V_{max} \times 0.637$ where V_{avg} = average value (in volts) V_{max} = peak value (in volts) 0.637 = constant	$V_{rms} = V_{max} \times 0.707$ where V_{rms} = rms value (in volts) V_{max} = peak value (in volts) 0.707 = constant

CONDUCTIVE LEAKAGE CURRENT

$$I_L = \frac{V_A}{R_I}$$

where
I_L = leakage current (in microamperes)
V_A = applied voltage (in volts)
R_I = insulation resistance (in megohms)

TEMPERATURE CONVERSIONS

Convert °C to °F	Convert °F to °C
$°F = (1.8 \times °C) + 32$	$°C = \frac{(°F - 32)}{1.8}$

FLOW RATE

$$Q = \frac{Q \times V_d}{231}$$

where
Q = flow rate (in gpm)
N = pump drive speed (in rpm)
V_d = pump displacement (in cu in./rev)
231 = constant

BRANCH CIRCUIT VOLTAGE DROP

$$\%V_D = \frac{V_{NL} - V_{FL}}{V_{FL}} \times 100$$

where
$\%V_D$ = percent voltage drop (in volts)
V_{NL} = no-load voltage drop (in volts)
V_{FL} = full-load voltage drop (in volts)
100 = constant

GLOSSARY

A

abbreviation: A letter or combination of letters that represents a word.

absolute pressure: Pressure measured from perfect vacuum, which is the zero point on the pressure scale.

absolute zero: The lowest theoretical temperature that a substance can reach, which corresponds to –273.16° on the Celsius scale and –459.6° on the Fahrenheit scale.

acronym: A word formed from the first letters of a compound term.

AC voltage: Voltage in a circuit that has current that reverses its direction of flow at regular intervals.

aerial cable: Cable suspended in the air on poles or other overhead structures.

alternation: One half of a cycle.

ammeter: A test instrument that measures the amount of current in an electrical circuit.

ampere: The number of electrons passing a given point in 1 sec.

amplification: The process of taking a small signal and increasing the size of the signal (gain).

amplitude: The distance that a vibrating object moves from a position of rest during vibration.

analog display: An electromechanical device that indicates readings by the mechanical motion of a pointer.

analog multimeter: A meter that can measure two or more electrical properties and displays the measured properties along calibrated scales using a needle.

arc blast: An explosion that occurs when the air surrounding electrical equipment becomes ionized and conductive.

arc flash: An extremely high-temperature discharge produced by an electrical fault in the air.

attenuation test: A test that measures the attenuation (power loss) of each cable pair.

audio spectrum: The part of the frequency spectrum that humans can hear (20 Hz to 20 kHz).

average value (V_{ave}): The mathematical mean of all instantaneous voltage values in a sine wave.

B

backbone cables: Conductors (copper or fiber optic) used between telecommunication closets, or floor distribution terminal and central offices (equipment rooms) within a building.

bandwidth: The width of a range of frequencies that have been specified as performance limits within which a meter can be used.

bar graph: A graph composed of segments that function as an analog pointer.

bel: The logarithm of an electric, acoustic, or other power ratio.

bench oscilloscope: A test instrument that displays the shape of a voltage waveform and is used mostly for bench testing electrical and electronic circuits.

bench testing: Testing performed when equipment being tested is brought to a designated service area.

board: A group of electronic components that are connected together on a printed circuit (PC) board and that perform a set task.

break-in period: The time just after the installation of a new piece of equipment in which defects resulting from defective parts, poor manufacturing quality, contamination, or environmental stress appear.

British thermal unit (Btu): The amount of heat required to raise 1 lb of water 1°F.

brownout: The reduction of the voltage level by a power company to conserve power during times of peak usage or excessive loading of the power distribution system.

building grounding: The connection of an electrical system to earth ground by using grounding electrodes, the metal frame of the building, concrete-encased electrodes, a ground ring, or an underground metal water pipe.

C

cable: Two or more conductors grouped together within a common protective cover and used to connect individual components.

cable fault finder (wire map tester or cable map tester): A test instrument used to find open circuits, short circuits, and improper cable wiring.

cable height meter: A test instrument that is used to measure the height of a power cable in feet and inches or in meters.

cable length meter: A handheld test instrument that measures the length of cables that are still on rolls, a cut length of cable, or an already installed cable.

cable length test: A test that measures the length of conductors (in feet or meters).

calorie: The amount of heat required to raise 1 g of water 1°C.

capacitance measurement mode: A DMM mode used to measure capacitance or test the condition of a capacitor.

capacitive leakage current: Leakage current that flows through conductor insulation due to a capacitive effect.

capacitor: An electronic device used to store an electric charge.

carrier frequency: The frequency that controls the number of times the solid-state switches in the inverter section of a pulse width modulated (PWM) variable frequency drive turn ON and turn OFF per second.

cascaded amplifier: Two or more amplifiers connected to obtain a required gain.

caution: A signal word that indicates a potentially hazardous situation which, if not avoided, may result in minor or moderate injury.

chip: An integrated circuit that is packaged in an insulating polymer case and is typically used to create digital circuits.

circuit analyzer: A receptacle plug and meter that determines circuit wiring faults (reverse polarity or open ground), tests for proper operation of GFCIs and arc fault breakers (AFCIs), and displays important circuit measurements (hot/neutral/ground voltages, impedance, and line frequency).

circuit tracer: A two-piece test instrument that includes a transmitter that is plugged into a receptacle, and a receiver that provides an audible indication of which circuit the transmitter is connected to.

cladding: A layer of glass or other transparent material surrounding the fiber core of fiber optic cable that acts as insulation.

clamp-on ammeter: A test instrument that measures current in a circuit by measuring the strength of the magnetic field around a single conductor.

clamp-on leakage current ammeter: An ammeter that can measure currents as low as a few milliamps (mA).

coaxial cable: A cable that has a center conductor surrounded by an insulating layer with a braided metal jacket around the outside of the insulating layer.

coaxial cable tester: A handheld test instrument that is used to test for open or short circuits on coaxial cable.

code: A regulation or minimum requirement.

common mode noise: Noise produced between ground and hot lines.

complementary metal-oxide semiconductor (CMOS) ICs: A group of ICs that employ MOS transistors.

component: An individual device used on a board or module.

compression: An area of increased pressure in a sound wave produced when a vibrating object moves outward.

conductive leakage current: The small amount of current that normally flows through the insulation of a conductor.

conductivity (G): The ability of a substance or material to conduct electric current.

conductor: A low resistance metal that carries electricity to various parts of a circuit.

confined space: An area that has limited openings for entry and/or unfavorable natural ventilation that could cause a dangerous air contamination condition.

contact medium voltage meter: A voltmeter specifically designed to take voltage measurements on high voltage cables.

contact tachometer: A test instrument that measures the rotational speed of an object through direct contact of the tachometer tip with the object being measured.

contact temperature probe: A probe that measures temperature at a single point by direct contact with the area being measured.

continuity tester: A test instrument that tests for a complete path for current to flow.

cord: Two or more very flexible conductors grouped together and used to deliver power to a load by means of a plug.

crosstalk: Any unwanted reception of signals induced on a communication line from another communication line or from an outside source.

current: The amount of electrons flowing through a conductor, component, or circuit.

current unbalance (imbalance): The unbalance that occurs when current on each of the three power lines of a three-phase power supply to a 3ϕ motor or to other 3ϕ loads is not equal.

cutoff region: The point at which a transistor is turned OFF and no current flows.

cycle: One complete wave of alternating positive and alternating negative voltage or current.

D

danger: A signal word that indicates an imminently hazardous situation which, if not avoided, will result in death or serious injury.

DC voltage: Voltage in a circuit that has current that flows in one direction only.

decibel (dB): 1. An acoustical unit used to measure the intensity (volume) of sound. 2. An electrical unit used for expressing the ratio of the magnitudes of two electric values such as voltage or current.

decibel scale: A scale that indicates the comparison (ratio) between two or more signal powers or voltages.

delay skew test: A test that determines the difference in propagation delay between cable pairs.

dielectric absorption test: A test that verifies the absorption characteristics of insulation in good working order.

differential pressure: The difference between two pressures where the reference pressure can be any pressure (not necessarily zero).

digital display: An electronic device that displays measurements as numerical values.

digital logic probe: A special test instrument (DC voltmeter) that detects the presence or absence of a high or low signal.

digital multimeter: A meter that can measure two or more electrical properties and displays the measured properties as numerical values.

digital signal: A signal represented by one of two states.

diode: An electronic device that allows current to flow in one direction only.

diode test mode: A DMM mode used to test the condition of a diode.

direct light: Light that travels directly from a lamp to the surface being illuminated.

displacement power factor (DPF): The power factor of the fundamental frequency (60 Hz) only.

double-insulated: A term used to describe an electrical product designed so that a single ground fault cannot cause a dangerous electrical shock through any exposed parts of the product that can be touched by an electrician.

dual-trace oscilloscope: An oscilloscope that displays two signal traces simultaneously.

E

efficacy: The ratio of lumens generated by a lamp to the watts consumed by the lamp.

electric arc: A discharge of electric current across an air gap.

electrical measurement: A measurement taken when a test instrument is briefly connected to a circuit or to a component and then removed.

electrical power system: A system that produces, transmits, distributes, and delivers electrical power to satisfactorily operate electrical loads designed for connection to the system.

electrical warning: A warning used to indicate a high voltage location and conditions that could result in death or serious personal injury from an electrical shock if proper precautions are not taken.

electromagnetic interference (EMI): An unwanted signal (noise) induced on electrical power or VDV cables through magnetic coupling of adjacent wires or magnetic field-producing devices such as motors and transformers.

electronic grounding: The connection of electronic equipment to earth ground to reduce the chance of electrical shock by grounding the equipment and all non-current-carrying exposed metal.

electronic leak detector: A leak detector that detects the presence of gases.

electrostatic discharge (ESD): The movement of static electricity (electricity at rest) from the surface of one object to another object.

equipment grounding: The connection of an electrical system to earth ground to reduce the chance of electrical shock by grounding all non-current-carrying exposed metal.

equipment grounding conductor (EGC): An electrical conductor that provides a low-impedance grounding path between electrical equipment and enclosures within the distribution system.

Ethernet: A cabling technology used for network connectivity.

existing grounding electrode: A standard part of a facility that includes metal underground water pipes, the metal frame of the building (if effectively grounded), and the reinforcing bars in concrete foundations.

explosion warning: A warning used to indicate locations and conditions where exploding parts may cause death or serious personal injury if proper precautions and procedures are not followed.

F

face shield: Any eye and face protection device that covers the entire face with a plastic shield, and is used for protection from flying objects.

failure analysis: The testing of a circuit using test instruments to identify the actual faulty component in the circuit.

fault current: Any current that travels a path other than the normal operating path for which a system was designed.

fiber optic cable DMM test attachment: A DMM accessory used to test fiber optic cable and fittings.

field testing: Testing performed when the test instrument is taken to the location of the equipment to be tested.

fixed leak detector: A stationary leak detector system with sensors and controllers to detect one specific type of refrigerant.

flash protection boundary: The distance at which personal protective equipment (PPE) is required to prevent burns when an arc occurs.

flat-topping: A condition that occurs when a sine wave has lower peaks from current drawn only at the peaks of the voltage.

floating input: A digital input signal that is too high or too low at times.

flow chart: A diagram that indicates a logical sequence of steps for a given set of conditions.

fluorescent leak detector: A leak detector that uses a UV light to detect fluorescent dye that is added to a system.

footcandle (fc):A unit of illumination equal to 1 lm per sq ft.

frequency: The number of cycles per second of an AC sine wave.

frequency meter (frequency counter): A test instrument that is used to measure the frequency of an AC signal.

frequency spectrum: Range of all possible frequencies.

fundamental frequency: The frequency of the voltage used to control motor speed.

G

gain: A ratio of the amplitude of the output signal to the amplitude of the input signal.

gate turn-on: A method of turning on a thyristor that occurs when the proper signal is applied to the gate at the correct time.

gauge pressure: Pressure relative to atmospheric pressure (barometric pressure).

ground: A low resistance conducting connection between electrical circuits, equipment, and the earth.

ground bond test: A test that verifies that a grounding circuit (ground conductor and all grounding parts) of an electrical device is of sufficient size to carry a fault current to ground.

ground continuity test: A test that verifies that a low impedance (resistance) path exists between all exposed conductive metal parts and the ground (green) conductor.

grounded conductor: A conductor such as a neutral conductor that has been intentionally grounded.

grounding: The connection of all exposed non-current-carrying metal parts to the earth.

grounding conductor: A conductor that does not normally carry current, except during a fault (short circuit).

grounding electrode conductor (GEC): A conductor that connects grounded parts of a power distribution system (equipment grounding conductors, grounded conductors, and all metal parts) to the NEC®-approved earth grounding system.

ground loop: A circuit that has more than one grounding point connected to earth ground, with a voltage potential difference between the grounding points high enough to produce a circulating current in the grounding system.

H

halide leak detector: A leak detector that uses a torch flame that changes color depending on the type of gas pulled across a copper element.

handheld oscilloscope: A test instrument that displays the shape of a voltage waveform and is typically used for field testing.

hardware: The physical components in a system.

hard-wired unit: A unit that has conductors connected to terminal screws, as with a programmable controller that has hundreds of hard-wired connections.

harmonic: A frequency that is an integer (whole number) multiple (second, third, fourth, fifth, etc.) of the fundamental frequency.

hasp: A multiple lockout/tagout device.

heat: Thermal energy.

heat sink: A device that conducts and dissipates heat away from an electrical component.

hertz (Hz): The international unit of frequency; equal to one cycle per second.

hipot test: A test performed on a product or circuit with a hipot tester to ensure that there is no chance of an electrical shock or that the component was not damaged during installation.

hipot tester: A test instrument that measures insulation resistance by measuring leakage current.

horizontal control: The control on an oscilloscope that adjusts the left-to-right position of the displayed voltage trace.

humidity: The amount of moisture (water vapor) in the air as a gas.

hypothesizing probable faulty functions: The theoretical analysis of what could be causing a machine to malfunction.

I

illumination: The effect that occurs when light falls on a surface.

indirect light: Light that is reflected from one or more objects to the surface being illuminated.

impedance (Z): The total opposition that any combination of resistance (R), inductive reactance (X_L) and capacitive reactance (X_C) offers to the flow of alternating current.

impedance test: A test that measures the impedance of each cable pair.

improper phase sequence (phase reversal): The changing of the sequence of any two phases in a three-phase system or circuit.

impulse transient voltage: A transient voltage commonly caused by a lightning strike that results in a short, unwanted voltage being placed on a power distribution system.

infrared meter: A meter that measures heat energy by measuring the infrared energy emitted by a material and displays the temperature as a numerical value.

infrared temperature probe: A noncontact temperature probe that senses the infrared energy emitted by a material.

installed grounding electrode: Grounding electrode installed specifically to provide a grounding system; includes rods, metallic plates, buried copper conductors, or pipes in the ground installed for grounding.

instrumentation: The use of gauges and other measurement mechanisms to determine the value of electrical, environmental, physical, and operating conditions.

insulation spot test: A test that verifies the integrity of insulation on electrical devices such as stator winding, load conductors, cables, switchgear, heaters, and transformers.

insulation step voltage test: A test that creates electrical stress on internal insulation cracks to reveal aging or damage not found during other motor or conductor insulation tests.

intensity: The level of trace brightness.

interface: An electronic device that allows communication between various components that are being used together in a system.

International Electrotechnical Commission (IEC): An organization that develops international safety standards for electrical equipment.

invisible light: The portion of the electromagnetic spectrum on either side of the visible light spectrum.

L

lamp: An electrical system output device that converts electrical energy into light.

laser tachometer: A test instrument that uses a laser light to measure the speed of an object without directly contacting the object.

leakage current: Current that leaves the normal path of current flow (hot to neutral) and flows through a ground wire.

leak detector: A device that is used to detect refrigerant or other gas leaks in pressurized air conditioning, refrigeration, or process systems.

light: The portion of the electromagnetic spectrum that produces radiant energy.

light-emitting diode (LED): A diode that emits light when forward current is applied.

light meter: A test instrument that measures the amount of light in footcandles (fc) or lumens (lm).

linear load: Any load in which current increases proportionately as voltage increases and current decreases proportionately as voltage decreases.

linear scale: A scale that is divided into equally spaced segments.

linear speed: The distance traveled per unit of time in a straight line.

load: Any electrical device that converts electrical energy into some other form of energy, such as light, heat, sound, or mechanical.

locating the faulty circuit: A troubleshooting process that requires that a circuit be tested along various points for the presence or absence of a signal, depending on the type of circuit.

locating the faulty function: The process where the determination is made as to which section of a multifunction machine is malfunctioning or what part of a single-function machine is actually at fault.

lockout: The process of removing the source of electrical power and installing a lock that prevents the power from being turned ON.

lumen (lm): A unit used to measure the total amount of light produced by a light source

lux: The unit of illumination equal to 1 lm per sq m.

M

main bonding jumper (MBJ): A connection in a service panel that connects the equipment grounding conductor (EGC), the grounding electrode conductor (GEC), and the grounded conductor (neutral conductor).

malfunction: The failure of a system, equipment, or component to operate properly.

measurement hold mode (HOLD): A DMM mode that captures the measurement, beeps, and locks the meter measurement on the display for later viewing.

mechanical switch: Any switch that uses silver contacts to start and stop the flow of current in a circuit.

medium voltage phasing tester: A test instrument specifically designed to identify the three phases of a three-phase distribution system as L1, L2, and L3.

megohmmeter: A high-resistance ohmmeter used to measure insulation deterioration on various wires by measuring high resistance values during high voltage test conditions.

medium voltage detector checker (piezo verifier): A test instrument used to verify that a high voltage detector is operating properly before any actual high voltage measurements are taken.

medium voltage test probe: A DMM accessory used to increase the voltage measurement range above the DMM listed range.

micro-ohmmeter: A test instrument that accurately measures resistance using microhms (µΩ).

minimum holding current: The minimum current that ensures proper operation of a solid-state switch.

MIN MAX recording mode: A DMM mode that captures and stores the lowest and highest measurements for later display.

module: A group of electronic and electrical components that are housed in an enclosure and that perform a set task.

momentary power interruption: A decrease to 0 V on one or more power lines lasting from 0.5 cycles up to 3 sec.

multifunction test instrument (multimeter): A device that is capable of measuring two or more electrical properties.

multimeter: A portable test instrument that is capable of measuring two or more electrical properties.

N

neon test light: A bulb that is filled with neon gas and uses two electrodes to ionize the gas (excite the atoms).

neutral conductor: A current-carrying conductor that carries current from the load back to the power source and is intentionally grounded.

NEXT test: A test that checks cable for near end cross-talk (in dB).

noncontact medium voltage ammeter: A test instrument specifically designed to take current measurements on high voltage cables.

noncontact temperature probe: A device used for taking temperature measurements on energized circuits or on moving parts.

noncontact voltage detector: A test instrument that indicates the presence of voltage without displaying the actual amount of voltage present.

nonlinear load: Any load in which the instantaneous load current is not proportional to the instantaneous voltage.

nonlinear scale: A scale that is divided into unequally spaced segments.

NPN transistor: A transistor that has a thin layer of P-type material placed between two pieces of N-type material.

O

off-site troubleshooting: Troubleshooting from a location other than where the hardware is installed.

ohmmeter: A test instrument that measures resistance.

Ohm's law: The relationship between voltage, current, and resistance in an electrical circuit.

on-site troubleshooting: Troubleshooting at the location where the hardware is installed.

open circuit transition switching: A process in which power is momentarily disconnected when switching a circuit from one voltage supply or level to another voltage supply or level.

open wire: Any conductor (cable or wire) that does not have continuity (low resistance path) between the two ends of the conductor.

optical time domain reflectometer (OTDR): A test instrument that is used to measure cable attenuation.

oscillatory transient voltages: Transient voltages commonly caused by turning OFF high inductive loads and by switching OFF large utility power factor correction capacitors.

oscilloscope: A test instrument that provides a visual display of voltages.

overvoltage: An increase of voltage of more than 10% above the normal rated line voltage for a period of time longer than 1 min.

P

peak value (V_{max} or V_p): The maximum instantaneous value of either the positive or negative alternations of a sine wave.

period: The time required to produce one complete cycle of a waveform.

permanent test instrument: A device that is installed in a process or at a bench to continually measure and display quantities and is powered by a 115 V receptacle.

personal protective equipment (PPE): Clothing and/or equipment worn by a technician to reduce the possibility of injury in the work area.

pH: The degree of acidity or alkalinity of a solution or substance.

phase sequence tester: A test instrument used to determine which of the three-phase power lines are powered and which power line is phase A, which is phase B, and which is phase C.

phase unbalance (imbalance): The unbalance that occurs when three-phase power lines are more or less than 120° out of phase.

phasor rotation: The order in which waveforms from each phase (phase A, phase B, and phase C) cross zero.

photo tachometer: A test instrument that uses light beams to measure the speed of an object without directly contacting the object.

PNP transistor: A transistor that has a thin layer of N-type material placed between two pieces of P-type material.

portable test instrument (meter): A device that is used to take measurements.

power factor (PF): The ratio of true power used in an AC circuit to apparent power delivered to the circuit.

Power over Ethernet (PoE): A network system technology that is used to safely pass electric power over Ethernet cabling.

pressure: The force per unit of area.

preventive maintenance: The work performed to keep machines, assembly lines, production operations, and plant operations running with little or no downtime.

preventive maintenance program: A combination of unscheduled and scheduled maintenance work required to maintain equipment in peak operating condition.

primary division: A division of an analog scale with a listed value.

process: An operation or sequence of operations in which the substance being treated is changed.

propagation delay test: A test that measures the time (in ns) required for a signal to travel the length of a cable pair.

pull-up resistor: A electronic component (resistor) that prevents a floating input condition.

R

radio frequency interference (RFI): An unwanted signal (noise) induced on electrical power and VDV cables from transmitted radio frequencies such as AM radio and analog cellular.

rarefaction: An area of reduced pressure in a sound wave produced when a vibrating object moves inward.

real-time (single-shot) bandwidth: The highest frequency an oscilloscope can capture in a single pass.

receptacle tester: A device that is plugged into a standard receptacle to determine if the receptacle is properly wired and energized.

relative humidity (% RH): The amount of moisture in the air compared to the amount of moisture the air can hold at saturation.

relative mode (REL): A DMM mode that records a measurement and displays the difference between that reading and subsequent measurements.

remote return path: A separate pathway located at a distance away from an adjacent pathway in an electrical circuit.

repair and service record form: An example of documentation used for the troubleshooting and repair of a particular piece of equipment.

resistance (R): The opposition to the flow of electrons in a circuit.

resolution: The degree of precise measurement a test instrument is capable of taking.

return loss (RL) test: A test that measures the signal loss due to signal reflection (return loss) in a cable.

reversed wire pair: A common cable pair in which the tip and ring conductors are reversed with each other.

ring conductor: The second wire in a pair of wires.

root-mean-square (effective) value (V_{rms}): The value of an AC sine wave that produces the same amount of heat in a pure resistive circuit as DC of the same value.

rotational speed: The distance traveled per unit of time in a circular direction.

rubber insulating matting: A personal protective device that provides electricians protection from electrical shock when working on energized electrical circuits.

S

safety glasses: An eye protection device with special impact-resistant glass or plastic lenses, reinforced frames, and side shields.

sample rate: The speed with which an oscilloscope takes a "picture" of an incoming signal.

satellite finder meter (satellite strength meter): A test instrument that is used to locate a satellite signal and measure and display the signal strength.

saturation region: Transistor operating condition where maximum current is flowing through the transistor.

secondary division: A division of an analog scale that divides primary divisions into halves, thirds, fourths, fifths, etc.

separately derived system (SDS): A system that supplies electrical power derived or taken from transformers, storage batteries, solar photovoltaic systems, or generators.

shorted wire (short): Any conductor (cable or wire) that has an unwanted low resistance path between two conductors or to ground.

signal (function) generator: A test instrument that provides a known input signal to a component, circuit, or system for testing purposes.

silicon controlled rectifier (SCR): A thyristor that is triggered into conduction in only one direction, and is suited for DC current use.

single-function test instrument: A device capable of measuring and displaying only one electrical property, such as an ammeter.

single phasing: The operation of a three-phase load on two phases due to one phase being lost.

sinusoidal waveform: A waveform that is consistent with a pure sine wave.

software: The programs and procedures that allow hardware to operate.

solid-state switch: A switch with no moving parts (contacts).

sound: Energy that consists of pressure vibrations in the air.

sound frequency (f): The number of air pressure fluctuation cycles produced per second.

sound intensity (volume): A measure of the amount of energy flowing in a sound wave.

standard: An accepted reference or practice.

static turn-on: A method of turning on a thyristor that occurs when a fast-rising voltage is applied across the terminal of a triac.

strobe tachometer: A test instrument that uses a flashing light to measure the speed of a moving object.

subdivision: A division of an analog scale that divides secondary divisions into halves, thirds, fourths, fifths, etc.

substitution: The replacement of a malfunctioning piece of equipment or component.

surface leakage current: Current that flows from areas on conductors where insulation has been removed to allow electrical connections.

surge protection device: A device that limits the intensity of voltage surges that occur on the power lines of a power distribution system.

sustained power interruption: A decrease in voltage to 0 V on all power lines for a period of more than 1 min.

sweep: The movement of the displayed trace across the oscilloscope screen.

symbol: A graphic element that represents a quantity, unit, or component.

symptom elaboration: The process of obtaining a more detailed description of the machine malfunction.

symptom recognition: The action of recognizing a malfunction in electronic equipment and/or systems.

system: A combination of equipment, components, and/or modules that are connected to perform work or meet a specific need.

system analysis: The breakdown of system requirements and components performed when designing, implementing, maintaining, or troubleshooting a system.

systems analyst: An individual who troubleshoots at the system level.

T

tachometer: A test instrument that measures the speed of a moving object.

tagout: The process of placing a danger tag on the source of electrical power, which indicates that the equipment may not be operated until the danger tag is removed.

tap: One of many connection points along a transformer coil that are commonly provided at 2.5% increments.

telephone line tester: A test instrument that is used to simulate the telephone of a caller so the telephone equipment and line can be tested.

telephone test: A test that is used to test the voice carrying capability of circuits and to identify circuits.

temperature: The measurement of the intensity of heat.

temperature probe: The part of a temperature test instrument that measures the temperature of liquids, gases, surfaces, and pipes.

temporary power interruption: A decrease to 0 V on one or more power lines lasting for more than 3 sec up to 1 min.

test instrument: A device used to measure electrical properties such as voltage, current, resistance, frequency, power, and conductivity.

test light: A test instrument with a bulb that is connected to two test leads to give a visual indication when voltage is present in a circuit.

thermal conductivity: The ability of a material to conduct heat in the form of thermal energy.

thermal imaging camera: A meter that measures heat energy by measuring the infrared heat energy emitted by a material and displays the temperature as a color-coded thermal picture.

thermal turn-on: A method of turning on a thyristor that occurs when heat levels exceed the limit of the thyristor (typically 230°F or 110°C).

thermocouple: A device that produces electricity when two different metals that are joined together are heated.

three-prong Category I power cord device: A device that has three conductors extending from it, one hot, one neutral, and one ground (green ground wire).

three-wire solid-state switch: A solid-state switch that has three terminals for connecting wires (exclusive of ground).

thyristor: A solid-state switching device that switches current ON by using a quick pulse of control current.

time/division (time per division) control: The control on an oscilloscope that selects the width of the displayed waveform.

time domain reflectometer (TDR) test: A test used for measuring cable length and locating faults (short or open circuits, poor connections) on cables (twisted pair or coaxial).

tip conductor: The first wire in a pair of wires.

total demand distortion (TDD): The ratio of the current harmonics to the maximum load current.

total harmonic distortion (THD): The amount of harmonics on a line compared with the fundamental frequency of 60 Hz.

total power factor (power factor or PF): The power factor equal to the total difference between true power and apparent power (VA) in a circuit.

tone and probe test: A test that identifies specific pairs of wires within a system.

trace: A reference point or line that is visually displayed on the face of the oscilloscope screen.

transient voltage (voltage spike): A temporary, undesirable voltage in an electrical circuit.

transistor: A three-element device made of semiconductor material.

transistor-transistor logic (TTL) ICs: A broad family of ICs that employ a two-transistor arrangement.

transverse mode noise: Noise produced between hot and neutral lines.

triac: A thyristor that is triggered into conduction in either direction and is suited for AC current use.

troubleshooting: A logical, step-by-step process used to find a problem in an electrical power system or process as quickly and easily as possible.

troubleshooting by knowledge and experience: A method of finding a malfunction in a system or process by applying information acquired from past malfunctions.

troubleshooting procedure: A logical step-by-step process used to find a malfunction in a system or process as quickly and easily as possible.

twisted conductors: Conductors that are intertwined at regular intervals.

two-prong Category II power cord device: A device that has only two conductors extending from it, one hot and one neutral.

two-wire solid-state switch: A solid-state switch with two terminals for connecting wires (not including a ground).

U

ultrasonic leak detector: A leak detector that listens for the sounds created by a leak.

undervoltage: A drop in voltage of more than 10% (but not to 0 V) below the normal rated line voltage for a period of time longer than 1 min.

ungrounded conductor: A current-carrying conductor that is connected to loads through fuses, circuit breakers, and switches.

unit: An individual component that performs a specific task by itself.

useful life: The period of time after the break-in period when most equipment operates properly.

V

vacuum: An absolute pressure, expressed in inches of mercury, starting at atmospheric pressure and increasing in value (maximum is 29.92 in. Hg) as pressure drops to a perfect vacuum.

velocity: The speed at which air, liquids, or solids travel through a system.

vibration: A continuous periodic change in displacement with respect to a fixed reference.

visible light: The portion of the electromagnetic spectrum that the human eye can perceive.

voltage breakover turn-on: A method of turning on a thyristor that occurs when the voltage across the thyristor terminals exceeds the maximum voltage rating of the device.

voltage fluctuation: An increase or decrease in the normal line voltage within the range of +5% to –10%.

voltage indicator: Test instrument that indicates the presence of voltage when the test tip touches, or is near, an energized hot conductor or energized metal part.

voltage regulator (stabilizer): Device that provides precise voltage control to protect equipment from voltage sags (voltage dips) and voltage swells (voltage surges).

voltage sag: A voltage drop of more than 10% (but not to 0 V) below the normal rated line voltage that lasts from 0.5 cycles up to 1 min.

voltage surge: A higher-than-normal voltage that temporarily exists on one or more power lines.

voltage swell: A voltage increase of more than 10% above the normal rated line voltage lasting from 0.5 cycles up to 1 min.

voltage tester: An electrical test instrument that indicates the approximate voltage amount and type of voltage (AC or DC) in a circuit by the movement of a pointer (and vibration on some models).

voltage unbalance (imbalance): The unbalance that occurs when voltages at the terminals of a motor or other 3ϕ load are not equal.

voltmeter: A test instrument that measures voltage.

volts/division (volts per division) control: The control on an oscilloscope that selects the height of the displayed waveform.

W

warning: A signal word that indicates a potentially hazardous situation which, if not avoided, could result in death or serious injury.

wavelength: The distance covered by one complete cycle of a sound wave as the wave passes through air.

wear-out period: The period of time after the useful life period of a piece of equipment ends and typical equipment failures occur.

wire: An individual conductor.

wireless communication: The sending and receiving of voice, data, and video signals without using wires or cables.

wire map test: A basic test for testing circuits for open or short circuits, crossed pairs, reversed wires, and split pairs.

INDEX

Numbers in italic refer to figures.

D

E

F

G

H

I

K

L

M

N

O

P

R

S

T

U

V

W

Z